BMW
3 Series

Service Manual

M3, 318i, 323i, 325i, 328i
Sedan, Coupe and Convertible
1992, 1993, 1994, 1995, 1996, 1997, 1998

BentleyPublishers®
.com

BENTLEY PUBLISHERS® | Automotive Reference™

Bentley Publishers, a division of Robert Bentley, Inc.
1734 Massachusetts Avenue
Cambridge, MA 02138 USA
800-423-4595 / 617-547-4170

Information that makes the difference®

BentleyPublishers®.com

Technical contact information

We welcome your feedback. Please submit corrections and additions to our BMW technical discussion forum at:

http://www.BentleyPublishers.com

Updates and corrections

We will evaluate submissions and post appropriate editorial changes online as updates or tech discussion. Appropriate updates and corrections will be added to the book in future printings. Check for updates and corrections for this book before beginning work on your vehicle. See the following web address for additional information:

http://www.BentleyPublishers.com/updates/

WARNING—Important Safety Notice

Do not use this manual for repairs unless you are familiar with basic automotive repair procedures and safe workshop practices. This manual illustrates the workshop procedures for some maintenance and service work. It is not a substitute for full and up-to-date information from the vehicle manufacturer or for proper training as an automotive technician. Note that it is not possible for us to anticipate all of the ways or conditions under which vehicles may be serviced or to provide cautions as to all of the possible hazards that may result.

We have endeavored to ensure the accuracy of the information in this manual. Please note, however, that considering the vast quantity and the complexity of the service information involved, we cannot warrant the accuracy or completeness of the information contained in this manual.

FOR THESE REASONS, NEITHER THE PUBLISHER NOR THE AUTHOR MAKES ANY WARRANTIES, EXPRESS OR IMPLIED, THAT THE INFORMATION IN THIS MANUAL IS FREE OF ERRORS OR OMISSIONS, AND WE EXPRESSLY DISCLAIM THE IMPLIED WARRANTIES OF MERCHANTABILITY AND OF FITNESS FOR A PARTICULAR PURPOSE, EVEN IF THE PUBLISHER OR AUTHOR HAVE BEEN ADVISED OF A PARTICULAR PURPOSE, AND EVEN IF A PARTICULAR PURPOSE IS INDICATED IN THE MANUAL. THE PUBLISHER AND AUTHOR ALSO DISCLAIM ALL LIABILITY FOR DIRECT, INDIRECT, INCIDENTAL OR CONSEQUENTIAL DAMAGES THAT RESULT FROM ANY USE OF THE EXAMPLES, INSTRUCTIONS OR OTHER INFORMATION IN THIS MANUAL. IN NO EVENT SHALL OUR LIABILITY WHETHER IN TORT, CONTRACT OR OTHERWISE EXCEED THE COST OF THIS MANUAL.

Before attempting any work on your BMW, read the **Warnings and Cautions** beginning on page vii, and any **WARNING** or **CAUTION** that accompanies a procedure in the manual. Review the **WARNINGS** and **CAUTIONS** each time you prepare to work on your BMW.

Your common sense and good judgment are crucial to safe and successful service work. Read procedures through before starting them. Think about whether the condition of your car, your level of mechanical skill, or your level of reading comprehension might result in or contribute in some way to an occurrence which might cause you injury, damage your car, or result in an unsafe repair. If you have doubts for these or other reasons about your ability to perform safe repair work on your car, have the work done at an authorized BMW dealer or other qualified shop.

Part numbers listed in this manual are for identification purposes only, not for ordering. Always check with your authorized BMW dealer to verify part numbers and availability before beginning service work that may require new parts.

Special tools required to perform certain service operations are identified in the manual and are recommended for use. Use of improper tools may be detrimental to the car's safe operation as well as the safety of the person servicing the car.

The vehicle manufacturer will continue to issue service information updates and parts retrofits after the editorial closing of this manual. Some of these updates and retrofits will apply to procedures and specifications in this manual. We regret that we cannot supply updates to purchasers of this manual.

This manual is prepared, published and distributed by Bentley Publishers, 1734 Massachusetts Avenue, Cambridge, Massachusetts 02138 USA. All information contained in this manual is based on the information available to the publisher at the time of editorial closing. BMW has not reviewed and does not vouch for the accuracy or completeness of the technical specifications and work procedures described and given in this manual.

ISBN 978-0-8376-1709-1 **Job code: B398-13** editorial closing 01 / 1999.

Library of Congress Cataloging-in-Publication Data
BMW 3 Series Service Manual M3, 318i, 323i, 325i, 328i, Sedan, Coupe and Convertible: 1992, 1993, 1994, 1995, 1996, 1997, 1998.
p. cm.
Includes index.
ISBN 978-0-8376-0326-1 (alk. paper)
1. BMW 3 Series automobile--Maintenance and repair--Handbooks, manuals, etc.
I. Robert Bentley, inc.

TL215.B25B56 1999
629.28'722--dc21
99-17179

 The paper used in this publication is acid free and meets the requirements of the National Standard for Information Sciences–Permanence of Paper for Printed Library Materials. ∞

Manufactured in the United States of America.

202601R002

Selected Books and Repair Information From Bentley Publishers

Motorsports

Alex Zanardi: My Sweetest Victory
Alex Zanardi and Gianluca Gasparini
ISBN 978-0-8376-1249-2

The Unfair Advantage *Mark Donohue and Paul van Valkenburgh*
ISBN 978-0-8376-0069-7

Driving Forces: The Grand Prix Racing World Caught in the Malestrom of the Third Reich
Peter Stevenson
ISBN 978-0-8376-0217-2

Equations of Motion – Adventure, Risk and Innovation
William F. Milliken
ISBN 978-0-8376-1570-7

The Racing Driver – The Theory and Practice of Fast Driving
Denis Jenkinson
ISBN 978-0-8376-0201-1

Engineering

The Hack Mechanic Guide to European Automotive Electrical Systems *Rob Siegel*
ISBN 978-0-8376-1751-0

Physics for Gearheads *Randy Beikmann*
ISBN 978-0-8376-1615-5

Bosch Automotive Handbook
Updated 9th Edition *Robert Bosch GmbH*
ISBN 978-0-8376-1732-9

Bosch Fuel Injection and Engine Management *Charles O. Probst, SAE*
ISBN 978-0-8376-0300-1

Maximum Boost: Designing, Testing, and Installing Turbocharger Systems
Corky Bell ISBN 978-0-8376-0160-1

Supercharged! Design, Testing and Installation of Supercharger Systems
Corky Bell ISBN 978-0-8376-0168-7

Race Car Aerodynamics *Joseph Katz*
ISBN 978-0-8376-0142-7

Audi Repair Manuals

Audi A4 Service Manual: 2002-2008, 1.8L Turbo, 2.0L Turbo, 3.0L, 3.2L
Bentley Publishers
ISBN 978-0-8376-1574-5

Audi A4 Service Manual: 1996-2001, 1.8L Turbo, 2.8L *Bentley Publishers*
ISBN 978-0-8376-1675-9

Audi TT Service Manual: 2000-2006, 1.8L turbo, 3.2 L *Bentley Publishers*
ISBN 978-0-8376-1625-4

Audi A6 (C5 platform) Service Manual: 1998-2004 *Bentley Publishers*
ISBN 978-0-8376-1670-4

BMW

BMW 3 Series (F30, F31, F34) Service Manual: 2012-2015 *Bentley Publishers*
ISBN 978-0-8376-1752-7

BMW 3 Series (E30) Service Manual: 1984-1990 *Bentley Publishers*
ISBN 978-0-8376-1647-6

BMW X5 (E53) Service Manual: 2000-2006 *Bentley Publishers*
ISBN 978-0-8376-1643-8

BMW 5 Series (E39) Service Manual: 1997-2003 *Bentley Publishers*
ISBN 978-0-8376-1672-8

BMW Z3 Service Manual: 1996-2002
Bentley Publishers
ISBN 978-0-8376-1617-9

BMW 7 Series (E32) Service Manual: 1988-1994 *Bentley Publishers*
ISBN 978-0-8376-1619-3

BMW 7 Series (E38) Service Manual: 1995-2001 *Bentley Publishers*
ISBN 978-0-8376-1618-6

Memoirs of a Hack Mechanic
Rob Siegel ISBN 978-0-8376-1720-6

Porsche

Porsche 911 (Type 996) Service Manual: 1999-2005 *Bentley Publishers*
ISBN 978-0-8376-1710-7

Porsche 911 Carrera Service Manual: 1984-1989 *Bentley Publishers*
ISBN 978-0-8376-1696-4

Porsche 911 SC Service Manual: 1978-1983 *Bentley Publishers*
ISBN 978-0-8376-1705-3

Porsche 911 (964): Enthusiast's Companion *Bentley Publishers*
ISBN 978-0-8376-0293-6

Porsche Boxster. Boxster S (986) Service Manual: 1997-2004
Bentley Publishers
ISBN 978-0-8376-1645-2

Porsche: Excellence Was Expected
Karl Ludvigsen ISBN 978-0-8376-0235-6

Ferdinand Porsche – Genesis of Genius
Karl Ludvigsen ISBN 978-0-8376-1557-8

Porsche – Origin of the Species
Karl Ludvigsen ISBN 978-0-8376-1331-4

Volkswagen

Volkswagen Jetta, Golf, GTI Service Manual: 1999-2005 *Bentley Publishers*
ISBN 978-0-8376-1678-0

Volkswagen Jetta Service Manual: 2005-2010 *Bentley Publishers*
ISBN 978-0-8376-1616-2

Volkswagen Passat Service Manual: 1998-2005 *Bentley Publishers*
ISBN 978-0-8376-1669-8

Volkswagen Jetta, Golf, GTI: 1993-1999, Cabrio: 1995-2002 Service Manual *Bentley Publishers*
ISBN 978-0-8376-1660-5

Volkswagen GTI, Golf, Jetta Service Manual: 1985-1992 *Bentley Publishers*
ISBN 978-0-8376-1637-7

MINI Repair Manuals

MINI Cooper Service Manual: 2002-2006 *Bentley Publishers*
ISBN 978-0-8376-1639-1

MINI Cooper Service Manual: 2007-2013 *Bentley Publishers*
ISBN 978-0-8376-1730-5

Featured Repair Manuals

BMW X3 (E83) Service Manual: 2004-2010
Bentley Publishers
ISBN 978-0-8376-1731-2

BMW 3 Series (E90, E91, E92, E93) Service Manual: 2006-2011
Bentley Publishers
ISBN 978-0-8376-1723-7

BMW 5 Series (E60, E61) Service Manual: 2004-2010
Bentley Publishers
ISBN 978-0-8376-1689-6

BMW 3 Series (E46) Service Manual: 1999-2005
Bentley Publishers
ISBN 978-0-8376-1657-5

Bentley Publishers®
Automotive Reference

Bentley Publishers has published service manuals and automobile books since 1950. For more information, please contact Bentley Publishers at 1734 Massachusetts Avenue, Cambridge, MA 02138 USA, or visit our web site at
`BentleyPublishers.com`

Foreword

This service manual covers 1992 through 1998 BMW 3 Series models, and is specifically designed to cover only those models built for sale in the United States.

BMW, and most accessories and parts sellers specializing in BMWs, refer to the 3 Series cars covered by this manual as the E36 model. The code "E36" is BMW's internal code for the basic platform shared by all of the cars covered by this manual. Except for the engine and a few other components, the models covered by this manual share many of same components such as suspension parts, body panels, and the interior pieces. When ordering parts, especially body parts, knowing the E36 code may be helpful.

BMW also uses an internal code for the engines used in the 3 Series models. Consult the text in the **100 Engine–General** repair group of this manual for a complete listing of engine codes. Engine code information is primarily used internally by BMW, but may be helpful if you have access to BMW technical information.

For the BMW owner with basic mechanical skills, this manual gives detailed maintenance and repair information. In addition, the BMW owner who has no intention of working on his or her own car will find that owning and reading this manual will make it possible to be better informed and to discuss repairs more intelligently with a professional technician. This manual has been prepared from the repair information that BMW provides to its factory-trained technicians and has been developed primarily with the do-it-yourself BMW owner in mind. The aim throughout has been clarity and understanding with practical descriptions, step-by-step procedures, and accurate specifications.

The BMW owner intending to do maintenance and repair should have a set of tools including a set of metric wrenches and sockets, screwdrivers, a torque wrench, and feeler gauges, since these basic tools will be used to do the majority of the maintenance and repair procedures described in this manual. This manual includes detailed information on these basic tools and other tips for the beginner in the first section of the manual, entitled **010 Fundamentals for the Do-It-Yourself Owner**. For some of the repairs described in this manual, BMW technicians use special tools. The text will note when a repair requires these special tools and, where possible, will recommend practical alternatives.

We have endeavored to ensure the highest degree of accuracy possible. When the vast array of data presented in the manual is taken into account, however, no claim to infallibility can be made. We therefore cannot be responsible for the result of any errors that may have crept into the text. The publisher encourages comments from the readers of this manual with regard to any errors and, also, suggestions for improvement in the presentation of technical material. These communications have been and will be carefully considered in the preparation of future printings of this and other manuals. Please contact Robert Bentley, Inc. using the contact information on the copyright page at the beginning of this manual.

BMW offers extensive warranties, especially on components of the fuel delivery and emissions control systems. Therefore, before deciding to repair a BMW that may still be covered wholly or in part by any warranties issued by BMW of North America, consult your authorized BMW dealer. You may find that he can make the repair either free or at minimum cost.

Regardless of its age and whether or not it is still protected by warranty, your BMW is an easy car to get serviced. So if at any time a repair is needed that you feel is too difficult to do yourself, a trained BMW technician is ready to do the job for you. Each authorized BMW dealer service department has made a significant investment in service and diagnostic test equipment, specials tools, and BMW original parts. It is also the best source of the most up-to-date repair and service techniques, which includes factory training and technical literature. Servicing your BMW through an authorized BMW dealer will insure that your investment will be protected while maintaining the highest degree of service standards.

Robert Bentley

Please read these warnings and cautions before proceeding with maintenance and repair work.

WARNING—

- See also Cautions on page viii

- Some repairs may be beyond your capability. If you lack the skills, tools and equipment, or a suitable workplace for any procedure described in this manual, we suggest you leave such repairs to an authorized BMW dealer service department or other qualified shop.

- Do not re-use any fasteners that are worn or deformed in normal use. Many fasteners are designed to be used only once and become unreliable and may fail when used a second time. This includes, but is not limited to, nuts, bolts, washers, self-locking nuts or bolts, circlips and cotter pins. Always replace these fasteners with new parts.

- Never work under a lifted car unless it is solidly supported on stands designed for the purpose. Do not support a car on cinder blocks, hollow tiles or other props that may crumble under continuous load. Never work under a car that is supported solely by a jack. Never work under the car while the engine is running.

- If you are going to work under a car on the ground, make sure that the ground is level. Block the wheels to keep the car from rolling. Disconnect the battery negative (–) terminal (ground strap) to prevent others from starting the car while you are under it.

- Never run the engine unless the work area is well ventilated. Carbon monoxide kills.

- Finger rings, bracelets and other jewelry should be removed so that they cannot cause electrical shorts, get caught in running machinery, or be crushed by heavy parts.

- Tie long hair behind your head. Do not wear a necktie, a scarf, loose clothing, or a necklace when you work near machine tools or running engines. If your hair, clothing, or jewelry were to get caught in the machinery, severe injury could result.

- Do not attempt to work on your car if you do not feel well. You increase the danger of injury to yourself and others if you are tired, upset or have taken medication or any other substance that may keep you from being fully alert.

- Illuminate your work area adequately but safely. Use a portable safety light for working inside or under the car. Make sure the bulb is enclosed by a wire cage. The hot filament of an accidentally broken bulb can ignite spilled fuel or oil.

- Catch draining fuel, oil, or brake fluid in suitable containers. Do not use food or beverage containers that might mislead someone into drinking from them. Store flammable fluids away from fire hazards. Wipe up spills at once, but do not store the oily rags, which can ignite and burn spontaneously.

- Always observe good workshop practices. Wear goggles when you operate machine tools or work with battery acid. Gloves or other protective clothing should be worn whenever the job requires working with harmful substances.

- Greases, lubricants and other automotive chemicals contain toxic substances, many of which are absorbed directly through the skin. Read the manufacturer's instructions and warnings carefully. Use hand and eye protection. Avoid direct skin contact

- Disconnect the battery negative (–) terminal (Ground strap) whenever you work on the fuel system or the electrical system. Do not smoke or work near heaters or other fire hazards. Keep an approved fire extinguisher handy.

- Friction materials (such as brake pads or shoes or clutch discs) contain asbestos fibers or other friction materials. Do not create dust by grinding, sanding, or by cleaning with compressed air. Avoid breathing dust. Breathing any friction material dust can lead to serious diseases and may result in death.

- Batteries give off explosive hydrogen gas during charging. Keep sparks, lighted matches and open flame away from the top of the battery. If hydrogen gas escaping from the cap vents is ignited, it will ignite gas trapped in the cells and cause the battery to explode.

- Connect and disconnect battery cables, jumper cables or a battery charger only with the ignition switched off, to prevent sparks. Do not disconnect the battery while the engine is running.

- Do not quick-charge the battery (for boost starting) for longer than one minute. Wait at least one minute before boosting the battery a second time.

- Do not allow battery charging voltage to exceed 16.5 volts. If the battery begins producing gas or boiling violently, reduce the charging rate. Boosting a sulfated battery at a high charging rate can cause an explosion.

- The air-conditioning system is filled with chemical refrigerant, which is hazardous. The A/C system should be serviced only by trained technicians using approved refrigerant recovery/recycling equipment, trained in related safety precautions, and familiar with regulations governing the discharging and disposal of automotive chemical refrigerants.

- Do not expose any part of the A/C system to high temperatures such as open flame. Excessive heat will increase system pressure and may cause the system to burst.

- Some aerosol tire inflators are highly flammable. Be extremely cautious when repairing a tire that may have been inflated using an aerosol tire inflator. Keep sparks, open flame or other sources of ignition away from the tire repair area. Inflate and deflate the tire at least four times before breaking the bead from the rim. Completely remove the tire from the rim before attempting any repair.

- Cars covered by this manual are equipped with a supplemental restraint system (SRS), that automatically deploys an airbag(s) in the event of a frontal impact. The airbag(s) is inflated by an explosive device. Handled improperly or without adequate safeguards, can be accidently activated and cause serious injury.

- The ignition system produces high voltages that can be fatal. Avoid contact with exposed terminals and use extreme care when working on a car with the engine running or the ignition switched on.

- Place jack stands only at locations specified by manufacturer. The vehicle lifting jack supplied with the vehicle is intended for tire changes only. A heavy duty floor jack should be used to lift vehicle before installing jack stands. See **010 Fundamentals for the Do-it-Yourself Owner**.

- Battery acid (electrolyte) can cause severe burns. Flush contact area with water, seek medical attention.

- Aerosol cleaners and solvents may contain hazardous or deadly vapors and are highly flammable. Use only in a well ventilated area. Do not use on hot surfaces (engines, brakes, etc.).

- Do not remove coolant reservoir or radiator cap with the engine hot. Danger of burns and engine damage.

continued on next page

Please read these warnings and cautions before proceeding with maintenance and repair work.

CAUTION—

- See also Warnings on page vii

- If you lack the skills, tools and equipment, or a suitable workshop for any procedure described in this manual, we suggest you leave such repairs to an authorized BMW dealer or other qualified shop.

- BMW is constantly improving its cars and sometimes these changes, both in parts and specifications, are made applicable to earlier models. Therefore, part numbers listed in this manual are for reference only. Always check with your authorized BMW dealer parts department for the latest information.

- Before starting a job, make certain that you have all the necessary tools and parts on hand. Read all the instructions thoroughly, do not attempt shortcuts. Use tools appropriate to the work and use only replacement parts meeting BMW specifications. Makeshift tools, parts and procedures will not make good repairs.

- Use pneumatic and electric tools only to loosen threaded parts and fasteners. Never use these tools to tighten fasteners, especially on light alloy parts. Always use a torque wrench to tighten fasteners to the tightening torque specification listed.

- Be mindful of the environment and ecology. Before you drain the crankcase, find out the proper way to dispose of the oil. Do not pour oil onto the ground, down a drain, or into a stream, pond or lake. Dispose of in accordance with Federal, State and Local laws.

- If battery power is lost or the battery has been disconnected, the power windows must be re-initialized. Both one touch up/down and pinch-protection will be inactive until windows are re-initialized. See **512 Door Windows**.

- The control module for the anti-lock brake system (ABS) cannot withstand temperatures from a paint-drying booth or a heat lamp in excess of 203°F (95°C) and should not be subjected to temperatures in excess of 185°F (85°C) for more than two hours.

- Before doing any electrical welding on cars equipped with ABS, disconnect the battery negative (–) terminal (ground strap) and the ABS control unit connector.

- On cars equipped with anti-theft radios, make sure you know the correct radio activation code before disconnecting the battery or removing the radio. If the wrong code is entered into the radio when power is restored, that radio may lock up and be rendered inoperable, even if the correct code is then entered.

- Always make sure ignition is off before disconnecting battery.

- Label battery cables before disconnecting. On some models, battery cables are not color coded.

- Disconnecting the battery may erase fault code(s) stored in control module memory. Using special BMW diagnostic equipment, check for fault codes prior to disconnecting the battery cables. If the Check Engine light is illuminated, see **100 Engine–General** for On-Board Diagnostics (OBD) fault code information. If any other system faults have been detected (indicated by an illuminated warning light), see an authorized BMW dealer.

- If a normal or rapid charger is used to charge battery, the battery must be disconnected and removed from the vehicle in order to avoid damaging paint and upholstery.

- Do not quick-charge the battery (for boost starting) for longer than one minute. Wait at least one minute before boosting the battery a second time.

- Connect and disconnect a battery charger only with the battery charger switched off.

- Sealed or "maintenance free" batteries should be slow-charged only, at an amperage rate that is approximately 10% of the battery's ampere-hour (Ah) rating.

- Do not allow battery charging voltage to exceed 16.5 volts. If the battery begins producing gas or boiling violently, reduce the charging rate. Boosting a sulfated battery at a high charging rate can cause an explosion.

010 Fundamentals for the Do-It-Yourself Owner

TABLES

GENERAL

Although the BMW is a sophisticated and complex machine, basic maintenance can be accomplished by an interested owner with mechanical skills and the right information. Most of the preventive maintenance that is required in the lifetime of the average BMW is well within the capabilities of the do-it-yourselfer.

> ***WARNING —***
> *Do not use this manual unless you are familiar with basic automotive repair procedures and safe workshop practices. This manual illustrates the workshop procedures required for most service work; it is not a substitute for full and up-to-date information from the vehicle manufacturer or for proper training as an automotive technician. Note that it is not possible for us to anticipate all of the ways or conditions under which vehicles may be serviced or to provide cautions as to all of the possible hazards that may result.*

> ***WARNING —***
> *Your common sense and good judgment are crucial to safe and successful service work. Read procedures through before starting them. Think about whether the condition of your car, your level of mechanical skill, or your level of reading comprehension might result in or contribute in some way to an occurrence that might cause you injury, damage your car, or result in an unsafe repair. If you have doubts for these or other reasons about your ability to perform safe repair work on your car, have the work done at an authorized BMW dealer or other qualified shop.*

This section of the manual is intended to help the beginner get started. To begin with there is a discussion on **How To Use This Manual**. Tips on mechanic's skills and workshop practices that can help the beginner do a faster and more thorough job can be found under **Getting Started.** The basic tools needed to do most of the procedures in this manual are found under **Tools**. The section ends with a quick reference guide to **Emergencies**, including basic troubleshooting and information on how to gauge the seriousness of a problem.

HOW TO USE THIS MANUAL

The manual is divided into nine sections:

0 GENERAL DATA AND MAINTENANCE
1 ENGINE
2 TRANSMISSION
3 SUSPENSION, STEERING AND BRAKES
4 BODY
5 BODY EQUIPMENT
6 ELECTRICAL SYSTEM
7 EQUIPMENT AND ACCESSORIES
ELECTRICAL WIRING DIAGRAMS

0 GENERAL DATA AND MAINTENANCE covers the recommended maintenance schedules and service procedures needed to perform BMW scheduled maintenance work. Also within this section is the **010 Fundamentals for the Do-It-Yourself Owner** section, which contains basic instructions, tips and helpful hints for do-it-yourself maintenance and repair.

The next seven sections (1 through 7) are repair based and are further broken down into three digit repair groups. Each major section begins with a **General** repair group, e.g. **100 Engine–General.** These "00" (double zero) groups are mostly descriptive in nature, covering topics such as theory of operation and troubleshooting. The remainder of the repair groups contain the more involved repair information. The last major section contains detailed electrical wiring diagram schematics.

A master listing of the 9 major sections and the corresponding individual repair groups can be found on the inside front cover.

Each repair group begins with a Table of Contents listing the major subject headings within the group. Page numbers throughout the manual are organized according to the repair group system. For example, you can expect to find repair information on brakes (Repair Group 340) beginning on page 340-1. A comprehensive index can be found at the back of the manual.

Warnings, Cautions and Notes

Throughout this manual are many passages with the headings **WARNING**, **CAUTION**, or **NOTE**. These very important headings have different meanings.

WARNING —
The text under this heading warns of unsafe practices that are very likely to cause injury, either by direct threat to the person(s) performing the work or by increased risk of accident or mechanical failure while driving.

CAUTION —
A caution calls attention to important precautions to be observed during the repair work that will help prevent accidentally damaging the car or its parts.

NOTE —
A note contains helpful information, tips that will help in doing a better job and completing it more easily.

Please read every **WARNING**, **CAUTION**, and **NOTE** at the front of the manual and as they appear in repair procedures. They are very important. Read them before you begin any maintenance or repair job.

Some **WARNING**s and **CAUTION**s are repeated wherever they apply. Read them all. Do not skip any. These messages are important, even to the owner who never intends to work on the car.

GETTING STARTED

Most of the necessary maintenance and minor repair that an automobile will need can be done with ordinary tools, even by owners with little or no experience in car repair. Below is some important information on how to work safely, a discussion of what tools will be needed and how to use them.

Safety

Although an automobile presents many hazards, common sense and good equipment can help ensure safety. Many accidents happen because of carelessness. Pay attention and stick to these few important safety rules.

WARNING —
- *Never run the engine in the work area unless it is well-ventilated. The exhaust should be vented to the outside. Carbon monoxide (CO) in the exhaust kills.*
- *Remove all neckties, scarfs, loose clothing, or jewelry when working near running engines or power tools. Tuck in shirts. Tie long hair and secure it under a cap. Severe injury can result from these things being caught in rotating parts.*
- *Remove rings, watches, and bracelets. Aside from the dangers of moving parts, metallic jewelry conducts electricity and may cause shorts, sparks, burns, or damage to the electrical system when accidentally contacting the battery or other electrical terminals.*
- *Disconnect the battery negative (–) cable whenever working on or near the fuel system or anything that is electrically powered. Accidental electrical contact may damage the electrical system or cause a fire.*

WARNING —
- *Never work under a lifted car unless it is solidly supported on jack stands that are intended for that purpose. Do not support a car on cinder blocks, bricks, or other objects that may shift or crumble under continuous load. Never work under a car that is supported only by the lifting jack.*
- *Fuel is highly flammable. When working around fuel, do not smoke or work near heaters or other fire hazards. Keep an approved fire extinguisher handy.*
- *Illuminate the work area adequately and safely. Use a portable safety light for working inside or under the car. A fluorescent type light is best because it gives off less heat. If using a light with a normal incandescent bulb, use rough service bulbs to avoid breakage. The hot filament of an accidentally broken bulb can ignite spilled fuel or oil.*
- *Keep sparks, lighted matches, and any open flame away from the top of the battery. Hydrogen gas emitted by the battery is highly flammable. Any nearby source of ignition may cause the battery to explode.*
- *Never lay tools or parts in the engine compartment or on top of the battery. They may fall into confined spaces and be difficult to retrieve, become caught in belts or other rotating parts when the engine is started, or cause electrical shorts and damage to the electrical system.*
- *The fuel system is designed to retain pressure even when the ignition is off. When working with the fuel system, loosen the fuel lines slowly to allow the residual pressure to dissipate gradually. Take precautions to avoid spraying fuel.*

Lifting the Car

For those repairs that require raising the car, the proper jacking points should be used to raise the car safely and avoid damage. There are four jacking points from which the car can be safely raised. The jack supplied with the car by BMW can only be used at the four side points—just behind the front wheel or just in front of the rear wheel. See Fig. 1.

WARNING —
- *When raising the car using a floor jack or a hydraulic lift, carefully position the jack pad to prevent damaging the car body. A suitable liner (wood, rubber, etc.) should be placed between the jack and the car to prevent body damage.*
- *Watch the jack closely. Make sure it stays stable and does not shift or tilt. As the car is raised, the car may roll slightly and the jack may shift.*

Fig. 1. 3-Series jacking points (**arrows**).

Raising car safely

1. Park car on flat, level surface.
2. Remove the round cover from jack mount using a screwdriver. See Fig. 2.

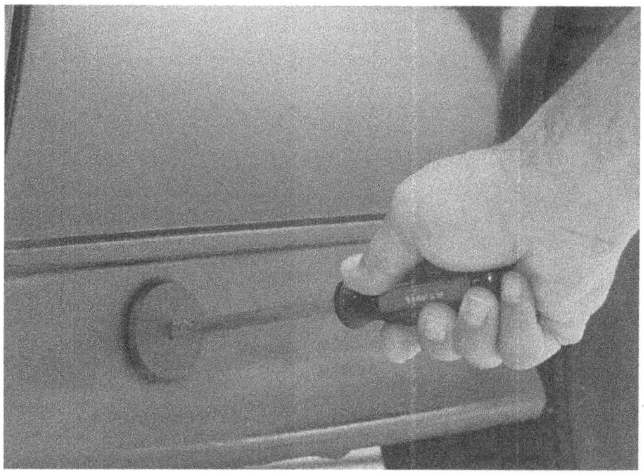

0013126

Fig. 2. Jack mount cover being removed from rocker panel.

3. Place jack fully into position. See Fig. 3. Make sure jack is resting on flat, solid ground. Use a board or other support to provide a firm surface for the jack, if necessary.
4. Raise car slowly while constantly checking position of jack and car.
5. Once car is raised, block the wheel that is opposite and farthest from jack to prevent car from unexpectedly rolling.

0013127

Fig. 3. BMW supplied jack correctly positioned in front jacking point.

> ***WARNING —***
> - *Do not rely on the transmission or the emergency brake to keep the car from rolling. They are not a substitute for positively blocking the opposite wheel.*
> - *Never work under a car that is supported only by a jack. Use jack stands that are properly designed to support the car. See* **Tools**.

Working under car safely

1. Disconnect negative (–) cable from battery so that no one else can start the car. Let others know what you will be doing.

> ***CAUTION —***
> *Prior to disconnecting the battery, read the battery disconnection cautions given at the front of this manual on page viii.*

2. Raise car slowly as described above.

3. Use at least two jack stands to support the car. A jack is a temporary lifting device and should not be used alone to support the car while you are under it. Use jack stands designed for the purpose of supporting a car. For more information on jack stands, see **Tools** below.

> ***WARNING —***
> *Do not use wood, concrete blocks, or bricks to support a car. Wood may split. Blocks or bricks, while strong, are not designed for that kind of load, and may break or collapse.*

4. Place jack stands on firm, solid surface. If necessary, use a flat board or similar solid object to provide a firm footing.

5. Lower car slowly until its weight is fully supported by jack stands. Watch to make sure that the jack stands do not tip or lean as the car settles on them.

6. Observe all jacking precautions again when raising car to remove jack stands.

ADVICE FOR THE BEGINNER

The tips in the paragraphs that follow are general advice to help any do-it-yourself BMW owner perform repairs and maintenance tasks more easily and more professionally.

Planning Ahead

Most of the repairs and maintenance tasks described in this manual can be successfully completed by anyone with basic tools and abilities. To prevent getting in too deep, know what the whole job requires before starting. Read the procedure thoroughly, from beginning to end, in order to know just what to expect and what parts will have to be replaced.

Cleanliness

Keeping things organized, neat, and clean is essential to doing a good job. When working under the hood, fender covers will protect the finish from scratches and other damage. Make sure the car is relatively clean so that dirt under the cover does not scratch the finish.

Any repair job will be less troublesome if the parts are clean. For cleaning old parts, there are many solvents and parts cleaners commercially available.

For cleaning parts prior to assembly, commercially available aerosol cans of parts cleaner or brake cleaner are handy to use, and the cleaner will evaporate completely.

> ***WARNING —***
> *Virtually all solvents used for cleaning parts are highly flammable, especially in aerosol form. Use with extreme care. Do not smoke. Do not use these products near any source of heat, sparks or flame.*

Let any solvent or cleaning product dry completely. Low-pressure, dry compressed air is helpful if available. Also, use only lint-free rags for cleaning and drying.

> *WARNING —*
> *Avoid getting tools or clothing near the battery. Battery electrolyte is a corrosive acid. Be careful with brake fluid, as it can damage the car's paint. Finally, keep rubber parts such as hoses and belts free from oil or gasoline, as they will cause the material to soften and fail prematurely.*

Non-reusable Fasteners

Many fasteners used on the cars covered by this manual must be replaced with new ones once they are removed. These include but are not limited to: bolts, nuts (self-locking, nylock, etc.), cotter pins, studs, brake fittings, roll pins, clips and washers. Genuine BMW parts should be the only replacement parts used for this purpose.

Some bolts are designed to stretch during assembly and are permanently altered rendering them unreliable once removed. These are known as torque-to-yield fasteners. Always replace fasteners where instructed to do so. Failure to replace these fasteners could cause vehicle damage and personal injury. See an authorized BMW dealer for applications and ordering information.

Tightening Fasteners

When tightening the bolts or nuts that attach a component, it is always good practice to tighten the bolts gradually and evenly to avoid misalignment or over stressing any one portion of the component. For components sealed with gaskets, this method helps to ensure that the gasket will seal properly and completely.

Where there are several fasteners, tighten them in a sequence alternating between opposite sides of the component. Fig. 4 shows such a sequence for tightening six bolts attaching a typical component. Repeat the sequence until all the bolts are evenly tightened to the proper specification.

For some repairs a specific tightening sequence is necessary, or a particular order of assembly is required. Such special conditions are noted in the text, and the necessary sequence is described or illustrated. Where no specific torque is listed, **Table a** can be used as a general guide for tightening fasteners.

> *WARNING —*
> **Table a** *is a general reference only. The values listed in the table are not intended to be used as a substitute for torques specifically called out in the text throughout this manual.*

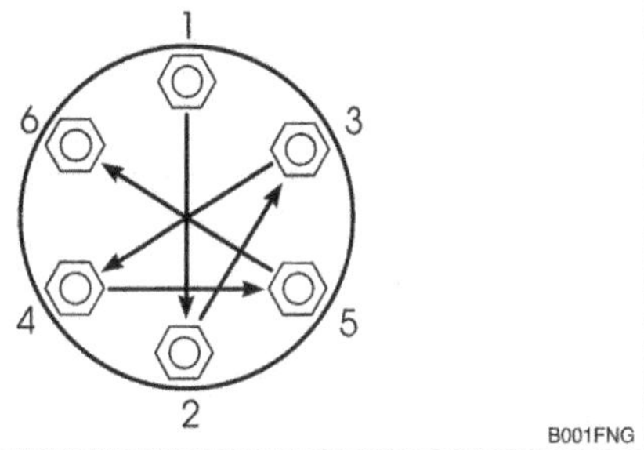

Fig. 4. General sequence for alternately tightening multiple fasteners.

> *NOTE —*
> - *Metric bolt classes or grades are marked on the bolt head.*
> - *Do not confuse wrench size with bolt diameter size. For a listing of the common wrenches used on various bolt diameters, see* **Basic Tool Requirements***.*

Table a. General Bolt Tightening Torques in Nm (max. permissible)

Bolt diameter	Bolt Class (according to DIN 267)					
	5.6	5.8	6.8	8.8	10.9	12.9
M5	2.5	3.5	4.5	6	8	10
M6	4.5	6	7.5	10	14	17
M8	11	15	18	24	34	40
M10	23	30	36	47	66	79
M12	39	52	62	82	115	140
M14	62	82	98	130	180	220
M16	94	126	150	200	280	340
M18	130	174	210	280	390	470

Gaskets and Seals

The smoothest metal mating surfaces still have imperfections that can allow leakage. To prevent leakage at critical joints, gaskets of soft, form-fitting material are used to fill in the imperfections.

To be most effective, gaskets are designed to crush and become thinner as the mating parts are bolted together. Once a gasket has been used, it is no longer capable of making as good a seal as when new, and is much more likely to leak. For this reason, gaskets should not be reused. Always plan to use new gaskets for any reassembly. Some gaskets—such as head gaskets—are directional. Make sure that these are installed correctly. This same logic applies to any part used for sealing, including rubber O-rings and copper sealing washers.

In places where a shaft must pass through a housing, flexible lip seals are used to keep the lubricating oil or grease from leaking out past the rotating shaft. Seals should never be reused once they have been removed. When removing a seal, be careful not to scratch or otherwise damage the metal surfaces. Even minor damage to sealing surfaces can cause seal damage and leakage.

The key to seal installation is to get the seal in straight without damaging it. Use a seal driver that is the same diameter as the seal housing to gently and evenly install into place. If a proper size seal driver is not available, a socket of the right size will do.

When installing a seal, it a good idea to coat the seal with oil to aid installation. Some seals are directional and special installation instructions apply. Make sure it is installed with the lip facing the correct way. Normally the lip faces the inside. Note the installation direction of the old seal before removing it.

Electrical Testing

Many electrical problems can be understood and solved with only a little fundamental knowledge of how electrical circuits function.

Electric current only flows in a complete circuit. To operate, every electrical device in the car requires a complete circuit including a voltage source and a path to ground. The positive (+) side of the battery is the original voltage source, and ground is any return path to the negative (–) side of the battery, whether through the wiring harness or the car body. Except for portions of the charging system, all electrical current in the car is direct current (DC) and flows from positive (+) to negative (–).

Switches are used to turn components on or off by completing or interrupting the circuit. A switch is "open" when the circuit is interrupted, and "closed" when the circuit is completed. Fig. 5 shows a basic circuit schematic. See **600 Electrical System–General** for electrical troubleshooting.

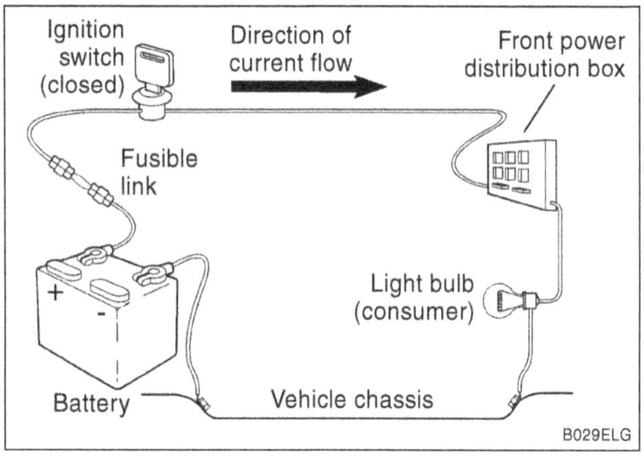

Fig. 5. Schematic representation of simple circuit for light bulb. Ignition switch is shown closed, making circuit complete.

Wire Repairs

Repairs to a wiring harness require special care to make the repair permanent. The wire ends must be clean. If frayed or otherwise damaged, cut off the end. If the wire is too short, splice in a new piece of wire of the same size and make two connections.

Use connectors that are designed for the purpose. Crimped-on or soldered-on connectors are best. Crimp connectors and special crimping pliers are widely available. If soldering, use needlenose pliers to hold the wire near the solder joint and create a "heat dam". This keeps the heat and the solder from traveling up the wire. Always use a solder made specifically for electrical work (rosin core).

NOTE —
Twisting wires together to make a repair is not recommended. Corrosion and vibration will eventually spoil the connection and may lead to irreparable damage to sensitive electronic components.

Insulate the finished connection. Electronics stores can supply heat-shrinkable insulating tubing that can be placed onto the wire before connecting, slid over the finished joint, and shrunk to a tight fit with a heat gun or hair dryer. The next best alternative is electrical tape. Make sure the wire is clean and free of solder flux or other contamination. Wrap the joint tightly to seal out moisture. See **600 Electrical–General** for more information.

BUYING PARTS

Many of the maintenance and repair tasks in this manual call for the installation of new parts, or the use of new gaskets and other materials when reinstalling parts. Most often, the parts that will be needed should be on hand before beginning the job. Read the introductory text and the complete procedure to determine which parts will be needed.

NOTE —
For some bigger jobs, partial disassembly and inspection are required to determine a complete parts list. Read the procedure carefully and, if necessary, make other arrangements to get the necessary parts while your car is disassembled.

Genuine BMW Parts

Genuine BMW replacement parts from an authorized BMW dealer are designed and manufactured to the same high standards as the original parts. They will be the correct material, manufactured to the same specifications, and guaranteed to fit and work as intended by the engineers who designed the car. Some genuine BMW parts have a limited warranty.

Many independent repair shops make a point of using genuine BMW parts, even though they may at times be more expensive. They know the value of doing the job right with the right parts. Parts from other sources can be as good, particularly if manufactured by one of BMWs original equipment suppliers, but it is often difficult to know.

BMW is constantly updating and improving their cars, often making improvements during a given model year. BMW may recommend a newer, improved part as a replacement, and your authorized dealer's parts department will know about it and provide it. The BMW parts organization is best equipped to deal with any BMW parts needs.

Non-returnable Parts

Some parts cannot be returned for credit, even if they are the wrong parts for the car. The best example is electrical parts, which are almost universally considered non-returnable because they are so easily damaged internally.

Buy electrical parts carefully, and be as sure as possible that a replacement is needed, especially for expensive parts such as electronic control units. It may be wise to let an authorized BMW dealer or other qualified shop confirm your diagnosis before replacing an expensive part that cannot be returned.

Information You Need to Know

Model. When ordering parts it is important that you know the correct model designation for your car. Models covered in this manual are E36 3 Series in both 4- and 6-cylinder configurations.

Model Year. This is not necessarily the same as date of manufacture or date of sale. A 1997 model may have been manufactured in late 1996, and perhaps not sold until early 1997. It is still a 1997 model. Model years covered by this manual are 1992 to 1998.

Date of Manufacture. This information is helpful when ordering replacement parts or determining if any of the warranty recalls are applicable to your car. The label on the driver's door below the door latch will specify the month and year that the car was built.

Vehicle Identification Number (VIN). This is a combination of letters and numbers that identify the particular car. The VIN appears on the state registration document, and on the car itself. One location is in the rear of the engine compartment, another in the lower left corner of the windshield.

The National Highway Traffic Safety Administration (NHTSA) requires passenger cars with a high theft rate to have the VIN marked on specific parts of the car when manufactured. On BMW cars, these parts are identified by an adhesive label.

Original body panels and other large components are identified by a label bearing the VIN and two BMW roundel logos. Replacement parts have a similar label, bearing one BMW roundel logo and the letters DOT-R. See Fig. 6. Parts or assemblies bearing the label are the engine, transmission, front and rear bumpers, front fenders, rear quarter panels, hood, trunk lid and doors. These labels should not be removed as they will tear apart.

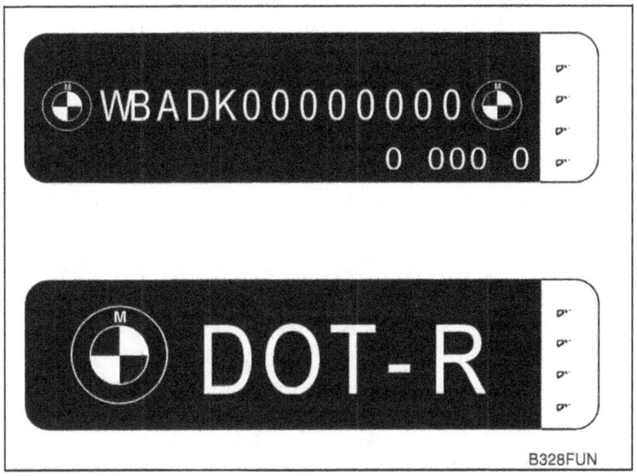

Fig. 6. Labels used to identify parts. Original equipment label with VIN number and roundel logos (top) and replacement part label with one roundel logo.

Engine. 3-Series cars covered in this manual are powered by either a 4- or 6-cylinder engine. For information on engine codes and engine applications, see **100 Engine–General**.

Transmission. The transmission type with its identifying code may be important when buying clutch parts, seals, gaskets, and other transmission-related parts. For information on transmission codes and applications, see **200 Transmission–General**.

SERVICE

BMW dealers are uniquely qualified to provide service for BMW cars. Their authorized relationship with the large BMW service organization means that they are constantly receiving new tools and equipment, together with the latest and most accurate repair information.

The BMW dealer's service technicians are highly trained and very capable. Unlike most independent repair shops, authorized BMW dealers are intensely committed to supporting the BMW product. They share the owner's interest in BMW value, performance, and reliability. On the other hand, there are many independent shops that specialize in BMW service and are capable of doing high quality repair work. Checking with other BMW owners for recommendations on service facilities is a good way to learn of reputable BMW shops in your area.

TOOLS

Most maintenance can be accomplished with a small selection of the right tools. Tools range in quality from inexpensive junk, which may break at first use, to very expensive and well-made tools for the professional. The best tools for most do-it-yourself BMW owners lie somewhere in between.

Many reputable tool manufacturers offer good quality, moderately priced tools with a lifetime guarantee. These are your best buy. They cost a little more, but they are good quality tools that will do what is expected of them. Sears' Craftsman® line is one such source of good quality tools.

Some of the repairs covered in this manual require the use of special tools, such as a custom puller or specialized electrical test equipment. These special tools are called out in the text and can be purchased through an authorized BMW dealer. As an alternative, some special tools mentioned may be purchased from the following tool manufacturers and/or distributors:

- Assenmacher Specialty Tools
  6440 Odell Place, Boulder, CO 80301
  (303) 530-2424
- Baum Tools Unlimited. Inc.
  P.O. Box 87, Longboat Key, FL 34228
  (800) 848-6657
- Schley Products Inc.
  5350 E. Hunter Ave., Anaheim Hills, CA 92807
  (714) 693-7666
- Zelenda Machine and Tool Corp.
  66-02 Austin Street, Forest Hills, NY 11375
  (718) 896-2288

Basic Tool Requirements

The basic hand tools described below can be used to accomplish most of the simple maintenance and repair tasks.

Screwdrivers. The common flat-blade type and the Phillips type will handle almost all screws used on BMWs. Two or three different sizes of each type will be required, since a screwdriver of the wrong size will damage the screw head. See Fig. 7.

A complete set of screwdrivers should also include Torx® type screwdrivers.

Wrenches. Wrenches come in different styles for different uses. Fig. 8 shows several. The basic open-end wrench is the most widely used, but grips on only two sides. It can spread apart and slip off more easily. The box-end wrench has better grip, on all six sides of a nut or bolt.

A 12-point box-end can loosen a nut or bolt where there is less room for movement, while a 6-point box-end provides better grip. For hex fasteners on fluid lines, like brake lines and fuel lines, a flare-nut wrench offers the advantages of a box-end wrench with a slot that allows it to fit over the line.

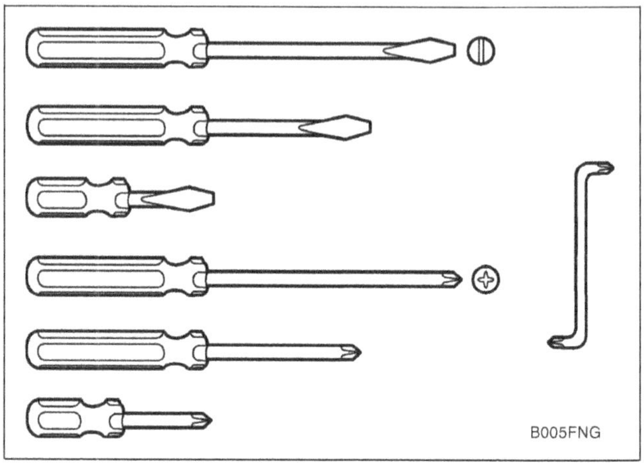

Fig. 7. Common flat-blade (top) and Phillips (bottom) screwdrivers. Offset screwdriver (right) is used for screws with limited access.

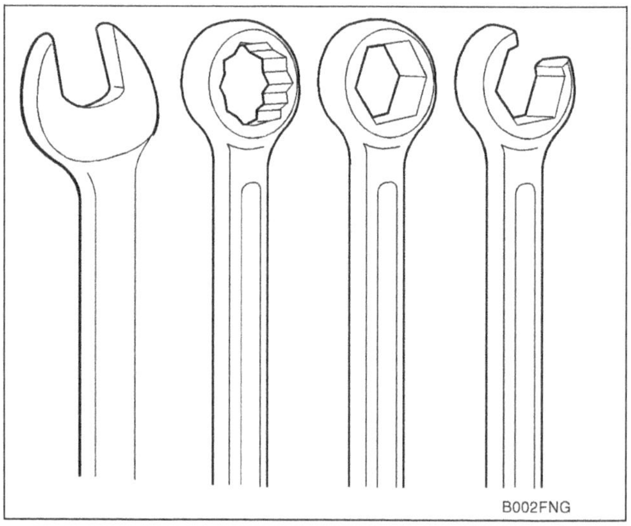

Fig. 8. Types of wrench heads. From left, open-end, 12-point box-end, 6-point box-end, flare nut.

The combination wrench is the most universal. It has one open-end and one box-end. 10mm and 13mm wrenches are the most common sizes needed. A more complete set of wrenches would include 6mm through 19mm sizes.

Sockets. Sockets perform the same job as box-end wrenches, but offer greater flexibility. They are used with a ratchet handle for speed and convenience and can be combined with extensions and universal joints (swivels) to reach fasteners more easily. Sockets come with different size connections to drive handles or extensions, called the drive size. The most common drive sizes are ¼ in., 3/8 in., and ½ in.

Sockets come in 6-point and 12-point styles. For use with a ratchet, the 6-point offers a better grip on tight nuts and bolts. 6mm to 19mm sockets are the most needed sizes. Below is a list of typical bolt diameters and the corresponding wrench sizes.

Bolt Diameter and Wrench Size

- M5 . . . . 8 mm
- M6 . . . . 10 mm
- M8 . . . . 12 mm or 13 mm
- M10 . . . . 17 mm
- M12 . . . . 19 mm
- M14 . . . . 22 mm

Spark Plug Socket. A special socket for spark plugs is the correct size, is deep enough to accommodate a spark plug's length, and includes a rubber insert to both protect the spark plug from damage and grip it for easier removal. See Fig. 9.

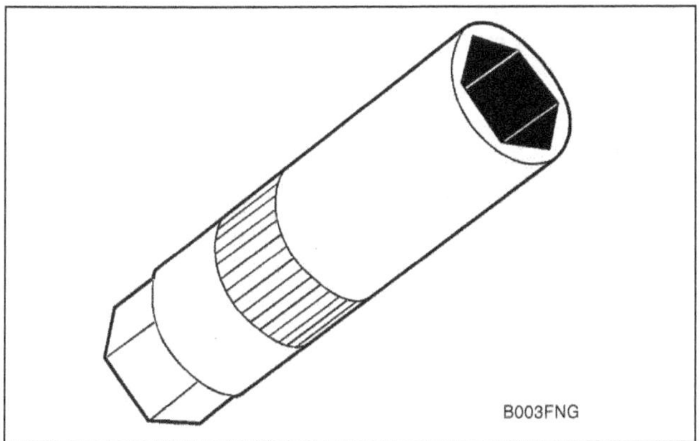

Fig. 9. Spark plug socket.

Pliers. A few of the many types of pliers are shown in Fig. 10. Most are used for holding irregular objects, bending, or crimping. Some have special applications.

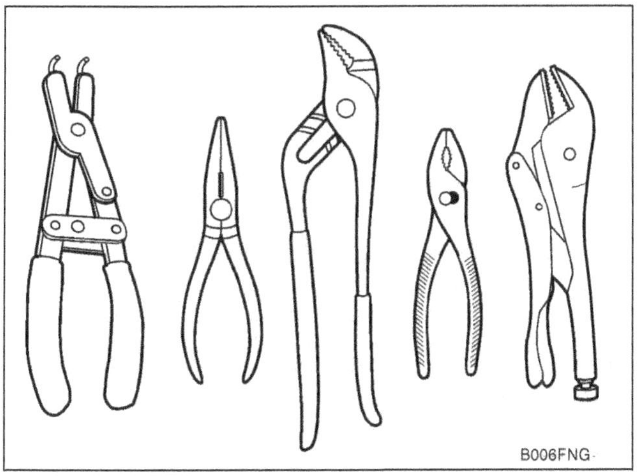

Fig. 10. Pliers. From left, snap-ring, needlenose, adjustable-joint (Channellock®), slip-joint, and locking (Vise-Grip®).

A needlenose plier is used for gripping small and poorly accessible objects, and is useful for wiring and other electrical work. A locking plier such as the Vise-Grip® is useful because of its tight grip.

Snap-ring and circlip pliers with special tipped jaws are used to remove and install snap-rings or circlips. A Channel-lock® or water pump plier has adjustable jaws that can be quickly changed to match the size of the object being held to give greater leverage.

An adjustable wrench can be a useful addition to a small tool kit. See Fig. 12. It can substitute in a pinch, if two wrenches of the same size are needed to remove a nut and bolt. Use extra care with adjustable wrenches, as they tend to loosen, slip, and damage fasteners.

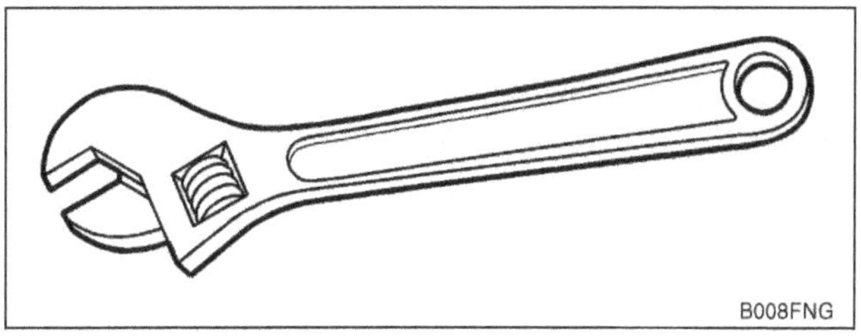

Fig. 12. Adjustable wrench.

Compared to a wrench of the correct size, an adjustable wrench is always second best. They should only be used when the correct size wrench is not available. Choose one of average size range, about 6 to 8 inches in length.

Jack Stands

Strong jack stands are extremely important for any work that is done under the car. Use only jack stands that are designed for the purpose. Blocks of wood, concrete, bricks, etc. are not safe or suitable substitutes.

> ***WARNING —***
> *A jack should be used only to raise the vehicle and should not be used to support the car for a long period. Always use jack stands to support a raised vehicle.*

Jack stands are available in several styles. A typical jack stand is shown in Fig. 12. The best ones are made of heavy material for strength, have a wide base for stability, and are equipped to positively lock in their raised positions. Get the best ones available.

Oil Change Equipment

Changing engine oil requires a 17mm socket or wrench to loosen and tighten the drain plug and a drain pan (at least 8 qt. capacity). An oil filter wrench is not required. These items are shown in Fig. 13. A wide, low drain pan will fit more easily under the car. Use a funnel to pour the new oil into the engine.

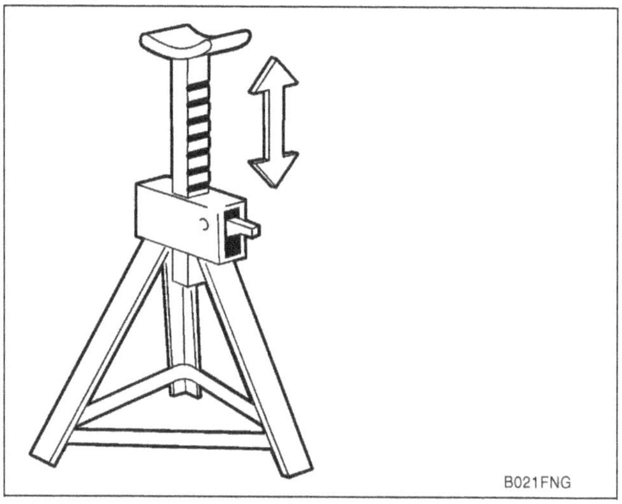

Fig. 13. Jack stand for safely supporting car to work underneath.

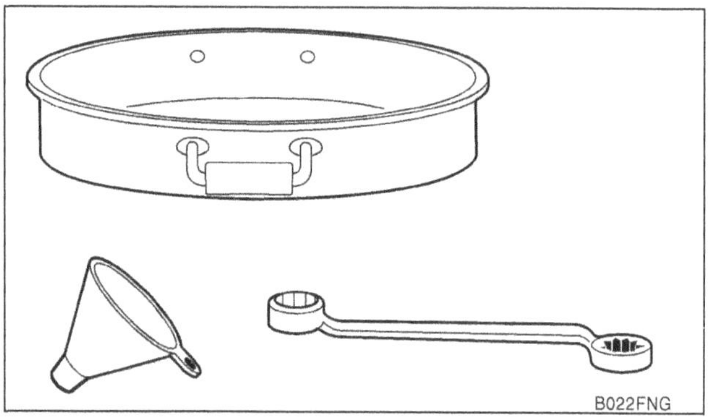

Fig. 14. Oil change equipment includes drain plug wrench (17mm), 8 qt. drain pan, and funnel.

Torque Wrench

A torque wrench is used to precisely tighten threaded fasteners to a predetermined value. Many of the repair procedures in this manual include BMW-specified torque values in Newton-meters (Nm) and the equivalent values in foot-pounds (ft-lb).

Several types of torque wrenches are available. They all do the same job, but offer different convenience features at different prices. Two typical torque wrenches are shown in Fig. 14. The most convenient ones have a built-in ratchet, and can be preset to indicate when a specific torque value has been reached. Follow the wrench manufacturer's directions for use to achieve the greatest accuracy.

A torque wrench with a range up to about 250 Nm (185 ft-lb) has adequate capacity for most of the repairs covered in this manual. For recommended torque values of 10 Nm or below, the English system equivalent is given in inch-pounds (in-lb). These small values may be most easily reached using a torque wrench calibrated in inch-pounds. To convert inch-pounds to foot-pounds, divide by 12.

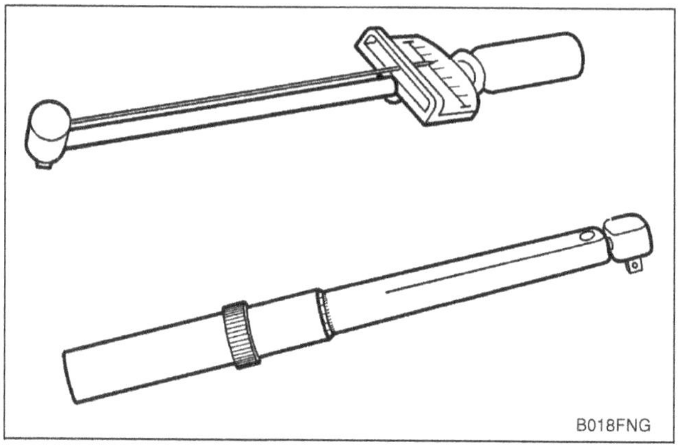

Fig. 15. Torque wrenches. Inexpensive beam-type (top) is adequate but must be read visually. Ratchet-type (bottom) can be preset to indicate (click) when torque value has been reached.

Feeler Gauges

Feeler gauges are thin metal strips of precise thickness, used to measure small clearances. They are normally available as a set, covering a range of sizes. See Fig. 16.

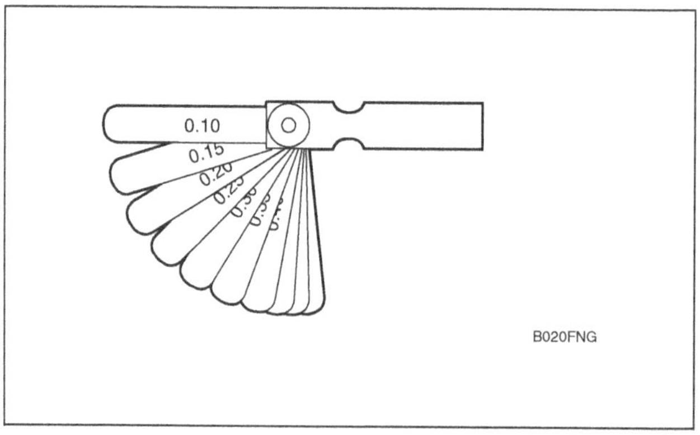

Fig. 16. Feeler gauge set.

Digital Multimeter

Many of the electrical tests in this manual call for the measurement of resistance (ohms) or voltage values. For safe and accurate tests of ignition, fuel injection, and emission control systems, the multi-meter, shown in Fig. 16, should be digital, with high (at least 10,000 ohms) input impedance. Some meters have automotive functions such as dwell and pulse width that are useful for troubleshooting ignition and fuel injection problems.

> ***CAUTION —***
> *The DME system, central body electronics, and other electronic systems may be damaged by the high current draw of a test light with a normal incandescent bulb. As a general rule, use a high impedance digital multimeter or an LED test light for all electrical testing.*

0012223

Fig. 17. Multimeter or Digital Volt/Ohmmeter (DVOM).

BMW Special Tools

Some of the more challenging repairs covered in this manual call for the use of BMW special tools. This, however, does not automatically mean that the job is too complicated or out of reach of the novice.

Many of the BMW special tools mentioned in this manual are simply the best thing to use to do the job correctly. In these cases, the tool is identified with a BMW part number. See your authorized BMW dealer parts department for information on how to order special tools.

There are some jobs for which expensive special tools are essential, and not a cost-effective purchase for one-time repair by the do-it-yourself owner. This manual includes such repairs for the benefit of those with the necessary experience and access to tools. For the do-it-yourselfer, the need for special tools is noted in the text, and whether or not BMW dealer service is recommended.

EMERGENCIES

Changing a Tire

Stop the car on as flat a surface as possible, in a place where you can be easily seen by other drivers. Avoid stopping just over the crest of a hill. Turn on the emergency flashers, and set out flares or emergency markers well behind the car. Chock the wheel (wheel chock located in trunk) diagonally opposite to the one being changed. Passengers should get out of the car and stand well away from the road. Remove the spare tire from the spare tire storage tray, as described later.

> ***WARNING —***
> *If a tire goes flat while driving, pull well off the road. Changing a tire on a busy street or highway is very dangerous. If necessary, drive a short distance on the flat tire to get to a safe place. It is much better to ruin a tire or rim than to risk being hit.*

Take the jack and tools from the tool area beneath the trunk mat. Remove the spare tire from the tire storage tray.

Loosen the wheel bolts while the car is on the ground, but leave them a little snug. Place the jack in the lifting point nearest the wheel being changed, lifting points are shown in Fig. 1 and Fig. 2). Use a board to provide a firm footing for the jack if the ground is soft. Raise the car only far enough so that the wheel is fully off the ground and then remove the wheel nuts and the wheel.

Install the spare wheel. Install the wheel nuts and tighten them by hand, then lower the car. With all wheels on the ground, fully tighten the nuts in a crisscross pattern. Torque the wheel nuts when installing the wheel. Check the inflation pressure of the spare tire.

Tightening torques
- Wheel to wheel hub . . . . . 100 ± 10 Nm (74 ± 7 ft-lb)

Car Will Not Start

If the engine turns over slowly or not at all, especially on cold mornings, the battery may not be sufficiently charged. Jump-starting the battery from another car may help.

> ***WARNING —***
> *On cars with manual transmission, push starting (or tow starting) a car is not recommended by BMW.*

NOTE —
Be sure to read the cautions under **Jump Starting Car** *prior boosting a low battery. Failure to follow the cautions may result in damage to the electronic components in the car.*

If the engine is turning over at normal speed with the starter motor, the battery and starter are fine. Check to make sure that there is fuel in the tank. Do not rely on the fuel gauge, it may be faulty. Instead, remove the gas filler cap and rock the car. If there is gas in the tank, you should hear a sloshing sound at the filler neck. If there is plenty of fuel in the tank, see **100 Engine–General** for in-depth diagnostics and troubleshooting procedures.

Jump Starting Car

Cars with discharged or dead batteries can be jump-started using the good battery from another car. When jump-starting the engine, always note the following warnings.

> ***WARNING —***
> - *Battery acid (electrolyte) can cause severe burns, and will damage the car and clothing. If electrolyte is spilled, wash the surface with large quantities of water. If it gets into eyes, flush them with water for several minutes and call a doctor.*
> - *Batteries produce explosive and noxious gasses. Keep sparks and flames away. Do not smoke near batteries.*
> - *Do not jump-start the engine if you suspect that the battery is frozen. Trapped gas may explode. Allow the battery to thaw first.*
> - *Do not quick-charge the battery (for boost starting) for longer than one minute, and do not exceed 15 volts at the battery with the boosting cables attached. Wait at least one minute before boosting the battery a second time.*
> - *Some electrical system "comfort" features may not function correctly once battery power is restored. For example, the front window one-touch-up function may be disabled. Some electrical systems will need to be reinitialized. Consult the appropriate repair group for more specific information.*

1. Place cars close together, but do not allow them to touch each other. Turn off the engine of the car with the good battery.

2. Connect one end of the positive (+) cable to the positive (+) post of the good battery. Remove the cover from the positive (B+) junction post in the engine compartment of the dead car and connect the other end of the positive (+) cable to the junction post. See Fig. 18.

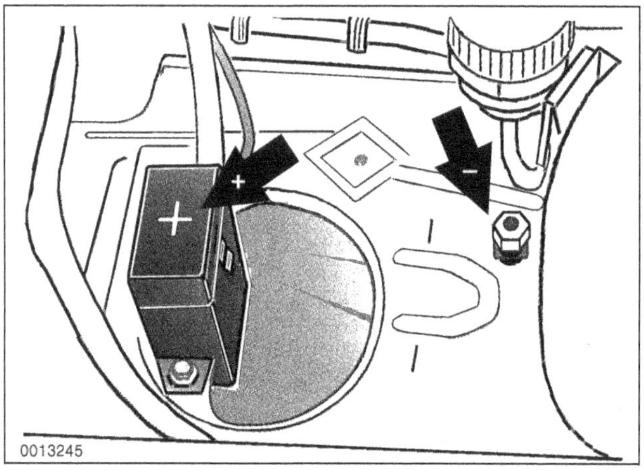

Fig. 18. Battery jump starting posts (**arrows**) behind right front strut tower. Lift cover to access positive post.

3. Connect one end of the negative (–) cable to the negative (–) battery post of the good battery. Connect opposite end of the negative cable (–) to the engine block of the car with the dead battery.

4. Start the car with the good battery and run the engine at about 2,000 rpm, then start the car with the dead battery.

5. With the engine at idle, switch on the headlights, blower motor and rear window defogger to avoid damaging the cars electrical system. Carefully disconnect the jumper cables, starting with the negative cable on the engine block. Turn all electrical consumers off.

> ***NOTE —***
> *The engine should be run for at least an hour to recharge the battery.*

Overheating

If the temperature gauge needle goes into the red band on the gauge face, the coolant temperature is too high. Find a safe place to stop and turn the engine off. Open the hood and allow the engine to cool until the temperature gauge needle is at the lower third of the scale. Continuing to drive an overheated car can cause extensive engine damage.

> ***WARNING —***
> *Do not remove the coolant reservoir or radiator cap with the engine hot. Undoing either could spray hot coolant and cause burns or damage the engine.*

> ***CAUTION —***
> *Do not add cold water or coolant to a hot engine. Engine damage could result from the sudden temperature change.*

> ***NOTE —***
> *If the engine cannot be safely turned off, make sure the air conditioner is off and turn the heater to high. This will help cool the engine until a safe stopping place can be reached.*

Overheating may be caused by low coolant level or a damaged engine drive belt. Visually check the coolant level and engine belts as described in **020 Maintenance Program**. If the coolant level is low, check the filler cap, hoses, clamps and radiator for signs of leakage. Check for leaks at the water pump on the front of the engine.

If no leaks are found, add coolant after the engine has cooled. The car can be driven, but have the cooling system thoroughly checked as soon as possible. If replacement coolant is not available, then plain water can be used, but the coolant should later be drained and refilled with the proper mixture of anti-freeze and water.

If steam is coming from the engine compartment then there is most likely a burst coolant hose or a large leak in the cooling system. To find the leak, look for signs of coolant leakage on hoses, at hose connections, or on the radiator. Let the engine cool thoroughly, then add coolant or water to fill the system and start the engine. If a great deal of water or coolant flows out of the hole, then the car should not be driven until repairs are made. If there is slight seepage, then it may be possible to drive a short distance, adding coolant as needed.

Low Oil Pressure

If the red engine oil pressure warning light illuminates, immediately stop the car and turn off the engine. Have the vehicle flat-bedded or towed to an authorized service center and have the engine oil pressure checked.

> ***NOTE —***
> *Detailed information on checking the oil pressure is covered in* **119 Lubrication System**.

Brake Fluid Level

The brake fluid level warning light may be an indicator of brake fluid loss. Problems with the brake system should be checked and repaired immediately. See **340 Brakes** for more information.

Check Engine Warning Light

If the "CHECK ENGINE" warning light comes on or flashes, it indicates that an emissions-related fault has occurred. Faults such as a failed oxygen sensor or a faulty fuel injector can turn the light on, causing the exhaust or evaporative emissions to exceed a specified limit. The engine can be safely driven with the light on, although the emission systems should be checked as soon as possible. See **100 Engine–General** for more information on the Check Engine Light and the On-board diagnostic system.

Dim Lights

Headlights and instrument panel illumination that are dim or gradually getting dimmer generally indicate a problem with the battery or charging system. The battery charge indicator light may come on as the lights are dimming. In either case, the engine and accessories are running off of the battery alone, and will soon discharge it altogether.

If possible, do not stop the engine unless you have the capability to jump start it. There may not be enough power in the starting system to restart the engine. Instead, turn off as many electrical consumers as possible. This will reduce the current drain and will allow the car to be driven farther before you lose all battery power.

With the engine and ignition off, check for loose or corroded battery cables or wires at the battery or the alternator. Disconnecting, cleaning, and reinstalling corroded wires and connectors may solve the problem. Also check drive belt tension as described in **020 Maintenance Program**.

Towing

The cars covered by this manual should be towed with a tow truck using wheel lift or flat bed equipment. Do not tow the car on all four wheels except for very short distances to move it to a safe place.

A towing eye is provided in the luggage compartment tool kit. The towing eye can be screwed into the front or rear bumper. See Fig. 18 and Fig. 19.

0013128

Fig. 19. Pry open rear tow lug access panel.

When towing the car from the front with the rear wheels on the ground, a maximum distance of 20 miles is acceptable. If the car needs to be towed further, have the rear wheels placed on dollies.

> ***CAUTION —***
> - *Do not tow with sling-type equipment. The front spoilers and bumper covers may sustain damage.*
> - *Towing a BMW with an automatic transmission with the rear wheels on the ground can result in transmission damage due to lack of lubrication and should be avoided.*

Fig. 20. Install rear tow lug into threaded hole.

If absolutely necessary, cars with automatic transmissions can be towed with the rear wheels on the ground, but the tow should not exceed 25 miles (40 km), at speeds at or below 30 mph (35 km/h). Be sure the transmission fluid has been topped off before starting the tow. Always tow the car with the transmission lever in "N" (neutral). If the tow must exceed 25 miles (40 km), add one quart of ATF to the transmission, or better yet, remove the driveshaft. Be sure to drain or pump out the added fluid once the tow has been completed.

NOTE —
- *ATF draining and filling is covered in* **240 Automatic Transmission**.
- *Driveshaft removal is covered in* **260 Driveshaft**.

Spare Parts Kit

Carrying a basic set of spare parts can prevent a minor breakdown from turning into a major annoyance. Many of the following items won't allow you to do major repair work on the car, but they will help in the event of the failure of something that can disable the car or compromise its safety.

Spare Parts Kit - Basic Contents
- Drive belt (V-belt or poly-ribbed)
- Engine oil (one or two quarts)
- Engine coolant (1 gallon of premixed 50/50 anti-freeze and water)
- Fuse assortment (7.5A, 10A, 15A, 20A)
- Radiator hoses (upper and lower)

Spare Parts Kit - Additional Contents
- Exterior lighting bulbs (headlight, brakelight, turn signal, and taillight)
- Wiper blades
- Brake fluid (new unopened bottle, DOT 4 specification)
- Main relay for DME system

020

020 Maintenance Program

TABLES

GENERAL

The information given in this repair group includes the routine checks and maintenance steps that are both required by BMW under the terms of the vehicle warranty protection and recommended by BMW to ensure long and reliable vehicle operation.

NOTE —
Aside from keeping your car in the best possible condition, proper maintenance plays a role in maintaining full protection under BMW's new-car warranty coverage. If in doubt about the terms and conditions of your car's warranty, an authorized BMW dealer should be able to explain them.

BMW has taken a unique approach to establishing maintenance intervals. BMW's Service Indicator System computes maintenance intervals based not only on elapsed mileage, but also on such inputs as engine speed, engine temperature, number of starts, and length of trips. At the appropriate time, the system indicates, through lights in the instrument cluster, when the next routine maintenance is due.

NOTE —
BMW is constantly updating their recommended maintenance procedures and requirements. The information contained here is as accurate as possible at the time of publication. If there is any doubt about what procedures apply to a specific model or model year, or what intervals should be followed, remember that an authorized BMW dealer has the latest information on factory-recommended maintenance.

BMW Service Indicator

The BMW Service Indicator notifies the driver when maintenance is required. The service indicator consists of multiple light emitting diodes (LEDs)—typically five green, one yellow, and one red—as well as **OIL SERVICE** and **INSPECTION** indicators.

When the ignition is turned on, the green LEDs come on. They go out when the engine is started. After the Service Indicator is reset, all green LEDs will be illuminated with the key on. As the car is driven in normal use, fewer and fewer green LEDs will be illuminated before start-up, indicating that the next maintenance interval is approaching.

When the car has accumulated sufficient use to require the next maintenance interval, the yellow LED will come on along with either the OIL SERVICE indicator or the INSPECTION indicator. These LEDs will stay on after the engine is started. If maintenance service is delayed, the red LED will also illuminate as a reminder that maintenance service is overdue.

An OIL SERVICE interval will always be followed by an INSPECTION interval, which will then be followed by an OIL SERVICE interval, and so on. See Fig. 1.

Fig. 1. Service indicator display in lower section of speedometer.

Oil Service

The Oil Service indicator signals the need for the most basic level of routine maintenance. BMW's required oil service specifies changing the engine lubricating oil and the engine oil filter after the engine has been warmed up. BMW-recommended additional maintenance for this service is listed in **Table a**.

NOTE —
- *For reference, the BMW Oil Service requirements are approximately equivalent to the maintenance that other European manufacturers specify at intervals with a maximum of every 7,500 miles or 6 months.*

CAUTION —
Be sure to follow the instructions for resetting the Oil Service indicator light. The Inspection indicator can be accidentally reset rendering its function inaccurate.

Inspection I and Inspection II

The Inspection indicator signals the need for more comprehensive maintenance and inspection. There are two sets of inspection requirements, Inspection I and Inspection II. These inspections alternate throughout a car's maintenance history. If the last inspection interval was Inspection I, the next inspection interval (following an Oil Service) will be Inspection II, the next after that will be Inspection I, and so on.

Inspection I tasks are listed in **Table b**. Inspection II includes most of the tasks from Inspection I with additional Inspection II tasks. A complete listing of Inspection II tasks are listed in **Table c**.

NOTE —
For reference, the BMW Inspection I and Inspection II requirements are approximately equivalent to the maintenance requirements that other European manufacturers specify. Inspection I is normally due at intervals with a maximum of 15,000 miles or 12 months. Inspection II is normally due at intervals with a maximum of 30,000 miles or 24 months.

Resetting Service Indicator

When the specified maintenance has been carried out, the service indicator memory should be reset.

The service indicator is reset using a special tool. The tool is plugged into the diagnostic connector in the engine compartment. See Fig. 2.

0011910

Fig. 2. Service indicator reset tool installed in diagnostic connector (aftermarket tool shown).

> ***CAUTION —***
> *Follow the manufacturer's directions when resetting the service indicator. If the reset procedures are done incorrectly, the reset tool or the electronic service indicator may be damaged.*

Aftermarket reset tools can be purchased from one of the following manufacturers:

Assenmacher Specialty Tools
Boulder, CO 80301
(303) 530-2424

Peake Research, Automotive Products Division
P.O. Box 28776
San Jose, CA 95159
(408) 369-0406

MAINTENANCE TABLES

Table a, **b**, **c**, and **d** on the following pages list the routine maintenance tasks specified by BMW. The intervals for most of these tasks are determined by the Service Indicator. The maintenance intervals for a few additional items are based on either elapsed mileage or time. These intervals are indicated in separate tables.

Except where noted, the maintenance items listed apply to all models and model years covered by this manual. The columns on the right side of each table give quick-reference information about the job. The bold text references the repair groups where additional repair information can be found.

Table a. Oil Service

Maintenance item	Tools required	New parts required	Warm engine required	Dealer service recommended
Engine compartment maintenance				
Change oil and oil filter		*	*	
Check fluid levels (coolant, steering, brake/clutch, ATF)		*	*	
Under car maintenance				
Check steering rack and steering linkages for leaks and excess play. **Repair Group 320**				
Inspect brake system. Check brake calipers, rotors, and brake pad thickness. Inspect for damaged hoses and lines, leaks or damage. Check hand brake, adjust cable if necessary. **Repair Group 340**				
Check and adjust tire pressures, including spare	*			
Body and interior maintenance				
Check operation of exterior and interior lights, including glove box light, engine compartment light, trunk light, instrument panel lights, horns, headlight flasher and dimmer switch. Check headlight and driving light aiming. **6 Electrical System**				
Check function/condition of seat belts. **Repair Group 720**				
Check windshield washer fluid level and concentration, add as necessary. Check operation of washer system. Check condition of wiper blades. **Repair Group 611**		*		
Road Test				
Check braking performance, steering, heating and ventilation, manual or automatic transmission, and mirrors				
Service Interval Reset				
Reset service indicator using special reset tool.	*			

Table b. Inspection I Service

Maintenance item	Tools required	New parts required	Warm engine required	Dealer service recommended
Engine compartment maintenance				
Change oil and oil filter.	*	*	*	
Check brake/clutch fluid level		*		
Check battery acid level and correct as necessary. **Repair Group 121**				
Check engine coolant level and anti-freeze protection and add as necessary. Inspect for cooling system leaks. **Repair Group 170**		*		
Lubricate throttle linkage		*		
Under car maintenance				
Check transmission for external leaks.				
Check manual transmission oil level and add as necessary. **Repair Group 230**	*	*		
Check final drive lubricant and add as necessary.	*	*		
Check fuel tank, fuel lines, and all connections for leaks.				
Inspect exhaust system. **Repair Group 180**				
Check steering rack and steering linkages. Check power steering system for leaks. Check fluid level. **Repair Group 320**		*		*
Inspect brake system for damaged hoses and lines, leaks or damage. Remove calipers and check brake pad wear. Check brake rotors for scoring damage. **Repair Group 340**	*			
Check parking brake operation, adjust cable if necessary. **Repair Group 340**	*			
Inspect front and rear suspension, including wheel bearings, for play. **Repair Group 320/330**				*
Inspect wheels and tires, including spare, check tire pressure and condition				
Body and Interior Maintenance				
Lubricate door hinges and hood latch, hood hinges				
Check headlight and driving light aiming and adjust as necessary.	*			*
Check operation of headlights, parking lights, back-up lights, license plate lights, interior lights, glove box light, engine compartment light, trunk light, instrument panel lights, turn signals, emergency flashers, stop lights, horns, headlight flasher and dimmer switch				
Replace ventilation/A/C micro-filter. Check function of air conditioning. **Repair Group 640**	*			
Check function of seat belts. **Repair Group 720**				
Check tightness of mounting screws on airbag front crash sensors (Airbag I). **Repair Group 721**				
Check windshield washer fluid level and concentration, add as necessary. Check operation of washer system. Check condition of wiper blades. **Repair Group 611**		*		
Tighten nuts and bolts for door locks and striker plates				
Replace batteries for alarm/remote control in master key (where applicable).		*		
Road Test				
Check braking performance, steering, HVAC, manual or automatic transmission, and mirrors				
Service Interval Reset				
Reset service indicator using special reset tool.	*			

Table c. Inspection II Service

Maintenance item	Tools required	New parts required	Warm engine required	Dealer service recommended
Engine compartment maintenance				
Inspect V-belt /Poly-ribbed drive condition. Check V-belt tension, if applicable.	*			
Lubricate throttle linkage.		*		
Replace air filter.	*	*		
Change oil and oil filter.	*	*	*	
Replace spark plugs. **Repair Group 120**	*	*		
Check brake/clutch fluid level.		*		
Check battery acid level and correct as necessary. **Repair Group 121**				
Check engine coolant level and anti-freeze protection and add as necessary. Inspect for cooling system leaks. **Repair Group 170**	*			
Under car maintenance				
Change automatic transmission fluid. **Repair Group 240**	*	*	*	
Change manual transmission oil. **Repair Group 230**	*	*	*	
Check clutch plate for wear. **Repair Group 210**	*			
Check drive axle boots for leaks. **Repair Group 330**				
Change final drive fluid (1995 and later models: change synthetic oil every 2nd Inspection II).	*		*	
Replace fuel filter.	*	*		
Check fuel tank, fuel lines, and all connections for leaks.				
Inspect exhaust system. **Repair Group 180**				
Check steering gear and steering linkages. Check power steering system for leaks. Check fluid level. **Repair Group 320**		*		*
Inspect brake system for damaged hoses and lines, leaks or damage. Remove calipers and check brake pad wear. Check brake rotors for scoring damage. **Repair Group 340**				
Check parking brake operation, adjust cable if necessary. Check thickness of parking brake linings. **Repair Group 340**		*		
Inspect front and rear suspension, including front and rear wheel bearings, for play. **Repair Group 310/330**				*
Inspect wheels and tires, including spare, check tire pressure and condition.				
Body and Interior Maintenance				
Lubricate door hinges and hood latch, hood hinges.		*		
Check headlight and driving light aiming and adjust as necessary.	*			*
Check operation of headlights, parking lights, back-up lights, license plate lights, interior lights, glove box light, engine compartment light, trunk light, instrument panel lights, turn signals, emergency flashers, stop lights, horns, headlight flasher and dimmer switch.				
Check active check control panel for faults (where applicable). **6 Electrical System**				

Continued on next page

Table c. Inspection II Service (continued)

Maintenance item	Tools required	New parts required	Warm engine required	Dealer service recommended
Body and Interior Maintenance (cont.)				
Replace ventilation/A/C micro-filter. Check function of air conditioning. **Repair Group 640**	*	*		
Check function of seat belts. **Repair Group 720**				
Check windshield washer fluid level and concentration, add as necessary. Check operation of washer system. Check condition of wiper blades **Repair Group 611**		*		
Tighten nuts and bolts for door locks and striker plates.	*			
Replace batteries for alarm/remote control in master key (where applicable).		*		
Check ground connection (mounting screws) of SRS (airbag) front sensor to body, if applicable. **Repair Group 721**	*			*
Road Test				
Check braking performance, steering, HVAC, manual or automatic transmission, and mirrors				
Service Interval Reset				
Reset service indicator using special reset tool.	*			

Table d. Scheduled Maintenance—Time/Mileage Intervals

Maintenance item	every 12 months	every 24 months	every 36 months	every 50,000 miles	every 60,000 miles	every 100,000 miles	New parts required	Dealer service recommended
Replace oxygen sensor. **Repair Group 180**								
1992-1994				*			*	
1995- on						*	*	
Flush and replace brake fluid/clutch fluid. **Repair Group 340**		*					*	
Drain and flush cooling system and replace coolant. **Repair Group 170**								
1992-1996			*				*	
1997-1998			*				*	
Anti-corrosion warranty inspection.	*							*

FLUID AND LUBRICANT SPECIFICATIONS

The fluids and lubricants specified by BMW are listed in **Table e**. Shown below is engine oil viscosity (SAE grade) vs. operating temperature range for the BMW engines covered in this manual.

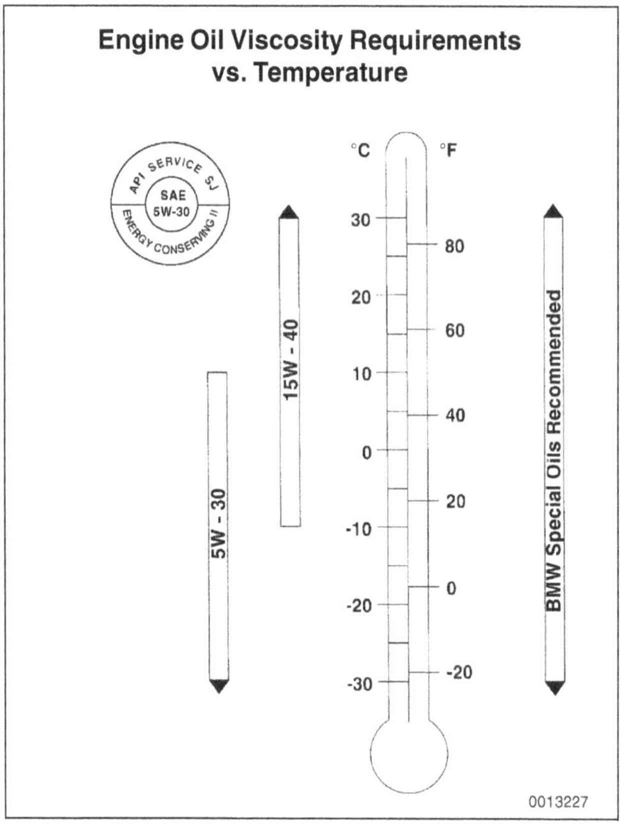

> ***WARNING —***
> *The use of fluids that do not meet BMW's specifications may impair performance and reliability, and may void warranty coverage.*

Table e. Fluids and Lubricants

Fluid	Approximate capacity	Specification
Engine oil with filter change		
M42 engine	5.0 l (5.3 US qt.)	API service SH or SJ
M44 engine	5.0 l (5.3 US qt.)	
M50 engine	6.5 l (6.9 US qt.)	
M52 engine	6.5 l (6.9 US qt.)	
S50US engine	6.5 l (6.9 US qt.)	
S52US engine	6.0 l (6.3 US qt.)	
Manual transmission oil		
M42 engine	1.1 l (1.2 US qt.)	See **230 Manual Transmission** for fluid specification information
M44 engine	1.1 l (1.2 US qt.)	
M50 engine	1.1 l (1.2 US qt.)	
M52 engine	1.2 l (1.3 US qt.)	
S50US engine	1.2 l (1.3 US qt.)	
S52US engine	1.2 l (1.3 US qt.)	

Continued

Table e. Fluids and Lubricants (continued)

Fluid	Approximate capacity	Specification
Automatic transmission fluid (ATF) (drain and fill) (additional fluid required when installing a dry torque converter)		
A4S 310R A4S 270R A5S 310Z	See **240 Automatic Transmission** for fluid specification information	
Final drive (drain and fill) (Lifetime fluid; no fluid change required)		
4-cylinder	1.1 l (1.2 US qt.)	BMW SAF-XO synthetic oil (ex. limited slip) BMW SAF-XLS synthetic oil (limited slip)
6-cylinder	1.7 l (1.8 US qt.)	
Power steering fluid		
All models	permanently sealed no drain plug	Dexron III® ATF
Brake fluid		
All models	–	SAE Dot 4
Engine coolant		
M42 engine	6.5 l (6.9 US qt.)	50% BMW anti-freeze/ 50% water
M44 engine	6.5 l (6.9 US qt.)	
M50 engine	10.5 l (11.1 US qt.)	
M52 engine	10.5 l (11.1 US qt.)	
S50US engine	10.75 l (11.4 qt.)	
S52US engine	10.75 l (11.4 qt.)	

> ***CAUTION —***
> *Multi-viscosity engine oils should not be used in the manual transmission. Use of such an oil could shorten the service life of the transmission.*

Manual Transmission Fluid

The manual transmissions installed in the E36 models are normally filled with automatic transmission fluid (ATF), although alternative synthetic lubricants may have been used from the factory, depending on transmission type and model year. Consult **230 Manual Transmission** for additional information on identifying the type of fluid installed.

Automatic Transmission Fluid

The automatic transmissions installed in the E36 models are normally filled with Dexron III automatic transmission fluid (ATF), although alternative fluids have been installed from the factory, depending on transmission type and model year. Some transmission are filled with a special BMW "life-time" automatic transmission fluid, which does not require periodic fluid changes. Consult **240 Automatic Transmission** for additional fluid specification information.

> ***NOTE —***
> *The transmission lubricant type can be found on the "type-plate" on the side of the transmission. Consult an authorized BMW dealer for alternate fluid use and the most-up-to-date information regarding transmission operating fluids.*

Final Drive Gear Oil

BMW recommends using only a specially formulated synthetic gear oil that is available through an authorized BMW dealer parts department. For additional information on this lubricant and any other lubricants that may be compatible, contact an authorized BMW dealer service department. See **331 Final Drive**.

Brake Fluid

Brake fluid absorbs moisture easily, and moisture in the fluid affects brake performance and reliability. This is why brake fluid should be flushed from the system every two years. When replacing or adding brake fluid, use only new fluid from previously unopened containers. Do not use brake fluid that has been bled from the system, even if it is brand new. Use only DOT 4 brake fluid. See **340 Brakes** for more information.

Engine Coolant (Anti-freeze)

BMW recommends coolant that is a 50/50 mixture of distilled water and phosphate/nitrate free anti-freeze containing ethylene glycol. Anti-freeze raises the boiling point and lowers the freezing point of the coolant. It also contains additives that help prevent cooling system corrosion.

Power Steering Fluid

The power steering fluid is Dexron III® ATF, or equivalent. The system is permanently filled and does not have a drain. Routinely adding ATF is not required unless the system is leaking.

> ***WARNING —***
> *Be sure to check the fluid specifications on the power steering reservoir to confirm the fluid type installed in the system. In some instances, the power steering system may be filled with an alternative hydraulic fluid designated CHF 7.1.*

ENGINE OIL SERVICE

Engine oil, checking level

The engine oil level is checked with a dipstick in the engine block. Check the oil level with the car on a level surface, after the engine has been stopped for at least a few minutes. Check the level by pulling out the dipstick and wiping it clean. Reinsert it all the way and withdraw it again. The oil level is correct if it is between the two marks near the end of the stick. See Fig. 3.

Add oil through the filler cap on the top of the cylinder head. See Fig. 4. Add only the amount needed to bring the oil level to the **MAX** mark on the dipstick, using an oil of the correct viscosity and grade. Too much oil can be just as harmful as too little.

0012208

Fig. 3. Engine oil level dipstick location on 4-cylinder engine (**arrow**). Dipstick is marked with minimum and maximum level marks.

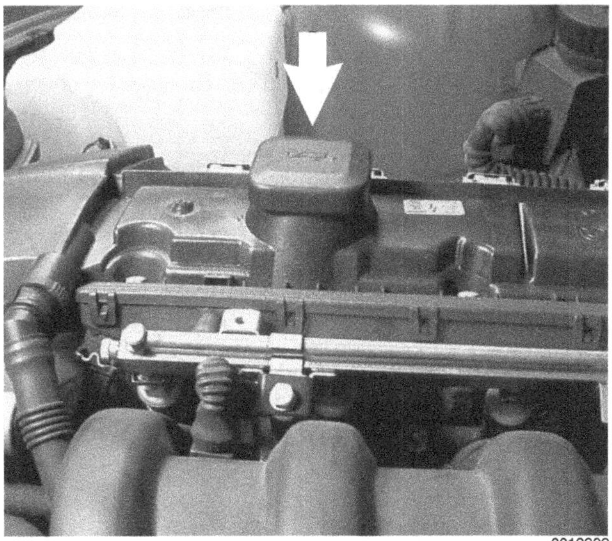

0012202

Fig. 4. Oil filler cap (**arrow**). 6-cylinder engine shown.

Engine oil and filter, changing

A complete oil change requires new oil (see **Table e**), a new oil filter insert kit, and a new drain plug sealing washer. The tools needed, a drain plug socket or box wrench (17 mm or 19 mm) and a drain pan (8-10 US qt. capacity), are described in **010 Fundamentals for the Do-It-Yourself Owner**.

> ***NOTE —***
> *If using a "fast-lube" service facility for oil changes, make sure the technician hand-starts and torques the engine oil drain plug using hand-tools. Power tools can strip the threads of the plug and the oil pan.*

1. Run engine for a few minutes to warm engine oil. Shut engine off.

2. With car on level ground, place drain pan under oil drain plug. See Fig. 5.

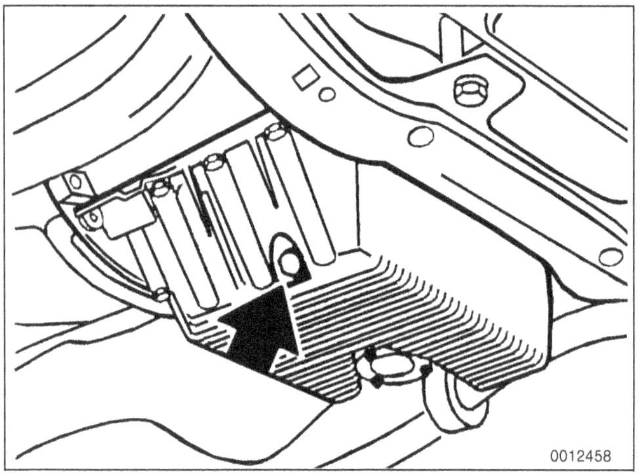

3. Remove filter housing cover. Remove filter cartridge and discard any O-rings. See Fig. 6.

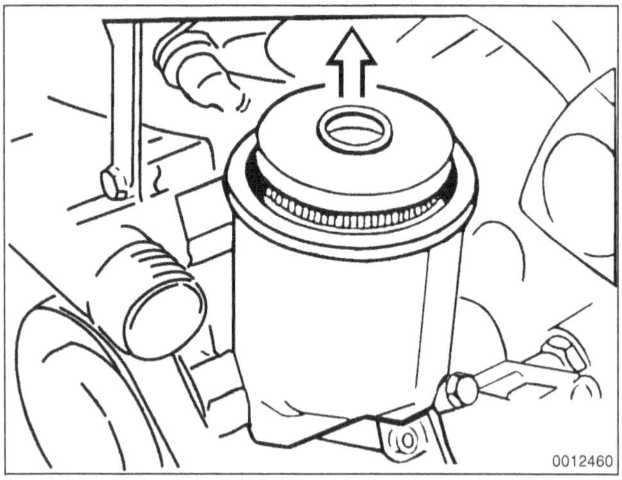

Fig. 6. Engine oil filter cartridge. Note O-ring locations when removing cover and filter.

4. Using a socket or box wrench, loosen drain plug at oil drain pan. By hand, remove plug and let oil drain into pan.

> ***CAUTION—***
> *Pull the loose plug away from the hole quickly to avoid being burned by hot oil. It will run out quickly when the plug is removed. If possible, use gloves to protect your hands.*

5. When oil flow has diminished to an occasional drip, reinstall drain plug with a new metal sealing washer and torque plug.

Tightening Torques
- Engine oil drain plug
  - M12 bolt (17mm wrench) . . . . 25 Nm (18 ft-lb)
  - M22 bolt (19mm wrench) . . . . 60 Nm (44 ft-lb)

6. Lubricate and install new O-rings. Install a new filter cartridge and housing cover. Tighten cover. See Fig. 7.

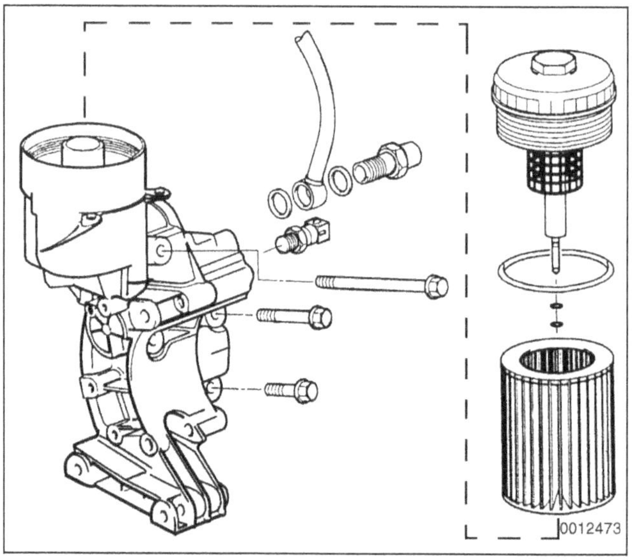

Fig. 7. Oil filter housing on 6-cylinder engine. 4-cylinder engine is similar.

Tightening Torque
- Engine oil filter cover to oil filter housing. 25 Nm (18 ft-lb)

7. Refill crankcase with oil. Approximate oil capacity is listed in **Table e**. Use dipstick to check correct oil level.

8. Start engine and check that oil pressure warning light immediately goes out.

9. Allow engine to run for a few minutes to circulate new oil, then check for leaks at drain plug and oil filter. Stop engine and recheck oil level.

ENGINE COMPARTMENT MAINTENANCE

The information under this heading describes routine maintenance—other than oil change—done in the engine compartment. It is not necessary for the car to be raised and supported off the ground. Information on oil change is given earlier under **Engine Oil Change**.

Air filter, replacing

The specified maintenance intervals for the air filter are based on normal use. If the car is operated primarily in dusty conditions, the air filter should be serviced more frequently.

To replace the air filter on 6-cylinder engines, unclip and slide the filter cartridge assembly out of the air box and remove the filter from the frame assembly. See Fig. 8.

Fig. 8. To remove air filter cartridge on 6-cylinder engines, release filter cartridge at sides (**arrows**) and then remove filter element insert from cartridge.

To replace the air filter element on 4-cylinder engines, separate the upper air filter housing section from the lower section by unfastening the spring clips. Remove the air filter element from the air filter housing. See Fig. 9. Wipe the inside of the air filter housing clean and install the new element. Reinstall the upper air filter housing, making sure that the two halves are mated correctly. Refasten the spring-clips.

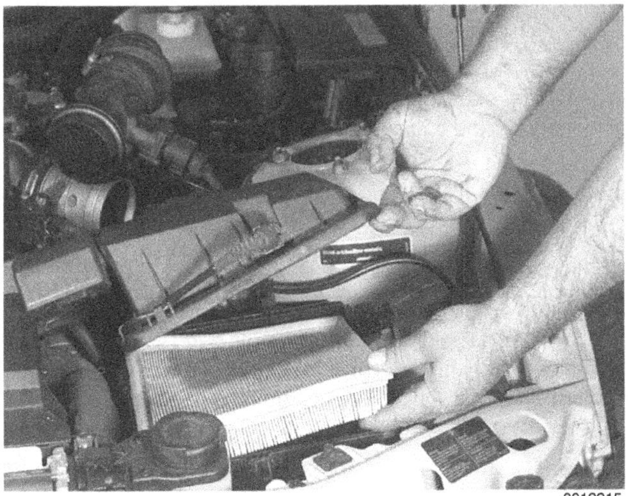

0012215

Fig. 9. Air filter element being removed on 4-cylinder engine.

Spark plugs, replacing

To access the spark plugs on 4-cylinder engines, remove the cover from atop the spark plugs. See Fig. 10. Using the special spark plug wire removal tool (attached to rear of cylinder head beneath engine cover), gently pull the spark plug wire from the spark plug. See Fig. 11. Blow away any dirt from around the base of the plug to prevent it from entering the engine. Remove and install the spark plugs one at a time.

> ***NOTE —***
> *Spark plug applications are often updated and superseded from the original plug specification. Consult an authorized BMW parts department for the latest parts information. The spark plug listed in the owner's manual may not be the latest recommend spark plug, especially on older cars.*

0012631

Fig. 10. To remove engine cover above spark plugs on 4-cylinder engine, turn retainers 90° counterclockwise and lift cover off.

0012630

Fig. 11. To remove spark plug wire from spark plug on 4-cylinder engine, slide special removal tool over spark plug wire end (**arrow**) and pull straight off. Use a screwdriver for added leverage. Special tool is stored at rear of cylinder head below top cover.

6-cylinder engines use a coil-on-plug configuration, and removal of the top engine cover (between the camshafts) and ignition coils are necessary. See Fig. 12. Blow away any dust or dirt around the ignition coils, and then remove the coils. Blow away any dust or dirt around the spark plugs. Remove the spark plugs.

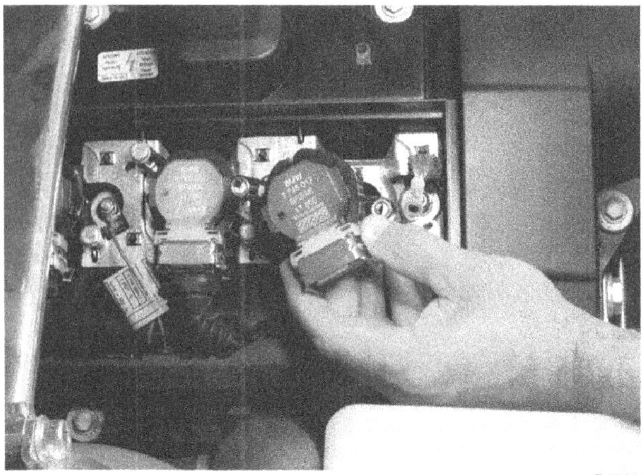

B11004

Fig. 12. Ignition coil being removed from spark plug on M52 engine.

> ***NOTE —***
> *If necessary, see* **120 Ignition System** *for more details on coil removal.*

Lightly lubricate the new spark plug threads with a small amount of copper anti-seize paste. Thread the plugs into the cylinder head by hand to prevent cross-threading.

Tightening Torque
- Spark plug to cylinder head . . . . 25 Nm (18 ft-lb)

Idle Speed

Engine idle speed can change due to a number of factors, including normal wear. The idle speed is electronically adaptive and non-adjustable. See **130 Fuel Injection** for more information.

Battery, checking and cleaning

The battery is located in the right side of the luggage compartment. Simple maintenance of the battery and its terminal connections will ensure maximum starting performance, especially in winter when colder temperatures reduce battery power.

> ***NOTE —***
> *Design characteristics of the convertible body cause vibrations in the trunk area. Therefore, E36 convertibles require a special battery designed for constant vibration. When replacing the battery, be sure the replacement is designed specifically for the convertible.*

Battery cables should be tight. The terminals, the cable clamps, and the battery case should be free of the white deposits that indicate corrosion and acid salts. Even a thin layer of dust containing conductive acid salts can cause battery discharge.

To remove battery corrosion, begin by disconnecting the cables. Disconnect the negative (–) cable first. Clean the terminal posts and the cable clamps with a wire brush. Clean the main chassis ground terminal next to the battery. Corrosion can be washed away with a baking soda and water solution that will neutralize the acid. Apply the solution carefully, though, since it will also neutralize the acid inside the battery. Reconnect the cable clamps, positive (+) cable first. Lightly coat the outside of the terminals, hold down screws, and clamps with petroleum jelly, grease, or a commercial battery terminal corrosion inhibitor.

> ***WARNING —***
> - *Prior to disconnecting the battery, read the battery disconnection cautions given at the front of this manual on page viii.*
> - *Battery acid is extremely dangerous. Take care to keep it from contacting eyes, skin, or clothing. Wear eye protection. Extinguish all smoking materials and do not work near any open flames.*

Battery electrolyte should be maintained at the correct level just above the battery plates and their separators. The correct level is approximately 5 mm (¼ in.) above the top of battery plates or to the top of the indicator marks (if applicable). The battery plates and the indicator marks can be seen once the filler caps are removed. If the electrolyte level is low, replenish it by adding distilled water only.

Battery, replacing

Batteries are rated by ampere hours (Ah), the number of hours a specific current drain can be sustained before complete discharge, or by cold cranking amps (CCA), the number of amps available to crank the engine in cold weather conditions. In general, replacement batteries should always be rated equal or higher than the original battery.

> ***CAUTION —***
> *Prior to disconnecting the battery, read the battery disconnection cautions given at the front of this manual on page viii.*

The battery is held in place by a single hand screw and plate. A secure battery hold-down is important in order to prevent vibrations and road shock from damaging the battery.

> ***NOTE —***
> - *Always disconnect the negative (–) cable first, and connect it last. While changing batteries, clean away any corrosion in or around the battery tray.*
> - *Design characteristics of the convertible body cause vibrations to oscillate in the trunk area. Therefore, the convertible model uses a special battery designed for this constant vibration.*

Accelerator and Throttle Linkage

The accelerator and throttle linkage should be lubricated periodically. Use a general purpose oil on the joints and bearings of the linkage. Use a multipurpose grease on the bearing points of the throttle plate.

Engine Drive Belts

Drive belts and pulleys transfer power from the engine crankshaft to various accessories. See Fig. 13. Depending on model and model year, engine driven accessories are driven by V-belts, poly-ribbed (serpentine) belts, or a combination of the two. For example, early 4-cylinder engines, use two V-belts and one poly-ribbed belt.

Inspect drive belts with the engine off. If the belt shows signs of wear, cracking, glazing, or missing sections, it should be replaced immediately.

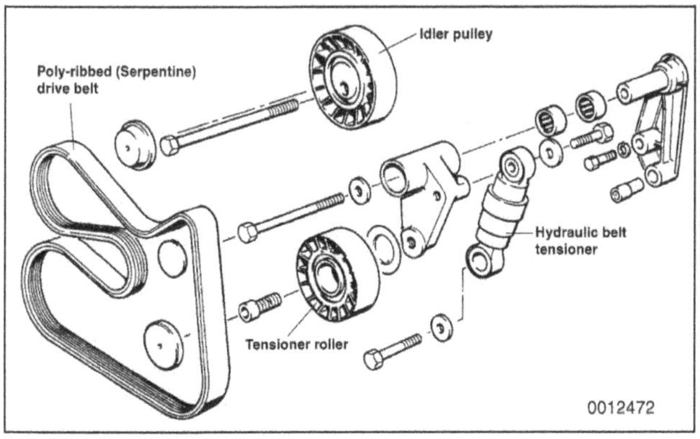

Fig. 13. Drive belt and tensioner assembly for late 4-cylinder engine with poly-ribbed drive belt.

V-belt, replacing (4-cylinder engines built up to 1/94)

V-belts should be inspected during regular maintenance intervals. Incorrect tension can decrease the life of the belt and the component it drives. The V-belt tension is adjusted through a toothed-rack mechanism.

Inspect belts with the engine off. Twist the belt to inspect its sidewalls and bottom. Belt structural damage, glazed or shiny sidewalls caused by a loose belt, or separation are all reasons to replace a belt.

Replacement of the V-belts every four years is recommended. Loosen the mounting bolts and adjust the belt-driven unit until the belt tension is as loose as possible. Remove the belt by slipping it over the pulleys. In some cases it may be necessary to remove one V-belt to get to another. Use a torque wrench and a crowfoot wrench to turn the tensioning gear bolt and set belt tension. Hold the wrench steady and tighten the locknut on the rear of the tensioning gear bolt. Tighten all other mounting nuts.

NOTE —
- *For the most accurate check, V-belt tension should be checked using a V-belt tension gauge.*
- *Belt squealing is normally caused by loose belt tension. Belt dressings should not be used.*

Tightening Torques
- Tensioning gear (alternator belt tensioning) . . . . 7 Nm (62 in-lb)
- Tensioning gear (power steering belt tensioning) . . . . 8 to 8.5 Nm (71 to 75 in-lb)
- Tensioning gear locking nut . . . . 24 Nm (18 ft-lb)

NOTE —
When belts are replaced with new ones, store the old set in the luggage compartment for emergency use.

When installing a new belt, gently pry it over the pulleys. Too much force may damage the belt or the accessory. Tension the belt(s), run the engine for a few minutes (at least 1500 rpm), then recheck the belt tension.

CAUTION —
Do not over tighten the V-belts. Overtightening will cause shaft bearings to fail prematurely.

Poly-ribbed belt, replacing (4-cylinder engines built from 1/94 and all 6-cylinder engines)

An automatic belt tensioner is used to keep the belts tensioned properly. Unless the tensioner mechanism malfunctions, the poly-ribbed belts do not require tension adjustment.

To reduce the chance of belt failure while driving, replacement of the belts every four years is recommended. Mark drive belt rotation direction if removing and reusing an old belt.

To remove the belt, it is necessary to retract the self-tensioner. Using a breaker bar or other long socket assembly, rotate the tensioner to loosen the belt. See Fig. 14 or Fig. 15. On 4-cylinder engines, it will be necessary to remove the A/C compressor belt to remove the main drive belt.

WARNING —
Observe care when replacing the belt. Personal injury could result if the tensioner springs back into position uncontrollably.

With the belt removed from the pulleys, inspect the pulleys for wear or damage that may cause early failure of the new belt. This is also a good opportunity to inspect the belt-driven accessories, checking for bearing wear and excess play. Drive belt routing is shown in Fig. 16 and Fig. 17.

Fig. 14. Engine drive belt tensioner being released on late 4-cylinder engine. Pry off tensioner cover and then lever tensioner in clockwise direction (as facing engine) and slip belt off pulleys.

Fig. 15. Engine drive belt tensioner being released on 6-cylinder engine. Pry off tensioner cover and then lever tensioner in clockwise direction (as facing engine) and slip belt off pulleys.

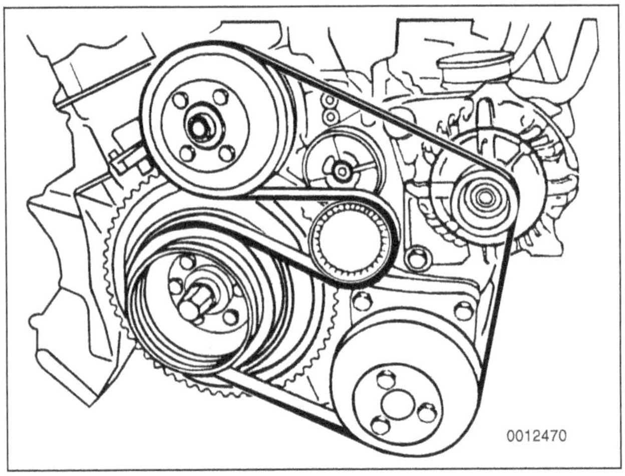

Fig. 16. Engine drive belt routing for 6-cylinder engine. A/C compressor drive belt not shown.

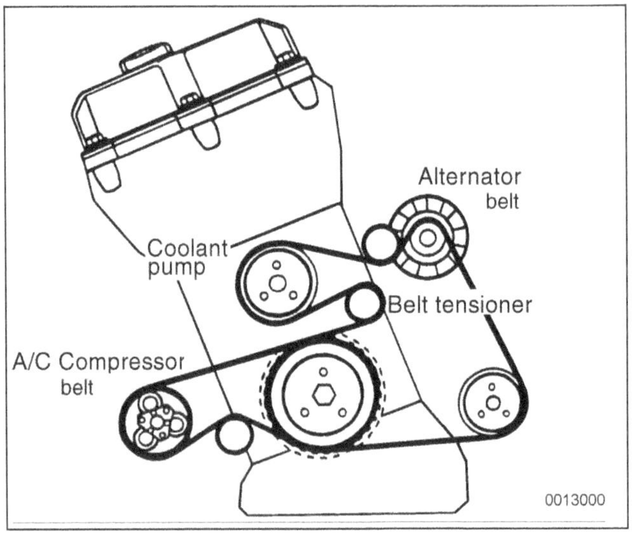

Fig. 17. Engine drive belt routing for late 4-cylinder engine with poly-ribbed drive belts.

Cooling System Service

Cooling system maintenance consists of maintaining the coolant level, inspecting hoses. Because the coolant's anti-corrosion and anti-freeze additives gradually lose their effectiveness, replacement of the coolant every two years up to 1996 models, and every three years from 1996 is recommended. As a preventive measure, replacement of the cooling system hoses every four years is also recommended.

> ***CAUTION—***
> *Use only BMW approved phosphate-free anti-freeze when filling the cooling system. Use of anti-freeze containing phosphates is considered to be harmful to the cooling system.*

A translucent expansion tank, or overflow reservoir, provides monitoring of coolant level. Because the expansion tank is translucent, the coolant level can be checked visually without opening the system. Always check the coolant level with the engine cold. The coolant level should be at the maximum mark on the expansion tank, as shown in Fig. 18.

Hose connections should be tight and dry. Coolant seepage indicates either that the hose clamp is loose, that the hose is damaged, or that the connection is dirty or corroded. Dried coolant has a chalky appearance. Hoses should be firm and springy. Replace any hose that is cracked, that has become soft and limp, or has been contaminated by oil. See Fig. 19.

Power steering fluid, checking level

Check the power steering fluid level in the fluid reservoir. Park the car on level ground with the engine off. The level is correct if it is between the **MIN** and **MAX** marks on the dipstick. See Fig. 20. If the level is below the **MIN** mark, start the engine and add fluid to the reservoir to bring the level up. Stop the engine and recheck the level. Hand-tighten the reservoir cap.

0011976

Fig. 18. Fill mark on coolant expansion tank. Coolant level should be at mark when engine is cold.

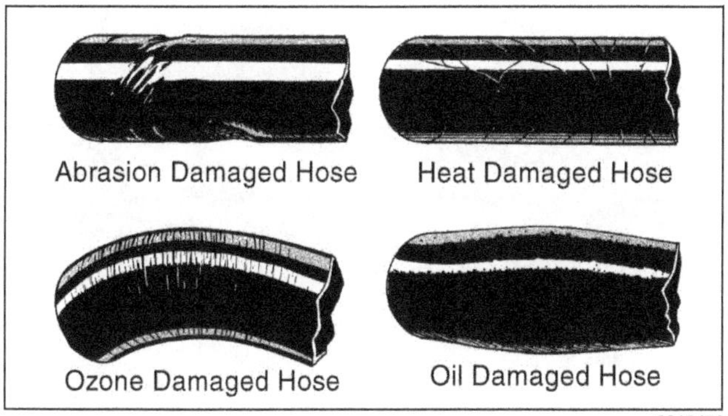

0012476

Fig. 19. Examples of damage to coolant hoses. Any of conditions shown is cause for replacement. Courtesy of Gates Rubber Company, Inc.

Specification

- Power steering fluid . . . . Dexron III® ATF

Oxygen Sensors

1996 and later engines are equipped with multiple oxygen sensors. See Fig. 21. A regulating sensor is mounted before each catalytic converter and a monitoring sensor downstream of each converter. The regulating sensor monitors engine combustion efficiency and helps to control the fuel injection system and exhaust emissions. The monitoring sensor is used by the On-board diagnostic system to monitor the function of the catalytic converter.

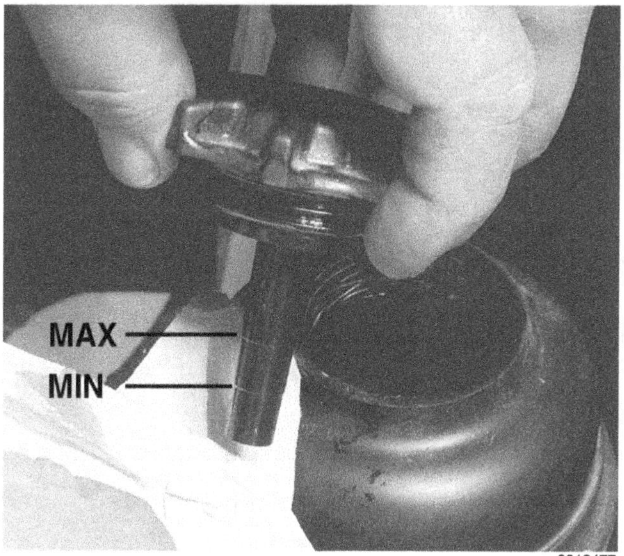

0012477

Fig. 20. Power steering fluid dipstick showing MIN and MAX marks.

OBD II enhanced emission standards require the engine control module (ECM) to monitor the oxygen content in the exhaust both before and after the catalytic converter. This allows for tighter control of the tail pipe emissions and also allows the ECM to diagnose converter problems. If the DME detects that catalytic converter or oxygen sensor efficiency has degraded past a certain pre-programmed limit, it will turn on the Check Engine light, and store a diagnostic trouble code (DTC) in the ECM.

Replacement of oxygen sensors at the specified intervals ensures that the engine and emission control system will continue to operate as designed. Extending the replacement interval may void the emission control warranty coverage.

1. To replace oxygen sensor, detach electrical harness connector, then unscrew sensor from exhaust pipe.

> ***WARNING —***
> *Allow exhaust system to cool thoroughly.*

NOTE —

Special sockets for replacing the oxygen sensor are available from most automotive parts stores. The socket has a groove cut down one side to allow the sensor to be installed without damaging the wire harness.

Tightening Torque

- Oxygen sensor to exhaust system. . . . . 55±5 Nm (40±4 ft-lb)

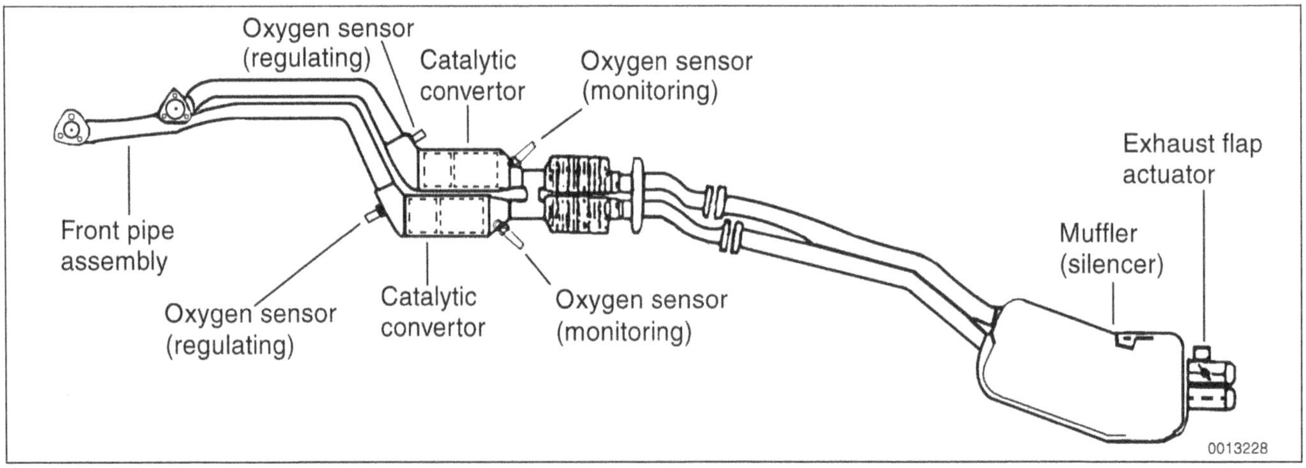

Fig. 21. Typical 1996 and later exhaust system showing oxygen sensor locations.

UNDER-CAR MAINTENANCE

Fuel filter, replacing

On early 4-cylinder cars, the fuel filter is mounted under the car on the left front bulkhead. On early 6-cylinder cars, the fuel filter is mounted to the front left motor mount in the engine compartment. See Fig. 22.

Fig. 22. Fuel filter location (**arrow**) on bulkhead under left side of vehicle. Filter for M42 engine shown.

On later cars, the fuel filter is located beneath the center of the car, approximately under the driver's seat. See Fig. 23. A protective cover is installed over the fuel filter, which must be removed to access the filter.

NOTE—
Drain the fuel filter from the inlet side into a container and inspect the drained fuel. Check for rust, moisture and contamination.

Fig. 23. Fuel filter location under center of car (**arrow**). Filter for M52 engine shown.

When replacing the fuel filter, disconnect the battery negative (–) cable and clamp the filter inlet and outlet hoses to lessen fuel spillage. Loosen the center clamping bracket and the two hose clamps on either end of the filter. Note the arrow or markings indicating direction of flow on the new filter. Install the filter and use new hose clamps.

WARNING—
Fuel will be expelled when the filter is removed. Do not smoke or work near heaters or other fire hazards. Keep a fire extinguisher handy.

CAUTION—
Clean thoroughly around the filter connections before removing them.

Tires, checking inflation pressure

Correct tire pressures are important to handling and stability, fuel economy, and tire wear. Tire pressures change with temperature. Pressures should be checked often during seasonal temperature changes. Correct inflation pressures can be found on the driver's door pillar and in the owner's manual. Note that tire pressures should be higher when the car is more heavily loaded.

> ***WARNING —***
> *Do not inflate any tire to a higher pressure than the tire's maximum inflation pressure listed on the sidewall. Use care when adding air to warm tires. Warm tire pressures can increase as much as 4 psi (0.3 bar) over their cold pressures.*

Tires, rotating

BMW does not recommend tire rotation. Due to the car's suspension design, the front tires begin to wear first at the outer shoulder and the rear tires begin to wear first at the middle of the tread or inner shoulder. Rotating the tires may adversely affect road handling and tire grip.

NOTE —
The main purpose of tire rotation is to promote even wear and maximum tire life. Tire life may be decreased slightly if the tires are not rotated.

Wheels, aligning

BMW recommends checking the front and rear alignment once a year and whenever new tires are installed. See **320 Steering and Wheel Alignment** for a more detailed discussion of alignment requirements and specifications.

Brake system, inspecting

Routine maintenance of the brake system includes maintaining the brake fluid in the reservoir, checking brake pads for wear, checking hand brake function, and inspecting the system for fluid leaks or other damage.

Check that brake hoses are correctly routed to avoid chafing or kinking. Inspect unions and brake calipers for signs of fluid leaks. Inspect rigid lines for corrosion, dents, or other damage. Inspect flexible hoses for cracking. Replace faulty hoses or lines.

> ***WARNING —***
> *Incorrect installation or overtightening hoses, lines, and unions may cause chafing or leakage. This can lead to partial or complete brake system failure.*

Brake pad/rotor wear, checking

All E36 cars are fitted with disc brakes at all four wheels. See Fig. 24. Although the brakes are equipped with a brake pad warning system, the system only monitors one wheel per axle. It is recommended that pad thickness should be checked whenever the wheels are off or brake work is being done.

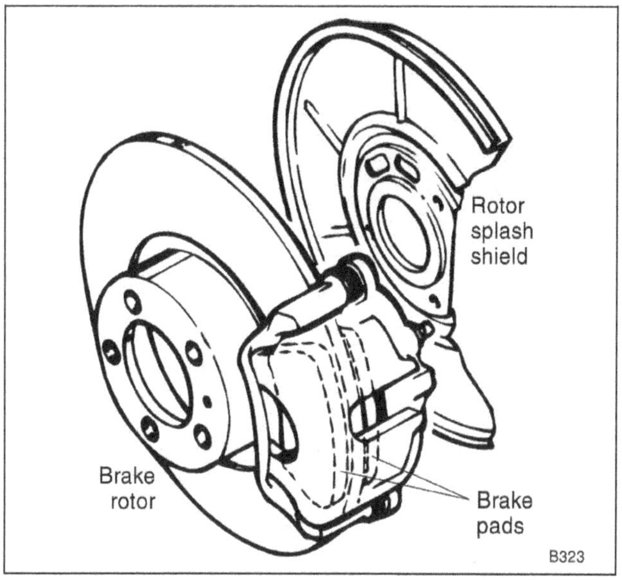

Fig. 24. Disc brake pad wear can checked through opening in caliper.

The brake caliper should be unbolted from the suspension strut to properly inspect the brake pads and the rotors. See Fig. 25. Brake caliper removal and installation procedures are given in **340 Brakes**.

Brake Pad Lining Minimum Thickness (Dimension A)
- Front and rear disc brake pads. . . . 3.0 mm (0.12 in)

Brake fluid, replacing

BMW strictly recommends replacing the brake fluid every two years. This will help protect against corrosion and the effects of moisture in the fluid. See **340 Brakes** for brake fluid flushing procedures.

Parking brake, checking

The parking brake system is independent of the main braking system and may require periodic adjustment depending on use. Adjust the parking brake if the brake lever can be pulled up more than 8 clicks. Check that the cable moves freely. A description of the parking brake and parking brake adjustment can be found in **340 Brakes**.

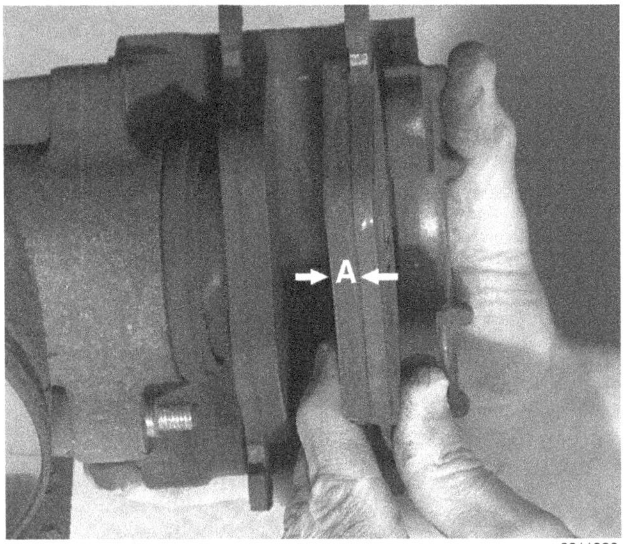

Fig. 25. Inspect brake pads with the caliper removed. Minimum brake pad thickness shown by dimension (**A**).

NOTE —
The parking brake may lose some of its effectiveness if it is not used frequently. This is due to corrosion build-up on the parking brake drum. To remove corrosion, apply the parking brake just until it begins to grip, then pull the lever up one more stop (click). Drive the car approximately 400 meters (1,300 ft.) and release the brake. To recheck the adjustment of the parking brake see **340 Brakes**.

Exhaust system, inspecting

Exhaust system life varies widely according to driving habits and environmental conditions. If short-distance driving predominates, the moisture and condensation in the system will not fully dry out. This will lead to early corrosion damage and more frequent replacement.

Scheduled maintenance of the exhaust system is limited to inspection. Check for restrictions due to dents or kinks. Check for weakness or perforation due to rust. Check to see that all the hangers are in place and properly supporting the system and that the system does not strike the body. Alignment of the system and the location of the hangers are described in **180 Exhaust System**.

Manual Transmission Service

Manual transmission service consists of inspecting for leaks and checking and changing the fluid.

Evidence of transmission leaks is likely to be seen around the driveshaft mounting flange and at the bottom of the bellhousing. For more information on identifying oil leaks and their causes, see **230 Manual Transmission** and **210 Clutch**.

Manual transmission fluid, checking and filling

The manual transmission fluid level should be checked at specified intervals. Check and fill the transmission with the car on a level surface. Transmission fluid level checking and replacement procedures are covered in **230 Manual Transmission**.

Automatic Transmission Service

The automatic transmission is not equipped with a dipstick. Therefore, checking the ATF level is an involved procedure, which includes measuring and maintaining a specified ATF temperature during the checking procedure.

For more complete ATF service, including checking ATF level and ATF filter replacement procedures, see **240 Automatic Transmission**.

Front suspension and steering linkages, inspecting

Inspection of the front suspension and steering includes a check of all moving parts for wear and excessive play. Also inspect the rubber seals and boots for cracks or tears that could allow the entry of dirt, water, and other contaminants. See **310 Front Suspension**.

Rear suspension, inspecting

Final drive and rear drive axle service consists of checking and changing the gear oil, inspecting for leaks, and checking the rear drive axle rubber boots for damage.

The areas where leaks are most likely to occur are around the drive shaft and drive axle mounting flanges. For more information on identifying oil leaks and their causes, see **330 Rear Suspension**.

Final drive oil level, checking

Check the lubricant level with the car level. Remove the oil filler plug. The level is correct when the fluid just reaches the edge of the filler hole. Install and tighten the oil filler plug when the oil level is correct. See Fig. 26.

The final drive should be filled with a special BMW lubricant available through an authorized BMW dealer. In addition, the lubricant type varies depending on whether or not a limited-slip differential is fitted.

Tightening Torque
- Final drive filler plug to final drive housing .... 70 Nm (52 ft-lb)

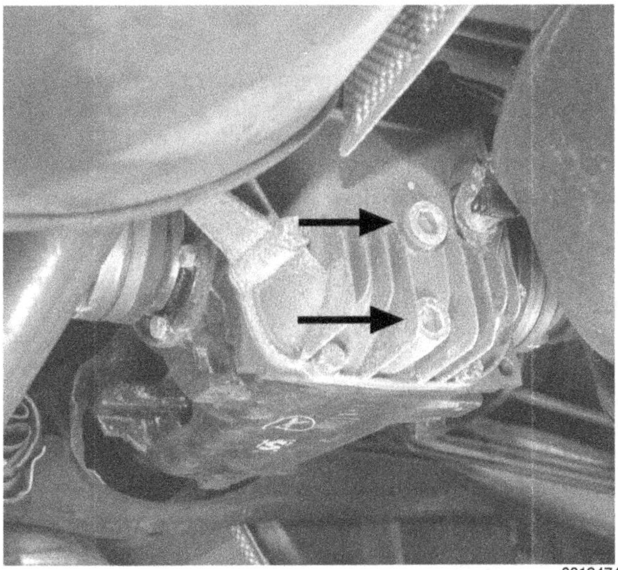

0012474

Fig. 26. Final drive oil filler and oil drain plugs.

NOTE —

- *Use a 14 mm or 17 mm allen bit socket to remove the drain plug.*
- *If the car is raised in the air, it should be level.*

Fuel tank and fuel lines, inspecting

Inspect the fuel tank, fuel lines, and fuel system for damage or leaks. Check for fuel leaks in the engine compartment or fuel odors in the passenger compartment. Check for faulty fuel lines by bending them. If any leaks are present, fuel should be expelled. Check for any evaporative emissions hoses that may have become disconnected, checking carefully at the charcoal canister and evaporative emissions purge system. See **130 Fuel Injection** and **160 Fuel Tank and Fuel Pump** for component locations and additional information.

> ***WARNING —***
> *When checking for fuel leaks, the engine must be cold. A hot exhaust manifold or exhaust system could cause the fuel to ignite or explode causing serious personal injury. Ventilate the work area and clean up spilled fuel immediately.*

Clutch fluid, checking

The hydraulic clutch and the brake system share the same reservoir and the same brake fluid. Clutch fluid level and brake fluid level are checked at the same time. See **340 Brakes** for more information. See **210 Clutch** for information on the clutch and the hydraulic clutch operating system.

Drive axle joint boots, inspecting

The protective boots must be closely inspected for cracks and any other damage that will allow contaminants to get into the joint. If the rubber boots fail, the water and dirt that enter the joint will quickly damage it. Replacement of the drive axle joint boots and inspection of the joints are described in **330 Rear Suspension**.

BODY AND INTERIOR MAINTENANCE

Windshield Wiper Blade Maintenance

Common problems with the windshield wipers include streaking or sheeting, water drops after wiping, and blade chatter. Streaking is usually caused when wiper blades are coated with road film or car wash wax. Clean the blades using soapy water. If cleaning the blades does not cure the problem then they should be replaced. BMW recommends replacing the wiper blades twice a year, before and after the cold season. On older cars, check the tension spring that holds the wiper to the glass. Replace the wiper arm if the springs are weak.

Drops that remain behind after wiping are caused by oil, road film, or diesel exhaust coating the windshield. Use an alcohol or ammonia solution, or a non-abrasive cleanser to clean the windshield.

Wiper blade chatter may be caused by dirty or worn blades, by a dirty windshield, or by bent or twisted wiper arms. Clean the blades and windshield as described above. Adjust the wiper arm so that there is even pressure along the blade, and so that the blade is perpendicular to the windshield at rest. Lubricate the wiper linkage with a light oil. The linkage is located under the hood on the driver's side. If the problem persists, the blades are excessively aged or worn and should be replaced. See **611 Wipers and Washers**.

Body and hinges, lubricating

The door locks and lock cylinders can be lubricated with an oil that contains graphite.

The body and door hinges, the hood latch, and the door check rods should be lubricated with SAE 30 or SAE 40 engine oil. Lubricate the seat runners with multipurpose grease. Do not apply any oil to rubber parts. If door weatherstrips are sticking, lubricate them with silicone spray or talcum powder. The hood release cable should be lubricated as well.

The use of winter lock de-icer sprays should be kept to an absolute minimum, as the alcohol in the de-icer will wash the grease out of the lock assemblies, and may cause the locks to corrode internally, or become difficult to operate.

Seat Belts

Dirt and other abrasive particles will damage seat belt webbing. If it is necessary to clean seat belts, use a mild soap solution. Bleach and other strong cleaning agents may weaken the belt webbing and should be avoided.

> *WARNING —*
> *Do not clean the seat belt webbing using dry cleaning or other chemicals. Allow wet belts to dry before allowing them to retract.*

The condition of the belt webbing and the function of the retractor mechanisms should be inspected. See **720 Seat Belts** for seat belt inspection information.

Exterior Washing

The longer dirt is left on the paint, the greater the risk of damaging the glossy finish, either by scratching or by the chemical effect dirt particles may have on the painted surface.

Do not wash the car in direct sunlight. If the engine hood is warm, allow it to cool. Beads of water not only leave spots when dried rapidly by the sun or heat from the engine, but also can act as small magnifying glasses and burn spots into the finish. Wash the car with a mixture of lukewarm water and a car wash product. Rinse using plenty of clear water. Wipe the body dry with a soft cloth towel or chamois to prevent water-spotting.

Waxing

For a long-lasting, protective, and glossy finish, apply a hard wax after the car has been washed and dried. Use carnauba or synthetic based products. Waxing is not needed after every washing. You can tell when waxing is required by looking at the finish when it is wet. If the water coats the paint in smooth sheets instead of forming beads that roll off, a new coat of wax is needed. Wax should not be applied to black trim pieces, rubber, or other plastic parts.

Polishing

Use paint polish only if the finish assumes a dull look after long service. Polish can be used to remove tar spots and tarnish, but afterwards a coat of wax should be applied to protect the clean finish. Do not use abrasive polish or cleaners on aluminum trim or accessories.

Washing Chassis

Periodic washing of the underside of the car, especially in winter, will help prevent accumulation of road salt and rust. The best time to wash the underside is just after the car has been driven in wet conditions. Spray the chassis with a powerful jet of water. Commercial or self-service car washes may not be best for this, as they may recycle the salt-contaminated water.

Special Cleaning

Tar spots can be removed with a bug and tar remover. Never use gasoline, kerosene, nail polish remover, or other unsuitable solvents. Insect spots also respond to tar remover. A bit of baking soda dissolved in the wash water will facilitate their removal. This method can also be used to remove spotting from tree sap.

Interior Care

Dirt spots can usually be removed with lukewarm soapy water or a dry foam cleaner. Use spot remover for grease and oil spots. Do not pour the liquid directly on the carpet or fabric, but dampen a clean cloth and rub carefully, starting at the edge of the spot and working inward. Do not use gasoline, naptha, or other flammable substances.

Leather Upholstery and Trim

Leather upholstery and trim should be periodically cleaned using a slightly damp cotton or wool cloth. The idea is to get rid of the dirt in the creases and pores that can cause brittleness and premature aging. On heavily soiled areas, use a mild detergent (such as Woolite®) or other specially formulated leather r cleaners. Use two tablespoons to one quart of cold water. Dry the trim and upholstery completely using a soft cloth. Regular use of a good quality leather conditioner will reduce drying and cracking of the leather.

100 Engine–General

TABLES

GENERAL

There are various engine configurations used in the 1992-1998 E36 cars. See **Table a**.

Table a. Engine Specifications

Model	Engine code	No. of cylinders	Displacement liters (cu. in.)	Compression ratio	Horsepower SAE net @ rpm
318i/is/iC 1992–1995 1996–1998	 M42 M44	 4 4	 1.8 (109.6) 1.9 (115.6)	 10.0:1 10.0:1	 138 @ 6000 138 @ 6000
323is/iC 1998	 M52	 6	 2.5 (152.2)	 10.5:1	 168 @ 5,500
325i/is/iC 1992-1995	 M50	 6	 2.5 (152.2)	 10.5:1	 189 @ 5,900
328i/is/iC 1996-1998	 M52	 6	 2.8 (170.4)	 10.2:1	 190 @ 5,300
M3 1995 1996-1998	 S50US S52US	 6 6	 3.0 (182.5) 3.2 (192.3)	 10.5:1 10.5:1	 240 @ 6,000 240 @ 6,000

Cylinder Block and Crankshaft

On both four- and 6-cylinder engines, the cylinder block is cast iron with integral cylinders. The cylinders are exposed on all sides to circulating coolant.

The fully counterweighted crankshaft rotates in replaceable split-shell main bearings. Oilways drilled into the crankshaft provide bearing lubrication. Oil seals pressed into alloy seal housings are installed at both ends of the crankshaft.

Connecting Rods and Pistons

The forged connecting rods use replaceable split-shell bearings at the crankshaft end and solid bushings at the piston pin end. The pistons are of the three-ring type with two upper compression rings and a lower one-piece oil scraper ring. Full-floating piston pins are retained with circlips.

Cylinder Head and Valvetrain

The aluminum cylinder head uses chain-driven double overhead camshafts and four valves per cylinder. See Fig. 1. The cylinder head employs a crossflow design for greater power and efficiency. Intake air enters the combustion chamber from one side while exhaust gasses exit from the other. Oilways in the head provide lubrication for the camshaft(s) and valvetrain.

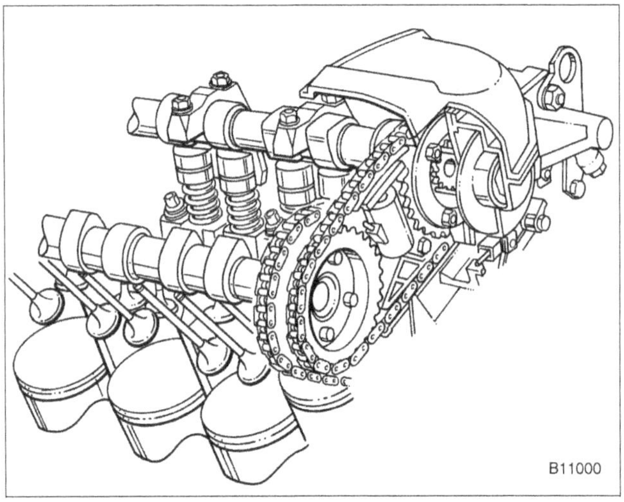

Fig. 1. M52 twin-cam, 4-valve-per-cylinder engine with hydraulic lifters.

On all engines except the M44 engine, valve clearance is by self-adjusting hydraulic lifters. On M44 engines, instead of hydraulic lifters, hydraulic pedestals are used in combination with roller rocker arms to actuate the valves. Hydraulic pedestals have the same function as hydraulic lifters, which is to maintain zero valve clearance, reduce valve noise, and eliminate routine adjustment. See Fig. 2.

VANOS (Variable Valve Timing)

1993 and later 6-cylinder engines are equipped with a variable intake valve timing system, known as VANOS (from the German words Variable Nockenwellen Steuerung). The VANOS system electro-hydraulically adjusts intake valve timing for enhanced mid-range performance. The VANOS system is controlled by the engine control module (ECM), using engine speed, engine load and engine temperature as the primary inputs.

At low speeds, the intake valves open late to ensure smooth engine operation. At mid-range speeds, the valves open early (valve timing advanced, VANOS actuated) for increased torque, improved driveability, and reduced emissions. And at high speeds, the valves again open late for optimum power and performance.

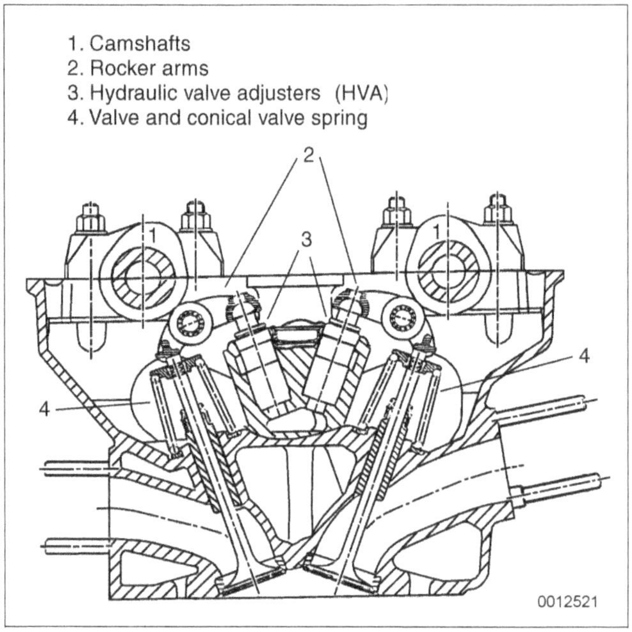

Fig. 2. Cross section of M44 twin-cam, 4-valve-per-cylinder head. Note function of hydraulic pedestal in combination with rocker arm (with roller bearing for reduced friction).

The main components of the VANOS system are the piston housing with integral spool valve and solenoid, and the modified intake camshaft and sprocket assembly. See Fig. 3.

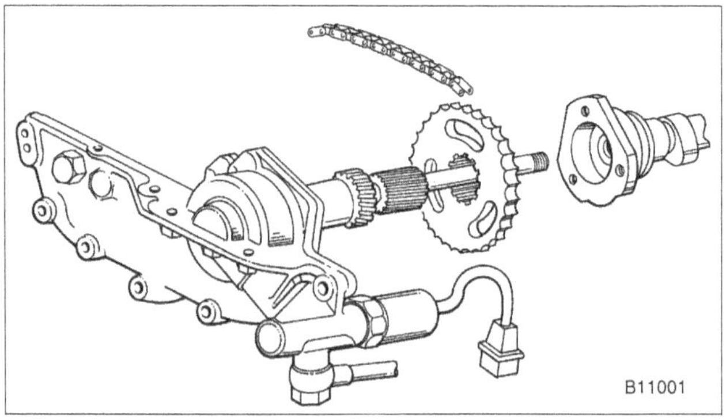

Fig. 3. VANOS (variable intake valve timing) system used on M52 engine. When solenoid is actuated, oil pressure is directed to front side of gear cup piston. This forces gear cup into camshaft to advance intake valve timing.

When the engine is running, the piston housing is supplied with pressurized engine oil via the solenoid-actuated spool valve. Depending on the position of the spool valve, oil is directed to either the front or back side of the gear cup piston.

When the solenoid is in the off position, engine oil is directed to the back side of the piston. This holds the gear cup forward and valve timing is maintained at the normal "late" position. When the solenoid is energized, the spool valve is moved forward and oil pressure is directed to the front side of the piston. This in turn moves the gear cup further into the camshaft secondary drive, causing the camshaft to "advance" 12.5°. The helical gears are cut so that forward motion of the gear cup is translated into rotational motion of the camshaft. See **117 Camshaft Timing Chain** for testing and repair information on the VANOS system.

DISA (Dual Resonance Intake System)

DISA, used on 4-cylinder engines, is a dual intake runner system that effectively provides the advantages of both short and long intake runners within the same engine. For best performance, long intake runners are most beneficial at low- and mid-engine speeds (below 4,200 rpm), and short intake runners enhance torque at high engine speeds (above 4,200 rpm).

> ***NOTE —***
> *The term DISA comes from the German words Differenzierte Sauganlage, and can roughly be translated as a differing intake manifold configuration.*

The DISA solenoid valve is controlled by the DME control module, using engine speed as the primary input. The main components of the system are the modified twin-section intake manifold with change-over valve, the twin-barrel throttle body, and the electrical/pneumatic actuating components. See Fig. 4.

The DISA system electro-pneumatically changes the intake runner length through the twin-section intake manifold and a change-over butterfly valve. The change-over valve is located at a point in the intake manifold where four pipes come into two. When the change-over valve is closed, the 4-way primary intake manifold and the 2-way ram air manifold are connected to make the "long" configuration. When the change-over valve opens, intake air flow is redirected through only the short 4-way primary intake manifold.

The change-over valve is held in the normally open position by a spring. This allows for an open valve in the event of system failure. During low and mid engine speeds, the DME control module supplies power to the solenoid valve, which in turn allows vacuum to be applied to the vacuum diaphragm. This causes the valve to close (long-pipe configuration). When engine speed reaches approximately 4,800 rpm, the DME control module electrically signals the solenoid valve and the valve opens, creating the short pipe configuration.

Further details on DISA can be found in **130 Fuel Injection**.

Engine Management System

All engines covered by this manual use an advanced engine management system called Digital Motor Electronics (DME). In the DME system, advanced On-Board Diagnostics (OBD), fuel injection, ignition, and other functions, are combined under the control of the Engine Control Module (ECM). See Fig. 5.

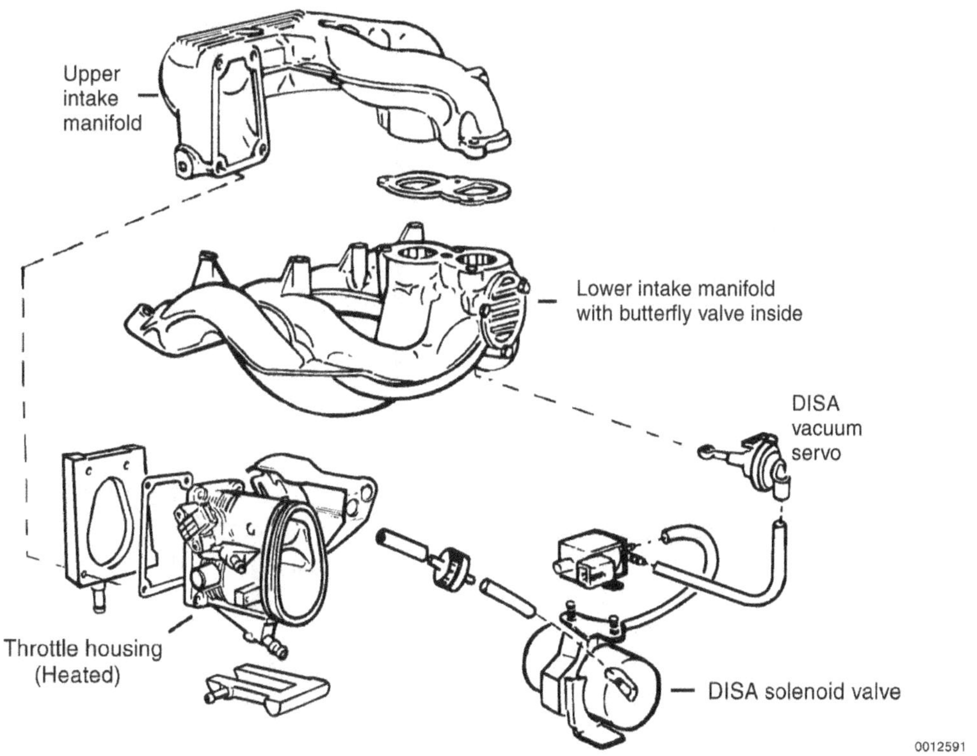

Fig. 4. DISA system components. DISA changes the intake runner length based on engine speed.

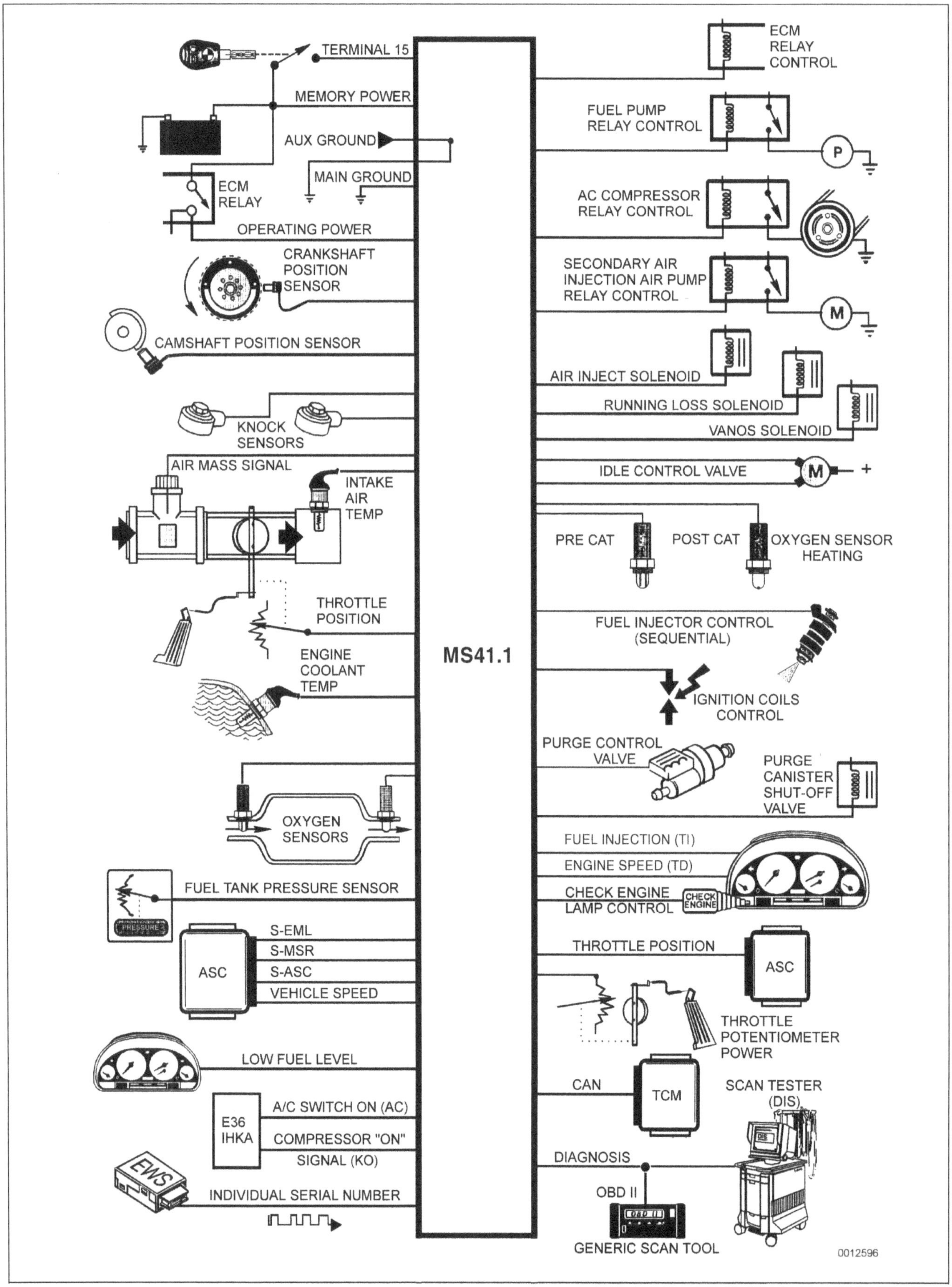

Fig. 5. Siemens MS 41.1 OBD II engine management system used on 1996 and later M52 engines.

Various versions of DME systems are used on the cars covered by this manual. See **Table b**. Each system is highly adaptive to compensate for things such as engine wear and vacuum leaks.

Table b. Engine Management Systems

Engine	System
4-cylinder engines	
M42 (1992-1995)	Bosch DME M1.7
M44 (1996-1998)	Bosch DME M5.2 (OBD II)
6-cylinder engines	
M50 (1992)	Bosch DME M3.1
M50 VANOS (1993-1995)	Bosch DME M3.3.1
M52 (1996-1998)	Siemens MS41.1 (OBD II)
S50US (1995)	Bosch DME M3.3.1
S52US (1996-1998)	Siemens MS41.1 (OBD II)

Ignition

Both the 4-cylinder and 6-cylinder engines use a distributorless ignition system with individual ignition coils for each cylinder.

Fuel Delivery

Pressurized fuel from the in-tank fuel pump is injected via solenoid-type fuel injectors. The ECM controls the opening and closing of the injectors by switching the ground side of each injector circuit. The exact amount of fuel injected is determined by the amount of time the injectors are open.

Cooling System

Whenever the engine is running, a coolant pump circulates coolant through the engine and, if either heater control valves are open, through the heater core in the passenger compartment. The coolant absorbs excess heat and carries it to the radiator where it is transferred into the passing airstream. A thermostat controls the flow of coolant through the radiator based on engine temperature.

Lubrication System

The lubrication system is pressurized whenever the engine is running. The oil pump draws oil through a pickup in the bottom of the oil pan, then forces it through a replaceable oil filter and into the engine oil passages.

On 4-cylinder engines, the oil pump is mounted to the front engine cover. On 6-cylinder engines, the chain-driven oil pump is bolted to the bottom of the cylinder block.

A pressure relief valve limits the maximum system pressure. A bypass valve prevents the oil filter from bursting and insures engine lubrication should the filter become plugged. See **119 Lubrication System** for additional information.

MECHANICAL TROUBLESHOOTING

When troubleshooting an engine that fails to start or runs poorly, first check its mechanical condition—particularly if the engine has high mileage.

Warnings and Cautions

For personal safety, as well as the protection of sensitive electronic components, the following warnings and cautions must be adhered to during all troubleshooting, maintenance, and repair work.

WARNING —

- *The ignition system produces high voltages that can be fatal. Avoid contact with exposed terminals and use extreme caution when working on a car with the ignition switched on or the engine running.*
- *Do not touch or disconnect any high voltage cables from the coil, distributor, or spark plugs while the engine is running or being cranked by the starter*
- *Connect and disconnect the engine management system wiring and test equipment leads only when the ignition is switched off.*
- *Gasoline is highly flammable and its vapors are explosive. Do not smoke or work on a car near heaters or other fire hazards when diagnosing and repairing fuel system problems. Have a fire extinguisher available in case of an emergency.*
- *Disconnecting the battery may erase fault code(s) stored in control module memory. Using special BMW diagnostic equipment, check for fault codes prior to disconnecting the battery cables. If the Check Engine light is illuminated, see* **On-Board Diagnostics (OBD)** *for fault code information. If any other system faults have been detected (indicated by an illuminated warning light), see an authorized BMW dealer.*

CAUTION —

- *Prior to disconnecting the battery, read the battery disconnection cautions given at the front of this manual on page viii.*
- *Do not connect any test equipment that delivers a 12-volt power supply to terminal 15 (+) of the ignition coil. The current flow may damage the ECM. In general, connect test equipment only as specified by BMW, this manual, or the equipment maker.*
- *Do not disconnect the battery with the engine running. Do not run the engine with any of the spark plug wires disconnected.*

Cylinder compression, checking

A compression tester is needed to make a compression test. To obtain accurate test results, the battery and starter must be capable of cranking the engine at 250–300 rpm, and the engine should be at normal operating temperature. Use compressed air to clean around the spark plugs before removal.

1. Disable the ignition system by removing the engine management system main relay and the fuel pump relay. See Fig. 6.

> ***WARNING —***
> *The ignition system produces high voltages that can be fatal. Avoid contact with exposed terminals and use extreme caution when working on a car with the ignition switched on or the engine running.*

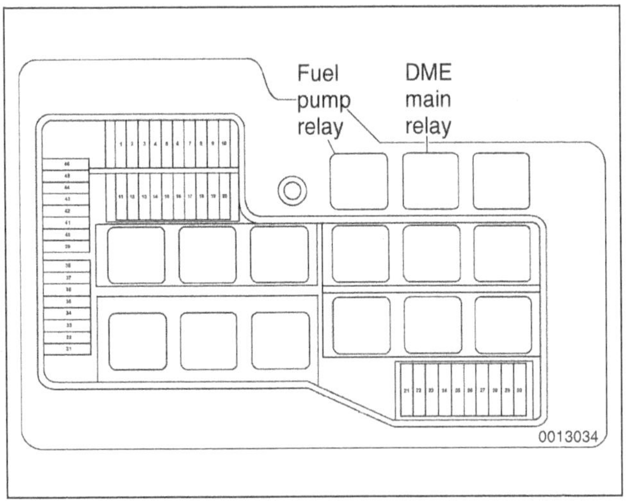

Fig. 6. Engine management relays in power distribution box in left rear of engine compartment.

> ***CAUTION —***
> - *On OBD II cars, making a compression test may cause a fault to set in the ECM and may also illuminate the Check Engine light. The light can only be turned out using special scan tool equipment, using either the BMW special service tool or using a "generic" OBD II scan tool. Disconnecting the battery will not erase the fault memory nor turn out the light.*
> - *Failure to remove the main relay or attempting to disable the ignition system by other methods may result in damage to the engine control module.*

2. On 4-cylinder engine: Remove plastic engine cover from top of cylinder head. Disconnect spark plug wires and remove spark plugs. See Fig. 7.
3. On 6-cylinder engine: Remove top engine cover by prying off nut covers and removing mounting nuts. See Fig. 8.

Fig. 7. Remove spark plug wires from spark plugs on 4-cylinder engine using special tool. Tool is stored under cover at rear of cylinder head (**arrow**).

Fig. 8. Remove engine cover on 6-cylinder engine by prying off nut covers and removing nuts (**A**). Be careful not to let rubber insulators fall off as cover is removed (**arrows**).

4. On 6-cylinder engines, disconnect the harness connectors from the ignition coils by lifting the retaining clip. Remove the coil mounting nuts and remove the six coils. See Fig. 9. Remove the spark plugs.

> ***NOTE —***
> *Used spark plugs should be reinstalled in the same cylinder from which they were removed.*

5. Install the compression gauge in the first cylinder's spark plug hole, tight enough to form a good seal.

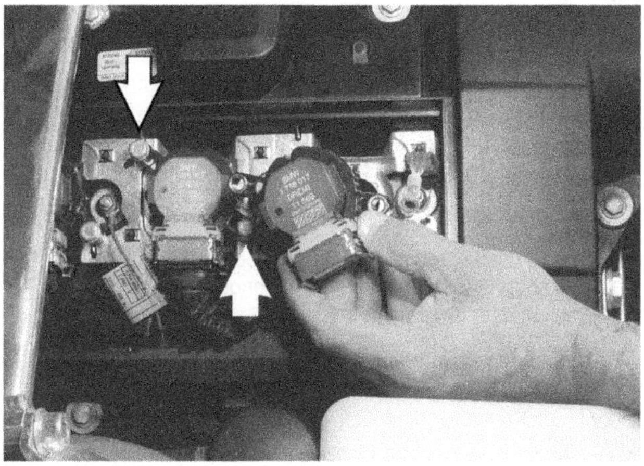

B11004

Fig. 9. Remove ignition coils on 6-cylinder engine by disconnecting harness connector and removing mounting bolts (**arrows**).

6. With the parking brake set, the transmission in Park or Neutral, and the accelerator pedal pressed to the floor, crank the engine with the starter. Record the highest value indicated by the gauge.

NOTE —

- *The compression gauge reading should increase with each compression stroke and reach near its maximum reading in about 4-6 strokes.*
- *All cylinders should reach maximum compression in the same number of strokes. If a cylinder needs significantly more strokes to reach maximum compression, there is a problem.*

7. Release the pressure at the compression gauge valve, then remove the gauge from the spark plug hole. Repeat the test for each of the other cylinders and compare the results with the values given below.

Compression Pressure

- Minimum. . . . . 10-11 bar (142-156 psi)
- Maximum difference between cylinders. . . . . 0.5 bar (7 psi)

Reinstall the spark plugs and spark plug wires or ignition coils. The remainder of installation is the reverse of removal. Be sure to reinstall all wires disconnected during the test, especially ground wires at the coils and cylinder head cover (where applicable).

Tightening Torque

- Spark plug to cylinder head . . . . 25 Nm (18 ft-lb)

Low compression indicates a poorly sealed combustion chamber. Relatively even pressures that are below specification normally indicate worn piston rings and/or cylinder walls. Erratic values tend to indicate valve leakage. Dramatic differences between cylinders are often the sign of a failed head gasket, burned valve, or broken piston ring.

Engine Mechanical Troubleshooting Table

Table c lists the symptoms of common engine mechanical problems, their probable causes and the suggested corrective actions. The bold type indicates the repair groups where applicable test and repair procedures can be found.

Table c. Engine Mechanical Troubleshooting

Symptom	Probable cause and corrective action
1. Engine will not start or run. Starter cranks engine at normal speed.	a. No fuel reaching engine. Check for fuel in tank. Fuel pump not operating. Repair Group **160** b. No spark. Check ignition system. Repair Group **120** c. Incorrect valve timing. Check timing chain. Repair Group **117**
2. Check Engine warning light illuminated.	a. DME engine management system self-diagnostic fault detected. See an authorized BMW dealer or other qualified shop for OBD II fault code diagnosis. See also **On-board Diagnostics (OBD)**, given later in this repair group.
3. White exhaust smoke (steam).	a. Failed cylinder head gasket (probably accompanied by low compression readings). Replace gasket and resurface cylinder head if necessary. Repair Group **113** b. Warped or cracked cylinder head. Resurface or replace cylinder head. Repair Group **113** c. Cracked cylinder block. Replace engine or short block.
4. Blue-gray exhaust smoke and oily spark plugs. Indicates oil burning in combustion.	a. Contaminated or gasoline-diluted engine oil. Change engine oil and replace oil filter. Repair Group **020** b. Faulty valve guide seals or valve guides. Replace valve guide oil seals or overhaul cylinder head. Repair Group **116** c. Worn piston rings. Overhaul or replace engine. d. Failed cylinder head gasket. Replace gasket. Repair Group **113**
5. Pinging or rattling when under load, traveling uphill, or accelerating, especially from low speeds. Indicates pre-ignition or detonation.	a. Ignition timing too advanced. Ignition knock sensors faulty (Check Engine light illuminated). Repair Group **120** b. Fuel octane level too low. c. Engine running too hot or overheating. Check cooling system. Repair Group **170** d. Air/fuel mixture too lean (Check Engine light illuminated). Repair Group **130**
6. Light metallic tapping that varies directly with engine speed. Oil pressure warning light **not** illuminated.	a. Valve clearances excessive. Check hydraulic valve adjusters. Repair Group **116** b. Low oil pressure or possibly defective warning light circuit. Check oil pressure and warning light circuit. Repair Group **119** c. Dirty oil. Change oil. Repair Group **020**.
7. Light metallic knock that varies directly with engine speed. Oil pressure warning light blinking or fully illuminated; may be most noticeable during hard stops or cornering.	a. Low oil level. Check and correct oil level. b. Low oil pressure/worn or faulty oil pump. Oil pump pickup restricted/clogged. Check oil pressure. Repair Group **119**
8. Screeching or squealing under load that goes away when coasting.	a. Loose, worn, or damaged drive belt. Inspect belt(s). Repair Group **020**
9. Growling or rumbling that varies with engine rpm. Indicates a bad bearing or bushing in an engine-driven accessory.	a. Remove drive belt(s) to identify faulty component. Check for play, bearing roughness, and loose mountings. Refer to appropriate repair group for repair procedures.

DRIVEABILITY TROUBLESHOOTING

On-Board Diagnostics is incorporated into all engine management systems used on the cars covered by this manual. Therefore, checking for fault codes should be the first step in troubleshooting a driveability problem. For additional information on engine management and repair see **130 Fuel Injection**.

On-Board Diagnostics (OBD)

Two generations of OBD are used on the cars covered by this manual:

On-Board Diagnostics

- 1992-1995 models .... OBD I
- 1996-1998 models .... OBD II

The OBD system detects emissions-related engine management malfunctions and turns on the Check Engine light in the instrument cluster. The Check Engine light will stay on until the code(s) are called up and the memory is erased. When faults are detected, the OBD system stores a Diagnostic Trouble Code (DTC) in the system ECM. In addition, the Check Engine warning light will come on if an emissions-related fault is detected.

NOTE —

- *On cars with OBD II, specialized OBD II scan tool equipment must be used to access the DTCs, either using the BMW special tool or a "generic" OBD II scan tool. See* **130 Fuel Injection**.
- *The OBD II fault memory (including an illuminated Check Engine light) can only be reset using the special scan tool. Removing the connector from the ECM or disconnecting the battery will not erase the fault memory.*

On cars with OBD I, fault codes can be read by turning the ignition key on and fully depressing the accelerator pedal to the floor five times within five seconds. The Check Engine light will then remain on for five seconds, blink off, come on for 2.5 seconds, then go off for 2.5 seconds. At this point, the fault codes will begin to flash. See **Table d**.

If more than one fault has been detected, each code will be separated by a 2.5 second pause. When all fault codes have been displayed, there will be 0.5 second flash and then the light will remain off. To read the codes again, turn the key off and then on again. Depress the accelerator pedal to the floor five times quickly (within 5 seconds). The codes will begin to repeat.

To erase the fault code memory, first make sure the fault code 1000 (short blink and then light goes out for long period) is present, then depress the throttle fully for at least 10 seconds. Read the fault codes as described earlier and check for the 1444 code (no faults stored).

Table d. OBD I Fault (Blink) Codes (1992-1995 models only)

Graphic representation of flashing Check Engine light fault code: 1221

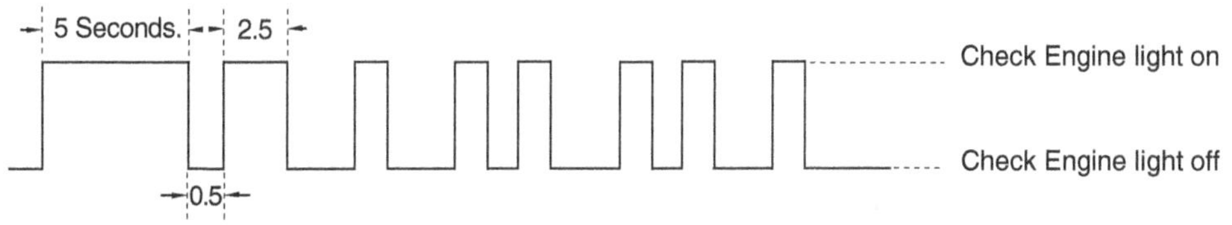

Fault code and meaning	Corrective action
Code 1211: DME Control Module	DME control module may be faulty. This code appears if the control module fails the self test. Check inputs to control module. Repair Group **130**
Code 1215: Mass air flow sensor	Check air flow or mass air flow sensor and wiring to sensor. Repair Group **130**
Code 1216: Throttle potentiometer	Test throttle potentiometer resistance and wiring. Repair Group **130**
Code 1218: Output stage, group #1 (DME 3.3.1only)	Test DME control module input/outputs. Repair Group **130**
Code 1219: Output stage, group #2 (DME 3.3.1 only)	Test DME control module input/outputs. Repair Group **130**
Code 1221: Oxygen sensor #1	Check sensor output signal to DME control module. Repair Group **130**
Code 1222: Oxygen sensor lean/rich control stop	Check for intake air leaks or reasons for rich mixture. See Driveability Troubleshooting given earlier in this repair group
Code 1223: Coolant temperature sensor (NTC)	Test coolant temperature sensor. Repair Group **130**
Code 1224: Air temperature sensor (NTC)	Test intake air temperature sensor. Repair Group **130**
Code 1225: Knock sensor #1 (DME 3.3.1 only)	Check knock sensor and sensor wiring. Repair Group **120**
Code 1226: Knock sensor #2 (DME 3.3.1 only)	Check knock sensor and sensor wiring. Repair Group **120**
Code 1231: Battery voltage monitor	Test battery voltage and battery, check charging system and starter. Repair Group **121**
Code 1234: Speedometer "A" signal (DME 3.3.1 only)	Check wiring between instrument cluster and DME control module. **Electrical Wiring Diagrams**
Code 1237: A/C compressor cutoff (DME 3.3.1 only)	Test DME control module inputs/outputs from A/C system. Repair Group **130**
Code 1242: A/C compressor signal (DME 3.3.1 only)	Test DME control module inputs/outputs from A/C system. Repair Group **130**

Table d. OBD I Fault (Blink) Codes (continued)
(1992-1995 models only)

Graphic representation of flashing Check Engine light fault code: 1221

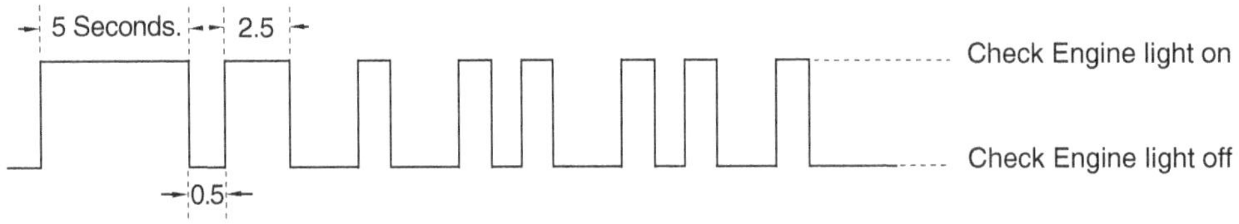

Fault code and meaning	Corrective action
Code 1243: Crankshaft position sensor (DME 3.3.1 only)	Test crankshaft position/rpm sensor and wiring from sensor to DME control module. Repair Group **120**
Code 1244: Camshaft position sensor (DME 3.3.1 only)	Test camshaft position sensor and wiring from sensor to DME control module. Repair Group **130**
Code 1245: Electronic transmission control intervention (DME 3.3.1 only)	Check wiring between DME control module and auto. transmission control module
Code 1247: Ignition secondary monitor (DME 3.3.1 only)	Check secondary voltage to ignition coils. Check wiring at ignition coils. Repair Group **120**
Code 1251: Fuel injector #1 (DME 3.1/3.3.1 only)	Test injector operation/signal to injector. Repair group **130**
Code 1252: Fuel injector #2 (DME 3.1/3.3.1 only)	Test injector operation/signal to injector. Repair group **130**
Code 1253: Fuel injector #3 (DME 3.1/3.3.1 only)	Test injector operation/signal to injector. Repair group **130**
Code 1254: Fuel injector #4 (DME 3.1/3.3.1 only)	Test injector operation/signal to injector. Repair group **130**
Code 1255: Fuel injector #5 (DME 3.1/3.3.1 only)	Test injector operation/signal to injector. Repair group **130**
Code 1256: Fuel injector #6 (DME 3.1/3.3.1 only)	Test injector operation/signal to injector. Repair group **130**
Code 1261: Fuel pump control	Test fuel pump relay and fuel pump circuit. Repair group **160**
Code 1262: Idle speed control	Test idle air control valve and signal to valve. Repair Group **130**
Code 1263: Fuel tank evaporative (EVAP) system	Test EVAP purge valve. Repair Group **160**
Code 1264: Oxygen sensor heater	Test oxygen sensor heater and heater relay. Repair group **130**
Code 1265: Check engine lamp (DME 3.3.1 only)	Test for faulty bulb or wiring. **Electrical Wiring Diagrams**
Code 1266: VANOS (DME 3.3.1 only)	Test VANOS solenoid. Check for signal to VANOS solenoid.
Code 1267: Air pump relay control (DME 3.3.1 only)	Test air pump relay and wiring (where applicable) **Electrical Wiring Diagrams**
Code 1271: Ignition coil #1 (DME 3.3.1 only)	Test ignition coil and wiring to ignition coil. Repair Group **120**
Code 1272: Ignition coil #2 (DME 3.3.1 only)	Test ignition coil and wiring to ignition coil. Repair Group **120**
Code 1273: Ignition coil #3 (DME 3.3.1 only)	Test ignition coil and wiring to ignition coil. Repair Group **120**
Code 1274: Ignition coil #4 (DME 3.3.1 only)	Test ignition coil and wiring to ignition coil. Repair Group **120**
Code 1275: Ignition coil #5 (DME 3.3.1 only)	Test ignition coil and wiring to ignition coil. Repair Group **120**
Code 1276: Ignition coil #6 (DME 3.3.1 only)	Test ignition coil and wiring to ignition coil. Repair Group **120**
Code 1281: DME control module memory supply (DME 3.3.1 only)	Check voltage supply from battery to DME control module. **Electrical Wiring Diagrams**
Code 1282: Fault code memory (DME 3.3.1 only)	Check DME control module inputs/outputs. Control module may be faulty. Repair Group **130**
Code 1283: Fuel injector output stage (DME 3.3.1 only)	Check DME control module inputs/outputs. Control module may be faulty. Repair Group **130**

Table d. OBD I Fault (Blink) Codes (continued)
(1992-1995 models only)

Graphic representation of flashing Check Engine light fault code: 1221

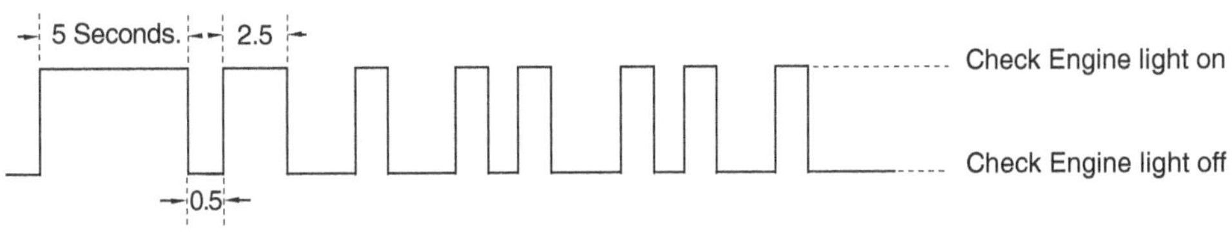

Fault code and meaning	Corrective action
Code 1286: Knock control test pulse (DME 3.3.1 only)	Check DME control module inputs/outputs. Control module may be faulty. Repair Group **130**
Code 1000 (light remains off):	End of fault code output—all fault codes have been displayed. No corrective action necessary. Repeat test if necessary
Code 1444:	No more faults. No corrective action necessary. This code must be present to erase fault memory

Basic Requirements

The following list contains basic checks that should be made when experiencing driveability problems.

1. Check intake (induction) system for leaks. Check for cracked, loose, or disconnected hoses and duct work. Check that all hose clamps are tight.

   NOTE —
   An air leak allows unmeasured air to enter the engine, often resulting in an in overly lean fuel mixture and causing driveability problems which will cause the Check Engine light to come on.

2. Check that the battery is in good condition. Check that the cables are tight and free of corrosion at both ends. Check that all related ground points are firmly connected and in good condition. Check all harness connectors for damage and corrosion.

3. Check for power and ground at the Engine Control Module (ECM). Check the main grounds for the ECM. See **130 Fuel Injection**.

4. Check the fuses. Check for sufficient fuel in the tank. If the engine ran out of fuel, it will take a little time to restore fuel pressure. See **160 Fuel Tank and Fuel Pump**.

5. Check for spark at the spark plugs. If the tachometer needle bounces while the engine is cranked by the starter then the ignition system is probably working correctly. See **120 Ignition System**.

6. Check for any faults through the On-Board Diagnostics system. See **On-Board Diagnostics (OBD)**.

Preventive Maintenance

The condition of the fuel, ignition and emission control system components has a direct effect on engine performance and driveability. BMW specifies maintenance of certain parts at regular intervals to keep the engine in proper tune. Extending maintenance intervals beyond the time or mileage recommended by the manufacturer can adversely affect the way the engine runs. When troubleshooting driveability problems, a good starting point is to perform a major engine service, particularly if one is overdue. For maintenance schedules, major engine service and other driveability-related maintenance procedures, see **020 Maintenance Program**.

Basic Engine Settings

Idle speed, idle mixture (%CO), and ignition timing are not adjustable. The adaptive engine management system is designed to automatically compensate for changes in engine operating conditions, although the adaptive range is limited. Once these limits are exceeded, driveability problems usually become noticeable.

NOTE —
If the DME adaptive limits are exceeded, the Check Engine light will come on, indicating an emission related fault. See **130 Fuel Injection**.

Oxygen Sensors

A high oxygen level in the engine exhaust indicates a lean air-fuel mixture and a low oxygen level indicates a rich mixture. The oxygen sensor, shown in Fig. 10, measures the oxygen content in the exhaust gas and generates a variable voltage signal. Using that feedback signal as an input, the DME control module fine tunes the air-fuel mixture.

NOTE —

The oxygen sensor signal is ignored until the engine reaches a specified minimum temperature. Therefore, when troubleshooting cold engine driveability problems, the oxygen sensor can be ruled out as a possible cause.

As the oxygen sensor ages, its ability to react quickly to changing conditions deteriorates, and it may eventually cease to produce any signal at all. To check the operation of the oxygen sensor, see **130 Fuel Injection**.

NOTE —

On 1996 and later cars, OBD II enhanced emission standards require the engine control module (ECM) to monitor the oxygen content in the exhaust both before and after the catalytic converter. This allows for tighter control of the tail pipe emissions and also allows the ECM to diagnose converter problems. If the DME detects that catalytic converter or oxygen sensor efficiency has degraded past a certain pre-programmed limit, it will turn on the Check Engine light, and store a diagnostic trouble code (DTC) in the ECM.

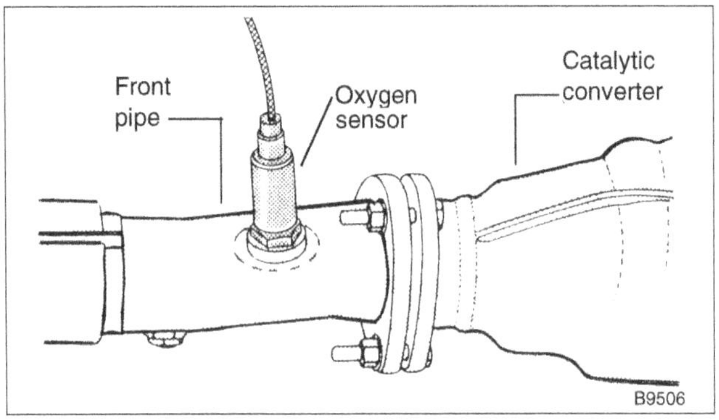

Fig. 10. Regulating oxygen sensor measures oxygen content of exhaust gas as indicator of air-fuel ratio and combustion efficiency. Monitoring oxygen sensor (not shown) monitors catalytic converter operation.

Replacement of oxygen sensors at the specified intervals ensures that the engine and emission control system will continue to operate as designed.

Air Flow Measurement and Vacuum Leaks

The primary input used by the fuel injection system to determine how much fuel should be delivered is the signal from the mass air flow sensor in the intake air ductwork.

Because proper fuel metering depends on accurate measurement of the incoming air, any unmeasured air that enters the engine downstream of the mass air flow sensor will cause a lean air-fuel mixture and possibly affect driveability.

NOTE —

If a large air leak is causing driveability problems, the Check Engine light will usually be illuminated and a fault will be stored in the fault code memory. See **On Board Diagnostics (OBD)**.

There are many places for unmeasured air to enter the engine. First, inspect all hoses, fittings, ducts, seals, and gaskets in the intake air tract for cracks or looseness. It may be necessary to remove parts that cannot be fully checked in their installed positions. Also make sure all of the intake manifold and throttle body mounting nuts are tight.

In addition to air leaks, air restrictions can also cause driveability problems. Remove the air filter and hold it up to a strong light source. If the filter does not pass light, it is restricted and should be replaced. Also, remove the intake air ductwork and check for oily deposits at the throttle plate. Buildup in this area can cause an erratic idle. Clean away any deposits using a cloth wetted with carburetor cleaner and reinstall the ductwork.

CAUTION —

Do not spray carburetor cleaner directly in the throttle area. Spraying cleaner may force it into the throttle position switch or sensor and cause damage.

Battery Voltage

All of the electrical components in the fuel injection and ignition systems require a minimum voltage to operate properly. When troubleshooting engine driveability problems, one of the first checks should be to make sure the battery is fully charged and capable of delivering all its power to the electrical system.

To make a quick check of battery charge, measure the voltage across the battery terminals with all cables attached and the ignition off. A fully charged battery will measure 12.6 volts or slightly more, compared to 12.15 volts for a battery with a 25% charge.

Even a fully charged battery cannot deliver power unless it is properly connected to the electrical system. Check the battery terminals for corrosion and loose cable connections. If the battery does not maintain the proper voltage, the charging system may be at fault. See **121 Battery, Starter, Alternator**.

If a battery cable connection has no visible faults, but is still suspect, measure the voltage drop across the connection. A large drop indicates excessive resistance, meaning the connection is corroded, dirty, or damaged. Clean or repair the connection and retest.

> ***NOTE —***
> *For instructions on conducting a voltage drop test, and other general electrical troubleshooting information, see* **600 Electrical System—General**.

Wiring and Harness Connections

The electronic fuel injection and ignition systems operate at low voltage and current values, making them sensitive to small increases in resistance. The electrical system is routinely subjected to corrosion, vibration and wear, so faults or corrosion in the wiring harness and connectors are common causes of driveability problems.

Visually inspect all wiring, connectors, switches and fuses in the system. Loose or damaged connectors can cause intermittent problems, especially the small terminals in the ECM connectors. Disconnect the wiring harness connectors to check for corrosion, and use electrical cleaning spray to remove contaminants. Often, simply disconnecting and reconnecting a dirty connector several times will clean the terminals and help to reestablish good electrical contact.

If a wiring harness connector has no visible faults, but is still suspect, measure the voltage drop across the connector. A large drop indicates excessive resistance, meaning the connector is corroded, dirty or damaged. Clean or repair the connector and retest.

Ground Connections

For any electrical circuit to work, it must make a complete path, beginning at the positive (+) battery terminal and ending at the negative (–) terminal. The negative (–) battery cable is attached to the car's chassis. Therefore, any wire or metal part attached to the chassis provides a good ground path back to the negative (–) battery terminal.

Poor ground connections are a major source of driveability problems. If any of the main ground connections for the ignition system or the fuel injection system are faulty, the increased resistance in that circuit will cause problems.

Visually inspect all ground wires and connections for breaks, looseness or corrosion. Be careful because wires sometimes break internally or in areas not easily visible. The main grounds for the DME system are shown in Fig. 11, Fig. 12, and Fig. 13. Also check the main fuel pump ground in the center console, below the emergency brake handle.

If a ground connection has no visible faults, but is still suspect, measure the voltage drop across the connection. A large drop indicates high resistance, meaning the connection is corroded, dirty or damaged. Clean or repair the connection and retest.

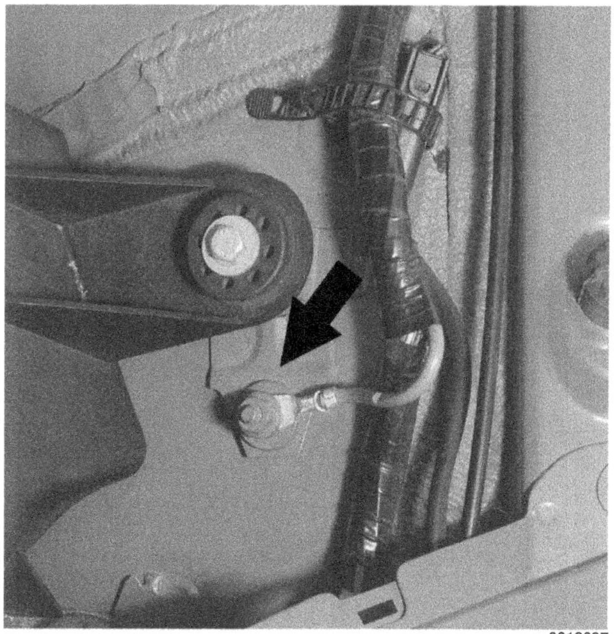

0012697

Fig. 11. Main ground (**arrow**) for engine management system. Grounds are in right rear of engine compartment.

0013130

Fig. 12. Main ground for ignition coils on MS 41.1 engine management system (**arrow**).

0013131

Fig. 13. Main chassis ground (**arrow**) in left front of engine compartment.

Fuel Supply

For the engine to start and run properly, the injection system must deliver fuel in precise proportion to the amount of air entering the engine. To do this, the injection system requires an unrestricted supply of fuel from the fuel pump.

If the fuel pump is not working, the engine will not run. If the fuel filter or a fuel line is restricted, the engine may run poorly. If the restriction is severe enough the engine will not start. If fuel delivery problems are suspected, perform the tests described in **160 Fuel Tank and Fuel Pump**.

The fuel pressure created by the fuel pump is controlled by a pressure regulator that returns excess fuel to the tank. Any change in fuel pressure will cause a change in the base air-fuel mixture delivered to the engine. If the fuel pressure is too low, the base air-fuel mixture will be lean. If the fuel pressure is too high, the base mixture will be rich. Fuel pressure tests are described in **160 Fuel Tank and Fuel Pump** and **130 Fuel Injection**.

> ***NOTE —***
>
> *Fuel pressure tests require a pressure gauge. If this tool is not available, the tests can be performed by an authorized BMW dealer or other qualified shop.*

Properly operating fuel injectors play a major role in fuel delivery. The DME ECM switches the injectors on and off at the negative (–) or ground side of the connectors. Positive (+) battery voltage is always present at the connectors when the engine is running. An injector that fails or loses power will not open, creating a lean air-fuel mixture and causing the engine to run poorly when cold or stumble on acceleration. An injector that shorts to ground will remain open constantly when the engine is running, creating a rich air-fuel mixture that can dilute engine oil, foul the spark plugs, cause a rough idle, and damage the catalytic converter.

Table e lists additional symptoms of common engine driveability problems, their probable causes, and the suggested corrective actions. The entries in bold type in the corrective action column indicate the repair groups where applicable test and repair procedures can be found.

> ***NOTE —***
>
> *Most of the symptoms listed in* **Table e** *will also cause the Check Engine light to come on. If the light is on, check for any stored faults as the first step in troubleshooting driveability complaints.*

Table e. Engine Driveability Troubleshooting

SYMPTOMS

a. Engine fails to start
b. Engine starts but stops immediately
c. Erratic engine idle
d. Poor engine response on acceleration
e. Erratic engine operation in all speed ranges
f. Excessive fuel consumption
g. Poor engine power, fails to rev up
h. CO content too low
i. CO content too high

a	b	c	d	e	f	g	h	i	CAUSES	CORRECTIVE ACTION
a									Fuel pump faulty	Test fuel pump. Repair Group **160**
a				e					Ignition system faulty	Test ignition system. Repair Group **120**
a									Main relay or fuel pump relay faulty	Test main relay. Repair Group **130**
a									Crankshaft/rpm position sensor faulty	Test crankshaft/rpm sensor. **120**
	b	c					h	i	Idle speed control valve faulty	Test idle speed control valve.
		c	d	e					Throttle position sensor faulty or idle signal to DME control module missing.	Test throttle position sensor.
		c	d				h	i	Mass air flow sensor faulty	Test mass air flow sensor Repair Group **130**
a	b	c	d	e			h		Large air intake system leak	Check for major intake air leaks downstream of mass air flow sensor.
a		c	d	e		g	h		Fuel pressure too low	Test fuel pressure. Repair Group **130**
		c			f			i	Fuel pressure too high	Test fuel pressure. Repair Group **130**
			d	e		g	h		Fuel pump delivery volume too low	Test fuel pump delivery volume. Repair Group **160**
a		c		e	f		h	i	Coolant temperature sensor faulty	Test coolant temperature sensor.
		c			f				Oxygen sensor faulty	Test oxygen sensor.
		c		e	f			i	Fuel injectors leaking	Check fuel injectors and replace leaking injectors. Repair Group **130**
		c							Throttle plate binding or incorrectly adjusted	Check throttle plate and adjust if necessary.
a		c							Electrical connections loose, broken, or corroded	Visually inspect connectors and correct any faults. Repair Group **600**
a		c	d						Ground connections loose, broken, or corroded	Visually inspect ground connections and correct any faults. Repair Group **600**
		c		e	f				CO content too high	Test air flow or mass air flow sensor. Repair Group **130**
a	b	c	d	e		g			CO content too low	Test mass air flow sensor.
	b	c	d	e	f	g	h	i	Input signals to ECM missing, ECM in limp home mode	Make electrical tests at DME ECM relay. Repair Group **130**
a	b	c	d	e	f	g	h	i	ECM faulty	Test DME ECM inputs. If all inputs are correct, replace ECM.

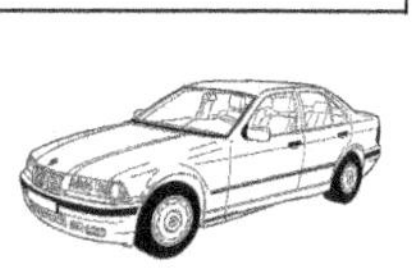

110 Engine Removal and Installation

GENERAL

Some special tools are required for removal and installation of the engine. Be sure to have the necessary equipment on hand before starting the job.

CAUTION—
- *Engine removal requires disconnecting the battery. This may erase any system fault code(s) that have been stored in control unit memories. Check for fault codes prior to disconnecting the battery cables.*
- *If the Check Engine light is illuminated with the engine running, see* **100 Engine–General** *for On-Board Diagnostics (OBD) fault code information.*
- *If any other system faults have been detected, as indicated by an illuminated ANTI LOCK, SRS or AST warning light, see the appropriate repair group in this manual or an authorized BMW dealer for more information on fault codes.*

ENGINE REMOVAL AND INSTALLATION

Engine, removing and installing (4-cylinder engines)

Cover all painted surfaces before beginning the removal procedure. As an aid to installation, label all components, wires, and hoses before removing them. Do not reuse gaskets, O-rings or seals during reassembly.

WARNING—
Due to risk of personal injury, be sure the engine is cold before beginning the removal procedure.

1. Disconnect negative (–) cable from battery in luggage compartment.

CAUTION—
Prior to disconnecting the battery, read the battery disconnection cautions given at the front of this manual on page viii.

2. Remove transmission from car. See **230 Manual Transmission** or **240 Automatic Transmission**.
3. Remove engine hood or place hood in service position. See **410 Fenders, Engine Hood**.

NOTE—
It is not necessary to remove the engine hood, but it is helpful and will make engine removal and installation easier.

4. Remove splash guard under engine, if applicable.
5. Unbolt ignition coil assembly from bracket on strut tower. Disconnect wiring to ignition coils, then place coil assembly on engine.
6. Disconnect all cables and harness connectors at throttle housing.
7. Unbolt and remove complete air cleaner housing with mass air flow sensor, disconnecting and labeling connectors, ducting and hoses as necessary. See Fig. 1.

0011969

Fig. 1. Intake air duct hose clamp (**A**), mass air flow sensor connector (**B**), and air cleaner to air flow sensor clips (**C**) on M44 engine.

8. Drain engine coolant and remove coolant hoses attached to cylinder head.
   - Drain radiator and engine block. See **170 Radiator and Cooling System**.
   - Disconnect hoses from thermostat housing at front of cylinder head.
   - Disconnect heater hoses at rear of engine. See Fig. 2.

NOTE —

- *The block drain plug is located on the exhaust side towards rear of engine.*
- *Remove small plastic lock clip to pull radiator drain plug out completely.*

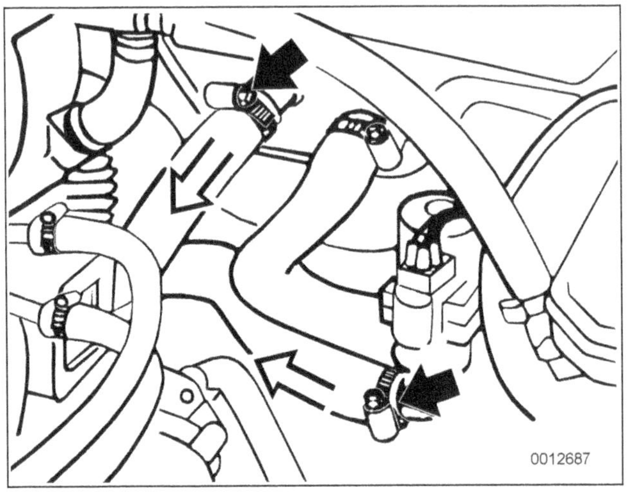

Fig. 2. Coolant hoses at heater valve and heater core to be disconnected (**arrows**).

9. Remove air shroud from top of radiator. See Fig. 3.

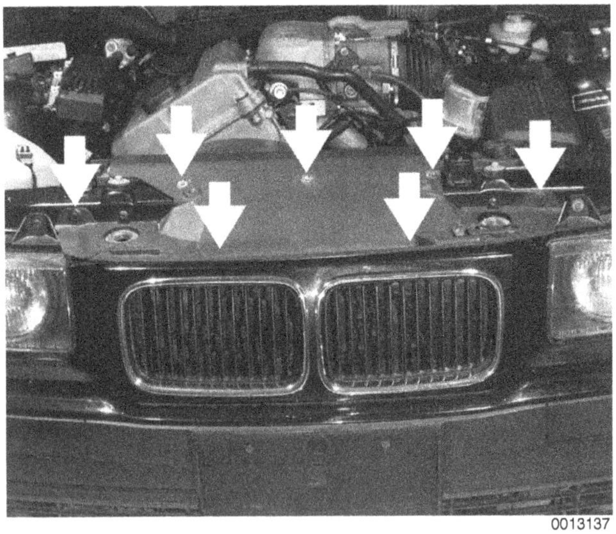

Fig. 3. Front air shroud mounting screws (**arrows**).

10. Remove radiator cooling fan and radiator as described in **170 Radiator and Cooling System**.

NOTE —

Some late 4-cylinder models use an electric primary cooling fan.

CAUTION —

On cars with viscous-type cooling fans, the radiator fan has left hand threads.

11. Remove upper intake manifold, unfasten cable duct from lower intake manifold, crankcase vent valve hose (M44 engine only)

12. Disconnect fuel supply and fuel return lines, main engine electrical connectors, and lower intake manifold as described in **113 Cylinder Head Removal and Installation**. See Fig. 4.

Fig. 4. Crankcase vent valve (**A**) on top of lower intake manifold of M44 engine. Fuel delivery and return lines shown at **B**. Engine harness and sensor connectors shown at **C**.

WARNING —

- *Fuel may be expelled under pressure. Do not smoke or work near heaters or other fire hazards. Keep a fire extinguisher handy. Before disconnecting fuel hoses, wrap a cloth around fuel hoses to absorb any leaking fuel. Plug all open fuel lines.*
- *Always unscrew the fuel tank cap to release pressure in the tank before working on the tank or lines.*

CAUTION —

Stuff clean rags into the open intake ports to prevent any parts from falling into the engine intake.

13. Disconnect vacuum hose from brake booster on bulkhead. Cover hole in booster and plug hose end.

14. Unbolt power steering fluid reservoir from its mounting bracket. Use stiff wire to hang reservoir to one side. Do not disconnect fluid lines.

15. Remove power steering pump drive belt and remove pump from its mounting bracket. Use stiff wire to hang pump from body.

16. On cars with automatic transmission remove front and rear brackets holding automatic transmission cooler lines to engine.

17. Remove A/C compressor from its mounting bracket without disconnecting any refrigerant lines. See Fig. 5.

Fig. 5. A/C compressor mounting bolts (**arrows**).

18. Move A/C compressor out of way without distorting or damaging any lines. Support compressor by hanging it from chassis using stiff wire.

19. Disconnect wiring from starter and alternator. Move wiring harness out of way. See **121 Battery, Starter, Alternator**.

20. Remove oil dipstick guide tube. Note O-ring at base of tube when removing.

21. Remove front exhaust pipe from exhaust manifold. See Fig. 6.

22. Install an engine lifting device and raise engine until its weight is supported. Remove nuts and ground strap from left and right engine mounts.

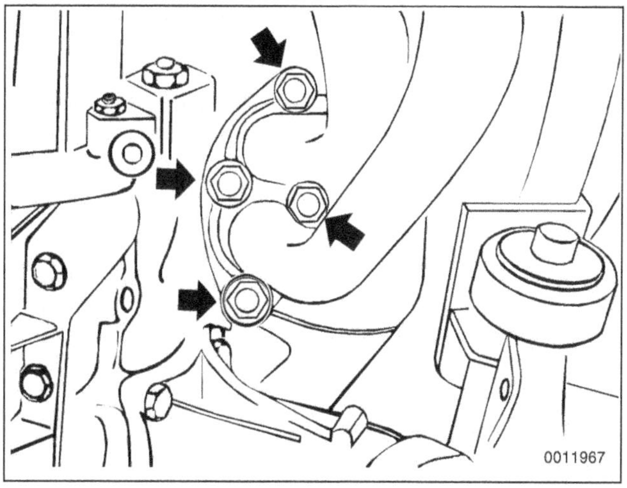

Fig. 6. Front exhaust pipe to exhaust mounting nuts (**arrows**), as viewed from below on 4-cylinder engine.

23. Carefully raise engine out of car, checking for any wiring harnesses, fuel lines, or mechanical parts that might become snagged as engine is removed.

24. Installation is reverse of removal, noting the following:
    - Replace all gaskets, O-rings and seals.
    - Change engine oil and filter and check all other fluid levels. See **020 Maintenance Program**.
    - Refill and bleed cooling system. See **170 Radiator and Cooling System**.
    - Install the front exhaust pipes using new gaskets and self-locking nuts. Use copper paste on threads. See **180 Exhaust System**.

Tightening Torques
- Coolant drain plug to cylinder block . 25 Nm (18 ft-lb)
- Engine mount to subframe
  - M8 . . . . 22 Nm (17 ft-lb)
  - M10 . . . . 45 Nm (33 ft-lb)
- Intake manifold to cylinder head
  - M7 . . . . 15 Nm (11 ft-lb)
  - M8 . . . . 22 Nm (16 ft-lb)
- Radiator cooling fan to coolant pump . . . . 40 Nm (30 ft-lb)
- Radiator drain screw to radiator . . . 2.5 Nm (22 in-lb)
- Wiring to alternator (M8 nut). . . . . 12 Nm (9 ft-lb)
- Wiring to starter
  - M6 nut . . . . .5 Nm (44 in-lb)
  - M8 nut . . . . 12 Nm (9 ft-lb)

Engine, removing and installing (6-cylinder engines)

Engine removal procedures for the various 6-cylinder engines are covered in this section. Most steps in the procedures are similar or the same for all engines. Specific differences that apply are noted at the beginning of each step.

Be sure to cover all painted surfaces before beginning the removal procedure. As an aid to installation, label all components, wires, and hoses before removing them. Do not reuse gaskets, O-rings or seals during reassembly.

> ***WARNING —***
> *Due to risk of personal injury, be sure the engine is cold before beginning the removal procedure.*

1. Disconnect negative (–) battery cable in luggage compartment.

> ***CAUTION —***
> *Prior to disconnecting the battery, read the battery disconnection cautions given at the front of this manual on page viii.*

2. Remove engine hood or place hood in service position. See **410 Fenders, Engine Hood**.

> ***NOTE —***
> *It is not necessary to remove the engine hood, but it is helpful and will make engine removal and installation easier.*

3. Remove splash guard under engine, if applicable.
4. Remove transmission from car. See **230 Manual Transmission** or **240 Automatic Transmission**.
5. Remove vacuum hose from brake booster on bulkhead. Cover hole in booster and plug hose end.
6. Remove intake air plenum panel in rear of engine compartment. See **640 Heating and Air Conditioning**.
7. Remove ground strap from timing case cover, if applicable.
8. Remove top engine covers and disconnect ignition coil harness connectors. Working at fuel injectors, pry open small wire clips at each injector. Remove main harness hold-down nuts and lift complete harness away. See Fig. 7.
9. Unbolt and remove complete air cleaner housing with mass air flow sensor, disconnecting and labeling connectors, ducting and hoses as necessary. See Fig. 8.

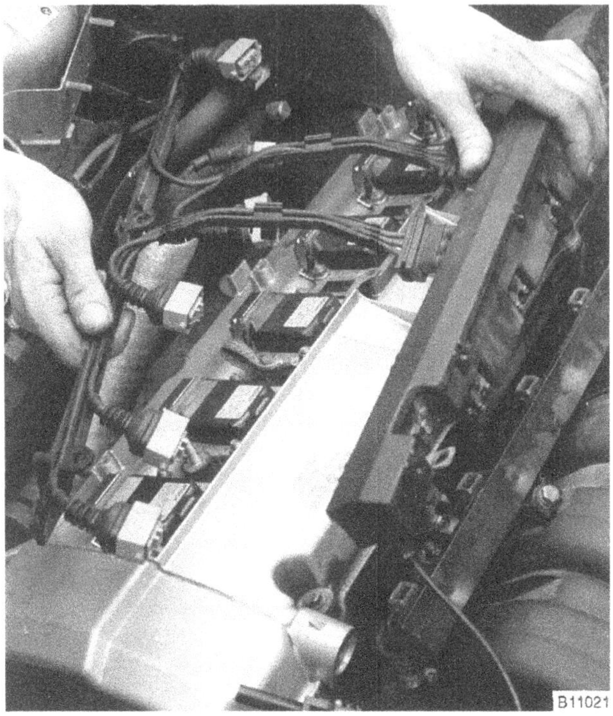

Fig. 7. Remove main harness from top engine and place at base of windshield.

0012703

Fig. 8. Mass air flow sensor connector, air duct hose clamp, and air cleaner housing mounting bolts (**arrows**).

10. On cars with automatic transmission, remove front and rear brackets holding transmission cooler lines to side of engine.

11. Drain engine coolant and remove coolant hoses attached to cylinder head.
    - Drain engine block by removing block drain.
    - Disconnect hoses from thermostat housing in front of cylinder head.
    - Disconnect heater hoses at rear of engine. See Fig. 9.
    - Drain radiator by removing threaded plastic plug from lower left corner.

NOTE —
- *The block drain plug is located on the exhaust side towards rear of engine.*
- *Remove small plastic lock clip to pull radiator drain plug out completely.*

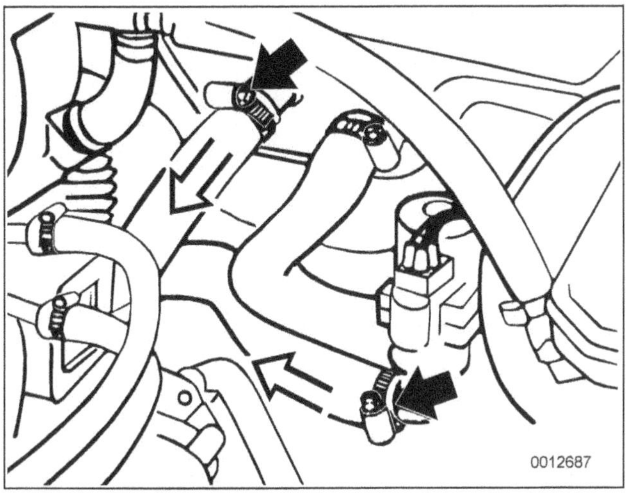

Fig. 9. Coolant hoses at heater valve and heater core to be disconnected (**arrows**).

12. Remove radiator cooling fan and radiator as described in **170 Radiator and Cooling System**.

CAUTION —
Radiator fan has left hand threads.

13. Label and disconnect coolant hoses, cables, and harness connectors from throttle housing. See Fig. 10.

14. Working at rear of intake manifold, label and then remove fuel supply and fuel return lines. Disconnect fuel lines from support bracket on manifold. See Fig. 11.

NOTE —
On M52 and S52US engines, special fittings are used to retain the fuel lines to the fuel rail. Use BMW special tool no. 16 1 050 to compress the locking clip inside the end of the fuel line fittings to release the line.

15. Unbolt power steering fluid reservoir from its mounting bracket. Use stiff wire to hang reservoir to one side. Do not disconnect fluid lines.

Fig. 10. Throttle housing assembly on M52 engine. Disconnect hoses, cables and harness connectors at and around housing.

Fig. 11. Special fuel line fitting at fuel rail on M52/S52US engine. Use BMW special tool 16 1 050 to release locking fittings at ends of fuel lines.

WARNING —
- *Fuel may be expelled under pressure. Do not smoke or work near heaters or other fire hazards. Keep a fire extinguisher handy. Before disconnecting fuel hoses, wrap a cloth around fuel hoses to absorb any leaking fuel. Plug all open fuel lines.*
- *Always unscrew the fuel tank cap to release pressure in the tank before working on the tank or lines.*

16. Remove intake manifold from engine by removing seven mounting nuts from above and two mounting bracket bolts from below. See Fig. 12.

CAUTION —
Stuff clean rags into the open intake ports to prevent any parts from falling into the engine intake.

Fig. 12. Intake manifold mounting nuts on 6-cylinder engine. Also remove two bolts from support brackets (**arrows**).

17. Working below intake manifold, disconnect harness connectors. Cut wire ties holding engine harness to bulkhead and disconnect main engine harness connector at rear of engine compartment. See Fig. 13.

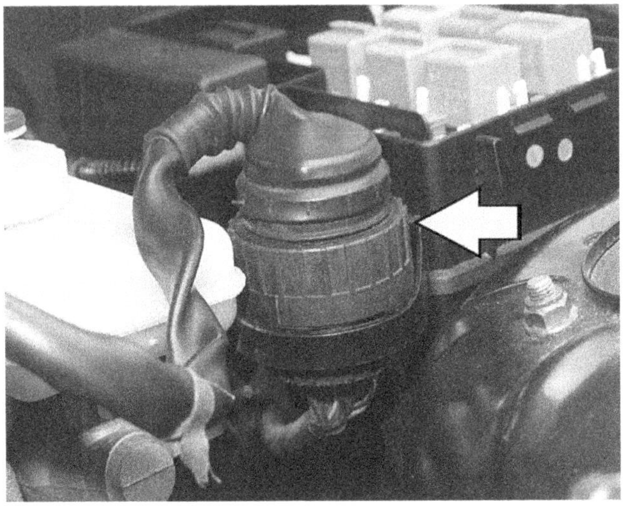

0013132

Fig. 13. Main engine harness connectors in rear of engine compartment (**arrow**).

18. Remove poly-ribbed drive belt(s).

19. Remove engine drive belt. See **020 Maintenance Program**.

20. Remove power steering pump from its mounting bracket. See Fig. 14. Use stiff wire to hang pump from body.

Fig. 14. Power steering pump mounting bolts (**arrows**). M52 engine shown.

21. Remove A/C compressor from its mounting bracket without disconnecting any refrigerant lines. See Fig. 15.

Fig. 15. A/C compressor mounting bolts (**arrows**).

22. Move A/C compressor out of the way without distorting or damaging any lines. Support compressor by hanging it from chassis using stiff wire.

23. Disconnect wiring from starter and alternator. Move wiring harness out of the way. See **121 Battery, Starter, Alternator**.

24. Remove front exhaust pipe from exhaust manifold. See Fig. 16.

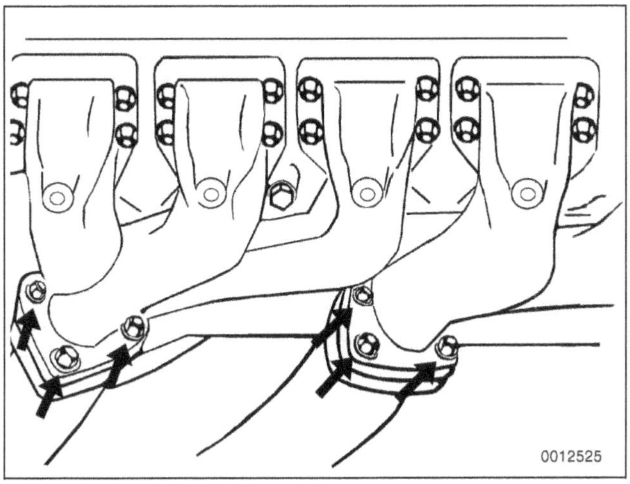

Fig. 16. Front exhaust pipe to exhaust manifolds mounting nuts on 6-cylinder engine (**arrows**).

25. Install an engine lifting device and raise engine until its weight is supported. Remove nuts and ground strap from left and right engine mounts.

26. Carefully raise engine out of car, checking for any wiring, fuel lines, or mechanical parts that might become snagged as engine is removed.

27. Installation is reverse of removal, noting the following:
    - Replace all gaskets, O-rings and seals.
    - Change engine oil and filter and check all other fluid levels. See **020 Maintenance Program**.
    - Refill and bleed cooling system. See **170 Radiator and Cooling System**.
    - Check that engine drive belts properly engage the pulley grooves.
    - Install the front exhaust pipes using new gaskets and self-locking nuts. Use copper paste on threads. See **180 Exhaust System**.

Tightening Torques

- Coolant drain plug to cylinder block . 25 Nm (18 ft-lb)
- Engine mount to subframe
  M10 . . . . 45 Nm (33 ft-lb)
  M8 . . . . 22 Nm (16 ft-lb)
- Intake manifold to cylinder head
  M7 . . . . 15 Nm (11 ft-lb)
  M8 . . . . 22 Nm (16 ft-lb)
- Radiator cooling fan to
  coolant pump . . . . 40 Nm (30 ft-lb)
- Radiator drain screw to radiator . . . 2.5 Nm (22 in-lb)
- Wiring to alternator (M8 nut). . . . . 12 Nm (9 ft-lb)
- Wiring to starter
  M6 nut . . . . .5 Nm (44 in-lb)
  M8 nut . . . . 12 Nm (9 ft-lb)

113 Cylinder Head Removal and Installation

GENERAL

The information given in this repair group assumes that the engine is installed in the engine bay. For cylinder head, camshaft, and valvetrain disassembly and reconditioning information, see **116 Cylinder Head and Valvetrain**.

In the procedures given below, engines are referred to by engine code. If necessary, see **100 Engine–General** for engine code and application information.

NOTE —
If a head gasket problem is suspected, a compression test will usually detect the fault. See **100 Engine–General** *for compression testing information.*

Special BMW service tools are required to properly remove and install the cylinder head on both the 4-cylinder and 6-cylinder engines. The special tools are used to time the valvetrain to the crankshaft and to remove the Torx (E12) head bolts. Read the procedure through before beginning the job.

Cover all painted surfaces before beginning the removal procedure. As an aid to installation, label all components, wires, and hoses before removing them. Do not reuse gaskets, O-rings or seals during reassembly.

> *WARNING —*
> *Due to risk of personal injury, be sure the engine is cold before beginning the removal procedure.*

CYLINDER HEAD, 4-CYLINDER

Cylinder head, removing (4-cylinder engine)

For ease of use this removal procedure is broken into subsections, as follows:

- **General**
- **Cooling system, draining**
- **Cylinder head cover, removing**
- **Intake manifold, removing**
- **Timing chains, uncoupling**
- **Cylinder head assembly, removing**

> *WARNING —*
> *Allow engine to cool before beginning this procedure.*

General

1. Disconnect negative (–) cable from battery.

> *CAUTION —*
> - *Prior to disconnecting the battery, read the battery disconnection cautions given at the front of this manual on page viii.*
> - *Disconnecting the battery may erase fault code(s) stored in control module memory. Check for fault codes using special BMW diagnostic equipment.*

NOTE —
The battery is mounted in the luggage compartment behind the right side trim panel.

2. Working underneath vehicle, remove front exhaust pipe from exhaust manifold. See Fig. 1.

3. Disconnect vacuum hose fitting from brake booster on rear bulkhead. Plug opening in booster and hose end.

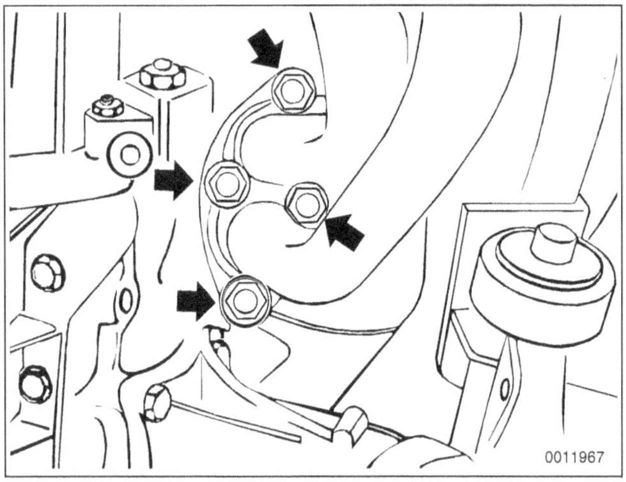

Fig. 1. Front exhaust pipe to exhaust manifold mounting nuts (**arrows**).

4. Remove air filter housing complete with mass air flow sensor and attached hoses and ducting. See Fig. 2.

Fig. 2. Air filter housing/air intake connections on M44 engine with traction control. Clamp at throttle housing (**A**); mass air flow sensor connector (**B**); and clips at upper air filter housing (**C**).

Cooling system, draining

5. Drain coolant from engine block and radiator. See **170 Radiator and Cooling System**.

NOTE —

- *The block drain is on the exhaust side of the engine, near cylinder no. 4.*
- *The radiator drain plug is on the bottom of the radiator on the driver's side. Remove small plastic lock clip to pull radiator drain plug out completely.*

6. Disconnect hoses from coolant flange/thermostat housing on front of cylinder head. Unbolt thermostat housing from cylinder head.

7. On M42 engines, disconnect hose from coolant pipe on left side of engine. See Fig. 3.

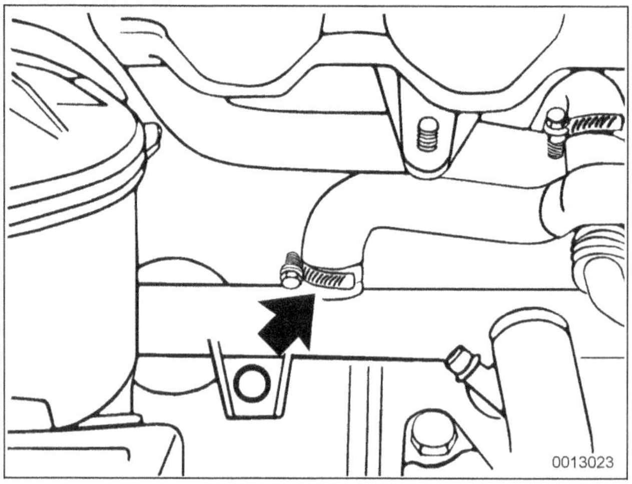

Fig. 3. Coolant hose connection beneath intake manifold on M42 engine (**arrow**).

8. Disconnect heater hoses from coolant flange on rear of cylinder head.

Cylinder head cover, removing

9. Remove ignition coil pack from right strut tower. See Fig. 4.

Fig. 4. Cylinder head cover showing location of engine breather hose (**A**), wiring harness duct (**B**), and ignition coil pack (**C**). M44 engine shown. Cover on M42 engine is similar.

10. Disconnect engine breather hose from cylinder head cover.

11. Detach wiring harness duct at rear bulkhead panel and pull complete duct forward and up to allow access to rear of cylinder head cover.

12. Remove spark plugs and spark plug wire loom.
    - Remove plastic cover from top of cylinder head.
    - Disconnect spark plug wires from spark plugs and remove spark plugs.
    - Unbolt spark plug cable harness and heat shield from right side of cylinder head cover and set aside.

13. Unscrew and remove cylinder head cover cap screws and remove cover. See Fig. 5.

> ***CAUTION —***
> *Two of the cover hold-down screws are under the spark plug cable duct.*

NOTE —
Make note of the arrangement of cap screws, washers and rubber grommets holding the cylinder head cover in place.

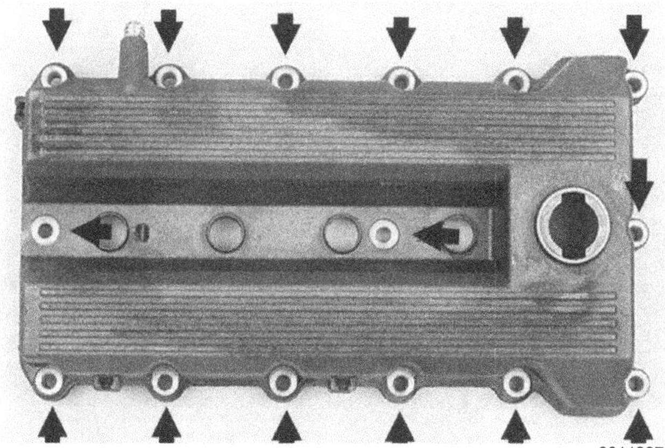
0011997

Fig. 5. Cylinder head cover mounting points (**arrows**).

Intake manifold, removing

14. On late engines, disconnect injector air shrouding hose from fitting at center of upper intake manifold.

15. Remove upper section of intake manifold.
    - Disconnect all cables harness connectors from throttle housing and intake manifold.
    - Disconnect fuel tank vent hose from vent valve on lower side of manifold.
    - Remove manifold support brackets. See Fig. 6.
    - Remove upper-to-lower manifold fasteners (1 bolt, 2 nuts) and separate upper manifold from lower manifold.

NOTE —
Note locating dowel sleeves on lower to upper manifold mounting studs. Remove dowels and set aside. See Fig. 8.

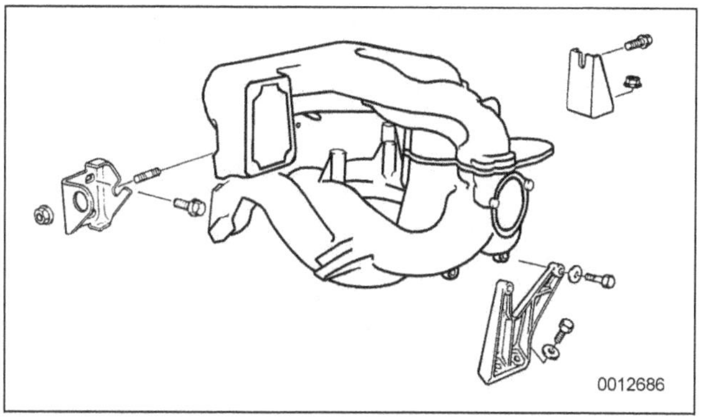

Fig. 6. Intake manifold support brackets and mounting hardware. Manifold for M44 engine shown.

16. Disconnect oil dipstick tube bracket from lower manifold section.

17. On M44 engine, relieve fuel pressure using compressed air (maximum 3 bar pressure) at schroeder valve on fuel rail. Briefly apply air pressure to force fuel back into fuel tank. See Fig. 7.

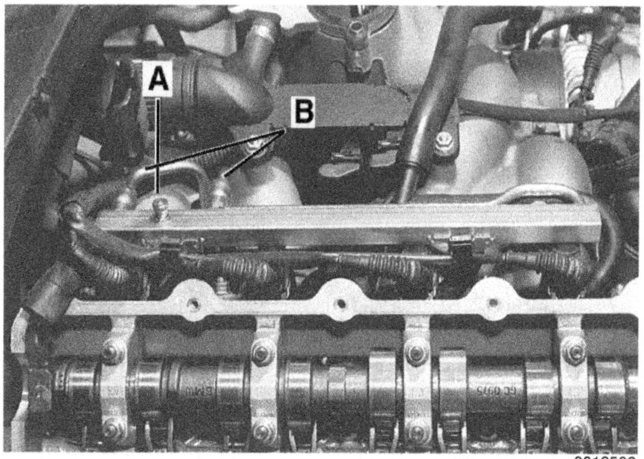

0012503

Fig. 7. M44 fuel rail air connection (**A**) and fuel supply and return lines (**B**).

18. Disconnect fuel supply and return lines from fuel rail.

> ***WARNING —***
> - *The fuel system retains fuel pressure when the engine is off (up to 75 psi). Use care when disconnecting fuel lines. Unscrew the fuel tank cap to release pressure in the tank. Wrap a clean shop towel around the fitting when loosening.*
> - *Fuel is highly flammable. When working around fuel, do not disconnect any wires that could cause electrical sparks. Do not smoke or work near heaters or other fire hazards. Place a fire extinguisher in the vicinity of the work area.*

19. Disconnect coolant hoses from vent valve on lower intake manifold. Remove crankcase vent valve from manifold.

20. Disconnect main engine wiring harness connector. Label and disconnect harness connectors for crankshaft and camshaft sensors. See Fig. 8.

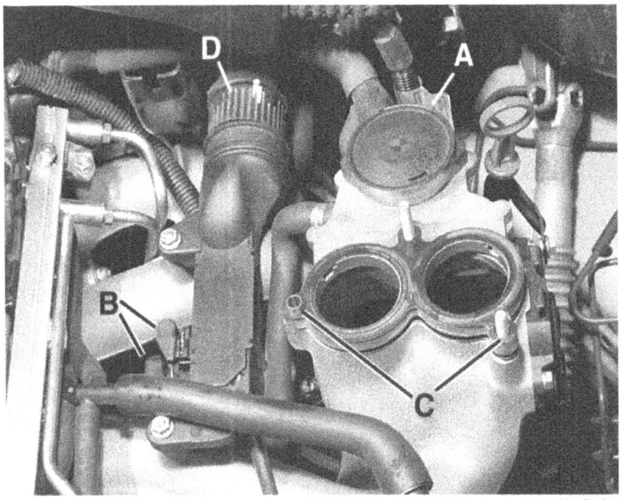

0012504

Fig. 8. Crankcase vent valve (**A**), camshaft and crankshaft sensor harness connectors (**B**), locating bushings (**C**), main engine wiring harness connector (**D**). M44 engine shown.

NOTE—

Early engines do not have a main disconnect at the lower intake manifold harness. If necessary, remove the cable ducting bolts and set harness ducting aside.

21. Remove lower section of intake manifold.
    - Remove Y-shaped support bracket from underside of manifold. (See Fig. 6.)
    - Unbolt support brackets from alternator and starter.
    - Remove nuts holding manifold to cylinder head and pull manifold back until harness connectors for oil pressure switch and coolant temperature sensor are accessible. Label and disconnect connectors, then remove manifold and set aside.

Timing chains, uncoupling

22. On M42 engines, remove thermostat housing from upper timing chain cover.

23. On late M44 engines (January 1997 and later), remove secondary air injection check-valve from front of engine and lay aside.

24. Remove cylinder identification sensor from upper timing chain cover. Remove upper timing chain cover. See Fig. 9.

25. Set engine to approximate TDC by rotating in normal operating direction until camshaft lobes at cylinder no. 1 are facing each other. See Fig. 10.

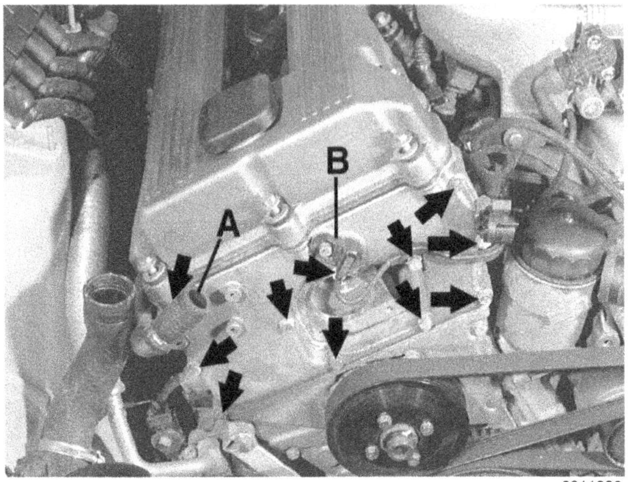

0011996

Fig. 9. Upper timing chain cover bolts (**arrows**). Remove secondary air injection check valve from hose (**A**) on M44 engine. Remove cylinder identification sensor from chain cover (**B**).

0011998

Fig. 10. Engine set to approximate TDC. Cylinder no. 1 camshaft lobes face in and arrows on sprockets point up (**arrows**).

26. Lock crankshaft in TDC position by inserting BMW special tool no. 11 2 300 through bore and into hole in flywheel or drive plate. See Fig. 11.

NOTE—

- *If necessary, remove plug from locating bore in lower edge of bell housing to insert locking tool.*
- *Confirm that the locking tool has been correctly installed by trying to rotate the crankshaft.*

27. Lock camshafts at TDC by mounting BMW special tool no. 11 3 240 over square ends of camshafts. See Fig. 12.

NOTE—

Check that tool is squarely seated on cylinder head gasket surface. If necessary, rotate camshafts slightly using an open-end wrench at hex on camshafts.

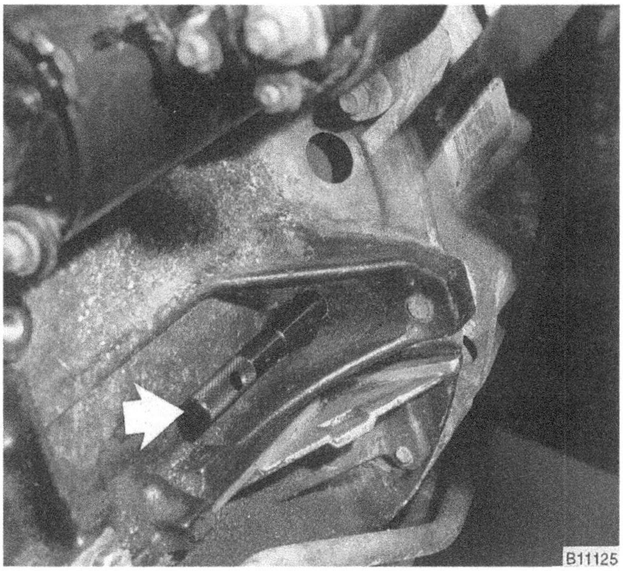

Fig. 11. BMW special tool no. 11 2 300 installed through bellhousin and into flywheel (**arrow**) to lock crankshaft at TDC.

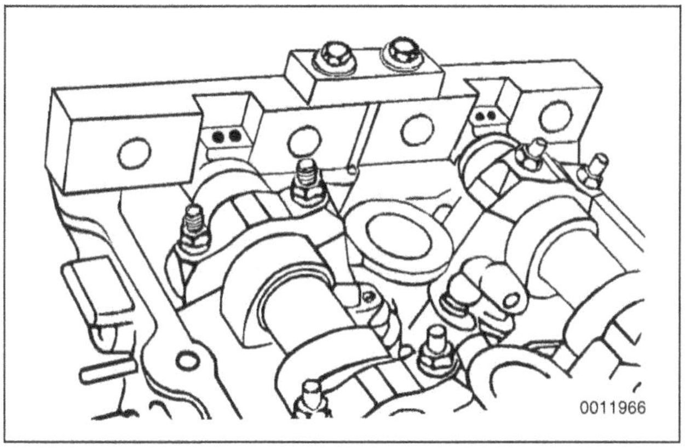

Fig. 12. BMW special tool no. 11 3 240 installed on rear of camshafts to hold them at TDC. (M44 engine shown. Other engines are similar.)

> ***CAUTION—***
>
> *The camshafts must be locked in the TDC position using the special service tool, or equivalent, before removing the timing chain. The tool holds the camshafts parallel to each other and perpendicular to the valve cover gasket surface.*

28. Remove hydraulic chain tensioner from right side of cylinder head. See Fig. 13.

29. Remove mounting bolts from left and right camshaft sprockets. Then remove camshaft sprockets from camshafts. On M44 engines, note orientation of sensor wheel.

30. Remove upper bolt from left chain guide. Loosen chain guide adjusting sleeve approximately two turns. See Fig. 14.

Fig. 13. Hydraulic chain tensioner (**arrow**).

Fig. 14. Chain guide and guide adjusting sleeve (**arrow**). Allen-head retaining screw has been removed. Use slotted screwdriver to loosen adjusting sleeve.

NOTE—

It is not necessary to remove the cam sprockets from the chain unless these parts are being replaced. Wire tie each sprocket to the chain to keep the sprockets timed to each other. Keeping the sprockets timed will simplify installation.

Cylinder head assembly, removing

31. Remove crankshaft locking tool (BMW special tool no. 11 2 300). Using the crankshaft vibration damper bolt, turn engine opposite its normal direction of rotation (counterclockwise as viewed from the front) approximately 45°. This prevents accidental contact between valves and pistons during installation.

32. Using Torx E12 socket, remove cylinder head bolts in the sequence indicated. See Fig. 15. Remove cylinder head and head gasket. Discard head bolts.

Fig. 15. Cylinder head bolt loosening sequence. Use Torx E12 socket.

Cylinder head, installing (4-cylinder engine)

> ***CAUTION—***
> *If the camshafts were removed/installed while the cylinder head was off, a minimum waiting time is required after installing the camshafts. When the camshafts are removed, the hydraulic lifters can expand. This expansion can cause increased valve lift, resulting in open valves and piston contact.*

Cylinder head installation waiting times

- At or above 68°F (20°C). . . . . 10 minutes
- 50-68°F (10-20°C) . . . . 30 minutes
- 32-50°F (0-10°C) . . . . 75 minutes

1. Clean cylinder head and gasket surfaces of cylinder block and all timing chain covers.

   - Remove all foreign matter and any liquid from bolt holes, then clean with a thread chaser.

> ***CAUTION—***
> *Do not use a metal scraper or wire brush to clean the aluminum cylinder head or pistons. If necessary, use a hard wooden or plastic scraper. Also available are abrasive discs to be used in conjunction with an electric drill. Be sure to use the correct disc for the type of metal being cleaned.*

2. Check gasket surface of cylinder head and cylinder block for warpage using a straight edge.
3. On M42 engine: Be sure oil feed check valve and spacer are correctly positioned in block. See Fig. 16.

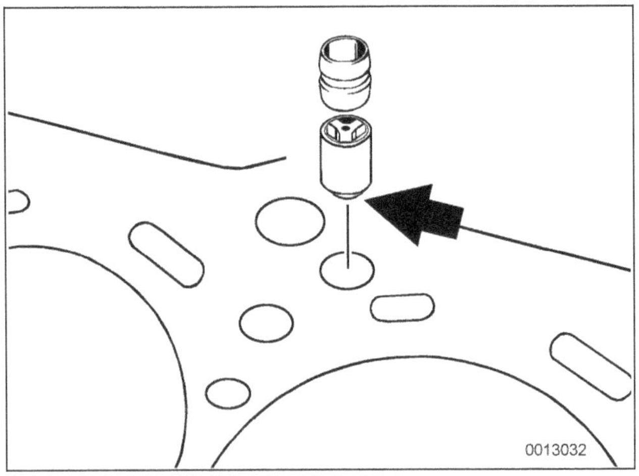

Fig. 16. Oil feed check valve and spacer used on M42 engine. Install check valve with collar (**arrow**) facing down.

> ***CAUTION—***
> *If the check valve is not installed correctly, oil flow to the cylinder head would be restricted.*

4. Place new cylinder head gasket on surface of cylinder block. Check that two locating dowels are correctly positioned in block and not damaged.

> ***CAUTION—***
> *The word "OBEN", printed on the gasket, should face up.*

5. Place new rubber profile gasket in groove of front cover.
6. With camshafts locked in TDC position with BMW special tool as described earlier, set cylinder head in position.

> ***CAUTION—***
> - *The camshafts must be locked in the TDC position at the rear of the cylinder head with BMW special tool no. 11 3 240 before installing the cylinder head.*
> - *Make sure the crankshaft, which had been rotated approximately 45° opposite the direction of engine rotation, is still in that position before lowering the cylinder head into position.*
> - *Torx-type cylinder head bolts should not be reused. They are stretch-type bolts and must always be replaced whenever loosened.*

7. Lightly lubricate new cylinder head bolts. Loosely install bolts and their washers, then thread them in until they are finger tight.

NOTE —

Check that all washers for the cylinder head bolts are in place before installing the bolts.

8. Tighten cylinder head bolts in correct sequence. See Fig. 17.

Fig. 17. Cylinder head bolt tightening sequence for 4-cylinder engines. Note oil supply tube seal (**arrow**) on M44 engine.

WARNING —

The bolts should be tightened in three stages as listed below. The final stages require the use of a special tool (BMW tool No. 11 2 110) or a suitable protractor to tighten the bolts to a specified torque angle. See Fig. 18.

Tightening Torques

- Cylinder head to engine block (Torx E12 M10 bolts)
  - Stage 1 .... 30 Nm (22 ft-lb)
  - Stage 2 .... +90°
  - Stage 3 .... +90°

9. Carefully rotate crankshaft in direction of engine rotation and lock it at TDC by inserting BMW special tool no. 11 2 300 through transmission bellhousing. See Fig. 19.

CAUTION —

Be sure timing chain does not bind on crank gear when turning crankshaft.

Fig. 18. Protractor and pointer being used to tighten bolts to specified torque angle (6-cylinder engine shown).

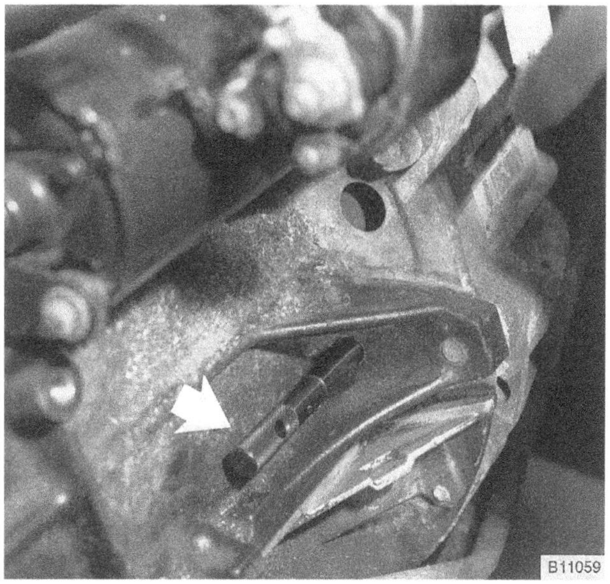

Fig. 19. BMW special tool 11 2 300 inserted through bellhousing hole and into flywheel (**arrow**) to lock crankshaft at TDC.

10. Turn left chain guide adjusting sleeve in until it just contacts the cylinder head (See Fig. 14.). Insert chain guide retaining screw and tighten.

11. Place chain sprockets with chains on camshafts so that arrows on sprockets point up and elongated holes in sprockets are centered to tapped holes in camshafts. See Fig. 20.

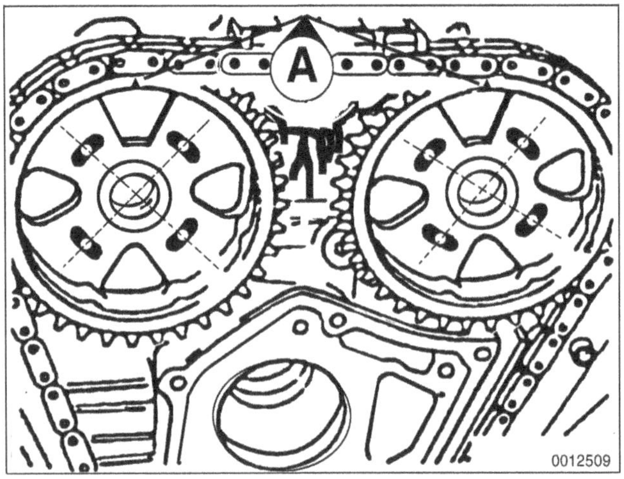

Fig. 20. Camshaft chain sprockets installed with arrows (**A**) pointing up and camshaft threaded holes centered in slotted holes.

12. On M44 engine: Place sensor wheel on intake camshaft sprocket so that arrow on sensor wheel points up.

13. Install and hand-tighten sprocket mounting bolts.

> ***CAUTION —***
> *Different hydraulic chain tensioners were used during manufacture, depending on engine type. See Fig. 21. Note that the late style tensioner can be retrofitted to the earlier engine, so be sure to identify the version of tensioner installed as installation instructions vary between the two.*

14. On M42 engine: Disassemble chain tensioner by striking outer sleeve against a solid object. This will release snap ring and tensioner will separate. See Fig. 22.

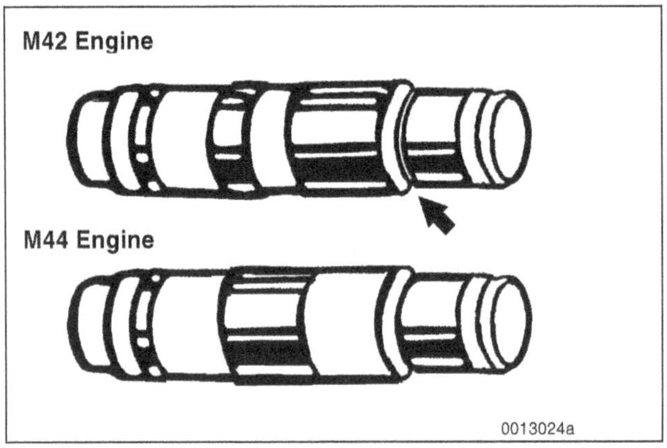

Fig. 21. Two versions of hydraulic chain tensioners. Original chain tensioner on M42 engine (**top**) can be disassembled. Chain tensioner on M44 engine (**bottom**) cannot be disassembled. Note differences in outer sleeve. Also, the M42 tensioner detent ring groove (**arrow**) is visible when tensioner is extended.

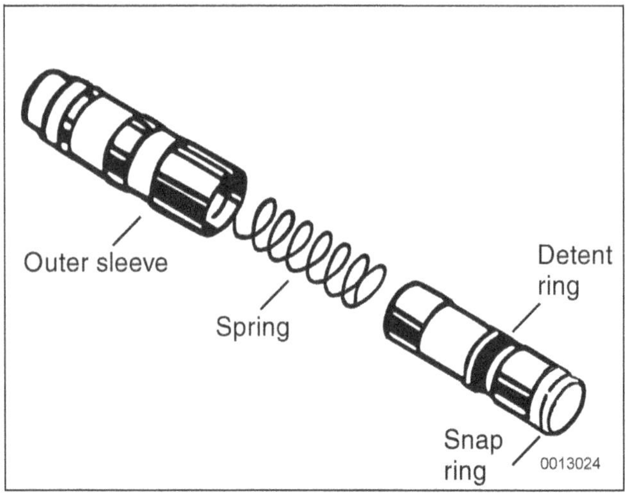

Fig. 22. Disassembled chain tensioner for M42 engine.

15. On M42 engine: Clamp tensioner piston in vice (soft jaws). Slowly compress piston, making sure detent ring fits into taper of outer sleeve. If necessary, press ends of detent ring together. Slowly continue pressing tensioner together until snap ring audibly clicks into outer sleeve. Measure overall length of tensioner to confirm correct assembly: 68.5 mm (2.7 in.).

> ***NOTE —***
> *The late-style tensioner on the M44 engine can be retrofitted to the M42 engine.*

Tightening Torque

- Chain tensioner plug to cylinder head . . . . 40 Nm (30 ft-lb)

---

16. On M44 engine: Clamp tensioner piston in vice (soft jaws) and squeeze oil from it. Slowly compress piston only up to end circlip. Repeat procedure twice to ensure all oil is expelled. See Fig. 23.

17. Install hydraulic chain tensioner to cylinder head. Use a new sealing washer on tensioner when installing.

18. On M42 engine: Unlock chain tensioner by pressing chain rail against tensioner until it expands out.

19. Check that chain is free of slack. Tighten sprocket mounting bolts.

Tightening Torque

- Chain sprocket to camshaft flange . . . 15 Nm (11 ft-lb)

---

20. Remove crankshaft locking tool from bellhousing and camshaft locking tool from rear of cylinder head.

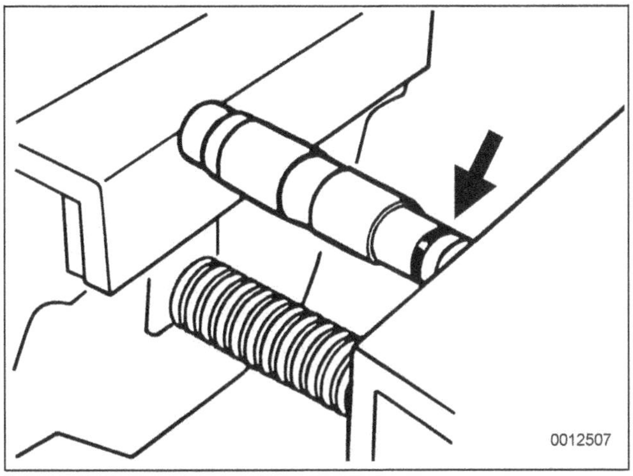

Fig. 23. When reinstalling chain tensioner on M44 engine, old oil in tensioner piston must be expelled by compressing piston in vise. When compressing piston, only compress up to end circlip (**arrow**).

21. Turn engine over by hand in direction of rotation at least two full rotations. Reinstall camshaft and crankshaft locking tools to check engine timing. Remove locking tools if no faults are found.

22. Install upper timing chain cover.
    - Install gaskets and install cover using two mounting bolts. Using a screwdriver, lever down on front cover until cover gasket surface is flush with cylinder head gasket surface. See Fig. 24. Tighten two bolts.
    - Install and tighten remainder of bolts.

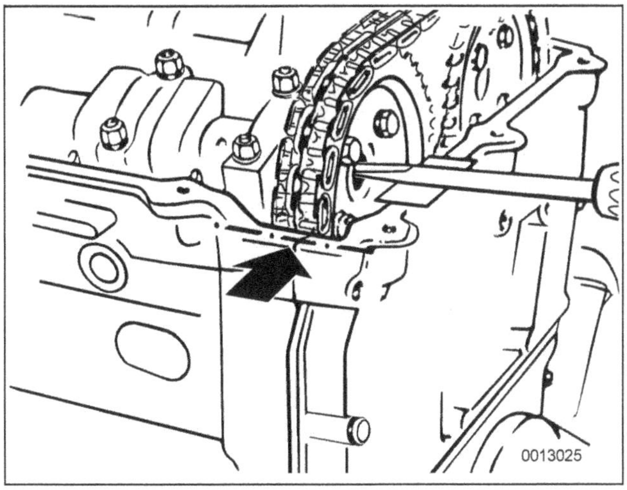

Fig. 24. Install upper timing case cover with two bolts, then press down on cover so that top surface is flush with cylinder head (**arrow**). Hold cover in position and tighten bolts.

23. Install cylinder head cover.
    - Clean old gasket glue from sealing surfaces. Use a small amount of 3-Bond®1209 adhesive or equivalent at corners of half-moon cut-outs at back of cylinder head. See Fig. 25.

NOTE —

On M44 engines, replace seal for oil supply tube. See Fig. 17.)

Fig. 25. Rear of cylinder head cover showing area where sealant (**arrows**) should be applied. Inset shows cylinder head cover mounting bolt and insulator assembly: M6 screw (**1**), rubber damper (**2**), washer (**3**).

Tightening Torque
- Cylinder head cover to cylinder head (M6 screws). . . . . . . 10 Nm (89 in-lb)

24. Installation of remaining parts is reverse of removal, noting the following:
    - Make sure all sealing surfaces are clean and free of old gasket material. Install new intake manifold gasket.
    - On M44 engine: Replace seals above and below crankcase vent valve.
    - When installing lower intake manifold, make sure hollow locating bushings are on two lower studs.
    - Reinstall harness connectors for oil pressure switch and coolant temperature sensor before installing intake manifold. Tighten manifold nuts in stages.
    - Refill cooling system as described in **170 Radiator and Cooling System**.
    - Change engine oil and filter as described in **020 Maintenance Program**.
    - If necessary, adjust accelerator cable.
    - Run engine at approx. 3,500 rpm for about 20 seconds to allow hydraulic chain tensioner to fill with oil.
    - Reconnect battery last.

Tightening Torques

- Coolant drain plug to cylinder block . . 25 Nm (18 ft-lb)
- Intake manifold support to manifold . . 15 Nm (11 ft-lb)
- Lower intake manifold to cylinder head . . . . 15 Nm (11 ft-lb)
- Lower intake manifold to upper manifold . . . . 15 Nm (11 ft-lb)
- Radiator cooling fan to coolant pump. 40 Nm (30 ft-lb)
- Radiator drain screw to radiator . . . .2.5 Nm (22 in-lb)

CYLINDER HEAD, 6-CYLINDER

Cylinder head, removing (6-cylinder engine)

For ease of use this procedure is broken into subsections, as follows:

- **General**
- **Cylinder system, draining**
- **Cylinder head cover, removing**
- **Intake manifold, removing**
- **Timing chains, uncoupling**
- **Cylinder head assembly, removing**

> *WARNING —*
> *Allow the engine to cool before beginning this procedure.*

General

1. Working inside luggage compartment, disconnect negative (–) battery cable.

> *CAUTION —*
> - *Prior to disconnecting the battery, read the battery disconnection cautions given at the front of this manual on page viii.*
> - *Disconnecting the battery may erase fault code(s) stored in control module memory. Check for fault codes using special BMW diagnostic equipment.*

2. Remove front exhaust pipes from manifolds. Discard nuts.

3. Remove air filter housing complete with mass air flow sensor and intake air ducting. Disconnect vacuum hoses from throttle housing.

4. On M52 and S52US engines: Remove secondary air injection check-valve from front of engine and set aside.

Cooling system, draining

5. Drain engine coolant and remove coolant hoses from thermostat housing. Drain engine block by removing block drain. See **170 Radiator and Cooling System**.

> *NOTE —*
> *The block drain is on the exhaust side of the engine, near cylinder no. 5.*

6. Remove radiator cooling fan and fan shroud. See **170 Radiator and Cooling System**.

> *NOTE —*
> *The radiator cooling fan nut (32 mm wrench) has left-hand threads.*

7. Disconnect heater hoses at rear of cylinder head. See Fig. 26.

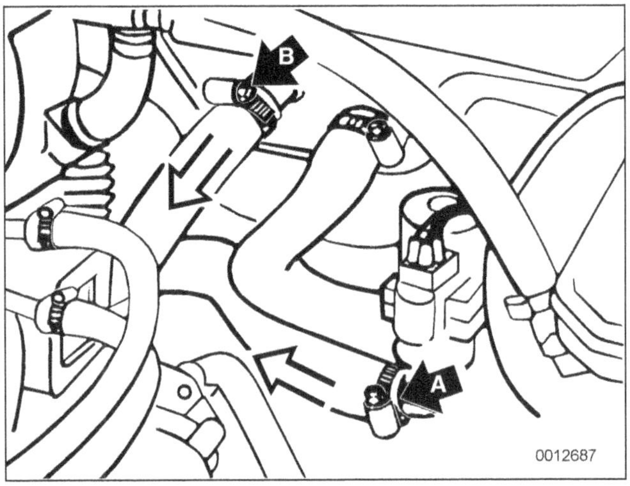

Fig. 26. Disconnect heater hoses at heater valve (**A**) and heater core (**B**).

8. Remove alternator cooling duct.

Cylinder head cover, removing

9. Remove air plenum from rear of engine compartment. See **640 Heating and Air Conditioning**.

10. Remove plastic cylinder head covers from above fuel injectors and ignition coils. Pull larger cover forward when removing. See Fig. 27.

11. Disconnect crankcase ventilation hose fitting from cylinder head cover.

12. Disconnect ignition coil harness connectors from coils. Remove ignition coils and spark plugs. See **120 Ignition System**.

Fig. 27. Remove top engine covers by prying out plugs and removing nuts (**A**). Make sure rubber insulators on either side of large cover do not fall off during cover removal (**B**).

13. Remove hold down nuts/bolts from fuel injection harness. Pry back injector connector retaining clips and lift entire harness off engine. See Fig. 28.

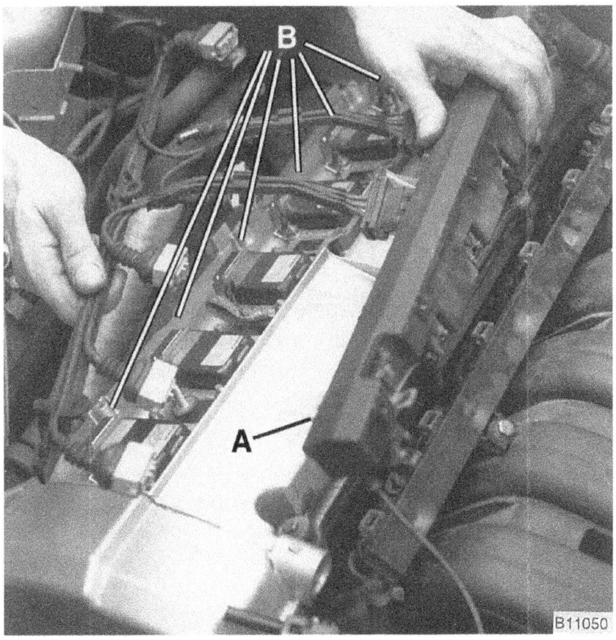

Fig. 28. Fuel injection harness being lifted off engine to be placed at base of windshield (**A**). Ignition coils are also visible (**B**). M50 engine shown, other 6-cylinder engines are similar.

14. Remove cylinder head cover mounting bolts and remove cylinder head cover.

15. Remove oil baffle cover from above intake camshaft. See Fig. 29.

Fig. 29. Camshaft oil baffle cover being removed.

NOTE —

The cylinder head cover mounting bolt insulators and gaskets should be reinstalled in the same order. Make note of their arrangement during removal.

Intake manifold, removing

16. Disconnect vacuum hose from brake system vacuum booster. Plug opening in booster.

17. Remove throttle housing mounting bolts and remove throttle housing. Set throttle housing aside without disconnecting cables or hoses. See Fig. 30.

Fig. 30. Remove 4 corner bolts (**arrows**) from throttle housing and unbolt complete assembly from intake manifold. Remove complete throttle assembly without disconnecting connectors, hoses or cables. (M50 engine shown.)

18. Working from underside of intake manifold, disconnect harness connectors, vent hoses, and air bypass hose(s).

19. Working at rear of intake manifold, label and disconnect fuel supply and fuel return lines. Remove fuel line hold down bracket to free lines. See Fig. 31.

> ***NOTE —***
> *On M52/S52US engines, a special fitting is used to retain the fuel lines to the fuel rail. Use BMW special tool no. 16 1 050 to expand the locking clip inside the end of the fuel line fittings.*

Fig. 31. Fuel lines at intake manifold on M52/S52 engine. Use BMW special tool 16 1 050 to release locking fittings at ends of fuel lines.

> ***WARNING —***
> - *The fuel system is designed to retain pressure even when the ignition is off. When working with the fuel system, loosen the fuel lines slowly to allow residual fuel pressure to dissipate gradually. Avoid spraying fuel.*
> - *Fuel is highly flammable. When working around fuel, do not disconnect any wires that could cause electrical sparks. Do not smoke or work near heaters or other fire hazards.*
> - *Always unscrew the fuel tank cap to release pressure in the tank before working on the tank or lines.*

20. Remove mounting screws and release wiring harness duct at rear bulkhead and move aside. Remove wiring harness brackets from bulkhead.

21. Detach oil dipstick guide tube from manifold. Disconnect vent hose at base of dipstick.

22. Remove intake manifold from engine by removing seven mounting nuts from above and two support bracket bolts from below. See Fig. 32.

Fig. 32. Intake manifold nuts and support bracket bolts (**arrows**).

> ***CAUTION —***
> *Stuff clean rags into the open intake ports to prevent any parts from falling into the engine.*

Timing chains, uncoupling

23. Remove VANOS unit from front of cylinder head. See VANOS section in **117 Camshaft Timing Chain**.

24. Set engine to approximate TDC by rotating in normal operating direction until camshaft lobes at cylinder no. 1 are facing each other. Arrows on camshaft sprockets should be pointing up.

> ***NOTE —***
> *To confirm that the crankshaft is set to the TDC position, check that the pointer on the front lower timing chain cover is aligned with the 0/T mark (0° TDC) vibration damper. See Fig. 33.*

25. Lock crankshaft at TDC.
    - Insert BMW special tool no. 11 2 300 through transmission bellhousing and into hole in flywheel or drive plate. See Fig. 34.

Fig. 33. Crankshaft is positioned at TDC when notch on front timing chain cover is aligned with 0/T mark on vibration damper.

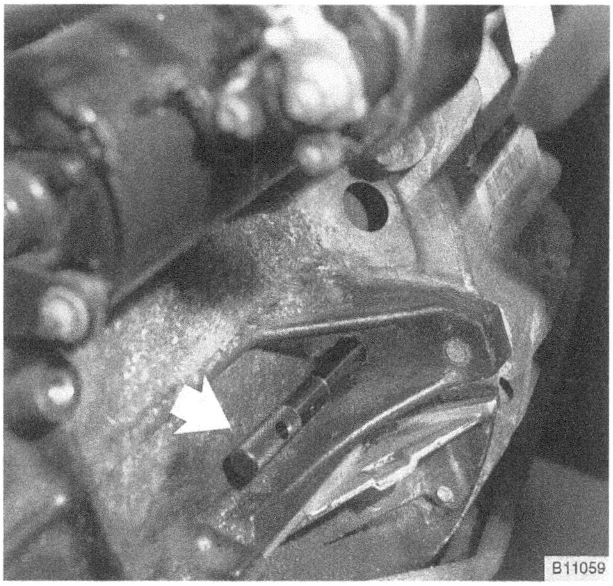

Fig. 34. Crankshaft locking tool (BMW special tool 11 2 300) inserted through bellhousing hole and into flywheel (**arrow**).

26. Lock camshafts at TDC.
   - Remove three cylinder head cover mounting studs from across rear of head, if applicable.
   - Mount BMW special tool no. 11 3 240 over square ends of camshafts. Check that tool is squarely seated on cylinder head gasket surface. If necessary, rotate camshafts slightly using a 24 mm wrench at hex on camshaft(s). See Fig. 35.

27. Depress secondary timing chain tensioner and lock it down by inserting BMW special tool no. 11 3 292. Alternatively, a stiff piece of wire can be used. See Fig. 36.

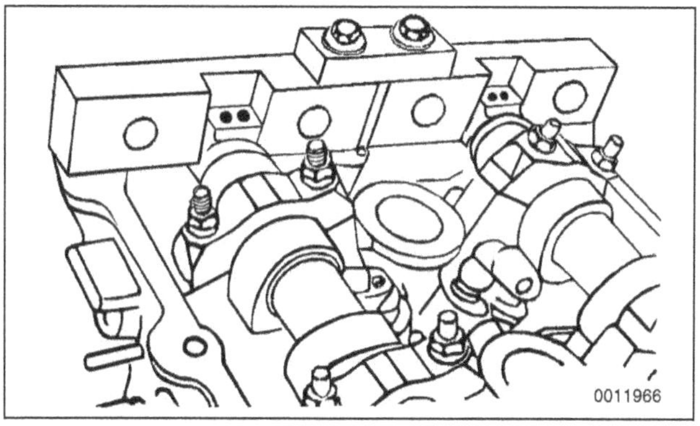

Fig. 35. Camshaft locking tool (BMW special tool no. 11 3 240) installed on rear of camshafts. (M44 engine shown. Other engines are similar.)

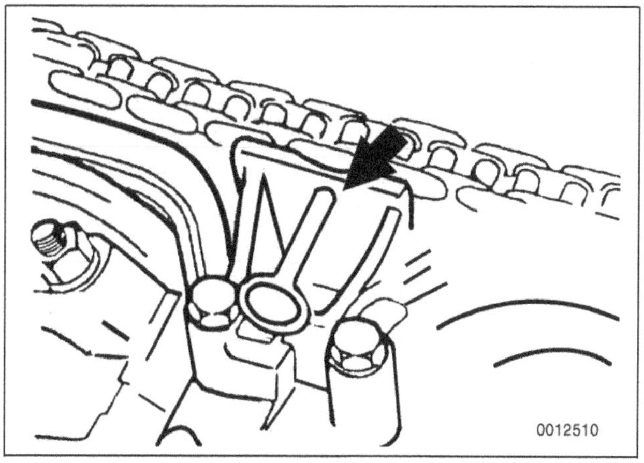

Fig. 36. Press secondary tensioner down fully and insert wire through holes (**arrow**) to lock. Timing chain tensioner varies depending on engine (M52 engine tensioner shown).

> ***CAUTION—***
> *The camshafts must be locked in the TDC position using the special service tool before removing the timing chain. The arrows on the sprockets should not be used to accurately set the engine to TDC. The tool holds the camshafts parallel to each other and perpendicular to the cylinder head cover gasket surface.*

28. Remove mounting nuts from front of intake camshaft sprocket. Remove camshaft sprocket, thrust washer(s), and where applicable, spring plate.

NOTE—
Where applicable, make note of thrust washer positions. The thicker (0.4 mm) washer is mounted outermost.

29. Remove mounting nuts from front of exhaust camshaft sprocket. Remove secondary sprockets together with timing chain from front of camshafts.

> ***CAUTION —***
> *The crankshaft must not be allowed to rotate once the camshaft timing chain is removed. The pistons can contact the valves.*

NOTE —
Do not remove the sprockets from the chain unless these parts are being replaced. Wire tie each sprocket to the chain to keep the sprockets timed to each other.

30. Unbolt secondary timing chain tensioner from top of cylinder head.

31. Remove primary timing chain tensioner from side of cylinder head. See Fig. 37.

B11132

Fig. 37. Primary timing chain tensioner (**arrow**).

> ***WARNING —***
> *Remove tensioner slowly—it is under spring pressure.*

32. Remove timing chain guide (Torx E8 socket). See Fig. 38.

33. Lift primary chain sprocket off exhaust camshaft. Remove sprocket from chain. Suspend chain using stiff wire.

Cylinder head assembly, removing

34. Remove two mounting bolts holding cylinder head to lower timing chain cover (Torx E8 socket). See Fig. 39.

Fig. 38. Timing chain guide mounting bolts (**arrows**). M50 engine shown.

Fig. 39. Cylinder head-to-timing chain cover mounting bolts (**arrows**). (Bolt **A** also retains secondary chain tensioner.)

35. Using BMW special tool no. 11 2 250 or an equivalent tool, loosen cylinder head bolts in several stages. See Fig. 40. Discard head bolts.

> ***CAUTION —***
> *Torx-type cylinder head bolts should not be reused. They are stretch-type bolts and must always be replaced whenever loosened.*

Fig. 40. Cylinder head bolt loosening sequence.

> ***NOTE —***
> *BMW special tool no. 11 2 250 is a thin-walled Torx E12 socket with an extended reach. The cylinder head bolts are recessed into the head below the camshaft towers with little working space.*

36. Remove crankshaft locking tool (BMW special tool no. 11 2 300). Using the crankshaft vibration damper bolt, turn engine opposite its normal direction of rotation (counterclockwise as viewed from the front) approximately 45°. This prevents accidental contact between valves and pistons during installation.

> ***NOTE —***
> *Guide cam chain to keep it from jamming when turning the crankshaft.*

Cylinder head, installing (6-cylinder engine)

> ***CAUTION —***
> *If the camshafts were removed and installed while the cylinder head was off, a minimum waiting time is required after installing the camshafts. When the camshafts are removed, the hydraulic lifters can expand. This expansion can cause increased valve lift when the camshafts are bolted down, resulting in open valves.*

Cylinder head installation waiting times

- 68°F (20°C) and higher .... 10 minutes
- 50-68°F (10-20°C) .... 30 minutes
- 32-50°F (0-10°C) .... 75 minutes

1. Clean cylinder head and gasket surfaces of cylinder block and all timing chain covers.

   - Remove all foreign matter and any liquid from bolt holes, then clean with a thread chaser.

> ***CAUTION —***
> *Do not use a metal scraper or wire brush to clean the aluminum cylinder head or pistons. If necessary, use a hard wooden or plastic scraper. Also available are abrasive discs to be used in conjunction with an electric drill. Be sure to use the correct disc for the type of metal being cleaned.*

2. Check gasket surface of cylinder head and cylinder block for warpage using a straight edge.

> ***NOTE —***
> *If the cylinder head has been machined, a special 0.3 mm thicker gasket should be installed. The thicker gasket is available from an authorized BMW dealer.*

3. Place new cylinder head gasket on cylinder block. Check that two locating dowels are correctly positioned in block and are not damaged.

> ***NOTE —***
> *The word "OBEN", printed on the gasket, should face up. The cylinder head gasket will fit correctly in only one orientation.*

4. With camshafts locked in TDC position with BMW special tools as described earlier, set cylinder head in position, guiding primary chain through cylinder head opening.

> ***CAUTION —***
> - *The camshafts must be locked in the TDC position at the rear of the cylinder head with BMW special tool no. 11 3 240 before installing the cylinder head.*
> - *Make sure the crankshaft, which had been rotated approximately 45° opposite the direction of engine rotation, is still in that position before lowering the cylinder head into position.*

5. Lightly lubricate new cylinder head bolts. Loosely install bolts and their washers, then thread them in until they are finger tight. Install cylinder head-to-lower timing chain cover bolts finger tight.

> ***CAUTION —***
> *Torx-type cylinder head bolts should not be reused. They are stretch-type bolts and must always be replaced whenever loosened.*

NOTE —
Check that all washers for the cylinder head bolts are in place before installing the bolts. Some of the washers may be staked to the cylinder head.

6. Tighten cylinder head bolts in correct sequence. See Fig. 41.

Fig. 41. Cylinder head bolt tightening sequence for M50 engine.

Fig. 42. Protractor and pointer being used to tighten bolts to specified torque angle.

WARNING —
The bolts should be tightened in three stages as listed below. The final stages require the use of a special tool (BMW tool No. 11 2 110) or a suitable protractor to tighten the bolts to a specified torque angle. See Fig. 42.

7. Install and tighten cylinder head-to-lower timing chain cover bolts.

Tightening Torques

- Cylinder head to engine block (Torx E12 M10 bolts)
  - Stage 1 . . . . 30 Nm (22 ft-lb)
  - Stage 2 . . . . +90°
  - Stage 3 . . . . +90°
- Cylinder head to lower timing chain cover . . . . 10 Nm (89 in-lb)

8. Carefully rotate crankshaft in direction of engine rotation and lock it at TDC by inserting BMW special tool no. 11 2 300 through transmission bellhousing. Guide primary timing chain to keep it from jamming on crank gear.

9. Mount primary chain sprocket to primary chain so that arrow on sprocket faces up. Mount sprocket to exhaust camshaft so tapped holes in camshaft are on left most side of elongated holes in sprocket. See Fig. 43.

Fig. 43. Primary chain sprocket correctly aligned to exhaust camshaft. Camshaft holes should be on left most side of sprocket holes.

10. Install timing chain guide and secondary chain tensioner. Tighten mounting bolts.

11. Install as an assembly, secondary chain sprockets and chain so that arrow(s) on sprocket(s) are pointing up. Center mounting holes in sprockets to bolt holes in camshafts. See Fig. 44.

12. Install thrust washers and spring plate (if applicable) to sprockets. See Fig. 45.

Fig. 44. Secondary chain sprockets and chain being installed. Attach sprockets with arrows pointing up and elongated holes centered.

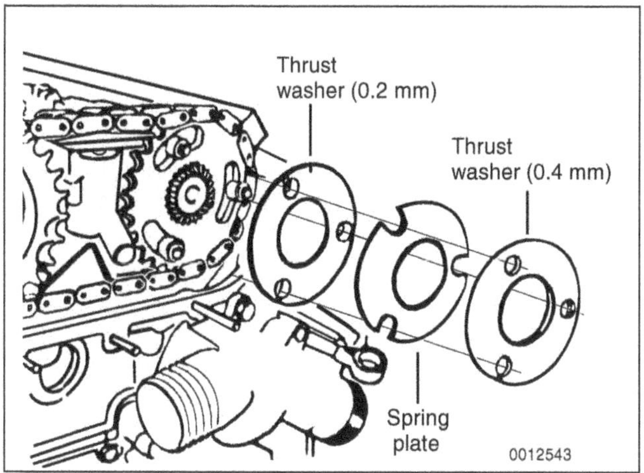

Fig. 45. VANOS spring plate and thrust washers on later engines.

> ***CAUTION —***
> *Mount intake camshaft sprocket so that collar side of the sprocket faces camshaft.*

13. Install and tighten exhaust camshaft sprocket mounting bolts finger tight only. Tighten intake camshaft sprocket mounting nuts to specified torque.

Tightening Torque

- Intake camshaft sprocket mounting nuts (M6) . . . . 10 Nm (89 in-lb)

---

14. Install BMW special tool no. 11 3 390 into primary timing chain tensioner sleeve. Thread tool in finger tight until slack is removed from chain.

NOTE —
BMW special tool no. 11 3 390 is a dummy primary chain tensioner and simulates the function of the tensioner.

15. Install VANOS unit as described in **117 Camshaft Timing Chain**.

> ***CAUTION —***
> *Special installation instructions apply when mounting the VANOS unit to the cylinder head. Be sure to follow the procedure found in* **117 Camshaft Timing Chain**.

16. Slowly unlock and release secondary chain tensioner.

17. Preload primary chain by tightening BMW special tool no. 11 3 390 (dummy chain tensioner) to 1.3 Nm (11 in-lb). Then tighten exhaust secondary camshaft sprocket mounting bolts in two stages.

Tightening Torque

- Secondary timing chain sprocket to exhaust camshaft (Torx bolts)
  - Stage 1 . . . . 5 Nm (44 in-lb)
  - Stage 2 . . . . 22 Nm (17 ft-lb)

---

18. Remove flywheel locking tool from transmission bellhousing. Remove camshaft locking tool from rear of cylinder head.

19. Turn engine over by hand at least two full crankshaft rotations. Reinstall camshaft locking tool and flywheel locking tool to make sure all adjustments are still correct. If settings are correct, remove locking tools.

20. Remove BMW special tool 11 3 390 (dummy chain tensioner) from side of cylinder head.

21. Install primary chain tensioner so that cutout in tensioner piston engages chain rail. See Fig. 46. Using new seals, install and tighten access plugs in VANOS control unit.

Tightening Torques

- Access plugs to VANOS control unit. . . . . 50 Nm (37 ft-lb)
- Primary camshaft timing chain tensioner to tensioner sleeve . . . . 50 Nm (37 ft-lb)

Fig. 46. Primary chain tensioner being installed. Piston cutout must engage chain rail.

22. Install intake cam cover and then install cylinder head cover. Check for correct seating of half-moons in back of cylinder head cover. Use a small amount of 3-Bond® 1209 or equivalent sealant at corners of half-moon cutouts.

Tightening Torque
- Cylinder head cover to cylinder head (M6 screws) . . . . 10 Nm (89 in-lb)

23. Install front exhaust pipes to manifolds using new gaskets and nuts. Coat manifold studs with copper paste prior to installing nuts.

24. Installation of remaining parts is reverse of removal. Refill cooling system as described in **170 Radiator and Cooling System**. Change engine oil and filter as described in **020 Maintenance Program**. Check adjustment of accelerator cable. Additional tightening torques are given below. Reconnect battery last.

25. Installation of remaining parts is reverse of removal, noting the following:
   - Make sure all sealing surfaces are clean and free of old gasket material. Install new intake manifold gasket.
   - Reinstall harness connectors for oil pressure switch and coolant temperature sensor before installing intake manifold.
   - Refill cooling system as described in **170 Radiator and Cooling System**.
   - Change engine oil and filter as described in **020 Maintenance Program**.
   - If necessary, adjust accelerator cable.
   - Reconnect battery last.

CAUTION—
To prevent damaging engine electronic systems, be sure to install all ground wires previously removed, including the ground wires at the cylinder head for the ignition coils.

Tightening Torques
- Coolant drain plug to cylinder block . . 25 Nm (18 ft-lb)
- Intake manifold to cylinder head. . . . . 15 Nm (11 ft-lb)
- Radiator cooling fan to coolant pump. 40 Nm (30 ft-lb)
- Radiator drain screw to radiator . . . 2.5 Nm (22 in-lb)

116 Cylinder Head and Valvetrain

116

GENERAL

This repair group covers cylinder head and valvetrain service and repair. Most of the repairs described here require that the cylinder head first be removed as described in **113 Cylinder Head Removal and Installation**.

NOTE —
- *For timing chain and VANOS repair information, see* **117 Camshaft Timing Chain**.
- *If it is determined that the cylinder head will require significant reconditioning work, a remanufactured cylinder head may be a good alternative. Remanufactured cylinder heads are available from an authorized BMW dealer.*

The information given in this repair group is organized according to engine code. For engine application information, see **100 Engine–General**.

CYLINDER HEAD, DISASSEMBLY AND ASSEMBLY

BMW special tools are required for most cylinder head service described in this repair group. Many of these tools are expensive and only available through an authorized BMW dealer. If the special tools are not available, one alternative is to remove the cylinder head and have it disassembled by an authorized BMW dealer. Be sure to read each procedure thoroughly before starting a job to determine which special tools and equipment will be necessary.

Camshafts, removing and installing (M42 engine)

The camshafts can be removed with the cylinder head mounted on the engine. The first step is to remove the cylinder head cover. See **113 Cylinder Head Removal and Installation**.

Special BMW service tools should always be used to remove and install the camshafts. Removal and installation of the camshafts without the special tools can result in camshaft and valve train damage.

> *WARNING —*
> *The process of evenly loosening the camshaft bearing cap nuts is NOT an acceptable method of removing the camshafts. Be sure to read the procedure through before starting the job.*

1. Disconnect negative (–) battery cable.

> *CAUTION —*
> *Prior to disconnecting the battery, read the battery disconnection cautions given at the front of this manual on page viii.*

2. Remove radiator cooling fan shroud and cooling fan. See **170 Radiator and Cooling System**.

> *CAUTION —*
> *Radiator fan has left hand threads.*

3. Remove timing chain sprockets and timing chains from camshafts. Remove upper timing chain guide from cylinder head. See **117 Camshaft Timing Chain**.

NOTE —

It is not necessary to remove the lower timing chain cover.

4. Remove spark plugs and install BMW special tool 11 3 260 to cylinder head. Tighten tool retaining bolts in spark plug holes. Do not over-tighten bolts. See Fig. 1.

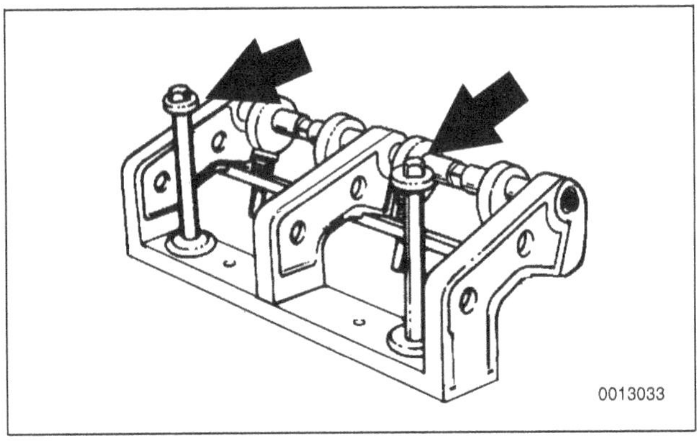

Fig. 1. Special BMW tool no. 11 3 260. Tool mounting bolts (**arrows**) thread into spark plug holes.

Tightening Torque

- BMW special tool 11 1 260 to cylinder head . . . . 23 Nm (17 ft-lb)

5. Turn eccentric shaft on special tool until pressure is applied to camshaft bearing caps. See Fig. 2.

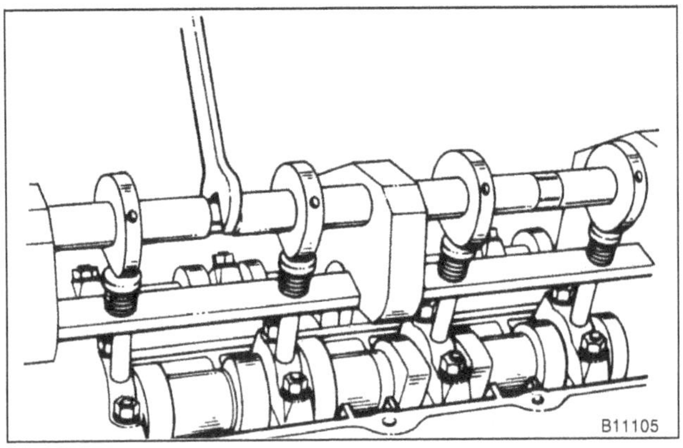

Fig. 2. Special tool (eccentric shaft) being turned to apply pressure to camshaft bearing caps.

6. Remove bearing cap mounting nuts.

7. Slowly release pressure on camshaft bearing caps by turning eccentric shaft. Remove bearing caps and camshaft. Repeat procedure on remaining camshaft.

> ***CAUTION —***
> *Camshaft bearing caps must be reinstalled in their original locations.*

NOTE —

- *The camshaft bearing caps are coded with a letter and a number. For example, "A1" is the exhaust camshaft bearing cap at the front of the engine, and "E1" is the intake camshaft bearing cap at the front of the engine.*
- *Camshafts are stamped with a code letter—"E" for the intake side and "A" for the exhaust side.*
- *"E" is for Einlaß, (German for intake) and "A" is for Auspuff, (German for exhaust).*
- *The front of the engine is at the front of the car.*

8. Camshaft installation is reverse of removal, noting the following:
   - Prior to installation, check all components for wear as described below under **Cylinder Head and Valvetrain, Reconditioning**.
   - Crankshaft must be at least 45° away from TDC before installing camshafts to cylinder head.
   - Lubricate camshafts prior to installation.
   - Install camshafts into cylinder head with lobes for cylinder no. 1 pointing up and towards each other.
   - After installing camshaft(s), observe specified waiting time for hydraulic cam followers to fully compress before bringing crankshaft back to TDC.

Camshaft installation waiting times

- At or above 68°F (20°C). . . . . 10 minutes
- 50-68°F (10-20°C) . . . . 30 minutes
- 32-50°F (0-10°C) . . . . 75 minutes

- Reinstall timing chain sprockets and tensioner as specifically described in **117 Camshaft Timing Chain**.
- Wait at least 10 minutes before cranking engine over with starter motor for the first time.

Tightening Torques

- Camshaft bearing caps to cylinder head (M7) . . . . 15 Nm (11 ft-lb)
- Cylinder head cover to cylinder head (M6) . . . . 10 Nm (89 in-lb)
- Primary timing chain tensioner plug to cylinder head . . . . 40 Nm (30 ft-lb) (see **117 Camshaft Timing Chain**)
- Spark plugs (M12) . . . . 23 Nm (17 ft-lb
- Timing chain cover (upper) to cylinder head . . . . 10 Nm (89 in-lb)
- Timing chain sprockets to camshaft
  M6 . . . . .10 Nm (7.5 ft-lb)
  M7 . . . . .15 Nm (11 ft.lb)
  (see **117 Camshaft Timing Chain**)

Camshafts, removing and installing (M44 engine)

The camshafts can be removed with the cylinder head mounted on the engine. The first step is to remove the cylinder head cover. See **113 Cylinder Head Removal and Installation**.

Special BMW service tools should always be used to remove and install the camshafts. Removal and installation of the camshafts without the special tools can result in camshaft and valvetrain damage.

> ***WARNING —***
> *The process of evenly loosening the camshaft bearing cap nuts is NOT an acceptable method of removing the camshafts. Be sure to read the procedure through before starting the job.*

1. Disconnect negative (–) battery cable.

> ***CAUTION —***
> *Prior to disconnecting the battery, read the battery disconnection cautions given at the front of this manual on page viii.*

2. Remove spark plugs. See **120 Ignition System**.

3. Rotate engine until a pair of camshaft lobes are pointing up. Using BMW special tool 11 5 130, depress each valve and valve spring enough to remove rocker arms. Rotate engine for the next pair of rocker arms. See Fig. 3.

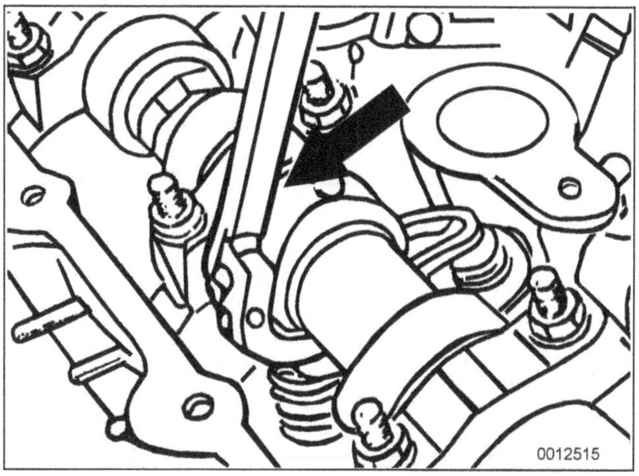

Fig. 3. BMW special tool no. 11 5 130 (**arrow**) being used to depress valve stem to remove rocker arm.

> ***WARNING —***
> *Removed rocker arms must be reinstalled in the same locations.*

4. Remove primary timing chain tensioner from right side of cylinder head.

5. Remove upper timing chain cover.

> ***NOTE —***
> *It is not necessary to remove the lower timing chain cover.*

6. Remove timing chain sprockets with timing chains from front of camshafts. See **117 Camshaft Timing Chain**. Remove upper timing chain guide from cylinder head.

7. Remove camshaft bearing caps and remove camshafts.

> ***CAUTION —***
> *Camshaft bearing caps must be reinstalled in their original locations.*

NOTE—
- *The camshaft bearing caps are coded with a letter and a number. For example, "A1" is the exhaust camshaft bearing cap at the front of the engine, and "E1" is the intake camshaft bearing cap at the front of the engine.*
- *Camshafts are stamped with a code letter—"E" for the intake side and "A" for the exhaust side.*
- *"E" is for Einlaß, (German for intake) and "A" is for Auspuff, (German for exhaust).*
- *The front of the engine is at the front of the car.*

8. Camshaft installation is reverse of removal, noting the following:
   - Prior to installation, check all components for wear as described below under **Cylinder Head and Valvetrain Reconditioning**.
   - Lubricate camshafts prior to installation.
   - Reinstall timing chain sprockets and tensioner and set valve timing **before** reinstalling rocker arms, as specifically described in **117 Camshaft Timing Chain**.
   - Be sure corresponding camshaft lobe is pointing up before installing rocker arms. Note that each rocker arm angles away from corresponding spark plug hole. See Fig. 4.
   - Wait at least 10 minutes for hydraulic cam followers to fully compress before cranking engine over with starter for the first time.

Tightening Torques
- Camshaft bearing caps to cylinder head (M7) . . . . 15 Nm (11 ft-lb)
- Cylinder head cover to cylinder head (M6) . . . . 10 Nm (89 in-lb)
- Primary timing chain tensioner plug to cylinder head . . . . 40 Nm (30 ft-lb) (see **117 Camshaft Timing Chain**)
- Spark plugs (M12) . . . . 23 Nm (17 ft-lb
- Timing chain cover (upper) to cylinder head
  M6 . . . . 10 Nm (7.5 ft-lb)
  M7 . . . . 15 Nm (11 ft-lb)
- Timing chain sprockets to camshaft
  M6 . . . . 10 Nm (7.5 ft-lb)
  M7 . . . . 15 Nm (11 ft.lb)
  (see **117 Camshaft Timing Chain**)

Camshafts, removing and installing (6-cylinder engine)

Camshafts on 6-cylinder engines can be removed with the cylinder head mounted on the engine. The first step is to remove the cylinder head cover. See **113 Cylinder Head Removal and Installation**.

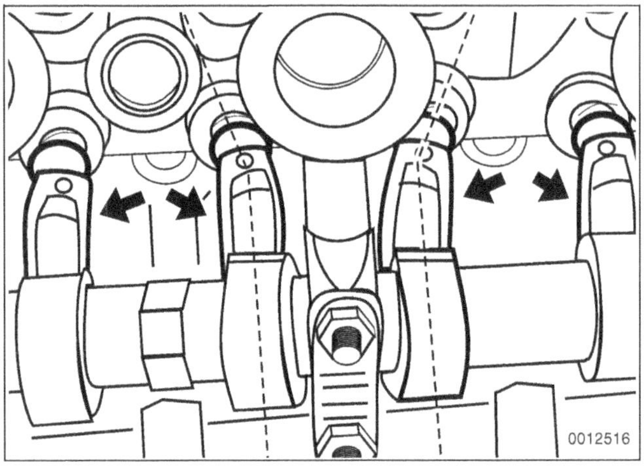

Fig. 4. Correct installation position of rocker arms (**arrows**) on M44 engine.

Special BMW service tools should always be used to remove and install the camshafts. Removal of camshafts without the special tools can result in personal injury and/or cylinder head damage.

WARNING—
The process of evenly loosening the camshaft bearing cap nuts is NOT an acceptable method of removing the camshafts. Special tools should always be used to remove the camshafts. Be sure to read the procedure through before starting a job.

1. Disconnect negative (–) battery cable.

CAUTION—
Prior to disconnecting the battery, read the battery disconnection cautions given at the front of this manual on page viii.

2. Set engine to TDC and lock into position by installing BMW special tool 11 2 300 through transmission bellhousing and into hole in flywheel or drive plate. See Fig. 5.

NOTE—
Arrows on camshaft sprockets should be pointing up. See **117 Camshaft Timing Chain**.

3. Remove VANOS control unit from cylinder head. Remove timing chain sprockets from camshafts. See **117 Camshaft Timing Chain**.

4. Remove spark plugs and install BMW special tool 11 3 260/11 3 270 to cylinder head. Tighten tool retaining bolts in spark plug holes. Do not over-tighten bolts. See Fig. 6.

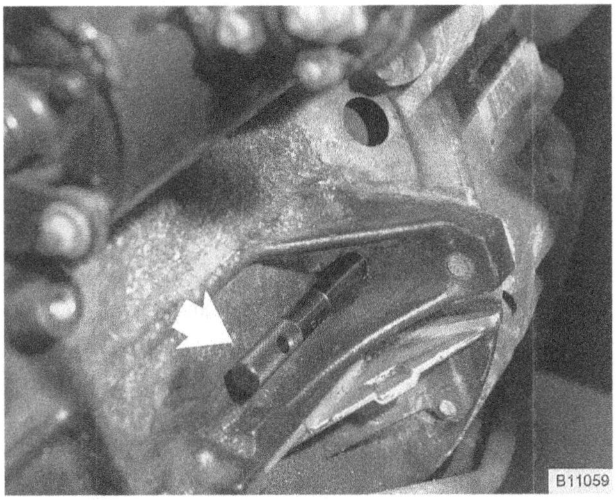

Fig. 5. BMW special tool 11 2 300 (**arrow**) inserted through bell-housing hole and into flywheel.

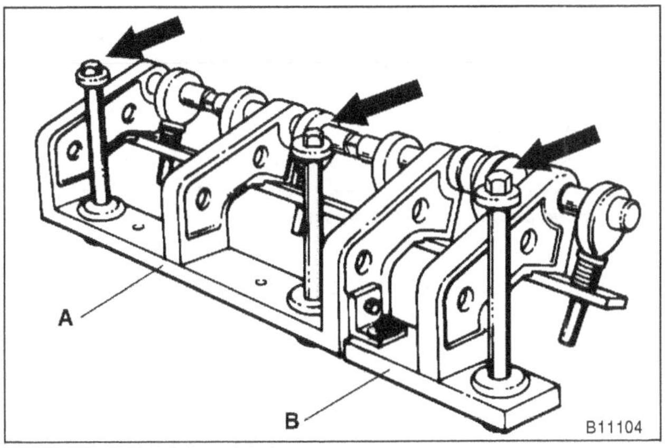

Fig. 6. Special BMW tool no. 11 3 260 (**A**) with adapter 11 3 270 (**B**). Tool mounting bolts (**arrows**) thread into spark plug holes.

Tightening Torque

- BMW special tool 11 1 260/11 3 270 to cylinder head . . . . 23 Nm (17 ft-lb)

---

5. Remove intake camshaft bearing cap no. 1.

6. Turn eccentric shaft on special tool until pressure is applied to camshaft bearing caps. See Fig. 7.

7. Remove remaining bearing cap mounting nuts.

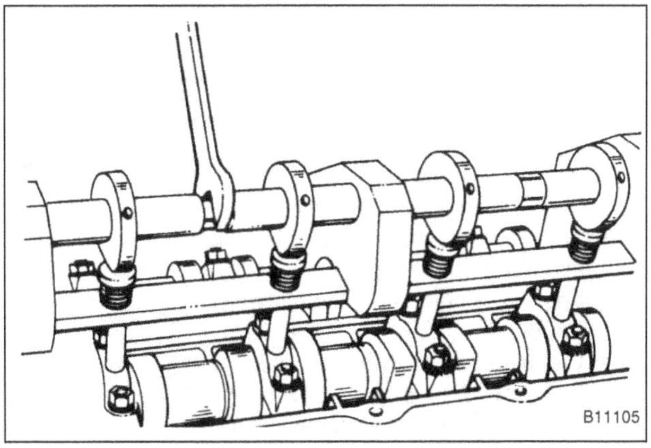

Fig. 7. Special tool (eccentric shaft) being turned to apply pressure to camshaft bearing caps.

8. Slowly release pressure on camshaft bearing caps by turning eccentric shaft. Remove bearing caps and camshaft. Repeat procedure on remaining camshaft.

> ***CAUTION—***
> *Camshaft bearing caps must be reinstalled in their original locations.*

NOTE—

- *The camshaft bearing caps are coded with a letter and a number. For example, "A1" is the exhaust camshaft bearing cap at the front of the engine, and "E1" is the intake camshaft bearing cap at the front of the engine.*
- *The camshafts are stamped with a code letter—"E" for the intake side and "A" for the exhaust side.*
- *"E" is for Einlaß, (German for intake) and "A" is for Auspuff, (German for exhaust).*
- *The front of the engine is at the front of the car.*

9. Camshaft installation is reverse of removal, noting the following:
   - Prior to installation, check all components for wear as described below under **Cylinder Head and Valvetrain Reconditioning**.
   - Crankshaft must be at least 45° away from TDC before installing camshafts to cylinder head.
   - Lubricate camshafts prior to installation.
   - Install camshafts into cylinder head with lobes for cylinder no. 1 pointing up and towards each other. See Fig. 8.
   - After installing camshaft(s), observe specified waiting time for hydraulic cam followers to fully compress before bringing crankshaft back to TDC.

Camshaft installation waiting times
- at or above 68°F (20°C) . . . . 10 minutes
- 50-68°F (10-20°C) . . . . 30 minutes
- 32-50°F (0-10°C) . . . . 75 minutes

   - Reinstall timing chain sprockets and tensioner as specifically described in **117 Camshaft Timing Chain**.
   - Wait at least 10 minutes before cranking engine over with starter motor for the first time.

Fig. 8. Camshaft correctly positioned for reassembly so that cylinder no. 1 lobes (**arrows**) face each other.

Tightening Torques
- Camshaft bearing caps to cylinder head (M7) . . . . 15 Nm (11 ft-lb)
- Cylinder head cover to cylinder head (M6) . . . . 10 Nm (89 in-lb)
- Primary timing chain tensioner to cylinder head—see **117 Camshaft Timing Chain**
- Spark plugs (M12) . . . . 23 Nm (17 ft-lb)
- Timing chain sprocket to camshaft flange (M6)—see **117 Camshaft Timing Chain**

Valves, removing and installing

1. Remove cylinder head as described in **113 Cylinder Head Removal and Installation**. Remove exhaust manifold(s).

2. Remove camshafts using special tools as described earlier.

3. On M42 and all 6-cylinder engines: Remove camshaft carriers by pulling them straight up off mounting studs.

> ***CAUTION —***
> - *Do not let the hydraulic cam followers fall out as the camshaft carrier is removed. Special suction cups are available from BMW to hold the hydraulic cam followers in place during carrier removal.*
> - *Hydraulic cam followers should be stored in an upright position.*

4. On M44 engines: Remove all hydraulic valve adjusters. See **Hydraulic valve adjusters, checking and replacing (M44 engine),** later in this section.

> ***CAUTION —***
> *Hydraulic valve adjusters should be stored in an upright position. If necessary, use a magnetic tool to aid in removal of the adjusters.*

5. Remove valves using a valve spring compressor.

> ***NOTE —***
> - *The M44 cylinder head has recessed valve stems. This requires the use of a special valve spring compressor tool.*
> - *Label each valve assembly as it is removed so it can be installed in its original position.*

6. Remove and discard valve stem oil seals from valve guides. See **Valve stem oil seals** later in this section.

7. Valve installation is reverse of removal.
   - Reinstall exhaust manifold(s) with new gaskets. Use Molycote HSC® or equivalent anti-seize thread compound on fasteners.

Tightening Torque
- Exhaust manifold to cylinder head M7 . . . . 20 Nm (15 ft-lb)

CYLINDER HEAD AND VALVETRAIN, RECONDITIONING

This section provides the specifications and special reconditioning information necessary to repair the cylinder heads covered by this manual.

If machine shop services are not readily available, one alternative is to install a remanufactured cylinder head from an authorized BMW dealer.

Cylinder head

The disassembled cylinder head should be inspected for warpage and cracks. Check the valve guides and valve seats for wear before machining a warped head. Always decarbonize and clean the head before inspecting it. A high-quality straight edge can be used to check for warpage.

Visually inspect the cylinder head for cracks. If a cracked cylinder head is suspected and no cracks are detected through the visual inspection, have the head further tested for cracks by an authorized BMW dealer. A cracked cylinder head must be replaced.

NOTE —
When replacing the cylinder head on engines with removable camshaft carriers, be sure to check the camshaft carrier bearing surfaces for warpage.

A warped cylinder head can be machined provided no more than 0.3 mm (0.012 in) of material is removed. If further machining is required, the head should be replaced. Removing more than this amount will reduce the size of the combustion chamber and adversely affect engine performance. A 0.3 mm thicker gasket is available from an authorized BMW parts department for machined heads.

Before machining the head to correct for warpage, measure the total height (thickness of the cylinder head). See Fig. 9. Minimum height specifications are given in **Table a**.

Table a. Cylinder Head Height

Engine	New	Minimum height (dimension A)
4-cylinder	140.0 mm (5.512 in)	139.55 mm (5.494 in)
6-cylinder	140.0 mm (5.512 in)	139.7 mm (5.500 in)

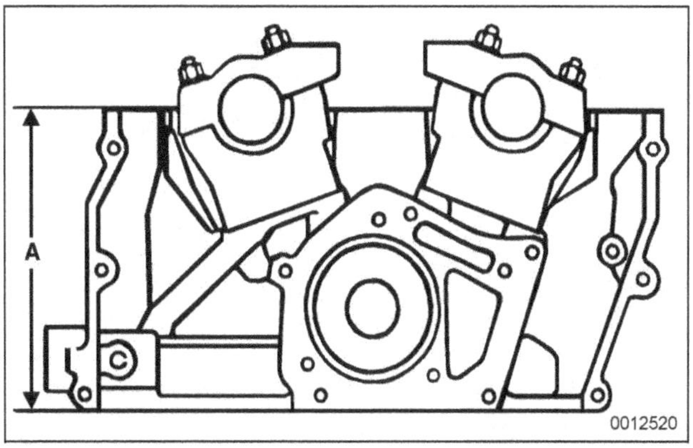

Fig. 9. Front view of cylinder head showing minimum resurfacing dimension (**A**).

Camshaft

Camshaft wear is usually caused by insufficient lubrication. Visually inspect the camshaft lobes and journals for wear. Camshaft wear specifications are given in **Table b**.

Table b. Camshaft Specifications

4-cylinder engines	
Axial play	
M42 engine	0.065-0.150 mm (0.0025-0.0059 in)
M44 engine	0.065-0.150 mm (0.0025-0.0059 in)
Radial play (Plastigage)	
M42 engine	0.020-0.061 mm (0.0008-0.0024 in)
M44 engine	0.040-0.082 mm (0.0016-0.0032 in)
6-cylinder engines	
Axial play	
All	0.150-0.330 mm (0.006-0.013 in)
Radial play (Plastigage)	
All	0.020-0.054 mm (0.0008-0.0021 in)

NOTE —
- *On M44 engines, check the oil sprayer tube in the top of the cylinder head cover and the tube seal in the cylinder head. A clogged sprayer tube or a defective seal will reduce oil flow to the camshaft. See Fig. 10.*
- *If the camshaft is being replaced due to cam lobe wear, the corresponding rocker arms or cam followers should also be replaced to avoid damaging the new camshaft.*

Fig. 10. Location of camshaft oil sprayer tube seal in M44 cylinder head (**arrow**).

Hydraulic valve adjusters, checking and replacing (M44 engine)

The M44 engine uses hydraulic valve adjusters (HVA) to keep valve clearances within a limited working range. See Fig. 11. HVAs are sealed units and require no maintenance.

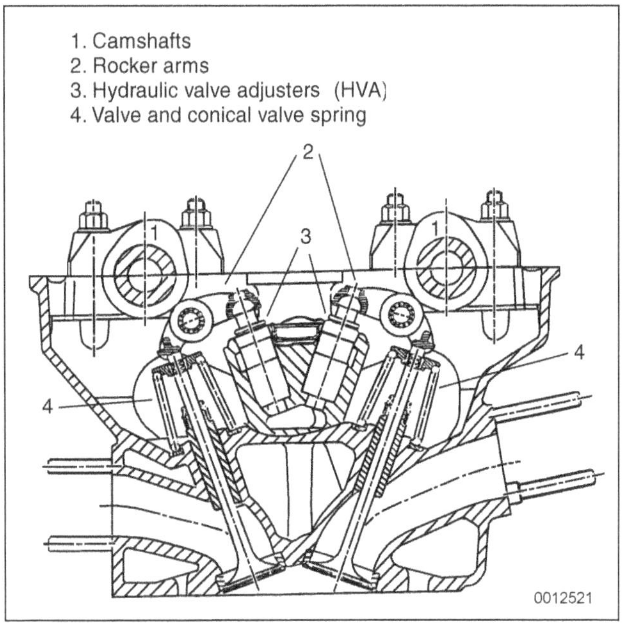

Fig. 11. Cutaway view of M44 cylinder head.

Under some circumstances, such as a cold start, the hydraulic units may become noisy. Hydraulic valve adjuster noise is usually a high-pitched tapping or chattering noise. In most instances, this is considered normal as long as the noise goes away in a few minutes (maximum 20 minutes). If the noise does not go away, either the hydraulic unit is faulty or the oil pressure to the adjuster(s) is low.

NOTE —
Before checking valve adjusters, make sure engine oil is fresh and at proper level.

1. Run engine until it reaches normal operating temperature.

2. Turn engine off and remove top engine cover, spark plugs, and cylinder head cover. See **113 Cylinder Head Removal and Installation**.

3. Use a plastic or wooden stick to press down on top of rocker arm end (side away from camshaft). If any play can be detected, the HVA is faulty and should be replaced.

NOTE —
When checking the HVA, make sure the camshaft lobe is facing up so that there is no valve spring pressure on the adjuster.

4. To remove an HVA, turn engine over until camshaft lobe points straight up. Using BMW special tool no. 11 5 130, depress valve spring and remove rocker arm from under camshaft. See Fig. 12.

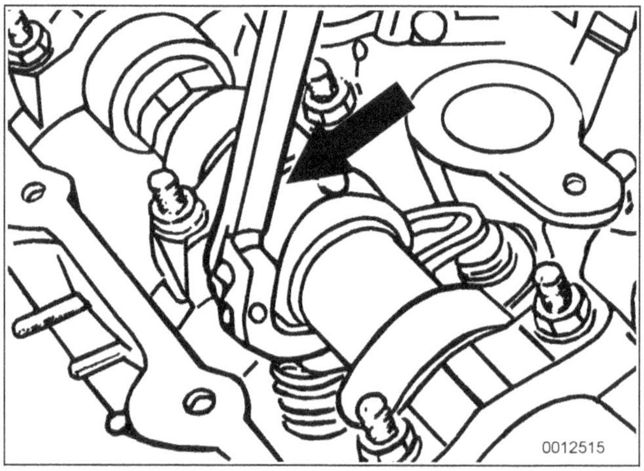

Fig. 12. BMW special tool no. 11 5 130 (**arrow**) being used to depress valve stem to remove rocker arm.

WARNING —
Removed rocker arms must be reinstalled in the same locations.

5. Pull hydraulic valve adjuster straight out of cylinder head. See Fig. 13.

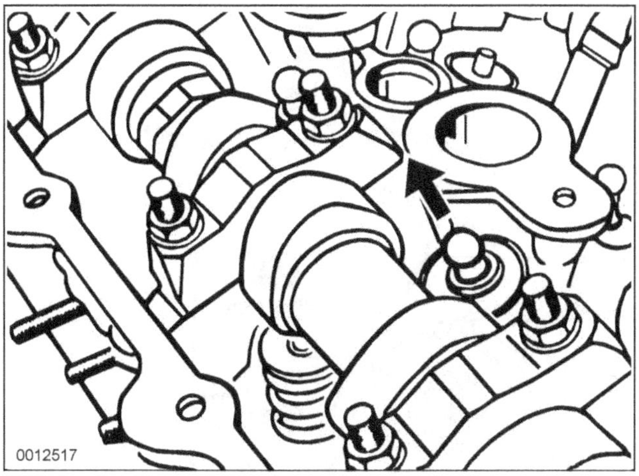

Fig. 13. Remove hydraulic valve adjuster by pulling straight out of cylinder head (**arrow**).

6. Installation is reverse of removal. Be sure to inspect HVA bore in the cylinder head for wear or scoring.

Tightening Torque

- Cylinder head cover to cylinder head/timing chain cover
  M6 bolt . . . . 10 Nm (89 in-lb)
  M7 bolt . . . . 15 Nm (11 ft-lb)

Hydraulic cam followers, checking and replacing (M42 and all 6-cylinder engines)

The M42 engine and all 6-cylinder engines use self-adjusting hydraulic cam followers to keep the valve clearances within a limited working range. The cam followers are sealed units and require no maintenance. See Fig. 14..

Under some circumstances, such as a cold start, the cam followers may become noisy. Hydraulic cam follower noise is usually a high-pitched tapping or chattering noise. In most instances, this is considered normal as long as the noise goes away in a few minutes (maximum 20 minutes). If the noise does not go away, either the follower is faulty or the oil pressure to the follower is low. Hydraulic cam follower replacement requires that the camshaft first be removed.

NOTE —

Before checking valve adjusters, make sure engine oil is fresh and at proper level.

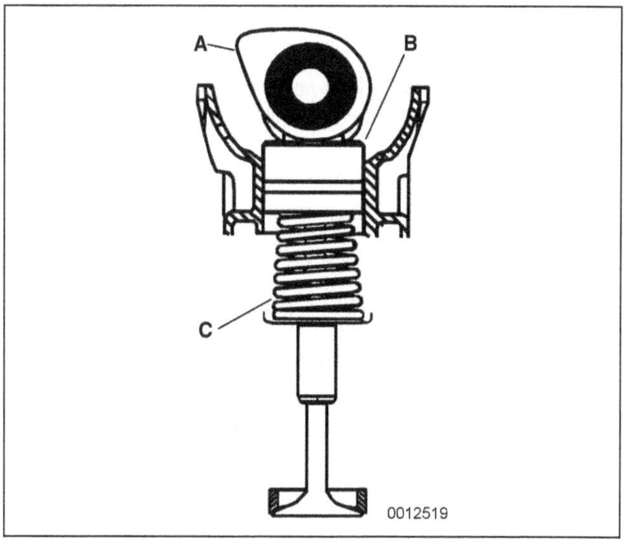

Fig. 14. Section view of camshaft (**A**), hydraulic cam follower (**B**), and valve with conical valve spring (**C**) used on M52 engine.

1. Run engine until it reaches normal operating temperature.

2. Turn engine off and remove top engine covers, ignition coils, and cylinder head covers. See **113 Cylinder Head Removal and Installation**.

3. Use a plastic or wooden stick to press down on top of cam follower. See Fig. 15. If it compresses noticeably, the follower is faulty and should be replaced.

B11107

Fig. 15. Hydraulic cam follower in M52 engine being checked. Note that camshaft lobe is pointing up.

NOTE —
When checking a cam follower, make sure the corresponding camshaft lobe is facing up so that there is no valve spring pressure on the follower.

4. To replace a cam follower, remove appropriate camshaft. Refer to camshaft removal procedures given earlier.

5. Once camshaft is out of way, withdraw faulty cam follower and replace with new one. Inspect cam follower bores for wear and scoring.

6. Camshaft, timing chain and cylinder head cover reassembly is reverse of disassembly.

Tightening Torques
- Camshaft bearing caps to cylinder head
  M7 . . . . 15 Nm (11 ft-lb)
- Cylinder head cover to cylinder head/timing chain cover
  M6 . . . . 10 Nm (89 in-lb)
- Timing chain hydraulic tensioner to cylinder head—see **117 Camshaft Timing Chain**
- Timing chain sprocket to camshaft flange (M6)—see **117 Camshaft Timing Chain**

Valve guides

BMW does not supply valve guides as replacement parts. If a valve guide is excessively worn, it should be reamed to accept oversized valve stems. Valve guide specifications are listed in **Table c**.

Valve guides should be checked for wear using a new valve. See Fig. 16. Be sure to thoroughly inspect the cylinder head to ensure that it can be reused before reworking the guides.

Fig. 16. Valve guide wear being checked with dial indicator. Insert new valve until stem end is flush with end of guide (**white arrow**). See **Table c** for wear limits.

NOTE —
- *International Organization for Standardization (ISO) tolerance classes are listed in* **Table c**. *ISO allowances are based on nominal sizes and should be used to determine proper fit. Most machine shops should have this information available*
- *Replacement valve guides may be available through aftermarket suppliers. Valve guide replacement requires special tools and a press. It is also necessary to heat the cylinder head and chill the valve guides when replacing the guides.*
- *BMW supplies oversize valves. See* **Valves** *for additional information.*

Table c. Valve Guide Specifications

Specifications	M42 engine (1992)	M42/M44 engines (1993-1998)	M50 engine (1992)	M50, M52, S50 US, S52 US engines (1993-1998)
Valve guide wear, maximum (valve tilt clearance measured with new valve)	0.5 mm (0.020 in)	0.5 mm (0.020 in)	0.5 mm (0.020 in)	0.5 mm (0.020 in)
Valve guide inside diameter, installed (tolerance per ISO allowance H7)				
Standard	7.0 mm (0.276 in)	6.0 mm (0.236 in)	7.0 mm (0.276 in)	6.0 mm (0.236 in)
Oversize 1	7.1 mm (0.279 in)	6.1 mm (0.240 in)	7.1 mm (0.279 in)	6.1 mm (0.240 in)
Oversize 2	7.2 mm (0.283 in)	6.2 mm (0.244 in)	7.2 mm (0.283 in)	6.2 mm (0.244 in)

Valves

Valve specifications are listed in **Table d**. Remove carbon deposits from the valves using a wire brush or wire wheel.

Table d. Valve Specifications

Specification	M42 engine (1992)	M42/M44 engines (1993-1998)	M50 engine (1992)	M50, M52 engines (1993-1998)	S50 US/S52 US engine (1995-1998)
Valve head dia. Intake Exhaust	 33.0 mm (1.299 in) 30.5 mm (1.201 in)	 33.0 mm (1.299 in) 30.5 mm (1.201 in)	 33.0 mm (1.299 in) 30.5 mm (1.201 in)	 33.0 mm (1.299 in) 30.5 mm (1.201 in)	 30.0 mm (1.181 in) 27.0 mm (1.063 in)
Valve stem dia. Standard Intake Exhaust	 $6.975^{-0.015}$ mm ($0.2746^{-0.0006}$ in) $6.960^{-0.015}$ mm ($0.2740^{-0.0006}$ in)	 $5.975^{-0.015}$ mm ($0.2352^{-0.0006}$ in) $5.960^{-0.015}$ mm ($0.2345^{-0.0006}$ in)	 $6.975^{-0.015}$ mm ($0.2746^{-0.0006}$ in) $6.960^{-0.015}$ mm ($0.2740^{-0.0006}$ in)	 $5.975^{-0.015}$ mm ($0.2352^{-0.0006}$ in) $5.960^{-0.015}$ mm ($0.2345^{-0.0006}$ in)	 $5.975^{-0.015}$ mm ($0.2352^{-0.0006}$ in) $5.960^{-0.015}$ mm ($0.2345^{-0.0006}$ in)
Oversize 1 Intake Exhaust	 $7.075^{-0.015}$ mm ($0.2785^{-0.0006}$ in) $7.060^{-0.015}$ mm ($0.2779^{-0.0006}$ in)	 $6.075^{-0.015}$ mm ($0.2392^{-0.0006}$ in) $6.060^{-0.015}$ mm ($0.2386^{-0.0006}$ in)	 $7.075^{-0.015}$ mm ($0.2785^{-0.0006}$ in) $7.060^{-0.015}$ mm ($0.2779^{-0.0006}$ in)	 $6.075^{-0.015}$ mm ($0.2392^{-0.0006}$ in) $6.060^{-0.015}$ mm ($0.2386^{-0.0006}$ in)	 $6.075^{-0.015}$ mm ($0.2392^{-0.0006}$ in) $6.060^{-0.015}$ mm ($0.2386^{-0.0006}$ in)
Oversize 2 Intake Exhaust	 $7.175^{-0.015}$ mm ($0.2825^{-0.0006}$ in) $7.160^{-0.015}$ mm ($0.2819^{-0.0006}$ in)	 $6.175^{-0.015}$ mm ($0.2431^{-0.0006}$ in) $6.160^{-0.015}$ mm ($0.2425^{-0.0006}$ in)	 $7.175^{-0.015}$ mm ($0.2825^{-0.0006}$ in) $7.160^{-0.015}$ mm ($0.2819^{-0.0006}$ in)	 $6.175^{-0.015}$ mm ($0.2431^{-0.0006}$ in) $6.160^{-0.015}$ mm ($0.2425^{-0.0006}$ in)	 $6.175^{-0.015}$ mm ($0.2431^{-0.0006}$ in) $6.160^{-0.015}$ mm ($0.2425^{-0.0006}$ in)

Valves, leak test

To test the valves for leakage, the camshafts and camshaft carriers (where applicable) should first be removed. Install the valve assemblies and the spark plugs in each cylinder. Place the cylinder head on a workbench with the combustion chamber facing upward. Fill each combustion chamber with a thin non-flammable liquid, such as a parts cleaning fluid. After fifteen minutes, check the level of the fluid. If the fluid level in any cylinder drops, that cylinder is not sealing properly.

Valve Stem Oil Seals

The purpose of the valve stem oil seal is to prevent excess oil from entering the combustion chamber. The sign of faulty valve stem oil seals is excessive oil consumption and smoke from the exhaust immediately after starting and during deceleration

NOTE —

Valve stem oil seals should not be reused. If valves are removed, new valve stem oil seals should be installed.

Valve stem oil seal replacement requires that the cylinder head be disassembled and the valves removed as described above under **Valves, removing and installing**.

NOTE —

BMW special tools are available to remove the valve stem oil seals. As an alternative, standard valve seal removal tools are available from most automotive parts stores. See Fig. 17.

Fig. 17. Valve stem oil seal being removed from valve guide.

Lubricate the new seals and install them using hand pressure only. Be sure to install the valve spring seat(s) before installing the seal.

Valve Seats

The valve seats should be resurfaced whenever new valves or valve guides are installed. Cutters are required to resurface the seats. Always check the valves for leaks after reconditioning a valve seat as described above. **Table e** lists valve seat dimensions.

> ***NOTE —***
> *Standard size replacement valve seats are not available from BMW. Replacement valve seats are only available from BMW in 0.4 mm oversize (oversized in both height and diameter). The manufacturer does not provide specifications for valve seat replacement for the engines covered by this manual.*

Valve Springs

The valve springs should be checked for fatigue. To quickly check the springs, line them up in a row. Place a straight edge across the top of the springs. Any spring that is significantly shorter than the others is worn and should be replaced.

> ***NOTE —***
> *Valve spring specifications and wear limits are not available from BMW.*

Table e. Valve Seat Dimensions

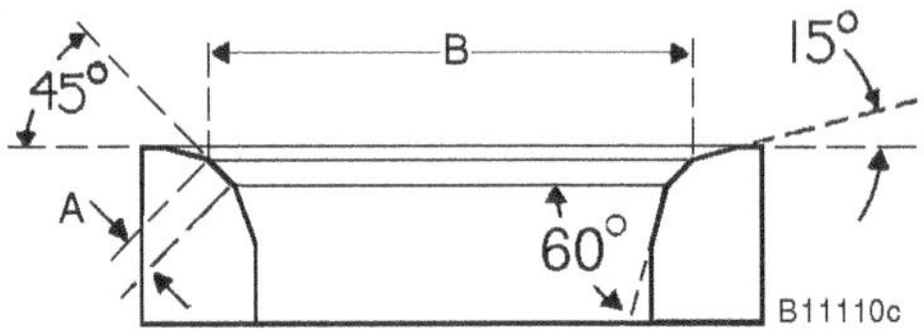

Specification	M42 engine (1992)	M42 engine (1993-1995)	M44 engine (1996-1998)	M50 engine (1992)	M50/M52 engines (1993-1998)	S50US/S52US engines (1995-1998)
Valve seat angle	45°	45°	45°	45°	45°	45°
Correction angle, outside	15°	15°	15°	15°	15°	15°
Correction angle, inside	60°	60°	60°	60°	60°	60°
Valve seat width (**A**)						
intake	1.65 ± 0.25 mm (0.065 ± 0.010 in)	1.65 ± 0.25 mm (0.065 ± 0.010 in)	1.65 ± 0.25 mm (0.065 ± 0.010 in)	1.4 - 1.9 mm (0.055-0.075 in)	1.65 ± 0.25 mm (0.065 ± 0.010 in)	1.65 ± 0.25 mm (0.065 ± 0.010 in)
exhaust	2.0 ± 0.25 mm (0.079 ± 0.010 in)	2.0 ± 0.25 mm (0.079 ± 0.010 in)	1.65 ± 0.25 mm (0.065 ± 0.010 in)	1.4 - 1.9 mm (0.055-0.075 in)	1.65 ± 0.25 mm (0.065 ± 0.010 in)	1.65 ± 0.25 mm (0.065 ± 0.010 in)
Valve seat outside dia. (**B**)						
intake	32.4 mm (1.276 in)	32.4 mm (1.276 in)	32.4 mm (1.276 in)	32.4 mm (1.276 in)	32.4 mm (1.276 in)	29.4 mm (1.157 in)
exhaust	30.1 mm (1.185 in)	30.1 mm (1.185 in)	30.1 mm (1.185 in)	30.1 mm (1.185 in)	29.4 mm (1.157 in)	26.4 mm (1.039 in)

117 Camshaft Timing Chain

GENERAL

This repair group covers timing chain repair information for the 4- and 6-cylinder engines.

4-cylinder engines use one double-row chain to drive the dual overhead camshafts. See Fig. 1.

Fig. 1. M44 engine valvetrain chain drive. M42 engine is similar.

6-cylinder engines use two chains to drive the camshafts. A primary chain is driven off the crankshaft and drives the primary sprocket on the exhaust camshaft. The secondary chain drives the intake camshaft. Each chain is tensioned by a hydraulic tensioner. See Fig. 2.

The timing chains are lubricated by engine oil and do not require maintenance. A worn timing chain and sprockets can lead to noisy operation and erratic valve timing. A faulty tensioner can also cause timing chain noises.

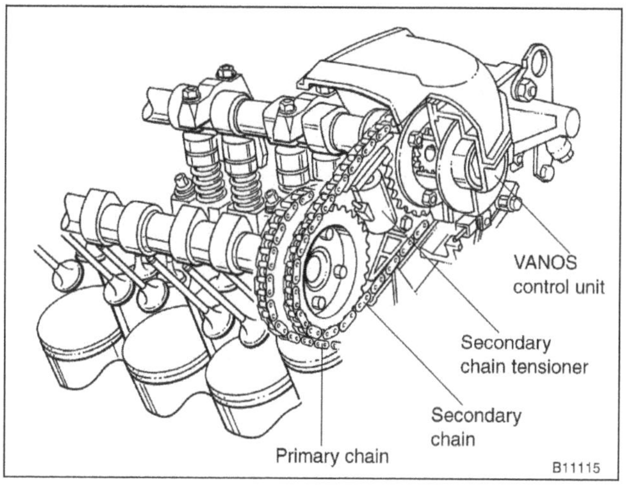

Fig. 2. Cutaway view of camshaft timing chains on 1993 and later 6-cylinder engine with VANOS.

NOTE —

See **100 Engine–General** *for engine code and application information.*

CAMSHAFT TIMING CHAIN, 4-CYLINDER

Special BMW service tools are needed for timing chain removal and installation procedures. The special tools assure proper timing of the valvetrain. Precise marks to set the timing on the camshafts are not provided for reassembly. Read the procedures through before beginning the job.

> ***CAUTION —***
> *If the camshafts are not properly timed, the pistons can contact the valves.*

Camshaft timing chain, removing (4-cylinder engines)

1. Disconnect negative (–) cable from battery.

> ***CAUTION —***
> *Prior to disconnecting the battery, read the battery disconnection cautions given at the front of this manual on page viii.*

2. Drain engine coolant. Drain engine block by removing block drain. Remove radiator cooling fan shroud and cooling fan. See **170 Radiator and Cooling System**..

> ***WARNING —***
> *Allow the engine to cool before opening or draining the cooling system.*

> ***CAUTION —***
> *On cars with viscous-type fan clutch, fan has left hand threads.*

NOTE —
The block drain plug is located on the exhaust side of the engine, below cylinder no. 4.

3. Remove cylinder head cover. See **113 Cylinder Head Removal and Installation.**

NOTE —
Make note of the arrangement of the rubber insulators when removing the cylinder head cover mounting bolts.

4. Loosen spark plugs. See **120 Ignition System**.
5. Disconnect coolant hoses from thermostat housing on front of cylinder head. Unbolt thermostat housing from cylinder head.
6. Remove cylinder identification sensor from upper timing chain cover. Remove upper timing chain cover. See Fig. 3.

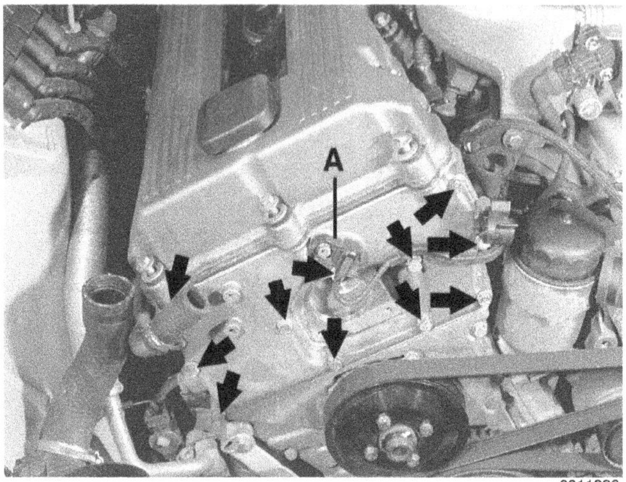

0011996

Fig. 3. Upper timing chain cover mounting bolts (**arrows**). Thermostat housing shown removed. Cylinder identification sensor (**A**) also shown.

7. Set engine to approximate TDC by rotating in normal operating direction until camshaft lobes at cylinder no. 1 are facing up and toward each other. See Fig. 4.

0011998

Fig. 4. Engine set to approximate TDC. Cylinder no. 1 camshaft lobes face in and sprocket arrows point up (**arrows**).

8. Remove air conditioning drive belt. Loosen and push aside air conditioning compressor. Remove air conditioning compressor mounting bracket. See **640 Heating and Air Conditioning**.
9. Remove engine drive belts and coolant pump pulley. See **020 Maintenance Program**.
10. Secure crankshaft vibration damper to prevent crankshaft from turning. Using a socket, loosen vibration damper hub center bolt.

NOTE —

The vibration damper center bolt is very tight. A heavy duty holding device, such as BMW special tool no. 11 2 150, should be used to hold the vibration damper hub stationary while the bolt is loosened.

11. Using loosened bolt at vibration damper, turn crankshaft to TDC and lock in position by inserting BMW special tool no. 11 2 300 through transmission bellhousing. See Fig. 5.

NOTE —

- *If necessary, remove plug from locating bore in lower edge of bell housing.*
- *Confirm that locking tool has been correctly installed by trying to rotate the crankshaft.*

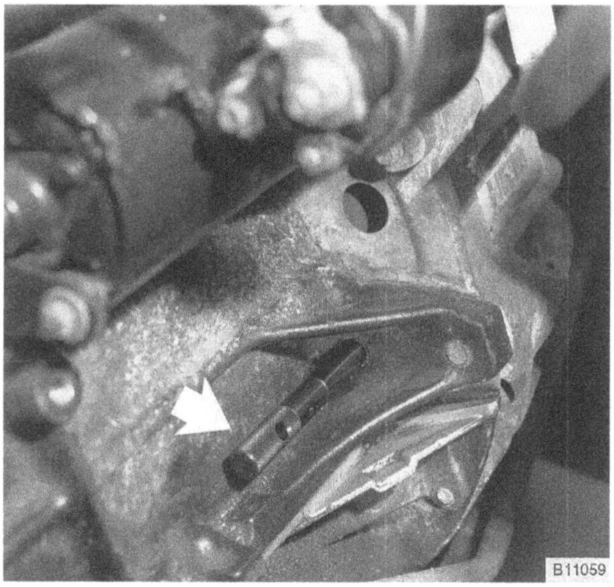

Fig. 5. BMW special tool no. 11 2 300 installed through bellhousing and into flywheel to lock crankshaft at TDC (**arrow**).

12. Lock camshafts at TDC by mounting BMW special tool no. 11 3 240 over square ends of camshafts. See Fig. 6.
    - Check that tool is squarely seated on cylinder head gasket surface. If necessary, rotate camshafts slightly using an open-end wrench at hex on camshafts.

CAUTION —
The camshafts must be locked in the TDC position using the special service tool before removing the timing chain. The tool holds the camshafts parallel to each other and perpendicular to the valve cover gasket surface.

13. Remove lower timing chain cover mounting bolts and carefully lift cover off. Clean sealing surfaces and gasket grooves. See Fig. 7.

14. Remove hydraulic chain tensioner from right side of cylinder head. See Fig. 8.

Fig. 6. BMW special tool no. 11 3 240 installed on rear of camshafts to position camshafts at TDC. (M44 camshafts shown. Other engines are similar.

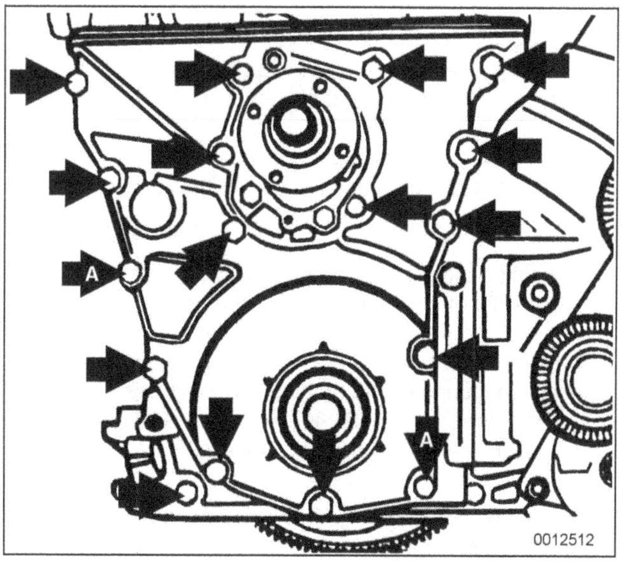

Fig. 7. Lower timing chain cover mounting bolts (**arrows**). Bolts A are installed through hollow locating dowels.

Fig. 8. Hydraulic chain tensioner plug (**arrow**).

15. Remove mounting bolts from left and right camshaft sprockets.
16. On M42 engines, remove upper bolt on right side chain guide and unbolt lower timing chain guide below crankshaft.
17. Remove camshaft sprockets from camshafts together with chain and crankshaft sprocket. Note Woodruff key when removing crankshaft sprocket.

> ***CAUTION —***
> *The crankshaft must not be allowed to rotate when the timing chain is removed. The pistons can contact the valves.*

Camshaft timing chain, installing (4-cylinder engines)

Inspect all sprockets for wear or damage. Inspect the chain guide and tensioner rails for grooves caused by chain contact. Replace any part that is worn. If any of the sprockets are worn, the chain and sprockets are replaced as an assembly.

The procedure outlined below assumes that the camshafts and the crankshaft are locked in the TDC installation position with special tools (shown earlier in Fig. 5 and Fig. 6).

1. Install timing chain to crankshaft sprocket, then slide sprocket on crankshaft while aligning woodruff key.

2. Place camshaft sprockets on chain, then place sprockets on camshafts so that arrows on sprockets point up and elongated holes in sprockets are centered to tapped holes in camshafts. See Fig. 9.

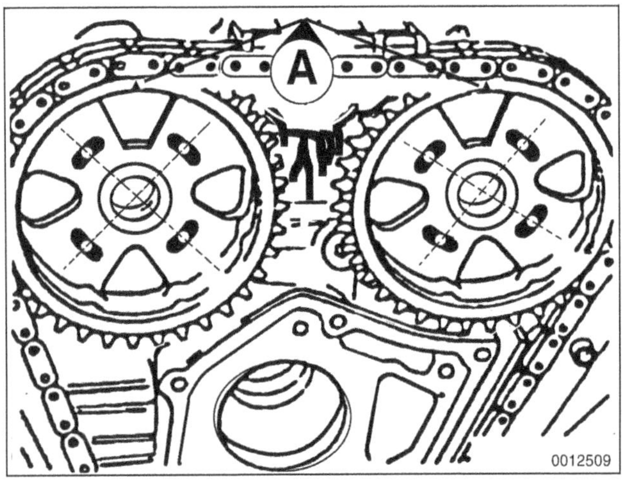

Fig. 9. Camshaft sprockets correctly installed with arrows pointing up and bores centered to tapped holes (M44 engine).

3. On M44 engine: Place cylinder position sensor plate on intake camshaft sprocket so that arrow on sensor plate points up.

4. Install and hand-tighten sprocket mounting bolts.

5. On M42 engine: Install lower chain guide.

6. On M44 engine: Remove allen-head retaining screw at left chain-guide. Turn slotted-head of adjusting sleeve so that sleeve contacts cylinder head. Then install and tighten allen-head screw fully. See Fig. 10.

0011999

Fig. 10. Left upper chain guide adjusting sleeve (**arrow**). Note camshaft position

> ***CAUTION —***
> *Different hydraulic chain tensioners were used during manufacture, depending on engine type. See Fig. 11. Note that the late style tensioner can be retrofitted to the earlier engine, so be sure to identify the version of tensioner installed as installation instructions vary between the two.*

7. On M42 engine: Disassemble chain tensioner by striking outer sleeve against a solid object. This will release snap ring and tensioner will separate. See Fig. 12.

8. On M42 engine: Clamp chain tensioner piston in vice (soft jaws). Slowly compress piston, making sure detent ring fits into taper of outer sleeve. If necessary, press ends of detent ring together. Slowly continue pressing tensioner together until snap ring audibly clicks into outer sleeve. Measure overall length of tensioner to confirm correct assembly: 68.5 mm (2.7 in.).

NOTE —
The late-style tensioner on the M44 engine can be retrofitted to the M42 engine.

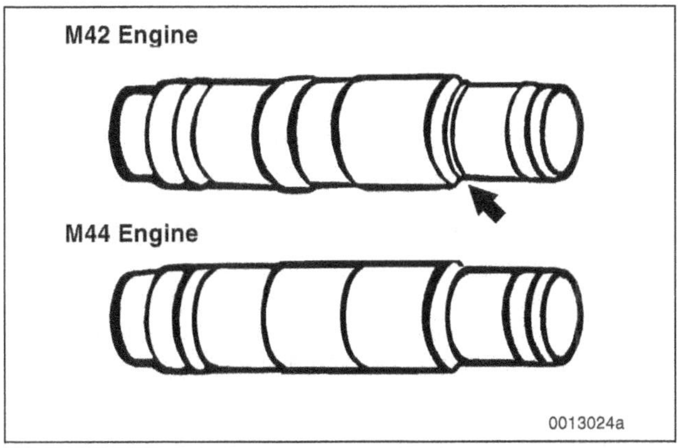

Fig. 11. Two versions of hydraulic chain tensioners. Original chain tensioner on M42 engine (**top**) can be disassembled. Chain tensioner on M44 engine (**bottom**) cannot be disassembled. Note differences in outer sleeve. Also, the M42 tensioner detent ring groove is visible when tensioner is extended (**arrow**).

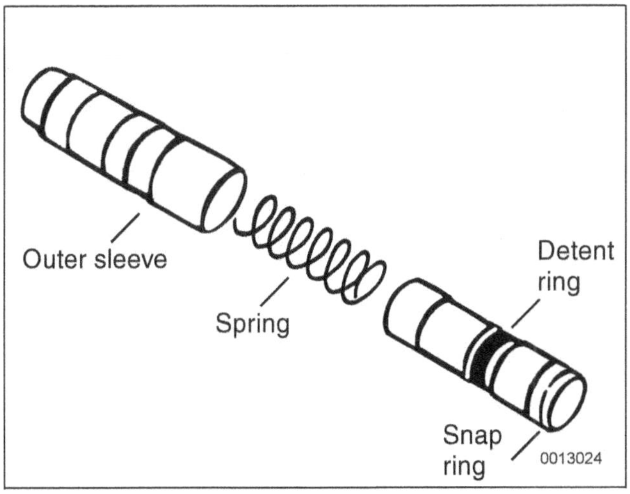

Fig. 12. Disassembled early-style chain tensioner used on M42 engine.

9. On M44 engines: Clamp tensioner piston in vice (soft jaws) and squeeze oil from it. Slowly compress piston only up to end circlip. Repeat procedure twice to ensure all oil is expelled. See Fig. 13.

10. Install hydraulic chain tensioner to cylinder head. Use a new sealing washer on tensioner plug when installing.

Tightening Torque
- Chain tensioner to cylinder head .... 40 Nm (30 ft-lb)

11. On M42 engines: Release early-style chain tensioner by pressing chain guide against tensioner until it unlocks and expands out.

12. Check that chain is now free of slack. Tighten down sprocket mounting bolts.

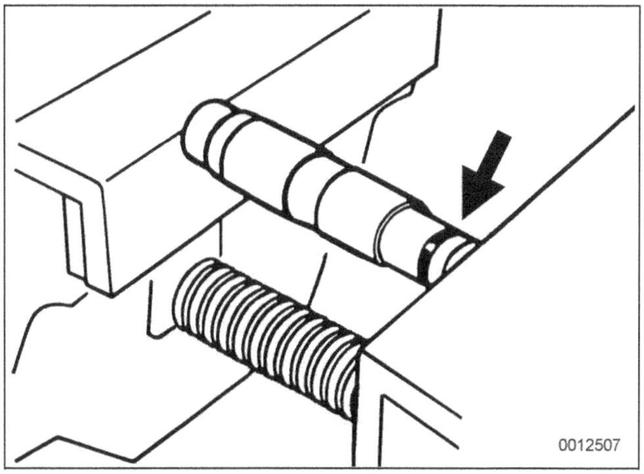

Fig. 13. When reinstalling chain tensioner, oil in tensioner piston must first be expelled by compressing piston in vise. When compressing piston, only compress up to end circlip (**arrow**).

Tightening Torque
- Chain sprocket to camshaft flange (M7) .... 15 Nm (11 ft-lb)

13. Remove crankshaft locking tool from bellhousing and camshaft locking tool from rear of cylinder head.

14. Install lower timing chain cover:
    - Install new crankshaft oil seal in cover.
    - Use new gaskets when installing cover.
    - Thoroughly clean profile gasket groove and all gasket sealing surfaces on lower chain cover.
    - Apply a small bead of silicon sealer (3-Bond 1209® or equivalent) to joint corners where chain cover meets head and block. See Fig. 14.
    - Tap cover into position to engage locating dowels.
    - Tighten cover mounting bolts alternately and in stages.

Fig. 14. Apply sealer (3-Bond 1209® or equivalent) to joint corners where chain cover meets head and block (**arrows**).

15. Install upper timing chain cover:
    - Thoroughly clean profile gasket groove and all sealing surfaces.
    - Apply a small bead of silicon sealer (3-Bond 1209® or equivalent) to upper and lower joint corners where chain cover meets cylinder head.
    - Install new profile gasket to groove in lower chain cover.
    - Protect top of profile gasket using thin sheet metal (BMW special tool no. 11 2 330 or equivalent). See Fig. 15.
    - Attach new gaskets to cover with light dabs of gasket sealant, hold gaskets in place using cover mounting screws.
    - Place upper timing cover into position. Snug down all screws, without tightening. Carefully slide out sheet metal.
    - Temporarily install cylinder head cover without gasket. Install and tighten at least two cylinder head cover mounting bolts to seat upper timing chain cover.
    - Tighten all upper timing cover screws.
    - Remove valve cover.

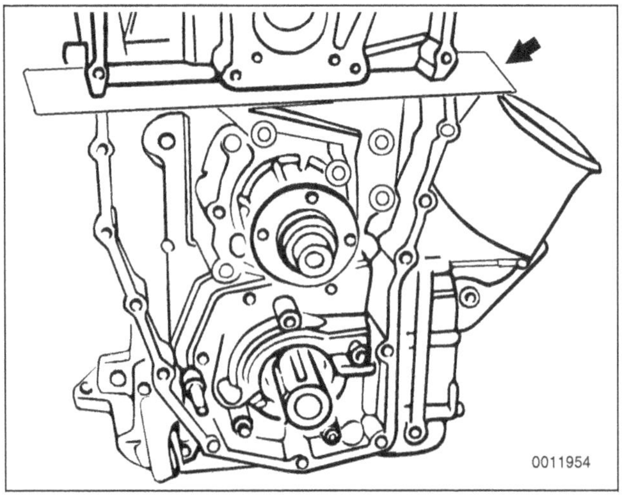

Fig. 15. When installing upper timing chain cover, use thin sheet metal (**arrow**) to protect profile gasket when installing upper cover.

Tightening Torque
- Upper timing chain cover to engine block or cylinder head (M6 screws) . . . . 10 Nm (89 in-lb)

16. Install vibration damper while aligning woodruff key. Hold crankshaft stationary and tighten mounting bolt.

Tightening Torques
- Pulley to vibration damper . . . . 22 Nm (17 ft-lb)
- Vibration damper to crankshaft M16 stretch bolt (always replace) . 330 Nm (244 ft-lb)

17. Remove crankshaft locking tool from transmission bell-housing and camshaft locking tool from rear of cylinder head.

18. Turn engine over by hand in direction of rotation at least two full rotations. Reinstall camshaft and crankshaft locking tools to check engine timing. Repeat cam timing procedure if tools do not fit easily into position. Remove locking tools.

19. Install cylinder head cover as described in **113 Cylinder Head Removal and Installation**.

NOTE —
On M44 engines, note oil supply O-ring and replace if necessary.

Tightening Torque
- Cylinder head cover to cylinder head (M6 screws). . . . . . . 10 Nm (89 in-lb)

20. Installation of remaining parts is reverse of removal. When installing thermostat, make sure arrow or vent hole faces up. Refill cooling system as described in **170 Radiator and Cooling System**.

Tightening Torques
- Coolant drain plug to cylinder block . . . . 25 Nm (18 ft-lb)
- Radiator cooling fan to coolant pump . . . . 40 Nm (30 ft-lb)
- Radiator drain screw to radiator . . . . 2.5 Nm (22 in-lb)

CAMSHAFT TIMING CHAINS, 6-CYLINDER

Camshaft timing chain removal on 6-cylinder engines requires that you remove the engine oil pan by raising the engine and/or lowering the front suspension crossmember.

Special BMW service tools are needed for timing chain removal and installation procedures. The special tools assure proper timing of the valvetrain. Precise marks to set the timing on the camshafts are not provided for reassembly. Read the procedures through before beginning the job.

CAUTION —
If the camshafts are not properly timed, the pistons can contact the valves.

Camshaft timing chains, removing (6-cylinder engines)

1. Disconnect negative (–) cable from battery.

> ***CAUTION —***
> *Disconnecting the battery may erase fault code(s) stored in control module memory. Check for fault codes using special BMW diagnostic equipment.*

2. Drain engine coolant. Drain engine block by removing block drain. Remove radiator cooling fan shroud and cooling fan. See **170 Radiator and Cooling System**..

> ***WARNING —***
> *Allow the engine to cool before opening or draining the system.*

> ***CAUTION —***
> *Radiator fan has left hand threads.*

NOTE —
The block drain plug is located on the exhaust side of the engine, below cylinder no. 4.

3. Drain engine oil and remove oil pan. See **119 Lubrication System**.
4. Remove alternator cooling duct.
5. Remove two large coolant hoses from thermostat housing. Remove engine lifting bracket and thermostat housing from front of cylinder head.
6. Loosen bolts for coolant pump pulley. Then remove drive belts for alternator and A/C compressor. See **020 Maintenance Program**. Remove coolant pump pulley.
7. Remove drive belt tensioner from front of engine. See Fig. 16.

NOTE —
On M50 engines with a two-roller drive belt tensioner, remove the upper roller to access the tensioner mounting bolts.

8. Remove engine speed sensor from lower timing chain cover, if applicable. See Fig. 17.
9. Remove cylinder head cover. See **113 Cylinder Head Removal and Installation**.

NOTE —
Make note of the arrangement of the rubber insulators when removing the cylinder head cover mounting bolts.

Fig. 16. Drive belt tensioner mounting bolts (**arrows**).

B11055

Fig. 17. 1992-1995 M50 engine speed sensor on timing cover (**arrow**).

10. Unclip and remove baffle cover from above intake camshaft. See Fig. 18.
11. On 1992 engines: Remove upper timing chain cover from front of cylinder head. See Fig. 19.
12. Loosen spark plugs. See **120 Ignition System**.
13. Set engine to approximate TDC by rotating in normal operating direction until camshaft lobes at cylinder no. 1 are facing each other. See Fig. 20.
14. Set engine to TDC by aligning “0/T” mark (0°TDC) on front vibration damper with cast boss on lower timing chain cover. See Fig. 21.
15. Remove vibration damper mounting bolts and remove vibration damper and pulley. See Fig. 22.

Fig. 18. Oil baffle cover being removed.

Fig. 19. Upper timing chain cover mounting nuts (**arrows**) on pre-VANOS engines.

16. Secure crankshaft hub to prevent crankshaft from turning. Using a socket, loosen but do not remove crankshaft hub center bolt.

NOTE —

The crankshaft hub center bolt is tightened to a torque of 410 Nm (300 ft-lb). A heavy duty holding device, such as BMW special tool no. 11 2 150, should be used to hold the crankshaft hub stationary while the bolt is loosened. Use a heavy duty 3/4-inch drive socket and breaker bar to break the bolt free.

Fig. 20. Engine set to approximate TDC. Camshaft lobes face in and arrows on sprockets point up (**arrows**).

Fig. 21. O/T (0°TDC) mark on vibration damper aligned with boss on timing chain cover (**arrow**).

17. Lock crankshaft in TDC position by inserting BMW special tool no. 11 2 300 through transmission bellhousing and into hole in flywheel or drive plate. See Fig. 23.

NOTE —

Confirm that the locking tool has been correctly installed by trying to rotate the crankshaft.

18. Remove crankshaft hub bolt and washer. Remove hub from crankshaft.

19. Remove cylinder head cover mounting studs at rear of head. Lock camshafts at TDC using BMW special tool no. 11 3 240 at ends of camshafts. See Fig. 24..

Fig. 22. Vibration damper with drive belt pulley being removed. Note hub locating dowel (**arrow**).

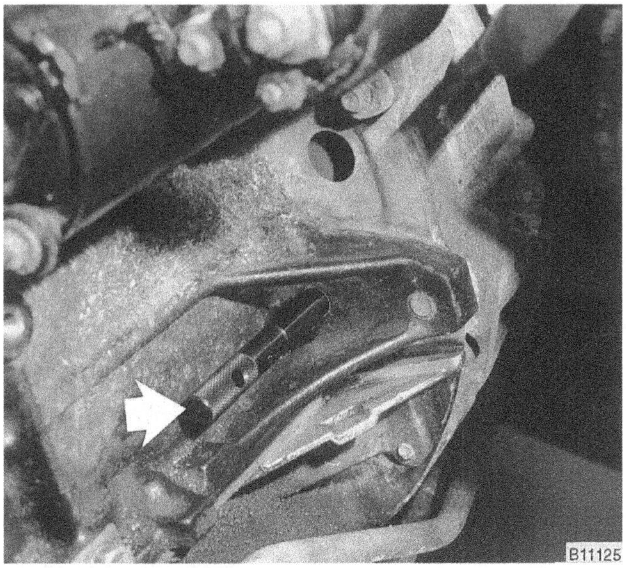

Fig. 23. BMW special tool no. 11 2 300 installed through bellhousing and into flywheel.

20. Remove primary chain tensioner plug and primary chain tensioner. See Fig. 25.

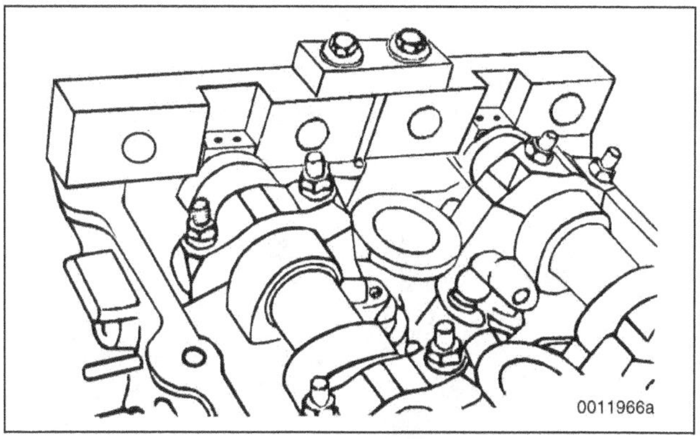

Fig. 24. BMW special tool no. 11 3 240 installed on rear of camshafts to position camshafts at TDC. (M44 camshafts shown. Other engines are similar.)

Fig. 25. Primary chain tensioner plug (**arrow**).

CAUTION—

- *Check that tool is squarely seated on cylinder head gasket surface. If necessary, rotate camshafts slightly using a 24 mm wrench*
- *On pre-VANOS engines (engines built up to 8-92), the camshaft locking tool may not fit correctly when the crankshaft is set to TDC. Do not assume the valve train is out of time. On some engines, valve overlap may have been reduced as a dealer service campaign to correct for an uneven idle. Check the timing by placing a 3mm drill bit under the left and right outside edges while pressing down on the center of the tool (tool center bolts loosened). If the tool now fits, the campaign has been done on the engine.*

21. On 1993 and later engines: Remove access plugs from VANOS control unit in front of exhaust camshaft sprocket. Remove 4 bolts from exhaust camshaft sprocket (use Torx E10 socket). See Fig. 26.

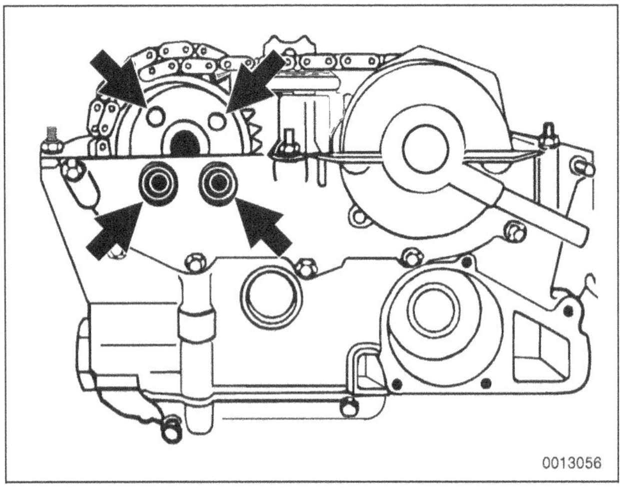

Fig. 26. Exhaust camshaft sprocket mounting bolts (**arrows**).

22. Depress secondary timing chain tensioner and lock down by inserting stiff wire into the openings in rear of tensioner. See Fig. 27.

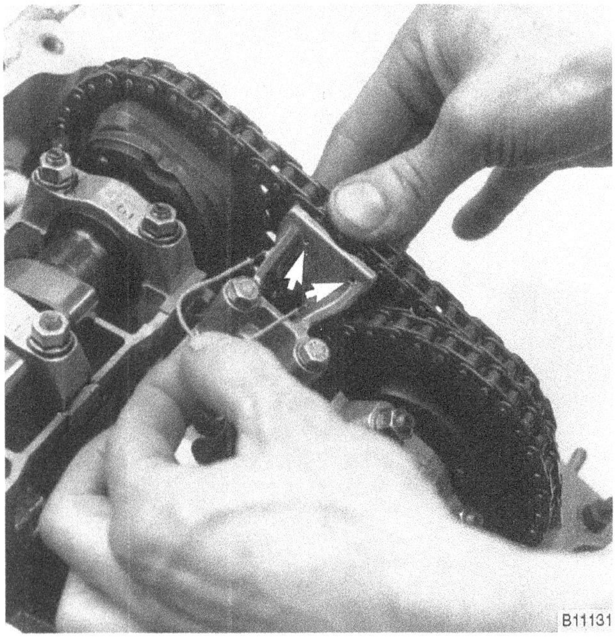

Fig. 27. Secondary timing chain tensioner being locked down on M50 engine. Press tensioner down and insert locking wire through holes (**arrows**). Tensioner on M52/S52 US engines vary slight-

23. Remove secondary chain tensioner and lift out primary chain guide.

24. On 1993 and later engines:
    - Remove VANOS control unit as described later in this repair group.
    - Remove intake camshaft sprocket mounting nuts.
    - Remove intake and exhaust sprockets together with secondary chain. Note location of thrust washers and spring plate (where applicable) on intake sprocket.

NOTE —

Hold the camshaft stationary (at hex on camshaft) when loosening or tightening sprocket mounting bolts.

CAUTION —

The crankshaft must not be allowed to rotate once the timing chain is removed. The pistons can contact the valves.

25. On 1992 engines:
    - Remove intake and exhaust sprocket mounting bolts.
    - Remove sprockets together with secondary chain.

26. Remove bolts holding lower timing chain cover to cylinder head (use Torx E8 socket). See Fig. 28.

Fig. 28. Cylinder head-to-timing chain cover mounting bolts (**arrows**).

27. Using a drift, drive two locating dowels in left and right sides of lower timing chain cover toward rear of car. Remove cover mounting bolts and carefully lift cover off. See Fig. 29.

CAUTION —

Use care when removing the cover from the cylinder head gasket. If the cover is stuck, use a sharp knife to separate it from the head gasket.

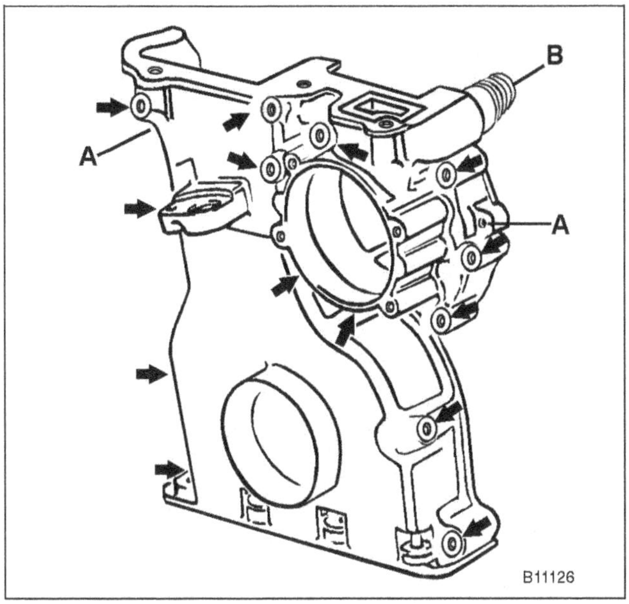

Fig. 29. Lower timing chain cover mounting bolts (**arrows**) and dowel locations (**A**). Coolant hose fitting **B** is replaced by coolant pipe pressure fitting in later models.

NOTE —

- *The timing chain cover can be removed with the coolant pump installed.*
- *Use care when removing the chain cover from the coolant pipe or hose at the rear of the cover.*

28. Remove remaining Torx bolt from timing chain guide. See Fig. 30.

Fig. 30. Timing chain guide mounting bolt (**A**).

29. Lift primary chain sprocket off exhaust camshaft and remove chain.

NOTE —

- *If any of the sprockets are being replaced due to wear, the chain should also be replaced.*
- *If the crankshaft sprocket requires replacement, the oil pump drive sprocket and chain must be removed as described under* **119 Lubrication System**.

Camshaft timing chains, installing (6-cylinder engines)

Inspect timing chain sprockets. Sprockets that have worn or missing teeth should be replaced. Inspect guide rail and tensioner rail for deep grooves caused by chain contact. Replace any part that is worn.

The procedure outlined below assumes that the camshafts and the crankshaft are locked in the TDC installation position with special tools (shown earlier in Fig. 23 and Fig. 24).

NOTE —

On pre-VANOS engines (engines built up to 8-92), the camshaft locking tool (Fig. 24) should be raised up on either side by 3mm (use feeler gauge or drill bits). Be sure the clamping bolts on tool are loose and the center of the tool is contacting the cylinder head surface. This service campaign helps improve idle characteristics by reducing valve overlap.

1. Install primary timing chain to crankshaft sprocket and tie it up through cylinder head opening.
2. Using new gaskets and coolant pipe O-ring, install lower timing chain cover:
   - Drive timing cover dowels in until they just protrude slightly from sealing surface.
   - Apply a small bead of silicon sealer (3-Bond 1209® or equivalent) to corners of cylinder head where timing cover meets cylinder head and engine block.
   - Tap cover into position to engage dowels.
   - Install mounting bolts hand tight, including two outer Torx bolts from above.
   - Drive dowels in flush to front of cover.
   - Tighten cover mounting bolts alternately and in stages.

Tightening Torque

- Lower timing cover to cylinder block

M7	15 Nm (11 ft-lb)
M8	20 Nm (15 ft-lb)

3. Install oil pan using a new gasket. See **119 Lubrication System**.

4. Mount primary timing chain sprocket to primary chain so that arrow on sprocket faces up. Mount sprocket to camshaft so that tapped holes in camshaft are on left side of elongated holes in sprocket. See Fig. 31.

Fig. 31. Primary timing chain sprocket correctly aligned to exhaust camshaft. Elongated sprocket holes should be aligned to left side of threaded camshaft holes.

5. On 1993 and later engines: Install BMW special tool no. 11 3 390 into primary timing chain tensioner sleeve. Thread tool in finger tight until slack is removed from chain.
   - Check that elongated holes in primary sprocket are centered to bolt holes in camshaft and chain is free of slack.

NOTE —

BMW special tool no. 11 3 390 is a dummy primary chain tensioner and simulates the function of the hydraulic tensioner.

6. Install primary timing chain guide and secondary chain tensioner. Tighten mounting bolts. (Tensioner remains locked down for now.)

7. Install cylinder identification trigger plate to front of intake camshaft. On 1993 and later engines, install thrust washer to front of intake camshaft.

8. Install secondary chain sprockets and chain as an assembly so that arrows on sprockets are pointing up. Center mounting holes in sprockets to bolt holes in camshafts. See Fig. 32.

9. On 1993 and later engines: Install thrust washers and spring plate (where applicable) to intake cam sprocket.

Fig. 32. Secondary chain sprockets being installed. Attach sprockets with arrows pointing up and elongated holes centered to threaded holes in shaft flange.

NOTE —

On engines built beginning from March 1995, the VANOS intake cam sprocket assembly was updated to include a spring plate and two thrust washers. See Fig. 33.

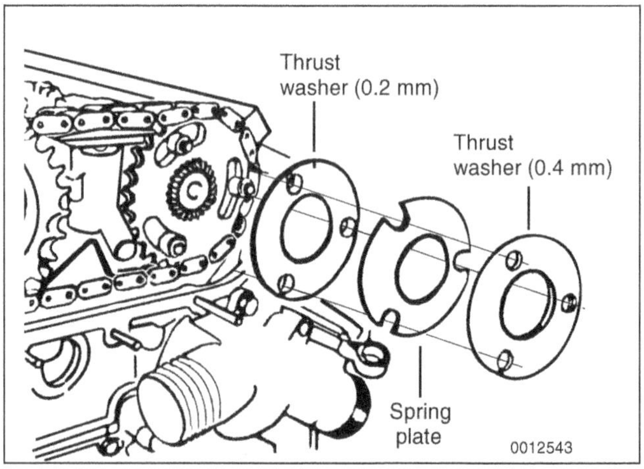

Fig. 33. VANOS spring plate and thrust washers on engines built 3/95 and later.

10. On 1993 and later cars, tighten nuts at intake camshaft sprocket.

Tightening Torque (1996 and later)
- Intake camshaft sprocket nut to stud (M6) . . . . 10 ± 1 Nm (89 ± 9 in-lb)

11. On 1992 engines: Install primary chain tensioner so that cutout in tensioner piston engages chain rail. See Fig. 34. Check that chain is free of slack.

Fig. 34. Primary timing chain tensioner being installed. Piston cutout must be in position shown to engage chain rail.

12. On 1993 and later engines: Install VANOS control unit as described later in this repair group.

13. Unlock and release secondary chain tensioner.

14. On 1992 engines: Tighten intake and exhaust timing chain sprocket bolts alternately in two stages, working at intake side first.

Tightening Torques
- Secondary timing chain sprockets to camshafts (M7 Torx bolts)
  Stage 1 . . . . 5 Nm (44 in-lb)
  Stage 2 . . . . 22 Nm (17 ft-lb)

15. On 1992 engines: Install upper timing chain cover to cylinder head using a new gasket. Apply a small amount of liquid gasket sealer (3-Bond 1209® or equivalent) around cover alignment dowels.

16. On 1993 and later engines: Preload primary chain by tightening BMW special tool no. 11 3 390 (dummy chain tensioner) to 1.3 Nm (11 in-lb). Then tighten exhaust camshaft sprocket bolts alternately in two stages.

Tightening Torques (1993 and later engines)
- Exhaust camshaft sprockets to camshafts (M7 Torx bolts)
  Stage 1 . . . . 5 Nm (44 in-lb)
  Stage 2 . . . . 22 Nm (17 ft-lb)

17. Remove flywheel locking tool from transmission bellhousing. Remove camshaft locking tool from rear of cylinder head.

18. Install crankshaft hub and washer while aligning woodruff key. Hold hub stationary and tighten mounting bolt. Install vibration damper and pulley, making sure that locating pin in hub engages hole in vibration damper.

NOTE —
When installing crankshaft hub mounting bolt, the stepped (shouldered) side of the washer should face the hub.

Tightening Torques
- Crankshaft hub to crankshaft (stretch bolt-always replace) . . . . 410 ± 20 Nm (300 ± 15 ft-lb)
- Vibration damper and pulley to crankshaft hub (M8) . . . . 22 Nm (17 ft-lb)

19. Turn engine over by hand in direction of rotation at least two full rotations. Reinstall camshaft and crankshaft locking tools to check engine timing. Remove locking tools if no faults are found.

NOTE —
On pre-VANOS engines, be sure to install the 3mm drill bits on left and right sides of tool.

20. On 1993 and later engines: Remove BMW special tool 11 3 390 (dummy tensioner). Install primary chain tensioner so that cutout in tensioner piston engages the chain rail. See Fig. 34. Install and tighten access plugs in VANOS control unit.

Tightening Torques
- Access plugs to VANOS control unit . . . . 50 Nm (37 ft-lb)
- Primary timing chain tensioner to tensioner sleeve
  M50/S50US engine . . . . 50 Nm (37 ft-lb)
  M52/S52US engine . . . . 40 Nm (30 ft-lb)

21. Install cylinder head cover mounting studs. Install oil baffle cover. Install cylinder head cover.

Tightening Torque
- Cylinder head cover to cylinder head . . . . 10 Nm (89 in-lb)

22. Installation of remaining parts is reverse of removal, noting the following:
    - When installing thermostat, make sure arrow or vent hole faces up, if applicable.
    - Refill cooling system as described in **170 Radiator and Cooling System**.
    - Install oil pan as described in **119 Lubrication System**.
    - Fill engine with oil and install a new oil filter as described in **020 Maintenance Program**.
    - Install ground wires at cylinder head cover mounting studs and at front of cylinder head and thermostat housing, where applicable.

Tightening Torques
- Coolant drain plug to cylinder block . . 25 Nm (18 ft-lb)
- Radiator cooling fan to coolant pump. 40 Nm (30 ft-lb)
- Radiator drain screw to radiator . . . . 2.5 Nm (22 in-lb)
- Upper timing chain cover to cylinder head
  M6 nut . . . . 10 Nm (89 in-lb)
  M8 bolt . . . . 22 Nm (17 ft-lb)
- VANOS control unit to cylinder head
  M6 nut . . . . 10 Nm (89 in-lb)
  M8 bolt . . . . 22 Nm (17 ft-lb)
- VANOS oil supply pipe to VANOS control unit
  banjo bolt . . . . 32 Nm (24 ft-lb)

VANOS (VARIABLE VALVE TIMING)

1993 and later 6-cylinder engines are equipped with a variable valve timing system called VANOS. This system is controlled by the engine management system and dynamically adjusts intake camshaft timing based on engine load, engine speed and engine temperature.

The main components of the VANOS system are the control unit with piston housing and integral spool valve, and the modified intake camshaft. See Fig. 35.

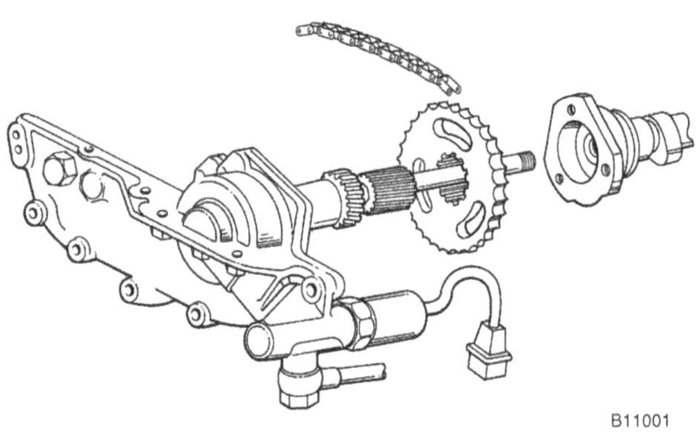

Fig. 35. VANOS control unit with modified intake camshaft.

When the engine is running, the piston housing is supplied with pressurized engine oil. At idle, the solenoid is in the off position (de-energized) and valve timing is maintained in the normal position. When the solenoid is energized, the gear cup piston moves forward to advance the camshaft by a maximum of 12.5°.

> *WARNING —*
> *Special BMW service tools are required to check and repair the VANOS system. Read the procedures through before beginning the job.*

VANOS system operation, testing

There are 3 special tools required to check VANOS operation; an electrical test lead (BMW special tool no. 12 6 410), an air line fitting (BMW special tool no. 11 3 450), and a crankshaft TDC locking tool (BMW special tool no. 11 2 300).

The test lead is used to power the solenoid, simulating the ground signal from the DME control unit. The air line fitting takes the place of the oil supply line fitting to simulate oil pressure. The locking tool positions and locks the crankshaft at TDC, cylinder no.1.

1. Remove alternator cooling air duct.

2. Disconnect crankcase ventilation hose fitting from cylinder head cover.

3. Remove top plastic engine covers. Disconnect ignition coil harness connectors from coils. Remove ignition coils.

4. Remove cylinder head cover mounting bolts and remove cylinder head cover. Unclip and remove oil baffle cover from above intake camshaft. See **113 Cylinder Head Removal and Installation**.

   NOTE —
   Note the arrangement of the cylinder head cover bolt insulators and gaskets during removal.

5. Set engine to approximate TDC by rotating in normal operating direction until camshaft lobes at cylinder no. 1 are facing each other. See Fig. 20.

6. Set engine to TDC by aligning "0/T" mark (0°TDC) on front vibration damper with cast boss on lower timing chain cover. See Fig. 21.

7. Lock crankshaft in position by inserting BMW special tool no. 11 2 300 through transmission bellhousing and into hole in flywheel or drive plate. See Fig. 36.

   NOTE —
   Confirm that the locking tool has been correctly installed by trying to rotate the crankshaft.

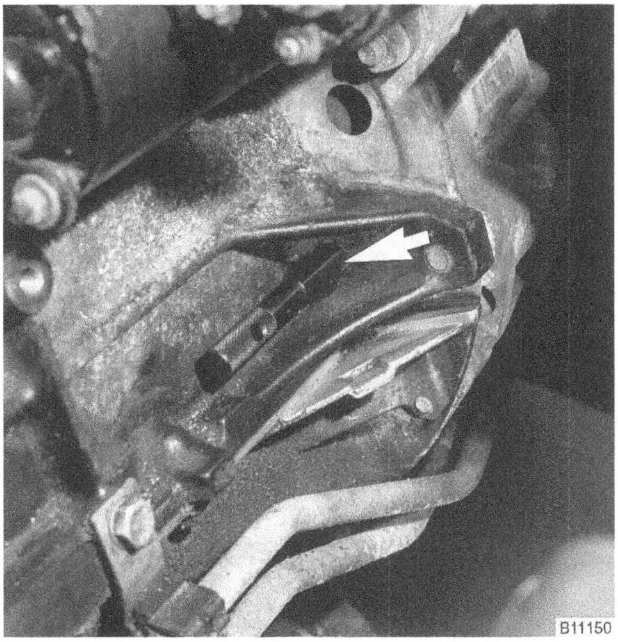

Fig. 36. BMW special tool no. 11 2 300 installed through bellhousing and into flywheel.

8. Remove oil line fitting from VANOS control unit.

NOTE —

Wrap the VANOS oil line fitting with a shop to absorb leaking oil.

9. Using hollow bolt and seals from oil supply line, install BMW special tool no. 11 3 450 (air line fitting) and connect a supply of compressed air (30-115 psi) to VANOS oil fitting.

10. Measure and record distance between trigger plate edge and side of secondary timing chain tensioner. See Fig. 37.

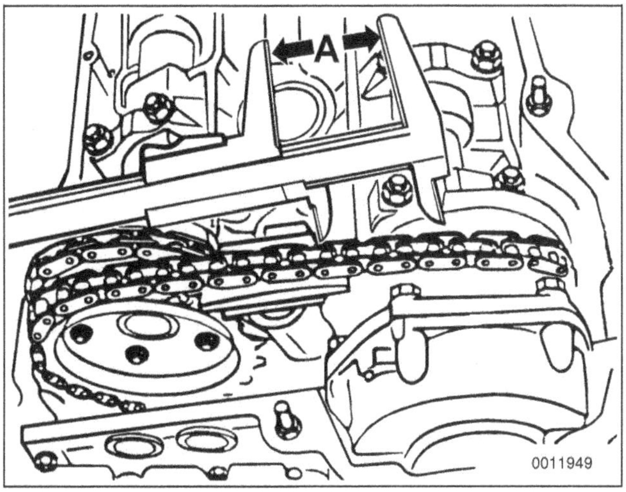

Fig. 37. VANOS reference measurement (dimension **A**) between trigger plate (sender gear) and side of secondary chain tensioner.

11. Disconnect harness connector from VANOS solenoid. Connect BMW special tool no. 12 6 410 (electrical test lead) to VANOS solenoid connector.
    - Connect positive (+) test lead to positive (+) terminal of battery. Connect negative (–) test lead to chassis ground. Solenoid should audibly click and intake camshaft should advance.

CAUTION —

- *Be sure to connect the test harness polarity correctly. If the polarity is reversed, the internal diode in the VANOS solenoid will be destroyed. Although the solenoid will still function with a faulty diode, a fault code may be set in the ECM memory.*
- *To confirm voltage polarity, turn the ignition on and check for positive (+) battery voltage at the terminal corresponding to the red/white wire in the main harness connector. Check that the positive terminal (+) in the main connector corresponds to the same terminal in the solenoid connector that is being connected to the battery.*

12. With solenoid actuated, measure and record new distance between trigger plate edge and side of secondary timing chain tensioner, as shown in Fig. 37.

13. Difference between first measurement (step 10) and second measurement (step 12) should be at least 8.5 mm (0.3346 in.). If any faults are found, solenoid may be faulty, or hydraulic control unit may be faulty or incorrectly installed.

14. Check VANOS solenoid by removing it from control unit. Check that solenoid plunger and control unit plunger move freely. Install solenoid using tightening torque given below.

NOTE —

- *The solenoid is available as a replacement part. If the control unit plunger is sticking, the complete control unit will have to be replaced.*
- *If the VANOS system does not advance correctly and no other faults can be found, it is possible that the VANOS control unit may have been installed incorrectly—especially if the camshaft sprockets have previously been removed for other repairs. Check the installation by removing and reinstalling the control unit as described later.*

15. Remove crankshaft locking tool. Remove special test equipment. Install oil supply line using new seals and reconnect solenoid connector. Remainder of installation is reverse of removal.

Tightening Torque
- Cylinder head cover to cylinder head. 10 Nm (89 in-lb)
- VANOS oil supply line to VANOS control unit (banjo bolt) . . . . 32 Nm (24 ft-lb)
- VANOS solenoid to VANOS control unit . . . . 30 Nm (22 ft-lb)

VANOS control unit, removing

Special BMW service tools are required to remove and install the VANOS control unit. Read the procedure through before starting the job.

1. Remove radiator cooling fan shroud and cooling fan. See **170 Radiator and Cooling System**.

> ***CAUTION—***
> *Radiator fan has left hand threads.*

2. Remove top plastic engine covers. Disconnect ignition coil harness connectors from coils. Remove ignition coils. Remove cylinder head cover. Unclip and remove oil baffle cover from above intake camshaft.See **113 Cylinder Head Removal and Installation**.

NOTE—
Make note of the mounting bolt insulator arrangement during removal of the cylinder head cover.

3. Set engine to TDC by rotating engine in normal operating direction until camshaft lobes at cylinder no. 1 are facing each other and "0/T" mark (0°TDC) on vibration damper lines up with cast boss on lower timing chain cover. See Fig. 38.

4. Lock crankshaft at TDC by inserting BMW special tool no. 11 2 300 through transmission bellhousing and into flywheel or drive plate. Refer to Fig. 36 given earlier.

NOTE—
Confirm that the locking tool has been correctly installed by trying to rotate the crankshaft.

5. Remove three cylinder head cover mounting studs from rear of head. Lock camshafts at TDC by mounting BMW special tool no. 11 3 240 at ends of camshafts. Check that tool is squarely seated on cylinder head gasket surface. If necessary, rotate camshafts slightly using a 24 mm wrench. Refer to Fig. 24 given earlier.

Fig. 38. Camshafts positioned at TDC—Cyl. No. 1 cam lobes face in Note 0/T mark (0°TDC) on vibration damper aligned with boss on front cover.

> ***CAUTION—***
> *The camshafts must be locked in the TDC position using the special service tool or an equivalent before removing the timing chain. The arrows on the sprockets should not be used to accurately set the engine to TDC. The special tool holds the camshafts parallel to each other and perpendicular to the valve cover gasket surface.*

6. Disconnect harness connector and oil line fitting from bottom of VANOS control unit.

NOTE—
Wrap the VANOS oil line fitting with a shop rag to absorb leaking oil.

7. Remove two access plugs from VANOS control unit in front of exhaust camshaft sprocket. Loosen 4 exhaust camshaft sprocket mounting bolts but do not remove (Torx E10 socket).

NOTE—
Removal of the access plugs allows the lower exhaust camshaft sprocket mounting bolts to be accessed.

8. Depress secondary timing chain tensioner and lock it in down position by inserting a stiff wire into rear of tensioner. See Fig. 27.

9. Remove VANOS control unit mounting nuts and bolt at front of cylinder head.

10. On cars built up to 3/95: Slide control unit off front of cylinder head, hand turning the exhaust sprocket clockwise (as viewed from front) as the unit is removed.

11. On cars built 3/95 and later: Use a spanner tool to rotate exhaust sprocket until stop on intake sprocket is contacted. See Fig. 39. Then slide control unit off front of cylinder head.

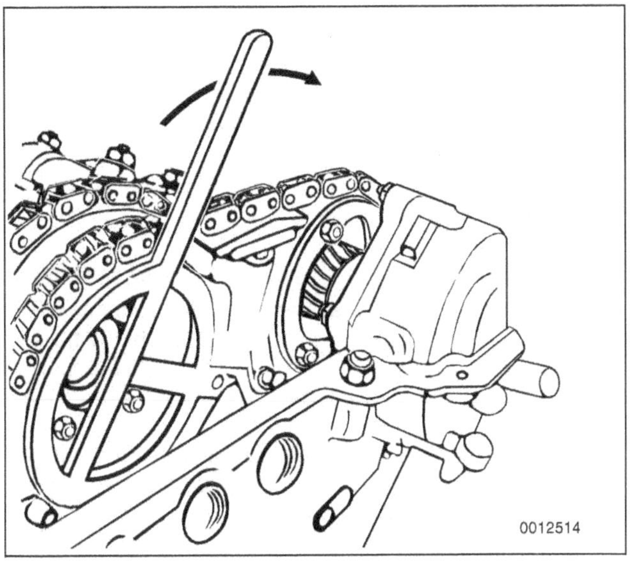

Fig. 39. On VANOS control unit with spring plate (cars built 3/95 and later), rotate exhaust camshaft sprocket clockwise until intake camshaft contacts stop. BMW special tool 11 5 490 shown.

VANOS control unit, installing

> ***CAUTION—***
> *Incorrect installation of the VANOS control may result in damage to the engine valvetrain.*

1. Lock camshafts and crankshaft in TDC position with BMW special tools as described above in removal procedure.

> ***CAUTION—***
> *The camshafts and crankshaft must be locked in the TDC position using BMW special tools no. 11 2 300 and 11 3 240. If the camshafts and crankshaft are not at TDC, the valves can contact the pistons when the engine is turned over.*

NOTE—

Be sure the secondary timing chain tensioner is locked down and the exhaust camshaft mounting bolts are loose before proceeding.

2. Remove primary timing chain tensioner from side of cylinder head. See Fig. 40.

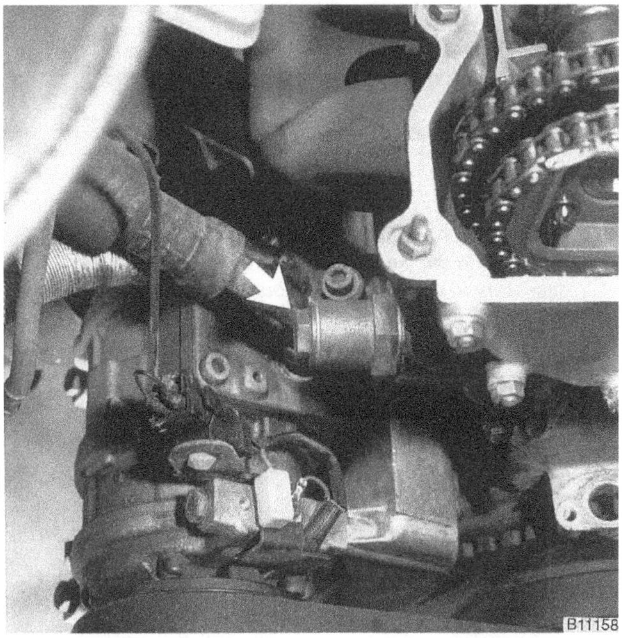

Fig. 40. Primary timing chain tensioner (**arrow**). Remove tensioner slowly to relieve spring tension beneath end plug.

3. Install BMW special tool no. 11 3 390 into primary timing chain tensioner sleeve. Thread tool in finger tight to remove chain slack.

NOTE—

BMW special tool no. 11 3 390 is a dummy chain tensioner and simulates the function of the tensioner.

4. Hand turn secondary sprockets clockwise (as viewed from front of engine) until sprockets contact stops.

5. Apply a small amount of liquid gasket sealer (3-Bond 1209® or equivalent) to VANOS housing alignment dowels on front of cylinder head. Install a new gasket.

6. Push VANOS piston gear cup into VANOS housing until it bottoms out in housing. Position VANOS control unit into place, but do not engage splines on gear cup.

7. On cars built up to 3/95:
   - Hand turn exhaust sprocket counterclockwise (as viewed from front of engine) **only enough** to mesh gear cup splines with camshaft gear.
   - Slowly push VANOS control unit in until fully seated, guiding chains in counterclockwise direction.

CAUTION —
It is important that the exhaust camshaft sprocket be turned only the minimum amount necessary to engage the gear cup. If this procedure is done incorrectly, the VANOS system may not fuly function.

NOTE —
As the control unit is installed, the secondary sprockets will rotate in the counterclockwise direction. It may be necessary to guide sprockets counterclockwise as the control unit is installed.

8. On cars built from 3/95: Use a spanner tool (See Fig. 39.), turn exhaust sprocket counterclockwise (as viewed from front of engine) **only enough** to mesh gear cup splines with camshaft gear. Slowly push VANOS control unit towards camshaft until it is fully seated, guiding chains in counterclockwise direction.

CAUTION —
Ensure that the first available tooth engages.

9. Install and tighten control unit nuts and bolt.

10. Connect harness connector to VANOS solenoid. Using new seals, install oil line fitting.

Tightening Torques
- VANOS control unit to cylinder head
  M6 nut . . . . 10 Nm (89 in-lb)
  M8 bolt . . . . 22 Nm (17 ft-lb)
- VANOS oil supply line to control unit . 32 Nm (24 ft-lb)

11. Slowly unlock and release secondary chain tensioner.

12. Preload primary chain by tightening BMW special tool no. 11 2 240 (dummy chain tensioner) to 1.3 Nm (11 in-lb). Then tighten exhaust camshaft sprocket mounting bolts alternately in two stages.

Tightening Torque
- Secondary timing chain sprockets to camshafts (Torx bolts)
  Stage 1. . . . . 5 Nm (44 in-lb)
  Stage 2. . . . . 22 Nm (17 ft-lb)

13. Remove flywheel locking tool from transmission bell-housing. Remove camshaft locking tool from rear of cylinder head.

14. Turn engine over by hand at least two full crankshaft rotations. Reinstall camshaft locking tool and flywheel locking tool to make sure all adjustments are correct. If settings are correct, remove locking tools.

15. Remove BMW special tool 11 2 420 (dummy chain tensioner) from side of cylinder head. Install primary chain tensioner so that cutout in tensioner piston engages chain rail. See Fig. 34. Using new seals, install and tighten access plugs in VANOS control unit.

Tightening Torque
- Access plugs to
  VANOS control unit. . . . . 50 Nm (37 ft-lb)
- Primary timing chain tensioner plug to tensioner sleeve
  M50/S50USengine . . . . 50 Nm (37 ft-lb)
  M52/S52US engine . . . . 40 Nm (30 ft-lb)

16. Install oil baffle cover above intake camshaft. Install three cylinder head cover mounting studs to rear of head. Install cylinder head cover.

Tightening Torque
- Cylinder head cover
  to cylinder head (M6 screws) . . . . . . 10 Nm (89 in-lb)

17. Installation of remaining parts is reverse of removal.

NOTE —
Be sure to install the ground wires at the cylinder head cover mounting studs and at the front of the cylinder head, if applicable.

119 Lubrication System

GENERAL

The oil pan and the oil pump can be removed with the engine installed, although engine lifting/support equipment is necessary.

> ***NOTE—***
> *Oil change procedure and oil filter replacement are covered in* **020 Maintenance Program**.

All engines are equipped with an oil pressure warning system to help prevent engine damage. Other safety features include:

- A filter bypass to provide lubrication should the oil filter become clogged.
- An oil pump pressure relief valve to prevent excessive system pressure.

TROUBLESHOOTING

The lubrication warning system consists of an oil pressure switch mounted in the oil circuit and an instrument panel warning light.

> ***CAUTION—***
> *If the red oil pressure warning light comes on or flashes on while driving, always assume that the oil pressure is low.*

Oil pressure, checking

To perform an oil pressure test, BMW specifies special tools that attach to the top of the oil filter housing. The following procedure works well using standard automotive oil pressure testing equipment attached to the oil pressure switch port in the engine. In some engines, however, access to this port may be extremely restricted.

1. Disconnect harness connector from oil pressure switch and remove switch. See Fig. 1.

> ***NOTE—***
> *Thoroughly clean around the oil pressure switch before removing it.*

Fig. 1. Oil pressure switch location on M50 engine (**arrow**).

Component Location
- Oil pressure switch
  all engines. . . . . on oil filter housing

> ***CAUTION —***
> *Some oil may drain out as the oil pressure switch is removed. Use a rag to soak up any spills.*

2. Install pressure gauge in place of switch.

3. With gauge installed, start engine and allow to reach operating temperature. Check oil pressure both cold and hot.

   NOTE —
   For the most accurate test results, the engine oil (and filter) should be new and of the correct grade.

Oil Pressure
- Idle (minimum) . . . . 0.5 bar (7 psi)
- Regulated pressure (elevated engine speed)
  4-cylinder engines . . . . 4.3 ± 0.2 bar (63 ± 3 psi)
  6-cylinder engines . . . . 4.0 bar (59 psi)

4. Remove pressure gauge and reinstall pressure switch.

If testing shows low oil pressure, one or more of the following conditions may be indicated:

- Worn or faulty oil pump.
- Worn or faulty engine bearings
- Severe engine wear.

All of these conditions indicate the need for major repairs.

Oil pressure warning system, testing

When the ignition is turned on, the oil pressure warning light comes on. When the engine is started and the oil pressure rises slightly, the oil pressure switch opens and the warning light goes out. Make sure the oil level is correct before making tests.

1. Turn ignition switch on.
   - Warning light on instrument panel must light up.

2. Remove connector from oil pressure switch.
   - Warning light on instrument panel must go out.

   NOTE —
   If the light does not go out, the wiring to the switch is most likely grounded somewhere between the switch terminal and the warning light. See **Electrical Wiring Diagrams** *at rear of manual for electrical schematics.*

3. If warning light does not light when ignition is on, remove connector from oil pressure switch and use a jumper wire to ground connector terminal to a clean metal surface.

   NOTE —
   If the warning light comes on, check the switch as described in the next step. If the warning light does not come on, the wiring to the instrument cluster or to the light itself is faulty.

4. To test switch, connect an ohmmeter between terminal in switch body and ground. With engine off, there should be continuity. With engine running, oil pressure should open switch and there should be no continuity. Replace a faulty switch.

> ***WARNING —***
> *Keep in mind that low oil pressure may be preventing the switch from turning the light out. If the light remains on while the engine is running, check the oil pressure as described earlier. Do not drive the car until the problem is corrected. The engine may be severely damaged.*

OIL PAN

The oil pan can be removed with the engine installed, although special engine support equipment will be needed.

Oil pan, removing and installing (4-cylinder engines)

1. Raise car and place securely on jackstands.

2. Remove splash shield(s) from under engine, where applicable.

3. Drain engine oil as described in **020 Maintenance Program**.

4. Disconnect vacuum hose adapter from vacuum brake booster at rear of engine compartment.

5. Remove oil dipstick guide tube mounting nut and pull guide tube from oil pan. See Fig. 2.

NOTE—
The guide tube is sealed in the pan with an O-ring. Check that the O-ring comes out with the tube. Use a new O-ring when installing the tube.

0011959

Fig. 2. Oil dipstick guide tube mounting nut (**arrow**) on M44 engine. Use new O-ring at base of tube during installation.

6. Install engine lifting equipment at front engine lifting point and raise engine approximately 5 mm (¼ inch) until engine weight is supported.

7. Working beneath car, separate steering column shaft from steering rack at universal joint.
   - Point wheels straight ahead before disconnecting shaft from rack. Mark steering column shaft joint to steering rack spindle. See **320 Steering and Wheel Alignment**.

CAUTION—
In order to avoid the need for front-end realignment, do not unbolt power steering rack from suspension crossmember.

8. Support suspension crossmember (subframe) from below using appropriate jacking equipment.

9. At left and right sides, unbolt control arm bushing carriers from body. Disconnect stabilizer bar links from control arms. See Fig. 3.

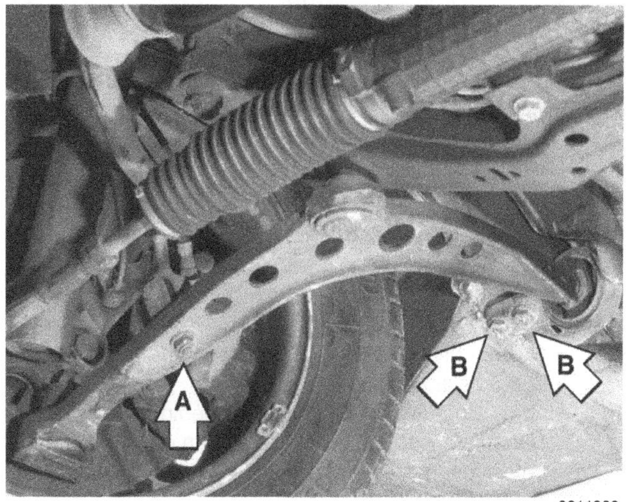

0011960

Fig. 3. Stabilizer bar link nut (**A**) and control arm bushing carrier bolts (**B**). Right side shown.

10. Remove bolts from left and right sides of suspension crossmember. Remove M10 nuts from bottom of left and right engine mounts. Lower crossmember as far as possible. See Fig. 4.

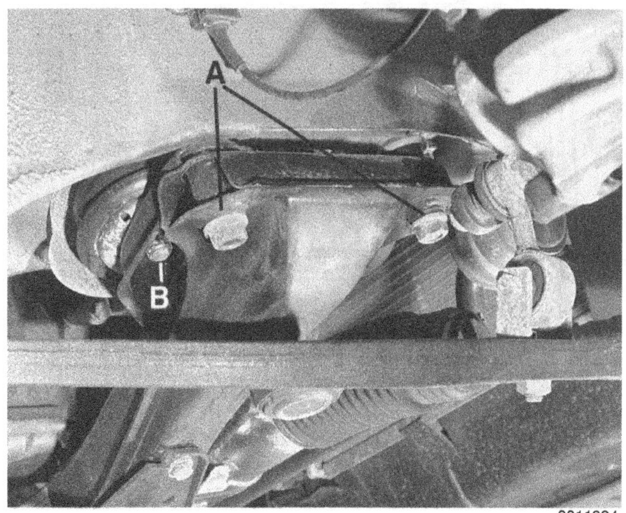

0011994

Fig. 4. Suspension crossmember bolts (**A**) and lower engine mounting nut (**B**). Right side shown.

11. Remove clamping brackets holding fuel lines to oil pan.

12. On cars with automatic transmission, remove ATF cooler line brackets from oil pan.

13. Remove engine drive belt from power steering pump. See **020 Maintenance Program**.

14. Unbolt power steering pump bracket and remove bracket with pump. Hang pump from chassis using wire.

15. Remove oil pan screws. Lower and remove oil pan toward rear.

> ***CAUTION—***
> *If the oil pan does not separate easily from the engine cylinder block, make sure all mounting bolts have been removed. If necessary, a few taps with a rubber mallet should break it free. Never pry the oil pan loose.*

16. Installation is reverse of removal.
    - Thoroughly clean all old gasket material from mating surfaces and use a new gasket.
    - Apply a small amount of non-hardening sealer (3-Bond®1209 or equivalent) to oil pan gasket directly below joints for end cover and front timing case cover. See Fig. 5.
    - When installing oil pan to engine, tighten forward screws first, then tighten rear.
    - Fill engine with oil as described in **020 Maintenance Program**.

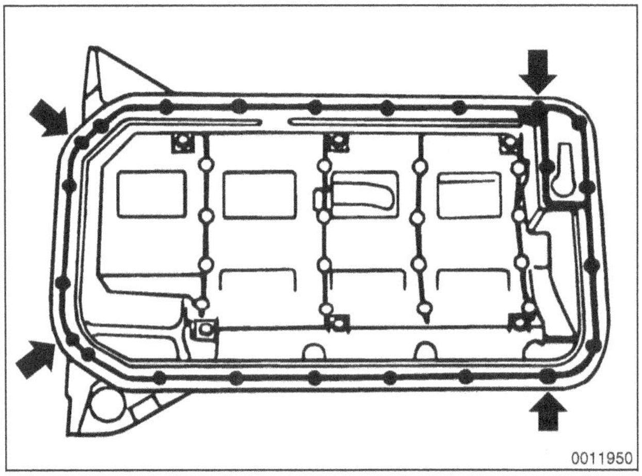

Fig. 5. Joint-mating areas at oil pan gasket where sealer should be applied (**arrows**). Apply a bead 3 mm wide by 2 mm high.

> ***WARNING—***
> *Always use new bolts when mounting the subframe to the body. The one-time only bolts should be replaced any time they are removed.*

NOTE—
The oil pickup is attached to the oil pan using self-tapping screws. For this reason no threading for the pickup is present on new oil pans. Replace the seal for the pickup and use the old screws to reattach the pickup to the pan.

Center the steering spindle to the steering rack before installing the steering column shaft. See **320 Steering and Wheel Alignment** for specific installation markings and procedures.

Tightening Torques

- Control arm bushing carrier to body (M10 bolt) . . . . 42 Nm (31 ft-lb)
- Engine mount to suspension crossmember (M10 nut) . 42 Nm (31 ft-lb)
- Front suspension crossmember to body . . . . See **310 Front Suspension**
- Oil pan to engine block (M6 bolt)
  - 8.8 grade . . . . 10 Nm (89 in-lb)
  - 10.9 grade . . . . 12 Nm (9 ft-lb)
- Power steering pump to bracket (self-locking nuts) . . . . 22 Nm (16 ft-lb)
- Power steering pump bracket to engine block or oil pan (self-locking nuts) . . . 22 Nm (16 ft-lb)
- Stabilizer bar link to control arm
  - M8 nut . . . . 22 Nm (17 ft-lb)
  - M10 nut . . . . 42 Nm (31 lb-ft)
- Steering column universal joint to steering rack spindle (M8 bolt) . . . . 19 Nm (14 ft-lb)

Oil pan, removing and installing (6-cylinder engine, 1992 models)

NOTE—
Cars built up to 9/92 (1992 models) use a different suspension crossmember than later production cars. Oil pan removal procedure on the earlier cars requires that the engine be raised and supported from above.

1. Raise car and place securely on jackstands.
2. Remove splash shield(s) from under engine, where applicable.
3. Drain engine oil as described in **020 Maintenance Program**.
4. Remove complete exhaust system. See **180 Exhaust System**.
5. Remove air filter housing complete with mass air flow sensor. See **113 Cylinder Head Removal and Installation**.
6. Remove alternator cooling duct from alternator and radiator support.
7. Remove radiator cooling fan and fan shroud. Remove radiator securing clips at top of radiator. See **170 Radiator and Cooling System**.

NOTE—
The radiator cooling fan nut (32 mm wrench) has left hand threads.

8. Remove air plenum from rear of engine compartment. See **640 Heating and Air Conditioning**.

9. Release drive belt tension and remove alternator drive belt. Remove A/C compressor drive belt. See Fig. 6.

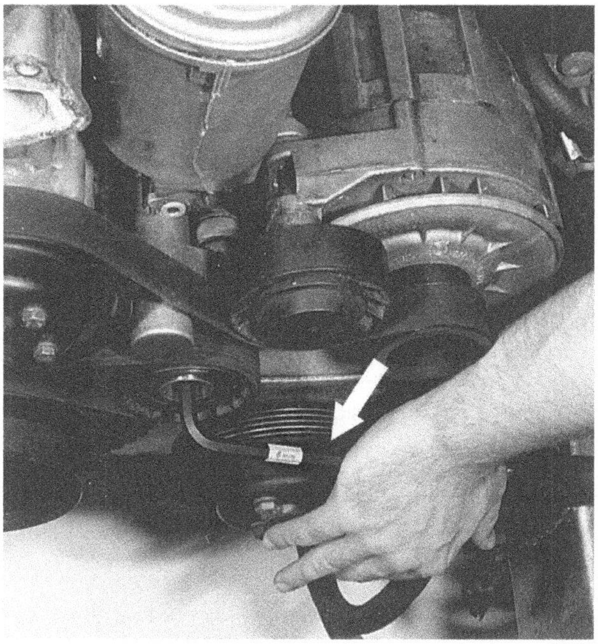

B11143

Fig. 6. To remove poly-ribbed drive belt, pry cover from front of tensioner. Then using 8 mm hex key, turn tensioner clockwise (**arrow**) to release tension and slip belt off pulleys.

10. Unbolt power steering reservoir and pull reservoir off engine mount bracket.

11. Without disconnecting fluid lines, remove power steering bracket (with pump) from oil pan and engine block. See Fig. 7. Disconnect fluid lines from bracket on engine mount. Hang pump from chassis using wire.

12. Without disconnecting refrigerant lines, remove A/C compressor from engine block. Hang compressor from chassis using wire. See **640 Heating and Air Conditioning**.

13. Remove oil dipstick guide tube mounting bolt and remove tube. See Fig. 8.

NOTE —
The guide tube is sealed in the oil pan with an O-ring. Check that the O-ring comes out with the tube. Use a new O-ring when installing the tube.

14. On cars with automatic transmission, remove brackets holding ATF cooler lines to oil pan and cylinder block.

Fig. 7. Power steering pump mounting bolts (**arrows**).

Fig. 8. Oil dipstick guide tube being removed. Use a new O-ring (**arrow**) during installation.

15. Install engine lifting equipment at front engine lifting point and raise engine just until its weight is supported.

16. Remove nuts at bottom of left and right engine mounts. Remove ground wire from right engine mount. Loosen nuts at top of left and right engine mounts (do not remove).

17. Raise engine as much as possible. Check carefully for obstructions, wiring harness clearance and pinched hoses or lines as engine is raised.

18. Remove oil pan mounting bolts and lower oil pan to subframe crossmember.

NOTE —
If the oil pan does not separate easily from the engine cylinder block, a few taps with a rubber mallet should break it free. Do not pry the oil pan loose.

19. Remove oil pump sprocket mounting nut. See Fig. 9. Lift sprocket off together with drive chain.

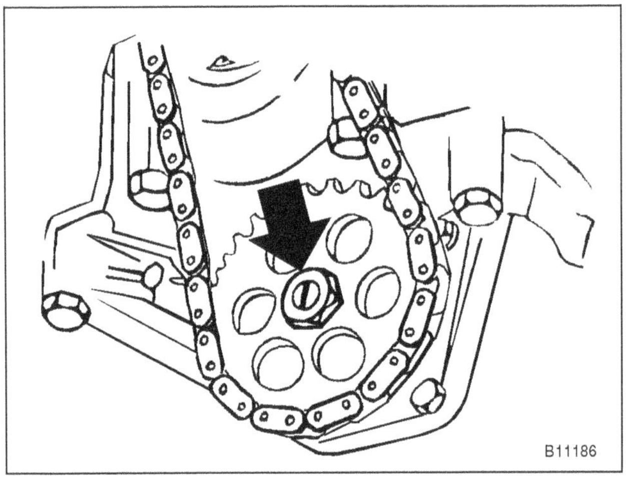

Fig. 9. Oil pump sprocket mounting nut (left-hand thread) on 6-cylinder engine (**arrow**).

NOTE —
The oil pump sprocket nut has left hand threads.

20. Unbolt oil pump pickup tube from oil pump. Unbolt oil pump from engine block. See **Oil pump, removing and installing (6-cylinder engine)**.

21. Remove oil pan from rear.

22. Installation is reverse of removal.
    - When installing oil pan, apply a bead of non-hardening sealing compound (3-Bond 1209® or equivalent) to front and rear end cover seam areas on block.
    - Position oil pump pickup tube and oil pan onto block and install pickup tube using a new gasket.
    - Be sure tab on gasket faces down towards intake of pickup tube.
    - Tighten forward oil pan screws first, then tighten rear.
    - Fill engine with oil as described in **020 Maintenance Program**.
    - After adding engine oil, start and run engine. Raise engine speed to 2,500 rpm until oil pressure warning lamp goes out (about 5 seconds).

Tightening Torques

- Engine mount to suspension crossmember(M10 nut) . . 42 Nm (31 ft-lb)
- Oil pan to engine block (M6 bolt)
  - 8.8 grade . . . . 10 Nm (89 in-lb)
  - 10.9 grade . . . . 12 Nm (106 in-lb)
- Oil pump to engine block . . . . 22 Nm (16 ft-lb)
- Oil pump sprocket to oil pump shaft (M10x1 left-hand thread) . . . . 25 Nm (18 ft-lb)
- Power steering pump to bracket (self-locking nuts) . . . . 22 Nm (16 ft-lb)
- Power steering pump bracket to engine block or oil pan (self-locking nuts) . . . 22 Nm (16 ft-lb)

Oil pan, removing and installing (6-cylinder engine, 1993 and later models)

NOTE —
Cars built after 9/92 use a different suspension crossmember than earlier production cars. Oil pan removal procedure on the later cars requires that the engine be supported from above and the front suspension crossmember be unbolted and lowered from the chassis.

1. Raise car and place securely on jackstands.

2. Remove splash shield from under engine, if applicable.

3. Drain engine oil as described in **020 Maintenance Program**.

4. Remove alternator cooling duct from alternator and radiator support.

5. Remove air filter housing complete with mass air flow sensor. See **113 Cylinder Head Removal and Installation**.

6. Remove oil dipstick guide tube mounting bolt. Disconnect oil separator hose from base of guide tube and remove tube from oil pan (where applicable). See Fig. 10.

NOTE—
The guide tube is sealed in the block using an O-ring. Check that the O-ring comes out with the tube. Use a new O-ring when installing the tube.

Fig. 10. Oil dipstick guide tube being removed. Use new O-ring (**arrow**) during installation.

7. M50/S50US engine: Using a clean syringe, remove power steering fluid from fluid reservoir. Disconnect power steering fluid lines from steering rack. See **320 Steering and Wheel Alignment**.

8. M52/S52US engine: Unbolt power steering reservoir from engine, then tie to chassis with wire.

9. Install engine lifting equipment at front engine lifting point and raise engine approximately 5 mm (¼ inch) until engine weight is supported. See Fig. 11.

10. Working beneath car, separate steering column shaft from steering rack at universal joint.

    - Mark steering column shaft joint to steering rack spindle. Point wheels straight ahead before disconnecting shaft from rack. See **320 Steering and Wheel Alignment**.

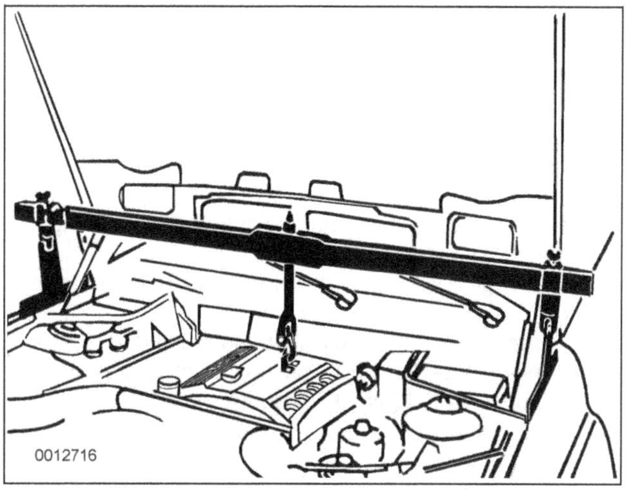

Fig. 11. Engine lifting equipment shown installed across engine.

CAUTION—
In order to avoid the need for front-end realignment, do not unbolt power steering rack from suspension crossmember.

11. Support suspension crossmember from below using appropriate jacking equipment.

12. Loosen nuts at top of left and right side engine mounts. Remove nuts from bottom of left and right side engine mounts.

13. At left and right sides, unbolt control arm bushing carriers from body. Disconnect stabilizer bar links from control arms. Refer to Fig. 3.

14. Remove bolts from left and right sides of suspension crossmember and lower front axle as far as possible.

15. Remove fuel line clamping brackets from oil pan. On cars with automatic transmission, remove ATF cooler line brackets from oil pan.

16. Remove oil pan screws. Lower and remove oil pan forward to remove.

CAUTION—
If the oil pan does not separate easily from the engine cylinder block, a few taps with a rubber mallet should break it free. Do not pry the oil pan loose.

17. Installation is reverse of removal.
    - Thoroughly clean all old gasket material from mating surfaces and use a new gasket.
    - Apply a small amount of non-hardening sealer (3-Bond 1209® or equivalent) to oil pan gasket directly below joints for end cover and front timing case cover. See Fig. 12.
    - Tighten oil pan bolts starting at front first, working toward back (transmission) end.
    - Fill engine with oil as described in **020 Maintenance Program**.
    - After adding engine oil, start and run engine. Raise engine speed to 2,500 rpm until oil pressure warning lamp goes out (about 5 seconds).

Fig. 12. Joint-mating areas at oil pan gasket where sealer should be applied (**arrows**). Apply a bead 3 mm wide by 2 mm high.

> ***WARNING —***
> - *Always use new bolts when mounting the subframe to the body. The one-time only bolts should be replaced any time they are removed.*
> - *Special installation instructions apply when installing the front suspension crossmember to the body. See* **310 Front Suspension** *for fastener specifications and tightening torques.*

Center the steering spindle to the steering rack before installing the steering column shaft. See **320 Steering and Wheel Alignment** for more specific procedures.

Tightening Torques
- Control arm bushing carrier to body (M10 bolt) . . . . 47 Nm (35 ft-lb)
- Engine mount to suspension crossmember(M10 nut) . . 42 Nm (31 ft-lb)
- Oil pan to engine block (M6 bolt)
  8.8 grade . . . . 10 Nm (89 in-lb)
  10.9 grade . . . . 12 Nm (106 in-lb)
- Stabilizer bar link to control arm (M10 nut). . . . . 42 Nm (31 lb-ft)
- Steering column universal joint to steering rack spindle (M8 bolt) . . . . 19 Nm (14 ft-lb)
- Suspension crossmember to body . . . . See **310 Front Suspension**

OIL PUMP

Oil pump, removing and installing (4-cylinder engine)

Oil pressure on M42 and M44 engines is generated by a gear-type pump mounted to the rear of the front engine cover (timing chain case). The pump is gear-driven off the front of the crankshaft.

> ***NOTE —***
> *Oil pump removal requires lowering of the front suspension and removal of the oil pan, as well as removal of the timing chain assembly.*

1. Disconnect negative (–) battery cable from battery.

> ***CAUTION —***
> *Prior to disconnecting the battery, read the battery disconnection cautions given at the front of this manual on page viii.*

2. Drain engine oil as described in **020 Maintenance Program**.

3. With engine cold, drain coolant. See **170 Radiator and Cooling System**.

4. Remove top cover from oil filter housing to allow engine oil to drain into oil pan. Remove oil pan as described earlier. Reinstall oil filter cover.

5. Remove alternator. Unbolt power steering pump from bracket, then remove alternator mounting bracket. See **121 Battery, Starter, Alternator**.

6. Remove crankshaft vibration damper and hub. Then remove upper and lower timing chain covers, complete timing chain, chain sprockets and chain guides. See **117 Camshaft Timing Chain**.

7. Unbolt timing chain case from front of engine. See Fig. 13.

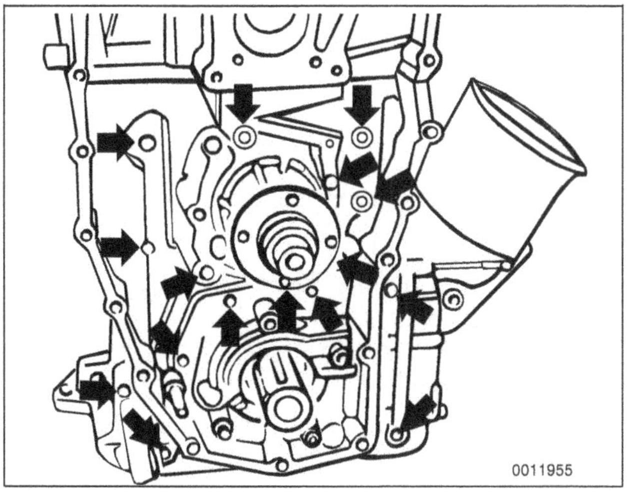

Fig. 13. Timing chain case mounting bolts (**arrows**) on M44 engine.

8. Remove cover from oil pump on rear of timing chain case. See Fig. 14. Inspect oil pump gears and oil pump cavity in timing chain case for wear and/or scoring.

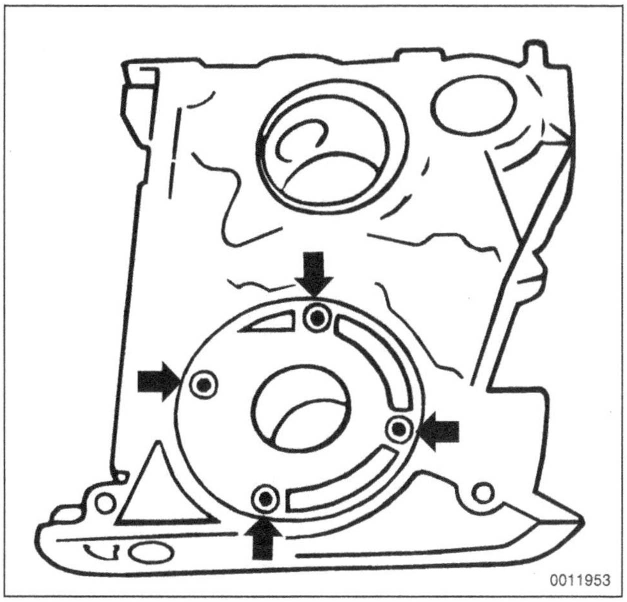

Fig. 14. Oil pump cover mounting bolts (**arrows**) on M44 engine.

9. Installation is reverse of removal.
   - Replace all gaskets and O-rings.
   - Install a new rubber seal (profile gasket) between top of timing chain case and cylinder head.
   - Be sure to thoroughly clean sealing surfaces.
   - Use 3-Bond 1209® or equivalent sealant on both sides of seal before installing.

10. Protect profile gasket using thin sheet metal (BMW special tool no. 11 2 330) when installing timing chain case. See Fig. 15.
    - Apply a thin coating of grease to both sides of sheet metal tool and to top surface of profile gasket.
    - Place sheet metal between gasket and cylinder head and carefully slide upper chain cover into position.
    - Tighten all mounting bolts and then carefully withdraw sheet metal. Install timing chain as described in **117 Camshaft Timing Chain**.

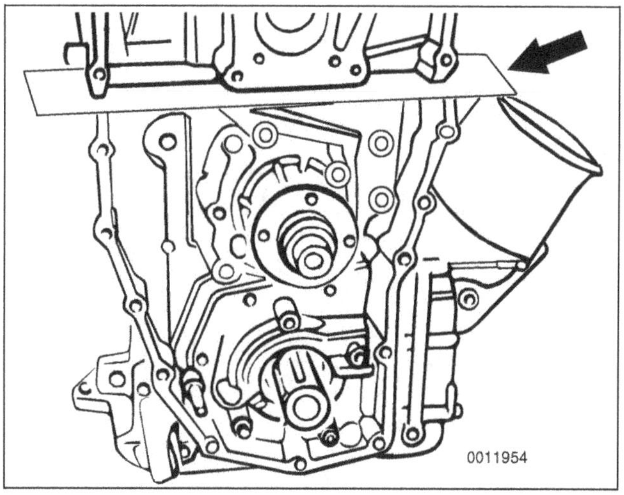

Fig. 15. Use thin sheet metal (**arrow**) to protect profile gasket during installation of timing chain case. Carefully withdraw sheet metal after tightening fasteners.

Tightening Torques

- Timing chain case to engine block
  M6 .... 10 Nm (7.5 ft-lb)
  M8 .... 22 Nm (16 ft-lb)
- Thermostat housing to upper timing chain cover(M6) .... 10 Nm (89 in-lb)
- Oil filter cover to oil filter housing (M8) .... 25 Nm (18 ft-lb)
- Oil pump cover to timing chain case (M6) .... 10 Nm (89 in-lb)
- Oil pan to engine block (M6 bolt)
  8.8 grade .... 10 Nm (89 in-lb)
  10.9 grade .... 12 Nm (106 in-lb)
- Alternator to alternator bracket ..... 43 Nm (32 ft-lb)

Oil pump, removing and installing (6-cylinder engine)

Oil pressure on the 6-cylinder engines is generated by a gear-type pump bolted to the bottom of the engine block. The oil pump is chain driven off the front of the crankshaft.

> ***NOTE —***
> *Oil pump removal requires raising the engine (cars built up to 9-92) or lowering the front suspension crossmember (cars built from 9-92) to remove the oil pan.*

1. Drain oil as described in **020 Maintenance Program**.
2. Remove oil pan as described earlier.
3. Remove oil pump sprocket mounting nut (left-hand thread). See Fig. 16. Lift sprocket off together with drive chain.

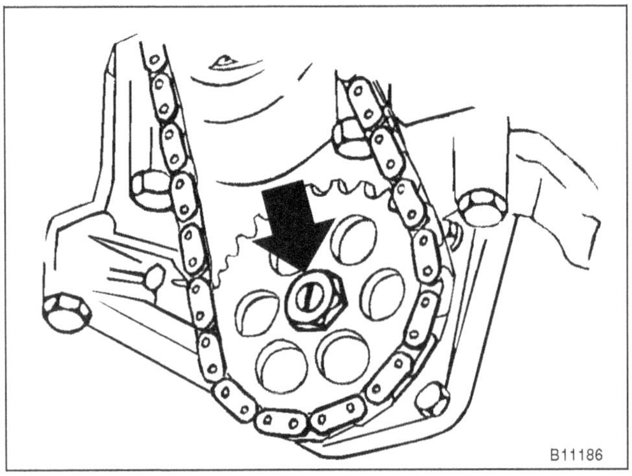

Fig. 16. Oil pump sprocket mounting nut (left-hand thread) on 6-cylinder engine (**arrow**).

4. Remove mounting bolts from oil pump and oil pump pickup tube. Withdraw pump.

> ***NOTE —***
> *• Note any spacers between pump and engine block. See Fig. 17.*
>
> *• Note position of locating dowels.*

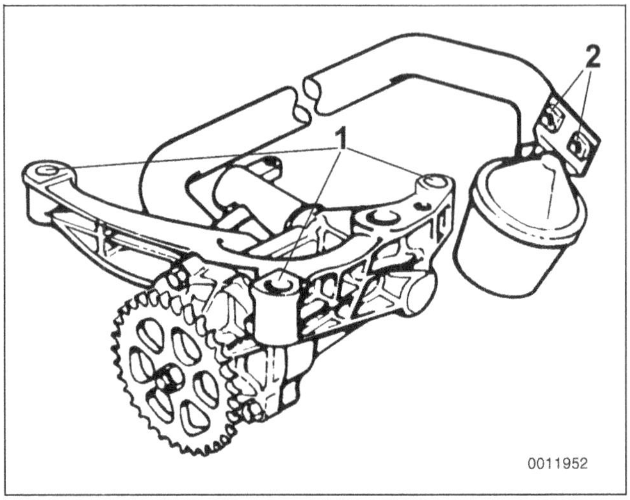

Fig. 17. Oil pump mounting points (**1**) and pickup tube mounting points (**2**).

5. Remove cover from oil pump and check for wear or scoring. Spin oil pump shaft and check that gears turn smoothly. Replace pump if gears spin with difficulty or any wear is present.
6. Installation is reverse of removal, noting the following:
   - Align sprocket splines to oil pump shaft splines before tightening sprocket nut.

Tightening Torques
- Oil pump to engine block (M8) ...... 22 Nm (16 ft-lb)
- Oil pan to engine block (M6 bolt)
  8.8 grade .... 10 Nm (89 in-lb)
  10.9 grade .... 12 Nm (106 in-lb)
- Oil pump sprocket to oil pump shaft
  (M10x1 left-hand thread) .... 25 Nm (18 ft-lb)

120 Ignition System

TABLES

GENERAL

This repair group covers repair and component replacement information for the ignition system.

NOTE—

- *Spark plug replacement is covered in* **020 Maintenance Program**.
- *For fuel related troubleshooting and testing, see* **130 Fuel Injection**.

All engines covered by this manual use an advanced engine management system. The engine management system incorporates on-board diagnostics, fuel injection, ignition and other advanced engine control functions.

Table a. Engine Management System Variants

Engine code	System
4-cylinder	
M42 (1.8 l)	Bosch DME M1.7
M44 (1.9 l)	Bosch DME M5.2 (OBD II)
6-cylinder	
M50	
1992 (2.5 l)	Bosch DME M3.1
1993-1995 (2.5 l)	Bosch DME M3.3.1
M52	
1996-1998 (2.8 l)	Siemens MS 41.1 (OBD II)
1998 (2.5 l)	Siemens MS 41.1 (OBD II)
S50US (3.0 l)	Bosch DME M3.3.1
S52US (3.2 l)	Siemens MS 41.1 (OBD II)

Troubleshooting and testing of the engine management system should be carried out using the BMW specialized service tester (scan tool).

All engines use a distributorless ignition system with individual ignition coils for each cylinder. There is no distributor cap or ignition rotor. Each coil can be selectively controlled by the engine control module on a cylinder-by-cylinder basis. On later systems, knock sensors are used to monitor and control ignition knock (ping) and adjust timing accordingly.

Ignition timing is electronically controlled and not adjustable. The engine control module (ECM) uses engine load, engine speed, coolant temperature, and intake air temperature as the basic inputs for timing control. Knock detection is also an input to the control module, where applicable.

NOTE—

All engines covered by this manual, with the exception of the 1992 M50, incorporate knock sensors as part of the engine management system.

The initial ignition point is determined by the crankshaft position/rpm sensor during cranking. Once the engine is running, timing is continually adjusted based on operating conditions. A characteristic ignition map is shown in Fig. 1. A map similar to the one shown is digitally stored in the engine control module.

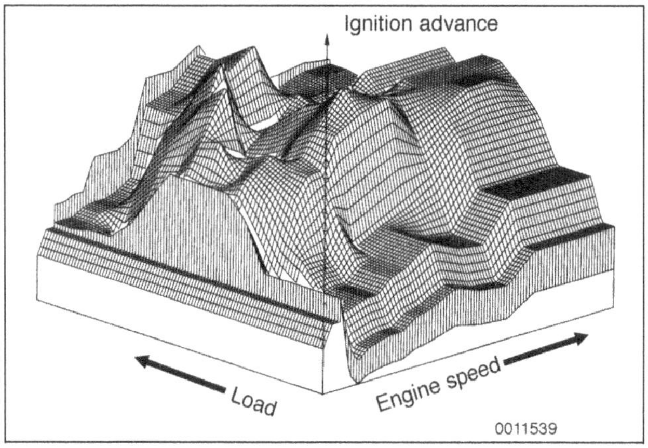

Fig. 1. Ignition characteristic map.

> ***WARNING —***
> *The ignition system is a high-energy system operating in a dangerous voltage range that could prove to be fatal if exposed terminals or live parts are contacted. Use extreme caution when working on a car with the ignition on or the engine running.*

Disabling Ignition System

The ignition system operates in a lethal voltage range and should therefore be disabled any time service or repair work is being done on the engine that requires the ignition to be switched on.

The engine management system can be disabled by removing the main relay. The relay is located in the power distribution box in the left rear of the engine compartment. See Fig. 2.

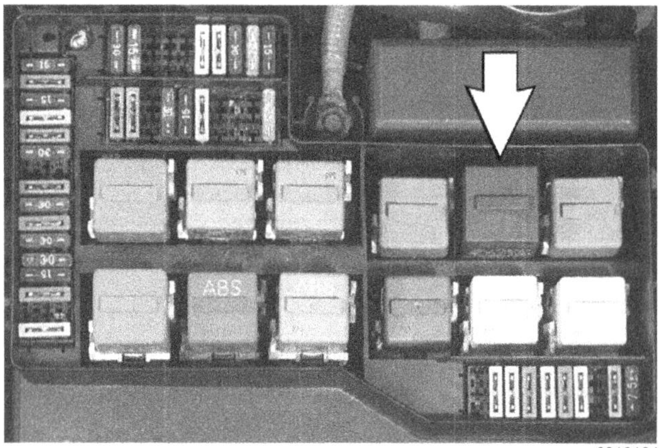

Fig. 2. Main system relay (**arrow**) in power distribution box (left rear of engine compartment).

> ***CAUTION —***
> *Relay locations may vary. Use care when identifying relays. The main relay has a large (4 or 6 mm^2) red wire at terminal 30 and at terminal 86 in the relay socket. Terminal 87 has a large (4 mm^2) red/white wire. See* **610 Electrical Component Locations**.

Warnings and Cautions

The engine management system contains sensitive electronic components. To protect the system and for general safety, the following warnings and cautions should be observed during ignition system troubleshooting, maintenance, or repair work.

> ***WARNING —***
> - *Do not touch or disconnect any of the high tension cables at the coils or spark plugs while the engine is running or being cranked by the starter. Fatal voltages are present.*
> - *Before operating the starter without starting the engine (for example when making a compression test) always disable the ignition.*

> ***CAUTION —***
> - *Prior to disconnecting the battery, read the battery disconnection cautions given at the front of this manual on page viii.*
> - *Do not attempt to disable the ignition by either removing the coil from the spark plugs (6-cylinder engines) or disconnecting the coil wires from the plugs (4-cylinder engines).*
> - *Connect or disconnect ignition system wires, multiple connectors, and ignition test equipment leads only while the ignition is off. Switch multimeter functions or measurement ranges only with the test probes disconnected.*
> - *Do not disconnect the battery while the engine is running.*
> - *Many of the tests of ignition system components require the use of high-impedance test equipment to prevent damage to the electrical components. A high impedance digital multimeter should be used for all voltage and resistance tests. An LED test light should be used in place of an incandescent-type test lamp.*
> - *In general, make test connections only as specified by BMW, as described in this manual, or as described by the instrument's manufacturer.*

IGNITION SYSTEM DIAGNOSTICS

Poor driveability may have a variety of causes. The fault may lie with the ignition system, the fuel system, parts of the emission control system, or a combination of the three. Because of these interrelated functions and their effects on each other, it is often difficult to know where to begin looking for problems.

For this reason, effective troubleshooting should always begin with an interrogation of the On-Board Diagnostic (OBD) system. The OBD system detects certain emissions-related engine management malfunctions. When faults are detected, the OBD system stores a Diagnostic Trouble Code (DTC) in the system ECM. In addition, the Check Engine warning light will come on if an emissions-related fault is detected.

Two generations of OBD are used on the cars covered by this manual. See **100 Engine–General** for OBD information.

On-Board Diagnostics

• 1992-1995 models	OBD I
• 1996 and later models	OBD II

NOTE —
- *On cars with OBD II, specialized OBD II scan tool equipment must be used to access DTCs, either using the BMW special tool or a "generic" OBD II scan tool.*
- *The OBD II fault memory (including an illuminated Check Engine light) can only be reset using the special scan tool. Removing the connector from the ECM or disconnecting the battery will not erase the fault memory.*

Basic Troubleshooting Principles

An engine that starts and runs indicates the ignition system is fundamentally working—delivering voltage to at least some of the spark plugs. A hard-starting or poor-running engine, however, may indicate ignition coil problems, cracked or deteriorated spark plug wires (4-cylinder engines only), and worn or fouled spark plugs.

> ***WARNING —***
> *Inefficient combustion (rich air/fuel mixture) can cause the catalytic converter to overheat and plug. An overheated catalytic converter can also be a fire hazard.*

IGNITION SYSTEM SERVICE

On 4-cylinder engines, an ignition coil pack is mounted to the passenger side strut tower in the engine compartment. The coil pack integrates 4 individual coils.

On 6-cylinder engines, an ignition coil is located directly above each spark plug.

Checking for Spark

> ***WARNING —***
> *If a spark test is done incorrectly, damage to the engine control module (ECM) or the ignition coil(s) may result.*

Checking for spark is difficult on engines with distributorless ignition systems.

Try removing the plugs and inspecting for differences between them. A poor-firing plug may be wet with fuel and/or black and sooty, but not always. If a coil is not operating, the engine management system will electrically disable the fuel injector to that cylinder. The key is to look for differences between cylinders.

Ignition coil, testing and replacing (4-cylinder engine)

1. Disconnect main harness connector from coils:
   - On M42 engine, remove plastic covering from coils and disconnect individual harness connectors.
   - On M44 engine, disconnect main harness connector at end of coil pack. See Fig. 3.

0012422

Fig. 3. Ignition coil pack for M44 engine (**arrow**). Coil harness connector shown at **1**.

2. Check for battery voltage to coils. If voltage is not present, check wire between terminal 15 and ignition switch. See **Electrical Wiring Diagrams**.
   - On M42 engine, connect digital multimeter between terminal 15(+) in each harness connector and ground. See Fig. 4.
   - On M44 engine, connect voltmeter between terminal 2 (15+) in connector and ground. Turn ignition on and check for voltage. See Fig. 5.

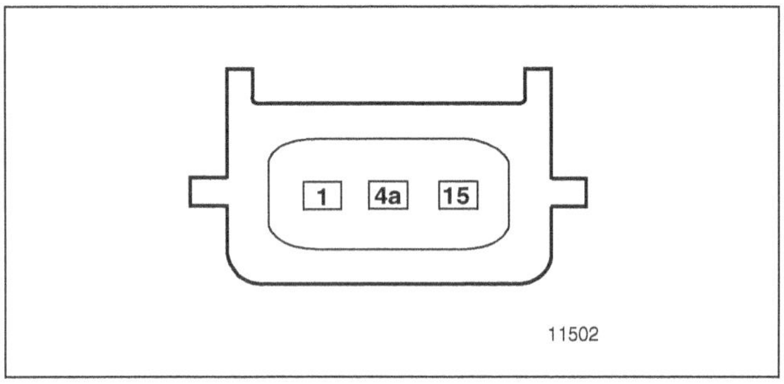

Fig. 4. Ignition coil harness connector terminal identification for M42 engine.

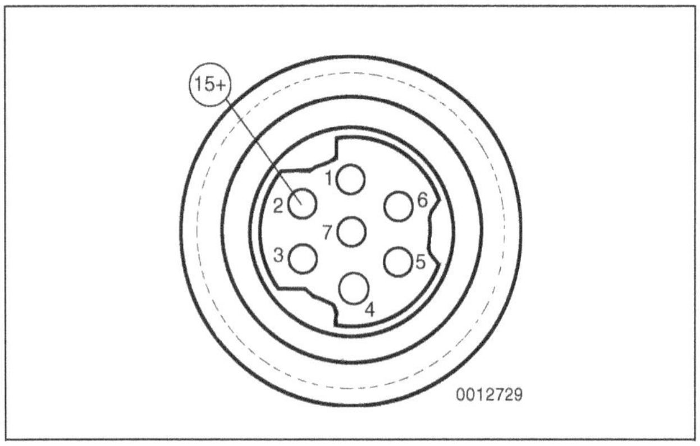

Fig. 5. Ignition coil harness connector terminal identification for M44 engine.

> ***CAUTION —***
> *The wiring to the coil (via the ignition switch 15+) is not fuse protected. Use care when testing this circuit.*

3. Turn ignition off.

4. Use a digital multimeter to test coil primary resistance.
   - On M42 engine, use **Table b** as a guide to check coil resistance. Make checks at each harness connector.
   - On M44 engine, use **Table c** as a guide to check coil primary resistance.

5. Inspect coil housing for hairline cracks or leaking casting material. See Fig. 7. A leaky ignition coil may indicate a faulty ECM (engine control module). Check the ECM before installing a new coil.

Table b. Ignition Coil Resistance (M42 Engine)

	Terminals (refer to Fig. 4.)	Resistance
Coil primary	1 (–) and 15 (+)	0.4-0.8 ohms
Coil secondary	N.A.	N.A.

Table c. Ignition Coil Resistance (M44 Engine)

	Terminals (refer to Fig. 5.)	Resistance
Coil #1 primary	1 (–) and 2 (+)	0.4-0.8 ohms
Coil #2 primary	6 (–) and 2 (+)	0.4-0.8 ohms
Coil #3 primary	7 (–) and 2 (+)	0.4-0.8 ohms
Coil #4 primary	5 (–) and 2 (+)	0.4-0.8 ohms
Coil secondary	N.A.	N.A.

> ***NOTE —***
> *If a single coil is faulty on M44 engines, the complete coil pack will have to be replaced.*

Ignition coil, testing and replacing (6-cylinder engine)

> ***CAUTION —***
> *Use a digital multimeter for the following tests.*

1. Remove plastic engine covers from top of engine by prying off nut covers and removing cover mounting nuts. See **113 Cylinder Head Removal and Installation**.

2. Disconnect harness connector from coil. Connect multimeter between terminal 15 (+) in connector and ground. See Fig. 6.

3. Turn ignition on and check for battery voltage.
   - If battery voltage is not present, check wire between terminal 15 and ignition switch. See **Electrical Wiring Diagrams**.

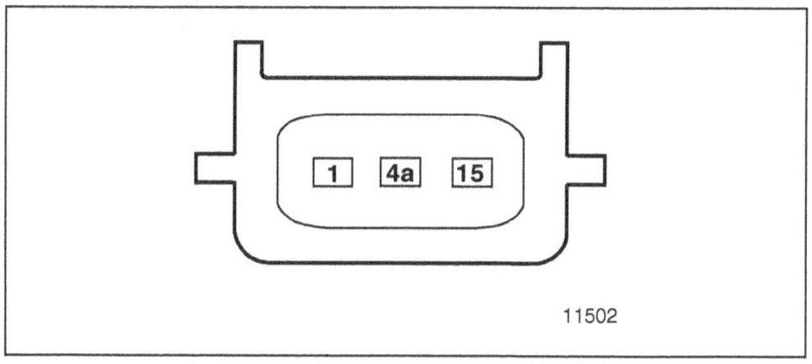

Fig. 6. Ignition coil harness connector terminal identification for 6-cylinder engine.

> ***CAUTION—***
> *The wiring to terminal 15 (+) of the coil (via the ignition switch) is not fuse protected. Use care when testing this circuit.*

4. Turn ignition off.

5. Use a multimeter to test coil primary resistance at coil terminals. See **Table d**.

Table d. Ignition Coil Resistance (6-cylinder Engine)

	Terminals (refer to Fig. 4.)	Resistance
Coil primary	1 (–) and 15 (+)	0.4-0.8 ohms
Coil secondary	N.A.	N.A.

6. Remove coil and inspect coil housing for hairline cracks or leaking casting material. See Fig. 7. A leaky ignition coil may indicate a faulty engine control module (ECM). Check ECM before installing a new coil.

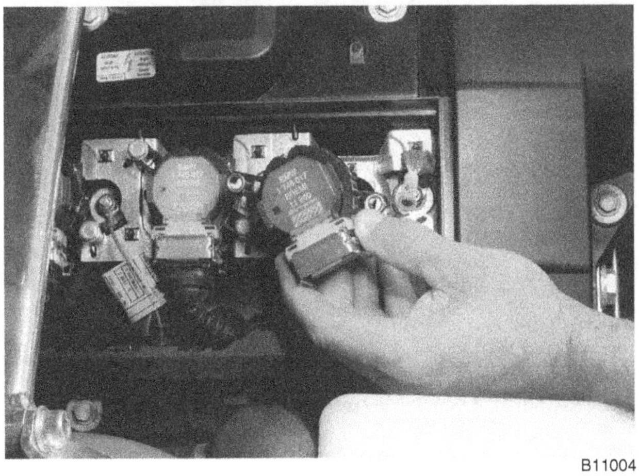

B11004

Fig. 7. Ignition coil being removed on 6-cylinder engine (M52 engine shown).

> ***CAUTION—***
> - *Note location of coil ground straps before coil removal; reinstall in the same location.*
> - *When replacing ignition coils, ensure that the replacement coil(s) are from the same manufacturer containing the same part/code numbers. If individual coils with the correct specifications are not available, all coils should be replaced.*

Crankshaft Position/rpm Sensor

If the engine control module (ECM) does not receive a crankshaft position signal during cranking, the engine will not start.

On 1992-1995 (pre-OBD II) cars, the crankshaft position/rpm sensor is mounted on the front engine cover and reads the toothed vibration dampener wheel. See Fig. 8.

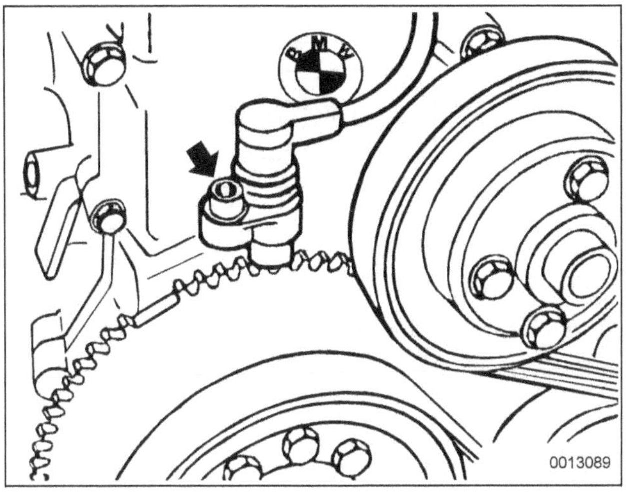

Fig. 8. Crankshaft position/rpm sensor mounted at front of engine on 1992-1995 cars. (**arrow**).

On 1996 and later cars (OBD II compliant), the crankshaft position/rpm sensor is mounted in the left rear side of the cylinder block. The sensor reads a toothed wheel mounted to the end of the crankshaft. See Fig. 9.

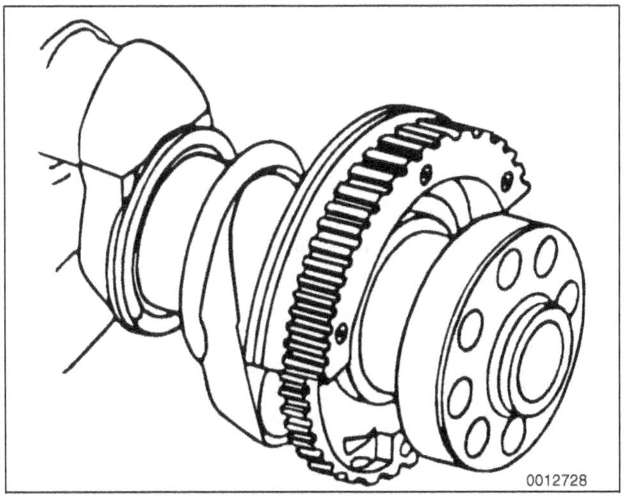

Fig. 9. Crankshaft position/rpm sensor toothed wheel mounted to rear of crankshaft OBD II compliant cars. The sensor is mounted in the left rear side of the cylinder block.

Crankshaft position/rpm sensor, testing and replacing (front mounted)

1. Disconnect sensor harness connector.

2. Using a digital multimeter, check resistance between terminals 1 and 2 in connector. See Fig. 10.

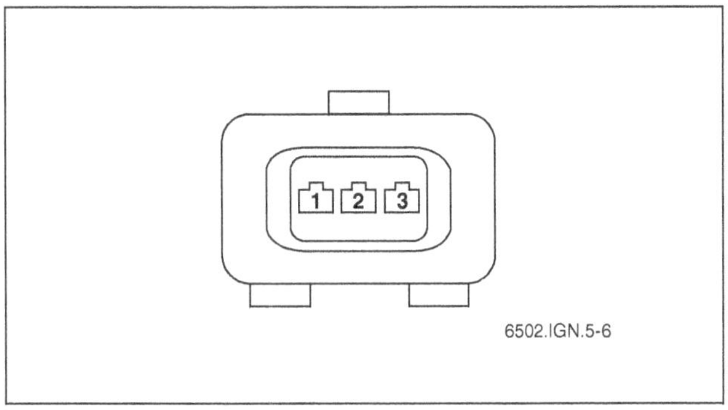

Fig. 10. Crankshaft position/rpm sensor connector.

Crankshaft position/rpm sensor specifications
- Coil resistance (approx.)
  terminals 1 and 2 . . . . 540 ± 10% ohms
- Air gap (sensor distance from toothed
  wheel) . . . . 1.0 ± 0.3 mm (0.04 ± 0.01 in.)

3. If the resistance is not correct, the sensor is faulty and should be replaced.

NOTE —
When installing the new sensor, be sure the wiring is rerouted in the same orientation. Secure the sensor using new wire ties.

Tightening Torque
- Crankshaft position/rpm sensor to
  mounting bracket . . . . 5 ± 1 Nm (62 ± 9 in-lb)

Crankshaft position/rpm sensor, replacing (rear mounted)

1. Disconnect sensor harness connector.

2. Locate sensor on rear left side of cylinder block. Remove sensor mounting bolt and remove sensor from cylinder block.

3. Installation is reverse of removal. Use a new O-ring when installing sensor. Be sure wiring is rerouted in same orientation. Secure sensor using new wire ties.

NOTE —
It may be easier to remove the sensor working from the underside of the vehicle.

Camshaft Position (CMP) Sensor

The camshaft position (CMP) sensor is used by the engine management system for sequential fuel injection and knock control.

Camshaft position (CMP) sensor, replacing (4-cylinder engine)

1. Remove CMP sensor from top timing cover, just above coolant thermostat housing. See Fig. 11.

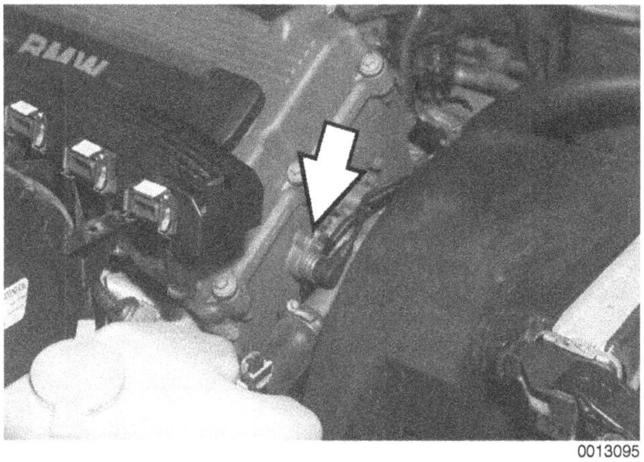

Fig. 11. Camshaft position sensor mounting bolt (**arrow**). M42 engine shown.

2. On M42 engines to 9/93: Disconnect CMP harness plug just above oil filter housing.

3. On M42 from 9/93 and all M44 engines: Remove upper intake manifold as described in **113 Cylinder Head Removal and Installation**. Then unplug CMP sensor harness connector below manifold.

4. Installation is reverse of removal. Use a new O-ring when installing sensor. Be sure wiring is rerouted in same orientation.

Camshaft position (CMP) sensor, replacing (6-cylinder engine)

1. Remove plastic cover from above fuel injectors.

2. Disconnect harness connector from VANOS solenoid and unscrew solenoid from VANOS control unit. Remove oil supply line from VANOS control unit. See **117 Camshaft Timing Chain**.

3. Remove camshaft position (CMP) sensor from left front of cylinder head, next to top of oil filter housing.

4. Disconnect CMP sensor harness from under intake manifold.

5. Installation is reverse of removal. Use a new O-ring when installing sensor. Be sure wiring is rerouted in same orientation.

Tightening Torques
- CMP sensor to cylinder head. . . . . 5 Nm (3.5 ft-lb)
- VANOS oil supply pipe to VANOS control unit . . . . 32 Nm (24 ft-lb)
- VANOS solenoid to VANOS control unit . . . . 30 Nm (22 ft-lb)

Knock sensors, replacing

Knock sensors are used on engines with Bosch DME M1.7, Bosch DME M3.3.1, Bosch DME M5.2, and Siemens MS 41.1 engine management systems. The knock sensors are bolted to the left side of the cylinder block and monitor the combustion chamber for engine-damaging knock. If engine knock is detected, the ignition point is retarded accordingly via the engine control module.

On 6-cylinder engines, the knock sensors are accessible after removing the intake manifold. On 4-cylinder engines, the upper section of the intake manifold should be removed to access the sensors. See Fig. 12.

Intake manifold removal and installation procedures are covered in **113 Cylinder Head Removal and Installation**.

CAUTION—
- *Label knock sensor harness connectors before disconnecting them. The connectors must not be interchanged. Engine damage may result if the sensors are monitoring the wrong cylinders.*
- *Note the installed angle of the knock sensor on the block before removing it. Reinstall the sensor in the same position. Be sure to use a torque wrench when tightening the sensor mounting bolt.*

NOTE—
- *On M52 engines, a single harness connects the two sensors to the main harness connector. The shorter cable lead is for the knock sensor for cylinders 4, 5, and 6.*
- *Clean contact surface on engine block before installing knock sensors.*

Tightening Torque
- Knock sensor to cylinder block. . . . . . 20 Nm (15 ft-lb)

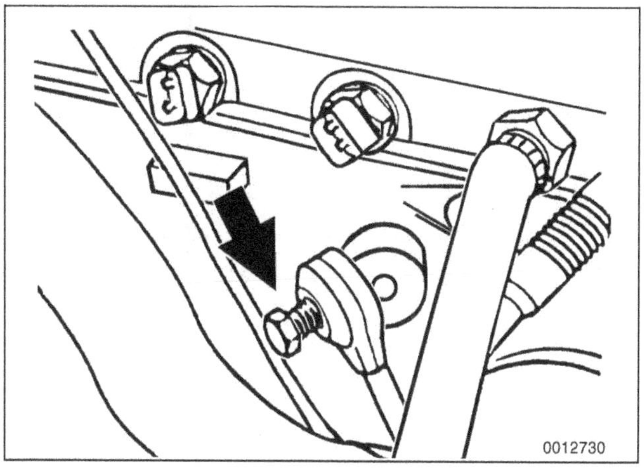

Fig. 12. Knock sensor mounting bolt on front left side of cylinder block.

NOTE—
Knock sensor function is monitored by the On-Board Diagnostic system. If a knock sensor is detected to be faulty, an appropriate diagnostic trouble code (DTC) may be stored in memory. See **100 Engine–General** *for information on retrieving DTCs.*

Ignition Firing Order

NOTE—
- *Cylinder no. 1 is at the front of the engine.*

Ignition Firing Order
- 4-cylinder engines 1-3-4-2
- 6-cylinder engines 1-5-3-6-2-4

IGNITION SYSTEM SCHEMATICS

Fig. 13 through Fig. 17 show ignition system schematics used on the engines covered by this manual.

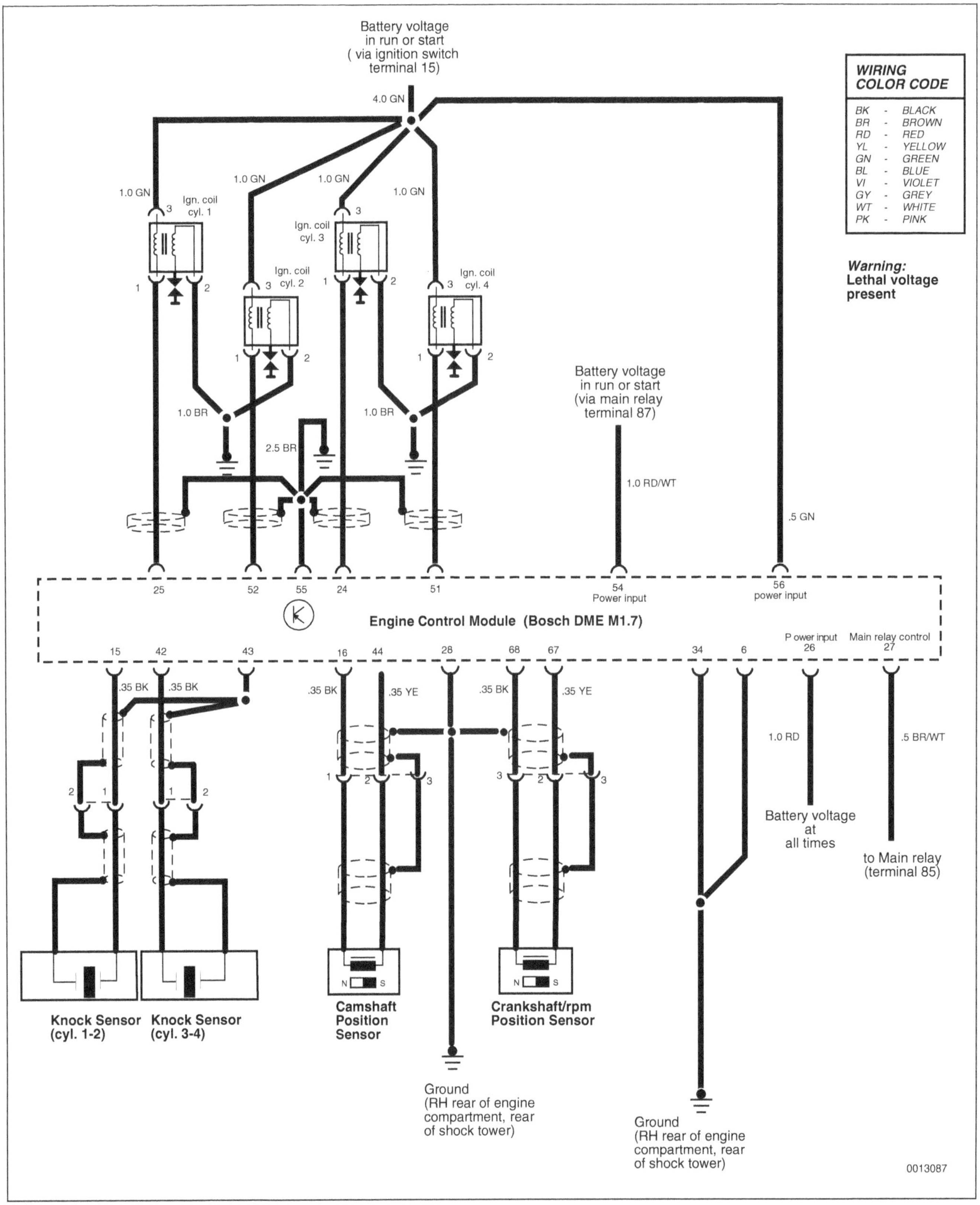

Fig. 13. Bosch DME M1.7 ignition system circuit for M42 engine.

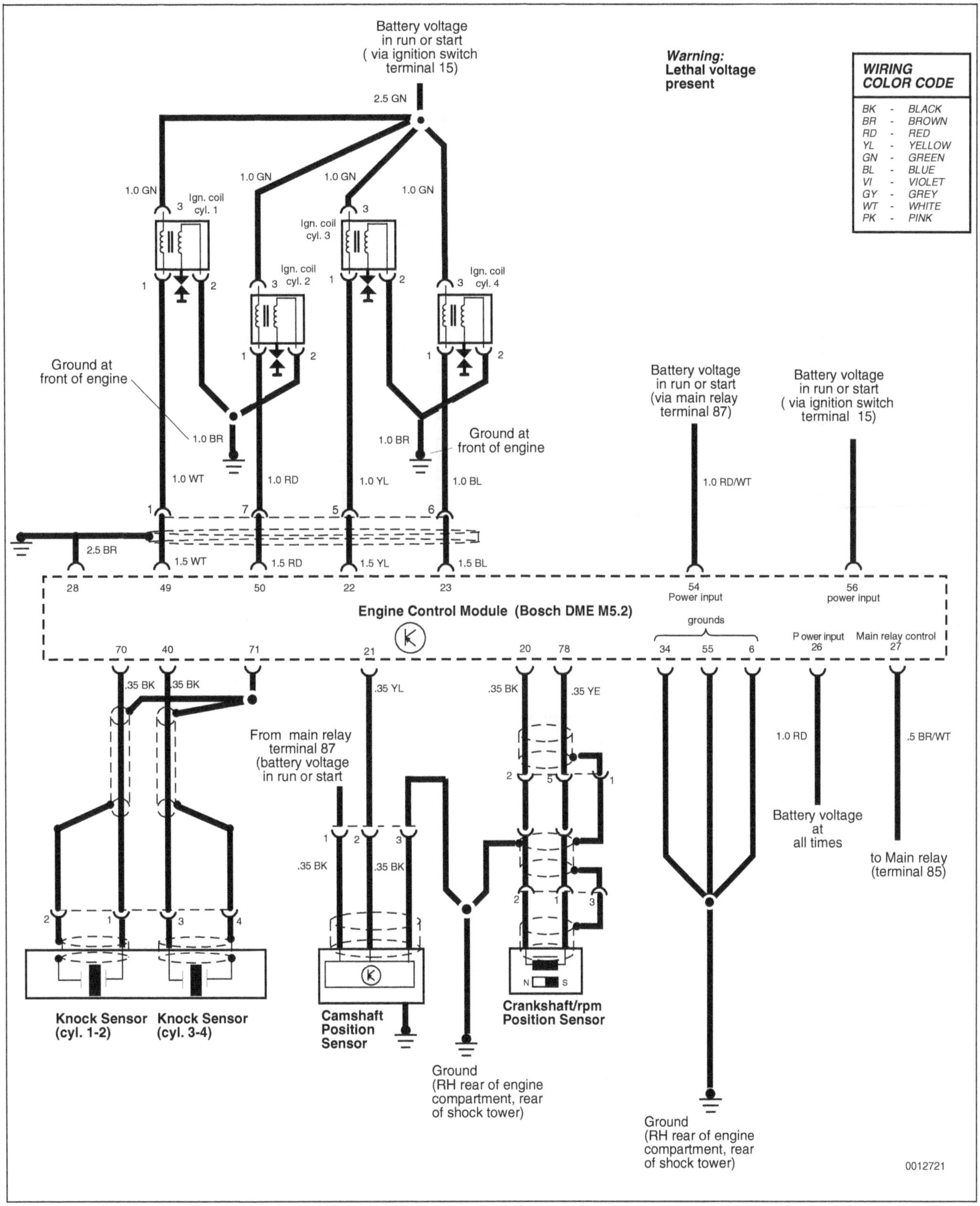

Fig. 14. Bosch DME M5.2 ignition system circuit for M44 engine.

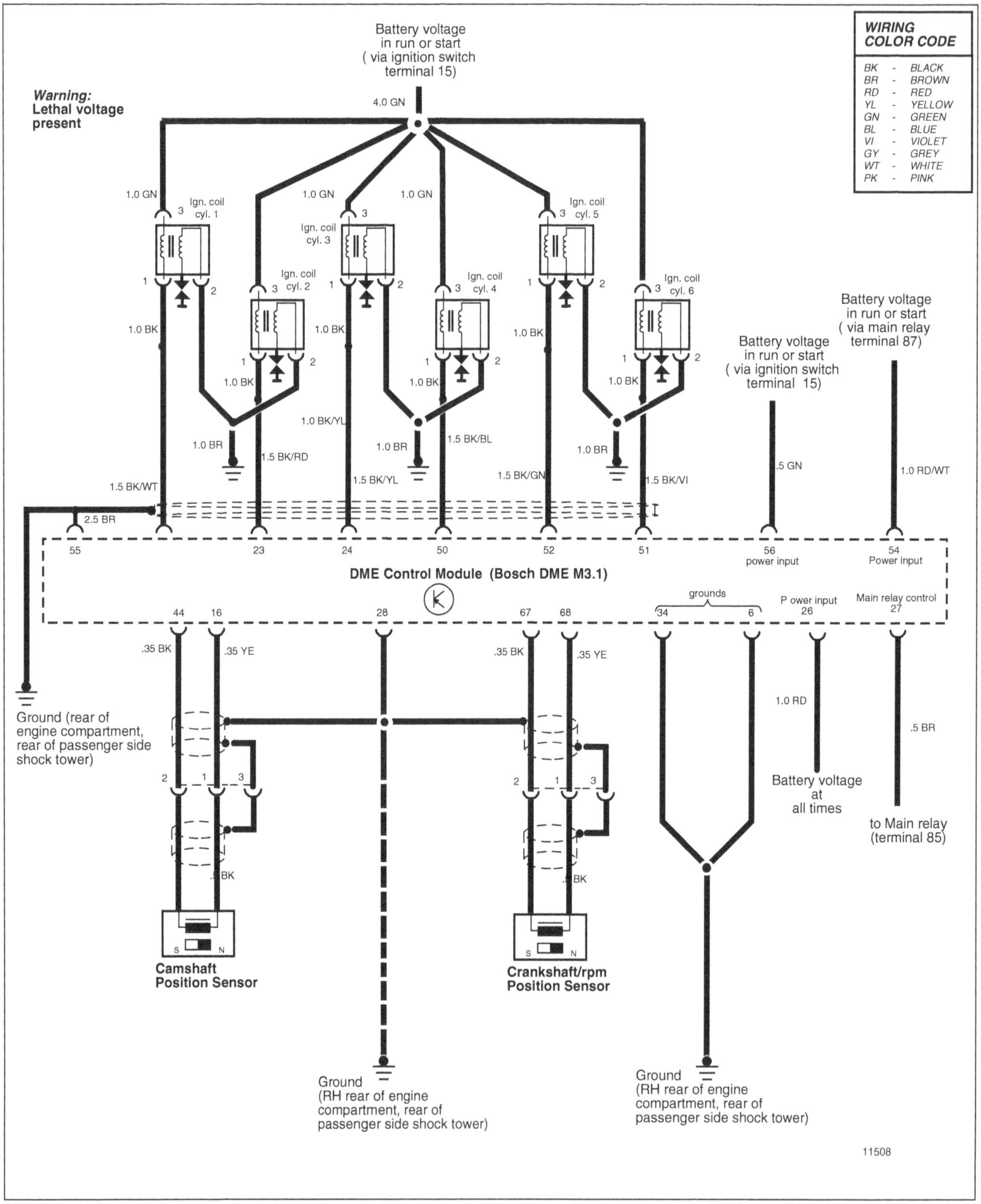

Fig. 15. Bosch DME M3.1 used on 1992 M50 engines (without VANOS).

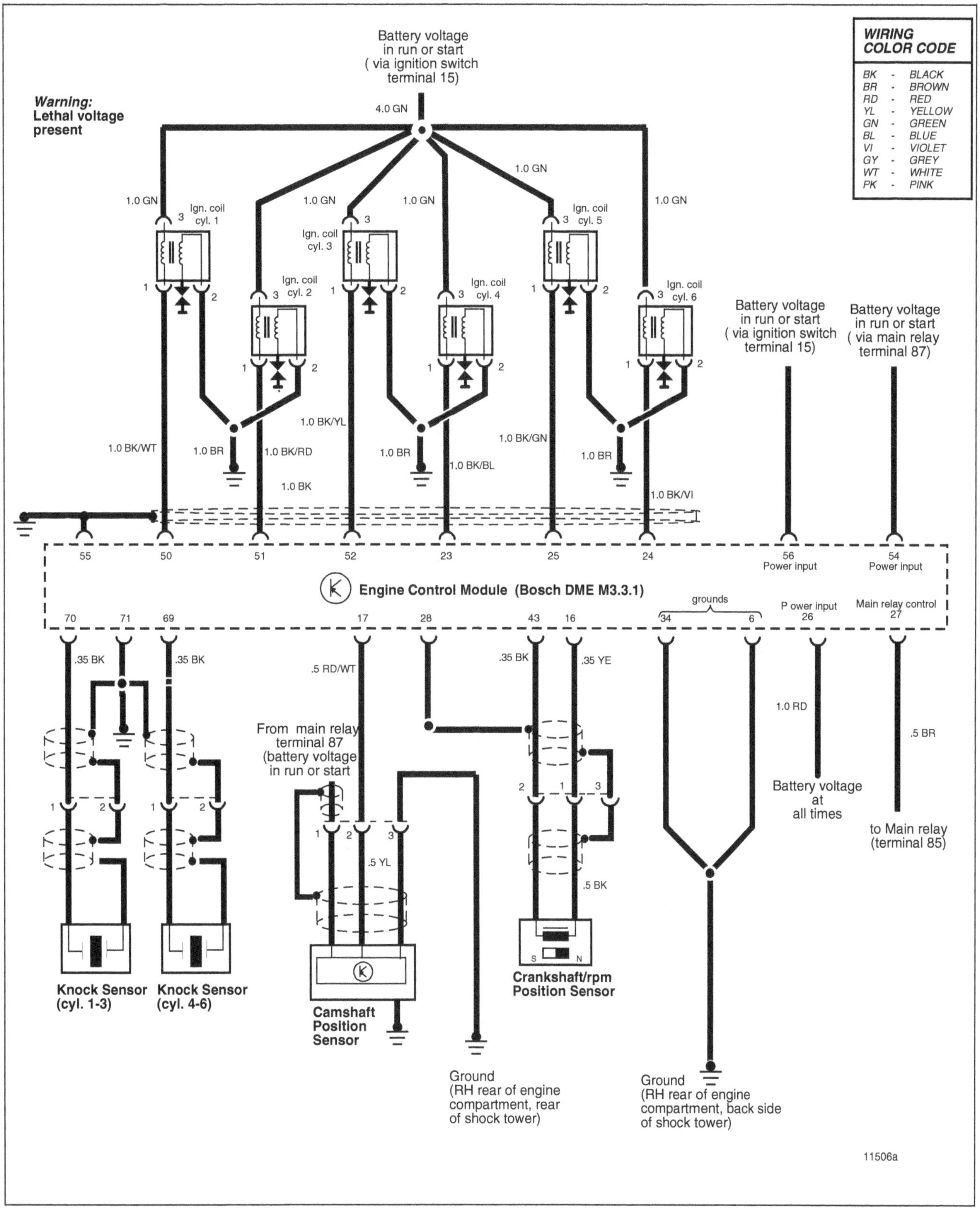

Fig. 16. Bosch DME M3.3.1 ignition system circuit for 1993-1995 M50 engines with VANOS.

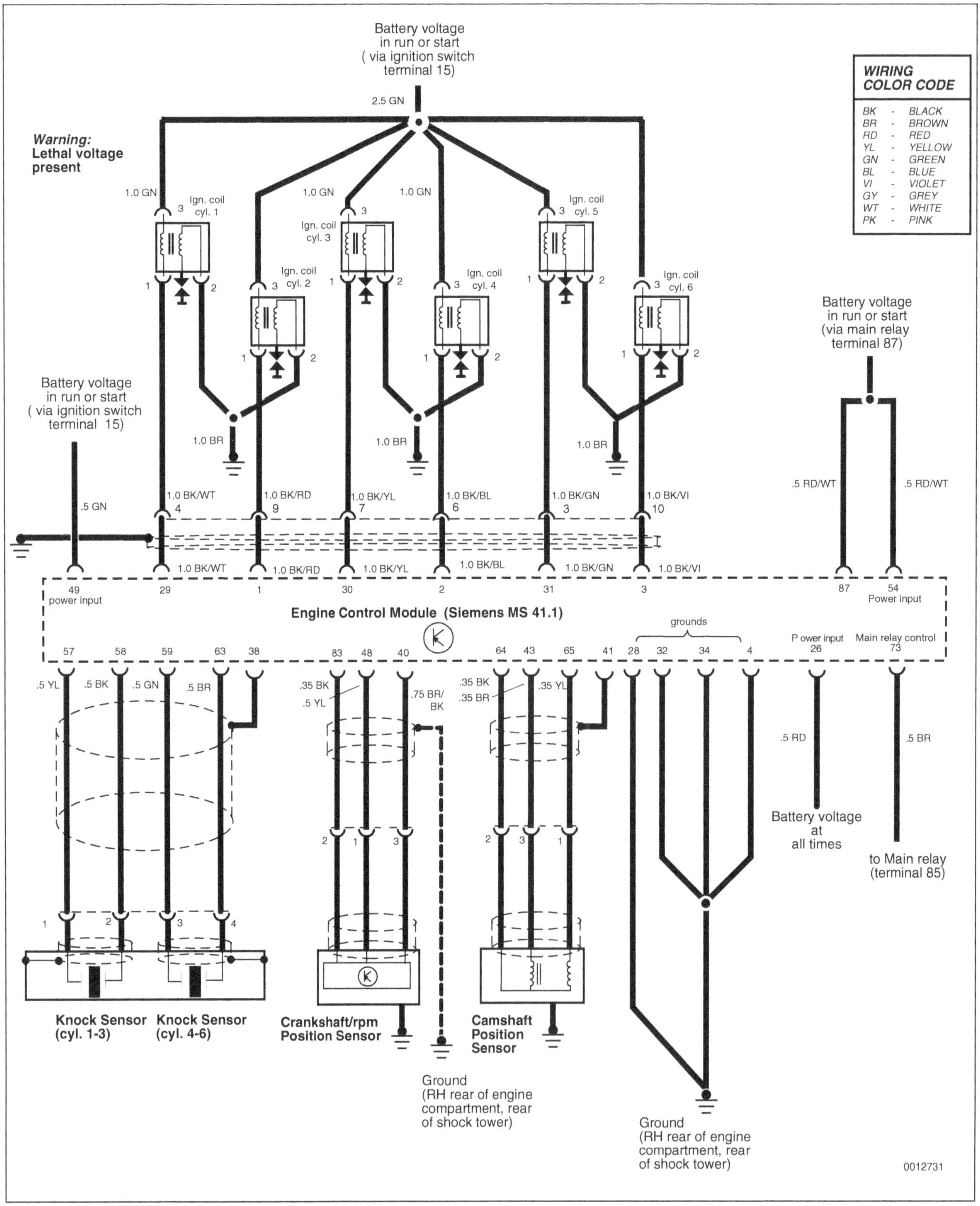

Fig. 17. Siemens MS 41.1 ignition system used on 1996 and later 6-cylinder engines.

121 Battery, Starter, Alternator

TABLES

GENERAL

The charging system consists of a belt-driven alternator with integral voltage regulator and a battery mounted in the luggage compartment.

Various versions of alternators, voltage regulators, starters, and batteries are used in the E36 cars. It is important to replace components according to the original equipment specification. Check with an authorized BMW dealer for specific application and parts information.

WARNING —
- *Wear goggles, rubber gloves, and a rubber apron when working around batteries and battery acid (electrolyte).*
- *Battery acid contains sulfuric acid and can cause skin irritation and burning. If acid is spilled on your skin or clothing, flush the area at once with large quantities of water. If electrolyte gets into your eyes, flush them with large quantities of clean water for several minutes and call a physician.*
- *Batteries that are being charged or are fully charged give off explosive hydrogen gas. Keep sparks and open flames away. Do not smoke.*

CAUTION —
- *Prior to disconnecting the battery, read the battery disconnection cautions given at the front of this manual on page viii.*
- *Disconnecting the battery cables may erase fault codes stored in control unit memory.*
- *Always disconnect the negative (–) battery cable first and reconnect it last. Cover the battery post with an insulating material whenever the cable is removed.*
- *After reconnecting the battery, the power window motors must be reinitialized. See* **511 Door Windows**.
- *Never reverse the battery cables. Even a momentary wrong connection can damage the alternator or other electrical components.*
- *Battery cables may be the same color. Label cable before removing.*

CHARGING SYSTEM TROUBLESHOOTING

Charging system diagnostics requires special test equipment. If the test equipment is not available, charging system fault diagnosis can be performed by an authorized BMW dealer or other qualified repair shop. A general troubleshooting guide is given in **Table a**.

Charging System Quick-Check

As a quick-check, use a digital multimeter to measure voltage across the battery terminals with the key off and then again with the engine running. The battery voltage should be about 12.6 volts with key off and approximately 14.0 volts with the engine running. If the voltage does not increase when the engine is running, there is a fault in the charging system.

> ***NOTE —***
> *The regulated voltage (engine running) should be between 13.5 and 14.5, depending on temperature and operating conditions. If the voltage is higher than 14.8, the voltage regulator is most likely faulty.*

Check for clean and tight battery cables. Check the ground cable running from the negative (–) battery terminal to the chassis and the ground cable running from the engine to the chassis. Check the alternator drive belt condition and tension.

Static current draw, checking

If the battery discharges over time, there may be a constant drain or current draw on the battery. A small static drain on the battery is normal, but a large drain will cause the battery to quickly discharge. Make a static current draw test as the first step when experiencing battery discharge.

1. Make sure ignition and all electrical accessories are switched off.

2. Disconnect battery negative (–) cable.

> ***CAUTION —***
> *Prior to disconnecting the battery, read the battery disconnection cautions given at the front of this manual on page viii.*

3. Connect a digital ammeter between negative battery post and negative battery cable to measure current. See Fig. 1. Wait at least one minute to get an accurate reading.

A range of about 0 to 100 milliamps is normal, depending on the number of accessories that need constant power. A current of 400 milliamps (0.4 amp) or more may indicate a problem.

Table a. Battery, Starter and Charging System Troubleshooting

Symptom	Probable Cause	Corrective Action
1. Engine cranks slowly or not at all, solenoid clicks when starter is operated.	a. Battery cables loose, dirty or corroded. b. Battery discharged. c. Body ground strap loose, dirty or corroded. d. Poor connection at starter motor terminal 30. e. Starter motor or solenoid faulty.	a. Clean or replace cables. See **020 Maintenance Program**. b. Charge battery, test and replace if necessary. c. Inspect ground strap, clean, tighten or replace if necessary. d. Check connections, test for voltage at starter. Test for voltage at neutral safety or clutch interlock switch. e. Test starter.
2. Battery will not stay charged more than a few days.	a. Short circuit draining the battery. b. Short driving trips and high electrical drain on charging system does not allow battery to recharge. c. Drive belt(s) worn or damaged. d. Battery faulty. e. Battery cables loose, dirty or corroded. f. Alternator or voltage regulator faulty.	a. Test for excessive current drain with everything electrical in the vehicle off. b. Evaluate driving style. Where possible, reduce electrical consumption when making short trips. c. Inspect or replace multi-ribbed belt(s). See **020 Maintenance Program**. d. Test battery and replace if necessary. e. Clean or replace cables. See **020 Maintenance Program**. f. Test alternator and voltage regulator.
3. Battery losing water.	a. Battery overcharging.	a. Test voltage regulator for proper operation.
4. Lights dim, light intensity varies with engine speed.	a. Drive belt(s) worn or damaged. b. Alternator or voltage regulator faulty. c. Body ground straps loose, dirty or corroded.	a. Inspect or replace multi-ribbed belt(s). See **020 Maintenance Program**. b. Test alternator and voltage regulator. c. Inspect ground straps, clean, tighten or replace as necessary.

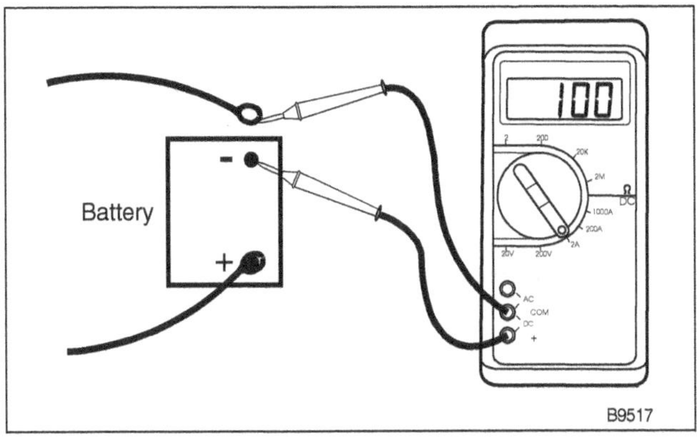

Fig. 1. Electrical system static current draw being measured.

To determine the circuit or component causing the problem, remove one fuse at a time until the current drops to a normal range.

BATTERY SERVICE

The E36 uses a six-cell, 12-volt lead acid battery mounted in the luggage compartment. See Fig. 2.

> ***NOTE —***
> *E36 convertible models require a special battery which is designed for constant vibration. A battery not designed for this will fail much earlier.*

Battery capacity is determined by the amount of current needed to start the vehicle, and by the amount of current consumed by the electrical system.

BMW batteries are rated by ampere/hours (Ah) and cold cranking amps (CCA) rating. The Ah rating is determined by the average amount of current the battery can deliver over time without dropping below a specified voltage. The CCA is determined by the battery's ability to deliver starting current at 0° F (-18° C) without dropping below a specified voltage.

Battery Testing

Battery testing determines the state of battery charge. On conventional or low-maintenance batteries the most common method of testing the battery is that of checking the specific gravity of the electrolyte using a hydrometer. Before testing the battery, check that the cables are tight and free of corrosion. See Fig. 2.

Hydrometer Testing

The hydrometer consists of a glass cylinder with a freely moving float inside. When electrolyte is drawn into the cylinder, the level to which the float sinks indicates the specific

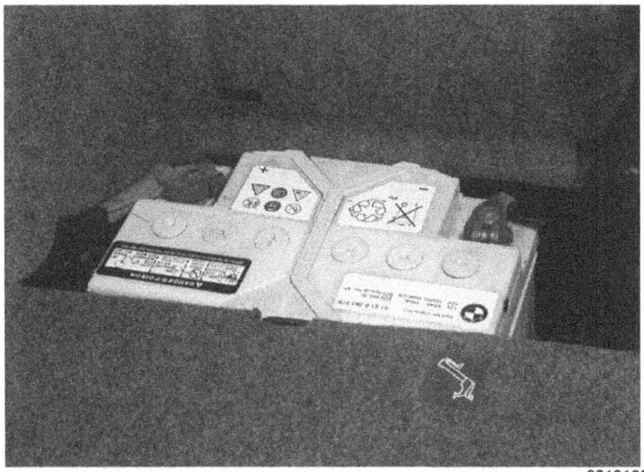

Fig. 2. Battery in right side of luggage compartment.

gravity of the electrolyte. The more dense the concentration of sulfuric acid in the electrolyte, the less the float will sink, resulting in a higher reading and indicating a higher state of charge).

> ***NOTE —***
> *Electrolyte temperature affects hydrometer reading. Check the electrolyte temperature with a thermometer. Add 0.004 to the hydrometer reading for every 10°F (6°C) that the electrolyte is above 80°F (27°C). Subtract 0.004 from the reading for every 10°F (6°C) that the electrolyte is below 80°F (27°C).*

Before checking the specific gravity of a battery, load the battery with 15 amperes for one minute. If the battery is installed in the vehicle, this can be done by turning on the headlights without the engine running. **Table b** lists the percentage of charge based on specific gravity values.

Table b. Specific Gravity of Battery Electrolyte at 80°F (27°C)

Specific gravity	State of charge
1.265	Fully charged
1.225	75% charged
1.190	50% charged
1.155	25% charged
1.120	Fully discharged

The battery is in satisfactory condition if the average specific gravity of the six cells is at least 1.225. If the specific gravity is above this level, but the battery lacks power for starting, determine the battery's service condition with a load voltage test, as described below. If the average specific gravity of the six cells is below 1.225, remove the battery from the luggage compartment and recharge. If, after recharging, the specific gravity varies by more than 0.005 between any two cells, replace the battery.

Battery Open-Circuit Voltage Test

Before making the test, load the battery with 15 amperes for one minute with a battery load-tester or turn on the headlights for about one minute without the engine running.Then disconnect the battery negative (–) cable and connect a digital voltmeter across the battery terminals. Open-circuit voltage levels are given in **Table c.**

If the open-circuit voltage is OK but the battery still lacks power for starting, make a load voltage test. If the open-circuit voltage is below 12.4 volts, recharge the battery and retest.

Table c. Open-Circuit Voltage and Battery Charge

Open-circuit voltage	State of charge
12.6 V or more	Fully charged
12.4 V	75% charged
12.2 V	50% charged
12.0 V	25% charged
11.7 V or less	Fully discharged

Battery Load Voltage Test

A battery load tester is required for a load voltage test. The test is made by applying a high resistive load to the battery terminals and then measuring battery voltage. The battery should be fully charged for the most accurate results. The battery cables must be disconnected before making the test. If the voltage is below that listed in **Table d**, the battery should be replaced.

> ***WARNING —***
> *Always wear protective goggles and clothing when performing a load test.*

Table d. Battery Load Test–Minimum Voltage (apply 200 amp load for 15 seconds)

Ambient temperature	Voltage
80°F (27°C)	9.6 V
60°F (16°C)	9.5 V
40°F (4°C)	9.3 V
20°F (-7°C)	8.9 V
0°F (-18°C)	8.5 V

Battery Charging

Discharged batteries can be recharged using a battery charger. The battery should be removed from the luggage compartment during charging.

Prolonged charging causes electrolyte evaporation to a level that can damage the battery. It is best to use a low-current charger (6 amperes or less) to prevent battery damage caused by overheating.

> ***WARNING —***
> *Hydrogen gas given off by the battery during charging is explosive. Do not smoke. Keep open flames away from the top of the battery, and prevent electrical sparks by turning off the battery charger before connecting or disconnecting it.*

> ***CAUTION —***
> - *Battery electrolyte (sulfuric acid) can damage the car. If electrolyte is spilled, clean the area with a solution of baking soda and water.*
> - *Always allow a frozen battery to thaw before attempting to recharge it.*
> - *Always disconnect both battery cables and remove battery from vehicle during battery charging. Do not exceed 16.5 volts at the battery.*

ALTERNATOR SERVICE

Before checking the alternator and regulator, make sure the battery is fully charged and capable of holding a charge. Check that the battery terminals are clean and tight and the alternator drive belt is properly tensioned and not severely worn.

Charging system, checking

> ***CAUTION —***
> *Do not disconnect the battery while the engine is running,. Damage to the alternator and/or engine electronic systems may result.*

1. Turn ignition key on. Check that the charge warning lamp comes on.

> ***NOTE —***
> *If the warning light does not come on, repair any wiring or bulb faults before continuing to check the charging system.*

2. Remove cooling duct or terminal cover from rear of alternator.

3. Check for battery voltage between ground and terminal **B+** at back of alternator. If voltage is not present, check wiring for faults. See Fig. 3.

Fig. 3. Terminal **B+** is supplied battery voltage directly from the battery. Terminal **D+** is ground for charge warning bulb when the key is on. With engine running, **D+** becomes alternator output voltage (approx. 14 volts) and charge warning bulb goes out.

4. If no faults are found up to this point, test alternator output using a load tester.

5. If a load tester is not available, a crude output test can be done by running engine at about 2000 rpm and turning on electrical loads (fans, lights and rear window defroster, wipers). With all accessories on, battery voltage should be above 12.0 VDC.

Alternator, removing and installing (4-cylinder engine)

A replacement alternator should have the same rating as the original. Alternator manufacturer and ampere rating are normally marked on the alternator housing.

1. Disconnect negative (–) battery cable.

> ***CAUTION—***
> *Prior to disconnecting the battery, read the battery disconnection cautions given at the front of this manual on page viii.*

2. Remove air cleaner upper section with mass air flow sensor.

3. Remove terminal cover from rear of alternator. Disconnect wiring.

4. Remove drive belt from alternator pulley. See **020 Maintenance program**.

> ***NOTE—***
> *If reusing drive belt, mark direction of rotation on belt before removing.*

5. On M44 engine with hydraulic belt tensioner: Remove tensioner idler pulley (upper roller) from alternator bracket.

6. Remove upper and lower mounting bolts and lift out alternator. See Fig. 4.

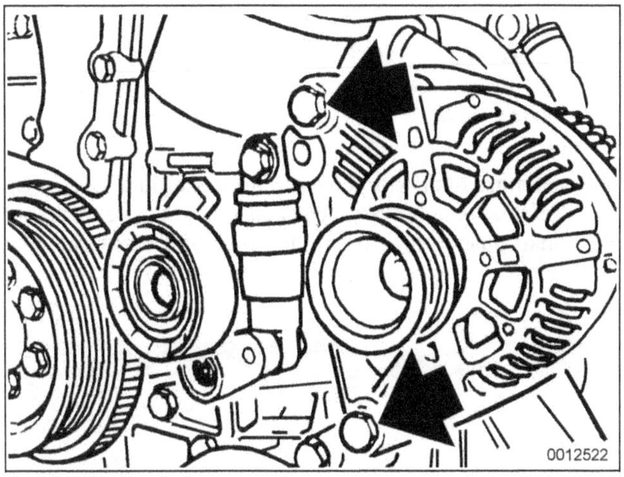

Fig. 4. Alternator mounting bolts (**arrows**).

7. Installation is reverse of removal. Install drive belt as described in **020 Maintenance Program**.

Tightening Torques

- D+ wire to alternator (M6 nut) . . . . 7 Nm (53 in-lb)
- B+ wire to alternator (M8 nut). . . . . . . 13 Nm (10 ft-lb)
- Pulley to alternator (M16 nut) . . . . 60 Nm (44 ft-lb)

Alternator, removing and installing (6-cylinder engine)

A replacement alternator should have the same rating as the original. Alternator manufacturer and ampere rating are normally marked on the alternator housing.

1. Disconnect negative (–) battery cable.

> ***CAUTION —***
> *Prior to disconnecting the battery, read the battery disconnection cautions given at the front of this manual on page viii.*

2. Remove alternator cooling duct and air filter housing assembly from car. Disconnect wiring from rear of alternator.

3. Remove radiator cooling fan and fan shroud. See **170 Radiator and Cooling System**.

> ***NOTE —***
> *The radiator cooling fan nut (32 mm wrench) has left-hand threads.*

4. Remove alternator drive belt from alternator pulley. On engines with hydraulic belt tensioner, remove tensioner idler pulley (upper roller) from alternator bracket. See Fig. 5.

Fig. 5. Pry off cover from tensioner and then lever tensioner clockwise to release belt tension. Remove upper roller (**A**) once belt is removed.

> ***NOTE —***
> *If reusing drive belt, mark direction of rotation on belt before removing.*

5. Remove upper and lower mounting bolts and lift out alternator.

6. Installation is reverse of removal. Note locating notches on tensioner idler pulley (upper roller) alternator bracket when installing pulley.

Tightening Torques
- D+ wire to alternator (M6 nut) .... 7 Nm (53 in-lb)
- B+ wire to alternator (M8 nut)....... 13 Nm (10 ft-lb)
- Pulley to alternator (M16 nut)....... 60 Nm (44 ft-lb)

Voltage regulator, removing and installing

1. Disconnect negative (–) cable from battery.

2. Remove alternator as described earlier.

3. Remove cooling duct from rear of alternator. See Fig. 6.

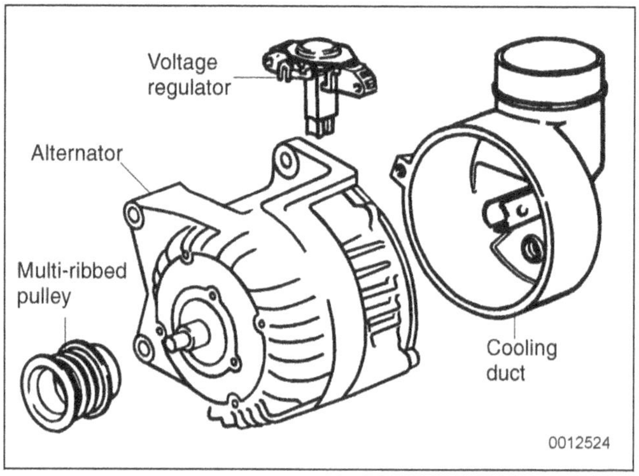

Fig. 6. Alternator assembly and related parts.

4. Remove voltage regulator mounting screws and remove regulator from alternator.

5. Installation is reverse of removal. Clean brush contact surfaces in alternator and check brush length as described later.

Alternator brushes, inspecting and replacing

Regulator brushes are not available as replacement parts from BMW. Replacement brushes may be available from aftermarket sources, however.

1. Remove voltage regulator as described above.

2. Clean brush contact surfaces and measure brush protrusion. See Fig. 7.

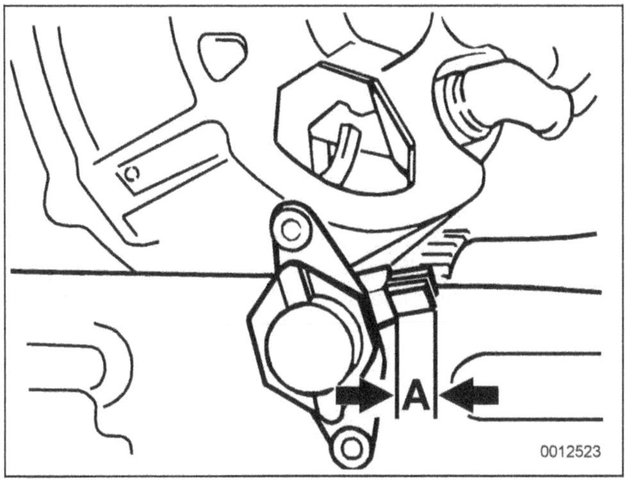

Fig. 7. Regulator brush protrusion (**A**).

Voltage Regulator
- Brush protrusion (minimum). . . . . 5 mm (¼ in.)

3. To replace brushes, carefully and as quickly as possible, unsolder brush lead from brush holder terminal, withdrawing brush from holder at same time.

4. Remove any traces of solder from brush holder terminal using solder wick.

5. Fit spring into brush holder and insert new brush.

6. Guide brush lead into terminal and solder into place. Check for free movement of brushes when solder cools.

7. Check brush slip rings in alternator for wear. Lightly clean slip rings using fine abrasive cloth.

8. Reinstall regulator and alternator.

STARTER SERVICE

Starter Troubleshooting

If the starter turns the engine slowly or fails to operate when the ignition is in the start position, check the battery first. Inspect the starter wires, terminals, and ground connections for good contact. In particular, make sure the ground connections between the battery, the body and the engine are completely clean and tight. If no faults can be found, the starter may be faulty and should be replaced.

NOTE —
- *Starting in 1/94, a factory-installed drive-away protection system, also referred to as EWS or EWS II, was used on all E36 cars. This system prevents operation of the starter when the system is engaged. See* **515 Central Locking and Anti-theft**.
- *On cars with automatic transmission, a starter relay and a neutral safety switch are used to prevent the engine from starting in gear positions other than park or neutral. If voltage is not present at terminal* ***50*** *with the key in the start position, check these components*

To make the most accurate check of the battery cables and starter wiring, make a voltage drop test on the cables and wiring as described in **600 Electrical System–General**.

Check for battery voltage at terminal **50** of the starter motor with the key in the start position. See Fig. 8. If voltage is not present, check the wiring between the ignition switch and the starter terminal. If voltage is present and no other visible wiring faults can be found, the problem is most likely internal in the starter motor.

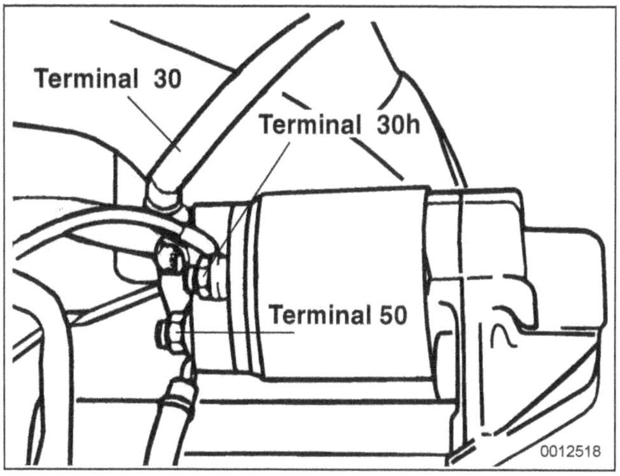

Fig. 8. Typical starter wiring terminal identification. Large wire at terminal **30** is direct battery voltage. Smaller wire at terminal **50** operates starter solenoid via ignition switch.

If the solenoid audibly clicks but the motor does not turn, switch on the lights and turn the key to the start position. If the lights go out while attempting to start, the battery cable may be loose or the starter may have a short circuit. If the lights stay on, the solenoid is most likely at fault.

Starter, removing and installing (4-cylinder engine)

1. Disconnect negative (–) cable from battery.

> ***CAUTION —***
> *Prior to disconnecting the battery, read the battery disconnection cautions given at the front of this manual on page viii.*

2. Remove guide tube for oil dipstick.

3. Raise vehicle.

> ***WARNING —***
> *Make sure the car is stable and well supported at all times. Use a professional automotive lift or jack stands designed for the purpose. A floor jack is not adequate support.*

4. Disconnect wiring from starter.

5. Remove top and bottom starter bolts. Bottom bolt must be removed from below car. Remove starter support bracket, if applicable.

6. Pull starter downward and turn until solenoid is at top, then remove from car.

7. Check starter pinion gear and flywheel teeth for damage.

8. Installation is reverse of removal.

Tightening Torques
- Starter to engine block (M10 bolt) . . . 50 Nm (37 ft-lb)
- Wire to terminal 50 (M6 nut) . . . . 6 Nm (53 in-lb)
- Wire to terminal 30 (M8 nut) . . . . 12 Nm (9 ft-lb)

Starter, removing and installing (6-cylinder engine with manual transmission)

The starter on a 6-cylinder car with manual transmission is removed from below.

1. Disconnect negative (–) battery cable.

2. Raise vehicle.

3. Remove reinforcing cross brace from under transmission, if applicable.

4. Remove cover from fuel filter and fuel lines on left side underneath car, if applicable. Detach fuel lines and harness connectors from retaining brackets, as necessary.

5. Disconnect wiring from starter.

6. Loosen and remove bolts and nuts fastening starter to transmission bell housing and/or engine block. Remove starter support bracket.

7. Remove starter from below.

8. Check starter pinion gear and flywheel teeth for damage.

9. Installation is reverse of removal.

Tightening Torques
- Reinforcing cross brace to chassis (M10). . . . . 42 Nm (31 ft-lb)
- Starter to engine block (M10 bolt) . . . 50 Nm (37 ft-lb)
- Support bracket to starter (M5 nut) . . . 5 Nm (44 in-lb)
- Support bracket to engine block (M10 bolt) . . . . 47 Nm (35 ft-lb)
- Wire to terminal 50 (M6 nut). . . . . 6 Nm (53 in-lb)
- Wire to terminal 30 (M8 nut). . . . . 12 Nm (9 ft-lb)

Starter, removing and installing (6-cylinder engine with automatic transmission)

The starter on a 6-cylinder car with automatic transmission is removed from above. It is necessary to remove the intake manifold and disconnect a number of electrical harness connectors.

1. Disconnect negative (–) battery cable.
2. Remove air plenum from rear of engine compartment. See **640 Heating and Air Conditioning**.
3. Remove top engine covers. See Fig. 9.

Fig. 9. Remove top engine cover by prying out plugs and removing nuts (**A**). Make sure rubber insulators (**B**) on either side of large cover do not fall off during cover removal.

4. Remove ignition coil harness connectors. Remove ground connection at front of cylinder head.
5. Remove two retaining bolts and lift off fuel injector harness connector strip. Carefully fold away all harnesses toward right side of engine compartment and place at base of windshield. See Fig. 10.
6. Remove intake manifold. See **113 Cylinder Head Removal and Installation**.
7. Disconnect wiring from starter.

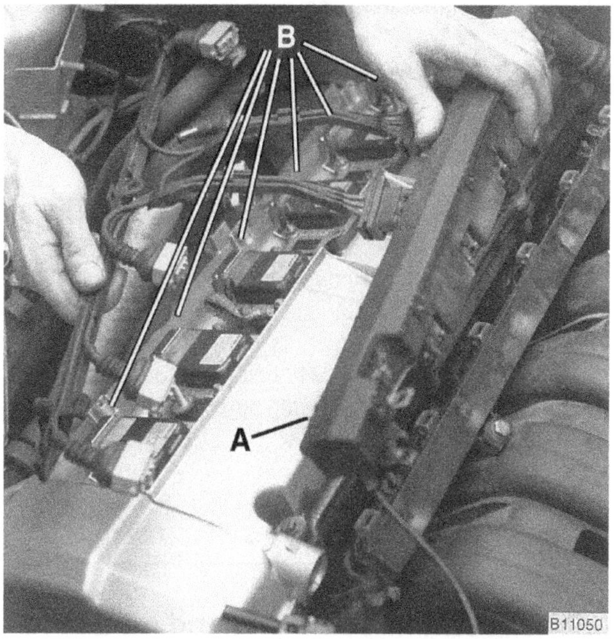

Fig. 10. Fuel injection harness (**A**) and ignition coil harness connectors (**B**) being lifted off engine to be placed at base of windshield. M50 engine shown. Other 6-cylinder engines are similar.

8. Loosen and remove bolts and nuts fastening starter to transmission bell housing and/or engine block. Remove starter support bracket.

Fig. 11. Starter wiring terminals and support bracket (**arrow**). Intake manifold has been removed.

9. Remove starter from above.

10. Check starter pinion gear and flywheel teeth for damage.

11. Installation is reverse of removal.

Tightening Torques

- Intake manifold to cylinder head
  M7 nut. . . . . 15 Nm (11 ft-lb)
  M8 nut/bolt . . . . 22 Nm (16 ft-lb)
- Starter to engine block (M10 bolt) . . . 50 Nm (37 ft-lb)
- Support bracket to starter (M5 nut) . . . 5 Nm (44 in-lb)
- Support bracket
  to engine block (M10 bolt) . . . . 47 Nm (35 ft-lb)
- Wire to terminal 50 (M6 nut) . . . . 6 Nm (53 in-lb)
- Wire to terminal 30 (M8 nut) . . . . 12 Nm (9 ft-lb)

Solenoid switch, removing and installing

1. Remove starter as described above.

2. Remove cover from solenoid switch.

3. Disconnect field winding strap between starter motor and solenoid switch.

   NOTE —
   The condition of the field winding strap is critical. If it is damaged, burned or partially melted through, a new or rebuilt starter motor is needed.

4. Remove solenoid switch mounting screws, and separate solenoid from starter. See Fig. 12.

   CAUTION —
   When installing field winding strap to starter, position it so that it does not contact the starter body.

B11184

Fig. 12. Starter solenoid mounting screws (**arrows**) are tight. Use an impact screwdriver to loosen the screws.

5. Installation is reverse of removal. Lubricate solenoid piston with light grease.

Tightening Torque

- Field winding strap to starter (M8) . . . . 12 Nm (9 ft-lb)

130 Fuel Injection

TABLES

130

GENERAL

This repair group covers fuel injection system component testing and repair. Special equipment is necessary for some of the procedures given in this repair group. If you do not have the equipment required to do the job, it is recommended that these repairs be left to an authorized BMW dealer. The BMW dealer is equipped with sophisticated diagnostic test equipment that is capable of quickly pinpointing hard-to-find fuel injection problems.

NOTE—
- *Wiring diagrams for the engine management system, can be found at the rear of the manual under* **Electrical Wiring Diagrams**.
- *For ignition system repair information, see* **120 Ignition System**.
- *For fuel supply system testing and repair, see* **160 Fuel Tank and Fuel Pump**.

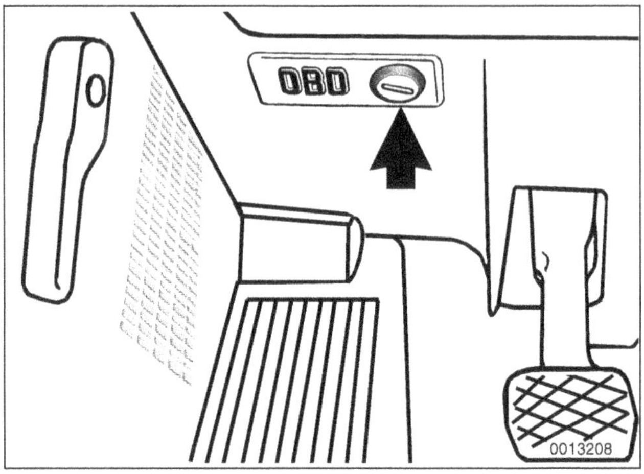

Fig. 1. OBD II diagnostic connector location.

Principles Of Operation

There are five versions of engine management systems used on the E36 cars. Each has the same basic components and operating principles. The most notable difference is that 1996 and later cars use a sophisticated OBD II-compliant system. See **Table a**.

Table a. Engine Management System Variants

Engine code/year	System
4-cylinder	
M42 (1.8 l) 1992-1995	Bosch DME M1.7
M44 (1.9 l) 1996-1998	Bosch DME M5.2 (OBD II)
6-cylinder	
M50	
1992 (2.5 l)	Bosch DME M3.1
1993-1995 (2.5 l)	Bosch DME M3.3.1 (VANOS)
M52	
1996-1998 (328i - 2.8 l)	Siemens MS 41.1 (OBD II)
1998 (323i - 2.5 l)	Siemens MS 41.1 (OBD II)
M-Power	
S50US (M3 - 3.0 l) 1995	Bosch DME M3.3.1
S52US (M3 - 3.2 l) 1997-1998	Siemens MS 41.1 (OBD II)

NOTE—
- *Descriptions and procedures in the first part of this repair group refer to all the various engine management systems.*
- *Particulars of each fuel injection system are treated in separate sections in the second part of this repair group.*
- *The 16-pin OBD II diagnostic connector is located on the lower left dash panel. See Fig. 1.*

The engine control module (ECM) uses electrical signals from the mass air flow sensor, the air and coolant temperature sensors, the crankshaft position/rpm sensor, the knock sensors and the oxygen sensors as the primary inputs to electronically control fuel delivery and ignition timing.

The fuel injection system is completely electronic in operation. Air flow is measured electronically via a mass air flow sensor and additional sensors supply information about engine operating conditions. The ECM calculates the amount of fuel needed for the correct air-fuel ratio and actuates the fuel injectors accordingly. The amount of fuel metered to the engine is determined by how long the injectors are open.

Air intake. Air entering the engine passes through a pleated paper air filter in the air cleaner. Intake air volume or mass is then measured by a mass air flow (MAF) sensor. In all except the vane type sensor (DME M1.7), a reference current is used to heat a thin wire or film in the sensor when the engine is running. The current used to heat the wire/film is electronically converted into a voltage measurement corresponding to the mass of the intake air.

Table b. Mass Air Flow Sensor Variants

System	Air flow sensor type
Bosch DME M1.7	Vane (volume sensor)
Bosch DME M3.1	Hot wire (mass sensor)
Bosch DME M3.3.1	Hot film (mass sensor)
Bosch DME M5.2	Hot film (mass sensor)
Siemens MS 41.1	Hot film (mass sensor)

NOTE—
On cars equipped with traction control, an additional throttle valve is controlled by an electronic throttle actuator (motor). This valve is used for engine speed intervention. Repair information for this system is not covered here due to the special electrical testing equipment required to service it.

Fuel metering. The ECM meters fuel by changing the opening time (pulse width) of the fuel injectors. To ensure that injector pulse width is the only factor that determines fuel metering, fuel pressure is maintained by a fuel pressure regulator. The injectors are mounted to a common fuel supply called the fuel rail.

The ECM monitors engine speed to determine the duration of injector openings. Other signals to the ECM help determine injector pulse time for different operating conditions. A temperature sensor signals engine temperature for mixture adaption. A throttle position sensor signals throttle position. The exhaust oxygen sensor(s) signal information about combustion efficiency for control of the air-fuel mixture. 1992 to 1995 engines are equipped with a single sensor. 1996 and later (OBD II) engines are equipped with an oxygen sensor before and one after each catalytic converter. For example, the M52 engine is equipped with four oxygen sensors.

Idle speed control. Idle speed is electronically controlled via the idle speed control valve, which maintains idle speed by bypassing varying amounts of air around the closed throttle valve. Idle speed is not adjustable.

Knock (detonation) control. Knock sensors monitor and control ignition knock through the ECM. The knock sensors function like microphones and are able to convert mechanical vibration (knock) into electrical signals. The ECM is programmed to react to frequencies that are characteristic of engine knock and adapt the ignition timing point accordingly. See **120 Ignition System** for further details.

> ***NOTE —***
> *The 1992 M50 engine is not equipped with knock sensors. All other engines are equipped with two knock sensors.*

Basic Engine Settings

Idle speed, idle mixture (%CO), and ignition timing are not adjustable. The adaptive engine management system is designed to automatically compensate for changes in engine operating conditions, although the adaptive range is limited. Once these limits are exceeded, driveability problems usually become noticeable.

> ***NOTE —***
> *If the system adaptive limits are exceeded, the Check Engine light will most likely come on, indicating an emissions- related fault. For Check Engine light diagnostics, see* **100 Engine–General**.

The engine management system compensates automatically for changes in the engine due to age, minor wear or small problems, such as a disconnected vacuum hose. As a result, idle speed and mixture do not need to be adjusted as part of routine maintenance.

> ***NOTE —***
> *Poor driveability may be encountered when the battery is disconnected and reconnected. When the battery is disconnected, the adaptive memory is lost. The system will readapt after about ten minutes of driving.*

DISA (Dual Resonance Intake System)

The E36 4-cylinder engine is equipped with a dual intake runner system, termed DISA. DISA offers the advantages of both short and long intake pipes. Long intake runners are most useful at low to medium engine rpm for producing good torque characteristics. Short intake runners produce higher horsepower at higher engine speeds.

> ***NOTE —***
> *The term DISA comes from the German words Differenzierte Sauganlage, and can roughly be translated as "differing intake manifold configuration." See* **100 Engine-General** *for additional information on DISA operation.*

Manifold construction. The intake manifold is a two-piece metal construction, with a pair of runners in the top section and four runners in the lower section. A butterfly valve is installed in the lower section, enabling the DISA solenoid to isolate one pair of runners from the other pair. See Fig. 2.

Operation. With the DISA butterfly valve closed, the pipes in the top half of the manifold act together with the ram air pipes in the lower half to produce a single, long air intake pipe for each cylinder. See Fig. 3. The column of air oscillating in this combined pipe significantly increases engine torque in the medium rpm range.

Above approximately 4,800 rpm, the butterfly valve between the intake air pipes for the two cylinder groups is opened. The shorter pipes in the lower manifold section now become the main suppliers of ram air to the cylinders, yielding greater power at the upper end of the engine rpm range. See Fig. 4.

Control components. The DISA butterfly valve is actuated electro-pneumatically via the engine control module (ECM). The valve begins to open as engine speed rises above 4,840 rpm and closes below 4,760 rpm. The action of the valve is deliberately delayed to prevent it from opening and closing repeatedly within a short time.

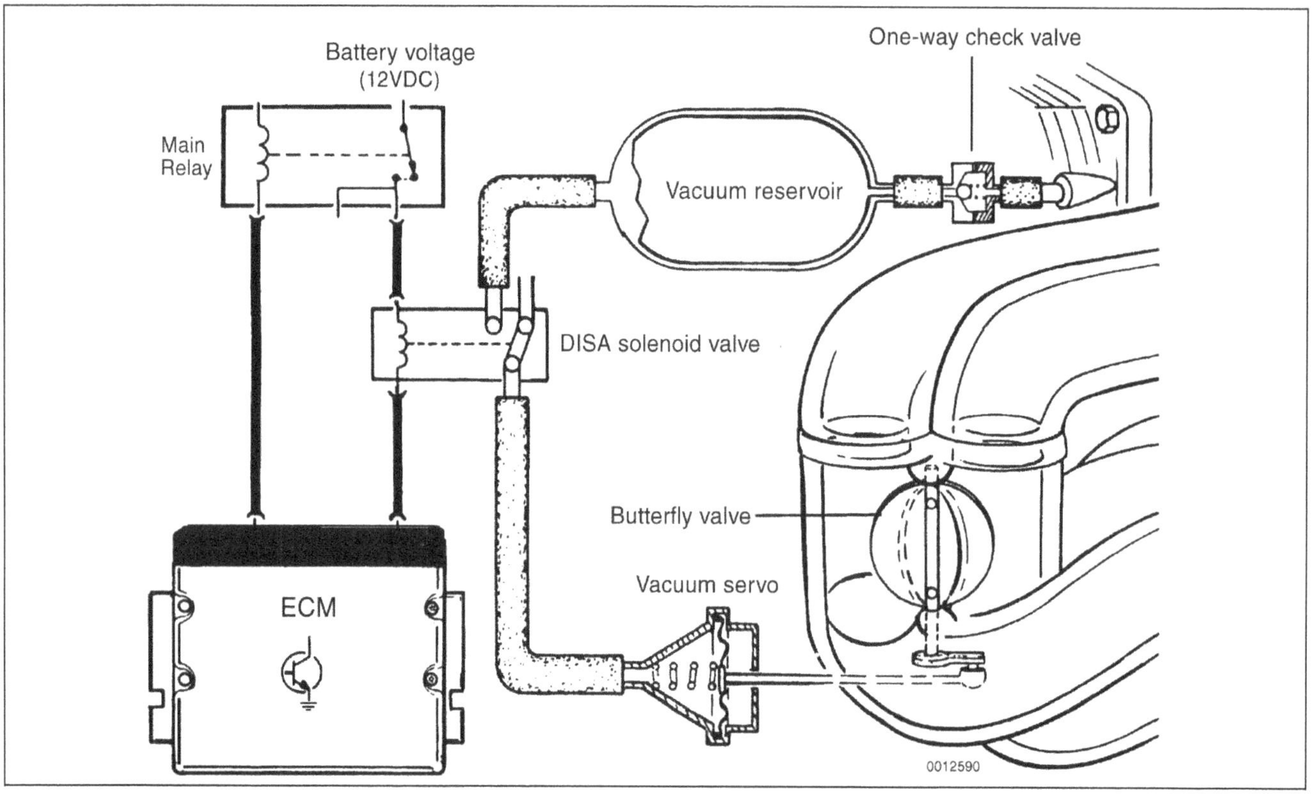

Fig. 2. DISA system components. DISA is controlled via the engine control module (ECM).

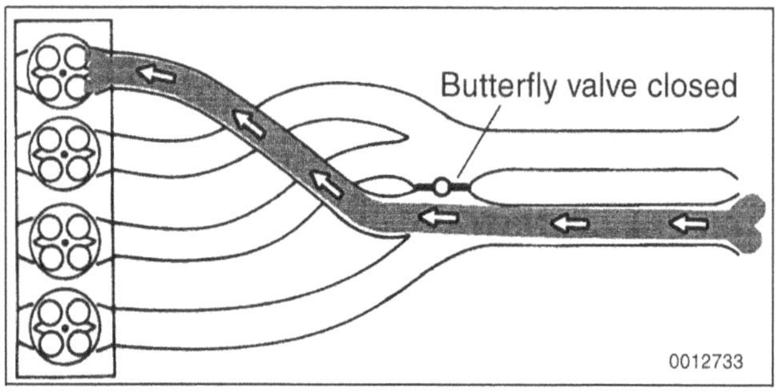

Fig. 3. DISA manifold runners with butterfly valve closed for low rpm-high torque operation.

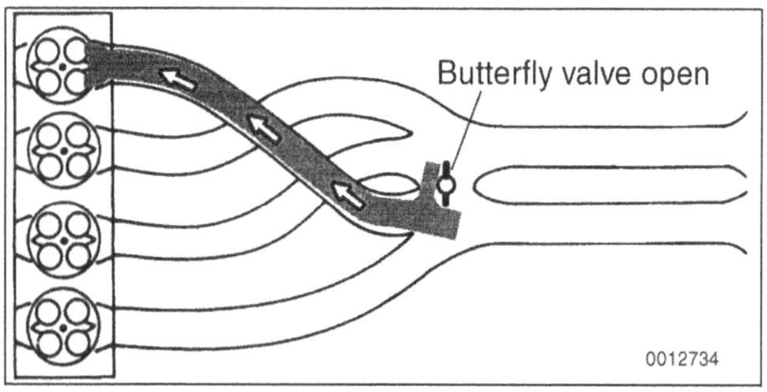

Fig. 4. DISA manifold runners with butterfly valve open for high rpm-high power operation.

On-Board Diagnostics (OBD) and Fault Diagnosis

Poor driveability may have a variety of causes. The fault may lie with the ignition system, the fuel system, parts of the emission control system, or a combination of the three. Because of these interrelated functions and their effects on each other, it is often difficult to know where to begin looking for problems.

For this reason, effective troubleshooting should always begin with an interrogation of the On-Board Diagnostic (OBD) system.

The OBD system detects certain emissions-related engine management malfunctions. When faults are detected, the OBD system stores a Diagnostic Trouble Code (DTC) in the system ECM. In addition, the Check Engine warning light will come on if an emissions-related fault is detected.

Two generations of OBD are used on the cars covered by this manual. See **100 Engine–General** for OBD I information and fault codes.

On-Board Diagnostics
- 1992-1995 models . . . . OBD I
- 1996-1998 models . . . . OBD II

The fuel injection systems used on the 1996 and later cars covered by this manual are OBD II compliant. This system is incorporated into both the Bosch M5.2 (M44 engine) and the Siemens MS 41.1 (M52/S52US engine) engine management systems. OBD II systems use sophisticated diagnostic software capable of recognizing and electronically storing hundreds of DTCs in the system ECM. DTCs can only be accessed using special scan tool test equipment. The BMW dealer is equipped with the specialized OBD II scan tool to quickly and efficiently locate engine management problems. Alternately, a "generic" scan tool can be used to access OBD II fault information.

NOTE —

- *At the time this manual went to press, generic scan tools were not widely available for BMW vehicles. The generic scan tool is a specialized tool that plugs into a standardized OBD II connector on 1996 and later passenger vehicles built for sale in the US.*
- *The OBD II fault memory (including an illuminated Check Engine light) can only be reset using the special scan tool. Removing the connector from the ECM or disconnecting the battery will not erase the fault memory.*
- *The 16-pin OBD II diagnostic connector is located on the lower left dash panel. Refer to Fig. 1.*

Secondary Air Injection

All OBD II 6-cylinder engines are equipped with a secondary air injection system. In addition, beginning in January 1997, the M44 engine was classified as a TLEV (Transitional Low Emission Vehicle) engine and is also equipped with secondary air injection.

Typical components of the secondary air injection system are depicted in Fig. 5.

Secondary air injection uses an electric pump to pump fresh air into the exhaust system upstream of the catalytic converter during engine warm-up. By providing extra oxygen to the unburned fuel in the exhaust, hydrocarbons oxidize and carbon monoxide combines with oxygen to form carbon dioxide and water. The air injection pump stops within a specified timed interval.

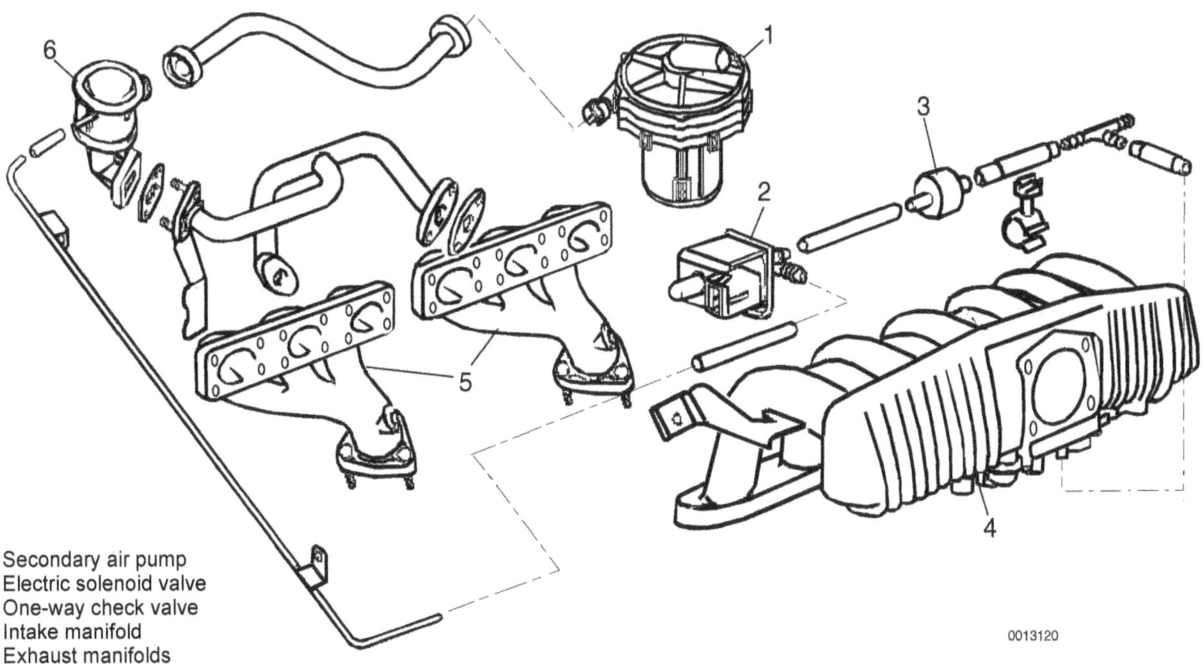

1. Secondary air pump
2. Electric solenoid valve
3. One-way check valve
4. Intake manifold
5. Exhaust manifolds
6. Diverter valve

Fig. 5. Components of 6-cylinder secondary air injection system. 4-cylinder system is similar.

Warnings and Cautions

For personal safety, as well as the protection of sensitive electronic components, the following warnings and cautions should be adhered to when working on the engine management system.

> ***WARNING —***
>
> - *The ignition system produces high voltages that can be fatal. Avoid contact with exposed terminals. Use extreme caution when working on a car with the ignition switched on or the engine running.*
> - *Do not touch or disconnect any high voltage cables from the coils or spark plugs while the engine is running or being cranked by the starter.*
> - *Connect and disconnect the DME system wiring and test equipment leads only when the ignition is switched off.*
> - *Gasoline is highly flammable and its vapors are explosive. Do not smoke or work on a car near heaters or other fire hazards when diagnosing and repairing fuel system problems. Have a fire extinguisher available in case of an emergency.*
> - *When working on an open fuel system, wear suitable hand protection, as prolonged contact with fuel can cause illnesses and skin disorders.*
> - *Renew fuel system hoses, clamps and O-rings any time they are removed.*
> - *Before making any electrical tests that require the engine to be cranked using the starter, disable the ignition system as described in* **120 Ignition System**.

> ***CAUTION —***
>
> - *Prior to disconnecting the battery, read the battery disconnection cautions given at the front of this manual on page viii.*
> - *Do not connect any test equipment that delivers a 12-volt power supply to terminal 15 (+) of the ignition coil. The current flow may damage the ECM. In general, connect test equipment only as specified by BMW, this manual, or the equipment maker.*
> - *Only use a digital multimeter for electrical test.*
> - *Only use an LED test light for quick tests.*
> - *Disconnecting the battery may erase fault code(s) stored in memory. Check for fault codes prior to disconnecting the battery cables. If the Check Engine light is illuminated, see* **100 Engine–General** *for DME fault code information. If any other system faults have been detected (indicated by an illuminated warning light), see an authorized BMW dealer. Additional systems with self-diagnostic capabilities include; ABS (Anti-lock brakes), SRS (Airbags), EML and ASC+T and AST (Traction Control).*
> - *Do not run the engine with any of the spark plug wires disconnected. Catalytic converter damage may result.*
> - *Always wait at least 40 seconds after turning off the ignition before removing the engine control module (ECM) connector. If the connector is removed before this time, residual power in the system relay may damage the control module.*
> - *Cleanliness is essential when working on an open fuel system. Thoroughly clean fuel line connections and surrounding areas before loosening. Avoid moving the car. Only install clean parts.*
> - *Fuel system cleaners and other chemical additives other than those specifically recommended by BMW may damage the catalytic converter, the oxygen sensor or other fuel supply components.*

ELECTRICAL CHECKS AND COMPONENT TESTING

Main relay, testing

The main relay is energized via the engine control module and supplies plus (+) power to many of the engine management components and subsystems, including the fuel pump relay. If this relay is faulty, the engine will not start.

1. With ignition off, remove main relay. See Fig. 6.

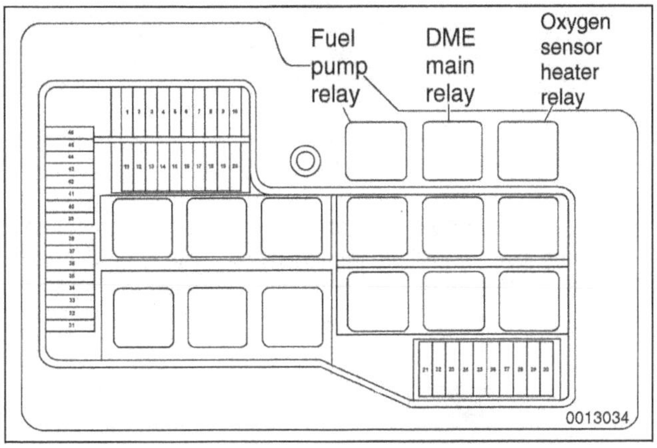

Fig. 6. Engine management relays in the power distribution box, left rear of engine compartment.

> ***CAUTION—***
> *Relay positions can vary. Be sure to confirm relay position by identifying the wiring in the socket using the wiring diagrams found at the rear of this manual.*

2. Check for voltage at terminal 30 and terminal 86 of relay socket. See Fig. 7.
   - If battery voltage is present continue testing
   - If battery voltage is not present at either point, check large red wire in relay socket. See **Electrical Wiring Diagrams.**

> ***NOTE—***
> *The wiring to terminals 85 and 86 in the main relay may be reversed. Electrically, the relay functions either way. If voltage is not present at terminal 86, check for voltage at terminal 85.*

3. Reinstall relay and turn ignition on. Gain access to underside of relay socket and check for ground at terminal 85 (brown wire).
   - If ground is present continue testing.
   - If ground is not present, signal from ECM is missing. Check wire between ECM and relay. If no faults can be found, ECM may be faulty.

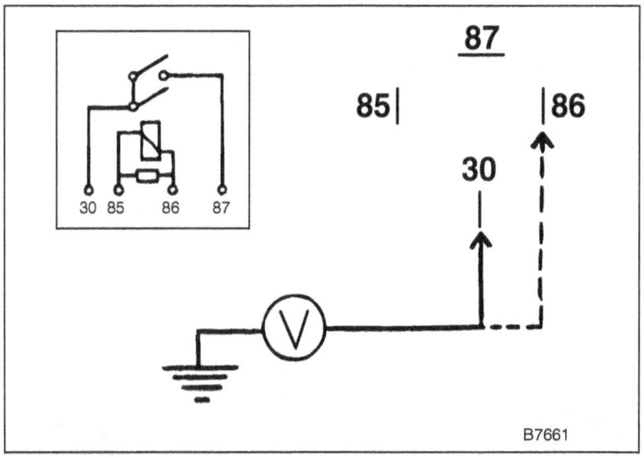

Fig. 7. Check for battery voltage at main relay terminals 30 and 86. Voltage should be present at all times.

4. With ignition on and relay installed, check for battery voltage at terminals 87.
   - If battery voltage is present, relay has energized and is functioning correctly.
   - If battery voltage is not present and all earlier tests are OK, relay is faulty and should be replaced.

Fuel pump relay, testing

The fuel pump relay is located in the power distribution box. Refer to Fig. 6.

The ECM supplies power to the coil side of the fuel pump relay. During starting, the fuel pump runs as long as the ignition switch is in the start position and continues to run once the engine starts. If the relay is faulty the fuel pump will not run.

1. Remove fuel pump relay from its socket.

2. With key off, check for battery voltage at socket terminal 30. See Fig. 8.

3. With ignition key in start position check for battery voltage at terminal 86.

4. With ignition key, use digital multimeter to check for ground at terminal 85.

> ***NOTE—***
> *The ground at terminal 85 is switched by the ECM. The ECM harness must be connected to check the switched ground connection.*

5. If no faults are found up to this point, test fuel pump operation as described in **160 Fuel Tank and Fuel Pump**.

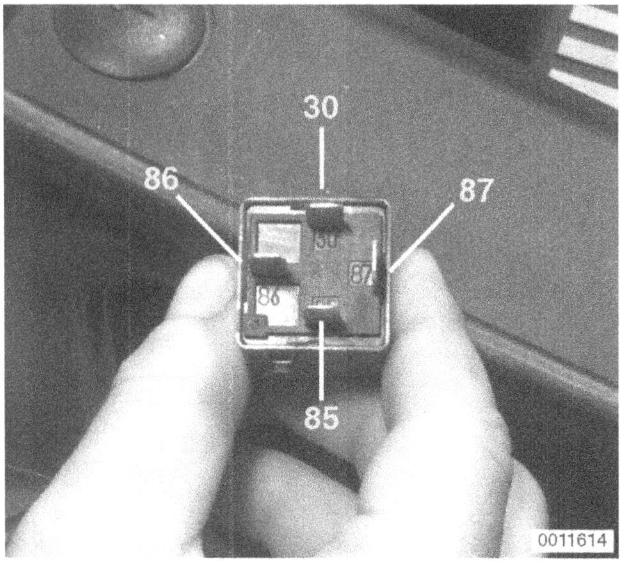

Fig. 8. Fuel pump relay terminal identification.

Oxygen sensor, testing (Bosch systems only)

> ***CAUTION —***
> - *The information in this section applies to Bosch engine management systems only.*
> - *The Siemens MS 41.1 system uses resistive-type oxygen sensors. This sensor uses a 5-volt reference signal input and the output to the ECM varies from 1-5 volts, in stepped increments based on oxygen content in the exhaust.*

NOTE —
- *On pre-OBD II (1992-1995) cars, there is one oxygen sensor mounted upstream of the catalytic converter(s) in the exhaust pipe. See Fig. 9.*
- *On OBD II cars (1996-1998) cars, there is one oxygen sensor before and one after each catalytic converter.*
- *On M52 and S52US engines, the front pair of oxygen sensors are installed in the exhaust manifolds.*

The oxygen sensor provides an input voltage signal (0-1 VDC) to the ECM based on the oxygen content in the exhaust gas. To generate voltage, the sensor temperature must exceed 575°F (300°C). Therefore it is electrically heated.

NOTE —
- *The test given below is not a conclusive test of oxygen sensor efficiency and does not test how quickly the oxygen sensor can react to changing conditions.*
- *Pin numbers and wire colors can vary. Always check the wiring diagrams to confirm wire color and pin assignment.*

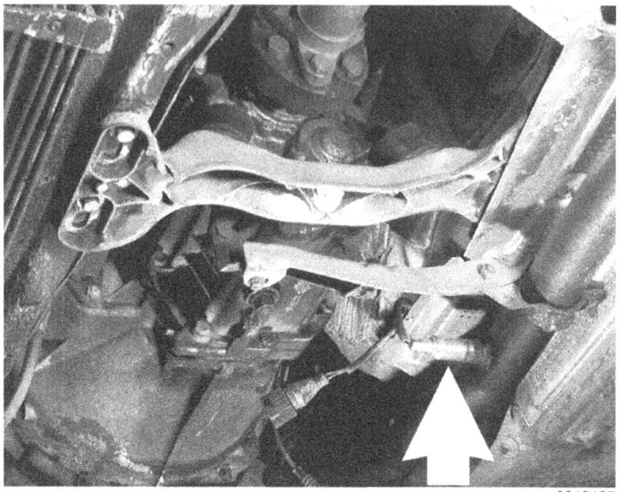

Fig. 9. Oxygen sensor location (**arrow**) ahead of catalytic converter. 318i model with M42 engine shown.

1. Peel back rubber boot on oxygen sensor electrical harness connector (car wiring side). With connector attached to sensor, connect digital voltmeter to pins 1 and 2 in rear of connector. See Fig. 10.

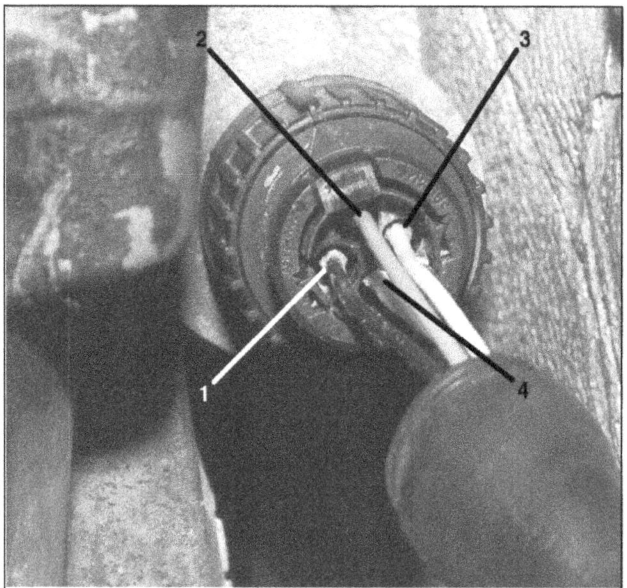

1. Black (ground)
2. Grey (signal)
3. White (heater circuit)
4. White (heater circuit)

0013189

Fig. 10. Oxygen sensor connector terminal identification (sensor side). Terminals numbers are molded into connector.

2. Start engine. Oxygen sensor should start to output a fluctuating voltage within a short period. If voltage is incorrect, turn engine off and check preheater circuit as described below.

> ***WARNING —***
> *Exhaust manifolds and pipes can be hot enough to cause serious burns. Wear suitable heavy gloves and other appropriate protection.*

Oxygen Sensor
- Voltage at idle. . . . . 0.2 to 0.8 VDC, fluctuating

NOTE —
To check sensor response to lean and rich mixtures, create an air leak, or pull vacuum hose off fuel pressure regulator to increase fuel pressure.

3. Separate sensor harness connector from sensor. Check for battery voltage between terminals 3 and 4 (green wire and brown wire) in main wiring harness side of connector with engine running. If voltage is not present, check oxygen sensor heater relay. See **610 Electrical Component Locations**.

4. Check heater element resistance between terminals 3 and 4 in sensor side of connector. If element is electrically open (no continuity), replace sensor.

NOTE —
The oxygen sensor heater relay is mounted in the main power distribution box in the left rear of the engine compartment. Refer to Fig. 6. The heater relay is energized with positive (+) battery voltage from the main relay and a switched ground from the ECM. See **Electrical Wiring Diagrams**.

5. If oxygen sensor doesn't produce a fluctuating voltage and preheater circuit is OK, replace sensor.

NOTE —
If not already applied, coat the oxygen sensor threads with an anti-seize compound before installation. Do not get the compound on the sensor tip.

Tightening Torque
- Oxygen sensor to exhaust pipe . . . . . 55 Nm (41 ft-lb)

FUEL DELIVERY TESTS

Checking fuel delivery is a fundamental part of troubleshooting and diagnosing the engine management system. Fuel pressure directly influences fuel delivery. An accurate fuel pressure gauge will be needed to make the tests.

There are three significant fuel delivery values to be measured:

- **System pressure**—created by the fuel pump and maintained by the pressure regulator.
- **Fuel delivery volume**—created by the fuel pump and affected by restrictions, such as clogged fuel filter.
- **Residual pressure**—the pressure maintained in the closed system after the engine and fuel pump are shut off.

Procedures for measuring the first two quantities are described in **160 Fuel Tank and Fuel Pump**. Residual fuel pressure is checked using the procedure detailed later in this group.

Operating fuel pump for tests

To operate the fuel pump for testing purposes without having to run the engine, the fuel pump relay can be bypassed to power the pump directly. Fuel pump relay location is shown in Fig. 6.

To run the fuel pump, remove the fuel pump relay and connect the socket for relay terminal 30 to the socket for relay terminal 87 with a fused jumper wire. After completing the tests, remove the jumper wire.

CAUTION —
- *Relay locations may vary. Use care when identifying relays and making electrical checks at the fuse/relay panel. See* **610 Electrical Component Locations** *for additional relay information.*
- *The fuel pump relay has a 1.5 mm² red wire at terminal 30 in the relay socket. Terminal 87 has a 1.5 mm² green/violet wire. See* **Electrical Wiring Diagrams** *for additional wiring information.*

NOTE —
The jumper wire should be 1.5 mm² (14 ga.) and include an in-line fuse holder with a 15 amp fuse. To avoid fuse/relay panel damage from repeated connecting and disconnecting, also include a toggle switch. A heavy-duty jumper, BMW tool no. 61 3 050, is also available from an authorized BMW dealer.

Relieving fuel pressure and connecting fuel pressure gauge

WARNING —
- *Gasoline is highly flammable and its vapors are explosive. Do not smoke or work on a car near heaters or other fire hazards when diagnosing and repairing fuel system problems. Have a fire extinguisher available in case of an emergency.*
- *When working on an open fuel system, wear suitable hand protection. Prolonged contact with fuel can cause illnesses and skin disorders.*

To prevent fuel from spraying on a hot engine, system fuel pressure should be relieved before disconnecting fuel lines. One method is to tightly wrap a shop towel around a fuel line fitting and loosen or disconnect the fitting.

Cleanliness is essential when working with fuel circuit components. Thoroughly clean the unions before disconnecting fuel lines.

NOTE —
- *On M44 engines, a Schræder valve is integrated in the fuel rail. Compressed air pressure can be applied at the valve to force the fuel in the system back into the tank. See Fig. 11.*

0012503

Fig. 11. Fuel rail showing location of Schræder valve fitting on M44 engine (**arrow**).

NOTE —
BMW specifies a special gauge and special connecting adapters to measure fuel pressure. If the special tools are not available, a length of fuel line and a T-fitting can be installed to the inlet fuel line and connected to a fuel gauge. On some vehicles, this is not possible without cutting into a fuel line. In these instances, it is recommended that fuel pressure be checked at the output side of the fuel pump. See **160 Fuel Tank and Fuel Pump**.

On OBD II 6-cylinder engines, the fuel lines use a special locking fitting. Use BMW special tool 16 1 050 to release the fittings and connect the fuel gauge. See Fig. 12.

0012699

Fig. 12. Fuel lines at rear of intake manifold on M52 engine. Use BMW special tool 16 1 050 to release fittings.

WARNING —
- *Gasoline is highly flammable and its vapors are explosive.*
- *The fuel pressure gauge must be securely connected to prevent it from coming loose under pressure.*

NOTE —
- *The fuel pressure gauge should have a range of 0 to 5 bar (0 to 75 psi).*
- *On 6-cylinder engine, the top left-side engine cover will have to be removed to access the fuel rail.*

Residual fuel pressure, testing

For quick restarts and to avoid vapor lock when the engine is hot, the fuel injection system is designed to retain fuel pressure after the engine has been turned off. This residual pressure is primarily maintained by a check valve at the fuel pump outlet and the fuel pressure regulator. The fuel pump check valve is not serviceable as an individual part.

1. Relieve fuel pressure and connect a pressure gauge as described earlier.

WARNING —
Fuel will be discharged. Wrap a shop towel around the fuel line fitting when disconnecting the fuel line. Do not smoke or work near heaters or other fire hazards. Keep an approved fire extinguisher handy.

2. Operate fuel pump for approximately one minute by bypassing fuel pump relay as described earlier. Observe fuel pressure.

3. Observe fuel pressure gauge after 20 minutes. The pressure should not drop off more than 0.5 bar from system pressure listed in **Table c**.

4. When finished testing, disconnect pressure gauge and reconnect fuel line.

If the fuel system does not maintain pressure, check visually for leaks in fuel lines or at unions. Leaks can also be due to a leaking injector or a faulty fuel pump check valve. Check the pump check valve by repeating the test, but before turning the fuel pump off, pinch off the return line at the fuel rail. If the pressure is now maintained, the fault is most likely the fuel pump check valve.

Table c. Fuel Pressure Specifications

Engine	Fuel pressure
4-cylinder	3.0 ± 0.2 bar (43.5 ± 2.9 psi)
6-cylinder M50/S50US engine M52/S52US engine	 3.0 ± 0.2 bar (43.5 ± 2.9 psi) 3.5 ± 0.2 bar (51 ± 2.9 psi)

Fuel pressure regulator response to engine load, testing

1. With fuel pressure gauge connected, reinstall fuel pump relay. Start engine and allow it to idle.

NOTE —
The fuel pressure should be lower by 0.4-0.7 bar from the specifications listed in **Table c**. *Engine vacuum acting on the fuel pressure regulator diaphragm reduces the fuel pressure.*

2. With engine idling, remove vacuum hose from regulator. Pressure should increase.

3. Reconnect hose and check that pressure decreases.

4. When finished testing, disconnect pressure gauge and reconnect fuel line. Tighten hose clamp.

If fuel pressure does not drop with the vacuum hose connected and no faults are found with fuel the pressure regulator vacuum hose, the fuel pressure regulator is faulty.

NOTE —
A cracked or leaking pressure-regulator vacuum hose may cause an erratic idle.

Fuel rail and injectors, checking

The fuel injectors are switched on and off (opened and closed) by the ECM. The injectors are connected to a common fuel supply, called the fuel rail.

The fuel injectors are removed by first removing the complete fuel rail assembly and then unclipping the injectors from the fuel rail.

CAUTION —
Use only a digital multimeter or an LED injector tester when testing injectors or wiring. Use of an analog VOM or incandescent test light may damage the engine control module.

NOTE —
- *On 4-cylinder cars pry up the injector wiring duct and then pry off the cover. See Fig. 13.*
- *On 6-cylinder cars it is necessary to remove the top engine cover(s) to access the fuel rail and injectors.*

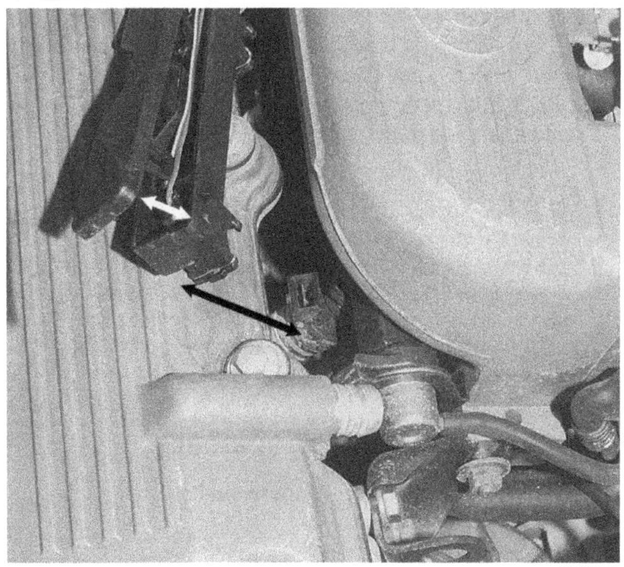

0013190

Fig. 13. For electrical tests, 4-cylinder fuel injector harness duct can be accessed between intake manifold and cylinder head cover.

1. Check that ECM signal is present at injector connector.
   - Back probe injector harness connector using digital voltmeter. See Fig. 14.
   - Operate starter or run engine and check for pulsed voltage signal.
   - Repeat for each injector.

NOTE —
To quick-check if an injector functioning, place a screwdriver or stethoscope on the injector with the engine running. If the injector is operating, there should be an audible buzz.

Fig. 14. Voltmeter connected across fuel injector connector with engine running. M50 engine shown.

2. If no signal is present, check for b+ power to injectors. There should be battery voltage (+) at red/white wire of each injector connector with the ignition key on. If voltage is not present, check main relay output. Injector b+ power comes from main relay (terminal 87).

NOTE—

If there is positive (+) battery voltage at injectors but there was no pulsed response at voltmeter, check the wire(s) from the ECM to the injectors. If no wiring faults can be found, the pulsed ground signal from the control module may be missing. Check the outputs from the ECM.

Fuel rail and injectors, replacing

1. Disconnect negative (–) battery cable.

CAUTION—

Prior to disconnecting the battery, read the battery disconnection cautions given at the front of this manual on page viii.

2. M42 and M50/S50US engines: Working at end of fuel rail, disconnect small vacuum hose from fuel pressure regulator.

3. M50/S50US engines: Remove two nuts from injector duct. Carefully pry injector wiring duct off injectors. See Fig. 15.

4. On 4-cylinder cars, remove upper intake manifold to access the fuel-rail and injectors. See **113 Cylinder Head Removal and Installation.**

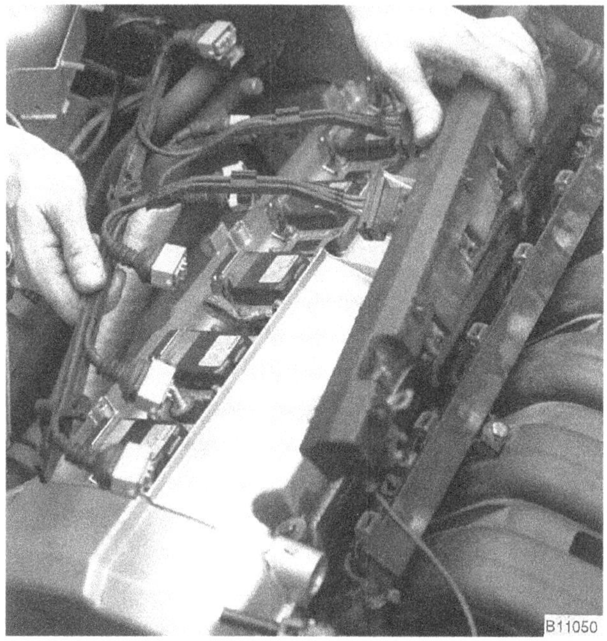

Fig. 15. Injector wiring harness duct being removed from M50 engine.

NOTE—

Late-model M42 and all M44 fuel injectors are air-shrouded. See Fig. 16. When removing the fuel rail it is necessary to remove the air shroud feeder hose from the intake manifold ducting.

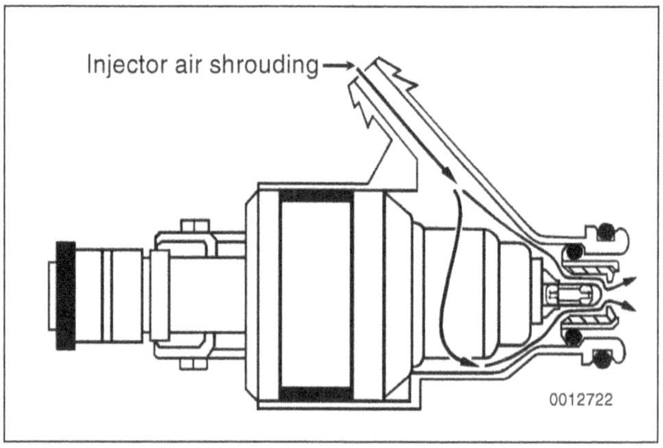

Fig. 16. Cutaway view of fuel injector used on M42 or M44 engine.

5. Disconnect fuel supply line and fuel return line from fuel rail ends.

WARNING—

- *Clamp off the fuel lines and then wrap a clean shop towel around the lines before removing them. Residual fuel pressure is present in the lines.*
- *Fuel will be discharged. Do not disconnect any wires that could cause electrical sparks. Do not smoke or work near heaters or other fire hazards. Keep an approved fire extinguisher handy.*

6. Remove fuel rail mounting bolts. See Fig. 17.

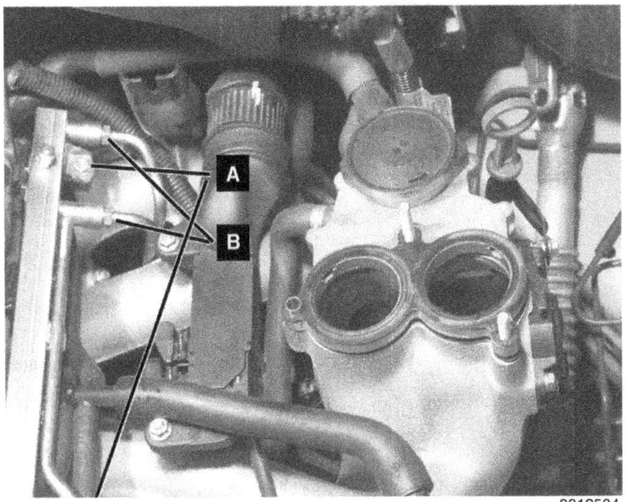

0012504

Fig. 17. Location of fuel rail mounting bolts (**A**) and fuel inlet and return lines (**B**) on M44 engine. Upper manifold has been removed.

7. Pull up on fuel rail and remove injector rail with injectors. Remove individual injectors by removing injector retaining clip and pulling injector from rail. See Fig. 18.

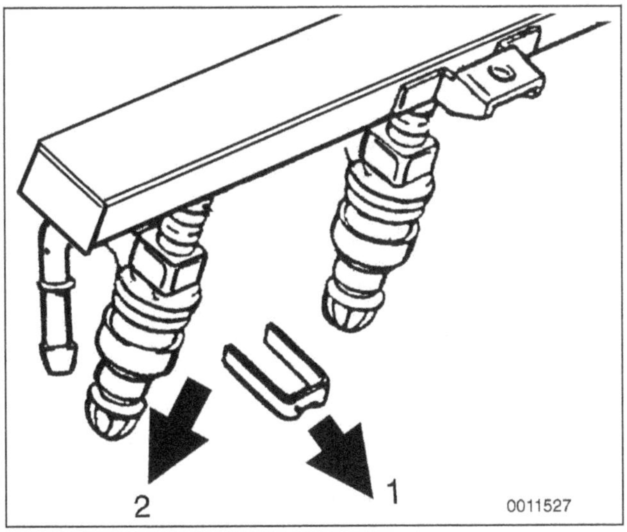

Fig. 18. Pry retaining clip from injector (**1**), then pull injector from rail (**2**).

8. Installation is reverse of removal.
   - Fit new O-rings when installing injectors. For ease of installation, lightly lubricate O-rings with SAE 90 or equivalent lubricant.
   - Check that injector electrical connections are correctly fitted and that injectors are fully seated prior to installing fuel rail mounting bolts.
   - Replace wire ties.

Tightening Torque
- Fuel rail to cylinder head .... 10 Nm (89 in-lb)

Fuel Pressure Regulator

The fuel pressure regulator on E36 models is in one of two locations:

Fuel Pressure Regulator Location	
• 4-cylinder engines	
1992- January 1997	on fuel rail
January 1997- on	under car
• 6-cylinder engines	
1992-1995	on fuel rail
1996- on	under car

Fuel pressure regulator, replacing (fuel rail mount)

> ***WARNING —***
> *Fuel will be discharged. Do not disconnect any wires that could cause electrical sparks. Do not smoke or work near heaters or other fire hazards. Keep an approved fire extinguisher handy.*

1. On 6-cylinder engines, remove top engine covers. Remove fuel rail mounting bolts and lift up fuel rail.

2. Disconnect vacuum hose from fuel pressure regulator. See Fig. 19.

0012732

Fig. 19. Fuel pressure regulator (**arrow**) on M42 engine. Vacuum hose shown at **A**.

3. Remove bolt from regulator clamp. Remove clamp. Note installed direction of regulator to fuel rail.

4. Wrap a shop rag around regulator, then remove regulator from fuel rail by pulling straight out.

NOTE —
Be sure to retrieve thrust washer behind fuel pressure regulator on 6-cylinder engine.

5. Installation is reverse of removal. Replace O-rings.

Fuel pressure regulator, replacing (under car mount)

WARNING —
Fuel will be discharged. Do not disconnect any wires that could cause electrical sparks. Do not smoke or work near heaters or other fire hazards. Keep an approved fire extinguisher handy.

On late 4- and 6-cylinder cars, the fuel pressure regulator is mounted beneath the left side of the car, under a protective cover. See Fig. 20.

0012726

Fig. 20. Fuel pressure regulator location underneath car (**arrow**). Vacuum hose to regulator is shown at **A**. (Protective cover has been removed.)

1. Working under car below driver's seat, remove protective cover from below fuel pressure regulator.

2. Remove vacuum hose from fuel pressure regulator.

3. Remove locking clip retaining fuel pressure regulator.

4. Wrap a shop rag around regulator, then remove regulator from 2/3 way valve by pulling straight out.

5. Installation is reverse of removal. Replace O-rings.

BOSCH DME M1.7 COMPONENT TESTS AND REPAIRS

CAUTION —
Use only a digital multimeter when testing components and wiring. Use of an analog VOM may damage the engine control module.

4-cylinder cars with M42 engines (1992 to 1995) use the Bosch DME M1.7 fuel injection system.

Electrical tests of the main and fuel pump relays and the DME engine control module (ECM) are covered earlier in this section. Fuel pump tests are covered in **160 Fuel Tank and Fuel Pump**.

Air flow sensor, testing and replacing

DME M1.7 fuel injection uses a volume air flow type sensor with integrated intake air temperature (IAT) sensor. The sensor provides a varying voltage signal to the ECM based on the position of the air vane. As the vane door swings open the potentiometer increases the voltage signal to the ECM. The IAT sensor adapts the output signal to the ECM based on intake air temperature.

1. Check ECM reference voltage to sensor:
   - Peel back rubber boot from air flow sensor harness connector.
   - Turn ignition key on.
   - Check for 5 volts between terminal 1 of harness connector and ground. See Fig. 21.
   - Turn ignition key off.
   - If voltage is not present or incorrect, check wring from ECM and check air flow sensor reference voltage output at ECM. See **Table h**.

2. Check air flow sensor potentiometer:
   - Remove intake air boot from sensor.
   - Connect a digital multimeter (ohms) across terminals 1 and 2. Swing air flow sensor vane through its travel range. Resistance should change steadily without interruption.
   - If any faults are found, the air flow sensor is faulty and should be replaced.

3. Check IAT sensor resistance:
   - With harness connector disconnected at air flow sensor, check resistance across sensor terminals 4 and 5 of air flow sensor. Compare tests results to values in **Table d** given later. If any faults are found, the air flow sensor should be replaced.

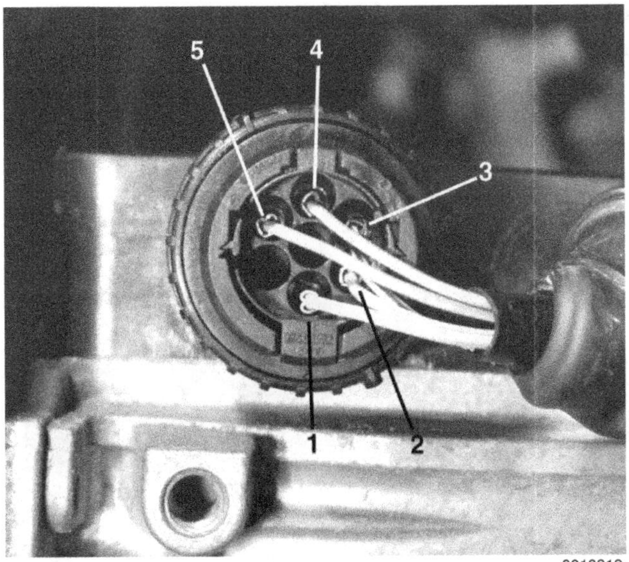

Fig. 21. Air flow sensor harness connector terminal identification.

4. Replacing air flow sensor:
   - Remove large band clamp attaching air boot to sensor.
   - Remove air filter top cover, then remove air flow sensor mounting nuts inside cover.
   - Reconnect air boot and harness connector.

Engine coolant temperature (ECT) sensor, testing and replacing

The engine coolant temperature (ECT) sensor sends continuous engine coolant temperature information to the ECM. As temperature increases, sensor resistance decreases.

The ECT sensor is located under number one intake runner. See Fig. 22.

1. Check ECM reference voltage to sensor:
   - Disconnect harness connector from ECT sensor.
   - Turn ignition key on.
   - Check for 5 volts between supply voltage (brown/red) wire of harness connector and ground.
   - Turn ignition key off.
   - If voltage is not present or incorrect, check wiring from ECM and check ECT sensor reference voltage output at ECM. See **Table h.**
2. Check ECT sensor resistance:
   - With harness connector disconnected, check resistance across sensor terminals.
   - Compare tests results to values in **Table d**.

Fig. 22. M42 engine coolant temperature (ECT) sensor (**arrow**).

Table d. Engine Coolant Temperature Sensor or Intake Air Temperature Sensor Test Values (DME 1.7)

Test temperatures	Resistance (k ohms)
14 ± 2°F (-10 ± 1°C)	7 - 11.6
68 ± 2°F (20 ± 1°C)	2.1 - 2.9
176 ± 2°F (80 ± 1°C)	0.27 - 0.40

NOTE —
The test values listed represent only three test points from a continuous resistance NTC sensor. Check the full linear response to increasing temperature as the engine warms up.

3. If ECT sensor fails these tests, it is faulty and should be replaced. If no faults are found, reconnect electrical harness.

NOTE —
Use a new copper sealing washer when installing sensor. Replace any lost coolant.

WARNING —
Do not replace the ECT sensor unless the engine is cold. Hot coolant can scald.

Tightening Torque
- ECT sensor to cylinder head . . . . 13 Nm (10 ft-lb)

Throttle position sensor (TPS), testing and replacing

The throttle position sensor (TPS) is mounted on the side of the throttle housing and is directly connected to the throttle valve shaft. The ECM sends a voltage signal to the potentiometer-type sensor and monitors the voltage that comes back.

Check TPS function by disconnecting the harness connector and checking reference voltage and sensor resistance. See **Table e** and Fig. 23. If voltage is not present, check the output voltage signal from the ECM and check the wiring between the sensor and the ECM. If the sensor resistance is incorrect, replace the throttle position sensor.

Fig. 23. Throttle position sensor terminal identification on M42 engine.

NOTE —
The throttle position sensor is not adjustable. If test results are incorrect, the sensor should be replaced.

Table e. Throttle Position Sensor Tests (DME 1.7)

Test conditions	Terminals	Test value
Harness connector disconnected, ignition on	1 and ground in harness connector	5 VDC (approx.)
Harness connector disconnected, ignition off	1 and 3 at sensor terminals	4 k ohms (approx.)
Connector disconnected, ignition off. Throttle rotated from idle to full position	1 and 2 at sensor terminals	Continuously variable from 1– 4 k ohms (approx.) without interruption

Idle speed control valve, testing and replacing

Idle speed is maintained by the ECM via the idle speed control valve. See Fig. 24. Idle speed is adaptive through the ECM and no idle speed adjustments can be made. Before testing the valve, confirm that the throttle position sensor (TPS) is working correctly.

Fig. 24. Idle speed control valve (**arrow**) on M42 engine.

NOTE —
- *The tests given below are electrical checks only. They do not check the mechanical operation of the valve or if the valve is sticking or worn. If the valve is suspect, substituting a known good valve is the best way to check for a mechanical fault.*

1. Check battery (+) voltage to valve:
   - Disconnect harness connector from valve.
   - Check for battery voltage at terminal 2 (red/white wire).
   - If voltage is not present check wiring between valve and main relay (terminal 87).
2. Check that ECM signal is reaching valve:
   - With engine running, check that idle speed control valve is audibly buzzing.
   - If valve is not working, disconnect wiring harness connector.
   - Connect 12V probe light across connector terminals.
   - Turn ignition key on; probe should light. If probe does not light, check the wiring from the ECM (pin 29) to the valve. See **Table h**. If probe does light but idle quality is poor, the valve is most likely sticking and or worn and should be replaced.

BOSCH DME M3.1 AND M3.3.1 COMPONENT TESTS AND REPAIRS

Consult **Table a** for engine application information for the Bosch DME 3.1 and 3.3.1 systems.

The DME 3.1 and DME 3.3.1 systems are similar in operation, with knock control and VANOS operation being the key differences. DME 3.1 engines are not equipped with VANOS or knock detectors, while the DME 3.3.1 system is.

> ***CAUTION—***
> *Use only a digital multimeter when testing wiring. Use of an analog VOM may damage the engine control module.*

Electrical tests of the main and fuel pump relays and the DME engine control module (ECM) are covered earlier in this section. Fuel pump tests are covered in **160 Fuel Tank and Fuel Pump**.

Mass Air Flow Sensor

There are two types of mass air flow sensors used on the cars covered by this section. Testing procedures vary depending on type installed. The air flow sensor is not adjustable and must only be tested with a digital multimeter.

Mass Air Flow Sensor Variants

- 1992 M50 engine
  DME M3.1 . . . . mass air flow sensor—hot wire
- 1993-1995 M50 and S50US
  DME M3.3.1) . . . . mass air flow sensor—hot film

Mass air flow sensor (hot wire), testing and replacing

When the engine is running, a current is used to heat a thin wire in the center of the sensor. See Fig. 25. The current in the wire is regulated to maintain a temperature of 100°C more than the air passing over it. The current used to heat the wire is electronically converted into a voltage measurement corresponding to the mass of intake air.

To keep the wire clean, it is heated to a temperature of about 1,000°C (1,830°F) for one second. This "burn-off" cycle takes place automatically, four seconds after the engine is turned off.

If the hot wire breaks or if there is no output from the air flow sensor, the ECM automatically switches to a "limp-home" mode and turns on the Check Engine light. The engine can usually be started and driven. The air flow sensor has no internal moving parts and cannot be serviced.

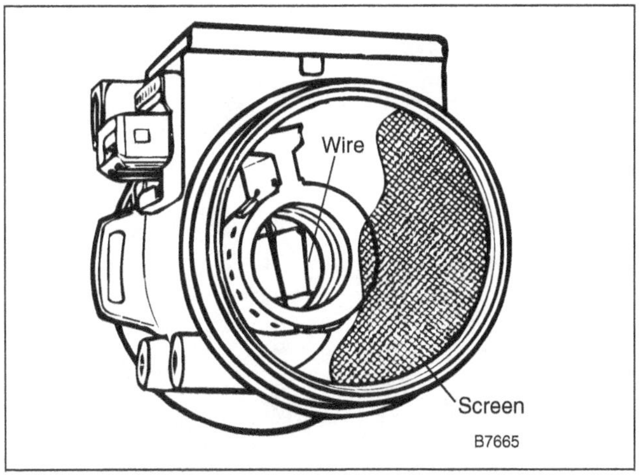

Fig. 25. Mass air flow sensor. Hot wire sensor used on cars with DME 3.1 (1992 6-cylinder models).

> ***CAUTION—***
> *Use only a digital multimeter when checking the mass air flow sensor. An analog meter can damage the air flow sensor.*

1. Disconnect air flow sensor from air cleaner only. Leave it connected to duct leading to intake manifold and leave wiring harness connected.

2. Start engine and run it to normal operating temperature.

3. Rev engine to at least 2,500 rpm, then shut it off. Look through meter at hot wire. After approximately four seconds wire should glow brightly for about one second.

> ***NOTE—***
> *If the wire glows as specified, then the air flow meter and ECM are probably operating correctly. If the wire does not glow, continue testing.*

4. If the wire does not glow, remove air flow sensor and look through it to see if wire is broken. If wire is broken, meter will have to be replaced.

5. Reinstall air flow sensor and harness connector. Peel back rubber boot from harness connector. Working from rear of connector, connect digital voltmeter across terminals 1 and 4. See Fig. 26.

6. Start and rev engine to at least 2,500 rpm, then shut it off. After about 4 seconds, voltage should rise to about 4 volts for about one second. If voltage is present, but wire does not glow, air flow sensor is faulty and should be replaced.

7. If voltage is not present in step 6, turn ignition key on and check for voltage and ground at sensor. There should be ground at pin 4. There should be positive (+) battery voltage at pin 2.

Fig. 26. Mass air flow sensor terminal identification on cars with DME M3.1 fuel injection.

8. With ignition off, disconnect harness connector from air flow sensor. Using a digital multimeter, check resistance at terminals listed.

Air Flow Sensor Test Values (DME M3.1)	
• Terminals 5 and 6	3-4 ohms

If any faults are found, check wiring to and from the ECM. Make ECM pinout test as listed in See **Table i**. Main power to air flow sensor comes from DME main relay.

Mass air flow sensor (hot film), testing and replacing

On cars with DME M3.3.1 a hot film mass air flow sensor is used. When the engine is running, a current is used to heat a thin film in the center of the sensor. This current is electronically converted into a voltage measurement corresponding to the mass of intake air.

NOTE —
A burn-off cycle is not used on hot film sensors.

If the hot film breaks or if there is no output from the air flow sensor, the ECM automatically switches to a "limp-home" mode and turns on the Check Engine light. The engine can usually be started and driven. The air flow sensor has no internal moving parts and cannot be serviced or adjusted.

> ***CAUTION —***
> *Use only a digital multimeter when checking the mass air flow sensor. An analog meter can damage the air flow sensor.*

1. Disconnect harness connector from air flow sensor.

2. Turn ignition on and check for voltage and ground at connector. There should be ground at pin 1. There should be positive (+) battery voltage at pin 3. If any faults are found, check wiring to and from ECM. Make ECM pinout test. See **Table j**.

NOTE —
Positive (+) battery voltage to air flow sensor comes from DME main relay when the ignition is switched on or engine is running.

Engine coolant temperature (ECT) sensor, testing and replacing

The engine coolant temperature (ECT) sensor sends continuous engine coolant temperature information to the ECM. As temperature increases sensor resistance decreases.

ECT Sensor Location	
• M50/S50US engine	left side of cylinder head under intake manifold

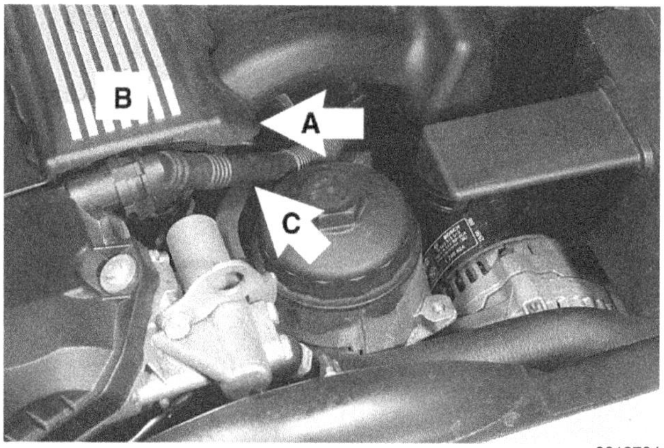

Fig. 27. Engine coolant temperature (ECT) sensor (**A**) is located beneath top engine cover (**B**) and crankcase vent hose (**C**). M52 engine shown. M50 is similar.

1. Check ECM reference voltage to sensor:
   - Disconnect harness connector from ECT sensor.
   - Turn ignition key on.
   - Check for 5 volts between supply voltage wire (brown/red) wire of harness connector and ground.
   - Turn ignition key off.
   - If voltage is not present or incorrect, check wiring from ECM and check reference voltage output at ECM (pin 78). See **Table i** (DME 3.1) or **Table j** (DME 3.3.1)

2. Check ECT sensor resistance:
   - With harness connector disconnected, check resistance across sensor terminals.
   - Compare tests results to values in **Table f**.

Table f. Engine Coolant Temperature Sensor or Intake Air Temperature Sensor Test Values (DME 3.1/3.3.1)

Test temperatures	Resistance (k ohms)
14 ± 2°F (-10 ± 1°C)	7 - 11.6
68 ± 2°F (20 ± 1°C)	2.1 - 2.9
176 ± 2°F (80 ± 1°C)	0.27 - 0.40

NOTE —
The test values listed represent only three test points from a continuous resistance NTC sensor. Check the full linear response to increasing temperatures as the engine warms up.

3. If ECT sensor fails these tests, it is faulty and should be replaced. If no faults are found, reconnect electrical harness.

WARNING —
Do not replace the ECT sensor unless the engine is cold. Hot coolant can scald.

NOTE —
- *Use a new copper sealing washer when installing sensor. Replace any lost coolant.*

Tightening Torque
- Engine coolant temperature sensor to cylinder head . . . . 13 Nm (10 ft-lb)

Intake air temperature (IAT) sensor, testing and replacing

The intake air temperature (IAT) sensor signal is used as a correction factor for fuel injection and ignition timing. This sensor is mounted in the intake manifold behind the throttle position switch. See Fig. 28.

1. Check that ECM reference voltage is reaching IAT sensor:
   - Disconnect IAT sensor harness connector.
   - Turn ignition key on.
   - Check for 5 volts between supply voltage wire of harness connector and ground.
   - Turn ignition key off.

IAT Sensor Supply Voltage
- M50/S50US engines . . . . grey wire and ground

Fig. 28. Intake air temperature sensor location on M50/S50US engine (**arrow**).

If voltage is not present or incorrect, check wiring from ECM and check reference voltage signal at ECM (pin 77). See **Table i** or **Table j**.

2. Check IAT sensor resistance:
   - With harness connector disconnected, check resistance across sensor terminals. Compare tests results to values in **Table f** given earlier.
   - If IAT sensor fails this test it is faulty and should be replaced.
3. If no faults are found, reconnect electrical harness.

Throttle position sensor (TPS), testing and replacing

The throttle position sensor (TPS) is mounted on the side of the throttle housing and is directly connected to the throttle valve shaft. The ECM sends a voltage signal to the potentiometer-type sensor and monitors the voltage that comes back.

Check TPS function by disconnecting the harness connector and testing continuity across the terminals while changing the throttle position. Resistance test values are listed below. If the results are incorrect, replace the throttle position sensor. See Fig. 29.

NOTE —
The throttle position sensor is not adjustable. If test results are incorrect, the sensor should be replaced.

Fig. 29. Throttle position sensor terminal identification on M50 engine.

Table g. Throttle Position Sensor Tests (DME 3.1/3.3.1)

Test conditions	Terminals	Test value
Harness connector disconnected, ignition on	3 and ground in harness connector	5 VDC (approx.)
Harness connector disconnected, ignition off	1 and 3 at sensor terminals	4 k ohms (approx.)
Throttle plate rotated from idle to full throttle position	1 and 2 at sensor terminals	Variable from 1 – 4 k ohms (approx.) without interruption

NOTE —

On cars with traction control, do not confuse the throttle position sensor on the main throttle body with the throttle position switch on the secondary throttle body, where applicable.

Idle speed control valve, testing

Idle speed is maintained by the ECM through the idle speed control valve. The idle control function compensates for engine load and engine operating conditions. Idle speed is adaptive through the ECM and no idle speed adjustments can be made.

Before testing the valve, confirm that the throttle position sensor is working correctly.

NOTE —

• The tests given below are electrical checks only. They do not check the mechanical operation of the valve. If the valve is suspected of causing poor idle, substituting a known good valve is the best way to check for a mechanical fault.

1. With engine running, check that idle speed control valve is buzzing.
2. Turn on A/C or shift car into drive. Idle should remain steady or increase slightly.
3. If valve is not buzzing, or if idle decreases in step 2, stop engine and disconnect harness connector from valve. Check resistance of valve across its terminals. See Fig. 30. Test values are listed below.

NOTE —

If you suspect an intermittent fault, lightly tap the valve while testing resistance.

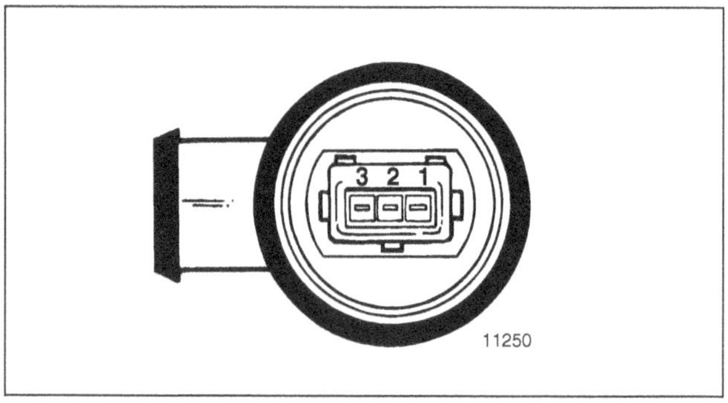

Fig. 30. Idle speed control valve terminal identification.

Idle Speed Control Valve Coil Resistance Values

- M50/S50US engine
  Terminals 1 and 2 . . . . 20 ± 5 ohms
  Terminals 2 and 3 . . . . 20 ± 5 ohms
  Terminals 1 and 3 . . . . 40 ± 5 ohms

4. With valve harness connector disconnected, check for battery voltage at red/white wire in connector with ignition turned on.
   - If there is no voltage, check wiring between connector and main relay terminal 87. See **Electrical Wiring Diagrams**.

NOTE —

The idle speed control valve receives positive (+) battery voltage from the main relay.

5. If voltage is present as described above, check wiring between ECM and valve. If no wiring faults are found, check ECM signal to valve.

NOTE —

- *The idle speed control valve signal can be checked using a duty cycle meter (or dwell meter). Connect the meter to the valve following the manufacturer's instructions. Turn on the A/C or create a large air leak and check that the meter reading reacts while the idle speed remains steady.*
- *There are some additional inputs to the control module that affect idle speed (i.e. throttle position, A/C-on signals, and A/T Drive position). Check these signals if idle problems persist. See* **Table i** *or* **Table j**.

Idle speed control valve, replacing

The idle speed control valve is mounted beneath the intake manifold. Accessing the valve is best accomplished by first removing the intake manifold.

NOTE —

It is recommended that the intake manifold gaskets be replaced whenever the intake manifold is removed.

1. Remove top engine covers from engine. See Fig. 31.

Fig. 31. Remove top engine covers by prying out caps and removing nuts (**A**). Note rubber gaskets at cover edge (**arrows**).

2. Remove two hold down nuts from fuel injector wiring duct. Pry back injector connector retainers and lift off wiring duct.
3. Disconnect intake air boot from throttle housing. Where applicable disconnect accelerator cable and cruise control cable from their brackets.
4. Unbolt intake manifold and move it out of way enough to access idle speed control valve.

NOTE —

The intake manifold is held in place with 7 nuts from above and 2 support bracket bolts from underneath. On DME 3.1 cars, carefully disconnect the idle air hose from the intake manifold when lifting the manifold off. For more information, see **113 Cylinder Head Removal and Installation**.

5. Disconnect all necessary hoses and remove idle speed control valve.
6. Installation is reverse of removal. Use new gaskets when installing intake manifold.

NOTE —

Poor driveability may be noticed after installing a replacement idle speed control valve. After about 10 minutes of driving, the system will adapt the base setting of the valve and the idle speed should return to normal.

Tightening Torque

- Intake manifold to cylinder head (M7) . . . . 15±2 Nm (11±1 ft-lb)

BOSCH DME M5.2 COMPONENT REPLACEMENT

On-Board Diagnostics II (OBD II) is incorporated into the Bosch DME M5.2 engine management systems used on the M44 engine. OBD II is capable of detecting a full range of faults. When faults are detected, a Diagnostic Trouble Code (DTC) is stored in the system ECM. The Check Engine warning light will also come on if an emissions-related fault is detected.

The most efficient way to diagnose the Bosch OBD II system is by using a specialized scan tool. The OBD II system is capable of storing hundreds of faults, making diagnostics with a scan tool the only viable method. Therefore, system diagnostics is not covered here.

NOTE —

- *The OBD II fault memory (including an illuminated Check Engine light) can only be reset using the special scan tool. Removing the connector from the ECM or disconnecting the battery will not erase the fault memory.*
- *The 16-pin OBD II diagnostic connector is located on the lower left dash panel. Refer to Fig. 1.*
- *BMW special scan tool or a "generic" OBD II scan tool must be used to access DTCs.*
- ***Table k*** *lists engine control module (ECM) pinouts for the DME M5.2 system*

Mass air flow sensor, replacing

1. Disconnect harness connector and intake air boot from air flow sensor. See Fig. 32.

0011969b

Fig. 32. To detach air flow sensor, disconnect harness connector (**A**); unclip retaining clips to air filter housing (**B**); and remove large hose clamp at air intake bellows (**C**). M44 engine with traction control shown.

2. Detach two large clips at air filter housing and work air flow sensor out of rubber seal in filter housing.

3. Installation is reverse of removal.
   - For ease of installation, lubricate large rubber seal in air filter housing with silicone lubricant or equivalent.
   - No adjustment to air flow sensor is possible.
   - Carefully check that all hoses and seals are seated properly.

Engine coolant temperature (ECT) sensor, replacing

The engine coolant temperature (ECT) sensor is a dual temperature sensor. One circuit provides coolant temperature information to the ECM while the other circuit provides coolant temperature information to the instrument cluster.

> ***WARNING —***
> *Do not replace the ECT sensor unless the engine is cold. Hot coolant can scald.*

1. Disconnect harness connector from ECT sensor. See Fig. 33.

2. Unscrew temperature sensor from cylinder head and remove

0011984a

Fig. 33. M44 engine coolant temperature (ECT) sensor located above oil filter (**arrow**).

3. Installation is reverse of removal.
   - Use a new copper sealing washer when installing sensor.
   - Replace any lost coolant.

Tightening Torque
- Engine coolant temperature sensor to cylinder head . . . . 13 Nm (10 ft-lb)

Intake air temperature (IAT) sensor, replacing

The intake air temperature (IAT) sensor on cars with M44 engine is mounted in the top section of the intake air filter housing. See Fig. 34.

1. Remove electrical harness connector from IAT sensor.

2. Unclip temperature sensor and remove from air filter housing.

3. Installation is reverse of removal.

Throttle position sensor (TPS), replacing

The throttle position sensor is located on the side of the throttle housing. See Fig. 35.

1. Unplug harness connector from sensor.

Fig. 34. Intake air temperature (IAT) sensor in top section of air filter housing.

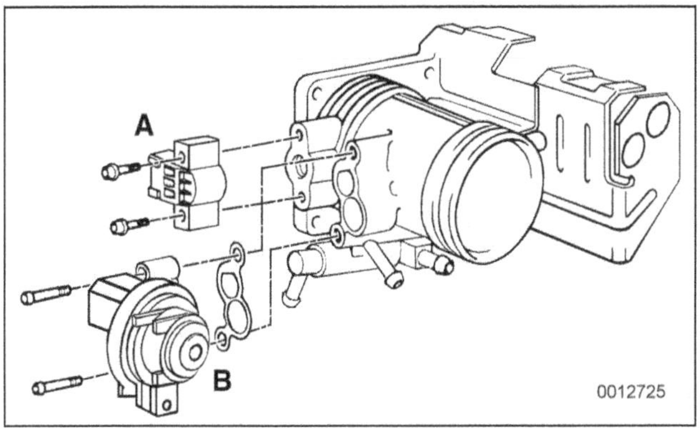

Fig. 35. M44 engine: Throttle housing showing (**A**) throttle position sensor (TPS) and (**B**) idle speed control valve.

2. Remove two mounting screws holding sensor to throttle housing.

3. Installation is reverse of removal. No adjustment is necessary.

Idle speed control valve, replacing

The idle speed valve on the M44 engine is located on the side of the throttle body. See Fig. 35.

1. Disconnect harness connector from idle valve.

2. Disconnect vacuum hose and retaining screws.

3. Pull idle speed control valve from its fittings.

4. Installation is reverse of removal. Always replace O-rings and seals.

NOTE —
Poor driveability may be noticed after installing a replacement idle control valve. After about 10 minutes of driving, the idle speed should return to normal.

SIEMENS MS 41.1 COMPONENT REPLACEMENT

On-Board Diagnostics II (OBD II) is incorporated into the Siemens MS 41.1 engine management systems used in the M52 and S52US engine. Refer to Fig. 36.

OBD II is capable of detecting a full range of faults. When faults are detected, a Diagnostic Trouble Code (DTC) is stored in the system ECM. The Check Engine warning light will also come on if an emissions-related fault is detected. The most efficient way to diagnose the Siemens OBD II system is by using a specialized scan tool. The OBD II system is capable of storing hundreds of faults, making diagnostics with a scan tool the only viable method. Therefore, system diagnostics is not covered here.

NOTE —
- *The OBD II fault memory (including an illuminated Check Engine light) can only be reset using the special scan tool. Removing the connector from the ECM or disconnecting the battery will not erase the fault memory.*
- *The 16-pin OBD II diagnostic connector is located on the lower left dash panel. Refer to Fig. 1.*
- *BMW special scan tool or a "generic" OBD II scan tool must be used to access DTCs.*
- ***Table I** lists engine control module (ECM) pinouts for the MS 41.1 systems.*

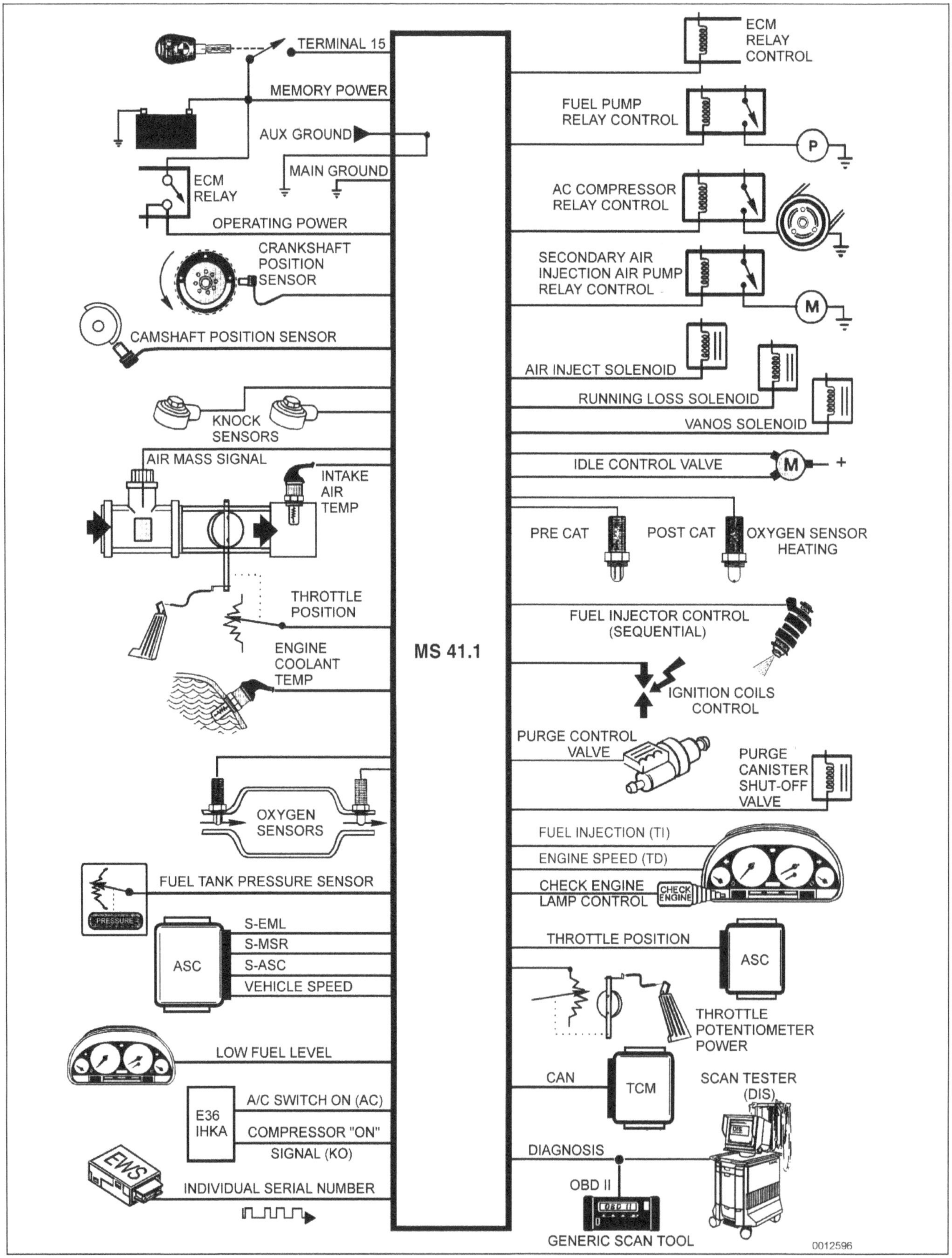

Fig. 36. Siemens MS 41.1 OBD II engine management system used on M52 engine. Bosch M5.2 system used on M44 engine is similar.

SIEMENS MS 41.1 COMPONENT REPLACEMENT

Mass air flow sensor, replacing

1. Disconnect harness connector and intake air boots from air flow sensor. See Fig. 37.

Fig. 37. Mass air flow sensor (**arrow**) on M52 engine.

2. Installation is reverse of removal.
   - No adjustment to air flow sensor is possible.
   - Check intake hoses for cracks and vacuum leaks.

Engine coolant temperature (ECT) sensor, replacing

The ECT sensor is a dual temperature sensor. One circuit provides coolant temperature information to the ECM while the other circuit provides coolant temperature information to the instrument cluster.

1. Remove left-side top engine cover.
2. Unclip crankcase venting hose from cylinder head cover. See Fig. 38.
3. Under intake manifold, disconnect harness connector from sensor.
4. Remove sensor.
5. Installation is reverse of removal.
   - Use a new copper sealing washer when installing new sensor.
   - Replace any lost coolant.

Tightening Torque

- Engine coolant temperature sensor to cylinder head . . . . 13 Nm (10 ft-lb)

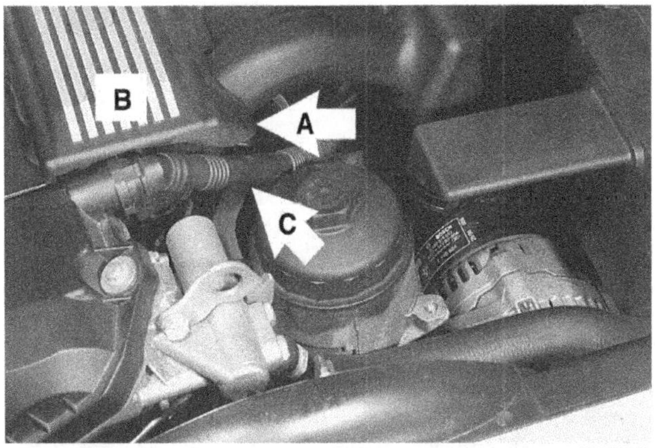

Fig. 38. M52 engine: Engine coolant temperature (ECT) sensor (**A**) is located beneath top engine cover (**B**) and crankcase vent hose (**C**).

Intake air temperature (IAT) sensor, replacing

The intake air temperature (IAT) sensor for the M52/S52US engine is mounted on the bottom of the intake manifold. See Fig. 39.

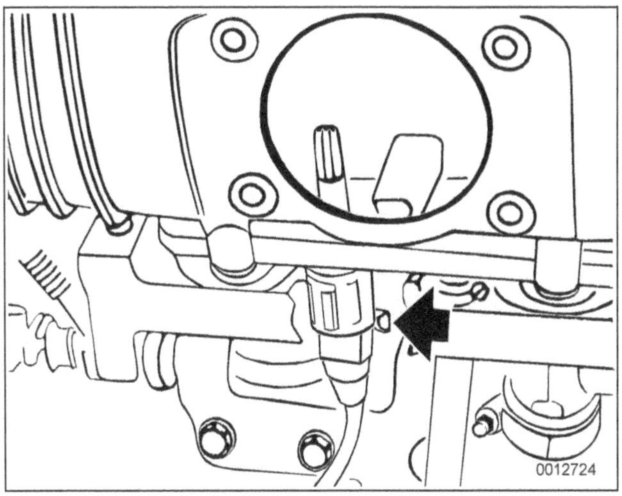

Fig. 39. M52/S52US engine: Intake air temperature sensor location in bottom of intake manifold (**arrow**). Throttle housing shown removed.

1. Disconnect intake air boot from throttle housing. Unbolt throttle housing and lay aside. (It is not necessary to disconnect throttle cable or electrical harness connectors from throttle housing.)
2. Remove electrical harness connector from IAT sensor.
3. Unclip temperature sensor and remove from intake manifold.
4. Installation is reverse of removal. Use a new gasket at the throttle housing.

Throttle position sensor (TPS), replacing

The throttle position sensor is located on the side of the throttle housing. See Fig. 40.

Fig. 40. M52 engine: Throttle position sensor (**arrow**).

1. Unplug harness connector from sensor.

NOTE —
On cars with traction control, do not confuse the throttle position sensor on the main throttle body with the throttle position switch on the secondary throttle body, where applicable.

2. Remove two mounting screws holding sensor to throttle housing.

3. Installation is reverse of removal.

Idle speed control valve, replacing

The idle speed control valve is mounted on the underside of the intake manifold adjacent to the dipstick tube bracket. Accessing the valve is best accomplished by first removing the throttle housing.

1. Disconnect intake air boot from throttle housing. Unbolt throttle body and lay aside. (Do not disconnect throttle cable or electrical harness connectors to throttle body.)

2. Unbolt dipstick tube bracket from intake manifold.

3. Working under intake manifold, disconnect electrical harness connector from idle speed control valve.

4. Remove two bolts retaining idle speed control valve to underside of intake manifold. Remove valve from intake manifold. Remove hose clamp and disconnect hose.

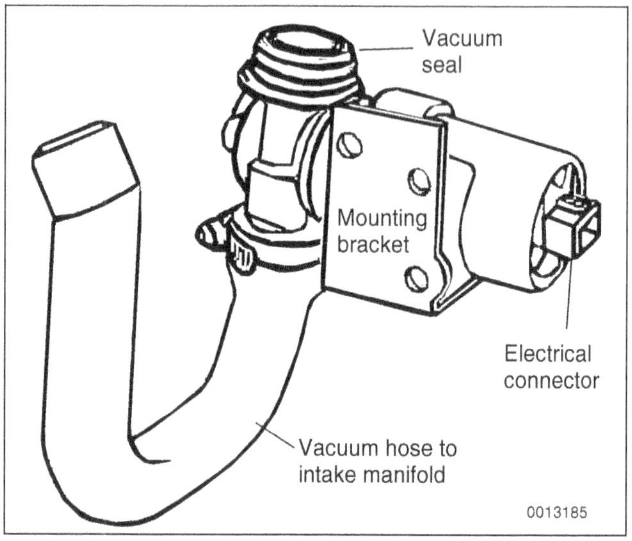

Fig. 41. M52 engine: Idle speed control valve (located under intake manifold).

5. Installation is reverse of removal. Use new gaskets when installing.

NOTE —
Poor driveability may be noticed after installing a replacement idle speed control valve. After about 10 minutes of driving the idle speed should return to normal.

ECM PIN ASSIGNMENTS

Engine control module (ECM), accessing

1. Disconnect negative (–) battery cable. Wait approximately one minute.

CAUTION —
Prior to disconnecting the battery, read the battery disconnection cautions given at the front of this manual on page viii.

2. Remove engine control module (ECM) compartment cover from right side of engine compartment rear bulkhead. Cover is retained with four captive screws. See Fig. 42.

3. Disconnect control module harness connector by releasing fastener and pivoting connector up and off ECM.

4. Remove ECM from retaining brackets and pull ECM from its holder.

5. Installation is reverse of removal.

Fig. 42. Engine control module (ECM) located in compartment in right-rear of engine compartment (**arrow**).

NOTE —

Replacement ECMs must be coded with application information (i.e. engine code, transmission type, etc.) prior to installation. Consult an authorized BMW dealer before replacing the ECM.

ECM pin assignments are given in **Tables h** through **Table l**. This information can be helpful when diagnosing faults to or from the ECM. If all inputs and wiring are OK but operational problems still exist, the ECM itself may be faulty.

Generally, absence of voltage or continuity means there is a wiring or connector problem. Test results with incorrect values do not necessarily mean that a component is faulty. Check for loose, broken or corroded connections and wiring before replacing components. If the results are still incorrect, test the component itself. For engine management system electrical schematics, see **Electrical Wiring Diagrams**.

CAUTION —

- *Always wait at least 40 seconds after turning off the ignition before removing the connector from the engine control module (ECM). If the connector is removed before this time, residual power in the system relay may damage the ECM.*
- *Always connect or disconnect the control module connector and meter probes with the ignition off.*

When making checks at the ECM itself, a breakout box should be used to allow tests to be made with the connector attached to the ECM. This also prevents damage to the small terminals in the connector. As an alternative, the harness connector housing can be separated so that electrical checks can be made from the back of the connector. ECM pin numbering is shown in Fig. 43.

NOTE —

On cars with traction control, do not confuse the throttle position sensor on the main throttle body with the throttle position switch on the secondary throttle body.

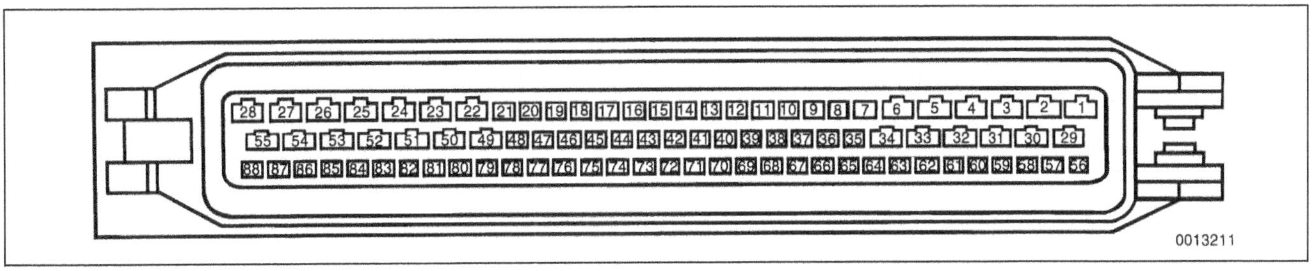

Fig. 43. ECM pin numbers as seen from back of ECM connector.

Table h. ECM Pin Assignment—Bosch DME M1.7

Pin	Signal	Component/function	Signal
1	Output	Fuel pump relay control	Fuel pump relay, terminal 85
2	Not used		
3	Output	Cyl. 2 and cyl. 4 fuel injection valve control	Cyl. 2 and cyl. 4 fuel injection valves
4	Not used		
5	Not used		
6	Ground	Ground, fuel injection valve output stages	Ground point
7	Not used		
8	Output	"Check engine" indicator control	Instrument cluster
9	Not used		
10	Not used		
11	Output	DKV potentiometer signal	Transmission control module (EGS)
12	Input	Throttle position sensor	Throttle position signal
13	Not used		
14	Input	Volume air flow sensor	Volume air flow sensor (signal ground)
15	Input	Cyl. 1-2 knock sensor	Cyl. 1-2 knock sensor
16	Input	Camshaft position sensor (cylinder identification)	Camshaft position sensor (cylinder identification)
17	Output	ti measurement signal (fuel consumption)	Instrument cluster
18	Output	Intake air resonance changeover valve (DISA)	Intake air resonance changeover valve
19	Not used		
20	Not used		
21	Not used		
22	Not used		
23	Not used		
24	Output	Cyl. 3 ignition coil control	Cyl. 3 ignition coil
25	Output	Cyl. 1 ignition coil control	Cyl. 1 ignition coil
26	Input	Battery voltage	B + junction point
27	Output	Engine control module relay control	Engine control module relay, terminal 85
28	Ground	Ground for electronics and shielding of sensors	Ground point
29	Output	Idle speed control valve control	Idle speed control valve
30	Not used		
31	Not used		
32	Output	Cyl. 1 and cyl. 3 fuel injection valve control	Cyl. 1 and cyl. 3 fuel injection valves
33	Not used		
34	Ground	Ground, output stages (except for ignition and fuel injection valves)	Ground point
35	Not used		
36	Output	Evaporative emission valve control	Evaporative emission valve
37	Output	Oxygen sensor heater control	Oxygen sensor relay, terminal 85
38	Not used		
39	Not used		
40	Not used		
41	Input	Volume Air Flow sensor	Volume air flow sensor (voltage varies with engine load)
42	Input	Cyl. 3-4 knock sensor	Cyl. 3-4 knock sensor
43	Ground	Ground for sensors	Engine coolant temperature sensor, cyl. 1-2 knock sensor, cyl. 3-4 knock sensor, and throttle position sensor
44	Input	Camshaft position sensor (cylinder identification)	Camshaft position sensor (cylinder identification)

Table h. ECM Pin Assignment—Bosch DME M1.7

Pin	Signal	Component/function	Signal
45	Not used		
46	Not used		
47	Not used		
48	Output	A/C compressor cut-out	Compressor control relay, terminal 85
49	Not used		
50	Not used		
51	Output	Cyl. 4 ignition coil control	Cyl. 4 ignition coil
52	Output	Cyl. 2 ignition coil control	Cyl. 2 ignition coil
53	Not used		
54	Input	Battery voltage from engine control module	Engine control module relay, terminal 87a
55	Ground	Ground, ignition	Ground point
56	Input	Ignition switch, terminal 15	Ignition switch, terminal 15
57	Not used		
58	Not used		
59	Output	Throttle position sensor and volume air flow sensor	Reference voltage (5 VDC)
60	Input	Programming voltage	Data link connector
61	Not used		
62	Not used		
63	Not used		
64	Input	Ignition timing intervention	Transmission control module (EGS)
65	Input	Drive range P/N	Automatic transmission range switch
66	Not used		
67	Input	Crankshaft position RPM sensor	Crankshaft position RPM sensor
68	Input	Crankshaft position RPM sensor	Crankshaft position RPM sensor
69	Not used		
70	Input	Oxygen sensor signal	Heated oxygen sensor
71	Ground	Oxygen sensor signal ground	Heated oxygen sensor
72	Not used		
73	Input	Vehicle speed signal	Instrument cluster
74	Output	Engine speed signal	Instrument cluster
75	Not used		
76	Input	Volume air flow sensor	Volume air flow sensor
77	Input	Intake air temperature sensor (IAT)	Intake air temperature sensor (IAT)
78	Input	Engine coolant temperature	Engine coolant temperature sensor
79	Not used		
80	Not used		
81	Input	Drive-away protection signal (code)	On-board computer
82	Not used		
83	Not used		
84	Not used		
85	Input	A/C pressure	Pressure switch
86	Input	A/C switch	Integrated climate regulation control module
87	Output	RxD diagnosis data line	Data link connector
88	Output-Input	TxD diagnosis data line	Data link connector

Table i. ECM Pin Assignment—Bosch DME M3.1

Pin	Signal	Component/function	Signal
1	output	Fuel pump relay control	Fuel pump relay switches with engine running or cranking (crankshaft position signal must be present for relay switchover)
2	output	Idle speed control valve	Pulsed ground– close signal (see also pin 29)
3	output	Fuel injector control, cyl. 1	Pulsed ground (injection pulse width in ms) with engine running
4	output	Fuel injector control, cyl. 3	Pulsed ground (injection pulse width in ms) with engine running
5	output	Fuel injector control, cyl. 2	Pulsed ground (injection pulse width in ms) with engine running
6	ground	Ground	Ground for fuel injector output stages
7	vacant	—	—
8	output	Check Engine	Check engine lamp control ground
9	vacant	—	—
10	vacant	—	—
11	output	Throttle valve position	Load signal to transmission control module
12	input	Throttle position sensor	Voltage varies with throttle position
13	output	Mass air flow sensor	Air flow sensor hot wire burn off (voltage for 0.5 seconds after shutdown)
14	ground	Mass air flow sensor	Ground for air flow sensor
15	vacant	—	—
16	input	Cylinder identification sensor	A/C voltage pulse per camshaft revolution (between pin 16 and 44)
17	output	Fuel consumption (ti)	Fuel consumption output (KVA signal) to instrument cluster
18	vacant	—	—
19	vacant	—	—
20	vacant	—	—
21	vacant	—	—
22	vacant	—	—
23	output	Ignition control (terminal 1), cyl. no. 2	Primary signal, ignition coil cyl. no. 2
24	output	Ignition control (terminal 1), cyl. no. 3	Primary signal, ignition coil cyl. no. 3
25	output	Ignition control (terminal 1), cyl. no. 1	Primary signal, ignition coil cyl. no. 1
26	input	Power supply (terminal 30)	Battery voltage (B+) at all times (terminal 30)
27	output	Main relay control	Main relay activation (to relay terminal 85)
28	ground	Ground	Ground for ECM and sensor shielding
29	output	Idle speed control valve	Pulsed ground– open signal (see also pin 2)
30	vacant	—	—
31	output	Fuel injector, cyl. no. 5	Pulsed ground (injection pulse width in ms), cyl. no. 5
32	output	Fuel injector, cyl. no. 6	Pulsed ground (injection pulse width in ms), cyl. no. 6
33	output	Fuel injector, cyl. no. 4	Pulsed ground (injection pulse width in ms), cyl. no. 4
34	ground	Ground	Ground for output stages
35	vacant		
36	output	Evaporative purge valve control	Pulsed ground with engine at normal temperature and varying engine load
37	output	Oxygen sensor heater relay control	Oxygen sensor heater relay activation (ground at terminal 85)
38	vacant	—	—
39	vacant	—	—
40	vacant	—	—
41	input	Mass air flow sensor	Voltage (+)
42	vacant	—	—
43	ground	Ground	Ground for temperature sensors (ECT sensor, IAT sensor, TP sensor)
44	input	Cylinder identification sensor	A/C voltage pulse per camshaft revolution (between pin 16 and 44)

Table i. ECM Pin Assignment—Bosch DME M3.1

Pin	Signal	Component/function	Signal
45	vacant	—	—
46	vacant	—	—
47	vacant	—	—
48	output	A/C compressor control	A/C compressor disabled via compressor control relay
49	vacant	—	—
50	output	Ignition control (terminal 1), cyl. no. 4	Primary signal, ignition coil cyl. no. 4
51	output	Ignition control (terminal 1), cyl. no. 6	Primary signal, ignition coil cyl. no. 6
52	output	Ignition control (terminal 1), cyl. no. 5	Primary signal, ignition coil cyl. no. 5
53	vacant	—	—
54	input	Power supply	Battery voltage (+) from main relay terminal 87
55	ground	Ground	Ground for ignition control
56	input	Power supply (terminal 15)	Battery voltage (+) with key on or engine running
57	vacant	—	—
58	vacant	—	—
59	output	Throttle position sensor	Throttle position sensor supply voltage (5 VDC)
60	input	Data link connector	Programming voltage
61	vacant	—	—
62	vacant	—	—
63	vacant	—	—
64	input	Ignition timing intervention	from A/T control module (only active during gearshift)
65	input	Automatic transmission (A/T) range switch	Transmission park or neutral signal
66	vacant	—	—
67	input	Engine speed/crankshaft position sensor	Crankshaft position/rpm (voltage (VAC) between pins 67 and 68)
68	input	Engine speed/crankshaft position sensor	Crankshaft position/rpm (voltage (VAC) between pins 67 and 68)
69	vacant	—	—
70	input	Oxygen sensor	Oxygen sensor signal (0–1 VDC fluctuating with engine running)
71	ground	Oxygen sensor	Oxygen sensor signal ground
72	vacant	—	—
73	input	Road speed	Road speed signal from instrument cluster
74	output	Engine speed (TD)	Engine speed (TD) signal to instrument cluster
75	vacant	—	—
76	vacant	—	—
77	input	Intake air temperature (IAT) sensor	Intake air temperature (0-5 V, temperature dependent)
78	input	Engine coolant temperature (ECT) sensor	Engine coolant temperature (0-5 V, temperature dependent)
79	vacant	—	—
80	vacant	—	—
81	input	On-board computer	Drive-away protection enable
82	vacant	—	—
83	vacant	—	—
84	vacant	—	—
85	input	A/C pressure switch	From Integrated climate control module via A/C pressure switch
86	input	A/C compressor on	From Integrated climate control module
87	input	Diagnostic connector (RxD)	Diagnostic RxD (receive) signal to pin 15 in Data link connector
88	input/output	Diagnostic connector (TxD)	Diagnostic TxD (transmit) signal to pin 20 in Data link connector

Table j. ECM Pin Assignment—Bosch DME M3.3.1

Pin	Signal	Component/function	Signal
1	output	Fuel pump relay control	Fuel pump relay switches with engine running or cranking (crankshaft position signal must be present for relay switchover)
2	output	Idle speed control valve	Pulsed ground—close signal (see also pin 29)
3	output	Fuel injector control, cyl. 5	Pulsed ground (injection pulse width in ms) cyl. 5
4	output	Fuel injector control, cyl. 6	Pulsed ground (injection pulse width in ms) cyl. 6
5	output	Fuel injector control, cyl. 4	Pulsed ground (injection pulse width in ms) cyl. 4
6	ground	Ground	Ground for fuel injector output stage
7	output	Camshaft actuator (VANOS solenoid) control	Camshaft actuator (VANOS solenoid)
8	output	Check Engine	Instrument cluster, Check Engine lamp
9	vacant	—	—
10	vacant	—	—
11	output	Throttle position	Throttle angle signal to A/T control module
12	vacant	—	—
13	input	Oxygen sensor	Oxygen sensor signal (0-1 VDC fluctuating with engine running)
14	input	Mass air flow sensor	Mass air flow sensor
15	ground	Ground	Ground
16	input	Crankshaft position/rpm sensor	Voltage pulse (VAC) between pin 16 and 43 (crank position/rpm sensor)
17	input	Camshaft position sensor	Hall effect camshaft sensor
18	vacant	—	—
19	vacant	—	—
20	vacant	—	—
21	vacant	—	—
22	vacant	—	—
23	output	Ignition coil control, cyl. 4	Ignition coil 4
24	output	Ignition coil control, cyl. 6	Ignition coil 6
25	output	Ignition coil control, cyl. 5	Ignition coil 5
26	input	Power supply (terminal 30)	Battery voltage (B+) at all times
27	input	Main relay control	Main relay activation (terminal 85)
28	ground	Ground	Ground for ECM and sensor shielding
29	output	Idle speed control valve	Pulsed ground—open signal (see also pin 2)
30	vacant	—	—
31	output	Fuel injector control, cyl. 3	Injection pulse width in ms—cyl. 3
32	output	Fuel injector control, cyl. 2	Injection pulse width in ms—cyl. 2
33	output	Fuel injector control, cyl. 1	Injection pulse width in ms—cyl. 1
34	ground	Ground	Ground for remaining output stages
35	vacant	—	—
36	output	Evaporative purge valve control	
37	vacant	—	—
38	output	Oxygen sensor heater relay control	Oxygen sensor heater relay switchover (terminal 85)
39	vacant	—	—
40	ground	Oxygen sensor	Oxygen sensor signal ground
41	input	Mass air flow sensor	Mass air flow voltage signal
42	input	Vehicle speed	Vehicle speed signal from instrument cluster
43	input	Crankshaft position/rpm sensor	Voltage pulse (VAC) between pin16 and 43
44	ground	Ground	Ground for intake air temp. sensor, engine coolant temp. sensor, throttle position sensor

Table j. ECM Pin Assignment—Bosch DME M3.3.1

Pin	Signal	Component/function	Signal
45	ground	Ignition circuit shield	Ground shield for ignition circuit monitoring
46	output	Fuel consumption (KVA signal)	To instrument cluster
47	output	Crankshaft rpm	Engine speed (TD) signal to instrument cluster
48	output	A/C compressor control	A/C compressor relay terminal 85
49	vacant	—	—
50	output	Ignition coil control, cyl. 1	Primary signal, ignition coil 1
51	output	Ignition coil control, cyl. 2	Primary signal, ignition coil 2
52	output	Ignition coil control, cyl. 3	Primary signal, ignition coil 3
53	vacant	—	—
54	input	Power supply	Battery voltage from main relay (terminal 87a)
55	ground	Ground	Ground for ignition control
56	input	Power supply (terminal 15)	Battery voltage with key on or engine running
57	input	Ignition timing intervention	From A/T control module
58	vacant	—	—
59	output	Throttle position sensor (TPS)	Voltage supply to TPS (5 VDC)
60	input	Programming voltage	Data link connector
61	vacant	—	—
62	vacant	—	—
63	vacant	—	—
64	input	A/C on signal	From integrated climate control module
65	input	A/C pressure signal	From integrated climate control module via A/C pressure switch
66	input	On-board computer	Drive-away protection enable (starter immobilization relay)
67	vacant	—	—
68	vacant	—	—
69	input	Knock sensor #2 (cyl. 4, 5, 6)	Knock sensor #2 signal
70	input	Knock sensor #1 (cyl. 1, 2, 3)	Knock sensor #1 signal
71	ground	Ground	Ground for knock sensors and shields
72	vacant	—	—
73	input	Throttle position sensor (TPS)	Throttle position signal
74	vacant	—	—
75	vacant	—	—
76	vacant	—	—
77	input	Intake air temperature	Intake air temperature sensor (0-5 VDC)
78	input	Engine coolant temperature	Engine coolant temperature sensor (0-5 VDC)
79	vacant	—	—
80	vacant	—	—
81	input	Automatic transmission gear position/neutral safety switch	A/T park or neutral position signal
82	vacant	—	—
83	vacant	—	—
84	vacant	—	—
85	vacant	—	—
86	vacant	—	—
87	input	Diagnostic connector (RxD)	Diagnostic RxD (receive) signal to pin 15 in Data link connector
88	input/output	Diagnostic connector (TxD)	Diagnostic TxD (transmit) signal to pin 20 in Data link connector

Table k. ECM Pin Assignment—Bosch DME M5.2

Pin	Signal	Component/function	Signal
1	output	Oxygen sensor (monitoring sensor) heater	Oxygen sensor heater control (switched ground)
2	output	Idle speed control valve	Pulsed ground—close signal (see also pin 29)
3	output	Fuel injector control, cyl. 1	Pulsed ground (injection pulse width in ms) cyl. 1
4	output	Fuel injector control, cyl. 4	Pulsed ground (injection pulse width in ms) cyl. 4
5	not used	—	—
6	ground	Ground	Ground for fuel injector output stage
7	not used	—	—
8	output	Check Engine	Instrument cluster, Check Engine lamp
9	not used	—	—
10	input	Electronic immobilizer control (EWS II)	Electronic immobilizer control (EWS II) module
11	output	Automatic climate control	Automatic climate control, to evaporator controller
12	not used	—	—
13	not used	—	—
14	not used	—	—
15	not used	—	—
16	input	Intake air temperature	Intake air temperature signal
17	input	Mass air flow meter	Intake air signal
18	not used	—	—
19	input	Oxygen sensor (monitoring sensor)	Oxygen sensor control
20	output	Crankshaft/rpm sensor	Crankshaft/rpm sensor control
21	input	Camshaft position sensor	Camshaft position signal
22	output	Ignition coil control, cyl. 3	Primary signal, ignition coil 3
23	output	Ignition coil control, cyl. 4	Primary signal, ignition coil 4
24	not used	—	—
25	not used	—	—
26	input	Power supply (terminal 30)	Battery voltage (B+) at all times
27	input	Main relay control	Main relay activation (terminal 85)
28	ground	Ground	Ground for ECM and sensor shielding
29	output	Idle speed control valve	Pulsed ground—open signal (see also pin 2)
30	output	Oxygen sensor (monitoring sensor) heater	Oxygen sensor heater control (switched ground)
31	output	Fuel injector control, cyl. 3	Pulsed ground (injection pulse width in ms) cyl. 3
32	output	Fuel injector control, cyl. 2	Pulsed ground (injection pulse width in ms) cyl. 2
33	not used	—	—
34	ground	Ground	Ground for ECM/ output stages
35	not used	—	—
36	input	A/C compressor relay	A/C compressor relay control
37	not used	—	—
38	not used	—	—
39	not used	—	—
40	input	Knock sensor #2 (cyl. 3,4)	Knock sensor #2 signal
41	not used	—	—
42	input	Vehicle speed	Vehicle speed signal from instrument cluster
43	not used	—	—
44	output	Throttle position sensor (TPS)	Throttle position reference signal

Table k. ECM Pin Assignment—Bosch DME M5.2 (continued)

Pin	Signal	Component/function	Signal
45	output	Mass air flow meter	Intake air signal
46	output	Oxygen sensor (monitoring sensor)	Oxygen sensor control
47	not used	—	—
48	not used	—	—
49	output	Ignition coil control, cyl. 1	Primary signal, ignition coil 1
50	output	Ignition coil control, cyl. 2	Primary signal, ignition coil 2
51	not used	—	—
52	not used	—	—
53	input	Throttle position sensor	Throttle position signal
54	input	Power supply	Battery voltage from main relay (terminal 87)
55	ground	Ground	Ground for ECM
56	input	Power supply (terminal 15)	Battery voltage with key on or engine running
57	output	Activate cooling fan (man. trans.) control	Normal speed relay
58	not used	—	—
59	not used	—	—
60	input	Programming voltage	Programming voltage via data link connector, pin 18
61	output	Evaporative emission valve	Evaporative emission valve control
62	vacant	—	—
63	output	Fuel pump relay	Fuel pump relay control
64	not used	—	—
65	not used	—	—
66	not used	—	—
67	not used	—	—
68	input	Signal above 80°C (man. tran.)	Double temperature switch
69	input	Automatic climate control	Automatic climate control
70	input	Knock sensor #1 (cyl. 1,2)	Knock sensor #1 signal
71	ground	Ground	Ground for analog signals and knock sensors
72	not used	—	—
73	not used	—	—
74	input	Engine coolant temperature sensor	Engine coolant temperature signal
75	not used	—	—
76	not used	—	—
77	output	Oxygen sensor (regulating sensor)	Oxygen sensor control
78	input	Crankshaft/rpm sensor	Crankshaft position/rpm signal
79	input	ABS or traction control	ABS or AST control
80	input	Engine speed	Engine speed signal
81	not used	—	—
82	not used	—	—
83	input	On-board computer	From On-board computer (terminal 4)
84	not used	—	—
85	not used	—	—
86	not used	—	—
87	input	Diagnostic connector (RxD)	Diagnostic RxD (receive) signal to pin 15 in Data link connector
88	output	Diagnostic connector (TxD)	Diagnostic TxD (transmit) signal to pin 17in Data link connector

Table I. ECM Pin Assignment—Siemens DME MS 41.1

Pin	Signal	Component/function	Signal
1	output	Ignition coil control, cyl. 2	Primary signal, ignition coil 2
2	output	Ignition coil control, cyl. 4	Primary signal, ignition coil 4
3	output	Ignition coil control, cyl. 6	Primary signal, ignition coil 6
4	ground	Ground	Ground
5	output	Fuel injector control, cyl. 2	Pulsed ground (injection pulse width in ms) cyl. 2
6	output	Fuel injector control, cyl. 1	Pulsed ground (injection pulse width in ms) cyl. 1
7	output	Mass air flow meter	Mass air meter signal
8	input	Mass air flow meter	Mass air meter signal
9	output	Instrument cluster	Fuel consumption signal
10	output	Engine coolant temperature (ECT) sensor	ECT signal
11	output	Fuel tank pressure sensor	Fuel tank pressure sensor control
12	input	Throttle position sensor (TPS)	Throttle position signal
13	—		
14	input	Intake air temperature (IAT) sensor	Intake air temperature signal
15		Traction control	AST module
16	input	Automatic climate control	Automatic climate control
17	—		
18	input	Electronic immobilizer control (EWS II)	Electronic immobilizer control (EWS II) module
19		Automatic climate control	Automatic climate control
20	—	Instrument cluster	Instrument cluster
21	output	Camshaft actuator (VANOS solenoid) control	Camshaft actuator (VANOS solenoid), switched ground
22	output	Fuel injector control, cyl. 3	Pulsed ground (injection pulse width in ms) cyl. 3
23		Fuel injector control, cyl. 6	Pulsed ground (injection pulse width in ms) cyl. 6
24		Fuel injector control, cyl. 4	Pulsed ground (injection pulse width in ms) cyl. 4
25	output	Oxygen sensor heater control	Oxygen sensor heater ground
26	input	Power supply (terminal 30)	Battery voltage (B+) at all times
27	output	Idle speed control valve	Pulsed ground—open signal (see also pin 53)
28	ground	Ground	Ground
29	output	Ignition coil control, cyl. 1	Primary signal, ignition coil 1
30	output	Ignition coil control, cyl. 3	Primary signal, ignition coil 3
31	output	Ignition coil control, cyl. 5	Primary signal, ignition coil 5
32	ground	Ground	Ground
33	output	Fuel injector control, cyl. 5	Pulsed ground (injection pulse width in ms) cyl. 5
34	ground	Ground	Ground
35	output	Secondary air injection	Secondary air injection pump relay control
36	output	Engine speed output	Engine speed signal
37	—		
38	ground	Knock sensor	Shielding for knock sensors
39	output	Intake air temperature sensor (IAT Sensor) Engine coolant temperature (ECT) sensor	Voltage supply to IAT sensor and ECT sensor
40	output	Crankshaft position sensor (Hall effect)	Crankshaft position sensor control
41	ground	Camshaft position (CMP) sensor	Shielding for CMP sensor
42	output	Throttle position sensor (TPS)	TPS ground
43	input	Camshaft position/rpm sensor	Camshaft position/rpm sensor control
44	output	Throttle position sensor (TPS)	Voltage supply to TPS (5 VDC)
45		Traction control	AST module
46	—	Instrument cluster	Instrument cluster

Table I. ECM Pin Assignment—Siemens DME MS 41.1 (continued)

Pin	Signal	Component/function	Signal
47	—		
48	input	Crankshaft position sensor (Hall effect)	Crankshaft position sensor control
49	input	Power supply (terminal 15)	Battery voltage with key on or engine running
50	output	Solenoid valve (running losses)	Running losses
51	output	Carbon canister valve	Carbon canister valve control
52	vacant	—	—
53	output	Idle speed control valve	Pulsed ground—close signal (see also pin 29)
54	input	Power supply	Battery voltage from main relay (terminal 87)
55	vacant	—	—
56	—		
57	input	Knock sensor (cyl. 1-3)	Knock sensor input signal
58	output	Knock sensor (cyl. 1-3)	Knock sensor control
59	input	Knock sensor (cyl. 4-6)	Knock sensor input signal
60	input/output	Diagnostic connector (TxD)	Diagnostic TxD (transmit) signal to pin 18 in Data link connector
61	output	Oxygen sensor heater (monitoring sensor)	Oxygen sensor heater ground
62	output	Secondary air injection	Secondary air injection control valve
63	output	Knock sensor (cyl. 4-6)	Knock sensor control
64	input	Camshaft position/rpm sensor	Camshaft position/rpm sensor control
65	input	Camshaft position/rpm sensor	Camshaft position/rpm sensor control
66	—		
67	output	Oxygen sensor	Oxygen sensor reference voltage
68	output	Evaporative purge valve control	Pulsed ground with engine at normal temperature and varying engine load
69	output	Fuel pump relay control	Fuel pump relay switches with engine running or cranking (crankshaft position signal must be present for relay switchover)
70	vacant	—	—
71	output	Oxygen sensor heater (regulating sensor)	Oxygen sensor heater ground
72	output	Oxygen sensor (monitoring sensor)	Oxygen sensor reference voltage
73	input	Main relay control	Main relay activation (terminal 85)
74	output	A/C compressor control	A/C compressor relay control
75	input	Oxygen sensor	Oxygen sensor signal
76	not used	—	—
77	input	Oxygen sensor (regulating sensor)	Oxygen sensor signal
78	input	Oxygen sensor (monitoring sensor)	Oxygen sensor signal
79	output	Oxygen sensor (regulating sensor)	Oxygen sensor reference voltage
80		Traction control	AST module
81		Traction control	AST module
82		Traction control	AST module
83	output	Crankshaft position sensor (Hall effect)	Crankshaft position sensor control
84	vacant	—	—
85	output	Automatic transmission	Automatic transmission control module
86	input	Automatic transmission	Automatic transmission control module
87	input	Power supply	Battery voltage from main relay (terminal 87)
88	input/output	Diagnostic connector (TxD)	Diagnostic TxD (transmit) signal to pin 17 in Data link connector

160 Fuel Tank and Fuel Pump

TABLES

GENERAL

This repair group covers service information specifically for the fuel supply system. Information on the fuel injection system is covered in **130 Fuel Injection**.

NOTE —
Fuel filter replacement is covered in **020 Maintenance Program**.

WARNING —
- *The fuel system is designed to retain pressure even when the ignition is off. When working with the fuel system, loosen the fuel lines slowly to allow residual fuel pressure to dissipate gradually. Avoid spraying fuel.*
- *Before beginning any work on the fuel system, place a fire extinguisher in the vicinity of the work area.*
- *Fuel is highly flammable. When working around fuel, do not disconnect any wires that could cause electrical sparks. Do not smoke or work near heaters or other fire hazards.*
- *Always unscrew the fuel tank cap to release pressure in the tank before working on the tank or lines.*
- *Do not use a work light near any fuel. Fuel may spray onto the hot bulb causing a fire.*
- *Make sure the work area is properly ventilated.*

The cautions and warnings on this page should be observed when servicing the fuel system.

CAUTION —
- *Prior to disconnecting the battery, read the battery disconnection cautions given at the front of this manual on page viii.*
- *Before making any electrical tests with the ignition turned on, disable the ignition system as described in* **120 Ignition System**. *Be sure the battery is disconnected when replacing components.*
- *To prevent damage to the ignition system or the electronic fuel system components, including the control unit, always connect and disconnect wires and test equipment with the ignition off.*
- *Cleanliness is essential when working with the fuel system. Thoroughly clean the fuel line unions before disconnecting any of the lines.*
- *Use only clean tools. Keep removed parts clean and sealed or covered with a clean, lint-free cloth, especially if completion of the repair is delayed.*
- *Do not move the car while the fuel system is open.*
- *Avoid using high pressure compressed air to blow out lines and components. High pressure can rupture internal seals and gaskets.*
- *Always replace seals and O-rings.*

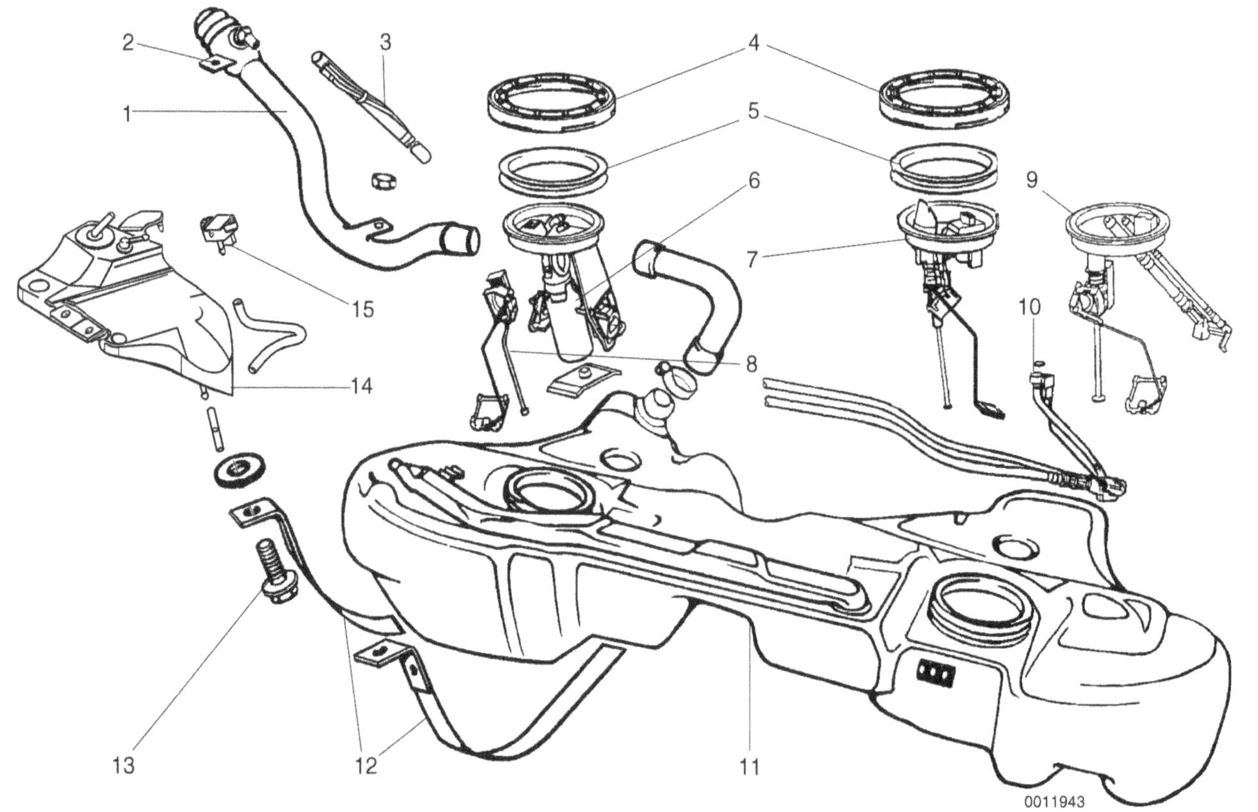

1. Filler pipe
2. Filler pipe grounding tab
3. Vent pipe
4. Threaded collar
5. Rubber ring seal
6. Fuel pump
7. Left side fuel level sender (models built from 1/95)
8. Right side fuel level sender
9. Left side fuel level sender with fuel compensating siphon pump (models built up to 1/95)
10. Fuel compensating siphon pump and hose assembly (models built from 1/95)
11. Fuel tank
12. Tank support strap
13. Bolt—tighten to 23 Nm (17 ft-lb)
14. Fuel expansion tank
15. Evap system pressure sensor (OBD II)

Fig. 1. E36 fuel tank assembly.

FUEL TANK AND LINES

The plastic fuel tank is mounted beneath the center of the car (beneath rear seat). Mounted in the fuel tank are the fuel pump and fuel level sending units. Connecting lines for the evaporative emission control system and expansion tank are also attached to the tank. When performing the procedures that follow, refer to Fig. 1.

Fuel tank capacity for E36 models

- All ex. 1996 M3 (3.0 L). . . . . 65 liters (17.0 gal)
- 1996 M3 (3.0 L) models . . . . .62 liters (16.4 gal)

Fuel tank, draining

The fuel tank should be drained into an approved storage unit or another car's fuel tank.

WARNING —

- *Before removing the tank, be sure that all hot components such as the exhaust system, are completely cooled down.*
- *Fuel may be spilled. Do not smoke or work near heaters or other fire hazards.*

1. Disconnect negative (–) cable from battery.

CAUTION —

Prior to disconnecting the battery, read the battery disconnection cautions given at the front of this manual on page viii.

2. Remove fuel tank filler cap.

3. Remove rear seat cushion and pull back insulation mats to expose fuel tank access covers. See Fig. 2.

0013140

Fig. 2. Right side fuel tank access cover under rear seat cushion.

4. Remove right and left access covers to expose fuel hoses and electrical connections at tank.

5. Label fuel hoses. Then disconnect hoses and harness connectors from fuel sending units and fuel pump. See Fig. 3.

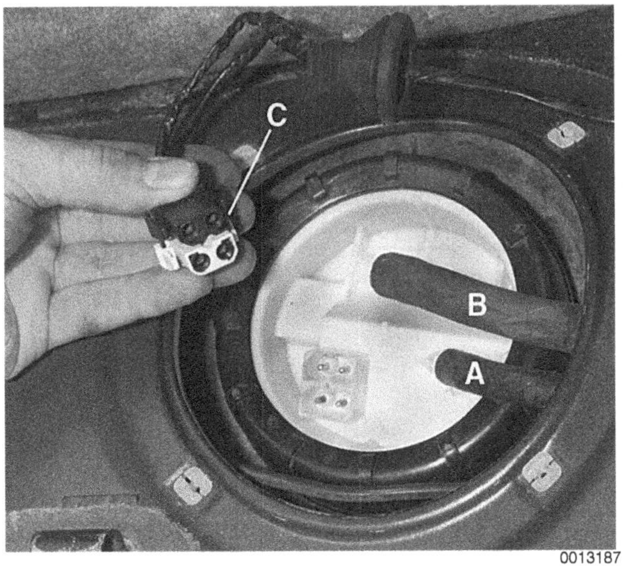

0013187

Fig. 3. Fuel pump/fuel level sender assembly in top of fuel tank (right side shown). Disconnect supply hose (**A**), return hose (**B**) and harness connector (**C**).

6. Unscrew threaded collars from fuel tank. Slowly withdraw fuel sender assemblies from tank, allowing fuel to drain off.

NOTE —

- *If necessary, push fuel level sender arm toward fuel pump assembly to facilitate removal.*
- *BMW special tool no. 16 1 020 should be used to remove and install the threaded collar. Damage to the collar may result if the special tool is not used.*
- *Pump out contents of each tank lobe using approved pumping/extraction equipment and flexible fuel-grade hose.*

> ***WARNING —***
> *Fuel may be spilled. Do not smoke or work near heaters or other fire hazards.*

7. Installation of sending units is reverse of removal, noting the following:
   - Use new sealing rings when installing pump/sending unit assemblies.
   - Be sure that fuel line connections point in same direction as they came out.
   - Fill tank and check for leaks by running engine.

> ***CAUTION —***
> *Before starting the engine, fill the fuel tank with at least 1.5 gallons (5 liters) of fuel. The pump will be damaged if you run it without fuel.*

Fuel tank, removing and installing

1. Disconnect negative (–) cable from battery.

> ***CAUTION —***
> *Prior to disconnecting the battery, read the battery disconnection cautions given at the front of this manual on page viii.*

2. Remove fuel tank filler cap and drain tank as described earlier.

3. Working in left tank access hole (under rear seat cushion), remove vent pipe from tank.

4. Working inside car, disconnect parking brake cable ends from parking brake lever. See **340 Brakes**.

5. Working underneath car, gently pry fuel lines away from retaining bracket in front of tank. Remove clamps and disconnect fuel hoses as shown in Fig. 4.

6. Remove exhaust system and heat shield as described in **180 Exhaust System**. See Fig. 5.

7. Remove driveshaft. See **260 Driveshaft**.

8. Pull parking brake cables backward away from bottom of fuel tank, disengaging them from brackets as needed.

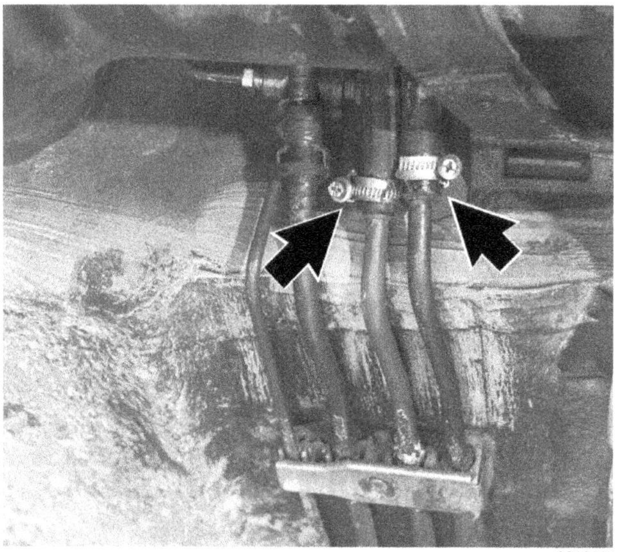

0013191

Fig. 4. Pinch off supply and return fuel hoses (**arrows**) before disconnecting from rigid metal lines.

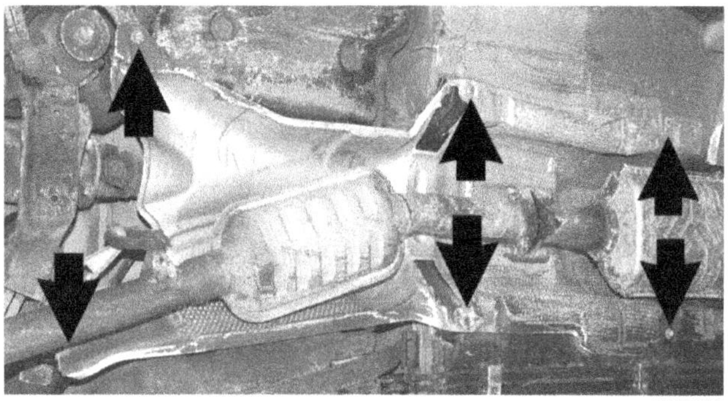

0013143

Fig. 5. After removing exhaust system, remove heat shield screws (**arrows**) and heat shield.

9. Working underneath car, loosen hose clamp and disconnect filler hose.

10. Support tank from below. Remove tank strap mounting bolts. Lower and remove tank, disconnecting right side vent and other hoses/electrical connectors as necessary.

11. Installation is reverse of removal.
    - Always use new seals, gaskets, O-rings, and hose clamps.
    - Inspect hoses and replace any that are chafed, dried out or cracked.
    - Inspect heat shield and replace if corroded.
    - Inspect rubber buffers and liners on fuel tank, support straps and on underside of body. Replace rubber parts that are hardened or damaged.

> ***CAUTION—***
> - *Before starting the engine, fill the fuel tank with at least 1.5 gallons (5 liters) of fuel to prevent damaging a dry fuel pump.*
> - *If the filler neck has been removed from the body, be sure to reattach the neck's grounding screw (where applicable). Check electrical resistance between the ground tab and wheel hub. The resistance should be no higher than 0.6 ohms.*

Tightening Torques
- Fuel tank to body . . . . 23 Nm (17 ft-lb)
- Hose clamps (10 mm-16 mm dia.) . . 2.0 Nm (18 in-lb)

FUEL TANK EVAPORATIVE CONTROL SYSTEM

The fuel tank evaporative control system is used to vent the fuel tank to a charcoal canister, which is then purged by engine vacuum during normal engine operation. This controls hydrocarbon (fuel) emissions to the atmosphere. See Fig. 1.

The overall operation of the system provides:

- An inlet vent, to an otherwise sealed fuel tank, for the entry of air to replace the fuel consumed during engine operation.
- An outlet vent with a storage canister to trap and hold fuel vapors that are produced by the expansion/evaporation of fuel in the tank.

A fuel expansion tank, located under the right rear fender, traps overflow fuel and incorporates extra fuel tank plumbing.

On most early cars (1992-1995), the charcoal canister is in the engine compartment. On later cars, the charcoal canister is under the spare tire in the trunk.

OBD II evaporative emissions (1996 - 1998)

OBD II standards require that the fuel tank evaporative control system be monitored for leaks. This is accomplished using a system pressure sensor, vent shut off valve, and evaporative purge valve. See Fig. 6.

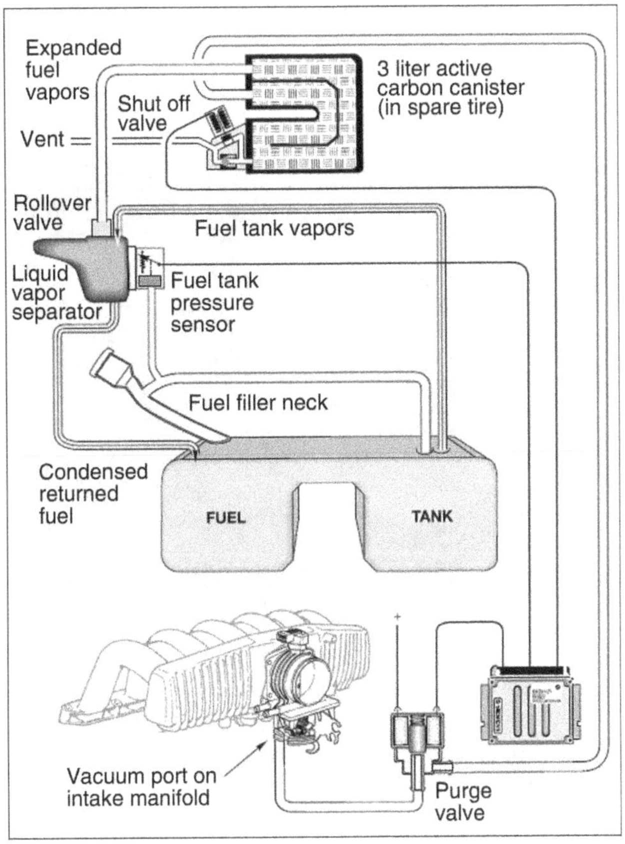

Fig. 6. Evaporative emissions system on 1996 and later models (OBD II).

Fuel Tank Pressure Sensor: The pressure sensor (transducer) mounted on the liquid / vapor separator provides fuel tank pressure input to the ECM. This is used by the ECM to check the fuel storage and evaporative (purge) system for leaks.

The pressure sensor receives a power supply from the ECM and produces a varying voltage input (0-5v) to the ECM, representing the amount of pressure in the fuel tank / evaporative system. If this input is defective, a fault code will be set and the CHECK ENGINE light will illuminate. See Fig. 7.

Active carbon cannister: As hydrocarbon vapors enter the canister, they are absorbed by the active carbon. The remaining air is vented to the atmosphere, allowing the fuel tank to breath. When the engine is running, the canister is purged using intake manifold vacuum to draw air through the canister, extracting the hydrocarbon vapors into the combustion chamber. The canister is located in the spare tire well. See Fig. 8.

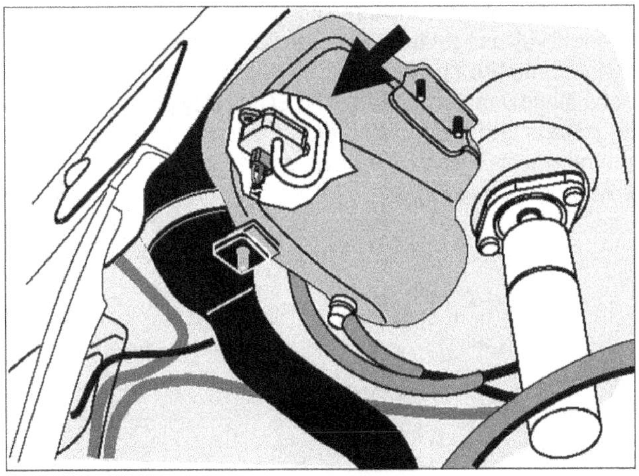

Fig. 7. OBD II evaporative emissions pressure sensor (**arrow**) mounted on liquid/vapor separator.

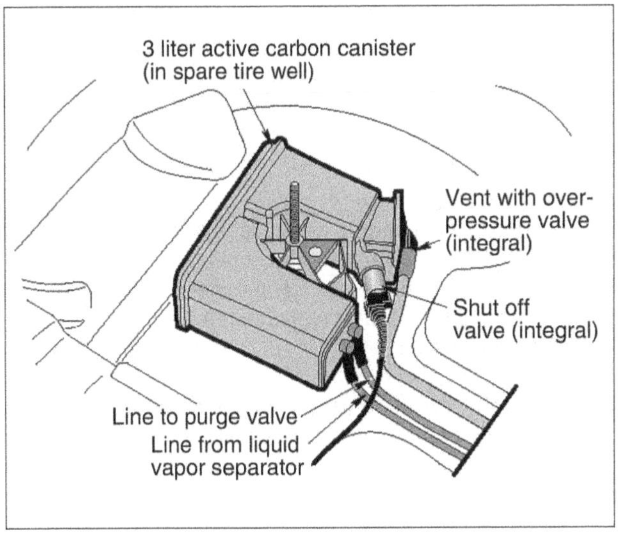

Fig. 8. Carbon canister located in spare tire well

Air inlet shut off valve: A shut-off valve is installed on the intake or vent side of the charcoal canister as part of the leak detection system. Operation of the valve is controlled by the ECM. When the shut-off valve and purge valve are closed, the entire fuel tank/evaporative system is sealed from the atmosphere, allowing the OBD pressure test to be executed.

If this valve is defective, a fault code will be set and the CHECK ENGINE light will illuminate when the OBD II criteria is achieved.

Evaporative emission purge valve: This ECM controlled solenoid valve regulates the purge flow from the active carbon canister into the intake manifold (located next to the hot film mass air flow meter (HFM)). The ECM relay provides operating voltage and the ECM controls the valve by regulating the ground circuit. The valve is powered open and closed by an internal spring. See Fig. 9.

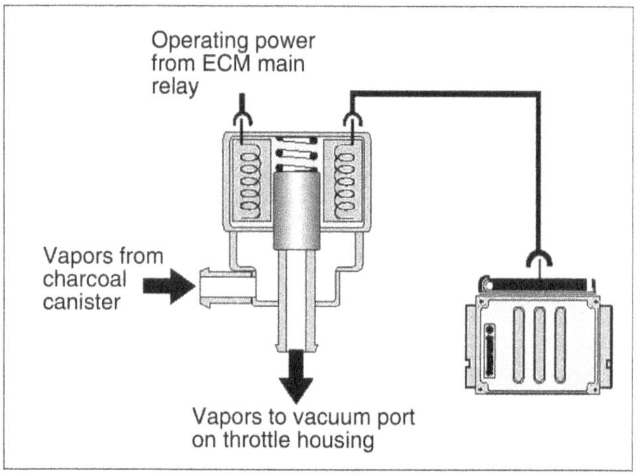

Fig. 9. Evap purge valve operation.

If the purge valve circuit is defective, a fault code will be set and the CHECK ENGINE light will illuminate when the OBD II criteria is achieved. If the valve is mechanically defective, a driveability complaint could be encountered and a mixture related fault code will be set.

The system hoses and components should also be checked if leak-related faults codes are present. This is best accomplished through the use of an automotive smoke machine.

FUEL LEVEL SENDERS

The E36 model uses a two-lobed fuel tank and each lobe of the tank has its own fuel level sending unit.

The right side fuel level sender is integrated with the fuel pump. Each sender has a float connected to a variable resistance for fuel level. When replacing the sender/pump assembly always replace hose clamps, gaskets and O-rings.

> ***WARNING —***
> *When removing the fuel level sender or the fuel pump, the fuel tank should be emptied.*

Fuel level senders, testing

The left and right fuel level sending units are wired in series. The fuel gauge receives a signal that is the average of the two input values.

1. Remove rear seat cushion, lift up insulation mats, and remove both fuel tank access covers.

2. Disconnect white plug on each sending unit. See Fig. 10.

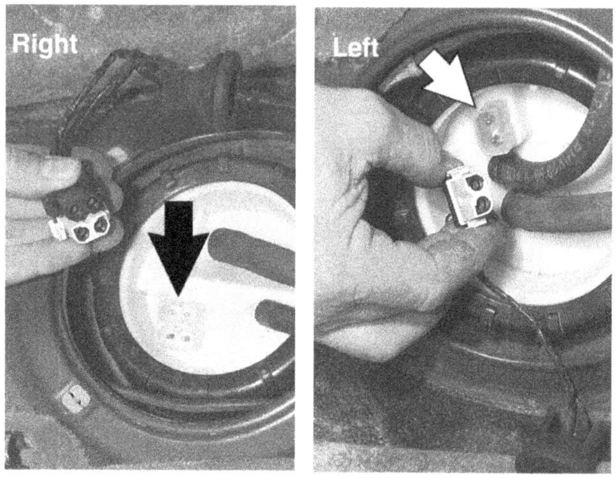

0013187 & 0013194

Fig. 10. Fuel pump/level sender terminal identification. Check resistance at terminals of white connectors (**arrows**).

3. Check resistance value of each sending unit. Values should be the same.

4. If values are different, remove sending units as described below, and check resistance values across terminals. **Table a** lists resistance values for senders at various float positions.

> ***NOTE —***
> *The compensating siphon pump (Fig. 1) maintains the fuel level between the right and left tank lobes. If the resistances are different between the left and right senders, remove the senders and visually inspect the level in each lobe. If the level is different, check the siphon compensating pump.*

Table a. Fuel Level Sender Resistances

Float position	Test result (ohms)
Right side empty	10 ohm ± 2
Right side full	250 ohm ± 5
Left side empty	10 ohm ± 2
Left side full	250 ohm ± 5

Fuel level sender (right side) and fuel pump, removing and installing

1. Disconnect negative (–) cable from battery.

> ***CAUTION —***
> *Prior to disconnecting the battery, read the battery disconnection cautions given at the front of this manual on page viii.*

2. Drain fuel from tank as described earlier.

> ***WARNING —***
> *Always unscrew the fuel tank cap to release pressure in the tank before working on the tank or lines.*

3. Remove rear seat cushion and pull back insulation pad under right side.

4. Remove screws holding right fuel tank access cover. Remove cover.

5. Label fuel hoses. Disconnect hoses and harness connectors from fuel sending unit and fuel pump.

6. Unscrew threaded collar from fuel pump/fuel level sender. Slowly withdraw assembly from tank, allowing fuel to drain off. See Fig. 11.
   - If necessary, push level sender arm toward fuel pump assembly to facilitate removal.

NOTE —
BMW special tool no. 16 1 020 should be used to remove and install the threaded collar. Damage to the collar may result if the special tool is not used

> ***WARNING —***
> *Fuel may be spilled. Do not smoke or work near heaters or other fire hazards.*

7. Using a new sealing ring, install pump and sending unit, being sure that fuel line connections point in same direction as they came out.

8. Reconnect harness connector and hoses. Fill tank and check for leaks by running engine. Install access cover and its mounting screws. Reconnect negative (–) cable to battery.

> ***CAUTION —***
> *Before starting the engine, fill the fuel tank with at least 1.5 gallons (5 liters) of fuel. The pump will be damaged if you run it without fuel.*

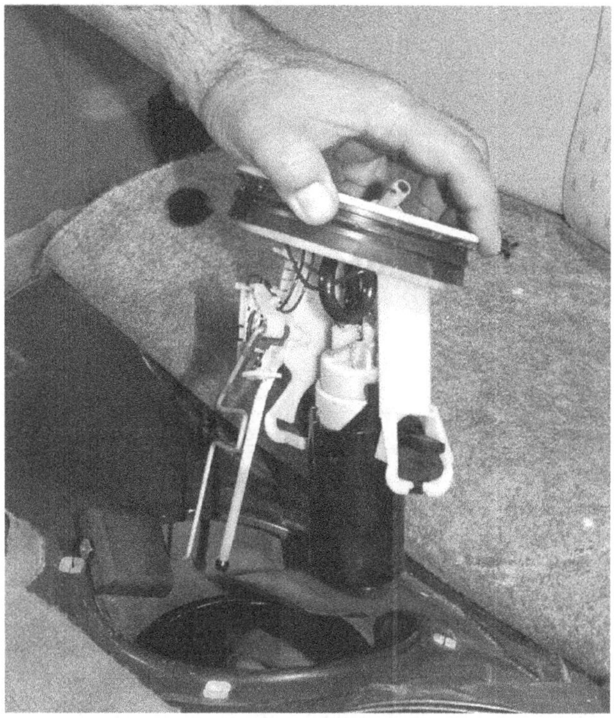

0013142

Fig. 11. Fuel sender and pump assembly.

Fuel level sender (left side), removing and installing

The procedure for removing and replacing the left side fuel level sender is similar to that used for the right side. There is no fuel pump on the left side.

FUEL PUMP

The electric fuel pump is mounted in the fuel tank in tandem with the right side fuel level sender. The fuel pump delivers fuel at high pressure to the fuel injection system. A pressure regulator maintains system pressure. The quantity of fuel supplied exceeds demand, so excess fuel returns to the fuel tank via a return line. See **130 Fuel Injection** for more information on system pressure and the fuel pressure regulator.

NOTE —
Fuel pump removal procedures are given earlier under **Fuel level sender (right side) and fuel pump, removing and installing**.

Fuel Pump Fuse and Relay

The fuel pump is operated by a fuel pump relay located in the power distribution box. The fuel pump circuit is protected by fuse no. 18 located in the power distribution box. See Fig. 12..

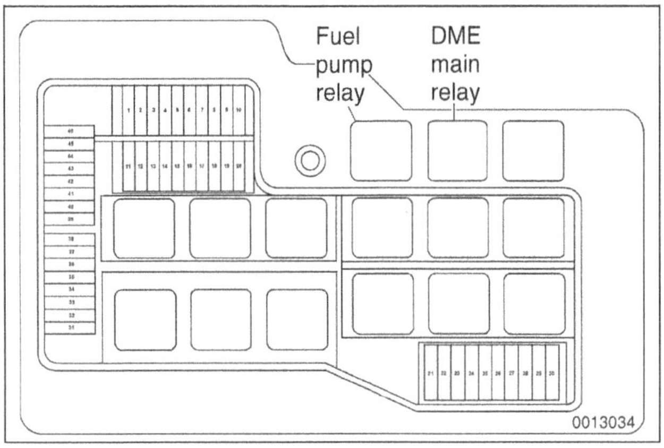

Fig. 12. Fuel pump relay location in power distribution box.

The DME engine control module (ECM) and the DME main relay supply power to the coil side of the fuel pump relay. During starting, the fuel pump runs as long as the ignition switch is in the start position and continues to run once the engine starts. If an electrical system fault interrupts power to the fuel pump, the engine will not run.

Operating fuel pump for tests

To operate the fuel pump for testing purposes without having to run the engine, the fuel pump relay can be bypassed to power the pump directly.

To run the fuel pump, remove the relay and connect the socket for relay terminal 30 to the socket for relay terminal no. 87 with a fused jumper wire. Relay terminal identification is shown in Fig. 13. After completing the tests, remove the jumper wire.

CAUTION —
Relay locations may vary. Use care when identifying relays and when troubleshooting the electrical system at the fuse/relay panel. The fuel pump relay is a four pin relay and has a 1.5 mm² red wire at terminal 30 in the relay socket. If in doubt, consult an authorized BMW dealer.

NOTE —
The jumper wire should be 1.5 mm² (14 ga.) and include an in-line fuse holder with a 16 amp fuse. To avoid fuse/relay panel damage from repeated connecting and disconnecting, also include a toggle switch. A heavy-duty jumper, BMW tool no. 61 3 050, is also available from an authorized BMW dealer.

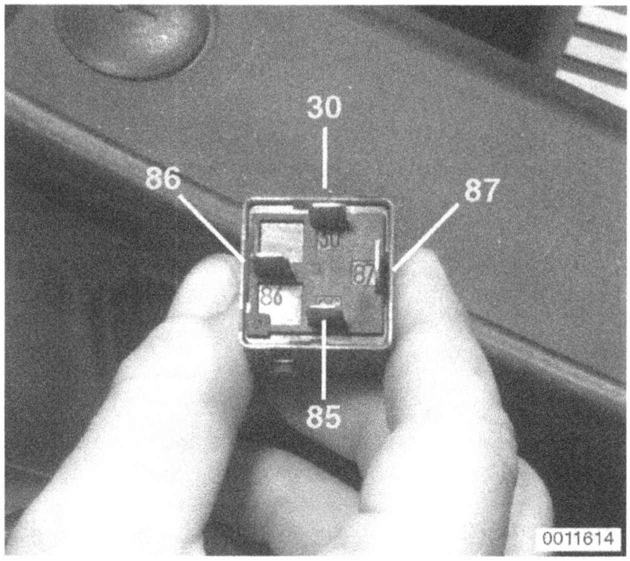

Fig. 13. Fuel pump relay terminal identification.

If the pump does not run with the jumper installed, the fault could be in the fuel pump or the wiring to the pump. Check the pump and its wiring as described below.

Fuel Pump Electrical Tests

Troubleshooting of any fuel pump fault should begin with checking the fuel pump fuse and the fuel pump relay. The DME main relay should also be checked.

NOTE —
Special tools are required for some of the tests described here.

Fuel pump electrical circuit, testing

The fuel pump electrical circuit diagram is shown in Fig. 14.

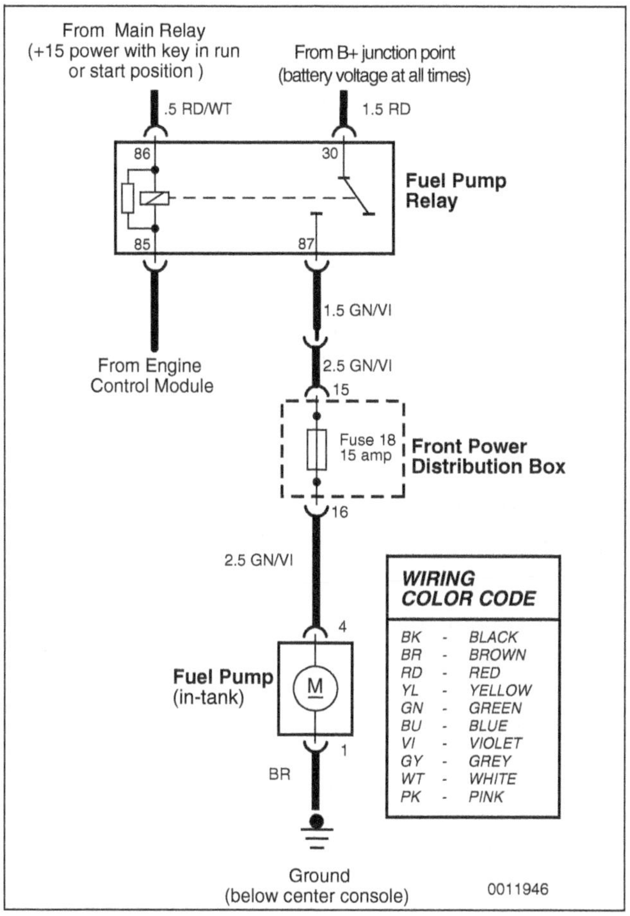

Fig. 14. Fuel pump electrical circuit.

> ***CAUTION—***
> *Fuse and relay locations may vary. Use care when troubleshooting the electrical system at the fuse/relay panel. To resolve problems in identifying a relay, see an authorized BMW dealer.*

1. Remove rear seat cushion, pull right side insulation mat back to expose fuel tank access cover. Remove cover to expose wiring connections.

2. Remove fuel pump relay and operate fuel pump as described under **Operating fuel pump for tests** earlier. Pump should run. Disconnect jumper wire when finished.

3. If fuel pump does not run, disconnect black harness connector from tank sender unit. With jumper wire connected as described in step 2 above, check for positive (+) battery voltage at harness connector terminals. See Fig. 15.

4. If voltage and ground are present, fuel pump is probably faulty. If there is no voltage, check wiring from fuel pump relay and make sure relay is functioning correctly.

Fig. 15. Voltage supply to fuel pump harness (black) connector being checked (**arrow**).

Fuel pump power consumption, testing

> ***NOTE—***
> - *To achieve accurate test results, the battery voltage at the connector should be approximately 13 volts. Charge the battery as necessary.*
> - *A higher than normal power consumption usually indicates a worn fuel pump, which may cause intermittent fuel starvation due to pump overheating and seizure. The only remedy is pump replacement. Be sure to check that the return line and the pump pickup are not obstructed before replacing the pump.*

1. Remove rear seat cushion, pull right side insulation mat back to expose fuel tank access cover. Remove cover to expose wiring connections.

2. Disconnect (black) harness connector from fuel pump.

3. Connect an ammeter and an insulated jumper wire between terminals in connector and corresponding pump terminals. See Fig. 16.

> ***CAUTION—***
> *Do not allow the test leads to short to ground.*

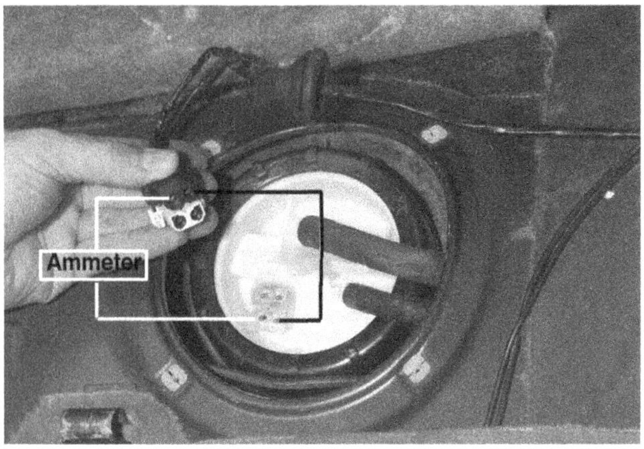

0013187

Fig. 16. Test fuel pump for current draw by attaching ammeter and jumper wire as shown.

> ***NOTE —***
> *See* **600 Electrical System–General** *for information on electrical tests using a digital multimeter (DMM).*

4. Run pump as described in **Operating fuel pump for tests**.
5. Compare ammeter reading with specification listed in **Table b**.

Table b. Fuel Pump Current

Maximum current consumption	5.0 amps

Fuel Delivery Tests

Checking fuel delivery is a fundamental part of troubleshooting and diagnosing the DME system. Fuel pressure directly influences fuel delivery. An accurate fuel pressure gauge will be needed to make the tests.

There are three significant fuel delivery values to be measured:

- **System pressure**—created by the fuel pump and maintained by the pressure regulator.
- **Fuel delivery volume**—created by the fuel pump and affected by restrictions, such as clogged fuel filter.
- **Residual pressure**—the pressure maintained in the closed system after the engine and fuel pump are shut off.

Procedures for measuring the first two quantities are described here.

Residual fuel pressure is checked using the procedure detailed in **130 Fuel Injection**.

Relieving fuel pressure and connecting fuel pressure gauge

> ***WARNING —***
> - *Gasoline is highly flammable and its vapors are explosive. Do not smoke or work on a car near heaters or other fire hazards when diagnosing and repairing fuel system problems. Have a fire extinguisher available in case of an emergency.*
> - *When working on an open fuel system, wear suitable hand protection. Prolonged contact with fuel can cause illnesses and skin disorders.*

> ***CAUTION —***
> *Cleanliness is essential when working with fuel circuit components. Thoroughly clean the unions before disconnecting fuel lines*

To prevent fuel from spraying on a hot engine, system fuel pressure should be relieved before disconnecting fuel lines. One method is to tightly wrap a shop towel around a fuel line fitting and loosen or disconnect the fitting.

Measuring fuel pressure requires special tools. Earlier cars use pressure hose with clamps for fuel delivery. These can be attached to a fuel pressure gauge below the intake manifold.

> ***NOTE —***
> *On 1996 and later 4-cylinder cars, use BMW special tool 13 6 051 (hose with quick release coupling) to attach pressure gauge to Shraeder valve fitting on top of the M44 fuel rail. See Fig. 17.*

0012503a

Fig. 17. Fuel rail showing location of Shrader valve fitting on M44 engine (**arrow**).

Later model 6-cylinder fuel systems require BMW special tool 16 1 050 to release fittings and connect the fuel gauge. See Fig. 18.

If the special tools are not available, a length of fuel line and a T-fitting can be installed to the inlet fuel line and connected to a fuel gauge.

Fig. 18. Fuel lines at rear of intake manifold on M52 engine. Use BMW special tool 16 1 050 to release fittings.

> ***NOTE —***
> - *Use BMW pressure gauge Part No. 13 3 060, or an equivalent. The fuel pressure gauge should have a range of 0 to 5 bar (0 to 75 psi) and must be securely connected to prevent it from coming loose under pressure.*
> - *On cars with 6-cylinder engine, the top left-side engine cover will have to be removed to access the fuel rail.*

System pressure, testing

System pressure is the pressure created by the fuel pump and maintained by the pressure regulator. See Fig. 19. System pressure is not adjustable.

> ***CAUTION —***
> *The fuel pump is capable of developing a higher pressure than that regulated by the pressure regulator. In the event the fuel pump check valve is faulty (stuck closed), make sure the fuel pressure does not rise above 6.0 bar (87 psi). Damage to the fuel lines or fuel system components could result.*

1. Remove fuel tank filler cap.

2. Connect in-line a T-fitting and fuel pressure gauge to outlet hose at fuel pump.

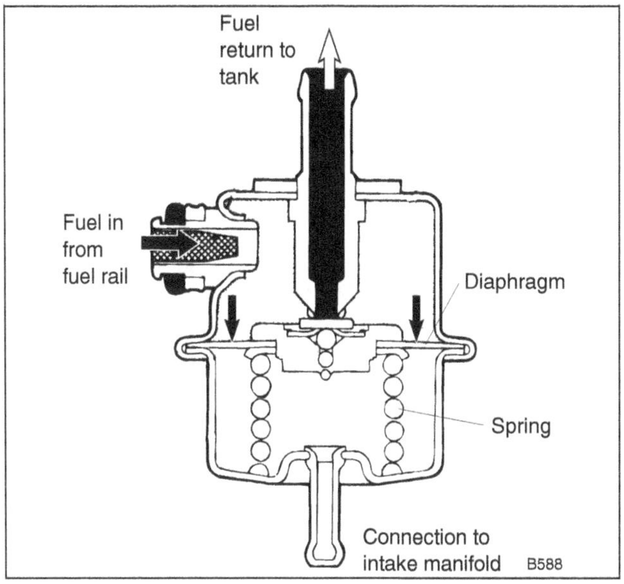

Fig. 19. Fuel pressure regulator. Fuel pressure deflects diaphragm to return fuel to tank when pressure reaches desired limit.

3. Operate fuel pump as described earlier under **Operating fuel pump for tests**. Check that fuel pressure corresponds to specifications listed in **Table c**.
   - If system pressure is low, repeat test while gradually pinching off return hose. Pressure should rise rapidly. If not, fuel pump is most likely faulty.
   - If system pressure is too high, check return line from pressure regulator to tank. Check for kinks in hose. Blow compressed air through line to check for blockages. If no faults can be found, pressure regulator is most likely faulty. **See 130 Fuel Injection** for testing and replacement procedures.

> ***WARNING —***
> *Fuel under pressure is present during the test. Use hose clamps at all connections.*

Table c. Fuel Pressure Specifications

Engine	Fuel pressure
4-cylinder M42/M44	3.0 ± 0.06 bar (43.5 ± 0.9 psi)
6-cylinder M50/S50US M52/S52US	 3.0 ± 0.06 bar (43.5 ± 0.9 psi) 3.5 ± 0.06 bar (51 ± 0.9 psi)

> ***CAUTION —***
> *Do not use compressed air above 40 psi to blow out lines or components. Internal damage to components may result.*

4. When finished, disconnect pressure gauge.

Fuel delivery volume, testing

1. Disconnect return line from fuel rail in engine compartment.

2. Connect a length of hose to fitting on fuel rail and place open end of hose in a suitable container for catching fuel (2-quart capacity).

3. Run fuel pump for exactly 30 seconds as described earlier under **Operating fuel pump for tests** and measure fuel collected. Refer to **Table d**.

Table d. Fuel Pump Delivery Specifications

Engine	Delivery rate (30 seconds @ 12V)
4-cylinder M42/M44	 0.875 liter (0.93 qt)
6-cylinder M50/S50US M52/S52US	 0.875 liter (0.93 qt) 1.12 liter (1.16 qt)

4. When finished testing, reconnect fuel line. Tighten all hose clamps.

NOTE —
It is common practice to replace the fuel filter any time the fuel pump unit is replaced.

170 Radiator and Cooling System

TABLES

GENERAL

This section covers repair and troubleshooting information for the engine cooling system. For heater core and related heating and air conditioning components, see **640 Heating and Air Conditioning**.

Coolant Pump and Thermostat

A centrifugal-type coolant pump is mounted to the front of the engine. The belt-driven pump circulates coolant through the system whenever the engine is running. A thermostat controls the flow of coolant into the radiator.

When the engine is cold the thermostat is closed so coolant bypasses the radiator, recirculating from the engine directly back to the pump. When the engine reaches operating temperature, the thermostat opens and coolant circulates through the whole system including the radiator.

Radiator and Expansion Tank

The radiator is a crossflow design. A translucent expansion tank provides for coolant expansion at higher temperatures and easy monitoring of the coolant level.

On 4-cylinder models, the radiator expansion tank is integral with the radiator. See Fig. 1.

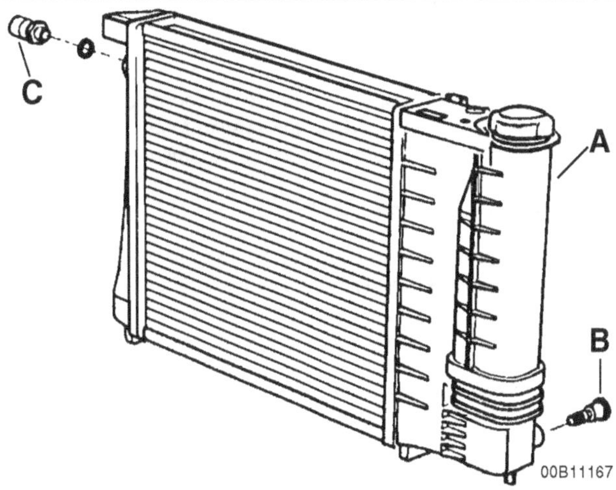

A. Expansion tank
B. Radiator drain screw
C. Dual temperature fan switch

Fig. 1. Radiator assembly with integral expansion tank (4-cylinder engine).

On 6-cylinder models, a stand alone expansion tank is used.

On cars with automatic transmission, ATF is circulated through an additional heat exchanger (ATF cooler).

Cooling Fans

Belt-driven cooling fan. The primary cooling fan is belt-driven. It is mounted to the front of the coolant pump through a fan clutch. The fan clutch is a viscous fluid coupling that controls the speed of the fan based on engine compartment temperature.

Electric cooling fan. Models with M44 engine and standard transmission substitute an electric fan for the belt-driven viscous fan. This is attached to the rear of the radiator and controlled via the DME 5.2 engine management system.

> ***NOTE —***
> *The electric cooling fan in these models is activated by the engine control module (ECM).*

Auxiliary cooling fan. In all models a two-speed electric auxiliary cooling fan is mounted behind the front grill and in front of the radiator. This fan is primarily used for the A/C system, but also operates when the coolant temperature exceeds a predetermined level.

Warnings and Cautions

The following warnings and cautions should be observed when working on the cooling system.

> ***WARNING —***
> - *At normal operating temperature the cooling system is pressurized. Allow the system to cool as long as possible before opening—a minimum of an hour—then release the cap slowly to allow safe release of pressure.*
> - *Releasing the cooling system pressure lowers the coolant's boiling point and the coolant may boil suddenly. Use heavy gloves and wear eye and face protection to guard against scalding.*
> - *Use extreme care when draining and disposing of engine coolant. Coolant is poisonous and lethal to humans and pets. Pets are attracted to coolant because of its sweet smell and taste. Consult a veterinarian immediately if coolant is ingested by an animal.*

> ***CAUTION —***
> - *Avoid adding cold water to the coolant while the engine is hot or overheated. If it is necessary to add coolant to a hot system, do so only with the engine running and coolant pump turning.*
> - *Prior to disconnecting the battery, read the battery disconnection cautions given at the front of this manual on page viii.*

TROUBLESHOOTING

Most cooling system faults can be grouped into one of three categories:

- Cooling system leaks
- Poor coolant circulation
- Radiator cooling fan faults

When investigating the cause of overheating or coolant loss, begin with a visual inspection. Be sure to check the condition and tension of the coolant pump drive belt. Check hoses for cracks or softness. Check clamps for looseness. Check the coolant level and check for evidence of coolant leaks from the engine.

Check that the radiator fins are not blocked with dirt or debris. Clean the radiator using low-pressure water or compressed air. Blow outward, from the engine side out.

Inspect the coolant pump by first removing the drive belt from the pump. Firmly grasp opposite sides of the pulley and check for play in all directions. Spin the pulley and check that the shaft runs smoothly.

> ***NOTE —***
> *The coolant provides lubrication for the pump shaft, so an occasional drop of coolant leaking from the pump is acceptable. If coolant drips steadily from the vent hole, the pump should be replaced.*

The cooling system becomes pressurized at normal operating temperature, which raises the boiling point of the coolant. Leaks may prevent the system from becoming pressurized, allowing the coolant to boil at a lower temperature. If visual evidence is inconclusive, a cooling system pressure test can help to pinpoint hard-to-find leaks.

If the cooling system is full of coolant and holds pressure, the next most probable cause of overheating are:

- Faulty radiator fan
- Loose or worn drive belt
- Failed thermostat or coolant pump
- Clogged/plugged radiator or coolant passages.

> ***NOTE —***
> - *Some early style coolant pumps were fitted with fiber/plastic type impellers. Over time, this impeller can wear away and result in overheating. The plastic impeller can also slip or free-wheel on the pump shaft. If the engine overheats and no other faults can be found, the old style impeller may be the cause of the problem.*
> - *Only pumps with the updated metal impeller should be used for replacement.*

Cooling System Pressure Test

A cooling system pressure test is used to check for internal leaks. Some of the common sources of internal leaks are a faulty cylinder head gasket, a cracked cylinder head, or a cracked cylinder block.

To do a cooling system pressure test, a special pressure tester is needed.

> ***WARNING —***
> *At normal operating temperature the cooling system is pressurized. Allow the system to cool before opening. Release the cap slowly to allow safe release of pressure.*

With the engine cold, install a pressure tester to the expansion tank. Pressurize the system to the specification listed below. Pressure should not drop more than 0.1 bar (1.45 psi) for at least two minutes. If the pressure drops rapidly and there is no sign of an external leak, the cylinder head gasket may be faulty. Consider a compression test as described in **100 Engine—General.**

The screw-on type expansion tank cap should also be tested using a pressure tester and the correct adapter.

Cooling System Test Pressure
- Radiator test pressure .... 1.5 bar (21.75 psi)
- Radiator cap test pressure........... 2 bar (29 psi)

> ***CAUTION —***
> *Exceeding the specified test pressure could damage the radiator or other system components.*

Carefully inspect the radiator cap for damage. Replace a faulty cap or a damaged cap gasket.

Thermostat Quick Check

To check if the thermostat is opening and coolant is circulating through the radiator, allow a cold engine to reach operating temperature (temperature gauge needle approximately centered). Shut off engine. Feel the top radiator hose. If the hose is hot to the touch, the coolant is probably circulating correctly. If there are any cool areas in the hose or radiator, coolant flow to the radiator is probably restricted. Check for a faulty thermostat or a plugged radiator.

> ***NOTE —***
> *A thermostat that is stuck open will cause the engine to warm up slowly and run below normal temperature at highway speed. A thermostat that is stuck closed will restrict coolant flow to the radiator and cause overheating.*

If the engine overheats and no other cooling system tests indicate trouble, the radiator may have some plugged passages that are restricting coolant flow.

Temperature Gauge Quick Check

The coolant temperature sensor is located on the intake manifold (left) side of the cylinder head, under the intake manifold runners. See Fig. 2.

0011984

Fig. 2. Temperature gauge sender on M44 engine. Temperature gauge sender location is similar on all engines.

In early models, the engine coolant temperature (ECT) sensor for the fuel injection and the coolant temperature gauge sender are located side by side.

In later models, the ECT sensor and the gauge sender are combined into one sender unit. For wire colors refer to **Table a**.

Table a. Coolant Temperature Sensor Wire Colors

Function	Sensor location	Terminal number	Wire colors
Two sensors:			
Temperature gauge sender	Rear	1 2	Brown/violet Brown/yellow
ECT sensor	Front	1 2	Brown/red Brown or Brown/black
One sensor:			
Temperature gauge sender ECT sensor	Dual sensor	1 2 3 4	Brown/yellow Brown/violet Brown/red Brown/black or Grey/black

A quick test at the coolant temperature gauge sender can determine if the gauge is functioning correctly.

If the gauge needle remains at the rest position with the engine warm, remove the harness connector from the sender and jumper the correct terminals in the connector to simulate a high engine temperature. See **Table a**. Turn the ignition on. If the gauge needle moves upward, the sender is faulty. If the gauge does not respond, the wiring to the gauge is broken (open circuit) or the gauge itself is faulty.

> ***WARNING —***
> *1996 and later models are OBD II compliant. Disconnecting electrical connectors with the ignition turned on may set fault codes in the ECM. It is recommended that you leave the diagnosis of faults in the coolant temperature sensor system to the BMW dealer service department which has specialized OBD II scan tool equipment.*

If the gauge needle reads too high when the engine is cold, remove the harness connector from the sender. Turn the ignition on. If the gauge needle position does not change, the wiring or the gauge is shorted to ground. If the gauge needle drops, the sender is faulty and should be replaced. When replacing a faulty coolant temperature sender, the gasket ring on the sender should also be replaced.

Tightening Torque
- Temperature gauge sender to engine 18 Nm (13 ft-lb)

Cooling fan, testing

> ***NOTE —***
> *On M44 engines with manual transmission, the primary electric cooling fan is mounted on the engine side of the radiator and is controlled by the engine control module (ECM). Troubleshooting this circuit should be left to an authorized BMW dealer with the proper diagnostic equipment.*

An otherwise sound cooling system may still overheat, particularly with prolonged idling, due to a failure of the cooling fan(s).

The belt-driven cooling fan is controlled by a temperature dependent viscous clutch. A failed fan clutch may affect air flow through the radiator resulting in overheating or possibly overcooling.

With the engine off, check the fan clutch by spinning the fan. The fan should spin on the clutch with some resistance. Check for signs of leaking fluid from the clutch. If the fan freewheels with no resistance, cannot be turned by hand, or there are signs of oil leakage, the clutch should be replaced.

The auxiliary cooling fan comes on when coolant temperature exceeds a predetermined level or whenever the air conditioning is on. A dual-range temperature switch for cooling fan control is mounted on the right side of the radiator. See Fig. 3.

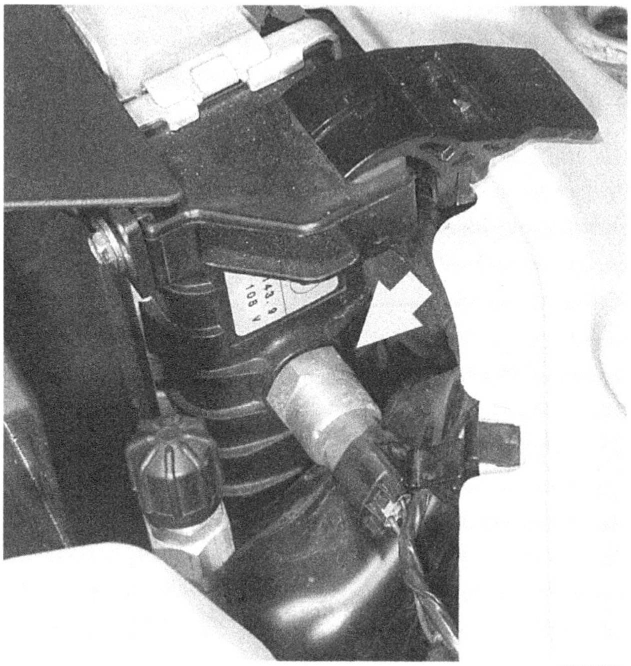

0012506

Fig. 3. Radiator cooling fan temperature switch (**arrow**).

> ***WARNING —***
> - *Use caution when testing the electric cooling fan(s) and coolant temperature switch. Keep hands and wires clear of the fan blades. The cooling fan(s) can run any time the ignition is ON.*
> - *For greatest safety, cooling fan and coolant temperature switch tests should be performed on a cold engine with the air conditioning off.*

If a faulty thermostat, trapped air, or a restriction in the system is not allowing the coolant to circulate through the radiator, the temperature switch will not close and the auxiliary cooling fan will not run. Before making the tests described below, make sure the thermostat is operating correctly as described earlier. The normal switching temperatures for the dual -speed switch are listed in **Table b**.

Table b. Auxiliary Cooling Fan Switching Temperatures

Speed	Switching temperature
Low speed	196°F (91°C)
High speed	210°F (99°C)

NOTE —
Some cars covered by this manual may have an alternate cooling fan switch with switching temperatures of 176°/190°F (80°/88°C). When replacing the switch check the switching specifications, which should be stamped on the switch body.

1. If coolant is circulating at normal operating temperature, but auxiliary cooling fan does not run, disconnect connector from radiator temperature switch and make tests listed in **Table c**.

Table c. Auxiliary Cooling Fan Temperature Switch Tests

Wires jumpered	Test conditions	Test results
Black/green (terminal 2) and brown (terminal 1)	Ignition ON	Fan runs on low speed
Black/gray (terminal 3) and brown (terminal 1)	Ignition ON	Fan runs on high speed

2. If fan runs only when powered directly by jumpered connector and hot coolant is circulating through radiator, radiator temperature switch is most likely faulty. Use a new sealing ring when replacing switch

Tightening Torque
- Temperature switch to radiator. . . . . . 15 Nm (11 ft-lb)

3. If auxiliary fan does not run when powered directly, check for battery voltage at temperature switch connector (black/green wire) with ignition on. If battery voltage is not present, check fuses. See **610 Electrical Component Locations**.

Auxiliary Cooling Fan Circuit Fuses
- Fuse 16 . . . . 5 amp
- Fuse 41 (ex. M44 w/man. trans) . . . . 30 amp
- Fuse 48 (M44 w/man. trans. only) . . . . 40 amp

4. If no faults are found, remove low speed relay and turn ignition ON. See Fig. 4.
   - Check for power at terminal 30 and terminal 86 of relay socket.
   - Reinstall low speed relay and repeat test at high speed relay socket. Fix any wiring faults found.

On early production cars (up to 9/92): If fan operates only on high speed and no electrical faults have been found up to this point, use an ohmmeter to check that fan resistor is not electrically open. Resistor is mounted on auxiliary cooling fan housing behind front grille. See Fig. 5. Wiring diagrams for the radiator cooling fan can be found under **Electrical Wiring Diagrams**.

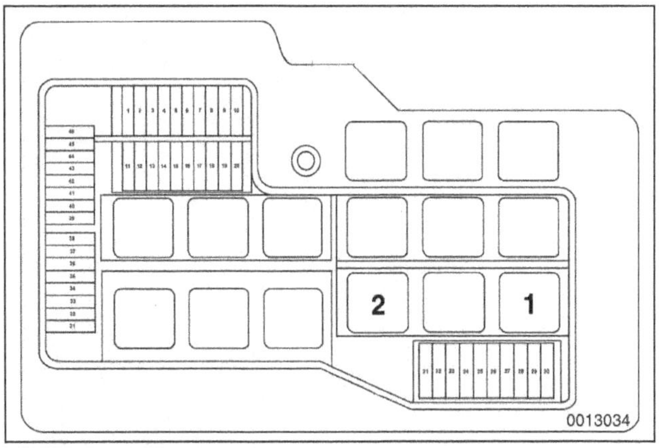

Fig. 4. Auxiliary radiator cooling fan low speed relay (**1**) and high speed relay (**2**) in power distribution box. (Relay locations may vary.)

Fig. 5. Auxiliary radiator fan resistor (**arrow**).

COOLING SYSTEM SERVICE

Coolant, draining and filling

1. Remove expansion tank cap. Set temperature controls to full warm.

WARNING —
Allow the cooling system to cool before opening or draining the cooling system.

2. Place 3-gallon pail beneath radiator drain plug and remove drain plug. See Fig. 6.

3. Place 3-gallon pail beneath rear of engine block. Loosen and remove engine block coolant drain plug.

Fig. 6. Radiator drain plug (**arrow**).

> ***WARNING —***
> *Coolant is poisonous. It is especially lethal to pets. Clean up spills immediately and rinse the area with water.*

NOTE —
The block drain plug is located on the exhaust side of the engine, near the rear of the engine.

4. Reinstall radiator and engine block drain plugs using new sealing washers. Leave heater controls on full warm.

5. Using a coolant mixture of 50% antifreeze and 50% distilled water, fill system slowly. On radiator with integral expansion tank, bleed cooling system as described below.

NOTE —
- *Tap water may cause corrosion of radiator, engine and coolant hoses.*
- *Coolant can often be reused provided it is clean and less than two years old. Do not reuse coolant when replacing damaged engine parts. Contaminated coolant may damage the engine or cooling system.*

Table d. Cooling System Capacities

Engine	Capacity
4-cylinder	6.5 liters (6.9 qt)
6-cylinder M50/M52 S50US/S52US	 10 liters (10.6 qt) 10.5 liters (11.1 qt)

Tightening Torques
- Radiator drain plug to radiator . . .2-3 Nm (18-27 in-lb)
- Engine block drain plug to block . . . . 25 Nm (18 ft-lb)

Cooling system, bleeding

Air may become trapped in the system during filling. Trapped air can prevent proper coolant circulation and overheating. Whenever the coolant is drained and filled, the system should be bled of trapped air.

> ***CAUTION —***
> *Always use genuine BMW coolant or its equivalent to avoid the formation of harmful, clogging deposits in the cooling system. Use of other antifreeze solutions may be harmful to the cooling system.*

1. With engine cold, loosen plastic bleed screw on coolant expansion tank. See Fig. 7.

NOTE —
- *On M3 models, the plastic bleed screw is located on the external expansion tank.*
- *On some models, the expansion tank may not have a plastic bleed screw. Check for a bleed screw on the thermostat housing and bleed the system by loosening screw as described here. If no bleed screw is available, the system is designed to be self bleeding via the vent hose on the expansion tank.*

Fig. 7. Cooling system bleed screw on radiator (**arrow**). Note cold level mark on expansion tank.

2. Set temperature controls in passenger compartment to full warm and turn ignition to on position (do *not* start engine).

3. Slowly add coolant until it spills from bleed screws. When coolant spilling from bleed screws is free of air bubbles, tighten screw.

NOTE —
Coolant level should be at or slightly above COLD (KALT) mark on tank.

4. Run engine until it reaches operating temperature. Run heater blower and check that hot air is present at vents (indicates coolant is circulating through system). Repeat bleeding procedure if necessary.

5. After engine has cooled, recheck coolant level and top up until level reaches COLD (KALT) mark on tank.

Tightening Torque
- Bleed screw (plastic) to expansion tank. 2 Nm (1.5 ft-lb)
- Bleed screw (metal) to thermostat housing) . . . . 8 Nm (71 in-lb)

Belt-driven cooling fan, replacing

1. Using a 32 mm wrench on fan clutch nut, turn wrench quickly in a clockwise direction (working from front of car) to loosen. Spin fan off pump. See Fig. 8.

NOTE —
- *The radiator cooling fan nut (32 mm wrench) has left-hand threads.*
- *The nut may be difficult to loosen. Use a tool to hold the coolant pump pulley stationary. BMW has a special tool for this purpose (BMW special tool no. 11 5 030).*

2. Remove expansion rivets holding shroud to radiator. See Fig. 9. Remove fan and shroud together.

NOTE —
Store the removed fan clutch assembly in an upright (installed) position to prevent loss of clutch fluid.

3. To replace fan clutch, remove fan mounting bolts and separate clutch from fan.

4. Installation is reverse of removal.

Fig. 8. Radiator cooling fan nut (**arrow**). Nut has left-hand threads.

0013203

Fig. 9. Fan shroud retaining rivet (**arrow**). Pry out center pin and remove rivet. Rivet design may vary depending on model and model year.

Tightening Torques
- Clutch nut to coolant pump (left-hand threads)
  Without BMW tool no. 11 5 040 . . . . . 40 Nm (29 ft-lb)
  With BMW tool no. 11 5 040. . . . . 30 Nm (22 ft-lb)
- Fan to viscous clutch . . . . 10 Nm (89 in-lb)

Electric cooling fan, replacing

On late 4-cylinder models (M44 engine) with manual transmission, the primary cooling fan is electrically operated and is mounted on the engine side of the radiator.

1. If necessary, remove cover from top of radiator. See Fig. 10.

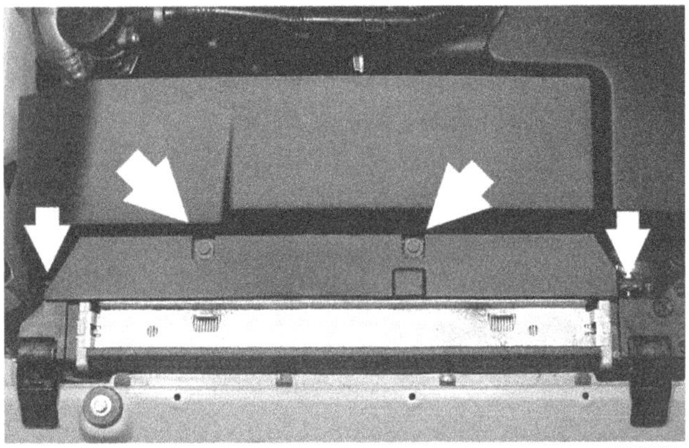

Fig. 10. Air cover in back of radiator. Unclip cover from fan shroud after removing mounting screws.

2. Disconnect fan harness connector at bottom left of radiator.

3. Remove fan shroud mounting screws at left and right side of radiator. Lift fan assembly straight up and off radiator.

4. Installation is reverse of removal. Be sure to align side tabs on fan housing with retaining tabs on radiator when reinstalling fan.

Auxiliary cooling fan, replacing

The auxiliary electric cooling fan is mounted behind the front bumper, in front of the A/C condenser. There are two versions of auxiliary fans installed, depending on model and engine installed.

▸Type 1 (up to 9/92)

1. Remove front radiator air shroud. See Fig. 11.

2. Remove fan mounting screws. See Fig. 12.

3. Tilt fan forward. Disconnect electrical harness connector before lifting fan up.

Fig. 11. Front radiator air shroud fasteners (**arrows**).

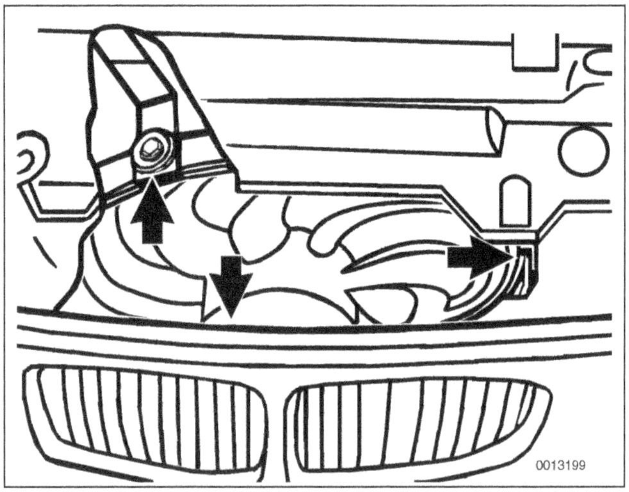

Fig. 12. Version 1 auxiliary fan mounting screws (**arrows**).

> ***NOTE —***
> *If necessary, remove front bumper or lower engine cover to access electrical harness connector.*

4. Installation is reverse of removal.

▸ Type 2 (from 9/92)

1. Remove front bumper and radiator grilles. See **510 Exterior Trim, Bumpers**.

2. Remove fan mounting screws. See Fig. 13.

3. Angle fan up and forward, disconnecting electrical harness connector behind it. Remove it through bumper opening.

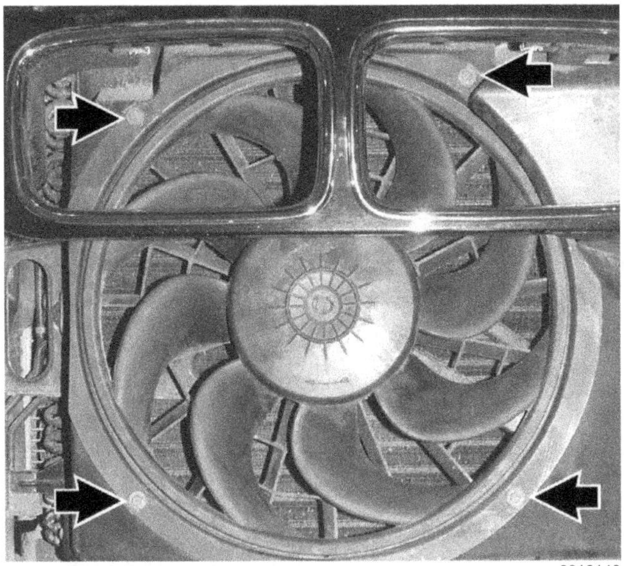

Fig. 13. Version 2 auxiliary fan mounting screws (**arrows**). Front bumper has been removed.

4. Installation is reverse of removal.

Thermostat, replacing

The coolant thermostat is installed in a housing at the front of the cylinder head. See Fig. 14.

Fig. 14. Coolant thermostat and housing.

> ***NOTE—***
> *On M44 engines, the thermostat is integrated into the thermostat housing. If the thermostat requires replacement, the complete thermostat housing must be replaced.*

1. Drain radiator as described above under **Coolant, draining and filling**.

> ***WARNING—***
> *Allow the cooling system to cool before opening or draining the system.*

2. Remove belt-driven cooling fan and fan shroud as described earlier.
3. On M52/S52US engines: Unclip wiring duct retaining clips and remove duct from above thermostat housing. Also remove engine front lifting bracket.
4. Remove thermostat housing from front of engine. See Fig. 15.

Fig. 15. Thermostat housing shown removed (6-cylinder engine).

5. Remove thermostat. Note thermostat direction and orientation before removing.
6. Install new thermostat and/or thermostat housing using new O-ring and gasket(s) as necessary.

> ***NOTE—***
> *Some thermostats have a direction arrow or vent hole near the edge. Install the thermostat so that either the arrow or hole are at the top.*

7. Installation is reverse of removal. Fill system with coolant as described under **Coolant, draining and filling.**

> ***CAUTION—***
> *Be sure to reconnect ground wire(s) at thermostat housing mounting bolt, where applicable.*

Tightening Torques

- Thermostat housing cover to housing or coolant pump . . . . 10 Nm (89 in-lb)
- Engine lifting bracket bolt to thermostat housing (M52/S52US engine)(M8) . . . . 22 Nm (17 ft-lb)

Coolant pump, replacing

The engine coolant pump is mounted in the front of the engine on the timing cover. See Fig. 16.

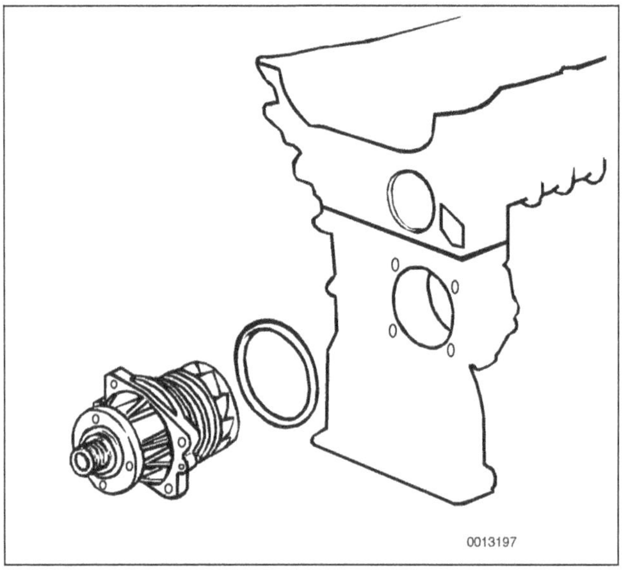

Fig. 16. Coolant pump mounted on timing cover.

NOTE —
Only replacement coolant pumps with metal impellers should be installed. Pumps with plastic impellers should not be used.

1. Drain cooling system as described earlier.

WARNING —
Allow cooling system to cool before opening or draining system.

2. Remove primary cooling fan shroud and cooling fan (belt-driven or electric) as described earlier.

3. Disconnect hoses from thermostat housing. Unbolt thermostat housing from front of cylinder head.

4. Remove coolant pump drive belt. See Fig. 17 or Fig. 18.

NOTE —
- *Mark direction of drive belt rotation if reusing belt.*
- *Several drive belt and tensioner configurations have been used in E36 models. Refer to* **020 Maintenance Program** *and* **640 Heating and Air Conditioning**.

5. Remove coolant pump pulley from pump.

6. Remove mounting fasteners from pump.

Fig. 17. Engine drive belt tensioner being released on M44 engine. Pry off tensioner cover and then lever tensioner in counterclockwise direction (as facing engine) and slip belt off pulleys.

Fig. 18. Drive belt tensioner being released on M52 engine. Pry off tensioner cover and then lever tensioner clockwise (as facing engine) to release belt tension.

NOTE —
- *The 4-cylinder coolant pump is mounted using two different length bolts. Note where the longer bolt fits during disassembly.*
- *The 6-cylinder coolant pump is mounted on studs and retained by nuts.*

7. Insert two screws (M6) in tapped bores and tighten uniformly until pump is free from timing chain cover. See Fig. 19.

Fig. 19. Coolant pump being removed. Thread two M6 bolts (**arrows**) in evenly to withdraw pump. (Thermostat and hoses have been removed for visual access.)

8. Installation is reverse of removal.
   - Be sure to replace O-ring and gaskets.
   - Coat O-ring with lubricant during installation.

Tightening Torque
- Coolant pump to timing chain cover
  M6 . . . . 10 Nm (89 in-lb)
  M8 . . . . 22 Nm (17 ft-lb)
- Coolant pump pulley
  to coolant pump . . . . 10 Nm (89 in-lb)

RADIATOR SERVICE

Radiator, removing and installing

1. Drain radiator as described earlier.

> ***WARNING —***
> *Allow cooling system to cool before opening or draining system.*

2. Remove primary cooling fan (belt-driven or electric) as described earlier.

3. Disconnect cooling fan and level sensor harness connectors from bottom of radiator, where applicable

4. Disconnect harness connector from auxiliary fan dual temperature switch. See Fig. 20.

5. Disconnect all coolant hoses from radiator.

Fig. 20. Cooling fan dual temperature switch (**arrow**) on right side of radiator.

6. Where applicable, disconnect automatic transmission fluid (ATF) cooler lines from radiator. See Fig. 21.

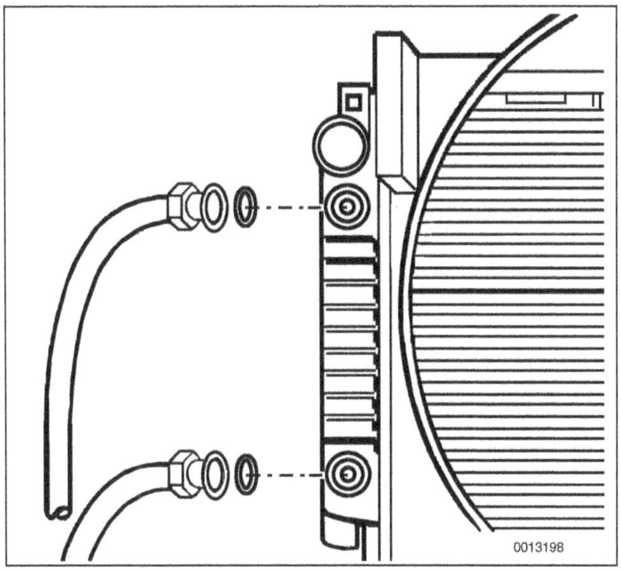

Fig. 21. Automatic transmission fluid (ATF) lines at radiator.

7. Carefully pry out radiator retaining clips from top of radiator. See Fig. 22.

8. Pull radiator up and out of car.

> ***NOTE —***
> *The radiator rests on two rubber mounts. Check that the mounts do not stick to the bottom of the radiator.*

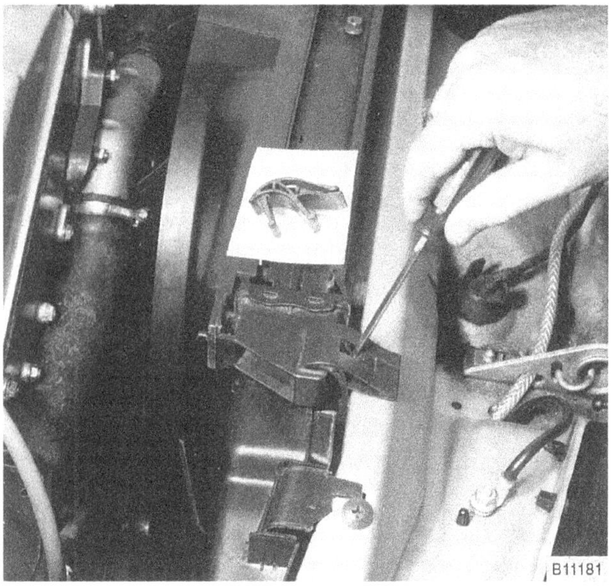

Fig. 22. Radiator retaining clip being removed. Push down and pull screwdriver forward to release clip.

9. Installation is reverse of removal.
   - Fill radiator and cooling system as described under **Coolant, draining and filling.**
   - Check ATF level and, if necessary, top up. See **240 Automatic Transmission**.

180 Exhaust System

GENERAL

The exhaust system is designed to be maintenance free, although regular inspection is warranted due to the harsh operating conditions. Under normal conditions, the catalytic converter does not require replacement unless it is damaged.

On 1996 and later cars, a pre- and post-catalytic converter oxygen sensor is used at each catalytic convertor. Be sure to take care when removing the system not to damage the sensors.

See **130 Fuel Injection** for information on testing oxygen sensors.

> ***NOTE —***
> *On M52 engines, two resonator pipes (short and long) are incorporated into the rear muffler assembly. See Fig. 1. An exhaust flap is integrated into the outlet side of the short pipe. The flap is operated by a vacuum actuator via the Siemens MS 41.1 engine control module). The flap is closed at engine speeds below 2,500 rpm. This allows exhaust gasses to flow through the long resonator pipe and reduce noise. Above 2,500 rpm, the exhaust flap is open (no vacuum applied to the actuator) and allows exhaust gasses to flow through the short pipe for maximum performance.*

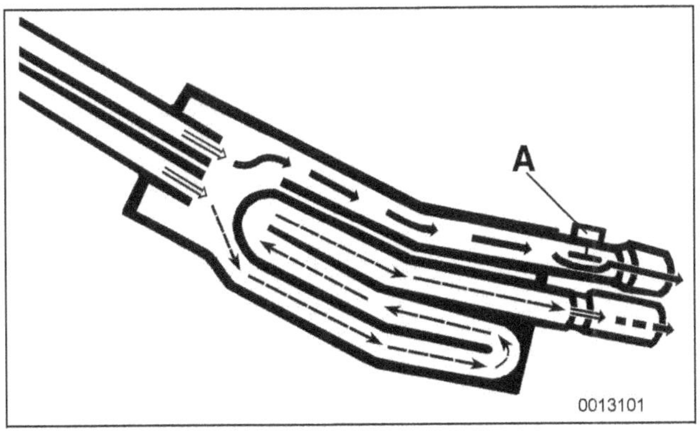

Fig. 1. Rear muffler (active silencer used on M52 engines). Vacuum operated flap (**A**) changes between short and long pipes, depending on engine speed.

> ***WARNING —***
> *Exhaust gases are colorless, odorless, and very toxic. Run the engine only in a well-ventilated area. Immediately repair any leaks in the exhaust system or structural damage to the car body that might allow exhaust gases to enter the passenger compartment.*

EXHAUST SYSTEM REPLACEMENT

Exhaust system components are detailed in Fig. 2 through Fig. 4.

New fasteners, clamps, rubber mounts, and gaskets should be used when replacing exhaust components. A liberal application of penetrating oil to the exhaust system nuts and bolts in advance may make removal easier.

> ***WARNING —***
> - *The exhaust system and catalytic converter operate at high temperatures. Allow components to cool before servicing. Wear protective clothing to prevent burns. Do not use flammable chemicals near a hot catalytic converter.*
> - *Old, corroded exhaust system components crumble easily and often have exposed sharp edges. To avoid injury, wear eye protection and heavy gloves when working with old exhaust parts.*

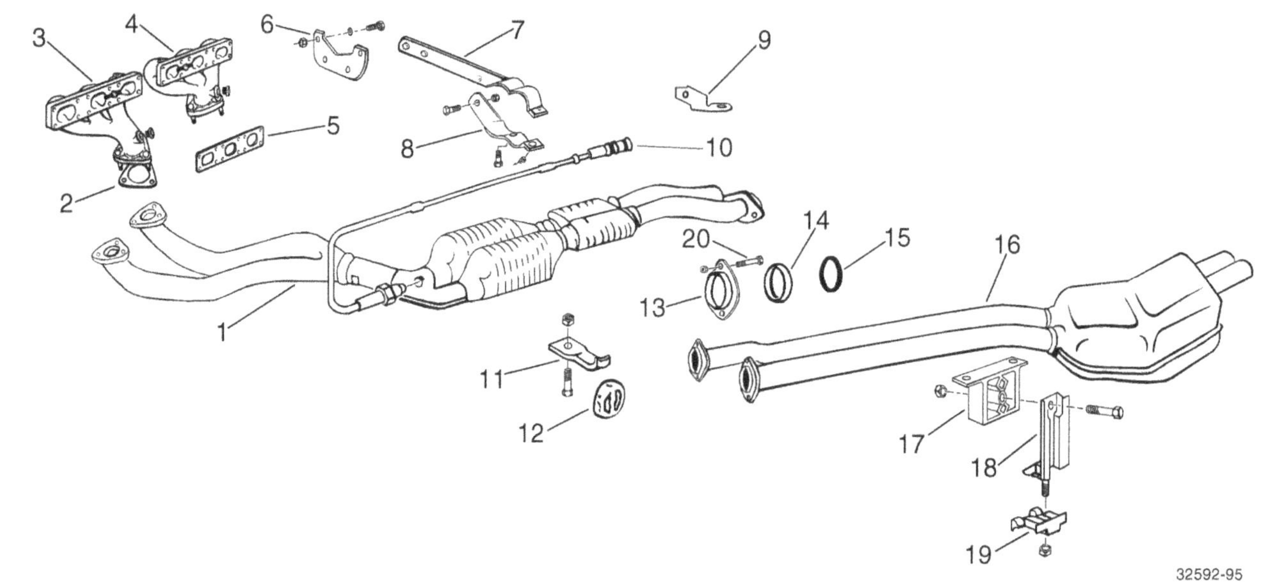

1. Front pipe with catalytic convertor
2. Gasket (2)
3. Exhaust manifold
4. Exhaust manifold
5. Manifold gasket
6. Bracket
7. Bracket (2)
8. Bracket
9. Bracket
10. Oxygen sensor
11. Bracket/clamp (2)
12. Rubber mounting ring (2)
13. Flange
14. Gasket (48mm)
15. Gasket (48mm)
16. Rear pipe with muffler
17. Rubber mount
18. Support
19. Bracket

Fig. 2. Exhaust system on 1992–1995 6-cylinder engines. 325i (M50) exhaust system shown.

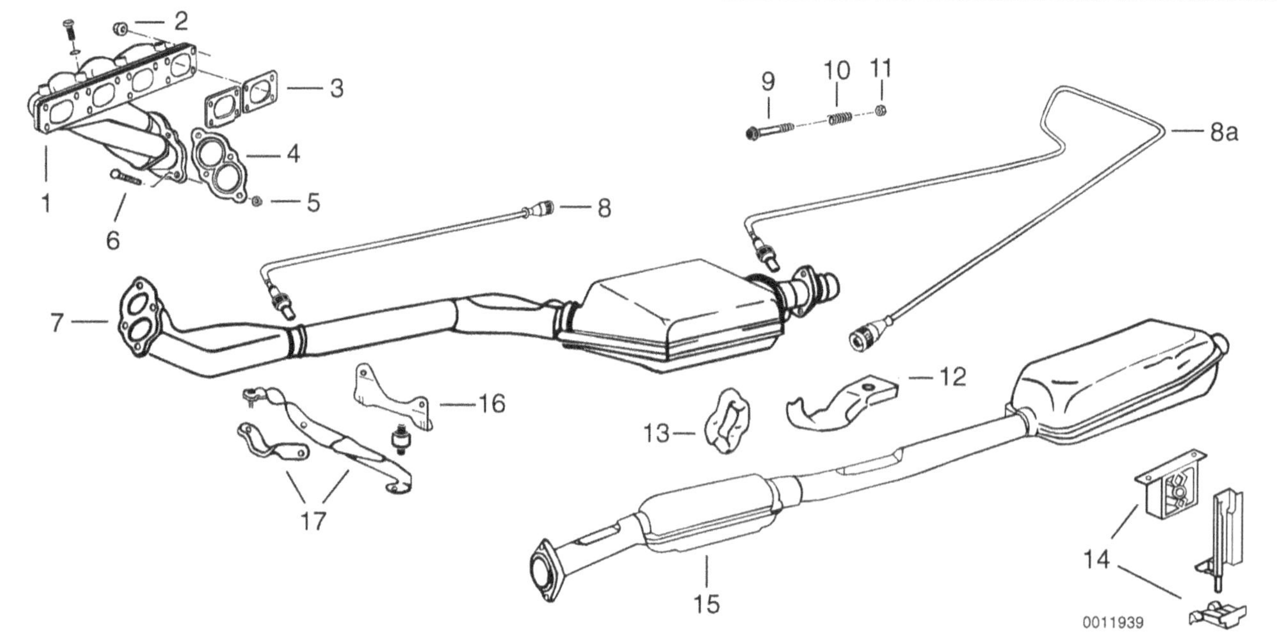

1. Exhaust manifold
2. Nut (M10)
3. Gasket (qty. 2)
4. Gasket
5. Nut
6. Bolt (M8X55)
7. Front pipe with catalytic convertor
8. Oxygen sensor (regulating) –tighten to 55 Nm (41 ft-lb)

8a. Oxygen sensor (monitoring) –tighten to 55 Nm (41 ft-lb)

9. Bolt (M8X90)
10. Compression spring
11. Self-locking nut (M8 -always replace
12. Bracket (rear pipe)
13. Rubber mounting ring (qty. 2)
14. Brackets (rear muffler)
15. Rear muffler and pipe
16. Bracket (transmission)
17. Brackets (transmission)

Fig. 3. Exhaust system on 1996–1998 4-cylinder (M44) engine.1992-1995 4-cylinder engine exhaust system is similar.

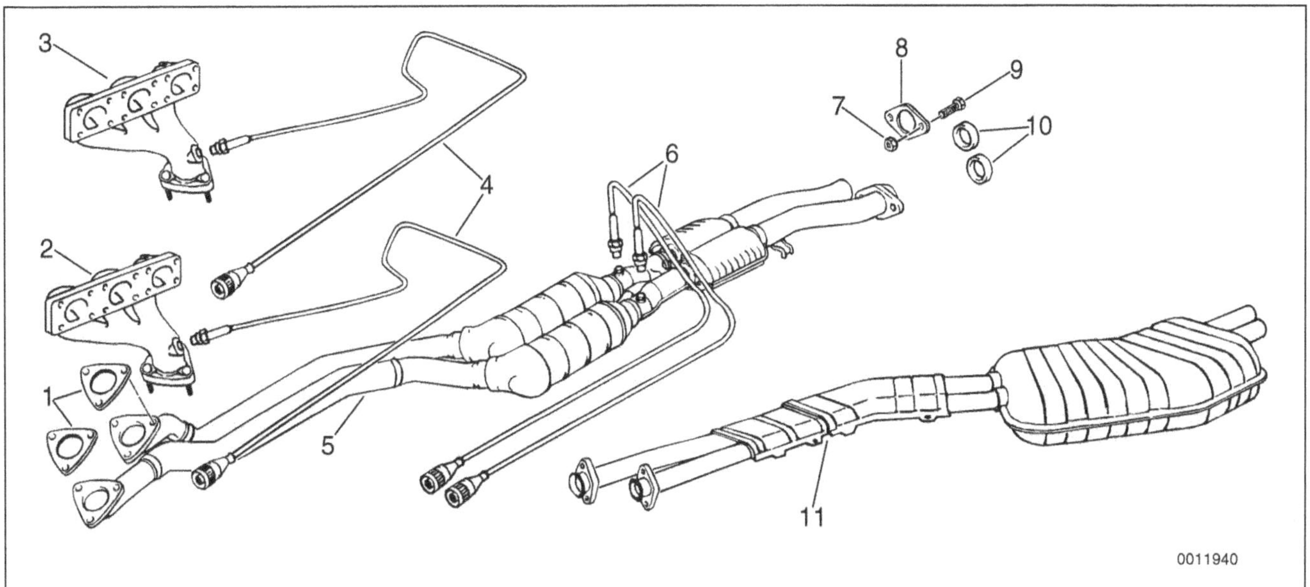

1. Gaskets
2. Exhaust manifold
3. Exhaust manifold
4. Oxygen sensor (regulating)
5. Front pipe with catalytic converter
6. Oxygen sensor (monitoring)
7. Nut (M8)
8. Flange
9. Bolt (M8x55)
10. Rubber mounting ring (48mm)
11. Rear pipe with mufflers

Fig. 4. Exhaust system on 1996–1998 6-cylinder engines. M3 (S50US) exhaust system shown.

Exhaust system, removing and installing

This section covers removing the exhaust system as a complete unit. Once the system is removed from the car, individual pipes and mufflers can be more easily replaced.

1. With exhaust system fully cold, raise and support car for access to exhaust system.

> ***WARNING —***
> *Do not work under a lifted car unless it is solidly supported on jack stands designed for that purpose. Never work under a car that is supported solely by a jack.*

2. Disconnect oxygen sensor connector(s).

NOTE —
On 1996 and later cars with multiple oxygen sensors, label the oxygen sensor connectors before disconnecting.

3. Loosen and remove bolts holding front exhaust pipes to exhaust manifolds.

4. Where applicable, remove support brace from transmission. See Fig. 5.

NOTE —
The automatic transmission will be supported by the rear crossmember once the brace is removed.

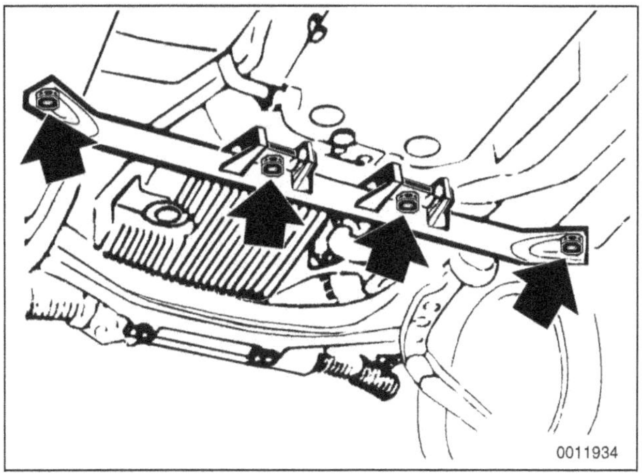

Fig. 5. Cross brace mounting bolts (**arrows**).

5. Disconnect exhaust support bracket assembly from transmission. See Fig. 6.

6. Support exhaust system from below and disconnect rubber supports/rubber rings from exhaust system. Remove exhaust system from below. See Fig. 7.

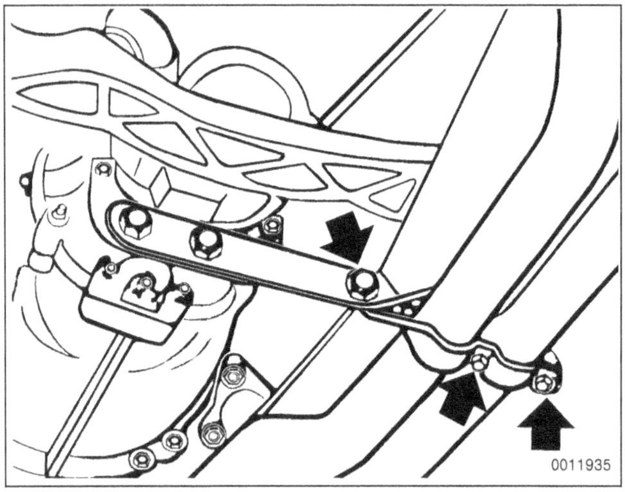

Fig. 6. Exhaust system support bracket on 6-cylinder engine. Remove bolts (**arrows**) to separate from exhaust pipe.

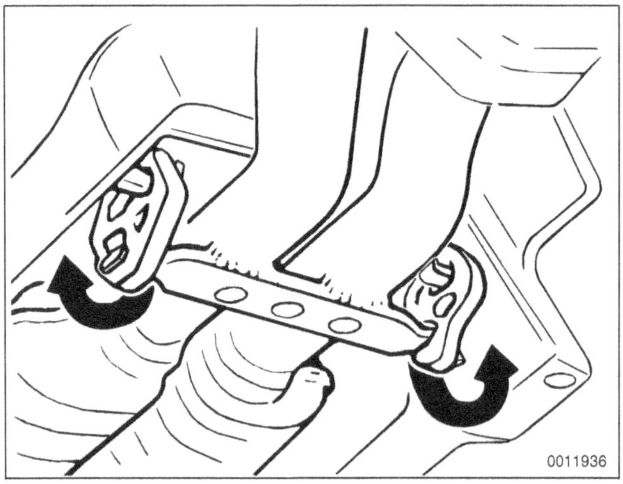

Fig. 7. Rubber exhaust system hangers. Remove hanger by prying off bracket (**arrows**).

7. Installation is reverse of removal.
   - Where necessary, transfer parts from old system to replacement system.
   - Coat manifold studs with copper paste before installing nuts.
   - Where applicable, tighten front pipe with semi-flexible compensator last when installing front pipe to manifolds.
   - Make sure there is at least 20 mm (¾ in.) of clearance between exhaust system and car body at every point.
   - Loosely install all exhaust system mounting hardware and hangers before tightening fasteners to their final torque.
   - On 4-cylinder engines: Tighten center mounting flange nuts evenly until correct dimension is obtained. See Fig. 8.

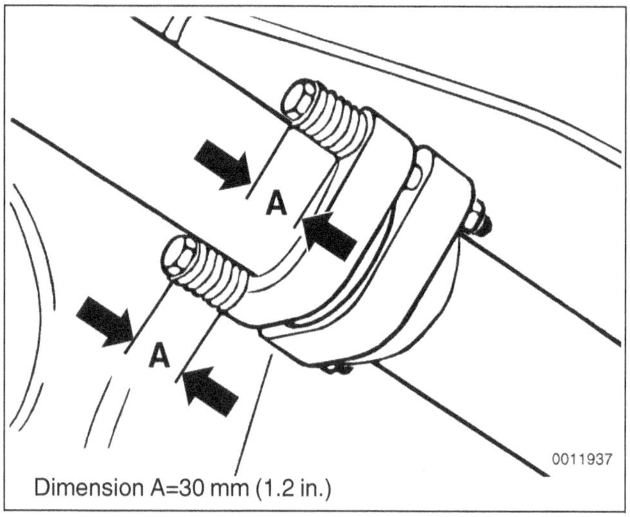

Fig. 8. On 4-cylinder engines, tighten nuts evenly so that springs are preloaded to **Dimension A**.

NOTE —
To prevent exhaust system rattles and vibration, preload the rear clamping brackets in the forward direction. See Fig. 9.

Tightening Torques
- Cross brace to chassis . . . . 21 Nm (17 ft-lb)
- Exhaust support bracket
  to transmission . . . . 21 Nm (17 ft-lb)
- Exhaust system mounting clamps . . . 15 Nm (11 ft-lb)
- Front exhaust pipe to exhaust manifold
  M10 nut . . . . 30 Nm (22 ft-lb)
- Front exhaust pipe to rear pipe
  M8 bolt/nut . . . . 22 Nm (16 ft-lb)
  with compression spring. . . . . .See Fig. 8
- Oxygen sensor to exhaust pipe . . . . . 55 Nm (41 ft-lb)

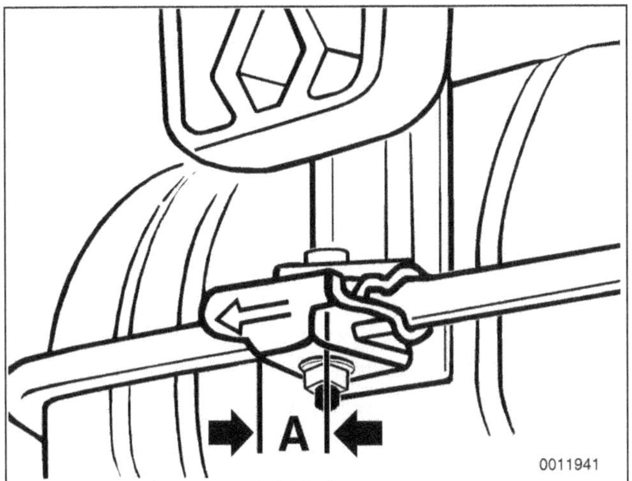

Dimension A=15mm (0.60 in.)

Fig. 9. Preload rear clamping bracket as specified to help prevent system rattling.

Exhaust manifolds, removing and installing

Always use new retaining nuts and gaskets when removing and installing the exhaust manifolds.

1. With exhaust system cold, raise and support car for access to exhaust system.

> ***WARNING —***
> *Do not work under a lifted car unless it is solidly supported on jack stands designed for that purpose. Never work under a car that is supported solely by a jack.*

2. On 1996 and later 6-cylinder engines, disconnect the secondary air check valve and pipe from the exhaust manifold.
3. Unbolt front exhaust pipe(s) from exhaust manifold(s).
4. Remove exhaust support bracket from transmission.
5. Loosen and remove nuts from exhaust manifolds and remove manifolds. Discard nuts and gaskets.

> ***NOTE —***
> *On cars with oxygen sensors mounted in the exhaust manifolds, use care when removing manifolds. To be safe, remove the sensors from the manifolds before removing manifolds.*

6. Installation is reverse of removal.
   - Coat exhaust manifold mounting studs with copper paste before installing new nuts.
   - Install manifold gaskets with arched side facing manifolds.

Tightening Torques
- Exhaust manifold to cylinder head
  4-cylinder engine (M7 nut) .... 15 Nm (11 ft-lb)
  6-cylinder engine (M7 nut) .... 20 Nm (15 ft-lb)

200 Transmission–General

TABLES

GENERAL

E36 models are equipped with a longitudinal drivetrain. The transmission is bolted directly to the rear of the engine. A driveshaft connects the output shaft of the transmission to the final drive. The final drive is mounted to the rear suspension and the body. Individual drive axles with integrated constant velocity joints transfer rotational power to the rear wheels.

Manual Transmission

Due to different power characteristics and performance requirements, four different manual transmissions are used in the models covered by this manual. Manual transmission applications are given in **Table a**.

For transmission gear ratio information and repair information, see **230 Manual Transmission**.

E36 cars with standard transmission use a single-disc clutch with dual-mass flywheel. For further information, see **210 Clutch**.

Table a. Manual Transmission Applications

Model	Year	Engine	Transmission
318i/is/iC	1992-1995	M42	Getrag S5D 200 G*
	1996-1998	M44	Getrag S5D 250 G
323is/iC	1998	M52	Getrag S5D 250 G
325i/is/iC	1992-1995	M50	Getrag S5D 250 G
328i/is/iC	1996-1998	M52	ZF S5D320 Z
M3	1995	S50US	ZF S5D310 Z
	1996-1998	S52US	ZF S5D320 Z

*The 1992 M42 engine may be fitted with either the Getrag S5D 200 G transmission or S5D 250 Getrag. The S5D 200G was discontinued in production in as of 9/92. This transmission is interchangeable with Getrag S5D 250 G.

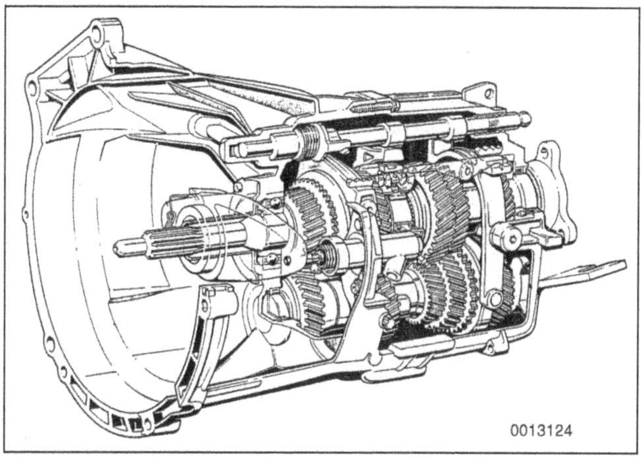

Fig. 1. ZF-manufactured manual transmission.

Automatic Transmission

Four different automatic transmissions are used, depending on model and model year and/or production date. All of the automatic transmissions are electrohydraulically controlled with either four or five forward speeds. Automatic transmission applications are given in **Table b**.

For automatic transmission repair information, see **240 Automatic Transmission**.

Table b. Automatic Transmission Applications

Model	Year	Engine	Transmission
318i/is/iC **323is/iC** **325i/is/iC**	1992-1995 1998 1992-1995	M42 M52 M50	A4S 310 R
318i/is/iC **328i/is/iC**	1996-1998 1996-1998	M44 M52	A4S 270 R
M3	1996-1998	S50US S52US	A5S 310 Z

Transmission ID Codes

On both manual and automatic transmissions, the code letter and transmission type numbers should be located on an ID plate attached to the side of the transmission. Do not rely on numbers cast on the transmission case for identification.

In addition, the transmission lubricant varies between transmissions and model years. Fluid application information should be clearly marked on a label on the transmission.

On manual transmissions, either ATF (orange label reads ATF-Oil!) or a special BMW lubricant (yellow label reads MTF-LT 1) are used. See **230 Manual Transmission**.

Automatic transmissions are filled with either Dexron III ATF (black trans. type plate reads ATF-Oil!), a special life-time oil (green trans. type plate reading Life-Time Öl). See **240 Automatic Transmission**.

> ***CAUTION —***
> *If in doubt as to the type of fluid installed in a particular transmission, consult an authorized BMW dealer service department for the latest in operating fluids.*

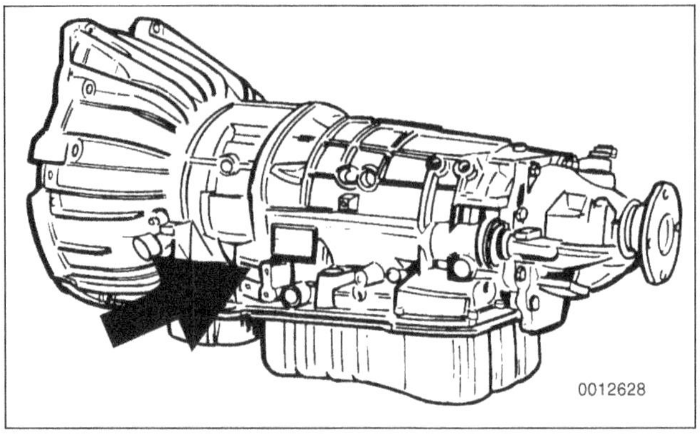

Fig. 2. Typical location of ID plate on automatic transmission housing (**arrow**). A4S 310R (THM-R1)

210 Clutch

GENERAL

This repair group covers replacement of the clutch mechanical and hydraulic components. Special tools may be required for some of the procedures. Read the procedure through before beginning a job.

> ***NOTE —***
> *See* **200 Transmission–General** *for transmission application information.*

The major components of the clutch system are shown in Fig. 1.

CLUTCH HYDRAULICS

The clutch is hydraulically actuated by the master and slave cylinders. Clutch disc wear is automatically taken up by the pushrod travel of the slave cylinder, making periodic adjustments unnecessary.

A soft or spongy feel to the clutch pedal, long pedal free-play, or grinding noises from the gears while shifting can all indicate problems with the clutch hydraulics. In these circumstances it is best to start with a clutch fluid flush, followed, if necessary, by replacement of the hydraulic parts.

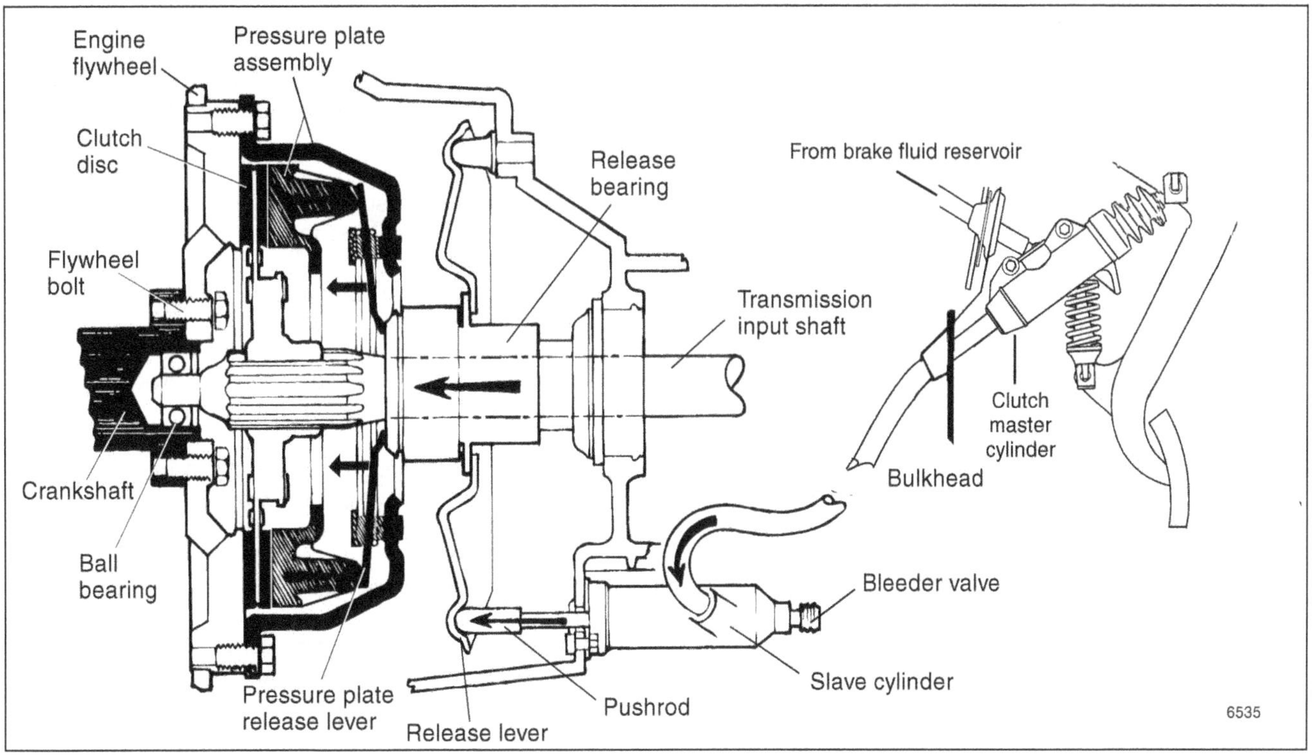

Fig. 1. Clutch assembly and hydraulics.

Clutch hydraulic system, bleeding and flushing

If the clutch/brake fluid is murky or muddy, or has not been changed within the last two years, the system should be flushed. Flushing the old fluid from the clutch lines is done using a brake system pressure bleeder.

1. Remove brake fluid reservoir cap. Using a clean syringe, remove brake fluid from reservoir. Refill reservoir with clean fluid.
2. Attach pressure brake bleeder to fluid reservoir.

> ***CAUTION —***
> *Do not exceed 2 bar (29 psi) pressure at the fluid reservoir when bleeding or flushing the hydraulic system.*

3. Connect a length of hose from clutch slave cylinder bleeder valve to a container. See Fig. 2.

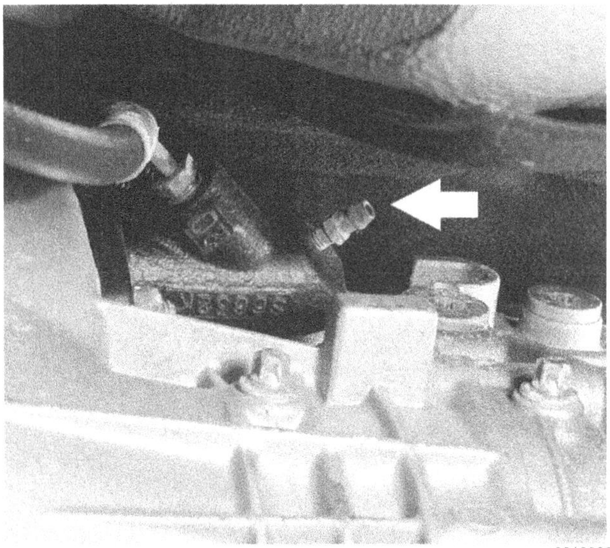

0012003

Fig. 2. Clutch slave cylinder bleeder valve on left side of transmission (**arrow**).

4. Open bleeder valve and allow brake fluid to expel until clean fluid comes out free of air bubbles.
5. Close bleeder valve and disconnect pressure bleeding equipment from fluid reservoir. Hose on bleeder valve remains connected.
6. Slowly operate clutch pedal about 10 times. Fill reservoir with clean fluid as necessary.
7. Unbolt slave cylinder from transmission.
8. Position slave cylinder so that bleeder valve is facing up (highest point).
9. Open bleeder valve and slowly push slave cylinder pushrod all the way in. Close bleeder valve and slowly release pushrod. Repeat procedure until fluid runs out clear and without bubbles.
10. Disconnect bleeder hose. Install slave cylinder to transmission. Add clean brake fluid to reservoir as necessary. Check clutch operation.

Tightening Torques
- Clutch slave cylinder to transmission . . . . 22 Nm (17 ft lb)

Clutch master cylinder, replacing

The clutch master cylinder is mounted to the pedal assembly, directly above the clutch pedal.

1. Disconnect negative (–) cable from battery.

> ***CAUTION —***
> *Prior to disconnecting the battery, read the battery disconnection cautions given at the front of this manual on page viii.*

2. Remove brake fluid reservoir cap. Using a clean syringe, remove brake fluid from reservoir.

> ***CAUTION —***
> *Brake fluid is poisonous, highly corrosive and dangerous to the environment. Wear safety glasses and rubber gloves when working with brake fluid. Do not siphon brake fluid with your mouth. Immediately clean away any fluid spilled on painted surfaces and wash with water, as brake fluid will remove paint.*

3. Disconnect fluid supply hose from master cylinder. Place a pan under hose to catch any excess fluid.
4. Working in engine compartment, disconnect fluid line fitting from master cylinder.
5. Working in passenger compartment, remove lower instrument panel trim above pedal cluster.
6. Disconnect clutch pedal from clutch master cylinder push rod by removing clevis pin locking clip and slide out clevis pin.
7. Remove master cylinder mounting bolts. See Fig. 3.
8. Installation is reverse of removal. Fill fluid reservoir with clean fluid. Bleed clutch hydraulics as described earlier. Connect negative cable to battery last.

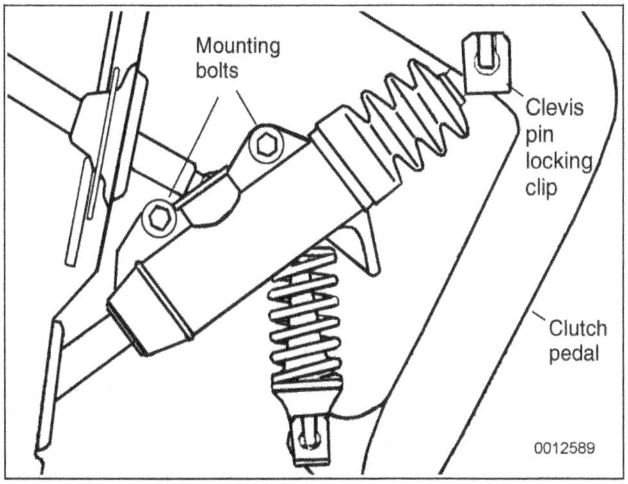

Fig. 3. Clutch master cylinder mounting bolts.

Tightening Torques
- Clutch master cylinder to bulkhead . . 22 Nm (17 ft-lb)
- Fluid line to master cylinder or slave cylinder . . . . 16 Nm (12 ft-lb)

Clutch slave cylinder, replacing

The clutch slave cylinder is mounted on the left side of the transmission.

1. Remove brake fluid reservoir cap. Using a clean syringe, remove brake fluid from reservoir.
2. Disconnect fluid hose from slave cylinder. Place a pan under hose to catch any excess fluid.
3. Unbolt slave cylinder from transmission. See Fig. 4.
4. Installation is reverse of removal, noting the following:
   - Check for wear on pushrod. Any other wear except on tip is caused by misalignment of clutch components.
   - Lightly coat pushrod tip with molybdenum disulfide grease (Molykote® Longterm 2 or equivalent).
   - During installation be sure pushrod tip engages recess in clutch release lever.
   - Fill fluid reservoir with clean fluid.
   - Bleed clutch hydraulics as described earlier.

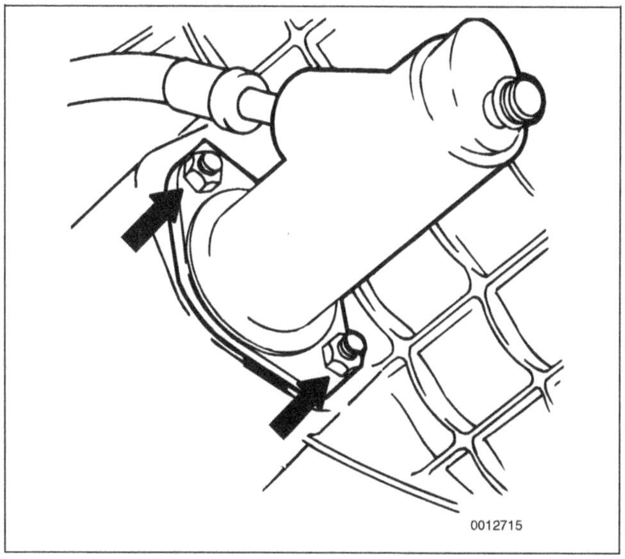

Fig. 4. Clutch slave cylinder mounting bolts (**arrows**).

Tightening Torques
- Clutch slave cylinder to transmission . . . . 22 Nm (17 ft lb)
- Fluid line to slave cylinder . . . . 17 Nm (10-12 ft lb)

CLUTCH MECHANICAL

The transmission must be removed from the engine to access the clutch mechanical components. It is recommended that the clutch disc, pressure plate and release bearing be replaced during a clutch overhaul. Be sure to check the bottom of the bellhousing for oil. If engine oil is found, check for a faulty rear crankshaft oil seal.

Table a. Clutch Disc Diameter

Model	Diameter
318i/is/iC	215 mm (8.46 in)
325i/is/iC 323is/iC	228 mm (8.98 in)
328i/is/iC M3	240 mm (9.45 in)

Clutch, removing

1. Remove transmission from engine. See **230 Manual Transmission**.

2. Remove release bearing from transmission input shaft.

3. Remove clutch release lever by sliding it out from under spring clip. See Fig. 5.

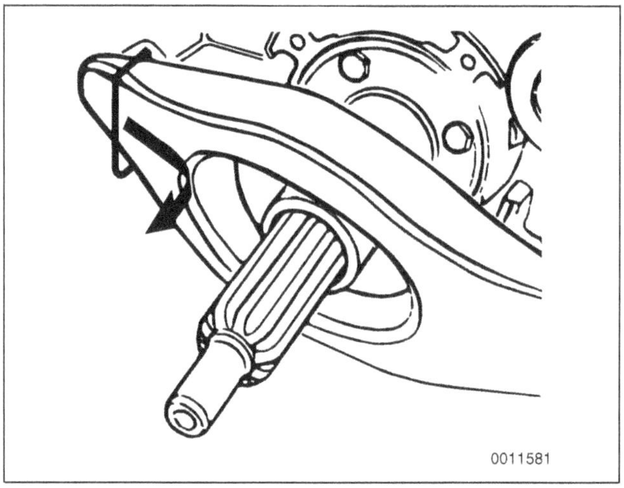

Fig. 5. Remove clutch release lever by pulling away from spring clip in direction of arrow.

4. Lock flywheel in position using a suitable locking tool (BMW Special tool no. 11 2 170, or equivalent). See Fig. 6.

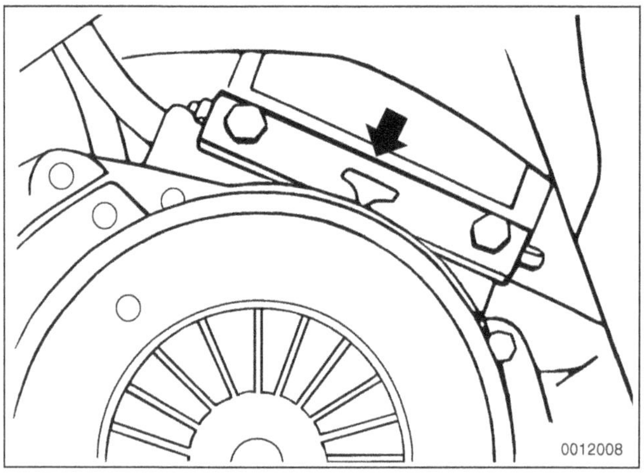

Fig. 6. Flywheel locking tool (**arrow**) installed to rear of cylinder block.

5. Install a clutch centering tool through clutch assembly to prevent clutch disc from falling out when removing pressure plate. See Fig. 7.

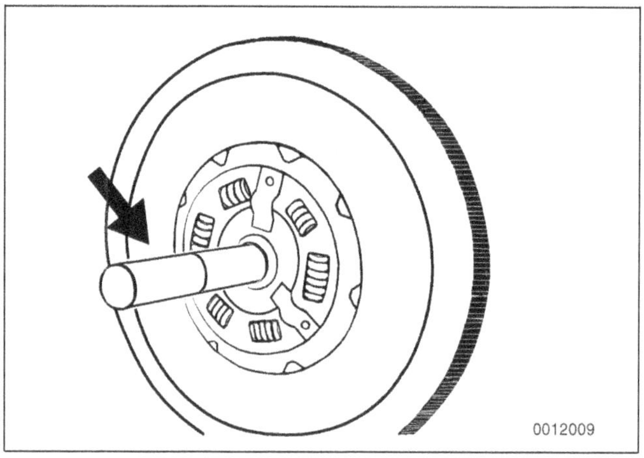

Fig. 7. Clutch centering tool (**arrow**) shown installed through clutch disc and flywheel.

6. Remove six bolts from perimeter of pressure plate. Remove pressure plate and clutch disc. See Fig. 8.

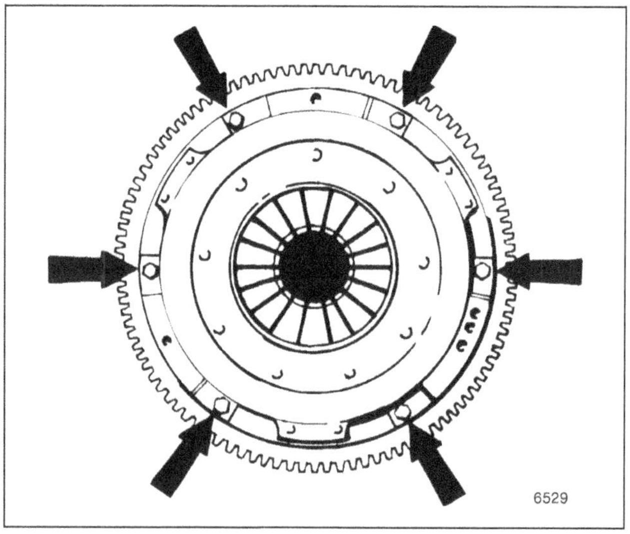

Fig. 8. Clutch pressure plate mounting bolts (**arrows**).

Clutch, inspecting and installing

1. Inspect clutch disc for wear, cracks, loose rivets, contamination or excessive runout (warping). Replace if necessary.

2. Inspect flywheel for scoring, hot spots, cracks or loose or worn guide pins. Replace flywheel if any faults are found.

> ***WARNING —***
> *If flywheel is removed from vehicle, use new bolts during installation. The old stretch-type bolts should not be reused.*

3. Inspect transmission pilot bearing in end of crankshaft. The bearing should rotate smoothly without play. If necessary, replace as described later.

4. Inspect and clean release bearing lever. Apply a thin coat of grease to release bearing lever contact points. See Fig. 9. Also, lightly lubricate clutch disc splines and transmission input shaft splines.

5. Install pressure plate and clutch disc. Do not fully tighten bolts.

6. Center clutch disc on flywheel using centering tool (BMW special tool no. 21 2 130 or equivalent). Refer to Fig. 7.

NOTE —
*Be sure clutch disc is facing the correct way. The new disc should be marked **engine side** or **transmission side**. See Fig. 10.*

7. Tighten each bolt one turn at a time until pressure plate is fully seated, and then torque to specification.

8. Remove clutch disc centering tool. Remove flywheel locking tool.

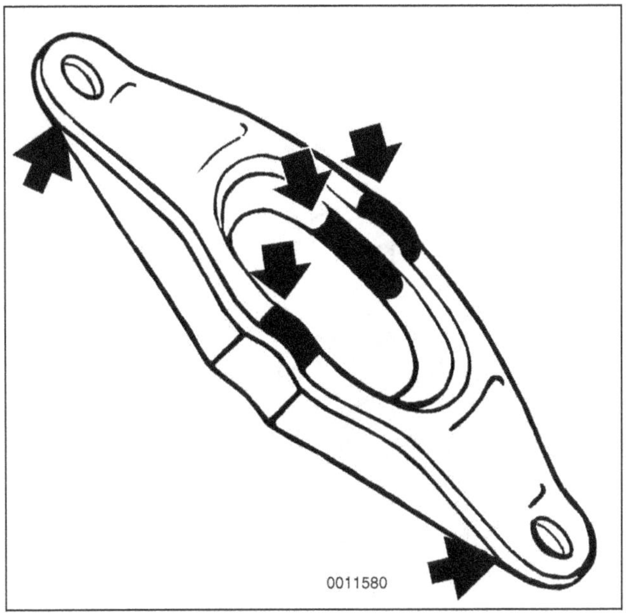

Fig. 9. Clutch release lever lubrication points (**arrows**).

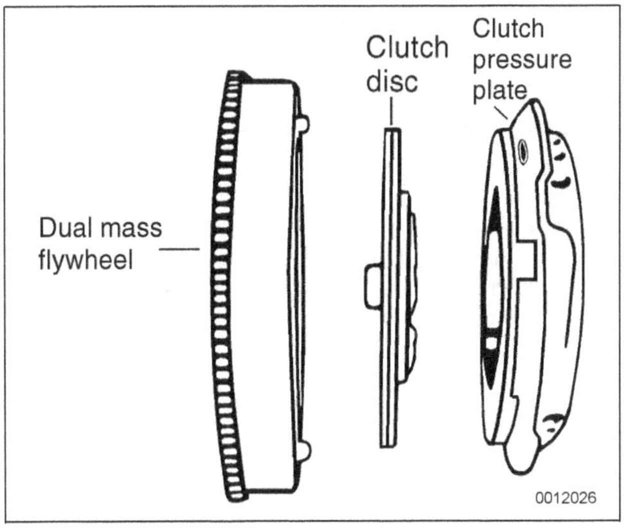

Fig. 10. Clutch assembly. Note clutch disc hub position during installation.

9. Clean and inspect release bearing guide sleeve on transmission. Install release lever and release bearing. See Fig. 11.

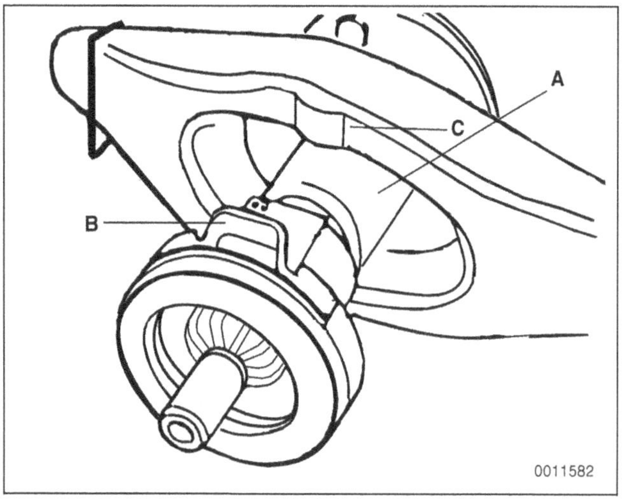

Fig. 11. Clutch release bearing correctly installed on guide sleeve (**A**). Bearing tabs (**B**) align with contact points (**C**) on release lever.

10. Install transmission. See **230 Manual Transmission**.

Tightening Torques

- Flywheel to crankshaft (use new bolts)
  M42/M44 engines . . . . 120 Nm (89 ft-lb)
  All others . . . . 105 Nm (77 ft-lb)
- Pressure plate to flywheel
  M8 8.8 grade . . . . 24 Nm (18 ft-lb)
  M8 10.9 grade . . . . 34 Nm (25 ft-lb)

Transmission pilot bearing, replacing

1. Remove clutch as described earlier.

2. Remove transmission pilot bearing from end of crankshaft using BMW special tool no. 11 2 340, or equivalent.

3. Press new bearing into place using BMW Special Tool no. 11 2 350, or equivalent.

230 Manual Transmission

230

GENERAL

This repair group covers external transmission service, including removal and installation of the transmission unit. Internal transmission repair is not covered. Special press tools and procedures are required to disassemble and service the internal geartrain.

Manual transmissions used are based on engine application. Consult **200 Transmission–General** for transmission application information. **Table a** lists gear ratio specifications.

NOTE —
Do not rely on forged casting numbers for transmission code identification.

TRANSMISSION FLUID SERVICE

The manual transmissions installed in the E36 models are normally filled with automatic transmission fluid (ATF), although alternative lubricants may have been used from the factory, depending on transmission type and model year. For example, transmissions fitted to the M3 models are normally filled with a “lifetime” manual transmission fluid (BMW MTF LT-1) and do not require periodic fluid changes.

NOTE —
The transmission lubricant type can be found on a label on the side of the transmission. Consult an authorized BMW dealer for alternate fluid use and the most-up-to-date information regarding transmission operating fluids.

Table a. Manual Transmission Gear Ratios

Transmission	Getrag S5D 200 G	Getrag S5D 250 G	ZF S5D 310 Z	ZF S5D 320 Z
Gear Ratios:				
1st	4.23	4.23	4.20	4.20
2nd	2.52	2.52	2.49	2.49
3rd	1.66	1.66	1.66	1.66
4th	1.22	1.22	1.24	1.24
5th	1.00	1.00	1.00	1.00
reverse	4.04	4.04	3.89	3.89

Transmission fluid level, checking

Checking the transmission oil level involves simply removing the side fill plug and inserting a finger into the hole to check the fluid level. If the fluid level is up to the bottom of the fill hole (finger is wetted by transmission fluid), the level is correct.

Transmission fluid, replacing

NOTE —
Manual transmission fluid type should be printed on a label on the side of the transmission. If in doubt, consult an authorized BMW dealer for the up-to-date information on operating fluids.

1. Drive vehicle for a few miles to warm transmission.

2. Raise and safely support vehicle to access drain plug.

WARNING —
- *When raising the car using a floor jack or a hydraulic lift, carefully position the jack pad to prevent damaging the car body. A suitable liner (wood, rubber, etc.) should be placed between the jack and the car to prevent body damage.*
- *Watch the jack closely. Make sure it stays stable and does not shift or tilt. As the car is raised, the car may roll slightly and the jack may shift.*

3. Place a drain pan under transmission and remove drain plug at bottom of transmission. See Fig. 1.

Fig. 1. Manual transmission fill (**1**) and drain plugs (**2**).

4. Install and torque drain plug.

5. Remove fill plug from side of transmission. Slowly fill transmission with fluid until fluid overflows fill hole.

6. Install and torque fluid fill plug.

Fluid Capacity
- Getrag transmission . . . . 1.0 liter (1.1 US qts)
- ZF transmission . . . . 1.2 liters (1.3 US qts)

Tightening Torque
- Transmission drain or fill plug . . . . 50 Nm (37 ft-lb)

TRANSMISSION SERVICE

Back-up light switch, replacing

1. Raise and safely support vehicle to access back-up light switch.

NOTE —
- *The back-up light switch on Getrag transmissions is in the left rear of the transmission.*
- *The back-up light switch on ZF transmission is in the right front of the transmission.*

2. Unscrew switch from transmission.

3. Install new switch.

Tightening Torque
- Back-up light switch to transmission . . . . 20 Nm (15 ft-lb)

4. Check transmission fluid level before lowering vehicle.

Selector shaft seal, replacing (transmission installed)

1. Raise and safely support vehicle to access drain plug.

2. Support transmission with transmission jack.

3. Remove driveshaft. See **260 Driveshaft**.

4. Remove transmission crossmember and tilt down rear of transmission.

> ***CAUTION—***
> *Tilting the engine to lower the transmission can lead to damage to various components due to lack of clearance.*
> *-On cars with AST (traction control) remove throttle body before tilting engine.*
> *-Remove brake fluid reservoir if necessary.*

5. Working at rear of selector shaft, disconnect shift linkage from selector shaft. See Fig. 2.

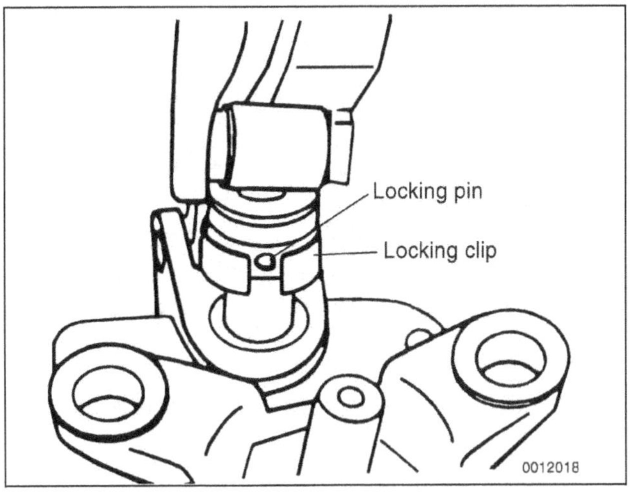

Fig. 2. Disconnect shift linkage from transmission selector shaft by prying back locking clip and pushing locking pin out.

6. Carefully pry seal out with a narrow seal remover or small screwdriver. See Fig. 3.

7. Coat new seal with transmission fluid. Drive new seal in flush with housing. Use BMW special seal installation tool no. 23 2 260 or equivalent and a soft-faced (plastic) hammer. See Fig. 4.

8. Reinstall shift linkage to selector shaft.

9. Raise transmission and reinstall crossmember.

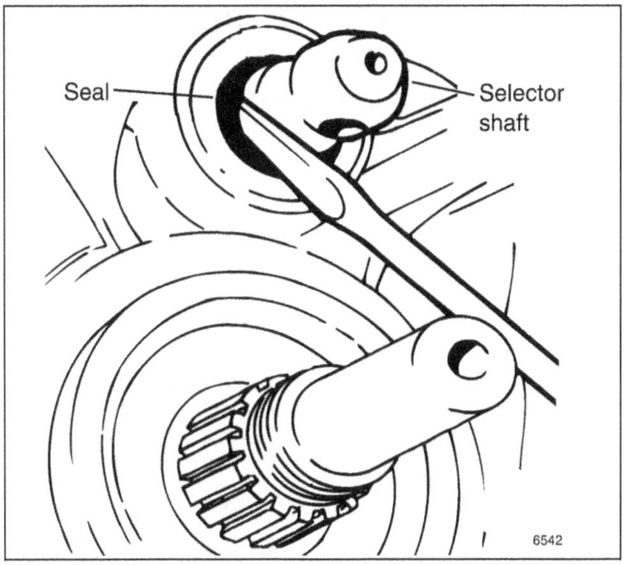

Fig. 3. Selector shaft oil seal being pried out with screwdriver.

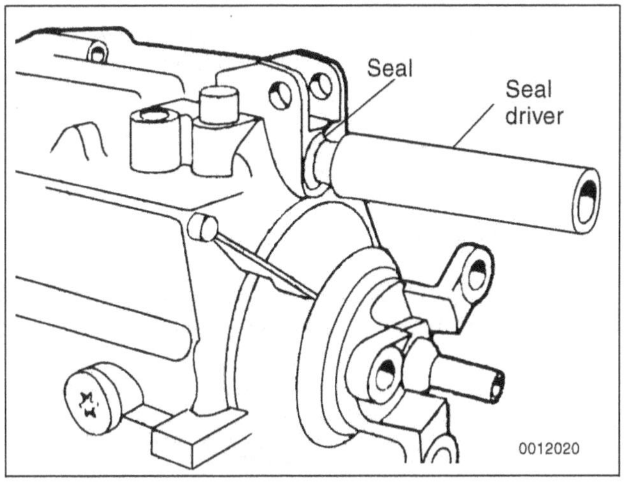

Fig. 4. Selector shaft seal being installed.

10. Install driveshaft using new nuts. See **260 Driveshaft**.

11. Check transmission oil level, topping up as necessary.

Tightening Torques
- Driveshaft to flex disc/ final drive flange . . . . See **260 Driveshaft**
- Transmission crossmember to chassis
  M8 . . . . 22-24 Nm (16-17 ft-lb)
  M10 . . . . 42 Nm (31 ft-lb)

Output shaft seal, replacing (transmission installed)

1. Raise and safely support vehicle. Support transmission with transmission jack.

> ***WARNING —***
> - *When raising the car using a floor jack or a hydraulic lift, carefully position the jack pad to prevent damaging the car body. A suitable liner (wood, rubber, etc.) should be placed between the jack and the car to prevent body damage.*
> - *Watch the jack closely. Make sure it stays stable and does not shift or tilt. As the car is raised, the car may roll slightly and the jack may shift.*

2. Remove driveshaft. See **260 Driveshaft.**

3. Remove transmission crossmember and lower transmission at rear.

> ***CAUTION —***
> *Tilting the engine to lower the transmission can lead to damage to various components due to lack of clearance.*
> *-On cars with AST (traction control) remove throttle body before tilting engine.*
> *-Remove brake fluid reservoir if necessary.*

4. Bend back and remove collar nut lockplate.

5. Hold flange steady and remove collar lock nut (use 30mm thin-walled deep socket.) See Fig. 5.

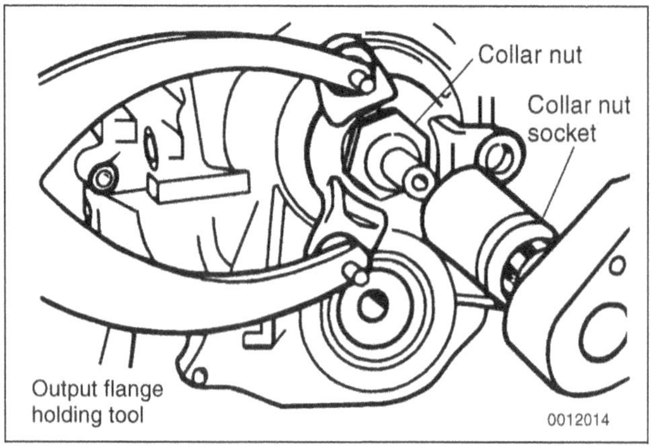

Fig. 5. Use holding tool when loosening output shaft collar nut.

6. Remove flange. If necessary, use a puller.

7. Use a seal puller to remove seal from transmission housing. See Fig. 6.

8. Coat new seal with transmission fluid and drive into position until it is flush with housing. See Fig. 7.

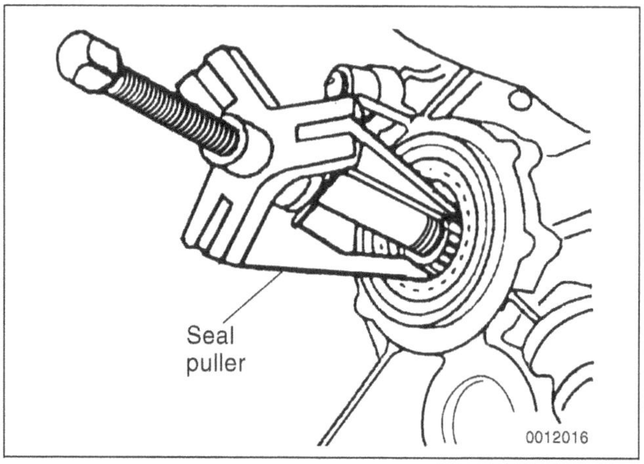

Fig. 6. Puller used to remove transmission seal.

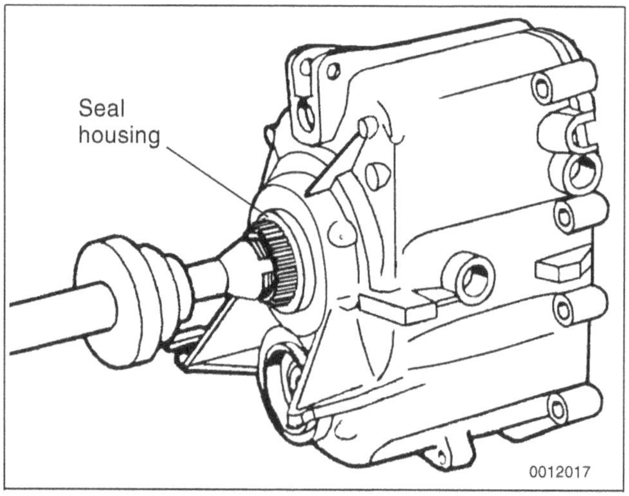

Fig. 7. Output shaft seal being installed using seal driver (BMW special tool no. 23 2 300).

9. Reinstall output flange (and shims, if applicable) to output shaft.

> ***NOTE —***
> *On Getrag S5D 200G and S5D 250G transmissions BMW recommends heating the output flange to about 176°F (80°C) to aid in installation. This can be done by placing the flange in hot water.*

10. Coat bearing surface of collar nut with sealer and install nut. Tighten collar nut in two stages. Install new lockplate. Bend tabs into flange grooves.

> ***NOTE —***
> *BMW recommends the use of a sealer such as Loctite® 242 when installing the flange collar nut to prevent oil from leaking past the threads.*

11. Raise transmission and reinstall transmission crossmember.

12. Install driveshaft using new nuts. See **260 Driveshaft**.

13. Check transmission oil level, topping up as necessary.

Tightening Torques
- Transmission crossmember to chassis
  M8 .... 22-24 Nm (16-17 ft-lb)
  M10 .... 42 Nm (31 ft-lb)
- Transmission output flange to output shaft
  Stage I .... 190 Nm (140 ft-lb)
  Stage II (after loosening) .... 120 Nm (89 ft-lb)

Input shaft seal, replacing (transmission removed)

Replacement of the input shaft seal requires the removal of the transmission from the vehicle as described later in this repair group.

1. Remove clutch release bearing and release lever from inside bellhousing.

2. Remove bolts for clutch release-bearing guide sleeve, noting bolt lengths. Remove sleeve and any spacers (shims) under it. See Fig. 8.

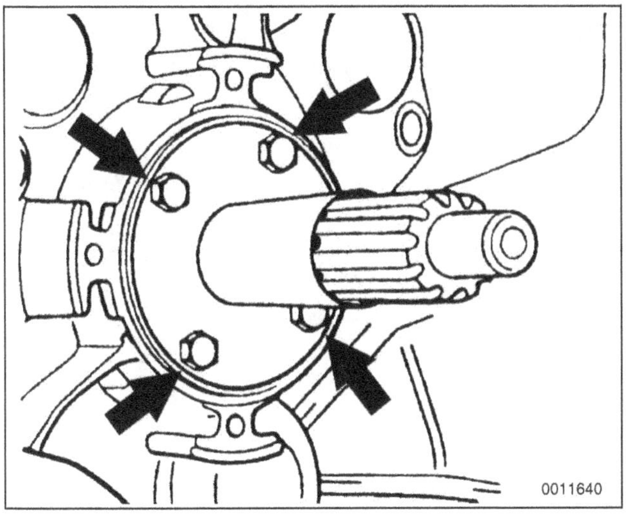

Fig. 8. Clutch release bearing guide sleeve bolts. (Getrag transmission shown).

3. With transmission on a workbench, remove input shaft seal cover from inside bell-housing.

4. Locate two indents in seal housing. Use a punch or scribe to punch holes in housing at indents. Thread two sheet metal screws into holes and pull seal out using screws. See Fig. 9.

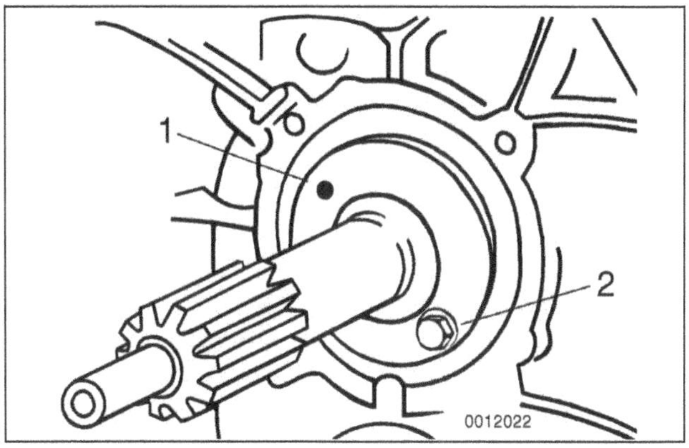

Fig. 9. Indent at oil seal housing (**1**). Punch holes in seal housing at indents and use sheet metal screw (**2**) to pull seal from transmission.

NOTE —
Use care not to scratch or damage input shaft. Use a protective sleeve or tape the shaft when removing and installing the seal.

5. Lubricate new seal with transmission oil and drive into place using BMW Special Tool no. 23 2 430 or equivalent.

6. Thoroughly clean guide sleeve mounting bolts, sealing surfaces, and threads in case. Apply sealer (Loctite® 242 or equivalent) to guide sleeve sealing surface and bolts. Reinstall guide sleeve and spacer(s).

Tightening Torque
- Guide sleeve to transmission
  M6x12 bolt .... 10 Nm (89 in-lb)

TRANSMISSION REMOVAL AND INSTALLATION

Removal and installation of the transmission is best accomplished on a lift using a transmission jack. The engine must be supported from above using appropriate support equipment. This allows the engine to pivot on its mounts to access the upper Torx-head bolts at the bellhousing.

WARNING —
- *Make sure the car is stable and well supported at all times. Use a professional automotive lift or jack stands designed for the purpose. A floor jack is not adequate support.*
- *The removal of the transmission may upset the balance of the vehicle on a lift.*

Transmission, removing and installing

1. Disconnect negative (–) cable from battery.

> ***CAUTION—***
> *Prior to disconnecting the battery, read the battery disconnection cautions given at the front of this manual on page viii.*

2. Install engine support across engine bay. Raise engine so that weight of engine is supported. See Fig. 10.

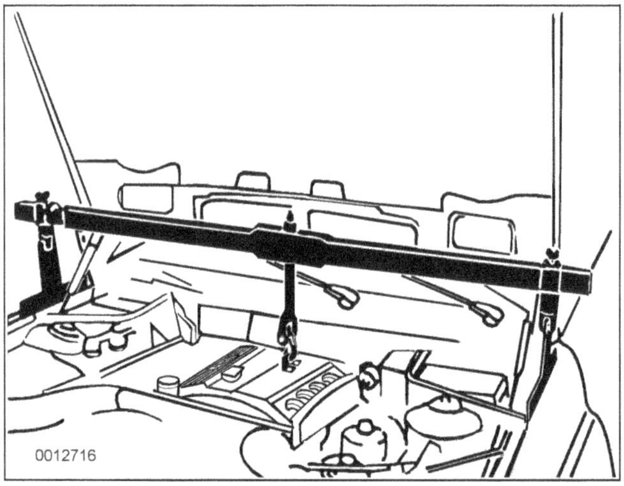

Fig. 10. Engine support equipment used to support engine from above before removing transmission.

3. Raise vehicle to gain access to underside of car.

4. Support transmission with transmission jack. If applicable, remove reinforcing cross brace from below engine/transmission.

5. Disconnect harness connector from reverse light switch on transmission.

6. Remove complete exhaust system and heat shield. See **180 Exhaust System**.

> ***NOTE—***
> *Disconnect oxygen sensor harness connector(s) before lowering exhaust system.*

7. Remove driveshaft. See **260 Driveshaft**.

8. Disconnect shift rod from transmission selector shaft. See Fig. 11.

9. Disconnect shift console from top of transmission. See **250 Gearshift Linkage**.

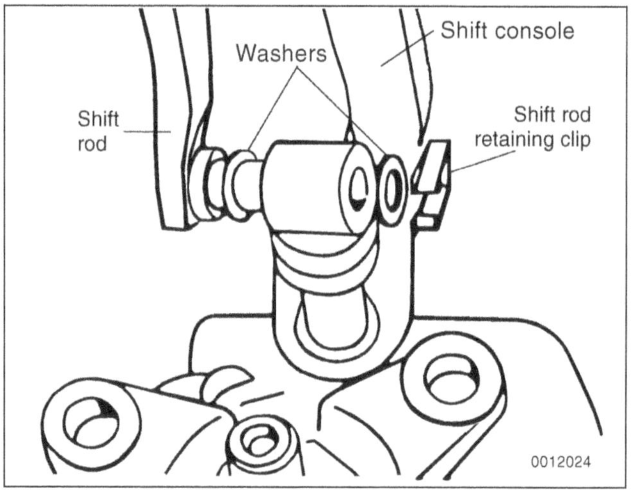

Fig. 11. Shift lever to transmission connection.

10. Unbolt clutch slave cylinder from side of transmission. Do not disconnect fluid hose. Suspend slave cylinder from chassis using stiff wire.

> ***CAUTION—***
> *Do not operate clutch pedal with slave cylinder removed from transmission.*

11. Support transmission from below with jack. Remove rear support crossmember from transmission. See Fig. 12.

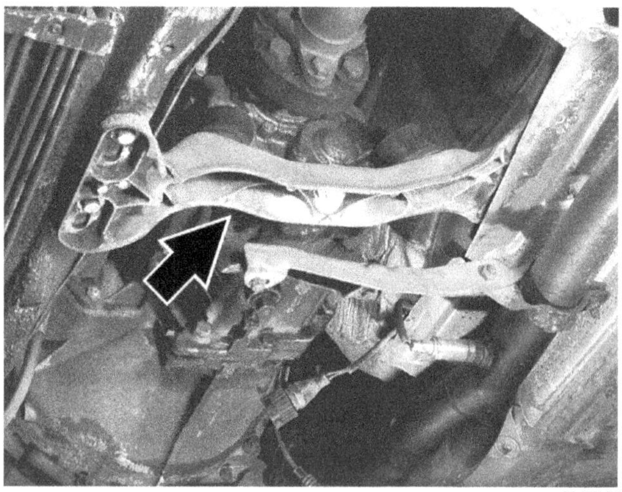

Fig. 12. Rear transmission crossmember (**arrow**).

12. Lower transmission/engine assembly until it rests on front suspension crossmember.

> ***CAUTION—***
> *Tilting the engine to lower the transmission can lead to damage to various components due to lack of clearance at rear of engine*
> *-On cars with AST remove throttle body before tilting engine.*
> *-Remove brake fluid reservoir if necessary.*

13. Remove transmission mounting bolts (Torx-head). Note length and location of bolts. See Fig. 13.

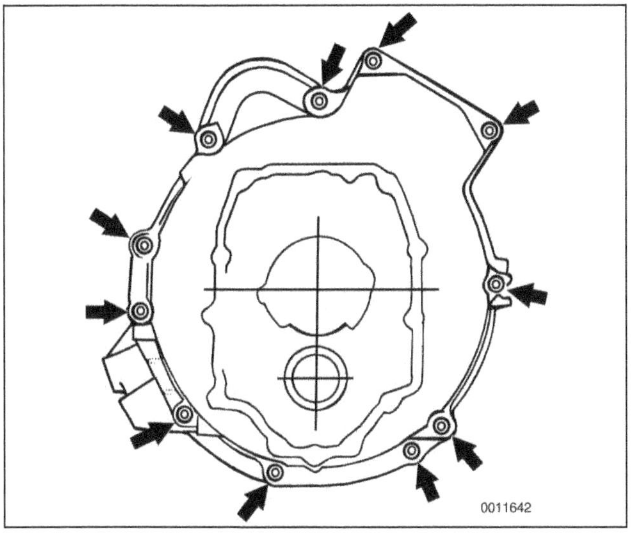

Fig. 13. Transmission mounting bolts.

> ***CAUTION —***
> *At no time should the weight of the transmission be supported by the transmission input shaft.*

14. Remove transmission by pulling backward until the transmission input shaft clears the clutch disc splines, then pull downwards. Lower jack and remove transmission.

15. Installation is reverse of removal, keeping in mind the following:
    - When installing exchange transmission, transfer parts from old unit to exchange unit
    - Thoroughly clean input shaft and clutch disc splines. Lightly lubricate transmission input shaft before installing.
    - Be sure bellhousing dowels are correctly located. See Fig. 14.
    - Center rear of transmission in driveshaft tunnel before tightening transmission support bracket.

NOTE —
Washers should always be used with Torx-head bolts. Installing bolts without washers will make them difficult to loosen.

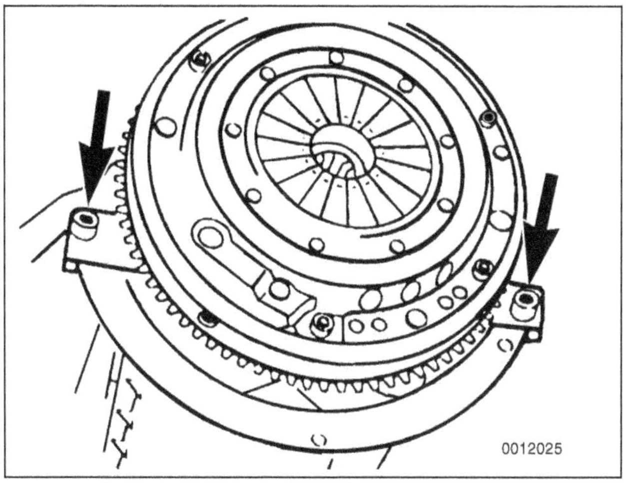

Fig. 14. Bellhousing alignment dowels (**arrows**).

16. Install driveshaft and preload center bearing bracket. Use new nuts when mounting driveshaft to transmission/flex disc and final drive. See **260 Driveshaft.**

17. Refill transmission with appropriate lubricant before starting or towing the car. See **Transmission Fluid Service** earlier in this repair group.

Tightening Torques

- Reinforcing cross brace to chassis (M10) . . . . 42 Nm (31 ft-lb)
- Rubber mount to transmission or bracket nut (M8) . . . . 21 Nm (16 ft-lb)
- Slave cylinder to transmission . . . . . . 22 Nm (16 ft-lb)
- Transmission crossmember to chassis
  - M8 . . . . 22-24 Nm (16-17 ft-lb)
  - M10 . . . . 42 Nm (31 ft-lb)
- Transmission to engine (Torx-head)
  - M8 . . . . 22 Nm (16 ft-lb)
  - M10 . . . . 43 Nm (32 ft-lb)
  - M12 . . . . 72 Nm (53 ft-lb)
- Transmission drain/fill plugs . . . . 50 Nm (37 ft-lb)

240 Automatic Transmission

GENERAL

Automatic transmissions are identified by code letters found on a data plate located on the side of the transmission case, behind the manual valve lever. See **200 Transmission–General** for application information. For gear ratio specifications, see **Table a**.

> ***CAUTION—***
> *ATF does not circulate unless the engine is running. When towing an automatic transmission vehicle, use a flat-bed truck or raise the rear wheels off the ground.*

The electronic transmission control module (TCM) monitors transmission operation for faults and alerts the driver by illuminating the transmission fault indicator on the instrument panel. On-board diagnostic codes stored in the TCM must be downloaded and interpreted by trained technicians using special BMW diagnostic equipment.

NOTE—
Internal repairs to the automatic transmission require special service equipment and knowledge. If it is determined that internal service is required, consult an authorized BMW dealer about a factory reconditioned unit or a transmission rebuild.

TRANSMISSION SERVICE

ATF level, checking

The automatic transmissions installed in the E36 models are normally filled with Dexron III automatic transmission fluid (ATF), although alternative fluids have been installed from the factory, depending on transmission type and model year. Some transmissions are filled with a special BMW "lifetime" automatic transmission fluid, which does not require periodic fluid changes.

> ***CAUTION—***
> *Mixing different types of transmission fluids will lead to transmission failure.*

NOTE—
The transmission lubricant installed can be found on a type plate on the side of the transmission. In addition, a sticker may be located on the fluid sump indicating the type of transmission lubricant used. If in doubt, consult an authorized BMW dealer for alternate fluid use and current information regarding transmission operating fluids.

The transmission is not equipped with a dipstick. Checking the ATF level requires measuring and maintaining a specified ATF temperature during the checking procedure. The checking temperature is not fully warm or cold (the ATF sump should be approximately warm to the touch).

Table a. Automatic Transmission Gear Ratios

Transmission	A4S 310R (THM-R1) (1992-1993)	A4S 310R (THM-R1) (1993-on)	A5S 310Z (5HP 18)	A4S 270R (THM-R1)
Gear ratios:				
1st	2.40	2.86	3.67	2.86
2nd	1.47	1.62	2.00	1.62
3rd	1.00	1.00	1.41	1.00
4th	0.72	0.72	1.00	0.72
5th	-	-	0.74	-
Reverse	2.00	2.00	4.10	2.00

> ***WARNING —***
> *If the appropriate temperature measuring equipment is not available, ATF level checking should be left to an authorized BMW dealer. The checking procedure should be done on a lift with the engine running in neutral. In addition, the transmission fluid temperature must be maintained within a narrow range during the checking procedure.*

1. Place vehicle on a lift.

> ***WARNING —***
> *To lift the vehicle safely, see* **010 Fundamentals for the Do-it-Yourself Owner**.

2. Warm up drivetrain by running engine until ATF reaches specified temperature. Apply brake and, while idling engine, shift through all gears several times. Turn engine off.

ATF Level Checking
- Fluid temperature . . . . 30-55°C (86-131°F)

3. Remove ATF fill plug. Place transmission in neutral and restart engine. See Fig. 1 or Fig. 2.

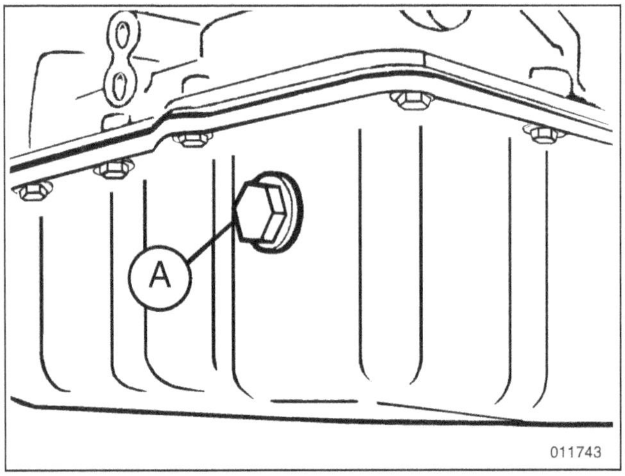

Fig. 1. ATF fill plug (**A**) on A4S 270 R/A4S 310R transmissions.

4. With ATF at specified temperature, add fluid until it spills from fill hole.

> ***WARNING —***
> *Hot ATF can scald. Wear eye protection and protective clothing and gloves during the check. If the transmission was overfilled, hot ATF will spill from the filler hole when the fill plug is removed.*

5. Reinstall fill plug using new sealing ring.

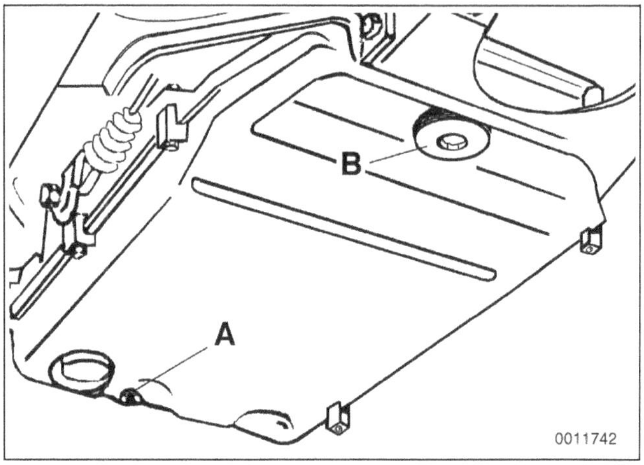

Fig. 2. ATF drain plug (**A**) and fill plug (**B**) on A5S 310Z transmission.

Tightening Torques
- ATF fill plug to ATF sump
  A4S 270R/A4S 310R . . . . 33 Nm (24 ft-lb)
  A5S 310Z (M30 plug) . . . . 100 Nm (74 ft-lb)

ATF, draining and filling

The procedure given here includes removal and installation of the ATF fluid strainer (filter).

> ***NOTE —***
> *It is recommended that you measure the drained ATF and begin the refill procedure by adding the same amount of fresh fluid as your starting point for checking the fluid level, described under* **ATF level, checking**.

1. Remove drain plug and drain fluid into container. See Fig. 3 or Fig. 2 given earlier.

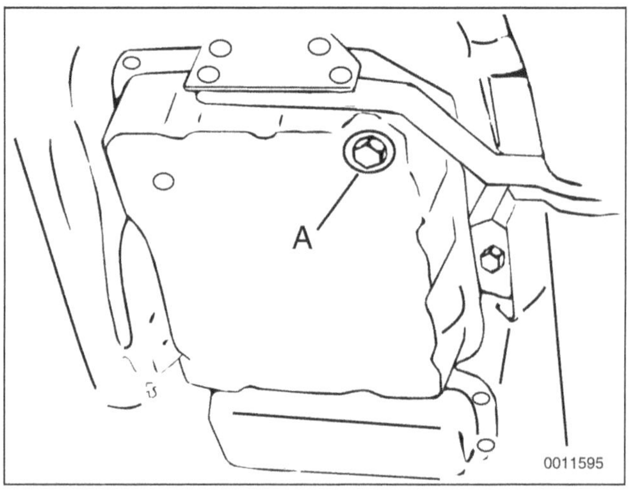

Fig. 3. ATF drain plug (**A**) on A4S 270 R/A4S 310R.

> ***WARNING —***
> *The ATF must not be hot when draining. Do not drain the ATF if the engine and/or transmission is hot. Hot ATF can scald. Wear eye protection, protective clothing and gloves.*

2. On A4S 270R/A4S 310R transmissions: Remove front and rear oil sump mounting bolts and remove sumps and sump gaskets.

NOTE —
The A4S 270 R/A4S 310 R transmissions may be fitted with ASE (inch) bolts.

3. On A5S 310Z transmission: Remove oil sump mounting bolts and brackets, noting position of brackets. Remove sump and sump gasket.

4. Remove ATF strainer mounting screws and remove strainer together with gasket or O-ring. See Fig. 4 or Fig. 5.

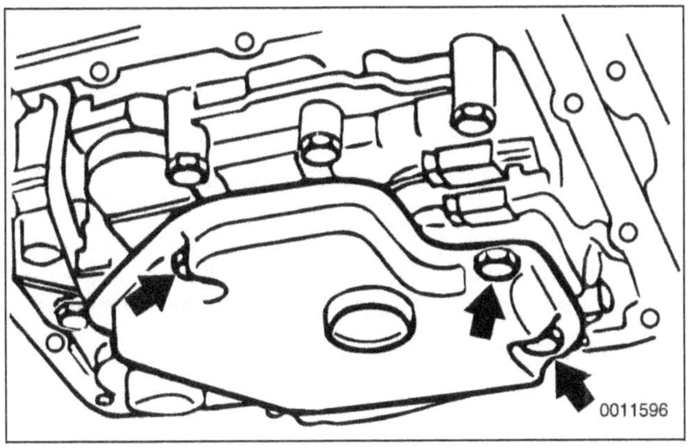

Fig. 4. A4S 270R/A4S 310R fluid strainer mounting bolts (**arrows**).

5. Installation is reverse of removal.
   - Always replace seal for drain plug and all gasket(s) and O-rings.
   - Clean sump and sump magnet(s) using a lint-free cloth.
   - On A4S 270R/A4S 310R transmissions: Use new bolts at sump and tighten bolts in-line, one after other (*do not* tighten bolts in a cross-wise pattern).
   - On A5S 310Z transmission, be sure to install rounded sump clamping brackets at corners of sump.
   - Fill transmission with approved fluid.
   - Check fluid level as described under **ATF level, checking**.

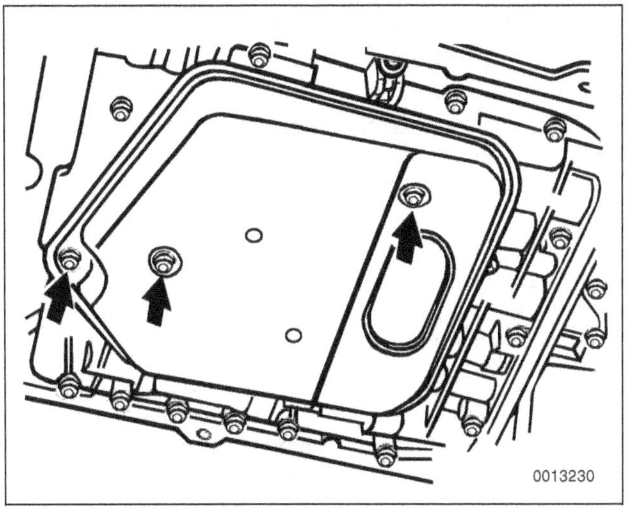

Fig. 5. A5S 310Z fluid strainer mounting bolts (**arrows**).

NOTE —
Use new sump bolts purchased from BMW. Alternatively, clean old bolts and coat with Loctite® thread locking compound or equivalent.

Automatic Transmission Fluid Capacity (refill only, not including torque converter draining)

- A4S 270R/A4S310R
  Removal of both sumps . . . . 7.8 liters (8.2 qt)
- A5S 310Z . . . . 3.3 liters (3.5 qt)

Tightening Torques

- ATF drain plug to ATF sump
  A4S 270R/A4S 310R . . . . 25 Nm (18 ft-lb)
  A5S 310Z (M10 plug) . . . . 16 Nm (12 ft-lb)
- ATF fill plug to ATF sump
  A4S 270R/A4S 310R . . . . 33 Nm (24 ft-lb)
  A5S 310Z (M30 plug) . . . . 100 Nm (74 ft-lb)
- ATF strainer to valve body
  A4S 270R/A4S 310R . . . . 20 Nm (15 ft-lb)
  A5S 310Z . . . . 6 Nm (53 in-lb)
- ATF sump to transmission
  A4S 270R/A4S 310R (M6) . . . . 12 Nm (9 ft-lb)
  A5S 310Z (M6) . . . . 6 Nm (53 in-lb)

TRANSMISSION REMOVAL AND INSTALLATION

Removal and installation of the transmission is best accomplished on a lift using a transmission jack. Use caution and safe workshop practices when working beneath car and lowering transmission.

> ***WARNING —***
> *To lift the vehicle safely, see* **010 Fundamentals for the Do-it-Yourself Owner**.

> ***CAUTION —***
> - *Be sure the vehicle is properly supported. The removal of the transmission may upset the balance of the vehicle on a lift.*
> - *Tilting the engine to remove the transmission can lead to damage to various components due to lack of clearance.*
>   *-On cars with AST remove throttle body.*
>   *-Remove brake fluid reservoir if necessary.*

Torx-head bolts are used to mount the transmission to the bellhousing. Be sure to have appropriate tools on hand before starting the job.

Transmission, removing and installing

1. Disconnect negative (–) cable from battery.

> ***CAUTION —***
> *Prior to disconnecting the battery, read the battery disconnection cautions given at the front of this manual on page viii.*

2. Install engine support across engine bay. Raise and safely support vehicle. See Fig. 6.

> ***WARNING —***
> - *When raising the car using a floor jack or a hydraulic lift, carefully position the jack pad to prevent damaging the car body. A suitable liner (wood, rubber, etc.) should be placed between the jack and the car to prevent body damage.*
> - *Watch the jack closely. Make sure it stays stable and does not shift or tilt. As the car is raised, the car may roll slightly and the jack may shift.*

> ***CAUTION —***
> *Removal of transmission will cause engine to tip unless engine support is used.*

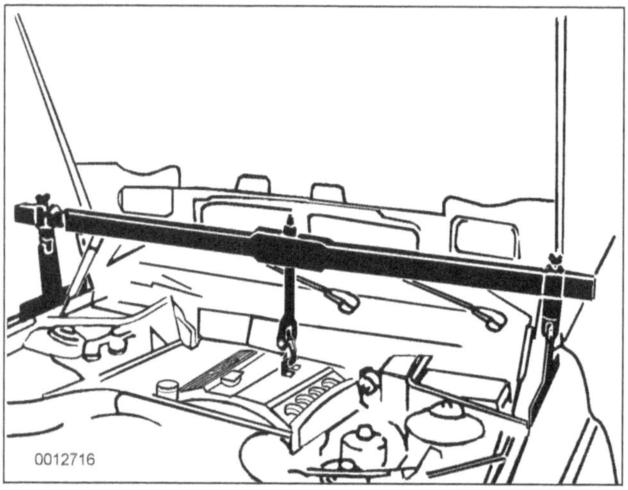

Fig. 6. Engine support equipment used to support engine from above before removing transmission.

3. Raise vehicle to gain access to underside of car.

4. Drain ATF from transmission as described earlier.

5. Disconnect oxygen sensor harness. Remove exhaust system, and exhaust heat shield. See **180 Exhaust System**.

6. Remove driveshaft. See **260 Driveshaft**.

7. With selector lever in **Park**, disconnect shift selector lever cable from lever and remove cable bracket. Hold linkage clamping bolt stationary when loosening clamping nut. See Fig. 7.

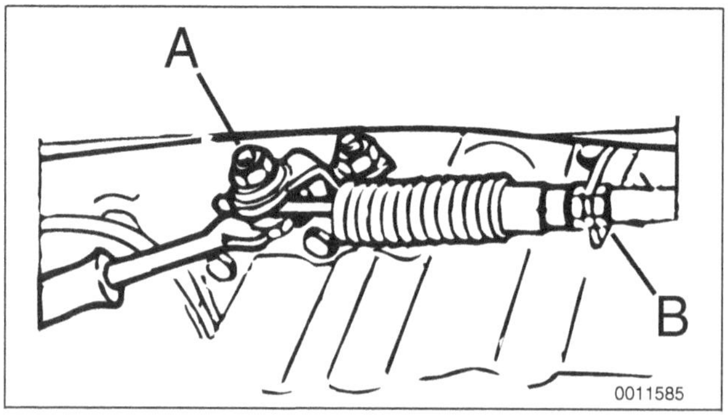

Fig. 7. Selector lever cable clamping nut (**A**). Loosen cable clamping nut (**B**) and remove cable from bracket. Always counterhold clamping bolt when loosening nut.

8. Disconnect main harness connector from transmission by turning its bayonet lock ring counterclockwise. Remove wiring harness from transmission housing. See Fig. 8.

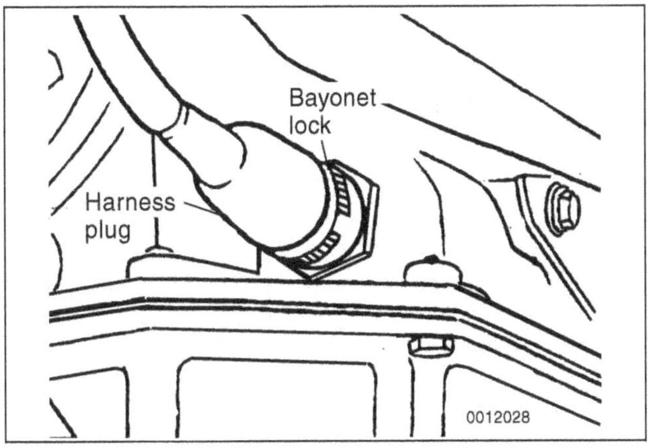

Fig. 8. Harness connector at automatic transmission.

9. Where applicable, disconnect harness connector from speed sensor at top of transmission.

10. Disconnect transmission cooler line clamps from engine. Disconnect lines from transmission.

11. Support transmission with transmission jack. Remove reinforcing cross brace (if applicable) and transmission support crossmember.

NOTE—
Note installation position of support crossmember.

12. Remove access plug in cover plate on right side of engine block and remove torque converter bolts. Turn crankshaft to access bolts. See Fig. 9.

13. Remove bellhousing-to-engine mounting bolts. See Fig. 10.

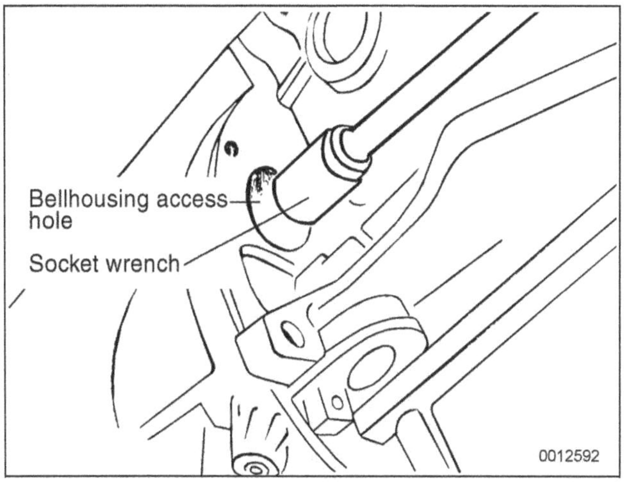

Fig. 9. Access and remove torque converter bolts through hole in bellhousing cover.

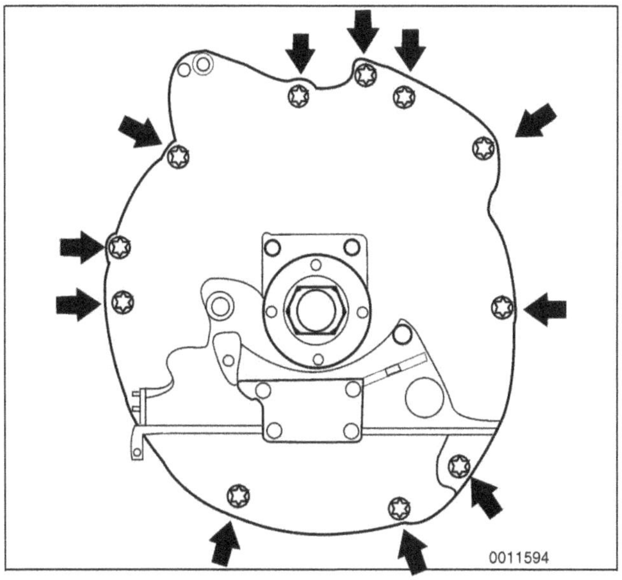

Fig. 10. Transmission-to-engine Torx-head mounting bolts.

14. Remove transmission by pulling back and down. Lower jack to allow transmission to clear vehicle.

> ***CAUTION —***
> *Do not allow the torque converter to fall off the transmission input shaft.*

15. Installation is reverse of removal, noting the following:

   - Blow out oil cooler lines with low-pressure compressed air and flush cooler with clean ATF twice before reattaching lines to transmission.

> ***CAUTION —***
> - *Wear safety glasses when working with compressed air.*
> - *Do not reuse ATF after flushing.*

   - Install new sealing washers on hollow bolts.
   - Install new O-rings on transmission cooler lines, where applicable.
   - Inspect drive plate for cracks or elongated holes. Replace if necessary.
   - Check to be sure torque converter is seated correctly in transmission during final installation.
   - When mounting transmission to engine, the three mounting tabs on torque converter must be aligned with indentations on driveplate. Use an alignment dowel to line up bolt holes, if necessary.
   - Fill transmission with clean ATF until oil level is even with fill hole. Then check fluid level as described earlier.
   - Adjust gearshift mechanism. See **250 Gearshift linkage**.

NOTE —
Torx-head mounting bolts should always be used with washers to prevent difficult removal.

Tightening Torques

- ATF drain plug to ATF sump
  A4S 270R/A4S 310R . . . . 25 Nm (18 ft-lb)
  A5S 310Z (M10 plug) . . . . 16 Nm (12 ft-lb)
- ATF fill plug to ATF sump
  A4S 270R/A4S 310R . . . . 33 Nm (24 ft-lb)
  A5S 310Z (M30 plug) . . . . 100 Nm (74 ft-lb)
- ATF cooler lines (A4S 270R/A4S 310R)
  Coupling nuts . . . . 20 Nm (15 ft-lb)
  Hollow bolts . . . . 37 Nm (28 ft-lb)
- Reinforcing cross brace
  to chassis (M10) . . . . 42 Nm (31 ft-lb)
- Torque converter
  to drive plate (M10) . . . . 45 Nm (33 ft-lb)
- Transmission support crossmember to chassis
  M8 . . . . 22-24 Nm (16-17 ft-lb)
  M10 . . . . 42 Nm (31 ft-lb)
- Transmission to engine (Torx-head with washer)
  M8 . . . . 21 Nm (15 ft-lb)
  M10 . . . . 42 Nm (31 ft-lb)
  M12 . . . . 72 Nm (53 ft-lb)

250 Gearshift Linkage

GENERAL

This repair group covers transmission gearshift and linkage service.

Special service tools are required for some of the procedures given here. To gain access to the complete gearshift mechanism it is necessary to remove the exhaust system and the driveshaft as described in **180 Exhaust System** and **260 Driveshaft**.

MANUAL TRANSMISSION GEARSHIFT

Gearshift lever, removing (manual transmission)

The manual transmission gearshift linkage is shown in Fig. 1. Use this illustration as a guide when removing and installing the linkage.

1. Pull straight up on shift knob to remove it from shift lever.

   NOTE —
   Removal of the shift knob will require about 90 lbs. of force.

2. Pry up on rear of shift boot to unclip, then remove boot from front retainers.

3. Raise vehicle to gain access to underside of vehicle.

   WARNING —
   Make sure the car is stable and well supported at all times. Use a professional automotive lift or jack stands designed for the purpose. A floor jack is not adequate support.

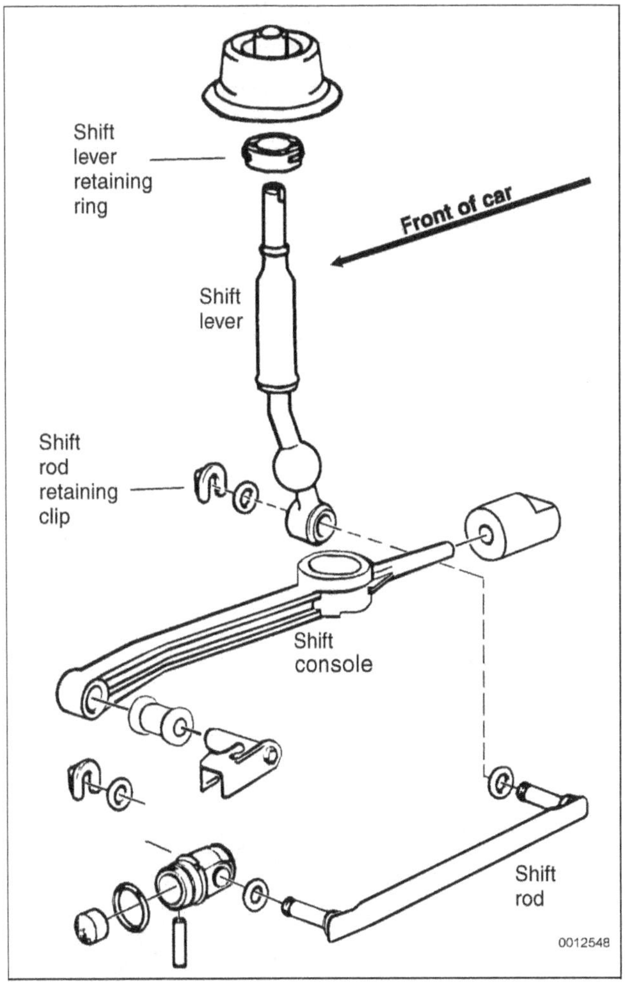

Fig. 1. Manual transmission gearshift linkage.

4. Disconnect oxygen sensor harness connectors. Remove complete exhaust system. See **180 Exhaust System**.

5. Remove driveshaft. See **260 Driveshaft**.

6. Support transmission with transmission jack. Remove crossmember from rear of transmission. See Fig. 2.

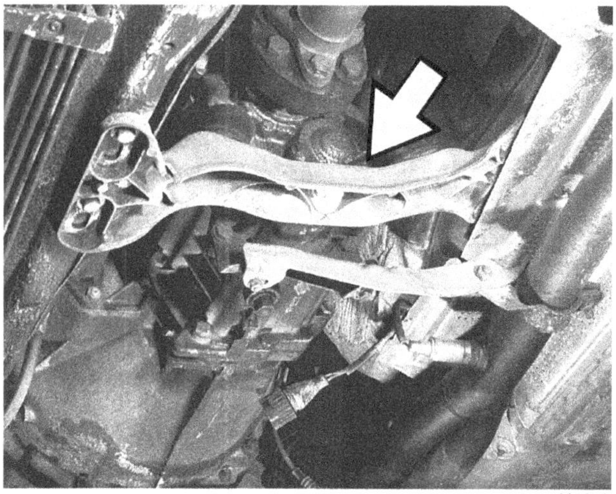

Fig. 2. Rear transmission crossmember (**arrow**).

7. Lower rear of transmission to access gearshift linkage.

> ***CAUTION —***
> *Tilting the engine to lower the transmission can lead to damage to various components due to lack of clearance.*
> *-On cars with AST remove throttle body.*
> *-Remove brake fluid reservoir if necessary.*

8. Working below transmission, disconnect shift rod from gearshift lever. Note washers on either side of shift rod end. See Fig. 3.

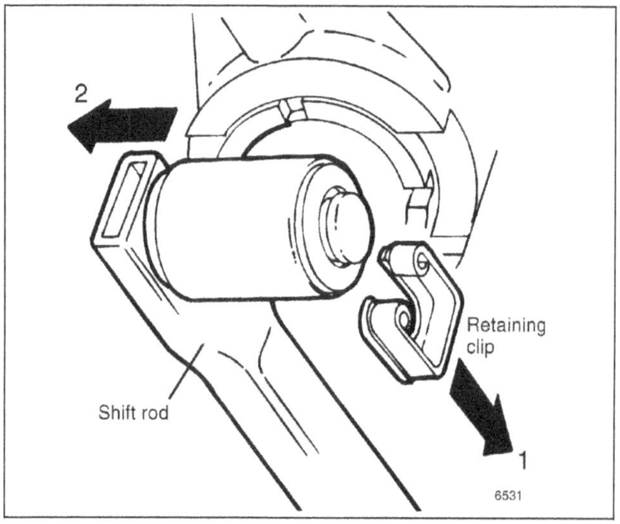

Fig. 3. Pull off shift rod retaining clip in direction of arrow (**1**); disengage shift rod from gearshift lever (**2**).

9. Release gearshift lever retaining ring from below by turning special tool 90° (¼ turn) counterclockwise. See Fig. 4.

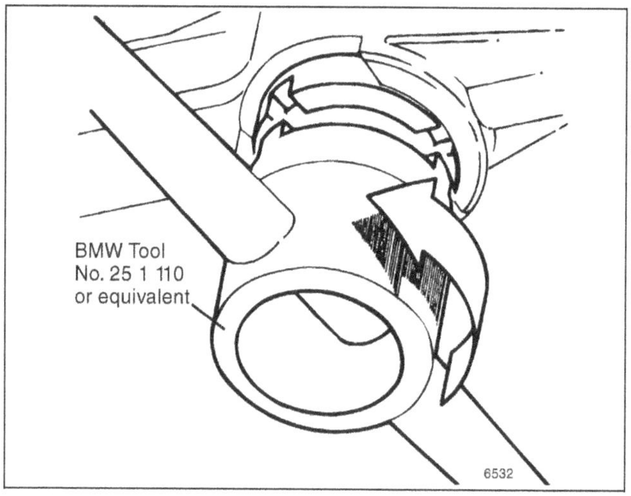

Fig. 4. Shift lever retaining ring being removed using BMW special tool 25 1 110.

10. Raise transmission and temporarily install transmission crossmember.

11. Lower vehicle. Working from inside passenger compartment, pull up on gearshift lever to remove it together with retaining ring and rubber grommet.

Gearshift lever, installing (manual transmission)

1. Install shift lever, aligning locking tabs with openings in shift console. Press down on retaining ring until it clicks into place. See Fig. 5.

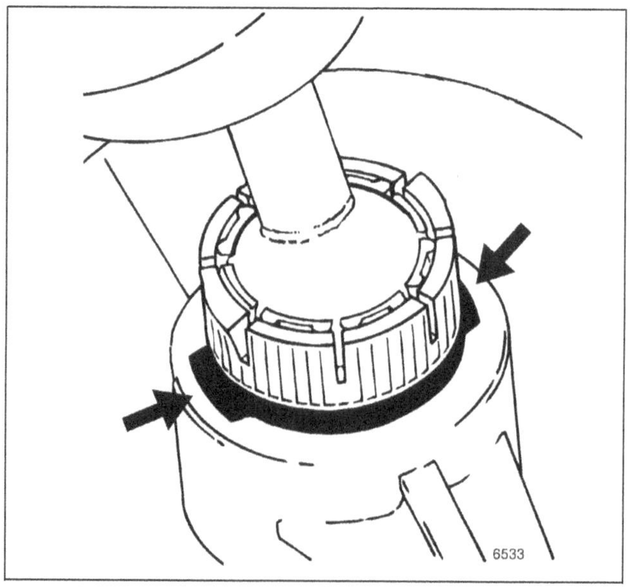

Fig. 5. Shift lever retaining ring. Fit plastic tabs into slots of shift console (**arrows**).

2. Install rubber grommet with arrow pointing forward. Install shift rod and shift boot.

NOTE —
Install rubber grommet correctly so that is seals out water.

3. Connect shift rod to shift lever. Install transmission crossmember. Lift transmission and tighten crossmember bolts.

NOTE —
Before connecting the shift rod to the lever, be sure the gearshift lever is facing the correct way as illustrated in Fig. 1.

4. Install driveshaft and heat shield. See **260 Driveshaft**.

5. Reinstall exhaust system. See **180 Exhaust System.** Connect oxygen sensor harness connectors.

6. Lower vehicle. Install shift boot cover. Push shift knob onto lever.

Tightening Torques
- Driveshaft to final drive flange (with U-joint) . . . . See **250 Driveshaft**
- Flex-disc to driveshaft or transmission flange. . . . . See **250 Driveshaft**
- Transmission crossmember to body (M8). 21 Nm (16 ft-lb)

AUTOMATIC TRANSMISSION GEARSHIFT

The automatic transmission shift mechanism is shown in Fig. 6. Use the illustration as a guide when servicing the gearshift mechanism, including shift cable replacement.

Gearshift mechanism, adjusting (automatic transmission)

1. Position selector lever in **Park**.

2. Raise vehicle to gain access to shift linkage.

WARNING —
Make sure the car is stable and well supported at all times. Use a professional automotive lift or jack stands designed for the purpose.

3. Loosen selector cable clamping nut. See Fig. 7.

4. Push shift lever forward toward engine (**Park** position) while applying light pressure on cable end. Tighten cable clamping nut.

NOTE —
Do not overtighten the nut so that it twists the cable.

Tightening Torque
- Shift cable clamping nut . . . . . . .10-12 Nm (7.5-9 ft-lb)

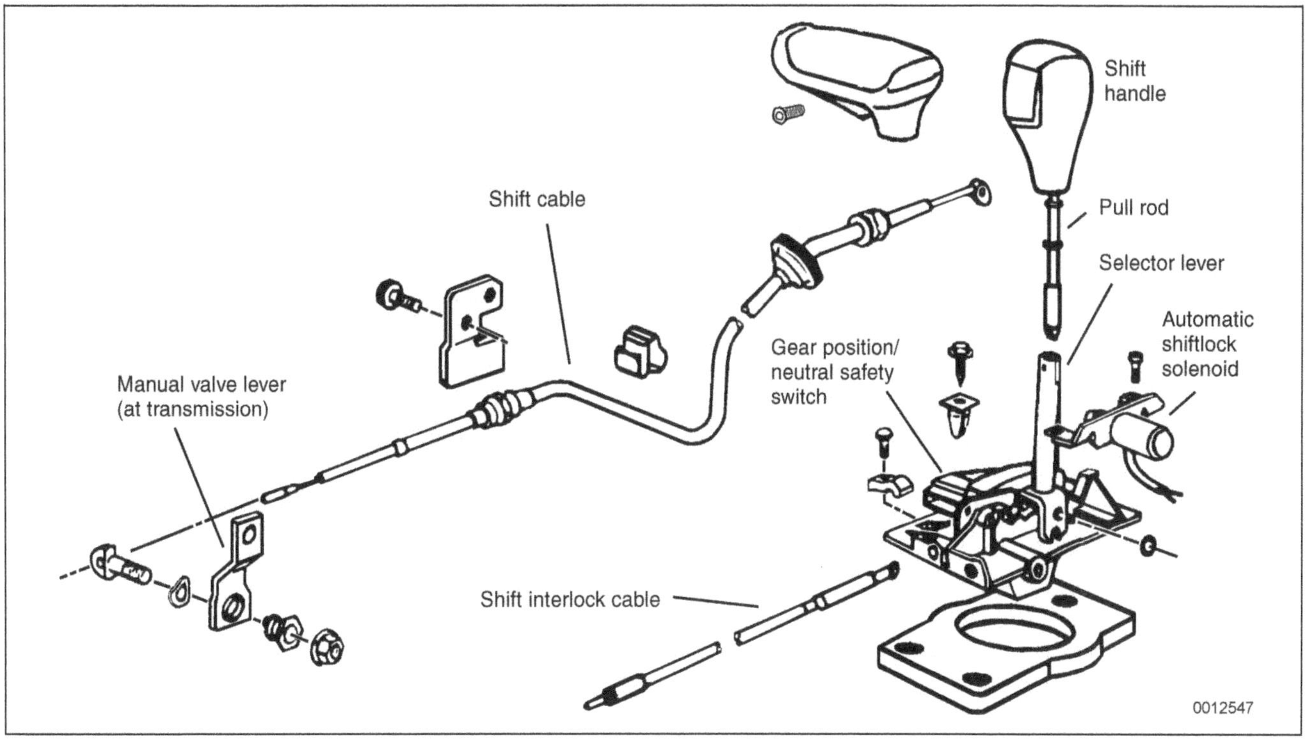

Fig. 6. Automatic transmission gearshift assembly.

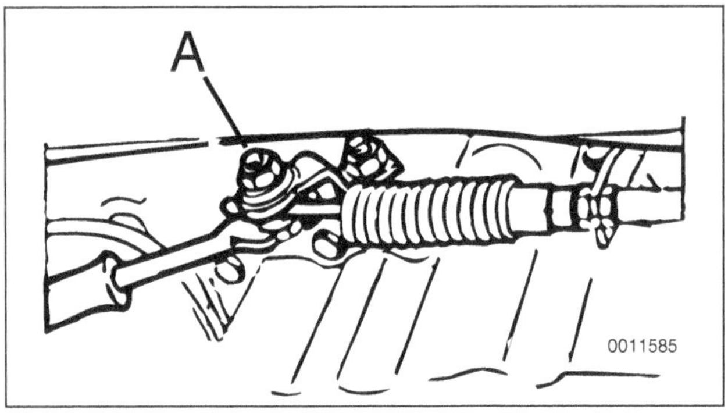

Fig. 7. Selector lever cable clamping nut (**A**). Counterhold clamping bolt when loosening nut.

Gear position/neutral safety switch, replacing (automatic transmission)

NOTE —

The gear-position/neutral safety switch is also sometimes referred to as the automatic transmission range switch.

Failure of the gear position/neutral safety switch can cause a no-start condition. In addition, since it signals gearshift position to the transmission control unit, a faulty switch can cause erratic transmission operation.

1. Remove hex-head screw (3mm allen key) from shift lever handle and remove handle. Carefully pry up cover from gearshift lever console.
2. Disconnect harness connector from gear position/neutral safety switch and remove mounting screws. See Fig. 8.
3. Remove gear position/neutral safety switch by moving selector until pin on lever aligns with slot on switch. Pull switch straight up and off. See Fig. 9.
4. Installation is reverse of removal, noting the following:
   - Before installing switch, align plastic contact slide inside switch with slot in switch. See Fig. 10.
   - Install switch so that contact slide on switch aligns with slot on switch. Then position gearshift lever pin to slot on switch and install switch.
   - When installing shift lever handle, engage pin on release button to hole in pull rod. See Fig. 11.

Tightening Torques
- Console to body . . . . 7 Nm (62 in-lb)
- Gear position/neutral safety switch to shift console . . . . .4.5 Nm (40 in-lb)

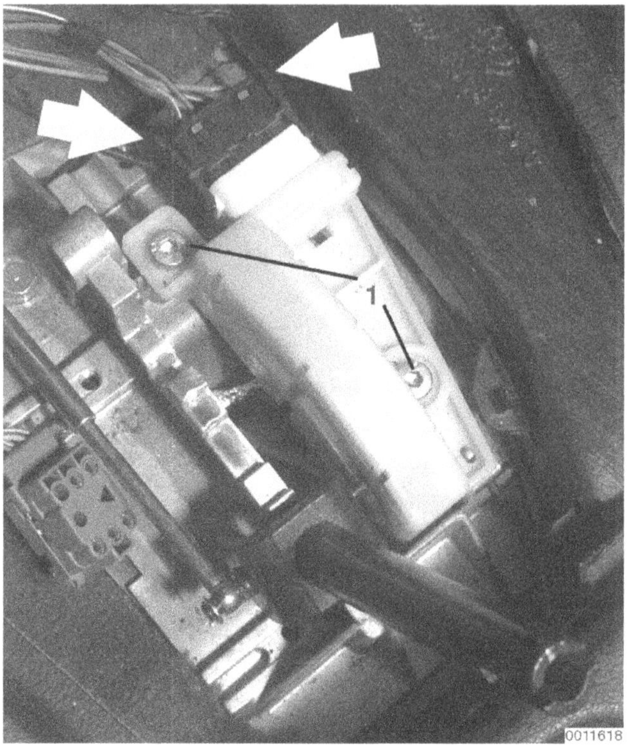

Fig. 8. Disconnect harness connector to gear position/neutral safety switch by pressing connector retaining clips together (**arrows**). Switch mounting screws shown at **1**.

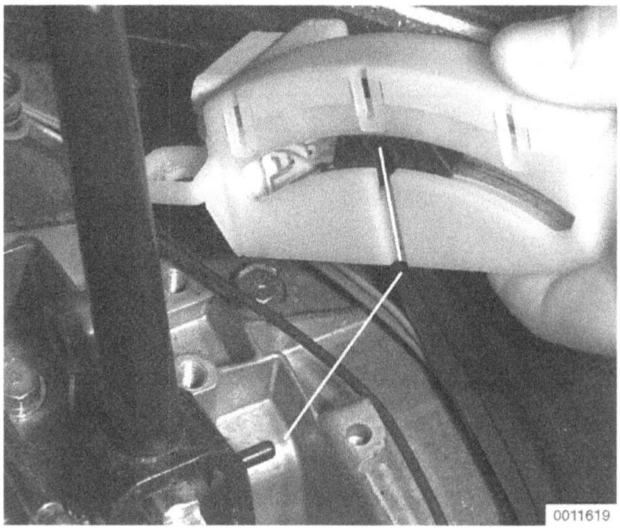

Fig. 9. Contact slide in switch aligns to slot on switch and pin on lever.

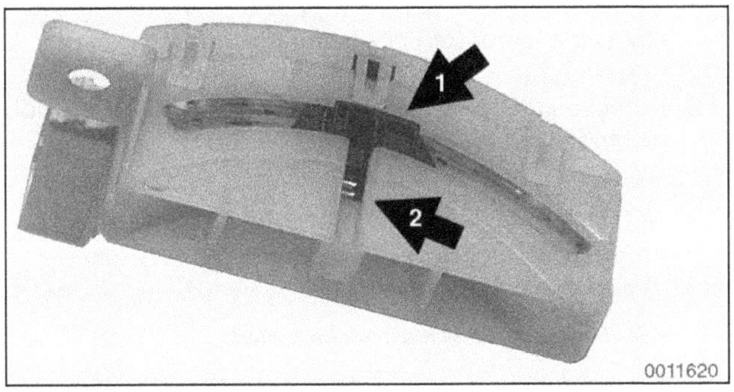

Fig. 10. Align contact slide (**1**) with slot in switch housing (**2**) before installing gear position/neutral safety switch.

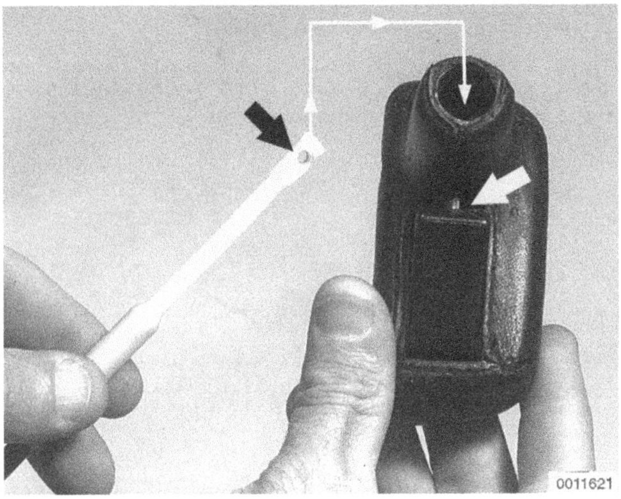

Fig. 11. Engage release button pin to hole in pull rod before installing shift lever handle.

Automatic shiftlock, checking function (automatic transmission)

The automatic shiftlock uses an electric solenoid to lock the selector lever in **P** or **N**. Depressing the foot brake with the ignition on energizes the solenoid, allowing the lever to be moved into a drive gear. The solenoid is energized only when the engine speed is below 2,500 rpm and the vehicle speed is below 3 mph. The solenoid is mounted in the right-hand side of the selector lever housing. See Fig. 12.

1. With engine running and car stopped, place selector lever in **P** or **N**.

2. Without depressing brake pedal, check that selector lever is locked in position **P** or **N**.

3. Depress brake pedal firmly. Solenoid should be heard to energize.

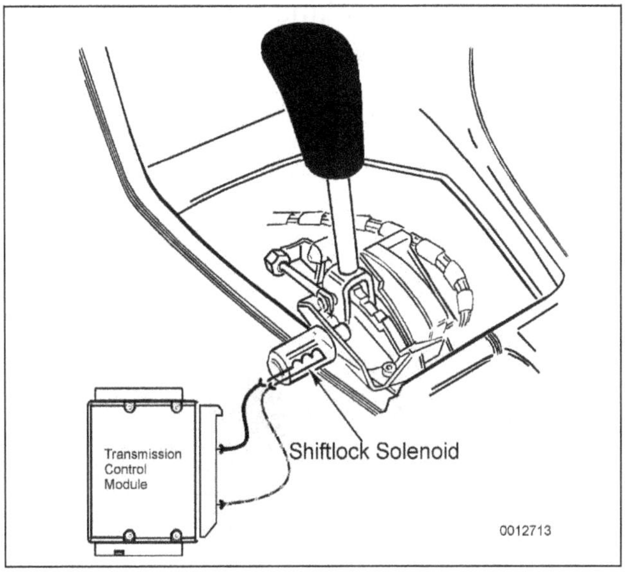

Fig. 12. Automatic shiftlock prevents drive gear selection until the brake pedal is depressed.

4. Check that selector lever can now be moved out of **P** or **N**.

> ***NOTE —***
> *The next test should be performed in an open area with the parking brake on and with extreme caution.*

5. With selector lever in **P** or **N** and brake pedal depressed, raise engine above 2,500 rpm. Check that selector lever cannot be moved out of **P** or **N**.

If any faults are found check the electrical operation of the shiftlock solenoid and check for wiring faults to or from the transmission control module (TCM). See **610 Electrical Component Locations** and **Electrical Wiring Diagrams**.

> ***NOTE —***
> *The solenoid is controlled via the TCM, using brake pedal position, engine speed, and road speed as controlling inputs.*

Shift interlock, checking function (automatic transmission)

The shift interlock uses a cable between the ignition switch and the shift lever to lock the shift lever in the park position when the key is in the off position or removed. This feature also prevents the key from being removed from the ignition lock until the selector lever is in **P**. Fig. 13.

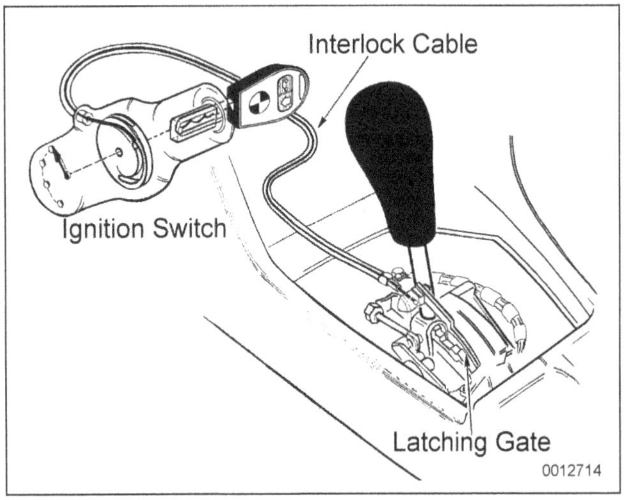

Fig. 13. Shift interlock serves two safety functions: 1) it locks shift lever in park (**P**) when the key is removed, and 2) it prevents removal of key until shift lever is moved into **P**.

1. Shift selector lever to park (**P**) position and turn ignition key to the "zero" (off) position.

2. Remove ignition key. Check that selector lever cannot be shifted out of **P** position.

   NOTE —
   - *It must only be possible to remove ignition key with selector lever in **P** position.*
   - *Shift interlock cable must not be kinked.*

3. Turn ignition key on and depress brake pedal. Check that selector lever moves freely from gear to gear.

4. With the selector lever in a gear position other than **P**, attempt to move the key to the zero position. The key should not go into the zero position.

If any faults are found check the cable for kinks and the cable attachment points for damage or faults.

260 Driveshaft

GENERAL

The two-piece driveshaft is joined in the center by a sliding splined coupling. This coupling compensates for fore and aft movement of the drive line. The driveshaft is connected to the transmission by a rubber flex-disc and to the final drive by a universal joint. It is supported in the middle by a center support bearing. The bearing is mounted in rubber to isolate vibration. See Fig. 1.

TROUBLESHOOTING

The source of driveline vibrations and noise can be difficult to pinpoint. Engine, transmission, rear axle, or wheel vibrations can be transmitted through the driveshaft to the car body. Noises from the rear of the car may be caused by final drive problems, or by faulty wheel bearings, drive axles, or even worn or improperly inflated tires.

> ***NOTE —***
> *For rear drive axle repair information, see* **330 Rear Suspension**.

Driveshaft noise or vibration may be caused by worn or damaged components. Check the universal joints for play. With the driveshaft installed, pull and twist the driveshaft while watching the joint. The BMW specification for play is very small, so almost any noticeable play could indicate a problem.

Check the torque of the fasteners at the flange connections. Check the rubber of the flex-disc and center bearing for deterioration or tearing and check for preload at the center bearing with a visual inspection. Check the driveshaft for broken or missing balance weights. The weights are welded tabs on the driveshaft tubes. In addition to inspecting for faulty driveshaft parts, the installed angles of the driveshaft should also be considered.

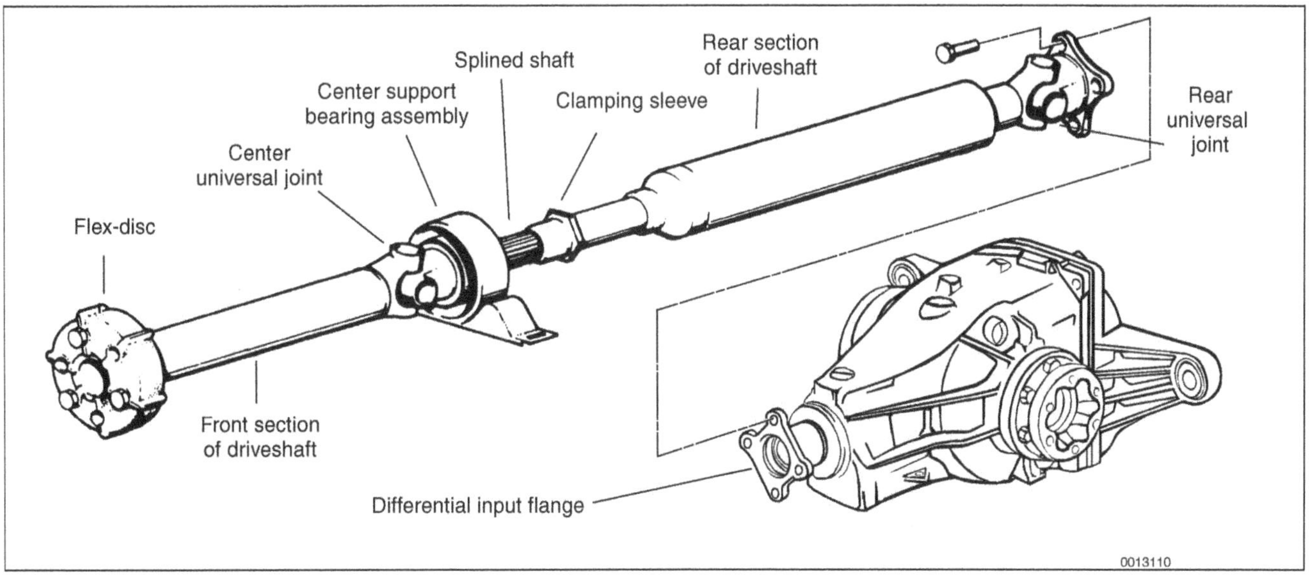

Fig. 1. Driveshaft and final drive assembly.

Further inspection requires removal of the driveshaft. Check the front centering guide on the transmission output flange for damage or misalignment. Also check runout at the transmission output flange and output shaft, and at the final drive input flange. Check the bolt hole bores in the flange for wear and elongation.

Driveshaft flange runout (max. allowable)
- Transmission output flange
  axial play .... 0.10 mm (0.004 in.)
  radial play.................... 0.07 mm (0.003 in.)
- Final drive input flange radial play (measured at driveshaft centering lip) .... 0.07 mm (0.003 in.)

Spin the center bearing and check for the smooth operation without play. Check that the splines of the sliding coupling move freely. Clean and lubricate the splines with molybdenum disulfide grease (Molykote® Longterm 2 or equivalent).

Check the universal joints for wear or binding. If any joint is difficult to move or binds, the driveshaft section should be replaced.

NOTE —
With the driveshaft installed, the actual amount that the universal joints pivot is limited. For the most accurate test, check universal joints in their normal range of movement.

If inspection reveals nothing wrong with the driveshaft, it may need to be rebalanced. This can be done by a speciality driveshaft repair shop. Also, check the alignment of the driveshaft as described below.

NOTE —
Minor driveshaft vibrations can often be corrected simply by disconnecting the driveshaft at the final drive, and repositioning the driveshaft 90°, 180° or 270° in relation to the final drive input flange.

Table a lists symptoms of driveshaft problems and their probable causes. Most of the repair information is contained within this repair group.

Table a. Driveshaft Troubleshooting

Symptom	Probable cause	Corrective action
1. Vibration when starting off (forward or reverse).	a. Incorrect preload of center bearing.	a. Check preload of center bearing. Readjust preload. See Fig. 6.
	b. Center bearing rubber deteriorated.	b. Inspect center bearing and rubber. Replace if necessary.
	c. Flex-disc damaged or worn.	c. Inspect flex-disc. Replace if necessary.
	d. Engine or transmission mounts faulty.	d. Inspect engine and transmission mounts. Align or replace, if necessary.
	e. Front centering guide worn, or driveshaft mounting flanges out of round.	e. Check front centering guide and replace if necessary. Check runout of driveshaft flanges.
	f. Universal joints worn or seized.	f. Check universal joint play and movement. Replace driveshaft if necessary.
	g. Sliding coupling seized.	g. Remove driveshaft and check movement of sliding coupling. Clean coupling splines and replace parts as necessary.
	h. Driveshaft misaligned.	h. Check driveshaft alignment.
2. Noise during on/off throttle or when engaging clutch.	a. Final drive components worn or damaged (excessive pinion-to-ring-gear clearance).	a. Remove final drive and repair.
	b. Drive axle or CV joint faulty.	b. Inspect drive axles and CV joints. Repair or replace as necessary.
	c. Sliding coupling seized.	c. Remove driveshaft and check movement of sliding coupling. Clean coupling splines and replace parts as necessary.
3. Vibration at 25 to 30 mph (40 to 50 km/h).	a. Front centering guide worn, or driveshaft mounting flanges out of round or damaged.	a. Check front centering guide and replace if necessary. Check runout of driveshaft mounting flanges.
	b. Universal joints worn or seized.	b. Check universal joint play and movement. Replace driveshaft if necessary.
	c. Flex-disc damaged or worn.	c. Inspect flex-disc. Replace if necessary.
	d. Center bearing rubber deteriorated.	d. Inspect center bearing. Replace if necessary.
	e. Sliding coupling seized.	e. Remove driveshaft and check movement of sliding coupling. Clean coupling splines and replace parts as necessary.
	f. Misaligned installed position.	f. Check driveshaft alignment.

Table a. Driveshaft Troubleshooting (continued)

Symptom	Probable cause	Corrective action
4. Vibration, audible rumble over 35 mph (60 km/h).	a. Front centering guide worn, or driveshaft mounting flanges out of round or damaged.	a. Check front centering guide and replace if necessary. Check run-out of driveshaft mounting flanges.
	b. Mounting flange bolts loose or holes worn.	b. Remove driveshaft and check transmission output flange and final drive input flange. Replace if necessary.
	c. Driveshaft unbalanced.	c. Check driveshaft for loose or missing balance weights. Have driveshaft rebalanced or replace if necessary.
	d. Universal joints worn or seized.	d. Check universal joint play and movement. Replace driveshaft if necessary.
	e. Sliding coupling seized.	e. Remove driveshaft and check movement of sliding coupling. Clean coupling splines or replace parts as necessary.
	f. Incorrect preload of center bearing.	f. Check preload of center bearing. Readjust if necessary.
	g. Center bearing faulty.	g. Replace center bearing.
	h. Final drive rubber mount faulty.	h. Inspect final drive rubber mount and replace if necessary.
	i. Driveshaft misaligned.	i. Check driveshaft alignment.

DRIVESHAFT SERVICE

Repair kits for the universal joints are not available. Worn or damaged universal joints usually require replacement of that section of the driveshaft.

The driveshaft is balanced to close tolerances. Whenever it is to be removed or disassembled, the mounting flanges and driveshaft sections should be marked with paint or a punch before proceeding with work. This will ensure that the driveshaft can be reassembled or installed in exactly the same orientation.

Universal joint play

- Maximum allowable . . . . 0.15 mm (0.006 in.)

Driveshaft, aligning

The alignment of the driveshaft does not normally need to be checked unless the engine/transmission or the final drive have been removed and installed. If, however, all other parts of the driveshaft have been inspected and found to be okay, but there is still noise or vibration, driveshaft alignment should be checked.

There are two important driveshaft alignment checks. The first is to make sure that the driveshaft runs straight from the transmission to the final drive, without any variation from side-to-side caused by misalignment of the engine/transmission in its mounts. Make a basic check by sighting along the driveshaft from back to front. Any misalignment should be apparent from the center bearing forward. To adjust the side-to-side alignment, loosen the transmission or engine mounts to reposition them, then retighten the mounts. The driveshaft should be exactly centered in the driveshaft tunnel. See Fig. 2.

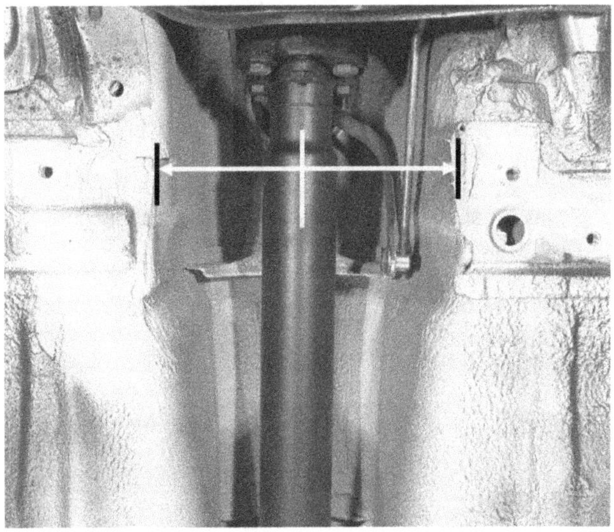

5326

Fig. 2. Driveshaft side-to-side alignment. Driveshaft should be centered in driveshaft tunnel.

The second important driveshaft alignment check is more complicated. It checks the amount the driveshaft is angled vertically at the joints. This angle is known as driveshaft deflection.

In general, there should be little deflection in the driveshaft between the engine, the center bearing, and the final drive. Precise checks require the use of a large protractor or some other means of measuring the angle of the engine and the final drive and comparing these angles to the angle of the driveshaft sections.

To change the deflection angle, shims can be placed between the center bearing and the body or between the transmission and its rear support. When using shims to change a deflection angle, keep in mind that the angle of adjacent joints will also change. Deflection angles should be as small as possible.

> ***CAUTION —***
> *The maximum allowable change in height of the center bearing or transmission support using shims is 3 mm (0.12 in.).*

Driveshaft, removing

> ***WARNING —***
> - *Be sure the wheels are off the ground before removing the driveshaft. Set the parking brake before removing the driveshaft.*
> - *Once the driveshaft has been removed, the vehicle can roll regardless of whether the transmission is in gear or not.*
> - *The driveshaft is mounted to the transmission and final drive with self-locking nuts. These nuts are designed to be used only once and should be replaced during reassembly.*

1. Disconnect oxygen sensor electrical connectors and remove complete exhaust system. See **180 Exhaust System**.
2. Remove exhaust heat shield and heat shield cross-member.
3. Matchmark front and rear driveshaft connections at transmission and final drive.
4. Remove driveshaft mounting bolts. See Fig. 3. Discard old nuts.

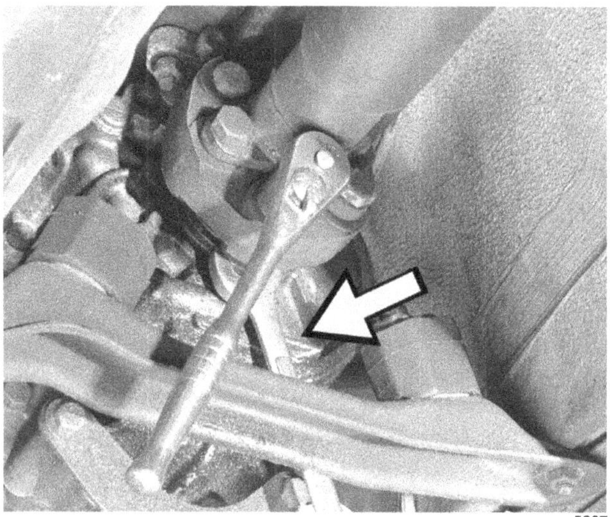

Fig. 3. Front driveshaft to transmission mounting bolt being removed. Note open-end wrench (**arrow**) being used to counterhold bolt.

5. Loosen threaded clamping sleeve on driveshaft a few turns. See Fig. 4.

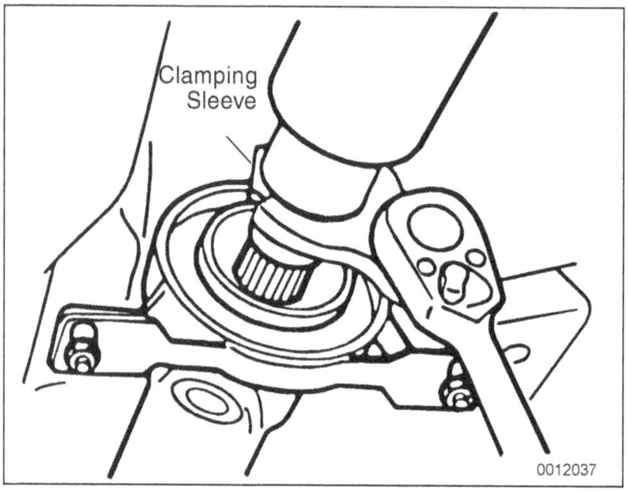

Fig. 4. Threaded clamping sleeve nut being loosened.

6. Support driveshaft and remove center support bearing mounting bolts.
7. Remove driveshaft but do not separate two halves. Pull down on center of driveshaft to facilitate removal.

> ***NOTE —***
> *If driveshaft halves were separate and not matchmarked, assemble driveshaft as shown in drawing. If a vibration occurs, disassemble driveshaft and rotate one section 180°. See Fig. 5.*

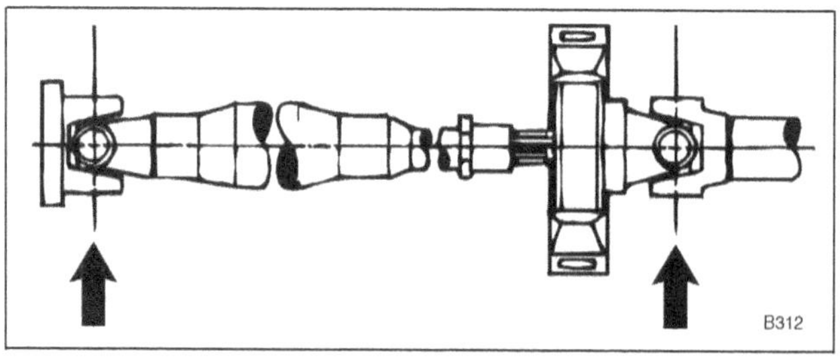

Fig. 5. Driveshaft alignment. Centerlines of universal joints (**arrows**) must be parallel or at 90° to each other.

Driveshaft, installing

1. Align driveshaft matchmarks at final drive and then at transmission flange. Start attaching nuts. Use new self-locking nuts.
2. Position center support bearing and start attaching nuts. Use new self-locking nuts.
3. Tighten flange nuts while preventing bolts from twisting in coupling. Tighten final drive flange first, then tighten coupling at transmission.
4. Preload center support bearing by pushing bearing forward 4-6 mm (0.16-0.24 in.) from center. See Fig. 6. Tighten attaching bolts.

5331

Fig. 6. Preload center bearing toward front 4-6 mm (**arrow**).

5. Tighten threaded sleeve on driveshaft to proper torque.

6. Install heat shield and heat shield crossmember.

7. Install exhaust system. See **180 Exhaust System**. Connect wiring harness to oxygen sensors.

8. Road test vehicle to check for noise or vibration.

> ***WARNING —***
> - *Do not reuse self-locking nuts. These nuts are designed to be used only once.*
> - *Avoid stressing the flex-disc when torquing the bolts. Do this by holding the bolts steady and turning the nuts on the flange side.*

Tightening Torques

- Driveshaft center mount to body .... 21 Nm (16 ft-lb)
- Driveshaft clamping sleeve .... 10 Nm (89 in-lb)
- Driveshaft to final drive flange
  - With CV joint (M8) .... 32 Nm (23 ft-lb)
  - Wth U-joint (M10 ribbed nut) .... 80 Nm (59 ft-lb)
  - With U-joint (M10 compressed nut) .. 60 Nm (44 ft-lb)
- Flex-disc to driveshaft or transmission flange
  - M10 (8.8 grade) .... 48 Nm (35 ft-lb)
  - M10 (10.9 grade) .... 64 Nm (47 ft-lb)
  - M12 (8.8 grade) .... 81 Nm (60 ft-lb)
  - M12 (10.9 grade) .... 100 Nm (74 ft-lb)
  - M12 (10.9 grade) (M3 models)..... 115 Nm (85 ft-lb)
- Transmission crossmember to body (M8).................... 21 Nm (16 ft-lb)

NOTE —
Bolt grade is marked on the bolt head. When replacing bolts, only use bolts of the same strength and hardness as the originals installed.

Flex-disc, replacing

The flex-disc between the front section of the driveshaft and the output flange of the transmission should be checked for cracks, tears, missing pieces, or distortion. Check for worn bolt hole bores in the flange.

1. Remove driveshaft as described in **Driveshaft, removing**.

NOTE —
It is possible to only partially remove the driveshaft, leaving it connected to the final drive. The driveshaft can be tilted down in the center and slid off the transmission flange once the clamping sleeve is loosened and the center bearing bracket is unbolted. Suspend the driveshaft using stiff wire in as close to the installed position as possible. If the driveshaft hangs unsupported, the rear universal joint may be damaged.

2. Unbolt flex-disc from driveshaft.

NOTE —
Removal and installation of the bolts may be made easier by placing a large hose clamp around the flex-disc, and tightening the clamp slightly to compress the coupling.

3. Install new flex-disc using new self-locking nuts. Molded arrows on coupling should face flange arms. See Fig. 7.

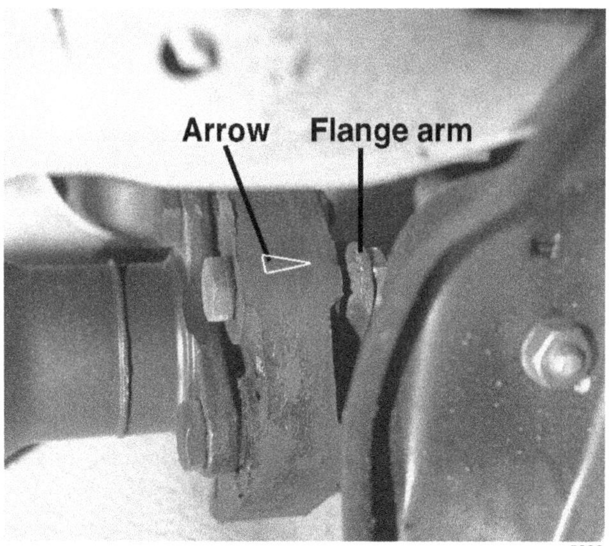

5332

Fig. 7. When attaching flex-disc, molded arrows must point toward flange arms.

4. Install driveshaft as described in **Driveshaft, installing**.

NOTE —
Torque only the nuts while holding the bolt heads. This will prevent damaging or fatiguing the rubber.

Tightening Torques

- Driveshaft center mount to body .... 21 Nm (16 ft-lb)
- Driveshaft clamping sleeve .... 10 Nm (89 in-lb)
- Driveshaft to final drive flange
  With CV joint (M8) .... 32 Nm (23 ft-lb)
  Wth U-joint (M10 ribbed nut) .... 80 Nm (59 ft-lb)
  With U-joint (M10 compressed nut) .. 60 Nm (44 ft-lb)
- Flex-disc to driveshaft or transmission flange
  M10 (8.8 grade) .... 48 Nm (35 ft-lb)
  M10 (10.9 grade.................. 64 Nm (47 ft-lb)
  M12 (8.8 grade) .... 81 Nm (60 ft-lb)
  M12 (10.9 grade) .... 100 Nm (74 ft-lb)
  M12 (10.9 grade) (M3 models)..... 115 Nm (85 ft-lb)
- Transmission crossmember
  to body (M8).................... 21 Nm (16 ft-lb)

Center bearing assembly, replacing

To replace the center bearing assembly, the driveshaft must be removed from the car. The center bearing assembly consists of a grooved ball bearing in a rubber mount. The bearing assembly is pressed onto the front section of the driveshaft and secured by a circlip. See Fig. 8.

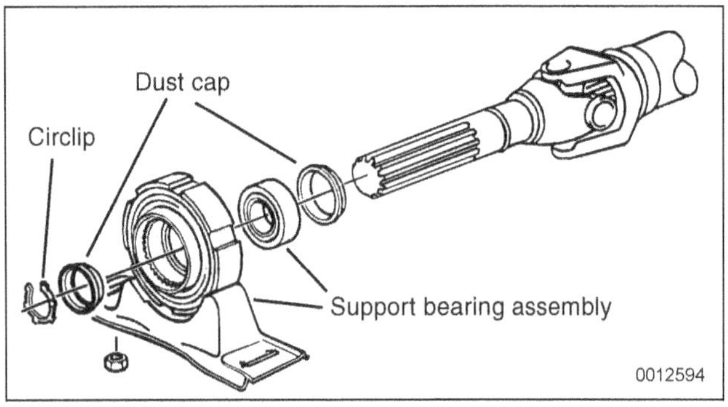

Fig. 8. Exploded view of driveshaft support bearing assembly.

1. Remove driveshaft. See **Driveshaft, removing**.

2. Make matching marks on front and rear driveshaft sections. See Fig. 9.

3. Loosen clamping sleeve fully and pull driveshaft sections apart. Remove rubber bushing, washer, and clamping sleeve from front section.

5333

Fig. 9. Before pulling apart driveshaft sections, make matching marks as shown.

4. Inspect condition of rubber bushing for splined coupling. Replace worn or damaged parts.

5. Remove center bearing circlip and dust guard. See Fig. 10. Remove bearing from driveshaft using puller.

5334

Fig. 10. Center bearing circlip (**arrow**) to be removed. Driveshaft shown installed.

NOTE —
Install puller so that it pulls on inner hub of bearing. Pulling on outer ring of mount may tear rubber, and entire bearing assembly will need to be replaced.

6. Before installation, make sure dust guard is on driveshaft, and then press center mount onto driveshaft flush with dust guard.

7. Place clamping sleeve, washer, and rubber bushing on front driveshaft section. Lubricate splines with molybdenum disulfide grease (Molykote® Longterm 2 or equivalent) and then reassemble driveshaft.

NOTE—
Do not retighten clamping sleeve until driveshaft is installed.

8. Install driveshaft. See **Driveshaft, installing**.

9. Tighten clamping sleeve.

Tightening Torque
• Clamping sleeve. . . . . 10 Nm (7.5 ft-lb)

Front centering guide, replacing

The front centering guide centers the driveshaft in relation to the transmission. The guide is press-fit into a cavity in the front of the driveshaft and slides onto the transmission output shaft.

No specifications are given for wear of the guide, but generally the guide should fit snugly on the transmission output shaft.

NOTE—
Some driveshafts have a dust cap installed on the end of the driveshaft, over the centering guide. The dust cap may become bent or distorted when the driveshaft is removed or installed. Damage to the dust cap should not affect the centering guide and should not be mistaken for guide wear.

1. Remove driveshaft. See **Driveshaft, removing**.

2. Pack cavity behind centering guide with heavy grease until grease is flush with bottom edge of guide.

3. Insert 14 mm (approximately ½ in.) diameter mandrel or metal rod into guide. Strike guide with hammer to force centering guide out.

NOTE—
The mandrel should fit snugly in the centering guide so that the grease cannot escape around the sides of the mandrel.

4. Remove old grease from driveshaft, lubricate new centering guide with molybdenum disulfide grease (Molykote® Longterm 2 or equivalent) and drive it into driveshaft.

NOTE—
The sealing lip of the guide should face outward and it should be driven into the driveshaft to a specified depth. See Fig. 11.

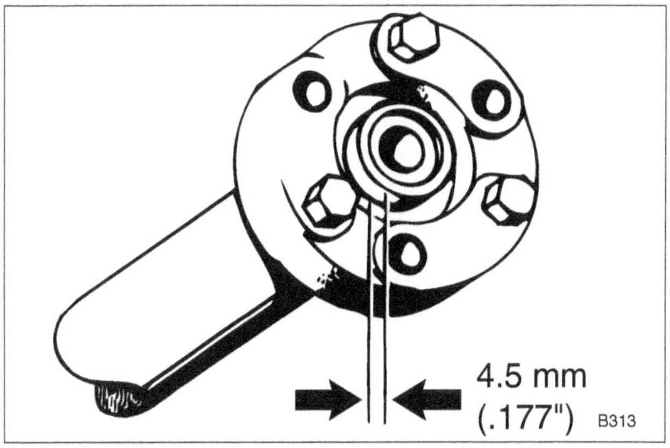

Fig. 11. When installing new driveshaft centering guide, drive guide in until its protrusion depth is as shown.

5. Install driveshaft. See **Driveshaft, installing**.

300 Suspension, Steering and Brakes–General

GENERAL

This section covers general information for front and rear suspension, the steering system, and the braking and traction control systems.

The BMW E36 suspension and steering systems are engineered to provide a compromise between taut, responsive handling and ride comfort. Although the front and rear suspension assemblies are independent subsystems, they work together to achieve BMW's overall combination of precise handling and comfort. See Fig. 1.

Front Suspension

The lower control arms connect the front suspension struts to mounting points on the subframe crossmember and the body. See Fig. 2.

Each front strut assembly includes a shock absorber cartridge inside the tubular strut housing and a large coil spring. The upper strut mount bearing allows the strut to turn with the wheels. The lower end of the strut, which includes the stub axle for the front wheel, connects to a ball-joint on the control arm.

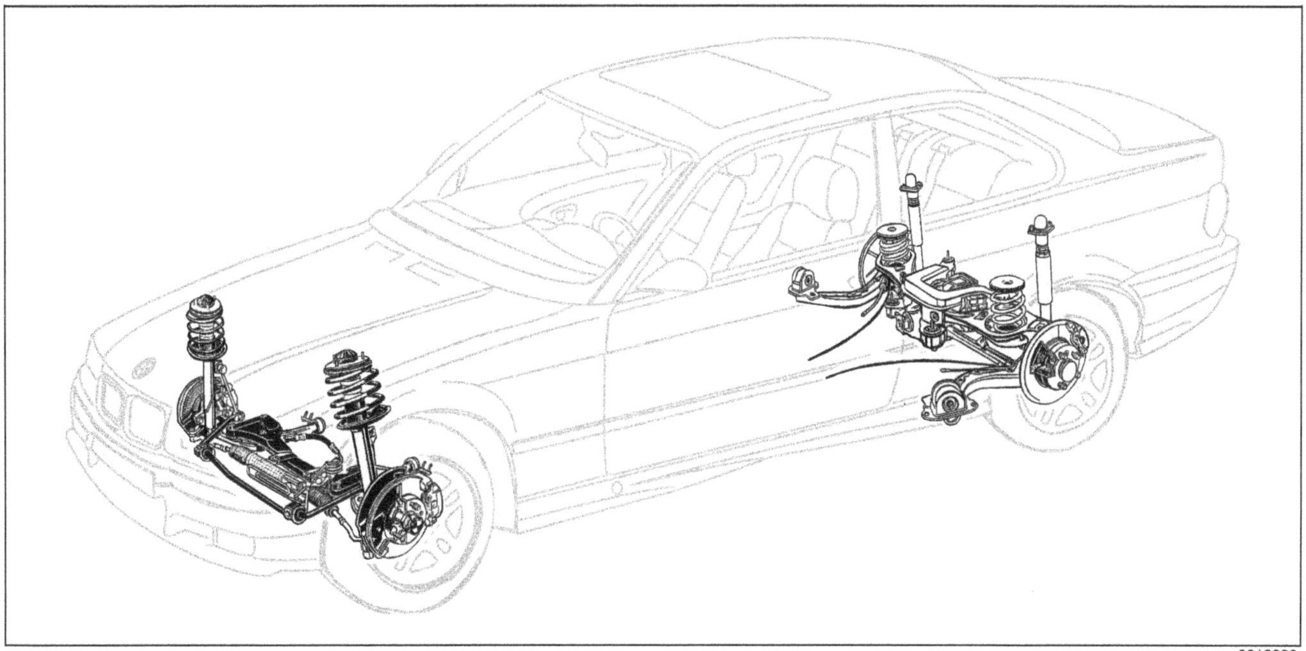

Fig. 1. BMW E36 front and rear suspension systems.

0012124

Fig. 2. Front suspension control arm (**arrow**).

The front suspension is designed with minimum positive steering offset. This geometry contributes to stability when traction is unequal from side to side. Suspension travel is limited by rubber bump stops.

The three point mounting of each L-shaped control arm precisely controls the front-to-rear and side-to-side position of the strut, while the flexibility of the joints and mounts also allows the movement necessary for suspension travel. The control arm mounting points are designed with anti-dive geometry. The suspension reduces the normal tendency for the front of the vehicle to dive under hard braking.

Control arm position is fixed, with no adjustment provisions on the control arms for altering front wheel alignment. A stabilizer bar mounted to both control arms helps to reduce body roll when cornering.

Steering

The variable-assist power steering system consists of an engine-driven hydraulic pump, a rack-and-pinion type steering gear, and connecting linkage to the road wheels. The E36 utilizes an engine-speed dependent variable effort steering system. At low speeds, maximum power assist is provided to ease parking and city driving. At high speeds, assist is reduced to ensure stability.

The steering linkage connects the rack-and-pinion unit through tie rods to the steering arms. The tie rod ends allow the wheels to pivot and react to suspension travel.

Rear Suspension

The rear axle carrier is the main mounting point for the final drive housing and the rear suspension components.

Trailing arms locate the rear wheels and anchor the springs, shocks and stabilizer bar. Drive axles with constant-velocity (CV) joints at both ends transfer power from the differential to the road wheels. The differential is mounted to the rear axle carrier through rubber mounts and bushings to help isolate drivetrain noise and vibration.

Brakes

E36 cars are equipped with power disc brakes with an integral antilock brakes (ABS). The parking brake is a dual-drum system integrated with the rear brake rotors. See Fig. 3.

Power assist is provided by a vacuum booster when the engine is running. The brake pedal pushrod is connected directly to the master cylinder, so failure of the vacuum booster does not normally result in total brake failure.

Each disc brake uses a caliper with a single hydraulic cylinder. Brake pads in the left front and right rear contain wear sensors. When the pads need replacement, the sensors illuminate a light on the dashboard.

Tires and Wheels

Tire size is critical to the proper operation of the E36 ABS or ABS/AST system. Several different styles of wheels, in 15, 16 and 17 inch diameters, are available from an authorized BMW dealer.

> ***NOTE—***
> *Aftermarket wheels should be selected with care. Improperly fitted wheels can contact and damage suspension, brake or body components and may adversely affect vehicle stability.*

INTEGRATED SYSTEMS

Antilock Brake System (ABS) is standard on all E36 cars. Standard on some models and installed as optional equipment on others, is All Season Traction (AST).

Antilock Brake System (ABS)

The electronically-controlled ABS maintains vehicle stability and control during emergency braking by preventing wheel lock-up. ABS provides optimum deceleration and stability dur-

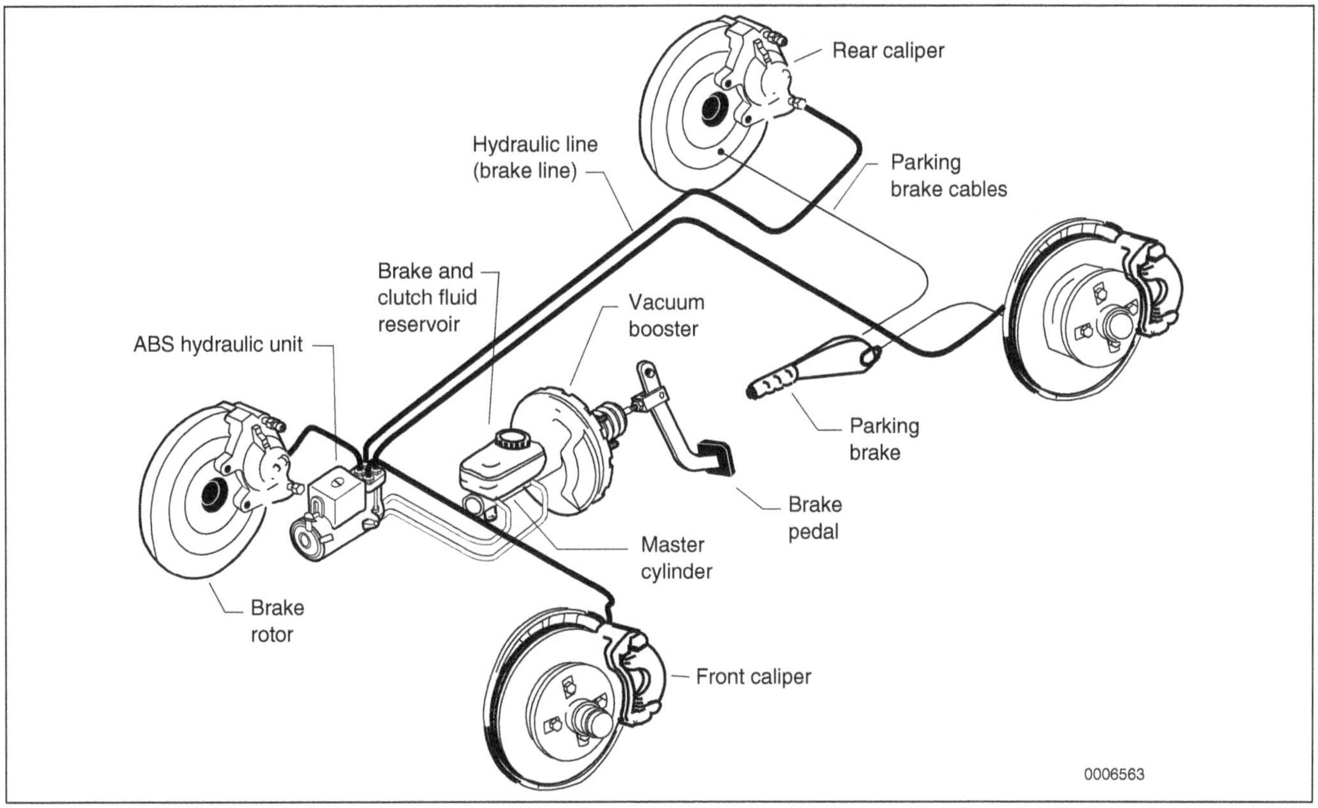

Fig. 3. Schematic view of dual circuit brakes with 4-channel ABS.

ing adverse conditions. It automatically adjusts brake system hydraulic pressure at each wheel to prevent wheel lock-up.

The system's main components are the wheel speed (pulse) sensors, the ABS control module, and the hydraulic control unit. See Fig. 4.

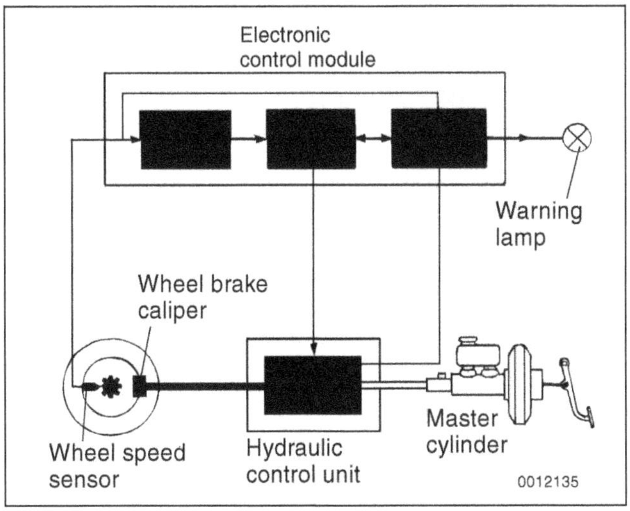

Fig. 4. Schematic representation of ABS.

The wheel speed sensors continuously send wheel speed signals to the control module. See Fig. 5. The control module compares these signals to determine, in fractions of a second, whether any of the wheels are about to lock. If any wheel is nearing a lock-up condition, the module signals the hydraulic unit to maintain or reduce pressure at the appropriate wheel(s). Pressure is modulated by electrically-operated solenoid valves in the hydraulic unit.

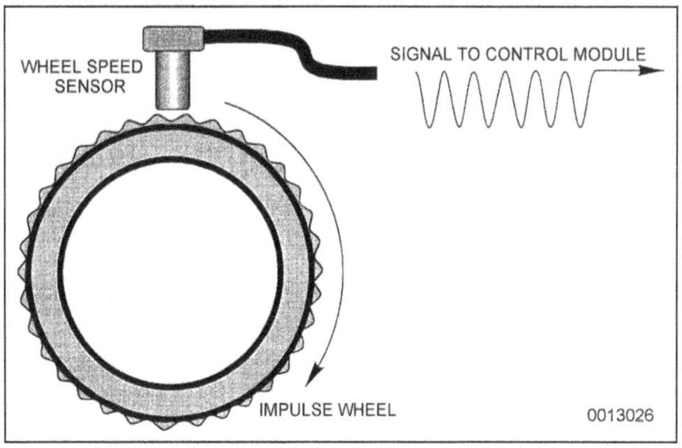

Fig. 5. ABS wheel speed sensor.

Three-channel ABS is used on all E36 cars equipped with standard ABS. Each front wheel has a separate solenoid valve to modulate hydraulic pressure to that brake caliper. The rear wheels are modulated as a pair, controlled by a single solenoid valve.

E36 cars equipped with ABS/AST utilize a four channel system, which functions exactly as the three channel system above, except that the rear wheels are modulated individually.

All Season Traction (AST)

The traction control system works in conjunction with ABS and the engine management system to enhance vehicle control. The main function of the AST system is to maintain the rolling contact between the tires and the road surface under all driving conditions. This is achieved through exact application and management of braking and drivetrain forces.

The throttle control system used in AST is illustrated in Fig. 6.

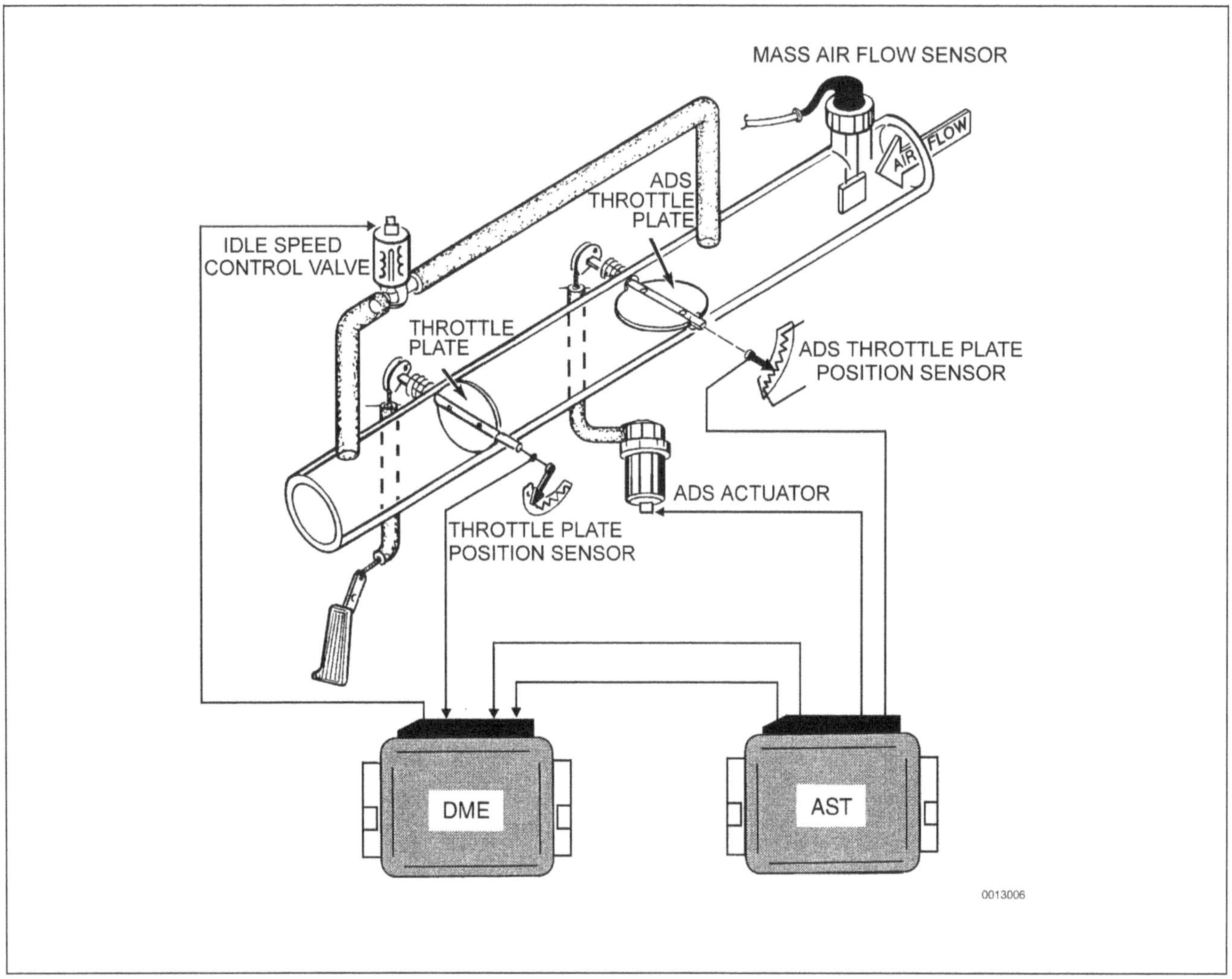

Fig. 6. All Season Traction (AST) works with other drivetrain systems to enhance vehicle control.

NOTE—
The traction control system referred to as AST (all season traction) may also be referred to as ASC (Automatic Stability Control) and ASC+T (Automatic Stability Control+Traction).

The AST system improves traction by electronically applying the rear brakes when the rear drive wheels are spinning at a faster rate than the front wheels. The combined ABS/AST control module, operating through the ABS hydraulic control unit, modulates braking force at the rear wheels.

In addition, AST uses retarded ignition timing and an auxiliary engine throttle plate to reduce engine torque and maintain vehicle stability.

The auxiliary throttle plate is held open by spring pressure. The AST system activates the auxiliary throttle position motor (ADS) to close the AST throttle as needed. This reduces the volume of engine intake air. Due to the throttle closing very rapidly during AST operation the driver cannot increase the engine power output regardless of how far the accelerator pedal is pushed to the floor.

The auxiliary throttle plate is placed ahead of the conventional throttle plate in the throttle body.

Traction control also comes into operation during deceleration. Decelerating on snowy or icy road surfaces can lead to rear wheel slip. If a rear wheel starts to drag or lock up when the throttle is closed rapidly or during downshifting, the AST system can limit the problem by advancing the ignition timing.

A switch on the center console is used to toggle the AST on or off.

The AST system is designed to be maintenance free. There are no adjustments that can be made. Repair and troubleshooting of the AST system requires special test equipment and knowledge and should be performed only by an authorized BMW dealer. **Table a** lists the conditions indicated by the AST indicator light in the instrument cluster

Table a. AST Indicator Lamp Function

Indicator lamp	Condition	Comment
Light on	Normal AST start-up	Automatic AST self-test
Light off	AST monitoring mode	Automatic AST operation
Press AST button, light comes on	AST off (disabled)	Rocking the car to get out of snow or other loose surface Driving with snow chains
Press AST button: light goes out	AST monitoring mode	Automatic AST operation
Light flashes	AST active mode	Normal AST operation as it controls wheel speed
Light stays on after start up or comes on while driving	Defect in AST	Consult BMW dealer for diagnosis/repair (Vehicle operation remains normal)

WARNING—
Even a car with AST is subject to the normal physical laws. Avoid excessive speeds for the road conditions encountered.

CAUTION—
- *If the tires on the car are of different makes, the AST system may over-react. Only fit tires of the same make and tread pattern,*
- *In adverse conditions, such as trying to rock the car out of deep snow or another soft surface, or when snow chains are fitted, it is advisable to switch off AST and allow the car's driveline to operate conventionally.*

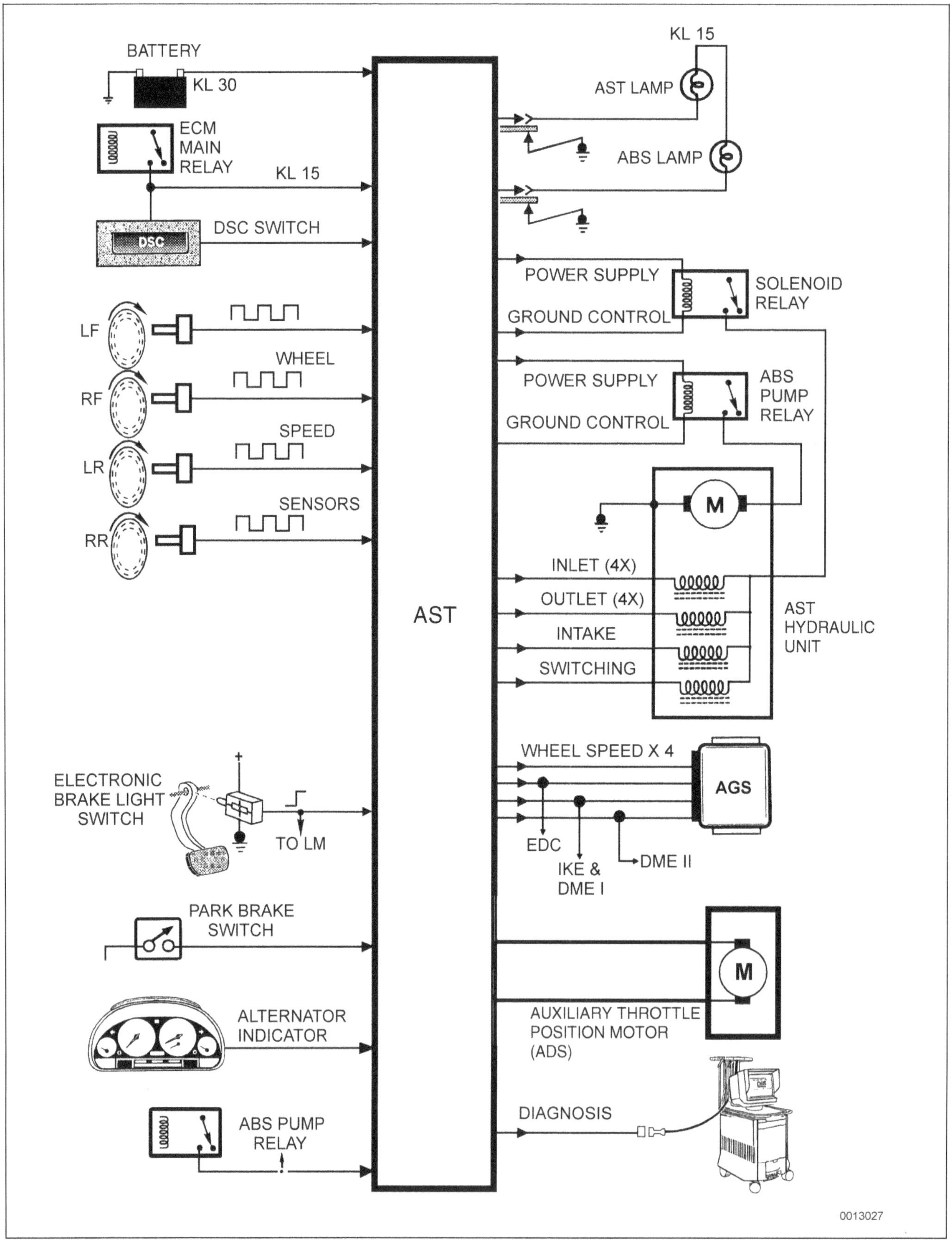

Fig. 7. Schematic of All Season Traction (AST) system.

M3 SUSPENSION AND BRAKES

The E36 M3 was introduced in late 1994 as a 1995 model. The design of the M3 suspension, steering and brakes is a modification of existing E36 systems which accommodates increased power output and road-handling demands.

The steering geometry is modified by changing the caster/camber angles slightly. This is accomplished by repositioning the upper strut mounting holes in the front strut towers and through different front and rear control arms.

M3 Front Suspension Improvements
- Reinforced spring plates, control arms, control arm bushings, steering arms.
- Recalibrated struts, firmer springs.
- Stiffer stabilizer bar, larger stabilizer bar bushings.
- Front stabilizer linked to struts.

M3 Rear Suspension Improvements
- Reinforced trailing arms.
- Recalibrated shock absorbers, firmer springs.

M3 Brake System Improvements
- Vented directional front rotors: 315 mm dia. x 28 mm thick.
- Vented directional rear rotors: 312 mm dia. x 20 mm thick.
- Larger brake pads and calipers.
- Modified Teves ABS system.

NOTE—
M3 brake rotors are directional, having a different part number for each side. Rotors with an odd part number go on the left while rotors with an even part number go on the right.

M3 Wheels/Tires
- Front wheels 1992-1998. . . . . . 7.5J x 17 in. light alloy
- Rear wheel 1992-1997. . . . . . . 7.5J x 17 in. light alloy
- Rear wheel (option) . . . . .8.5 J x 17 in. light alloy
- Front and rear tire 1992-1997 . . . . 235/40 x 17 in.
- Front tire 1998 only. . . . . 225/45 x 17 in.
- Rear tire 1998 only. . . . . 245/40 x 17 in.

CAUTION—
In an emergency, an M3 front wheel/tire assembly may be mounted on the rear axle. However a rear wheel/tire assembly must ***not*** *be mounted on the front axle.*

SUSPENSION AND STEERING TROUBLESHOOTING

Stable handling and ride comfort both depend on the integrity of the suspension and steering components. Any symptom of instability or imprecise road feel may be caused by worn or damaged suspension components.

When troubleshooting suspension and steering problems, also consider the condition of tires, wheels, and their alignment. Tire wear and incorrect inflation pressures can dramatically affect handling. Subtle irregularities in wheel alignment angles also affect stability. Mixing different types or sizes of tires, particularly on the same axle, can affect alignment and may unbalance a car's handling.

Table b lists the symptoms of common suspension and steering problems and their probable causes, and suggests corrective actions. Bold type indicates the repair groups where applicable test and repair procedures can be found.

Table b. Suspension and Steering Troubleshooting

Symptom	Probable cause	Repairs (Repair Groups shown in bold)
Car pulls to one side, wanders	a. Incorrect tire pressure b. Incorrect wheel alignment c. Faulty brakes (pulls only when braking)	a. Check and correct tire pressures. **020** b. Check and adjust wheel alignment. **320** c. Check for sticking/damaged front caliper. **340**
Steering heavy, poor return-to-center	a. Worn upper strut bearing(s) b. Incorrect tire pressure c. Power steering system faulty	a. Replace strut bearings. **310** b. Check and correct tire pressures. **020** c. Check power steering fluid level. Test system pressure. **320**
Front-end or rear-end vibration or shimmy	a. Worn strut cartridges or shock absorbers b. Worn suspension bushings (control arm or trailing arm) c. Worn front suspension ball joints (control arm, steering arm or steering tie-rod end) d. Unbalanced or bent wheels/tires e. Loose wheel lug bolts	a. Replace strut cartridges. **310**, **330** b. Replace worn bushings. **310**, **330** c. Replace worn ball joints. **310** d. Balance tires. Check tires for uneven wear patterns. Check wheels for damage. e. Tighten lug bolts to proper torque.
Poor stability, repeated bouncing after bumps, suspension bottoms out easily	a. Worn strut cartridges or shock absorbers	a. Replace strut cartridges. **310**, **330**
Suspension noise (especially over bumps)	a. Worn front upper strut bearings b. Loose front strut cartridge c. Worn suspension bushings (control arm or trailing arm) d. Worn stabilizer bar rubber mounts e. Loose suspension crossmember	a. Replace upper strut bearings. **310** b. Check/tighten strut cartridge collar nut. **310** c. Replace worn bushings. **310**, **330** d. Replace stabilizer bar rubber mounts. **310**, **330** e. Check crossmember for damage. Tighten mounting bolts.
Uneven ride height	a. Incorrect coil springs b. Bent or damaged suspension components c. Sagging coil springs	a. Measure ride height and replace springs as required. **310**, **330** b. Inspect, repair/replace as necessary. **310**, **330** c. Measure ride height and replace springs as required. **310**, **330**
Wheel noise, continuous growling, may be more noticeable when turning	a. Worn wheel bearing	a. Replace wheel bearing. **310**, **330**
Steering loose, imprecise	a. Incorrect tire pressure b. Loose steering gear mounting bolt(s) c. Worn tie rod end(s) d. Faulty front wheel bearing e. Worn or damaged steering gear f. Worn tires	a. Check and correct tire pressures. **020** b. Inspect and tighten bolts. **320** c. Replace tie rod(s) and align wheels. **320** d. Replace wheel bearing. **310** e. Adjust or replace steering gear. **320** f. Replace tires. **010**

310 Front Suspension

310

GENERAL

The E36 front suspension is shown in Fig. 1. For a general description of the front suspension and components, see **300 Suspension, Steering and Brakes–General**.

Special service tools are required for most of the work described in this repair group. Read the procedures through before beginning any job.

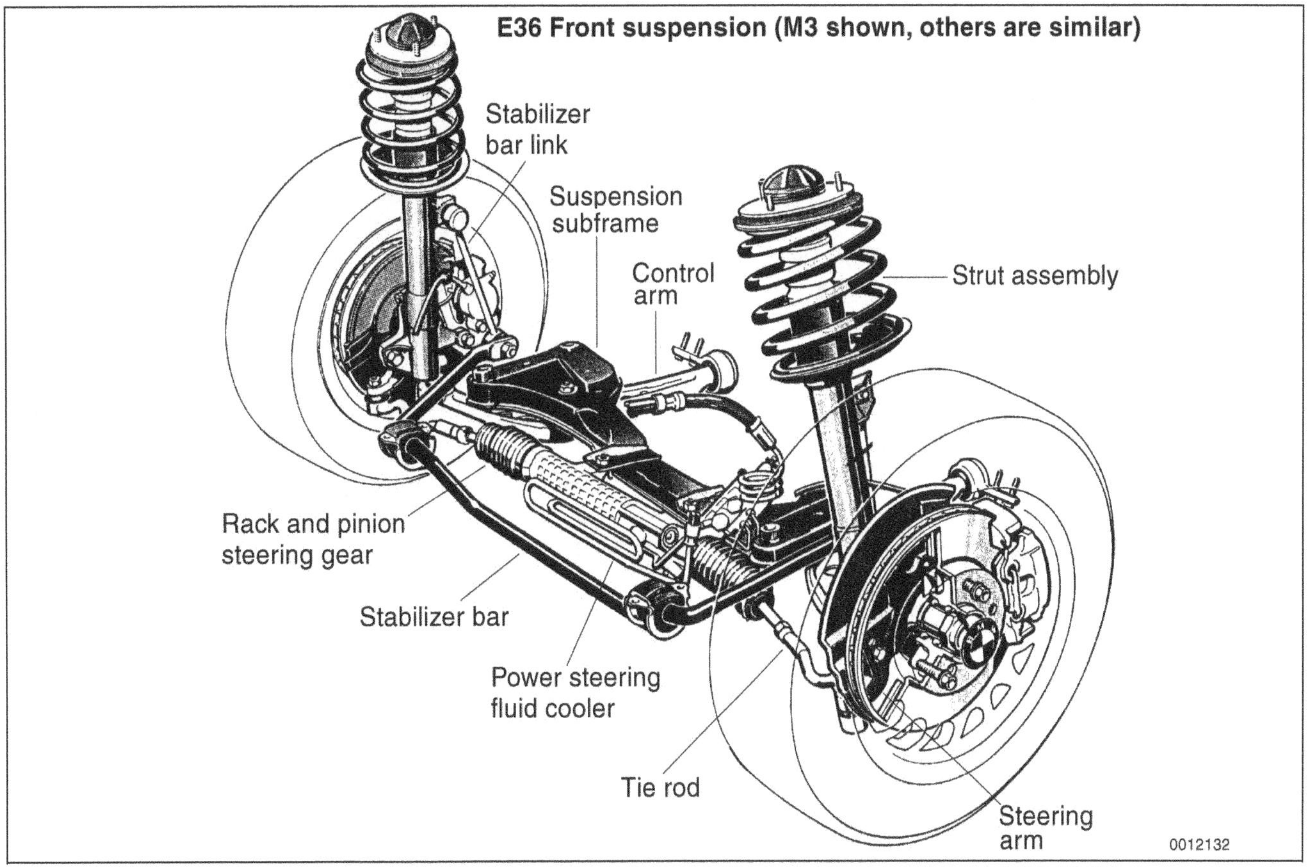

Fig. 1. BMW E36 front suspension. (M3 suspension shown. Others are similar.)

> ***WARNING—***
> - *Physical safety could be impaired if procedures described here are undertaken without the proper service tools and equipment. Be sure to have the right tools on hand before beginning any job.*
> - *Do not reuse self-locking nuts or bolts. They are designed to be used only once and may fail if reused. Always replace self-locking fasteners any time they are loosened or removed.*
> - *Do not install bolts and nuts coated with undercoating wax, as the correct tightening torque cannot be assured. Always clean the threads with solvent before installation, or install new parts.*
> - *Do not attempt to weld or straighten any suspension components. Replace damaged parts.*

SHOCK ABSORBERS AND SPRINGS

The complete front strut assembly is shown in Fig. 2. Most strut assembly components are available as replacement parts.

> ***NOTE—***
> *Struts and/or springs should always be replaced in pairs.*

Strut assembly, removing and installing

1. Raise car and remove front wheel.

> ***WARNING—***
> *Make sure that the car is firmly supported on jack stands designed for the purpose. Place the jack stands beneath a structural chassis point. Do not place jack stands under suspension parts.*

2. Unbolt brake caliper assembly, keeping brake hose connected. Detach brake hose from strut assembly. Suspend brake caliper from chassis using stiff wire. Remove brake rotor. See **340 Brakes**.

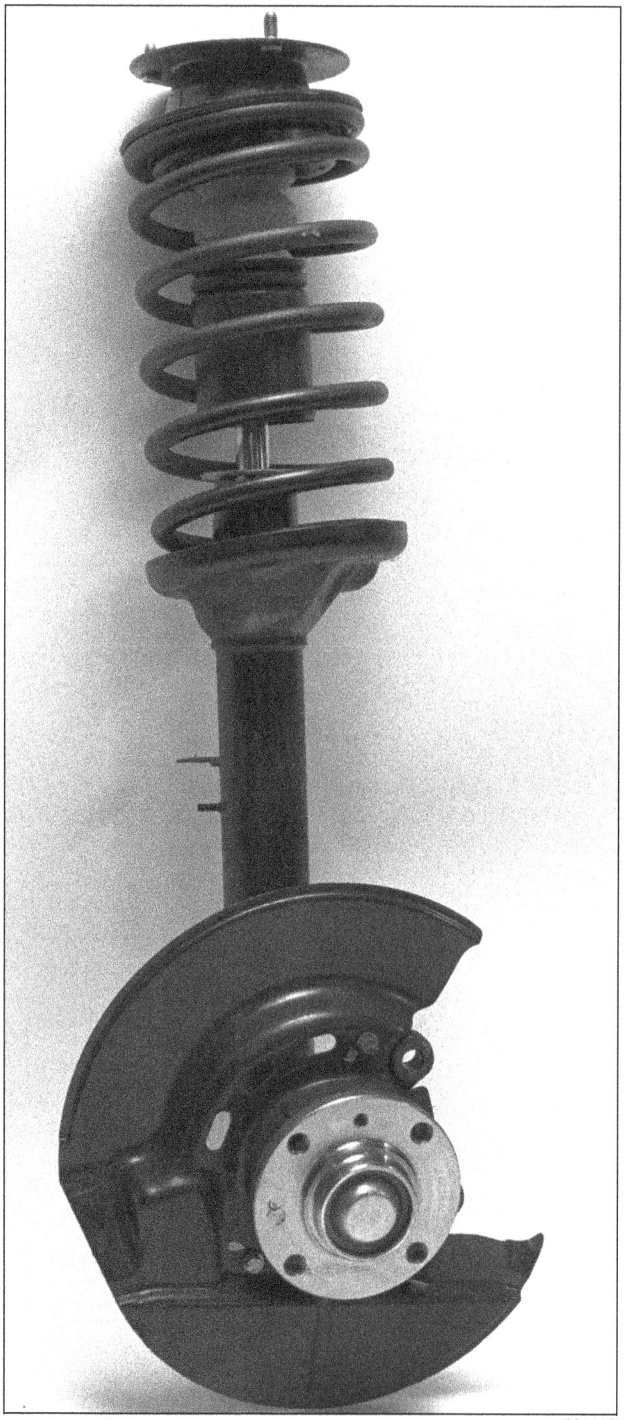

0011191

Fig. 2. Front strut assembly with steering arm, front hub, front spring and strut upper mount.

3. Remove ABS wheel speed sensor. See Fig. 3.

4. Disconnect ABS sensor wire harness and brake pad wear sensor wire harness (left side only) from strut housing.

5. On M3 models: Disconnect stabilizer bar link from strut.

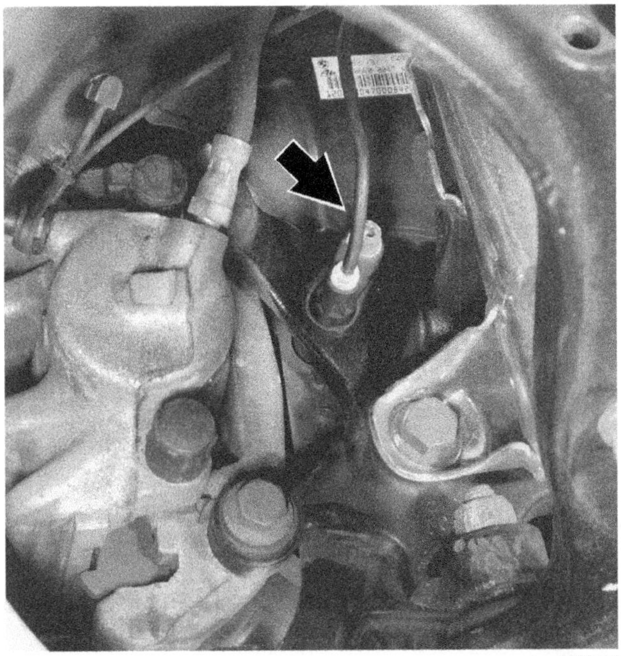

0012117

Fig. 3. ABS wheel speed sensor mounting bolt (**arrow**).

NOTE—
Note flats on stabilizer bar link ball joint for placement of open end wrench.

6. Unbolt splash shield and steering arm from strut. Suspend steering arm assembly from chassis using stiff wire. See Fig. 4.

> ***CAUTION—***
> *The steering arm ball joint can be damaged if the arm is allowed to hang unsupported.*

7. Support strut assembly from below. From above, remove three strut mounting nuts on strut tower.

> ***CAUTION—***
> *Do not remove center strut retaining nut.*

8. Lower strut assembly from car.

NOTE—
Store the removed strut assembly upright. Do not allow it to lie horizontally or upside-down for any length of time. If incorrect storage has occurred, place the strut assembly upright with the rod fully extended for at least 24 hours.

Strut and steering arm assembly

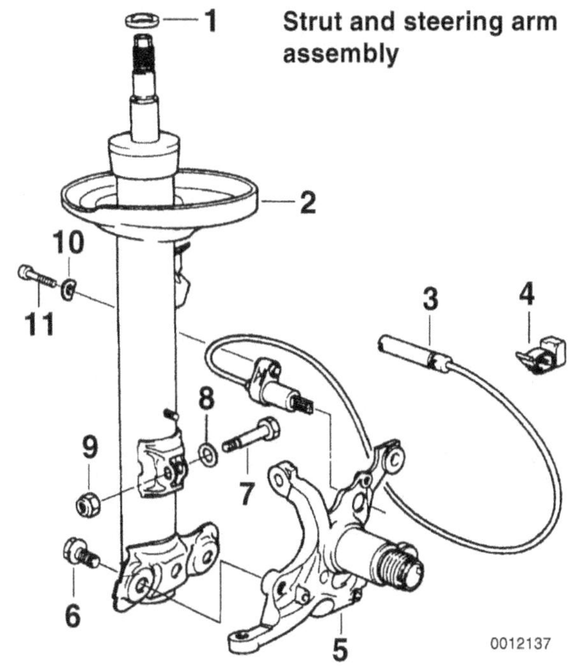

1. Nut
2. Strut
3. ABS wheel speed sensor
4. Cable holder
5. Steering arm
6. Bolt
7. Bolt
8. Flat washer
9. Self-locking nut
10. Washer
11. Bolt

Fig. 4. Strut and steering arm assembly.

9. Installation is reverse of removal. Be sure to use new self-locking nuts and bolts.
   - Use new steering arm mounting bolts, or clean bolts and use Locktite® 270 or equivalent thread-locking compound.
   - On M3 models: When attaching stabilizer link to strut, use a thin wrench to hold flats on link parallel to strut.
   - Have car professionally aligned when job is complete.

Tightening Torques
- Brake caliper to steering arm . . . . . . 110 Nm (81 ft-lb)
- Road wheel to hub . . . . 100±10 Nm (74 ±7 ft-lb)
- Stabilizer bar link to strut (M3) . . . . . . 59 Nm (44 ft-lb)
- Steering arm to strut housing
- (encapsulated bolts-always replace) 107 Nm (78 ft-lb)
- Strut assembly to strut tower (self-locking nuts) . . . . 24 Nm (16 ft-lb)

Strut assembly, disassembling and assembling

NOTE—
Springs and/or struts should always be replaced in pairs.

Replacing the upper strut mount, spring and strut requires that the strut first be disassembled. For a guide to the parts during strut replacement, see Fig. 5.

> *WARNING—*
> *Do not attempt to disassemble the struts without a spring compressor designed specifically for this job.*

1. Remove strut assembly as described earlier. Place in secure support.

2. Using spring compressor, compress spring until spring force on upper mount is relieved.

> *WARNING—*
> *Make sure the spring compressor grabs the spring fully and securely before compressing it.*

3. Pry protective cover off top of strut assembly, then remove top (center) nut and washer while holding strut shaft stationary.

4. Remove upper strut mount and related components.

5. Release spring compressor carefully and evenly, allowing spring to expand slowly.

6. Assembly is reverse of disassembly, noting the following:
   - Be sure upper spring pad is correctly installed to upper spring plate and spring end is correctly seated in upper and lower spring plates.
   - Use a new self-locking top (center) nut. Tighten nut fully before releasing spring compressor.
   - Have car professionally aligned when job is complete.

Tightening Torques
- Upper strut mount to strut shaft (self-locking nut)
  shaft with external hexagon . . . . 65 Nm (48 ft-lb)
  shaft with internal hexagon. . . . . 44 Nm (32 ft-lb)
- Road wheel to hub . . . . .100±10 Nm (74 ±7 ft-lb)

Ride Height

Ride height is measured from the lower edge of the wheel arch to the bottom edge of the wheel rim. See Fig. 6.

If the ride height is outside the specification listed, new springs should be installed. Be sure to have the old spring code number on hand when ordering new spring.

NOTE—
The spring part number is stamped near the end of the spring coils.

Table a lists front suspension ride height specifications. These specifications apply to a car in a normally loaded position. When checking ride height or installing suspension components that require the car to be "normally loaded," load the car as follows:

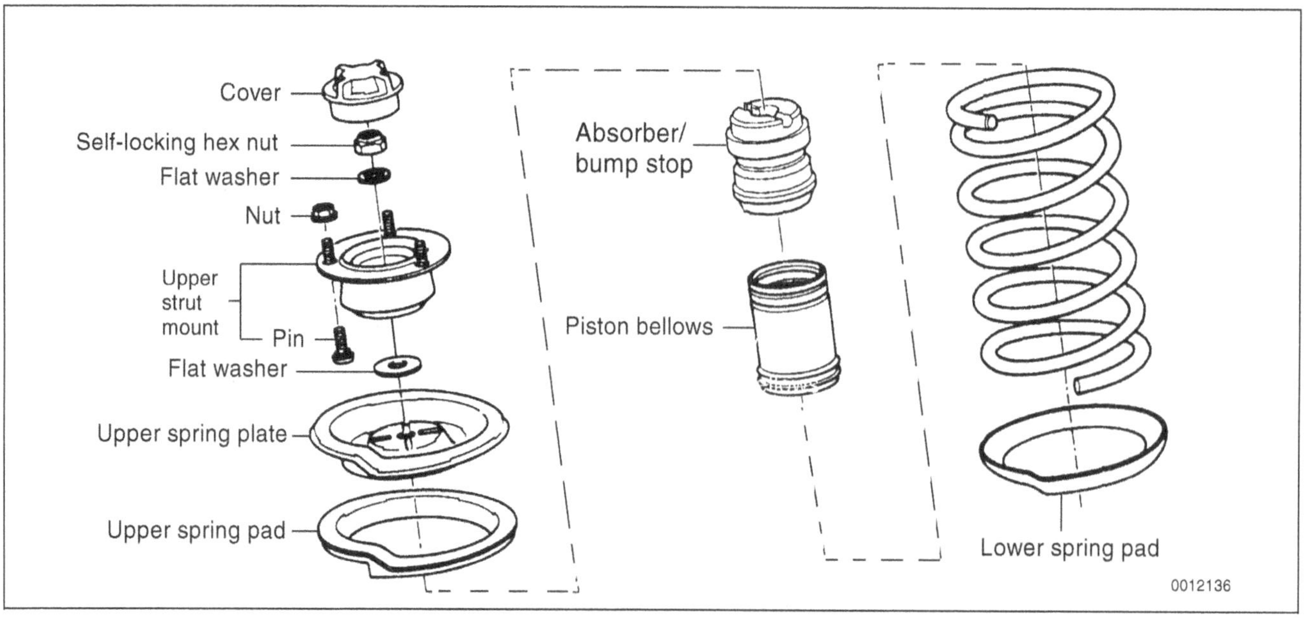

Fig. 5. Front spring, spring plate and related parts.

Fig. 6. Ride height measurement (**A**) is taken from center of wheel arch to bottom of wheel rim.

Normal loaded position

- Each front seat.................... 68 Kg (150 lbs)
- Rear seat (center) .... 68 Kg (150 lbs)
- Trunk .... 21 Kg (46 lbs)
- Fuel tank .... full

Table a. Front Ride Height Specifications

Wheel size	318, 323, 325, 328 models	M3 models
15 inch		
std. suspension	576 mm (22.67 in)	555 mm (21.85 in)
sport suspension	561 mm (22.09 in)	530 mm (20.87 in)
16 inch		
std. suspension	589 mm (23.19 in)	570 mm (22.44 in)
sport suspension	574 mm (23.52 in)	545 mm (21.46 in)
17 inch		
std. suspension	604 mm (23.77 in)	585 mm (23.03 in)
sport suspension	589 mm (23.19 in)	560 mm (22.05 in)
Maximum variation between sides: 10 mm (0.4 in)		
Maximum deviation from specifications: 10 mm (0.4 in)		

CONTROL ARMS

On all models except M3s, the outer mounting bushing and the outer ball joint on the control arm are available as replacement parts. If the control arm center ball joint is worn, the complete control arm will need to be replaced. See Fig. 7.

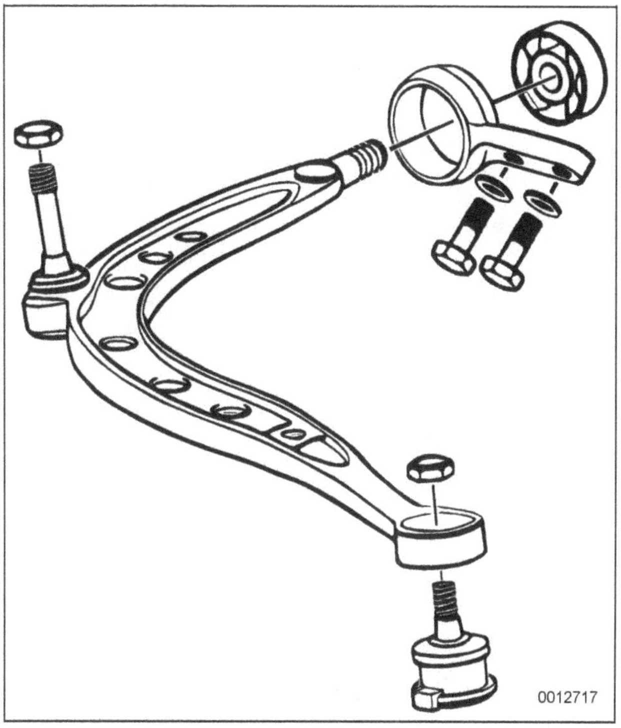

Fig. 7. Control arm. Outer ball joint and bushing are available as replacement parts (except M3). Center ball joint is integral with control arm.

Special press tools are required to remove the control arms and to replace the control arm bushings. Read the procedure through before beginning the job.

Control arm, removing and installing

1. Raise car and remove wheel.

> ***WARNING—***
> *Make sure that the car is firmly supported on jack stands designed for the purpose. Place the jack stands beneath a structural chassis point. Do not place jack stands under suspension parts.*

2. Remove stabilizer bar link mounting nut and separate link from control arm as described later in this repair group.

> ***NOTE—***
> *On M3 models, the stabilizer bar link attaches to the strut.*

3. Working at steering arm, remove control arm ball joint nut. See Fig. 8. Press ball joint out of steering arm.

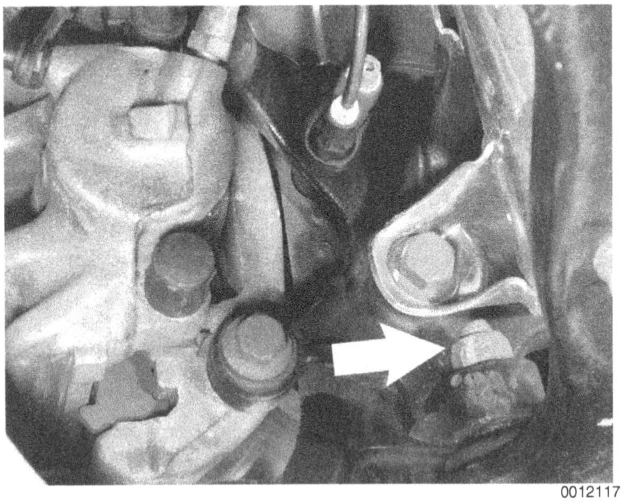

Fig. 8. Control arm ball joint nut at steering arm (**arrow**).

4. Remove control arm to subframe crossmember ball joint nut. Separate ball joint from crossmember. Working at other end of control arm, remove bolts attaching control arm bushing carrier to body. See Fig. 9.

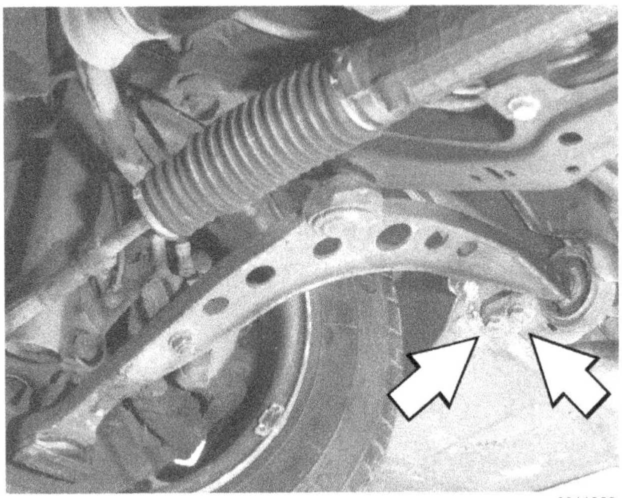

Fig. 9. Control arm bushing carrier bolts (**arrows**).

5. Installation is reverse of removal. Make sure all thread bores, bolts, nuts and mating surfaces are clean. Use new self-locking nuts or bolts, where applicable.

> ***WARNING —***
> - *Do not reuse self-locking nuts or bolts. They are designed to be used only once and may fail if re-used.*
> - *Bolts and nuts coated with a locking compound from the factory should not be reinstalled. Always replace "micro-encapsulated" fasteners such as the steering arm-to-strut bolts.*

Tightening Torques
- Control arm ball joint to subframe crossmember . . . . 85 Nm (62 ft-lb)
- Control arm ball joint to steering arm . 65 Nm (48 ft-lb)
- Control arm bushing carrier to body . . 47 Nm (34 ft-lb)
- Stabilizer bar link to control arm (M8). 42 Nm (31 ft-lb)
- Road wheel to hub . . . . 100±10 Nm (74±7 ft-lb)

Outer control arm ball joint, replacing (except M3 models)

On all models except M3s, the outer ball joint is available as a replacement part. The ball joint can be replaced with the control arm installed.

Special tools are required to replace the ball joint. Read the procedure through before beginning the job.

1. Disconnect stabilizer bar link from control arm as described later in this repair group.

> ***NOTE —***
> *On M3 models, the stabilizer bar link attaches to the strut. It is not necessary to disconnect the stabilizer bar link on M3 models for this repair procedure.*

2. Loosen nut for ball joint until it contacts steering arm.

3. Using a suitable ball joint tool, separate joint from steering arm. Remove ball joint mounting nut and pull ball joint out of steering arm.

4. Mark position of ball joint guide to control arm. See Fig. 10.

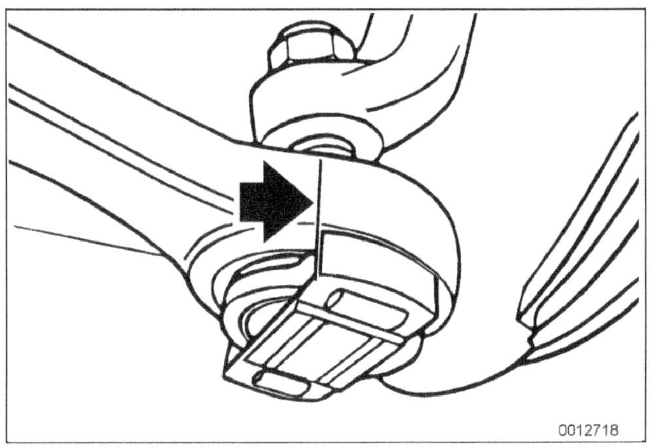

Fig. 10. Mark installed position of ball joint guide to control arm (**arrow**).

5. Move suspension strut out of way and press ball joint out of control arm.

6. Installation is reverse of removal.
   - Make sure control arm ball joint taper bore is clean and free of any corrosion.
   - Press new ball joint into position, making sure to align ball joint guide with mark made earlier.
   - Verify that locking clip on new ball joint faces towards center of vehicle.

Tightening Torques
- Control arm ball joint to steering arm. 65 Nm (48 ft-lb)
- Stabilizer bar link to control arm (M8) 42 Nm (31 ft-lb)
- Road wheel to hub......... 100±10 Nm (74±7 ft-lb)

Control arm bushing, replacing

Control arm bushings should always be replaced in pairs and the two bushings and bushing carriers should have the same markings, indicating same manufacturer.

1. Raise car.

> ***WARNING—***
> *Make sure that the car is firmly supported on jack stands designed for the purpose. Place the jack stands beneath a structural chassis point. Do not place jack stands under suspension parts.*

2. Unbolt control arm bushing carrier from body.
3. Mark installed position of bushing to control arm. See Fig. 11.

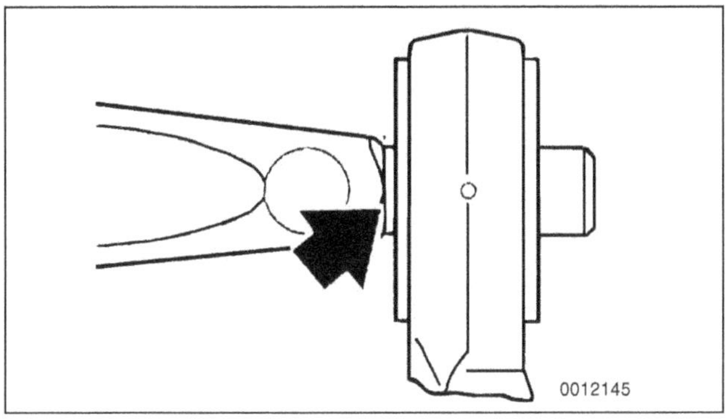

Fig. 11. Before removing bushing from control arm, mark installed position of old bushing to control arm. Install new bushing in same position.

4. Using appropriate press tools, pull control arm bushing assembly off control arm.
5. Coat end of control arm with BMW special lubricant part no. 81 22 9 407 284.

NOTE—
The kerosene-like special lubricant is used only as an aid in installing the bushing to the control arm stub. Work must be carried out in as short a time as possible. Within 30 minutes the control arm bushing will be "glued" in position on the control arm.

> ***CAUTION—***
> *BMW part numbers are provided for reference only and are subject to change. Always rely on an authorized BMW dealer parts department for the most up-to-date information.*

6. Press new rubber control arm bushing assembly onto the end of control arm to proper depth (as marked).

NOTE—
Be sure to reinstall the bushing carrier to the control arm correctly. The carrier has centering bores on side. This side must face up towards body. See Fig. 12.

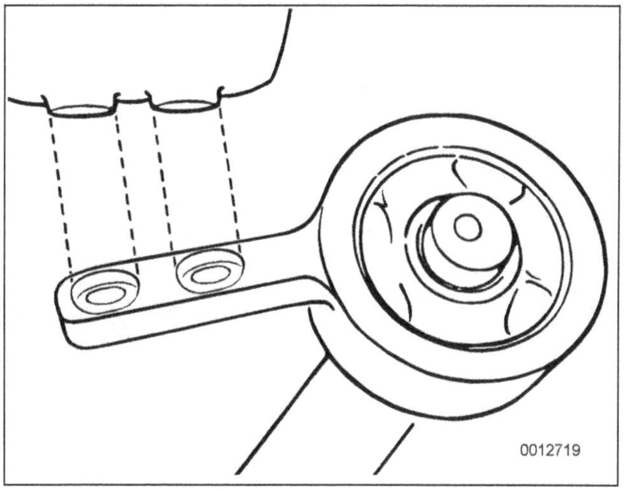

Fig. 12. Control arm bushing carrier centering bores must face up.

7. Immediately bolt control arm bushing carrier to body, torque bolts and lower car to ground.

Tightening Torques
- Control arm bushing carrier to body.. 47 Nm (34 ft-lb)

8. With car lowered to ground and loaded as described earlier, bounce suspension a few times, then leave car undisturbed for a minimum of 30 minutes.

> ***CAUTION—***
> *Failure to follow the above procedure may impair the car's handling and stability.*

FRONT WHEEL BEARINGS

The front wheel bearings are permanently sealed and require no maintenance. The bearing is integral with the wheel hub and pressed onto the steering arm stub axle.

Special press tools are required to replace the front wheel bearings. Read the procedure through before beginning the job.

Front wheel bearing, replacing

1. Raise car and remove wheel.

> ***WARNING—***
> *Make sure that the car is firmly supported on jack stands designed for the purpose. Place the jack stands beneath a structural chassis point. Do not place jack stands under suspension parts.*

2. Pry off dust cap from center of wheel hub. Bend back staked part of wheel hub (axle) collar nut. See Fig. 13.

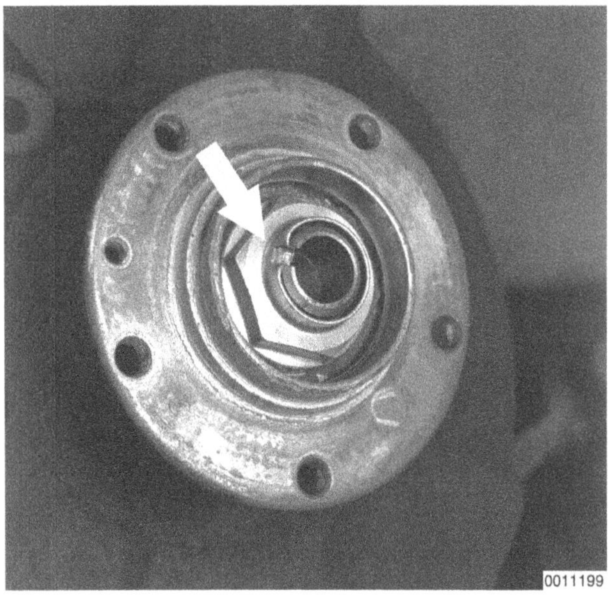

Fig. 13. Staked part of wheel hub collar nut to be bent back (**arrow**).

3. Remount wheel and lug bolts.

4. Lower car to ground. With an assistant applying brakes, loosen collar nut. Do not remove completely.

> ***NOTE—***
> *The wheel hub collar nut is tightened to a torque of 290 Nm (214 ft-lb). Make sure the car is firmly on the ground.*

5. Raise car and remove wheel.

6. Remove ABS wheel speed sensor. Refer to Fig. 3.

7. Remove brake caliper assembly and brake rotor as described in **340 Brakes**. Leave brake hose connected to caliper. Suspend caliper assembly from chassis using stiff wire.

8. Remove wheel hub collar nut.

9. Remove wheel hub with integral wheel bearing from steering arm using a puller. See Fig. 14.

0011200

Fig. 14. Wheel hub with integral wheel bearing being removed using a puller.

> ***NOTE—***
> *If bearing inner race stays on steering arm, remove splash guard and use two-jaw puller to remove race.*

10. Install splash guard and new dust shield behind bearing. Press new wheel hub/bearing assembly on using appropriate press tools.

> ***NOTE—***
> *When pressing the wheel hub/bearing on, apply force only to the bearing inner race.*

11. Install new collar nut. Do not tighten nut to its final torque at this time.
    - Install brake rotor and brake caliper. See **340 Brakes**.
    - Mount wheel and lug bolts. Lower car to ground to gain leverage.

12. With an assistant applying brakes, tighten collar nut and then stake nut to axle.

13. Raise car and remove wheel.

14. Install a new grease cap, using Loctite® 638 sealant or equivalent. Install wheel and lower car.

Tightening Torque

- Brake rotor to wheel hub .... 16 Nm (12 ft-lb)
- Collar nut to stub axle .... 290 Nm (214 ft-lb)
- Brake caliper to steering arm...... 110 Nm (81 ft-lb)
- Road wheel to hub......... 100±10 Nm (74±7 ft-lb)

STABILIZER BAR

Stabilizer bar, removing and installing

Stabilizer bar link arrangements differ slightly among models. In most models the links attach to the control arms near the outboard end. On M3 models the links attach to the strut assemblies.

1. Raise car and remove both front wheels.

> ***WARNING —***
> *Make sure that the car is firmly supported on jack stands designed for the purpose. Place the jack stands beneath a structural chassis point. Do not place jack stands under suspension parts.*

2. Remove stabilizer bar connecting link from stabilizer bar on left and right sides. See Fig. 15.

Fig. 15. Disconnect stabilizer bar connecting link at bar (**arrow**). (318i model shown. M3 stabilizer link attaches to strut.)

3. Remove two stabilizer bar bushing brackets on left and right sides. Remove bar.

4. Installation is reverse of removal. Use new self-locking nuts on connecting links.
   - On M3 models: When attaching stabilizer link to strut or to bar, use a thin wrench to hold flats on link parallel to strut.

NOTE —
Installation of the stabilizer bar is easiest with the car level (front wheels at the same height), and as near to normal ride height as possible.

Tightening Torques

- Stabilizer bar bushing brackets to subframe crossmember .... 22 Nm (16 ft-lb)
- Stabilizer bar link bracket to control arm .... 42 Nm (31 ft-lb)
- Stabilizer bar link to stabilizer bar (M10) .... 42 Nm (31 ft-lb)
- Stabilizer bar link to strut or bar (M3) . 59 Nm (44 ft-lb)

SUBFRAME CROSSMEMBER

The subframe crossmember provides rigid mounting points for the engine, suspension, and steering components. The crossmember is not normally subject to wear and should only be replaced if structurally damaged.

Removing the subframe crossmember requires engine lifting equipment to support the weight of the engine from above so that the subframe can be removed from below.

> ***WARNING —***
> *Removal or replacement of the subframe crossmember may affect suspension and steering geometry, including front wheel alignment. Make appropriate matching marks during removal and have the front end aligned once repairs are complete.*

Subframe crossmember, removing and installing

1. Using engine support equipment, raise engine until weight of engine is supported. See **110 Engine Removal and Installation**. See Fig. 16.

2. Raise car and remove front wheels.

> ***WARNING —***
> *Make sure that the car is firmly supported on jack stands designed for the purpose. Place the jack stands beneath a structural chassis point. Do not place jack stands under suspension parts.*

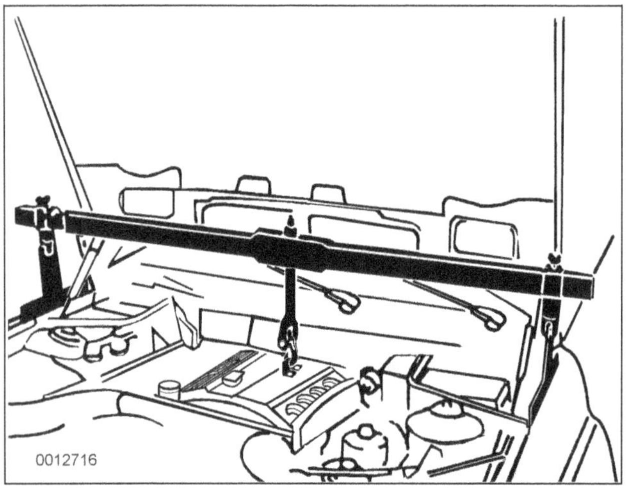

Fig. 16. Support equipment used to support engine from above.

3. Working beneath car, remove left and right nuts from engine mounts.

4. Remove control arm bushing carrier from body. Refer to Fig. 9.

5. Remove control arm ball joint from crossmember. Separate joint with a plastic hammer. Suspend control arm from chassis using stiff wire.

> ***WARNING —***
> *Do not allow the control arm to hang from the ball joint. This can damage the ball joint.*

6. Unbolt steering rack from subframe. See **320 Steering and Wheel Alignment**. Suspend steering rack from chassis using stiff wire. Using a transmission jack or equivalent, support crossmember from below.

7. Remove reinforcing brace below oil pan, where applicable.

8. Unbolt crossmember from body on both sides. See Fig. 17.

9. Slowly lower crossmember, making sure all electrical leads, suspension components and heat shields are clear during removal.

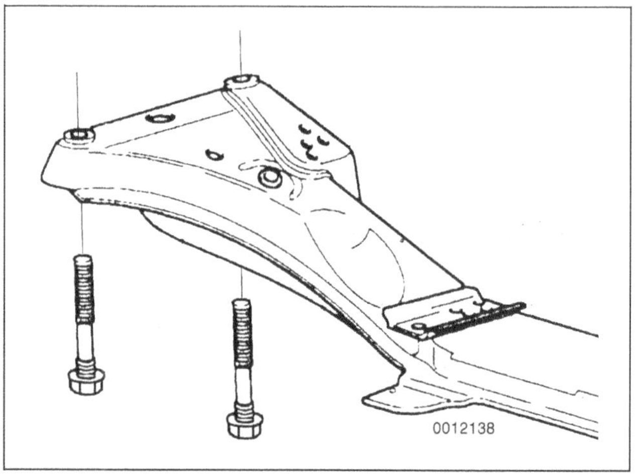

Fig. 17. Right side crossmember-to body mounting bolts.

10. Installation is reverse of removal, noting the following:
    - Make sure all bolts, bolt holes, and mating surfaces are clean to ensure proper tightening and alignment. Use new self-locking nuts or bolts, where applicable.
    - **On vehicles produced up 9-92 only:**
      - -Remove paint from contact surfaces before installing crossmember.
      - -Install 4 new toothed lock washers between subframe and chassis (install new washers regardless of whether washers were originally fitted).
      - -Replace all 4 mounting bolts.
      - -Lubricate suspension crossmember bolts with LM48 paste or equivalent anti-seize compound before installing.
    - Lower engine onto engine mounts, allowing it to settle fully before tightening engine mount bolts.
    - When the job is completed have front end professionally aligned.

Tightening Torques

- Subframe crossmember to body
  M12-8.8 bolts . . . . 77 Nm (56 ft-lb)
  M12-10.9 bolts . . . . 110 Nm (81 ft-lb)
  M12-12.9 bolts . . . . 105 Nm (77 ft-lb)
- Steering gear to subframe crossmember
  (M10-10.9 self-locking bolts) . . . . 42 Nm (31 ft-lb)
- Control arm ball joint to
  subframe crossmember . . . . 85 Nm (62 ft-lb)
- Control arm bushing carrier
  to body . . . . 47 Nm (34 ft-lb)

320 Steering and Wheel Alignment

TABLES

320

GENERAL

This repair group covers servicing of the mechanical and hydraulic components of the steering system, including wheel alignment information.

The variable-assist power steering system consists of an engine-driven hydraulic pump, a rack-and-pinion steering gear assembly with an integral hydraulic control valve, and connecting linkage to the road wheels.

At low speeds, maximum power assist is provided to ease parking and city driving. At high speeds, assist is reduced to ensure stability. The power steering system varies assist based on engine speed.

> ***WARNING—***
> - *Do not reuse self-locking nuts. They are designed to be used only once and may fail if reused. Always replace them with new locking nuts.*
> - *Do not install bolts and nuts coated with undercoating wax, as correct tightening torque cannot be assured. Always clean the threads with solvent before installation, or install new parts.*
> - *Do not attempt to weld or straighten any steering components. Always replace damaged parts.*

NOTE—
For information on steering column-mounted switches and the ignition lock, see **612 Switches and Electrical Accessories***.*

STEERING WHEEL

The BMW E36 is equipped with an SRS airbag mounted in the steering wheel. Improper handling of the airbag could cause serious injury. The airbag is an explosive device and should be treated with extreme caution. Always follow the airbag removal procedure as outlined in **721 Airbag System (SRS)**.

> ***WARNING—***
> - *The BMW Supplemental Restraint System (SRS) is complex and special precautions must be observed when servicing. Serious injury may result if system service is attempted by persons unfamiliar with the BMW SRS and its approved service procedures. BMW specifies that all inspection and service should be performed by an authorized BMW dealer.*
> - *BMW Supplemental Restraint Systems (SRS) are equipped with a back-up power supply inside the SRS control module. A 10 minute waiting period should be observed after the battery cable has been disconnected. This will allow the reserve power supply to discharge.*

Steering wheel, removing and installing

1. Center steering wheel. Make sure front wheels are pointed straight ahead.

2. Disconnect negative (–) cable from battery and cover terminal with insulating material.

> ***CAUTION—***
> *Prior to disconnecting the battery, read the battery disconnection cautions given at the front of this manual on page viii.*

3. Carefully remove airbag from front of steering wheel. See **721 Airbag System (SRS)**. Store airbag unit in a safe place.

> ***WARNING —***
> *Improper handling of the airbag could cause serious injury. Store the airbag with the horn pad facing up. If stored facing down, accidental deployment could propel it violently into the air, causing injury.*

4. Remove steering wheel center bolt. Lightly punch mark or scribe position of steering wheel to steering column shaft. See Fig. 1.

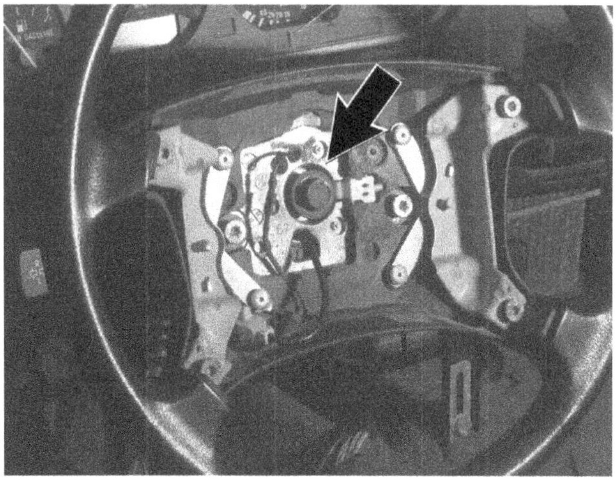

0012687

Fig. 1. Steering wheel center bolt (**arrow**).

5. Unlock steering wheel by turning ignition key on. Remove steering wheel.

> ***WARNING —***
> *The SRS contact reel is mounted to the rear of the steering wheel hub. The contact reel is a wound coil of wire that ensures continuous electrical contact for the air bag unit. Once the steering wheel mounting nut or bolt is removed the contact reel is locked in the center position and its position must not be altered.*

6. Install steering wheel while aligning matching marks. Make sure airbag contact ring locking pin engages cut out in contact reel. Install steering column center nut. Do not over-torque. See Fig. 2.

7. The remainder of installation is reverse of removal. Carefully install airbag to front of steering wheel. See **721 Airbag System (SRS)**.

Tightening Torque
- Steering wheel to steering column spindle . . . . 63 Nm (46 ft-lb)

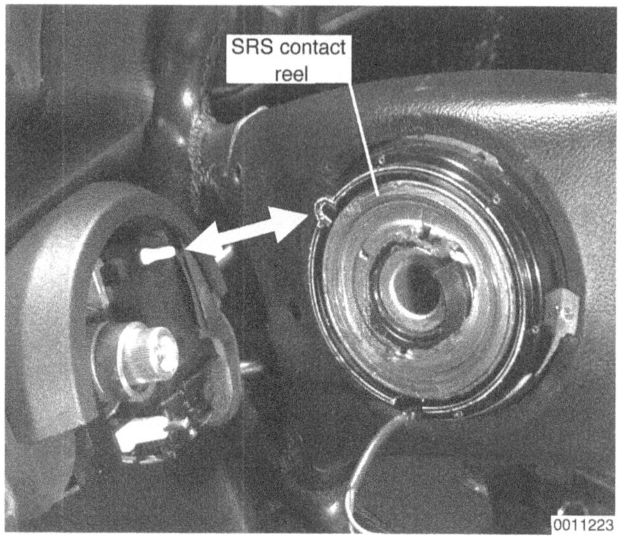

Fig. 2. Install steering wheel so that locking pin in column aligns with cutout in contact reel (**arrow**). Contact reel on late model steering wheel may vary from that shown.

POWER STEERING SYSTEM

Power assist is provided by a belt-driven pump on the lower left front of the engine, just below the alternator. The power steering fluid reservoir is located on the lower left front of the engine.

Power steering pump, removing and installing

1. Empty power steering fluid reservoir using clean syringe. Do not reuse fluid.

2. Raise front of car.

> ***WARNING —***
> *Make sure the car is firmly supported on jack stands designed for the purpose. Place the jack stands beneath a structural chassis point. Do not place jack stands under suspension parts.*

3. Remove fluid hoses from pump. Plug openings in pump and in hose ends.

4. Remove drive belt from power steering pump. See **020 Maintenance Program**.

5. Remove pump mounting bolts and remove pump. See Fig. 3.

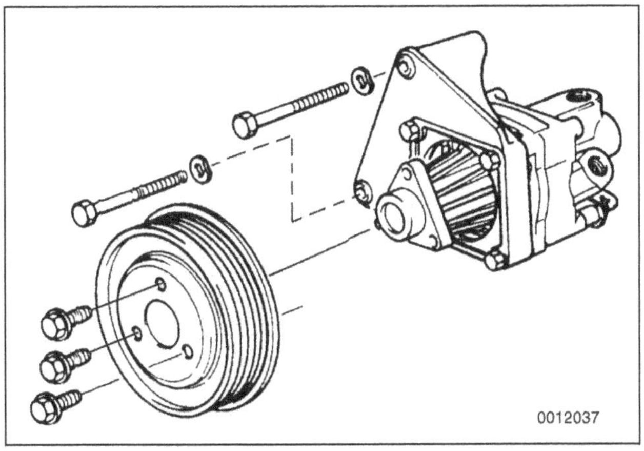

Fig. 3. Power steering pump and pulley mounting bolts.

6. Installation is reverse of removal, noting the following:
   - Make sure all thread bores, bolts, nuts, fluid couplings and mating surfaces are clean, and that inside surface of multi-ribbed-belt is free of grease and dirt.
   - Use new sealing washers when reattaching power steering pressure lines.
   - Make sure hoses have adequate clearance from chassis. Do not over-torque banjo bolts.
   - Fill and bleed power steering system as described below.

Tightening Torques
- Hydraulic hoses to power steering pump
  M10 banjo bolt . . . . 12 Nm (9 ft-lb)
  M14 banjo bolt . . . . 35 Nm (26 ft-lb)
  M16 banjo bolt . . . . 40 Nm (30 ft-lb)
- Power steering pump to bracket
  (self-locking nuts). . . . . 22 Nm (16 ft-lb)
- Power steering pump bracket to engine
  block or oil pan (self-locking nuts) . . . 22 Nm (16 ft-lb)

Power steering system, bleeding and filling

1. With engine off, fill power steering fluid reservoir with clean fluid. Fill level to **MAX** mark on dipstick.

2. Start engine. Slowly turn steering wheel from lock to lock a minimum of two times.

3. Turn engine off and check fluid level, adding fluid if necessary.

STEERING GEAR

A cutaway of the power-assisted rack-and-pinion power steering gear is shown in Fig. 4. The rack is designed to be maintenance free.

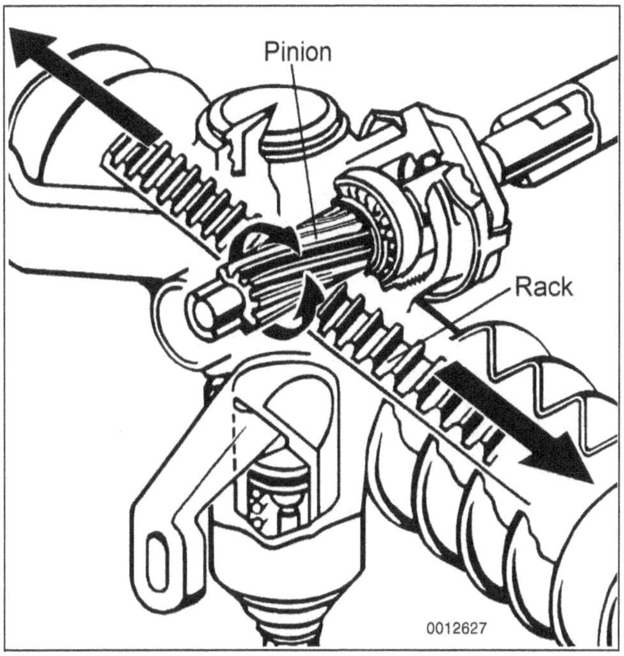

Fig. 4. Cutaway of rack-and-pinion steering gear.

The steering gear and linkage require no maintenance other than alignment and a periodic inspection for worn components.

Steering gear rack, removing and installing

1. Empty power steering fluid reservoir using clean syringe. Do not reuse fluid.

2. Disconnect negative (–) cable from battery and cover terminal with insulating material.

> ***CAUTION—***
> *Prior to disconnecting the battery, read the battery disconnection cautions given at the front of this manual on page viii.*

3. Carefully remove airbag from front of steering wheel. See **721 Airbag System (SRS)**. Store airbag unit in a safe place.

4. Make sure front wheels are pointed straight ahead.

5. Remove outer tie rod ball joint nuts, then separate outer tie rod ends from steering arms using appropriate press tools.

6. Disconnect power steering hose from fluid reservoir.

7. Remove fluid line banjo bolt from steering gear. See Fig. 5.

> **CAUTION—**
> *Plug the openings in the pump and hose ends to prevent dirt from entering the hydraulic system.*

NOTE—
It may be necessary to remove other components to gain access to the steering gear.

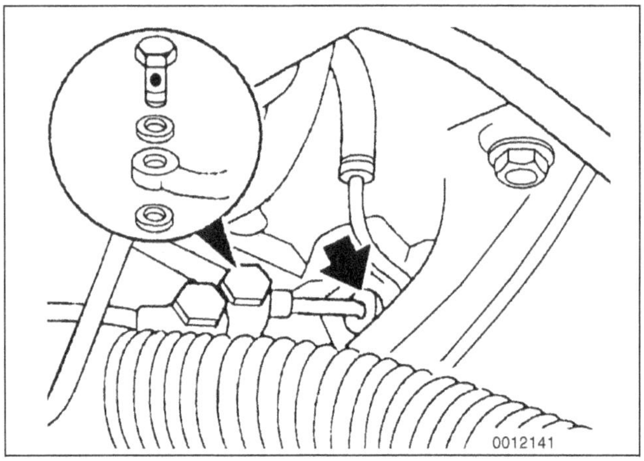

Fig. 5. Fluid hoses and banjo bolts at steering gear (**arrows**).

8. Remove lower steering shaft universal joint pinch bolt, then pull universal joint half way off steering gear shaft. Mark relationship between shaft and universal joint, then remove universal joint from steering gear. See Fig. 6.

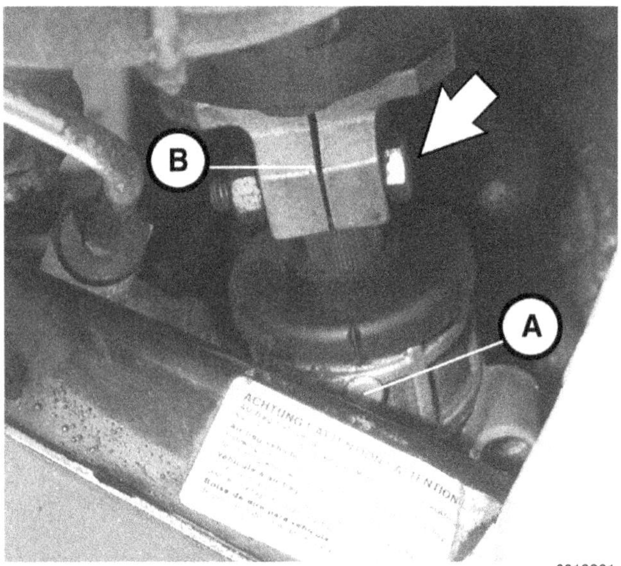

0013231

Fig. 6. Steering column universal joint pinch bolt (**arrow**). Mark U-joint to shaft before disconnecting. With wheels in straight ahead position, mark on steering gear (**A**) should align with split seam in universal joint (**B**).

9. Remove steering gear mounting bolts. See Fig. 7. Remove steering gear from subframe crossmember by pulling it forward.

Fig. 7. Steering gear mounting bolts (**arrows**).

10. Installation is reverse of removal, noting the following:
    - Make sure all thread bores, bolts, nuts, splines and mating surfaces are clean. Use new self-locking nuts wherever applicable.
    - Use marks made during removal to reassemble steering column universal joint to steering stub shaft. Make sure wheels are straight ahead.
    - Use a new sealing washer on banjo fitting. Do not over-torque banjo bolt.
    - Install airbag to the steering wheel as described in **721 Airbag System (SRS)**.
    - Fill and bleed power steering system as described earlier.
    - Have the car professionally aligned.

Tightening Torques
- Outer tie rod end to steering arm . . . . 45 Nm (33 ft-lb)
- Steering column universal joint to steering gear shaft (self-locking nut) . . . . 19 Nm (14 ft-lb)
- Steering gear to subframe crossmember
  M10 bolt . . . . 42 Nm (32 ft-lb)
- Steering fluid lines
  M10 banjo bolt . . . . 12 Nm (9 ft-lb)
  M14 banjo bolt . . . . 35 Nm (26 ft-lb)
  M16 banjo bolt . . . . 40 Nm (30 ft-lb)

Outer tie rod end, replacing

1. Raise car, remove outer tie rod ball joint nut, and press out outer tie rod end ball joint as described above.

2. Make a reference measurement of outer tie rod end to tie rod. See Fig. 8. Record measurement.

NOTE—
Accurate measuring of the tie rod end in reference to the tie rod will help approximate correct wheel alignment when new parts are installed.

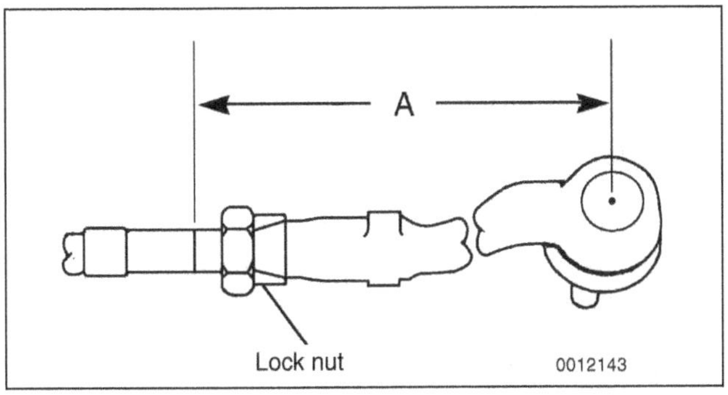

Fig. 8. Measuring tie rod end position (for reference only).

3. Loosen outer tie rod end lock nut. See Fig. 9.

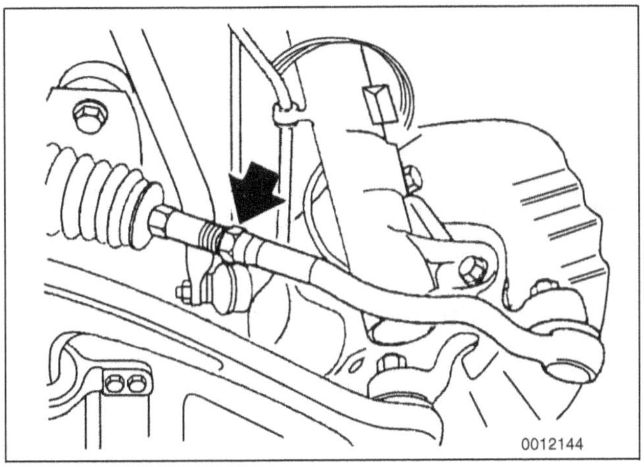

Fig. 9. Outer tie rod end lock nut (**arrow**).

4. Unscrew tie rod end from steering gear.

5. Installation is reverse of removal. Make sure all threaded parts are clean. Have the car professionally aligned.

Tightening Torques
- Outer tie rod end to steering arm.... 45 Nm (33 ft-lb)
- Outer tie rod end lock nut.......... 35 Nm (26 ft-lb)

Tie rod, replacing

1. Raise car, remove outer tie rod end ball joint nut, and press out outer tie rod end ball joint as described above.

2. Make a reference measurement of outer tie rod end to tie rod. See Fig. 8. Record measurement

NOTE—
Accurate measuring of the tie rod end in reference to the tie rod will help to approximate wheel alignment when new parts are installed.

3. Loosen rack boot bellows clamp and slide bellows back. Inspect boot for any sign of damage. Replace if necessary.

4. Unlock inner tie rod end lockplate using pliers.

5. Using special tool, unscrew inner tie rod end from steering rack. See Fig. 10.

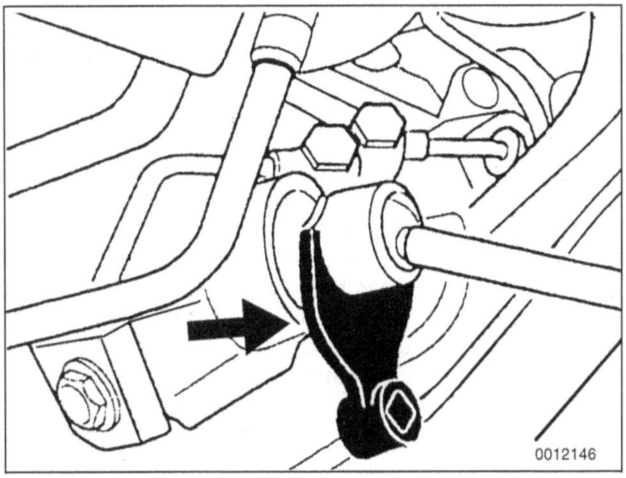

Fig. 10. Special tool (**arrow**) installed on inner tie rod end.

6. Installation is reverse of removal, noting the following:
   - Make sure all threaded parts are clean. Replace self-locking nuts and inner tie rod lockplate.
   - Install outer tie rod end to new tie rod using reference measurement recorded earlier.
   - Have car professionally aligned.

Tightening Torques
- Outer tie rod end to steering arm.... 45 Nm (33 ft-lb)
- Outer tie rod end lock nut.......... 35 Nm (26 ft-lb)
- Inner tie rod to steering gear .... 71 Nm (52 ft-lb)

WHEEL ALIGNMENT

Proper handling, stability, tire wear, and driving ease depend upon the correct alignment of all four wheels. The front axle is aligned in relation to the rear axle, then the front wheels are aligned in relation to one another. This is known as a four-wheel or thrust-axis alignment.

The BMW E36 uses a sophisticated multi-link suspension at the front and rear of the car. Proper alignment requires computerized alignment equipment.

Caster and Camber

Front and rear caster and camber are both fixed by the design of the car. Any deviations are usually the result of worn or damaged suspension or body parts.

Front Toe-in and Toe Difference Angle

Toe-in is the difference in the distance between the front of the front wheels and the rear of the front wheels. It is adjusted by altering the length of the outer tie rods.

Toe-in adjustments should only be made on alignment equipment with the car in normal loaded position as described below under **Wheel Alignment Specifications**.

Toe difference angle determines the progressively different paths taken by the front wheels as the car is turning. It is fixed by the steering gear, but depends on accurate setting of the toe-in.

Rear Toe-in

There is no provision for routine rear wheel toe-in adjustment. If any alignment angles deviate from specifications, carefully inspect the rear trailing arms, rear suspension subframe and all associated bushings and flexible mounts for wear or damage. After replacing any parts that appear worn or damaged, re-check the toe measurements.

Wheel Alignment Specifications

Alignment specifications for the E36 are listed in **Table a** and **Table b**. The specifications only apply under the following conditions:

- Correct wheels and tires are installed, in good condition, and are at the correct inflation pressures.
- All steering and suspension parts and bushings are undamaged and show no signs of abnormal wear.
- Wheel bearings are in good condition.
- Ride height is in accordance with specifications. See **310 Front Suspension** and **330 Rear Suspension**.
- Car is in normal loaded position.

Normal loaded position

- Each front seat . . . . 68 Kg (150 lb)
- Rear seat (center) . . . . 68 Kg (150 lb)
- Trunk. . . . . 21 Kg (46 lb)
- Fuel tank. . . . . full

Table a. Front Wheel Alignment Specifications

Model	318/325	318/325 sport suspension	M3 3.0 liter	M3 3.2 liter
Toe angle (total)	0° 18' ±8'	0° 18' ±8'	0° 10' ±5'	0° 10' ±5'
Camber				
To 1995	–0° 40' ±30'	–0° 58' ± 30'	–0° 55' ±30'	—
1995 on	–0° 30' ±30'	–0° 51' ±30'	–0° 55' ±30'	–0° 46' ±30'
Caster				
@ 10° wheel lock	3° 41' ±30'	3° 50' ±30'	6° 38' ±30'	9° 35' ±30'
@ 20° wheel lock	3° 52' ±30'	3° 57' ±30'	6° 52' ±30'	7° 50' ±30'
Front Wheel Displacement	0° ±15'	0° ±15'	0° ±15'	0° ±15'

Table b. Rear Wheel Alignment Specifications

Model	318/325	318/325 sport suspension	M3 3.0 liter	M3 3.2 liter
Toe angle (total)	0° 24' ±6'	0° 24' ±6'	0° 30' ±6'	0° 30' ±6'
Camber	–1° 40' ±15'	–2° 0' ±15'	–1° 45' ±10'	–1° 45' ±10'
Maximum allowable deviation between sides	0° ±3'	0° ±3'	0° ±3'	0° ±3'

330 Rear Suspension

GENERAL

Special service tools are required for some of the work described in this repair group. Read the procedures through before beginning any job.

NOTE —
A general description of the rear suspension and a troubleshooting guide can be found in **300 Suspension, Steering and Brakes—General**.

RIDE HEIGHT

Rear suspension ride height is controlled by the rear springs. If the rear ride height is not within the specifications listed, the rear springs should be replaced.

Ride height is measured from the lower edge of the wheel arch to the bottom edge of the wheel rim. See Fig. 1.

Table a lists rear suspension ride height specifications. These specifications apply to a car in a normally loaded position. When checking ride height or installing suspension components, load the car as follows:

Normal loaded position
- Each front seat. . . . . 68 kg (150 lb)
- Rear seat (center) . . . . 68 kg (150 lb)
- Trunk . . . . 21 kg (46 lb)
- Fuel tank . . . . full

NOTE —
M3 ride height is measured with the car unladen.

Fig. 1. Ride height measurement (**A**) is taken from center of wheel arch to bottom of wheel rim.

Table a. Rear Ride Height Specifications

Wheel size	318, 323, 325, 328 models	M3 models
15 inch std. suspension sport suspension	 518 mm (20.39 in) 507 mm (19.96 in)	
16 inch std. suspension sport suspension	 531 mm (20.91 in) 520 mm (20.47 in)	

Continued

Table a. Rear Ride Height Specifications

Wheel size	318, 323, 325, 328 models	M3 models
17 inch std. suspension sport suspension	 546 mm (21.69 in) 535 mm (21.06 in)	 540 mm (21.26 in)
Maximum variation between sides: 10 mm (0.4 in) **Maximum deviation from specifications: 10 mm (0.4 in)**		

SHOCK ABSORBERS AND SPRINGS

The BMW E36 independent rear suspension is shown in Fig. 2. Refer to this illustration while performing repairs on components covered in this repair group.

NOTE —
Shock absorbers and/or springs should always be replaced in pairs

WARNING —
- *Do not reuse self-locking nuts. They are designed to be used only once and may fail if reused. Always replace them with new self-locking nuts.*
- *Personal injury may result if procedures described here are undertaken without the proper service tools and equipment. Be sure to have the right tools on hand before beginning the job.*
- *Do not install bolts and nuts coated with undercoating wax, as correct tightening torque cannot be assured. Always clean the threads with solvent before installation, or install new parts.*
- *Do not attempt to weld or straighten any suspension components. Replace damaged parts.*

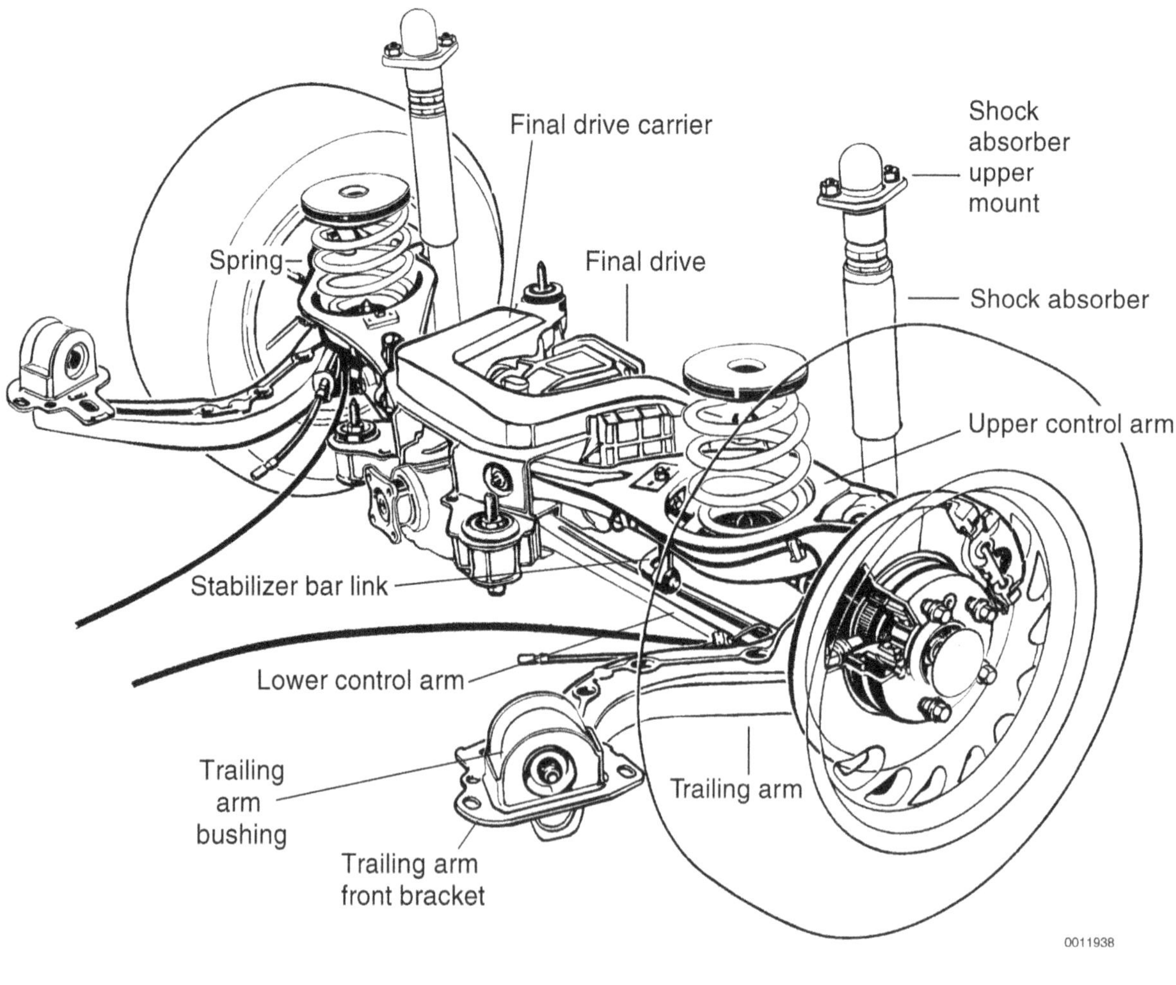

Fig. 2. BMW E36 rear suspension.

Rear shock absorber, removing and installing

1. Raise car and remove rear wheels.

> ***WARNING —***
> *Make sure that the car is firmly supported on jack stands designed for the purpose. Place the jack stands beneath a structural chassis point. Do not place jack stands under suspension parts.*

2. Pry out luggage compartment liner retaining clips and peel back liner to gain access to upper shock absorber mounting nuts. See Fig. 3.

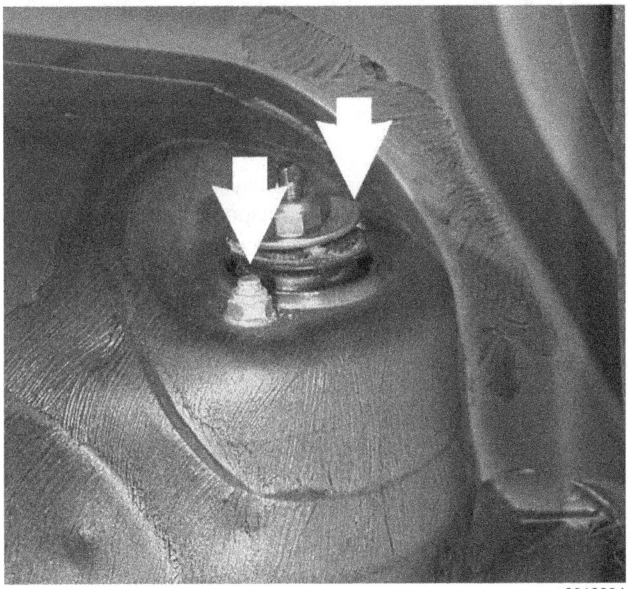

Fig. 3. Rear shock upper mounting nuts (**arrows**) in luggage compartment. Liner has been peeled back.

> ***NOTE —***
> *On convertible models, the upper shock absorber mount is in the convertible top compartment behind the rear seat.*

3. Support trailing arm from below using a transmission jack or equivalent. See Fig. 4. Remove shock absorber upper mounting nuts.

> ***CAUTION —***
> *The shock absorber prevents the drive axle and trailing arm from drooping too far, so always support the trailing arm before removing the shock. Damage to drive axle CV joints can result.*

4. While supporting shock absorber, carefully remove lower mounting bolt. Lower shock absorber out of wheel well.

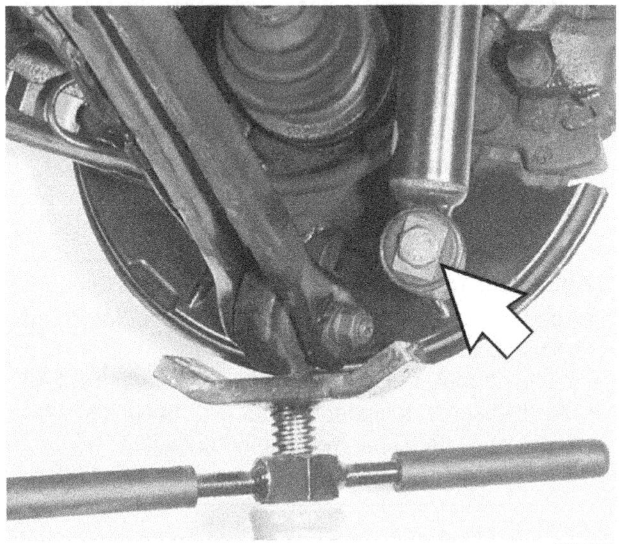

Fig. 4. Support rear trailing arm from below before removing upper or lower (**arrow**) shock mounting.

5. Transfer shock top mounting plate, dust cover and related components to new shock absorber. See Fig. 5.

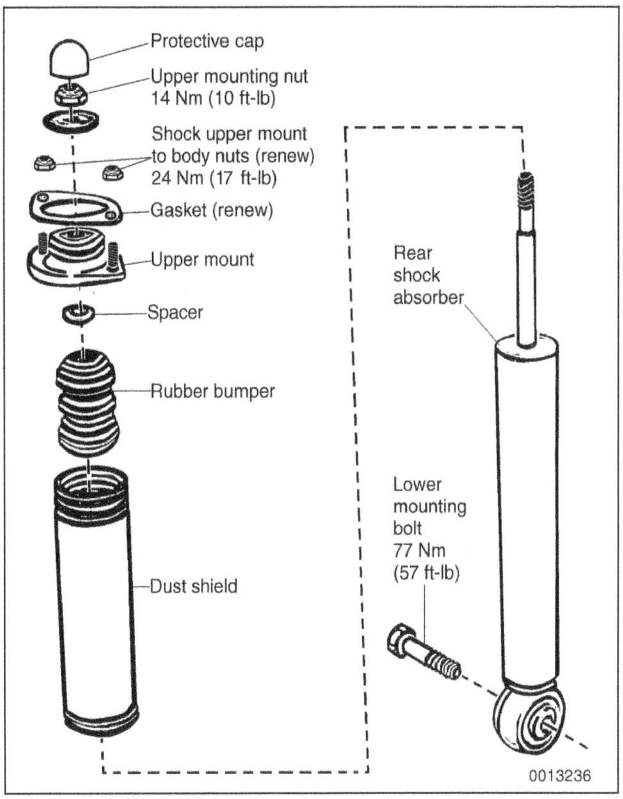

Fig. 5. Rear shock absorber assembly.

6. Installation is reverse of removal, noting the following:
   - Make sure all threaded bolts, nuts and mating surfaces are clean.
   - Install shock absorber into shock tower using a new upper mounting gasket and new self-locking nuts.
   - Tighten lower bolt to its final torque once car is on ground.

Tightening Torques
- Road wheel to hub . . . . 100±10 Nm (74±7 ft-lb)
- Shock absorber to trailing arm (car in normal loaded position). . . . . . 77 Nm (57 ft-lb)
- Shock absorber to upper mount. . . . . 14 Nm (10 ft-lb)
- Shock absorber upper mount to body (M8 self-locking nuts). . . . . 24 Nm (17 ft-lb)

Coil spring, removing and installing

WARNING —
- *The coil spring is under compressive force and is extremely dangerous until compression is removed.*
- *It is recommended that a restraining chain be installed between the bottom of the coil spring and the upper control arm to retain the spring in case of accidental release. Personal injury can result if the compressed coil spring is not released slowly and carefully.*
- *Make sure that the car is firmly supported on jack stands designed for the purpose. Place the jack stands beneath a structural chassis point. Do not place jack stands under suspension parts.*

1. Raise car and remove rear wheel.

2. Support suspension arm from below using a transmission jack or equivalent.

3. Remove drive axle from final drive. See Fig. 6. Suspend drive axle from chassis using stiff wire.

4. Disconnect shock absorber from trailing arm.

5. Lower suspension slowly and carefully until the compressed coil spring is fully unloaded.

CAUTION —
The shock absorber prevents the drive axle and trailing arm from drooping too far, so always support the trailing arm before removing the shock. Damage to drive axle CV joints can result.

6. Remove restraining chain and remove coil spring.

0012103

Fig. 6. Drive axle to final drive mounting bolts (**arrows**).

7. If spring is to be reused, carefully inspect it for any surface damage or corrosion. Inspect spring mounts for any signs of damage. Replace any parts showing evidence of wear or damage.

8. Fit coil spring to spring mounts and install restraining chain.

9. Slowly lift suspension back into position, making sure coil spring is correctly seated in upper and lower coil spring mounts.

10. When suspension has been lifted sufficiently, install shock absorber to trailing arm. Remove restraining chain.

11. The remainder of installation is reverse of removal.

Tightening Torques
- Drive axle to final drive flange
  - M8 Torx bolt. 64 Nm (47 ft-lb)
  - M10 Torx bolt. . . . . 83 Nm (62 ft-lb)
  - M10 ribbed bolt . . . . 100 Nm (74 ft-lb(
- Road wheel to hub . . . . 100±10 Nm (74±7 ft-lb)
- Shock absorber to trailing arm . . . . . . 77 Nm (57 ft-lb)

REAR WHEEL BEARINGS

The rear wheel bearing is a unitized assembly and is not repairable separately. See Fig. 7.

Special press tools, to be used with the trailing arm attached to the car, are required to replace the wheel bearings. Read the procedure through before beginning the job.

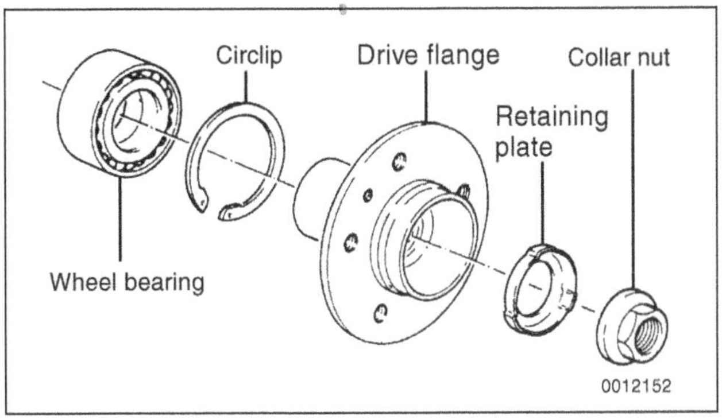

Fig. 7. Rear wheel bearing assembly.

Rear wheel bearing, replacing

1. Raise car and support safely. Remove rear wheel.
2. Remove drive axle as described later.
3. Remove brake caliper assembly and rotor as described in **340 Brakes**. Leave brake hose connected to caliper. Suspend caliper assembly from chassis using stiff wire.
4. Remove ABS pulse sensor. See Fig. 8.

Fig. 8. ABS pulse sensor being removed.

5. Remove drive flange from bearing assembly. See Fig. 9.

NOTE —
If the inner bearing race remains attached to the drive flange, use BMW special tool nos. 00 7 500 and 33 1 312 or equivalents to separate inner race from flange.

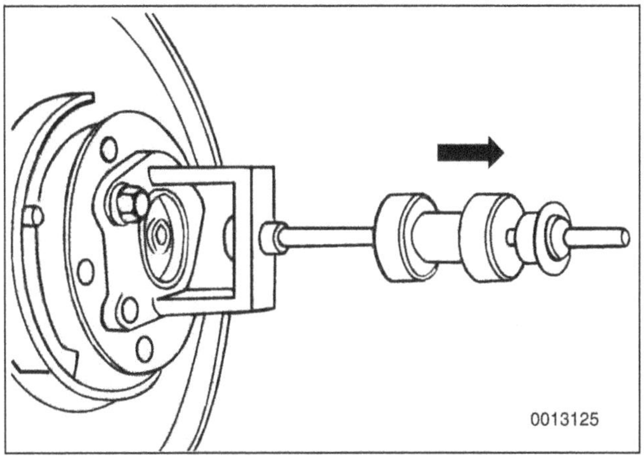

Fig. 9. Drive flange being removed from bearing assembly using impact style puller.

6. Remove bearing retainer circlip from trailing arm.
7. Using appropriate press tools, pull bearing assembly out of trailing arm bearing housing.
8. Inspect bearing housing for any damage or contamination. Clean housing bore thoroughly before installing new bearing. Make sure all thread bores, bolts, nuts and mating surfaces are clean
9. Install new bearing assembly using press tools. Apply force only to outer race when installing bearing.

WARNING —
M3 installation: red sealing ring on rear wheel bearing must face outward. Otherwise ABS will not function.

10. Install new circlip. The old circlip should never be re-used.

CAUTION —
Make sure that the bearing is pressed in far enough to contact the shoulder at the back side of the housing and that the circlip is fully seated in its groove.

11. Draw drive flange into bearing using appropriate press tools. Support bearing inner race when pressing in drive flange.

NOTE —
BMW specifies special tools to pull the drive flange through the wheel bearing into position. If using alternative tools, be sure to support the bearing inner race when pressing or pulling the drive flange into place.

12. Install brake caliper assembly and rotor as described in **340 Brakes**.

13. Install drive axle as described below.

Tightening Torques
- Brake rotor to drive flange . . . . 16 Nm (12 ft-lb)
- Brake caliper to trailing arm . . . . 67 Nm (50 ft-lb)
- Drive axle collar nut to drive flange. . . . . 250 Nm (184 ft-lb)
  M3 . . . . 300 Nm (221 ft-lb)
- Road wheel to hub . . . . 100±10 Nm (74±7 ft-lb)

DRIVE AXLES

The drive axles use constant-velocity (CV) joints on both ends. To replace a CV joint or dust boot, the drive axle must be removed from the car. The components of the drive axle are illustrated in Fig. 10.

Drive axle, removing and installing

1. Raise rear of car. Remove rear wheel.

> ***WARNING —***
> *Make sure that the car is firmly supported on jack stands designed for the purpose. Place the jack stands beneath a structural chassis point. Do not place jack stands under suspension parts.*

2. Pry off dust cap from center of wheel hub.

3. Remount wheel and lug bolts.

4. Lower car to ground. With an assistant applying brakes, loosen collar nut. Do not remove completely. See Fig. 11.

> ***WARNING —***
> *The drive flange collar nut is tightened to a torque of 250 Nm (184 ft-lb). Make sure the car is firmly on the ground.*

NOTE —
The collar nut may be staked or it may be locked in place using a lock plate, which must be bent away.

5. Raise car and remove wheel.

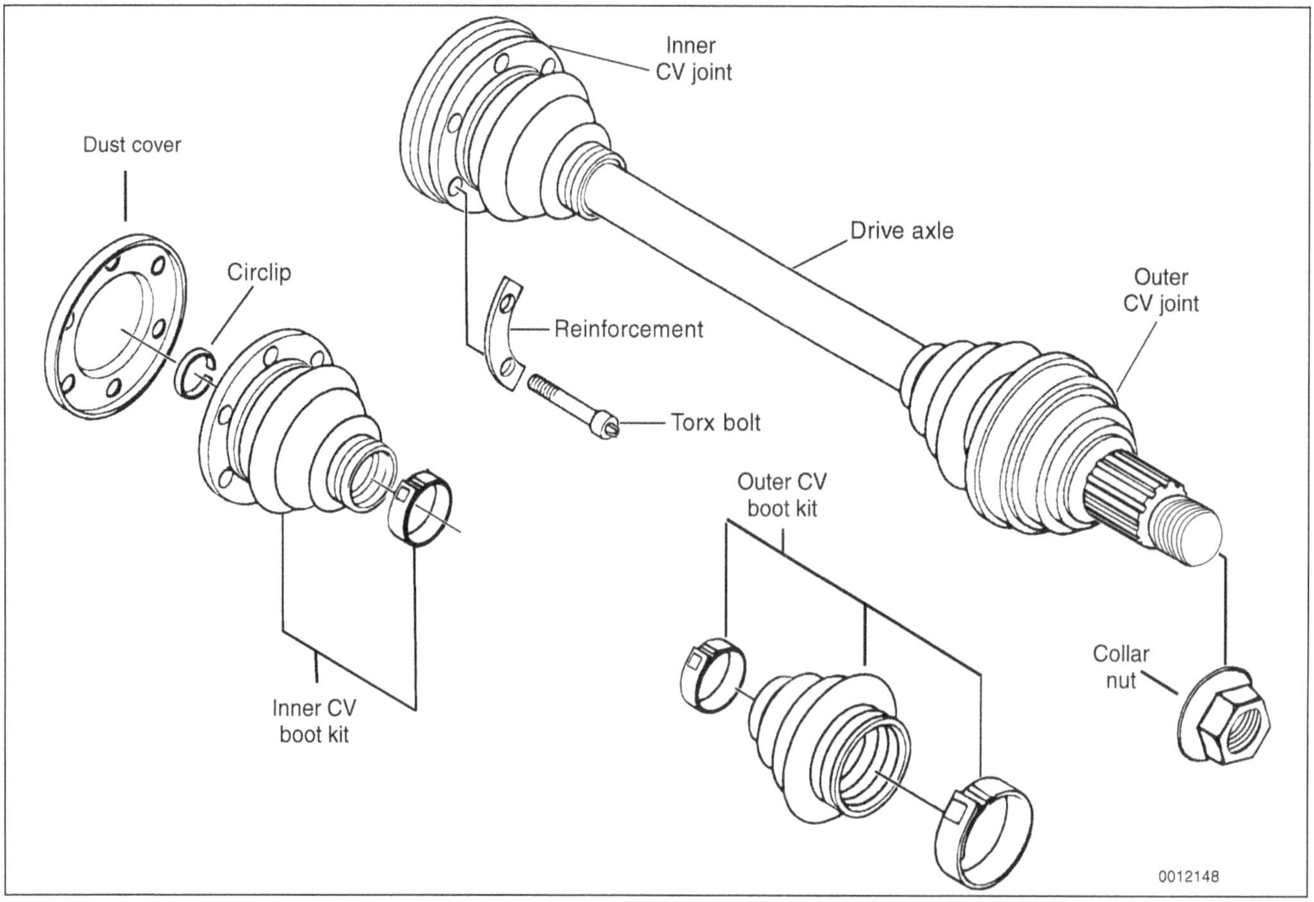

Fig. 10. Drive axle assembly.

0012111

Fig. 11. Drive flange collar nut (**arrow**).

6. Remove drive axle from drive flange. See Fig. 12. Suspend drive axle from chassis using stiff wire.

0012103

Fig. 12. Drive axle to final drive mounting bolts (**arrows**).

7. Support trailing arm from below, using a transmission jack or equivalent. Remove shock absorber to trailing arm bolt.

8. Lower trailing arm sufficiently to gain clearance for removing drive axle. If necessary, use an appropriate puller to remove drive axle from wheel bearing housing. See Fig. 13.

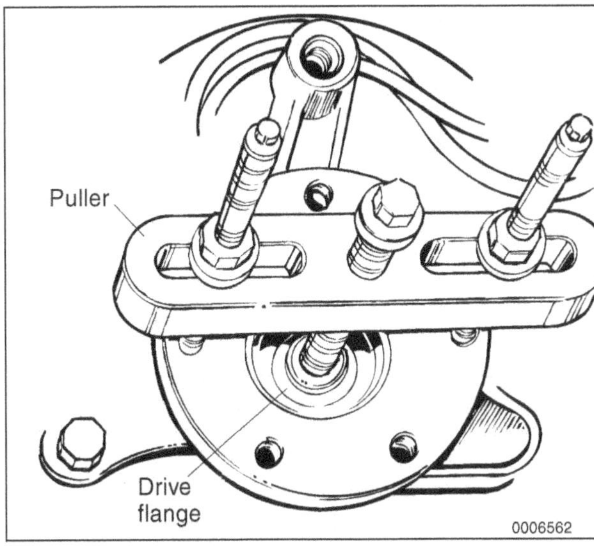

Fig. 13. Press drive axle from wheel bearing housing using puller.

> ***WARNING —***
> *It is recommended that a restraining chain be installed between the bottom coil of the coil spring and the upper control arm to retain the spring in case of accidental release. Personal injury can result if the compressed coil spring is not released slowly and carefully.*

9. Installation is reverse of removal.
   - Apply a light coating of oil to contact face of collar nut, and install it loosely.
   - Install road wheel and lower car to ground.
   - With an assistant applying brakes, tighten drive axle collar nut to its final torque.
   - Install a new collar nut retaining plate, if applicable.

Tightening Torques
- Brake caliper to trailing arm . . . . 67 Nm (50 ft-lb)
- Brake rotor to drive flange . . . . 16 Nm (12 ft-lb)
- Drive axle collar nut to
  drive flange. . . . . 250 Nm (184 ft-lb)
  M3 . . . . 300 Nm (221 ft-lb)
- Drive axle to final drive unit mounting flange . . . .
  M8 Torx bolt . . . . 64 Nm (47 ft-lb)
  M10 Torx bolt . . . . 83 Nm (62 ft-lb)
  M10 ribbed bolt. . . . . 100 Nm (74 ft-lb)
  Road wheel to hub . . . . 100±10 Nm (74±7 ft-lb)
- Shock absorber to
  trailing arm . . . . 77 Nm (57 ft-lb)

CV boot, replacing

NOTE —

- *The outer CV joint cannot be removed from the axle shaft. In order to replace the outer CV boot, it is necessary to remove the inner joint and boot first.*
- *If the CV joints are worn or defective, a complete rebuilt axle shaft is available from an authorized BMW dealer parts department.*

1. Remove drive axle from car as described above.
2. Remove inner and outer CV boot clamping bands and cut old boots off joints and shaft. Clean grease from joints.
3. Lift off dust cover from inner CV joint and remove circlip retaining joint inner hub to axle shaft. See Fig. 14.

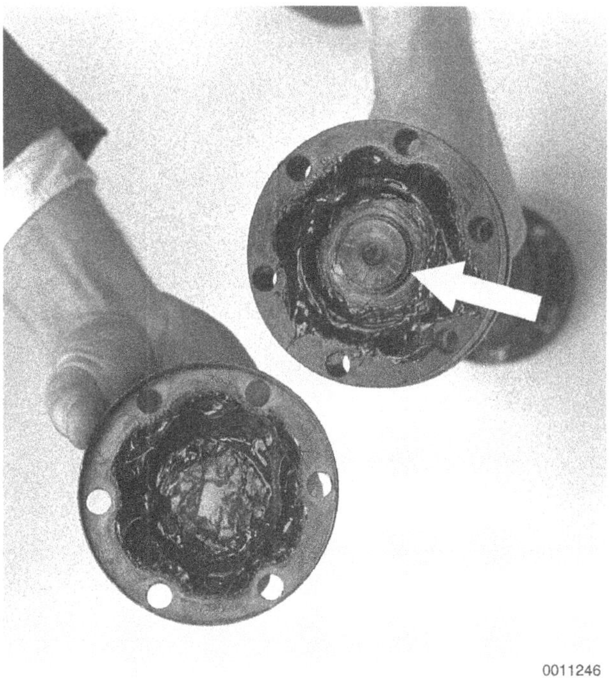

Fig. 14. Dust cover shown removed from inner CV joint. Clean away grease and remove circlip (**arrow**) from shaft end.

4. While supporting inner hub, press axle shaft out of joint. See Fig. 15.

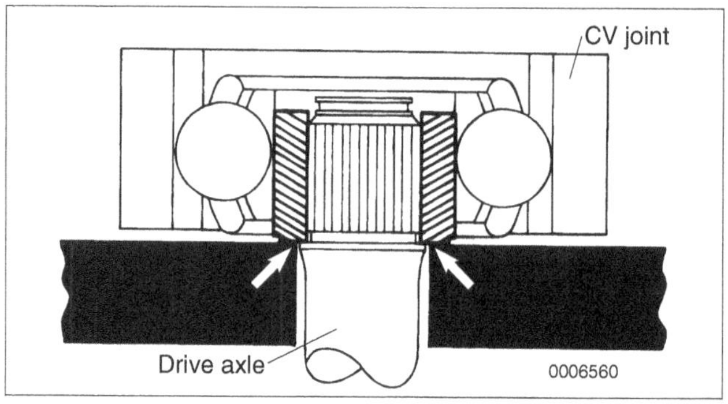

Fig. 15. Support CV joint at inner hub during pressing operation (**arrows**).

5. Clean all old lubricant off shaft splines and inner joint's splines.

NOTE —

To inspect a CV joint, clean away the grease and look for galling, pitting and other signs of wear or physical damage. Polished surfaces or visible ball tracks alone are not necessarily cause for replacement. Discoloration due to overheating indicates lack of lubrication.

6. Place new clamping bands and CV boots over drive axle.

NOTE —

When replacing CV joint boots, use complete CV joint boot repair kits. A kit will include a new boot, clamping bands, special lubricant, and a new inner CV joint circlip. The kit is available from an authorized BMW dealer parts department.

7. Apply Loctite® 270 or an equivalent heavy-duty locking compound to drive axle splines. Position new CV joint on shaft so that raised or taller side of hub is facing shaft.

WARNING —

Do not let the locking compound contact the balls in the joint. Apply only a thin coat to cover the splines.

8. While supporting axle shaft, press inner hub of CV joint onto shaft. Install new circlip.

NOTE —

Do not let the ball hub pivot more than 20° in the outer ring of the joint. The balls will fall out if the hub is pivoted too far.

9. Pack each CV joint and rubber boot with specified amount of lubricant supplied. Apply adhesive to large end of boot and mount it on joint. Secure boot with clamps. Apply sealer to inner CV joint dust cover and install.

NOTE —
- *Before installing each small boot clamp be sure to "burp" the boot by flexing the CV joint as far over as it will go. A small screw-driver inserted between the boot and the axle-shaft will help the process.*
- *BMW recommends Bostik®1513 or Epple®4851 adhesive, and Epple®39 or Curil®T sealer.*

CV Joint Lubricant Capacity

- Wheel hub end
  ex. M3 . . . . 80 gram (2.8 oz.)
  M3 . . . . 100 gram (3.5 oz.)
- Final drive end . . . . 85 gram (3.0 oz.)

10. Install drive axle as described earlier.

REAR SUSPENSION ARMS

The trailing arms, control arms and their mounting bushings control the position of the rear wheels. A damaged suspension arm or worn bushings will change the rear wheel alignment and may adversely affect handling and stability.

WARNING —
Do not attempt to straighten a damaged suspension arm. Bending or heating may weaken the original part. If the suspension arm shows any signs of damage or excessive corrosion, it must be replaced.

If the rear brake line is disconnected to remove a trailing arm, the complete braking system must be bled as part of the installation procedure.

When performing repairs to rear suspension components, refer to Fig. 16.

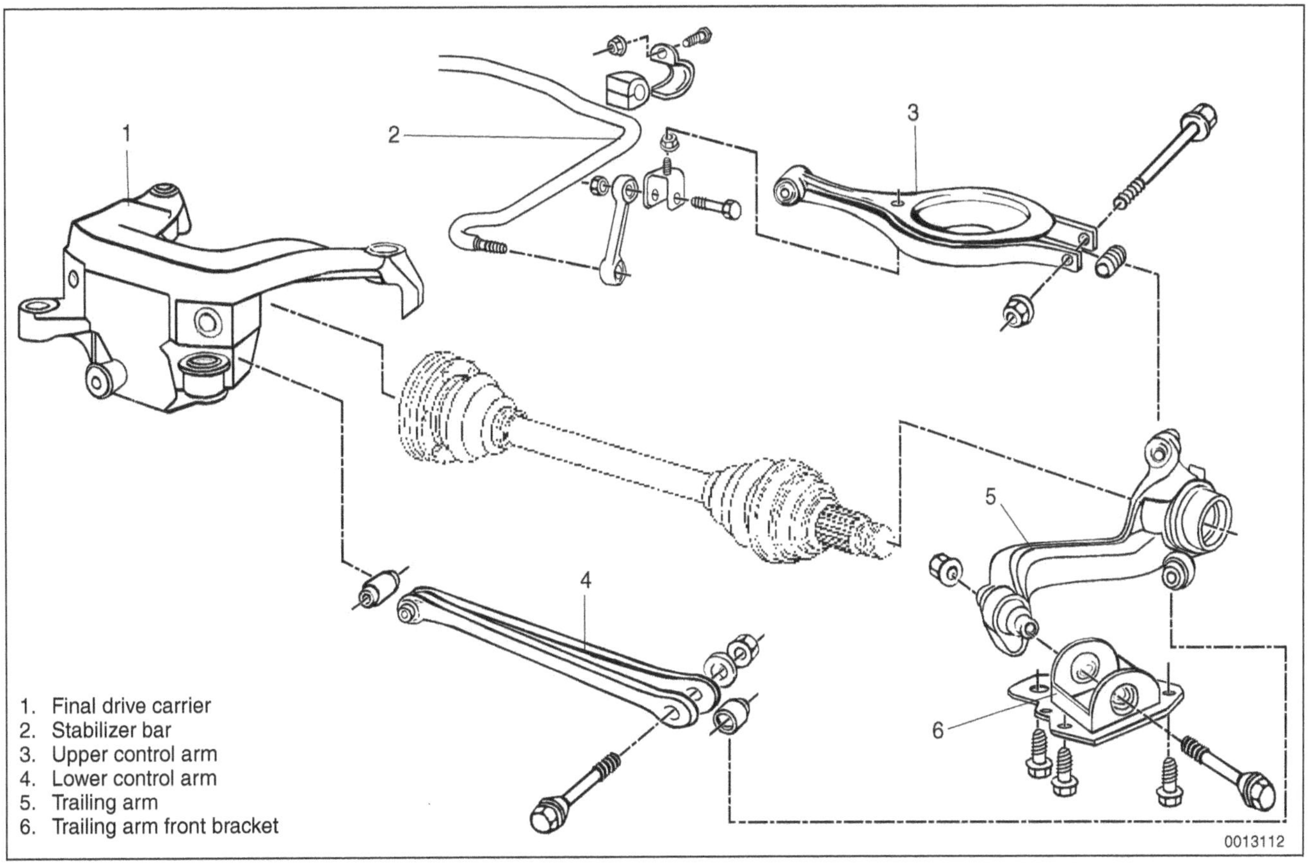

Fig. 16. Rear trailing arms, control arms and final drive carrier.

Trailing arm, removing and installing

1. Raise rear end of car and remove wheel.

> ***WARNING —***
> *Make sure that the car is firmly supported on jack stands designed for the purpose. Place the jack stands beneath a structural chassis point. Do not place jack stands under suspension parts.*

2. Remove drive axle as described earlier.
3. Remove brake rotor and caliper as described in **340 Brakes**. Do not remove brake line from caliper. Hang caliper from body with wire.
4. Remove parking brake cable from brake shoe expander. See **340 Brakes**.
5. Remove rear brake line bracket from trailing arm. See Fig. 17.

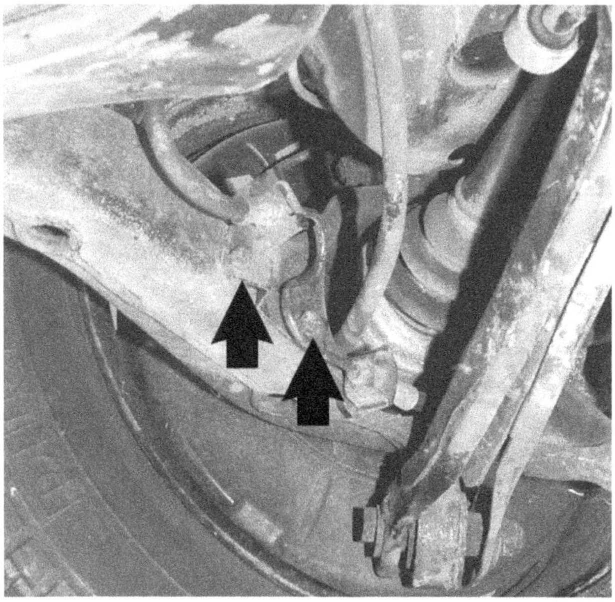

0013144

Fig. 17. Brake line bracket bolts on trailing arm (**arrow**).

6. Support trailing arm from below with a transmission jack or equivalent.

> ***WARNING —***
> *It is recommended that a restraining chain be installed between the bottom coil of the coil spring and the upper control arm to retain the spring in case of accidental release. Personal injury can result if the compressed coil spring is not released slowly and carefully.*

7. Remove shock absorber to trailing arm bolt. See Fig. 18. Slowly lower suspension until coil spring can be safely removed.

0013220

Fig. 18. Support trailing arm before removing lower shock bolt (**arrow**).

8. Unbolt upper and lower control arms from trailing arm, as described later.
9. Remove ABS wheel speed sensor from rear wheel bearing housing. Unclip sensor wiring from trailing arm. If removing right trailing arm, separate brake pad wear sensor connector. See Fig. 19.

0012104

Fig. 19. Brake pad wear sensor and ABS wheel speed sensor connectors (**arrow**).

10. Remove three bolts holding trailing arm front bracket to body. See Fig. 20. Remove trailing arm.

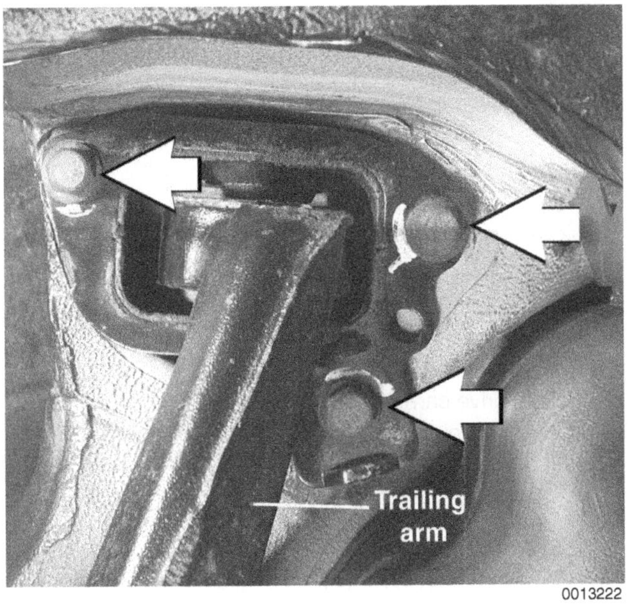

Fig. 20. Trailing arm front bracket bolts (**arrows**).

11. If trailing arm is being replaced, transfer trailing arm front bracket. Transfer attached brake system components to new arm as described in **340 Brakes**. Install a new wheel bearing as described earlier under **Rear Wheel Bearings**.

NOTE —

BMW-supplied replacement trailing arms come with the bushings installed. A new wheel bearing will have to be installed.

12. Installation is reverse of removal.
    - Always use new self-locking nuts.
    - Have car professionally aligned when job is complete.

Tightening Torques
- Drive axle to final drive unit mounting flange
  - M8 Torx bolt . . . . 64 Nm (47 ft-lb)
  - M10 Torx bolt . . . . 83 Nm (62 ft-lb)
  - M10 ribbed bolt . . . . 100 Nm (74 ft-lb)
- Road wheel to wheel hub. . . . 100±10 Nm (74±7 ft-lb)
- Shock absorber to trailing arm . . . . . . 77 Nm (57 ft-lb)
- Trailing arm to front bracket . . . . 110 Nm (81 ft-lb)
- Trailing arm bracket to body . . . . 77 Nm (57 ft-lb)
- Trailing arm to upper or lower control arm (M12 bolt) . . . . 110 Nm (81 ft-lb)

Trailing arm bushing, replacing

NOTE —

When replacing any bushing in the rear trailing arm, measure and record the orientation and protrusion of the bushing from its boss. Press in new bushing to match the position of the old one.

1. Remove trailing arm as described above.
2. Press bushing out of trailing arm using appropriate press tools.
3. Coat new bushing with Circolight® lubricant or equivalent.
4. Using appropriate press tools, draw new bushing into trailing arm until previously measured protrusion is achieved.
5. If front bracket bushing was replaced, align front bracket before tightening bolt. See Fig. 21.

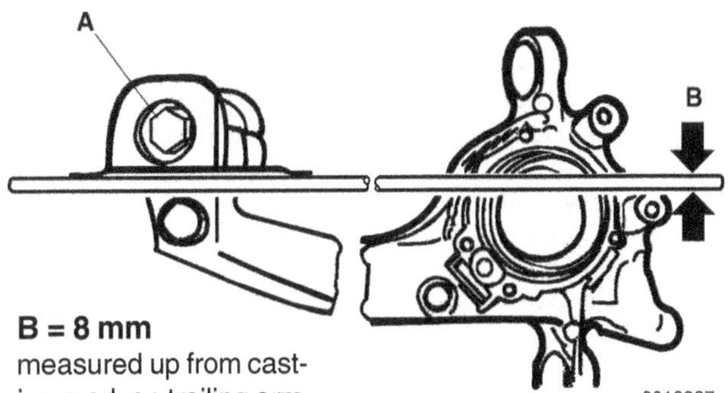

Fig. 21. Tighten trailing arm front mounting bolt (**A**) by lining up casting mark 8 mm (**B**) below level of trailing arm mounting bracket.

6. Installation is reverse of removal.
    - Have car professionally aligned when job is complete.

Tightening Torques
- Trailing arm to front bracket . . . . 110 Nm (81 ft-lb)
- Trailing arm bracket to body . . . . 77 Nm (57 ft-lb)

Upper control arm, removing and installing

1. Raise rear end of car and remove wheel.

> ***WARNING —***
> *Make sure that the car is firmly supported on jack stands designed for the purpose. Place the jack stands beneath a structural chassis point. Do not place jack stands under suspension parts.*

2. Remove drive axle as described earlier.

3. Remove stabilizer bar link from upper control arm.

4. Support trailing arm from below with a transmission jack or equivalent.

> ***WARNING —***
> *It is recommended that a retaining chain be installed between the bottom coil of the coil spring and the upper control arm to retain the spring in case of accidental release. Personal injury can result if the compressed coil spring is not released slowly and carefully.*

5. Remove lower shock absorber bolt. Refer to Fig. 18. Slowly lower trailing arm until coil spring is fully released. Remove restraining chain and remove spring.

6. Remove both upper control arm mounting bolts. Remove control arm.

7. Installation is reverse of removal.
   - Always use new self-locking nuts.
   - Have car professionally aligned when job is complete.
   - Install drive axle as described earlier.

Tightening Torques
- Drive axle to final drive flange
  M8 Torx bolt .... 64 Nm (47 ft-lb)
  M10 Torx bolt .... 83 Nm (62 ft-lb)
  M10 ribbed bolt................. 100 Nm (74 ft-lb)
- Shock absorber to trailing arm...... 77 Nm (57 ft-lb)
- Upper control arm
  to final drive carrier (M12 bolt)...... 77 Nm (57 ft-lb)
- Upper control arm
  to trailing arm (M12 bolt) .... 110 Nm (81 ft-lb)

Lower control arm, removing and installing

1. Raise rear end of car and support it securely on jackstands. Remove rear wheel.

2. Remove both lower control arm mounting bolts.

3. Use a soft hammer to tap control arm out of its mounting points.

4. Installation is reverse of removal.

> ***NOTE —***
> *Use a new bolt and washer assembly at the final drive carrier mounting point for the control arm.*

Tightening Torque
- Lower control arm
  to final drive carrier (M12 bolt) ...... 77 Nm (57 ft-lb)
- Lower control arm
  to trailing arm (M12 bolt) .... 110 Nm (81 ft-lb)

FINAL DRIVE CARRIER

Final drive carrier, removing and installing

1. Raise rear end of car and support it securely on jack stands. Remove rear wheels.

2. Remove driveshaft. See **260 Driveshaft**.

3. Remove both drive axles as described earlier.

4. Disconnect stabilizer bar from upper control arms. Remove stabilizer bar mounts.

5. Disconnect speedometer harness connector on final drive (where applicable).

6. Remove rear section of exhaust system. See **180 Exhaust System**.

> ***NOTE —***
> *It may be easier to remove complete exhaust as one piece.* **See 180 Exhaust System**.

7. Remove upper and lower control arms as described earlier.

8. Support final drive carrier with transmission jack. Remove mounting bolts and nuts at final drive carrier bushing mounts. See Fig. 22.

9. Slowly lower final drive and final drive carrier, detaching wire harness and connectors as necessary.

10. Final drive may now be unbolted from final drive carrier.

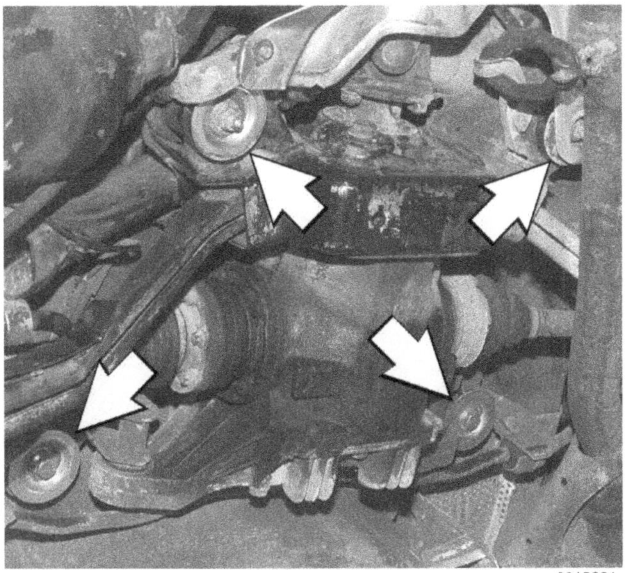
0013221

Fig. 22. Final drive carrier mounts (**arrows**).

11. Installation is reverse of removal.
    - Check and adjust parking brake as described in **340 Brakes**.
    - Always use new self-locking fasteners.
    - Have car professionally aligned when job is complete.

Tightening Torques

- Drive axle to final drive flange
  M8 Torx bolt . . . . 64 Nm (47 ft-lb)
  M10 Torx bolt . . . . 100 Nm (74 ft-lb)
- Driveshaft to final drive flange . . . See **260 Driveshaft**
- Final drive carrier to body . . . . 77 Nm (57 ft-lb)
- Final drive to final drive carrier (M12 bolts)
  Front mount. 95 Nm (70 ft-lb)
  Rear mounts . . . . 77 Nm (57 ft-lb)
- Road wheel to wheel hub. . . . 100±10 Nm (74±7 ft-lb)
- Upper or lower control arm
  to final drive carrier (M12 bolts) . . . . . 77 Nm (57 ft-lb)

331 Final Drive

GENERAL

This repair group covers repair operations that do not require complicated disassembly of the final drive. Internal repairs and final drive disassembly are not covered in this manual.

FINAL DRIVE SERVICE

All final drive work requires some method of raising the car and supporting it securely while the work is performed. Jack stands and a floor jack can easily be used, but use extreme caution when working beneath the car. See **010 Fundamentals for the Do-It-Yourself Owner**.

> ***NOTE—***
> *Removal of final drive carrier is covered in* **330 Rear Suspension**.

Final drive oil, draining and filling

1. Drive car to warm final drive fluid.

2. Raise car and support safely.

> ***WARNING—***
> *Make sure that the car is firmly supported on jack stands designed for the purpose. Place the jack stands beneath a structural chassis point. Do not place jack stands under suspension parts.*

3. Place a drain pail below final drive and remove drain and fill plug from final drive. See Fig. 1.

> ***NOTE—***
> - *Use a 14 mm allen bit socket to remove the drain plug. Alternatively, cut approximately 30 mm (1.2 in) from an allen key and use a box end wrench on the key stub.*

4. Install and tighten drain plug.

5. Fill final drive with appropriate type and quantity of lubricant. Install and tighten fill plug.

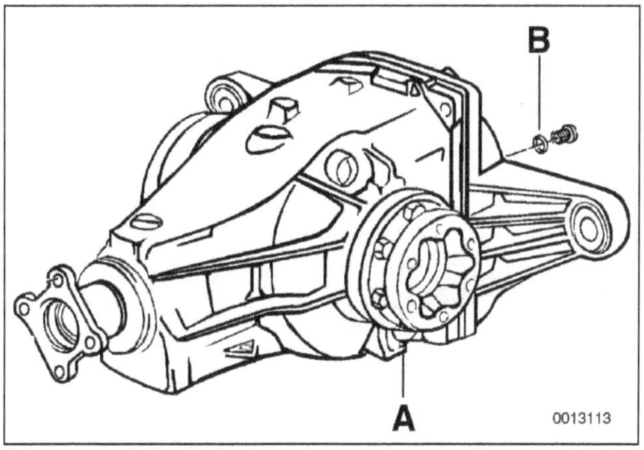

Fig. 1. Final drive drain plug (**A**) and fill plug (**B**).

> ***NOTE—***
> *The final drive fluid level is correct when the fluid begins to spill from the fill plug.*

Final Drive Drain and Fill
- Oil specifications
  - w/o limited slip . . . . BMW SAF-XO Synthetic Oil
  - with limited slip . . . . BMW SAF-XLS Synthetic Oil
- Final drive oil capacity
  - 4-cylinder. . . . . 1.1 liters (1.2 qt)
  - 6-cylinder. . . . . 1.7 liters (1.8 qt)

Final Drive Oil Seals

Low oil level caused by faulty oil seals may be the cause of noisy final drive operation or limited-slip chatter. The drive flange (side) and input shaft (front) oil seals can be replaced while the final drive is installed.

> ***NOTE—***
> *Do not mistake leaking CV joints for flange seal leaks. It may be helpful to degrease the final drive to pinpoint the source of the leak prior to replacing seals.*

Final drive flange oil seal, replacing

1. Raise car and support safely.

> ***WARNING —***
> *Make sure that the car is firmly supported on jack stands designed for the purpose. Place the jack stands beneath a structural chassis point. Do not place jack stands under suspension parts.*

2. Detach drive axle(s) from final drive. See **330 Rear Suspension**.

> ***CAUTION —***
> *Suspend the detached drive axle from the car body with a stiff wire hook to prevent damage to the outer CV joint.*

3. Pry drive flange from final drive. See Fig. 2.

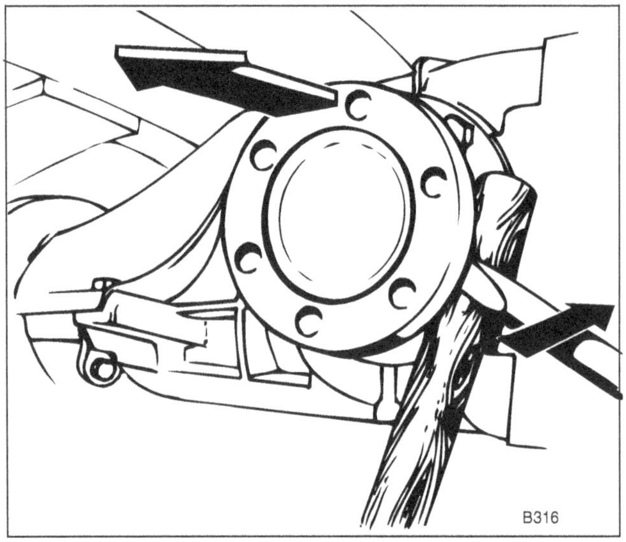

Fig. 2. Drive flange being pried off. For leverage, use a wooden dowel as shown.

4. Remove wire (snap) ring from final drive opening. See Fig. 3.

> ***NOTE —***
> *Inspect flange at the point where the oil seal rides on the shaft. Replace the flange assembly if there is a groove worn in the shaft.*

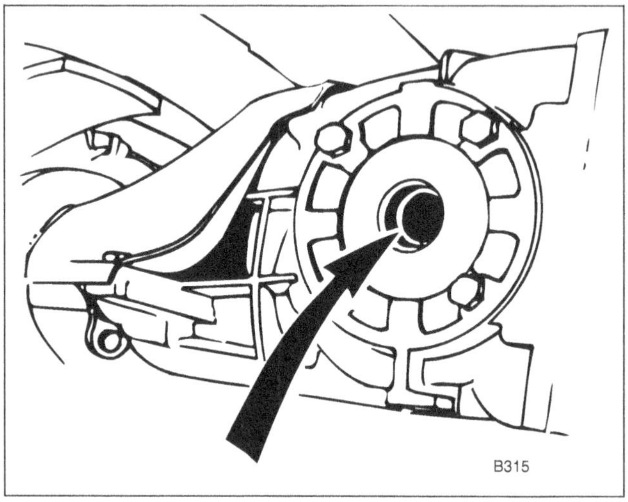

Fig. 3. Drive flange snap ring (**arrow**) to be removed from final drive housing.

5. Pry old oil seal from its recess using a hooked seal removal tool, or a large screwdriver. Use pan to catch draining fluid.

> ***CAUTION —***
> *Be careful not to mar the final drive housing when removing the seal.*

6. Dip new seal in final drive lubricant and drive seal into place until fully seated.

7. Install new snap ring in groove of final drive housing. Make sure both ends of ring are fully seated in groove.

8. Install drive flange by pressing it in by hand until snap ring engages. It may be necessary to turn drive flange slightly while pushing.

9. Attach drive axle and tighten bolts.

10. Top off final drive with oil.

Tightening Torque
- Drive axle to final drive flange
  - M8 Torx bolt. 64 Nm (47 ft-lb)
  - M10 Torx bolt. . . . . 100 Nm (74 ft-lb)

Final drive input shaft oil seal, replacing

1. Raise car and support safely.
2. Drain final drive oil as described earlier.
3. Remove driveshaft from final drive input shaft flange. See **260 Driveshaft**.
4. Make matching marks on input shaft and input shaft collar nut. See Fig. 4.

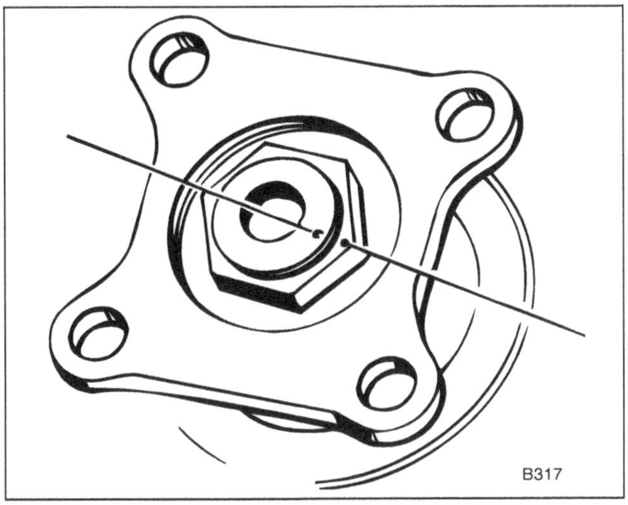

Fig. 4. Make matching marks on input shaft collar nut and flange.

5. Pry lockplate from nut. Hold input flange and remove collar nut. If necessary, use a puller to remove input flange.
6. Remove final drive front dust cover.
7. Pry faulty seal from its recess using a hooked seal removal tool or a large screwdriver. Dip new seal in final drive lubricant and drive it into position.

> ***CAUTION—***
> *Be careful not to mar the housing when removing the seal.*

8. Install new final drive front dust cover.
9. Lightly lubricate input shaft and press input flange back on. Install collar nut and slowly tighten until matching marks line up, coming as close as possible to specified torque.

Tightening Torques
- Final drive input shaft flange to pinion with 4 bolt side covers. . . . . 175 Nm (129 ft-lb) (or until match marks line up)
- Final drive input shaft flange to pinion with 6 bolt side covers. . . . . 185 Nm (137 ft-lb) (or until match marks line up)

> ***CAUTION—***
> *If the input flange nut torque is exceeded, or the nut is tightened past the marks, the crushable collar sleeve behind the flange will need to be replaced. This operation requires disassembly of the final drive unit.*

10. Install a new lockplate and refill final drive with lubricant.
11. Remaining assembly is reverse of disassembly.

Tightening Torques
- Driveshaft to final drive flange . . . See **260 Driveshaft**
- Flex-disc to driveshaft or transmission flange . . . . See **260 Driveshaft**
- Final drive drain and fill plugs. . . . . . . 70 Nm (52 ft-lb)

Final drive unit, removing and installing

1. Raise rear end of car and support it securely on jack stands. Remove rear wheels.
2. Remove driveshaft from final drive input flange. See **260 Driveshaft**.
3. Remove both drive axles from final drive unit. Suspend axle using stiff wire. See **330 Rear Suspension**.
4. Disconnect stabilizer bar from upper control arms. Remove left and right stabilizer bar mounts.
5. Disconnect speedometer harness connector on final drive (where applicable).
6. Support final drive with transmission jack. Remove mounting bolts at front and rear. See Fig. 5.
7. Slowly lower final drive unit and remove towards rear.

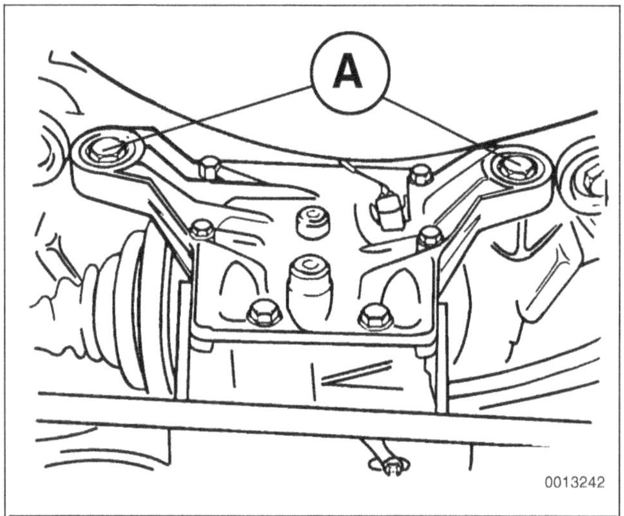

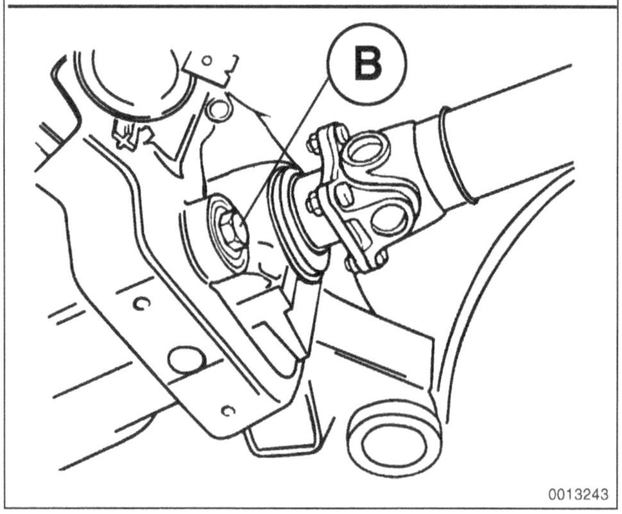

Fig. 5. Final drive to final drive carrier rear (**A**) and front mounting bolts (**B**).

8. Installation is reverse of removal.
   - Check and adjust parking brake as described in **340 Brakes**.
   - Always use new self-locking fasteners.
   - Have car professionally aligned when job is complete.

Tightening Torques

- Drive axle to final drive flange
  M8 Torx bolt . . . . 64 Nm (47 ft-lb)
  M10 Torx bolt . . . . 100 Nm (74 ft-lb)
- Driveshaft to final drive flange . . . See **260 Driveshaft**
- Final drive carrier to body. . . . . 77 Nm (57 ft-lb)
- Final drive to final drive carrier (M12 bolts)
  Front mount . . . . 95 Nm (70 ft-lb)
  Rear mounts. 77 Nm (57 ft-lb)
- Road wheel to wheel hub. . . . 100±10 Nm (74±7 ft-lb)
- Upper or lower control arm
  to final drive carrier (M12 bolts) . . . . .77 Nm (57 ft-lb))

340 Brakes

TABLES

340

GENERAL

WARNING —

- *Although semi-metallic and metallic brake friction materials in brake pads or shoes no longer contain asbestos, they produce dangerous dust.*
- *Brake fluid is poisonous, highly corrosive and dangerous to the environment. Wear safety glasses and rubber gloves when working with brake fluid. Do not siphon brake fluid with your mouth. Immediately clean away any fluid spilled on painted surfaces and wash with water, as brake fluid will remove paint.*
- *Always use new brake fluid from a fresh, unopened container. Brake fluid will absorb moisture from the air. This can lead to corrosion problems in the braking system, and will also lower the brake fluid's boiling point. Dispose of brake fluid properly.*
- *Do not reuse self-locking nuts, bolts or fasteners. They are designed to be used only once and may fail if reused. Always replace them with new self-locking fasteners.*

BMW E36 models are equipped with vacuum power-assisted four-wheel disc brakes with an integral Antilock Brake System (ABS). Single-piston calipers act on solid or vented front rotors and solid rear rotors. A brake pad wear sensor for each axle indicates when brake pads need replacement. The dual drum-type parking brake system is integrated with the rear brake rotors.

NOTE —

M3 models have vented directional brake rotors on the rear axle as well as on the front.

TROUBLESHOOTING

Brake performance is mainly affected by three things: the level and condition of the brake fluid, the system's ability to create and maintain hydraulic pressure, and the condition of the friction components.

Air in the brake fluid will make the brake pedal feel spongy during braking or will increase the brake pedal force required to stop. Fluid contaminated by moisture or dirt can corrode the system. Inspect the brake fluid inside the reservoir. If it is dirty or murky, or is over a year old, the fluid should be replaced.

Visually check the hydraulic system starting at the master cylinder. To check the function of the master cylinder hold the brake pedal down hard with the engine running. The pedal should feel solid and stay solid. If the pedal slowly falls to the floor, either the master cylinder is leaking internally, or fluid is leaking externally. If no leaks can be found, the master cylinder is faulty and should be replaced. Check all brake fluid lines and couplings for leaks, kinks, chafing and corrosion.

Check the brake booster by pumping the brake pedal approximately 10 times with the engine off. Hold the pedal down and start the engine. The pedal should fall slightly. If not, check for any visible faults before suspecting a faulty brake booster. Check for strong vacuum at the vacuum hose fitting at the booster, and check the non-return valve for one-way flow.

Worn or contaminated brake pads will cause poor braking performance. Oil-contaminated or glazed pads will cause stopping distances to increase. Inspect the rotors for glazing, discoloration and scoring. Steering wheel vibration while braking at speed is often caused by warped rotors, but can also be caused by worn suspension components.

When troubleshooting, keep in mind that tire inflation, wear and temperature can affect braking and suspension. See **310 Front Suspension** for more information on front suspension parts inspection.

Table a lists symptoms of brake problems, their probable causes, and suggested corrective actions.

Table a. Brake System Troubleshooting

Symptom	Probable cause	Repairs
Brake squeal	a. Incorrectly installed brake pads or parking brake shoes b. Brake pad carriers dirty or corroded c. Brake pad anti-rattle springs faulty or missing d. Brake pads heat-glazed or oil-soaked e. Wheel bearings worn (noise most pronounced when turning)	a. Check component installation. b. Remove brake pads and clean calipers. c. Install/replace anti-rattle springs. d. Replace brake pads. Clean rotors. Replace leaking calipers as required. e. Replace worn bearings. See **310 Front Suspension** or **330 Rear Suspension**.
Pedal goes to floor when braking	a. Brake fluid level low due to system leaks b. Master cylinder faulty	a. Check fluid level and inspect hydraulic system for signs of leakage. Fill and bleed system. b. Replace master cylinder.
Low pedal after system bleeding	a. Master cylinder faulty	a. Replace master cylinder.
Pedal spongy or brakes work only when pedal is pumped	a. Air in brake fluid b. Master cylinder faulty (internal return spring weak) c. Leaking line or hose unions	a. Bleed system. b. Replace master cylinder. c. Repair or replace lines and hoses. Bleed system.
Excessive braking effort	a. Brake pads wet b. Brake pads heat-glazed or oil-soaked c. Vacuum booster or vacuum hose connections to booster faulty	a. Use light pedal pressure to dry pads while driving. b. Replace brake pads. Clean rotors. Replace leaking calipers. c. Inspect vacuum lines. Test vacuum booster and replace as required. Test vacuum non-return valve for one-way air flow.
Brakes pulsate, chatter or grab	a. Warped brake rotors b. Brake pads worn c. Brake pads heat-glazed or oil-soaked	a. Resurface or replace rotors. b. Replace brake pads. c. Replace brake pads. Clean rotors. Replace leaking calipers.
Uneven braking, car pulls to one side, rear brakes lock	a. Incorrect tire pressures or worn tires b. Brake pads on one side of car heat-glazed or oil-soaked c. Caliper or brake pads binding d. Worn suspension components	a. Inspect tire condition. Check and correct tire pressures. b. Replace brake pads. Clean rotors. Replace leaking calipers. c. Clean and recondition brakes. d. Inspect for worn or damaged suspension components. See **310 Front Suspension** or **330 Rear Suspension**.
Brakes drag, bind or overheat	a. Brake caliper or brake pads binding b. Master cylinder faulty	a. Clean or replace caliper. b. Replace master cylinder.

> ***WARNING —***
> *On cars with All Season Traction (AST), special BMW service equipment is required to properly bleed the ABS/AST system. For safety reasons, the brake system on cars with ABS/AST must not be bled using the procedures described in this repair group.*

BLEEDING BRAKES

Brake bleeding is usually done for one of two reasons: Either to replace old brake fluid as part of routine maintenance or to expel trapped air in the system that resulted from opening the brake hydraulic system during repairs.

Always use new brake fluid from an unopened container. It is important to bleed the entire system when any part of the hydraulic system has been opened. On cars not equipped with traction control (AST), brake system bleeding should be done with a pressure bleeder. On cars with AST, brake bleeding should be done by an authorized BMW dealer.

> ***WARNING —***
> *On cars with All Season Traction (AST), special BMW service equipment is required to properly bleed the ABS/AST system. For safety reasons, the brake system on cars with ABS/AST must not be bled using the procedures described in this repair group.*

When bleeding the brakes, start at the wheel farthest from the master cylinder and progress in the following order:

- right rear brake
- left rear brake
- right front brake
- left front brake

Pressure bleeding brakes (except cars with AST)

1. Top off brake fluid in reservoir and connect pressure bleeder to reservoir. Connect bleeder hose and bottle to right rear caliper bleeder screw. Pressurize system to approximately 1 bar (14.5 psi).

> ***CAUTION—***
> *Do not exceed a pressure of 2 bar (29 psi) when pressure bleeding the brake system. Excessive pressure will damage the brake fluid reservoir.*

2. Have a helper hold brake pedal down.

3. Open bleeder screw. See Fig 1. Have helper slowly pump brakes about 10 times with bleeder screw open, holding pedal down on the last pump. When escaping fluid is free of air bubbles, close bleeder screw..

> ***CAUTION—***
> *Bleeder hose must always remain submersed in the clean brake fluid whenever the bleeder valve is open.*

0012113

Fig. 1. Brake caliper bleeder screw (**arrow**).

4. Close bleeder screw and release brake pedal. Refill brake fluid reservoir and proceed to rear left wheel.

5. Proceed with the remaining wheels using the order listed earlier.

Tightening Torques
- Bleeder screws:
  7 mm screw . . . . 5 Nm (4 ft-lb)
  9 mm screw . . . . 6 Nm (5 ft-lb)

BRAKE PADS, CALIPERS, AND ROTORS

The E36 front brake caliper is shown in Fig. 2. The rear brake caliper is shown in Fig. 7.

Brake pads can be replaced without disconnecting the brake fluid hose from the caliper or having to bleed the brakes. The rotors can be replaced without disassembling the wheel hub and bearing. Always machine or replace rotors in pairs. Replace pads in sets.

> ***WARNING—***
> - *Although semi-metallic and metallic brake friction materials in brake pads or shoes no longer contain asbestos, they produce dangerous dust.*
> - *Treat all brake dust as a hazardous material.*
> - *Do not create dust by grinding, sanding, or cleaning brake friction surfaces with compressed air.*
> - *Breathing any brake dust can cause serious diseases such as cancer, and may result in death.*

Brake pads, replacing

This procedure is applicable to both front and rear brakes. Front and rear brake assemblies are basically the same, except that the rear brake rotors house the parking brake mechanism

1. Raise car and remove wheels.

> ***WARNING—***
> *Make sure the car is firmly supported on jack stands designed for the purpose. Place the jack stands beneath a structural chassis point. Do not place jack stands under suspension parts.*

2. Remove plastic caps from guide bolts and then remove guide bolts. See Fig. 3.

3. If applicable, disconnect brake pad wear sensor connector and remove wiring from its holder.

4. Remove anti-rattle clip from caliper. See Fig. 4.

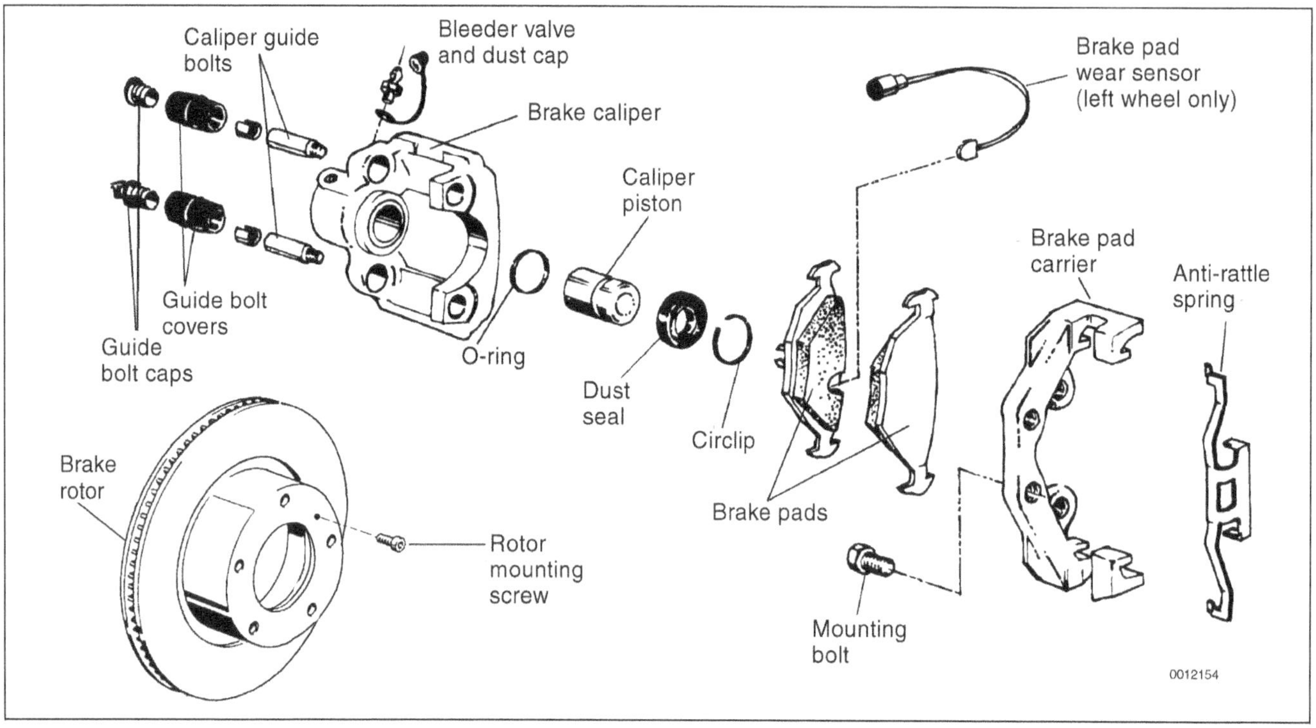

Fig. 2. Front brake caliper assembly.

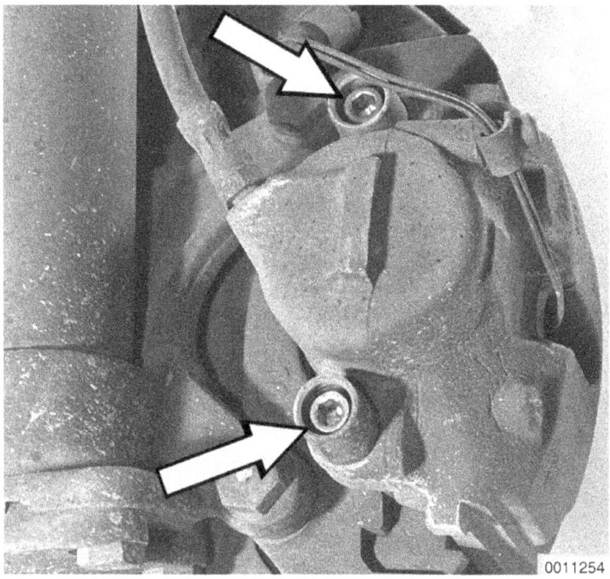

Fig. 3. Brake caliper guide bolts (**arrows**).

5. Pull caliper straight off rotor.

> ***CAUTION —***
> *Do not let the brake caliper hang from the brake hose. Suspend it from the chassis using stiff wire.*

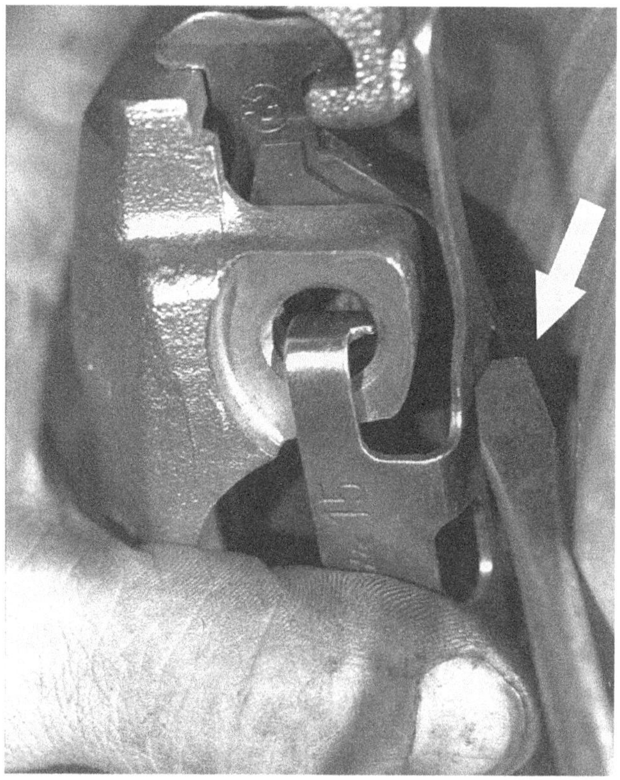

Fig. 4. Removing anti-rattle clip by unhooking at top and bottom. Use screwdriver as an aid (**arrow**).

> ***NOTE —***
> *If there is a ridge on the rotor edge, the caliper piston will have to be pushed back into the caliper before the caliper can be removed.*

6. Remove brake pads from caliper. Where applicable, carefully pry pad wear sensor from pad. See Fig. 5.

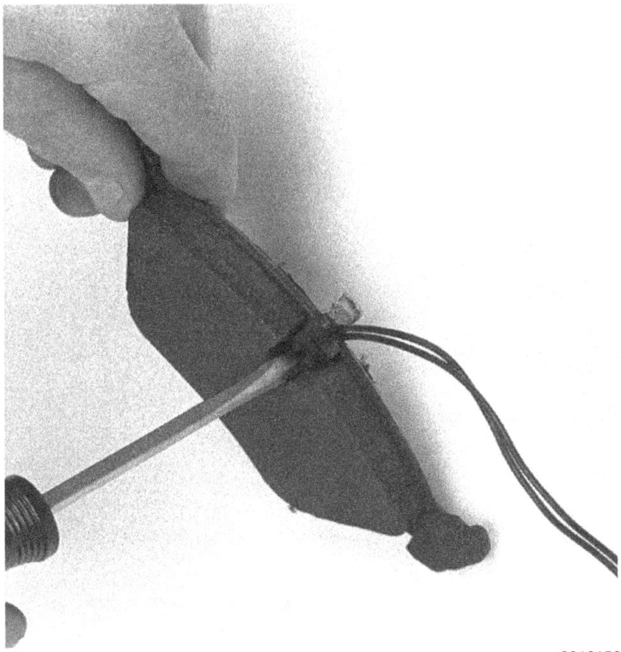

0012156

Fig. 5. Removing brake pad wear sensor.

7. Inspect brake caliper for signs of leakage. Check that the caliper piston slides smoothly into caliper. Replace caliper if any faults are found.

> ***CAUTION—***
> *With the pads removed, residual hydraulic pressure may cause the caliper piston to slide out. Use a wooden block to hold the piston in place.*

8. Insert brake pad wear sensor into cutout in new pad where applicable.

NOTE—
If the brake lining indicator light illuminated prior to brake pad replacement, the wear sensor should be replaced.

9. Slowly press piston back into caliper. Use care not to damage piston dust seal. See Fig. 6.

NOTE—
- *Open caliper bleeder screw when pressing piston back into caliper. Catch expelled fluid in appropriate container. This procedure is highly recommended in the case of cars with ABS.*
- *Open the bleeder screw only when applying force to the piston. Do not allow air to be drawn in through the bleeder screw.*
- *Pressing the caliper piston in may cause the brake fluid reservoir to overflow. To prevent this, use a clean syringe to first remove some fluid from the reservoir.*

0012157

Fig. 6. Brake caliper piston being pressed into caliper to make room for new pads. Use a rag to protect caliper piston and seals.

10. Thoroughly clean inner and outer surfaces of brake rotor. Inspect rotor for cracks, signs of overheating and scoring.

11. Measure thickness of rotor. See **Table b**. If rotor does not pass minimum thickness requirements, or is damaged, replace as described later.

12. Thoroughly clean all contact points on caliper and brake pad carrier. Clean guide bolts and make sure they slide freely.

13. Lightly coat pad contact points with brake anti-squeal paste. Place outer brake pad onto caliper. Apply a small amount of grease to the top and bottom contact points where the caliper contacts the pad carrier.

14. Place inner brake pad (with locating spring) onto brake caliper piston. Route pad wear sensor wiring through caliper opening and bleeder dust cap.

15. The remainder of installation is reverse of removal.
    - Do not lubricate guide bolts.
    - Once brake assembly is installed depress brake pedal several times before driving vehicle.

Tightening Torques
- Brake caliper to brake pad carrier (guide bolt) .... 30 Nm (22 ft-lb)
- Road wheel to hub .... 100±10 Nm (74±7 ft-lb)

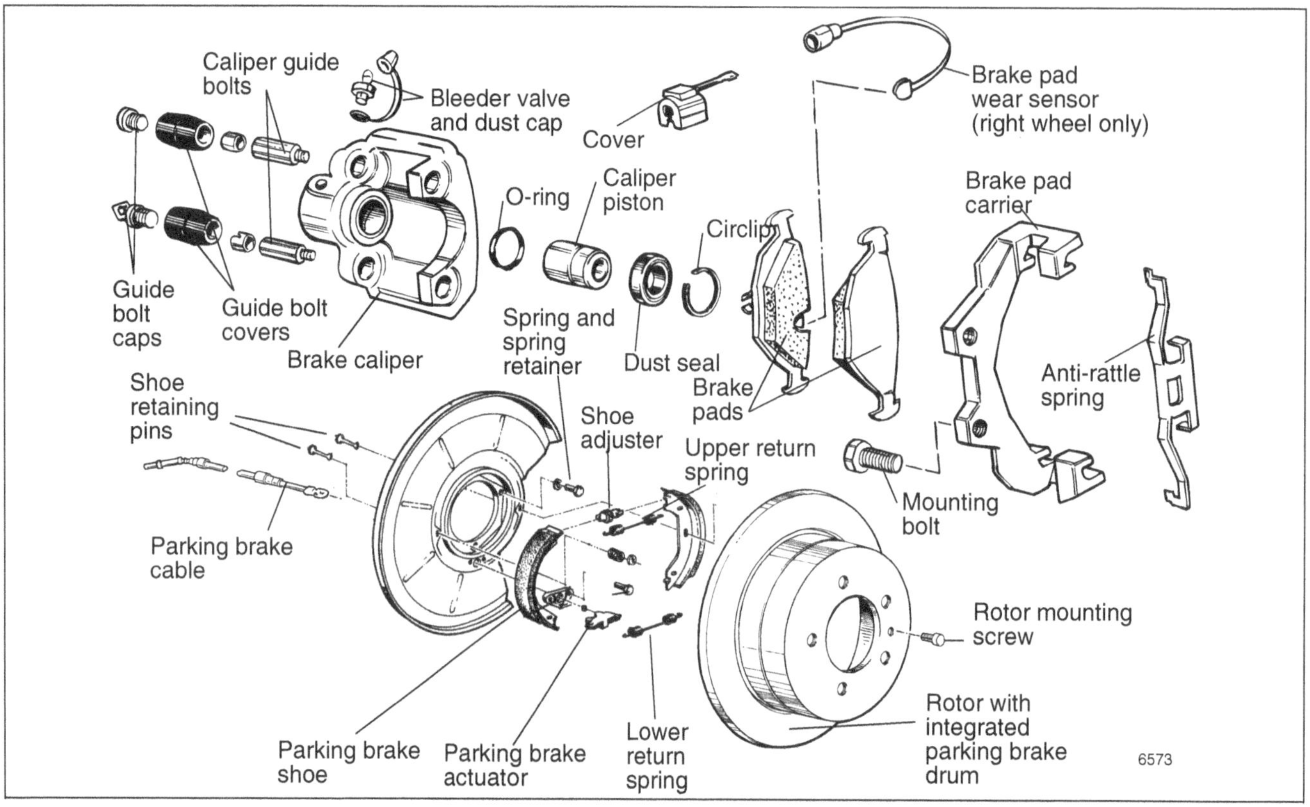

Fig. 7. Rear brake caliper assembly.

Brake caliper, removing and installing

1. Raise car and remove wheel.

> ***WARNING —***
> *Make sure the car is firmly supported on jack stands designed for the purpose.*

2. Loosen brake hose fitting at caliper. Do not remove.

3. Remove caliper mounting bolts from steering arm (front caliper) or from trailing arm (rear caliper). See Fig. 8.

4. Press piston back into caliper and slide caliper off brake rotor. Spin caliper off brake hose.

5. Inspect brake caliper for signs of leakage. Check that caliper piston slides smoothly into caliper. Replace caliper if any faults are found.

6. Installation is reverse of removal. Bleed entire brake system before driving car. See **Bleeding Brakes**.

Tightening Torques
- Front brake caliper to steering arm . . 110 Nm (81 ft-lb)
- Rear brake caliper to trailing arm. . . . 67 Nm (50 ft-lb)
- Brake fluid hose to caliper . . . . 17-19 Nm (13-14 ft-lb)
- Road wheel to hub . . . . 100±10 Nm (74±7 ft-lb)

Fig. 8. Front brake pad carrier mounting bolts (**arrows**). Rear caliper mounting bolts are similar.

Brake rotor, removing and installing

Brake rotors should always be replaced in pairs.

NOTE —
M3 models use directional brake rotors and must be installed on the proper side. Part numbers are cast into the rotors. A part that ends in an odd number is a left rotor. One that ends in an even number is a right rotor.

1. Remove caliper as described above and hang from chassis using stiff wire.

NOTE —
Do not loosen caliper brake hose.

2. Remove mounting screw holding rotor to wheel hub. See Fig. 9.

0011259

Fig. 9. Brake rotor mounting screw (**arrow**).

3. Inspect rotor for excessive grooving, cracks, and warping. Check rotor thickness. See **Table b**.

NOTE —
On original equipment rotors, the minimum allowable thickness is stamped on the rotor hub. Measure the rotor braking surface with a micrometer at eight to ten different points and use the smallest measurement recorded.

4. Installation is reverse of removal.
   - Clean rotor with brake cleaner before installing.
   - If brake fluid hose was removed or loosened, brake system must be bled as described above under **Bleeding Brakes**.

5. Once brake assembly is installed, depress brake pedal several times to adjust caliper and pads.
   - Pedal should feel firm and be at proper height.
   - Check fluid level. Top up if necessary.

NOTE —
When installing new rear brake rotors, the parking brake should be adjusted. See **Parking Brake**.

Tightening Torques
- Brake rotor to wheel hub .... 16 Nm (12 ft-lb)
- Front brake caliper to steering arm... 110 Nm (81 ft-lb)
- Rear brake caliper to trailing arm.... 67 Nm (50 ft-lb)
- Road wheel to hub ...... 100 ± 10 Nm (74 ± 7 ft-lb)

Table b. Brake Rotor Reconditioning Specifications

	Front	Rear
Vented rotor wear limit (min. thickness)	20.4 mm (0.803 in.)	
Solid rotor wear limit (min. thickness)		8.4 mm (0.331 in.)
M3 models wear limit (min. thickness)	26.4 mm (1.039 in.)	18.4 mm (0.723 in.)
Axial runout (max.) rotor removed rotor installed	 0.05 mm (0.002 in.) 0.20 mm (0.008 in.)	 0.05 mm (0.002 in.) 0.20 mm (0.008 in.)

CAUTION —
M3 brake rotors must not be machined.

MASTER CYLINDER

The brake master cylinder is mounted to the front of the vacuum booster on the driver side bulkhead.

Master cylinder, removing and installing

1. Using a clean syringe, empty brake fluid reservoir.

WARNING —
Brake fluid is highly corrosive and dangerous to the environment. Dispose of it properly.

2. Disconnect brake fluid level sender connector from fluid reservoir cap.

3. Disconnect brake fluid lines from master cylinder. See Fig. 10. Disconnect hydraulic clutch system supply line.

4. Unscrew mounting nuts and remove master cylinder from brake booster.

5. Make sure all nuts, fluid couplings, thread bores, and mating surfaces are clean.

0012164

Fig. 10. Brake fluid lines at master cylinder (**arrows**).

6. If fluid reservoir was removed, install it carefully using new sealing grommets.

7. Mount master cylinder to brake booster using a new O-ring and new self-locking nuts.

> ***CAUTION—***
> *Use care not to over-torque the master cylinder mounting nuts. This could damage the brake booster and prevent proper vacuum build-up.*

8. Connect all brake fluid lines. Connect hydraulic clutch hose to brake fluid reservoir.

9. Reconnect fluid level sender, and bleed entire brake system as described earlier.

Tightening Torque
- Brake master cylinder to brake booster . . . . 26 Nm (18 ft-lb)
- Brake fluid hose to caliper . . . . 17-19 Nm (13-14 ft-lb)

BRAKE BOOSTER

The brake booster is mounted to the bulkhead on the driver side of the engine compartment, directly behind the brake master cylinder. See Fig. 11.

0013225

Fig. 11. Brake booster (**arrow**). **A** is vacuum hose from intake manifold. **B** is one-way valve. Intake manifold vacuum acts on a large diaphragm in the brake booster to reduce brake pedal effort.

Brake booster, removing and installing

1. Disconnect negative (–) cable from battery.

> ***CAUTION—***
> *Prior to disconnecting the battery, read the battery disconnection cautions given at the front of this manual on page viii.*

2. Using a clean syringe, empty brake fluid reservoir.

> ***WARNING—***
> *Brake fluid is highly corrosive and dangerous to the environment. Dispose of it properly.*

3. Remove brake fluid level sender connector from reservoir cap.

4. Disconnect brake fluid lines from master cylinder. Plug openings.

5. Disconnect engine vacuum hose from brake booster.

6. Disconnect ABS electrical connectors. Label and then disconnect brake lines from hydraulic brake unit. Plug openings. See Fig. 12.

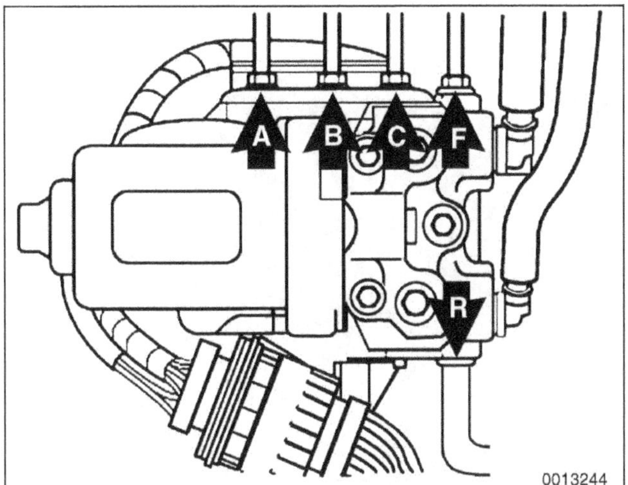

A. Right front
B. Left front
C. Rear
F. Front from master cylinder
R. Rear from master cylinder

Fig. 12. Brake fluid lines at ABS hydraulic unit.

> ***WARNING —***
> - *On cars with All Season Traction (AST), special BMW service equipment is required to properly bleed the Antilock Brake System (ABS). Removal of the ABS hydraulic unit is not recommended unless this equipment is available. For safety reasons, the brake system on cars with ABS must not be bled using the procedures described in this repair group.*
> - *Do not mix up the fluid lines at the ABS hydraulic unit. Label all connections before disconnecting.*

7. Remove ABS hydraulic unit from engine compartment.

8. Working inside car, remove panel(s) from beneath steering column to access pedal assembly. See **513 Interior Trim**.

9. Remove clip and clevis pin from brake booster pushrod and disconnect it from brake pedal. Remove brake booster mounting nuts. See Fig. 13.

10. Working in engine compartment, remove brake booster together with master cylinder.

11. If booster or master cylinder O-ring is being replaced, disconnect master cylinder from brake booster.

12. Make sure all nuts, fluid couplings, thread bores and mating surfaces are clean.

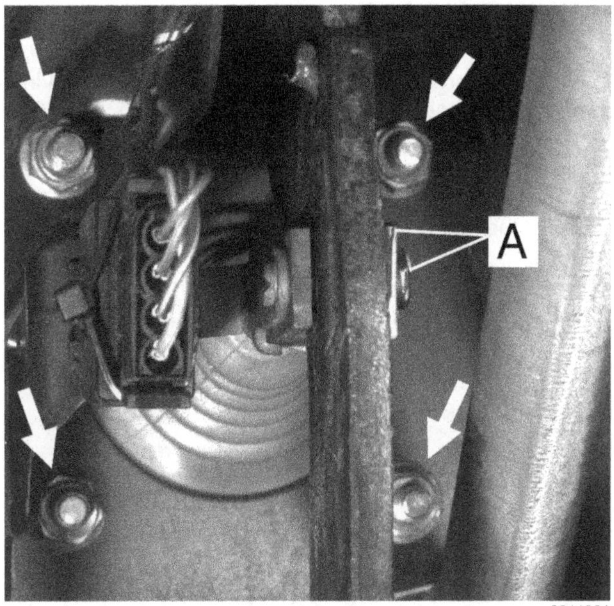

Fig. 13. Brake pedal clevis pin and retaining clip (**A**). Brake booster mounting nuts shown at **arrows**.

13. If necessary, mount master cylinder to vacuum booster using a new O-ring and new self-locking nuts.

> ***CAUTION —***
> *Do not over-torque the master cylinder mounting bolts. This could damage the brake booster and prevent proper vacuum build-up.*

14. Installation is reverse of removal.
    - Bleed brakes as described earlier.
    - Check and, if necessary, adjust brake light switch above brake pedal.

Tightening Torques
- Brake master cylinder to brake booster . . . . 26 Nm (18 ft-lb)
- Brake booster to bulkhead . . . . 22 Nm (16 ft-lb)
- Brake fluid line to master cylinder or ABS hydraulic unit . . . . 17-19 Nm (13-14 ft-lb)

> ***NOTE —***
> *When replacing the brake booster one-way valve or vacuum hose, install the valve so that the molded arrow is pointing toward the intake manifold. Use new hose clamps.*

PARKING BRAKE

The parking brake is a brake drum system integrated into the rear brake rotors. See Fig. 14.

Adjustment of the shoes may be necessary to compensate for wear. The parking brake should also be adjusted any time the cable, the rear brake rotor, or the parking brake shoes are replaced.

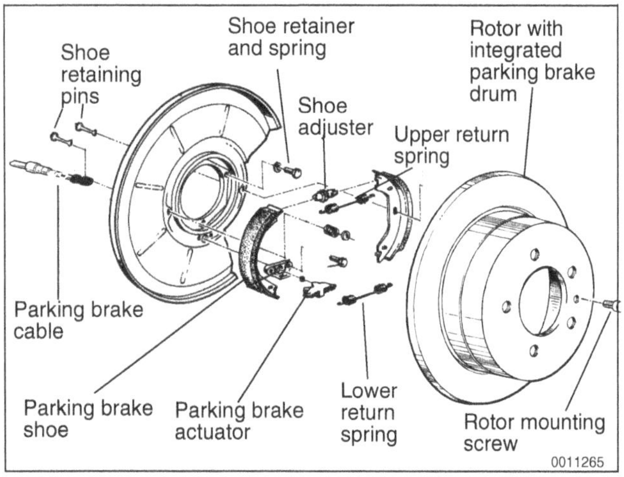

Fig. 14. Parking brake shoe assembly.

Parking brake, adjusting

The parking brake can be adjusted with the wheels installed, although the rear wheels will have to be raised off the ground.

1. Lift parking brake lever boot out of console. While holding cables stationary, loosen parking brake cable nuts until cables are completely slack. See Fig. 15.

2. Raise rear of car.

> ***WARNING —***
> *Make sure the car is firmly supported on jack stands designed for the purpose. Place the jack stands beneath a structural chassis point. Do not place jack stands under suspension parts.*

3. Remove one lug bolt from each rear wheel. Turn road wheel until lug bolt hole lines up with parking brake adjuster (approximately 65° to rear of wheel centerline). See Fig. 16.

4. Using a screwdriver, turn adjuster to expand brake shoes until road wheel can no longer turn, then back adjuster off. Repeat procedure on other rear wheel.

Fig. 15. Parking brake cable adjusting nuts (**arrows**).

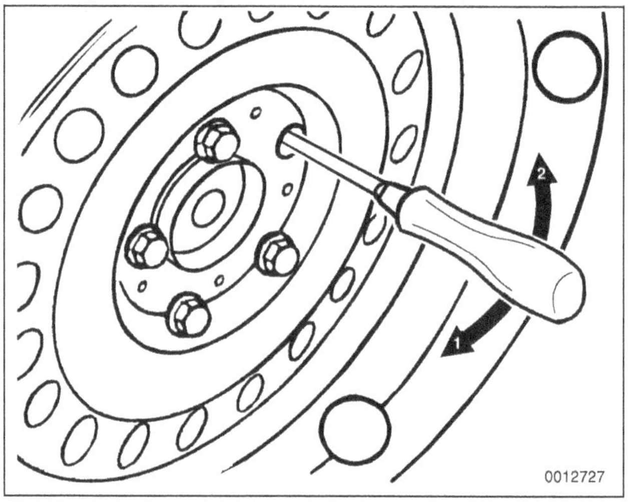

Fig. 16. Access parking brake shoe adjuster approximately 65° to rear of vertical. Use flat-bladed screwdriver to turn adjuster. On left wheel, turn adjuster in direction **1** to expand shoes. On right wheel, turn adjuster in direction **2** to expand shoes.

Parking Brake Adjusting (initial)
- Back off adjuster through wheel lug bolt hole
  - ex. M3 models . . . . 18 notches
  - M3 models . . . . 8 notches

5. Working inside car, set parking brake several times to seat cable. Then pull parking brake lever up four notches. Tighten cable adjusting nuts until it is just possible to turn rear wheels with slight resistance.

6. Release lever and make sure rear wheels turn freely.

7. Turn on ignition. Check that light comes on as soon as lever is pulled up.

8. Install parking brake lever boot. Install road wheel lug bolts.

Tightening Torque
- Road wheel to hub . . . . 100±10 Nm (74±7 ft-lb)

Parking brake shoes, removing and installing

1. Raise rear of car and remove road wheels.

> ***WARNING —***
> *Make sure the car is firmly supported on jack stands designed for the purpose. Place the jack stands beneath a structural chassis point. Do not place jack stands under suspension parts.*

2. Without disconnecting brake fluid hose, remove rear brake calipers from trailing arms. Remove rear brake rotors. See **Brake Pads, Calipers and Rotors**.

> ***CAUTION —***
> *Do not let the brake caliper assembly hang from the brake hose. Support caliper from chassis with strong wire.*

3. Unhook upper return spring from brake shoes. Remove shoe retainers by pushing them in and rotating ¼ turn. BMW special tool no. 34 4 000 can be used to remove retainers, if needed.

4. Spread shoes apart and lift them out.

NOTE —
When removing the brake shoes the shoe expander pivot pin in the brake actuator may fall out.

5. Inspect shoe expander to make sure it functions properly. Apply a thin coat of grease to sliding parts and pins.

6. Installation is reverse of removal. Be sure to adjust parking brake cables as described earlier.

Parking brake cable, replacing

1. Raise rear of car and remove road wheels.

2. Remove rear brake caliper and rotor. See **Brake Pads, Calipers and Rotors**.

3. Working inside car, lift parking brake lever boot out of console. Remove cable adjusting nut. Refer to Fig. 15.

4. Remove parking brake shoes as described above.

5. Disconnect cable from parking brake actuator and then remove parking brake cable from cable holder on trailing arm. See Fig. 17.

0013209

Fig. 17. Parking brake cable insertion point in rear trailing arm (**arrow**).

6. Slide cable out of housing.

NOTE —
On some models it may be necessary to remove rear muffler to access the brake cable(s).

7. Installation is reverse of removal. Adjust parking brake as described earlier.

ANTILOCK BRAKE SYSTEM (ABS)

Two versions of the Antilock Brake System (ABS) are installed on the cars covered by this manual. Early cars use the Teves Mark IV G ABS. Later cars use the Teves Mark 20-I ABS. The main difference between the two systems is that the Mark 20-I system integrates the ABS control module into the hydraulic unit as a single component.

ABS is designed to be maintenance free. There are no adjustments that can be made to the system. Repair and troubleshooting of the major ABS components requires special test equipment and knowledge and should be done by an authorized BMW dealer.

ABS is self-tested by the ABS diagnostic unit each time the car is started. Once the test is complete, the ABS dashboard light turns off. If the light remains lit or comes on at any time during driving, a system fault has occurred and ABS is electronically disabled. The conventional braking system remains fully functioning.

ABS Inspection

A visual inspection of the ABS may help to locate system faults. If no visual faults can be found and the ABS light remains on, have the system diagnosed by an authorized BMW dealer.

Carefully inspect the entire ABS wiring harness, particularly the pulse sensor harnesses and connectors near each wheel. Look for chafing or damage due to incorrectly routed wires.

Carefully remove the wheel speed sensors. Clean the sensor tips. Inspect the toothed wheel on the wheel hub. Check for missing, clogged or corroded teeth, or other damage that could alter the clearance between the sensor tip and the toothed wheel.

> ***NOTE —***
> *If a toothed ring is damaged, the entire wheel hub must be replaced. See* **310 Front Suspension** *or* **330 Rear Suspension**.

ABS wheel speed sensors, replacing

1. Raise applicable end of car.

> ***WARNING —***
> *Make sure the car is firmly supported on jack stands designed for the purpose. Place the jack stands beneath a structural chassis point. Do not place jack stands under suspension parts.*

2. Unscrew wheel speed sensor mounting bolt. Remove sensor from its housing. See Fig. 18. Remove harness from retaining mounts.

3. Front ABS sensor: Working in front wheel housing, disconnect and remove sensor. See Fig. 19.

4. Rear ABS sensor: Working in rear wheel housing, disconnect and remove sensor.

5. During installation, apply thin coat of Molykote® Longterm 2 or an equivalent grease to pulse sensor and housing.

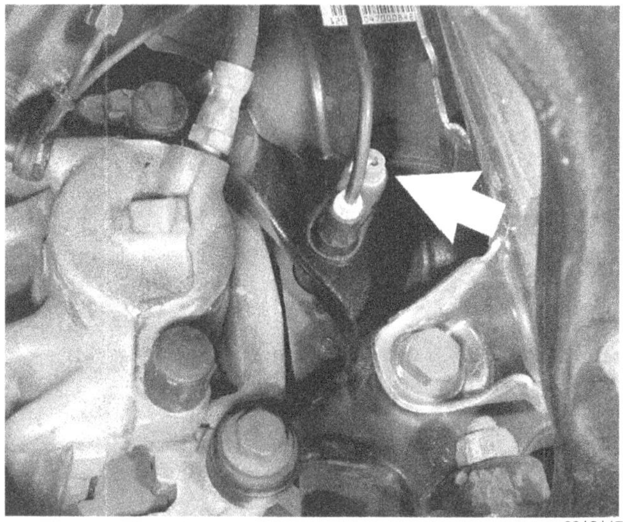

0012117

Fig. 18. Front ABS pulse sensor mounting bolt (**arrow**).

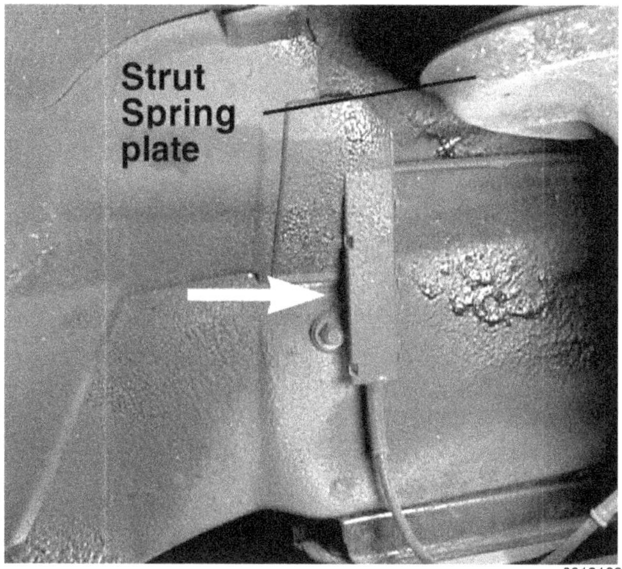

0012122

Fig. 19. Front ABS wheel speed electrical connector (**arrow**) in wheel well.

6. Installation is reverse of removal.

400 Body–General

GENERAL

This section covers system descriptions and general service information for the repair groups found in **4 Body** and **5 Body Equipment**.

When working or servicing cars equipped with SRS airbag(s), the following precautions must be observed to prevent injury.

WARNING —
- *Airbags are inflated by an explosive device. Handled improperly or without adequate safeguards, airbag units can be very dangerous. Special precautions must be observed prior to any work at or near the driver's side airbag, the passenger's side airbag (where applicable), or the side-impact airbags (where applicable). See* **721 Airbag System (SRS)**.
- *Always disconnect the battery and cover the negative (–) battery terminal with an insulator before starting diagnostic, troubleshooting or service work on cars fitted with SRS, and before doing any welding on the car.*

Fig. 1. The BMW E36 coupe.

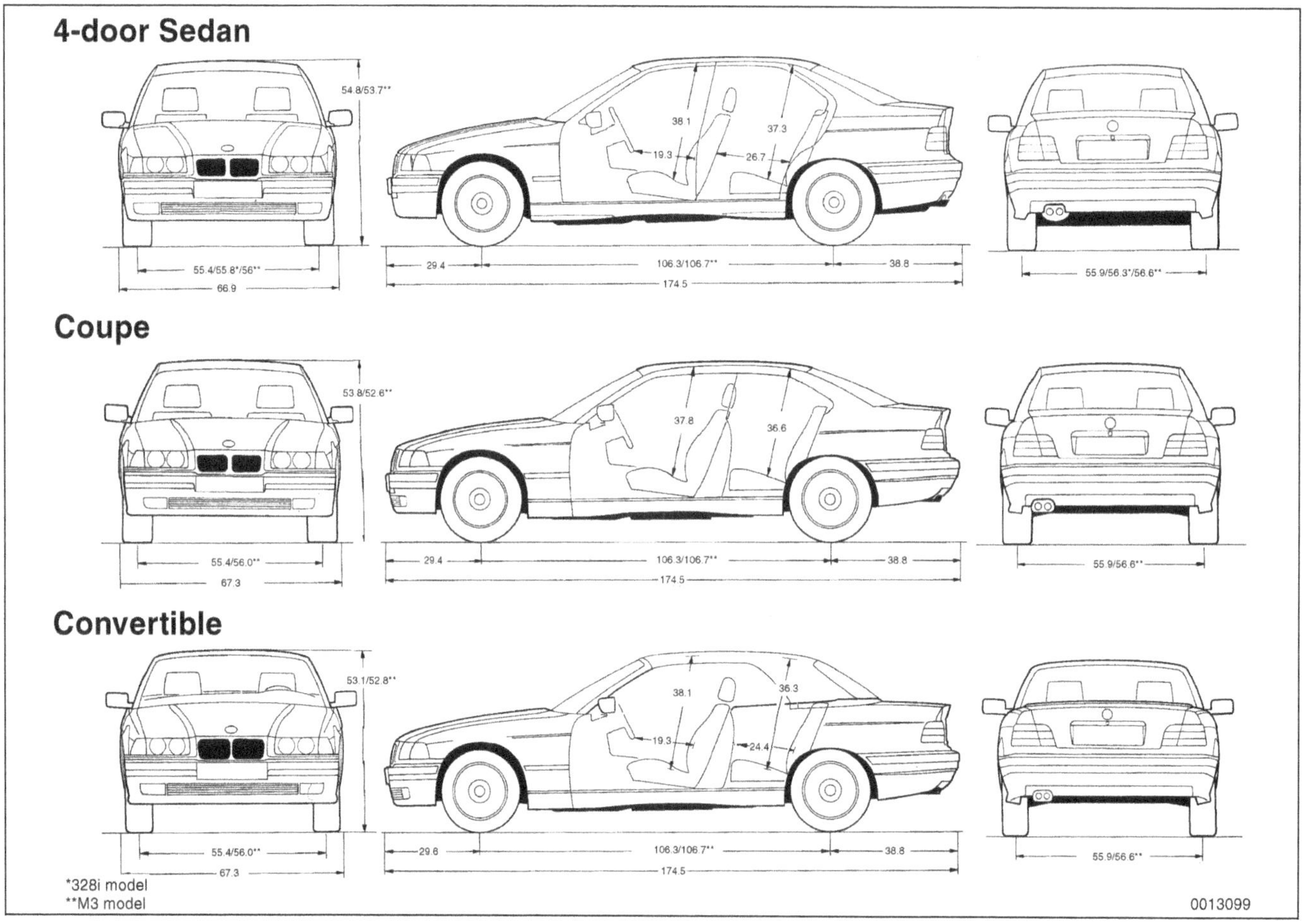

Fig. 2. E36 body dimensions.

BODY ASSEMBLY

The BMW 3 Series chassis is a unibody design constructed of high strength steel. Attaching parts, such as the front fenders, rear quarter panels, doors, trunk lid and hood are also constructed of steel.

The body styles of E36 cars covered by this manual are the 4-door sedan, 2-door coupe and convertible. Body dimensions vary slightly between models. See Fig. 2.

Exterior and Aerodynamics

The E36 can be immediately recognized as a typical BMW. With its traditional styling features, such as the BMW kidney grille, circular headlights behind a glass cover (for reasons of streamlining), and the "kick" in the rearmost roof pillar (C-pillar).

Windows bonded flush to the body and the guided flow of cooling air into the engine compartment provide a good coefficient of drag (C_D). The E36 cars have a C_D of approximately 0.30 for the sedan and coupe models, and 0.35 for the convertible models.

The exterior panels are corrosion-resistant zinc coated steel. The front fenders are bolted on and easily replaceable. See **410 Fenders, Engine Hood**.

The front and rear bumpers use an aluminum framework with an overlaid high-impact, deformable plastic cover. This bumper system keeps damage to a minimum in head-on collisions of up to 10 m.p.h. (15 km/h). See **510 Exterior Trim, Bumpers**.

Coupe Models

The E36 coupe is similar to the 4-door sedan, but has a more stretched and sleeker appearance. Thus, with the same wheelbase and identical length, it looks different but still familiar.

The 2-door variant differs from the 4-door sedan by a 3.12 inch (8 cm) longer front end, a flatter hood line with air slits, and a roof almost 1.2 inches (3 cm) shorter. Additional differences include wider doors with frameless windows, glass-covered door-posts (B-pillars) and a shorter and lower luggage compartment lid.

Convertible Models

The E36 convertible is an all-season vehicle with a high level of functionality, excellent interior comfort, and acoustic refinement. Interior materials and colors, the soft top and the light-alloy wheels have been carefully matched.

Various versions of manual and automatic soft tops have been used, with the fully automatic convertible roof being installed beginning in the 1997 model year.

A hard-top in body color with heated rear window is available as an option for the cold season.

The convertible body structure includes many passive safety refinements over its fixed-top counterparts. Some of the enhancements include:

- More rigid underbody.
- Transverse seat members.
- Reinforced windshield frame with bonded windshield, and tubular door reinforcements.
- Rollover protection system, which is tripped into position in a fraction of second. This is available as optional equipment (standard on the 1998 328iC) and may be fitted to earlier convertible models. With this feature, two bars fully independent of each other come up behind each rear-seat backrest when the vehicle is in danger of turning over.

A convertible body normally flexes and vibrates, transferring oscillations to the luggage compartment. In the E36 convertible, BMW utilizes a special battery in the luggage compartment, in conjunction with a unique battery tray, as a vibration dampening system to counteract this oscillation.

> ***CAUTION —***
> *The E36 convertible requires a special battery which is designed for constant vibration.*

Information for the convertible top and its related mechanisms can be found **541 Convertible Top**.

INTERIOR EQUIPMENT

E36 interiors have a unique character, with a combination of sporty features and innovative design. All body versions were made in a wide choice of interior and exterior colors. New colors with matching leather and/or fabric upholstery were introduced for each model year, including light-alloy wheels in new designs.

Driving comfort is enhanced by power steering, tilt steering wheel (optional), height-adjustable front seats, and electrically adjustable and heated rear-view mirrors. Inside the car there are reading lights at four seating positions.

Seats and Seat Belts

Anatomically-correct seats are constructed from polyurethane foam containing areas or “zones” of different firmness. The seat bases have steel springs for added support and strength. The power seats are provided with conveniently located switches along the outside edge of the seat. Heated seats are an option in all models. See **520 Seats** for more details.

The M3 sport seat is shown in Fig. 3.

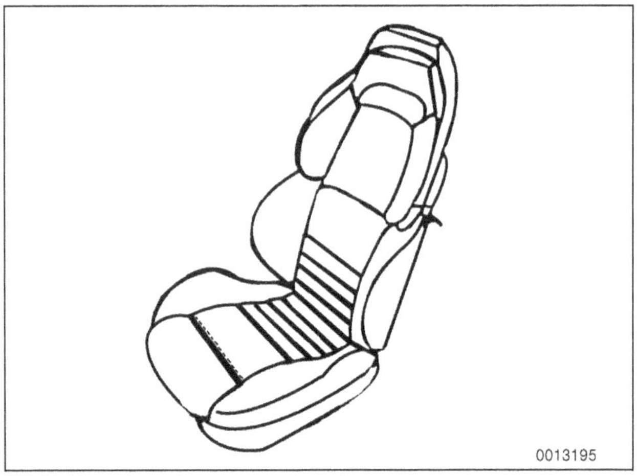

Fig. 3. M3 sport seat.

The early E36 cars are equipped with a single (driver side) airbag in the steering wheel. Starting with the 1994 models, a dual airbag Supplemental Restraint System has been standard equipment. Side-impact airbags in the front doors are optional in 1997 4-door cars and standard equipment in all 1998 models.

Through 1996, E36 seatbelts were equipped with BMW's original-design spring tensioners. In 1997, the seat base design was modified to accommodate a pyrotechnic (explosive charge) seatbelt tensioner. Both types of tensioners are designed to automatically tension the belts by about 2 inches (55 mm) in the event of a collision.

NOTE —
The seat base used on cars with pyrotechnic seatbelt tensioners is not interchangeable with the earlier seat base.

Instruments and Controls

Everything in the interior passenger compartment essential to the driver is logically grouped and easy to reach. All instruments and controls have been arranged ergonomically and are fully integrated into the overall design of the vehicle.

The padded dashboard houses the instrument cluster and the ventilation and heating system. The instrument cluster uses large easy-to-read analog instruments and is removable as a unit without removing the dashboard. The Service Interval Indicator (SII) will remind the driver when the car requires service. See **620 Instruments**.

The Multi-Information Display (MID) unit in the center of the dashboard comes in a variety of four configurations, the most advanced being the On-Board Computer (OBC). See Fig. 4.

All versions of the MID have a clock and calendar. In addition, different versions are equipped with a “check control,” display for fuel-economy information and system warnings. Consult the owner's manual for full operating instructions.

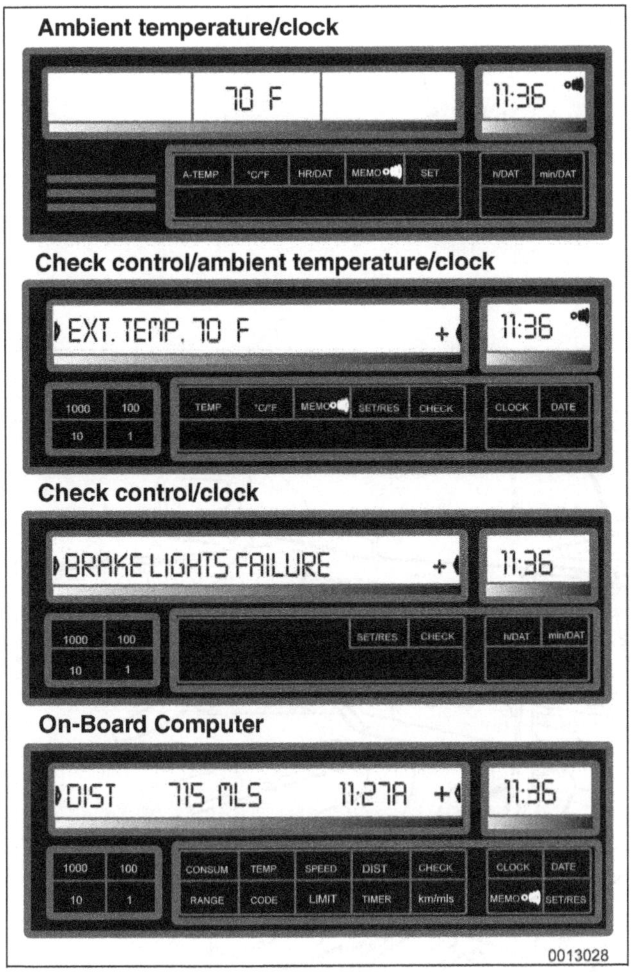

Fig. 4. Different versions of Multi-information Display (MID) installed in E36 cars.

Central Body Electronics

The E36 cars are equipped with a sophisticated centralized body electric/electronics plan. The system is self-diagnostic and incorporates many functions into a single control module. The consolidation of several systems into a single control module minimizes power requirements and the incorporation of the diagnostic link results in more efficient and accurate troubleshooting.

Self-diagnostic codes are accessible electronically through the data-link connector in the engine compartment. See **610 Electrical Component Locations**.

Two versions of central body electronics are used depending on model year.

1992 and 1993 models are equipped with a central locking module (ZVM) which incorporates control of the following systems:

- Interior lighting
- Central locking with double lock feature
- Output relays for power windows and sunroof
- Electronic power protection

E36 cars since the 1994 model year are equipped with ZKE IV. This is an electronic comfort and convenience system that includes the following additional features:

- One touch up/down power window operation
- Power window and sunroof closure from driver or passenger door lock
- Vehicle locking or unlocking from the trunk lock.

Other features of the system are discussed in **515 Central Locking and Anti-Theft**.

Electronic Immobilization System

Beginning in model year 1994, all E36 vehicles incorporate a driveaway protection system regardless of whether the vehicle is equipped with a BMW Alarm System or an On-Board computer. Two types of factory systems were installed, depending on vehicle production date.

On the early system (model year 1994 up to January 1995), the ignition and fuel injection functions are disabled whenever the vehicle is locked in the double lock mode. This is referred to as EWS or EWS I.

Beginning in January 1995, the driveaway protection system was upgraded to a more sophisticated system, called EWS II. This system uses a wireless communication link between a transponder chip in the ignition key and the ring antenna surrounding the ignition switch. The EWS II control module blocks the starting of the vehicle unless the correct coded ignition key is used.

For further information about drive away protection systems, see **515 Central Locking and Anti-Theft**.

Heating and Air Conditioning

The electronically controlled heating and air conditioning system uses an extra large, infinitely variable radial blower motor for good distribution of air. Fresh air enters through the grille below the engine hood and into the passenger compartment via the dashboard and footwell vents. See Fig. 5.

Repair information for the heating and air conditioning system is covered in **640 Heating and Air Conditioning**.

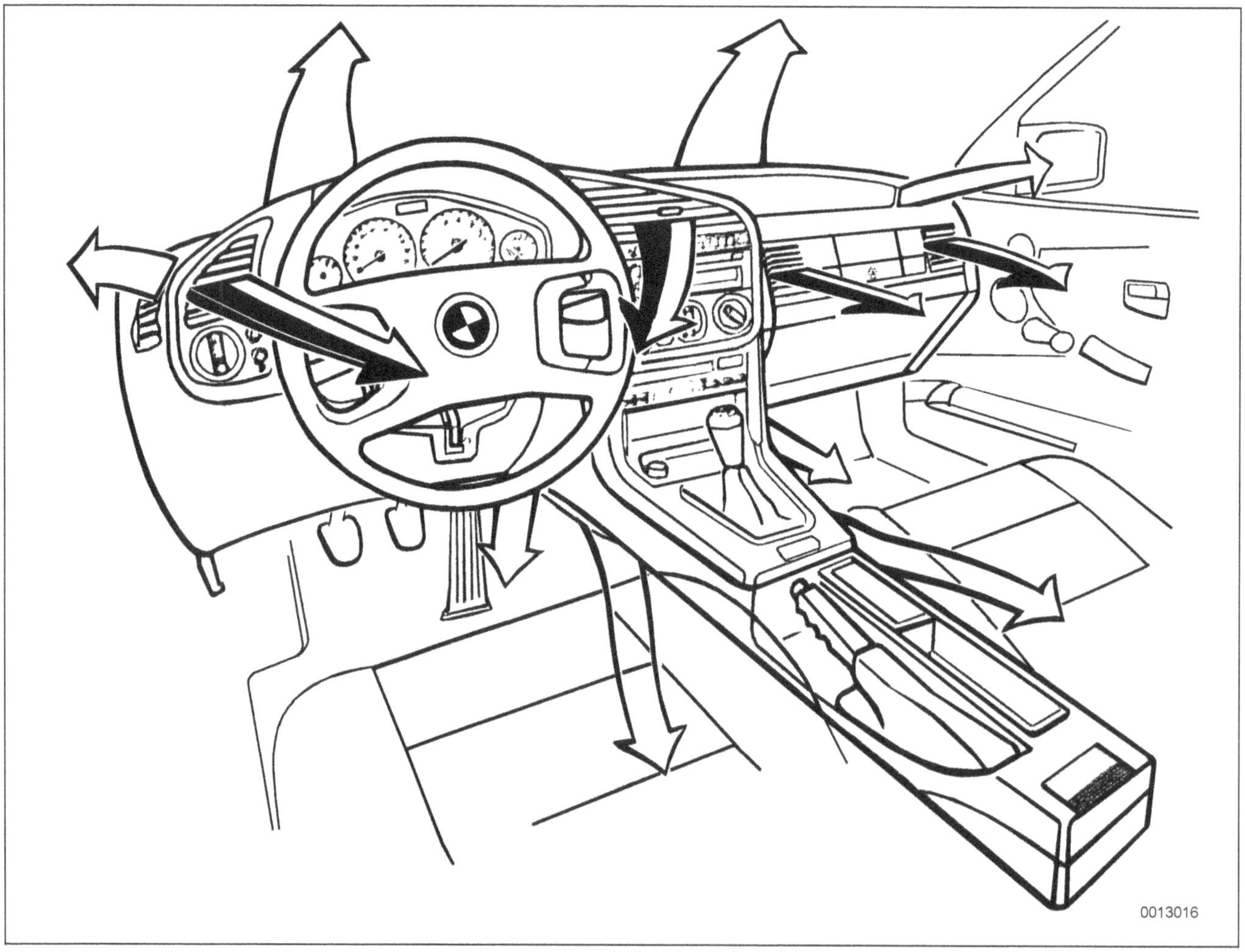

Fig. 5. Heating and air conditioning ventilation plan.

410 Fenders, Engine Hood

GENERAL

This repair group covers replacement of the front fenders and removal and installation of the engine hood.

> ***NOTE —***
> *The body is painted at the factory after assembly. Realignment of body panels may expose unpainted metal. Paint all exposed metal once the work is complete.*

FRONT FENDERS

Front fender, removing and installing

1. Raise hood. Raise and safely support front of vehicle.

> ***WARNING —***
> - *When raising the car using a floor jack or a hydraulic lift, carefully position the jack pad to prevent damaging the car body. A suitable liner (wood, rubber, etc.) should be placed between the jack and the car to prevent body damage.*
> - *Watch the jack closely. Make sure it stays stable and does not shift or tilt. As the car is raised, the car may roll slightly and the jack may shift.*

2. Remove front wheel. Remove inner plastic liner from wheel well.

3. Remove front bumper. See **510 Exterior Trim, Bumpers**.

4. Working at front corner of fender, remove front grille panel bolts (2 bolts at left and right). See Fig. 1.

0013145

Fig. 1. Front grille panel to front fender retaining bolts (**arrows**). Front bumper shown removed.

5. Remove front turn signal/lens assembly from applicable front corner. Carefully pry out side directional from fender (later cars only). See **630 Exterior Lighting**.

6. Remove fender attaching bolts at front underside of fender. See Fig. 2.

7. Remove bolts along top edge of fender. See Fig. 3.

8. Open front door and remove fender mounting bolts in door jamb. See Fig. 4.

9. Carefully remove fender from body.

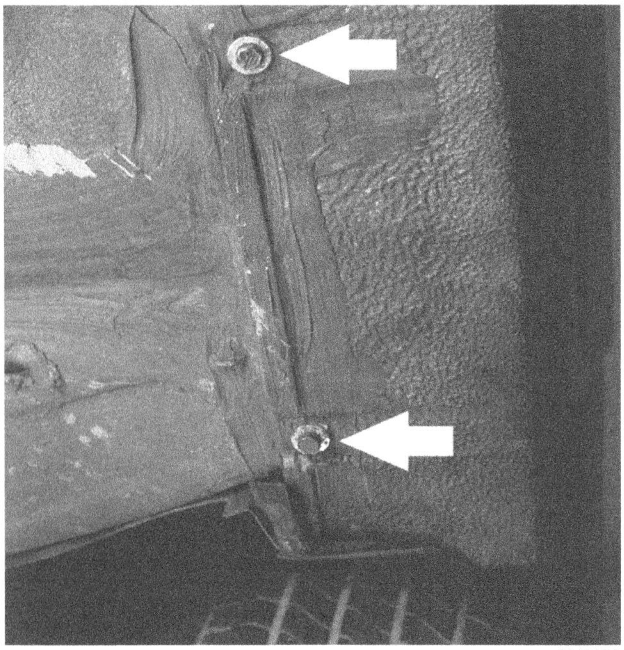

0013104

Fig. 2. Fender attaching bolts, lower front (**arrows**).

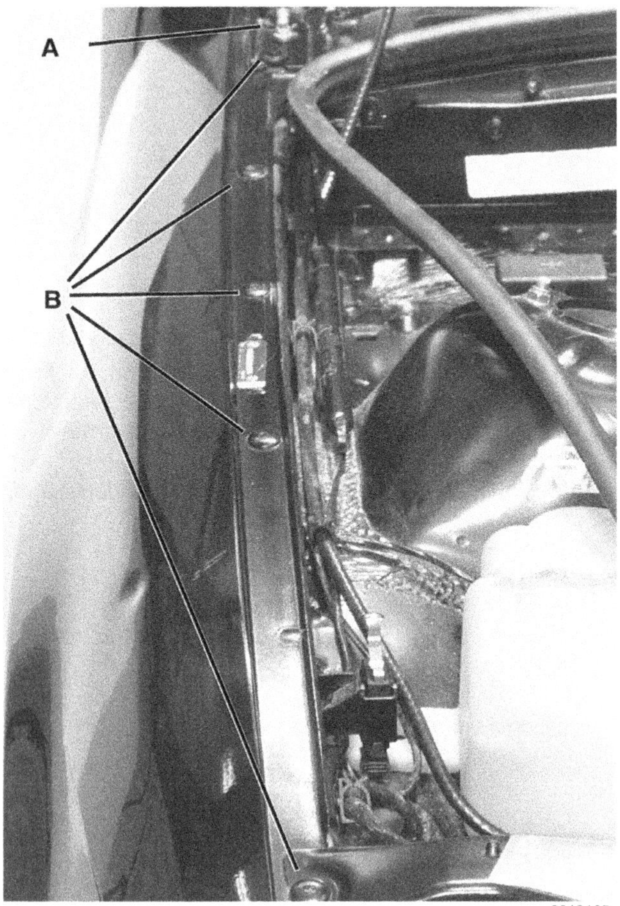

0013105

Fig. 3. Loosen bolt **A** at top of fender. Remove bolts **B**.

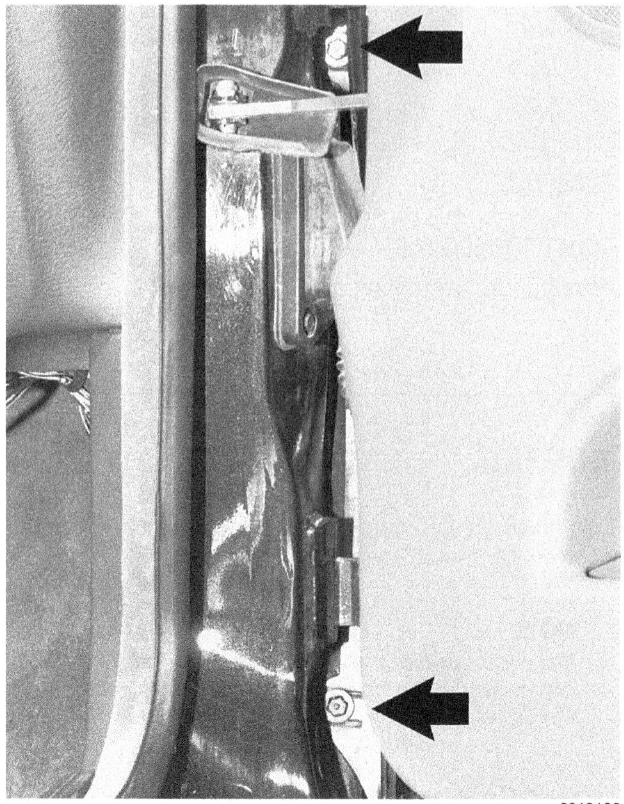

0013106

Fig. 4. Front fender attaching bolts as viewed through door jamb (**arrows**).

10. Installation is reverse of removal, noting the following:
    - Before installing new fender, clean old sealant and protective coating from mounting surfaces.
    - Position new fender and loosely install all mounting bolts. Align fender with door pillar and front panel, then tighten bolts.
    - Repair any paint damage and paint any exposed metal.
    - Reseal and apply protective coating to mounting surfaces.

Clearance Specification

- Fender to front hood or to door . . . .5.5 mm (0.216 in)

ENGINE HOOD

Hood, raising to service position

> ***CAUTION—***
> *Do not switch on the windshield wipers with the engine hood raised. As a precaution, remove the wiper motor fuse. See* **610 Electrical Component Locations**.

1. 4-door models: Open hood fully. Remove bolt from hood hinge behind hood supports (left and right). Raise hood to service position. See Fig. 5.

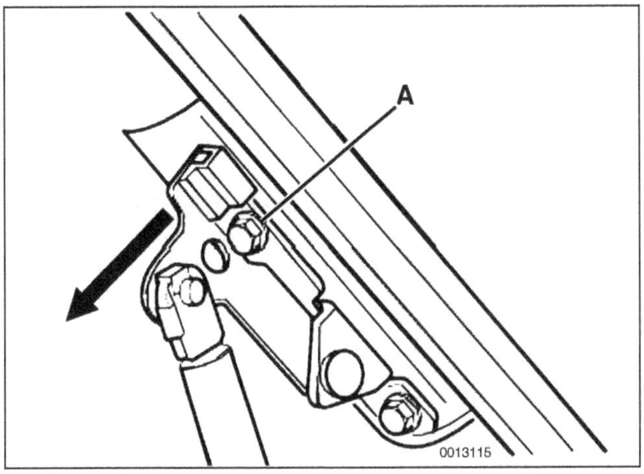

Fig. 5. Hood support bracket bolt to be removed (**A**) on 4-door cars. Remove bolts at left and right hinges. Swing bracket in direction of **arrow** to raise hood into service position.

2. 2-door models: Open hood fully.
   - Remove ground cable at right hood hinge.
   - While an assistant supports hood, detach both hood supports from hood.
   - Remove detent screws at back of hood hinges. See Fig. 6.
   - Lift hood while pulling hinge forward. Support hood at higher position.
3. When service work is complete, return hood to lower position and install screws or bolts.

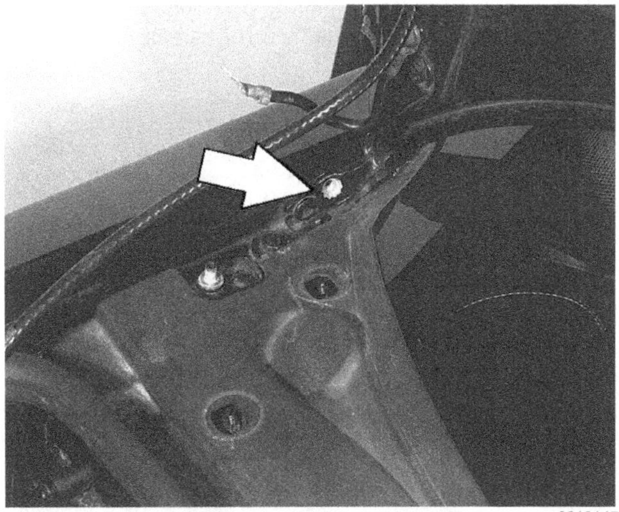

Fig. 6. Torx screw at base of hood hinge (**arrow**).

Hood, removing and installing

1. Raise hood. Where applicable, partially remove hood acoustic insulation to access rear of hood. See Fig. 7.

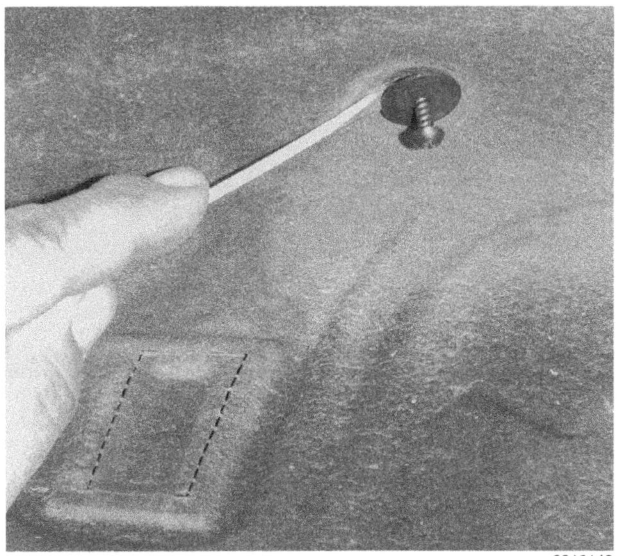

Fig. 7. Hood insulation retainer being pried out

2. Disconnect washer fluid hoses from washer nozzles. Where applicable, disconnect electrical harness connectors from heated nozzles.
3. Detach gas-charged hood supports from hood.

> ***CAUTION—***
> *The hood is heavy. Before removing the hood supports or hood retaining bolts, be sure to have an assistant help support the hood.*

4. With aid of assistant, remove hood hinge bolts from left and right sides of hood. Lift hood off carefully. See Fig. 8.

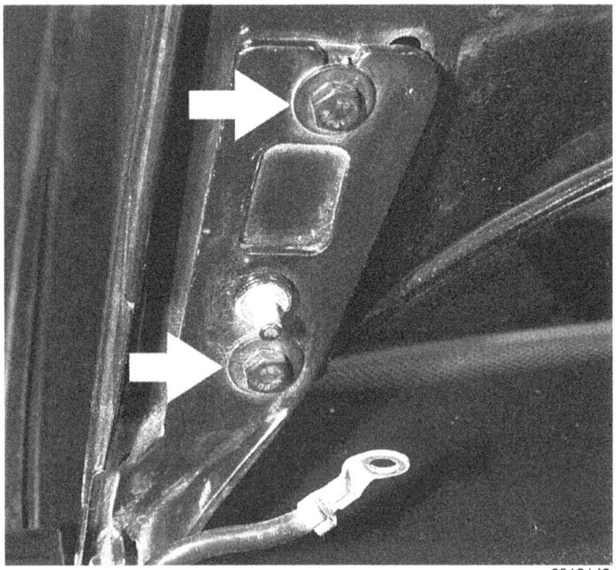

0013149

Fig. 8. Hood hinge bolts (**arrows**). Right side shown.

5. Installation is reverse of removal.
   - Repair any paint damage and paint any exposed metal.
   - Check hood alignment as described next.

NOTE —
The hood was fitted and attached to an unfinished body at the factory. Unfinished surfaces exposed by this procedure must be touched up with paint.

Hood, aligning

When installing the hood, align the hinges as close to the original painted surface as possible. Movement of the hood on its attaching hardware may require touch-up paint.

The hood should be aligned so that gaps on all sides are as even as possible. See Fig. 9.

Clearance Specification
- Hood edges to adjoining body panel (gap) . . . . 5.5 mm (0.216 in)

0013150

Fig. 9. Hood panel gaps should be even (approx. 5.5 mm) at all points (**arrows**).

Hood release cable and latches, adjusting

Before adjusting hood latches, be sure the hood is aligned evenly to fenders and front panel. The hood latch and cable assembly is shown in Fig. 10.

1. To adjust hood cables:
   - Remove left and right radiator grilles. See **510 Exterior Trim, Bumpers**.
   - Remove electric auxiliary cooling fan. See **170 Radiator and Cooling System**.
   - Adjust cable section between two hood latches so there is no play (cables must not be under tension either).
2. Loosen hood latches at hood on left and right sides. Loosen bolts only enough to allow movement of latch. See Fig. 11.
3. Lower (but do not fully latch) hood several times so latches on hood center themselves in lower locks.

CAUTION —
Do not let hood lock with latch bolts loose.

4. Tighten hood latch bolts when alignment is correct.
5. Remove one screw from each side, clean and coat with Loctite®270 or equivalent, and reinstall.

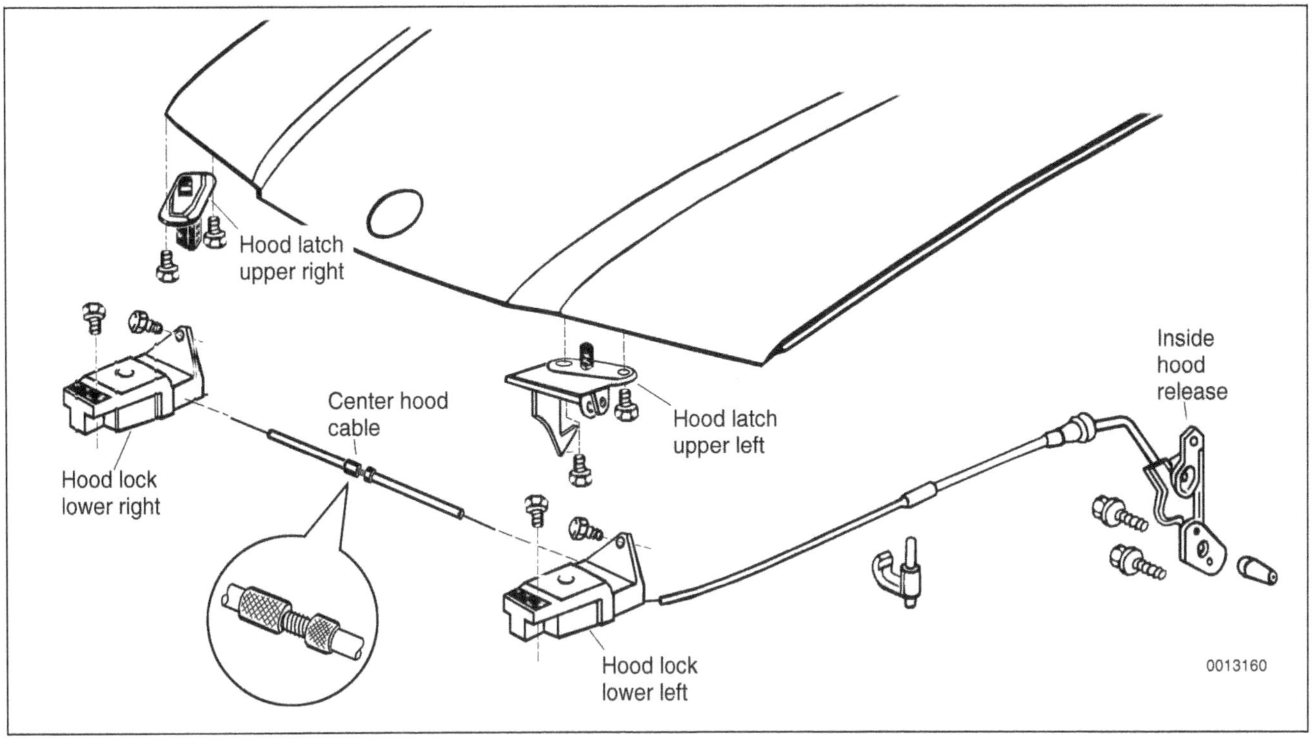

Fig. 10. Hood locking components. Cable adjusting section shown at inset.

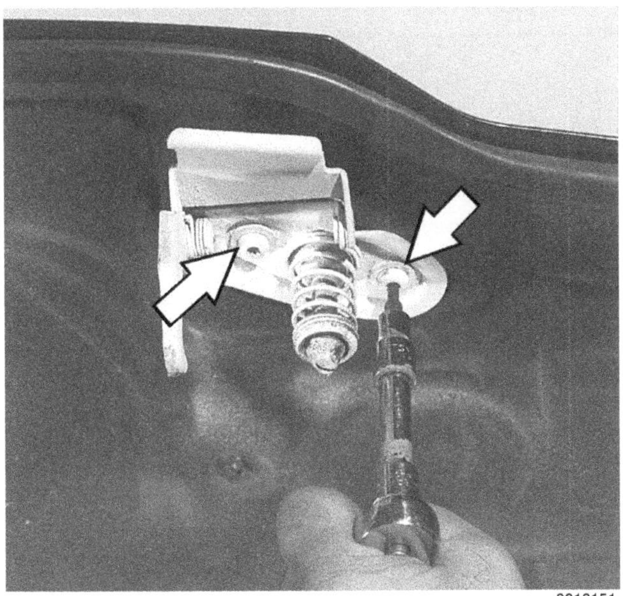

Fig. 11. Hood latch mounting bolts (**arrows**). Left latch shown.

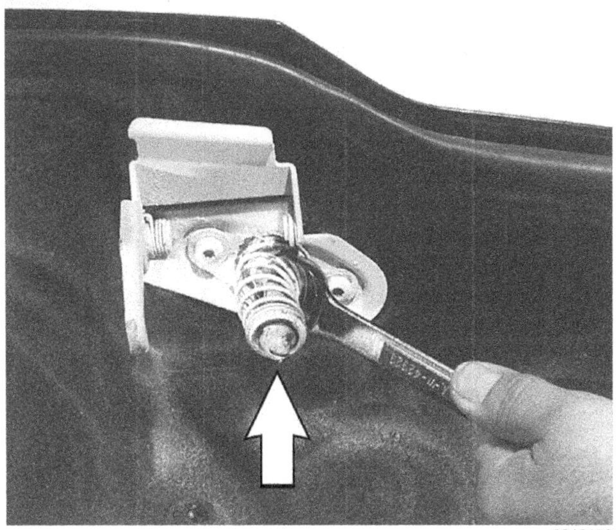

Fig. 12. To lengthen spring pin, loosen locknut with wrench inserted through spring, then turn pin (**arrow**) counterclockwise.

6. Test hood for correct closure and opening. If hood does not spring open, lengthen spring pins at latches. See Fig. 12.

411 Doors

GENERAL

This repair group covers front and rear door repair information, including removal and installation of the interior door panels.

> ***WARNING —***
> *Late M3 models have aluminum doors. Use extra care in handling aluminum parts in order to avoid damage from gouging or corrosion.*

NOTE —
- *For information on the power door windows and door glass replacement, see* **512 Door Windows**.
- *For information on the power door locking system, see* **515 Central Locking and Anti-Theft**.

DOORS

> ***WARNING —***
> *Some 1997 and all 1998 cars are fitted with side-impact airbags in the front doors. When servicing doors on cars with front side-impact airbags, always disconnect the negative (–) battery terminal. See* **721 Airbag System (SRS)** *for cautions and procedures relating to the airbag system.*

Front or rear door, removing and installing

1. If working on front door with side-impact airbag, disconnect negative (–) battery cable.

> ***CAUTION —***
> *Prior to disconnecting the battery, read the battery disconnection cautions given at the front of this manual on page viii.*

2. Unbolt main harness connector at door pillar and separate connector. See Fig. 1.

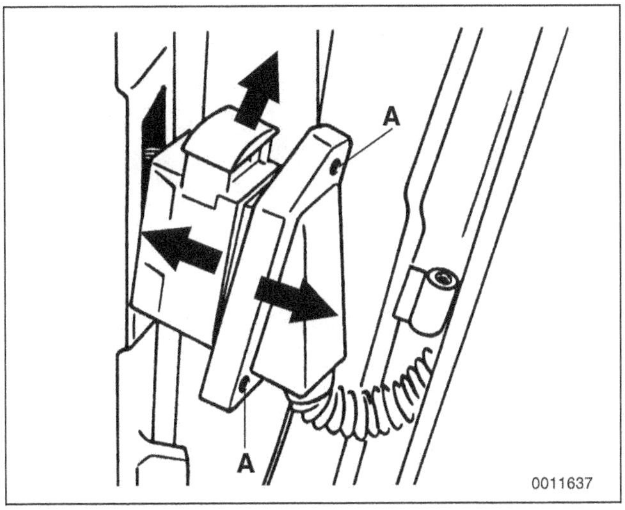

Fig. 1. Door harness connector being removed from door pillar. Remove mounting screws at **A**, then separate connector by pulling up on harness lock (**top arrow**).

3. With door fully open, remove pin bolts from top and bottom door hinges. See Fig. 2.

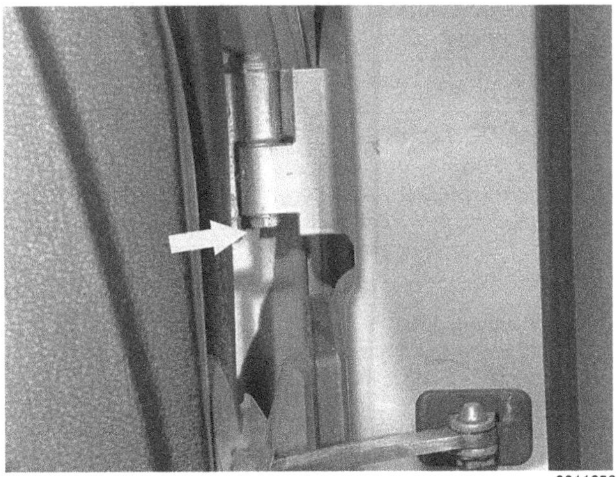

Fig. 2. Door hinge pin bolt (**arrow**). Top hinge shown. Also remove pin bolt from lower hinge.

4. Remove retainer clip (early models only) from door check pin on body pillar. Drive out door check pin. See Fig. 3.

Fig. 3. Door check pin being removed. Note retainer clip (**A**) used on early models.

5. Remove door by lifting up off lower hinge halves.

> ***CAUTION—***
> *Be careful not to damage door or other painted body surfaces. Make sure no load is placed on wiring harness.*

6. Installation is reverse of removal. Keep in mind the following:
   - Align door so that panel gaps are equal on either side, as described later. If necessary, adjust door hinges as described later.
   - Adjust door striker so that trailing edge of front door is slightly higher (1 mm) than leading edge of rear door. See **515 Central Locking and Anti-Theft**.
   - Repair any paint damage and paint any exposed metal.

> ***CAUTION—***
> - *In 2-door models, removing and installing the door may upset the adjustment and alignment of the window. Window adjustment, as described in* **512 Door Windows**, ***must*** *be carried out after installation to prevent damaging the glass.*
> - *If battery power is lost or the battery has been disconnected, the windows must be re-initialized. Both one touch up/down and pinch-protection will be inactive until windows are re-initialized. See* **512 Door Windows**.

Door Adjustment

- Door gap
  to fender/rear door . . . . approx. 5 mm (0.189 in)
- Permissible deviation from parallel. .0.5 mm (0.020 in)

Door check, replacing

> ***WARNING—***
> *Some 1997 and all 1998 cars are fitted with side-impact airbags in the front doors. When servicing doors on cars with front side-impact airbags, always disconnect the negative (–) battery terminal. See* **721 Airbag System (SRS)** *for cautions and procedures relating to the airbag system.*

1. Close door window completely.

2. If working on front door with side-impact airbag, disconnect negative (–) battery cable.

> ***CAUTION—***
> *Prior to disconnecting the battery, read the battery disconnection cautions given at the front of this manual on page viii.*

3. Remove interior door panel and vapor barrier as described later.

4. Where applicable, remove side-impact airbag from door. See **721 Airbag System (SRS)**.

> ***NOTE—***
> *In some models covered in this manual, it is not essential to remove the side-impact airbag to access internal door components. However, the procedure is recommended for safety reasons.*

5. Remove retainer clip (early models only) from door check pin on body pillar. Drive out door check pin. Refer to Fig. 3.

6. Remove bolts and rubber cover from door check lockplate on door.

7. Remove door check from inside door.

8. Installation is reverse of removal.
   - Lubricate door check before installing.
   - Use new mounting bolts when reinstalling the side-impact airbag to the door (where applicable).
   - If battery was disconnected, re-initialize window regulator motors as described in **512 Door Windows**.

Door Hinge Adjustment

If the installed door is uneven or out of parallel, shims can be used to correct its position. The shims are placed behind the hinge plate and are available in two different thicknesses (0.5 mm and 1.0 mm). See Fig. 4.

Tightening Torque
- Door hinge to door . . . . 21 Nm (15.5 ft-lb)
- Door striker to door . . . . 25 Nm (18 ft-lb)

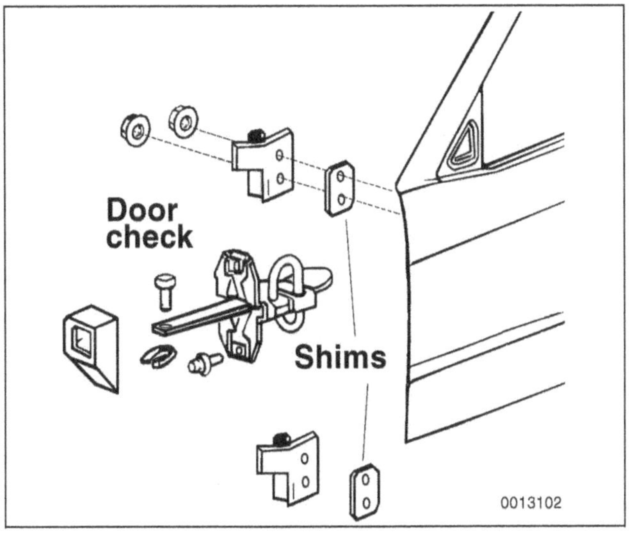

Fig. 4. Door hinge and door check components. Front and rear doors are similar.

DOOR PANELS

WARNING —
Some 1997 and all 1998 cars are fitted with side-impact airbags in the front doors. When servicing doors on cars with front side-impact airbags, always disconnect the negative (–) battery terminal. See **721 Airbag System (SRS)** *for cautions and procedures relating to the airbag system.*

Front door trim panel, removing and installing

1. On cars with side-impact airbags, disconnect negative (–) battery cable.

CAUTION —
Prior to disconnecting the battery, read the battery disconnection cautions given at the front of this manual on page viii.

2. Carefully pry out mirror adjustment switch and disconnect harness connector from switch. Fig. 5.

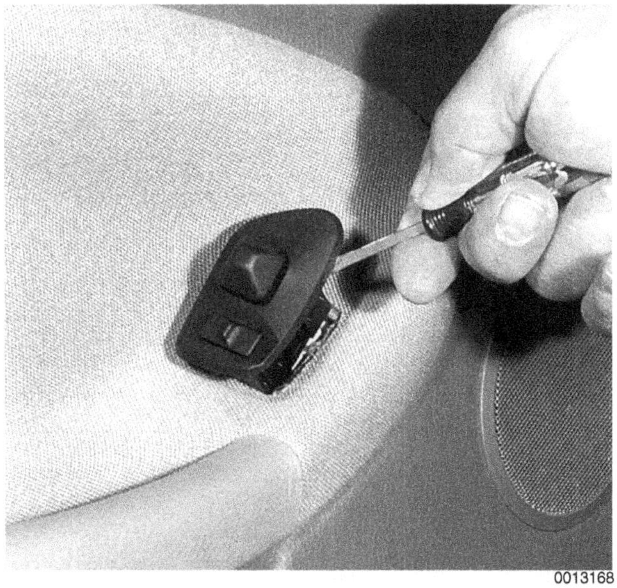

Fig. 5. Pry outside mirror switch from door armrest.

3. Unscrew door locking knob.

4. Pry off trim cover for inside door handle. Pry off concealing plugs from armrest. Remove door panel retaining screws. See Fig. 6.

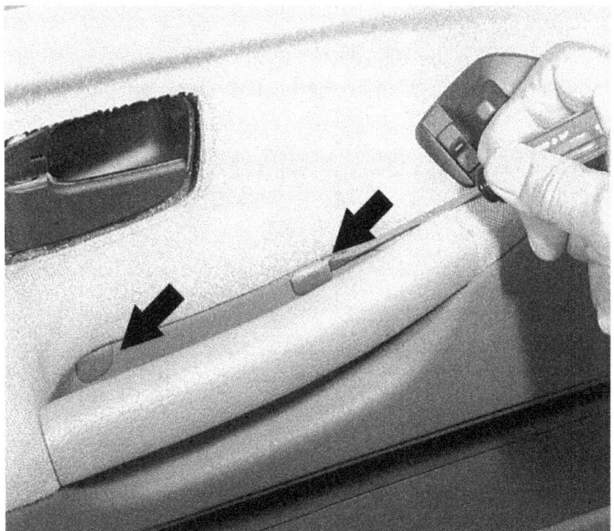

Fig. 6. Pry off plugs (**arrows**) and remove screws.

5. Unclip door panel from door perimeter. See Fig. 7.

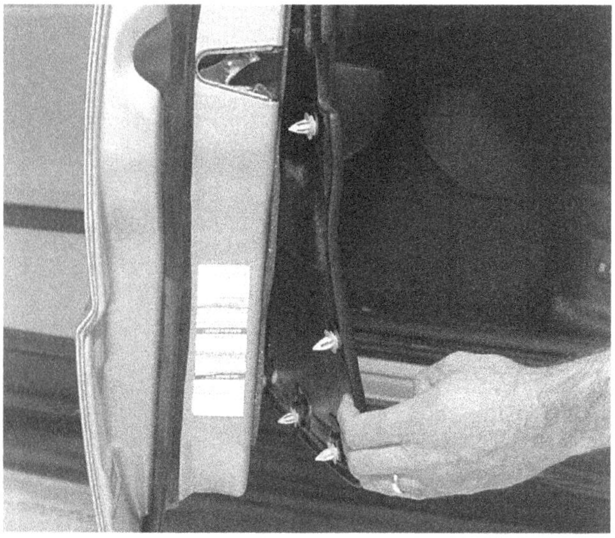

0011653

Fig. 7. Door panel being unclipped from door. Pull panel retaining clips off one at a time.

6. Disconnect radio speaker harness connector.
7. Installation is reverse of removal, noting the following:
   - Replace any damaged plastic parts or clips.
   - Renew vapor barrier if it is damaged.
   - Use new mounting bolts when reinstalling the side-impact airbag to the door (where applicable).
   - When installing door panel, align metal retainers on window sill with openings on top of door panel. See Fig. 8.
   - Check door-lock mechanism and window for ease of movement.
   - If battery was disconnected, re-initialize window regulator motors as described in **512 Door Windows**.

0013173

Fig. 8. Metal retainers at base of window should be in position to mate with openings in door trim panel.

Rear door trim panel, removing and installing

1. Open door and pry out power window switch. Disconnect harness connector.
2. Remove ash tray.
3. Pry off trim cover from inside door handle.
4. Pry off concealing plugs from armrest. Remove door panel retaining screws. See Fig. 9.

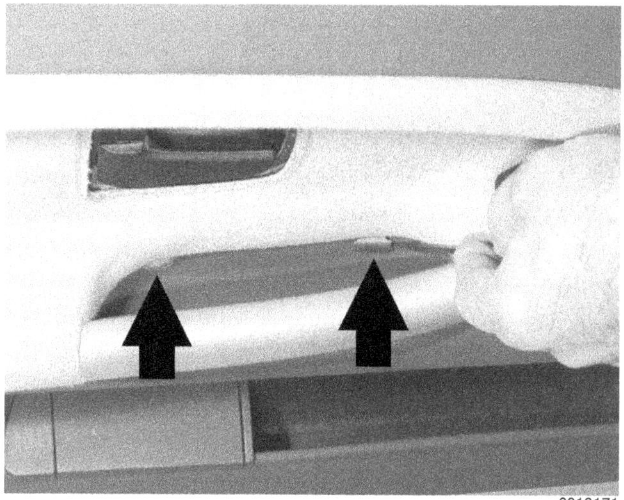

0013171

Fig. 9. Pry off plugs (**arrows**) and remove screws.

5. Unscrew door locking knob.
6. Unclip panel from door perimeter, starting at the top of the door.
7. Installation is reverse of removal.
   - Replace any damaged plastic parts or clips.
   - Renew vapor barrier if it is damaged.
   - Align metal retainers on window sill with openings in trim panel. Refer to Fig. 8.
   - Check door-lock mechanism and window for ease of movement.

Rear side trim panel, removing and installing (coupe models)

1. Remove rear seat cushion. Remove rear backrest side section. See **520 Seats**.
2. Pull off edge trim at bottom of rear side window. See Fig. 10.
3. Gently unclip trim panel and remove.
4. Installation is reverse of removal.
   - Replace any damaged plastic parts or clips.

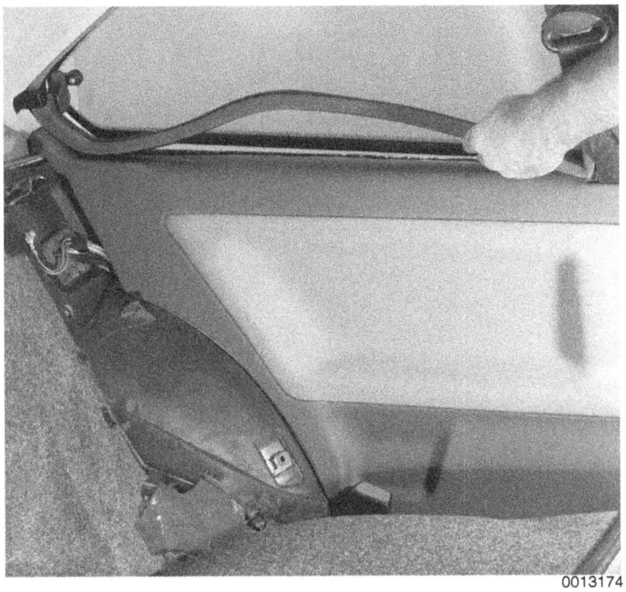

Fig. 10. Panel edge trim piece trim being removed.

Rear side trim panel, removing and installing (convertible models)

To remove the rear side panel in convertible models it is necessary to:

- Remove rear seat cushion and rear backrest side section.
- Lower side window; open convertible roof.
- Left side: Remove convertible top lid release lever trim.
- Remove power window switch and disconnect harness connector.
- Remove door panel inset by pulling forward.
- Remove corner protector rubber trim.
- Pull off side panel after removing screw fastener(s). Disengage from front shoulder belt.

412 Trunk Lid

GENERAL

This repair group covers trunk lid removal and installation. Also included here are replacement procedures for the gas-charged support struts that hold the trunk lid in the up position.

NOTE—
The body is painted at the factory after assembly. Realignment of body panels may expose unpainted metal. Paint all exposed metal once the work is complete.

TRUNK LID

Trunk lid, removing and installing

1. Raise trunk lid. Open tool kit and remove screws attaching tool kit to trunk lid.
2. Remove insulating liner from underside of trunk lid. See Fig. 1.

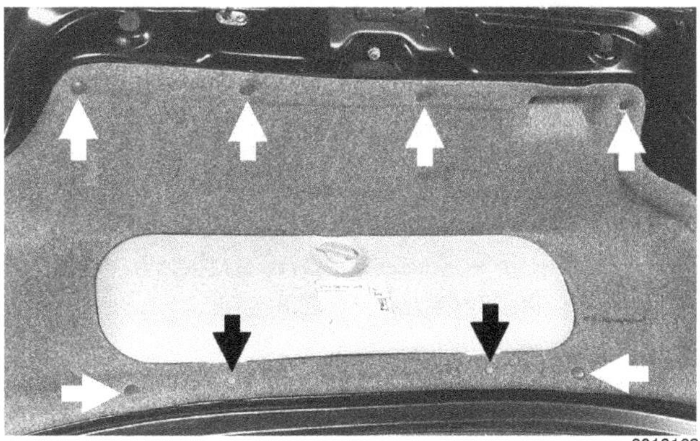

Fig. 1. Trunk lid liner retaining screws and clips (**arrows**).

3. Disconnect wiring from electrical components and remove complete wiring harness from trunk lid.
4. While supporting trunk lid, loosen top hinge bolts and remove lower bolts from left and right sides. Remove trunk lid with aid of helper. See Fig. 2.

NOTE—
Before loosening hinge bolts, mark hinge and hinge bolt locations for reinstallation.

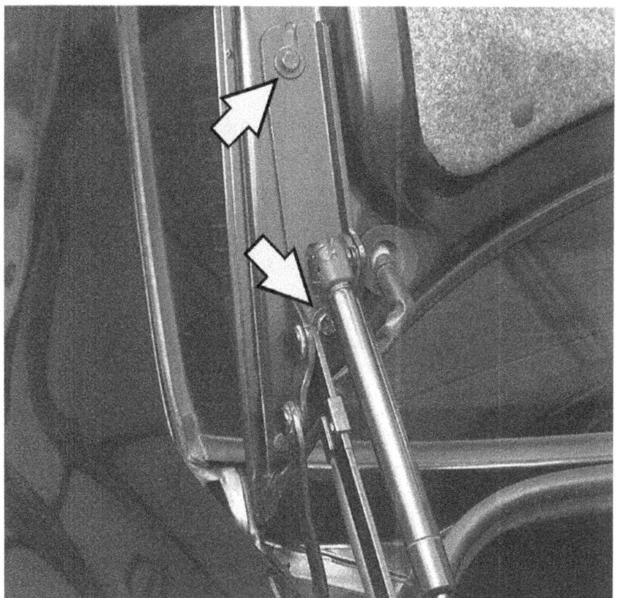

Fig. 2. Trunk lid hinge bolts (**arrows**). Loosen upper bolt and remove lower bolt.

5. Installation is reverse of removal. If necessary, align trunk lid to body as described next.

Trunk lid, aligning

1. Raise trunk lid.

2. Loosen bolts at trunk lid lock until it can just be moved. See Fig. 3.

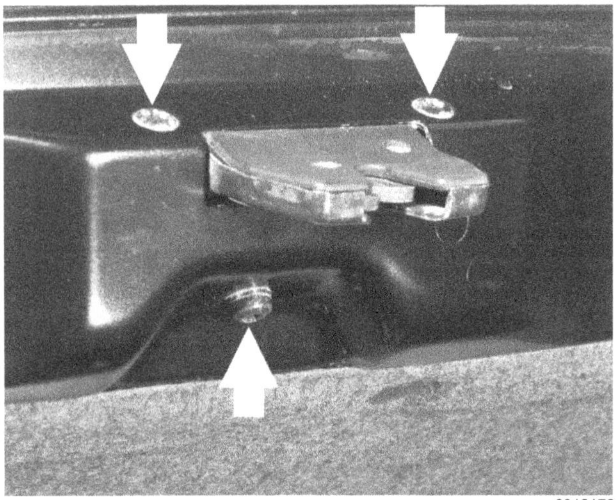

0013172

Fig. 3. Trunk lid lock mounting bolts (**arrows**).

3. Loosen bolts at left and right trunk hinges until lid can just be moved.

4. Working at lower corners of trunk lid, screw in rubber buffers on left and right sides fully.

5. Prop open trunk lid and remove gas pressure strut as described below.

> ***WARNING —***
> - *Make sure to support trunk lid before removing strut.*
> - *Trunk lid with gas pressure strut removed will drop without damping.*

6. Close trunk lid and align lid to body at rear and front. Open trunk lid and tighten hinge bolts.

7. Adjust lock so that closed trunk lid is approximately 1mm (0.04 in.) below top of rear fenders. Tighten lock bolts.

8. Open trunk lid and screw out rubber stops so that closed trunk lid is flush with top of rear fenders.

9. Reinstall gas pressure strut.

Seam Clearance Specification
- Trunk lid to rear fender or bumper . . . . 4.3 - 5.3 mm (0.169 - 0.209 in)
- Max. deviation from parallel . . . . . . .0.5 mm (0.019 in)

Trunk lid support strut, removing and installing

> ***WARNING —***
> *Make sure to support trunk lid before removing strut.*

1. Open trunk lid. Remove spring clip from support strut and remove strut from body end. See Fig. 4.

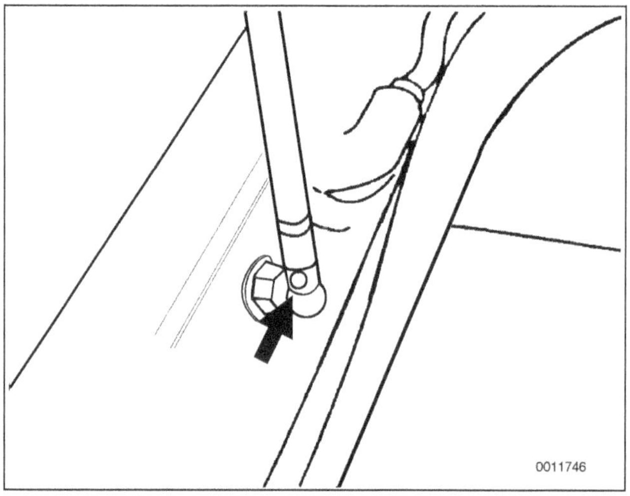

Fig. 4. Support strut spring clip (**arrow**) at body end.

2. Remove strut from trunk lid.

3. Installation is reverse of removal. Replace any trim retaining clips damaged during removal.

510 Exterior Trim, Bumpers

GENERAL

This repair group includes repair information for the outside rear view mirror, front and rear bumpers, and the easily removable exterior trim parts.

OUTSIDE REAR VIEW MIRRORS

The remote controlled mirrors come in two styles: heated and unheated. Many of the mirror components are separately available from an authorized BMW dealer, including the glass and outside plastic housing.

Outside mirror glass, replacing

> ***CAUTION —***
> *Mirror should be at or above room temperature before removal. Otherwise, small plastic parts or glass will break.*

1. Insert thin pry tool (wooden or tape-wrapped screwdriver) between bottom mirror edge and mirror housing and carefully pry out mirror glass from housing. See Fig. 1.

2. Heated mirror: remove heating element harness connectors from back of mirror glass.

3. Install new glass into position by pressing firmly until it snaps into place.

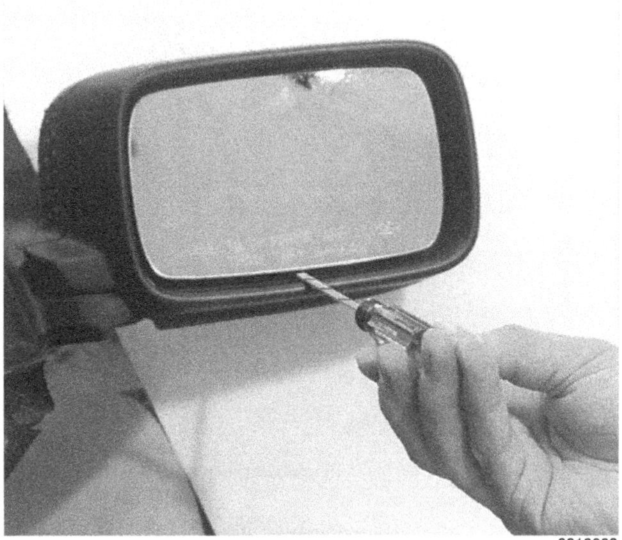

0013038

Fig. 1. Prying mirror glass out of mirror housing. Note screwdriver shaft wrapped with tape.

Outside mirror housing, removing and installing

1. Remove mirror glass as described above.

2. Remove housing retaining screws and lift off rear housing. See Fig. 2.

3. Tilt mirror housing forward and compress plastic retainers and lift off front housing. See Fig. 3.

4. Installation is reverse of removal. Check mirror function before installing covering parts.

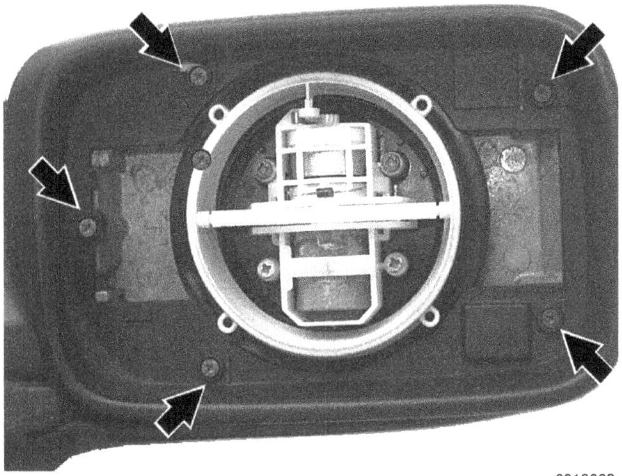

Fig. 2. Mirror housing retaining screws (**arrows**).

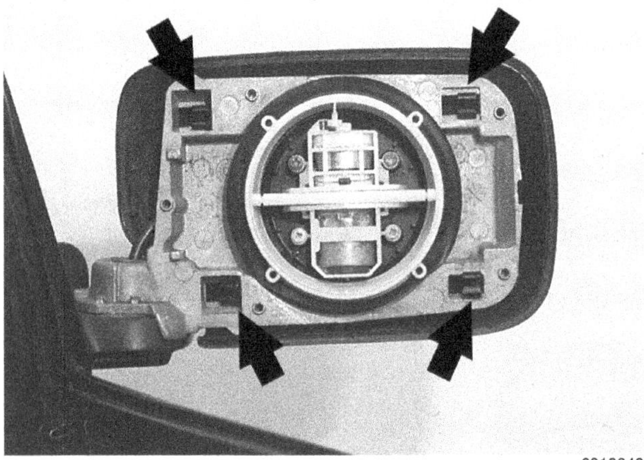

Fig. 3. Use needle-nose pliers to release retainers (**arrows**).

Outside mirror, removing and installing

1. Remove triangular trim plate by prying and lifting gently from door frame. See Fig. 4. Unhook harness splice connector from base of trim plate.

2. Separate harness connector. Support mirror and remove mirror mounting bolts. Lift mirror off door. See Fig. 5.

3. Installation is reverse of removal. Check mirror function before installing covering parts.

Tightening Torque
- Outside mirror to door (M6) . . . . 6 Nm (4.5 ft-lb)

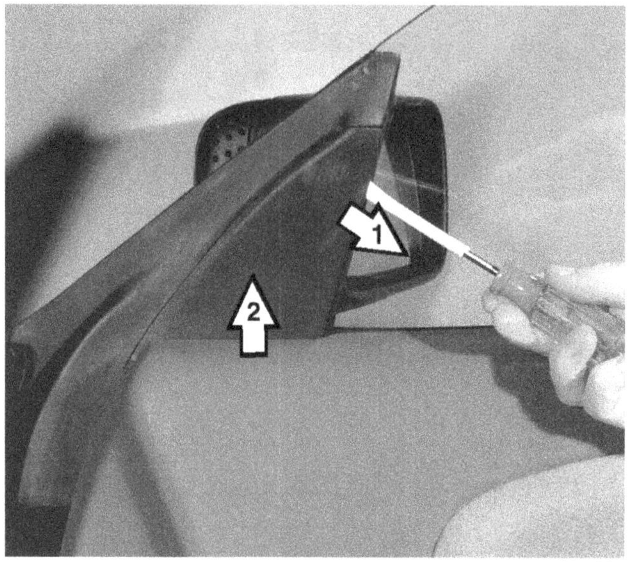

Fig. 4. Pry plastic trim in direction of **arrow 1**, then lift off in direction of **arrow 2**.

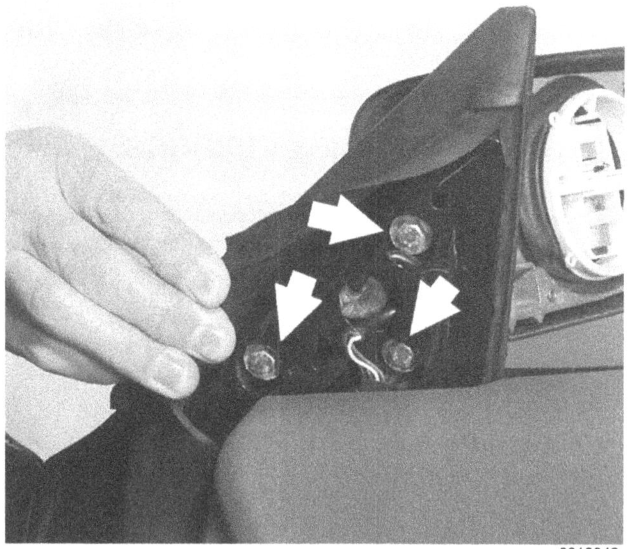

Fig. 5. Mirror mounting bolts (**arrows**). Two-door body style shown.

BUMPERS

Front bumper, removing and installing

Refer to Fig. 6. when removing and installing the front bumper.

1. Raise and properly support vehicle.

> ***WARNING —***
> *Make sure that the car is firmly supported on jack stands designed for the purpose. Place the jack stands beneath a structural chassis point. Do not place jack stands under suspension parts.*

2. Unclip left trim strip by prying gently under plastic trim at front of bumper. Similarly, unclip towing eye cover from right side.
3. Remove four mounting nuts from bumper mounts (two nuts on each mount). See Fig. 7.
4. Working underneath car, remove screws from left and right side corner trim, then remove trim.

Fig. 7. Front bumper mounting nuts (**arrows**). Left trim and tow eye cover on right have been removed.

5. Remove bumper cover retaining screws in front wheel well (left and right sides). See Fig. 8.
6. Disconnect harness connectors at foglight sockets (if applicable).

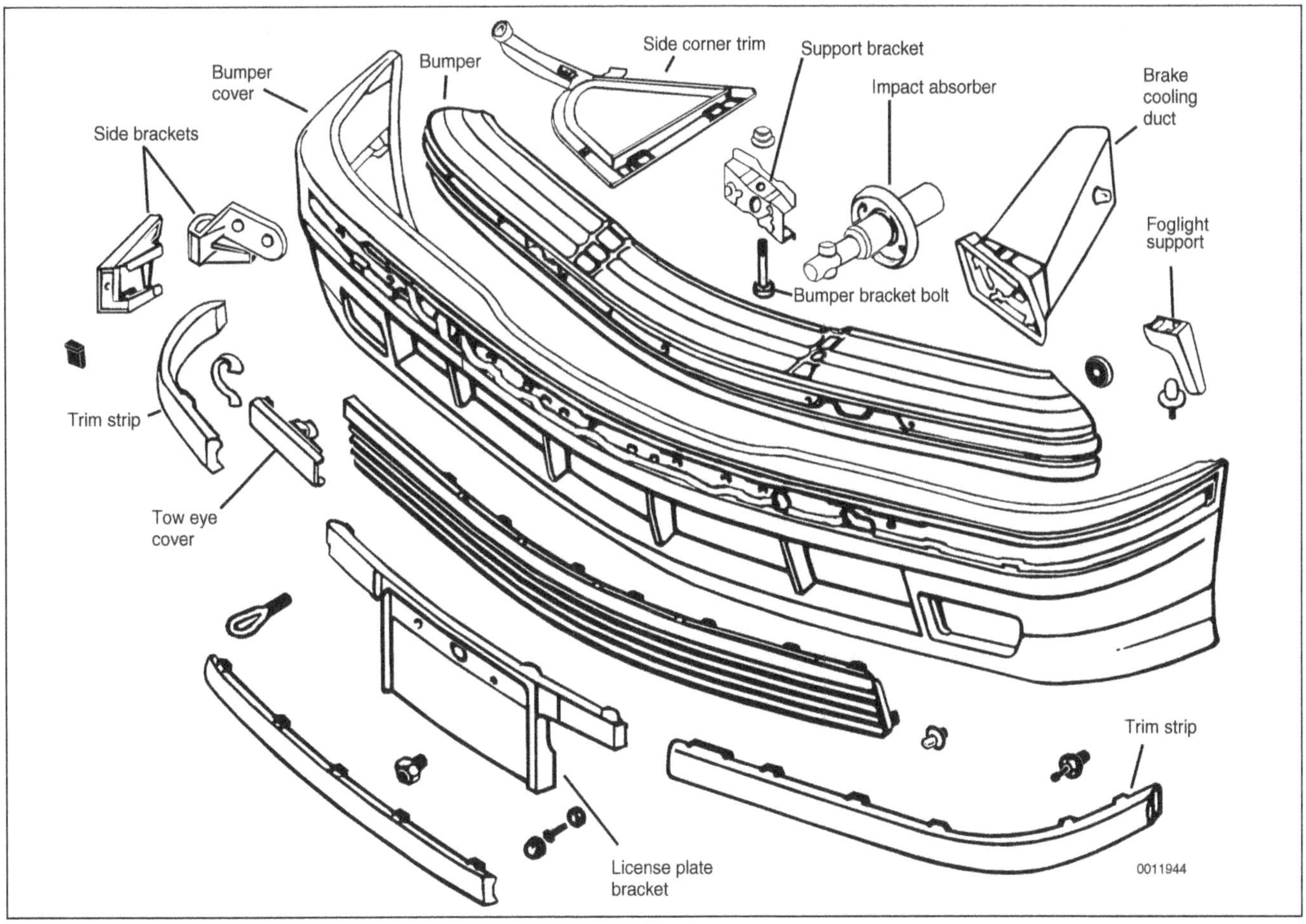

Fig. 6. Front bumper assembly.

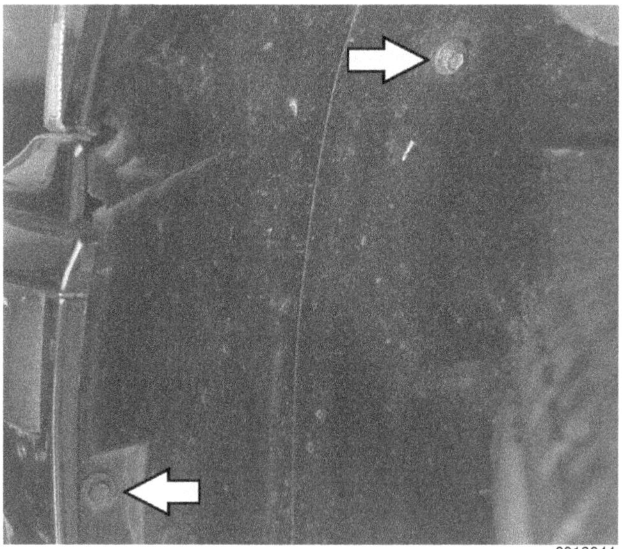

0013044

Fig. 8. Bumper cover retaining screws (**arrows**) in left wheel well.

7. Disconnect harness connectors at outside temperature sensor at rear of left side of bumper and temperature switch at rear of right side of bumper (if applicable).

8. Slide bumper straight off side brackets.

9. Installation is reverse of removal, noting the following:
   - Slide side brackets on bumper and body carefully together.
   - Make sure tabs on trim strip engage slots in bumper correctly.

Tightening Torque
- Bumper to bumper bracket (M8 nut) . . . . 22 Nm (16 ft-lb)

Front bumper impact absorber, replacing

The bumper mounting bracket and impact absorber mounting hardware are accessible with the front bumper removed, as described above. See Fig. 9.

NOTE —
Install the bumper bracket bolt with thread locking compound such as Loctite®270 or equivalent.

Tightening Torques
- Bumper bracket to impact absorber (M10 nut) . . . . 55 Nm (41 ft-lb)
- Impact absorber to chassis (M6 nut) . . . 9 Nm (7 ft-lb)

0013052

Fig. 9. Front bumper impact absorber mounting nuts (**A**) and bumper bracket hardware (**B**).

Rear bumper, removing and installing

When removing and installing the rear bumper or its components, refer to Fig. 10.

1. Raise and properly support vehicle.

> ***WARNING —***
> *Make sure that the car is firmly supported on jack stands designed for the purpose. Place the jack stands beneath a structural chassis point. Do not place jack stands under suspension parts.*

2. Working under bumper, remove screws or expansion rivets at corners and middle of lower bumper panel. Pull panel backward, uncoupling it from bumper cover. See Fig. 11.

3. Remove bumper cover expansion rivets and screws inside left and right rear wheel wells.

4. Remove left and right bumper bracket mounting bolts and remove bumper. See Fig. 12.

5. Installation is reverse of removal.

NOTE —
Install the bumper bracket bolt with thread locking compound such as Loctite®270 or equivalent.

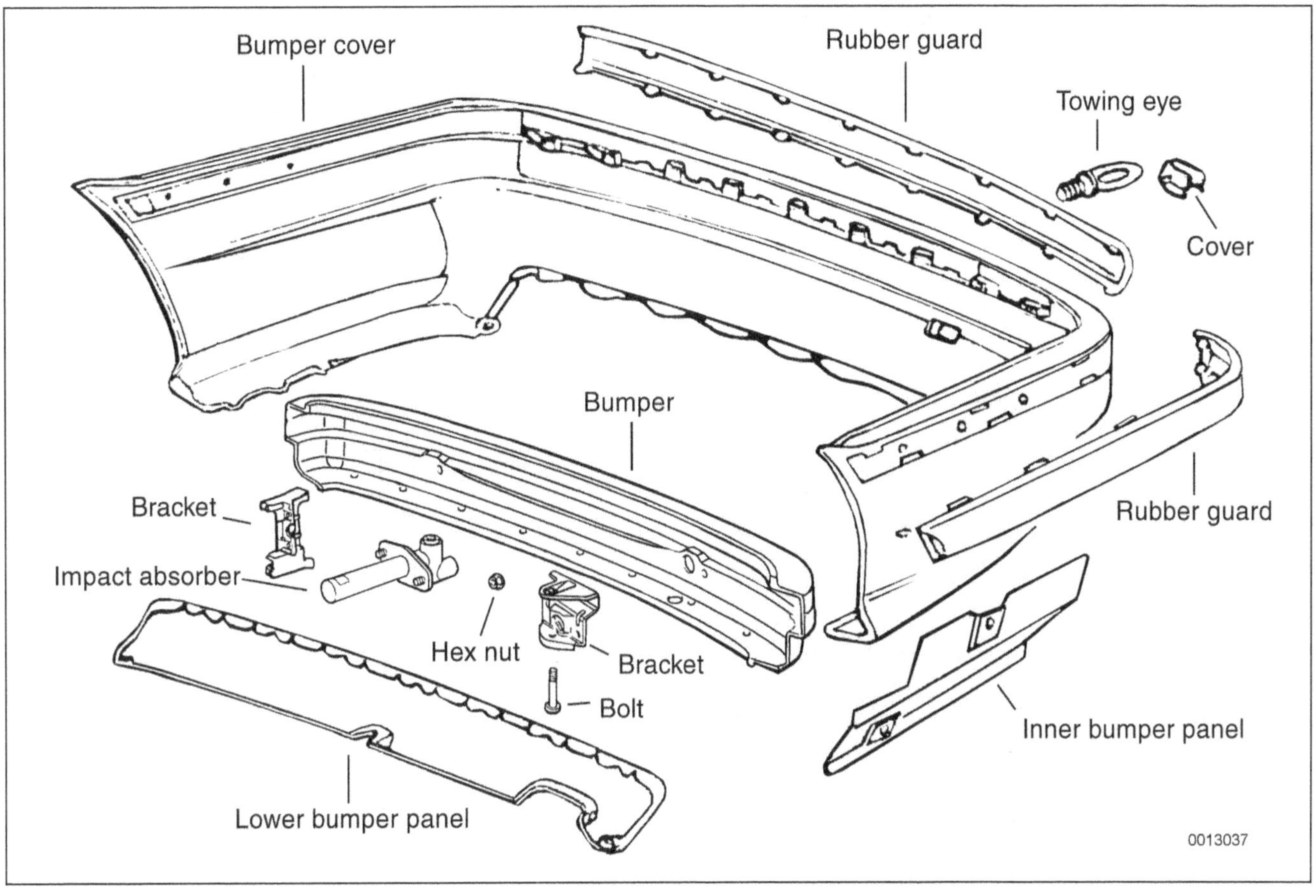

Fig. 10. Rear bumper assembly.

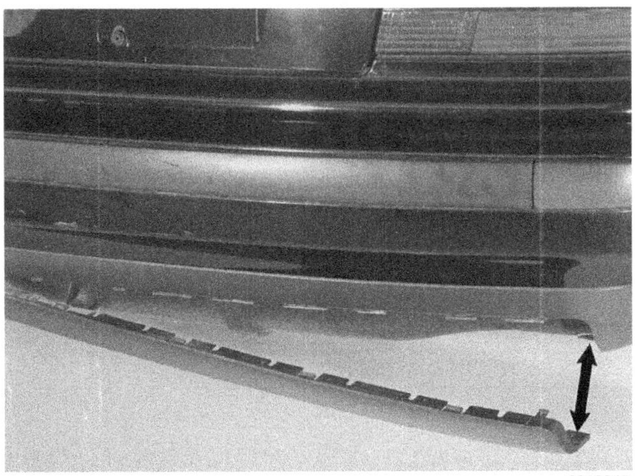

Fig. 11. Expansion rivet at corner of rear bumper lower panel (**arrow**). Some models use screws. Pull panel backward to disengage from bumper cover.

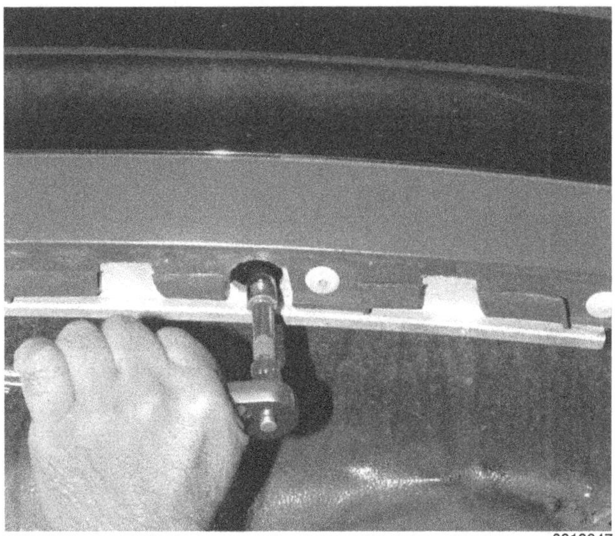

Fig. 12. Insert tool through rear bumper mount bores to remove mounting bolts (right side shown).

Tightening Torque

- Bumper bracket to impact absorber (M10 bolt) . . . . 42 Nm (31 ft-lb)

Rear bumper impact absorber, replacing

1. Remove bumper as described above.

2. Remove luggage compartment floor covering.

3. Right side: Remove battery and battery tray.

4. Remove fasteners from left or right luggage compartment floor panels and lift out panel. See Fig. 13. Impact absorber mounting nuts can now be accessed.

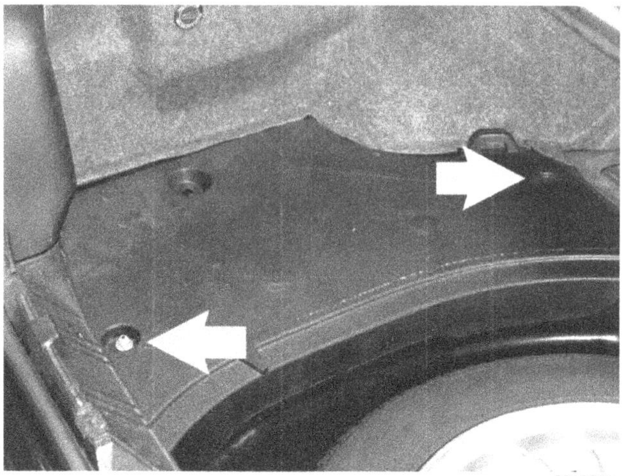

0013051

Fig. 13. Luggage compartment floor panel fasteners (**arrows**). (Left side shown. Right side is similar)

5. Installation is reverse of removal.

Tightening Torques
- Bumper bracket to impact absorber (M10 bolt) . . . . 42 Nm (31 ft-lb)
- Impact absorber to chassis (M8 nut) . . . . 22 Nm (16 ft-lb)

Bumper height, adjusting

The adjustment procedure given below applies to either the front or rear bumper. A 14mm allen socket is needed to make the adjustment.

1. Raise and properly support vehicle.

> ***WARNING —***
> *Make sure that the car is firmly supported on jack stands designed for the purpose. Place the jack stands beneath a structural chassis point. Do not place jack stands under suspension parts.*

2. Front bumper:
   - Remove bumper as described earlier.
   - Remove bumper bracket bolt. Refer to Fig. 9.

3. Rear bumper:
   - Remove lower bumper panel as described in **Rear bumper, removing and installing**.
   - Remove bumper bracket mounting bolt at impact absorber.

4. Using a 12-inch extension and a 14mm Allen socket, turn adjusting collar clockwise or counterclockwise as needed to change height of adjusting collar. See Fig. 14.

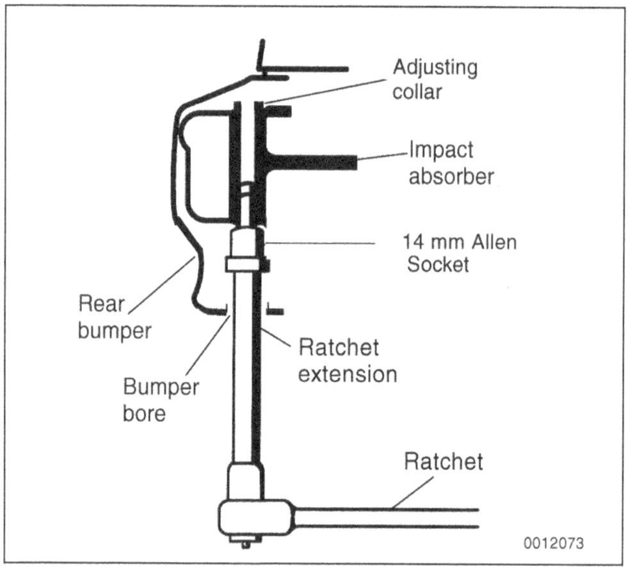

Fig. 14. Adjust bumper height at threaded adjusting collar in bumper impact absorber.

5. Reinstall mounting bolts and check bumper height. Refit front bumper.

Tightening Torques
- Front bumper to bumper bracket (M8 nut) . . . . 22 Nm (16 ft-lb)
- Front bumper bracket to impact absorber (M10 bolt) . . . . 55 Nm (41 ft-lb)
- Rear bumper bracket to impact absorber (M10 bolt) . . . . 42 Nm (31 ft-lb)

Exterior Trim

Much of the exterior trim is retained to the body by plastic clips and fasteners that may be damaged during removal. Be sure to have the necessary fasteners on hand when removing exterior trim pieces.

BMW emblem, removing and installing

The procedure given below applies to both front and rear emblems.

1. Wrap end of a screwdriver with tape.

> ***CAUTION —***
> *Protect hood paint by covering area around emblem with tape.*

2. Carefully pry out emblem. See Fig. 15.

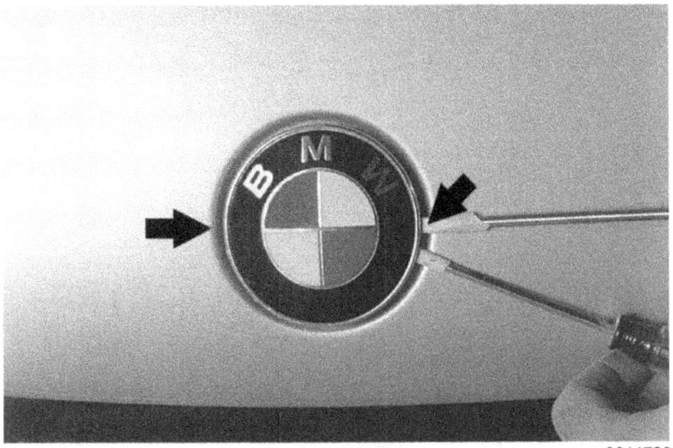

Fig. 15. BMW emblem being removed. Pry up emblem carefully on either side (**arrows**). Note tape on screwdriver tips.

3. Installation is reverse of removal.
   - Replace plastic inserts in body if damaged.
   - If emblem fits loosely into inserts, use a small amount of body molding tape or adhesive on rear of emblem before installing.

Body side molding, replacing

To remove body side moldings, carefully pry the moldings straight off. See Fig. 16.

Installation is the reverse of removal. Replace any clips or clip covering boots damaged during removal.

Radiator grille, removing and installing

1. Remove plastic front radiator shroud or air duct, as needed.

2. Lightly tap center of grille toward radiator and remove from panel opening. Chrome ring pulls out from front. See Fig. 17.

3. To replace, snap grille and chrome piece together. Push assembly into front panel, pressing at top and bottom until it snaps into place.

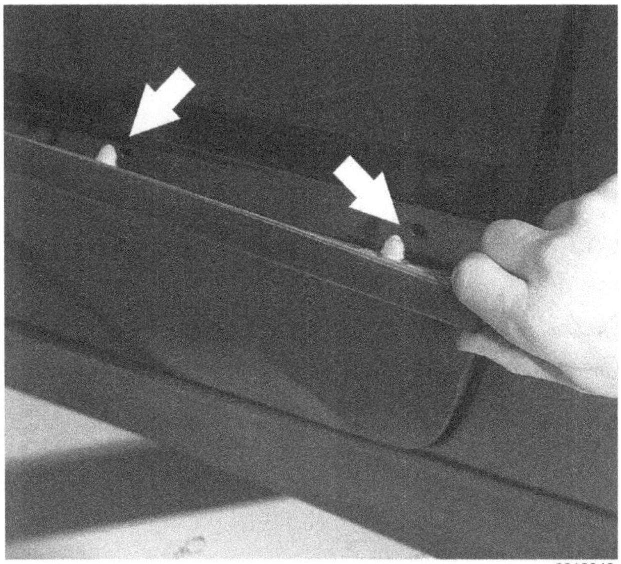

Fig. 16. Body side molding retaining clips (**arrows**).

Fig. 17. Radiator grille being tapped inward. Right side front panel top retaining screw is shown at **A**.

Front panel, removing and installing

The front panel is shown in Fig. 18.

1. Remove front bumper as described earlier.

2. Remove headlights. See **630 Exterior Lighting**.

3. Remove radiator grille as described earlier.

4. With engine hood open, remove front panel retaining screws at left and right sides of radiator. See Fig. 19.

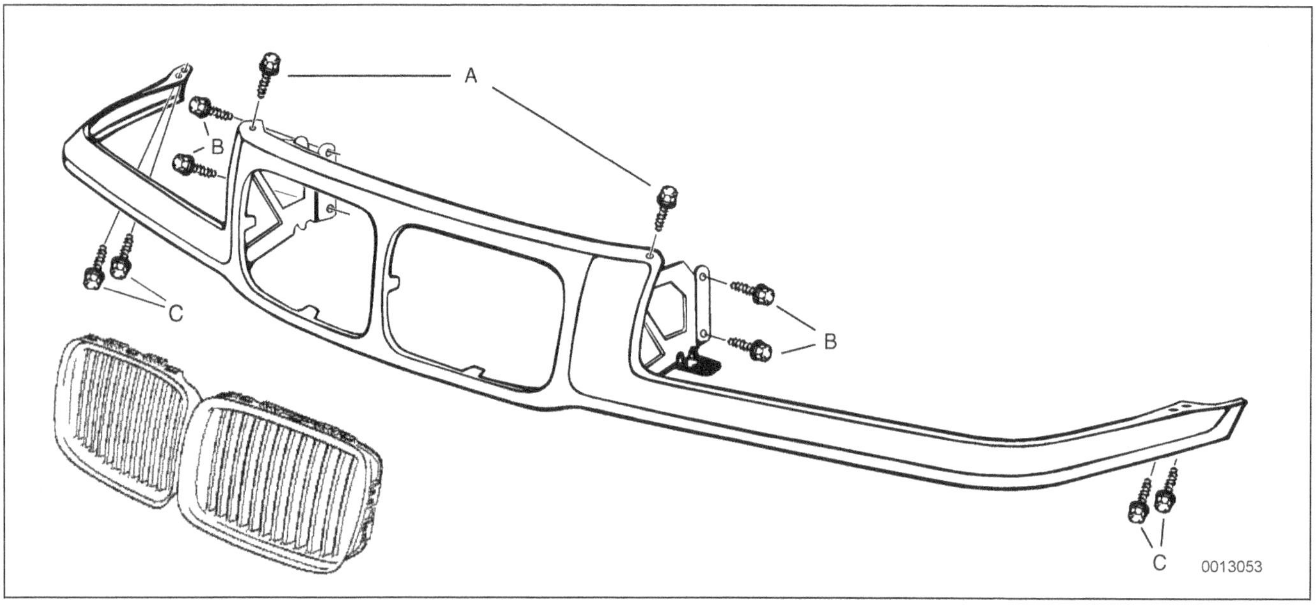

Fig. 18. Front panel assembly. Note top retaining screws (**A**), retaining screws at radiator support (**B**), and side retaining screws (**C**).

Fig. 19. Front panel retaining screws at radiator (**arrows**). Right side shown. To remove left side screws it may be necessary to remove air filter housing or duct.

5. Remove front panel top retaining screws near hood latches. Refer to Fig. 17.

6. Working under car, remove left and right side panel retaining screws. See Fig. 20.

7. Remove panel and repair or replace with painted part.

8. Installation is reverse of removal.

Fig. 20. Front bumper must be off to access far right and left front panel retaining screws from below (**arrows**). (Right side shown.)

512 Door Windows

GENERAL

This heading covers door glass, window regulator and power window motor repair information. Before proceeding with window repairs, specifically on late cars fitted with front side-impact airbags, read the procedure through to determine the scope of the repair.

The bonded windshield and rear glass are replaced using special adhesives and tools. It is recommended that bonded glass replacement be done by a professional glass installer.

> ***WARNING —***
> - *Always wear hand and eye protection when working with broken glass.*
> - *If a window is broken, all of the glass bits should be vacuumed out of the door cavity. Use a blunt screwdriver to clean out any remaining glass pieces from the window guide rails.*

> ***CAUTION —***
> *If battery power is lost or the battery has been disconnected, windows must be re-initialized. Both one touch up/down and pinch-protection will be inactive until windows are re-initialized. See below.*

Front door window, initializing

The front door windows incorporate a pinch-protection feature as well as one-touch up/down (1994 and later models). In order for these features to work correctly, the mechanical stop reference point must be re-initialized after repairs are carried out.

1. Close door and turn ignition key on. Fully open front window.
2. Fully close front window.
3. Hold window button in up position for at least 5 seconds after window is closed.
4. Repeat for remaining window.
5. The windows should now be initialized and the anti-pinch and one-touch up/down functions should be enabled.

512

WINDOW SERVICE, 4-DOOR MODELS

> ***WARNING —***
> *When servicing the door windows, the harness connector to the window regulator should always be disconnected to prevent pinching fingers in closing window.*

NOTE —
If the battery is disconnected, the front window regulator motors must be re-initialized after repairs are carried out. If the window motors are not reinitialized, the programmed one-touch up/down and anti-pinch functions will be disabled. See **Front door window, initializing**.

Front door window, removing and installing (4-door models)

1. Lower door window about 12 in. (300 mm).

2. Remove front door panel and vapor barrier as described in **411 Doors**.

> ***WARNING —***
> - *Some 1997 and all 1998 cars are fitted with side-impact airbags in the front doors. When servicing the door windows on cars with front side-impact airbags, always disconnect the negative (–) battery cable. See* **721 Airbag System (SRS)** *for cautions and procedures relating to the airbag system.*
>
> - *Prior to disconnecting the battery, read the battery disconnection cautions given at the front of this manual on page viii.*

3. On cars with side-impact airbags, disconnect negative (–) cable from battery. Then remove airbag unit from door. See **721 Airbag System (SRS)**.

NOTE —
In some models covered in this manual, it is not essential to remove the side-impact airbag to access internal door components. However, the procedure is recommended for safety reasons.

4. Disconnect harness connector from window motor to prevent accidental operation.

5. Carefully remove window inner seal and pry off metal retainers under sealing lip to prevent scratching glass. See Fig. 1.

6. Remove retaining clips from window slides. See Fig. 2.

Fig. 1. Lift off window seal from door. Note metal retainer (**arrow**). 2-door car shown.

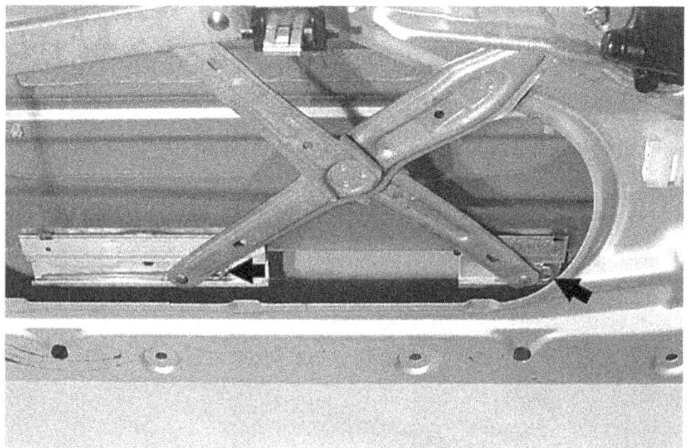

Fig. 2. Retaining clips for window slides (**arrows**).

7. While supporting window glass, pry window lifting arms out of left and right slides. See Fig. 3.

8. Pivot window up at rear and out of door.

9. Installation is reverse of removal, noting the following:
   - Use new mounting bolts when reinstalling the side-impact airbag to the door (where applicable). See **721 Airbag System (SRS)**.
   - Adjust window as described later.
   - Re-initialize window regulator motors as described earlier.

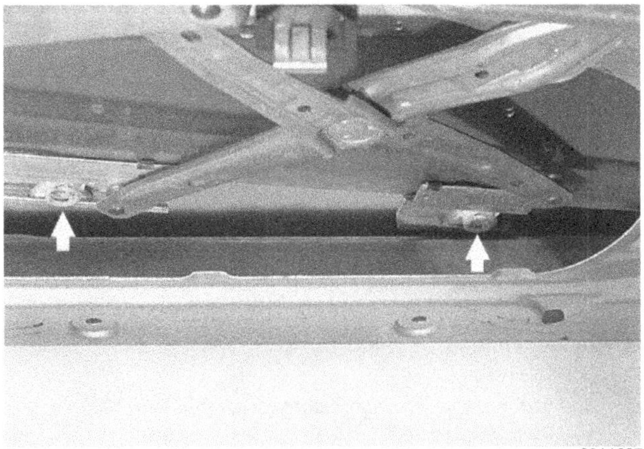

0011697

Fig. 3. Window lifting arms shown removed from slides (**arrows**).

Front door window, adjusting (4-door models)

Whenever the front window or window regulator is removed, window adjustment should be checked. The glass should contact the top of the window squarely and should seat against the window seal uniformly.

> ***NOTE —***
> *If the window does not contact the window seal uniformly, wind noise or water infiltration may result.*

Be sure to re-initialize the window motors as described earlier.

> ***WARNING —***
> *Some 1997 and all 1998 cars are fitted with side-impact airbags in the front doors. When servicing the door windows on cars with front side-impact airbags, always disconnect the negative (–) battery cable. See* **721 Airbag System (SRS)** *for cautions and procedures relating to the airbag system.*

> ***CAUTION —***
> - *In models with side-impact airbags, adjusting the front window requires that the battery cable be reconnected with the side-impact airbag disconnected. This will set a fault code in the SRS control module, turn on the SRS warning light, and disable the SRS system until the fault memory is cleared using special SRS test equipment.*
> - *Prior to disconnecting the battery, read the battery disconnection cautions given at the front of this manual on page viii.*

1. Raise door window to top of its travel.
2. On cars with side-impact airbags, disconnect negative (–) cable from battery.
3. Remove front door panel and vapor barrier as described in **411 Doors**.
4. Remove airbag unit from door. See **721 Airbag System (SRS)**.

> ***NOTE —***
> *In some models covered in this manual, it is not essential to remove the side-impact airbag to access internal door components. However, the procedure is recommended for safety reasons.*

5. Loosen regulator stop at window regulator support. See Fig. 4.

0013075

Fig. 4. Window regulator stop (**arrow**).

6. If glass is not squarely positioned in window frame, loosen support arm bolt and reposition arm as necessary. Tighten bolt. See Fig. 5.

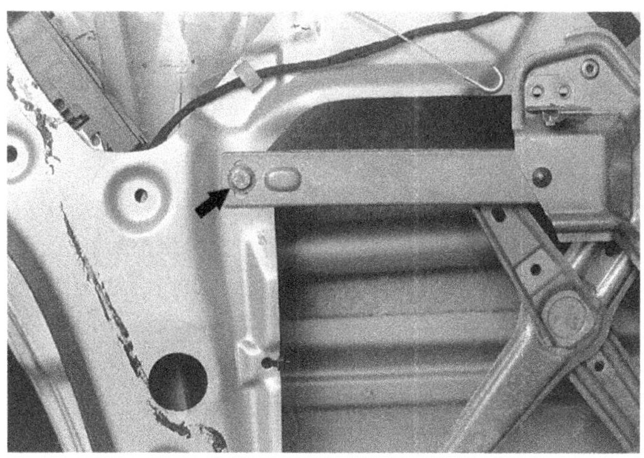

0011705

Fig. 5. Adjust window to its frame by loosening bolt (**arrow**) and repositioning support arm up or down.

7. Push regulator stop tightly against window regulator sector and tighten nut.
   - Check for tight, even fit at top of window frame.
   - If necessary, temporarily reconnect negative (–) cable to battery and raise window so it is pressing uniformly against top rubber seal before final tightening.

8. Reinitialize front window motors as described earlier.

Tightening Torques
- Regulator stop to window regulator . . . . 9 Nm (7 ft-lb)
- Window regulator bracket to door . . . . . 9 Nm (7 ft-lb)

Rear door window, removing and installing (4-door models)

1. Remove rear door panel and vapor barrier. See **411 Doors**.

2. Pull out rubber guide at front of window frame. Remove three screws and pull off exterior trim cover. See Fig. 6.

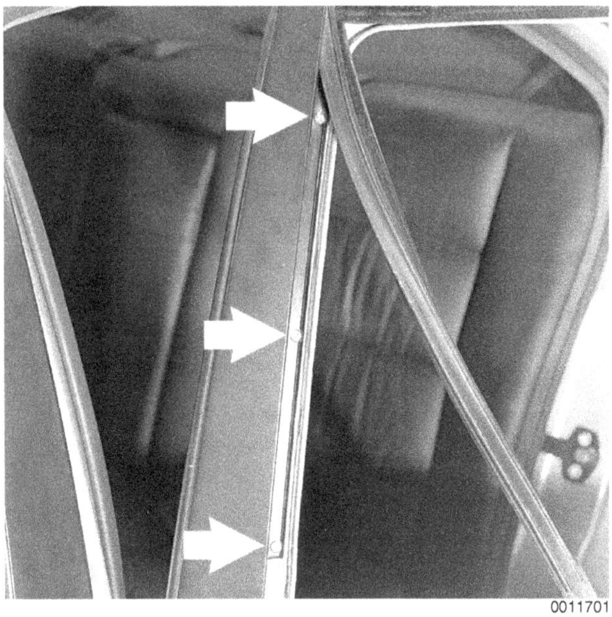

Fig. 6. Trim cover mounting screws (**arrows**). To remove cover, slide up and off door.

3. Lift up and remove window inner sealing strip. Remove clips at edge of window channel. Refer to Fig. 1.

4. Lower door window about 12 in. (300 mm). Remove window lifting arm retaining clip from window slide. Refer to Fig. 2.

> ***WARNING —***
> *Once the window is lowered, disconnect harness connector from power window motor to prevent accidental operation of the window.*

5. Pry lifting arm out of slide. Refer to Fig. 3.

NOTE —
Support window in door before disconnecting lifting arm from window slide.

6. Remove rubber channel from door frame perimeter.

7. Remove window glass from above.

8. Installation is reverse of removal.
   - Rear window adjustment is similar to front window adjustment, given earlier.
   - Reinitialize window motors as described earlier.

Rear door fixed glass, removing and installing (4-door models)

1. Remove window from rear door as described earlier.

2. Remove upper and lower bolts at guide rail. Lift up rubber seal to access upper bolt. See Fig. 7.

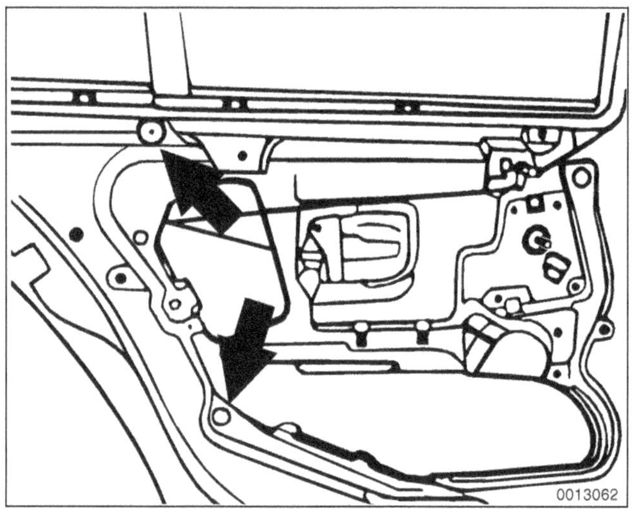

Fig. 7. Rear guide rail mounting bolts (**arrows**).

3. Remove guide rail screw at top and move rail aside enough to slide glass forward and out of door frame. See Fig. 8.

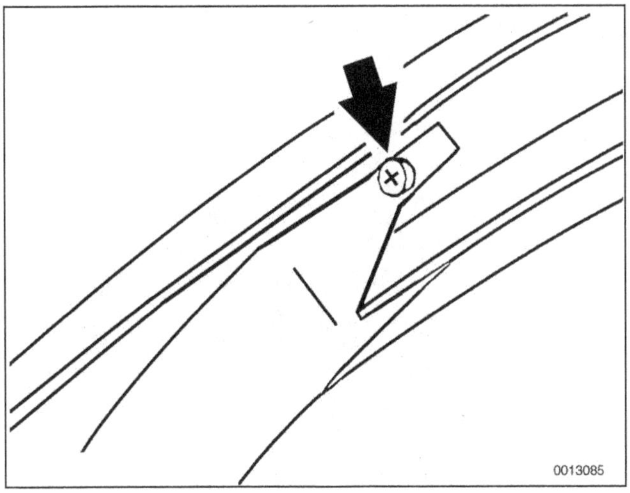

Fig. 8. Remove guide rail screw (**A**) and move rail aside.

4. Installation is reverse of removal.

WINDOW SERVICE, 2-DOOR MODELS

> ***WARNING —***
> *When servicing the door windows, the harness connector to the window regulator should always be disconnected to prevent pinching fingers in closing window.*

2-door models are equipped with the "close with door shut" feature. In this mode, the window glass moves down slightly when the door latch is activated, then goes up tightly against the weather-strip when the door is fully closed.

> ***NOTE —***
> *The front window regulator motors must be re-initialized after repairs are carried out. If the window motors are not reinitialized, the programmed one-touch up/down and anti-pinch functions will be disabled. See* **Front door window, initializing**.

Front door window, removing and installing (2-door models)

> ***WARNING —***
> *Some 1997 and all 1998 cars are fitted with side-impact airbags in the front doors. When servicing the front windows on cars with front side-impact airbags, always disconnect the negative (–) battery cable. See* **721 Airbag System (SRS)** *for cautions and procedures relating to the airbag system.*

> ***CAUTION —***
> - *In models with side-impact airbags, removing the front window requires that the battery cable be reconnected with the side-impact airbag disconnected. This will set a fault code in the SRS control module, turn on the SRS warning light, and disable the SRS system until the fault memory is cleared using special SRS test equipment.*
> - *Removing the front window will upset the adjustment and alignment of the window. Window adjustment, as described later,* ***must*** *be carried out after installation to prevent damaging the glass.*

1. On cars with side-impact airbags, disconnect negative (–) cable from battery.

> ***CAUTION —***
> *Prior to disconnecting the battery, read the battery disconnection cautions given at the front of this manual on page viii.*

2. Remove interior door panel and vapor barrier. See **411 Doors**.

3. Where applicable, remove side-impact airbag from door. See **721 Airbag System (SRS)**.

> ***NOTE —***
> *In some models covered in this manual, it is not essential to remove the side-impact airbag to access internal door components. However, the procedure is recommended for safety reasons.*

4. Pull off exterior door molding. See **510 Exterior Trim, Bumpers**.

5. Lower door window completely. If necessary, temporarily reconnect battery cable.

6. Lift up and remove window inner sealing lip. Remove all metal retainers under sealing lip. See Fig. 9.

7. Raise window to about 10 in. (250 mm). If necessary, temporarily reconnect battery cable. Remove guide bracket mounting nuts and remove bracket. See Fig. 10.

8. Lower window about 4½ in. (110 mm). If necessary, temporarily reconnect battery cable. Working through bore hole in door exterior, remove lower guide bracket through-bolt from glass. See Fig. 11.

9. Lower window about 2½ in. (65 mm). If necessary, temporarily reconnect battery cable. Repeat step 8 to remove upper through-bolt from guide bracket.

10. Raise window to about 7½ in. (190 mm) from top. If necessary, temporarily reconnect battery cable. Remove window rear limit stop bracket. See Fig. 12.

Fig. 9. Lift window inner sealing lip from door. Note metal retainer (**arrow**).

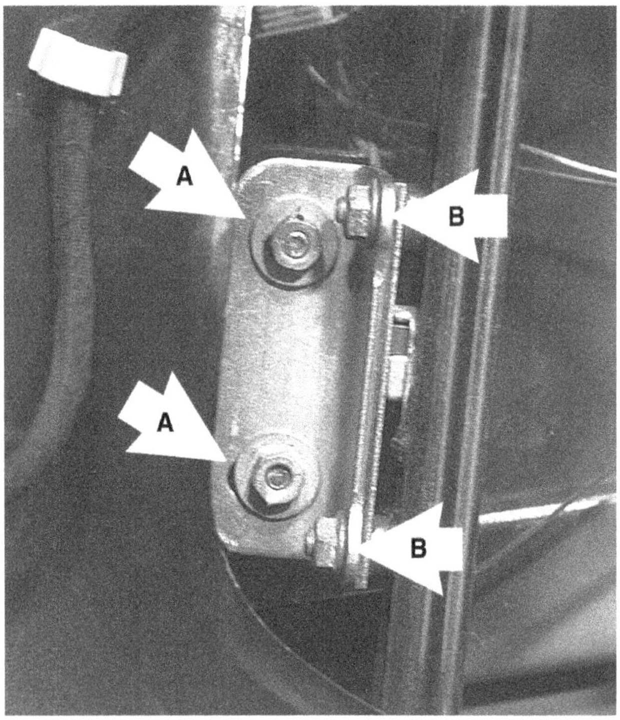

Fig. 10. Window rear guide bracket nuts **A** to be removed. Loosen nuts **B** to allow bracket room to swivel out of the way.

Fig. 11. Using 4mm Allen socket, reach through rear bore hole (**A**) in door to remove window guide bracket bolts. (Exterior door molding has been removed.)

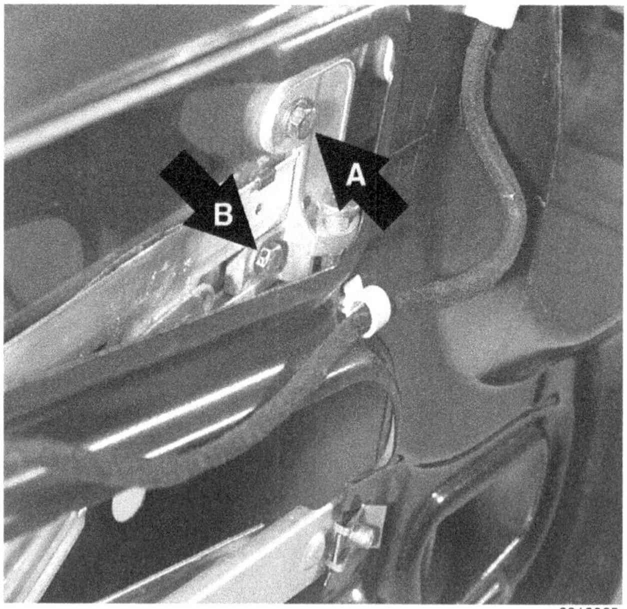

Fig. 12. To remove rear window limit stop bracket, remove bolt **A** and loosen clamping bolt **B**.

11. Lower window about 4½ in. (110 mm). If necessary, temporarily reconnect battery cable. While supporting window, remove retaining clips from window slides and disconnect lifting arms from slides. See Fig. 13.

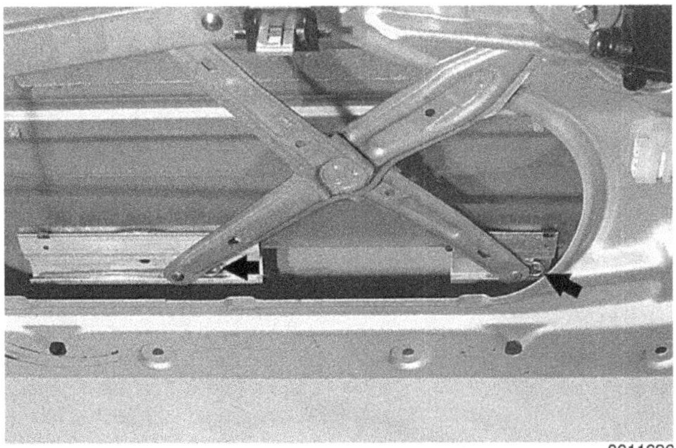

0011696

Fig. 13. Lifting arm retaining clips (**arrows**) at window slides.

12. Tip window so rear of glass is up out of door. While supporting window, remove window front limit stop bracket. See Fig. 14.

0013067

Fig. 14. With window glass shown partially tipped out of door, remove bolt **A** and loosen bolt **B** in order to remove front window limit stop bracket.

13. Take glass fully out of door.

14. Installation is reverse of removal, noting the following:
    - Guide front of window into front guide rail and rest it there while installing front window limit stop bracket.
    - When installing window rear guide bracket, tighten countersunk screws in glass first (through outside of door), then install guide rail bracket.
    - Insert rubber seal at top of door correctly. See Fig. 15.
    - Use new mounting bolts when reinstalling the side-impact airbag to the door (where applicable). Reset SRS warning light and clear SRS fault memory using special test equipment. See **721 Airbag System (SRS)**.
    - Adjust window as described later.
    - Re-initialize window regulator motors as described earlier.

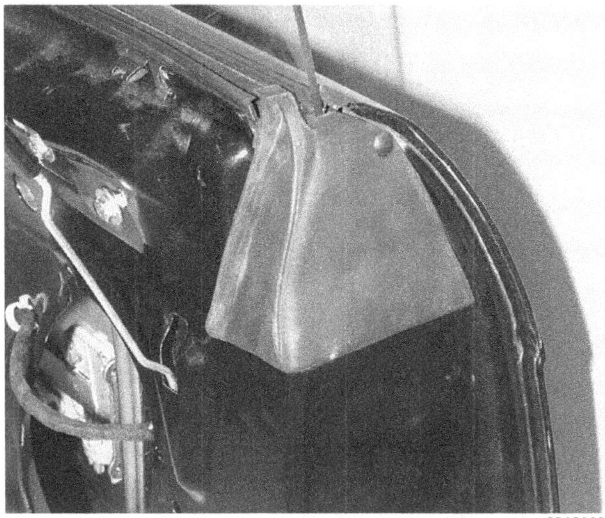

0013068

Fig. 15. Correct installation of door top rubber seal.

Tightening Torques
- Rear guide bracket to window . . . . 6 Nm (53 in-lb)
- Rear guide bracket to roller guide . . . . 9 Nm (7 ft-lb)

Front door window, adjusting (2-door models)

The "close with door shut" feature of 2-door cars, as well as the frameless window, makes window adjustment critical.

WARNING —
- *Some 1997 and all 1998 2-door cars are fitted with side-impact airbags in the front doors. When servicing the door windows on cars with front side-impact airbags, always disconnect the negative (–) battery cable. See* **721 Airbag System (SRS)** *for cautions and procedures relating to the airbag system.*
- *Adjusting the front window requires that the battery cable be reconnected with the side-impact airbag disconnected. This will set a fault code in the SRS control module, turn on the SRS warning light, and disable the SRS system until the fault memory is cleared using special SRS test equipment.*
- *There is risk of window glass breakage if correct adjustment procedures are not used. If in doubt, this procedure should be left to a trained BMW technician.*

NOTE —

To achieve an accurate window adjustment in 2-door models with "close with door shut" feature, simulate a closed door as follows: On early cars, manually turn the latch on the door to the closed position. See Fig. 16. On later cars, insert a wedge of wood into the striker on the body to close the striker switch. Be sure to open the rotary latch (by pulling up on the door handle) or remove the wedge before closing the door.

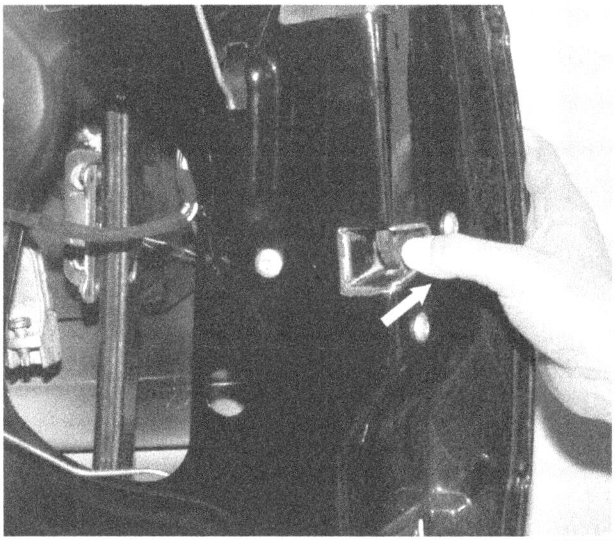

0013066

Fig. 16. To manually close door latch, push rotary latch in direction of **arrow**.

CAUTION —

Do not close the door with the rotary latch in closed position, or with a wedge in the striker.

When making adjustments to the front door window, see Fig. 17.

1. On cars with side-impact airbags, disconnect negative (–) cable from battery.

CAUTION —

Prior to disconnecting the battery, read the battery disconnection cautions given at the front of this manual on page viii.

2. Remove interior door panel and vapor barrier. See **411 Doors**.

3. Where applicable, remove side-impact airbag from door. See **721 Airbag System (SRS)**.

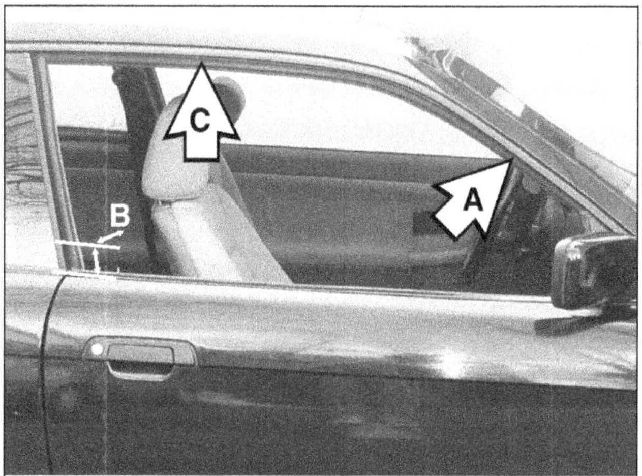

A Window insertion depth at A-pillar
B B-pillar preload gap (measure gap at 20 mm from top of door)
C Window insertion depth at roof line/window parallelism (make measurement at 3.5 in. and 16 in. from B-pillar edge)

0013076

Fig. 17. 2-door window adjustment positions.

NOTE —

- *In some models covered in this manual, it is not essential to remove the side-impact airbag to access internal door components. However, the procedure is recommended for safety reasons.*
- *After each adjustment step, fully lower and raise window and check adjustment before going on to the next step.*
- *If parts of the front window mechanism were replaced, make sure initial adjustments are in the center of their travel range. Once the glass is installed, position glass to the rear guide rail as shown in Fig. 18.*

4. To set window preload **B** (See Fig. 19.):
   - Measure up 20 mm (0.75 in.) from top of door and mark position on glass.
   - Close door just until top of glass contacts seal.
   - Measure gap between glass (at 20 mm mark) and window seal.
   - If adjustment is necessary, move bottom of window guide in or out to correct. See Fig. 20.
   - If further adjustment is needed, move bottom of window in or out by sliding guide bracket. Refer to Fig. 18.

Preload Gap

- **B** (Refer to Fig. 19.)

Coupe	8 mm (0.32 in.)
Convertible	3 mm (0.12 in.)

Fig. 18. Rear guide rail to rear guide bracket initial adjustment. Top should be set at 30 mm (1.2 in.) from glass to stud center; bottom should be set at 34 mm (1.3 in.) from glass to stud center.

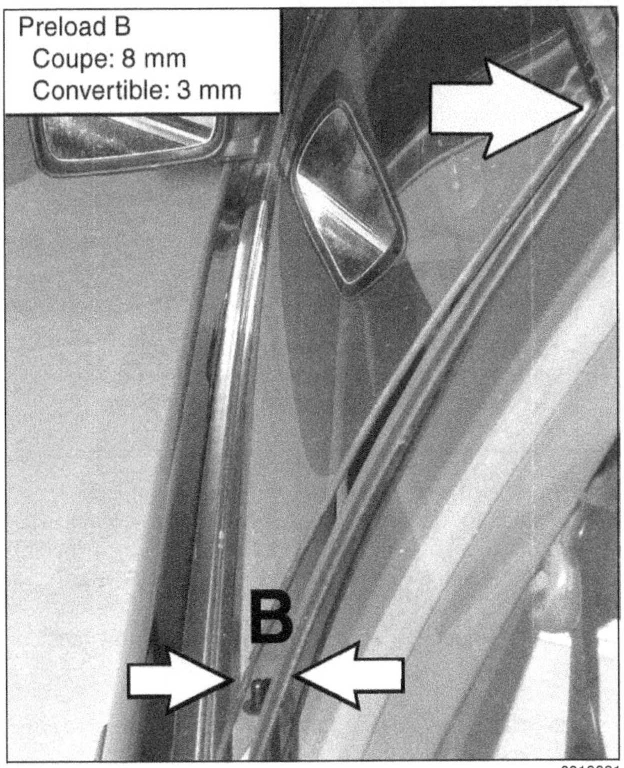

Fig. 19. With top of glass against door seal (**top arrow**), measure preload **B** at 20 mm (0.75 in.) above top of door.

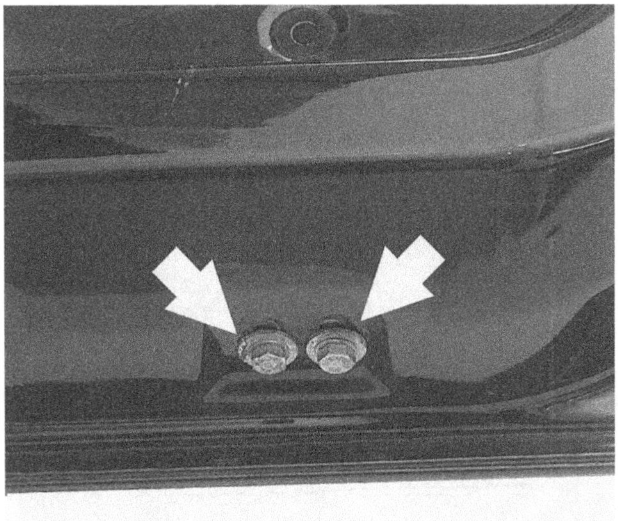

Fig. 20. Rear guide rail lower mounting bolts in bottom of door (**arrows**).

5. To set insertion depth **A** (Refer to Fig. 17.):
   - Lower window about 10 in. to access window bracket at rear guide. If necessary, temporarily reconnect battery cable.
   - Loosen guide rail bracket and/or rear guide rail mounting bolts and reposition window left to right. See Fig. 21. or Fig. 22.
   - Fully close and align window. Check that glass enters window seal to the specification listed.
   - When adjustment is correct, tighten loosened fasteners.

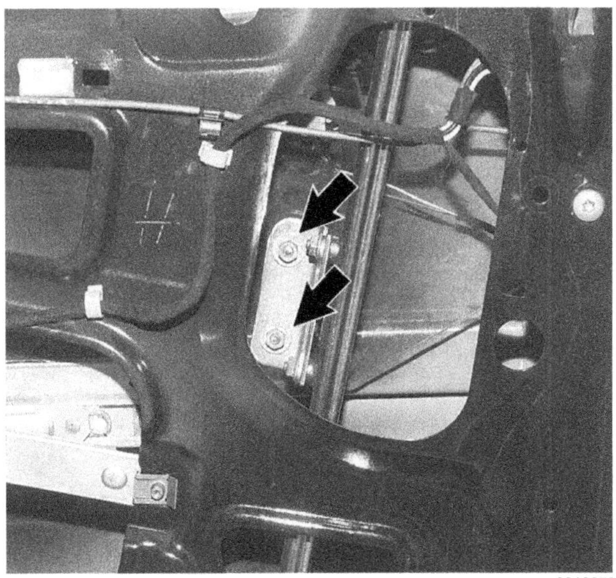

Fig. 21. Loosen nuts on rear guide bracket (**arrows**) and reposition glass fore/aft. Holes in bracket are elongated to allow for small amounts of adjustment.

0013078

Fig. 22. Move window fore/aft at rear guide rail top mounting bolts (**arrows**).

Insertion Depth

- **A** (See Fig. 17.) . . . . .1-1.5 mm (0.04-0.6 in.)

6. To set insertion depth **C** (Refer to Fig. 17.):
   - Lower window slightly. Loosen stop at window regulator motor. See Fig. 23.

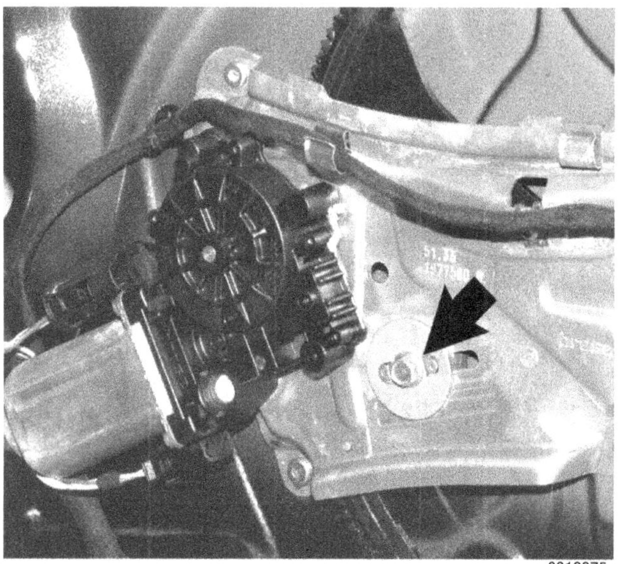

0013075

Fig. 23. Window regulator stop (**arrow**).

- Raise window. Loosen rear bolt at regulator support arm (5mm Allen wrench). See Fig. 24. Align window parallel to roof line by moving support arm up or down. Tighten arm bolt.
- Lower window and remove rubber plugs at bottom of door. If necessary, temporarily reconnect battery cable.
- Adjust front and rear limit stops so that window retracts into window seal as specified. See Fig. 25.
- Raise window to check adjustment.
- Push regulator stop fully forward and tighten bolt.
- Close door and check window operation.

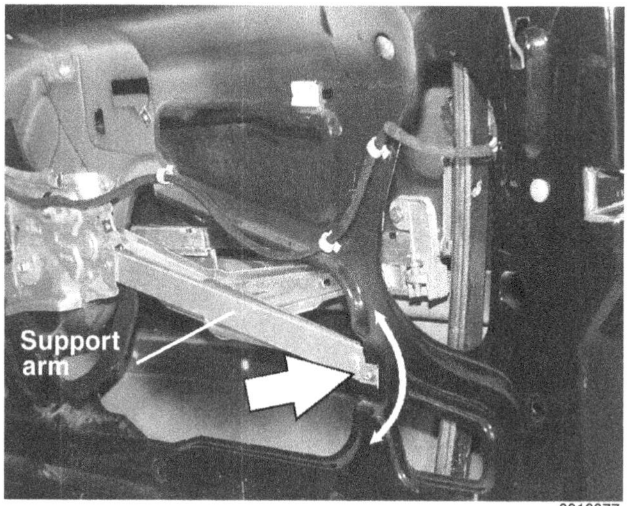

0013077

Fig. 24. Use 5mm Allen socket to loosen mounting bolt at support bracket (**arrow**) and slide bracket up or down to adjust window position at roof line. (Allen bolt head faces outside of door.)

CAUTION—

- *Recheck preload gap **B** (step 4). Excessive preload could cause window to shatter.*
- *The fully closed window must not contact the rain gutter when opening the door (in the event the "close with door shut" feature is faulty or the battery is dead).*

7. Use new mounting bolts when reinstalling the side-impact airbag to the door (where applicable). Reset SRS warning light and clear SRS fault memory using special test equipment. See **721 Airbag System (SRS)**.

8. Replace door panel, rubber door plugs and outside door molding.

9. Re-initialize window regulator motors as described earlier.

0013080

Fig. 25. Reach through door bottom to adjust window limit stops. Rear stop being adjusted.

Rear vent window or vent latch, replacing (2-door models)

1. Remove C-pillar (rear roof pillar) trim. See Fig. 26.

0013069

Fig. 26. C-pillar trim shown removed. Peel off window sealing trim and then unclip pillar trim piece. Unplug interior light socket (**arrow**).

2. Remove window latch screws at C-pillar. See Fig. 27.

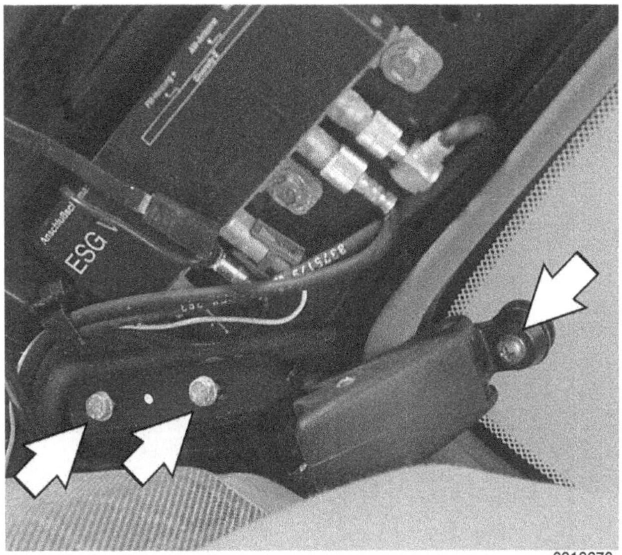

0013070

Fig. 27. Rear vent latch screws on C-pillar and on window (**arrows**).

3. Remove front seat belt sliding anchor trim at B-pillar (door post), and remove anchor nut. See Fig. 28.

0012672

Fig. 28. B-pillar seat belt anchor trim (**arrow**).

4. Pull off sliding anchor adjusting knob. Partially pull away door edge and vent window trim strips and remove top B-pillar plastic trim.

5. Remove seat belt sliding anchor. See Fig. 29.

6. While supporting window, remove its B-pillar mounting fasteners. See Fig. 30.

7. Installation is reverse of removal. Keep in mind the following:
   - Plastic sealing washers must be installed between glass and body on forward glass mounting studs.
   - Reinstall sealing washers on both sides of glass at rear latch attachment.

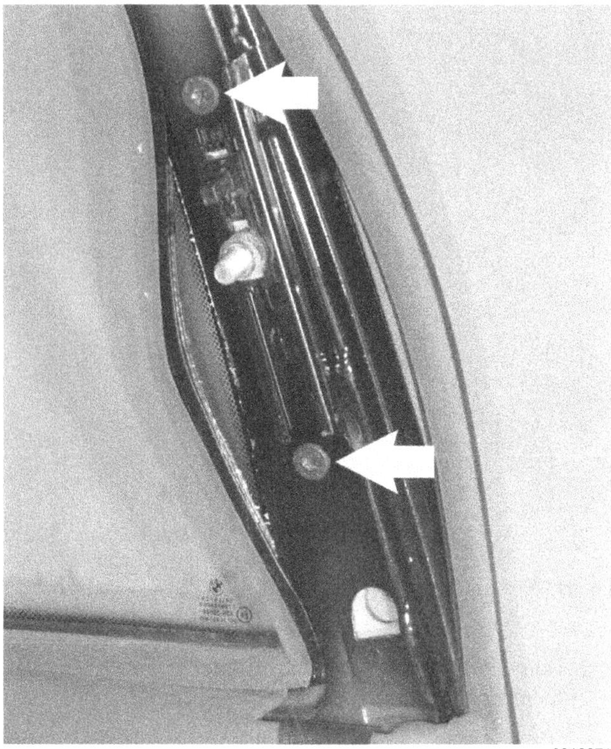

0013071

Fig. 29. Seat belt sliding anchor mounting bolts on B-pillar (**arrows**). Note door post and vent window trim strips peeled away.

0013072

Fig. 30. Vent window mounting nuts on B-pillar (**arrows**).

Tightening Torques

- Latch to vent window (M5 screw) . . . . . 6 Nm (53 in-lb)
- Seat belt anchor nut . . . . 31 Nm (23 ft-lb)
- Seat belt sliding anchor mounting bolts . . . . 24 Nm (17 ft-lb)
- Vent window to body (M6 nut) . . . . . 6.2 Nm (55 in-lb)

Window Regulator Service

Door window regulator and motor, removing and installing

This procedure applies to front or rear door windows.

1. Remove door panel and vapor barrier as described in **411 Doors**.

> *WARNING —*
> - *Some 1997 and all 1998 model year cars are equipped with front side-impact airbags in the front doors. Whenever servicing the doors or windows on cars with front side-impact airbags, always disconnect the negative (–) battery cable. See* **721 Airbag System (SRS)** *for cautions and procedures relating to the airbag system.*
> - *Prior to disconnecting the battery, read the battery disconnection cautions given at the front of this manual on page viii.*

2. Remove door glass as described earlier in this section.

3. Remove inside door handle. See Fig. 31.

4. Disconnect wiring connector to power window motor.

5. Remove regulator mounting bolts or drill out rivets. See Fig. 32.

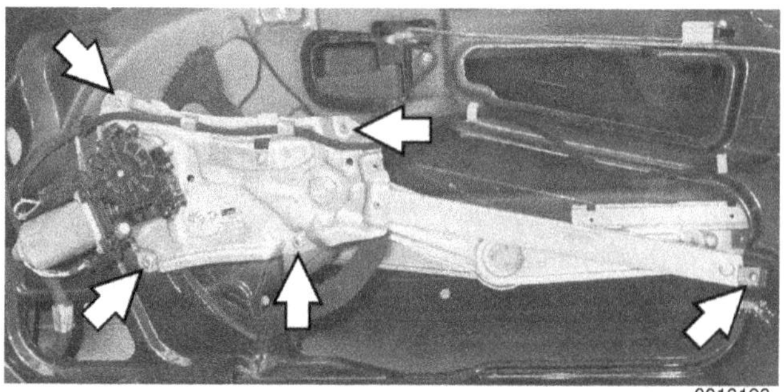

0013193

Fig. 32. Front window regulator mounting fasteners (**arrows**).

6. Remove regulator assembly from door.

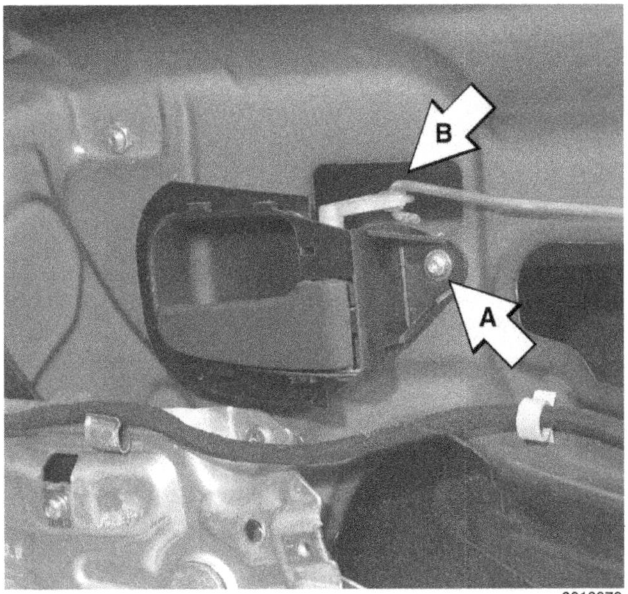

Fig. 31. Inside door handle. Remove screw **A**, then unhook from rod **B**.

7. Unbolt power window motor from rear of regulator.

NOTE —
Some of the window regulators are mounted to the doors with rivets. The rivets must be drilled out. During installation, M6 x 10 bolts and self-locking nuts can be substituted.

8. Installation is reverse of removal.

 • Adjust door window as described earlier.
 • Re-initialize window regulators as described earlier.

Tightening Torques
• Window regulator to door. . . . . 9 Nm (7 ft-lb)
• Window motor to regulator. . . . . 5 Nm (4 ft-lb)

513 Interior Trim

GENERAL

This repair group covers interior trim removal and installation procedures. For removal and installation of interior switches, including interior lighting equipment, see **612 Switches and Electrical Accessories**.

Most of the interior trim and finish panels are clipped or screwed into place. Many of the trim retaining clips are designed to be used only once. When removing trim that is held in place with clips, it is a good idea to have spares on hand before beginning the job.

CENTER CONSOLE

The front section center console houses the Multi-Information Display (MID) module, the front ashtray and cigarette lighter, shifter lever, power window switches and the hazard warning switch. The rear center console section houses the ashtray(s) and covers the emergency brake cable ends.

Center console, removing and installing

1. Disconnect negative (–) cable from battery and cover terminal with insulating material.

> ***CAUTION—***
> *Prior to disconnecting the battery, read the battery disconnection cautions given at the front of this manual on page viii.*

2. Remove shifter boot or selector lever cover. See **250 Gearshift Linkage**.

3. Remove rear ashtray retaining screws to take out ashtray. Remove rear console retaining screw under ashtray.

4. Remove Multi-Information Display module. See **620 Instruments**.

5. Remove hazard warning switch and power window switches by pushing up and out from below.

6. Remove lower left dashboard panel and glove compartment as described later in this repair group.

7. Remove retaining screw under hazard light switch. Pull rear console back and remove retaining nut at back of front console. See Fig. 1.

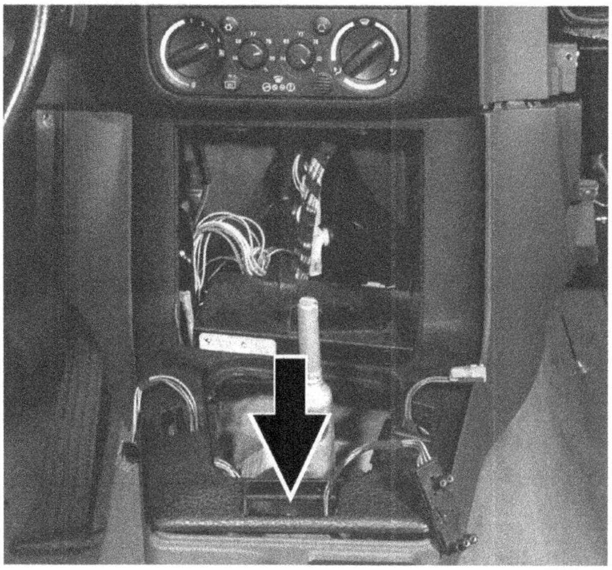

0013154

Fig. 1. Console retaining screw.

8. Take out rear console first, then front console.

513

DASHBOARD

Dashboard removal is a complex operation, involving disconnection and dismantling of a number of electrical and dash structural components.

The BMW E36 is equipped with SRS airbags mounted in the steering wheel, in the dashboard (1994 and later) on the passenger side, and in the front doors (1997 and later). Airbags units are pyrotechnic devices and should be treated with extreme caution. Improper handling of the airbags could cause serious injury.

> ***WARNING —***
> - *The BMW Supplemental Restraint System (SRS) is complex and special precautions must be observed when servicing. Serious injury may result if system service is attempted by persons unfamiliar with the BMW SRS and its approved service procedures. BMW specifies that all inspection and service should be performed by an authorized BMW dealer.*
> - *Before performing any work involving the steering wheel or interior trim in the vicinity of the airbags, disconnect the negative (–) battery cable. See* **721 Airbag System (SRS)**.
> - *SRS contains a back-up power supply within the SRS control module. A 10 minute discharge period should be allowed after the battery cable has been disconnected.*

Glove compartment, removing and installing

1. Open glove compartment door.
2. Remove glove compartment light and remove glove compartment mounting screws. See Fig. 2.
3. Pull glove compartment out, disconnecting electrical harness connectors as necessary.
4. Installation is reverse of removal.

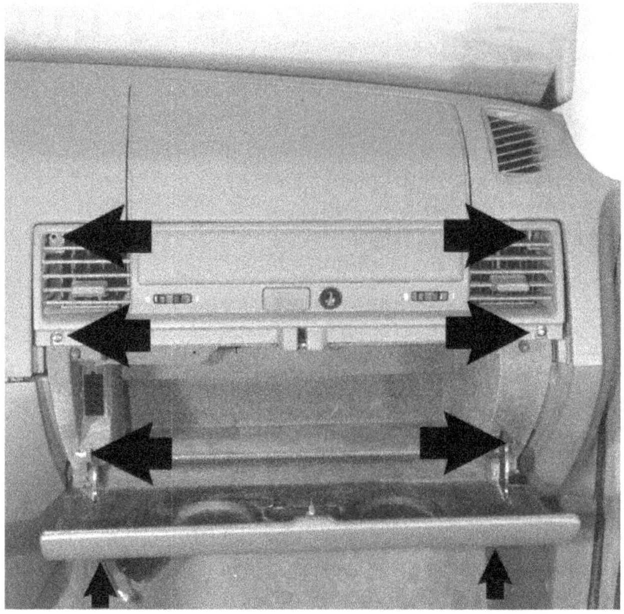

0013107

Fig. 2. Glove compartment mounting screws (**arrows**).

Lower left dash panel, removing and installing

1. Remove lower dash panel retaining screws. See Fig. 3.

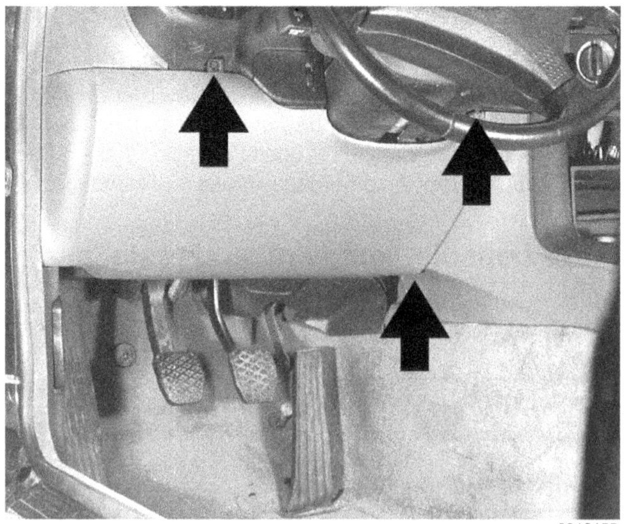

0013155

Fig. 3. Lower left dash panel retaining screws (**arrows**).

2. Pull lower dash panel left and back to disengage from driver's footwell retaining clips.
3. Installation is reverse of removal.

Dashboard, removing and installing

1. Disconnect negative (–) cable from battery and cover terminal with insulating material.

> ***CAUTION —***
> *Prior to disconnecting the battery, read the battery disconnection cautions given at the front of this manual on page viii.*

2. Remove center console as described earlier.

3. Remove glove compartment as described earlier.

4. Remove lower left dash trim piece as described above. Remove lower right dash trim panel. See Fig. 4.

0013158

Fig. 4. Lower right dash trim piece retaining screw (**arrow**).

5. Remove driver and passenger airbags. See **721 Airbag System (SRS)**.

> ***WARNING —***
> *Store the removed airbags in a safe place. Position the airbags facing up.*

6. Remove steering wheel. See **320 Steering and Wheel Alignment**.

7. Remove driver side knee bolster. See Fig. 5.

8. Remove instrument cluster. See **620 Instruments**.

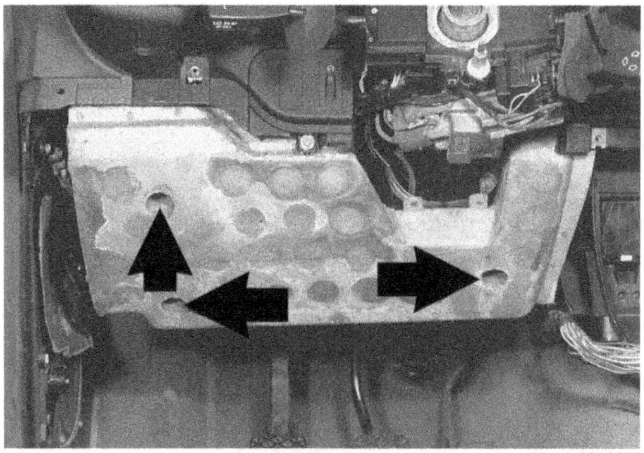

0012659

Fig. 5. Driver side knee bolster mounting bolts (**arrows**).

9. Remove headlight switch. Remove turn signal and windshield wiper/washer stalk switches. See **612 Switches and Electrical Accessories**.

10. Remove A/C control head and module. See **640 Heating and Air Conditioning**.

11. Remove radio. See **650 Radio**.

12. Remove A-pillar (windshield pillar) trim on left and right sides. See Fig. 6.

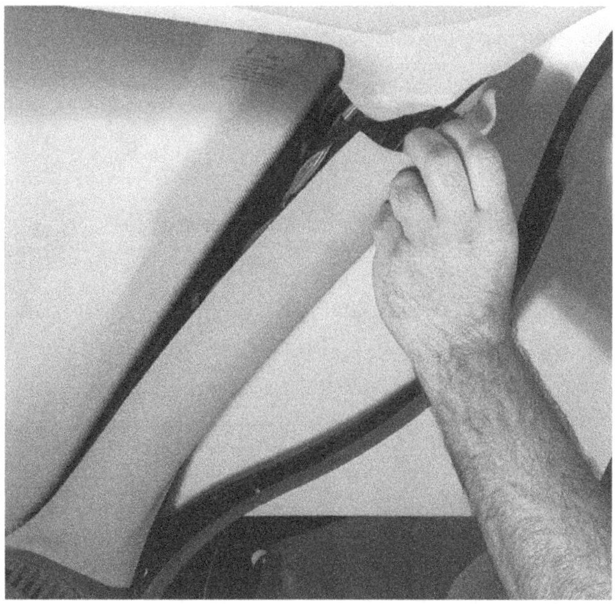

0013156

Fig. 6. A-pillar (windshield pillar) trim removal. Right side shown, left is similar.

13. Remove reinforcement bar below steering column and remove steering column shear bolts. Lower steering only slightly and support steering column. See Fig. 7.

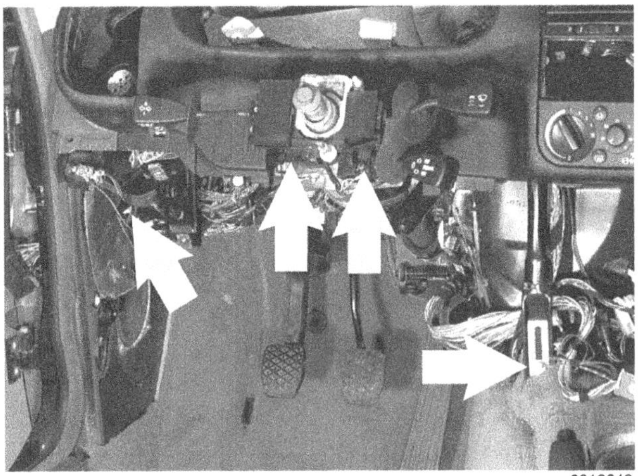

0012642

Fig. 7. Reinforcement bar bolts (**arrows**). Shear (headless) bolts at steering column will have to be drilled out and replaced during reinstallation.

> ***WARNING —***
> *Lower steering column only enough to facilitate removal of the dashboard. The steering column supports and swivel spindles can be damaged if the column is lowered too much.*

14. Remove glove compartment support frame. See Fig. 8.

0012643

Fig. 8. Glove compartment support frame bolts (**arrows**).

15. Carefully pry out defroster outlets (left and right sides) at base of windshield. Remove screw below each outlet. See Fig. 9.

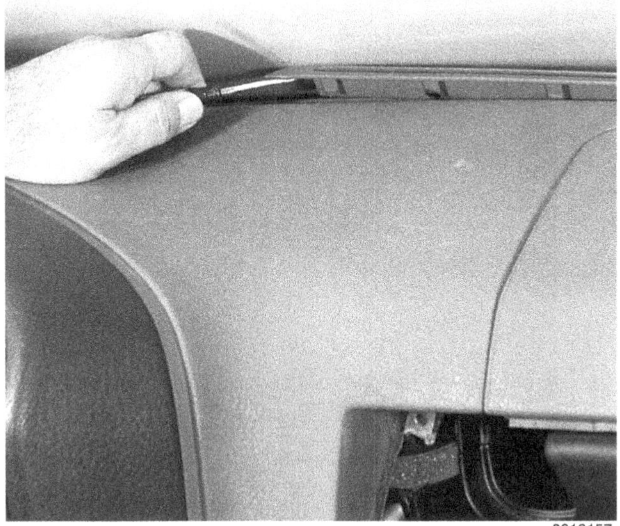

0013157

Fig. 9. Windshield defroster outlet being lifted off (**arrow**).

16. Remove mounting bolts on left and right ends of dashboard. See Fig. 10.

0013159

Fig. 10. Dashboard mounting bolt (**arrows**). Left side shown, right is similar. Some models may have may have two bolts on each side.

17. Lift instrument panel off slowly, making sure all harness connectors and wiring are disconnected.

18. Installation is reverse of removal.

515 Central Locking and Anti-Theft

GENERAL

This section covers repair information for the central locking and the anti-theft systems. Keep in mind that E36 cars are equipped with sophisticated and self-diagnostic electrical systems. When experiencing malfunctions relating to the central locking or the anti-theft system, it is recommended that the system be diagnosed using the BMW service tester (DIS). This advanced diagnostic tool can usually pinpoint electrical faults quickly and safely. Consult an authorized BMW dealer.

NOTE —
Additional general electrical information can be found in **610 Electrical Component Locations** *and* **Electrical Wiring Diagrams**.

DOOR HANDLES AND LOCKS

Front door lock, removing and installing

1. Remove front door panel and vapor barrier as described in **411 Doors**.

> ***WARNING —***
> *Some 1997 and all 1998 cars are fitted with side-impact airbags in the front doors. When servicing the door locks on cars with front side-impact airbags, always disconnect the negative (–) battery terminal. See* **721 Airbag System (SRS)** *for cautions and procedures relating to the airbag system.*

2. Raise door window completely.
3. On cars with side-impact airbags, disconnect negative (–) cable from battery, then remove airbag module.

> ***CAUTION —***
> *Prior to disconnecting the battery, read the battery disconnection cautions given at the front of this manual on page viii.*

NOTE —
In some models covered in this manual, it is not essential to remove the side-impact airbag to access internal door components. However, the procedure is recommended for safety reasons.

4. Disconnect harness connector from window motor to prevent accidental operation.
5. On 4-door cars: Peel back rubber window guide from rear guide rail. Remove guide rail mounting bolt at bottom. Unclip any wiring from guide rail and pull guide rail downward and remove from door. See Fig. 1.

NOTE —
The top of the guide rail is "hooked" to the door sheet metal.

6. Remove inside door handle retaining screw. Disconnect inside handle from linkage and remove. See Fig. 2.
7. Cut wire ties and remove door lock electrical harness connector. Remove door lock mounting screws. See Fig. 3.
8. Working inside door, disconnect outside door handle and lock cylinder linkage from door lock. Remove lock from inside door. See Fig. 4.

515

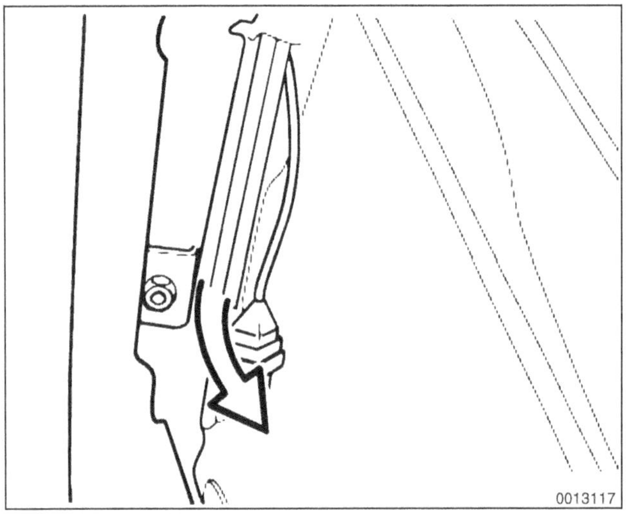

Fig. 1. 4-door cars: Remove lower retaining bolt at base of window rear guide rail and pivot rail out of door.

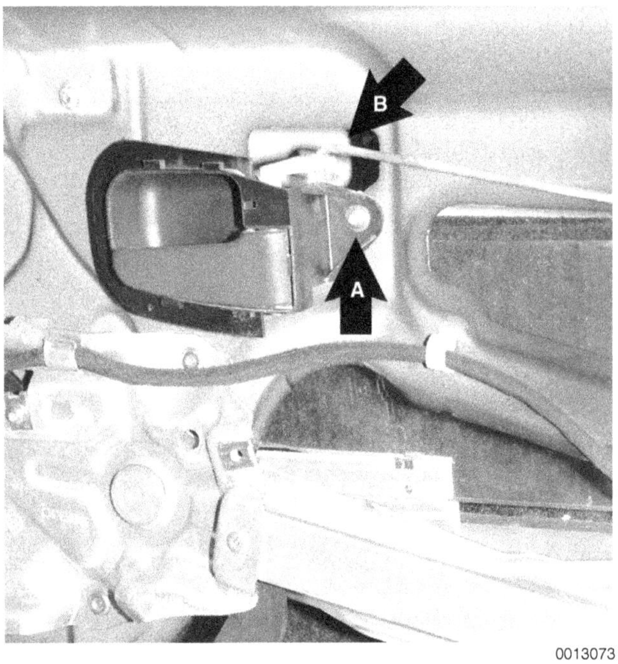

Fig. 2. Inside door handle retaining screw (**A**) Unhook linkage from handle at **B**.

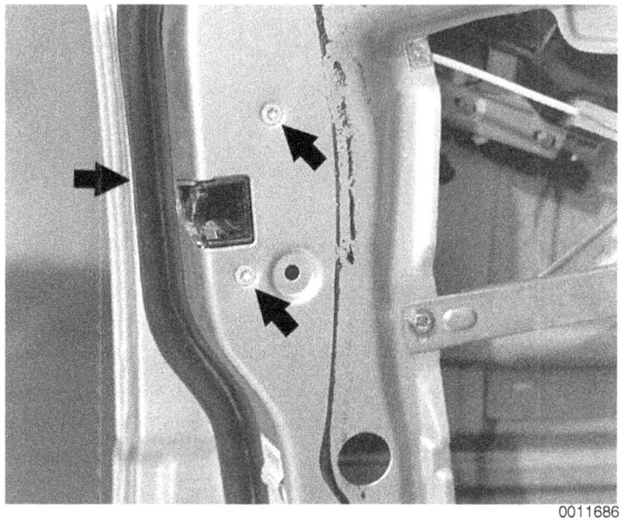

Fig. 3. Front door lock mounting screws (**arrows**).

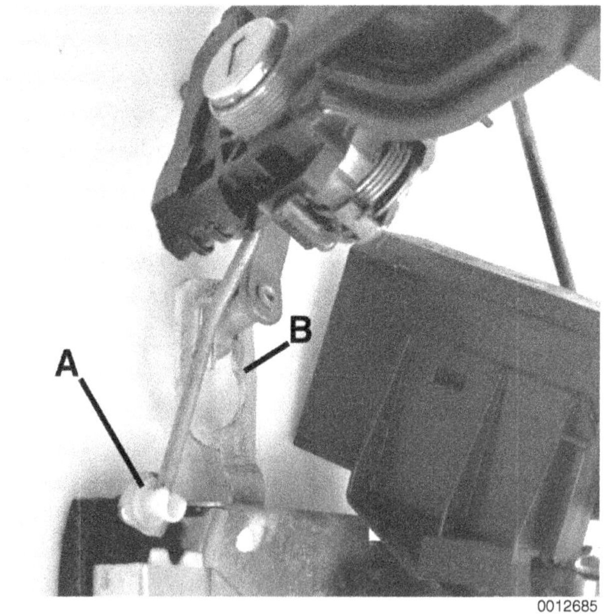

Fig. 4. Pry out lock linkage (**A**) from plastic retainer. Lift door lock mechanism slightly to remove latching pin from lock gate (**B**). Door lock mechanism shown removed from door.

9. Installation is reverse of removal, noting the following:
   - Install lock mounting bolts with Loctite® 270 or equivalent thread-locker.
   - Reconnect power window motor harness connector.
   - Use new mounting bolts when reinstalling the side-impact airbag to the door (where applicable).
   - Secure wire harnesses with wire ties.
   - Check function of door lock and window mechanism before reinstalling inner door panel.

Tightening Torques

- Door lock to door . . . . 9 Nm (80 in-lb)
- Window guide to door . . . . 10 Nm (89 in-lb)

Front door handle, removing and installing

1. Close door window completely.
2. Remove front door panel and vapor barrier as described in **411 Doors**.

> ***WARNING —***
> *Some 1997 and all 1998 cars are fitted with side-impact airbags in the front doors. When servicing the door handles on cars with front side-impact airbags, always disconnect the negative (–) battery terminal. See* **721 Airbag System (SRS)** *for cautions and procedures relating to the airbag system.*

3. On cars with side-impact airbags, disconnect negative (–) cable from battery, then remove airbag module.

> ***CAUTION —***
> *Prior to disconnecting the battery, read the battery disconnection cautions given at the front of this manual on page viii.*

> ***NOTE —***
> *In some models covered in this manual, it is not essential to remove the side-impact airbag to access internal door components. However, the procedure is recommended for safety reasons.*

4. Remove door lock as described earlier.
5. Remove window as described in **512 Door Windows**.

> ***CAUTION —***
> *On models with side-impact airbags, removing/installing the front window requires that the battery cable be reconnected with the side-impact airbag disconnected. This will set a fault code in the SRS control module, turn on the SRS warning light, and disable the SRS system until the fault memory is cleared using special SRS test equipment.*

6. Pry out access plug in end of door. Use screwdriver through access hole to slide exterior trim locking plate into released position. See Fig. 5. Remove door handle exterior trim piece.
7. Working inside door, remove locking clip for door handle retaining pin and push pin out.
8. Working inside door, disconnect electrical harness connector at door handle. Twist plastic connector bracket to remove from door handle.
9. Working at door exterior, remove handle collar nut while supporting handle. See Fig. 6.

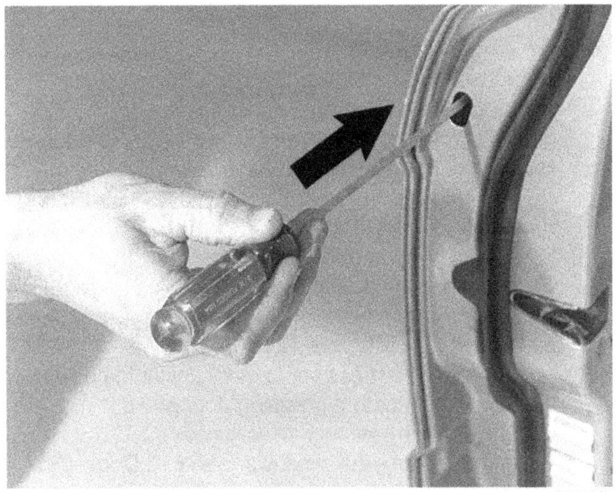

0011681

Fig. 5. Insert screwdriver into access hole and push locking plate forward (**arrow**) to release exterior handle trim.

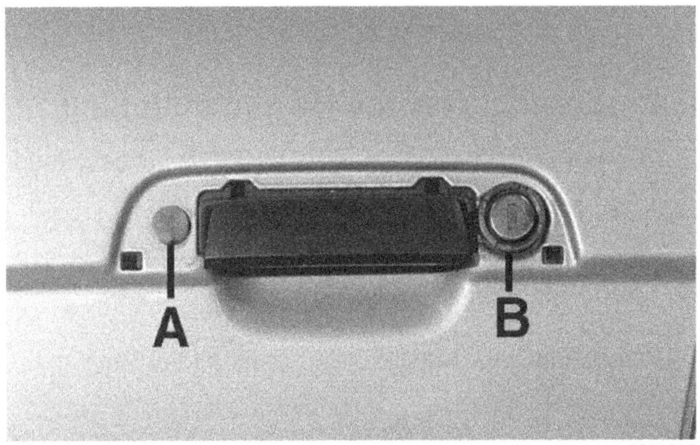

0011682

Fig. 6. Door handle retaining pin (**A**) and lock collar nut (**B**). Door handle trim shown removed.

10. Remove door handle from door by angling out of door cavity. Removed door handle assembly is shown in Fig. 7.

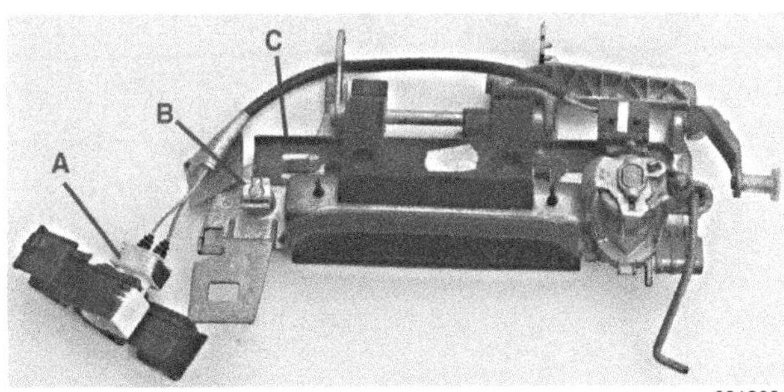

0012684

Fig. 7. Door handle harness connector (**A**); retaining pin locking clip (**B**); exterior trim locking plate (**C**). Handle assembly shown removed from door.

11. Installation is reverse of removal. Keep in mind the following:
    - Make sure rubber seal for handle is correctly positioned before tightening mounting fasteners.
    - Install exterior trim piece and then use a hooked tool to pull locking plate into locked position. Make sure door handle trim contacts body evenly before locking trim in place.
    - Use new mounting bolts when reinstalling the side-impact airbag to the door (where applicable). Reset SRS warning light and clear SRS fault memory using special test equipment. See **721 Airbag System (SRS)**.
    - Reconnect all harness connectors (including power window motor) before installing door panel.
    - Secure wire harnesses with wire ties.
    - Adjust and reinitialize windows. See **512 Door Windows**.

Tightening Torque
- Collar nut to door handle . . . . 10 Nm (89 in-lb)

Rear door lock, removing and installing

1. Remove rear door panel and vapor barrier as described in **411 Doors**.
2. Remove window from rear door as described in **512 Door Windows**.
3. Disconnect and remove inside door handle. See Fig. 8.

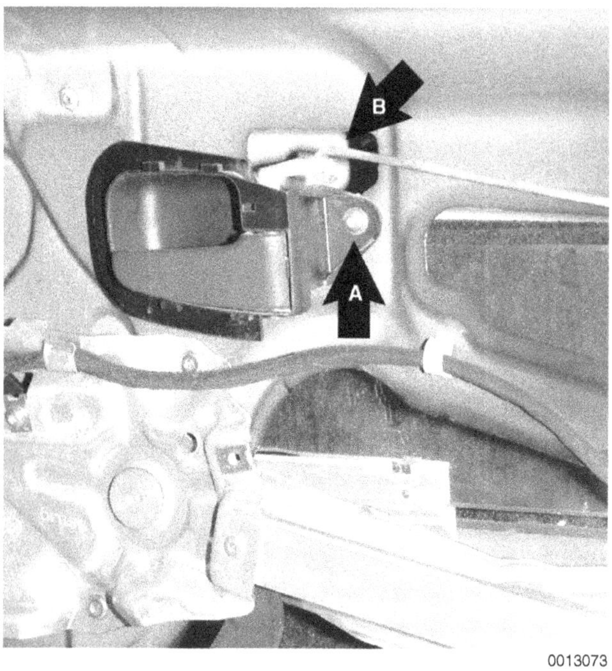

Fig. 8. Inside door handle retaining screw (**A**) Unhook linkage from handle at **B**.

4. Remove screws mounting bottom of window rear track to door. See Fig. 9.

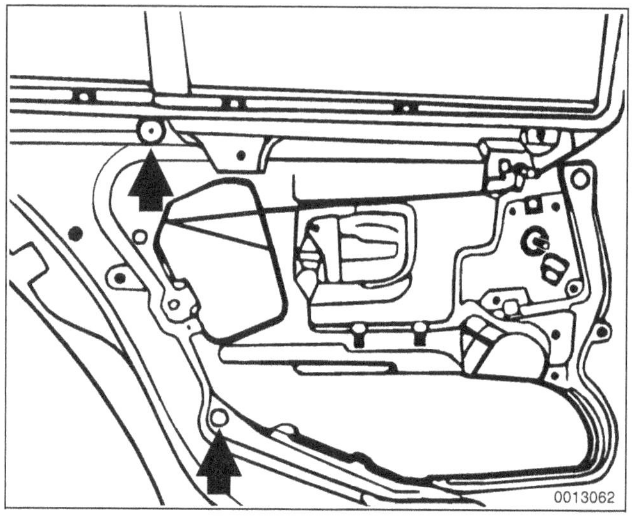

Fig. 9. Rear window track lower fasteners (**arrows**).

5. Remove window rear track top screw and move track aside. See Fig. 10.

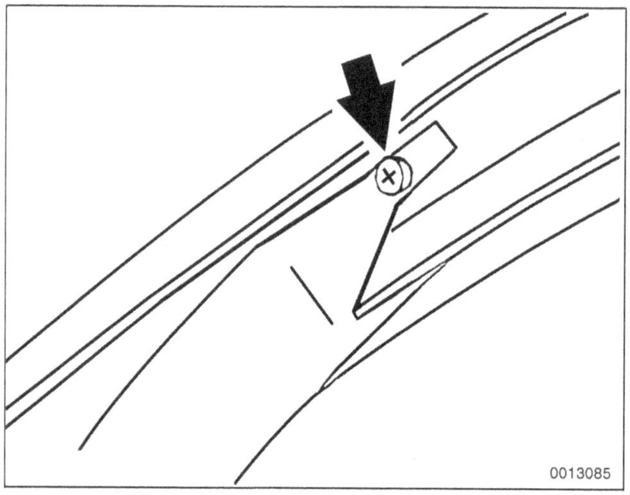

Fig. 10. Rear window track top screw (**arrow**).

6. Remove screw from lock button linkage rod retainer. Lift rod out of retaining clip. See Fig. 11.
7. Disconnect door lock harness connector.
8. Remove door lock screws. See Fig. 12.
9. Move lock up slightly to disengage from exterior handle linkage. Remove lock from inside door.

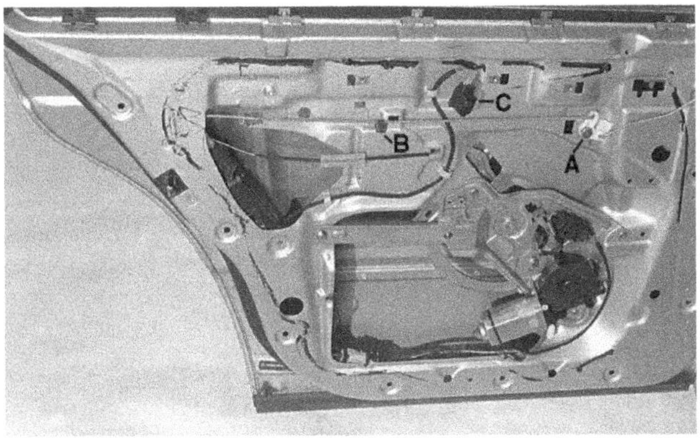

Fig. 11. Rear door lock button linkage rod retainer (**A**), rod retaining clip (**B**), and harness connector (**C**).

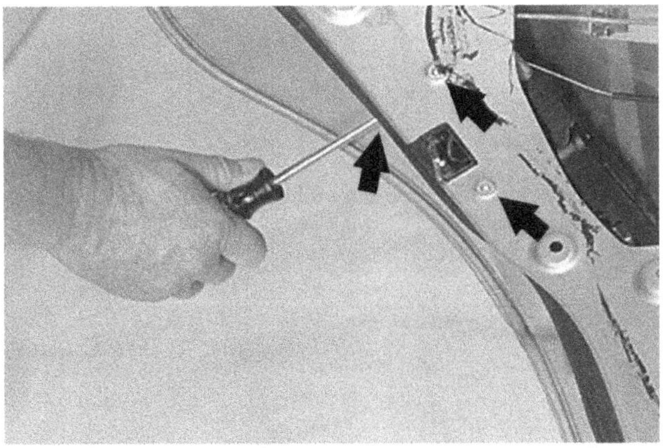

Fig. 12. Rear door lock mounting screws (**arrows**).

10. Installation is reverse of removal. Keep in mind the following:
    - Install lock mounting bolts with Loctite® 270 or equivalent thread-locker.
    - Reconnect power window motor harness connector before installing door panel.
    - Secure wire harnesses with wire ties.
    - Adjust window as described in **512 Door Windows**.

Tightening Torque

- Door lock to door . . . . 9 Nm (80 in-lb)

Rear door handle, removing and installing

1. Remove rear door panel and vapor barrier as described in **411 Doors**.
2. Remove window from rear door as described in **512 Door Windows**.
3. Remove door lock as described earlier.
4. Pry out access plug in end of door. Use screwdriver through access hole to slide exterior trim locking plate into released position. See Fig. 13. Remove door handle exterior trim piece.

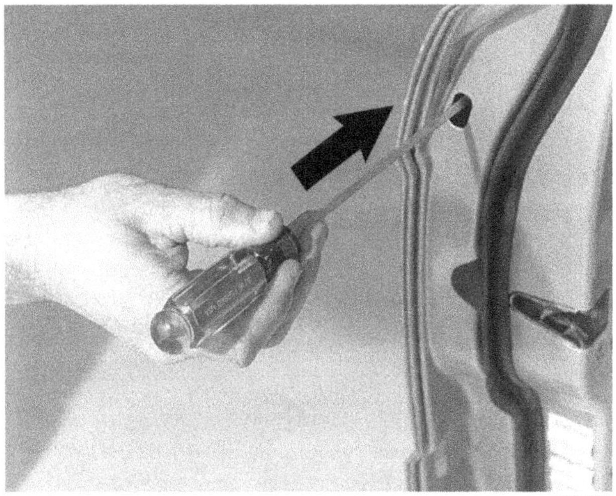

Fig. 13. Insert screwdriver into access hole and push locking plate forward (**arrow**) to release exterior handle trim.

5. Remove door handle mounting screws while supporting handle. See Fig. 14.

Fig. 14. Rear door handle mounting screws (**arrows**).

6. Remove door handle assembly from inside door.
7. Installation is reverse of removal. Keep in mind the following:
   - Make sure rubber seal for handle is correctly positioned before tightening mounting fasteners.
   - Install exterior trim piece and then use a hooked tool to pull locking plate into locked position. Make sure door handle trim contacts body evenly before locking trim in place.
   - Reconnect all harness connectors (including power window motor) before installing door panel.
   - Secure wire harnesses with wire ties.
   - Adjust window as described in **512 Door Windows**.

Tightening Torque
- Rear door handle mounting screws . . 10 Nm (89 in-lb)

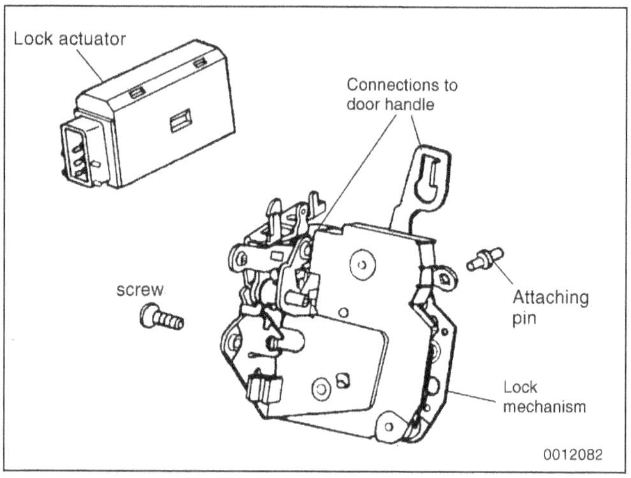

Fig. 15. Front door lock mechanism and related parts. Rear lock mechanism is similar.

Door striker, replacing

The door striker contains an electrical contact switch. If the switch is faulty, the striker must be replaced. To remove the striker, remove the mounting bolts. Disconnect the harness connector from the striker.

> ***NOTE—***
> *If the striker backup plate falls into the body cavity, it can be retrieved after removing the inside trim from the B-pillar (door-post pillar).*

Tightening Torque
- Striker plate to body pillar. . . . . 28 Nm (20 ft-lb)

LOCKING ACTUATORS

Door Locking Actuator

The front door lock mechanism and actuator are shown in Fig. 15. The rear door lock and actuator are similar. The lock must first be removed, as described earlier, to replace the actuator.

> ***NOTE—***
> - *If just one of the central locking drives does not operate, the locking linkage could be frozen or stiff enough to prevent actuation of the drive. Try remove the interior door panel and lubricating the locking linkage as the first step.*
> - *The actuator is not serviceable and must be replaced if defective.*

Luggage compartment locking actuator, removing and installing

1. Raise trunk lid. Remove tool kit and trunk lid trim panel. See Fig. 16.

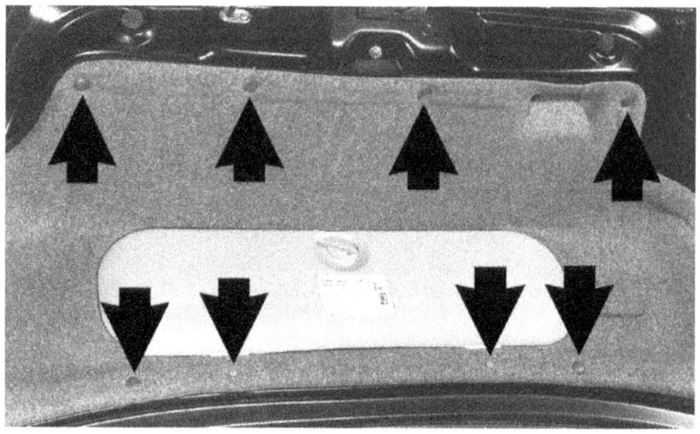

Fig. 16. Trunk lid trim and tool kit retaining screws and clips (**arrows**).

2. Disconnect actuating rod from lock. See Fig. 17.
3. Unplug two wiring connectors from actuator. See Fig. 18.

> ***NOTE—***
> *The actuator is adjustable. Matchmark the attaching screws before removal.*

4. Remove actuator attaching screws. Pull actuator backwards from behind metal shield to remove.
5. Installation is reverse of removal. Align actuator screws with matchmarks during installation.

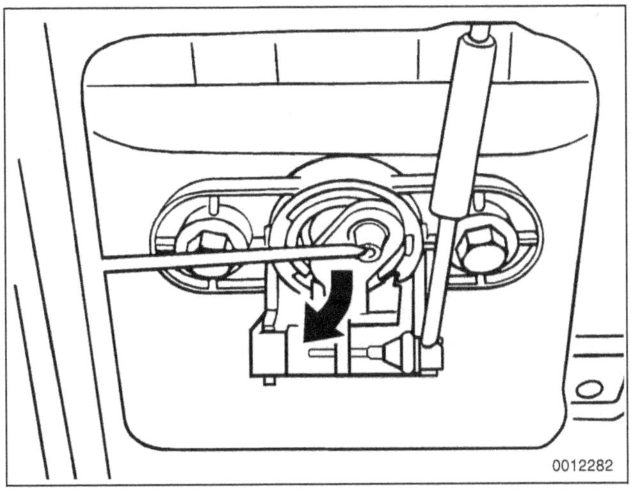

Fig. 17. Pull off luggage compartment lock actuating rod in the direction of **arrow**.

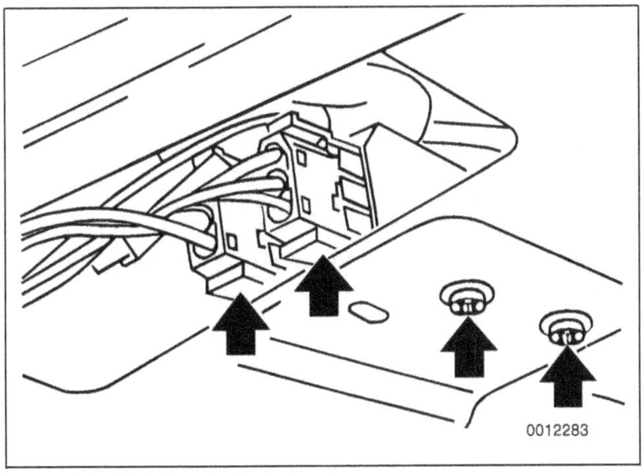

Fig. 18. Luggage compartment actuator wire connectors and attaching screws (**arrows**).

LOCKING SYSTEM AND ELECTRONIC IMMOBILIZATION

The central locking system uses electrical lock actuators at each door and at the trunk lid. The locking system can be actuated by either front door lock or by the luggage compartment lock. On some 1994 and later cars, an additional remote control key pad can also be used to activate the lock and anti-theft systems.

When the key is turned to the lock position, a micro-switch signals a central control module to lock all doors, luggage compartment, gas tank flap and, in later convertibles, the glove box lid. Where applicable, this also arms the anti-theft system.

NOTE —
The central locking system responds to accidents via an impact (inertia) switch. When this function is activated, the doors are automatically unlocked and the hazard warning lights and interior lights are turned on. The impact switches are mounted in the left and right footwells behind the speaker grilles.

Central Locking System (ZVM)

Different versions of the central locking control have been used in the E36 cars. In early production cars (model years 1992 and 1993), the locking system is controlled through the central locking module. This system is referred to as ZVM.

In this system, three microswitches are operated via the door lock cylinder at each front door. Turning the key approximately 45° (position **1**) operates the door locking microswitch. Turning the key approximately 90° (position **2**) actuates the double locking microswitch. Turning the key approximately 45° in the opposite direction through position **0** actuates the unlocking microswitch and deactivates the alarm system (position **3**). See Fig. 19.

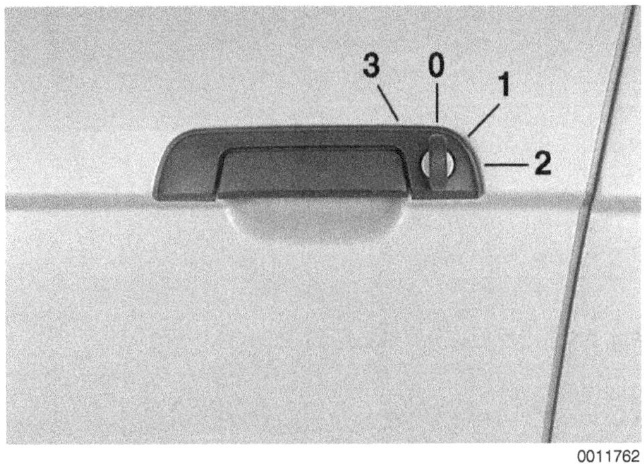

Fig. 19. ZVM door lock cylinder positions.

CAUTION —
Do not engage double locking position with passengers in the car unless the master key is available. The door cannot be opened from outside or inside without the master key.

NOTE —
In case of a discharged battery, the car can still be locked or unlocked using the key.

Central Body Electronics (ZKE IV)

Beginning with vehicles produced 9/93 (model year 1994), the central locking system was integrated with sunroof and window closure into the Central Body Electronics (ZKE IV) (generation four) system. The control module for this system is mounted in front of the glove compartment.

In this system two microswitches are operated via the door lock cylinder at both front doors. Turning the key approximately 45° (position **1**) operates the door locking microswitch and activates the alarm system. Holding the key in that position also closes any open windows and the sunroof.

Turning the key approximately 45° in the opposite direction through position **0** actuates the unlocking microswitch and deactivates the alarm system (position **2**). See Fig. 20.

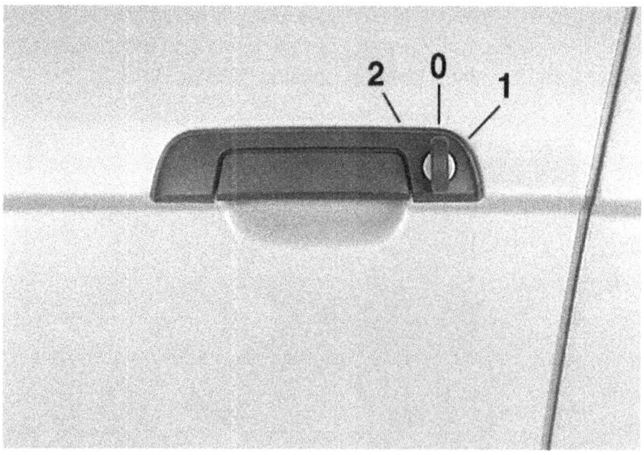

Fig. 20. ZKE IV door lock cylinder positions.

Electronic Immobilization System (EWS)

The electronic anti-theft system known as EWS was first introduced in January 1994. The early version of the system uses a starting inhibition module to interrupt the ignition, the fuel injection and the starter motor. This system is activated and deactivated by the central locking system. The control module for the system is installed under the left side of the dashboard.

Starting with models built since January 1995, E36 cars come equipped with a sophisticated coded electronic immobilization system called EWS II. For a schematic of the system, refer to Fig. 21.

NOTE —

The EWS II system is also sometimes referred to as the Driveaway Protection System or the electronic immobilization system.

On EWS II equipped cars, the ignition key is embedded with a computer chip and permanently encoded. A primary code is programmed into the key and into the vehicle itself. A secondary code is changed every time the vehicle is started. If the key code and EWS II control module code do not match, the engine management control module and the starter are disabled. EWS II ignition keys cannot be duplicated.

The system is designed to have up to ten keys and only an authorized BMW dealer can provide replacement keys.

NOTE —

It is possible to damage the electronic circuitry in the key, rendering it unusable. In that case, a new key should be purchased and initialized by an authorized BMW dealer.

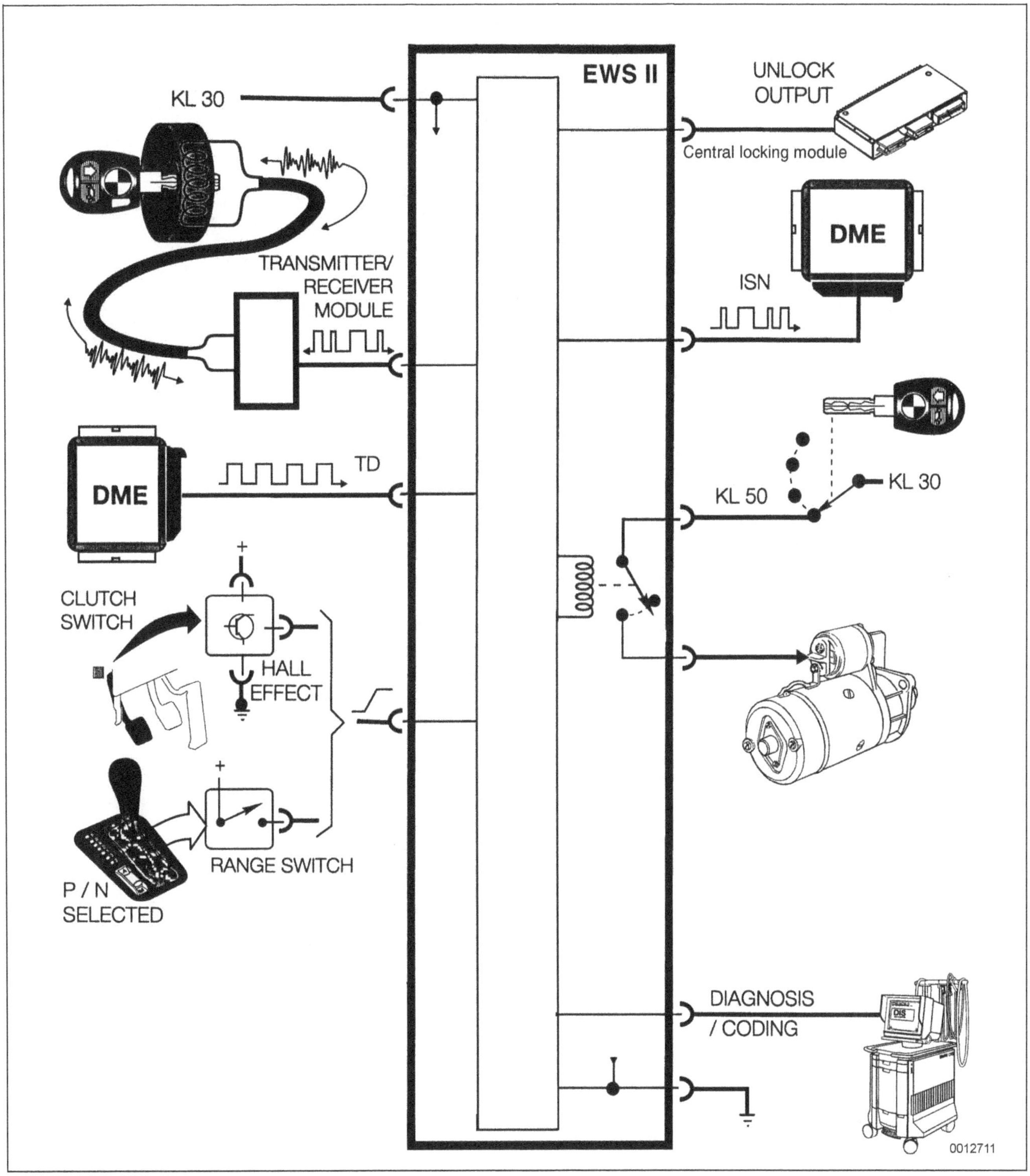

Fig. 21. EWS II system.

EWS II control module, replacing

1. Disconnect negative (–) cable from battery.

> *CAUTION—*
> *Prior to disconnecting the battery, read the battery disconnection cautions given at the front of this manual on page viii.*

2. Remove glove compartment as described in **513 Interior Trim**.

3. Remove EWS II control module harness connector. Pull module out of bracket. See Fig. 22.

NOTE—
The EWS II module should be identified with EWS II markings.

4. Installation is reverse of removal.

Ring antenna, replacing

1. Disconnect negative (–) cable from battery.

> *CAUTION—*
> *Prior to disconnecting the battery, read the battery disconnection cautions given at the front of this manual on page viii.*

2. Remove lower steering column cover. See Fig. 23.

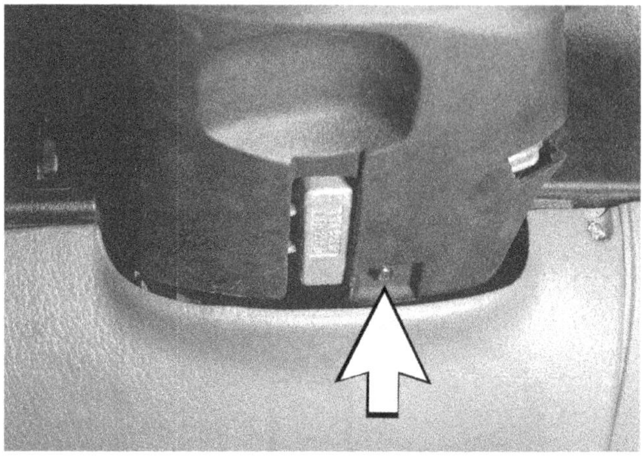

Fig. 23. Steering column cover retaining screw (**arrow**).

3. Remove lower left dash trim panel as described in **513 Interior Trim**.

4. Cut off wire ties holding wiring harness to steering column.

5. Gently pry off ring antenna from ignition switch.

6. Unplug ring antenna harness from EWS II transmitter/receiver module as described below.

7. Installation is reverse of removal.

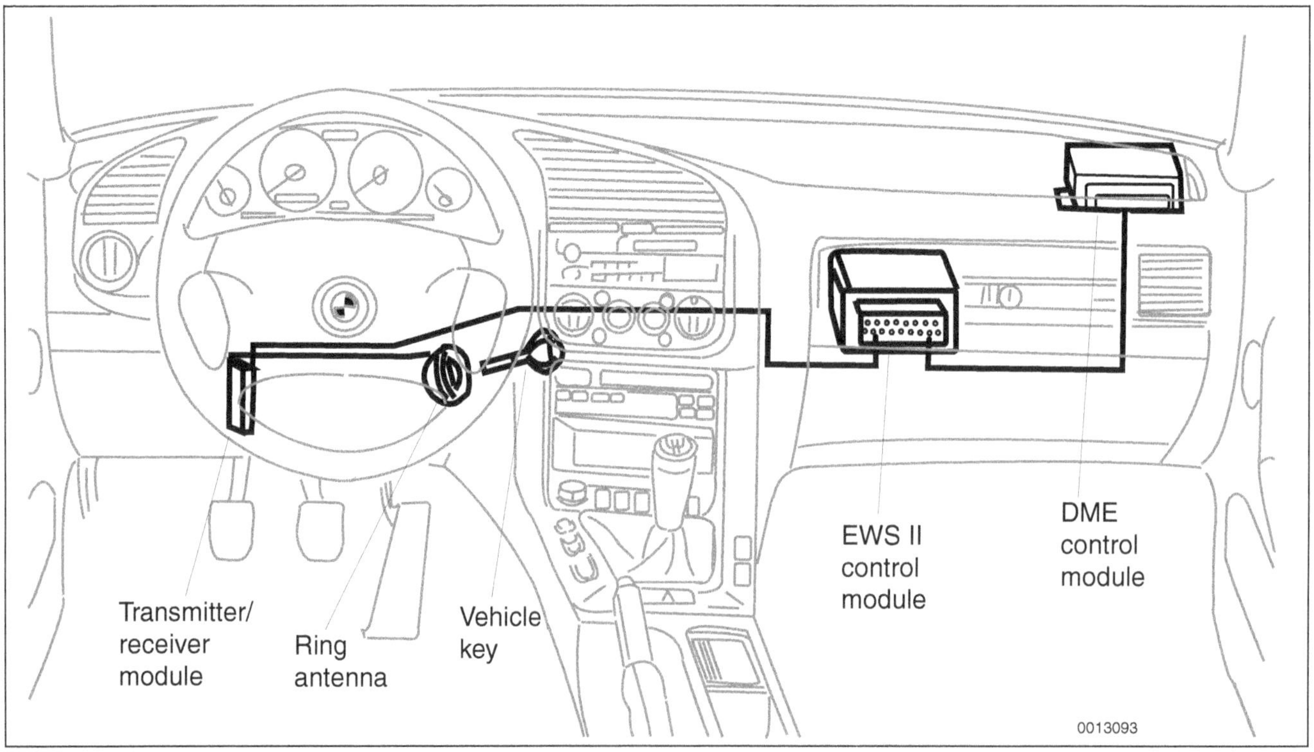

Fig. 22. Location of EWS II components.

EWS II transmitter/receiver module, replacing

1. Disconnect negative (–) cable from battery.

> ***CAUTION—***
> *Prior to disconnecting the battery, read the battery disconnection cautions given at the front of this manual on page viii.*

2. Remove lower left dash trim panel and knee bolster as described in **513 Interior Trim**.

3. Unclip auxiliary relay panel under steering column and lower it to access relays and harness connectors. See Fig. 24.

0013164

Fig. 24. Lowering the left side auxiliary relay panel.

4. Disconnect harness connector from transmitter/receiver module. Unclip module from panel. See Fig. 25.

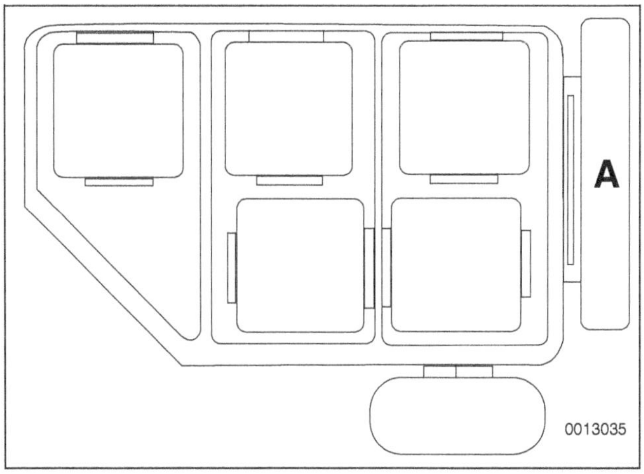

Fig. 25. Position of EWS II transmitter/receiver module (**A**) on left side auxiliary relay panel.

5. Installation is reverse of removal.

520 Seats

GENERAL

This repair group covers removal and installation of the front and rear seats.

FRONT SEATS

The front seats installed in E36 cars come in either manual or electrically powered versions. Cloth, vinyl and leather versions are offered. M3 sport seats have extra bolsters on the sides of the seat-back, a thigh support bolster, and extra controls for seat inclination.

Seat removal and installation is similar for all versions. Seat disassembly differs in details among the many models of seats.

Front seat, removing and installing

1. Move seat to full forward position and raise it as far as possible. Remove two mounting bolts at rear of seat rails. See Fig. 1.

2. Move seat to full rear position, keeping it raised.

3. Disconnect negative (–) cable from battery.

> ***CAUTION —***
> *Prior to disconnecting the battery, read the battery disconnection cautions given at the front of this manual on page viii.*

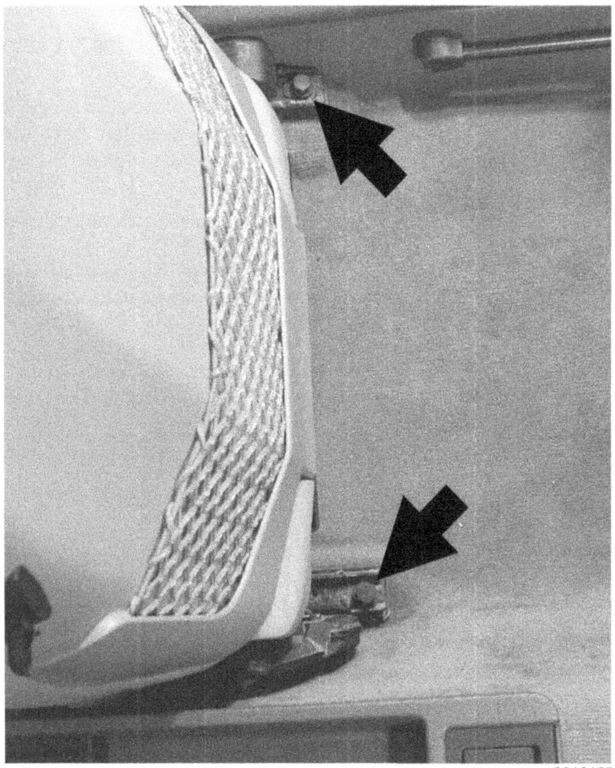

0013165

Fig. 1. Front seat rear mounting bolts (**arrows**).

4. Remove plastic caps from front mounting nuts, and then remove mounting nuts. See Fig. 2.

> ***WARNING —***
> *Deactivate front seat belt lock tensioners before proceeding to next step. See **720 Seat Belts**.*

NOTE —
Use a blanket to protect door sill from scuffing by seat rail during seat removal.

520

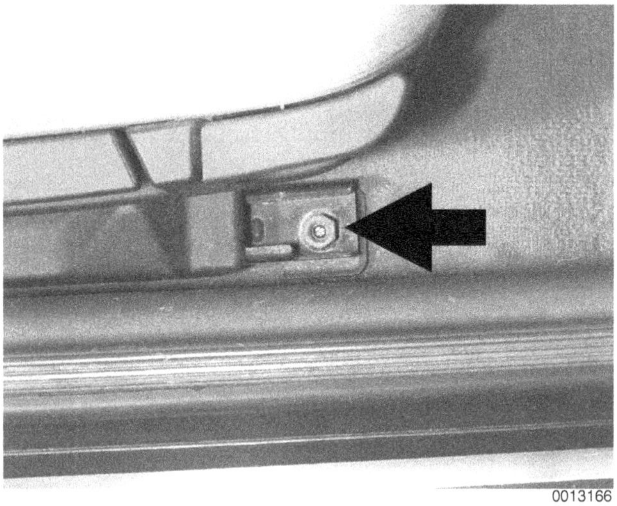

Fig. 2. Front seat front mounting nut (**arrow**).

5. On 2-door cars: unhook seat belt from guide on seat back.

6. Tilt seat back to access and unplug electrical harness connectors. Cut or untwist wire ties as necessary.

7. On 4-door cars: Unbolt seat belt from seat frame. See Fig. 3.

8. Remove seat from car.

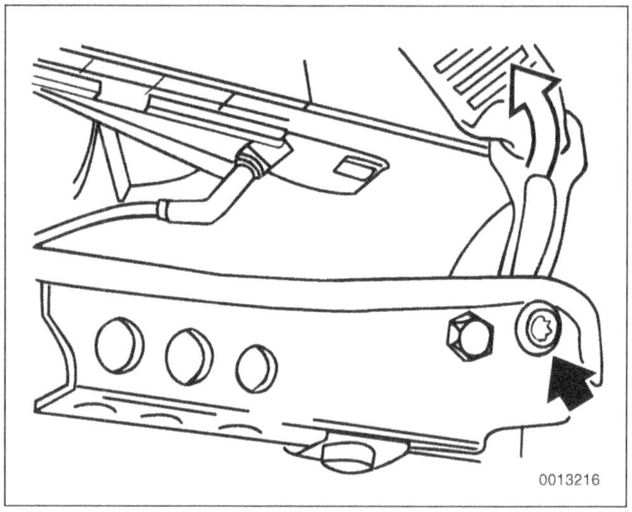

Fig. 3. Seat belt mounting bolt (**arrow**).

9. Installation is reverse of removal. Keep the following in mind:
   - Use wire ties or equivalent means to keep seat harness wiring from being exposed to fraying.
   - Reactivate front seat belt locks.
   - Start all seat mounting bolts.
   - Tighten seat mounting fasteners in this sequence: inner front; outer front; inner rear, outer rear.

Tightening Torque
- Front seat to floor . . . . 55 Nm (41 ft-lb)

Manual Front Seat Assembly

The non-power seat in E36 cars is shown disassembled in Fig. 4. Component replacement is possible once the seat has been removed from the vehicle as described earlier.

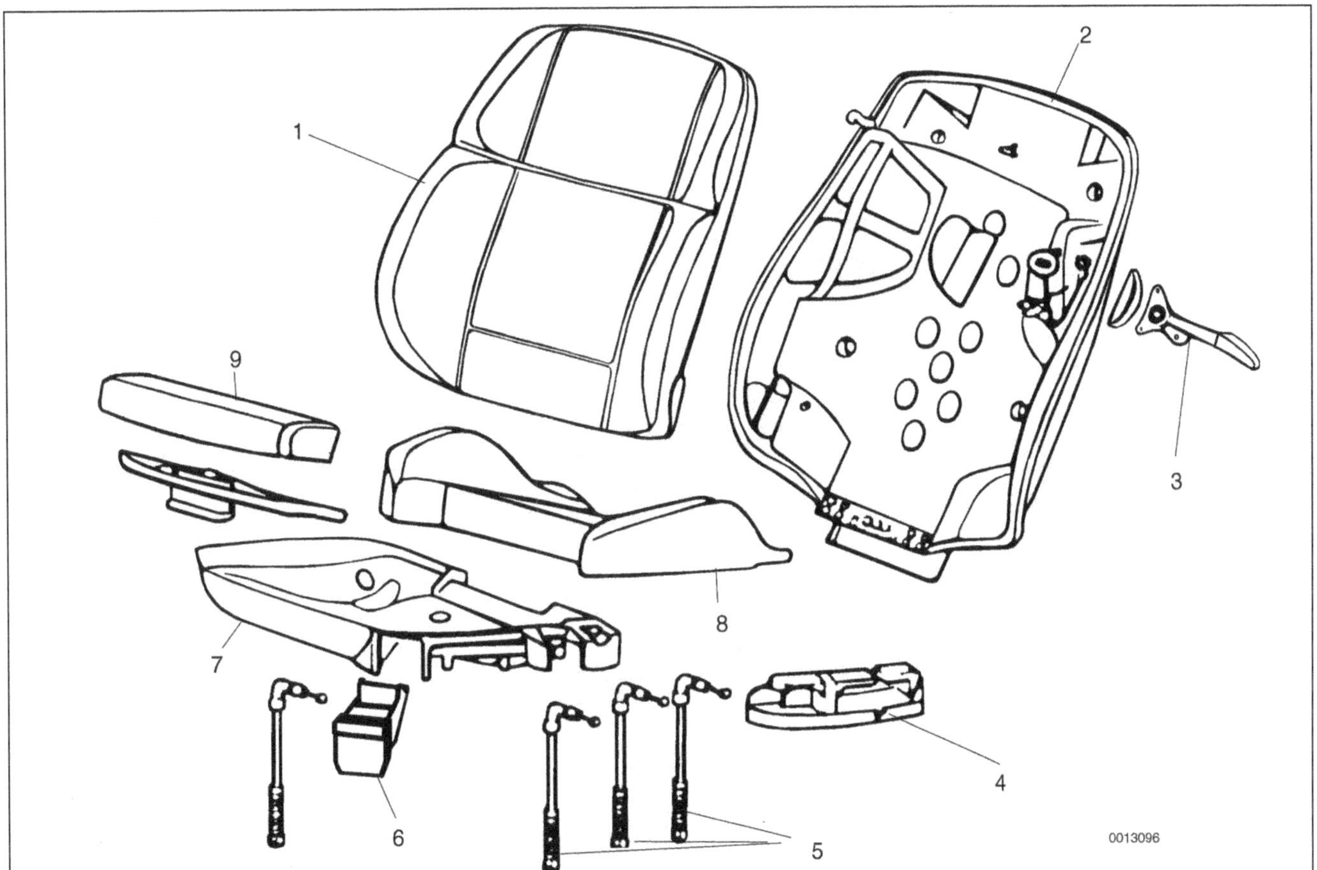

1. Backrest padding
2. Backrest frame
3. Backrest release lever (two-door)
4. Seat adjustment controls
5. Seat adjustment cables
6. Seat tilt control (M3 only)
7. Seat frame
8. Seat cushion
9. Knee bolster (M3 only)

Fig. 4. Construction of non-power front seat.

Power Front Seat Assembly

The power seat in E36 cars is shown disassembled in Fig. 5. Component, actuator and drive cable replacement is possible once the seat has been removed from the vehicle as described earlier.

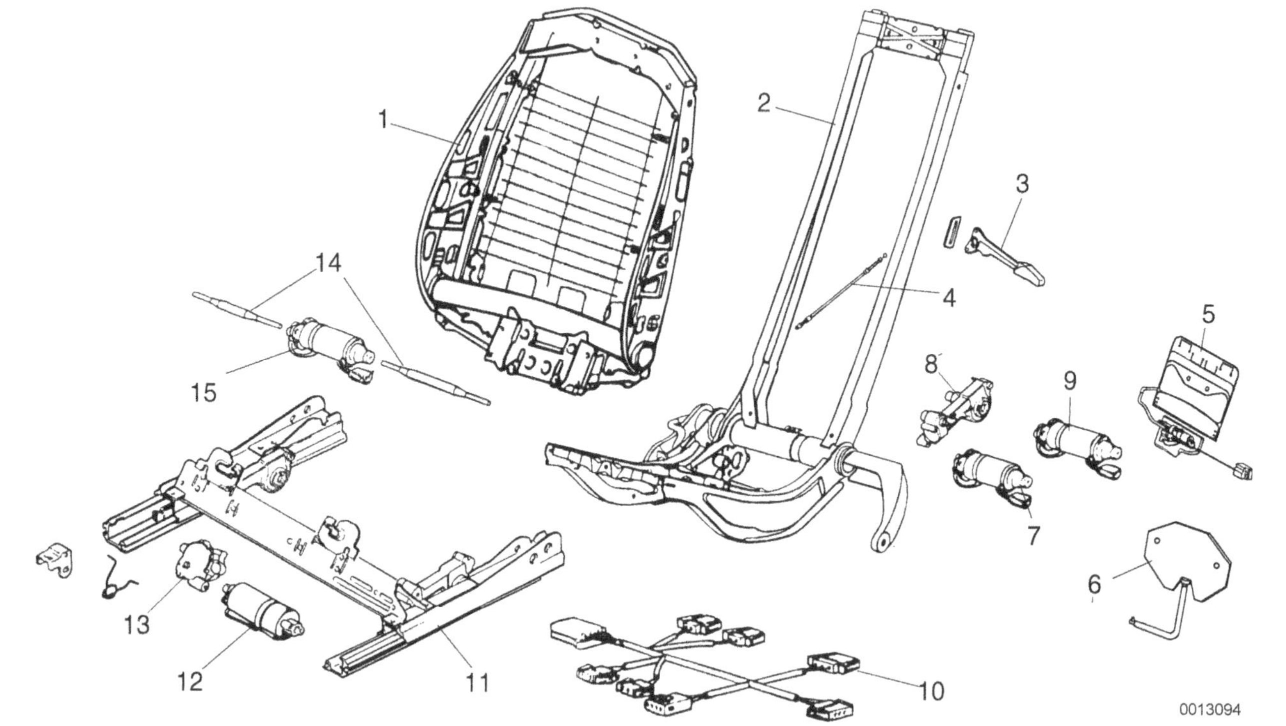

1. Backrest frame
2. Seat frame
3. Backrest release lever (two-door)
4. Backrest release lever cable
5. Lumbar support (from 9/94)
6. Lumbar support (to 9/94)
7. Vertical adjustment (rear)
8. Backrest adjustment gearbox
9. Backrest adjustment motor
10. Wiring harness
11. Seat track
12. Seat tilt motor
13. Tilt motor gearbox
14. Forward/backward driveshafts
15. Forward/backward motor

Fig. 5. Construction of power front seat.

REAR SEATS

Rear seat cushion and backrest, removing and installing (fixed seat back models)

1. Pull front of bottom cushion up and out of clips.

   NOTE —
   Remove rear headrests, if equipped.

2. Fold down center armrest and remove bolts from behind arm rest.

3. Slide seat back up off holders. Then slide to side to clear seat belts. See Fig. 6.

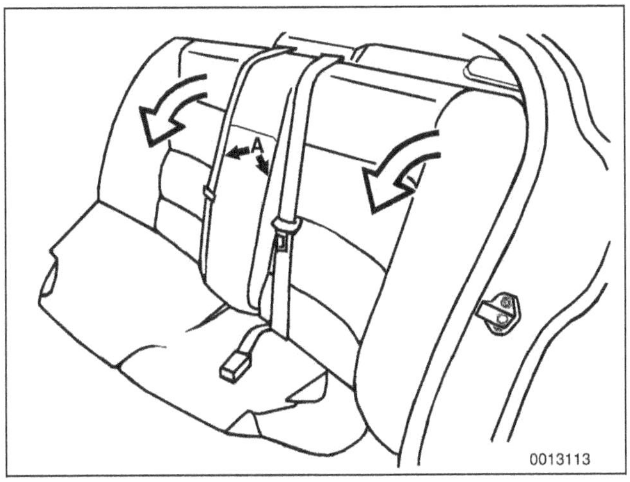

Fig. 6. To remove rear seat, remove bottom cushion. Remove bolts (**A**) behind center armrest. Slide seat back up off holders, then forward.

Rear seat cushion and backrest, removing and installing (fold-down seat back models)

1. Pull front of bottom cushion up and out of clips.

   NOTE —
   Remove rear headrests, if equipped.

2. Remove bottom seat belt anchor bolts. See Fig. 7.

   NOTE —
   Be sure to note relative positions of metal belt-end anchors in order to reassemble correctly.

0013167

Fig. 7. Rear seat belt anchor bolt under rear seat cushion (**arrow**).

3. Tilt backrests forward and unclip plastic cover. See Fig. 8.

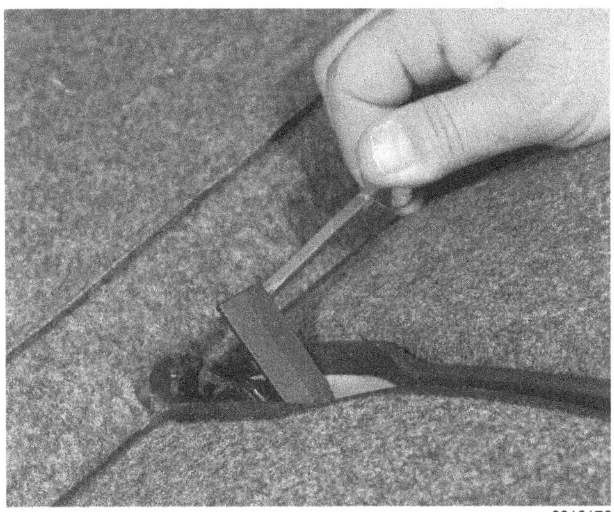

0013176

Fig. 8. Rear seat back mounting cover plate at base of seat back.

4. Push spring-loaded lock back and unclip seat back. See Fig. 9.

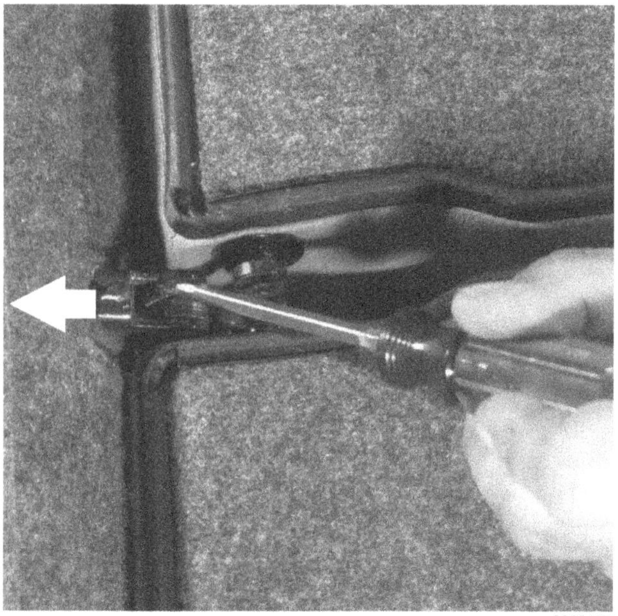

0013177

Fig. 9. Push spring-loaded clip back (**arrow**).

5. Installation is reverse of removal. Make sure backrest locks at top seat belt connection.

Rear seat backrest side section, removing and installing (fold-down seat back models)

1. Pull front of cushion up and out of clips.

2. Fold down rear seat backrest.

3. Lift backrest side section up and out. See Fig. 10.

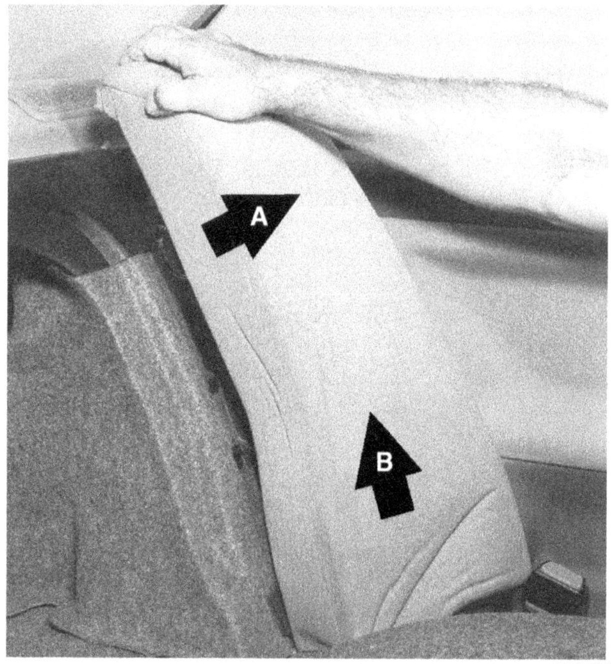

0013175

Fig. 10. Remove backrest side section by lifting out (**A**) and up (**B**).

4. Installation is reverse of removal.

540 Sunroof

GENERAL

The electrically-operated slide-tilt sunroof is controlled by a set of cables that move the sunroof panel along guide rails when the motor is operated. The sunroof can be adjusted without removing it from the car. Replacement of the sunroof liner or components such as the cable assembly require that the sunroof panel be removed.

The components of the slide-tilt sunroof are shown in Fig. 1.

NOTE —

- *Be sure to check the drains in the front corners of the sunroof carrier if water is entering the car through the headliner.*
- *Removal and repair of the sunroof and other components inside the roof cavity is beyond the scope of this manual.*

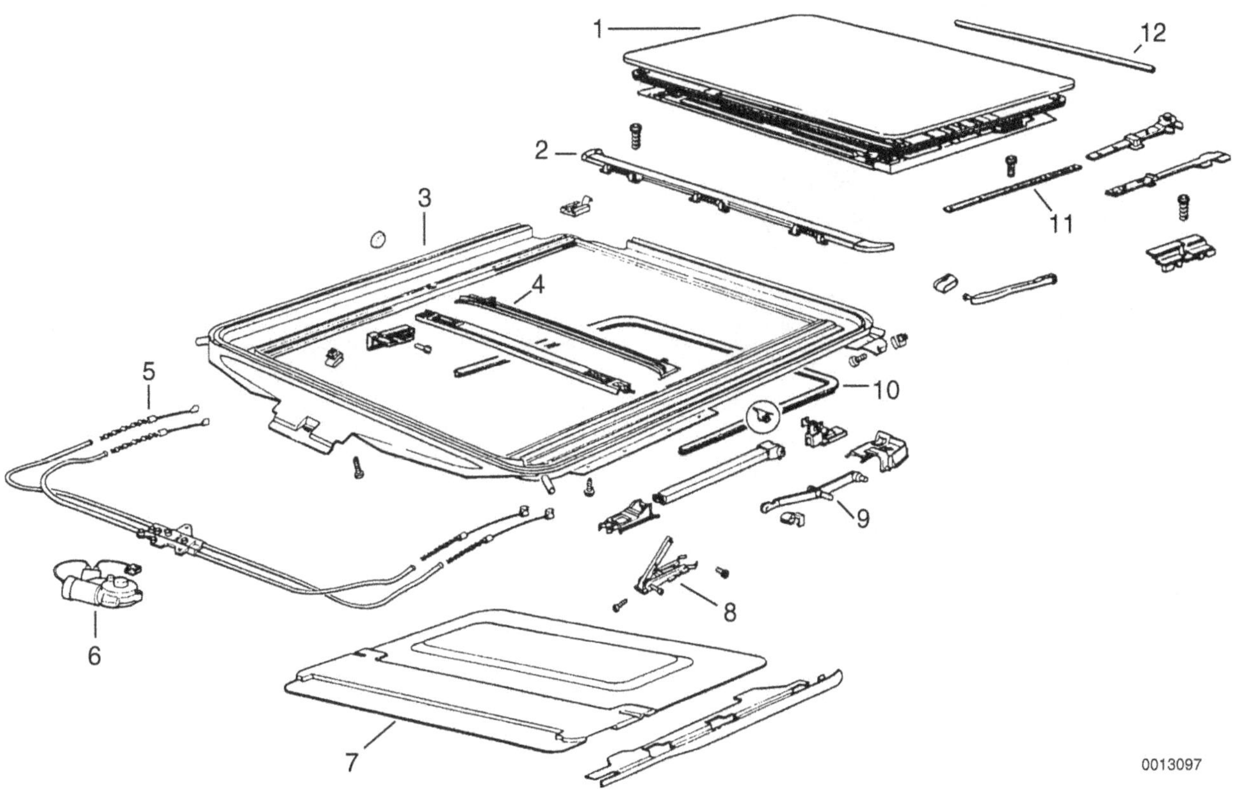

1. Sunroof panel
2. Wind deflector
3. Frame
4. Drain
5. Cable
6. Motor
7. Liner
8. Lever
9. Lever
10. Seal
11. Center rail
12. Seal

Fig. 1. Slide-tilt sunroof assembly.

SLIDE-TILT SUNROOF

Sunroof panel, emergency closing

1. Remove access panel from below sunroof motor. See Fig. 2.

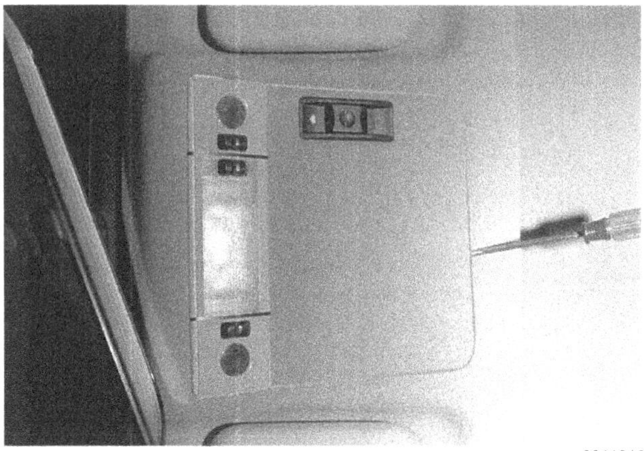

0011816

Fig. 2. Sunroof motor access cover being pried off.

2. Insert hex key into drive in sunroof motor and turn hex key to close panel. See Fig. 3.

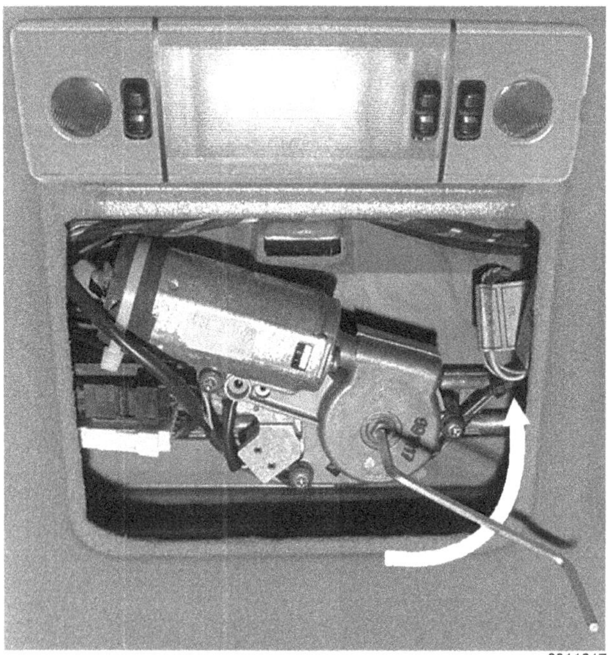

0011817

Fig. 3. Use hex key to manually close sunroof. Turn hex key to manually close sunroof.

NOTE —
The hex key for manually (emergency) closing the sunroof can be found in the tool kit located in the luggage compartment.

Sunroof panel, adjusting

The sunroof panel should be adjusted whenever the top of the closed sunroof becomes misaligned with the roof of the car, if it does not close squarely, if there are wind noises at speed, if there are water leaks, or if the sunroof has been removed.

1. Open sunroof slightly. Rear of sunroof should be 2 to 3 mm below surface of roof.

2. Slide sunroof liner slowly and carefully back into roof cavity as far as possible. See Fig. 4.

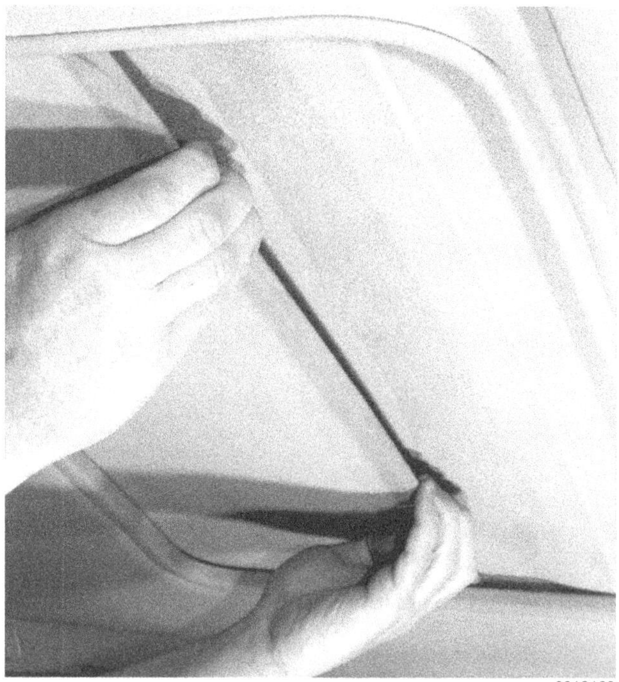

0013109

Fig. 4. Slide sunroof liner back into roof cavity.

3. Shut sunroof fully.

CAUTION —
Do not open sunroof with liner in this position. Damage to sunroof mechanism could result.

4. Rear adjustment: Loosen rear adjusting wedges and push in or out until rear of sunroof is aligned correctly with roof-top. See Fig. 5.

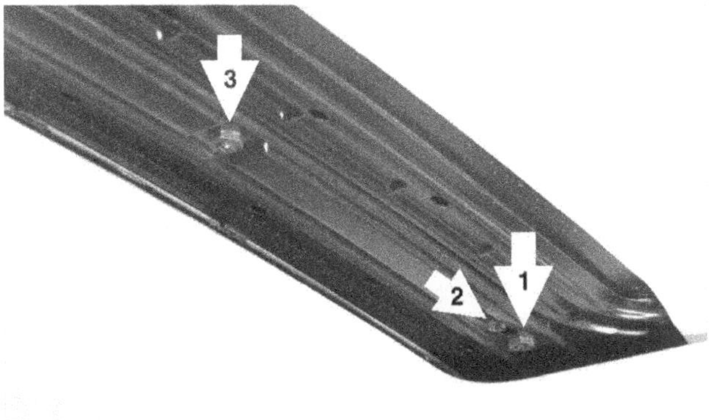

0013103

Fig. 5. Adjust rear of sunroof by loosening nuts **1** and **3**. Push adjusting wedges **2** in or out to achieve correct height.

> ***NOTE —***
> - *Push sunroof forward before retightening wedge fasteners.*
> - *Use Loctite® 270 or an equivalent thread sealing compound when reinstalling the retaining fasteners. Alternatively, install new fasteners.*

5. Front adjustment: Loosen retaining screws, then push front of sunroof panel up or down until it is aligned correctly with roof-top. Retighten screws. See Fig. 6.

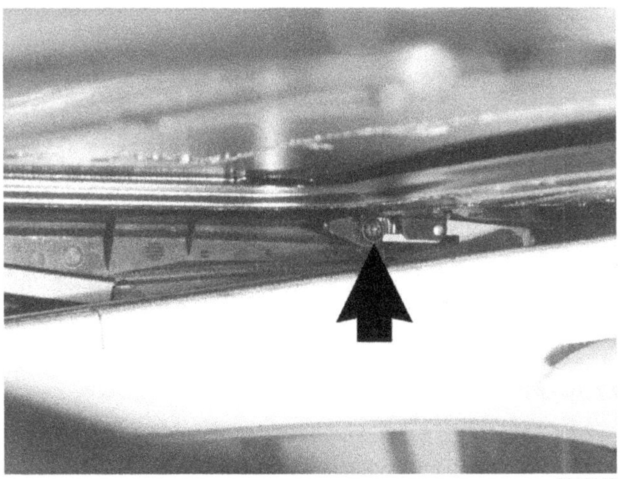

0013170

Fig. 6. Adjust front of sunroof by loosening screw (**arrow**). Push sunroof up or down to achieve correct height.

> ***NOTE —***
> *Use Loctite® 270 or an equivalent thread sealing compound when reinstalling the retaining screws. Alternatively, install new screws.*

6. Open sunroof slightly. Rear of sunroof should be 2 to 3 mm below surface of roof.

7. Slide sunroof liner slowly and carefully forward out of roof cavity as far as possible.

Sunroof Height Adjustment Specifications
- Rear . . . . flush/1 mm higher than roof top
- Front . . . . .flush/1 mm lower than roof top

Tightening Torque
- Front or rear adjusting fasteners. . . . . . 6 Nm (53 in-lb)

Sunroof motor, removing and installing

The sunroof should be in the closed position before removing or installing the sunroof motor. If necessary, manually close the sunroof panel as described under **Sunroof panel, emergency closing**.

1. With ignition switch off, remove access panel from below sunroof motor. Refer to Fig. 2.

2. Remove sunroof motor mounting screws. See Fig. 7.

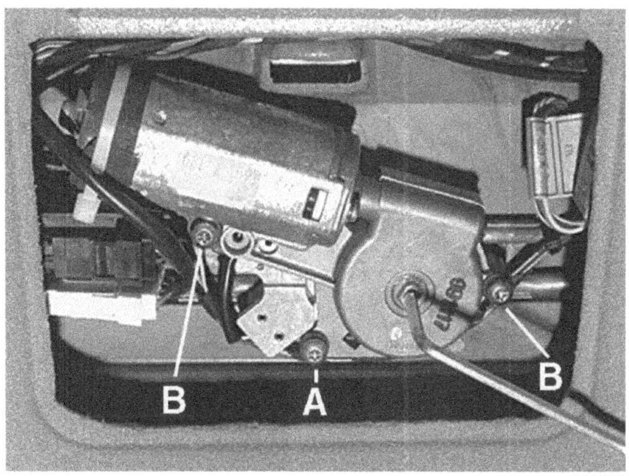

0011825

Fig. 7. Sunroof motor mounting screws (**A** and **B**). Screw at **A** is longer than screws at **B**.

3. Disconnect harness connector from motor and remove motor.

4. Installation is reverse of removal. Be sure the motor is in the "closed" position before installing it to the sunroof carrier. See Fig. 8.

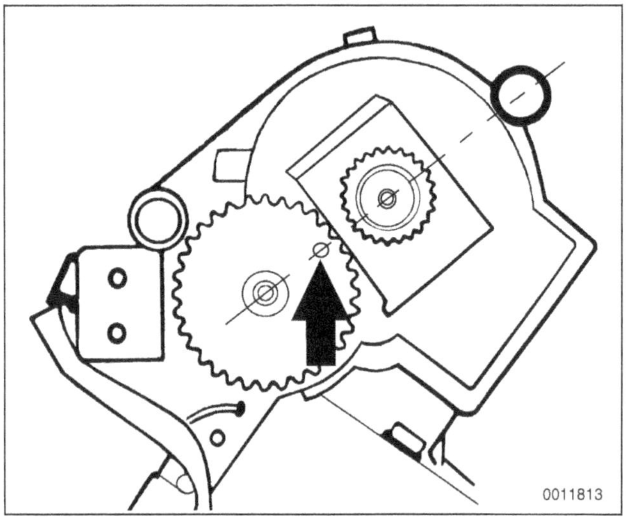

Fig. 8. Sunroof motor in "closed" position. Hole (**arrow**) in large gear aligns with shafts on both gears (to mid-1994).

NOTE —
On cars manufactured from mid-1994, the sunroof motor gear assembly is encased and cannot be seen. Before installing the motor in the roof, connect electrical harness connector, operate switch to rear until motor stops, then to front until motor stops. Then install motor in roof.

Tightening Torque
- Sunroof motor to motor carrier . . . . . 2.8 Nm (25 in-lb)

541 Convertible Top

GENERAL

The convertible top comes in three versions: a manually operated top, a semi-automatic power top and a fully automatic power top.

> ***NOTE—***
> *The convertible top control module (CVM) is located behind the left rear side trim panel. For the location of other convertible top electric components see* **610 Electrical Component Locations.**

The rear window in the convertible models is made of a strong, flexible vinyl which is available separately. Replacement of the rear window is not covered in this manual.

BMW has designed a special battery tray for the convertible which, in conjunction with the mass of the battery in the luggage compartment, serves to dampen convertible body flexing and vibration.

> ***CAUTION—***
> *The E36 convertible requires a special battery which is designed for constant vibration. A battery not designed for this will fail much earlier.*

MANUAL CONVERTIBLE TOP

Visor Latch Plate

The left and right visor latch plates can be accessed by removing the trim and moldings around the sides and top of the windshield. Each latch is secured to the top of the windshield using three bolts. Latch position is adjusted as follows:

- Loosen bolts
- Latch top
- Align top
- Retighten bolts. See Fig. 1.

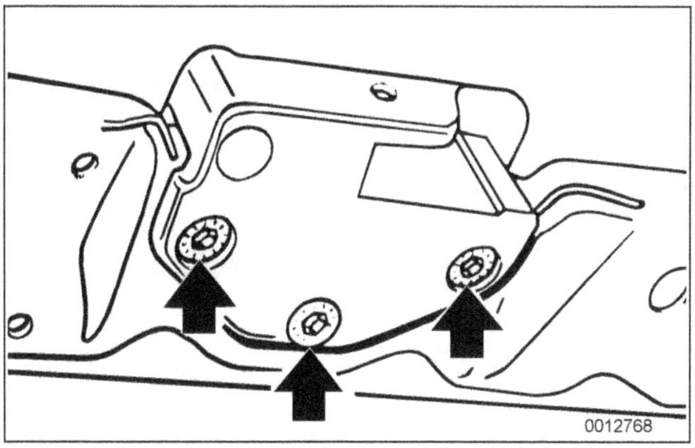

Fig. 1. Visor latch plate is held to top of windshield with three bolts (**arrows**).

Visor Latch

Check for ease of release and locking of convertible top front latches. Height of a latch can be changed by using the adjusting screw. To change latch tension, it is necessary to remove the trim on the front bow of the top. The ball-joint pull-rod length can then be adjusted. Lubricate latch if necessary. See Fig. 2.

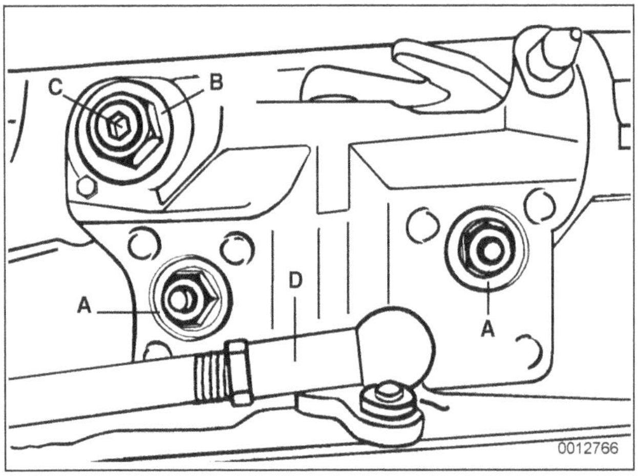

Fig. 2. Manually operated front latch. Loosen mounting nuts (**A**) and locknut (**B**), then adjust latch height with hex-bolt (**C**). (Fully automatic front latch height adjuster is similar.) Latch tension is adjusted at rod ball-joint end (**D**).

Convertible Top Lid Release Mechanism

The convertible lid release handle is situated in the trim panel behind the driver's seat. To remove the handle, pull its plastic insert forward and remove the trim insert. The handle is mounted with a single screw.

A bowden cable is routed from the handle to the left lid latch. Another cable is routed from the left to the right latch inside the luggage compartment. It is accessed by pulling down the trim lining off the front wall of the compartment.

Convertible top and frame, replacing

1. Open convertible top partially. Open convertible top lid. Open back windows.
2. Remove fasteners in compartments behind door posts and lift out complete top. Count number of shims on horizontal mounting bolts. See Fig. 3.
3. Installation is reverse of removal. Be sure to replace U-shaped shims on horizontal mounting bolts. See Fig. 4.
4. To adjust position and fit of top, remove or add U-shaped shims until top is properly aligned. Tighten nuts and screws.

POWER CONVERTIBLE TOP

The semi-automatic top, installed in 1996 convertibles, must first be manually released from the visor latch plates and pushed back past the "tension point," whereupon, the automatic retracting motor and linkage take over and pull it into the storage compartment behind the back seat.

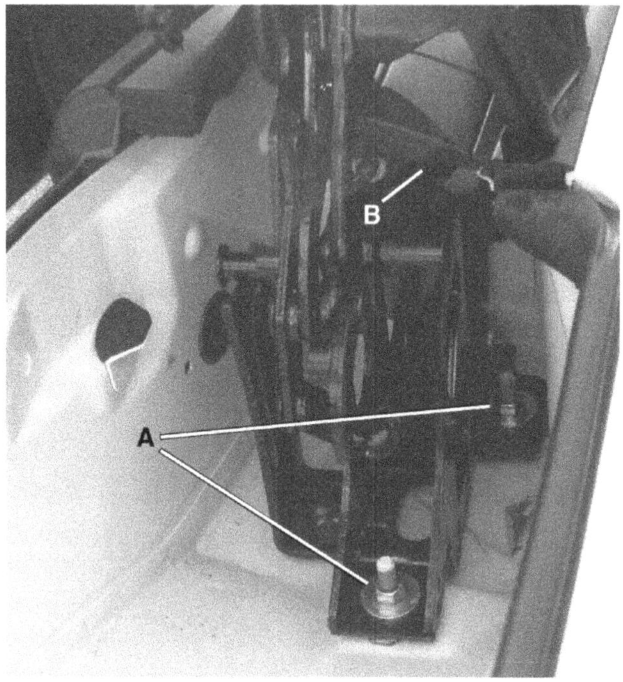

0012752

Fig. 3. Mounting nuts (**A**) for convertible top (two on each side). There is also a horizontal bolt with spacers on each side (**B**).

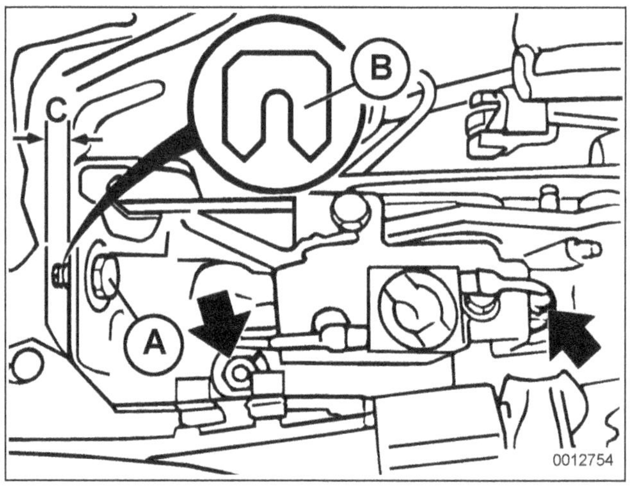

Fig. 4. Horizontal positioning bolt (**A**) and U-shaped adjusting shims (**B**) for convertible top. One of two main mounting nuts (**arrows**) also visible.

The fully automatic top, standard on 1997 and later convertibles, allows complete operation of the top, from fully open to fully closed, with the push of a toggle switch on the console.

The top switch incorporates a "top unlatched" warning LED.

A schematic representation of the fully automatic convertible top is shown in Fig. 5.

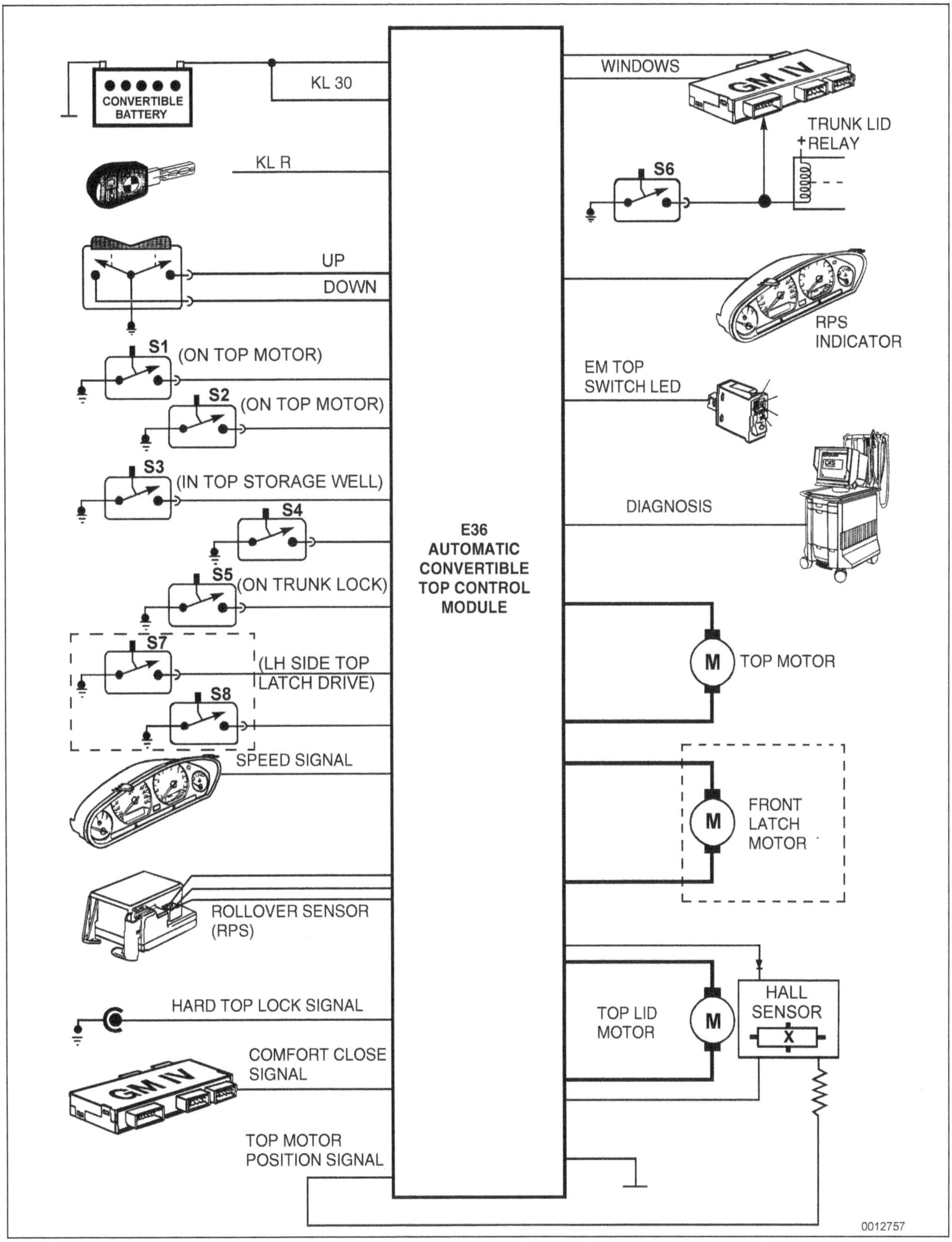

Fig. 5. Semi-automatic and fully automatic convertible top system. Main additional components in fully automatic top shown inside dashed lines.

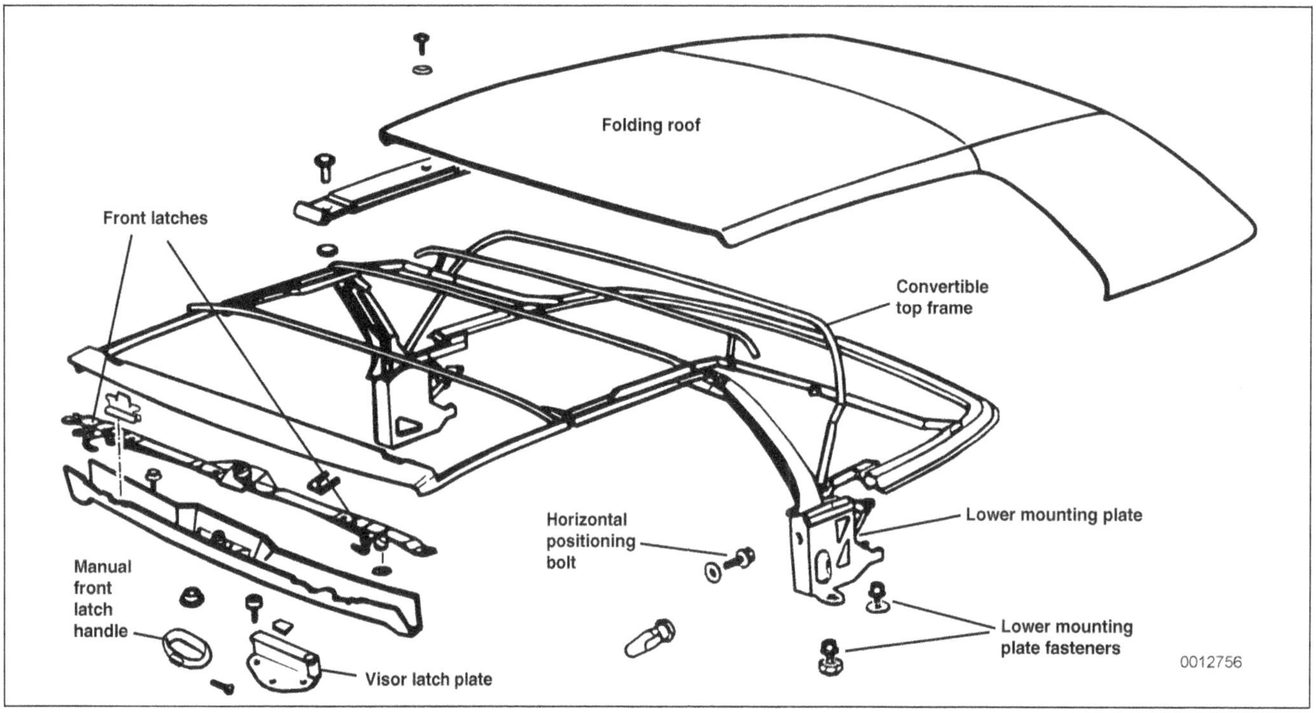

Fig. 6. Mechanical components of E36 convertible top. Manual, semi-automatic, and fully automatic tops have similar mechanical construction.

NOTE —
The electronic circuits used to control convertible top operation are complex. Diagnosis of the convertible top and the Rollover Protection System are beyond the scope of this book. Your authorized BMW dealer has the proper diagnostic equipment and tools to carry out these tasks.

Front Latching Mechanism

A close, tight fit between the front bow of the convertible roof and the top of the windshield is critical for preventing moisture and wind noise from intruding into the passenger compartment.

Visor Latch Plate

The left and right visor latch plates can be accessed by removing the trim and moldings around the sides and top of the windshield. Each latch is attached to the top of the windshield using three bolts. Latch position is adjusted as follows:

- Loosen bolts
- Latch top
- Align top
- Retighten bolts. See Fig. 7.

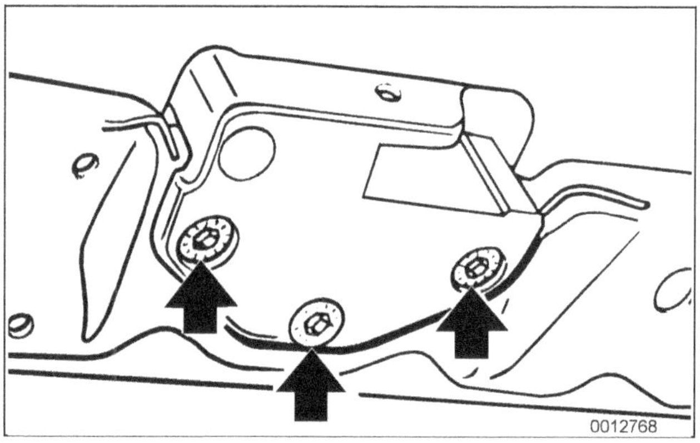

Fig. 7. Visor latch plate is held to top of windshield with three bolts (**arrows**).

Visor Latch

The semi-automatic and the fully automatic top front latches are of similar design. For the automatic latching mechanism an additional motor has been installed in the front bow of the convertible top. This releases the top from the visor latch plates and pulls it back far enough for the main retracting mechanism to complete the job.

The latches and the motor can be accessed by removing the front cover trim from the convertible top.

The left latch in the fully automatic version has two microswitches (S7 and S8) which signal open, closed and latched states. See Fig. 8.

The length of the dead center point rod is critical to correct opening and latching of the convertible top. See Fig. 9.

Dead Center Point Rod
- Dimension **A** .... 205.5 ± 1 mm (8.09 ± 0.04 in)

Visor Latch Motor

The fully automatic convertible top is latched and unlatched by one electric motor installed in the front roof bow. To replace, remove front cover trim from the roof. Remove electrical harness connectors from motor, and remove mounting screws. Slide the motor off the output shafts to the left and right latches.

During reinstallation, the output shafts must be turned until the S8 microswitch in the left latch is tripped.

Convertible top and frame, replacing

1. Open convertible top partially. Open convertible top lid. Open back windows.

2. Fully automatic power convertible top: Remove front cover trim from top. Disconnect electrical harness connectors from front latch motor. Disconnect connectors from S7 and S8 microswitches at left latch.

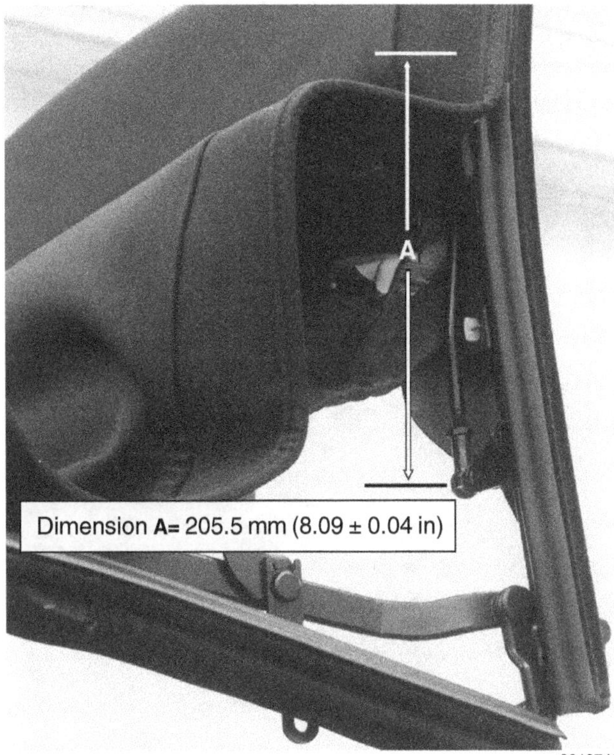

Fig. 9. Preset length of dead center point rod.

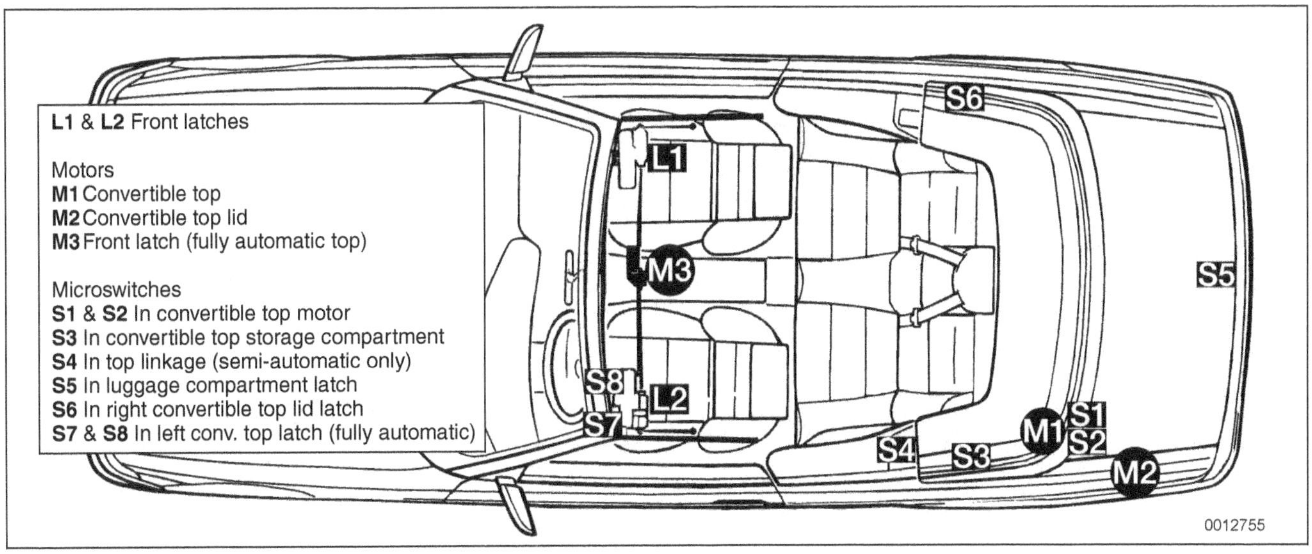

Fig. 8. Location of automatic and semi-automatic convertible latch mechanisms, motors and microswitches.

3. Remove headliner, starting from front and working backward. Note arrangement and lacing of tensioning cable so it can be reinstalled in its original configuration. See Fig. 10.

0012748

Fig. 10. Headliner tensioning cable and some of its retaining eyelets (**arrows**).

4. Disconnect electrical harness connector from S4 microswitch in middle linkage of convertible top.

5. Disconnect electric motor linkage rods in convertible top storage compartment behind left rear seat. See Fig. 11.

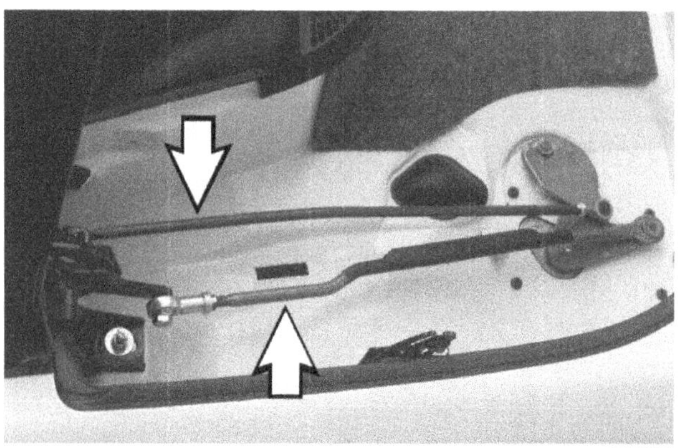

0012751

Fig. 11. Convertible top linkage rods (**arrows**) in compartment behind left rear seat.

6. Pull electrical harness through to disengage from top.

7. Remove fasteners in bottom of storage compartments behind door posts and lift out complete top. Note number of shims on horizontal mounting bolt. See Fig. 12.

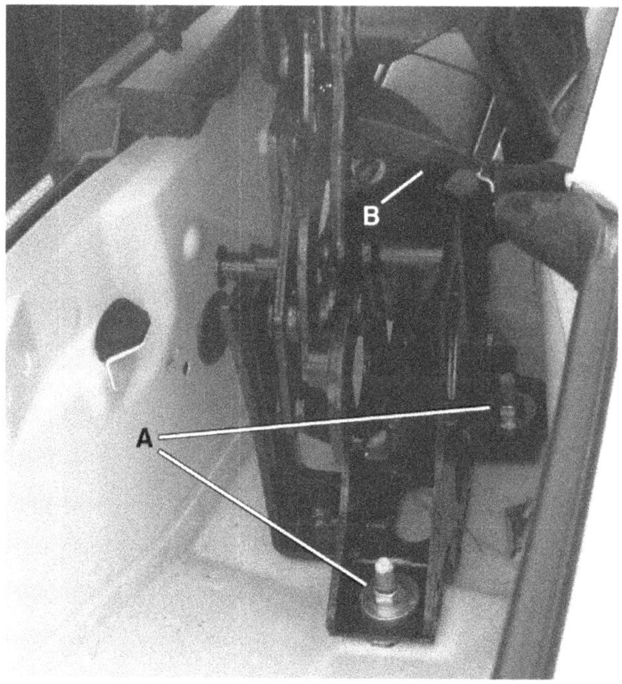

0012752

Fig. 12. Mounting nuts (**A**) for convertible top (two on each side). There is also a horizontal bolt with spacers on each side (**B**).

8. Installation is reverse of removal. Be sure to replace U-shaped shims on horizontal mounting bolts. See Fig. 13.

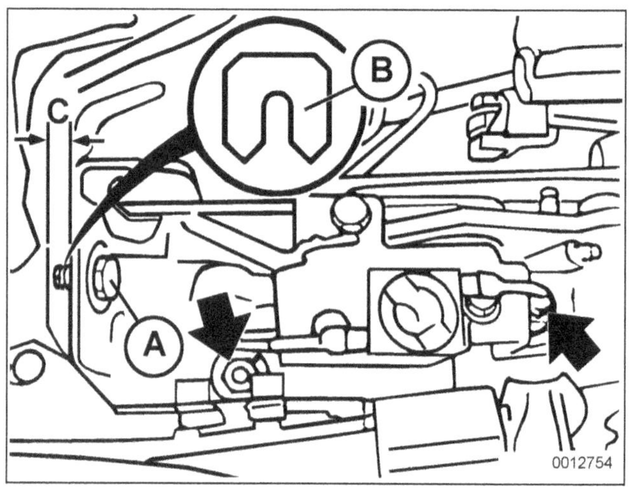

Fig. 13. Horizontal positioning bolt (**A**) and U-shaped adjusting shims (**B**) for convertible top. One of two main mounting nuts (**arrows**) also visible.

9. To adjust position and fit of top, remove or add U-shaped shims until top is properly aligned. Tighten nuts and screws.

Convertible top drive motor, replacing

1. Open convertible top partially, stopping when convertible top lid is fully open. Remove lid.

2. Release luggage compartment lock-out by pressing on microswitch in right convertible top lid latch. See Fig. 14.

0012745

Fig. 14. Right side convertible top lid latch. **Arrow** points to location of luggage compartment lock-out microswitch.

3. Open luggage compartment. Remove trim and inner lining from left side of compartment.

4. Disconnect electrical harness connectors at top motor. Release emergency release cable from lever. Push lever up to release motor. See Fig. 15.

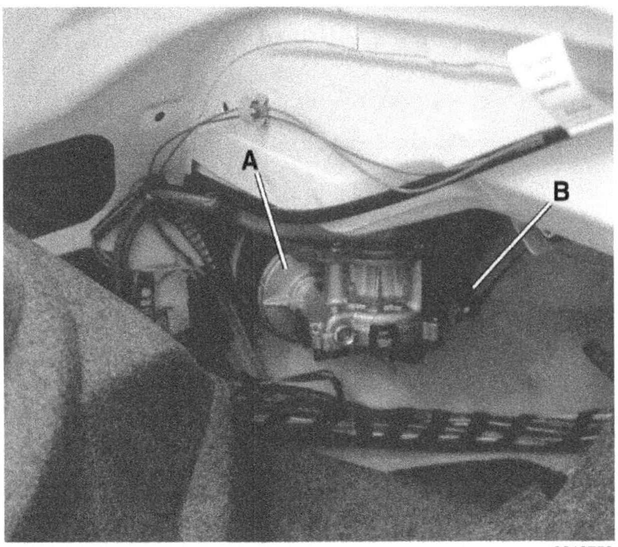

0012759

Fig. 15. Convertible top motor (**A**) and release lever (**B**) in luggage compartment.

5. Working in convertible top storage compartment behind driver's seat, remove convertible top linkage rods from top of motor. Refer to Fig. 11.

6. Remove four fasteners holding motor to body and remove motor through luggage compartment. See Fig. 16.

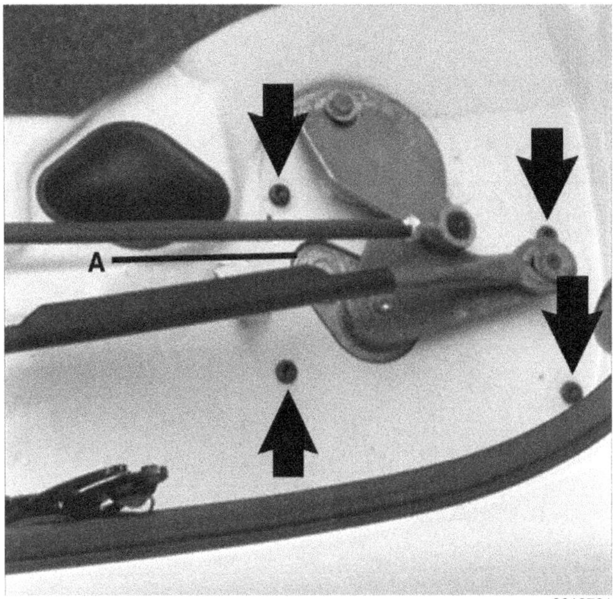

0012761

Fig. 16. Convertible top motor mounting screws (**arrows**). Gasket (A) must be renewed when motor is replaced.

7. Installation is reverse of removal. Keep the following in mind:
   - Replace sealing gasket between top of motor and body.
   - Secure motor mounting screws with Locktite® 270 or equivalent.

Tightening Torque
- Convertible top motor to body mounting screws. . . . . 10 Nm (7.5 ft-lb)

Convertible Top Emergency Operation

A malfunction in the electrical system or another fault in the convertible top mechanism can cause the automatic or semi-automatic top to be stuck in open, shut, or intermediate position. The following general procedures are suggested in order to close the top in an emergency situation. Resetting procedures for convertible top synchronization after emergency closure are beyond the scope of this manual.

Main Motor. To release the top linkage from the motor, lift the left corner of the rear seat to access emergency release handle. See Fig. 17.

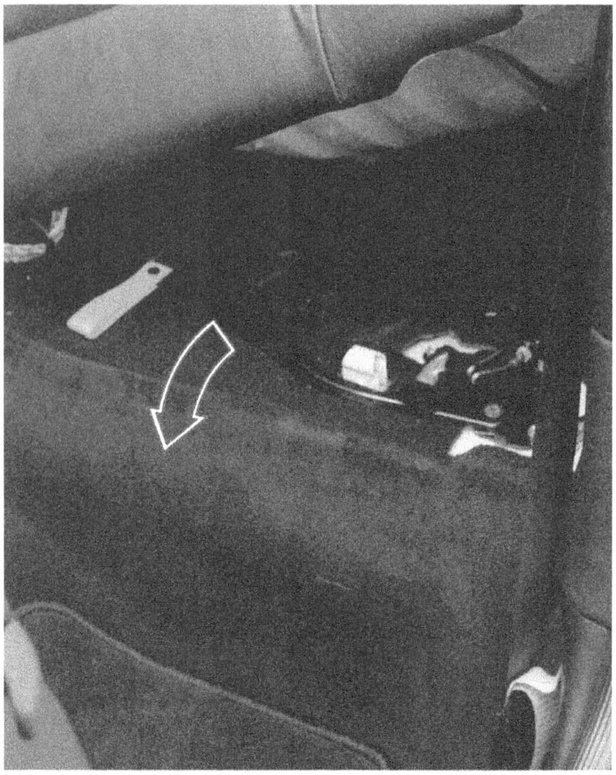

0012764

Fig. 17. Convertible top motor emergency release handle under rear seat. Pull in direction of **arrow** to release top.

> ***CAUTION —***
> *In case of failure of the automatic or semi-automatic convertible top function, only use the emergency procedure to close the top, never to open it. An authorized BMW dealer should then be consulted.*

It is then possible to push the top manually to a near closed position. In the case of the semi-automatic top, simply attach and lock the front latches.

Front Latches. Once the fully automatic convertible top has been released from the motor and manually pushed to near-closure, emergency latching of the top is possible by using the S-shaped hex key provided in the emergency tool kit.

First, remove the plastic cover over the latch motor in the front bow of the convertible top. See Fig. 18.

Next, use the hex key from tool box to crank the front latches shut. See Fig. 19.

Luggage Compartment Cover. A microswitch in the right convertible top lid latch prevents the luggage compartment from being opened while the lid is up. This lockout can be overridden by pressing on the microswitch. Refer to Fig. 14.

0012762

Fig. 18. Remove plastic cover from center of convertible roof front bow to access latching motor drive.

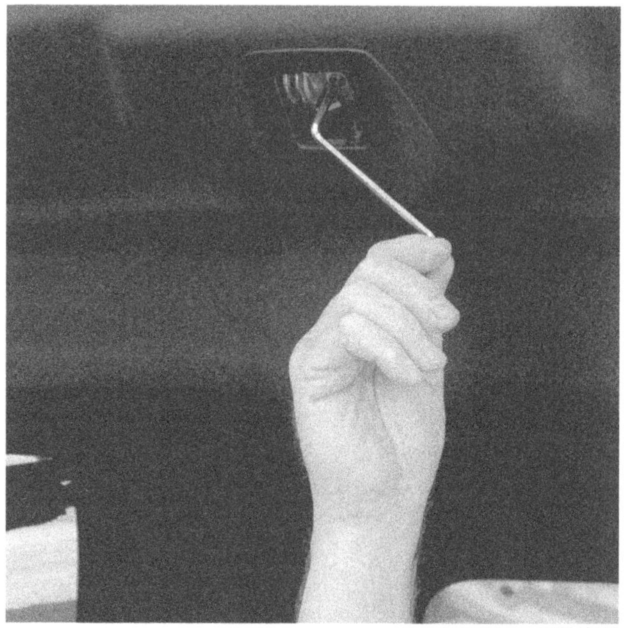

0012763

Fig. 19. Using emergency hex key to move front part of convertible top.

> ***CAUTION —***
> *The convertible top lid and the luggage compartment cover interfere with each other and cannot be open at the same time. To avoid damage to painted surfaces, an assistant should keep the lid down until the luggage compartment cover can be closed.*

Convertible Top Lid

The components of the convertible top lid are shown in Fig. 20.

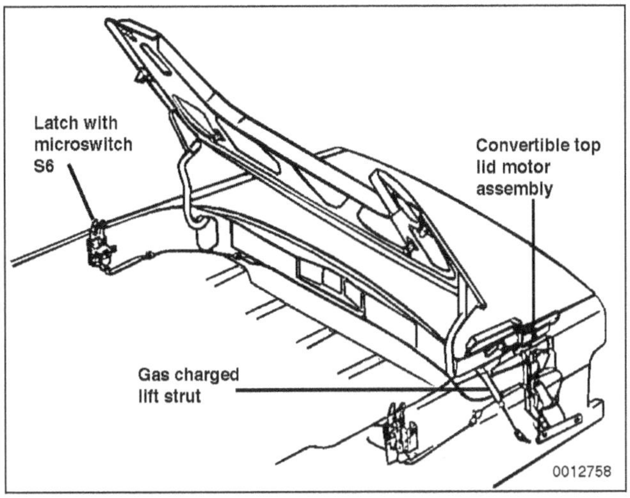

Fig. 20. Components of the convertible top lid.

The lid latches are cable operated by a drive motor located in the left side of the luggage compartment, behind the trim liner. The same motor is used to raise and lower the lid during automatic convertible top operation.

The latch cables must be adjusted so that the lid fits snugly against the body when locked.

The lid drive motor can be removed and replaced from the luggage compartment.

ROLLOVER PROTECTION SYSTEM

Two spring-loaded cassettes are mounted behind the rear seat. A single rollover sensor, mounted on the left cassette, signals the convertible top module (CVM) of an imminent rollover, whereupon the module triggers both cassettes within 3/10th of a second. See Fig. 21.

Once the car's ignition has been switched on, the rollover sensor performs a self-test lasting 6 seconds. During this time a yellow warning light is lit on the dashboard. If the warning light fails to go out after 6 seconds, this means that a fault has been detected in the system. The fault is stored in the CVM. Faults can be retrieved and diagnosed by special diagnostic equipment.

> ***NOTE —***
> *Diagnosis of the convertible top and the Rollover Protection System are beyond the scope of this book. Your authorized BMW dealer has the proper diagnostic equipment and tools to carry out these tasks.*

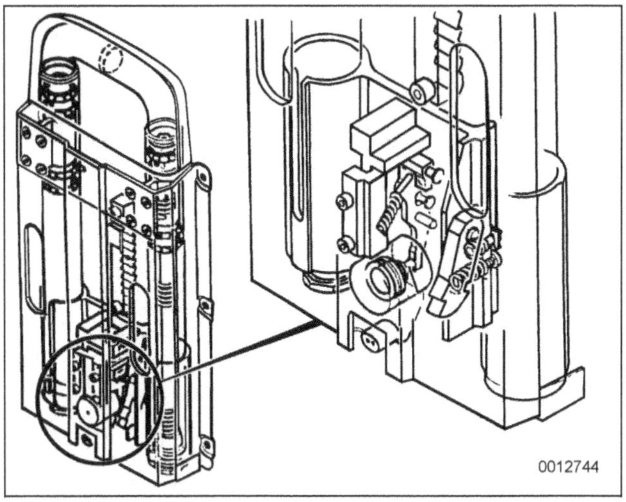

Fig. 21. Construction of the rollover protection cassette.

After deployment, the detent pawl in a rollover protection cassette can be retracted using the special tool in the tool kit, fitted to the screwdriver handle. The rear seat headrest needs to be raised for this procedure.

> ***WARNING —***
> *Ensure that the area above and adjacent to the rollover bars remains clear and unobstructed at all times.*

> ***CAUTION —***
> - *It is not possible to close the convertible top with the rollover bars extended.*
> - *If a hardtop is mounted, be sure to install the protective rollover bar covers provided with the hardtop to prevent damaging the rear window in case of deployment.*

600 Electrical System–General

GENERAL

A brief description of the principal parts of the electrical system is presented here. Also covered here are basic electrical system troubleshooting tips.

Voltage and Polarity

The vehicle electrical system is a 12-volt direct current (DC) negative-ground system. A voltage regulator controls system voltage at approximately the 12-volt rating of the battery. All circuits are grounded by direct or indirect connection to the negative (–) terminal of the battery. A number of ground connections throughout the car connect the wiring harness to chassis ground. These circuits are completed by the battery cable or ground strap between the body and the battery negative (–) terminal.

Wiring, Fuses and Relays

Nearly all parts of the wiring harness connect to components of the electrical system with keyed, push-on connectors that lock into place. Notable exceptions are the heavy battery cables and the starter wiring. The wiring is color-coded for circuit identification.

With the exception of the battery charging system, most electrical power is routed from the ignition switch or the battery through the main fuse/relay panel, located in the left rear corner of the engine compartment. Fuses are color coded to indicate current capacities.

The relays and control units/modules are mounted in various places throughout the vehicle. See **610 Electrical Component Locations**.

Electrical System Safety Precautions

Please read the following warnings and cautions before doing any work on your electrical system.

WARNING —

- *The cars covered by this manual are equipped with a Supplemental Restraint System (SRS) that automatically deploys one or more airbags. Each airbag unit houses an explosive powerful charge. Any work involving the SRS system should only be performed by an authorized BMW dealer. Making repairs without the proper knowledge and special test equipment may cause serious personal injury. See* **721 Airbag System (SRS)**.
- *The ignition system of the car operates at lethal voltages. People with pacemakers or weak hearts should not expose themselves to the ignition system. Extra caution must be taken when working on the ignition system or when servicing the engine while it is running or the key is on. See* **120 Ignition System** *for additional ignition system warnings and cautions.*
- *Before operating the starter without starting the engine (as when making a compression test), disable the ignition system as described in* **120 Ignition System**.
- *Keep hands, clothing and other objects clear of the electric radiator cooling fan when working on a warm engine. The fan may start at any time, even when the ignition is switched off.*

> ***CAUTION —***
> - *Always turn off the engine and disconnect the negative (–) cable from the battery before removing any electrical components. Disconnecting the battery may erase fault code(s) stored in control module memory. Check for fault codes using special BMW diagnostic equipment.*
> - *Prior to disconnecting the battery, read the battery disconnection cautions given at the front of this manual on page viii.*
> - *Connect and disconnect ignition system wires, multiple connectors, and ignition test equipment leads only while the ignition is off.*
> - *Do not disconnect the battery with engine running.*
> - *Do not quick-charge the battery (for boost starting) for longer than one minute, and do not exceed 16.5 volts at the battery with the boosting cables attached. Wait at least one minute before boosting the battery a second time.*
> - *Do not use a test lamp that has a normal incandescent bulb to test circuits containing electronic components. The high electrical consumption of these test lamps may damage the components.*
> - *Do not use an analog meter. Use only a digital multimeter.*
> - *Many of the solid-state modules are static sensitive. Static discharge will permanently damage them. Always handle the modules using proper static prevention equipment and techniques.*
> - *To avoid damaging harness connectors or relay panel sockets, use jumper wires with flat-blade connectors that are the same size as the connector or relay terminals.*
> - *Always switch a digital multimeter to the appropriate function and range before making test connections.*
> - *Do not try to start the engine of a car which has been heated above 176°F (80°C), (for example, in a paint drying booth). Allow it to cool to normal temperature.*
> - *Disconnect the battery before doing any electric welding on the car.*
> - *Do not wash the engine while it is running, or anytime the ignition is switched on.*

Electrical Test Equipment

Many of the electrical tests described in this manual call for measuring voltage, current or resistance using a digital multimeter (DMM). Digital meters are preferred for precise measurements and for electronics work because they are generally more accurate than analog meters. The numerical display is also less likely to be misread, since there is no needle position to be misinterpreted by reading at an angle.

An LED test light is a safe, inexpensive tool that can be used to perform many simple electrical tests that would otherwise require a digital multimeter. The LED indicates when voltage is present between any two test-points in a circuit.

> ***CAUTION —***
> - *Choose test equipment carefully. Use a digital multimeter with at least 10 megaohm input impedance, or an LED test light. An analog meter (swing-needle) or a test light with a normal incandescent bulb may draw enough current to damage sensitive electronic components.*
> - *An ohmmeter must not be used to measure resistance on solid state components such as control units or time delay relays.*
> - *Always disconnect the battery before making resistance (ohm) measurements on the circuit.*

WIRING DIAGRAMS

The wiring diagrams shown in **Electrical Wiring Diagrams** have been specially designed to enable quick and efficient diagnosis and troubleshooting of electrical malfunctions.

Wiring Codes and Abbreviations

A lot of information is included in each wiring diagram if you know how to read them. Wire colors in the diagrams are abbreviated. Combined color codes indicate a multi-colored wire. For example the code BLU/RED indicates a Blue wire with a Red stripe.

Many electrical components, connectors, fuses, and ground locations are identified using a unique number. Each of these numbers corresponds to a particular part in the circuit commonly found in **Electrical Wiring Diagrams**.

NOTE —
Sometimes the color of an installed wire may be different than the one on the wiring diagram. Don't be concerned. Just be sure to confirm that the wire connects to the proper terminals.

Wire color codes

- BLU .... Blue
- BRN .... Brown
- YEL .... Yellow
- GRN .... Green
- GRY .... Gray
- ORG .... Orange
- RED .... Red
- BLK .... Black
- VIO .... Violet
- WHT .... White

Most terminals are identified by numbers on the components and harness connectors. The terminal numbers for major electrical connections are shown in the diagrams. Though many terminal numbers appear only once, several other numbers appear in numerous places throughout the electrical system and identify certain types of circuits. Some of the most common circuit numbers are listed below in **Table a.**

Table a. Terminal and Circuit Numbers

Number	Circuit description
1	Low voltage switched terminal of coil
4	High voltage center terminal of coil
+X	Originates at ignition switch. Supplies power when the ignition switch is in the PARK, RUN, or START position
15	Originates at ignition switch. Supplies power when ignition switch is in RUN or START position
30	Battery positive (+) voltage. Supplies power whenever battery is connected. (Not dependent on ignition switch position, unfused)
31	Ground, battery negative (–) terminal
50	Supplies power from battery to starter solenoid when ignition switch is in START position only
+54	Originates at ignition switch. Supplies power when ignition switch is in the RUN position only
85	Ground side (–) of relay coil
86	Power-in side (+) of relay coil
87	Relay actuated contact
D	Alternator warning light and field energizing circuit

Additional abbreviations shown in the wiring diagrams are given below.

Abbreviations

- ABS .... antilock brakes
- A/C .... air conditioning
- AST/ASC+T .... all season traction
- CONV .... convertible
- DME .... digital motor electronics
- ECM .... electronic control module
- EWS/EWS II .... coded driveaway protection
- SRS .... supplemental restraint system-airbag
- TCM .... transmission control module
- ZKE (94-98 models) .... central body electronics
- ZVM (92-93 models)............... central locking

ELECTRICAL TROUBLESHOOTING

Four things are required for current to flow in any electrical circuit: a voltage source, wires or connections to transport the voltage, a load or device that uses the electricity, and a connection to ground. Most problems can be found using a digital multimeter (volt/ohm/amp meter) to check for voltage supply, for breaks in the wiring (infinite resistance/no continuity), or for a path to ground that completes the circuit.

Electric current is logical in its flow, always moving from the voltage source toward ground. Electrical faults can usually be located through a process of elimination. When troubleshooting a complex circuit, separate the circuit into smaller parts. The general tests outlined below may be helpful in finding electrical problems. The information is most helpful when used with the wiring diagrams.

Be sure to analyze the problem. Use the wiring diagrams to determine the most likely cause. Get an understanding of how the circuit works by following the circuit from ground back to the power source.

When making test connections at connectors and components, use care to avoid spreading or damaging the connectors or terminals. Some tests may require jumper wires to bypass components or connections in the wiring harness. When connecting jumper wires, use blade connectors at the wire ends that match the size of the terminal being tested. The small internal contacts are easily spread apart, and this can cause intermittent or faulty connections that can lead to more problems.

Voltage and Voltage Drops

The wires, connectors, and switches that carry current are designed with very low resistance so that current flows with a minimum loss of voltage. A voltage drop is caused by higher than normal resistance in a circuit. This additional resistance actually decreases or stops the flow of current. A voltage drop can be noticed by problems ranging from dim headlights to sluggish wipers. Some common sources of voltage drops are corroded or dirty switches, dirty or corroded connections or contacts, and loose or corroded ground wires and ground connections.

A voltage drop test is a good test to make if current is flowing through the circuit, but the circuit is not operating correctly. A voltage drop test will help to pinpoint a corroded ground strap or a faulty switch. Normally, there should be less than 1 volt drop across most wires or closed switches. A voltage drop across a connector or short cable should not exceed 0.5 volts.

NOTE —

- *A voltage drop test is generally more accurate than a simple resistance check because the resistances involved are often too small to measure with most ohmmeters. For example, a resistance as small as 0.02 ohms would results in a 3 volt drop in a typical 150 amp starter circuit. (150 amps x 0.02 ohms =3 volts).*
- *Keep in mind that voltage with the key on and voltage with the engine running are not the same. With the ignition on and the engine off (battery voltage), voltage should be approximately 12.6 volts. With the engine running (charging voltage), voltage should be approximately 14.0 volts. Measure voltage at the battery with the ignition on and then with the engine running to get exact measurements.*

Voltage, measuring

1. Connect digital multimeter negative lead to a reliable ground point on car.

NOTE —
The negative (–) battery terminal is always a good ground point.

2. Connect digital multimeter positive lead to point in circuit you wish to measure. See Fig. 1. If a reading is obtained, current is flowing through circuit.

NOTE —
The voltage reading should not deviate more than 1 volt from the voltage at the battery. If the voltage drop is more than this, check for a corroded connector or loose ground wire.

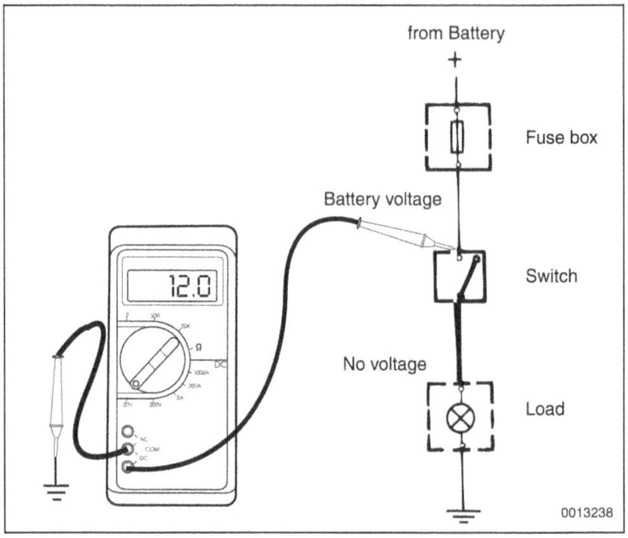

Fig. 1. Digital multimeter being used to test voltage.

Voltage drop, testing

Voltage drop can only be checked when current is running through the circuit, such as by operating the starter motor or turning on the headlights. A digital multimeter should be used to ensure accurate readings.

1. Connect digital multimeter positive lead to positive (+) battery terminal or a positive power supply close to battery source.
2. Connect digital multimeter negative lead to other end of cable or switch being tested. See Fig. 2.
3. With power on and circuit working, meter shows voltage drop (difference between two points). This value should not exceed 1 volt.

NOTE —
The maximum voltage drop in an automotive circuit, as recommended by the Society of Automotive Engineers (SAE), is as follows: 0 volts for small wire connections; 0.1 Volts for high current connections; 0.2 volts for high current cables; and 0.3 volts for switch or solenoid contacts. On longer wires or cables, the drop may be slightly higher. In any case, a voltage drop of more than 1.0 volt usually indicates a problem.

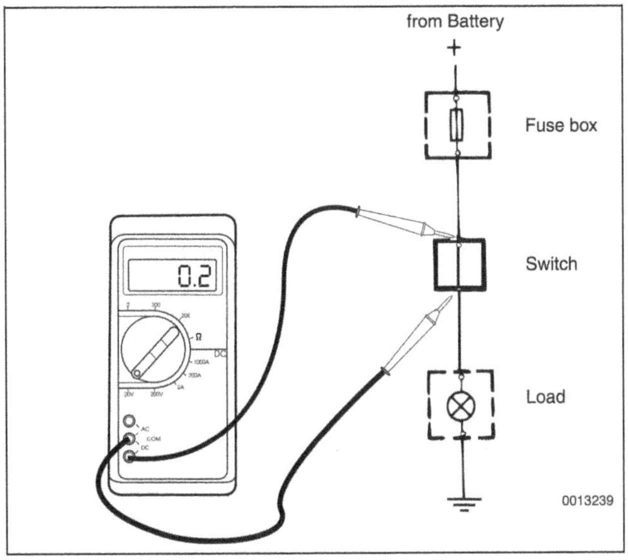

Fig. 2. Digital multimeter being used to check for voltage drop across a switch.

Continuity, checking

The continuity test can be used to check a circuit or switch. Because most automotive circuits are designed to have little or no resistance, a circuit or part of a circuit can be easily checked for faults using an ohmmeter. An open circuit or a circuit with high resistance will not allow current to flow. A circuit with little or no resistance allows current to flow easily.

> **CAUTION—**
> *Use only a high quality digital ohmmeter having high input impedance when checking electronic components. The internal power source used in most analog (swing-needle) meters can damage solid state components.*

When checking continuity, the ignition should be off. On circuits that are powered at all times, the battery should be disconnected. Using the appropriate wiring diagram, a circuit can be easily tested for faulty connections, wires, switches, relays, and engine sensors by checking for continuity. For a continuity check on a brake light switch, see Fig. 3.

Short Circuits

A short circuit is exactly what the name implies. The circuit takes a shorter path than it was designed to take. The most common short that causes problems is a short to ground where the insulation on a positive (+) wire wears away and the metal wire is exposed. When the wire rubs against a metal part of the car or other ground source, the circuit is shorted to ground. If the exposed wire is live (positive battery voltage), a fuse will blow and the circuit may possibly be damaged.

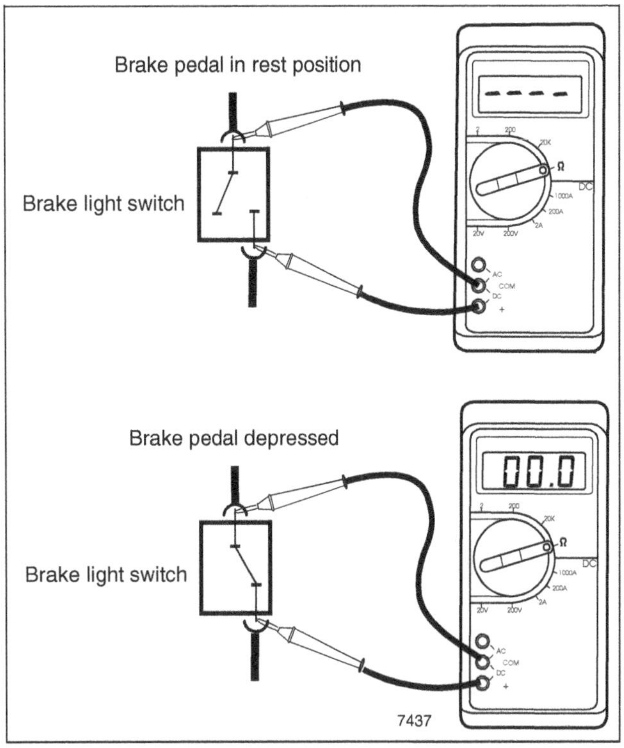

Fig. 3. Brake light switch being tested for continuity. With brake pedal in rest position (switch open) there is no continuity (infinite ohms). With pedal depressed (switch closed) there is continuity (zero ohms).

Shorts to ground can be located with a digital multimeter. Short circuits are often difficult to locate and may vary in nature. Short circuits can be found using a logical approach based on the current path.

> **CAUTION—**
> - *On circuits protected with high rating fuses (25 amp and greater), the wires or circuit components may be damaged before the fuse blows. Always check for damage before replacing fuses of this rating.*
> - *When replacing blown fuses, use only fuses having the correct rating. Always confirm the correct fuse rating printed on the fuse/relay panel cover.*

Short circuit, testing with ohmmeter

1. Remove blown fuse from circuit and disconnect cables from battery.

2. Disconnect harness connector from circuit's load or consumer.

3. Using an ohmmeter, connect one test lead to load side of fuse terminal (terminal leading to circuit) and the other test lead to ground. See Fig. 4.

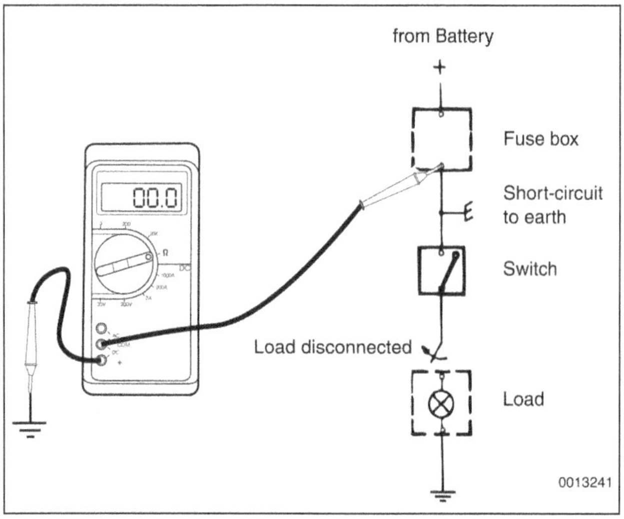

Fig. 4. Digital multimeter being used as an ohmmeter to find short circuit.

4. If there is continuity to ground, there is a short to ground.

5. If there is no continuity, work from wire harness nearest to fuse/relay panel and move or wiggle wires while observing meter. Continue to move down harness until meter displays a reading. This is the location of short to ground.

Visually inspect the wire harness at this point for any faults. If no faults are visible, carefully slice open the harness cover or the wire insulation for further inspection. Repair any faults found.

Short circuit, testing with voltmeter

1. Remove blown fuse from circuit.

2. Disconnect harness connector from circuit's load or consumer.

NOTE —
Most fuses power more than one consumer. Be sure all consumers are disconnected when checking for a short circuit.

3. Using a voltmeter, connect test leads across fuse terminals. See Fig. 5. Make sure power is present in circuit. If necessary turn key on.

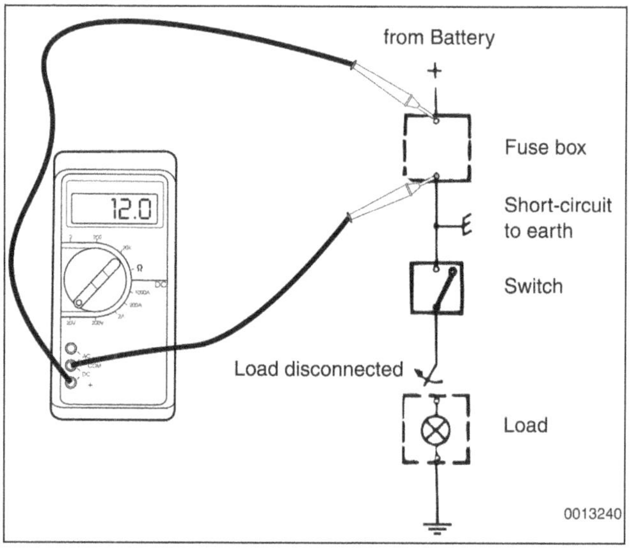

Fig. 5. Digital multimeter being used as a voltmeter to find short circuit.

4. If voltage is present at voltmeter, there is a short to ground.

5. If voltage is not present, work from wire harness nearest to fuse/relay panel and move or wiggle wires while observing meter. Continue to move down harness until meter displays a reading. This is the location of short to ground.

6. Visually inspect wire harness at this point for any faults. If no faults are visible, carefully slice open harness cover or wire insulation for further inspection. Repair any faults found.

610 Electrical Component Locations

TABLES

GENERAL

This repair group covers fuse, relay, and control module location information. Electrical equipment and accessories installed varies depending on model and model year. Always confirm that the proper electrical component has been identified by using the electrical wiring diagrams.

> ***WARNING —***
> *On cars equipped with airbags, special precautions apply to any electrical system testing or repair. Airbag units are explosive devices and must be handled with extreme care. Before starting any work on an airbag equipped car, refer to the warnings and cautions in* **721 Airbag System (SRS)**.

> ***CAUTION —***
> - *Prior to disconnecting the battery, read the battery disconnection cautions given at the front of this manual on page viii.*
> - *Relay and fuse positions are subject to change and may vary from car to car. If questions arise, an authorized BMW dealer is the best source for the most accurate and up-to-date information.*
> - *A good way to verify a relay position is to compare the wiring colors at the relay socket to the colors indicated on the wiring diagrams located at the rear of this manual.*
> - *Always switch the ignition off and remove the negative (–) battery cable before removing any electrical components.*
> - *Connect and disconnect ignition system wires, multiple connectors, and ignition test equipment leads only while the ignition is switched off.*
> - *Only use a digital multimeter for electrical tests.*

FUSE AND RELAY POSITIONS

The front power distribution box, an auxiliary relay panel, and two auxiliary splice panels contain most of the fuses and relays in E-36 vehicles. Refer to the component location later in this section, as well as the tables at the end of this repair group for electrical component location.

Fuse Positions

> ***CAUTION —***
> - *Replace fuses with those of the same rating. Installing a fuse with higher rating can lead to circuit failure and may also start a fire.*
> - *Fuse positions vary by model and equipment.*

Fuses number 1 through 46 are mounted in the front power distribution box. See Fig. 1.

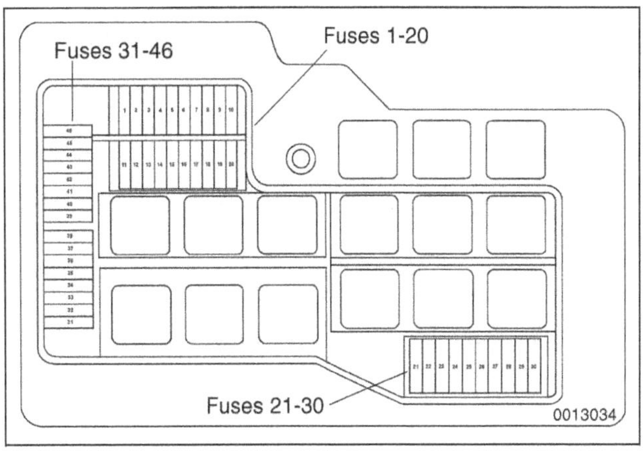

Fig. 1. Fuse positions 1–46 in front power distribution box.

Fuses number 47 and 50 are mounted on the left side splice panel. Refer to Fig. 6.

Fuse 48 is mounted on the auxiliary relay panel. Refer to Fig. 4.

Fuse 49 is mounted in the right side of the luggage compartment behind the trim cover.

Relay Positions

> ***CAUTION —***
> - *Relay positions vary by model and equipment.*

Mounted in the left-hand rear corner of the engine compartment, the power distribution box contains 46 fuses and 15 relay positions. See Fig. 2.

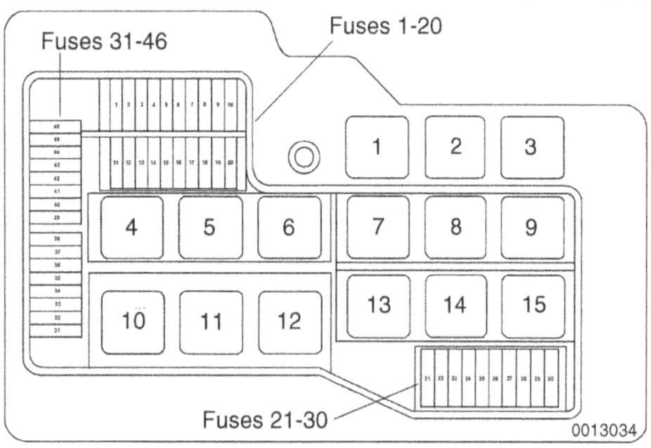

1. 1992 - 1997: Fuel pump relay
2. 1992 - 1997: System (main) relay
   1998: Fuel pump relay
3. Oxygen sensor heater relay
4. Horn relay
5. Front foglight relay
6. 1992 - 1993: High beam relay
   1994 - : Low beam relay
7. 1992 - 1993: Low beam relay
   1994 - : High beam relay
8. Hazard flasher relay
9. Blower relay
10. Rear defogger relay
11. ABS system relay
12. ABS pump relay
13. High speed fan relay
14. A/C compressor relay
15. Normal speed fan relay

Fig. 2. 318i, 323i, 325i, 328i front power distribution box.

M3 power distribution box relays are shown in Fig. 3.

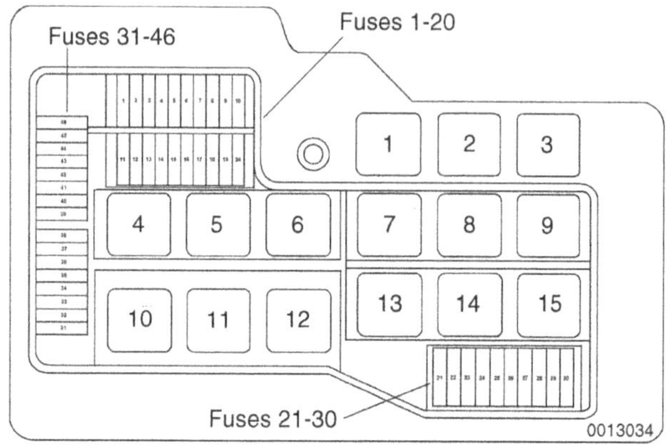

1. 1995 - 1996: Double relay
   1997: Secondary air injection relay
   1998: System (main) relay
2. 1995 - 1996: System (main) relay
   1997 - 1998: Fuel pump relay
3. 1995 - 1996: Fuel pump / start identification relay
   1997: System (main) relay
   1998: Secondary air injection relay
4. Horn relay
5. Front foglight relay
6. 1992 - 1993 : High beam relay
   1994 - : Low beam relay
7. 1992 - 1993: Low beam relay
   1994 - : High beam relay
8. Hazard flasher relay
9. Blower relay
10. Rear defogger relay
11. ABS system relay
12. ABS pump relay
13. High speed fan relay
14. A/C compressor relay
15. Low speed fan relay

Fig. 3. M3 front power distribution box.

Auxiliary Relay Panel

The auxiliary relay panel contains five relay positions. In later models an additional fuse holder (Fuse 48) and the EWS II (anti-theft) transmitter/receiver module are located on this panel. See Fig. 4.

1. Comfort relay (where applicable)
2. Crash control module
3. Park ventilation relay
4. Not used
5. Not used
6. EWS II transmitter / receiver module (from Jan. 1995)

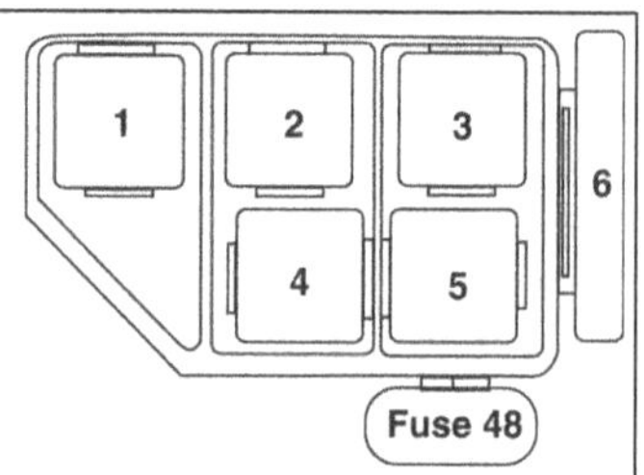

Fig. 4. Auxiliary relay panel under left side of dashboard.

The panel is located next to the steering column under the driver's side dashboard. To access this panel, remove the lower dash panel on driver's side. See **513 Interior Trim**. Then remove the driver's side knee bolster. Lower the relay panel by gently releasing the plastic retainers. See Fig. 5.

Fig. 5. Auxiliary relay panel under left side of dashboard (**arrow**).

During installation, be sure retainer tabs are engaged in relay panel.

Splice Panel, Left

The left splice panel contains four relay positions and an additional fuse holder (Fuses 47 and 50).

To access the left side electrical splice panel, remove the left side dash panel and knee bolster, as described earlier. The panel is located above and to the left of the clutch pedal. See Fig. 6.

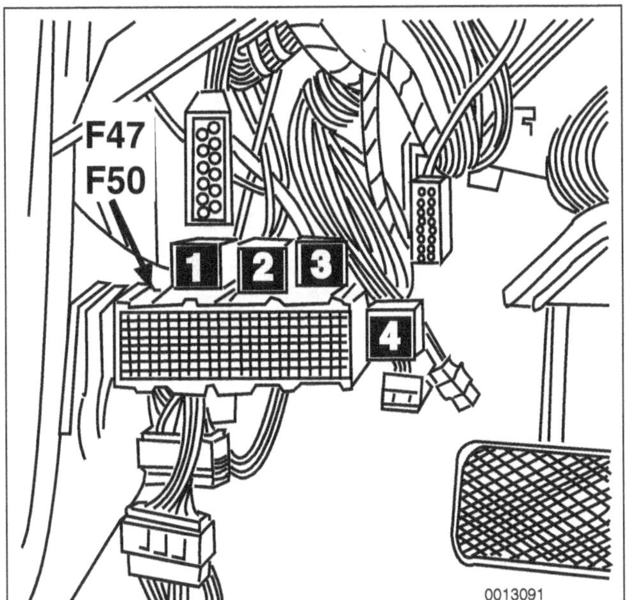

1. Unloader relay
2. Wiper relay
3. Wiper motor relay
4. Starter immobilization relay (1994 only)

F47 & **F50** Fuses 47 and 50

Fig. 6. Splice panel under left side of dash.

Splice Panel, Right

The right splice panel contains three relay positions. It is located under the right side of the dashboard.

To access the panel, remove the glove compartment as described in **513 Interior Trim**. See Fig. 7.

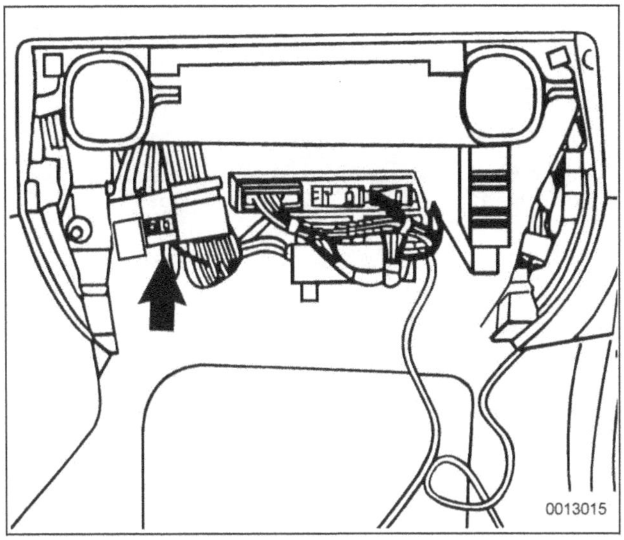

Fig. 7. Splice panel under right side of dash (**arrow**). Glove compartment and right side vents shown removed.

COMPONENT LOCATIONS

On the following pages are illustrations and photos showing the location of major electrical components in E36 vehicles.

> ***NOTE —***
> - *Every component is not installed in every car.*
> - *Due to changes in production, component locations may vary from what is illustrated. Consult your BMW dealer for the latest information.*
> - *The gear-position/neutral safety switch is also sometimes referred to as the automatic transmission range switch.*
> - *The EWS II system is also sometimes referred to as the Driveaway Protection System or the electronic immobilization system.*
> - *All-Season Traction (AST) is also sometimes referred to as ASC or ASC+T.*

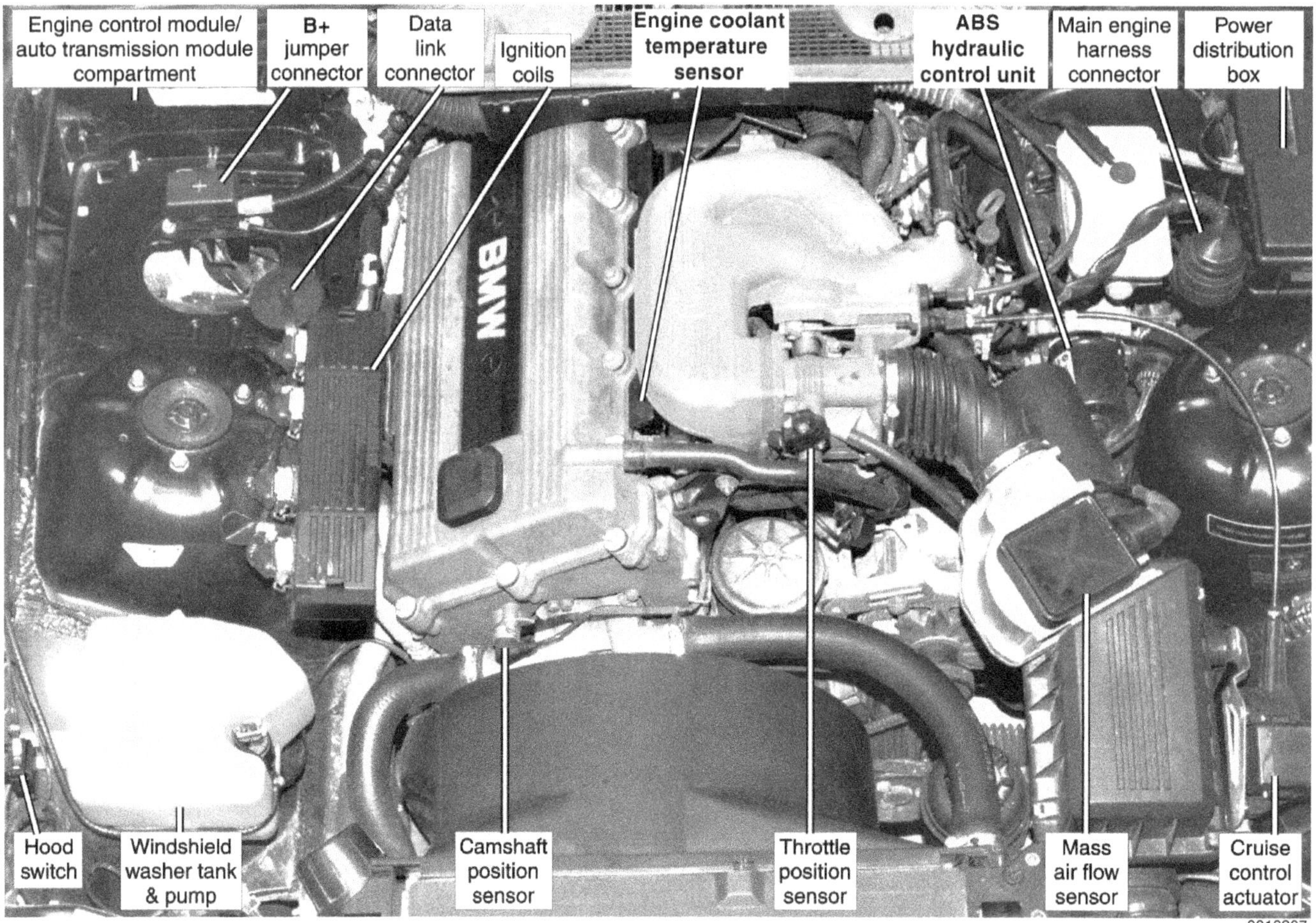

Fig. 8. E36 engine compartment layout. M42 engine compartment shown.

Fig. 9. Control modules under right side of dash. Glove compartment has been removed. For module identification, see Fig. 10.

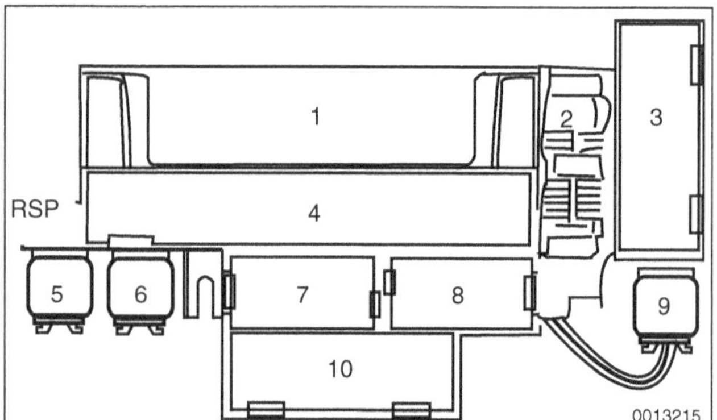

1. SRS (Airbag) control module (through 1993)
2. ABS/ASC+T control module (as applicable)
3. Central locking module (through 1993)
4. Anti-theft alarm module or
   ZKE IV control module (from 1994)
5. Rear window blower relay (convertible models)
6. Horn relay
7. Wiper/washer module
8. Rear power window relay (four-door models)
9. Trunk lid lock relay (convertible)
10. Cruise control module
    EWS II control module (from Jan. 1995)

RSP: Right splice panel

Fig. 10. Identification of control modules under right side of dash. Module utilization varies by year and model

1. Heater valves
2. ABS pedal travel sensor
3. ABS pump
4. Main engine harness connector
5. Power distribution box

0013202

Fig. 11. Left rear of engine compartment.

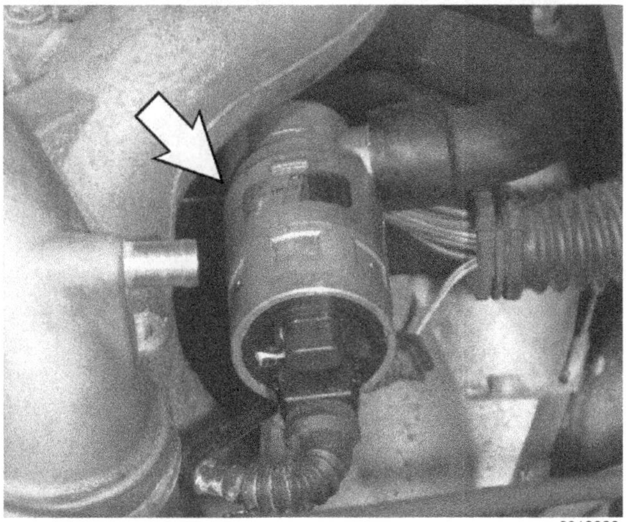

0013226

Fig. 12. M42 engine idle speed control valve (**arrow**) beneath intake manifold

0012421

Fig. 13. M44 engine intake air temperature (IAT) sensor in air filter housing

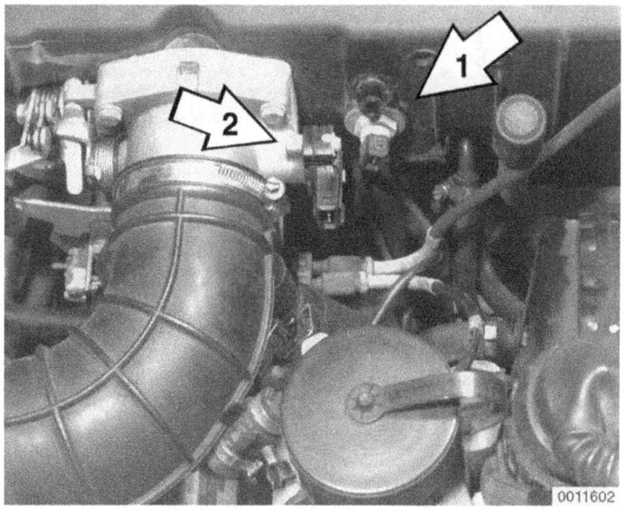

0011602

Fig. 14. M50/S50US engine intake air temperature (IAT) sensor (**1**) and throttle position sensor (**2**).

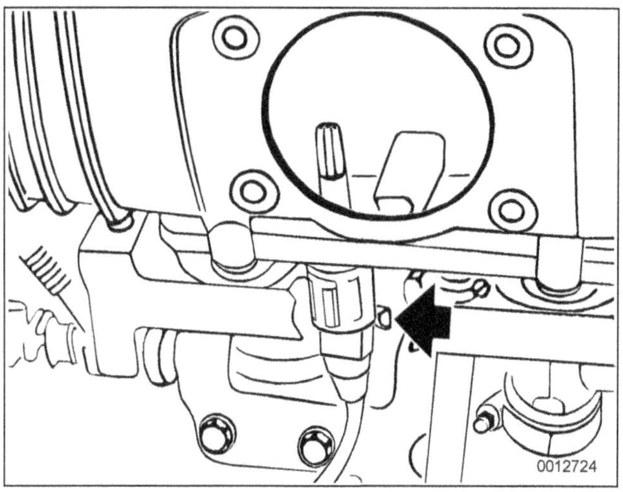

0012724

Fig. 15. M52/S52US engine intake air temperature (IAT) sensor in bottom of intake manifold (throttle housing shown removed).

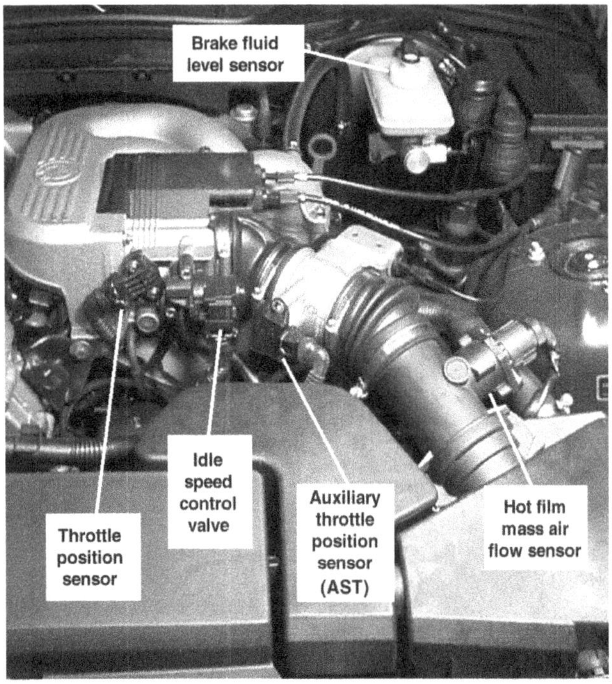

Fig. 16. M44 engine harness connectors at air intake

Fig. 17. A/C recharging ports (**arrows**), right rear of engine compartment

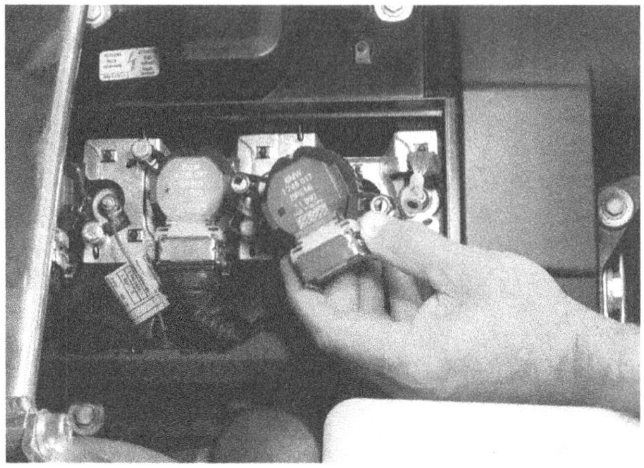

Fig. 18. Six-cylinder ignition coil in top of cylinder head (M52 engine).

Fig. 19. Cooling fan dual switch in radiator (**arrow**).

0011984

Fig. 20. M44 engine coolant temperature (ECT) sensor (**arrow**).

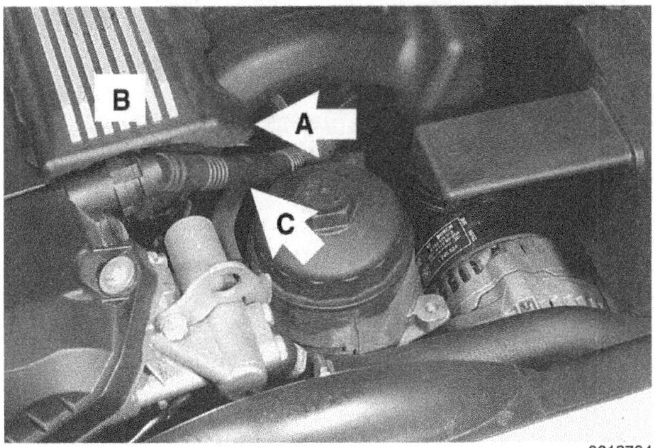

0012704

Fig. 21. M52 engine coolant temperature (ECT) sensor (**A**) is located beneath top engine cover (**B**) and crankcase vent hose (**C**).

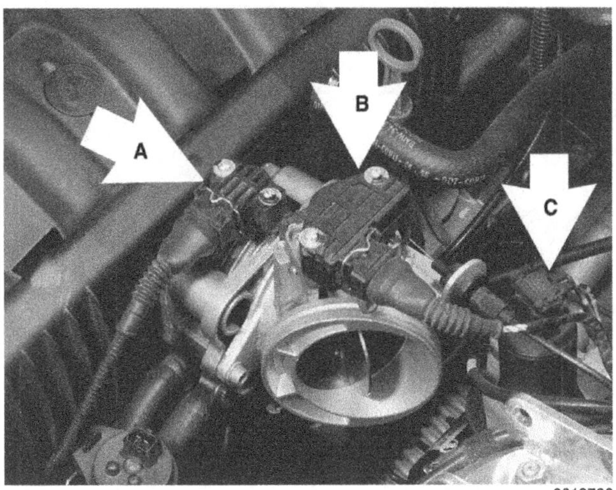

0012700

Fig. 22. M52 throttle position sensor (**A**), idle speed control valve (**B**), and secondary air pump (**C**).

0013131

Fig. 23. Headlights ground at left front of engine compartment (**arrow**).

0013130

Fig. 24. Main ground at right strut tower (**A**). **B** is negative jumper connection.

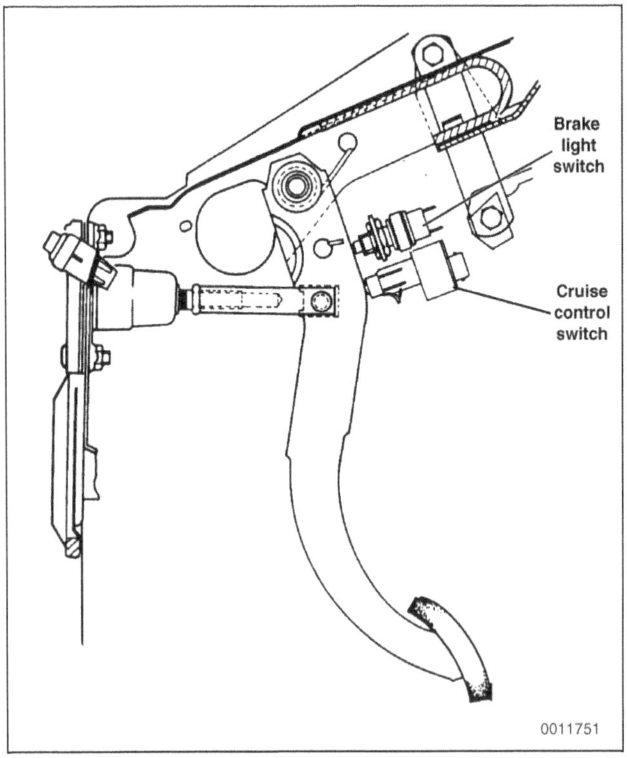

Fig. 25. Brake pedal switches.

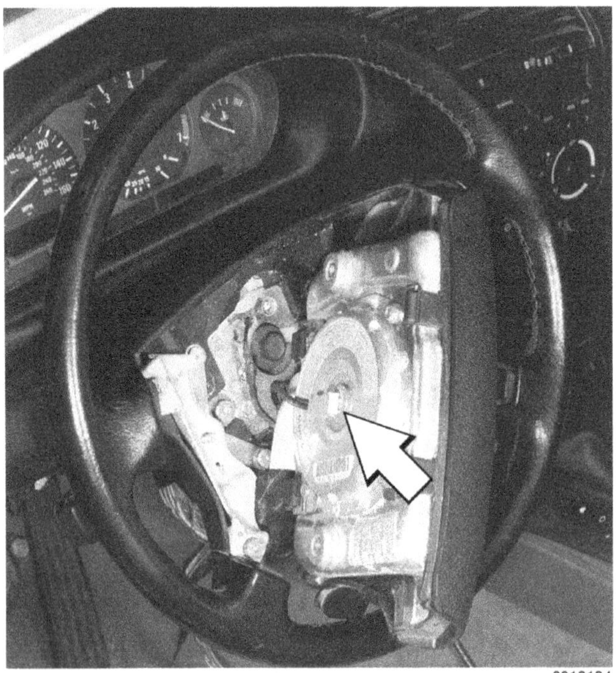

Fig. 26. Driver side airbag connector (**arrow**).

Fig. 27. Passenger side airbag connector (**arrow**) behind glove compartment.

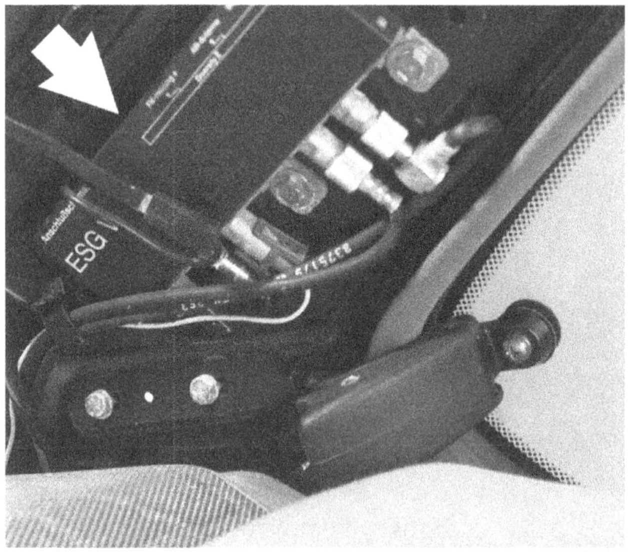

Fig. 28. Antenna diversity module (**arrow**) in left C-pillar.

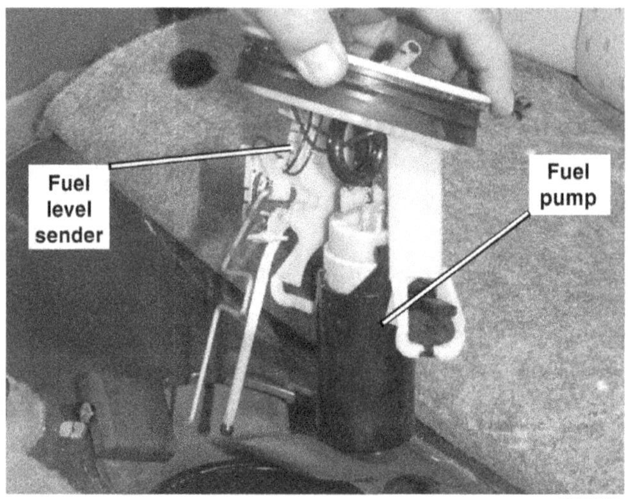

Fig. 29. Right side of fuel tank under rear seat cushion.

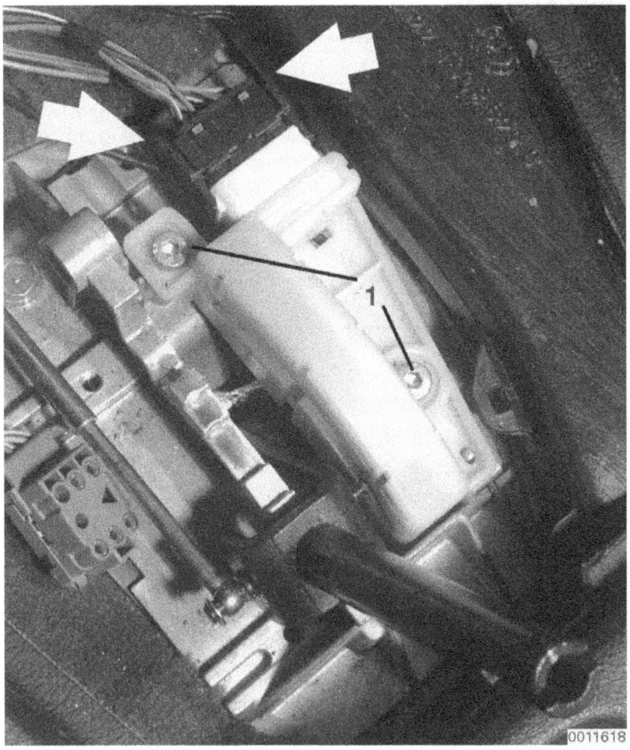

Fig. 30. Transmission gear-position/neutral safety switch under center console. Squeeze connector tabs (**arrows**) to disengage harness connector.

6263

Fig. 31. Back-up light switch (**arrow**) on transmission. (Switch may be found on right or left side of transmission, depending on application.)

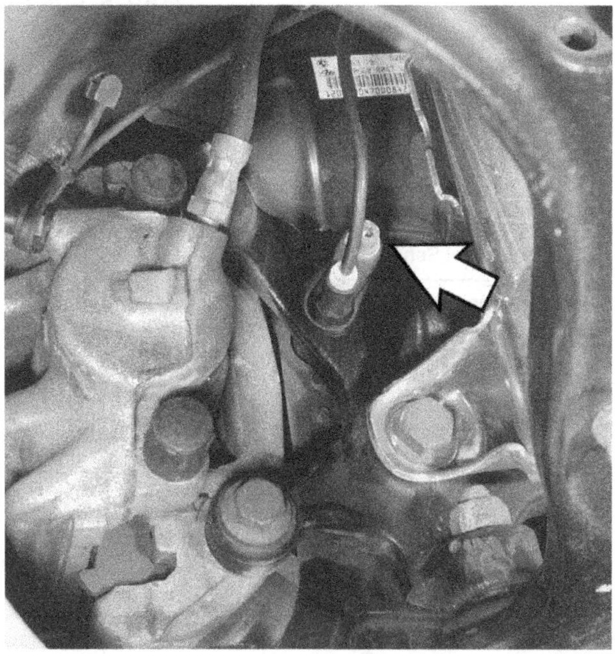

0012117

Fig. 32. ABS wheel speed sensor (**arrow**) in front steering arm.

Component Location Table

Table a is a cross-referenced listing of electrical components in E36 vehicles. In addition to location descriptions, where possible there is a reference to a figure showing the component.

NOTE —

- *Every component is not installed in every car.*
- *Due to changes in production, component locations may vary. Consult your BMW dealer for the latest information.*
- *The figure numbers are all within this section. Many other electrical components are illustrated throughout this book. Consult the main index.*
- *The gear-position/neutral safety switch is also sometimes referred to as the automatic transmission range switch.*
- *The EWS II system is also sometimes referred to as the Driveaway Protection System or the electronic immobilization system.*
- *All-Season Traction (AST) is also sometimes referred to as ASC or ASC+T.*

Table a. E36 Component Locations

Component	Model Year	Location	Refer to
ABS Hydraulic Control Unit	1992 - 1998	Left rear engine compartment under master cylinder	Fig. 8. Fig. 11.
ABS Relays	1992 - 1998	Power distribution box	Fig. 2. Fig. 3.
ABS Pedal Travel Sensor	1992 - 1998	On brake booster beside master cylinder	Fig. 11.
ABS/AST (Traction) Control Module	1992 - 1998	Behind glove compartment	Fig. 10.
ABS Speed Sensor, Left Front Wheel	1992 - 1998	Left front steering arm	Fig. 32.
ABS Speed Sensor, Left Rear Wheel	1992 - 1998	Left rear trailing arm	
ABS Speed Sensor, Right Front Wheel	1992 - 1998	Right front steering arm	
ABS Speed Sensor, Right Rear Wheel	1992 - 1998	Left rear trailing arm	
A/C Compressor	1992 - 1998	Right lower front of engine	
A/C Compressor Control Relay	1992 - 1998	Power distribution box	Fig. 2. Fig. 3.
A/C Control Module	1992 - 1998	Behind center console	
A/C Evaporator Temperature Sensor	1992 - 1998	Left side of heater box	
A/C Pressure Switch	1992 - 1998	Right front of engine compartment	
A/C Program Switch	1992 - 1998	Center of dash below radio	
A/C Recharging Ports	1992 - 1998	Right rear engine compartment	Fig. 17.
Air Flow Sensor	1992 - 1998	Between throttle housing & filter housing	Fig. 8.
Airbag Contact Slip Ring	1992 - 1998	Rear of steering wheel	
Airbag, Left Front	1992 - 1998	In center of steering wheel	Fig. 26.
Airbag, Right Front	1994 - 1998	Right front of dash above glove compartment	Fig. 27.
Airbag Right/Left Front Door	1997 - 1998	Behind right/left front door panel	
Airbag Control Module (Basic)	1992 - 1993	Behind glove compartment	Fig. 10.
Airbag Control Module (ZAE)	1994 - 1998	On center tunnel under rear seat cushion	
Airbag Crash Sensor, Airbag	1992 - 1993	On right/left front strut tower	
Airbag Crash Sensor, Right/Left Side Impact	1997 - 1998	Right/left door sill behind front seat	
Alternator (Generator)	1992 - 1998	Left lower front of engine	
Amplifier Hi-fi	1992 - 1998	Left front side of luggage compartment	
Antenna Amplifier AM/FM	1992 - 1998	In upper left side C-pillar	Fig. 28.
Anti-theft Control Module	1992 - 1998	Behind glove compartment	Fig. 10.

Continued

Table a. E36 Component Locations

Component	Model Year	Location	Refer to
Anti-theft Horn	1992 - 1998	Right rear engine compartment	
Anti-slip (Traction) Control Module (AST/ASC+T)	1994 - 1998	Behind glove compartment	Fig. 10.
Anti-slip (Traction) Control Switch	1994 - 1998	In center console	
Automatic Climate (A/C) Control Module	1992 - 1998	Behind center console	
Automatic Transmission Range Switch (Gear Position/Neutral Safety Switch)	1992 - 1998	Next to gear shift lever in center console	Fig. 30.
Auxiliary Cooling Fan Dual Switch	1992 - 1998	Right side radiator	Fig. 19.
Auxiliary Cooling Fan Motor	1992 - 1998	Behind front grill	
Auxiliary Fuse Box (Left Splice Panel)	1992 - 1998	Behind left front footwell kick-panel	Fig. 6.
Auxiliary Relay Panel	1992 - 1998	Left side of dash under steering column	Fig. 5.
Auxiliary Fuse Box In Trunk	1997 - 1998	Right side of trunk behind trim panel	
Auxiliary Throttle Position Motor (ADS)	1994 - 1998	Front of engine compartment	
B+ Jump Start Junction Point	1992 - 1998	Right rear of engine compartment	Fig. 8.
Back-Up Light Switch (Manual)	1992 - 1998	Right or left side top of transmission	Fig. 31.
Body Electronics Control Module (ZKE IV)	1994 - 1998	Behind glove compartment	Fig. 10.
Brake Fluid Level Sensor	1992 - 1998	Top of brake fluid reservoir at master cylinder	Fig. 16.
Brake Light Switch	1992 - 1998	On brake pedal support bracket	Fig. 25.
Brake Pad Sensor, Left Front	1992 - 1998	Left front brake caliper	
Brake Pad Sensor, Right Rear	1992 - 1998	Right rear brake caliper	
Camshaft Actuator (VANOS)	1993 - 1998	Front of engine compartment	
Camshaft Position Sensor (Cylinder Identification)	1992 - 1998	Front of engine	Fig. 8.
Carbon Canister Valve (TLEV)	1997 - 1998	In spare wheel well in trunk	
Central Locking Module (ZVM)	1992 - 1993	Behind glove compartment	
Check Control Module (if applicable)	1992 - 1998	Below left side of dash	
Child Protection Window Control (console)	1992 - 1998	Center console	
Chime Module	1992 - 1998	Below left side of dash in left dash trim panel	
Clock (Multi-Function)	1993 - 1998	Center console	
Clutch Pedal Switch	1992 - 1998	On clutch pedal support bracket	
Combination (Stalk) Switch	1993 - 1998	Left side steering column (turn signals/high-low bean switch)	
Convertible Top Actuating Switch	1994 - 1998	Center console	

Continued

Table a. E36 Component Locations

Component	Model Year	Location	Refer to
Convertible Top Control Module	1994 - 1998	In lower left C-pillar behind trim panel	
Convertible Top End Position Switch	1994 - 1998	In convertible top linkage	
Convertible Top Motor	1994 - 1998	In left side of trunk behind trim panel	
Convertible Top Position Microswitch	1994 - 1998	In convertible top linkage	
Convertible Top Storage Lid Lock Motor	1994 - 1998	In convertible top compartment left side	
Convertible Top Storage Lid Microswitch	1994 - 1998	In right side convertible top storage lid latch	
Convertible Top Storage Lid Motor	1994 - 1998	In left side of trunk behind trim panel	
Coolant Level Sensor (if applicable)	1992 - 1998	Bolted on radiator, left side	
Coolant Temperature Sensor	1992 - 1998	Left front of engine/cylinder head	Fig. 8. Fig. 20. Fig. 21.
Cooling Fan Dual Switch	1992 - 1998	Right side radiator	Fig. 19.
Cooling Fan Motor	1992 - 1998	Behind front grill	
Crankshaft Position/RPM Sensor	1992 - 1995	At front crankshaft pulley	
Crankshaft Position/RPM Sensor	1996 - 1998	Left rear of crankcase	
Crash Sensor, Airbag	1992 - 1993	On right/left front strut tower	
Crash Control Module	1992	Auxiliary relay panel	Fig. 4.
Crash Control Module	1993 - 1998	Driver's side footwell	
Crash Sensor, Right/Left Side Impact Airbag	1997 - 1998	Right/left door sill behind front seat	
Cruise Control Actuator	1992 - 1998	Left front of engine compartment	Fig. 8.
Cruise Control Module	1992 - 1998	Behind glove compartment	Fig. 10.
Data Link Connector	1992 - 1998	Right side of engine	Fig. 8.
Digital Motor Electronics (DME) Control Module	1992 - 1998	Right rear engine compartment, behind bulkhead panel	Fig. 8.
Door Lock Heater	1993 - 1994	Left front door handle	
Door Handle Switch	1994 - 1998	left front door	
Drive-away Protection (EWS II) Module	1/1995 - 1998	Behind glove compartment	Fig. 10.
Dual Coolant Temperature Sensor	1997 - 1998	Below intake manifold	Fig. 20. Fig. 21.
Electronic Immobilizer Control Module (EWS II)	1/1995 - 1998	Behind glove compartment	Fig. 10.
Engine Compartment Light Switch	1993 - 1996	Over right headlight assembly	Fig. 8.
Engine Control Module (ECM)	1992 - 1998	Right rear engine compartment, behind bulkhead panel	Fig. 8.

Continued

Table a. E36 Component Locations

Component	Model Year	Location	Refer to
Engine Control Module (System) Relay	1992 - 1998	Power distribution box	Fig. 2. Fig. 3.
Engine Coolant Level Sensor	1992 - 1998	Left front of engine compartment, radiator coolant reservoir	
Engine Coolant Temperature Sensor	1992 - 1998	Left front of engine under intake manifold	Fig. 8. Fig. 20. Fig. 21.
Evaporative Emission Valve	1992 - 1998	Left side strut tower area	
Evaporator Temperature Sensor	1992 - 1998	Left side of heater box	
EWS II Control Module	1/1995 - 1998	Behind glove compartment	Fig. 10.
EWS II Ring Antenna	1/1995 - 1998	At ignition cylinder	
EWS II Transmitter Module	1/1995 - 1998	Right side of steering column in auxiliary relay panel	Fig. 4.
Exhaust Gas Flap Valve	1996 - 1998	Left rear muffler area	
Flasher Relay	1992 - 1998	Power distribution box	Fig. 2. Fig. 3.
Foglight Relay	1992 - 1998	Power distribution box	Fig. 2. Fig. 3.
Fresh Air Flap Actuator	1992	Right side of heater box	
Fresh Air Flap Actuator	1993 - 1995	Below right side heater box	
Fresh Air Flap Actuator	1996 - 1998	Below left side heater box	
Fuel Injectors	1992 - 1998	Top left side of engine	
Fuel Level Sensor, Left	1992 - 1998	Under left side of rear seat in fuel tank	
Fuel Level Sensor, Right	1992 - 1998	Under right side of rear seat in fuel tank	Fig. 29.
Fuel Pump	1992 - 1998	Under right side of rear seat in fuel tank	Fig. 29.
Fuel Pump Relay	1992 - 1998	Power distribution box	Fig. 2. Fig. 3.
Fuel Pump/Start Identification Relay	1994 - 1996	Power distribution box	
Fuel Tank Pressure Sensor	1997 - 1998	Right rear wheel well behind trim liner	
Fuse and Relay Box (Power Distribution Box)	1992 - 1998	Left side of engine compartment	Fig. 2. Fig. 3. Fig. 8. Fig. 11.
Gas Filler Lock Motor	1992 - 1998	Right side of luggage compartment	
Gear Position/Neutral Safety Switch (Automatic Transmission Range Switch)	1992 - 1998	Next to gear shift lever in center console	Fig. 30.
Glove Compartment Lock Motor	1994 - 1998	Behind glove compartment	

Continued

Table a. E36 Component Locations

Component	Model Year	Location	Refer to
Grounds			
G 100	1995 - 1998	Front of left front fender (headlights)	Fig. 23.
G 101	1992 - 1998	Front of right front fender	
G 102	1992 - 1998	Left front strut tower	
G 103	1992 - 1998	Right front strut tower	
G 110	1992 - 1998	Left front of engine	
G 111	1992 - 1998	Near battery	
G 117	1992 - 1998	Right rear of engine	
G 119	1992 - 1998	Right front of engine	
G 123	1992 - 1998	Right side of safety wall	
G 200	1992 - 1998	Left kick panel	
G 201	1992 - 1998	Right side of I/P	
G 202	1992 - 1998	Left side of I/P	
G 203	1992 - 1998	Right kick panel	
G 230	1992 - 1998	Right kick panel	
G 301	1992 - 1998	Below right front seat	
G 302	1992 - 1998	Below center console	
G 303	1992 - 1998	Below right rear seat	
G 310	1992 - 1998	Behind right rear seat	
G 312	1992 - 1998	Behind left rear seat	
G 313	1992 - 1998	Right side of rear shelf	
G 314	1992 - 1998	Left side of rear shelf	
G 400	1992 - 1998	Left front side of trunk	
G 404	1992 - 1998	Left rear side of trunk	
Camshaft Position Sensor (Hall effect)	1992 - 1998	Front of engine	Fig. 8.
Hard Top, Connector I	1994 - 1998	In convertible top compartment left side	
Hard Top, Connector II	1994 - 1998	In convertible top compartment left side	
Hazard Flasher Relay	1992 - 1998	Power distribution box	Fig. 2. Fig. 3.
Headlight High Beam Relay	1998	Left hand side of dash behind instrument cluster	
Headlight High Beam Relay	1992 - 1997	Power distribution box	
Headlight Low Beam Relay	1992 - 1998	Power distribution box	
Heat Exchanger Temperature Sensors (left and right)	1993 - 1998	Center console	

Continued

Table a. E36 Component Locations

Component	Model Year	Location	Refer to
Heater Blower Motor	1992 - 1998	Behind fresh air cowl	
Heater Blower Motor Relay	1992 - 1998	Power distribution box	Fig. 2. Fig. 3.
Heat Rear Recirculation Flap Actuator	1997 - 1998	Right side of heater box	
Heater Recirculation Flap Actuator	1992 - 1998	Below left dash right side of steering column	
Heater Blower Motor Resistor Pack	1992 - 1998	Behind center of dashboard	
Heater Regulator Flap Actuator	1997 - 1998	Right side of heater box	
Heater Valve Assembly	1992 - 1998	Left rear engine compartment	
Heating/Ventilation Relay Box	1992 - 1993	Below left side of dash	
Heating/Ventilation Relay	1994 - 1998	Left front footwell	
High Speed Cooling Fan Relay	1992 - 1998	Power distribution box	Fig. 2. Fig. 3.
Horns, Right/Left	1993 - 1998	Front of engine compartment	
Horn Relay	1992 - 1998	Power distribution box	Fig. 2. Fig. 3.
Horn Relay (Alpine) Board Computer/Anti-Theft	1994 - 1998	Behind glove compartment	Fig. 10.
Idle Speed Control Valve	1992 - 1995	Top rear of engine behind intake manifold (M42 engine)	Fig. 12.
Idle Speed Control Valve	1992 -1995	Top left side of engine (M50 engine)	
Idle Speed Control Valve	1993 - 1996	Below intake manifold (M50/S50US engine)	
Idle Speed Control Valve	1996 - 1998	On throttle housing (M44 engine)	Fig. 16.
Idle Speed Control Valve	1997 - 1998	Left front of engine below intake manifold (M52/S52US engine)	
Ignition Coils	1992 - 1998	Right of engine under plastic cover (four-cylinder)	Fig. 8.
Ignition Coils	1992 - 1998	Top of engine under cover (six-cylinder)	Fig. 18.
Ignition Switch	1992 - 1998	Right side of steering column	
Intake Air Resonance Changeover Valve (DISA)	1992 - 1998	Left rear of engine compartment	
Intake Air Temperature Sensor	1992 - 1995	In mass air flow sensor (M42 engine)	
Intake Air Temperature Sensor	1996 - 1998	In top of air filter housing (M44 engine)	Fig. 13.
Intake Air Temperature Sensor	1992 - 1998	Left side of engine in intake manifold (six-cylinder)	Fig. 14. Fig. 15.
Kick-Down Switch (A/T)	1992 - 1998	Left footwell	
Knock Sensors	1992 - 1998	Left side of engine block	
Left Front Power Window Motor	1992 - 1998	Left front door	
Left Front Window Monitoring Switch	1997 - 1998	Left front door trim panel	
Left Front Window Retraction Switch	1992 - 1993	Left front door handle lock assembly	
Left Splice Panel	1992 - 1998	Under left side of dash at kick-panel	Fig. 6.

Continued

Table a. E36 Component Locations

Component	Model Year	Location	Refer to
Main Fuse and Relay Box (Power Distribution Box)	1992 - 1998	Left side of engine compartment	Fig. 2. Fig. 3.
Mass Air Flow Sensor	1992 - 1998	Left side of engine	Fig. 8. Fig. 16.
Mixing Actuators	1992 - 1998	Heater box	
Mixing Switch	1992 - 1994	Center console	
Multi-Function Clock	1993 - 1998	Center console	
Normal Speed Cooling Fan Relay	1992 - 1998	Power distribution box	Fig. 2. Fig. 3.
Neutral Safety Switch (Automatic Transmission Range Switch)	1992 - 1998	Next to gear shift lever in center console	Fig. 30.
Oil Pressure Switch	1992 - 1998	Left front of engine below intake manifold at oil filter housing	
On-Board Computer	1992 - 1998	Center console	
On-Board Computer Horn Relay	1992 - 1998	Behind glove compartment	Fig. 10.
On-Board Computer Temperature Sensor	1992 - 1998	Right front of engine compartment	
Outside Temperature Sensor	1992 - 1997	In right front brake cooling duct	
Outside Temperature Sensor	1998	Right rear of engine compartment behind fresh air cowl	
Oxygen Sensor Heater Relay	1992 - 1998	Power distribution box	Fig. 2. Fig. 3.
Oxygen Sensor(s)	1992 - 1998	On exhaust pipe ahead/behind catalytic converter (M42/M44 engine)	
Oxygen Sensor (x2)	1992 - 1995	On exhaust pipe ahead of catalytic converter (M50/S50US engine)	
Oxygen Sensors (x4)	1996 - 1998	On exhaust manifold/exhaust pipe behind catalytic converter (M52/S52US engine)	
Parking Brake Switch	1992 - 1998	On rear of parking brake lever	
Power Distribution Box	1992 - 1998	Rear left of engine compartment	Fig. 2. Fig. 3. Fig. 8. Fig. 11.
Radio Microswitch	1992 - 1996	Center console behind radio	
Radio Receiver	1997 - 1998	Right rear luggage compartment	
Relay And Fuse Box (Power Distribution Box)	1992 - 1998	Left side of engine compartment	Fig. 2. Fig. 3. Fig. 8. Fig. 11.
Rear Window Blower	1994 - 1998	Behind rear seat (convertible)	
Rear Window Blower Microswitch	1994 - 1994	At convertible top linkage	
Rear Window Blower Microswitch	1995 - 1998	Behind rear seat (convertible)	

Continued

Table a. E36 Component Locations

Component	Model Year	Location	Refer to
Rear Window Blower Relay	1994 - 1998	Behind glove compartment	Fig. 10.
Rear Window Defogger Relay	1992 - 1998	Power distribution box	Fig. 2. Fig. 3.
Reversing (Back-up) Light Switch	1992 - 1998	Right or left side top of transmission	Fig. 31.
Splice Panel, Right	1992 - 1998	Behind glove compartment	Fig. 7.
Rollover Protection Actuator	1994 - 1998	Behind rear seat (convertible)	
Rollover Sensor	1994 - 1998	In lower left C-pillar behind trim panel (Convertible)	
Seat Belt Switch	1992 - 1998	In front seat belt buckle	
Seat Cushion Tilt Motor	1992 - 1998	Under front seat	
Seat Heater	1993 - 1998	Under front seat	
Seat Height Motor Front	1992 - 1998	Under front seat	
Seat Movement Motor	1992 - 1998	Under front seat	
Seatback Heater	1993 - 1998	In upper front seatback	
Seatback Recliner Motor	1993 - 1998	Under front seat	
Seat Occupancy Detector	1994 - 1998	Under right front seat	
Secondary Air Pump	1997½ - 1998	Left engine compartment, below power distribution box (M44 engine)	
Secondary Air Pump	1996 - 1998	Front of engine compartment (M52/S52US engine)	
Secondary Air Pump Relay	1996 - 1998	Power distribution box	Fig. 3.
Secondary Air Pump Valve	1996 - 1998	Right side of engine	
Side Impact Airbag Right/Left	1997 - 1998	Behind right/left door trim panel	
Side Impact Airbag Crash Sensor, Right/Left	1997 - 1998	Right/left door sill behind front seat	
Shift Interlock	1992 - 1998	Center console beside shift selector	
Slip Control Module (ABS/AST)	1992 - 1998	Behind glove compartment	Fig. 10.
Splice Panel, Left Side	1992 - 1998	Behind footwell speaker grill	Fig. 6.
Splice Panel, Right Side	1992 - 1998	Behind glove compartment	Fig. 7.
Starter	1992 - 1998	Left rear of engine	
Starter Immobilization Relay	1994 - 1994	In left splice panel behind footwell speaker grill	
Starter Relay	1992 - 1993	Below left side of dash	
Starter Relay	1994 - 1998	Behind left footwell speaker grill	
Sunroof Control Assembly	1992 - 1998	In roof center panel	
Supplemental Restraint System (Airbag) Control Module	1992 - 1993	Behind glove compartment	Fig. 10.
Supplemental Restraint System (Airbag ZAE) Control Module	1994 - 1998	Under rear seat	
Switch, Power Window	1992 - 1998	Center console	

Continued

Table a. E36 Component Locations

Component	Model Year	Location	Refer to
Telephone Eject Box	1994 - 1996	Center console	
Telephone Transceiver	1992 - 1992	In luggage compartment	
Throttle Flap Heater	1994 - 1996	Left side of engine	
Throttle Position Sensor	1992 - 1998	Left side of engine in throttle housing	Fig. 8. Fig. 16. Fig. 22.
Transmission Control Module	1992 - 1998	Right rear engine compartment behind panel	Fig. 8.
Transmission Range Switch (Gear Position/Neutral Safety Switch)	1992 - 1998	Next to gear shift lever in center console	Fig. 30.
Transmission Valve Unit	1992 - 1993	In transmission pan	
Transmission Valve Unit	1994 - 1998	On left side of transmission	
Trunk Lid Microswitch	1994 - 1998	In trunk lid	
Trunk Lid Relay	1994 - 1998	Behind glove compartment (convertible)	Fig. 10.
Trunk Light Switch	1992 - 1998	Left underside of trunk lid	
Trunk Lock Motor	1992 - 1998	Rear of luggage compartment	
Unloader Relay	1992 - 1998	In left splice panel behind footwell speaker grill	Fig. 6.
Variable Camshaft Control Module (VANOS)	1994 - 1996	Rear right side of engine compartment	
Variable Camshaft Control (VANOS)	1993 - 1998	Front of engine	
Vehicle Speed Sensor	1992 - 1998	On final drive	
Volume Air Flow Sensor (M42 Engine)	1992 - 1995	Left side of engine compartment	Fig. 8.
Washer Fluid Level Sensor	1992 - 1998	Right front of engine compartment	
Washer Pump	1992 - 1998	Right side of engine compartment on windshield washer reservoir	Fig. 8.
Wiper Motor	1992 - 1998	Rear engine compartment	
Wiper Motor Relay	1992 - 1998	In left splice panel behind footwell speaker grill	Fig. 6.
Wiper/Washer Module	1992-1993 (early)	Left side of footwell under foot rest	
Wiper/Washer Module	1993 (late)-1998	Behind glove compartment	Fig. 10.
ZAE Airbag Control Module	1994 - 1998	On center tunnel under rear seat cushion	
ZKE IV Control Module	1994 - 1998	Behind glove compartment	Fig. 10.

FUSE POSITION TABLES

The following tables list fuse positions in the power distribution box and, where applicable, in the auxiliary relay panel, the left side splice panel and the luggage compartment. Refer to the beginning of this repair group for locations of these panels.

> ***CAUTION —***
> *Fuse designations and locations are subject to change, and may vary from car to car, depending on options. If questions arise, please remember that an authorized BMW dealer is the best source for the most accurate and up-to-date information.*

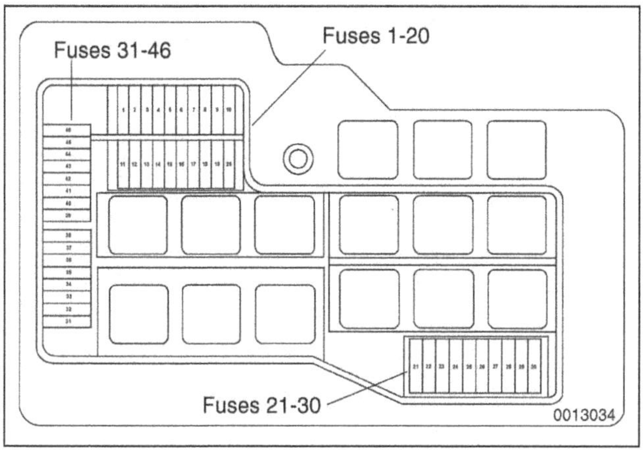

Fuse positions 1-46 in front power distribution box

Table b. 1992 E36 Fuse Positions

Fuse	Rating	Protected Circuit
1	30A	Power sunroof (see also fuse 13)
2		Not used
3	30A	Headlight washer
4	15A	Heated seats (see also fuse 23)
5	30A	Power seats (see also fuse 40)
6	20A	Rear window defogger (see also fuse 23)
7	5A	Anti-theft system (Alpine) (see also fuse 43) Central locking (ZVM II) (see also fuse 35, 43)
8	15A	Horn
9	20A	Sound system (see also fuse 44)
10	30A	ABS (see also fuse 21, 27, 38, 46)
11	7.5A	Crash control module (see also fuse 12, 25, 29, 34) Headlights/fog lights (see also fuse 12, 15, 22, 23, 25, 29, 30) On-Board Computer (see also fuse 12, 23, 27, 31, 33, 37, 45, 46)

Table b. 1992 E36 Fuse Positions

Fuse	Rating	Protected Circuit
12	7.5A	Crash control module (see also fuse 11, 25, 29, 34) Headlights/fog lights (see also fuse 11, 15, 22, 23, 25, 29, 30) On-Board Computer (see also fuse 11, 23, 27, 31, 33, 37, 45, 46)
13	5A	Power sunroof (see also fuse 1) Power windows (see also fuse 14)
14	30A	Power windows (see also fuse 13)
15	5A	Headlights/fog lights (see also fuse 11, 12, 22, 23, 25, 29, 30)
16	5A	Engine control module (see also fuse 28) Heating and A/C (IHKR E-36)(see also fuse 20, 23, 31, 37, 39, 41)
17		Not used
18	15A	Fuel pump
19	15A	Park ventilation (see also fuse 20)
20	30A	Heating and A/C (IHKR E-36)(see also fuse 16, 23, 31, 37, 39, 41) Park ventilation (see also fuse 19)
21	5A	ABS (see also fuse 10, 27, 38, 46)
22	5A	Headlights/fog lights (see also fuse 11, 12, 15, 23, 25, 29, 30) Instrument illumination (see also fuse 25, 33, 37) Park/tail lights (see also fuse 33, 37)
23	5A	Multi-function clock (see also fuse 27, 31, 45, 46) Headlights/fog lights (see also fuse 11, 12, 15, 22, 25, 29, 30) Heated seats (see also fuse 4) Heating and A/C (IHKR E-36)(see also fuse 16, 20, 31, 37, 39, 41) Instrument cluster (see also fuse 27, 31, 46) On-Board Computer (see also fuse 11, 12, 27, 31, 33, 37, 45, 46) Rear window defogger (see also fuse 6) Turn signals/hazard lights (see also fuse 34)
24	10A	Power mirrors
25	5A	Crash control module (see also fuse 11, 12, 29, 34) Headlights/fog lights (see also fuse 11, 12, 15, 22, 23, 29, 30) Instrument illumination (see also fuse 22, 33, 37)
26	15A	Back-up lights Electronic transmission control (A4S 310R) (see also fuse 28, 46)
27	5A	ABS (see also fuse 10, 21, 38, 46) Multi-function clock (see also fuse 23, 31, 45, 46) Instrument cluster (see also fuse 23, 31, 46) On-Board Computer (see also fuse 11, 12, 23, 31, 33, 37, 45, 46)

Table b. 1992 E36 Fuse Positions

Fuse	Rating	Protected Circuit
28	5A	Cruise control (see also fuse 46) Engine control module (see also fuse 16) Starting system Electronic transmission control (A4S 310R) (see also fuse 26, 46)
29	7.5A	Crash control module (see also fuse 11, 12, 25, 34) Headlights/fog lights (see also fuse 11, 12, 15, 22, 23, 25, 30)
30	7.5A	Headlights/fog lights (see also fuse 11, 12, 15, 22, 23, 25)
31	5A	Multi-function clock (see also fuse 23, 27, 45, 46) Heating and A/C (IHKR E-36)(see also fuse 16, 20, 23, 37, 39, 41) Instrument cluster (see also fuse 23, 27, 46) On-Board Computer (see also fuse 11, 12, 23, 27, 33, 37, 45, 46)
32	30A	Glove compartment light/cassette compartment/ashtray light/charging socket/cigar lighter (see also fuse 33, 37, 44)
33	10A	Cellular telephone (see also fuse 43) Glove compartment light/cassette compartment/ashtray light/charging socket/cigar lighter (see also fuse 32, 37, 44) Interior lights (also see fuse 43) License plate/luggage compartment lights (see also fuse 37) Instrument illumination (see also fuse 22, 25, 37) On-Board Computer (see also fuse 11, 12, 23, 27, 31, 37, 45, 46) Park/tail lights (see also fuse 22, 37)
34	15A	Crash control module (see also fuse 11, 12, 25, 29) Turn signals/hazard lights (see also fuse 23)
35	25A	Central locking (ZVM II) (see also fuse 7, 43)
36	30A	Wiper/washer (see also fuse 37, 44, 45)
37	10A	Glove compartment light/cassette compartment/ashtray light/charging socket/cigar lighter (see also fuse 32, 33, 44) Heating and A/C (IHKR E-36)(see also fuse 16, 20, 23, 31, 39, 41) Instrument illumination (see also fuse 22, 25, 33) License plate/luggage compartment lights (see also fuse 33) On-Board Computer (see also fuse 11, 12, 23, 27, 31, 33, 45, 46) Park/tail lights (see also fuse 22, 33) Wiper/washer (see also fuse 36, 44, 45)
38	30A	ABS (see also fuse 10, 21, 27, 46)
39	7.5A	Heating and A/C (IHKR E-36)(see also fuse 16, 20, 23, 31, 37, 41)
40	30A	Power seats (see also fuse 5)

Table b. 1992 E36 Fuse Positions

Fuse	Rating	Protected Circuit
41	30A	Heating and A/C (IHKR E-36)(see also fuse 16, 20, 23, 31, 37, 39) Radiator auxiliary fan
42	7.5A	Not used
43	5A	Anti-theft system (Alpine) (see also fuse 7) Cellular telephone (see also fuse 33) Central locking (ZVM II) (see also fuse 7, 35) Interior lights (also see fuse 33)
44	15A	Glove compartment light/cassette compartment/ashtray light/charging socket/cigar lighter (see also fuse 32, 33, 37) Sound system (see also fuse 9) Wiper/washer (see also fuse 36, 37, 45)
45	7.5A	Multi-function clock (see also fuse 23, 27, 31, 46) On-Board Computer (see also fuse 11, 12, 23, 27, 31, 33, 37, 46) Wiper/washer (see also fuse 36, 37, 44)
46	15A	ABS (see also fuse 10, 21, 27, 38) Brake lights Multi-function clock (see also fuse 23, 27, 31, 45) Cruise control (see also fuse 28) Electronic transmission control (A4S 310R) (see also fuse 26, 28) Instrument cluster (see also fuse 23, 27, 31) On-Board Computer (see also fuse 11, 12, 23, 27, 31, 33, 37, 45)

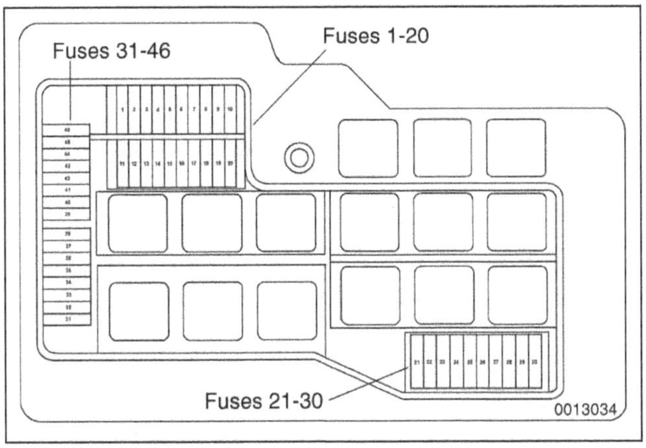

Fuse positions 1-46 in front power distribution box

Table c. 1993 E36 Fuse Positions

fuse	Rating	Protected Circuit
1	30A	Power sunroof
2		Not used
3	30A	Headlight washer
4	15A	Heated seats (see also fuse 23)
5	30A	Power seats (see also fuse 40)
6	20A	Rear window defogger (see also fuse 23)
7	5A	Central locking (ZVM) (see also fuse 35, 43)
8	15A	Anti-theft (Alpine) (see also fuse 43) Horn
9	20A	Sound system (see also fuse 44)
10	30A	ABS (see also fuse 21, 27, 38, 46)
11	7.5A	Headlights/fog lights (see also fuse 12, 15, 23, 25, 29, 30) On-Board Computer (see also fuse 12, 23, 27, 31, 33, 37, 45, 46)
12	7.5A	Headlights/fog lights (see also fuse 11, 15, 23, 25, 29, 30) On-Board Computer (see also fuse 11, 23, 27, 31, 33, 37, 45, 46)
13	5A	Not used
14	30A	Power windows (see also fuse 33)
15	5A	Headlights/fog lights (see also fuse 11, 12, 23, 25, 29, 30)
16	5A	Engine control module (see also fuse 28) Heating and A/C (IHKR E-36)(see also fuse 20, 23, 31, 37, 39, 41)
17		Not used
18	15A	Fuel pump
19	15A	Not used
20	30A	Heating and A/C (IHKR E-36)(see also fuse 16, 23, 31, 37, 39, 41) Park ventilation

Table c. 1993 E36 Fuse Positions

fuse	Rating	Protected Circuit
21	5A	ABS (see also fuse 10, 27, 38, 46)
22	5A	Instrument illumination (see also fuse 25, 33, 37) Park/tail lights (see also fuse 33, 37)
23	5A	Multi-function clock (see also fuse 27, 31, 45, 46) Headlights/fog lights (see also fuse 11, 12, 15, 25, 29, 30) Heated seats (see also fuse 4) Heating and A/C (IHKR E-36)(see also fuse 16, 20, 31, 37, 39, 41) Instrument cluster (see also fuse 27, 31, 46) On-Board Computer (see also fuse 11, 12, 27, 31, 33, 37, 45, 46) Rear window defogger (see also fuse 6) Turn signals/hazard lights (see also fuse 34)
24	10A	Power mirrors
25	5A	Headlights/fog lights (see also fuse 11, 12, 15, 23, 29, 30) Instrument illumination (see also fuse 22, 33, 37)
26	10A	Back-up lights Electronic transmission control (A4S 310R) (see also fuse 28, 46)
27	5A	ABS (see also fuse 10, 21, 38, 46) Multi-function clock (see also fuse 23, 31, 45, 46) Instrument cluster (see also fuse 23, 31, 46) On-Board Computer (see also fuse 11, 12, 23, 31, 33, 37, 45, 46)
28	5A	Cruise control (see also fuse 46) Engine control module (see also fuse 16) Starting system Electronic transmission control (A4S 310R) (see also fuse 26, 46)
29	7.5A	Headlights/fog lights (see also fuse 11, 12, 15, 23, 25, 30)
30	7.5A	Headlights/fog lights (see also fuse 11, 12, 15, 23, 25)
31	5A	Multi-function clock (see also fuse 23, 27, 45, 46) Heating and A/C (IHKR E-36)(see also fuse 16, 20, 23, 37, 39, 41) Instrument cluster (see also fuse 23, 27, 46) On-Board Computer (see also fuse 11, 12, 23, 27, 33, 37, 45, 46)
32	30A	Glove compartment light/cassette compartment/ashtray light/charging socket/cigar lighter (see also fuse 33, 37, 44)

Table c. 1993 E36 Fuse Positions

fuse	Rating	Protected Circuit
33	10A	Cellular telephone (see also fuse 43) Glove compartment light/cassette compartment/ashtray light/charging socket/cigar lighter (see also fuse 32, 37, 44) Interior lights (also see fuse 43) License plate/luggage compartment lights (see also fuse 37) Instrument illumination (see also fuse 22, 25, 37) On-Board Computer (see also fuse 11, 12, 23, 27, 31, 37, 45, 46) Park/tail lights (see also fuse 22, 37) Power windows (see also fuse 14)
34	15A	Crash control module Turn signals/hazard lights (see also fuse 23)
35	25A	Central locking (ZVM II) (see also fuse 7, 43)
36	30A	Wiper/washer (see also fuse 37, 44, 45)
37	10A	Glove compartment light/cassette compartment/ashtray light/charging socket/cigar lighter (see also fuse 32, 33, 44) Heating and A/C (IHKR E-36)(see also fuse 16, 20, 23, 31, 39, 41) Instrument illumination (see also fuse 22, 25, 33) License plate/luggage compartment lights (see also fuse 33) On-Board Computer (see also fuse 11, 12, 23, 27, 31, 33, 45, 46) Park/tail lights (see also fuse 22, 33) Wiper/washer (see also fuse 36, 44, 45)
38	30A	ABS (see also fuse 10, 21, 27, 46)
39	7.5A	Heating and A/C (IHKR E-36)(see also fuse 16, 20, 23, 31, 37, 41)
40	30A	Power seats (see also fuse 5)
41	30A	Heating and A/C (IHKR E-36)(see also fuse 16, 20, 23, 31, 37, 39) Radiator auxiliary fan
42	7.5A	Airbag (SRS) (see also fuse 43)
43	5A	Anti-theft system (Alpine) (see also fuse 8) Airbag (SRS) (see also fuse 42) Cellular telephone (see also fuse 33) Central locking (ZVM II) (see also fuse 7, 35) Interior lights (also see fuse 33)
44	15A	Glove compartment light/cassette compartment/ashtray light/charging socket/cigar lighter (see also fuse 32, 33, 37) Sound system (see also fuse 9) Wiper/washer (see also fuse 36, 37, 45)
45	7.5A	Multi-function clock (see also fuse 23, 27, 31, 46) On-Board Computer (see also fuse 11, 12, 23, 27, 31, 33, 37, 46) Wiper/washer (see also fuse 36, 37, 44)

Table c. 1993 E36 Fuse Positions

fuse	Rating	Protected Circuit
46	15A	ABS (see also fuse 10, 21, 27, 38) Brake lights Multi-function clock (see also fuse 23, 27, 31, 45) Cruise control (see also fuse 28) Electronic transmission control (A4S 310R) (see also fuse 26, 28) Instrument cluster (see also fuse 23, 27, 31) On-Board Computer (see also fuse 11, 12, 23, 27, 31, 33, 37, 45)

CAUTION —
Fuse designations and locations are subject to change, and may vary from car to car, depending on options. If questions arise, please remember that an authorized BMW dealer is the best source for the most accurate and up-to-date information.

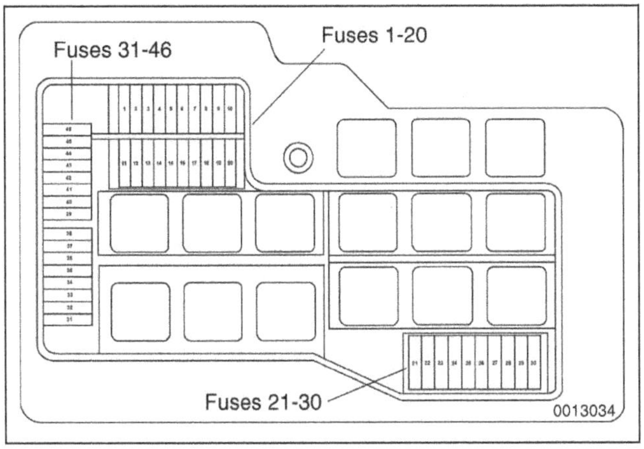

Fuse positions 1-46 in front power distribution box

Table d. 1994 E36 Fuse Positions

fuse	Rating	Protected Circuit
1	30A	Power sunroof
2	15A	Not used
3	30A	Headlight washer (also see fuse 37)
4	15A	Heated seats (see also fuse 23)
5	30A	Power seats (see also fuse 40)
6	20A	Rear window defogger/blower (see also fuse 23)
7	5A	Central body electronics (ZKE IV) (convertible only) (see also fuse 33, 43) Central locking (see also fuse 35, 43) Convertible roof (see also fuse 35, 43) Roll-over protection (see also fuse 35, 42, 43)
8	15A	Horn
9	20A	Sound system (see also fuse 44)
10	30A	ABS (see also fuse 21, 38, 46) AST (see also fuse 21, 38)
11	7.5A	Headlights/fog lights (see also fuse 12, 15, 23, 25, 29, 30) On-Board Computer (see also fuse 12, 23, 27, 31, 33, 37, 45, 46)
12	7.5A	Headlights/fog lights (see also fuse 11, 15, 23, 25, 29, 30) On-Board Computer (see also fuse 11, 23, 27, 31, 33, 37, 45, 46)
13	5A	Not used
14	30A	Front power windows
15	15A	Headlights/fog lights (see also fuse 11, 12, 23, 25, 29, 30)
16	5A	Engine control module (see also fuse 28) Heating and A/C (IHKR E-36)(see also fuse 20, 23, 31, 37, 39, 41)
17	10A	Not used

Table d. 1994 E36 Fuse Positions

fuse	Rating	Protected Circuit
18	15A	Fuel pump
19	30A	Rear power windows
20	30A	Blower motor Heating and A/C (IHKR E-36) (see also fuse 16, 23, 31, 37, 39, 41) Park ventilation
21	5A	ABS (see also fuse 10, 38, 46) AST (see also fuse 10, 38)
22	5A	Instrument illumination (see also fuse 25, 37) Park/tail lights (see also fuse 33, 37)
23	5A	Headlights/fog lights (see also fuse 11, 12, 15, 25, 29, 30) Heated seats (see also fuse 4) Heating and A/C (IHKR E-36)(see also fuse 16, 20, 31, 37, 39, 41) Instrument cluster (see also fuse 27, 31, 46) On-Board Computer (see also fuse 11, 12, 27, 31, 33, 37, 45, 46) Rear window defogger/blower (see also fuse 6) Turn signals/hazard lights (see also fuse 34)
24	10A	Power mirrors
25	5A	Headlights/fog lights (see also fuse 11, 12, 15, 23, 29, 30) Instrument illumination (see also fuse 22, 37)
26	10A	Back-up lights Electronic transmission control (A4S 310R) (see also fuse 26, 46)
27	5A	Instrument cluster (see also fuse 23, 31, 46) On-Board Computer (see also fuse 11, 12, 23, 31, 33, 37, 45, 46)
28	5A	Cruise control (see also fuse 46) Engine control module (see also fuse 16) Starting system Electronic transmission control (A4S 310R) (see also fuse 26, 46)
29	7.5A	Headlights/fog lights (see also fuse 11, 12, 15, 23, 25, 30)
30	7.5A	Headlights/fog lights (see also fuse 11, 12, 15, 23, 25)
31	5A	Multi-function clock Heating and A/C (IHKR E-36)(see also fuse 16, 20, 23, 37, 39, 41) Instrument cluster (see also fuse 23, 27, 46) On-Board Computer (see also fuse 11, 12, 23, 27, 33, 37, 45, 46)
32	30A	Cigar lighter/ashtray lights
33	10A	Central body electronics (ZKE IV) (see also fuse 7, 43) Charging socket Interior lights (also see fuse 43) License plate/luggage compartment lights (see also fuse 37) On-Board Computer (see also fuse 11, 12, 23, 27, 31, 37, 45, 46) Park/tail lights (see also fuse 22, 37) Telephone (see also fuse 43)

Table d. 1994 E36 Fuse Positions

fuse	Rating	Protected Circuit
34	15A	Crash control module Turn signals/hazard lights (see also fuse 23)
35	25A	Central locking (see also fuse 7, 43) Convertible roof (see also fuse 7, 43) Roll-over protection (see also fuse 7, 42, 43)
36	30A	Wiper/washer (see also fuse 37, 44, 45)
37	10A	Cassette compartment light Engine compartment light Heating and A/C (IHKR E-36)(see also fuse 16, 20, 23, 31, 39, 41) Instrument illumination (see also fuse 22, 25) License plate/luggage compartment lights (see also fuse 33) On-Board Computer (see also fuse 11, 12, 23, 27, 31, 33, 45, 46) Park/tail lights (see also fuse 22, 33) Wiper/washer (see also fuse 36, 44, 45)
38	30A	ABS (see also fuse 10, 21, 46) AST (see also fuse 10, 21)
39	7.5A	Heating and A/C (IHKR E-36) (see also fuse 16, 20, 23, 31, 37, 41)
40	30A	Power seats (see also fuse 5)
41	30A	Heating and A/C (IHKR E-36)(see also fuse 16, 20, 23, 31, 37, 39) Radiator auxiliary fan
42	7.5A	Airbag (SRS) (see also fuse 43) Roll-over protection (see also fuse 7, 35, 43)
43	5A	Anti-theft system (Alpine) Airbag (SRS) (see also fuse 42) Central body electronics (ZKE IV) (see also fuse 7, 33) Central locking (see also fuse 7, 35) Convertible roof (see also fuse 7, 35) Interior lights (also see fuse 33) Roll-over protection (see also fuse 7, 35, 42) Telephone (see also fuse 33)
44	15A	Glove compartment light Sound system (see also fuse 9) Wiper/washer (see also fuse 36, 37, 45)
45	7.5A	On-Board Computer (see also fuse 11, 12, 23, 27, 31, 33, 37, 46) Wiper/washer (see also fuse 36, 37, 44)
46	15A	ABS (see also fuse 10, 21, 38) Brake lights Multi-function clock (see also fuse 23, 27, 31, 45) Cruise control (see also fuse 28) Electronic transmission control (A4S 310R) (see also fuse 26, 28) Instrument cluster (see also fuse 23, 27, 31) On-Board Computer (see also fuse 11, 12, 23, 27, 31, 33, 37, 45)

CAUTION —

Fuse designations and locations are subject to change and may vary from car to car, depending on options. If questions arise, please remember that an authorized BMW dealer is the best source for the most accurate and up-to-date information.

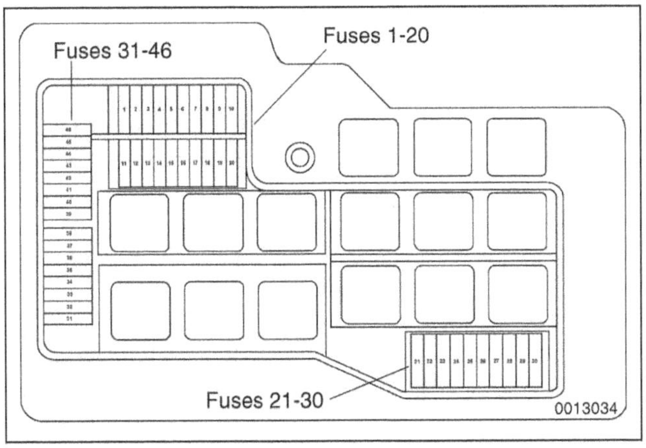

Fuse positions 1-46 in front power distribution box

Table e. 1995 E36 Fuse Positions

fuse	Rating	Protected Circuit
1	30A	Power sunroof
2	15A	Not used
3	30A	Headlight washer (also see fuse 37)
4	15A	Heated seats (see also fuse 23)
5	30A	Power seats (see also fuse 40)
6	20A	Rear window defogger/blower (see also fuse 23)
7	5A	Central body electronics (ZKE IV) (convertible only) (see also fuse 14, 31, 33, 35, 43) Central locking (see also fuse 26, 28, 35) Convertible roof (see also fuse 27, 35, 42, 43) Interior lights (also see fuse 33, 43) Roll-over protection (see also fuse 35, 42, 43)
8	15A	Horn
9	20A	Sound system (see also fuse 33, 44)
10	30A	ABS (see also fuse 21, 23, 27, 38, 46, 47) AST (see also fuse 21, 26, 27, 38, 46)
11	7.5A	Headlights/fog lights (see also fuse 12, 15, 17, 23, 25, 29, 30) On-Board Computer (see also fuse 12, 23, 27, 31, 33, 37, 45, 46)
12	7.5A	Headlights/fog lights (see also fuse 11, 15, 17, 23, 25, 29, 30) On-Board Computer (see also fuse 11, 23, 27, 31, 33, 37, 45, 46)
13	5A	Not used
14	30A	Central body electronics (ZKE IV) (see also fuse 7, 14, 31, 33, 35, 43) Front power windows
15	15A	Headlights/fog lights (see also fuse 11, 12, 17, 23, 25, 29, 30)

Table e. 1995 E36 Fuse Positions

fuse	Rating	Protected Circuit
16	5A	Engine control module (see also fuse 28, 39) Heating and A/C (IHKR E-36)(see also fuse 20, 23, 31, 37, 39, 41)
17	10A	Headlights/fog lights (see also fuse 11, 12, 15, 23, 25, 29, 30)
18	15A	Fuel pump
19	30A	Rear power windows
20	30A	Blower motor Heating and A/C (IHKR E-36) (see also fuse 16, 23, 31, 37, 39, 41) Park ventilation
21	5A	ABS (see also fuse 10, 23, 27, 38, 46, 47) AST (see also fuse 10, 26, 27, 38, 46)
22	5A	Instrument illumination (see also fuse 25, 33, 37) Park/tail lights (see also fuse 33, 37)
23	5A	ABS (see also fuse 10, 21, 27, 38, 46) Headlights/fog lights (see also fuse 11, 12, 15, 17, 25, 29, 30) Heated seats (see also fuse 4) Heating and A/C (IHKR E-36)(see also fuse 16, 20, 31, 37, 39, 41) Instrument cluster (see also fuse 27, 31, 46) Multi-function clock (see also fuse 26, 27, 31, 45, 46) On-Board Computer (see also fuse 11, 12, 27, 31, 33, 37, 45, 46) On-Board Display (see also fuse 26, 27, 31, 45, 46) Rear window defogger/blower (see also fuse 6) Turn signals/hazard lights (see also fuse 34)
24	10A	Power mirrors
25	5A	Headlights/fog lights (see also fuse 11, 12, 15, 17, 23, 29, 30) Instrument illumination (see also fuse 22, 33, 37)
26	10A	AST (see also fuse 10, 21, 27, 38, 46) Back-up lights Central locking (see also fuse 7, 28, 35) Electronic transmission control (A4S 310R/Z) (see also fuse 28, 46) Multi-function clock (see also fuse 23, 27, 31, 45, 46) On-Board Display (see also fuse 23, 27, 31, 45, 46) Starting system (see also fuse 28)
27	5A	ABS (see also fuse 10, 21, 23, 38, 46, 47) AST (see also fuse 10, 21, 26, 38, 46) Convertible roof (see also fuse 7, 35, 42, 43) Instrument cluster (see also fuse 23, 31, 46) Multi-function clock (see also fuse 23, 26, 31, 45, 46) On-Board Computer (see also fuse 11, 12, 23, 26, 31, 33, 37, 45, 46) On-Board Display (see also fuse 23, 31, 45, 46)

Table e. 1995 E36 Fuse Positions

fuse	Rating	Protected Circuit
28	5A	Central locking (see also fuse 7, 26, 35) Cruise control (see also fuse 46) Engine control module (see also fuse 16, 39) Electronic transmission control (A4S 310R/Z) (see also fuse 26, 46) Starting system (see also fuse 26)
29	7.5A	Headlights/fog lights (see also fuse 11, 12, 15, 17, 23, 25, 30)
30	7.5A	Headlights/fog lights (see also fuse 11, 12, 15, 17, 23, 25, 29)
31	5A	Central body electronics (ZKE IV) (see also fuse 7, 14, 33, 35, 43) Heating and A/C (IHKR E-36)(see also fuse 16, 20, 23, 37, 39, 41) Instrument cluster (see also fuse 23, 27, 46) Multi-function clock (see also fuse 23, 26, 27, 45, 46) On-Board Computer (see also fuse 11, 12, 23, 26, 27, 33, 37, 45, 46) On-Board Display (see also fuse 23, 27, 45, 46)
32	30A	Cigar lighter/ashtray lights
33	10A	Central body electronics (ZKE IV) (see also fuse 7, 14, 31, 35, 43) Charging socket Door lock heating Instrument illumination (see also fuse 22, 25, 37) Interior lights (also see fuse 7, 43) License plate/luggage compartment lights (see also fuse 37) On-Board Computer (see also fuse 11, 12, 23, 26, 27, 31, 37, 45, 46) Park/tail lights (see also fuse 22, 37) Sound system (see also fuse 9, 44) Telephone (see also fuse 43)
34	15A	Crash control module Turn signals/hazard lights (see also fuse 23)
35	25A	Central body electronics (ZKE IV) (see also fuse 7, 14, 31, 33, 43) Central locking (see also fuse 7, 26, 28) Convertible roof (see also fuse 7, 27, 42, 43) Roll-over protection (see also fuse 7, 42, 43)
36	30A	Wiper/washer (see also fuse 37, 44, 45)
37	10A	Cassette compartment light Engine compartment light Heating and A/C (IHKR E-36)(see also fuse 16, 20, 23, 31, 39, 41) Instrument illumination (see also fuse 22, 25, 33) License plate/luggage compartment lights (see also fuse 33) On-Board Computer (see also fuse 11, 12, 23, 26, 27, 31, 33, 45, 46) Park/tail lights (see also fuse 22, 33) Wiper/washer (Canada only) (see also fuse 36, 44, 45)
38	30A	ABS (see also fuse 10, 21, 23, 27, 46, 47) AST (see also fuse 10, 21, 26, 27, 46)

Table e. 1995 E36 Fuse Positions

fuse	Rating	Protected Circuit
39	7.5A	Engine control module (see also fuse 16, 28) Heating and A/C (IHKR E-36) (see also fuse 16, 20, 23, 31, 37, 41)
40	30A	Power seats (see also fuse 5)
41	30A	Heating and A/C (IHKR E-36)(see also fuse 16, 20, 23, 31, 37, 39) Radiator auxiliary fan
42	7.5A	Airbag (SRS) (see also fuse 43) Convertible roof (see also fuse 7, 27, 35, 43) Roll-over protection (see also fuse 7, 35, 43)
43	5A	Anti-theft system (Alpine) (see also fuse 47) Airbag (SRS) (see also fuse 42) Central body electronics (ZKE IV) (see also fuse 7, 14, 31, 33, 35) Central locking (see also fuse 7, 35) Convertible roof (see also fuse 7, 27, 35, 42) Interior lights (also see fuse 7, 33) Roll-over protection (see also fuse 7, 35, 42) Telephone (see also fuse 33)
44	15A	Glove compartment light Sound system (see also fuse 9, 33) Wiper/washer (see also fuse 36, 37, 45)
45	7.5A	Multi-function clock (see also fuse 23, 26, 27, 31, 46) On-Board Computer (see also fuse 11, 12, 23, 26, 27, 31, 33, 37, 46) On-Board Display (see also fuse 23, 27, 31, 46) Wiper/washer (Canada only) (see also fuse 36, 37, 44)
46	15A	ABS (see also fuse 10, 21, 23, 27, 38, 47) AST (see also fuse 10, 21, 26, 27, 38) Brake lights Cruise control (see also fuse 28) Electronic transmission control (A4S 310R) (see also fuse 26, 28) Instrument cluster (see also fuse 23, 27, 31) Multi-function clock (see also fuse 23, 26, 27, 31, 45) On-Board Computer (see also fuse 11, 12, 23, 26, 27, 31, 33, 37, 45) On-Board Display (see also fuse 23, 27, 31, 45)
47	15A	ABS (see also fuse 10, 21, 23, 27, 38, 46) Anti-theft system (Alpine) (see also fuse 43)
48	40A	Not used

CAUTION —

- *Fuse designations and locations are subject to change, and may vary from car to car, depending on options. If questions arise, please remember that an authorized BMW dealer is the best source for the most accurate and up-to-date information.*

- *To locate fuses 47-48, refer to* **Fuse and Relay Positions** *in this repair group.*

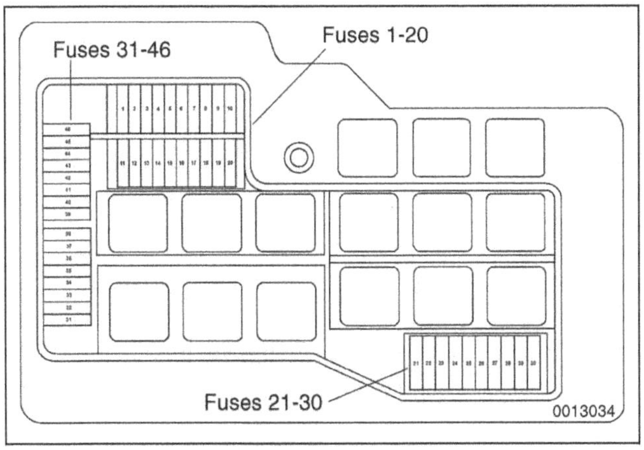

Fuse positions 1-46 in front power distribution box

Table f. 1996 E36 Fuse Positions

fuse	Rating	Protected Circuit
1	30A	Power sunroof (see also fuse 13)
2	15A	Not used
3	30A	Headlight washer (also see fuse 37)
4	15A	Heated seats (see also fuse 23)
5	30A	Power seats (see also fuse 40)
6	20A	Rear window defogger/blower (see also fuse 23)
7	5A	Central body electronics (ZKE IV) (convertible only) (see also fuse 14, 33, 35, 43) Central locking (see also fuse 26, 28, 35) Convertible roof (see also fuse 27, 35, 42, 43) Interior lights (also see fuse 33, 43) Roll-over protection (see also fuse 35, 42, 43)
8	15A	Horn
9	20A	Sound system (see also fuse 44)
10	30A	ABS (see also fuse 21, 23, 27, 38, 46, 47) AST (see also fuse 21, 26, 27, 38, 46)
11	7.5A	Headlights/fog lights (see also fuse 12, 15, 17, 23, 25, 29, 30) On-Board Computer (see also fuse 12, 23, 27, 31, 33, 37, 45, 46)
12	7.5A	Headlights/fog lights (see also fuse 11, 15, 17, 23, 25, 29, 30) On-Board Computer (see also fuse 11, 23, 27, 31, 33, 37, 45, 46)
13	5A	Power sunroof (see also fuse 1)
14	30A	Central body electronics (ZKE IV) (see also fuse 7, 14, 33, 35, 43) Front power windows
15	15A	Headlights/fog lights (see also fuse 11, 12, 17, 23, 25, 29, 30)
16	5A	Engine control module (see also fuse 28, 39) Heating and A/C (IHKR E-36)(see also fuse 20, 23, 31, 37, 39, 41, 48)
17	10A	Headlights/fog lights (see also fuse 11, 12, 15, 23, 25, 29, 30)
18	15A	Fuel pump
19	30A	Rear power windows
20	30A	Blower motor Heating and A/C (IHKR E-36) (see also fuse 16, 23, 31, 37, 39, 41, 48) Park ventilation
21	5A	ABS (see also fuse 10, 23, 27, 38, 46, 47) AST (see also fuse 10, 26, 27, 38, 46)
22	5A	Instrument illumination (see also fuse 25, 33, 37)
23	5A	ABS (see also fuse 10, 21, 27, 38, 46) Headlights/fog lights (see also fuse 11, 12, 15, 17, 25, 29, 30) Heated seats (see also fuse 4) Heating and A/C (IHKR E-36)(see also fuse 16, 20, 31, 37, 39, 41, 48) Instrument cluster (see also fuse 27, 31, 46) Multi-function clock (see also fuse 26, 27, 31, 45, 46) On-Board Computer (see also fuse 11, 12, 27, 31, 33, 37, 45, 46) On-Board Display (see also fuse 26, 27, 31, 45, 46) Rear window defogger/blower (see also fuse 6) Turn signals/hazard lights (see also fuse 34)
24	10A	Power mirrors
25	5A	Headlights/fog lights (see also fuse 11, 12, 15, 17, 23, 29, 30) Instrument illumination (see also fuse 22, 33, 37)
26	10A	Airbag (SRS) (see also fuse 43) AST (see also fuse 10, 21, 27, 38, 46) Back-up lights Central locking (see also fuse 7, 28, 35) Electronic transmission control (A4S 310R/Z) (see also fuse 28, 46) Multi-function clock (see also fuse 23, 27, 31, 45, 46) On-Board Display (see also fuse 23, 27, 31, 45, 46) Starting system (see also fuse 28)
27	5A	ABS (see also fuse 10, 21, 23, 38, 46, 47) AST (see also fuse 10, 21, 26, 38, 46) Convertible roof (see also fuse 7, 35, 42, 43) Instrument cluster (see also fuse 23, 31, 46) Multi-function clock (see also fuse 23, 26, 31, 45, 46) On-Board Computer (see also fuse 11, 12, 23, 26, 31, 33, 37, 45, 46) On-Board Display (see also fuse 23, 31, 45, 46)

Table f. 1996 E36 Fuse Positions

fuse	Rating	Protected Circuit
28	5A	Central locking (see also fuse 7, 26, 35) Cruise control (see also fuse 46) Engine control module (see also fuse 16, 39) Electronic transmission control (A4S 310R/Z) (see also fuse 26, 46) Driveaway protection (EWS II) (see also fuse 31, 46) Starting system (see also fuse 26)
29	7.5A	Headlights/fog lights (see also fuse 11, 12, 15, 17, 23, 25, 30)
30	7.5A	Headlights/fog lights (see also fuse 11, 12, 15, 17, 23, 25, 29)
31	5A	Driveaway protection (EWS II) (see also fuse 28, 46) Heating and A/C (IHKR E-36)(see also fuse 16, 20, 23, 37, 39, 41, 48) Instrument cluster (see also fuse 23, 27, 46) Multi-function clock (see also fuse 23, 26, 27, 45, 46) On-Board Computer (see also fuse 11, 12, 23, 26, 27, 33, 37, 45, 46) On-Board Display (see also fuse 23, 27, 45, 46)
32	30A	Cigar lighter/ashtray lights
33	10A	Central body electronics (ZKE IV) (see also fuse 7, 14, 35, 43) Charging socket Engine compartment light (also see fuse 37) Instrument illumination (see also fuse 22, 25, 37) Interior lights (also see fuse 7, 43) License plate/luggage compartment lights (see also fuse 37) On-Board Computer (see also fuse 11, 12, 23, 26, 27, 31, 37, 45, 46) Park/tail lights (see also fuse 37) Telephone (see also fuse 43)
34	15A	Crash control module Turn signals/hazard lights (see also fuse 23)
35	25A	Central body electronics (ZKE IV) (see also fuse 7, 14, 33, 43) Central locking (see also fuse 7, 26, 28) Convertible roof (see also fuse 7, 27, 42, 43) Roll-over protection (see also fuse 7, 42, 43)
36	30A	Wiper/washer (see also fuse 44, 45)
37	10A	Cassette compartment light Engine compartment light (also see fuse 33) Heating and A/C (IHKR E-36)(see also fuse 16, 20, 23, 31, 39, 41, 48) Instrument illumination (see also fuse 22, 25, 33) License plate/luggage compartment lights (see also fuse 33) On-Board Computer (see also fuse 11, 12, 23, 26, 27, 31, 33, 45, 46) Park/tail lights (see also fuse 33)
38	30A	ABS (see also fuse 10, 21, 23, 27, 46, 47) AST (see also fuse 10, 21, 26, 27, 46)

Table f. 1996 E36 Fuse Positions

fuse	Rating	Protected Circuit
39	7.5A	Engine control module (see also fuse 16, 28) Heating and A/C (IHKR E-36) (see also fuse 16, 20, 23, 31, 37, 41, 48)
40	30A	Power seats (see also fuse 5)
41	30A	Heating and A/C (IHKR E-36)(see also fuse 16, 20, 23, 31, 37, 39, 48) Radiator auxiliary fan
42	7.5A	Airbag (SRS) (see also fuse 26, 43) Convertible roof (see also fuse 7, 27, 35, 43) Roll-over protection (see also fuse 7, 35, 43)
43	5A	Anti-theft system (Alpine) (see also fuse 47) Airbag (SRS) (see also fuse 26, 42) Central body electronics (ZKE IV) (see also fuse 7, 14, 33, 35) Central locking (see also fuse 7, 35) Convertible roof (see also fuse 7, 27, 35, 42) Interior lights (also see fuse 7, 33) Roll-over protection (see also fuse 7, 35, 42) Telephone (see also fuse 33)
44	15A	Glove compartment light Sound system (see also fuse 9) Wiper/washer (see also fuse 36, 37, 45)
45	7.5A	Multi-function clock (see also fuse 23, 26, 27, 31, 46) On-Board Computer (see also fuse 11, 12, 23, 26, 27, 31, 33, 37, 46) On-Board Display (see also fuse 23, 27, 31, 46) Wiper/washer (see also fuse 36, 44)
46	15A	ABS (see also fuse 10, 21, 23, 27, 38, 47) AST (see also fuse 10, 21, 26, 27, 38) Brake lights Cruise control (see also fuse 28) Driveaway protection (EWS II) (see also fuse 28, 31) Electronic transmission control (A4S 310R) (see also fuse 26, 28) Instrument cluster (see also fuse 23, 27, 31) Multi-function clock (see also fuse 23, 26, 27, 31, 45) On-Board Computer (see also fuse 11, 12, 23, 26, 27, 31, 33, 37, 45) On-Board Display (see also fuse 23, 27, 31, 45)
47	15A	ABS (see also fuse 10, 21, 23, 27, 38, 46) Anti-theft (Alpine) (see also fuse 43)
48	40A	Heating and A/C (IHKR E-36)(see also fuse 16, 20, 23, 31, 37, 39, 41)

CAUTION —

- *Fuse designations and locations are subject to change and may vary from car to car, depending on options. If questions arise, please remember that an authorized BMW dealer is the best source for the most accurate and up-to-date information.*
- *To locate fuses 47-48, refer to* **Fuse Positions** *in this repair group.*

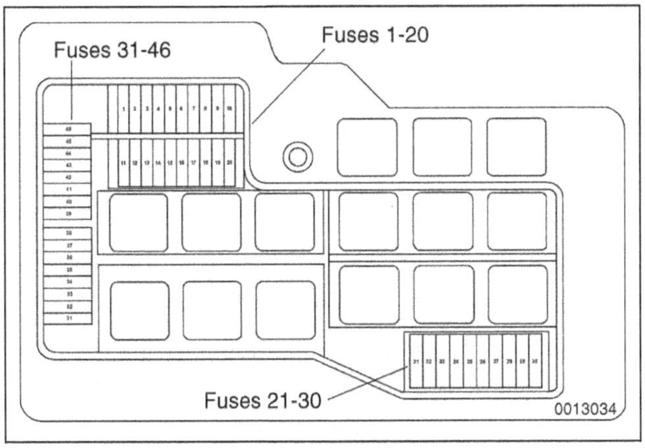

Fuse positions 1-46 in front power distribution box

Table g. 1997-1998 E36 Fuse Positions

fuse	Rating	Protected Circuit
1	30A	Power sunroof
2	5A	Engine controls DME M5.2 (see also fuses 16, 18, 39, 46)
3	30A	Wiper/washer module (see also fuse 36, 37, 44)
4	15A	Heated seats (see also fuse 23)
5	30A	Power seats (see also fuse 40)
6	20A	Rear window defogger/blower (see also fuse 23, 43) Antenna (see also fuse 23)
7	5A	Central locking (see also fuse 28, 31, 35, 45) Convertible roof (see also fuse 24, 27, 42, 43, 49)
8	15A	Horn
9	20A	Sound system (see also fuse 33, 43, 44)
10	30A	ABS (see also fuse 21, 27, 38, 46) AST (see also fuse 21,27, 38, 46)
11	7.5A	Headlights/fog lights (see also fuse 12, 15, 23, 25, 29, 30) On-Board Computer (see also fuse 12, 23, 26, 27, 31, 33, 37, 45, 46)
12	7.5A	Headlights/fog lights (see also fuse 11, 15, 23, 25, 29, 30) On-Board Computer (see also fuse 11, 23, 26, 27, 31, 33, 37, 45, 46)
13	5A	Not used
14	30A	Central body electronics (ZKE IV) (see also fuse 33, 35, 43) Power windows (see also fuse 19)
15	15A	Headlights/fog lights (see also fuse 11, 12, 23, 25, 29, 30)

Table g. 1997-1998 E36 Fuse Positions

fuse	Rating	Protected Circuit
16	5A	Engine control DME M5.2 (see also fuse 2, 18, 39, 46) Engine control DME MS 41.1(see also fuse 18, 39, 46, 50) Heating and A/C (IHKR E-36) (see also fuse 20, 23, 31, 39, 41, 48)
17	10A	Not used
18	15A	Engine control DME M5.2 (see also fuse 2, 16, 39, 46) Engine control MS 41.1(see also fuse 16, 39, 46, 50)
19	30A	Power windows (see also fuse 14
20	30A	Blower motor Heating and A/C (IHKR E-36) (see also fuse 16, 23, 31, 37, 39, 48) Park ventilation
21	5A	ABS (see also fuse 10, 27, 38, 46) AST (see also fuse 10, 27, 38, 46)
22	5A	Instrument illumination (see also fuse 25, 33, 37)
23	5A	Antenna (see also fuse 6) Headlights/fog lights (see also fuse 11, 12, 15, 25, 29, 30) Heated seats (see also fuse 4) Heating and A/C (IHKR E-36)(see also fuse 16, 20, 31, 39, 41, 48) Instrument cluster (see also fuse 27, 31,45, 46) Multi-function clock (see also fuse 26, 27, 31, 45, 46) On-Board Computer (see also fuse 11, 12, 27, 31, 33, 37, 45, 46) On-Board Display (see also fuse 26, 27, 31, 45, 46) Rear window defogger/blower (see also fuse 6, 43) Turn signals/hazard lights (see also fuse 34)
24	10A	Convertible roof (see also fuse 7, 27, 42, 43, 49) Power mirrors
25	5A	Headlights/fog lights (see also fuse 11, 12, 15, 23, 29, 30) Instrument illumination (see also fuse 22, 33, 37)
26	10A	Back-up lights Electronic transmission control (A4S 310R/Z) (see also fuse 28, 46) Electronic transmission control (A5S 310Z) (see also fuses 28, 46) Multi-function clock (see also fuse 23, 27, 31, 45, 46) On-Board Computer (see also fuse 11, 12, 23, 27, 31, 33, 37, 45, 46)

Table g. 1997-1998 E36 Fuse Positions

fuse	Rating	Protected Circuit
27	5A	Airbag (SRS) (see also fuse 42) ABS (see also fuse 10, 21, 38, 46) AST (see also fuse 10, 21, 38, 46) Convertible roof (see also fuses 7, 24, 42, 43, 49) Instrument cluster (see also fuse 23, 31, 45, 46) Multi-function clock (see also fuse 23, 26, 31, 45, 46) On-Board Computer (see also fuse 11, 12, 23, 26, 31, 33, 37, 45, 46) On-Board Display (see also fuse 23, 27, 31, 45, 46)
28	5A	Central locking (see also fuse 7, 31, 35, 45) Cruise control (see also fuse 46) Electronic transmission control (A4S 310R/Z) (see also fuse 26, 46) Electronic transmission control (A5S 310Z) (see also fuses 26, 46) Driveaway protection (EWS II) (see also fuse 31, 45) Starting system (see also fuse 31, 45)
29	7.5A	Headlights/fog lights (see also fuse 11, 12, 15, 23, 25, 30)
30	7.5A	Headlights/fog lights (see also fuse 11, 12, 15, 17, 23, 25, 29)
31	5A	Central locking (see also fuse 7, 28, 35, 45) Driveaway protection (EWS II) (see also fuse 28, 45) Heating and A/C (IHKR E-36) (see also fuse 16, 20, 23, 39, 41, 48) Instrument cluster (see also fuse 23, 27,45, 46) Multi-function clock (see also fuse 23, 26, 27, 45, 46) On-Board Computer (see also fuse 11, 12, 23, 26, 27, 33, 37, 45, 46) On-Board Display (see also fuse 23, 26, 27, 31, 45, 46) Starting system (see also fuses 28, 45)
32	30A	Cassette compartment (see also fuses 33, 37, 44) Charging socket (see also fuses 33, 37, 44) Cigar lighter/ashtray lights (see also fuses 33, 37,44) Glove compartment light (see also fuses 33, 37, 44)
33	10A	Ashtray light (see also fuses32, 37, 44) Cassette compartment (see also fuses 32, 37, 44) Cellular phone (see also fuse 43) Central body electronics (ZKE IV) (see also fuse 14, 35, 43) Charging socket (see also fuses 32, 37,44) Cigar lighter (see also fuses 32,37,44) Instrument illumination (see also fuse 22, 25, 37) Interior lights (see also fuse 43) Glove compartment light (see also fuse 32, 37, 44) License plate/luggage compartment lights (see also fuse 37) Luggage compartment light (see also fuse 37) On-Board Computer (see also fuse 11, 12, 23, 26, 27, 31, 37, 45, 46) Park/tail lights (see also fuse 37) Sound system (see also fuses 9, 43,44)
34	15A	Crash control module Turn signals/hazard lights (see also fuse 23)
35	25A	Central body electronics (ZKE IV) (see also fuse 14, 33, 43) Central locking (see also fuse 7, 28, 31,45)
36	30A	Headlight washers (see also fuses 3, 37) Wiper/washer (see also fuse 3, 37, 44)
37	10A	Ashtray light (see also fuses 32, 33, 44) Cassette compartment light (see also fuses 32, 33, 44) Charging socket (see also fuses 32, 33, 44) Cigar lighter (see also fuses 32, 33, 44) Glove compartment light (see also fuses 32, 33, 44) License plate/luggage compartment lights (see also fuse 33) Instrument illumination (see also fuses 22, 25, 33) On-Board Computer (see also fuse 11, 12, 23, 26, 27, 31, 33, 45, 46) Park/tail lights (see also fuse 33) Wiper washer module (see also fuses 3, 36, 44)
38	30A	ABS (see also fuse 10, 21, 27, 46) AST (see also fuse 10, 21, 27, 46)
39	7.5A	Engine control DME M5.2 (see also fuse 16,18, 46) Engine control DME MS 41.1(see also fuses 16, 18, 46, 50) Heating and A/C (IHKR E-36) (see also fuse 12, 20, 23, 31, 41, 48)
40	30A	Power seats (see also fuse 5)
41	30A	Heating and A/C (IHKR E-36) (see also fuse 16, 20, 23, 31, 39, 48) Radiator auxiliary fan
42	7.5A	Airbag (SRS) (see also fuse 27) Convertible roof (see also fuse 7, 24, 27, 43, 49) Roll-over protection (see also fuse 43)

Table g. 1997-1998 E36 Fuse Positions

fuse	Rating	Protected Circuit
43	5A	Anti-theft system (Alpine) (see also fuse 47) Cellular phone (see also fuse 33) Central body electronics (ZKE IV) (see also fuse 14, 33, 35) Convertible roof (see also fuse 7, 24, 27, 42, 49) Interior lights (also see fuse 33) Rear window defogger/blower (see also fuse 6, 23) Roll-over protection (see also fuse 42) Sound system (see also fuses 9, 33, 44)
44	15A	Ashtray light (see also fuses 32, 33, 37) Cassette compartment light (see also fuses 32, 33, 37) Charging socket (see also fuses 32, 33, 37) Cigar lighter (see also fuses 32, 33, 37) Glove compartment light (see also fuses 32, 33, 37) Sound system (see also fuse 9, 33, 43) Wiper/washer (see also fuse 3, 36, 37)
45	7.5A	Central locking (see also fuses 7, 28, 31, 35) Drive-away protection (EWS II) (see also fuse 28, 29, 31) Instrument cluster (see also fuses 23, 27, 31, 46) Multi-function clock (see also fuse 23, 26, 27, 31, 46) On-Board Computer (see also fuse 11, 12, 23, 26, 27, 31, 33, 37, 46) On-Board Display (see also fuse 23, 27, 31, 46) Starting system (see also fuses 28, 31)
46	15A	ABS (see also fuse 10, 21, 27, 38) AST (see also fuse 10, 21, 27, 38) Brake lights Cruise control (see also fuse 28) Electronic transmission control (A4S 310R) (see also fuse 26, 28) Electronic transmission control (A5S 310Z) (see also fuse 26, 28) Engine control DME M5.2 (see also fuses 2, 16, 18, 39) Engine control DME MS 41.1 (see also fuses 16, 18, 39, 50) Instrument cluster (see also fuse 23, 27, 31, 45) Multi-function clock (see also fuse 23, 26, 27, 31, 45) On-Board Computer (see also fuse 11, 12, 23, 26, 27, 31, 33, 37, 45)
47	15A	Anti-theft system (see also fuse 43)
48	40A	Heating and A/C (IHKR E-36)(see also fuse 16, 20, 23, 31, 39, 41)
49	20A	Convertible roof (see also fuses 7, 24, 27, 42, 43)
50	5A	Engine control DME MS 41.1 (see also fuses 16, 18, 39, 46)

CAUTION —

- *Fuse designations and locations are subject to change and may vary from car to car, depending on options. If questions arise, please remember that an authorized BMW dealer is the best source for the most accurate and up-to-date information.*
- *To locate fuses 47-50, refer to* **Fuse Positions** *in this repair group.*

611 Wipers and Washers

GENERAL

This repair group covers windshield wiper and washer system repair information. Please note that some procedures are different for 2- and 4-door models.

The windshield wiper/washer stalk switch assembly is covered in **612 Switches and Electrical Accessories**. Electrical wiring diagrams and relays for the wiper/washer system are covered in **Electrical Wiring Diagrams**.

WIPER BLADES

> ***CAUTION—***
> *Never turn on the wiper blades while the hood is open. Damage to the wiper system and hood may occur. To ensure safety during wiper system procedures, remove the windshield wiper fuse. See* **610 Electrical Component Locations.**

Wiper Blade Cleaning Problems

Common cleaning problems with the wipers include streaking or sheeting, water drops after wiping, and blade chatter.

Streaking is usually caused when wiper blades are coated with road film or car wash wax. Clean the blades using soapy water. If cleaning does not cure the problem, the blades should be replaced. BMW recommends replacing the wiper blades twice a year, before and after the cold season.

Drops that remain behind after wiping are usually caused by oil, road film, or diesel exhaust residue on the glass. Use an alcohol or ammonia solution or a non-abrasive cleaner (such as Bon-Ami®) to clean the windshield.

Wiper blade chatter may be caused by dirty or worn blades, or by wiper arms that are out of alignment. Clean the blades and windshield as described above. Adjust the wiper arm so that there is even pressure along the blade, and so that the blade at rest is perpendicular to the windshield. If problems persist, the blades and wiper arms should be replaced.

> ***NOTE—***
> *BMW has available a special tool/gauge (BMW special tool no. 00 9 210) to align the angle of the wiper arm to the windshield glass.*

The wiper blades can be renewed in one of two ways: either by replacing the complete blade or by replacing just the rubber insert.

Insert replacement is the most economical method, although over time the wiper blade itself will become worn. One rule of thumb would be to replace just the inserts every second time.

611

Wiper blades, replacing

1. Pivot wiper arm off windshield.

2. Position wiper blade approximately perpendicular to wiper arm.

3. Remove wiper blade from wiper arm by depressing retaining tab and sliding blade out of arm. See Fig. 1.

> ***NOTE—***
> *Some wiper blade versions may have two retaining tabs.*

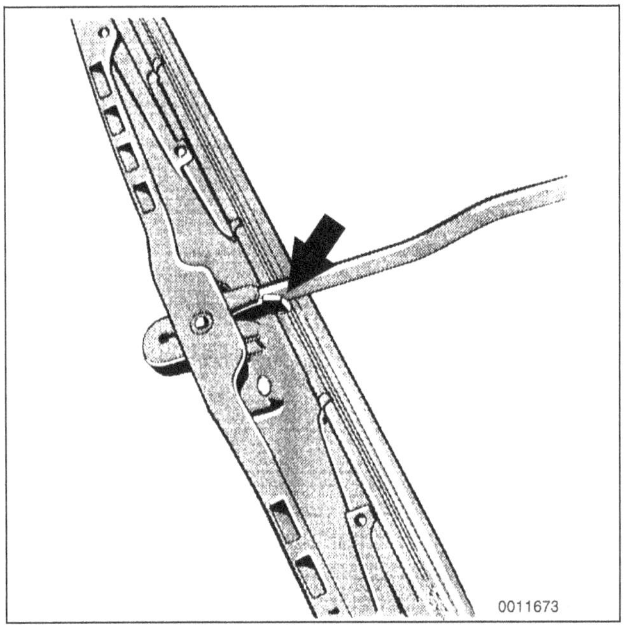

Fig. 1. Windshield wiper blade retaining tab. Press in on tab (**arrow**) to release blade from arm.

4. Installation is reverse of removal. Install wiper blade to wiper arm until retaining tab can be heard to click into position.

Wiper blade inserts, replacing

1. Remove wiper blade as described earlier.

2. Unhook end of wiper blade insert from wiper arm guides. See Fig. 2.

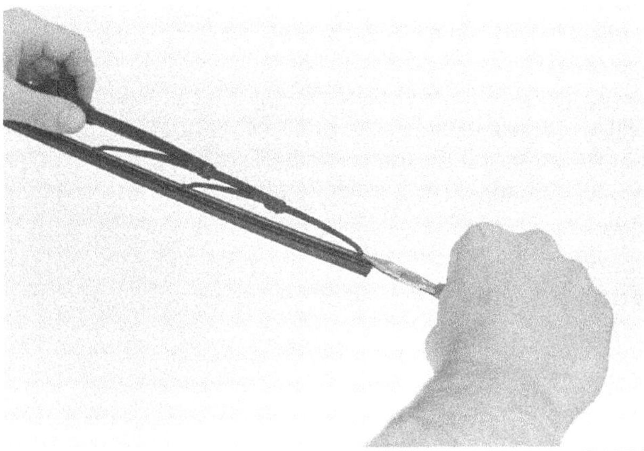

0011784

Fig. 2. Unhook wiper blade insert from wiper arm guide. If necessary, guide slightly using needle nose pliers.

3. Pull old insert from wiper arm guides, noting installation position.

4. Remove metal support strips from old insert and install into slots in new insert, noting installation direction of cutouts in support strips.

NOTE —
The notched cutouts in the retaining strips should engage the molded notches in the inserts.

5. Slide new insert through wiper blade guides. Lock insert in place at end guides.

Wiper arms, removing and installing

NOTE —
On 2-door models it may be necessary to raise the hood to the service position. See **410 Fenders, Engine Hood**.

CAUTION —
- *Use fender cover to protect windshield.*
- *Remove windshield wiper fuse to prevent accidents.*

1. With engine hood in closed position, remove both wiper blades.

2. Open hood. Remove cap covering wiper arm retaining nuts by inserting a narrow screwdriver blade in slot below cap and prying upward.

NOTE —
Marking position of wiper arm on shaft will aid installation.

3. Using a 13 mm socket, remove wiper arm retaining nut and remove wiper arm. See Fig. 3.

0012537

Fig. 3. Wiper arm retaining nut. (**arrow**).

NOTE —
It may be necessary to gently rock wiper arm back and forth to loosen arm from shaft.

4. Installation is reverse of removal. Recheck wiper arm tightening torque after waiting 15 minutes.

Tightening Torque
- Wiper arm to wiper shaft .... 25 Nm (18 ft-lb)

WINDSHIELD WIPER ASSEMBLY

The windshield wiper assembly (linkage and motor) is removed as a single unit. Once the assembly is removed, the wiper motor and other linkage parts can be repaired or replaced. For an illustration of wiper assembly components, see Fig. 4.

Windshield wiper assembly, removing and installing (4-door models)

NOTE —
The wiper arms should be in the parked position before removing the wiper assembly. To avoid damaging the wiper arms and pivots, do not manually slide or force the wiper arms across the windshield.

1. Disconnect negative (–) cable from battery.

CAUTION —
- *Prior to disconnecting the battery, read the battery disconnection cautions given at the front of this manual on page viii.*

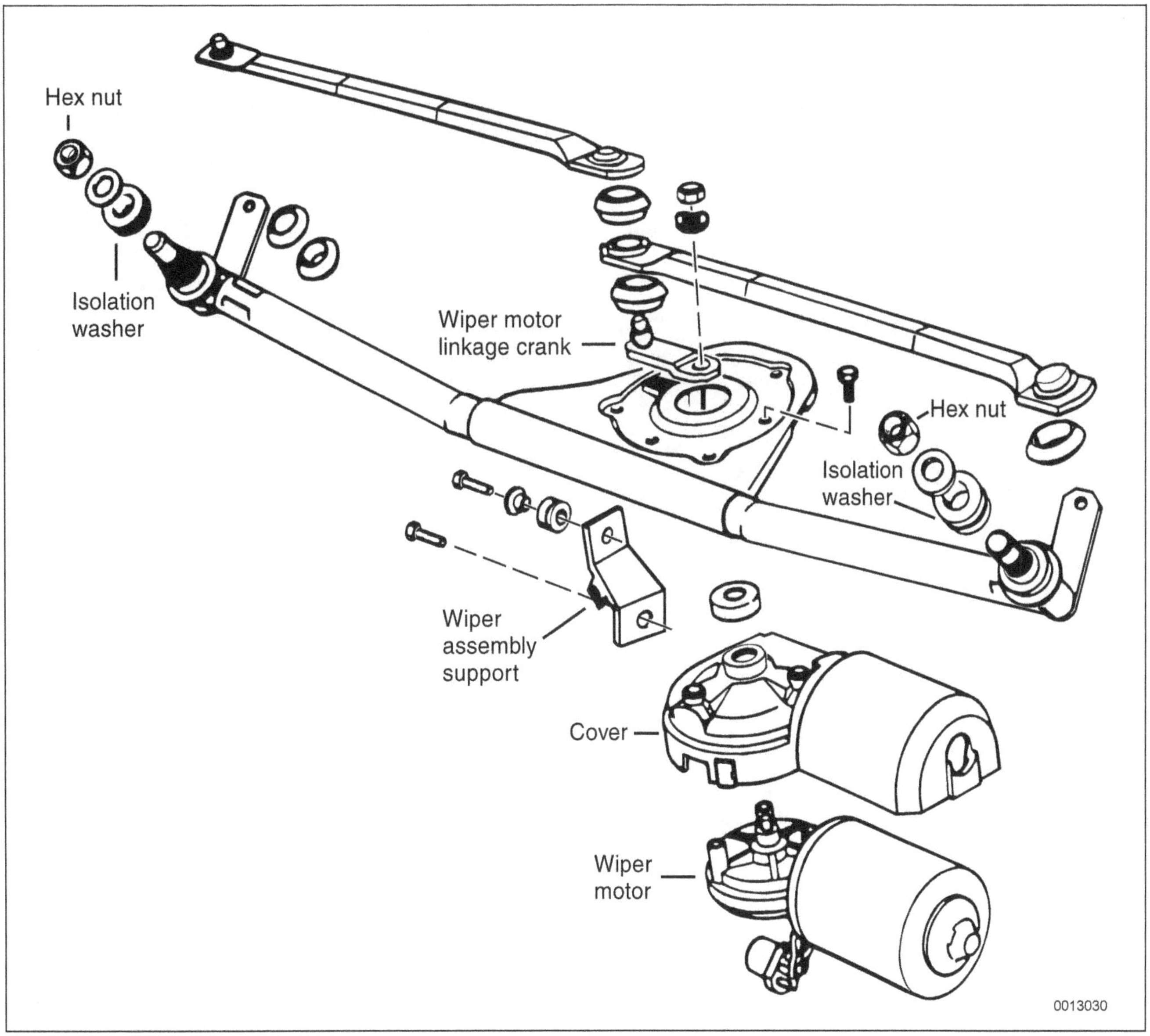

Fig. 4. Windshield wiper motor assembly and related parts.

2. Remove wiper arms as described earlier.

3. Place engine hood in service position. See **410 Fenders, Engine Hood**.

4. Unclip engine bulkhead plenum grille at base of windshield and pull it away from rubber gasket. See Fig. 5.

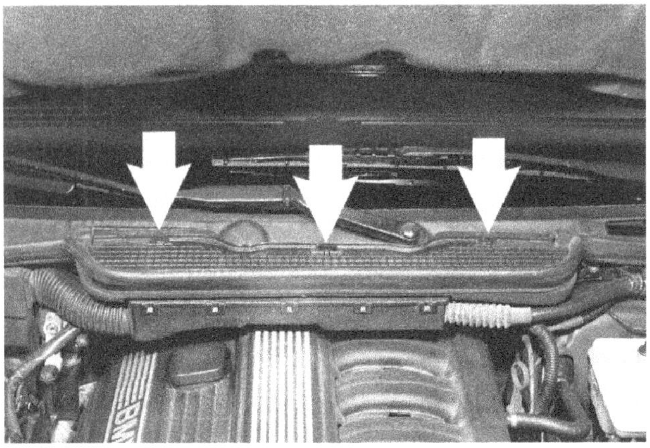

0012527

Fig. 5. Detach grill clips (**arrows**) and pull upward.

5. Remove wiring harness retaining clip from bulkhead in rear of engine bay. Remove screws retaining wiring harness protective cover. See Fig. 6.

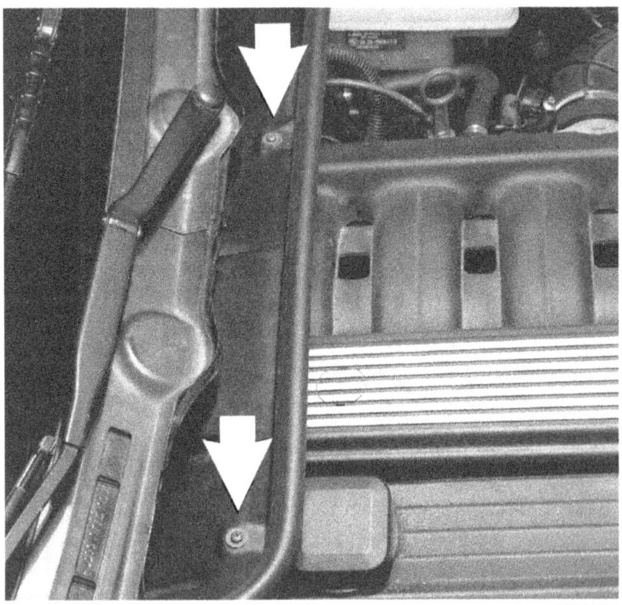

0012529

Fig. 6. Engine bulkhead wiring harness cover screws (arrows).

6. On passenger side of air plenum, push aside wiring harness and remove two screws retaining plenum housing. See Fig. 7.

0012635

Fig. 7. Right side plenum retaining screws (**arrows**).

7. On driver's side of air plenum, push aside wiring harness and remove screw retaining plenum. Pull plenum up and remove from engine compartment. See Fig. 8.

0012534

Fig. 8. Remove screw at **A** and pull plenum housing upward.

8. On 6-cylinder models: Remove plastic top engine covers. See **110 Engine Removal and Installation**.

NOTE —

Cover inside edges of cowl panel sheet metal with thick tape to prevent damage during wiper assembly removal.

9. Remove large nut and washer at base of wiper arm shafts cowl panel. See Fig. 9.

10. Remove wiper assembly support. See Fig. 10.

11. Separate harness at bulkhead connector. See Fig. 11.

12. Wrap protective tape around wiper shaft on driver's side to prevent damaging painted metal during removal.

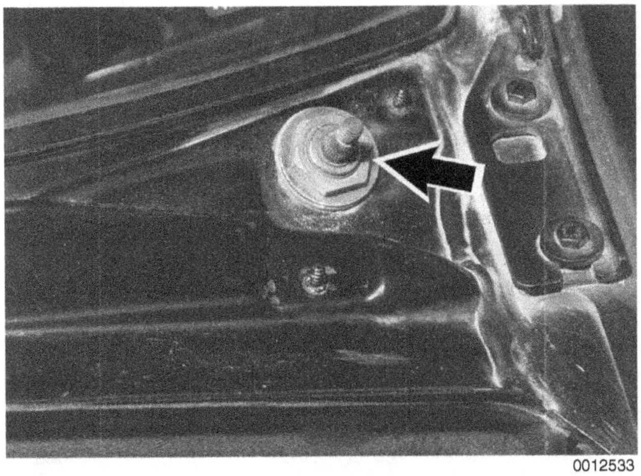

Fig. 9. Wiper shaft nut at cowl (**arrow**).

Fig. 10. Wiper assembly support bolts (**arrows**).

Fig. 11. Wiper assembly electrical connector (**arrow**).

13. Push wiper arm shaft down into cowl area and maneuver wiper assembly out of cowl.

14. Installation is reverse of removal, noting the following:
    - Install all fasteners finger-tight first.
    - Once installed, wiper motor should first be run to parked position and switched off.
    - Next, install wiper arms but not blades.
    - Close hood and fit wiper blades.
    - Check all electrical harness fittings and sealing grommets for correct reinstallation. Replace any wire ties cut off during removal.

Tightening Torques
- Wiper shaft nut at cowl .... 12 Nm (9 ft-lb)
- Wiper support to wiper assembly or to body .... 10 Nm (89 in-lb)
- Wiper arm to wiper shaft .... 25 Nm (18 ft-lb)

Windshield wiper assembly, removing and installing (2-door models)

NOTE —

The wiper arms should be in the parked position before removing the wiper assembly. To avoid damaging the wiper arms and pivots, do not manually slide or force the wiper arm across the windshield.

1. Disconnect negative (–) cable from battery.

CAUTION —
- *Prior to disconnecting the battery, read the battery disconnection cautions given at the front of this manual on page viii.*
- *Use fender cover to protect windshield.*

2. Remove wiper arms as described earlier.

3. Place engine hood in service position. See **410 Fenders, Engine Hood**.

4. Remove trim retainers at left and right side and remove air plenum grill. See Fig. 12.

5. Remove screws retaining wiring harness protective cover. See Fig. 13.

6. On passenger side of air plenum, push aside wiring harness and remove two screws retaining plenum housing. See Fig. 14.

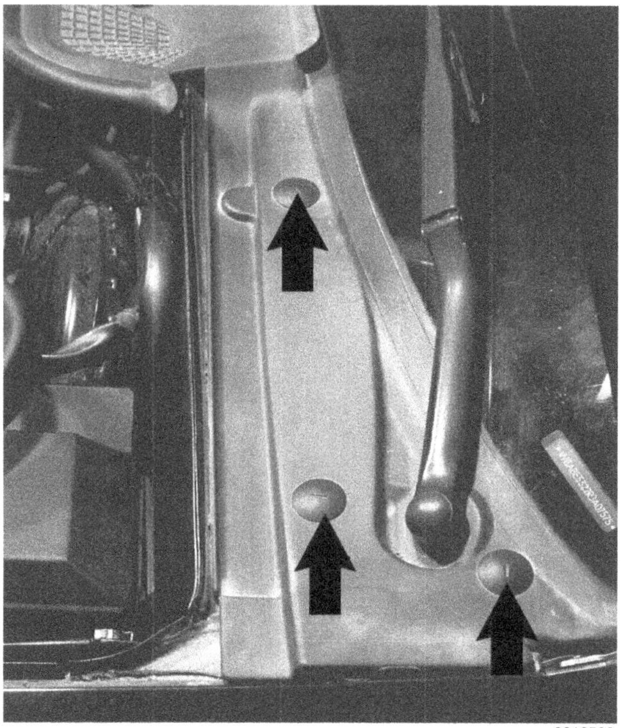

Fig. 12. Left side air plenum grill trim retainers (**arrows**).

Fig. 13. Engine bulkhead wiring harness cover screws (**arrows**).

Fig. 14. Right side plenum retaining screws (**black arrows**). Note positive leads to main (+) connector (**white arrow**).

7. On driver's side of air plenum, push aside wiring harness and remove screw retaining plenum. Pull plenum up and remove from engine bay. See Fig. 15.

Fig. 15. Remove screw at **A** and pull plenum upward.

8. On passenger side of engine compartment bulkhead, remove engine control module (ECM) compartment cover. See Fig. 16.

9. Disconnect harness connector(s) from ECM and automatic transmission control module (if applicable). Carefully pull harness and wire loom as far forward as possible. If necessary, disconnect positive leads at main (+) connector.

10. Remove wiper assembly mounting fasteners. See Fig. 17.

Fig. 16. Engine control unit (ECM) compartment cover screws (**arrows**).

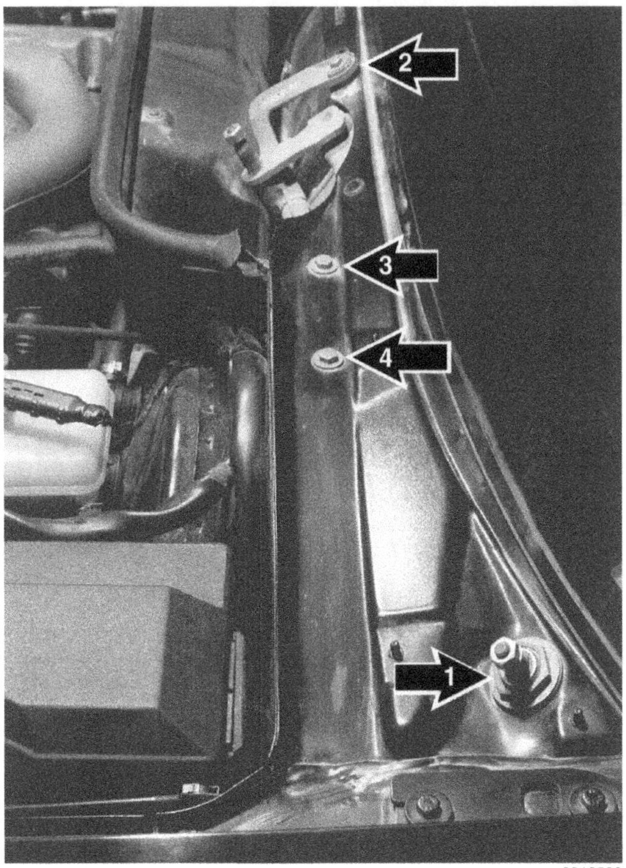

Fig. 17. Wiper assembly mounting fasteners (**arrows**).

11. Remove wiper assembly support. Refer to Fig. 10.

12. Separate harness at bulkhead connector. See Fig. 18.

Fig. 18. Wiper assembly electrical plug (**arrow**).

13. Cover inside edges of cowl panel sheet metal with thick tape to prevent damaging it during wiper assembly removal.

14. Wrap protective tape around wiper shaft on driver side to prevent damaging sheet metal during removal.

15. Push wiper arm shaft down into cowl area and maneuver wiper assembly out of cowl.

16. Installation is reverse of removal, keeping in mind:
    - Install all fasteners finger tight first.
    - Torque top fasteners in sequence indicated in Fig. 17. Then tighten wiper assembly support bolts inside air plenum.
    - Wire tie wiper harness to wiper assembly support inside plenum.
    - Once installed, wiper motor should first be run to parked position and switched off.
    - Next, install wiper arms but not blades.
    - Close hood and fit wiper blades.
    - Check all electrical harness fittings and sealing grommets for correct reinstallation. Replace all wire ties cut during removal.

> ***CAUTION—***
> *The gaskets and sealing grommets at the ECM compartment cover (passenger side of bulkhead) must be installed correctly to keep moisture out of critical electrical components.*

Tightening Torques

- Wiper shaft nut at cowl . . . . 12 Nm (9 ft-lb)
- Wiper support to wiper assembly or to body . . . . 10 Nm (89 in-lb)
- Wiper arm to wiper shaft . . . . 25 Nm (18 ft-lb)

Windshield wiper motor, removing and installing

1. Disconnect negative (–) cable from battery.

> ***CAUTION —***
> - *Prior to disconnecting the battery, read the battery disconnection cautions given at the front of this manual on page viii.*
> - *Use fender cover to protect windshield.*

2. Remove wiper assembly as described earlier.
3. Remove nut on wiper motor linkage crank and remove linkage crank from motor. See Fig. 19.

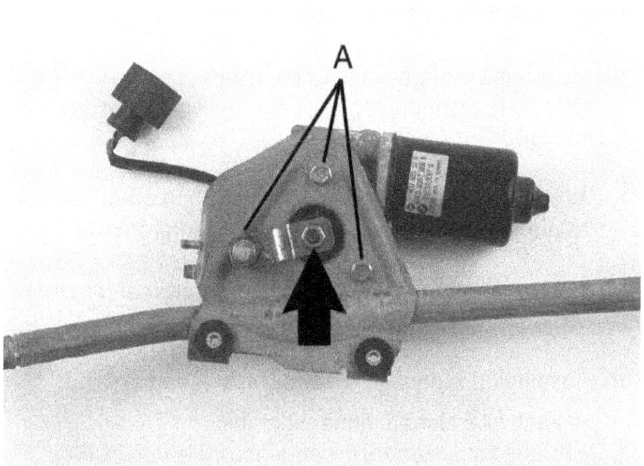

0012539

Fig. 19. Wiper linkage crank nut (**arrow**) and wiper motor mounting bolts (**A**).

4. Unscrew three motor to assembly mounting bolts and remove motor. Refer to Fig. 19.
5. Install new motor to wiper assembly. Do not install crank.
6. Connect motor to electrical harness. If necessary, reconnect battery temporarily.
7. Operate motor, then turn off switch. Motor will stop in "park" position.
8. Install wiper arm crank in straight line with linkage. See Fig. 20.

> ***NOTE —***
> *When installing wiper crank onto wiper motor, brace crank with a drift pin through the hole on crank into hole on wiper assembly.*

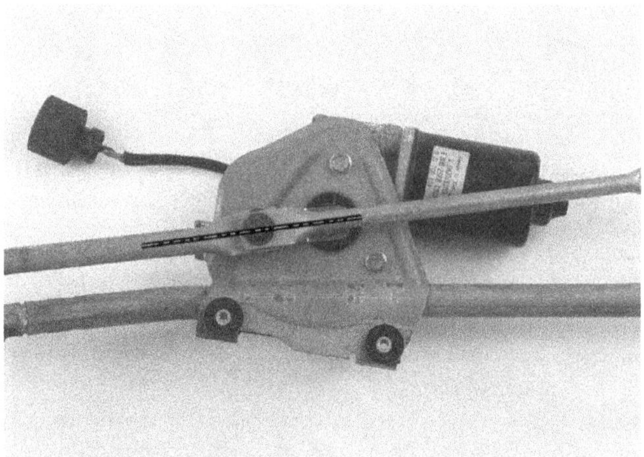

0012538

Fig. 20. Align wiper crank with linkage, as illustrated.

9. Remainder of installation is reverse of removal.

Tightening Torques
- Wiper motor to wiper assembly (use new bolts) . . . . . 10 Nm (89 in-lb)
- Wiper motor crank to wiper motor (M8 nut) . . . . 27 Nm (20 ft-lb)

WINDSHIELD CLEANING SYSTEM

The windshield cleaning system includes the spray nozzles in the engine hood, the washer fluid pump, and the washer fluid reservoir in the front passenger side of the engine compartment.

The windshield spray nozzles can be aimed by using a sewing needle or a similar diameter stiff piece of wire.

Windshield spray nozzle, removing and installing

1. Open hood and remove hood insulating pad retaining clips until bottom of nozzle can be fully accessed. See **410 Fenders, Engine Hood**.

2. Carefully disconnect hose from nozzle. Disconnect harness connectors for nozzle heater.

3. Push nozzle out of hood. See Fig. 21.

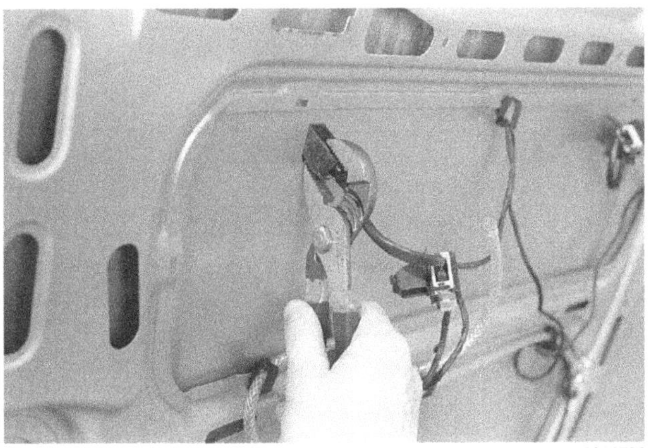

0011792

Fig. 21. First push nozzle forward then upward from underneath hood.

4. Installation is reverse of removal, noting the following:
    - Check that colors of harness and nozzle connector match.
    - Check and adjust nozzles before driving.

Washer fluid reservoir, replacing

1. Open hood and siphon out washer fluid reservoir.

2. Remove windshield washer pump and hose as described later.

3. Remove windshield washer fluid level sensor as described later.

4. Remove large plastic washer reservoir retainer at fender.

5. Remove reservoir.
6. Installation is reverse of removal, noting the following:
    - On installation check that washer fluid hose is not kinked.
    - Replace hose clamp.

Windshield washer fluid pump, replacing

The windshield washer fluid pump is mounted in the washer fluid reservoir.

1. Open hood and siphon out washer fluid reservoir.

2. Depress retaining tabs on electrical connector and unplug harness connector from pump.

3. Loosen hose clamp and remove washer fluid line from washer pump.

4. Pull pump out of tank. See Fig. 22.

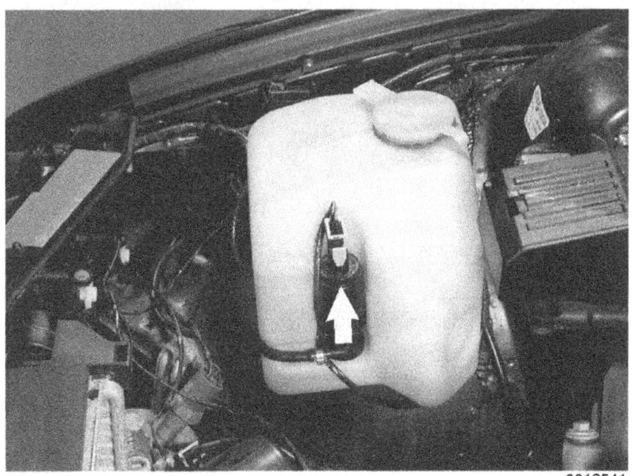

0012541

Fig. 22. Pull windshield washer pump up to remove (**arrow**).

5. Installation is reverse of removal. Check pump to tank sealing grommet for damage and correct fit. Replace hose clamp. Check that washer fluid hose is free of kinks.

Washer fluid level sensor, replacing (models with check control)

The washer fluid level sensor is mounted in the fluid reservoir, near the bottom of the container.

> ***NOTE —***
> *Check wire leads at sensor for worn insulation and corrosion. Check electrical contacts in harness connector for corrosion.*

1. Open hood and siphon out washer fluid reservoir.
2. Lift harness connector retaining lever and unplug harness connector to washer fluid level sensor. See Fig. 23.

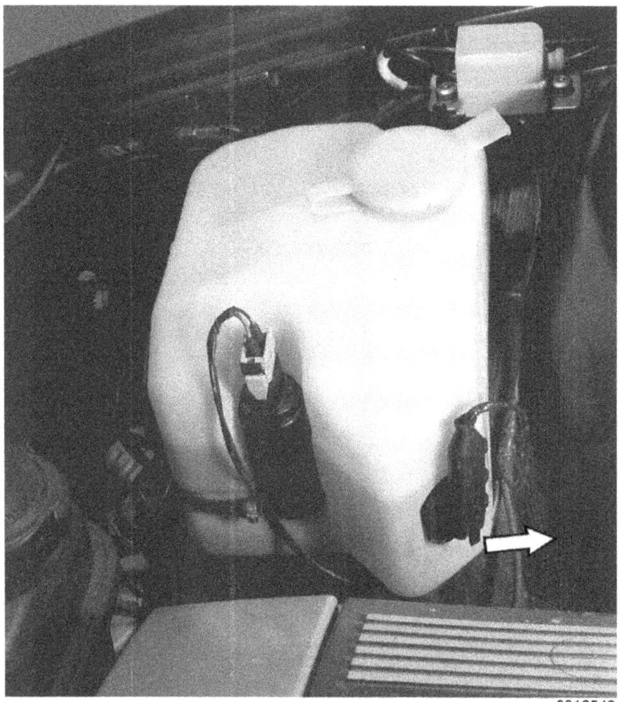

0012542

Fig. 23. Release electrical connector by pulling up on lever (**arrow**).

3. Pull fluid level sensor out from rubber grommet.
4. Install new sensor into reservoir. Check sealing grommet for correct fit. Replace if damaged.

612 Switches and Electrical Accessories

GENERAL

This section covers replacement of interior switches and accessories.

When working on electrical switches, always disconnect the negative (–) cable from the battery and insulate the cable end to prevent accidental re-connection.

> ***CAUTION —***
> *Prior to disconnecting the battery, read the battery disconnection cautions given at the front of this manual on page viii.*

NOTE —
- *Wiring diagrams can be found in* **Electrical Wiring Diagrams**.
- *Fuse, relay and electrical component locations can be found in* **610 Electrical Component Locations**.
- *Replacement information for the transmission gear position/neutral safety switch is covered in* **250 Gearshift Linkage**.
- *Testing and replacement information for the oil pressure switch is covered in* **119 Lubrication System**.

INTERIOR SWITCHES

NOTE —
To prevent marring interior trim, wrap the tips of screwdrivers and pliers with tape before prying out switches or electrical accessories.

Turn signal switch and wiper switch, replacing

The procedures for removing the turn signal stalk switch and the wiper system stalk switch are similar.

1. Disconnect negative (–) cable from battery and cover terminal with insulating material.
2. Remove airbag unit from steering wheel. See **721 Airbag System (SRS)**. Store airbag unit in a safe place with pad facing up.

> ***WARNING —***
> *The cars covered by this manual are equipped with a Supplemental Restraint System (SRS) that automatically deploys one or more airbags. Each airbag unit houses an explosive powerful charge. Any work involving the SRS system should only be performed by an authorized BMW dealer. Making repairs without the proper knowledge and special test equipment may cause serious personal injury. See* **721 Airbag System (SRS)**

3. With front wheels pointed straight ahead, remove steering wheel. See **320 Steering and Wheel Alignment**.

NOTE —
To help with reassembly, mark steering wheel and column before disassembly.

4. Remove trim panel from lower left side of instrument panel. See **513 Interior Trim**.

612

5. Remove lower steering column cover from steering column. See Fig. 1.

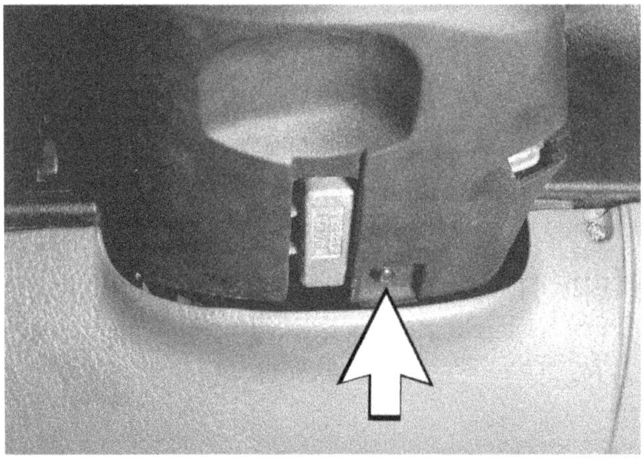

0012543

Fig. 1. Lower steering column cover retaining screw (**arrow**).

6. Remove stalk switch holder retaining screw. See Fig. 2.

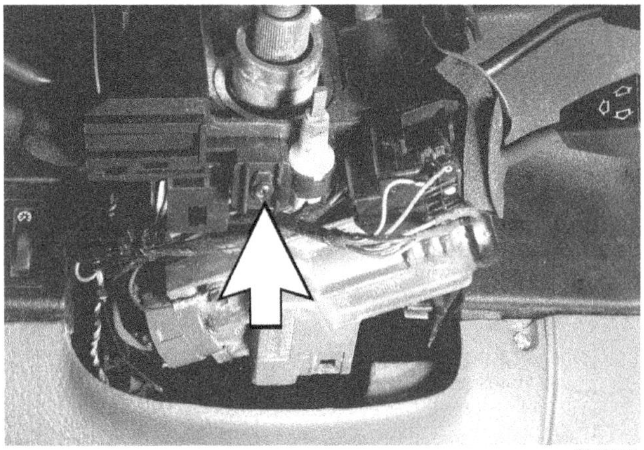

0012555

Fig. 2. Stalk switch holder retaining screw (**arrow**).

7. Depress retaining tabs to release stalk switch holder. Push stalk switch holder downwards and off steering column. See Fig. 3.

8. Depress catches on stalk switch and release switch from column. See Fig. 4.

9. Cut wire ties retaining harness to steering column. Disconnect switch electrical connector located at bottom of steering shaft.

10. Installation is reverse of removal, noting the following:
    - Place turn signal indicator in center position before installing.
    - Ensure that self cancelling cams on turn signal switch are not damaged during installation.

0012545

Fig. 3. Depress retaining tabs (**arrows**) and push holder downwards.

0012556

Fig. 4. Depress upper and lower catches simultaneously and unplug harness connector.

> ***WARNING —***
> *Once the airbag unit is installed and all other service procedures have been completed, start the engine and check that the SRS warning light goes out. If the warning light stays on, the SRS system will not function as designed. Have the system diagnosed and repaired by an authorized BMW dealer.*

Tightening Torques

- Steering wheel to steering column spindle (Bolt) . . . . 63 Nm (46 ft-lb)

Ignition Switch Assembly

The components of the ignition switch assembly are shown in Fig. 5. The ignition electrical switch and the ignition lock cylinder can be removed with the ignition switch assembly attached to the steering column.

Ignition electrical switch, replacing

1. Disconnect negative (–) cable from battery and cover terminal with insulating material.

> ***CAUTION—***
> *Prior to disconnecting the battery, read the battery disconnection cautions given at the front of this manual on page viii.*

2. Remove trim panel from lower left side of instrument panel. See **513 Interior Trim.**

3. Remove screw retaining lower steering column cover and remove cover. Refer to Fig. 1.

4. Clean away colored sealer from ignition switch set screws and remove slotted set screws on both sides of switch. See Fig. 6.

5. Release tabs and remove switch after separating electrical harness connector.

6. Installation is reverse of removal, noting the following:
   - Before mating ignition electrical switch to ignition lock, check that switch is in correct position relative to ignition lock.
   - Seal set screws using paint or lacquer.

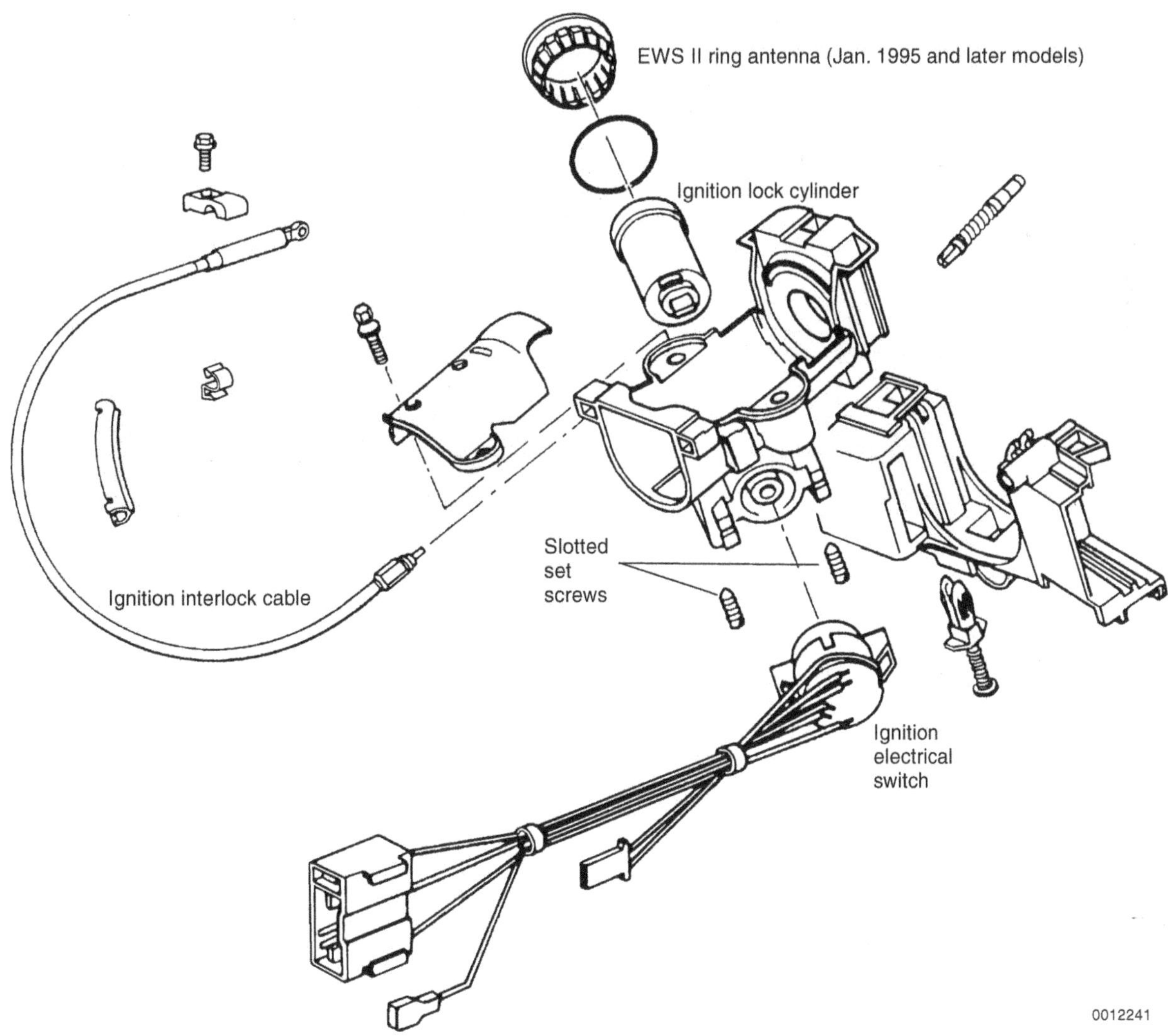

Fig. 5. Ignition switch assembly.

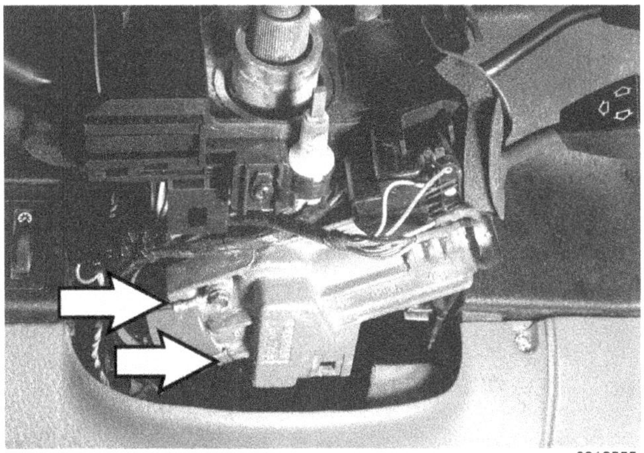
0012555

Fig. 6. Ignition switch set screws (**arrows**).

Ignition lock cylinder, replacing

NOTE—
On cars with EWS II (production date 1/95 and later), the ring antenna surrounding the ignition switch must be removed. See **515 Central Locking and Anti-theft**.

1. Insert master key into lock cylinder and turn 60° to "radio" position. See Fig. 7.

0011807

Fig. 7. With ignition key in "radio position" (60° from "off"), insert stiff wire into lock opening (**arrow**).

2. Insert a thin piece of stiff wire into opening in lock cylinder and pull lock cylinder out. See Fig. 8.

NOTE—
On early cars (up to production date 10/95), turn lock cylinder back 12° to release.

0011808

Fig. 8. Lock cylinder being released.

Headlight and Foglight Switch Assembly

The headlight and foglight switches are retained in the left dashboard insert. This also contains the driver's side face level air vents. To service either switch, remove insert by removing a single screw. See Fig. 9.

NOTE—
Access to electrical harness connectors behind dash insert may be easier if lower left dash trim and driver side knee bolster are removed. See **513 Interior Trim**.

0012557

Fig. 9. Remove retaining screw to pull out left dash insert (**arrow**).

Headlight switch, replacing

1. Remove left side dash insert and disconnect harness connector to headlight switch. See Fig. 10.

0012559

Fig. 10. Removing headlight switch electrical harness connector (**arrow**).

2. Pull off headlight switch knob. Wrapping adhesive tape around knob facilitates removal.

3. Remove nut mounting switch to dash insert. See Fig. 11.

0012558

Fig. 11. Headlight switch mounting nut (**arrow**).

4. Pull switch out from housing.

5. Installation is reverse of removal. Align new switch to dash insert. See Fig. 12.

Tightening Torque
- Headlight switch retaining nut .... 2 Nm (18 in-lb)

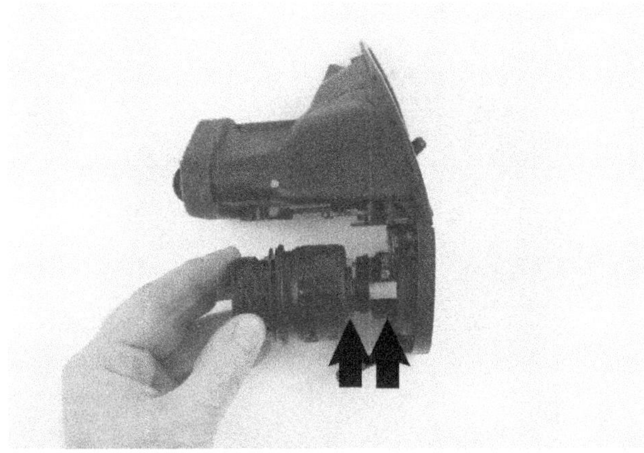

0012560

Fig. 12. Fit groove on headlight switch into switch bezel on dash insert (**arrows**).

Foglight switch, replacing

1. Remove left side dash insert, as described earlier, and remove foglight switch harness connector.

2. Slide foglight switch out of dash insert by squeezing spring tabs together.

3. Installation is reverse of removal.

Hazard warning switch, replacing

1. Unclip shift boot and lift up from shift lever console. Carefully remove insulation pad. See **250 Gearshift Linkage**.

2. Working underneath center console, push hazard warning switch up and out of console with your fingers.

3. Disconnect harness connector from switch.

4. Installation is reverse of removal.

Power window switch, replacing

1. Unclip shift boot and lift up from shift lever console. Carefully remove insulation pad. See **250 Gearshift Linkage**.

2. Working underneath center console, push window switch up out of console with your fingers. See Fig. 13.

3. Disconnect harness connector from switch.

4. Installation is reverse of removal.

0012561

Fig. 13. Removing window switch from console.

Outside mirror switch, replacing

1. Carefully pry mirror switch from arm rest. See Fig. 14.

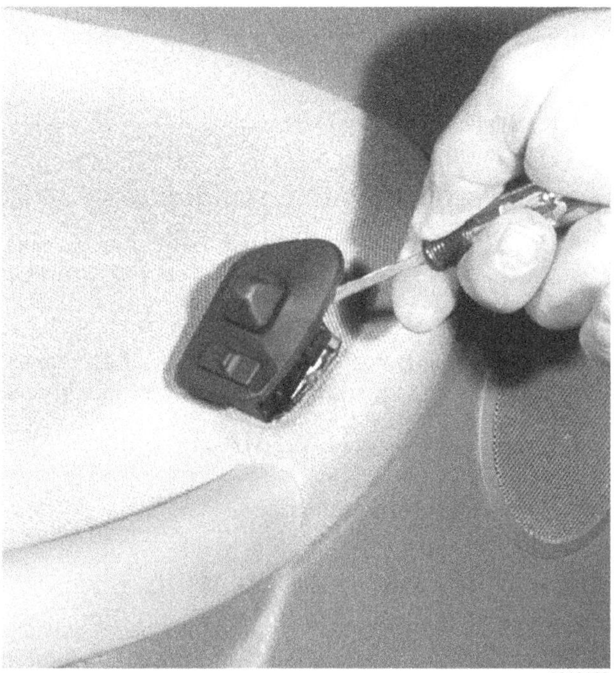

0013168

Fig. 14. Pry outside mirror switch from door arm rest.

2. Disconnect harness connector from switch.

3. Installation is reverse of removal.

INTERIOR LIGHTS

Dome light, rear map lights, trunk light, replacing bulbs

1. Pry dome light out from left side. See Fig. 14.

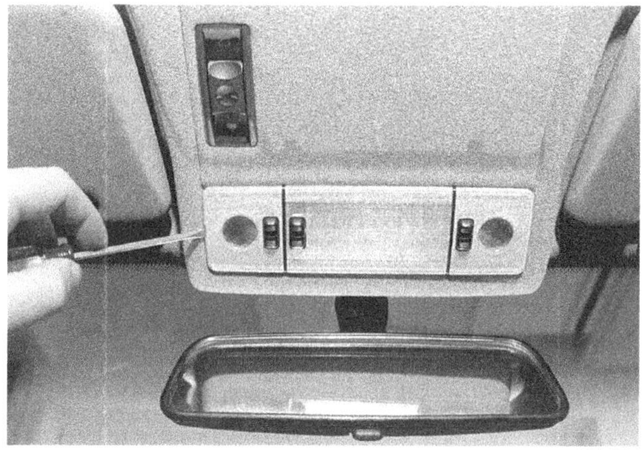

0011854

Fig. 15. Dome light assembly being removed.

2. Release locking blade and swing back reflector. Spread apart contact blades to replace bulb.

3. Pry rear map light out with small screwdriver, remove reflector and replace bulb. See Fig. 16.

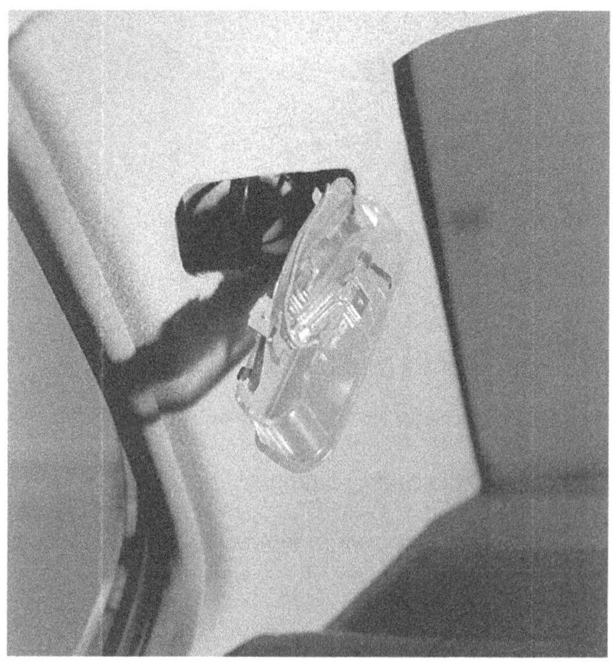

0011855

Fig. 16. Rear interior light assembly shown removed.

4. Use screwdriver to carefully pry off trunk light lens and replace bulb. See Fig. 17.

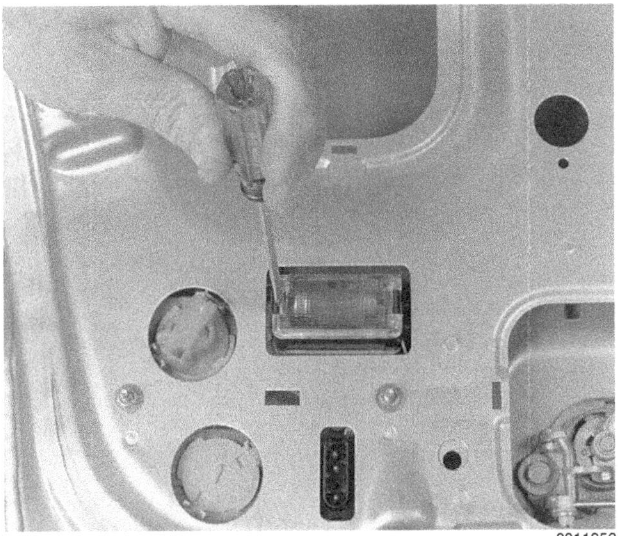

Fig. 17. Trunk light lens being removed.

Interior Light Bulb Wattages
- Dome light . . . . 10 watts
- Rear map light . . . . 5 watts
- Trunk light. . . . . 10 watts

620 Instruments

GENERAL

This repair group covers removal and installation of the instrument cluster and the Multi-information Display (MID) module. Bulb replacement information for these units is also included here.

> ***NOTE —***
> *A quick check of the temperature gauge and its electrical circuit is given in* **170 Radiator and Cooling System**. *Use this procedure to diagnose a faulty temperature gauge*

INSTRUMENT CLUSTER

When servicing the instrument cluster, always disconnect the negative (–) cable from the battery and insulate the cable end to prevent accidental re-connection.

Instrument cluster, self-testing

Several different versions of the basic instrument cluster have been fitted in E36 cars. It is possible to perform an instrument cluster test and display system test information.

Some of the information displayed by the instrument cluster during self-testing is used for internal BMW organization purposes, such as matching the correct instrument cluster components during parts replacement. Other functions test the various instrument modules. Self-test is also used to recode the instrument cluster after removal.

In later production vehicles (1996 and later) the test results can be downloaded, *via* the diagnostic plug, by an authorized BMW dealer technician.

1. Turn ignition to “radio” position.
2. Press and hold reset button on trip-odometer until trip-odometer display shows numerical codes.
3. With each subsequent press of the button, a different function is tested. Some functions will activate needle gauges.
4. Turn off ignition.

Instrument cluster, removing and installing

1. Disconnect negative (–) cable from battery and cover terminal with insulating material.

> ***CAUTION —***
> *Prior to disconnecting the battery, read the battery disconnection cautions given at the front of this manual on page viii.*

2. Remove airbag unit from steering wheel. See **721 Airbag System (SRS).** Store unit in a safe place.

> ***WARNING —***
> *The airbag is an explosive device and should be treated with extreme caution. Improper handling of the airbag could cause serious injury or death. Store the airbag with the horn pad facing up. If stored facing down, accidental deployment could propel it violently into the air, causing injury.*

3. With front wheels pointed straight ahead, remove steering wheel. See **320 Steering and Wheel Alignment**.
4. Remove instrument cluster retaining screws. See Fig. 1.
5. Place a towel over steering column. Carefully tilt top of cluster out of dash panel. Unlock and disconnect harness connectors from back of cluster. See Fig. 2.

620

0012562

Fig. 1. Instrument cluster retaining screws (**arrows**).

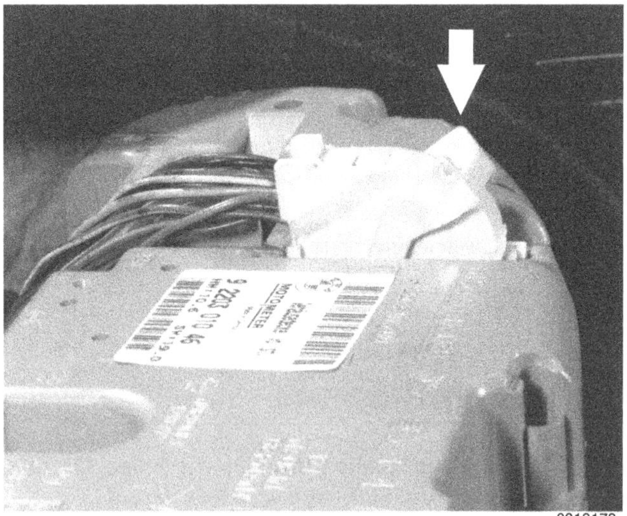

0013178

Fig. 2. Instrument cluster connector locking lever (**arrow**).

6. Installation is reverse of removal, noting the following:
   - Instrument cluster harness connector locking levers must be in up position before installing connector.
   - Install airbag to steering wheel as described in **721 Airbag System (SRS)**.
   - Reconnect negative (–) cable to battery last, once all harness connectors have been reconnected.
   - Recode cluster as described in **Instrument cluster, self-testing**.

Instrument cluster bulbs, replacing

1. Remove instrument cluster as described above.

2. Remove faulty bulb by turning bulb base 45° in its mount and remove from housing. See Fig. 3.

The back of the instrument cluster and its bulb locations are shown in Fig. 4.

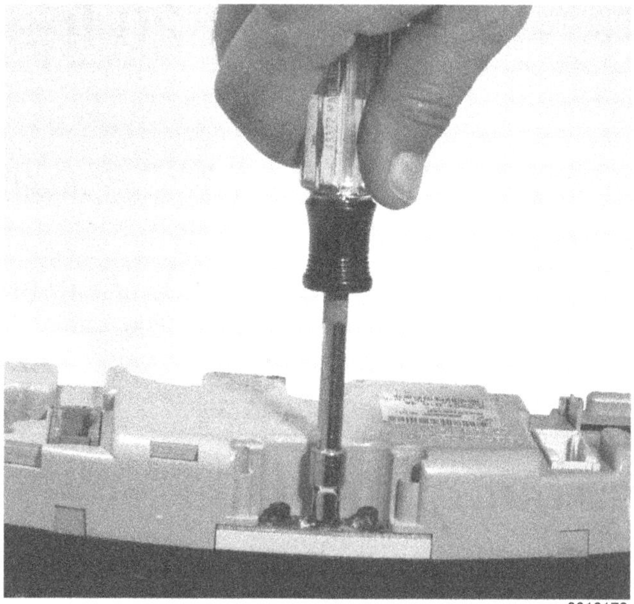

0013179

Fig. 3. Turn bulb base 45° in direction shown (**arrow**).

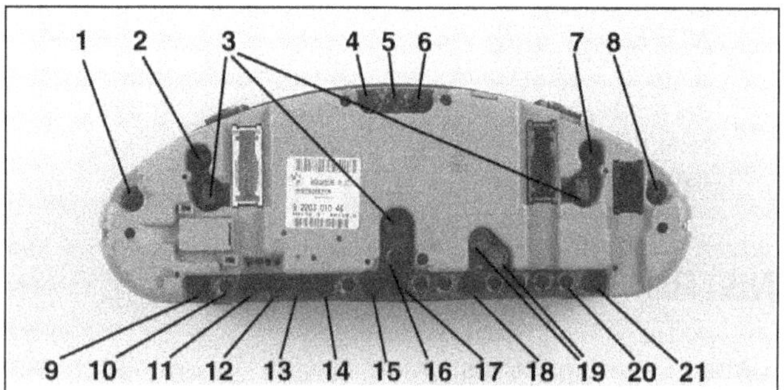

1. Temperature warning
2. Right turn signal
3. Instrument lighting
4. Oil pressure warning
5. High beam
6. Battery charging lamp
7. Left turn signal
8. Low fuel warning
9. Airbag warning
10. Catalytic converter warning
11. Emergency brake
12. Brake fluid warning
13. Brake pad wear warning
14. ABS warning
15. Seat belt warning
16. AST warning
17. Check control
18. Check engine
19. LCD lighting
20. Rear foglight (not used)
21. Front foglights

0012563

Fig. 4. Bulb designation chart.

> ***CAUTION —***
> *Note wattage of bulb. Excessive wattage could melt cluster housing. Replace defective bulbs with replacement bulbs of the same wattage.*

Odometer coding plug, replacing

The coding plug maintains a record of the vehicle mileage.

1. Record vehicle mileage.

2. Remove instrument cluster as described earlier.

3. Break open plastic retainer in back of instrument cluster and remove coding plug. See Fig. 5.

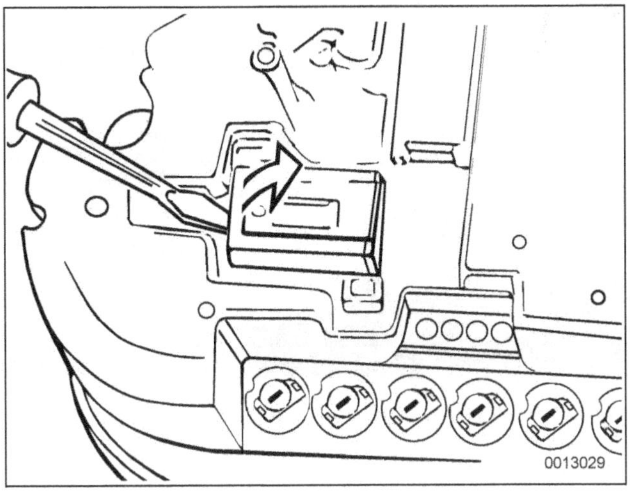

Fig. 5. Remove coding plug retainer with a screwdriver (**arrow**).

4. Install coding plug and new retainer. Reinstall instrument cluster.

5. Reset vehicle mileage.

> ***NOTE—***
> *Consult an authorized BMW dealer service department about resetting vehicle mileage using the special BMW Service tester.*

Vehicle speed sender, replacing

The vehicle speed sender supplies an electronic pulsed speed signal to the speedometer. The sender is mounted beneath the rear of the car, on the differential cover plate.

> ***NOTE—***
> *If the speedometer is malfunctioning, check the vehicle speed sensor harness connector on the rear differential. Water damage, frayed wires and corrosion of the connector contacts are often the cause of speedometer malfunction.*

1. Locate vehicle speed sender electrical connector at back of differential. Depress retaining levers on connector and remove. See Fig. 6.

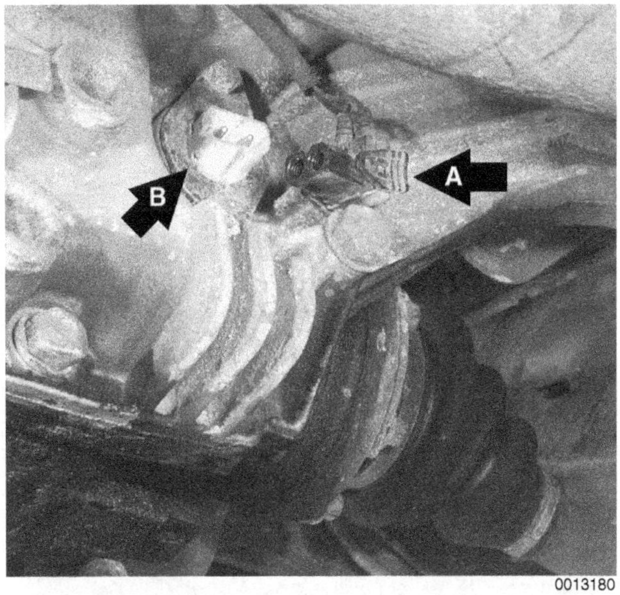

Fig. 6. Vehicle speed sender harness connector (**A**) and sender (**B**).

2. Remove vehicle speed sender retaining screws and remove sender from differential cover.

3. On installation, replace O-ring and coat with oil (90W). See Fig. 7.

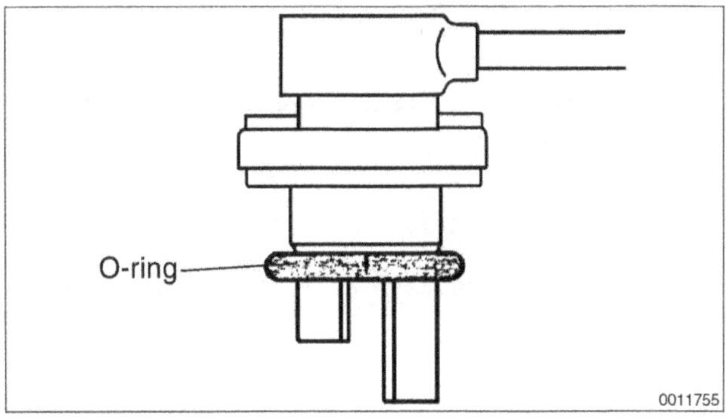

Fig. 7. O-ring correctly installed on vehicle speed sender.

4. Install sender into differential and tighten mounting bolts evenly in stages until the sender is firmly seated.

Tightening Torque
- Vehicle speed sender to differential cover . . . . 6.6-8.4 Nm (53-74 in-lb)

MULTI-INFORMATION DISPLAY (MID)

Several different versions of the Multi-Information Display, have been installed in E36 cars. The top-of-the-line model is called the On-Board Computer. See Fig. 8.

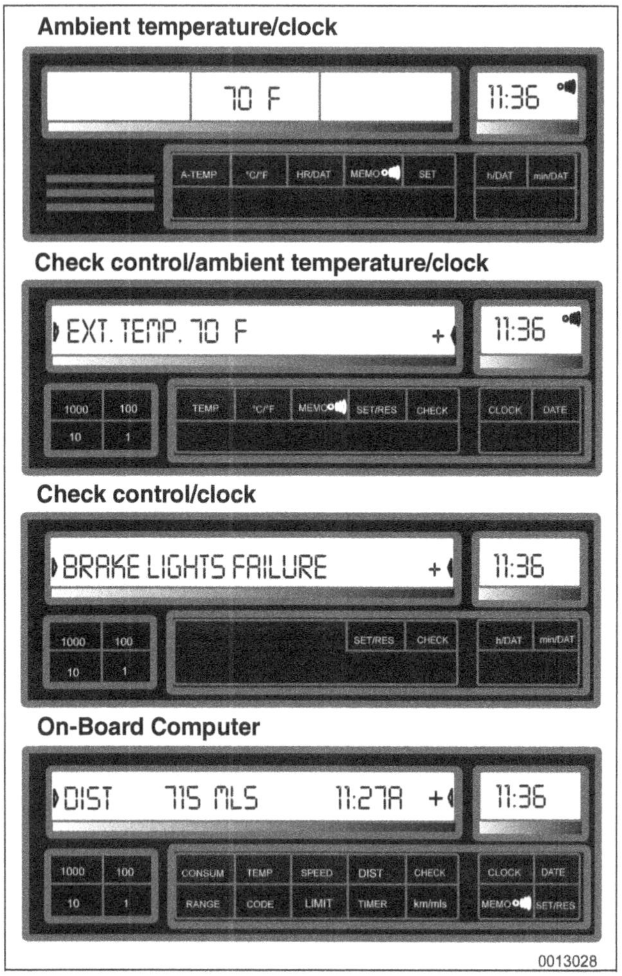

Fig. 8. Different versions of Multi-Information Display (MID) installed in E36 cars.

Multi-Information Display, removing and installing

The Multi-Information Display (MID) is illuminated by bulbs. It must be removed to replace the bulbs. Bulb replacement is included in this procedure.

1. Remove dashboard parcel shelf. Reach under MID module and push outward. See Fig. 9.

2. Slide harness retaining lever forward to release MID from wiring harness. See Fig. 10.

3. To remove a faulty bulb, use a narrow screwdriver to turn bulb base 45° counter clockwise. See Fig. 11.

Fig. 9. Removing the Multi-Information Display (MID) module.

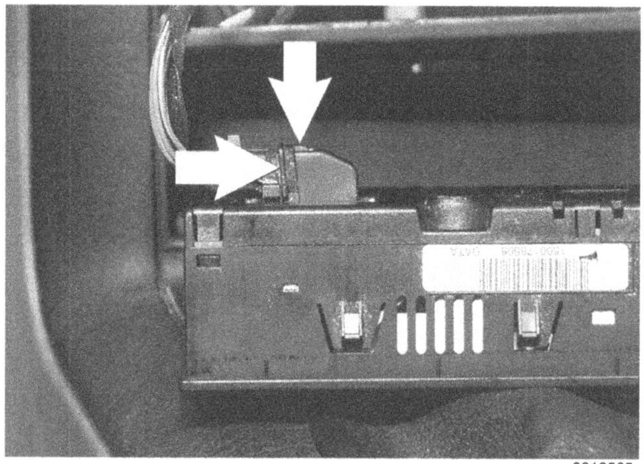

Fig. 10. MID wiring harness release lever (**arrows**).

Fig. 11. Turn bulb base 45° counterclockwise.

4. Installation is reverse of removal.

630 Exterior Lighting

GENERAL

The headlight and foglight assemblies use halogen bulbs. To ensure the longevity of halogen bulbs, the bulb glass should not be handled. Dirt and skin oils may cause a bulb to fail prematurely. If necessary wipe bulb using a clean cloth dampened with rubbing alcohol.

> ***NOTE —***
> *Use only original equipment replacement bulbs. Non-original equipment bulbs may cause false failure readings on the check control display (where applicable).*

> ***WARNING —***
> *When working on electrical systems, remove the fuse protecting the circuit under repair. See* **610 Electrical Component Locations**.

FRONT LIGHTING

Headlight bulb, replacing

1. Open engine hood.

2. Working in back of headlight assembly, grasp rear of bulb socket (harness connector) and turn counterclockwise 90°. Pull bulb out of headlight assembly. See Fig. 1.

3. Disconnect harness connector from bulb.

> ***CAUTION —***
> *Do not touch halogen bulbs with your bare hands. If necessary wipe bulb using a clean cloth dampened with rubbing alcohol.*

0012567

Fig. 1. To release bulb from headlight assembly, rotate socket counterclockwise to disengage retaining tabs (**arrows**). Right side bulb is shown in unlocked position. Note that bottom tab **A** is wider than other two.

4. Installation is reverse of removal. Align tabs on bulb base to cutouts on rear of lens assembly. Secure bulb by rotating clockwise 90°.

Bulb Applications
- Low beam headlight bulb H9006
- High beam headlight bulb H9005

630

Headlight assembly, removing and installing

> ***CAUTION —***
> *Following headlight assembly work, headlight beam aim must be adjusted using special equipment.*

1. Left side: Remove air filter housing, air intake duct, and air mass meter. See **100 Engine–General**.
2. Disconnect harness connectors from headlight bulbs.
3. Remove front turn signal. See **Front turn signal assembly, removing and installing** later in this repair group.
4. Remove radiator cover. See Fig. 2.

0012568

Fig. 2. Radiator cover screws (**arrows**).

5. Using a wrench, brace adjusting nut at top center of headlight. Remove three screws at top of assembly. See Fig. 3.

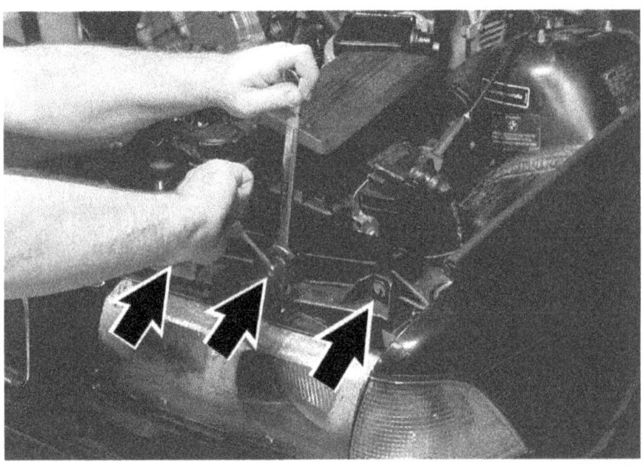

0012569

Fig. 3. Brace headlight adjusting nut as shown and remove top screws (**arrows**).

6. Remove two screws on bottom of assembly. See Fig. 4.

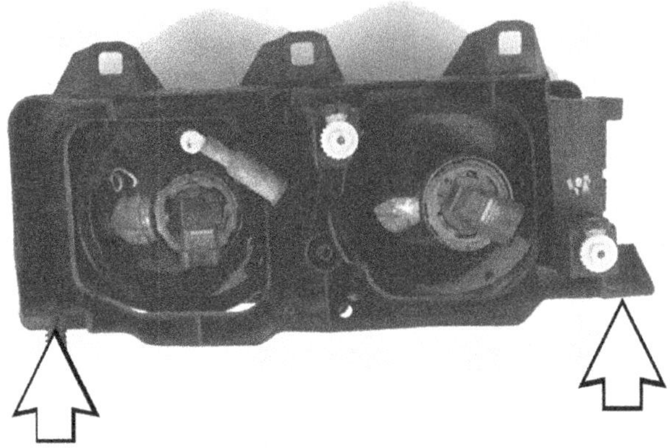

0012567a

Fig. 4. Headlight bottom retaining screws (**arrows**). Screws mount in threaded expanding rivets which act as headlight height adjusters.

7. Remove headlight assembly from front.
8. On installation, tighten lower screws first while holding expanding rivets with a wrench. There should be a 2.5 mm gap between bottom of lens and front panel. Use adjusting nut to raise or lower headlight unit. See Fig. 5.

0012681

Fig. 5. Measure gap **A** (2.5 mm) between bottom of headlight assembly and front panel.

9. Clearance between hood and headlight cover should be 1.5 mm. Adjust by turning upper headlight adjusting nuts, then tighten down center screw. See Fig. 6.

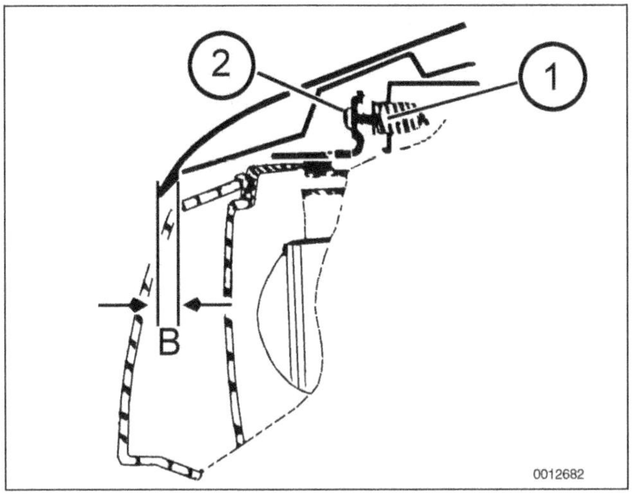

Fig. 6. Measure gap **B** (1.5 mm) between front of hood and headlight cover. Note horizontal headlight adjusting nut (**1**) and retaining screw (**2**).

Foglight assembly, removing and installing

The following procedure covers foglight lens and foglight bulb replacement. Foglight components are shown in Fig. 7.

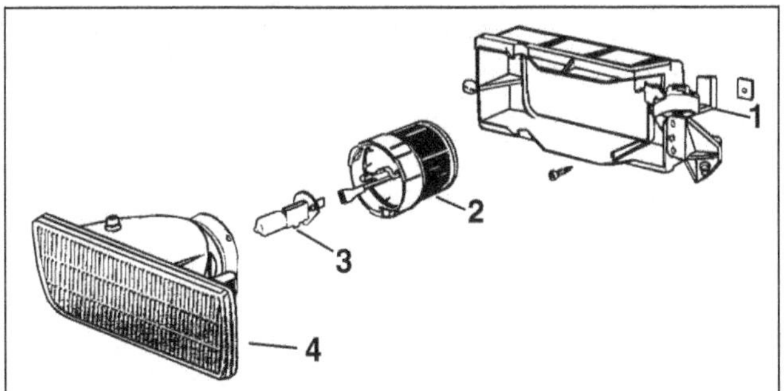

Fig. 7. Foglight assembly (removed).

1. Insert a screwdriver in upper air duct inlet of front lower valence and disengage foglight lens catch mechanism. Swing light outward and gently disengage hinged end. See Fig. 8.

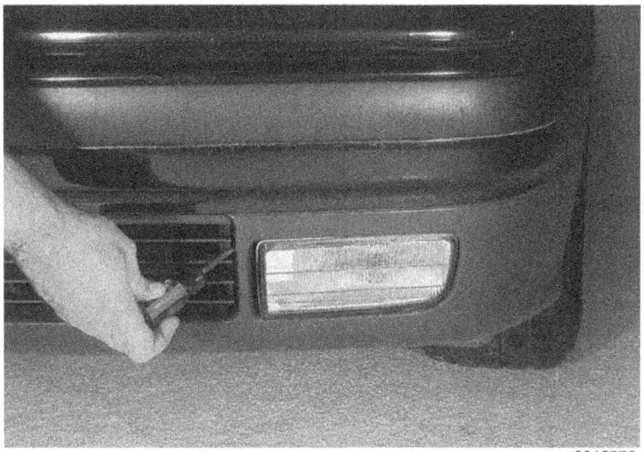

Fig. 8. Insert screwdriver in upper air duct inlet.

2. Remove rear protection cap and disengage steel wire foglight bulb clamp.

3. Remove bulb from rear of foglight assembly. Disconnect electrical harness from foglight bulb.

> ***CAUTION—***
> *Do not touch halogen bulbs with your bare hands. If necessary wipe bulb using a clean cloth dampened with rubbing alcohol.*

4. Installation is reverse of removal. Check electrical connector for corrosion. Spray with contact cleaner, if necessary.

Bulb Application (USA Models):

- Foglight bulb H1 (55w)

Front turn signal assembly, removing and installing

The procedure below also includes turn signal bulb replacement.

1. Working inside engine compartment, use a screwdriver to press turn signal assembly retaining clip away from retaining tab on headlight assembly frame. Push turn signal forward. See Fig. 9.

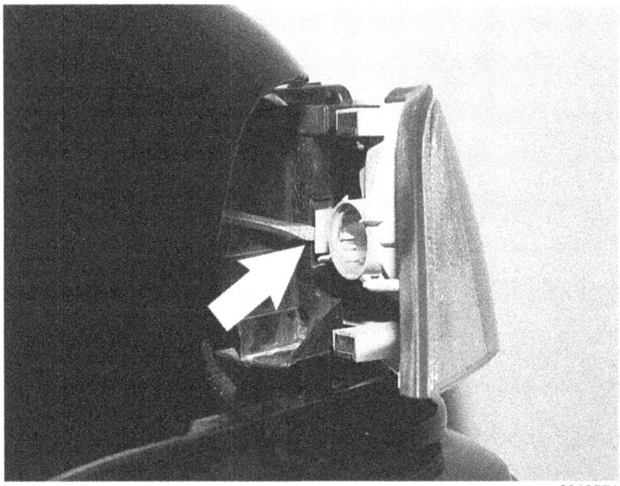

0012571

Fig. 9. Press on clip (**arrow**). Light slides straight out.

2. Depress tabs on bulb socket and remove from turn signal assembly. See Fig. 10.

0012572

Fig. 10. Depress locking tabs (**arrows**) to remove bulb and socket from turn signal housing.

Side Marker Lights

Side marker lights on later model E36 cars are snapped into the front fenders. To service, carefully pry the assembly out of the fender, then detach the electrical harness connector and bulb socket.

REAR LIGHTING

Taillight Bulbs

To access taillight bulbs, see Fig. 11. The location of the rear bulbs are shown in Fig. 12. Turn bulb socket ¼ turn to remove.

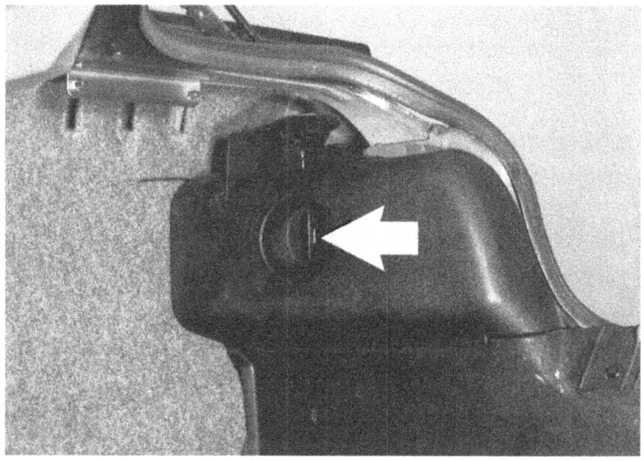

0012573

Fig. 11. To remove taillight bulb cover, turn retainer (**arrow**) ¼ turn counterclockwise.

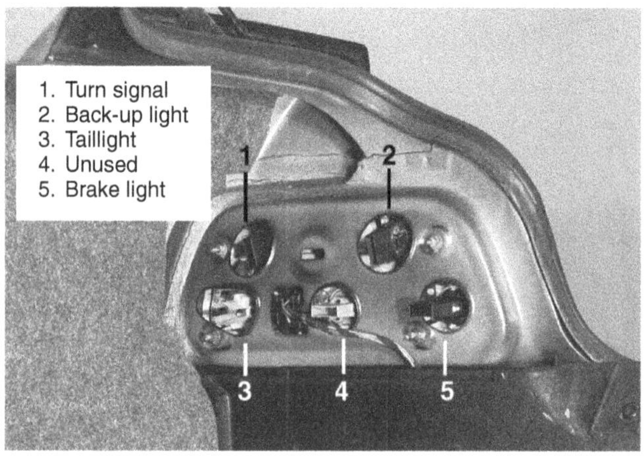

0012574

Fig. 12. Taillight bulb identification.

Taillight assembly, removing and installing

1. Open trunk lid. Remove cover from rear of taillight. Refer to Fig. 11.

2. Depress locking tabs on electrical harness connector and disconnect from rear of taillight lens assembly. Remove four taillight assembly retaining nuts. See Fig. 13.

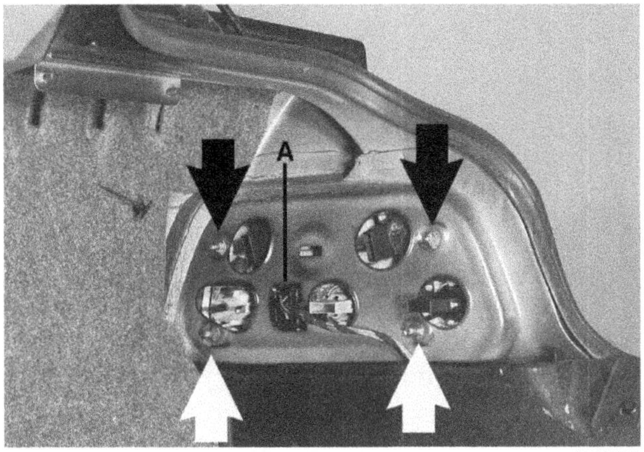

0012574

Fig. 13. Taillight mounting nuts (**arrows**) and harness connector (**A**).

3. Installation is reverse of removal.

Center Brake Light

On convertible models, it is necessary to open trunk lid and remove inner tailgate trim adjacent to license plate light assembly. Then remove the electrical connectors and four center brake light retaining nuts. Push center brake light out. Replace bulbs if necessary.

On 4-door and coupe models, open trunk lid. Working directly under center brake light, turn bulb socket counterclockwise ¼ turn and remove from center brake light assembly. Replace bulb if necessary. See Fig. 14.

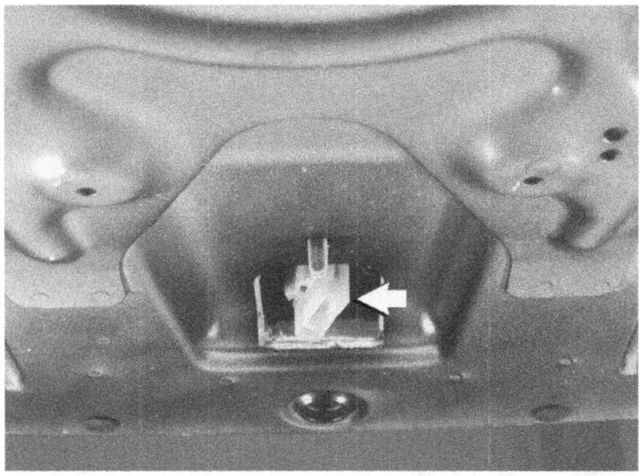

0012575

Fig. 14. Turn center brake light socket counterclockwise and remove bulb.

Center brake light housing, replacing (4-door and coupe models)

1. Remove rear window parcel shelf. See **513 Interior Trim**.

2. Depress tabs on center brake light electrical connector and disconnect.

3. Remove retaining screws and remove socket.

4. Replace bulb.

License plate light, removing and installing

1. With trunk lid raised, remove license plate light trim panel screws. See Fig. 15.

0012576

Fig. 15. Remove screws (**arrows**) and remove license plate light trim panel.

2. Using a flat tipped screw driver, pry gently on right side of license plate light assembly. See Fig. 16.

3. Lift license plate light assembly from sheet metal. Release harness connector locking tabs and separate harness from license plate light electrical connector.

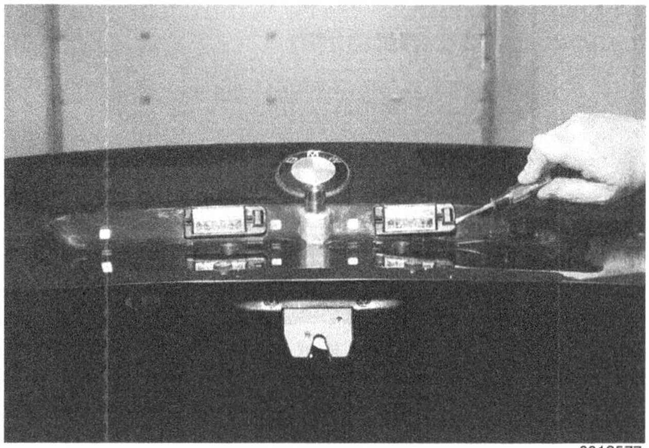

0012577

Fig. 16. Insert screwdriver (**arrow**).

NOTE —

Inspect bulb contact springs for damage and corrosion. Replace socket assembly if necessary.

4. Installation is reverse of removal.

640 Heating and Air Conditioning

TABLES

GENERAL

E36 cars are equipped with integrated automatic heating and air conditioning systems. Many of the procedures given in this repair group require that the A/C refrigerant charge be evacuated using specialized equipment. See **A/C System Warnings and Cautions** later in this repair group.

HEATER AND CONTROLS

Blower motor, removing and installing

NOTE —

On 6-cylinder cars, the plastic top engine cover and fuel rail cover must be removed. The fuel injector harness retaining screws must be removed and the harness set aside. The engine control module (ECM) cover must be removed, the control unit connectors unplugged and the harness pulled out of the way. See **130 Fuel Injection**.

1. Disconnect negative (–) cable from battery.

2. Open hood. Unclip engine bulkhead plenum grille at base of windshield and remove grille. See Fig. 1.

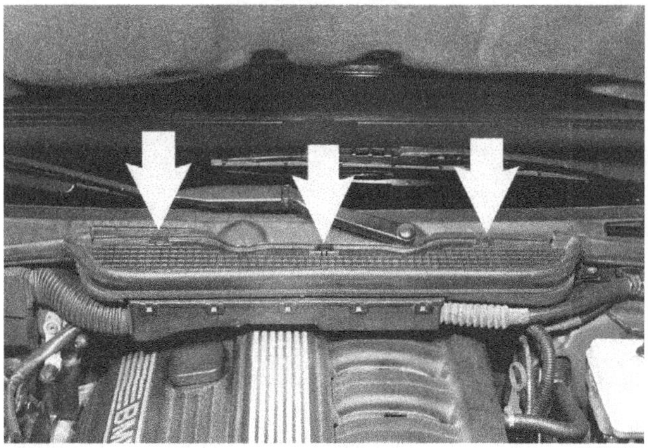

0012527

Fig. 1. Detach grille retaining clips (**arrows**) and pull grille upward and out of rubber gasket.

CAUTION —

Prior to disconnecting the battery, read the battery disconnection cautions given at the front of this manual on page viii.

640

3. Remove wiring harness retaining clip at bulkhead. Remove screws retaining wiring harness cover and pull cover forward. See Fig. 2.

0012529

Fig. 2. Wiring harness cover screws at base of windshield (**arrows**).

4. On passenger side, push aside wiring harness and remove two screws retaining plenum housing. See Fig. 3.

0012535

Fig. 3. Right side plenum retaining screws (**arrows**).

5. On driver side, push aside wiring harness and remove screw retaining plenum. Pull plenum up and remove from engine bay. See Fig. 4.

6. Unclip blower motor housing cover, disconnect control cable, and turn center retainer ¼ turn to remove cover. See Fig. 5.

7. Disconnect electrical connectors from blower motor.

8. Remove left upper section of blower housing by detaching spring clips at edge of housing. See Fig. 6.

0012534

Fig. 4. Remove screw at **A** and pull air plenum upward.

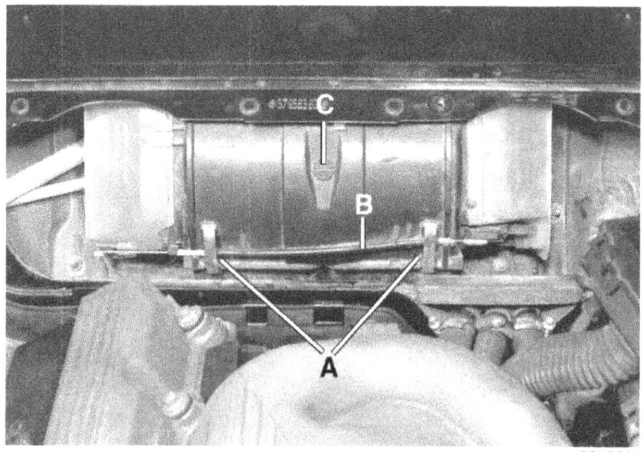

0012636

Fig. 5. Blower motor housing clips (**A**); control cable (**B**); and center retainer (**C**).

0012637

Fig. 6. Spring clips (**arrows**) on driver side of blower housing.

9. Release blower motor retaining strap and remove blower motor and fans from cowl opening.

NOTE —
Twist and rotate blower motor assembly to facilitate removal from opening. If the blower motor is to be reinstalled, use care not to damage fan wheels during removal. If the blower motor is to be replaced, cut off fan wheels at center to aid in removal.

10. Installation is reverse of removal. When installing a replacement blower motor, install motor into cowl area before installing fan wheels. Blower motor test specifications are listed in **Table a**.

NOTE —
On replacement blower motors from BMW, the fan wheels come unattached to aid in installation. Locate arrow on blower motor and install fan wheels at this position. Be sure color coding matches. Retain fan wheels using the new clips. See an authorized BMW parts department for the latest parts information.

Table a. Blower Motor Test Values

Testing voltage	13 VDC
Operating voltage	9–15 VDC
Max. current consumption (approx.)	12 amps
Maximum speed	3,600–4,000 rpm
Axial play (max. allowable)	1.0 mm (0.04 in.)

Blower motor resistor pack, replacing

The blower resistor pack is clipped into the upper left side of the heater box.

1. Remove lower left dash trim panel. See **513 Interior Trim**.

2. Working under left (driver's) side of dash, unclip and pull out resistor pack from side of heater box. Remove harness connector from resistor pack. See Fig. 7.

3. Installation is reverse of removal.

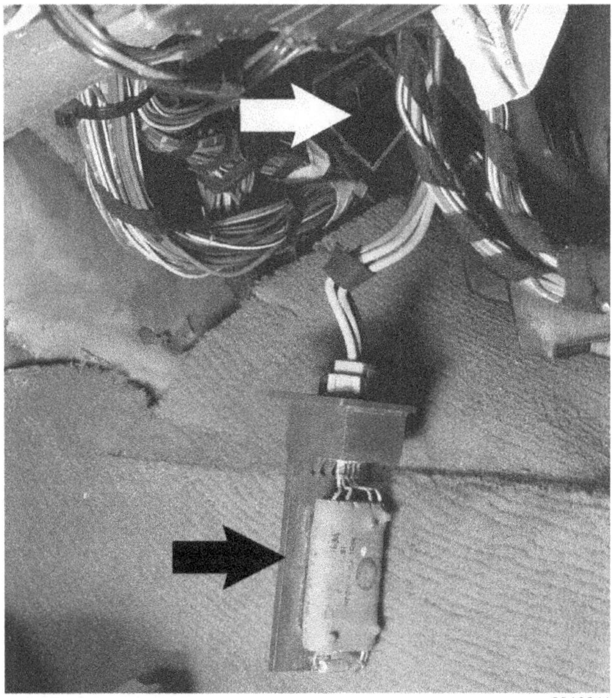

0012641

Fig. 7. Blower fan resistor pack (**black arrow**) and resistor pack receptacle (**white arrow**) below left side of dash.

Heater valve, replacing

The two-way heater valve is located in the rear of the engine compartment on the driver side. The solenoid operated valves control the coolant flow through the dual-chamber heater core.

1. Drain engine coolant. See **170 Radiator and Cooling System**.

WARNING —
Allow the cooling system to cool before opening or draining the cooling system.

2. Disconnect hoses and harness connector at heater valve. See Fig. 8.

0012638b

Fig. 8. Hoses (**A**) and harness connector (**B**) at heater valve.

3. Carefully pull heater valve out of rubber mounting and pull up and out of engine compartment.

4. Installation is reverse of removal. Use new hose clamps. Inspect hoses and replace as necessary.

Heater box, removing and installing

The heater box, located underneath the center of the dashboard, houses the heater core, the A/C evaporator, and associated sensors and actuators. Removal of the heater box requires that the A/C refrigerant be discharged.

> ***CAUTION —***
> - *Removing the heater box requires evacuating the A/C system. DO NOT attempt this procedure without proper tools and training.*
> - *Any person who services a motor vehicle air conditioner must, by law, be properly trained and certified, and use approved refrigerant recycling equipment. Technicians must complete an EPA-approved recycling course to be certified*

1. Drain engine coolant. See **170 Radiator and Cooling System**.
2. Remove complete dashboard. See **513 Interior Trim**.
3. Remove blower motor as described earlier.
4. Remove nut retaining water hose flange to rear bulkhead. Push flange to one side. See Fig. 9.

0012638b

Fig. 9. Nut retaining coolant hose flange at bulkhead (**arrow**).

> ***NOTE —***
> *Clear any remaining coolant from the heater core by blowing compressed air into the coolant feed hose.*

5. Remove nuts from heater box retaining bracket on engine bulkhead left side. See Fig. 10. Repeat procedure for right side nuts.

0012639

Fig. 10. Heater core mounting nuts at left side of bulkhead (**arrow**). Wiper assembly shown removed for clarity.

6. Cut wire tie holding wire harness to heater box and disconnect cable harness connector from heater box. See Fig. 11.

0012647

Fig. 11. Harness connectors (**A**) and cable tie (**B**) at heater box, under driver side dash.

7. Following manufacturer's instructions, connect approved refrigerant recovery/recycling/recharging unit to A/C system and discharge system.
8. Unbolt A/C refrigerant lines flange from engine bay bulkhead. See Fig. 12.
9. On models with rotary-knob A/C control head: Disconnect wiring harness connectors to A/C control module in center console and remove module. See Fig. 13.

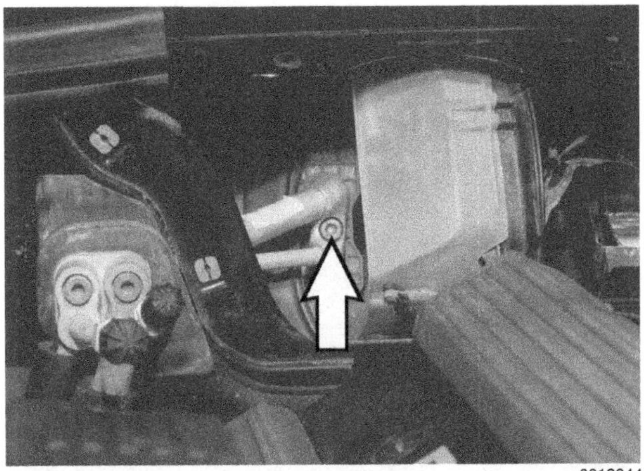

0012644

Fig. 12. Mounting bolt for A/C refrigerant line flange (**arrow**).

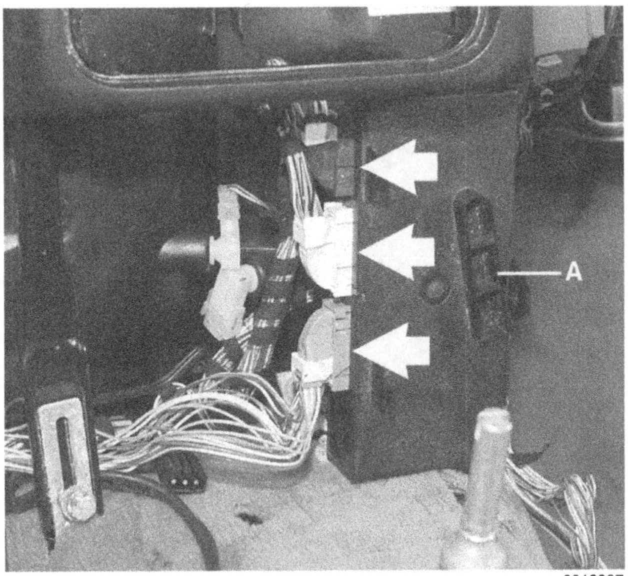

0013007

Fig. 13. A/C control module (**A**) behind center console. **Arrows** point to electrical harness connectors to be disconnected.

10. Unclip and remove heating and ventilation ducts on sides of heater box. See Fig. 14.

11. Remove heater box retaining bracket on center tunnel and remove heater box. See Fig. 15.

12. Installation is reverse of removal. Evacuate and recharge A/C system using specialized equipment. Fill and bleed cooling system. See **170 Radiator and Cooling System**.

> ***CAUTION—***
> *Replace heater and A/C flange O-rings and tighten fasteners to specified torques.*

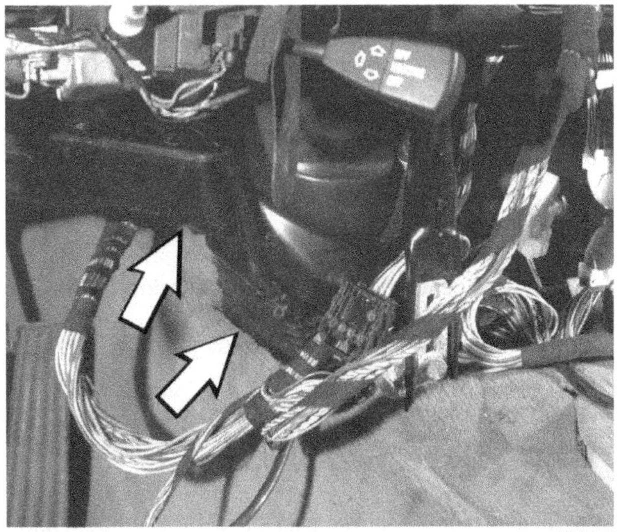

0013008

Fig. 14. Heating and ventilation ducts on left side of heater box (**arrows**). Right side ducts are similar.

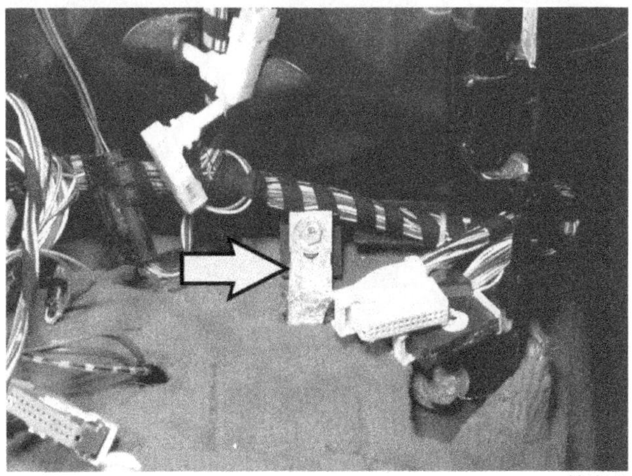

0012645

Fig. 15. Heater box retaining bracket in center console (**arrow**).

Tightening Torques

- A/C refrigerant lines flange to bulkhead.................... 3.5 Nm (31 in-lb)
- Water hose flange to bulkhead (M6 nut) .... 3.5 Nm (31 in-lb)
- Heater box retaining bracket to tunnel .... 4.5 Nm (40 in-lb)

Heater core, replacing

1. Drain engine coolant. See **170 Radiator and Cooling System**.

> ***WARNING —***
> *Allow the cooling system to cool before opening or draining the cooling system.*

2. Remove center console. See **513 Interior Trim**.
3. Remove Multi-Information Display (MID) module. See **620 Instruments**.
4. Remove lower dash panel and knee bolster on driver side. See **513 Interior Trim**.
5. Working under driver side of dash, remove coolant flange from heater core. See Fig. 16.

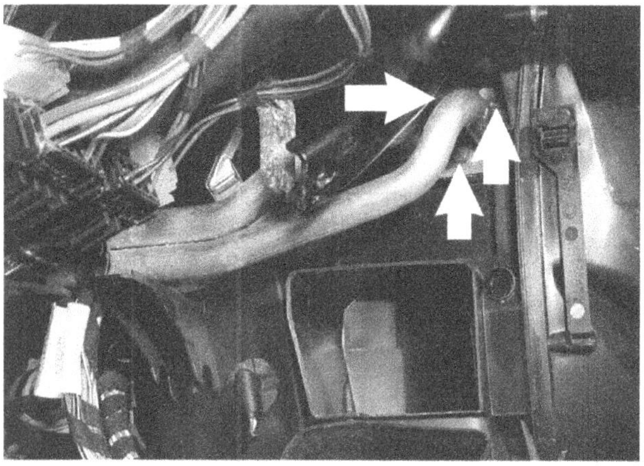

0012651

Fig. 16. Bolts at 3-way coolant flange under driver's side of dash (**arrows**).

> ***NOTE —***
> *Residual coolant in heater core will spill. Protect the carpet.*

6. Working at front of driveshaft tunnel, unclip wire harnesses and move to one side in front of heater box. See Fig. 17.
7. Remove heater core temperature sensors as described later in this repair group.
8. Remove front heater box cover. See Fig. 18.
9. On right side of heater box, unclip heater air duct actuating lever. Remove cross-rod (with flaps) from front of heater core. See Fig. 19.

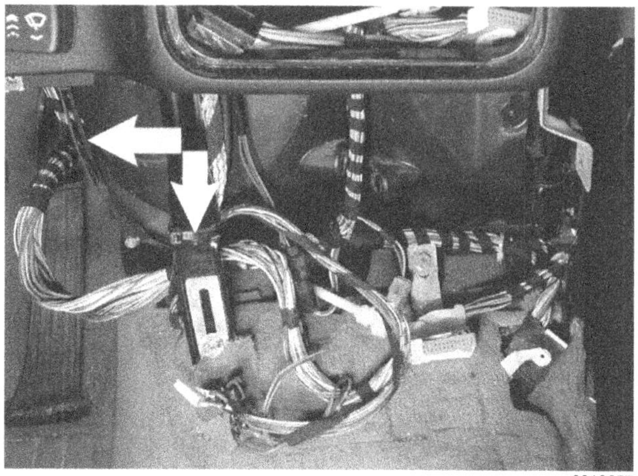

0012657

Fig. 17. Wire harness retaining clips and wire ties at heater box cover (**arrows**).

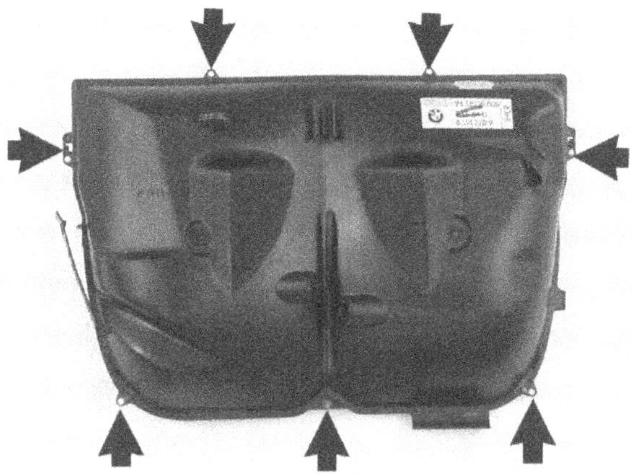

0012662

Fig. 18. Screws retaining front heater box cover (shown removed) to heater box (**arrows**).

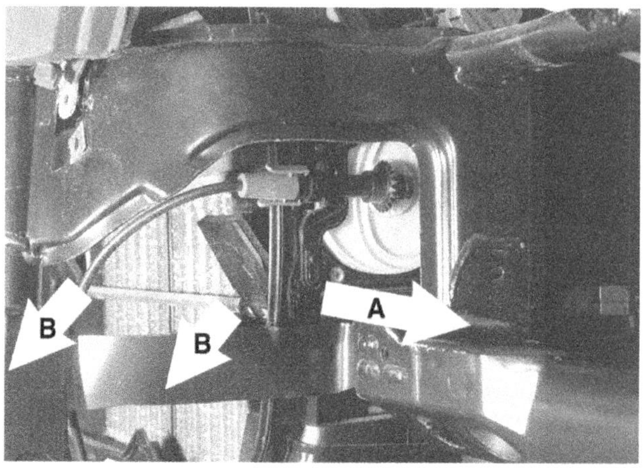

0012661

Fig. 19. Pry heater flap actuating lever in direction **A**. Pull off air flap cross-rod in direction **B**.

10. Pull heater core up and backward, away from heater box. If necessary, remove left and right reinforcing bracket bolts. See Fig. 20.

0012660

Fig. 20. Pull heater core up and toward back of car. Lower mounting bolts for steering column brace and glove compartment support frame may have to be removed (**arrows**).

11. Installation is reverse of removal. Fill and bleed cooling system. See **170 Radiator and Cooling System**.

> ***CAUTION —***
> *Replace coolant flange O-rings and tighten fasteners to specified torques.*

Tightening Torques

- Coolant flange to heater core .... 3 Nm (27 in-lb)

Heater core temperature sensor, replacing

The automatic heating and air conditioning system uses several sensors to control interior air temperature. Use the following procedure and the data in **Table b** to pinpoint problems with the heater core sensors.

1. Remove center console. See **513 Interior trim**.
2. Remove Multi-Information Display (MID) module. See **620 Instruments**.
3. Pull temperature sensors from heater box and disconnect harness connectors. See Fig. 21.
4. Installation is reverse of removal.

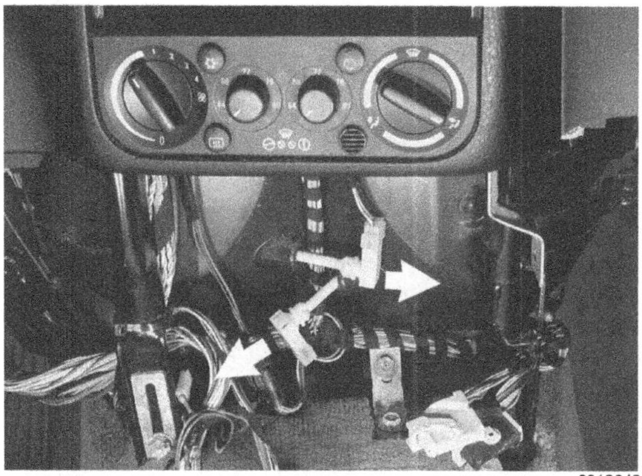

0012648

Fig. 21. Remove heater core temperature sensors in direction shown (**arrow**).

Table b. Heater Core Temperature Sensor Test Values

Test temperature	Resistance
4°F (-20°C)	84.39 - 109.61 k ohms
14°F (-10°C)	48.58 - 62.09 k ohms
32°F (0°C)	28.89 - 36.40 k ohms
68°F (20°C)	11.13 - 13.83 k ohms
140°F (60°C)	2.19 - 2.78 k ohms
212°F (100°C)	0.58 - 0.77 k ohms

Heater and A/C Control Head

Two types of heater and air conditioning control heads were installed in E36 cars. Early models have control heads with four rotary knobs and a separate control module. A flexible shaft, clipped to the right side of the control head, controls the heater flaps.

Later model E36 cars, starting in approximately mid-1995, have digital display control heads with buttons. The A/C control module is integrated with the digital control head.

Heater and A/C control head, removing and installing (rotary-knob)

1. Remove Multi-Information Display (MID) module from center of dash. See **620 Instruments**.
2. Pull off outer rotary knobs from A/C control head. Remove two screws on knob bezels. Remove control panel face plate by prying it off control head gently. See Fig. 22.
3. Carefully push control head back into dash and downward through MID opening. See Fig. 23.

0012649

Fig. 22. A/C control head faceplate screws on rotary knob version (**arrows**). Once screws are removed, faceplate must be unclipped from control head.

0012650

Fig. 23. Push control panel out of center console.

4. Remove harness connectors. Unfasten heater duct control flex-shaft on right side of control head by squeezing retaining tab. Pull shaft out of coupling in control head and remove head.
5. Installation is reverse of removal.

Heater and A/C control head, removing and installing (digital-display)

1. Remove Multi-Information Display (MID) module from center of dash. See **620 Instruments**.
2. Reach into dash through MID opening and push A/C control head outwards.
3. Disconnect harness connector and remove control head.
4. Installation is reverse of removal

Heater and A/C control module, removing and installing (rotary-knob)

1. Remove Multi-Information Display (MID) module from center of dash. See **620 Instruments**.
2. Remove center console. See **513 Interior Trim**.
3. Disconnect wiring harness connectors to A/C control module in center console and remove module. See Fig. 24.

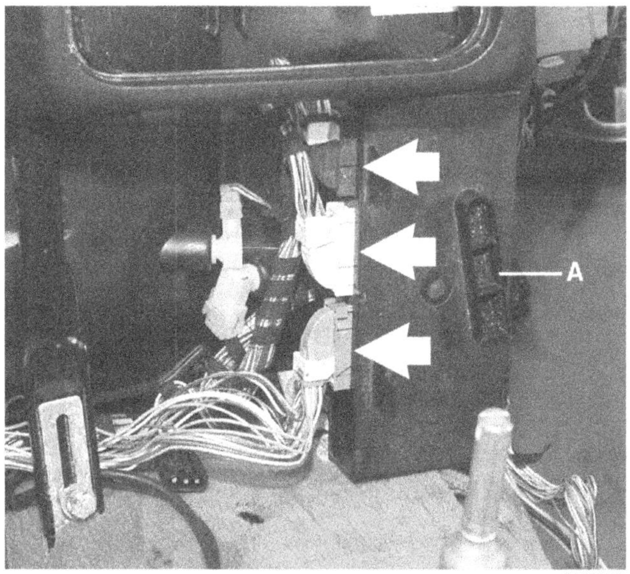

0013007

Fig. 24. A/C control module (**A**) in center console. **Arrows** point to electrical harness connectors at A/C module.

Heater and A/C control module, removing and installing (digital-display)

The A/C control module for cars with the digital display is mounted on the back of the control head. See Fig. 25.

1. Remove control head as described earlier and place on an anti-static mat.

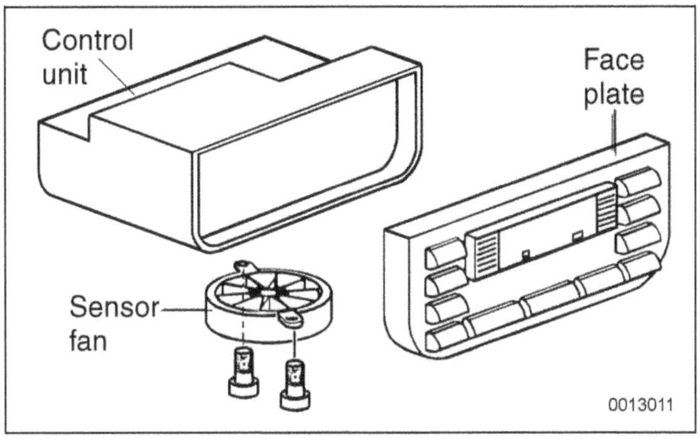

Fig. 25. Digital display A/C control head components.

2. Unscrew left and right screws retaining control head to control module. Press retaining clips and release front plate from control head. Remove A/C control module from control head housing.

Sensor fan for interior temperature, replacing

A small electric fan, installed in the A/C control head, draws cabin air past the climate control temperature sensor.

1. Remove A/C control head as described earlier.

2. Rotary-knob control head: Disconnect electrical connector from top of control head housing. Pry open plastic housing and remove sensor fan. See Fig. 26.

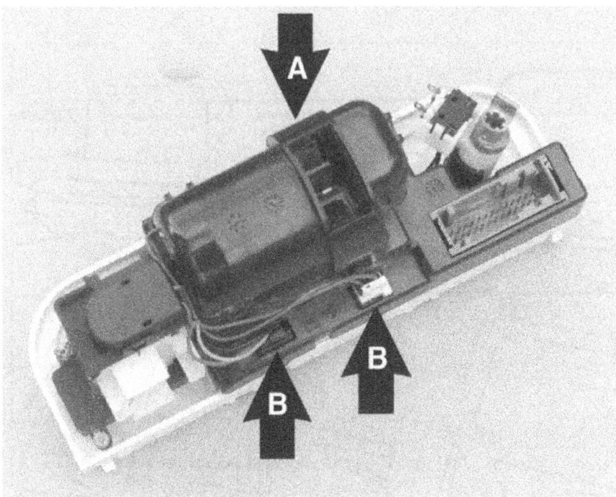

Fig. 26. Sensor fan (**A**) and harness (**B**) in back of A/C control head (rotary-knob control head shown).

3. Digital display models: Disconnect electrical harness connector from control head housing. Remove screws holding sensor fan to bottom of control module and remove. Refer to Fig. 25.

4. Installation is reverse of removal.

Heater and A/C Air Distribution Motors

The recirculation, fresh air and other vent-duct flaps are controlled electrically via a series of air distribution actuating motors. Cars with the rotary-knob control head use three electromechanical motors, each held in place via a small plastic clip. See Fig. 27.

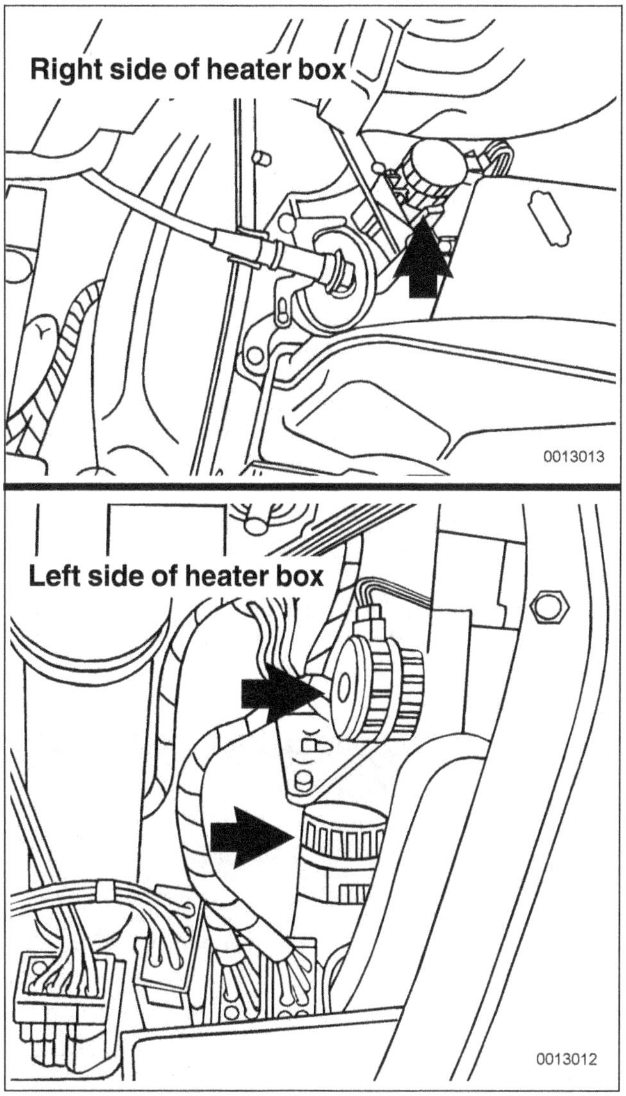

Fig. 27. Location of early production air distribution motors (**arrows**).

Later systems (with the digital display control head) use five distribution motors, four of which are digitally controlled. See Fig. 28.

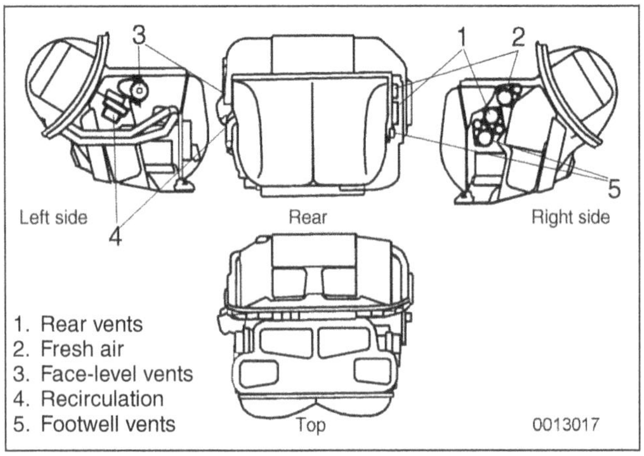

Fig. 28. Location of late production air distribution motors in heater box (shown in rear, side and top views).

Heater and A/C air distribution motor, replacing (rotary-knob)

1. To access left side motors (dash vent mixing or recirculation flap), remove lower dash panel and knee bolster on driver side. See **513 Interior Trim**.
2. To access right side motor (fresh air flap), remove glove compartment. See **513 Interior Trim**.
3. Each motor can be removed after its electrical wiring harness connector has been removed. Squeeze plastic clip to release motor. See Fig. 29.

Fig. 29. A/C air distribution motor plastic retaining clip (**arrow**). Release clip and tilt motor out of housing.

Rear Window Defogger

The electric rear window defogger is integrated with the radio antenna. See Fig. 30.

The rear window defogger switch is integrated with the heating and A/C control head.

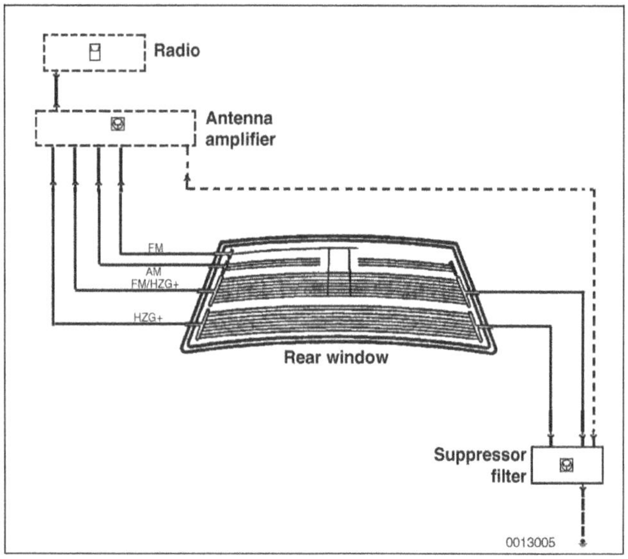

Fig. 30. Wiring diagram of rear window defogger and radio antenna.

Rear Window Blower (Convertible Models)

A blower fan, fitted into the rear seat back, defogs the rear window. It will only operate with the engine running and the top raised. A microswitch behind the rear seats provides the signal that the top is raised.

To access the blower fan motor, it is necessary to remove four screws from the center seat back. Remove the wiring harness connector and motor mounting bolts, then remove motor by pulling it down and forward. See Fig. 31.

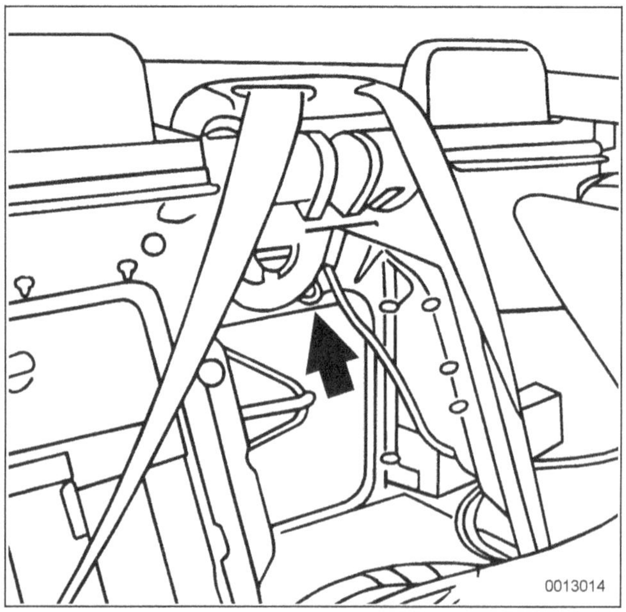

Fig. 31. Rear window blower in center of rear seat back (**arrow**).

To access the rear window blower relay, remove the glove compartment. See Fig. 32.

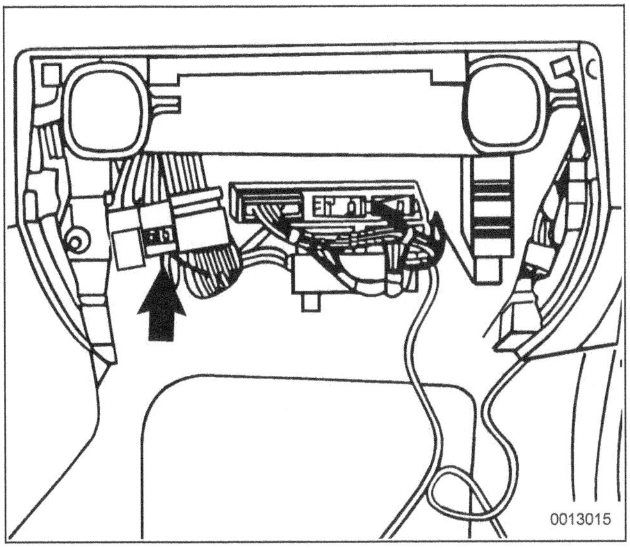

Fig. 32. Rear window blower relay located in front of glove compartment (**arrow**).

AIR CONDITIONING

This section covers removal and installation of air conditioning refrigerant components. A/C testing and diagnosis, refrigerant discharge, evacuation and recharge are not covered here.

The subassemblies and lines of the air conditioner are filled with either refrigerant R12 (1992 cars) or R-134a (1993 and later cars). The major components of the A/C system are shown in Fig. 33.

A/C System Warnings and Cautions

WARNING —
- *Always be careful that refrigerant does not come in contact with your skin. Always wear hand and eye protection (gloves and goggles) when working around the A/C system. If refrigerant does come in contact with your skin or eyes:*
—Do not rub skin or eyes;
—Immediately flush skin or eyes with cool water for 15 minutes;
—Rush to a doctor or hospital;
—Do not attempt to treat yourself.

- *Work in a well ventilated area. Avoid breathing refrigerant vapors. Exposure may irritate eyes, nose, and throat. Switch on exhaust/ventilation systems when working on the refrigerant system.*

- *Do not expose any component of the A/C system to high temperatures (Above 80°C/176°F) or open flames. Excessive heat will cause a pressure increase which could burst the system.*

WARNING —
- *Keep refrigerant away from open flames. Poisonous gas will be produced if it burns. Do not smoke near refrigerant gases for the same reason.*

- *The A/C system is filled with refrigerant gas which is under pressure. Pressurized refrigerant in the presence of oxygen may form a combustible mixture. Never introduce compressed air into any refrigerant container (full or empty), capped off A/C component, or piece of service equipment.*

- *Refrigerant gases are heavier than air, displace oxygen and may cause suffocation in areas of poor circulation, for example under a car. Switch on exhaust/ventilation systems when working on the refrigerant system.*

- *State and local governments may have additional requirements regarding air conditioning servicing. Always comply with state and local laws.*

- *Electric welding near refrigerant hoses causes R-134a to decompose. Discharge system before welding.*

CAUTION —
- *As of January 1, 1992, any person who services a motor vehicle air conditioner must, by law, be properly trained and certified, and use approved refrigerant recycling equipment. Technicians must complete an EPA-approved recycling course to be certified.*

- *It is recommended that all A/C system service be left to an authorized BMW dealer or other qualified A/C service facility.*

- *Do not top off a partially charged refrigerant system. Discharge system, evacuate and then recharge system.*

- *R-134a and R-12 refrigerants and refrigerant oils and system components, including hoses and sealing O-rings, are NOT compatible. Component damage and/or system contamination will result.*

- *The mixture of refrigerant oil (PAG oil) and refrigerant R-134a attacks some metals and alloys (for example, copper) and breaks down certain hose materials. Use only hoses and lines that are identified with a green mark (stripe) or the lettering "R-134a".*

- *Immediately plug open connections on A/C components to prevent dirt and moisture contamination.*

- *Do not steam clean condensers or evaporators. Use only cold water or compressed air.*

A/C Compressor, replacing

1. Following manufacturer's instructions, connect an approved refrigerant recovery/recycling/recharging unit to A/C system and discharge system.

> ***WARNING —***
> *Do not discharge/charge the A/C system without proper equipment and training. Damage to the vehicle and personal injury may result.*

2. Remove splash shield from under engine.

3. Mark A/C drive belt with direction of rotation. Remove A/C belt tensioner trim cap. See Fig. 34.

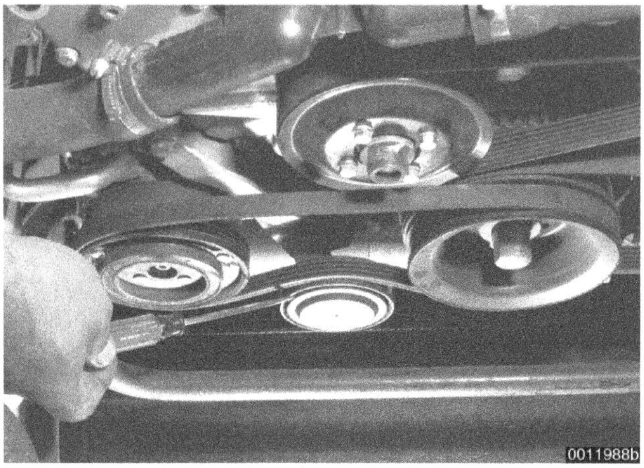

0011988

Fig. 34. To access tensioner retaining bolt, remove trim cap from drive-belt tensioner.

4. On 4-cylinder engines: Loosen A/C belt tensioner bolt under trim cap, then remove belt from A/C compressor. See Fig. 35.

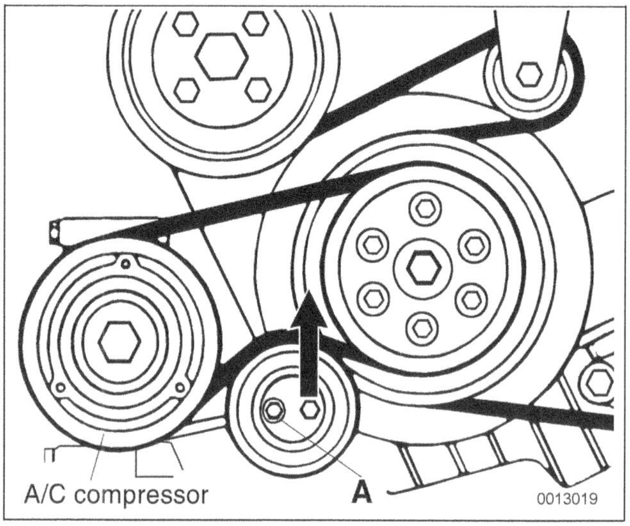

Fig. 35. 4-cylinder engines: Loosen retaining bolt **A** on tensioner to release drive-belt tension. To increase tension, rotate pulley in direction of arrow, then tighten bolt **A**.

5. On 6-cylinder engines: Release tension on A/C belt by turning tensioner pulley clockwise (against spring tension). See Fig. 36.

NOTE —
Detailed instructions for drive-belt removal and installation can be found in **020 Maintenance Program.**

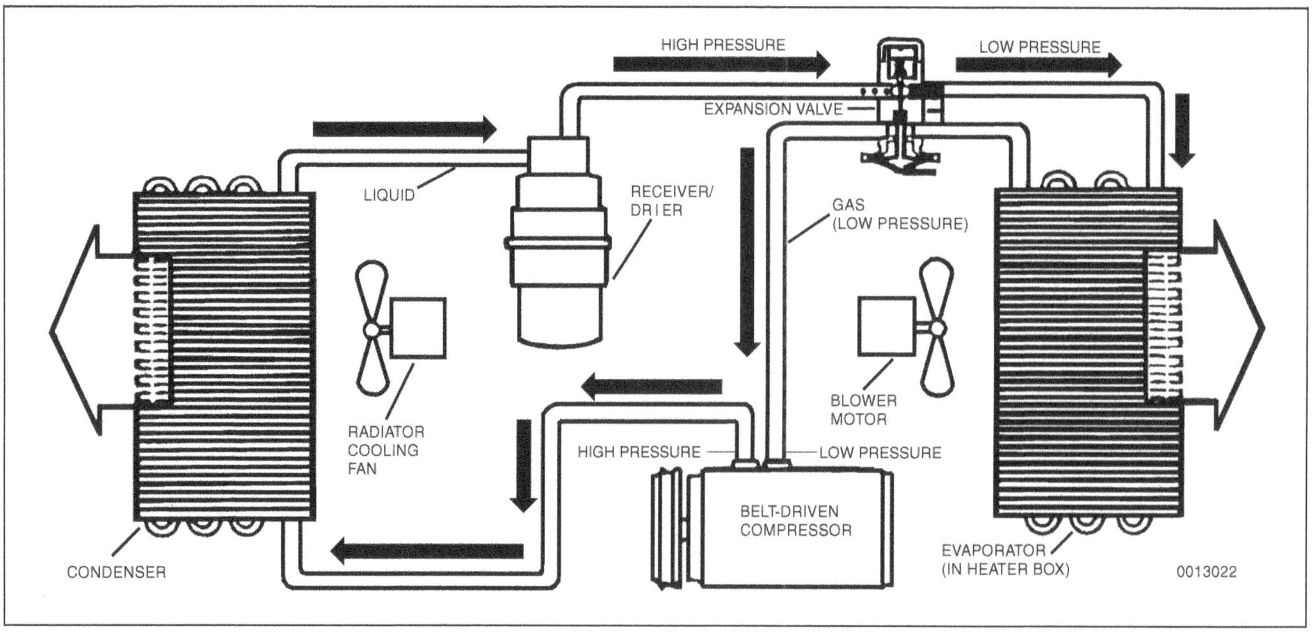

Fig. 33. A/C components. Arrows indicate direction of refrigerant flow.

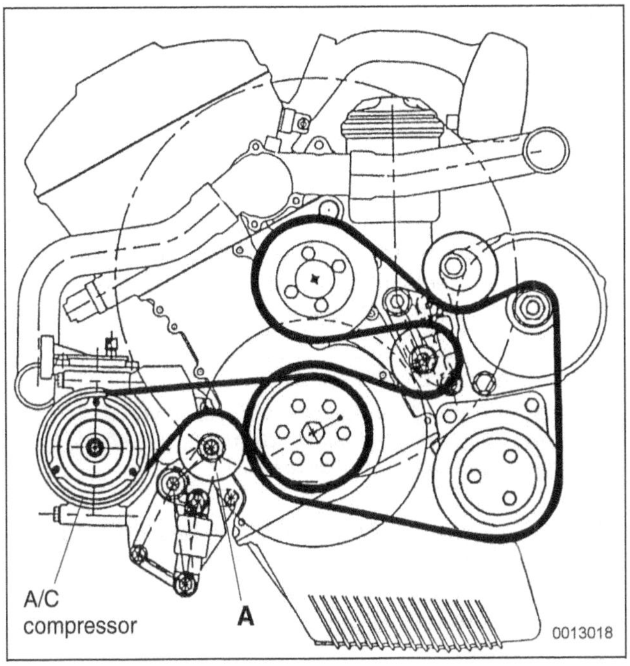

Fig. 36. 6-cylinder engine belt diagram: To remove A/C belt, lever tensioner (**A**) in clockwise direction using long-handled wrench or breaker-bar.

6. Disconnect electrical harness connector from compressor.

7. Remove A/C pressure hose and suction hose flange bolts from compressor. Plug hoses immediately. See Fig. 37.

Fig. 37. A/C compressor pressure hose (**A**) and suction hose (**B**). (6-cylinder engine shown. 4-cylinder engine is similar.)

8. Remove compressor mounting bolts. See Fig. 38. Remove compressor by pulling downward from underneath car.

Fig. 38. A/C compressor mounting bolts (**arrows**).

9. Installation is reverse of removal.

> ***CAUTION —***
> - *Always replace O-rings when reconnecting refrigerant lines.*
> - *When installing a new compressor always replace receiver/drier unit.*

10. Before recharging, evacuate system for at least 30 min. Recharge system following equipment manufacturer's instructions. See **Table d. A/C System Refrigerant Capacities**.

Tightening Torques

- A/C compressor to engine block .... 22 Nm (17 ft-lb)
- A/C line flange to A/C compressor
  - 5/8 in. line .... 20 Nm (15 ft-lb)
  - 3/4 in. line .... 39 Nm (29 ft-lb)
  - 7/8 in. line .... 42 Nm (31 ft-lb)
  - 11/16 in. line .... 48 Nm (35 ft-lb)

A/C Compressor Clutch

The compressor clutch can be replaced without disturbing the sealed refrigerant system, although clearances in the six-cylinder cars may be very tight. Detach the compressor from the compressor bracket to gain extra clearance.

Typical A/C compressor and clutch is shown in Fig. 39.

Remove drive plate and pulley to access clutch magnetic winding. See Fig. 40.

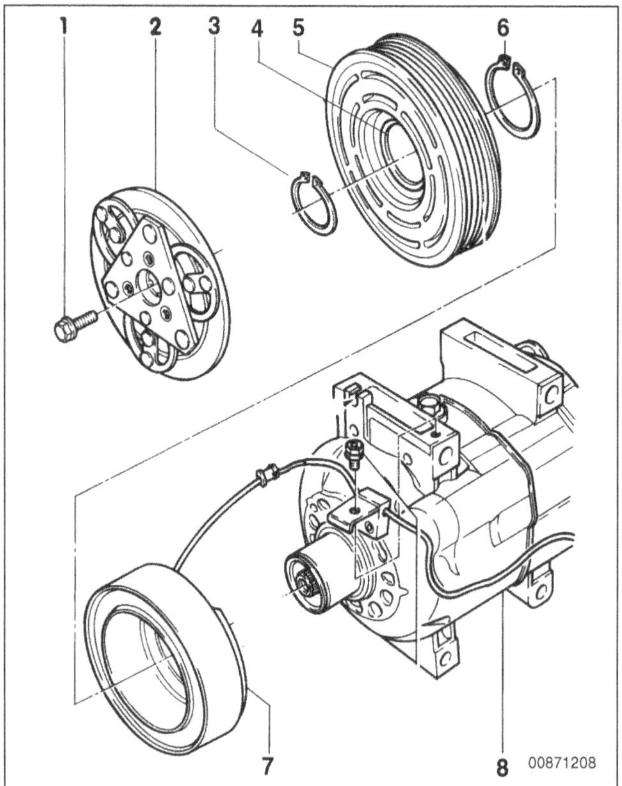

1. Clutch plate retaining bolt
2. Clutch plate
3. Pulley retaining circlip
4. Clutch bearing
5. A/C compressor pulley
6. Clutch retaining circlip
7. Clutch (magnetic winding)
8. Compressor

Fig. 39. A/C compressor and clutch components.

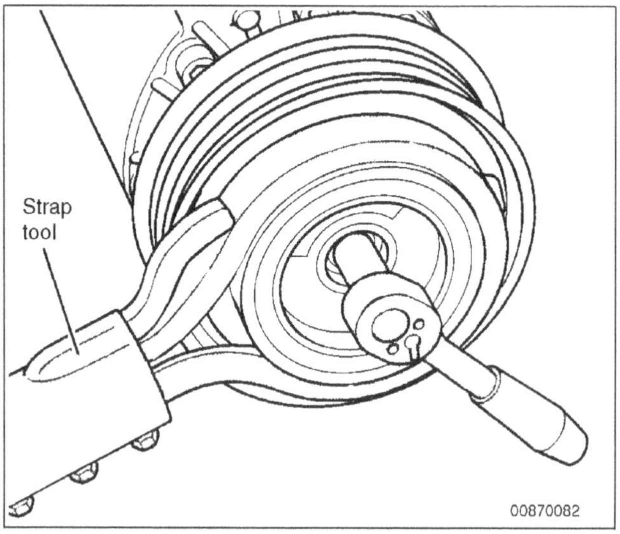

Fig. 40. A/C compressor drive plate can be held stationary with strap tool to allow removal of retaining bolt.

NOTE —
BMWs are equipped with compressors from various manufacturers. Special tools may be required to remove driveplate or pulley.

Tightening Torque
- Compressor clutch bolt to compressor shaft . . . . 20 Nm (15 ft-lb)

A/C condenser, replacing

1. Following manufacturer's instructions, connect an approved refrigerant recovery/recycling/recharging unit to A/C system and discharge system.

WARNING —
Do not discharge/charge the A/C system without proper equipment and training. Damage to the vehicle and personal injury may result.

2. Remove front bumper and radiator grilles. See **510 Exterior Trim, Bumpers**.
3. Remove auxiliary electric radiator fan. See **170 Cooling System**.
4. Detach A/C lines from lower right corner of condenser. Plug lines immediately. See Fig. 41.

Fig. 41. A/C flange bolts at condenser (**arrows**).

5. Remove condenser mounting bolts from front radiator support. See Fig. 42.
6. Lift up condenser to disengage from lower mounts and remove from below.
7. Installation is reverse of removal.

CAUTION —
- *Always replace O-rings when reconnecting refrigerant lines.*
- *When installing a new condenser always replace receiver/drier unit.*

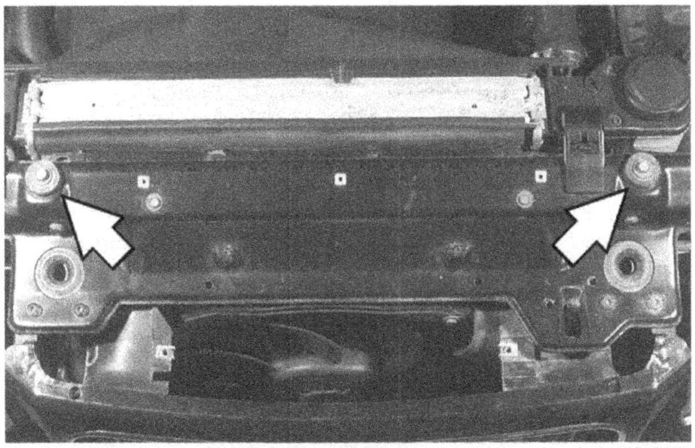

Fig. 42. Upper A/C condenser mounting bolts (**arrows**).

8. Before recharging, evacuate system for at least 30 min. Recharge system following equipment manufacturer's instructions. See **Table d. A/C System Refrigerant Capacities**.

A/C receiver/drier, replacing

The A/C receiver/drier is mounted in the front right corner of the engine compartment behind the headlight assembly.

> ***CAUTION—***
> *The receiver/drier must be replaced when*
> *–There is dirt in the A/C system*
> *–The compressor has seized or has been replaced for any reason*
> *–The condenser or evaporator are replaced for any reason*
> *–The A/C system is leaking and there is no more refrigerant*
> *–The A/C system was open for 24 hours or more.*

1. Following manufacturer's instructions, connect an approved refrigerant recovery/recycling/recharging unit to A/C system and discharge system.

2. Remove right headlight assembly. See **630 Exterior Lighting**.

3. Disconnect A/C electrical harness connectors. See Fig. 43.

4. Remove refrigerant pressure lines from receiver/drier. Plug lines immediately.

5. Remove receiver/drier mounting bolts.

6. Remove underbody splash shield on passenger side. See Fig. 44.

7. Remove passenger side brake cooling duct. Disconnect outside temperature sensor harness, if applicable.

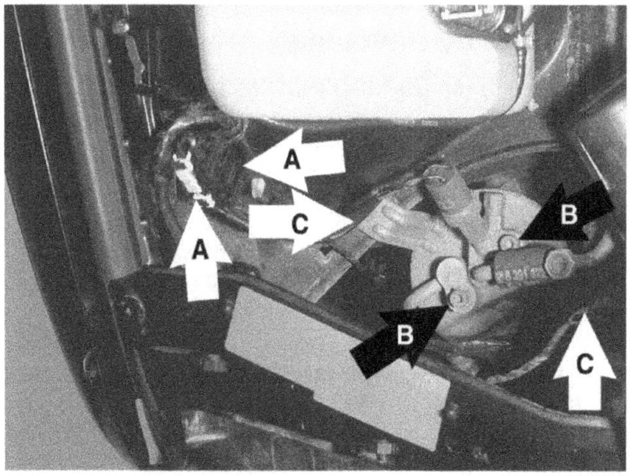

Fig. 43. A/C receiver/drier electrical harness connectors (**A**), pressure line retaining bolts (**B**), and mounting bolts (**C**).

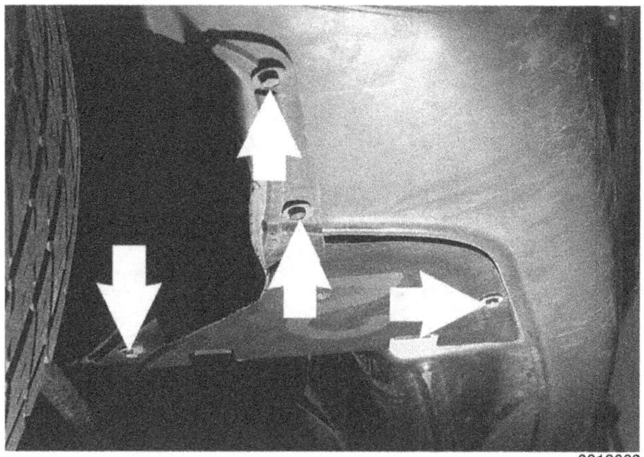

Fig. 44. Passenger side splash shield retaining screws (**arrows**).

8. Push receiver/drier down and remove from underneath car. Make sure that receiver/drier is clear of any electrical harnesses.

9. Installation is reverse of removal.

> ***CAUTION—***
> *Always replace O-rings when reconnecting refrigerant lines.*

10. Before recharging, evacuate system for at least 30 min. Recharge system following equipment manufacturer's instructions. See **Table d. A/C System Refrigerant Capacities**.

A/C expansion valve, removing and installing

1. Following manufacturer's instructions, connect an approved refrigerant recovery/recycling/recharging unit to A/C system and discharge system.

> ***WARNING —***
> *Do not discharge/charge the A/C system without proper equipment and training. Damage to the vehicle and personal injury may result.*

2. Remove glove compartment and glove compartment support frame. See **513 Interior Trim**.

3. Remove right side footwell vents.

4. Remove right side control module carrier screws and tilt carrier down. See Fig. 45.

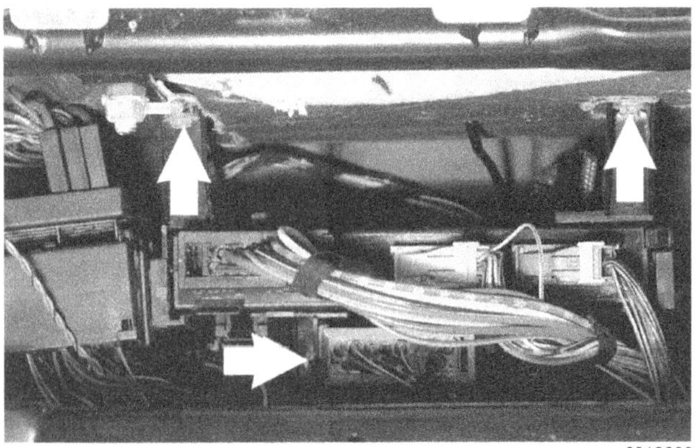

0012666

Fig. 45. Control module carrier retaining screws (**arrows**).

5. Remove expansion valve cover. See Fig. 46.

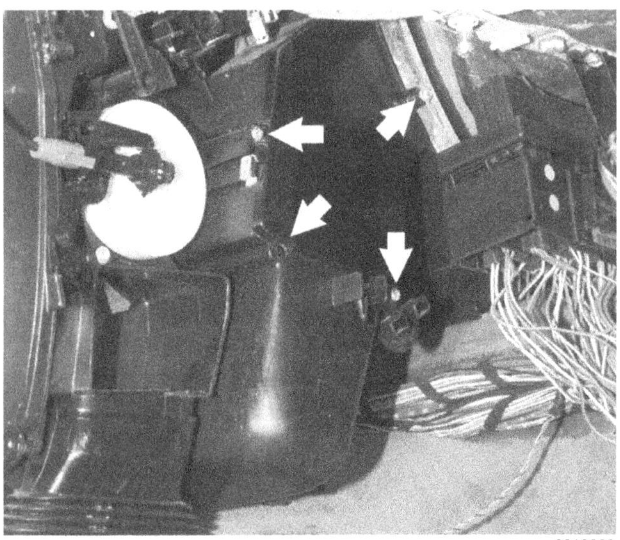

0012663

Fig. 46. Four screws retain expansion valve cover (**arrows**).

6. Remove A/C refrigerant line flange bolt(s) from expansion valve. Plug lines immediately. See Fig. 47.

> ***NOTE —***
> *If necessary, remove air plenum grille from rear of engine compartment as described earlier in* **Blower motor, removing and installing**. *Remove wiring harness cover screws and then remove right side plenum screws. Reaching into the plenum chamber, remove A/C line flange bolt to allow expansion valve room for removal.*

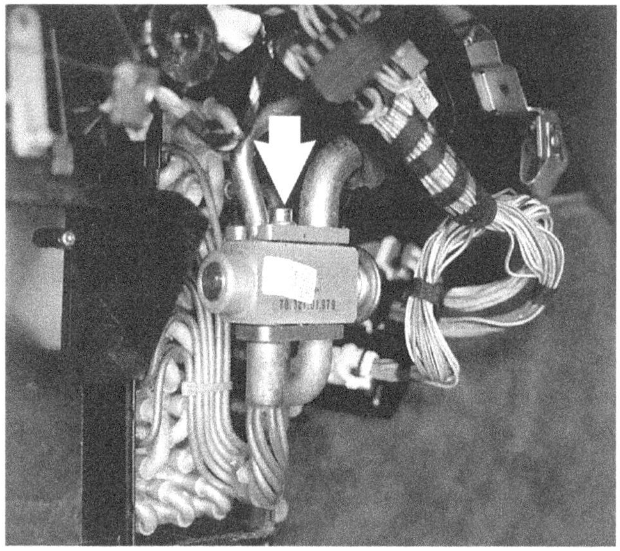

0011887

Fig. 47. Refrigerant line flange mounting bolt at expansion valve (**arrow**).

7. Remove screws from top of expansion valve and remove expansion valve from evaporator. See Fig. 48.

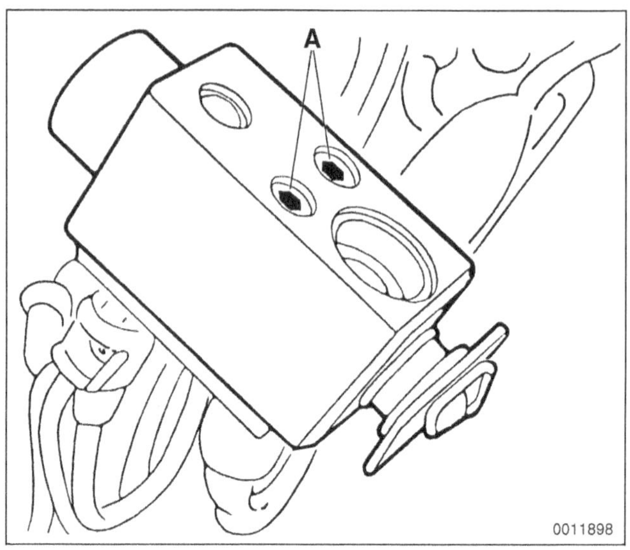

Fig. 48. Expansion valve mounting screws (**A**).

8. Installation is reverse of removal.

> **CAUTION—**
> *Always replace O-rings when reconnecting refrigerant lines.*

9. Before recharging, evacuate system for at least 30 min. Recharge system following equipment manufacturer's instructions. See **Table d. A/C System Refrigerant Capacities**.

A/C evaporator, removing and installing

1. Following manufacturer's instructions, connect an approved refrigerant recovery/recycling/recharging unit to A/C system and discharge system.

2. Remove complete heater box as described earlier.

3. Remove expansion valve as described earlier.

4. Unfasten evaporator sensor electrical harness connector and pull out sensor from left side of heater box. See Fig. 49.

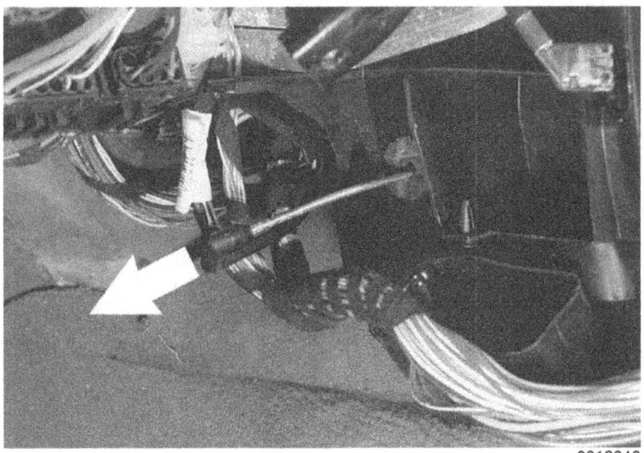

0012646

Fig. 49. Pull out evaporator sensor from housing in direction of **arrow**.

5. Remove rear heater box cover.

6. Remove clips retaining lower evaporator housing cover. Tilt out heater core, then pull evaporator out of housing.

7. Installation is reverse of removal.

> **CAUTION—**
> *Always replace O-rings when reconnecting refrigerant lines.*

8. Before recharging, evacuate system for at least 30 min. Recharge system following equipment manufacturer's instructions. See **Table d. A/C System Refrigerant Capacities**.

A/C evaporator temperature sensor, replacing

The evaporator temperature sensor is mounted in the left (driver) side of the heater box behind the center console.

1. Remove lower dash panel and knee bolster on driver side. See **513 Interior Trim**.

2. Remove left footwell ventilation duct by removing retaining rivet in bottom of duct. Pull it down and away from heater box. See Fig. 50.

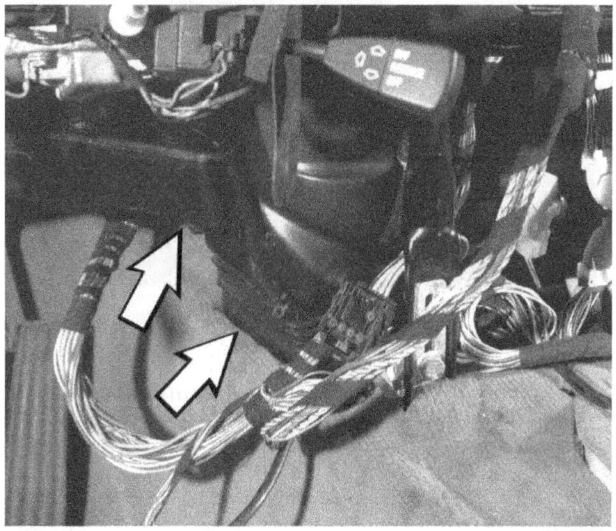

0013008a

Fig. 50. Left footwell ventilation duct (**arrow**).

3. Disconnect sensor electrical connector from heater box. Pull sensor straight out of heater box. Refer to Fig. 49.

4. Installation is reverse of removal. **Table c.** lists evaporator temperature sensor test values.

Table c. Evaporator Temperature Sensor Test Values

Test temp.	Resistance
23°F (-5°C)	11.4 – 11.9 k ohms
32°F (0°C)	8.8 – 9.2 k ohms
41°F (5°C)	6.8 – 7.2 k ohms
50°F (10°C)	5.3 – 5.6 k ohms
59°F (15°C)	4.2 – 4.5 k ohms
68°F (20°C)	3.3 – 3.6 k ohms
77°F (25°C)	2.6 – 2.9 k ohms
86°F (30°C)	2.1 – 2.3 k ohms
95°F (35°C)	1.7 – 1.9 k ohms

A/C microfilter, replacing

1. Remove glove compartment. See **513 Interior Trim**.

2. Remove right footwell ventilation duct by pushing down and to right. See Fig. 51.

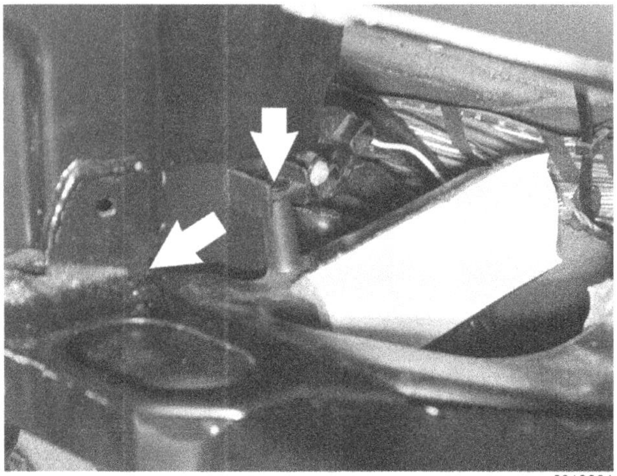

0012664

Fig. 51. Unfasten retaining rivets from right ventilation duct (**arrows**).

3. Remove right side control module carrier screws and tilt carrier down. See Fig. 52.

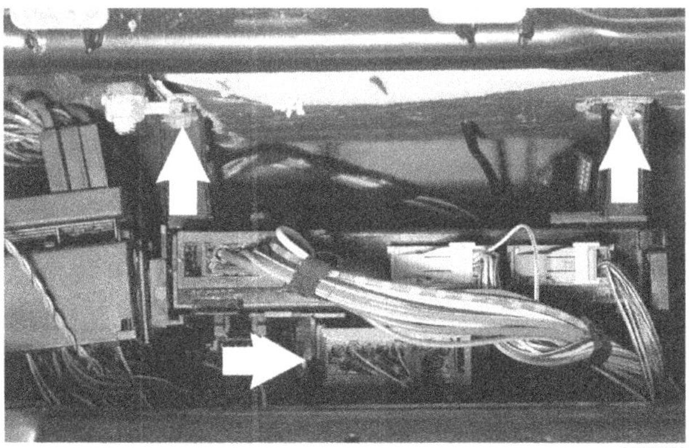

0012666

Fig. 52. Control module carrier retaining screws (**arrows**).

4. Rotate A/C microfilter housing release knob about 45° counterclockwise and remove filter housing cover. See Fig. 53.

5. Remove microfilter while moving module harnesses out of the way.

0012665

Fig. 53. A/C microfilter cover release knob (**arrow**) on passenger side of heater box.

6. Installation is reverse of removal.

Table d. A/C System Fluid Capacities (total)

Fluid	Quantity
R-12 refrigerant	1200g ± 25 g (2.65 lbs.± 0.05 lbs.)
R-134a refrigerant	1000 g ± 25 g (2.20 lbs.± 0.05 lbs.)
PAG oil SP-10 (R12 system)	200 ml ± 20 ml (6.8 oz. ± 0.7 oz.)
Refrigerant oil (R134a system)	Nippondeso 120ml Seiko Seiki 150ml

650 Radio

GENERAL

This section covers the BMW factory-installed sound system, including radio and speaker removal, radio connectors and terminal designations.

The antenna is incorporated into the rear window defogger.

0012159

Fig. 1. Radio amplifier (premium package) in luggage compartment behind left panel.

AUDIO SYSTEM

The BMW E36 is equipped with a three band radio/cassette player unit mounted in the middle of the center console. In addition to AM and FM bands, the radio will receive broadcasts from the National Weather Service.

The cassette deck is equipped with the Dolby™ C noise reduction system, for improved tape performance with reduced background tape noise.

A 6-speaker audio system was standard in the early years of the model line. Many additional features have been installed as optional or standard equipment during the vehicle production run.

- A 200-watt power amplifier with custom-contoured equalization and ten speaker output. See Fig. 1.
- A multi-disc compact disc changer, for which the vehicle and the radio are prewired. See Fig. 2.

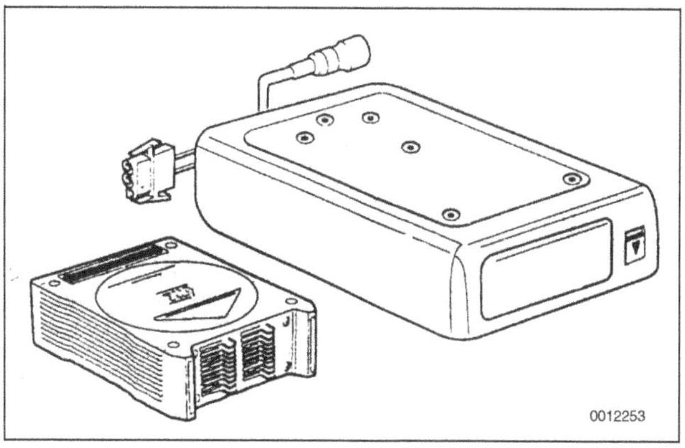

0012253

Fig. 2. Optional multi-disc compact disc changer and disc cartridge.

650

• A 320-watt Harmon-Kardon ten speaker system, including a separate high power multi-channel audio amplifier and two sub-woofers. In this system, the bass and volume outputs from the selected audio source automatically compensate for vehicle speed, making constant volume adjustments unnecessary. An additional feature of this sound system is the ability to switch from normal left-right stereo into Spatial Enhancement Mode, which produces a fuller, rounder sound. See Fig. 3.

The E36 radio is also equipped with a built-in anti-theft system. If power is removed from the radio circuitry by removing the radio electrical connectors, or by disconnecting the battery, the radio is rendered inoperative. The radio is not permanently damaged, and proper operation can be restored by entering the radio anti-theft code.

RADIO AND SPEAKER SERVICE

Radio, removing and installing

CAUTION —
- *Before beginning work on the radio or sound system, verify that the radio is an original equipment BMW radio and that the wiring harness has not been modified.*
- *Servicing of aftermarket sound equipment should be referred to an authorized agent of the equipment's manufacturer.*
- *Do not remove radio or disconnect the battery without first obtaining the anti-theft code.*

1. Carefully pry open small square covers at left and right sides of radio.

Fig. 3. Ten speaker sound system.

2. Using special hex key, turn radio mounting screws counterclockwise to loosen. See Fig. 4.

0011845

Fig. 4. Use special hex key or Torx driver to loosen/tighten radio mounting screws.

3. Pull radio out of instrument panel opening.

4. Remove connector(s) from rear of radio. See Fig. 5.

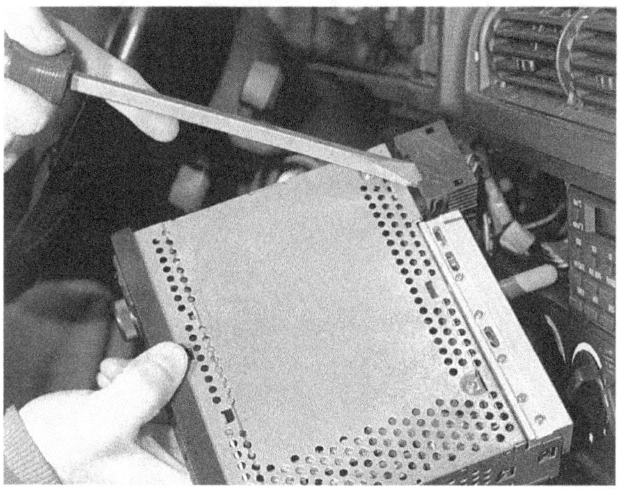

0011846

Fig. 5. Radio connector being removed.

5. Installation is reverse of removal.

Left footwell speaker, removing and installing

1. Remove screw from center of hood latch release lever. Remove lever. See Fig. 6.

0012578

Fig. 6. Hood release lever retaining screw (**arrow**).

2. Rotate plastic trim panel retainer ¼ turn. Pull off door seal from edge of speaker cover and pull speaker trim panel back toward rear of car. See Fig. 7.

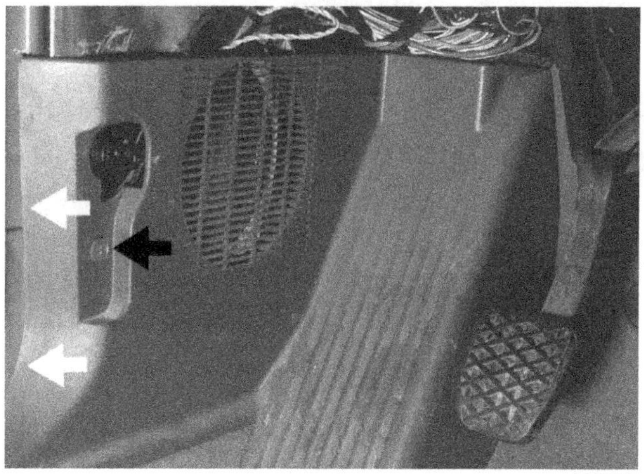

0012579

Fig. 7. Rotate plastic retainer (**black arrow**) and remove panel in direction shown (**white arrows**).

3. Remove speaker by removing mounting screws. See Fig. 8.

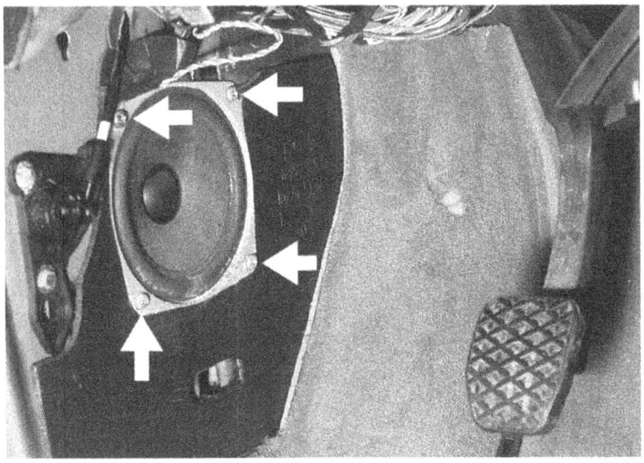

Fig. 8. Speaker mounting screws (**arrows**).

4. Disconnect harness connector(s) from speaker. See Fig. 9.

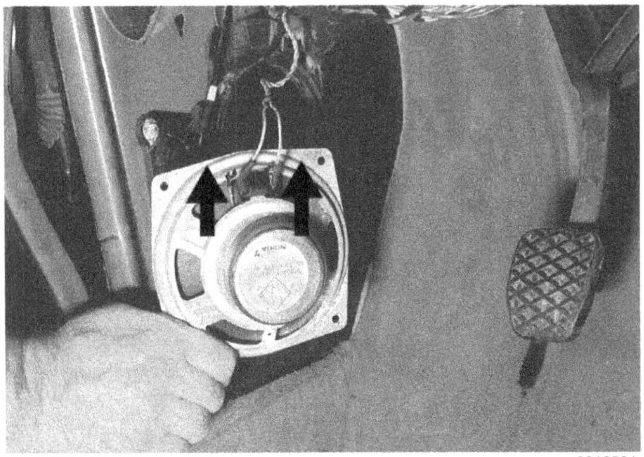

Fig. 9. Disconnect speaker electrical connector(s) in direction shown (**arrows**).

5. Installation is reverse of removal.

Right footwell speaker, removing and installing

1. Remove glove compartment lower trim. See Fig. 10.
2. Rotate speaker cover plastic retainer ¼ turn. Remove door seal at edge of speaker cover. See Fig. 11.
3. Pull speaker trim panel away from body and then pull panel back toward rear of car.

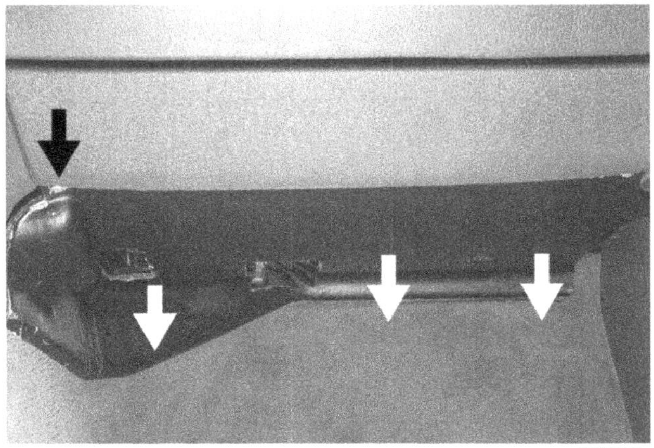

Fig. 10. Remove lower glove compartment trim panel retaining screw (**black arrow**). Pull trim in direction shown (**white arrows**).

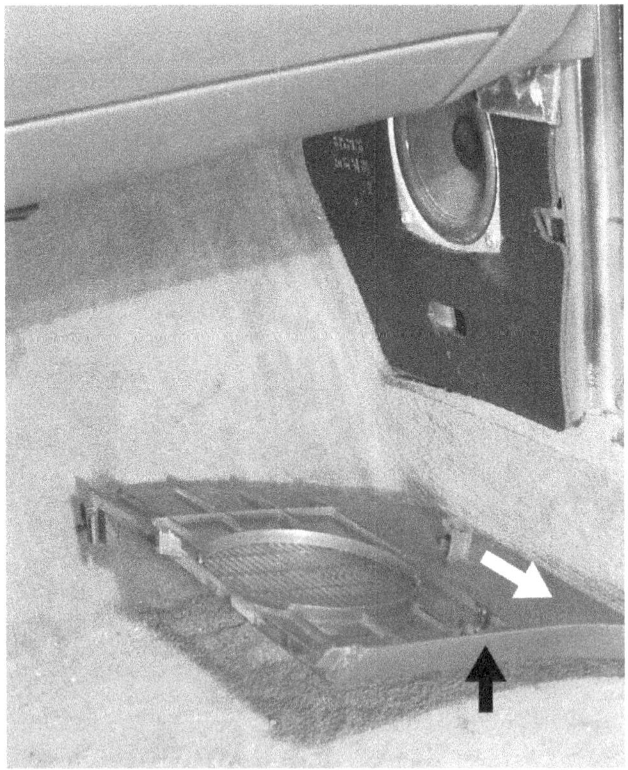

Fig. 11. Speaker cover retainer (**black arrow**). Pull trim in direction shown (**white arrow**).

4. Remove speaker by removing mounting screws. Disconnect harness connector(s) from speaker.
5. Installation is reverse of removal.

Rear speaker, removing and installing

NOTE —
Rear speakers in convertible models are installed under the rear side panels. For rear side panel removal procedure, see **411 Doors**.

1. Working at rear parcel shelf, pry speaker cover by inner edge and remove. See Fig. 12.

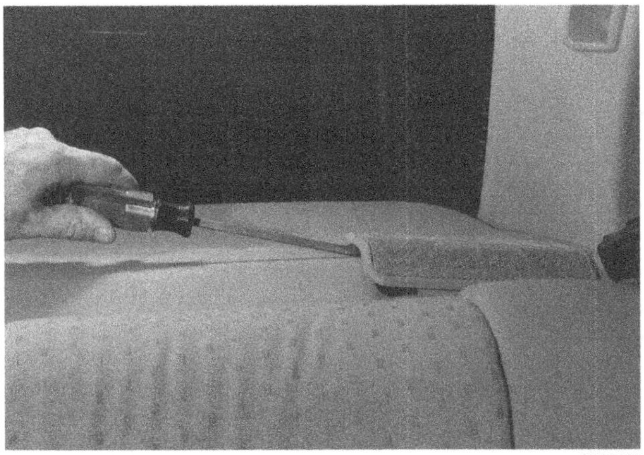

0012632

Fig. 12. Pry up inner edge of speaker cover off rear parcel shelf.

2. Remove mounting screws and depress speaker retaining clip. See Fig. 13. Lower speaker into trunk.

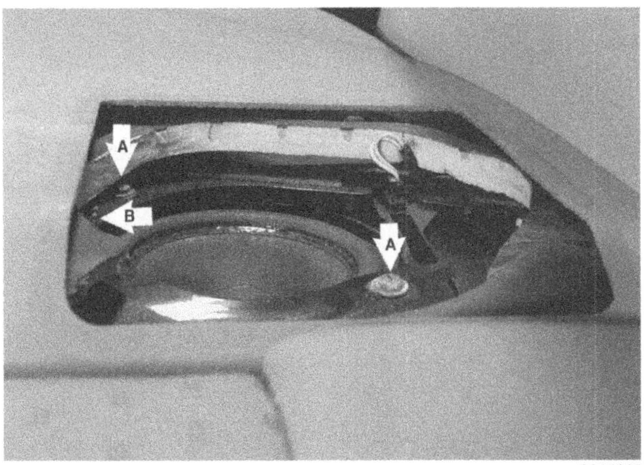

0012633

Fig. 13. Remove screws (**A**) and depress clip (**B**).

3. Disconnect speaker connector and remove speaker from box.

4. Installation is reverse of removal.

Door panel speaker, removing and installing

1. On cars with side-impact airbags, disconnect negative (–) cable from battery.

CAUTION —
Prior to disconnecting the battery, read the battery disconnection cautions given at the front of this manual on page viii.

2. Remove inner door panel. See **411 Doors**.

WARNING —
Some 1997 and all 1998 model year cars are equipped with front side-impact airbags in the front doors. When servicing the doors with front side-impact airbags, always disconnect the negative (-) battery terminal. See **721 Airbag System (SRS)** *for cautions and procedures relating to the airbag system.*

3. On back of door panel, remove nut on top of speaker retaining bracket and remove bracket from behind speaker. See Fig. 14.

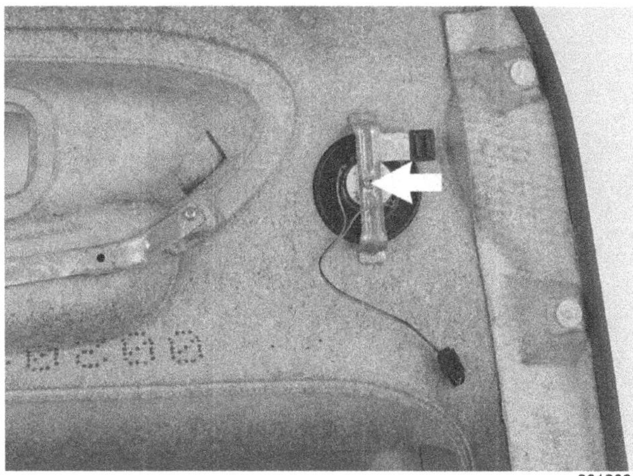

0012634

Fig. 14. Speaker retaining nut (**arrow**).

4. Disconnect speaker connector from harness connector and remove loudspeaker forward through door panel.

5. Installation is reverse of removal. Install side-impact airbag using new screws. See **721 Airbag System (SRS)**.

Radio Connector

Table a lists radio connector terminal designations. The pin numbers are identified in Fig. 15.

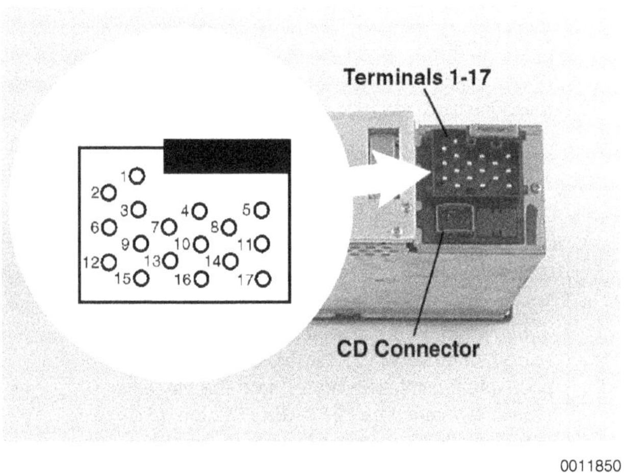

Fig. 15. Radio connector. See **Table a**.

Table a. Radio Connector Terminal Designation

Connector Terminal	Signal
1	Speaker, left front +
2	Speaker, right front +
3	Speaker, left rear +
4	Telephone mute feature (where applicable)
5	Positive (terminal R)
6	Front speaker, left
7	I-bus connection (diagnostics)
8	Speaker, left front
9	Constant positive, Terminal 30
10	Speedometer signal for speed-sensitive volume control (where applicable)
11	Speaker, right front
12	Speaker, left rear
13	Illumination (terminal 58G)
14	Speaker, right rear
15	Ground (terminal 31)
16	Automatic antenna or window antenna amplifier or premium sound system amplifier
17	Not assigned

720 Seat Belts

GENERAL

All E36 models are equipped with 3-point shoulder belts at four seating positions. On 4-door models, there is an additional lap belt in the center of the rear seat.

Automatic seat belt tensioners are integrated into the front seat belt locks. Through the 1996 model year cars, a mechanical spring tensioner is used. See Fig. 1. On 1997 and later cars, a pyrotechnic (explosive charge) tensioner is used. Automatic seat belt tensioners are designed to automatically retract and tension the seat belt by 55 mm (approximately 2 inches) in the event of an accident.

The seat belt reel should lock when driving quickly through curves, during severe braking or in case of a collision. The automatic reel does not require any servicing and should never be opened. Belts that do not retract and/or lock properly should be replaced promptly.

1. Seat belt lock bolt
2. Seat belt lock with mechanical spring tensioner
3. Seat belt anchor bar (two-door models)
4. Seat belt anchor bolt
5. Seat belt reel bolt
6. Seat belt reel
7. B-pillar guide nut
8. B-pillar sliding guide
9. Convertible seat belt reel
10. Convertible B-pillar guide

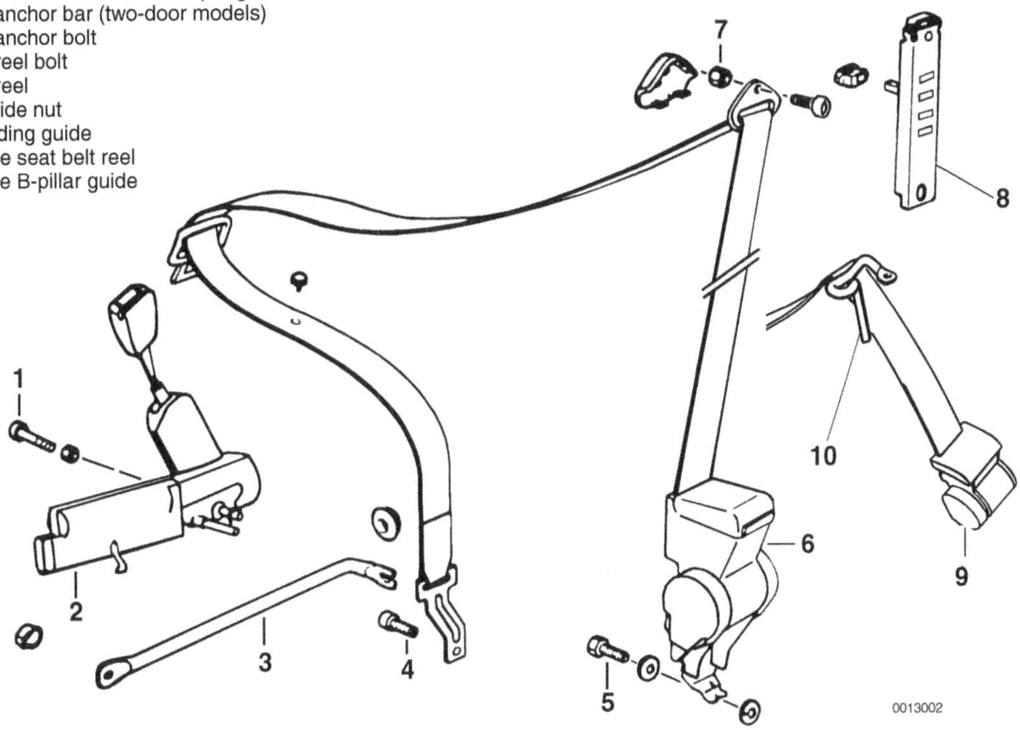

Fig. 1. Front seat belt components.

> ***WARNING —***
> - *For maximum protection from injury, seat belts should be replaced as a set (including all hardware), if they are subject to occupant loading in a collision.*
> - *Seat belts should not be modified or repaired. The seat belt anchorage points should not be changed or modified.*
> - *Do not bleach or dye seat belt webbing. Webbing that is severely faded or re-dyed will not meet the strength requirements of a collision and must be replaced.*
> - *Seat belts should be periodically inspected for webbing defects such as cuts or pulled threads.*

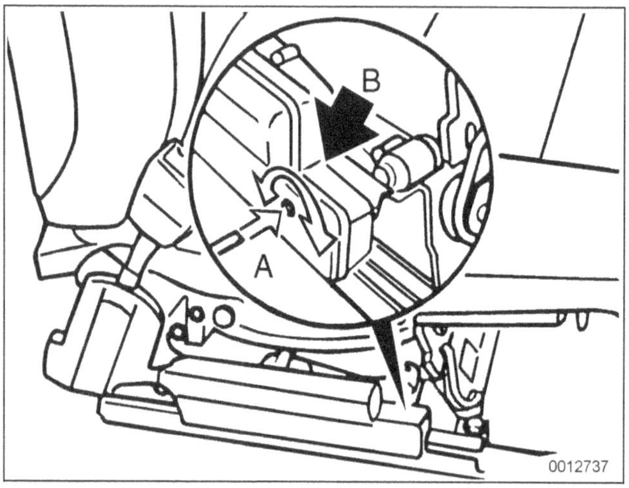

Fig. 2. Electrically adjustable front seat. Turn screw (**A**) to disarm mechanical tensioner. Display **B** turns from green to red.

FRONT SEAT BELT ASSEMBLY

Automatic Front Seat Belt Lock Tensioners

The procedure for disabling the mechanical spring tensioner used on early cars is covered below. On cars with the pyrotechnic (explosive device) tensioners, the orange harness connector should be removed from the tensioner any time the seat is removed.

> ***WARNING —***
> *Both the mechanical and the pyrotechnic tensioners are powerful devices and should be handled with extreme care.*

Mechanical spring tensioner, disarming

1. Remove front seat. See **520 Seats**.

2. Disarm tensioner–electrically adjustable seats:
   - Working at side of seat, turn activating/deactivating screw clockwise until display changes from green to red. See Fig. 2.
   - Before reinstalling seat, turn screw until display changes back to green.

   NOTE —
   Green is armed. Red is disarmed.

3. Disarm tensioner–manually adjustable seats:
   - Cut wire tie next to tab on bowden cable. See Fig. 3.
   - Turn and lift tab to disengage cable housing from seat bracket.
   - Disconnect cable end from seat lever.
   - When reinstalling, correctly engage tab end of cable into seat bracket and fasten in position using new wire tie.

Fig. 3. Manually adjustable front seat. Turn and lift tab (**arrow**) to disengage cable from bracket.

Front seat belt reel, removing and installing

1. On 4-door models: Remove seat belt anchor bolt from front seat. Refer to Fig. 1.
2. On 2-door models: Remove seat belt anchor bar from rear of door sill. See Fig. 4.

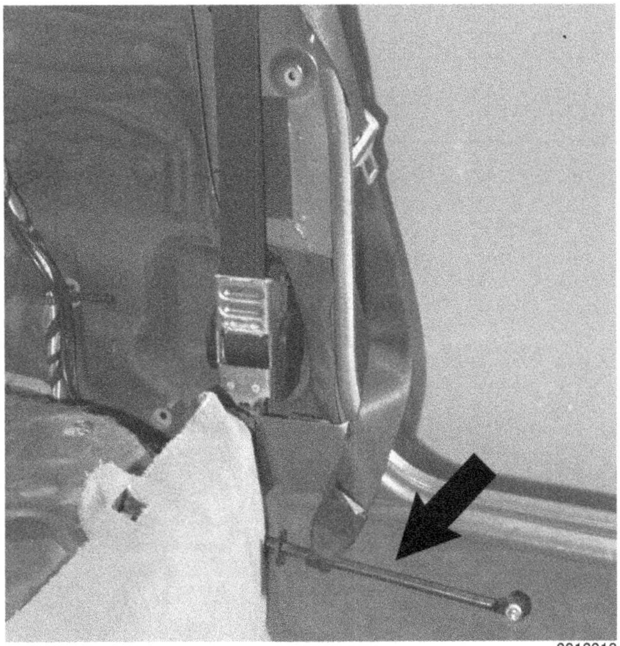
0013010

Fig. 4. Two-door seat belt anchor bar (**arrow**).

3. On 4-door and coupe models: Remove plastic trim cover from B-pillar (rear door jamb pillar) sliding seat belt guide. Remove nut and guide loop. See Fig. 5.
4. On 4-door models: Starting at upper corners, peel rubber door seals away from B-pillar trim. Remove trim from B-pillar.
5. On 2-door models: Remove rear side panel. See **411 Doors**.
6. Remove two bolts holding cross bracket to B-pillar. See Fig. 6.
7. Remove seat belt reel mounting bolt at base of B-pillar.

0012672

Fig. 5. B-pillar seat belt guide nut trim (**arrow**).

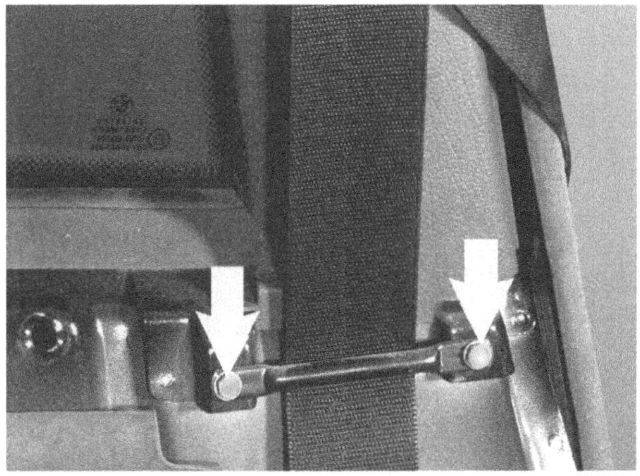
0012674

Fig. 6. B-pillar trim removed to show seat belt cross-bracket bolts (**arrows**).

8. Disconnect harness connector from reel assembly. Remove reel and seat belt.
9. Installation is reverse of removal.

Tightening Torque

- Front seat belt to seat (use thread locking compound) ..... 47 Nm (35 ft-lb)
- Front seat belt guide loop to B-pillar sliding anchor .... 31 Nm (23 ft-lb)
- Front seat belt reel to B-pillar .... 48 Nm (36 ft-lb)

REAR SEAT BELT ASSEMBLY

On 4-door E36 models, there are two 3-point shoulder belts and a center lap belt in back. On 2-door models, there is no center rear lap-belt.

In models with the stationary rear seat back, the shoulder belt reels are bolted to the middle of the parcel shelf. In models with fold-down rear seat back, the shoulder belt reels are attached to the folding seat back.

The rear seat belt assemblies are shown in Fig. 7.

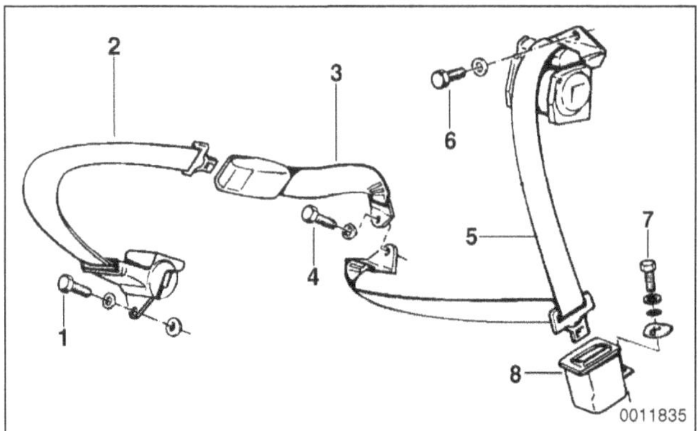

1. Lap-belt reel bolt/shoulder belt anchor bolt
2. Center lap-belt
3. Center lap-belt lock
4. Lap-belt lock bolt/shoulder belt anchor bolt
5. Shoulder belt
6. Shoulder belt reel bolt
7. Shoulder belt lock bolt
8. Shoulder belt lock

Fig. 7. Details of rear seat belt assembly.

Rear seat belt, removing and installing (fixed seat back models)

1. Remove rear seat cushion and backrest. See **520 Seats**.
2. Remove plastic trim cover over shoulder belt reels from rear window shelf. See Fig. 8.
3. Remove rear window shelf. See Fig. 9.
4. Unbolt and remove shoulder belt reels.
5. Remove shoulder belt anchor bolt from beneath rear seat.

NOTE —
The shoulder belt anchor bolts also hold the center lap-belt in place (where applicable). Be sure to note relative positions of metal belt-end anchors in order to reassemble correctly.

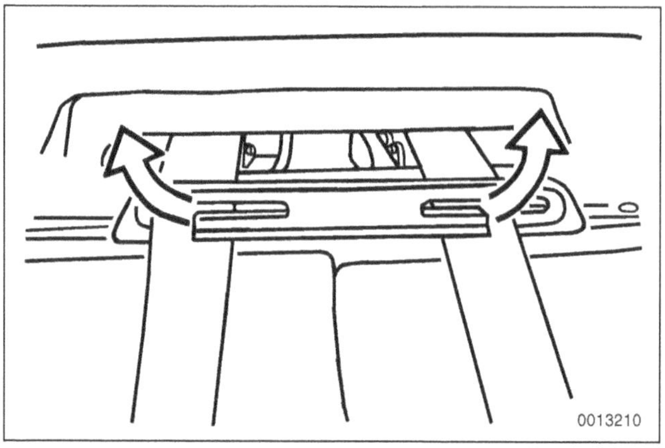

Fig. 8. Remove rear shoulder belt reel trim cover from rear window shelf. Unclip plastic, then slide off shoulder belts as shown by **arrows**.

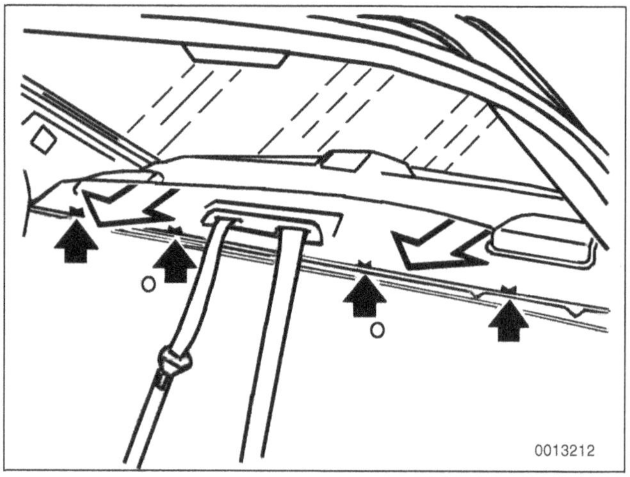

Fig. 9. Remove rear window shelf. Speaker grilles and third brake light socket will have to be removed. The front of the shelf lifts off its clips (**black arrows**) before the shelf slides forward.

6. On 4-door models: Unbolt shoulder belt lock bolt from body next to rear door frame.
7. On 2-door models: Remove rear side panel as described in **411 Doors**. Unbolt shoulder belt lock bolt from body next to quarter panel.
8. Installation is reverse of removal.

WARNING —
Alignment tabs on seat belt reels must be reinstalled in their original positions

Tightening Torques

- Rear seat belt reel bolt, anchor bolt, or lock bolt to body . . . . 48 Nm (36 ft-lb)

Rear seat belt, removing and installing (fold-down seat back models)

1. Remove rear seat cushion. See **520 Seats**.

2. Remove shoulder belt anchor bolt from beneath rear seat.

NOTE —
The shoulder belt anchor bolts also hold the center lap-belt in place (where applicable). Be sure to note relative positions of metal belt-end anchors in order to reassemble correctly.

3. On 4-door models: Unbolt shoulder belt lock bolt from body next to rear door frame.

4. On 2-door models: Remove rear side panel as described in **411 Doors**. Unbolt shoulder belt lock bolt from body next to quarter panel.

5. Unlock and flip down folding rear seat backrest. Remove trim plug from back of cushion (luggage compartment side).

6. Remove seat belt reel bolt and washer. See Fig. 10.

0012676

Fig. 10. Rear seat belt reel bolt (**arrow**).

7. Gently pry off plastic trim at top of backrest and pull out seat belt reel from seat.

NOTE —
The seat belt reel is held snugly by the backrest cushion. Moderate force will be needed to pull it out.

8. Installation is reverse of removal.

Tightening Torques
- Rear seat belt reel bolt, anchor bolt, or lock bolt to body . . . . 48 Nm (36 ft-lb)

Center rear lap-belt, removing and installing

1. Remove rear seat cushion. See **520 Seats**.

2. Remove shoulder belt anchor bolts from beneath rear seat.

NOTE —
The shoulder belt anchor bolts also hold the center lap-belt in place. Be sure to note relative positions of metal belt-end anchors in order to reassemble correctly.

3. Installation is reverse of removal.

WARNING —
Alignment tabs on seat belt reels must be reinstalled in their original positions.

Tightening Torques
- Lap-belt reel bolt or lock bolt to body . 48 Nm (36 ft-lb)

721 Airbag System (SRS)

GENERAL

This repair group covers only removal and installation of airbag components. Airbag system repair and fault diagnosis is not covered here. Diagnostics, component testing, and airbag system repair should be carried out by trained BMW service technicians.

> ***NOTE —***
> *Special test equipment is required to retrieve SRS fault codes, diagnose system faults, and reset/turn off the SRS indicator light. The SRS indicator light will remain on until any problem has been corrected and the fault memory has been cleared.*

When servicing cars equipped with SRS, the precautions on this page must be observed to prevent personal injury.

Airbag Systems

The airbag system installed on E36 cars through the 1993 model year consisted of a single airbag in the steering wheel. This system can be identified by the orange crash sensors located on the front wheel wells and the absence of a passenger side airbag.

The ZAE airbag system, starting in model year 1994 (production date 9/93 and later), provides both a driver side and a passenger side airbag. On ZAE vehicles, the crash sensors are integrated into the airbag control module, located under the center of the rear seat. The ZAE control module triggers deployment of the airbags when it senses the vehicle decelerating at a rate equivalent to a head-on collision of at least 13 mph (18 km/h).

In some 1997 4-door models and all 1998 models, side-impact airbags were installed in the front doors. The airbag units are mounted behind the door trim panels. Caution must be used when working on vehicle systems in the proximity of air bags, such as the steering column, the dashboard, or the doors.

> ***WARNING —***
> - *Airbag(s) are inflated by an explosive device. Handled improperly or without adequate safeguards, the system can be very dangerous. Special precautions must be observed prior to any work at or near the driver side airbag, the passenger side airbag (where applicable) or the side-impact airbags (where applicable).*
> - *The airbag is a vehicle safety system. Serious injury may result if system service is attempted by persons unfamiliar with the BMW SRS and its approved service procedures. BMW advises that all inspection and service be performed by an authorized BMW dealer.*
> - *Always disconnect the battery and cover the negative (–) battery terminal with an insulator before starting diagnostic, troubleshooting or service work not associated with SRS, and before doing any welding on the car.*
> - *If an airbag has been activated due to an accident, BMW specifies that SRS components be replaced. For more information on post-collision SRS service, see an authorized BMW dealer.*
> - *Do not allow the airbag unit to come in contact with cleaning solutions or grease. Never subject an airbag unit to temperatures above 212°F (100°C). When reconnecting the battery, no person should be inside the vehicle.*
> - *Always place an airbag unit that has been removed from its packaging with the padded side facing upward. Do not leave an airbag unit unattended.*
> - *If the airbag unit or airbag control module has been dropped from a height of ½ meter (1½ ft.) or more, the airbag unit should not be installed.*

Airbag system components are shown in Fig. 1.

721

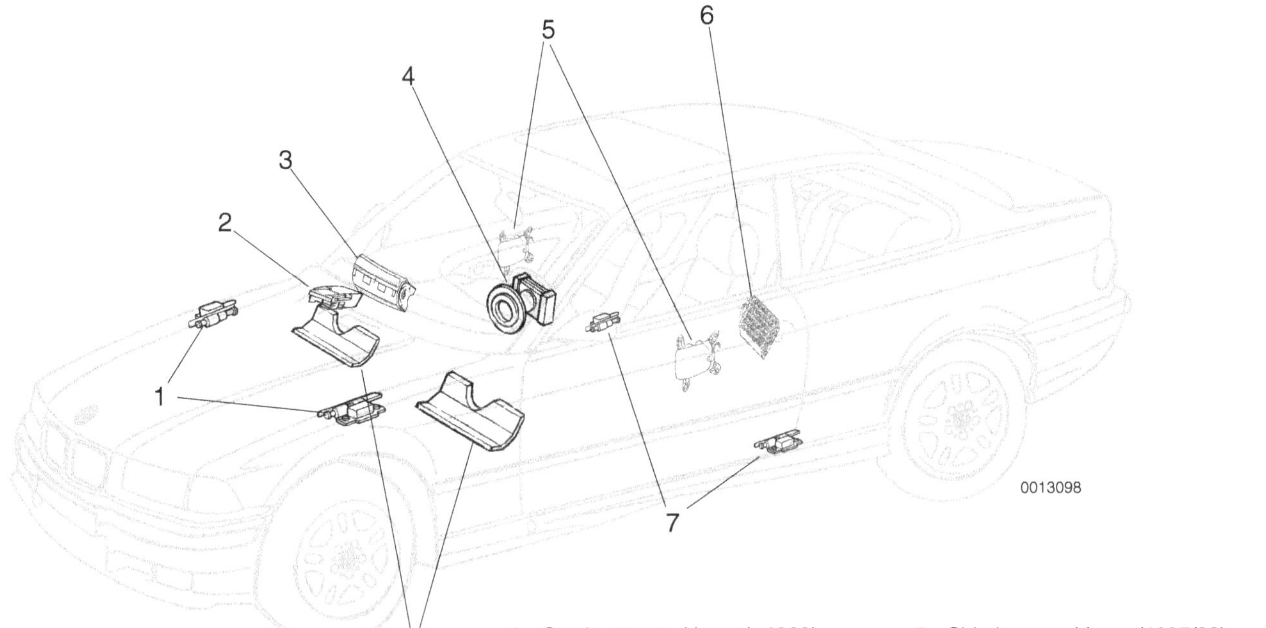

1. Crash sensors (through 1993)
2. SRS control module in front of glove compartment (through 1993)
3. Passenger side airbag unit in dashboard (ZAE system)
4. Driver side airbag unit in steering wheel
5. Side-impact airbags (1997/98)
6. ZAE control module under rear seat (from 1994)
7. Side-impact airbag crash sensors in door sills (1997/98)
8. Knee bolsters

Fig. 1. Airbag system components.

AIRBAG SENSORS AND ELECTRONICS

> ***CAUTION —***
> - *If an airbag is activated, the following components **must** be replaced:*
>   *–Airbag with contact ring;*
>   *–Pyrotechnic seat belt tensioners (if fitted);*
>   *–SRS Control module;*
>   *–Impact sensors;*
>   *–All airbags which did not deploy.*
> - *Consult your BMW dealer for parts and service information.*

The E36 airbag system consists of the following components: crash sensors (in early cars, through 1993), an electronic control/monitoring module (with integrated crash sensor in ZAE systems), an indicator light in the instrument panel, and inflatable airbags equipped with gas generators in the steering wheel and in the passenger side of the instrument panel (ZAE systems).

Some 1997 and all 1998 cars also have side-impact airbags in the front doors.

In addition to the airbags, the vehicles are equipped with knee bolsters for extra protection.

SRS Warning Light and Fault Display

The airbag system utilizes a self-diagnostic control module to detect and store system faults. If a fault is detected, the SRS indicator in the instrument cluster comes on and stays on.

When the ignition key is turned on, the indicator light will illuminate for approximately 4-6 seconds. The illuminated light indicates the self-test of the electronic control/monitoring module, the ignitors and crash-sensor circuits, and the system wiring. The airbag system should be inspected by an authorized BMW dealer if the indicator light does not come on, if it does not go out in approximately 4-6 seconds after switching on the ignition, or if the light comes on during driving. If the system has detected a fault, the light will come on and stay on.

Airbag contact ring, replacing

The airbag contact ring is mounted to the rear of the steering wheel and is made up of a wire ribbon that coils and uncoils as the steering wheel is turned. The contact ring assures continuous electrical contact to the driver side airbag unit.

1. Disconnect negative (–) cable from battery and cover negative terminal with insulating material.

> ***CAUTION—***
> *Prior to disconnecting the battery, read the battery disconnection cautions given at the front of this manual on page viii.*

2. Remove driver side airbag from steering wheel. See **Driver side airbag, removing and installing** in this repair group.

3. With wheels facing straight ahead and steering wheel centered, remove steering wheel. See **320 Steering and Wheel Alignment**.

4. Remove contact ring cover (flat plastic ring containing warning text) from steering wheel, if applicable.

5. Disconnect ground wire, if applicable

6. Lift out plastic lock from steering wheel, if applicable. See Fig. 2.

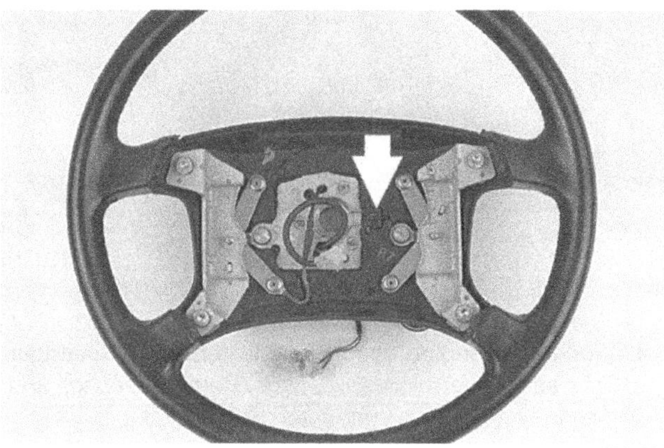

0011903

Fig. 2. Airbag contact ring lock (**arrow**). Pry out lock and remove spring from behind lock. Some airbag steering wheels are not equipped with contact ring lock.

> ***WARNING—***
> - *The contact ring lock holds the contact ring in the centered position. Once the lock is removed, the contact ring is free to rotate. Do not allow the contact ring to rotate once the lock is removed.*
> - *If the contact ring is accidentally rotated out of center, turn the ring either fully clockwise or counterclockwise, then turn in the opposite direction 3 complete revolutions until the alignment arrow can be seen through the lock opening.*

7. Remove mounting nuts or screws from contact ring. See Fig. 3. Disconnect wire lead(s) and remove contact ring from rear of steering wheel.

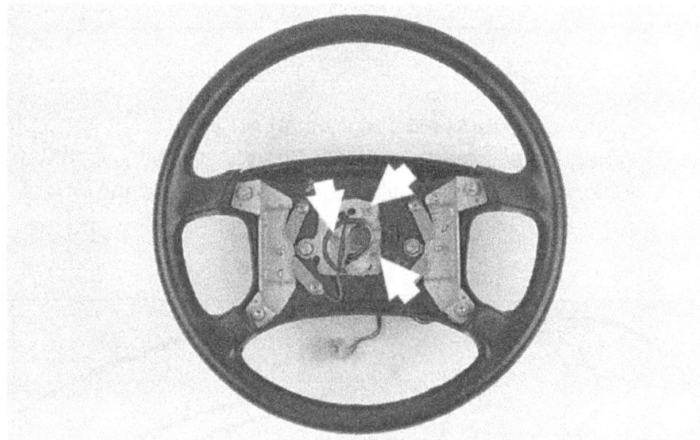

0011904

Fig. 3. Airbag contact ring mounting fasteners (**arrows**).

8. Installation is reverse of removal, noting the following:
   - Coat horn slip ring (electrical contact ring) with CRC® copper paste before installing steering wheel.
   - Press on lower lock until it is fully engaged.
   - When installing contact ring lock, make sure lock spring engages pins on lock and lock cover.
   - Special instructions apply when installing airbag equipped steering wheel. See **320 Steering and Wheel Alignment**.

Early cars (through 1993):

- Route wiring lead for horn through square opening in steering wheel.
- When installing a new contact ring from BMW, remove locking screw after installing the contact ring to steering wheel. See Fig. 4.
- Lock contact ring mounting fasteners in place using lacquer paint or varnish.

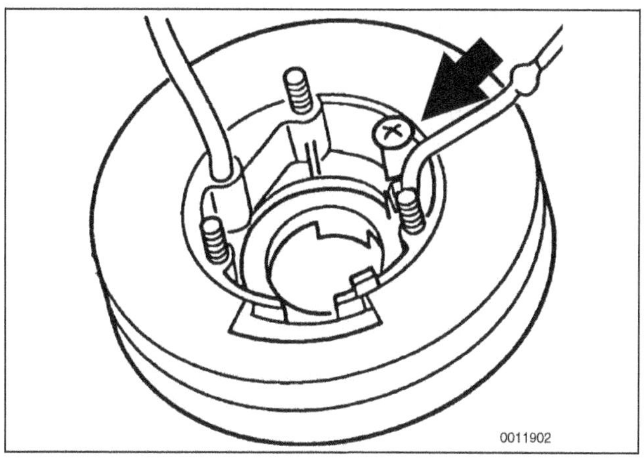

0011902

Fig. 4. Airbag contact ring locking screw (**arrow**), as installed on new replacement part from BMW (early production cars).

> **CAUTION —**
> *A new contact ring is held in center position with a screw. This locking device must be removed after contact ring is installed on steering wheel. See Fig. 4.*

Later cars (1994 on):

- Connect ground lead at contact ring.
- When installing a new contact ring from BMW, remove locking pin after installing contact ring to steering wheel. See Fig. 5.

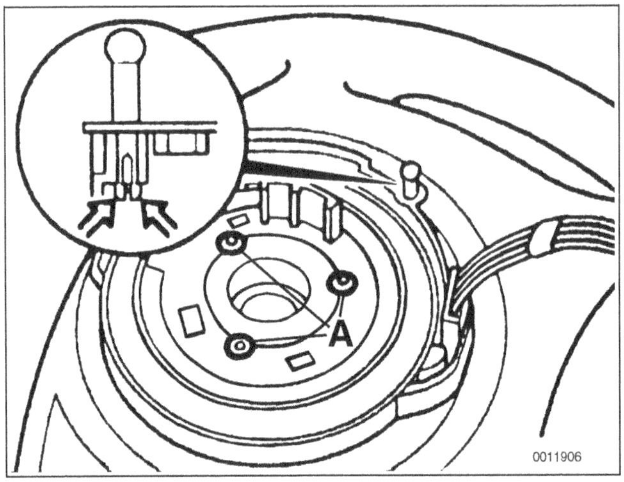

Fig. 5. Airbag contact ring locking pin (**arrows**), as installed on new replacement part from BMW (later production cars). Contact ring mounting screws shown at **A**.

> **CAUTION —**
> *A new contact ring is held in center position with a screw or pin. This locking device must be removed after contact ring is installed on steering wheel. See Fig. 5.*

Airbag crash sensor, replacing

The airbag system in cars produced through 1993 uses two crash sensors mounted in the engine compartment on the left and right wheel arches. The sensors are identified by their orange color.

> **WARNING —**
> *If an airbag is deployed, both sensors must be replaced.*

1. Disconnect negative (–) cable from battery and cover negative terminal with insulating material.

> **CAUTION —**
> *Prior to disconnecting the battery, read the battery disconnection cautions given at the front of this manual on page viii.*

2. Remove lower steering column trim mounting screw and remove lower trim. See Fig. 6.

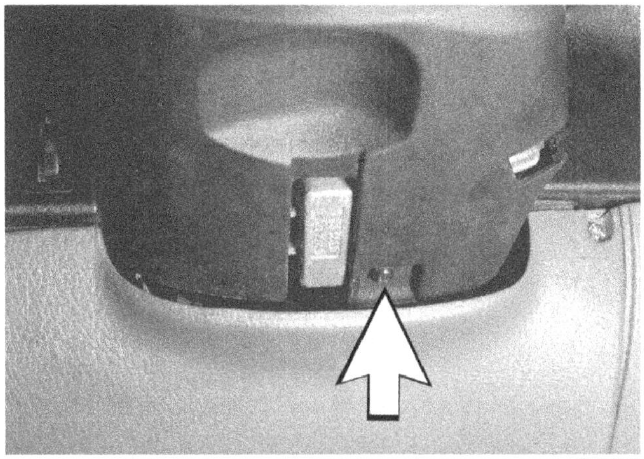

Fig. 6. Lower steering column retaining screw (**arrow**).

3. Remove orange SRS connector from its holder and carefully separate connector. See Fig. 7.

Fig. 7. Orange SRS connector below steering column (**arrow**).

4. Working in engine compartment, remove anti-tamper Torx screws from crash sensor. Remove sensor and disconnect harness connector. See Fig. 8.

5. Installation is reverse of removal. Make sure arrow on sensor faces forward when installing sensor.

> **WARNING —**
> *Once the airbag unit is installed and all other service procedures have been completed, start the engine and check that the SRS warning light goes out. If the warning light stays on, the SRS system will not function as designed. Have the system diagnosed and repaired by an authorized BMW dealer.*

0011897

Fig. 8. Airbag crash sensor mounting screws (**arrows**) on early cars. Note Torx screws with anti-tamper pin in screw-head.

ZAE control module, replacing

NOTE —
Some versions of early ZAE control modules are no longer available from BMW. When replacing control modules on cars built between September 1993 and June 1994, consult a BMW parts department for the latest information on retrofitting the later module.

1. Disconnect negative (–) cable from battery and cover negative terminal with insulating material.
2. Lift rear seat cushion. Unplug harness connectors and remove ZAE airbag control module, which is bolted to top of driveshaft tunnel.
3. Installation is reverse of removal.

Front side-impact airbag crash sensor, removing and installing

1. Disconnect negative (–) cable from battery and cover negative terminal with insulating material.
2. Remove front seat. See **520 Seats**.
3. Convertible: remove rear seat
4. Remove plastic door sill trim.
5. Fold up carpet in back of front seat.
6. Pull off connector from crash sensor, which is located under rug next to door sill.
7. Remove sensor by removing screws. Note direction arrow before removing sensor.
8. Installation is reverse of removal, noting direction arrow on sensor.

AIRBAGS

Driver side airbag, removing and installing

1. Disconnect negative (–) cable from battery and cover negative terminal with insulating material.

WARNING —
On cars with ZAE system: If the airbag system has detected a fault (SRS indicator illuminated), a waiting period of at least ten minutes must be observed before removing the airbag from the steering wheel. This will allow the system capacitor to discharge, reducing the possibility of a trigger pulse being generated to the airbag.

CAUTION —
Prior to disconnecting the battery, read the battery disconnection cautions given at the front of this manual on page viii.

2. Remove lower steering column trim mounting screw and remove lower trim. Refer to Fig. 6.
3. Remove orange SRS connector from its holder and carefully separate connector. Refer to Fig. 7.
4. Working behind steering wheel, completely loosen Torx screws (T30) while holding airbag in place. See Fig. 9. Support airbag unit to prevent it from falling out.

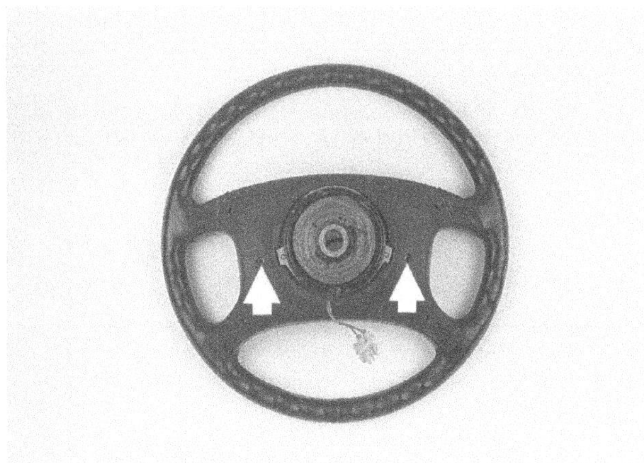

0011893

Fig. 9. Driver side airbag unit mounting screws (**arrows**).

5. Carefully lift airbag unit off steering wheel and disconnect orange harness connector from rear of airbag unit. See Fig. 10.

NOTE —
In some models, the airbag electrical connection to the steering wheel is via a plug in back of the airbag unit. There is no wire harness to disconnect.

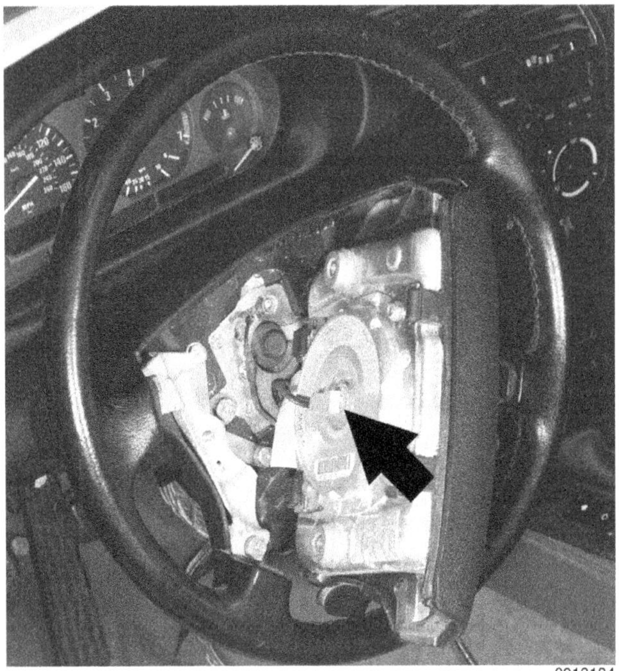

0013184

Fig. 10. Driver side airbag unit harness connector (**arrow**).

WARNING —

• Store the removed airbag unit with the horn pad facing up. If stored facing down, accidental deployment could propel it violently into the air, causing injury.

• Once an airbag is removed, the car must not be driven.

• Do not connect the battery with the airbag disconnected. A fault code will be stored, setting off the SRS warning light. Special tools are needed to reset the fault memory.

• Special instructions apply when installing an airbag equipped steering wheel. See **320 Steering and Wheel Alignment**.

• Once the airbag unit is installed and all other service procedures have been completed, start the engine and check that the SRS warning light goes out. If the warning light stays on, the SRS system will not function as designed. Have the system diagnosed and repaired by an authorized BMW dealer.

6. Installation is reverse of removal. Tighten right side screw on airbag first (as viewed from driver's seat), then left side screw.

NOTE —
If there is no electrical harness connector between the airbag and the steering wheel, be sure to fit the push-in electrical plug in back of the airbag into the proper receptacle in the steering wheel.

Tightening Torque
- Airbag unit to steering wheel . . . . 8 Nm (71 in-lb)

Passenger side airbag, removing and installing

1. Disconnect negative (–) cable from battery and cover negative terminal with insulating material.

2. Remove glove compartment. See **513 Interior Trim**.

3. Lift cover from top of passenger side airbag on dashboard. See Fig. 11.

0013108

Fig. 11. After removing glove compartment, lift out passenger side airbag cover. Note airbag mounting bolts (**A**) and airbag electrical connector (**B**).

4. Remove passenger side airbag mounting bolts. Disconnect electrical harness connector from rear of airbag unit and remove airbag.

5. Installation is reverse of removal. Make sure wiring harness is not pinched when installing airbag unit in dashboard.

Front side-impact airbag, removing and installing

1. Disconnect negative (–) cable from battery and cover negative terminal with insulating material.

> ***CAUTION —***
> *Prior to disconnecting the battery, read the battery disconnection cautions given at the front of this manual on page viii.*

2. Remove inside door panel as described in **411 Doors**. Remove vapor shield.

3. Disconnect electrical harness connector from door-mounted airbag unit. See Fig. 12.

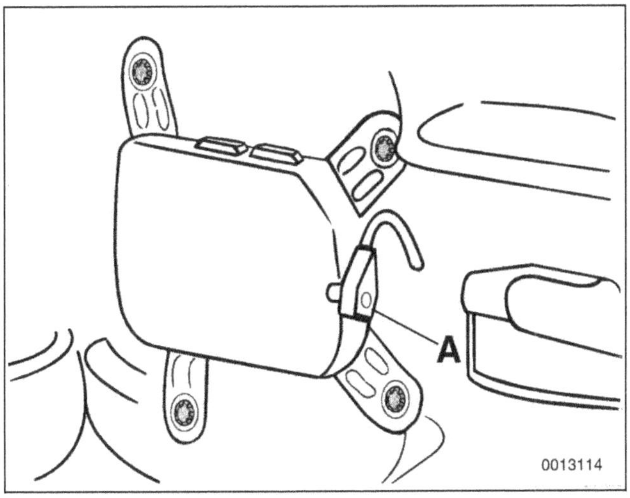

Fig. 12. Front side-impact airbag (inner door panel removed). **A** is electrical harness connector.

> ***WARNING —***
> *The plug connector inside the door is not marked in orange, unlike other SRS connectors in the car.*

> ***CAUTION —***
> *• The electrical harness connector of the side-impact airbag unit may be either on the side of the unit or in the back. When removing and installing the unit, pay attention to the routing of the harness to avoid kinks or breaks in the wire.*
>
> *• Store the removed airbag unit with the soft pad facing up. If stored facing down, accidental deployment could propel it violently into the air, causing injury.*

4. Remove mounting screws and remove airbag from door.

5. Installation is reverse of removal.

> ***WARNING —***
> *The side-impact airbag mounting screws must always be replaced any time they are removed. Use new screws with factory-applied locking sealer.*

Tightening Torque
- Side-impact airbag to door . . . . 8.5 Nm (75 in-lb)

Electrical Wiring Diagrams

GENERAL

This section contains wiring diagrams for 1992 through 1998 BMW 3 Series models.

> ***WARNING —***
> *On cars equipped with airbags, special precautions apply to any electrical testing or repair. The airbag unit is an explosive device and must be handled with exterme care. Before starting any work on an airbag equipped car, refer to the warnings and cautions in* **720 Seatbelts** *and* **721 Airbag Systems (SRS)**.

NOTE —
- *The gear position/neutral safety switch is also sometimes referred to as the automatic transmission range switch.*
- *The EWS II system is also sometimes referred to as the Driveaway Protection System or the electronic immobilization system.*
- *Generator = alternator.*

> ***CAUTION —***
> - *Always switch the ignition off and disconnect the negative (–) battery cable before removing any electrical components.*
> - *Prior to disconnecting the battery, read the battery disconnection cautions given at the front of this manual on page viii.*
> - *Connect and disconnect ignition system wires, multiple connectors and ignition test equipment leads only while ignition is switched off.*
> - *Always switch a test meter to the appropiate function and range before making test connections.*

- *All-Season Traction (AST) is also referred to as ASC or ASC+T.*

Power Distribution

1992-93 318i Only (1 of 7)

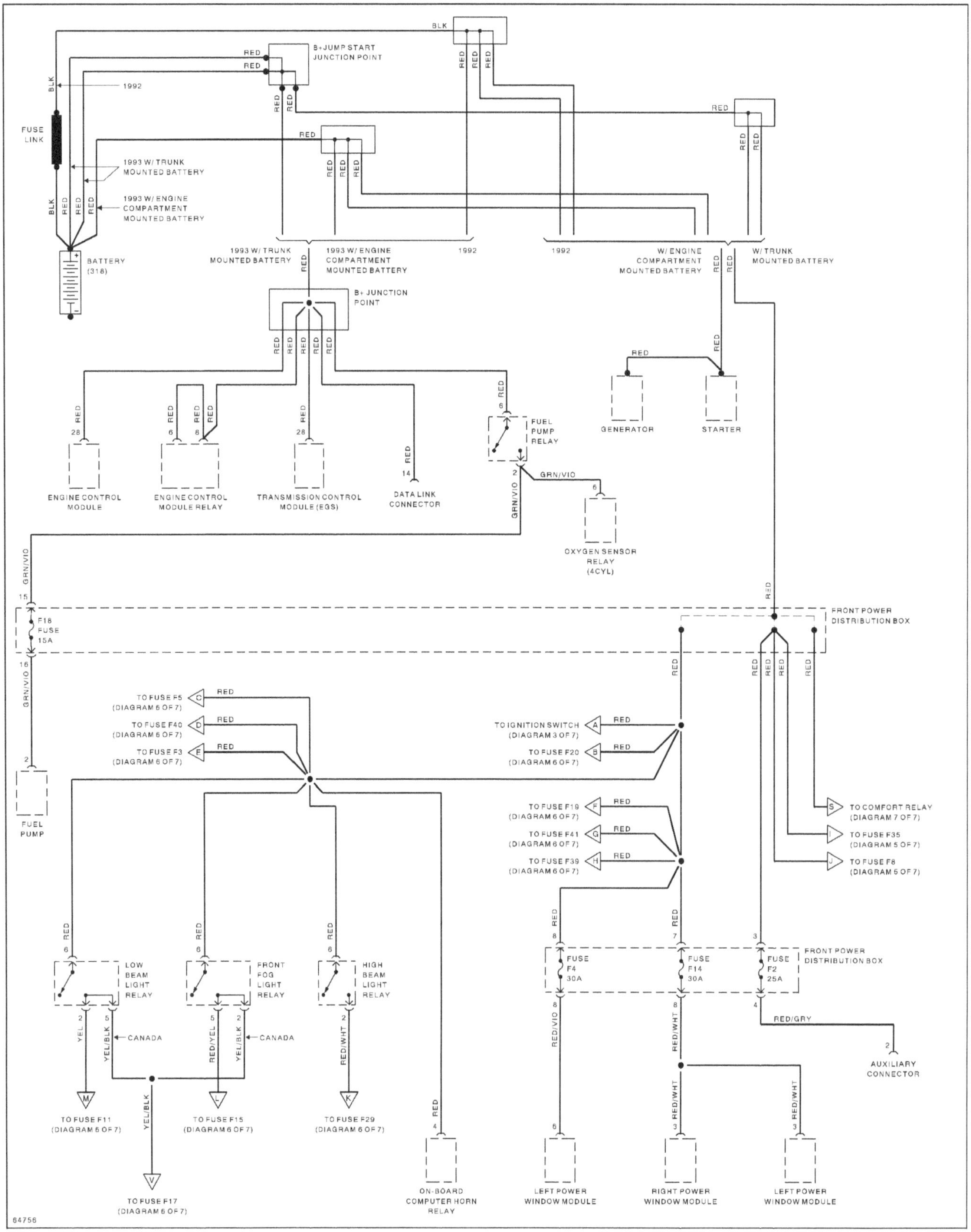

Power Distribution

1992-93 325i Only (2 of 7)

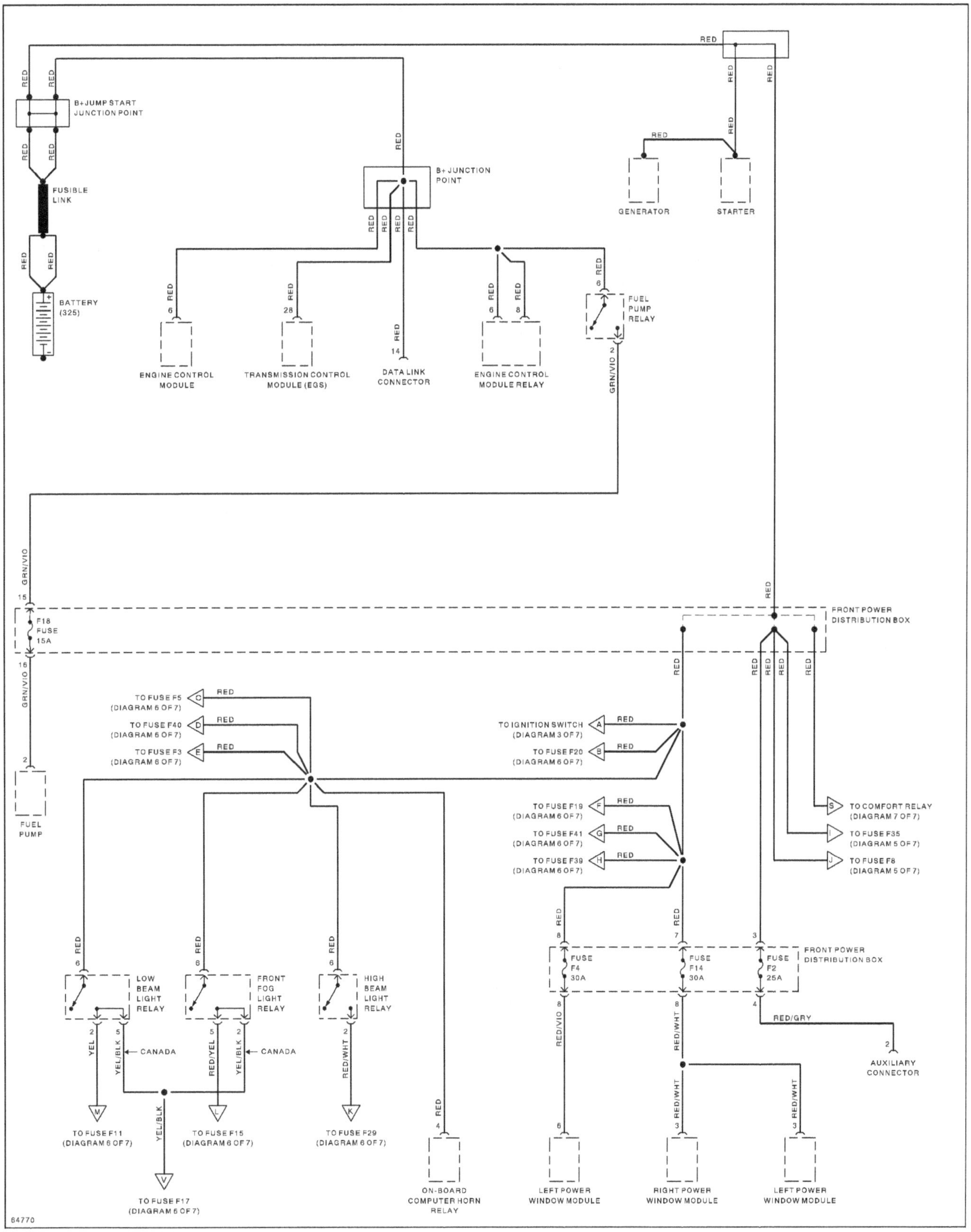

Power Distribution

1992-93 (3 of 7)

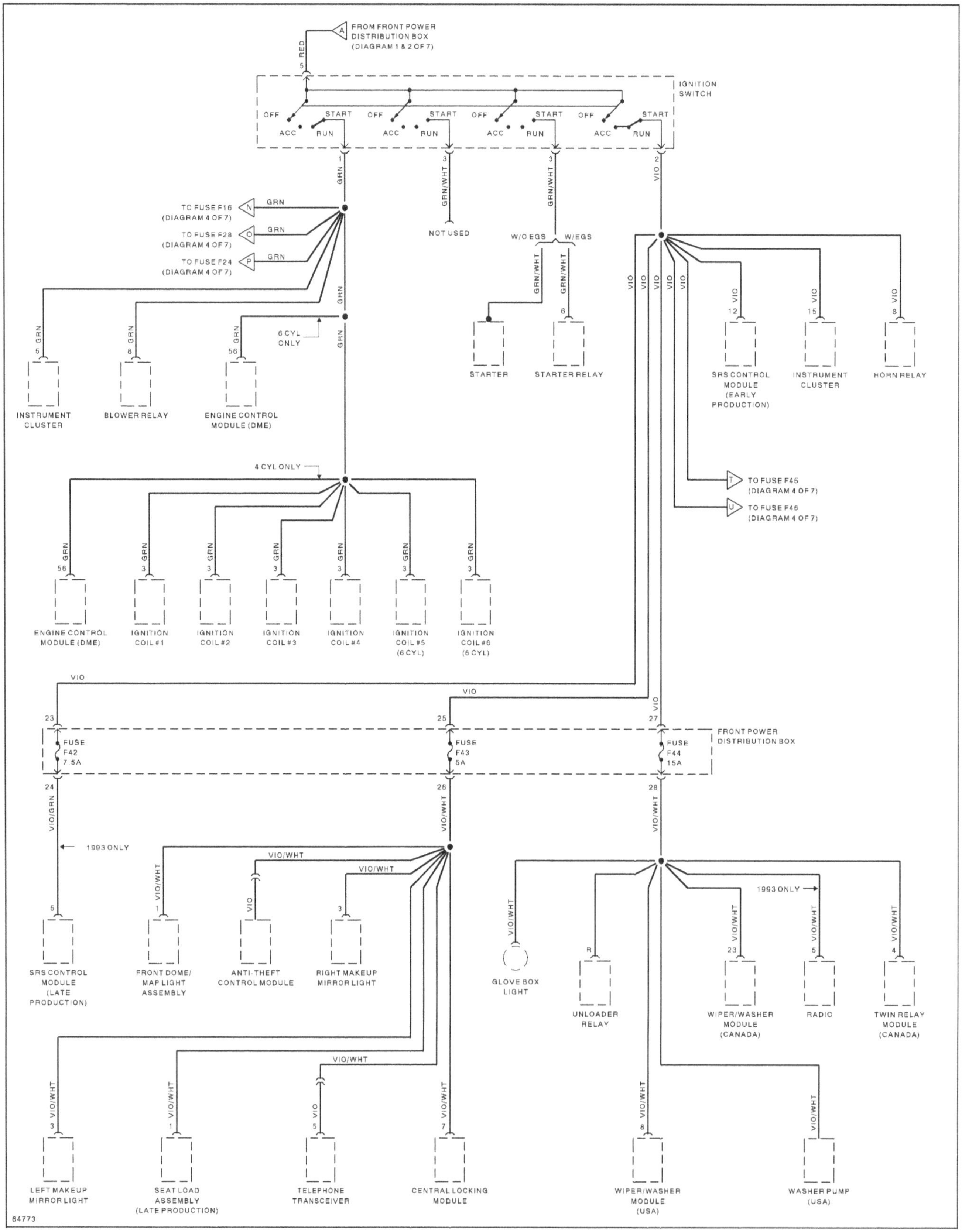

Power Distribution

1992-93 (4 of 7)

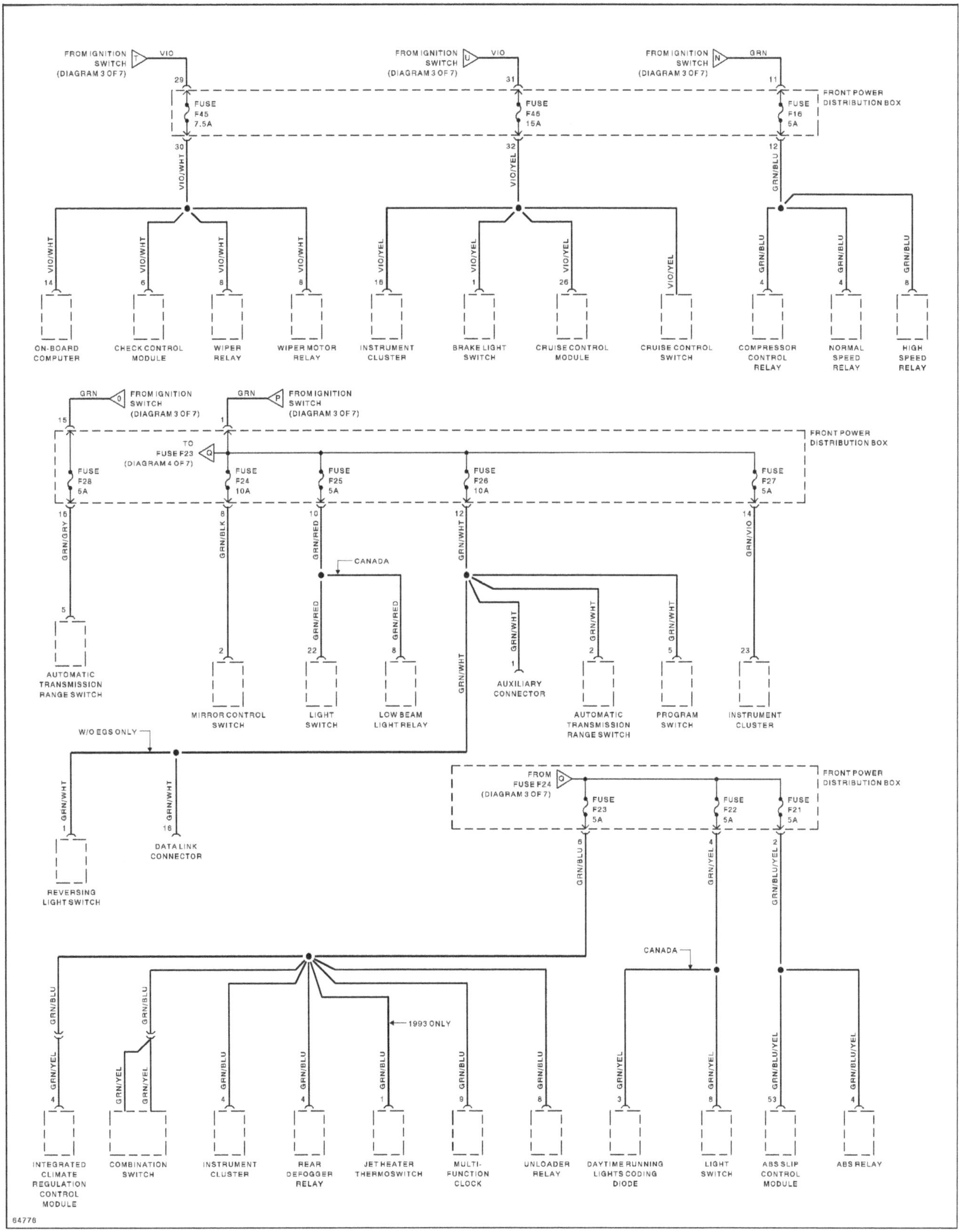

Power Distribution

1992-93 (5 of 7)

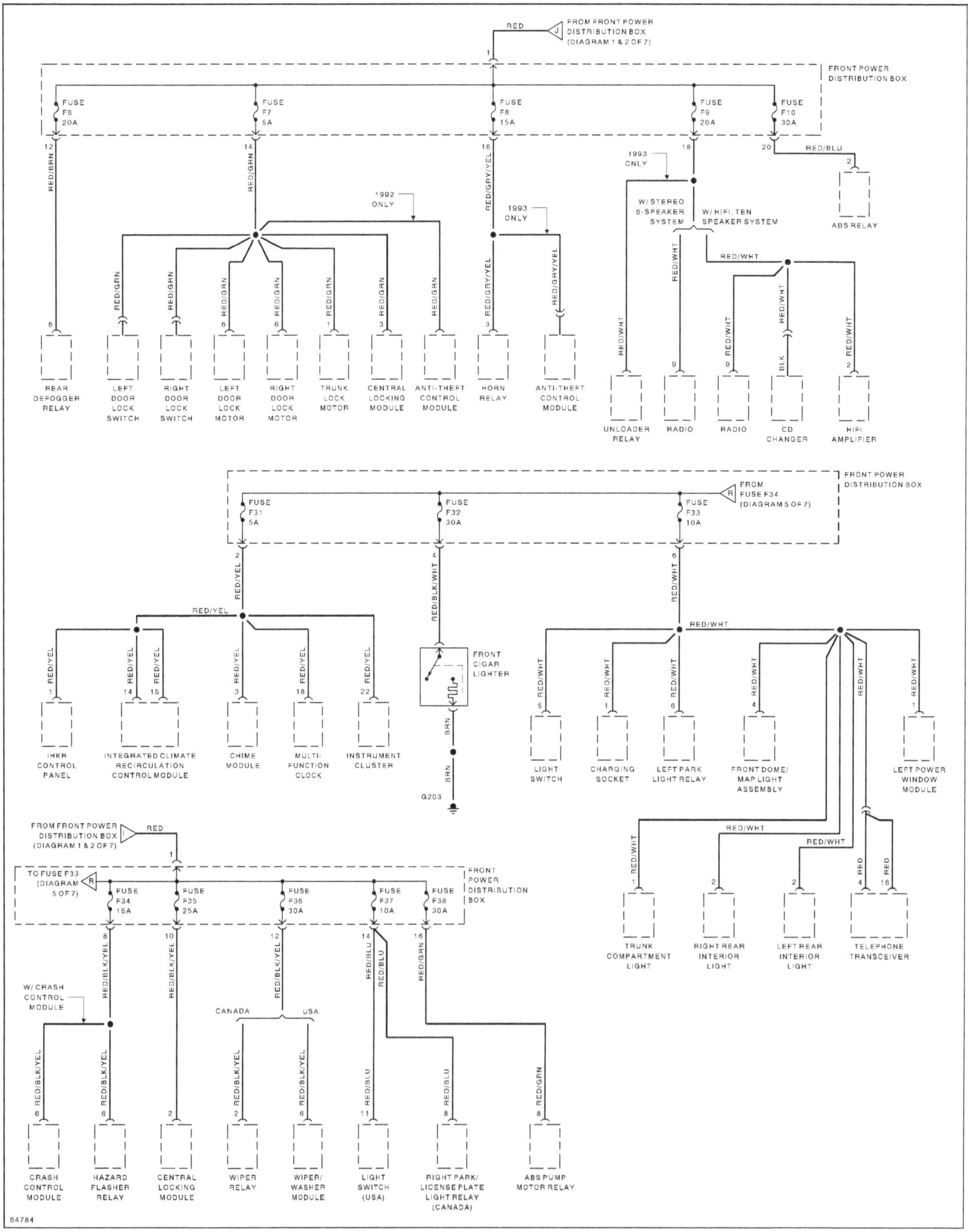

Power Distribution
1992-93 (6 of 7)

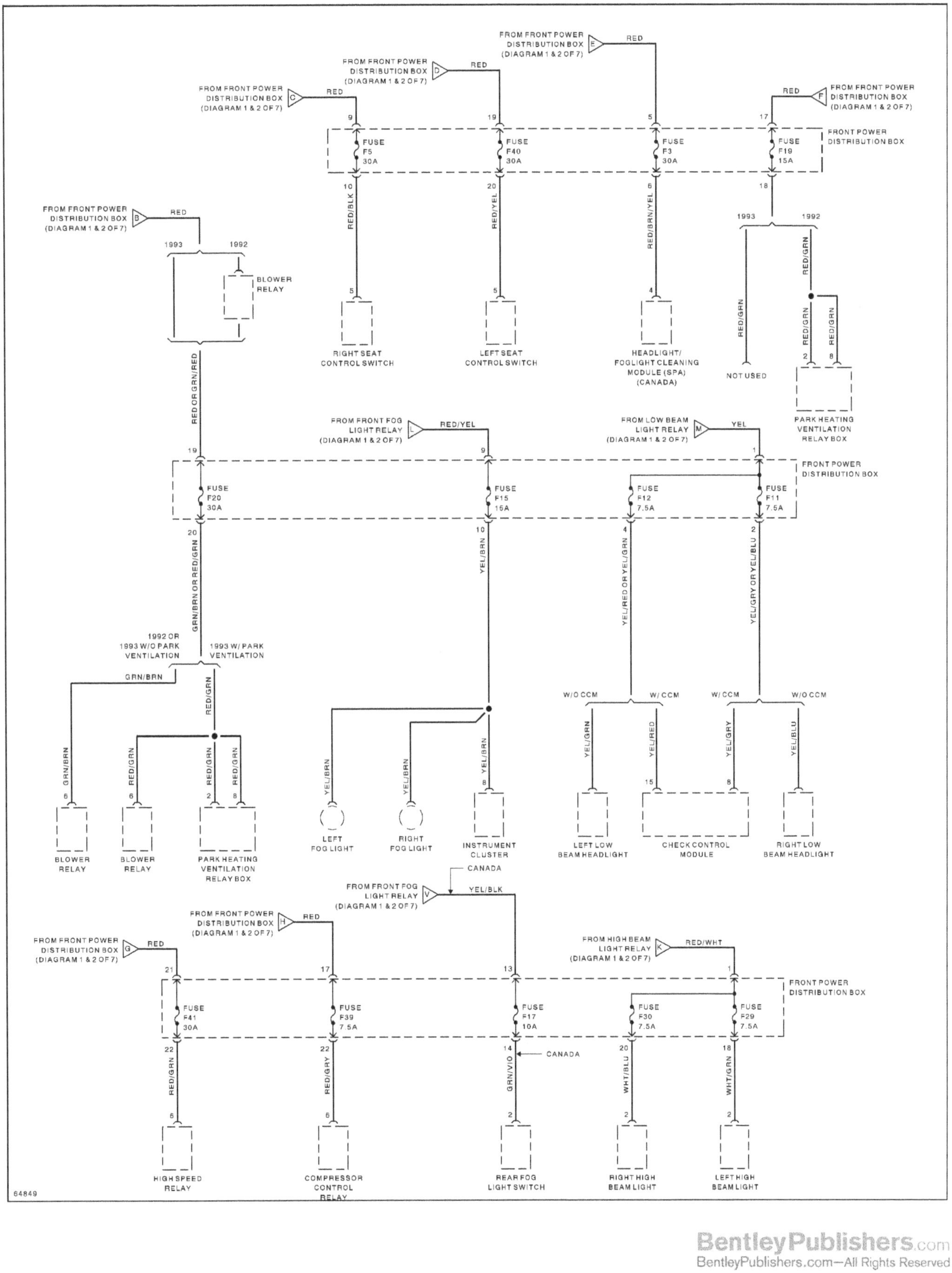

Power Distribution

1992-93 (7 of 7)

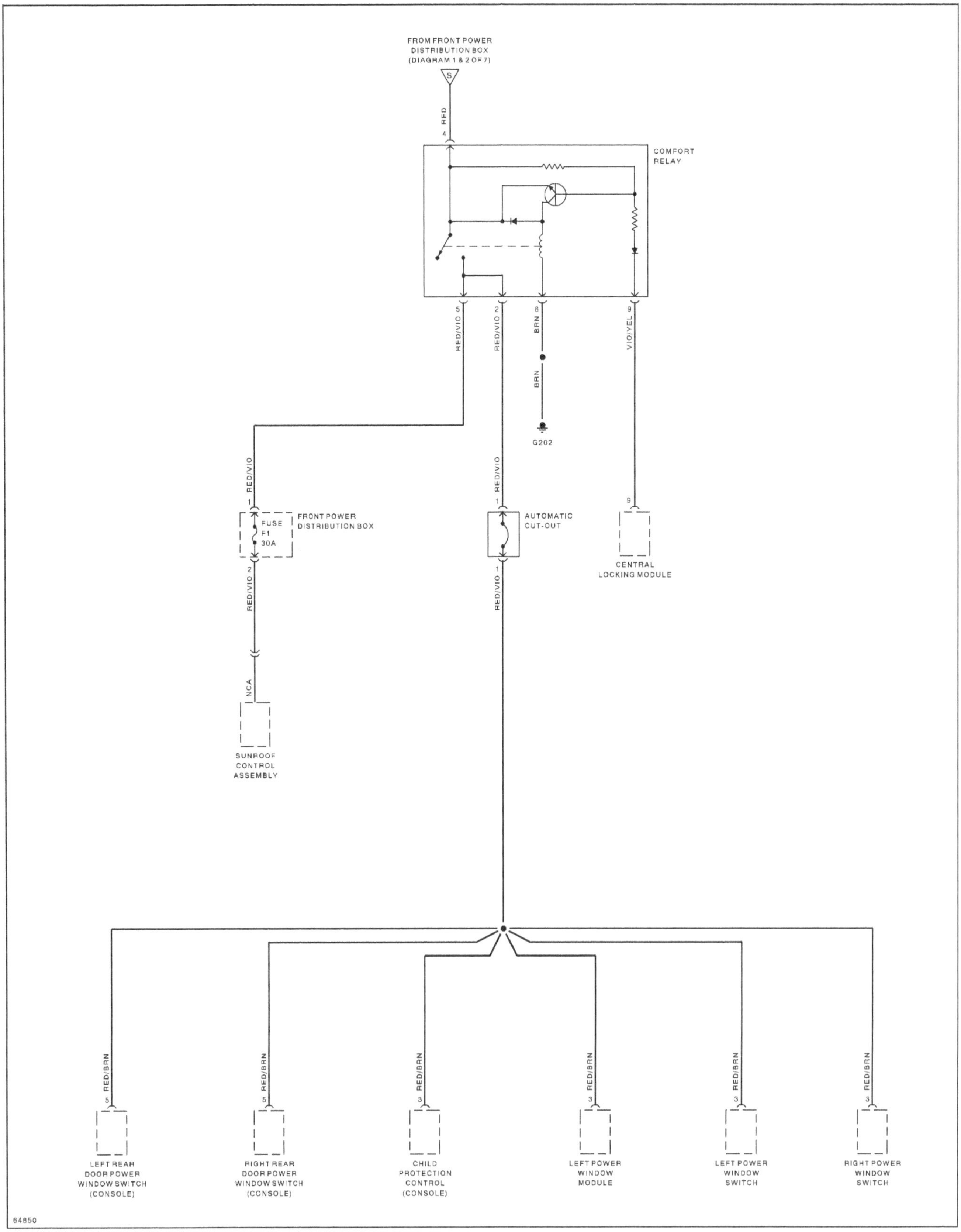

Power Distribution

1994-95 Except 318ti (1 of 5)

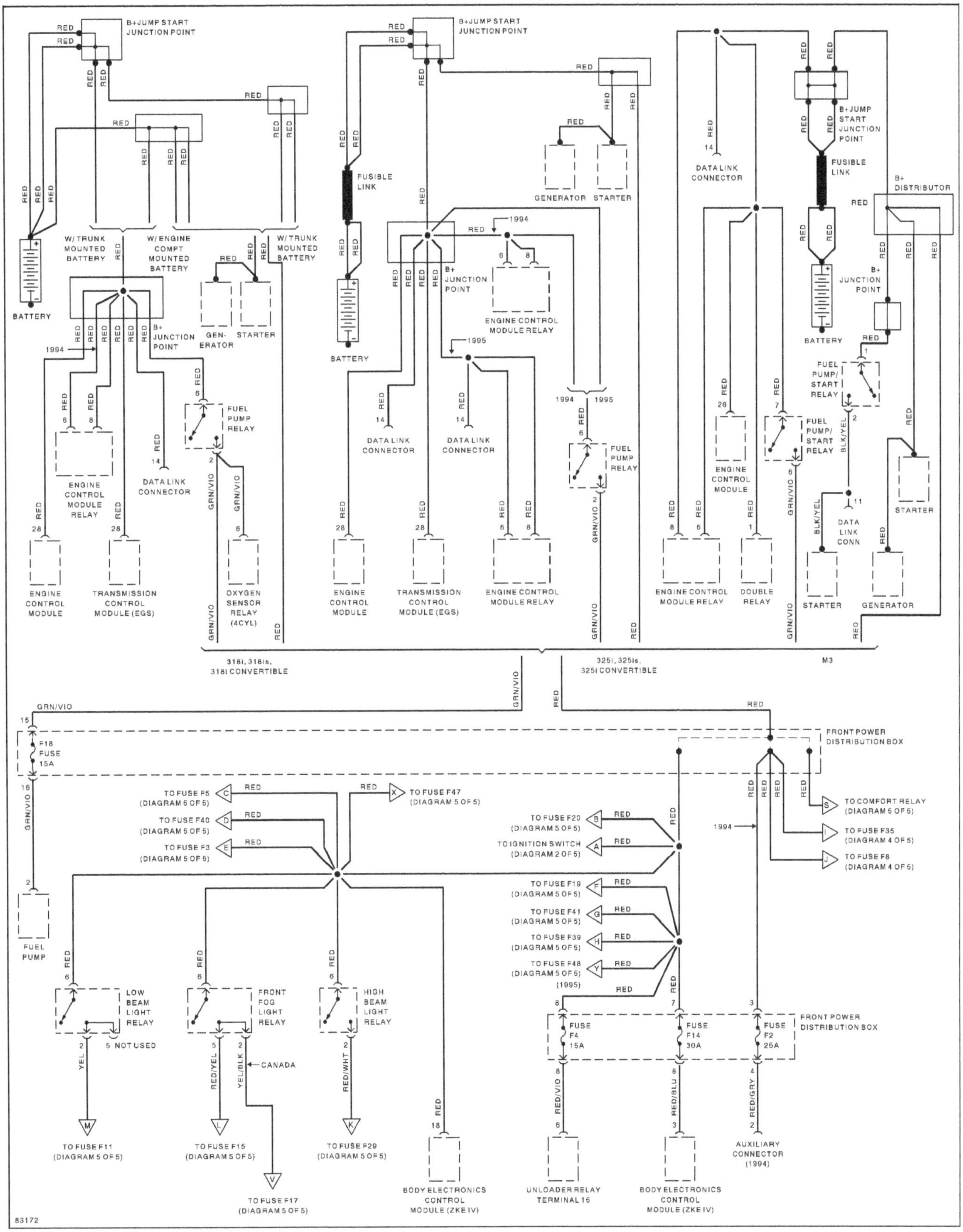

Power Distribution

1994-95 Except 318ti (2 of 5)

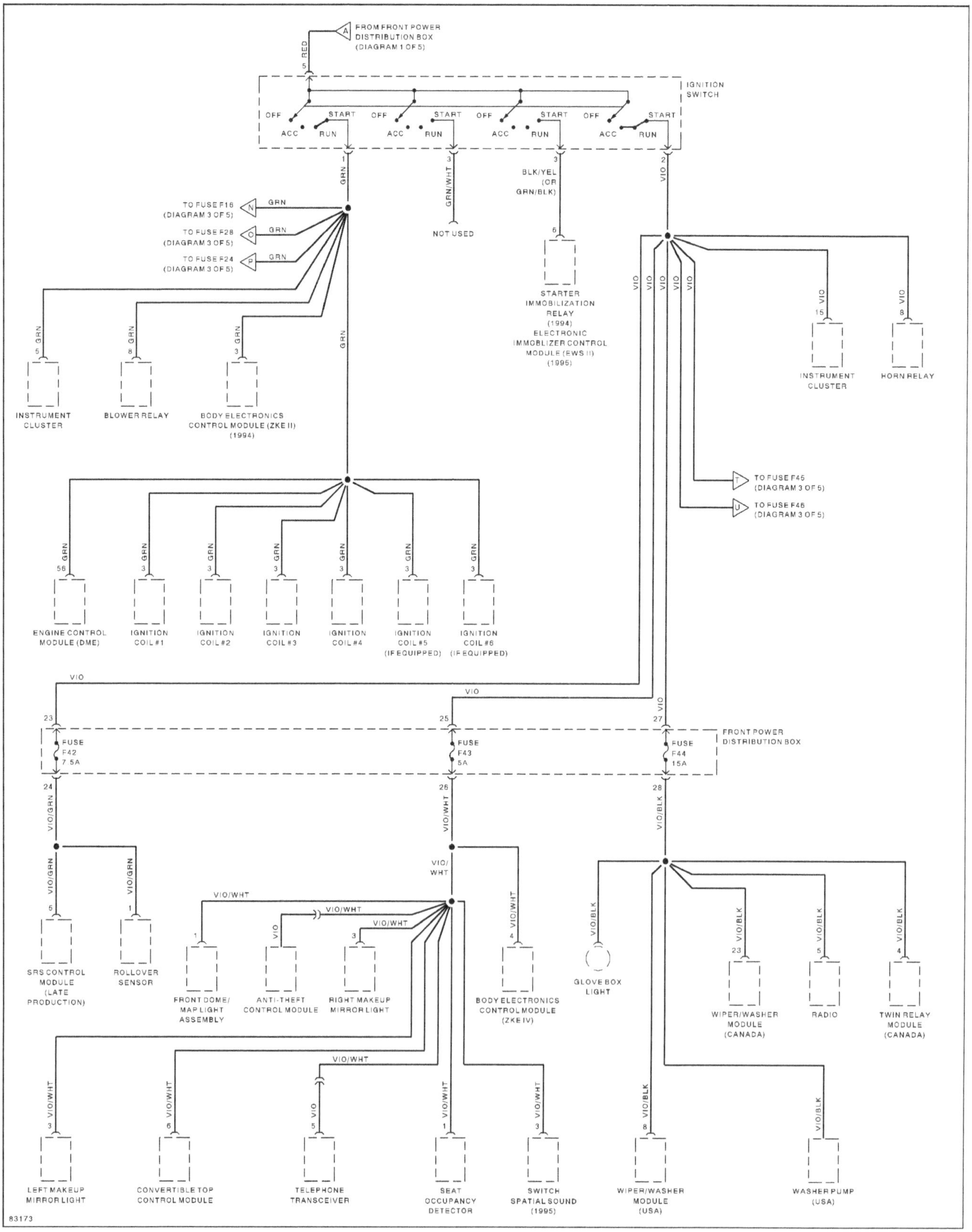

Power Distribution

1994-95 Except 318ti (3 of 5)

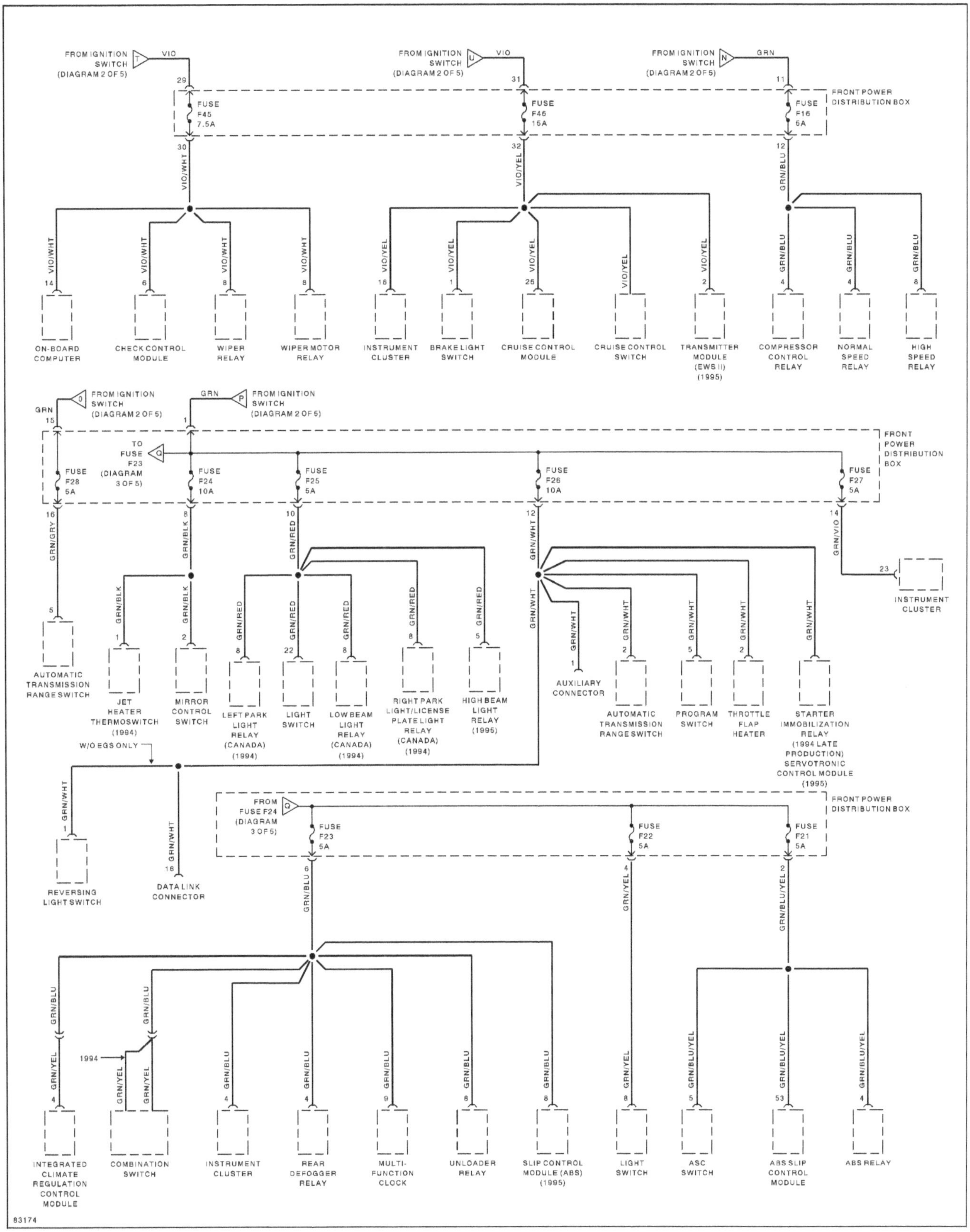

Power Distribution

1994-95 Except 318ti (4 of 5)

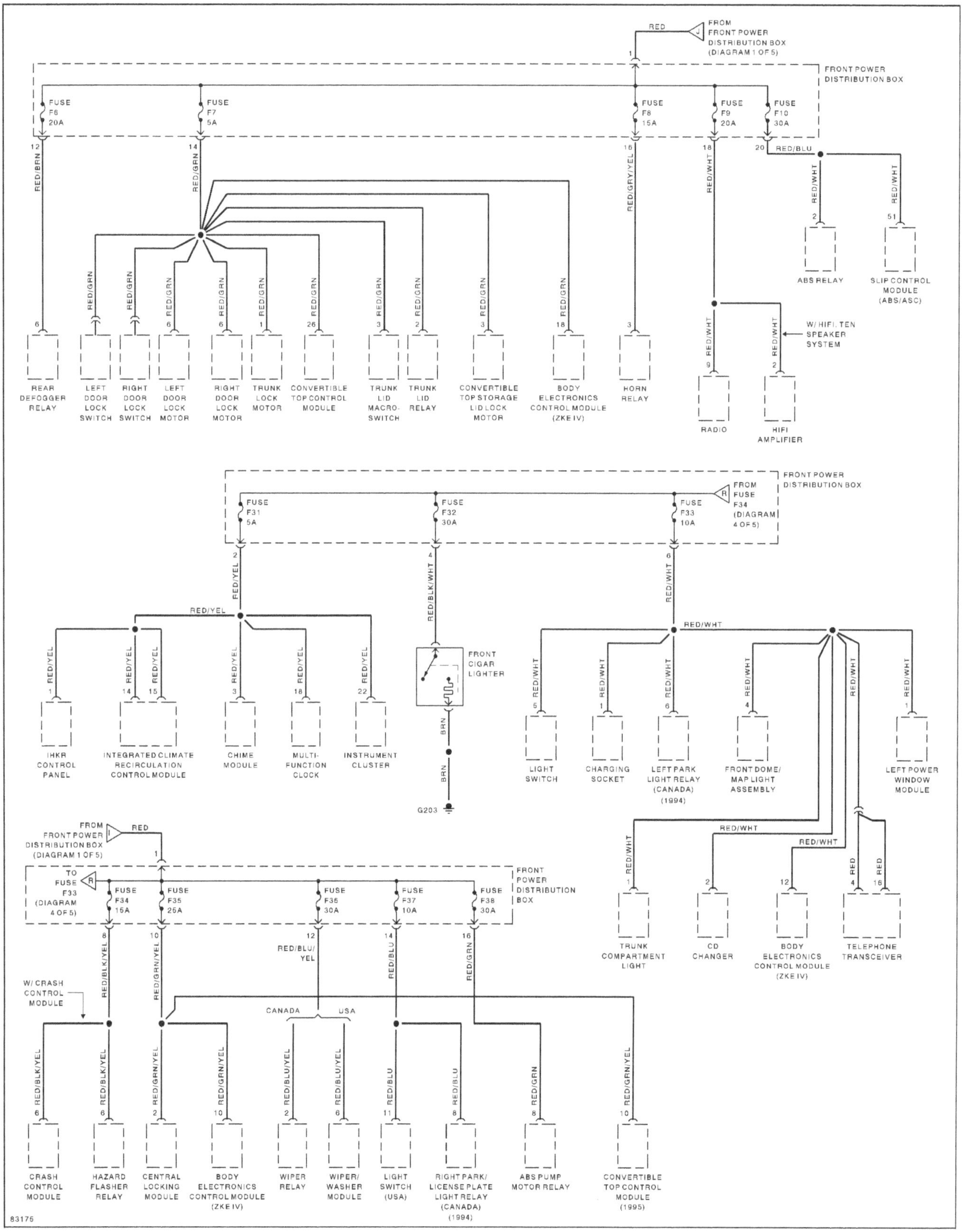

Power Distribution

1994-95 Except 318ti (5 of 5)

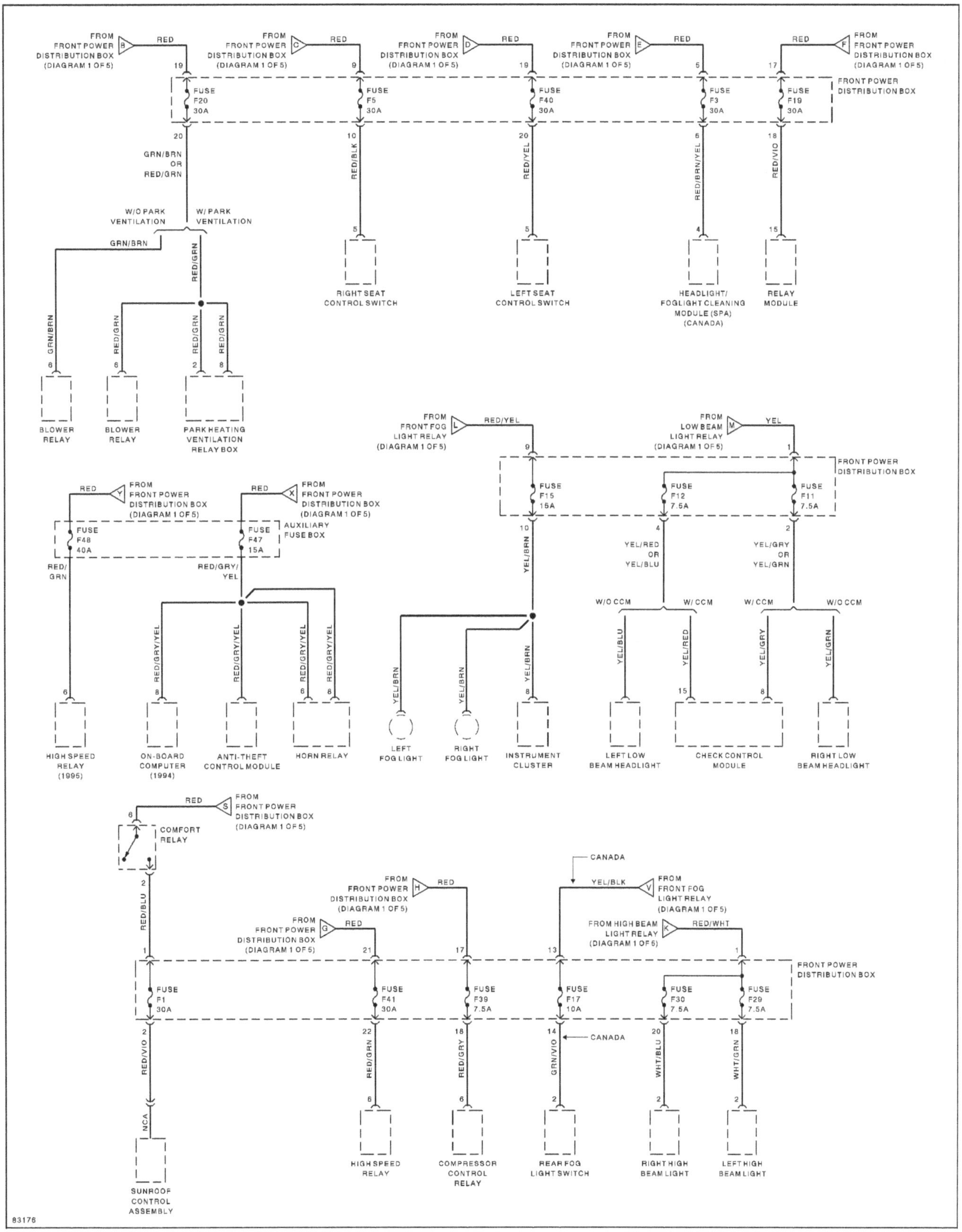

Power Distribution

1996-97 Except 318ti (1 of 5)

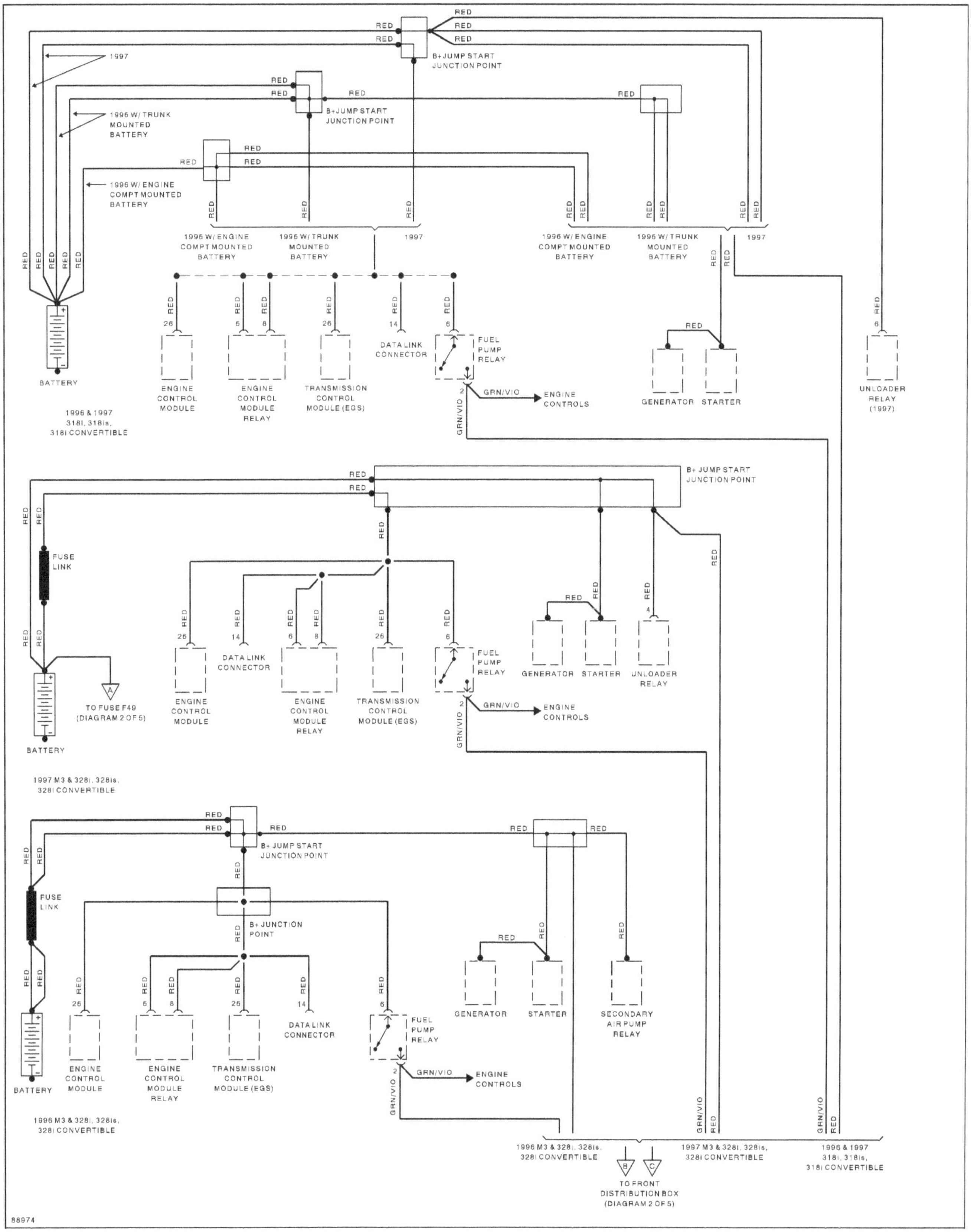

Power Distribution

1996-97 Except 318ti (2 of 5)

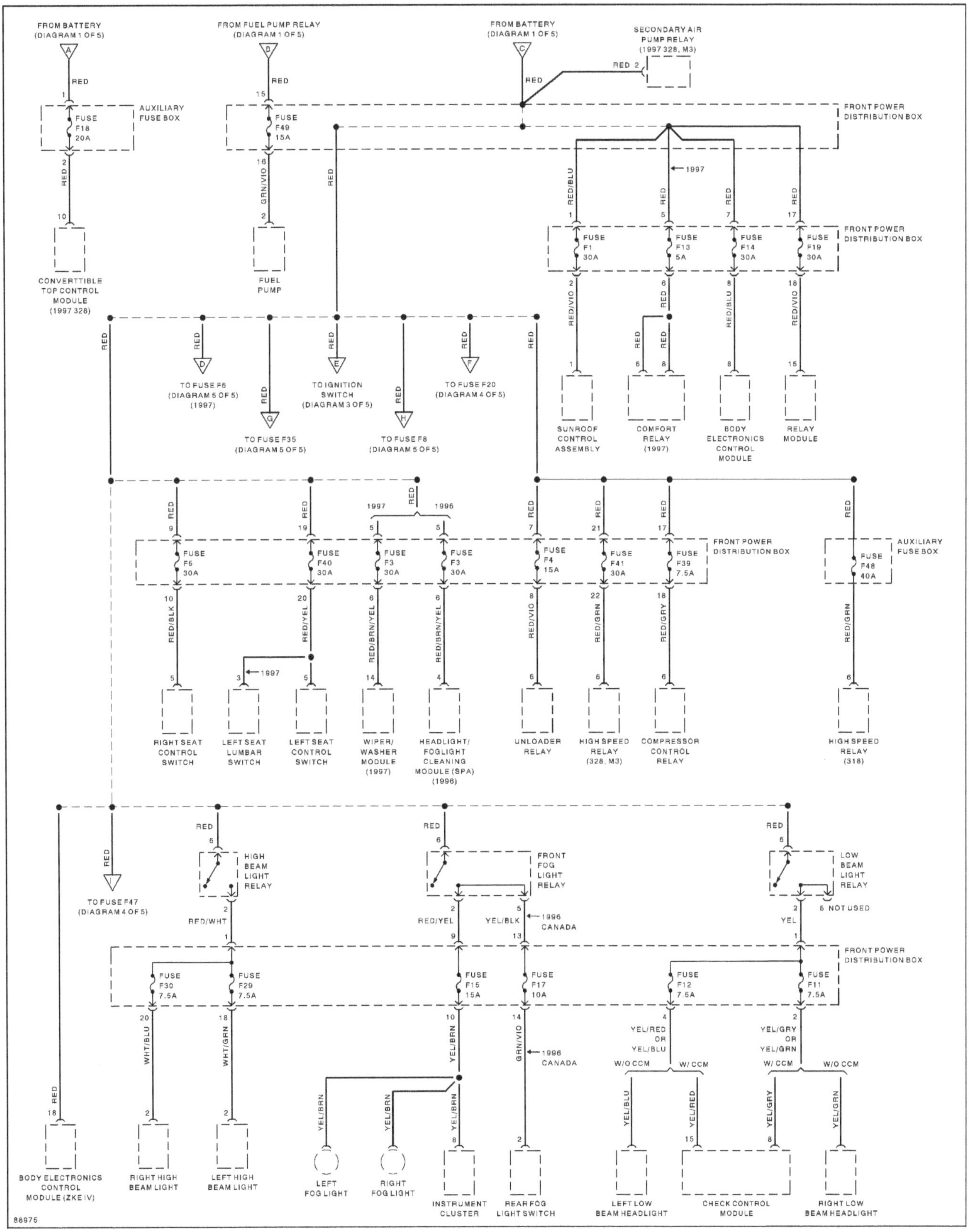

Power Distribution

1996-97 Except 318ti (3 of 5)

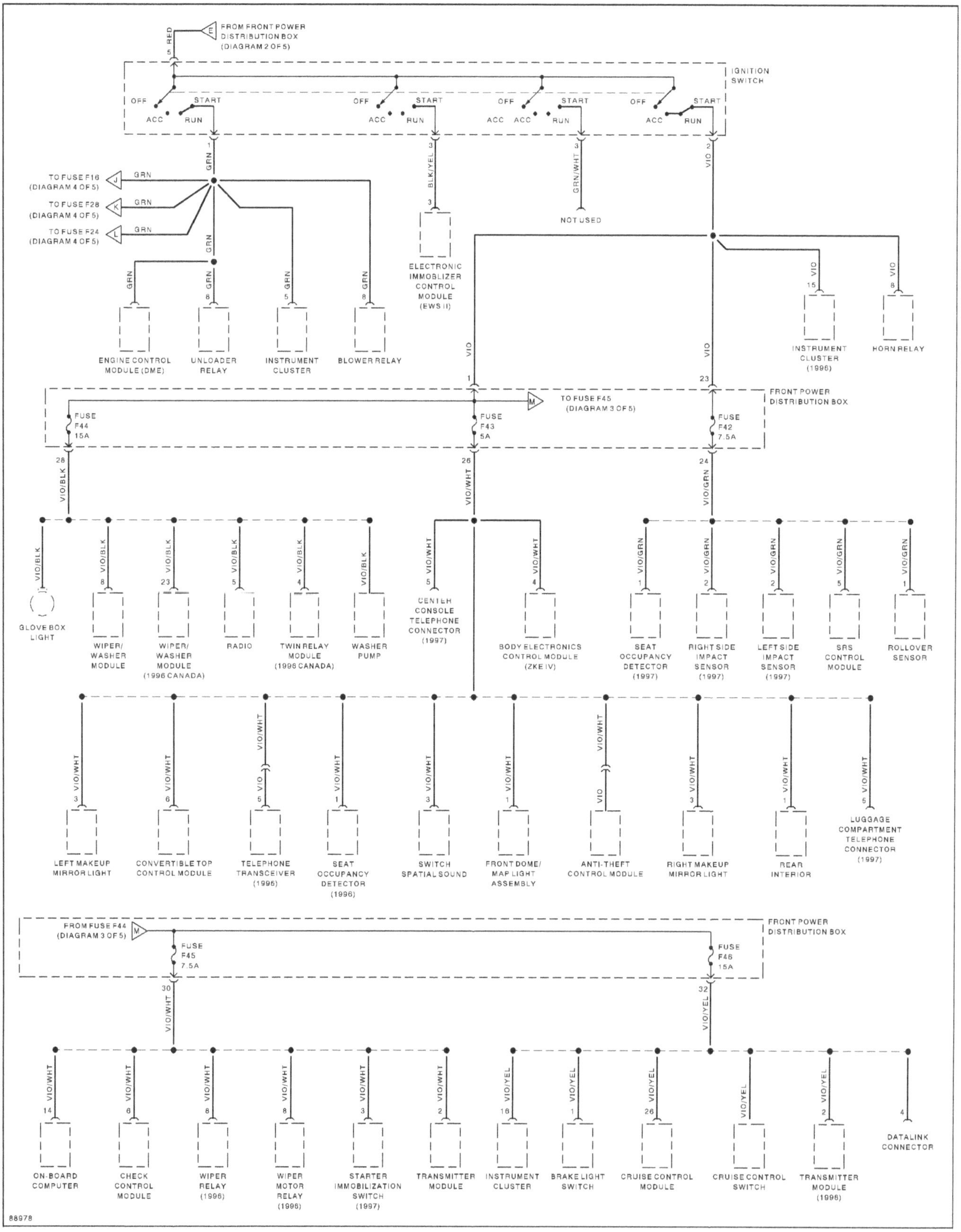

Power Distribution

1996-97 Except 318ti (4 of 5)

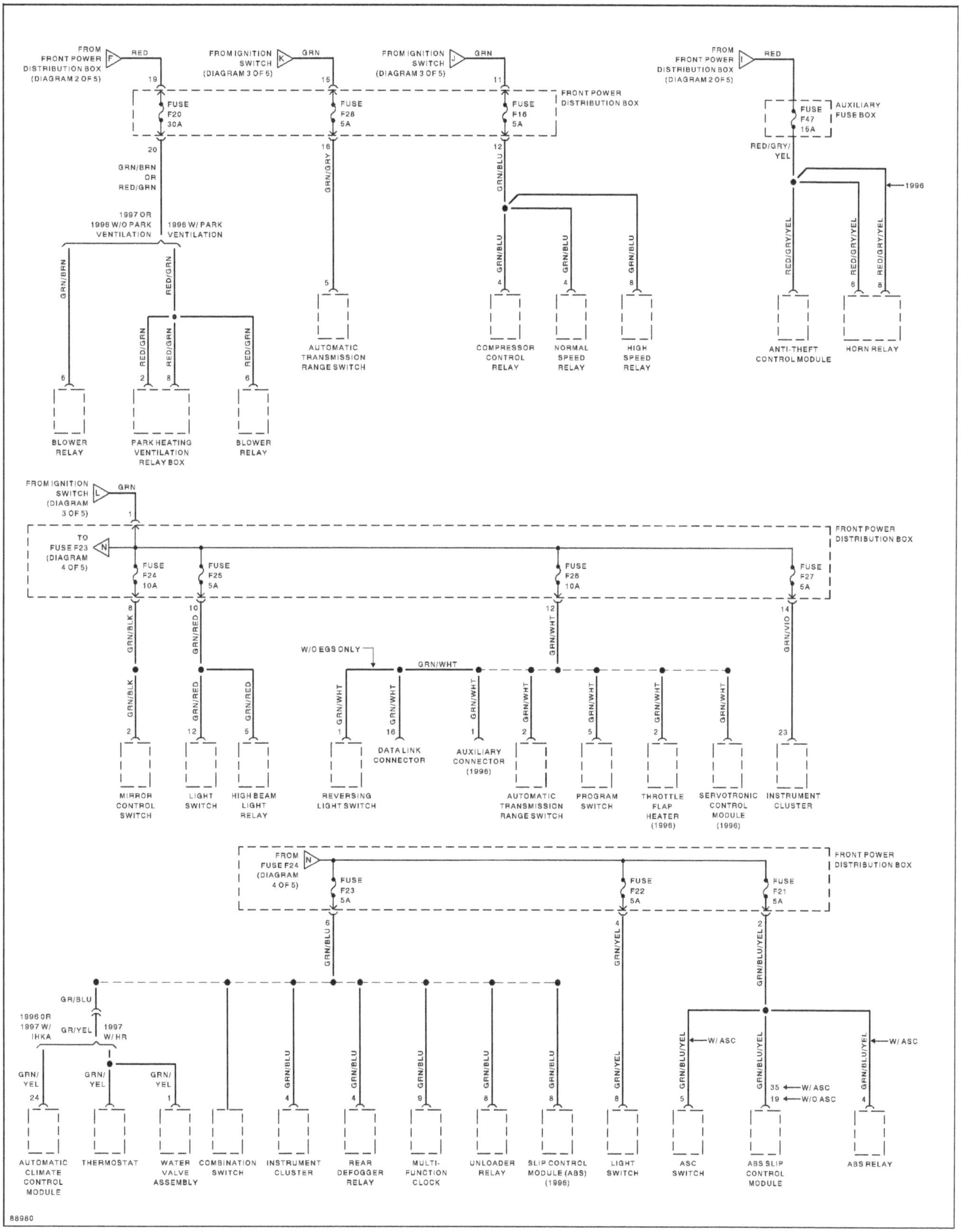

Power Distribution

1996-97 Except 318ti (5 of 5)

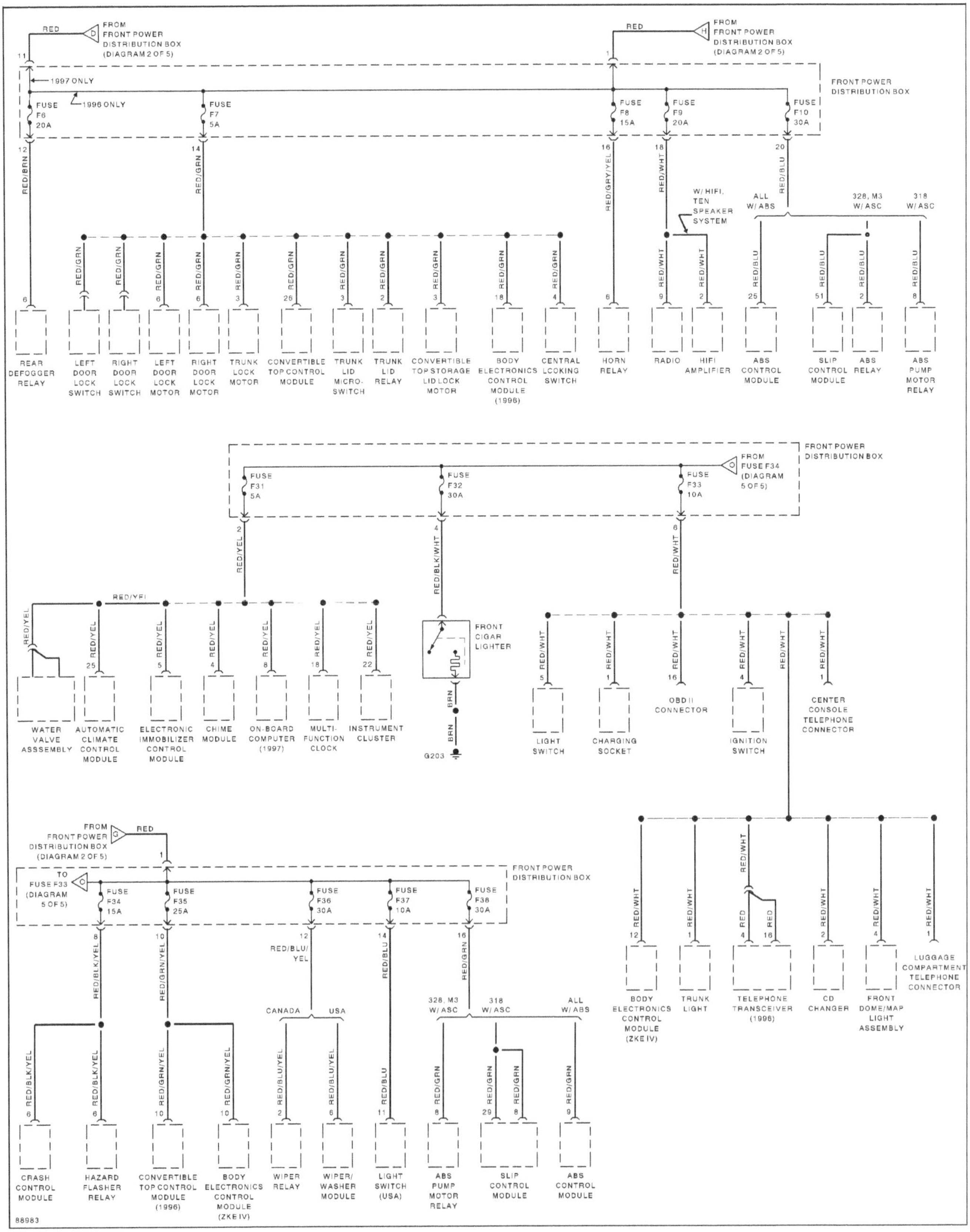

Power Distribution
1998 Except 318ti (1 of 5)

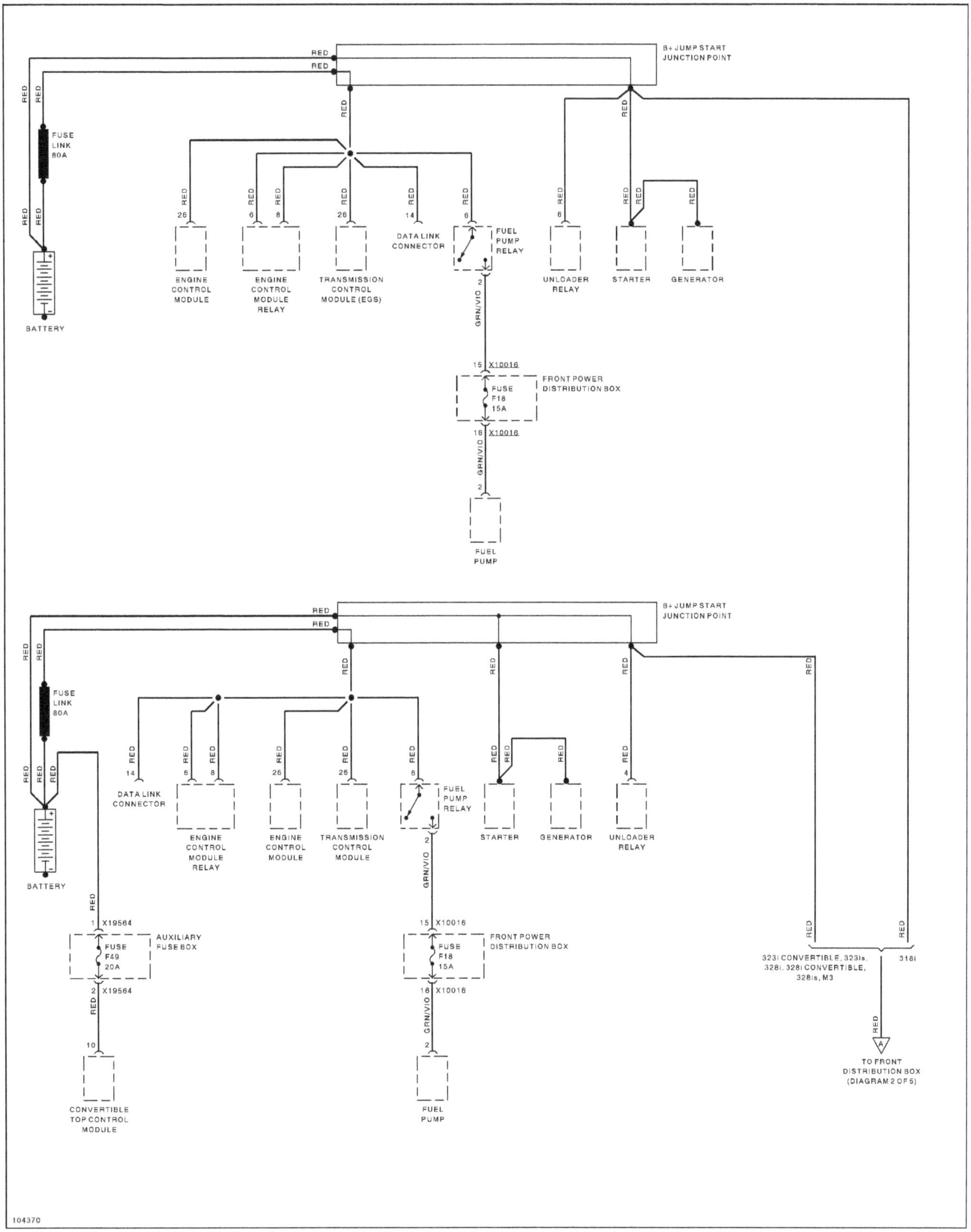

Power Distribution

1998 Except 318ti (2 of 5)

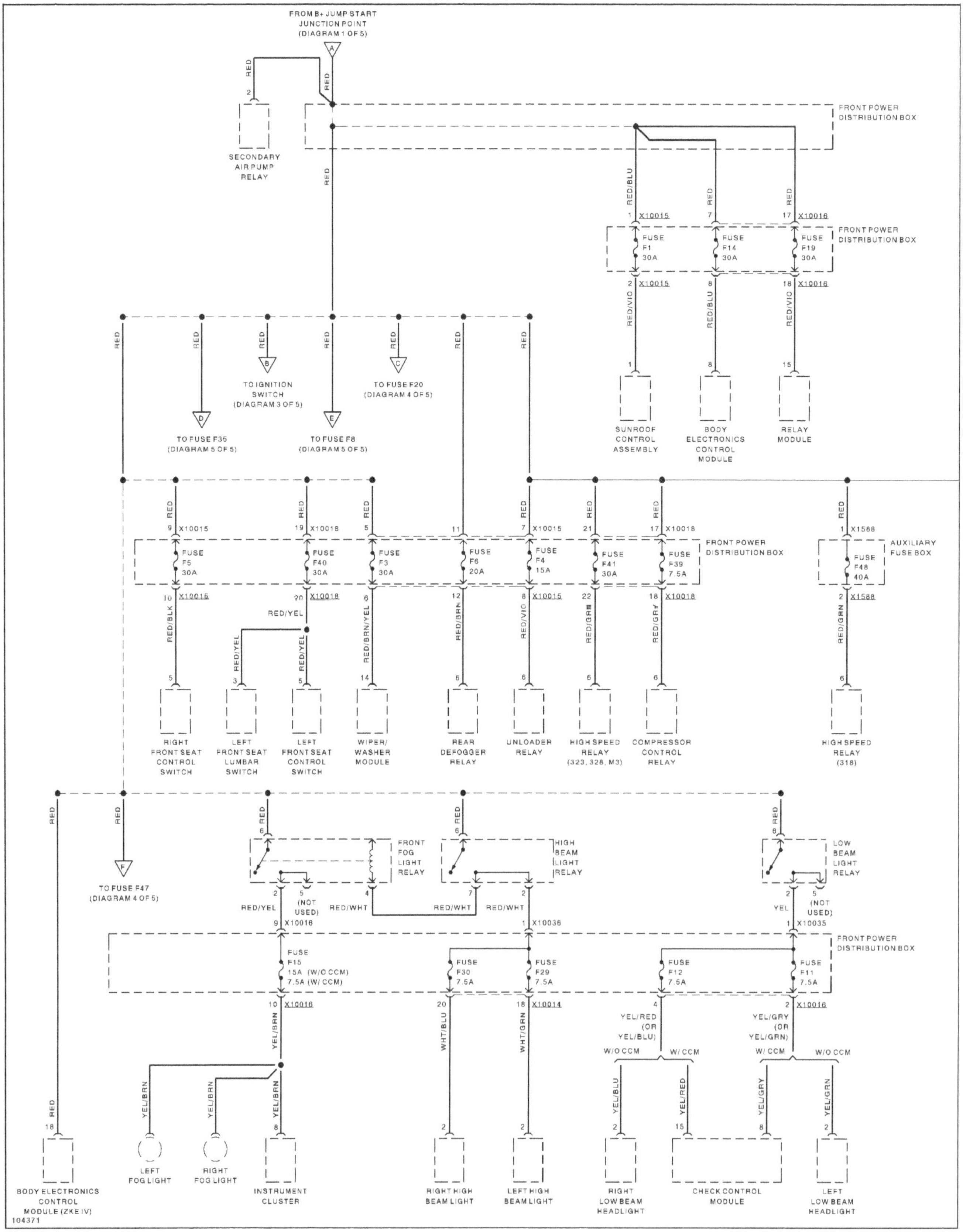

Power Distribution

1998 Except 318ti (3 of 5)

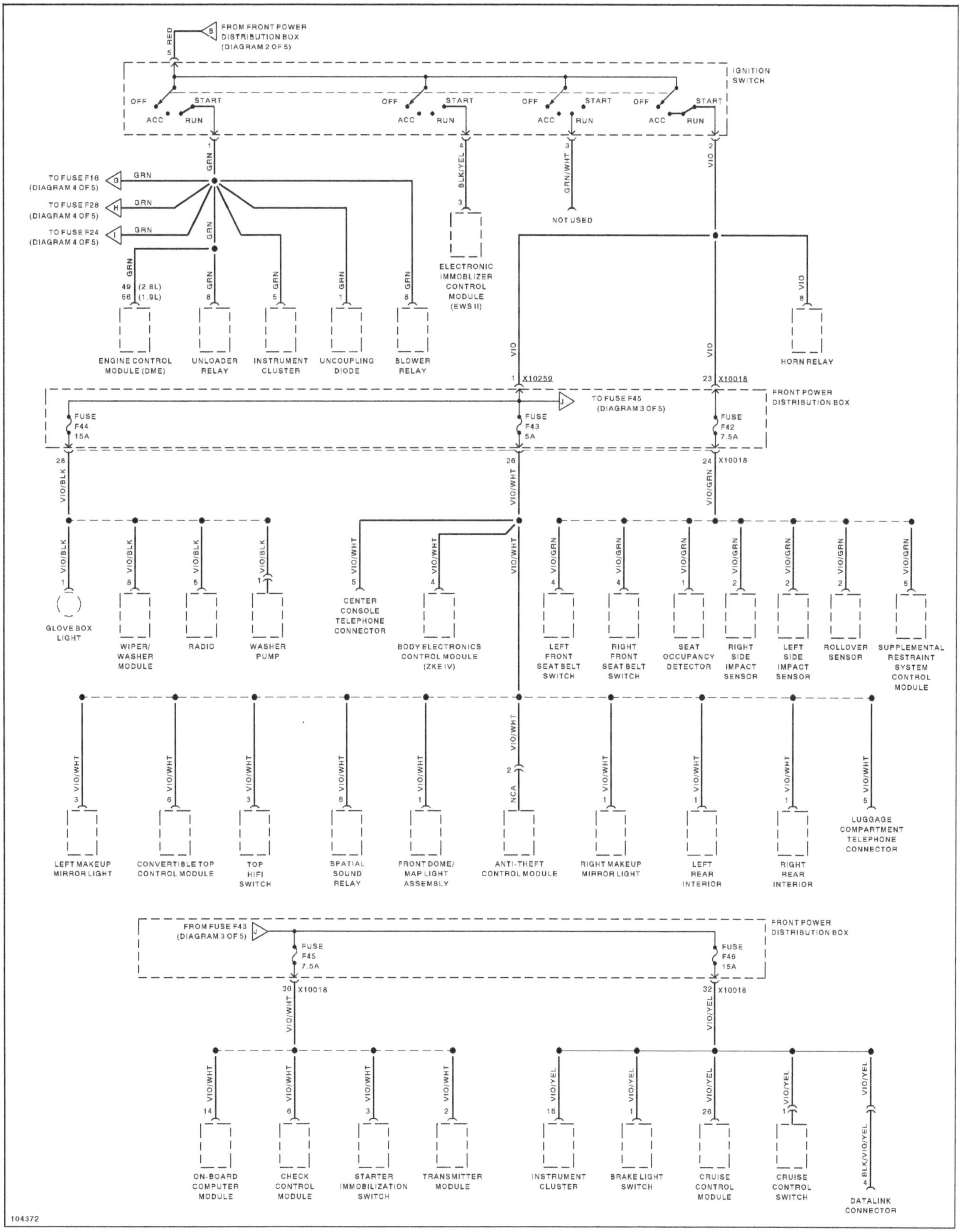

Power Distribution

1998 Except 318ti (4 of 5)

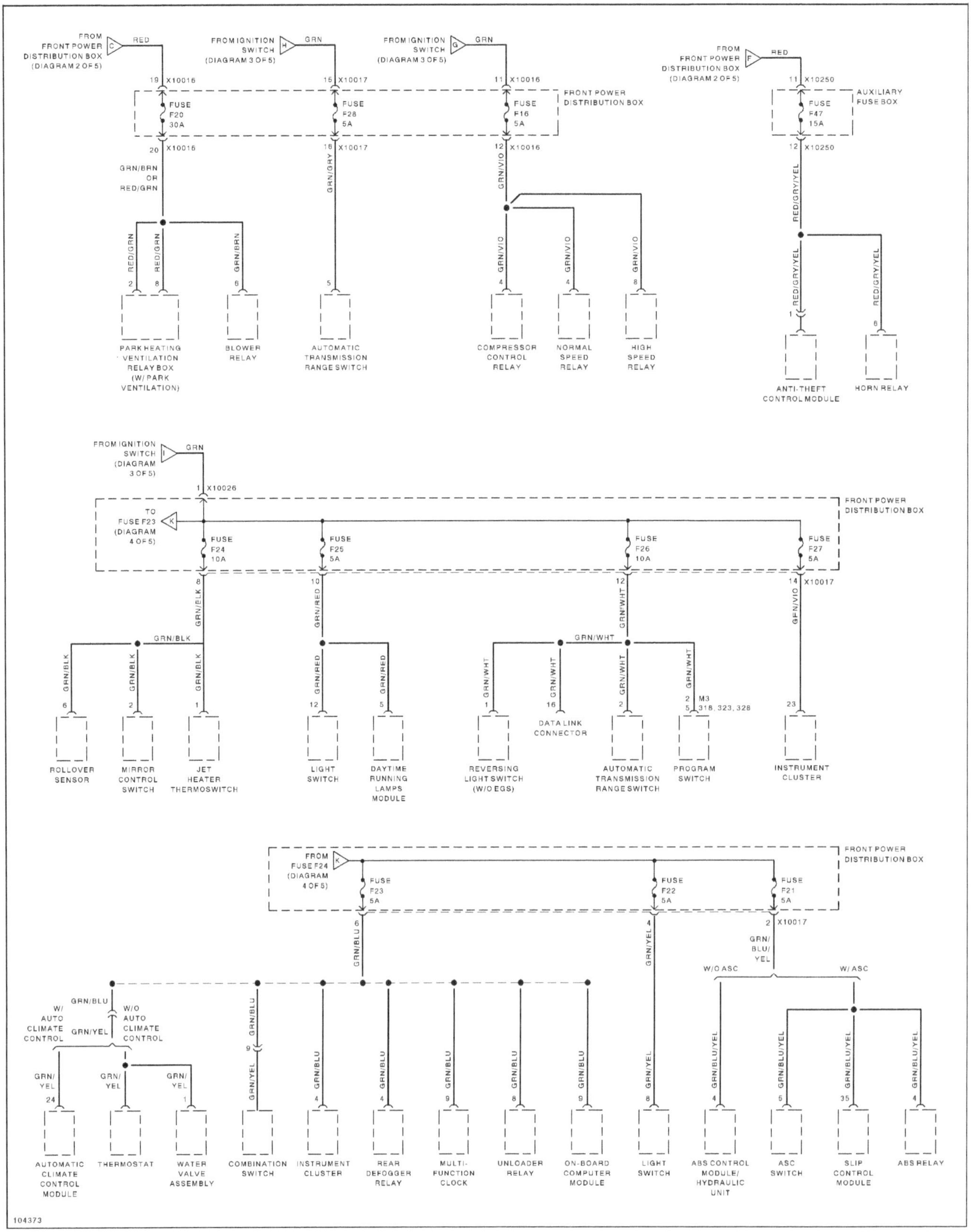

Power Distribution

1998 Except 318ti (5 of 5)

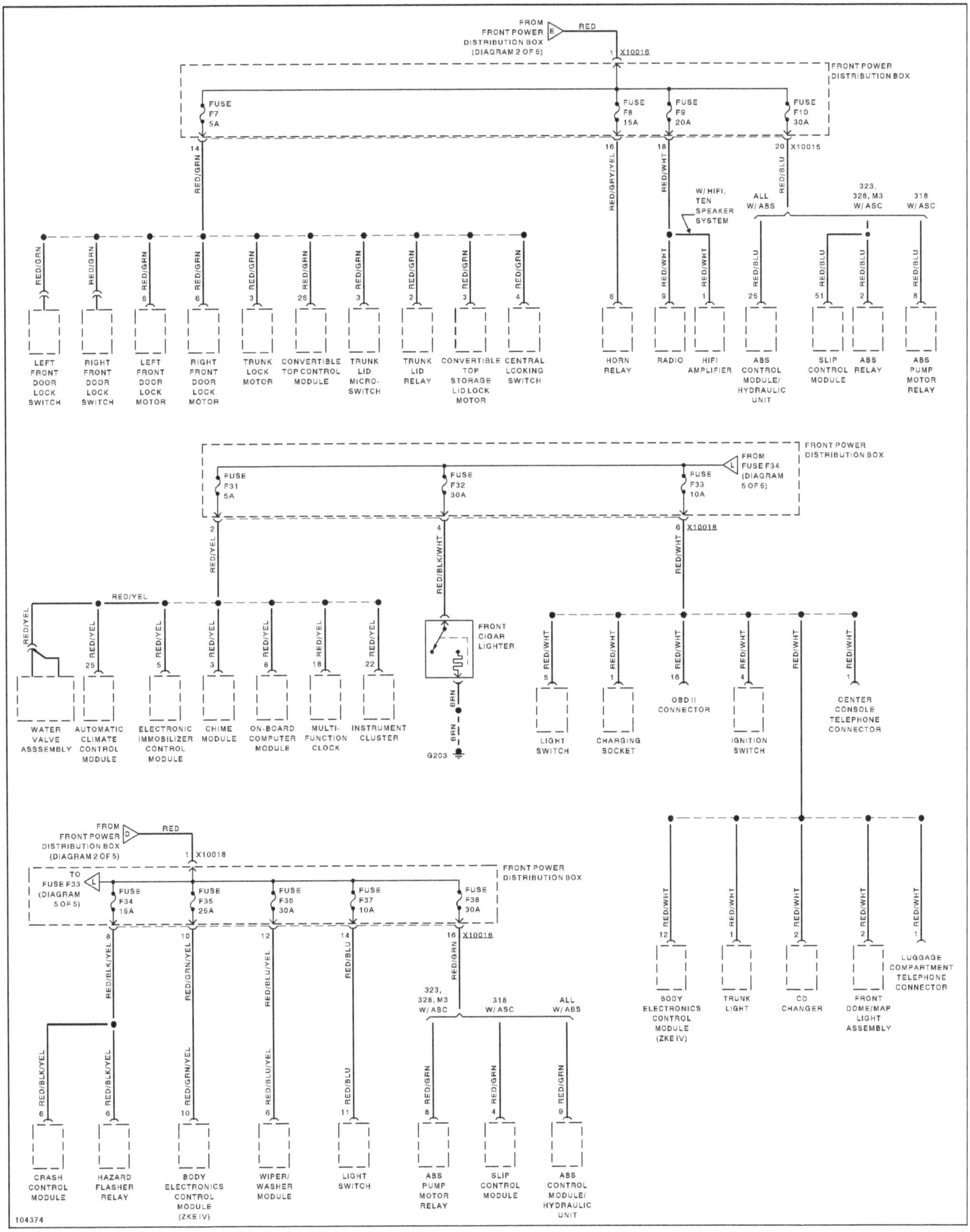

Power Distribution

1995-96 318ti (1 of 5)

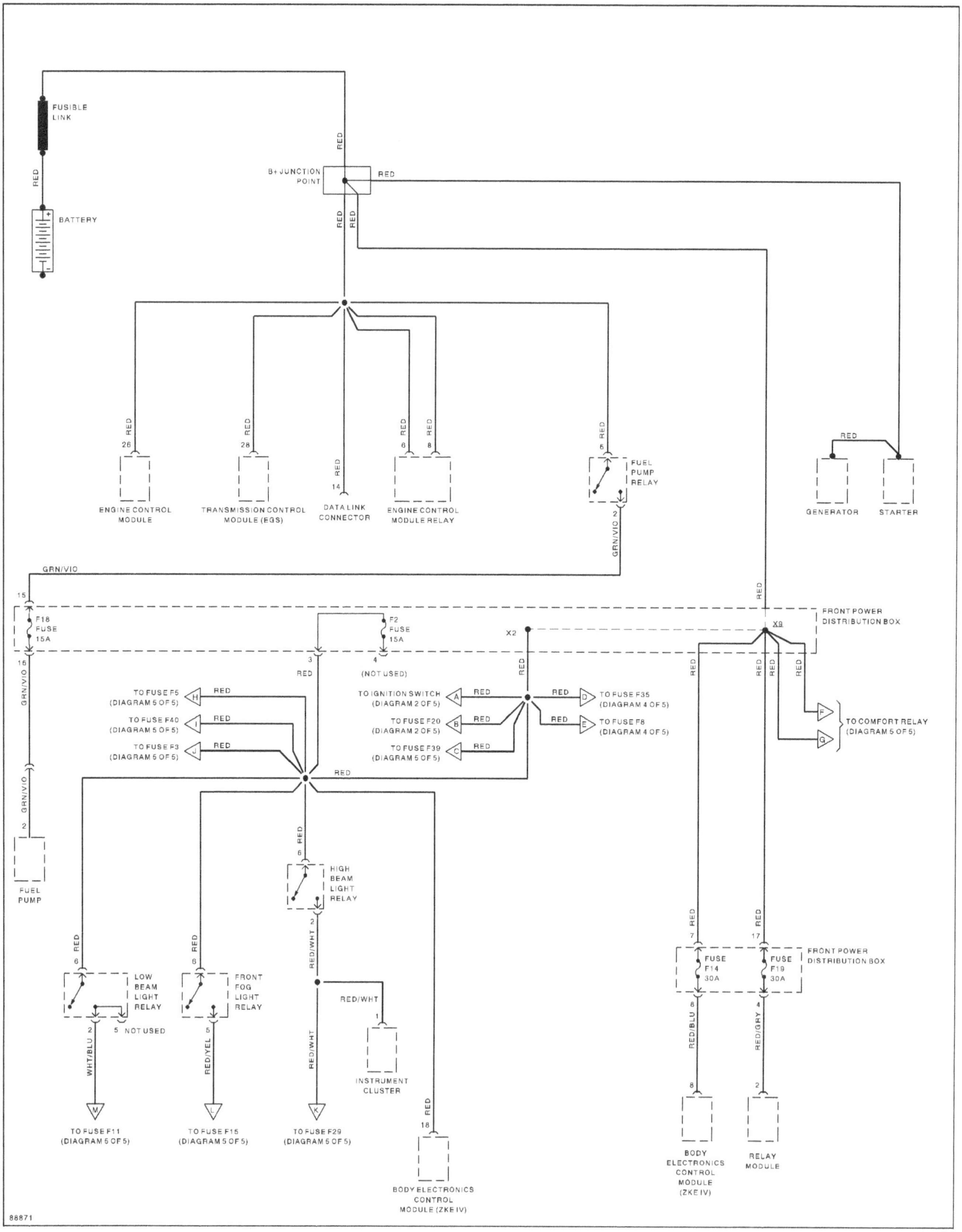

Power Distribution

1995-96 318ti (2 of 5)

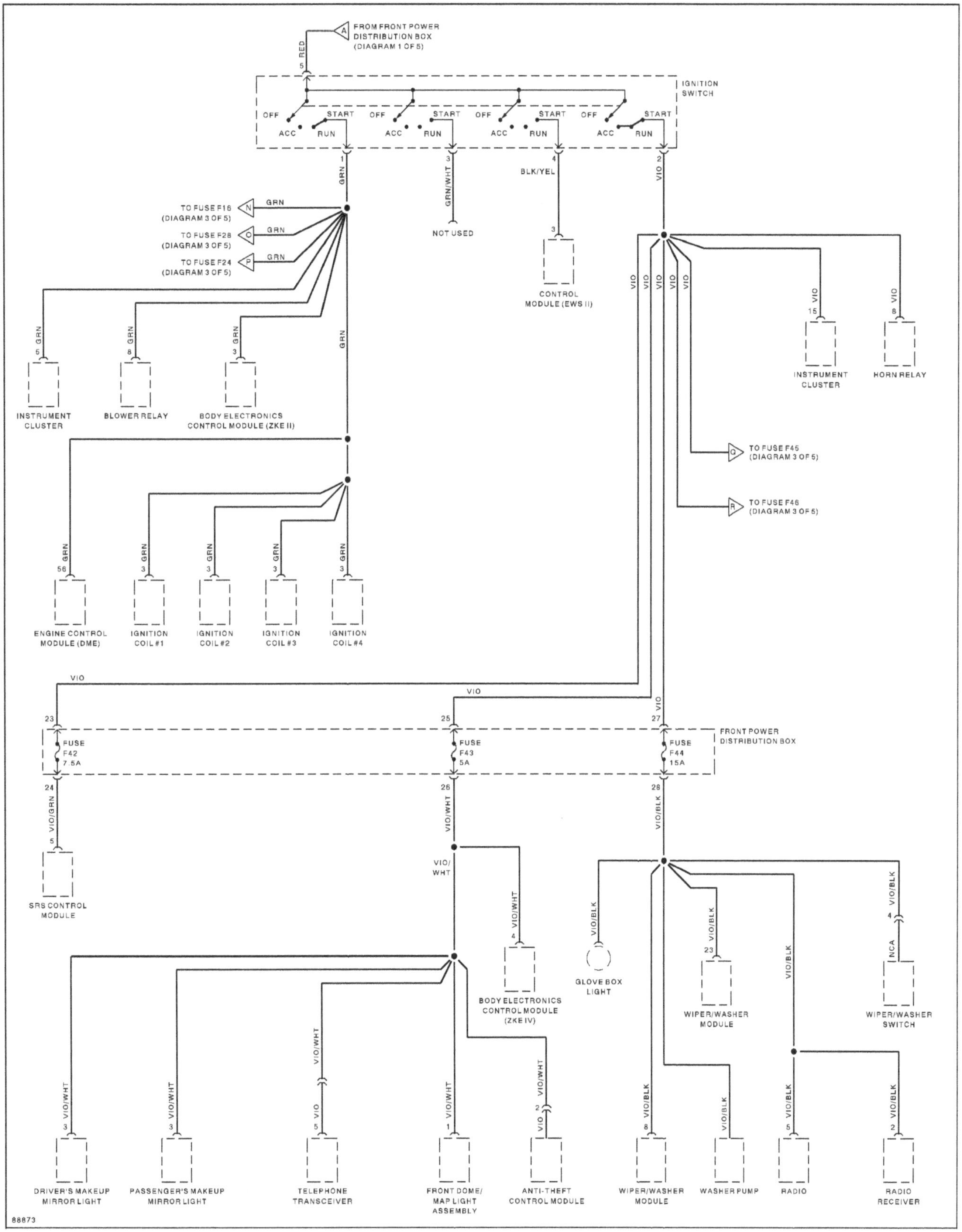

Power Distribution

1995-96 318ti (3 of 5)

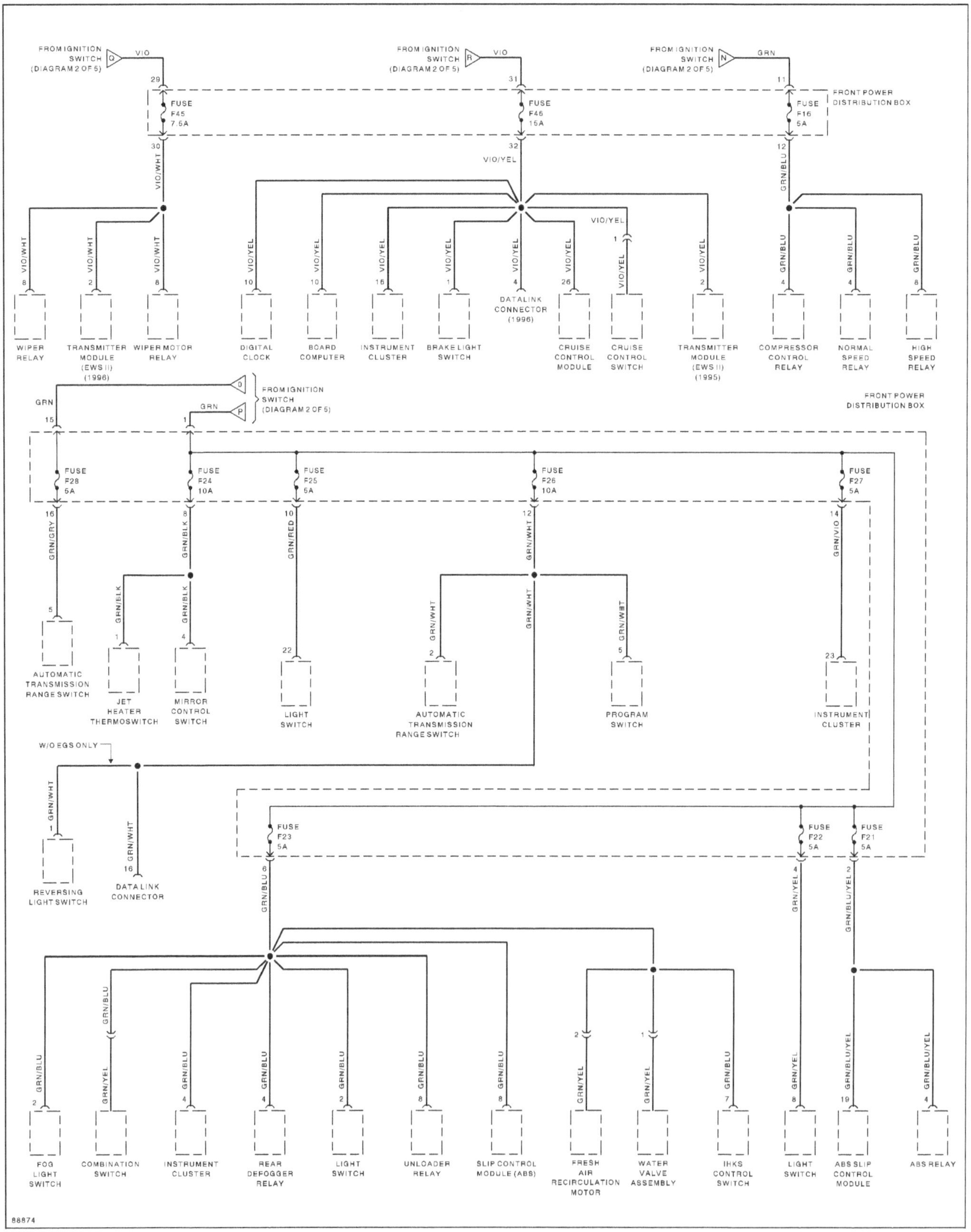

Power Distribution

1995-96 318ti (4 of 5)

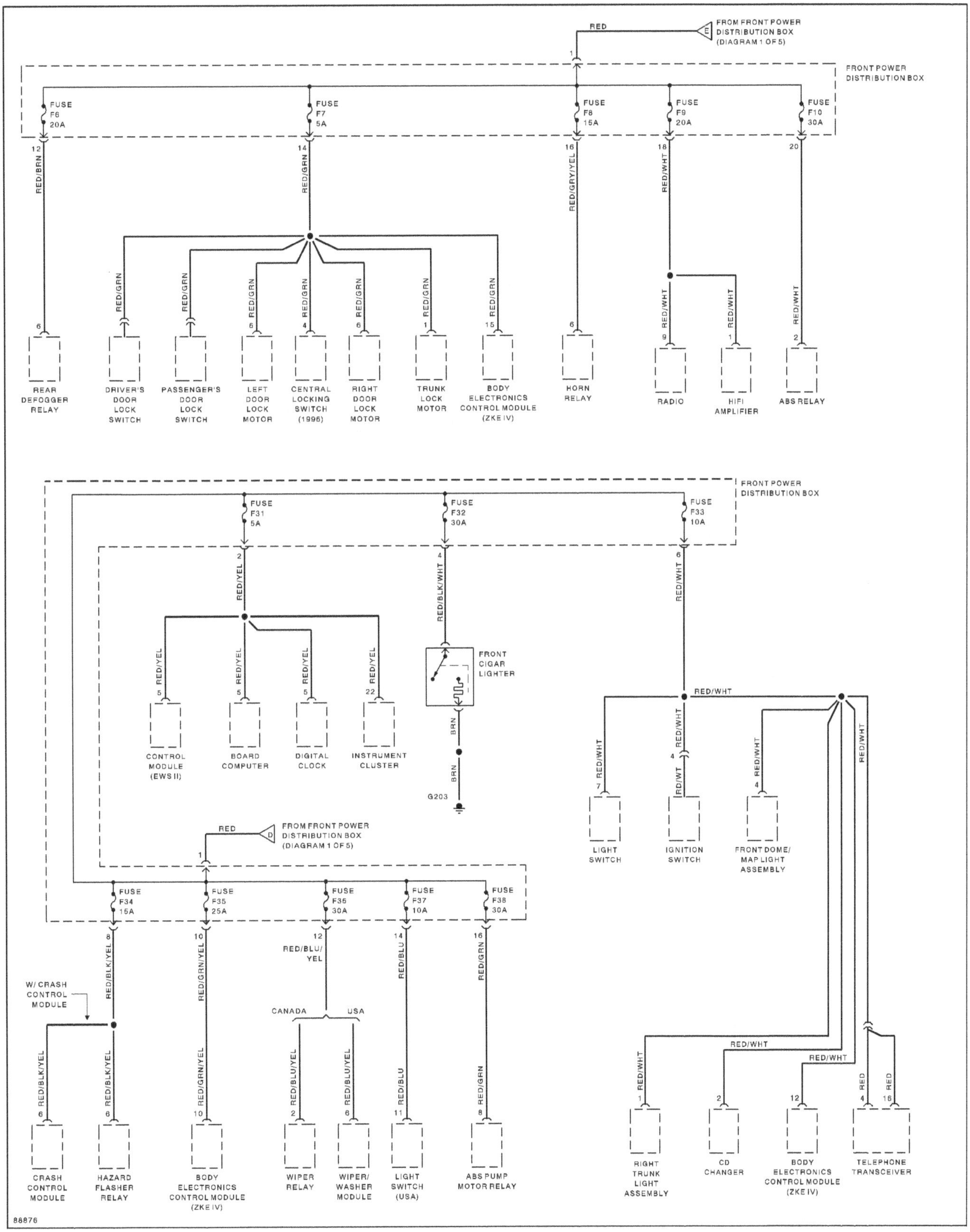

Power Distribution

1995-96 318ti (5 of 5)

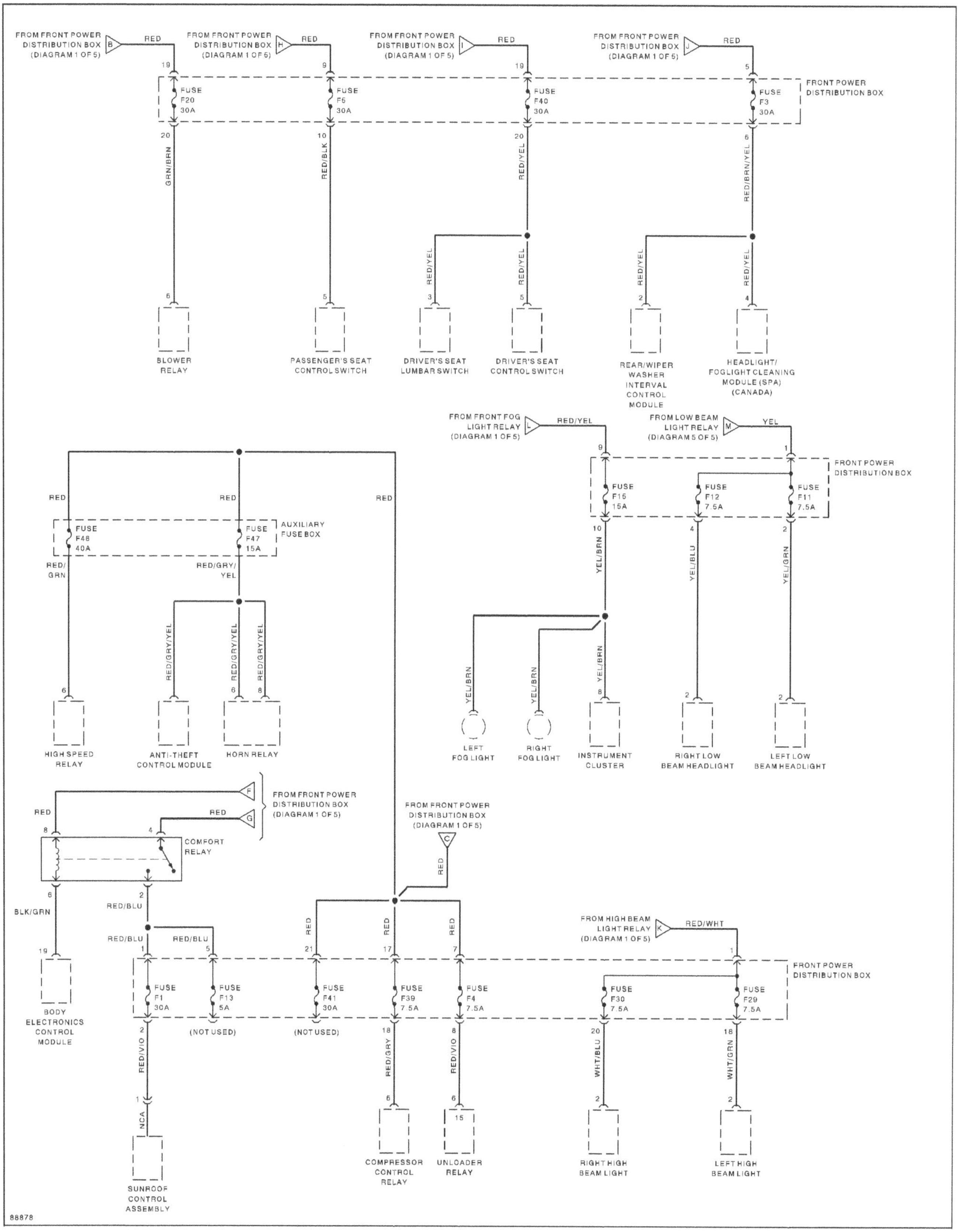

Power Distribution

1997 318ti (1 of 3)

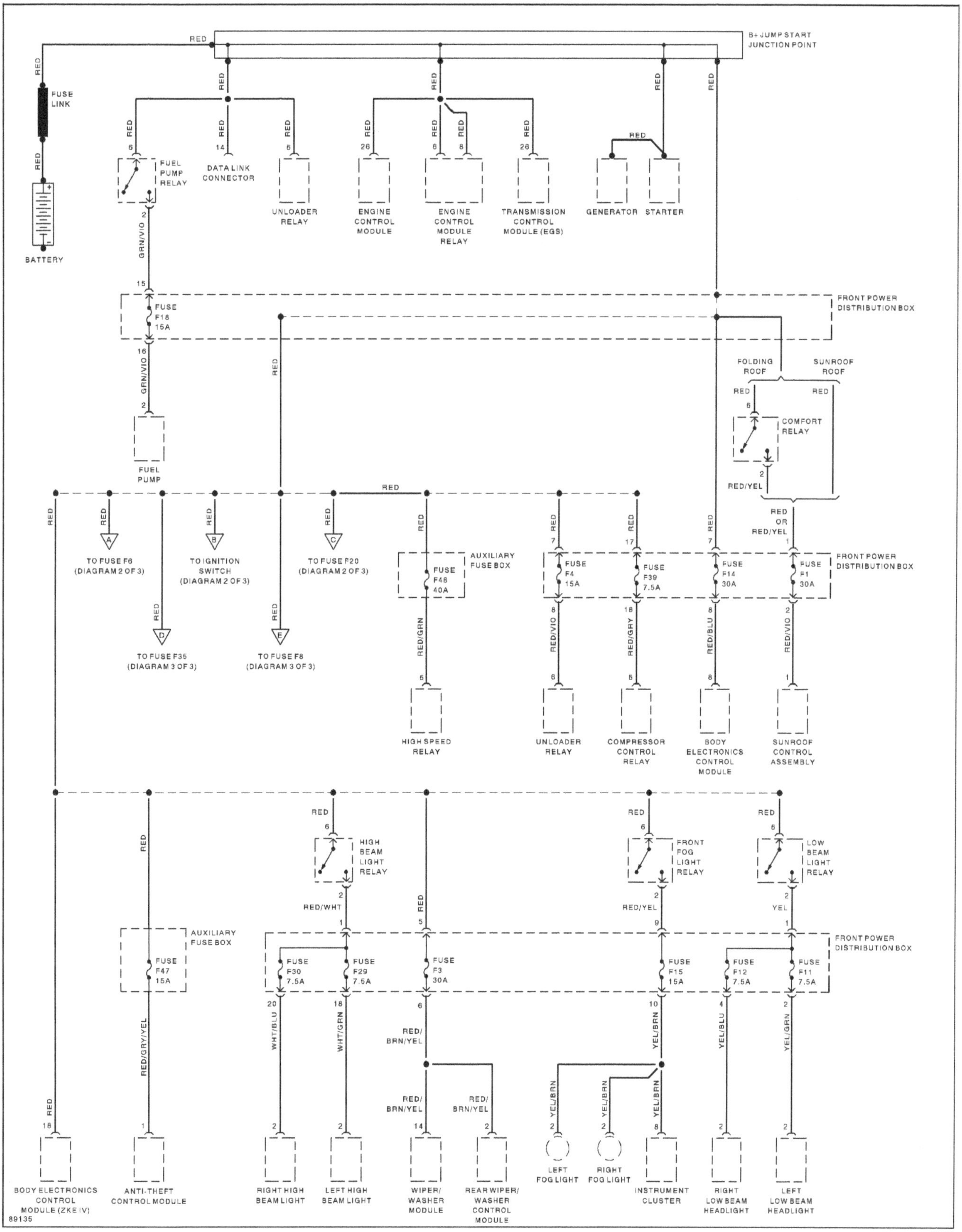

Power Distribution

1997 318ti (2 of 3)

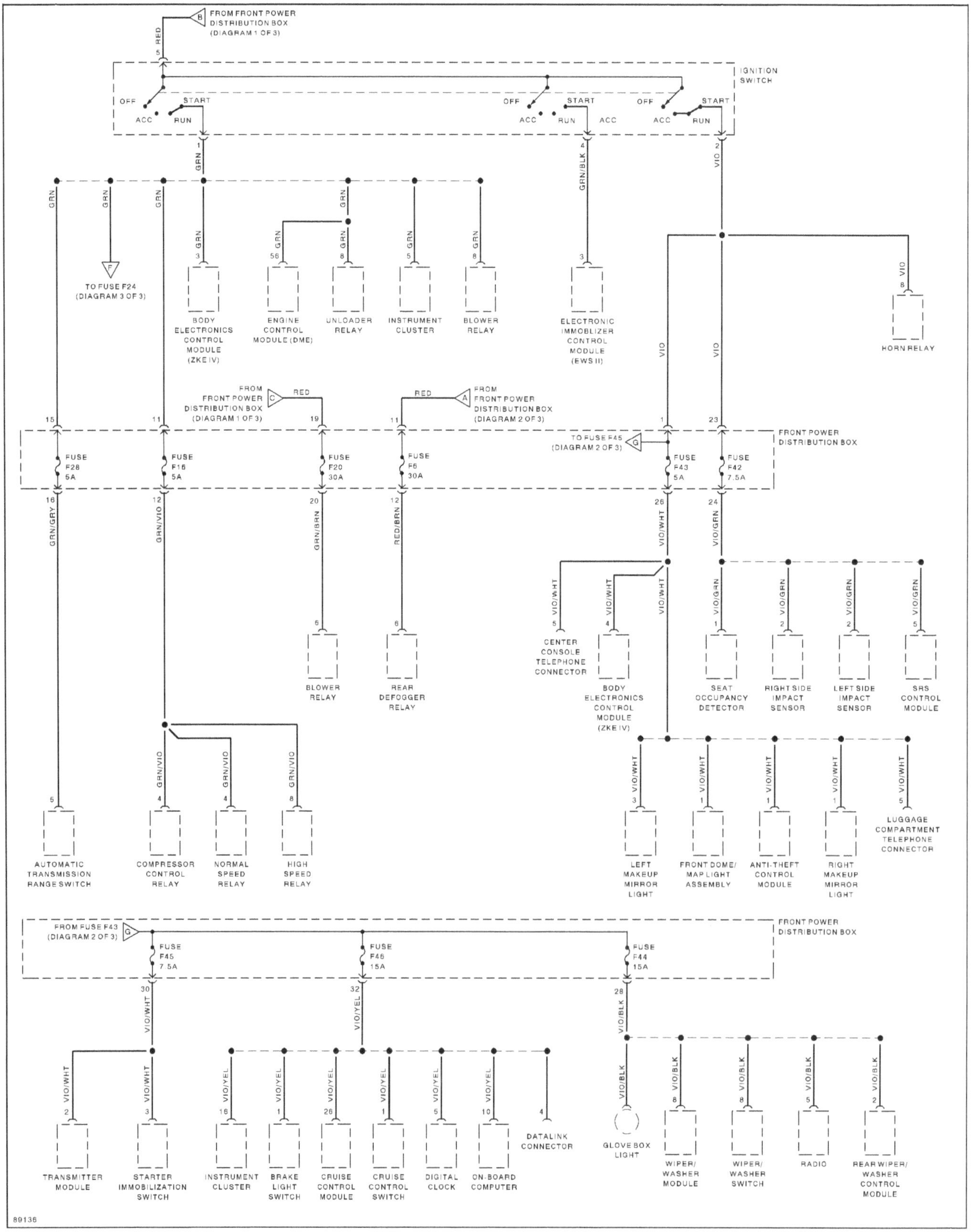

Power Distribution

1997 318ti (3 of 3)

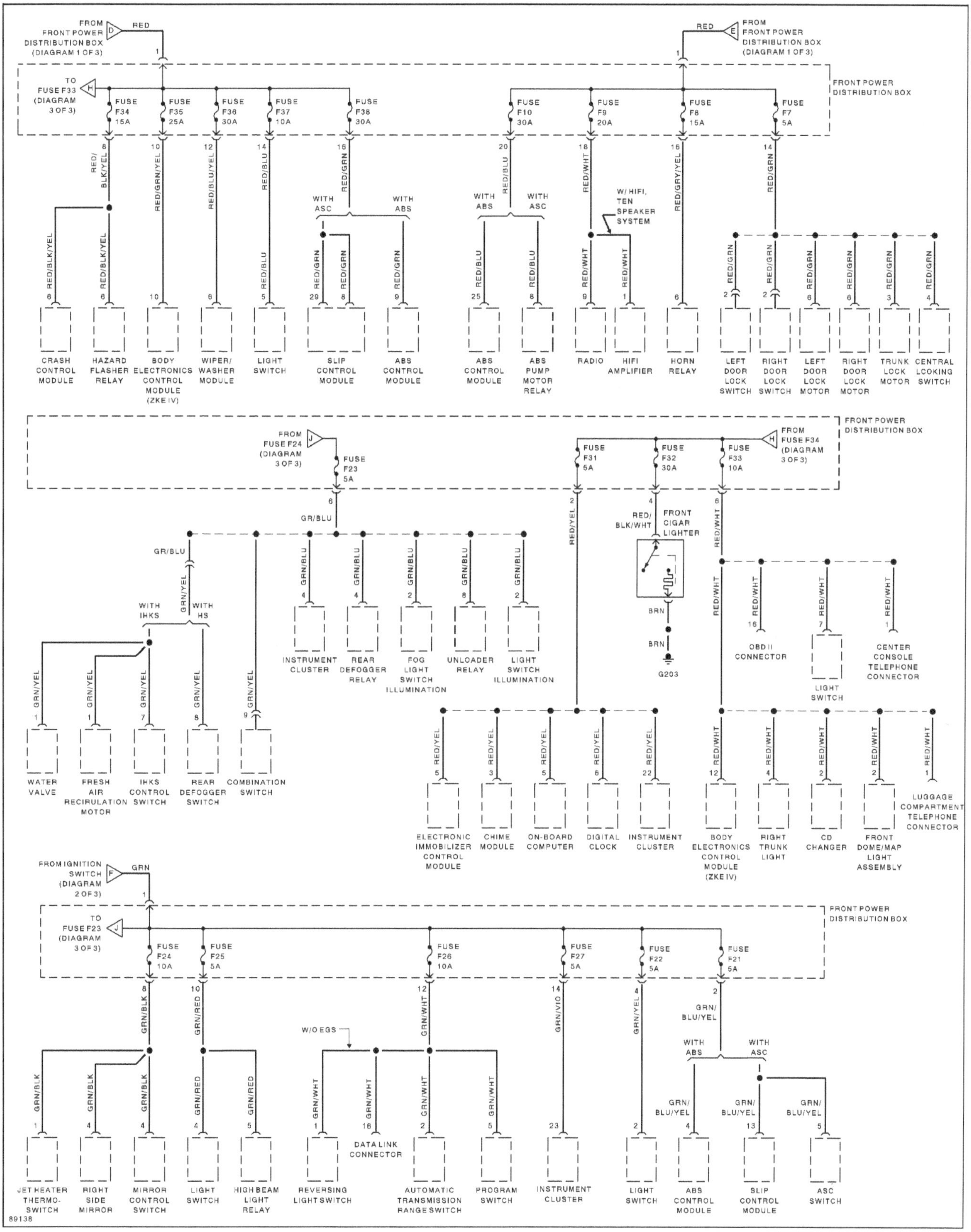

Power Distribution

1998 318ti (1 of 3)

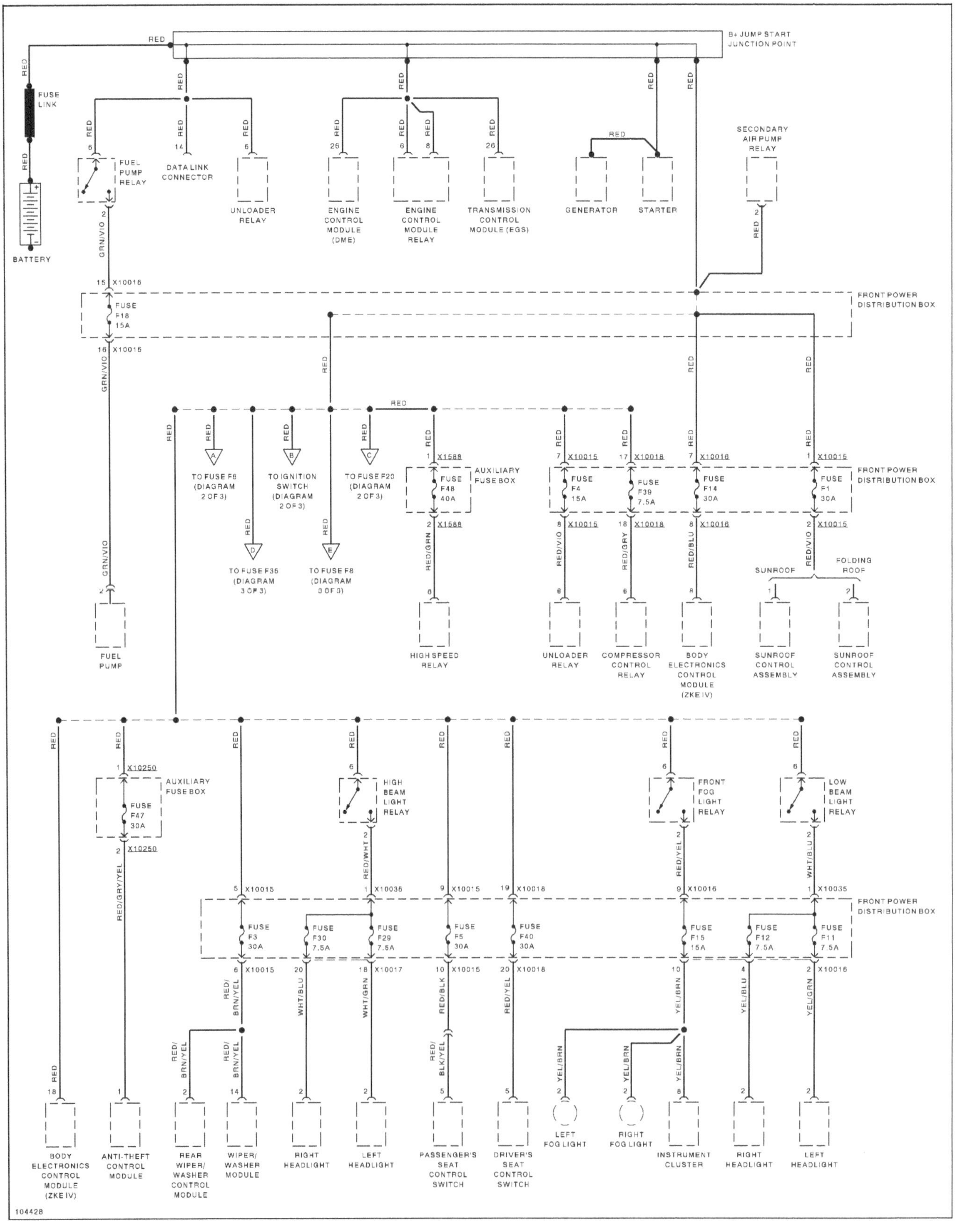

Power Distribution

1998 318ti (2 of 3)

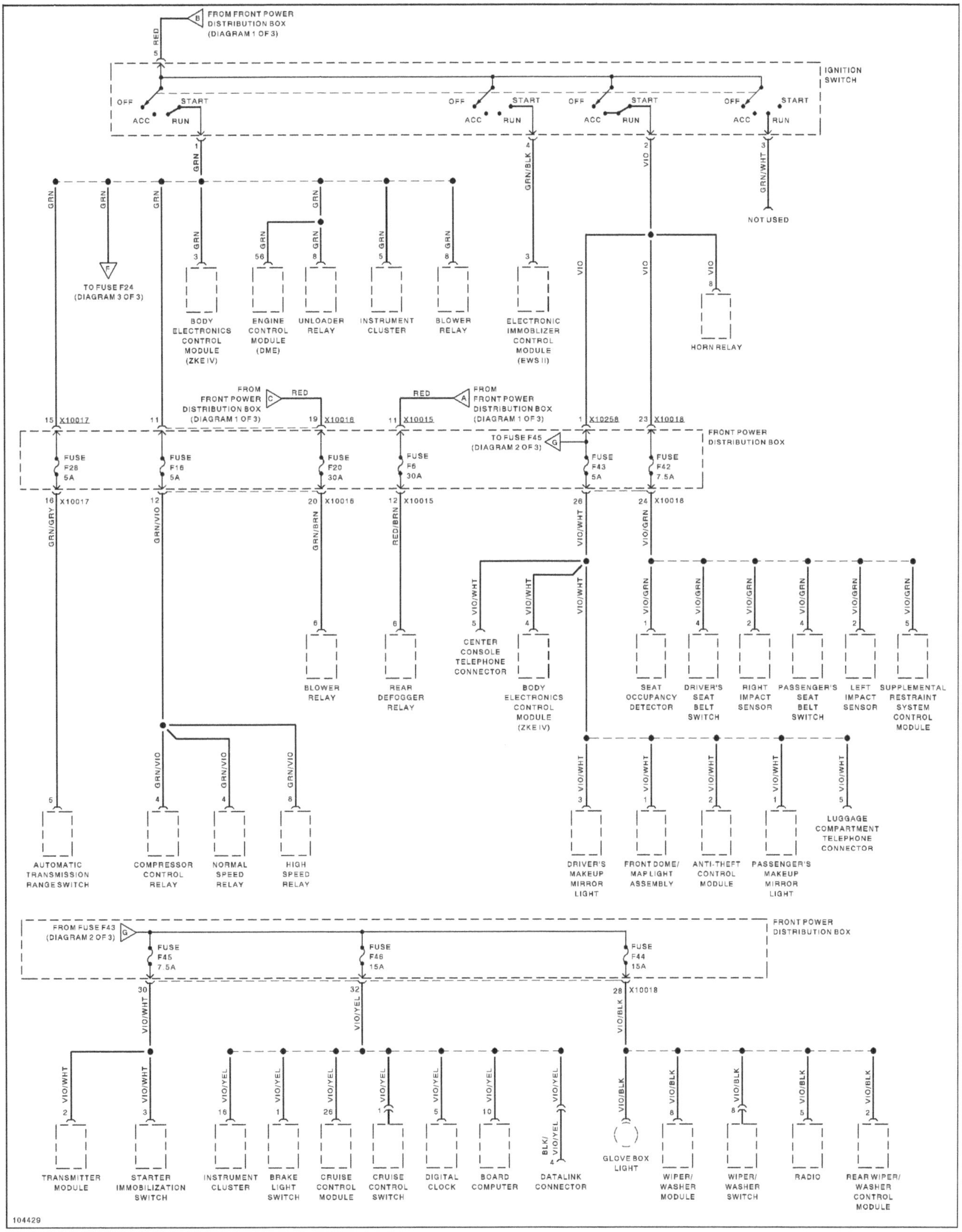

Power Distribution

1998 318ti (3 of 3)

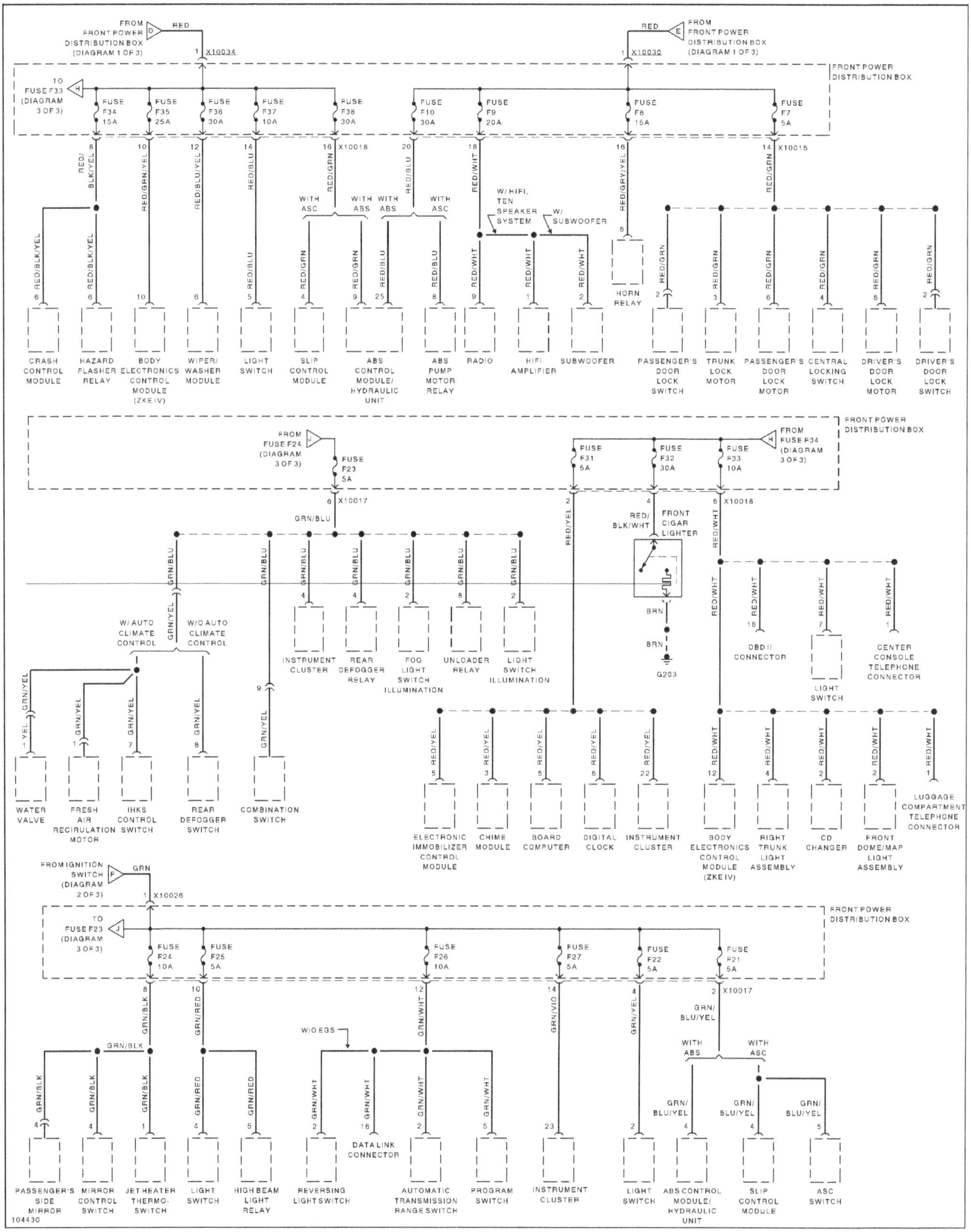

Ground Distribution

1992-93 (1 of 5)

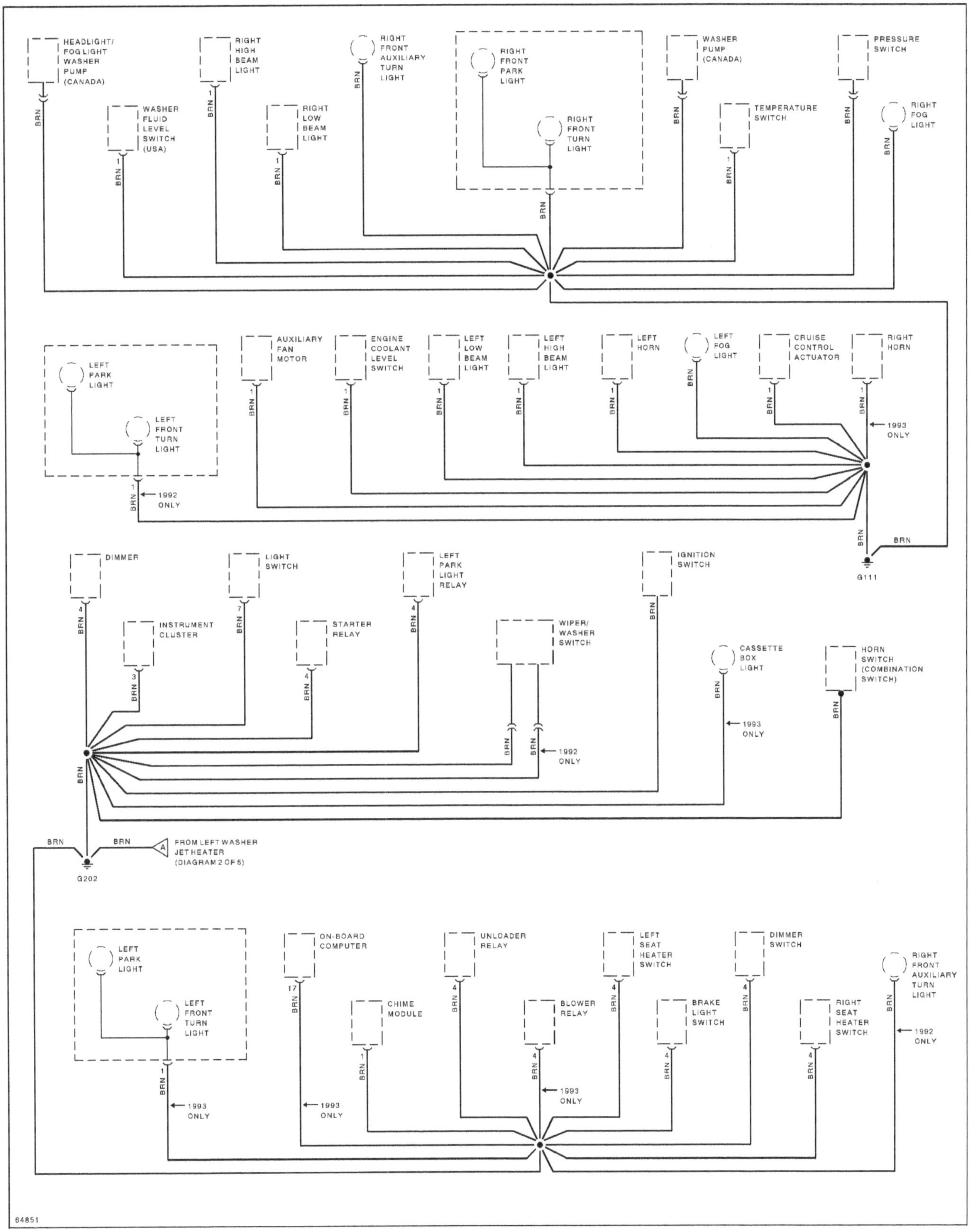

Ground Distribution

1992-93 (2 of 5)

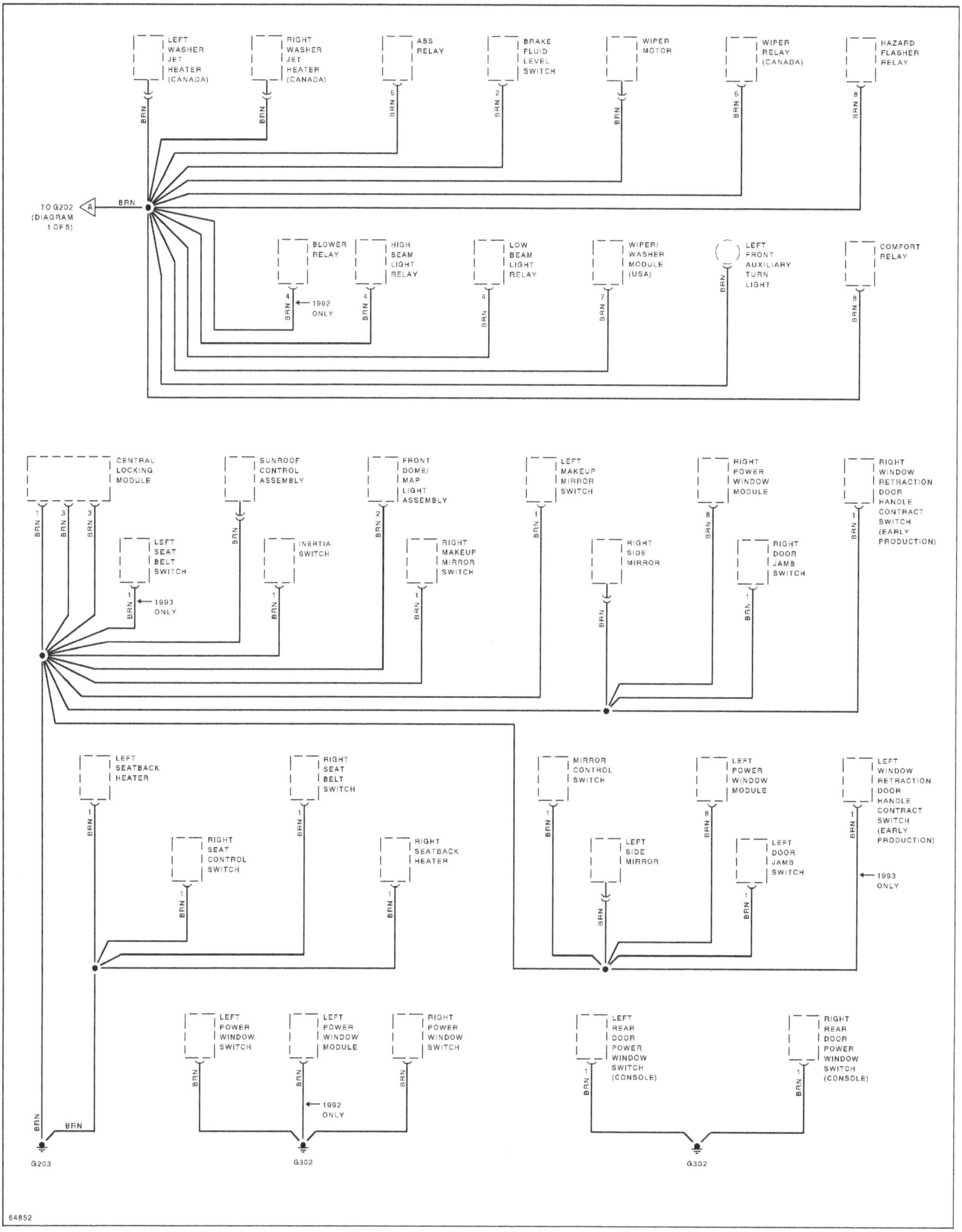

Ground Distribution

1992-93 (3 of 5)

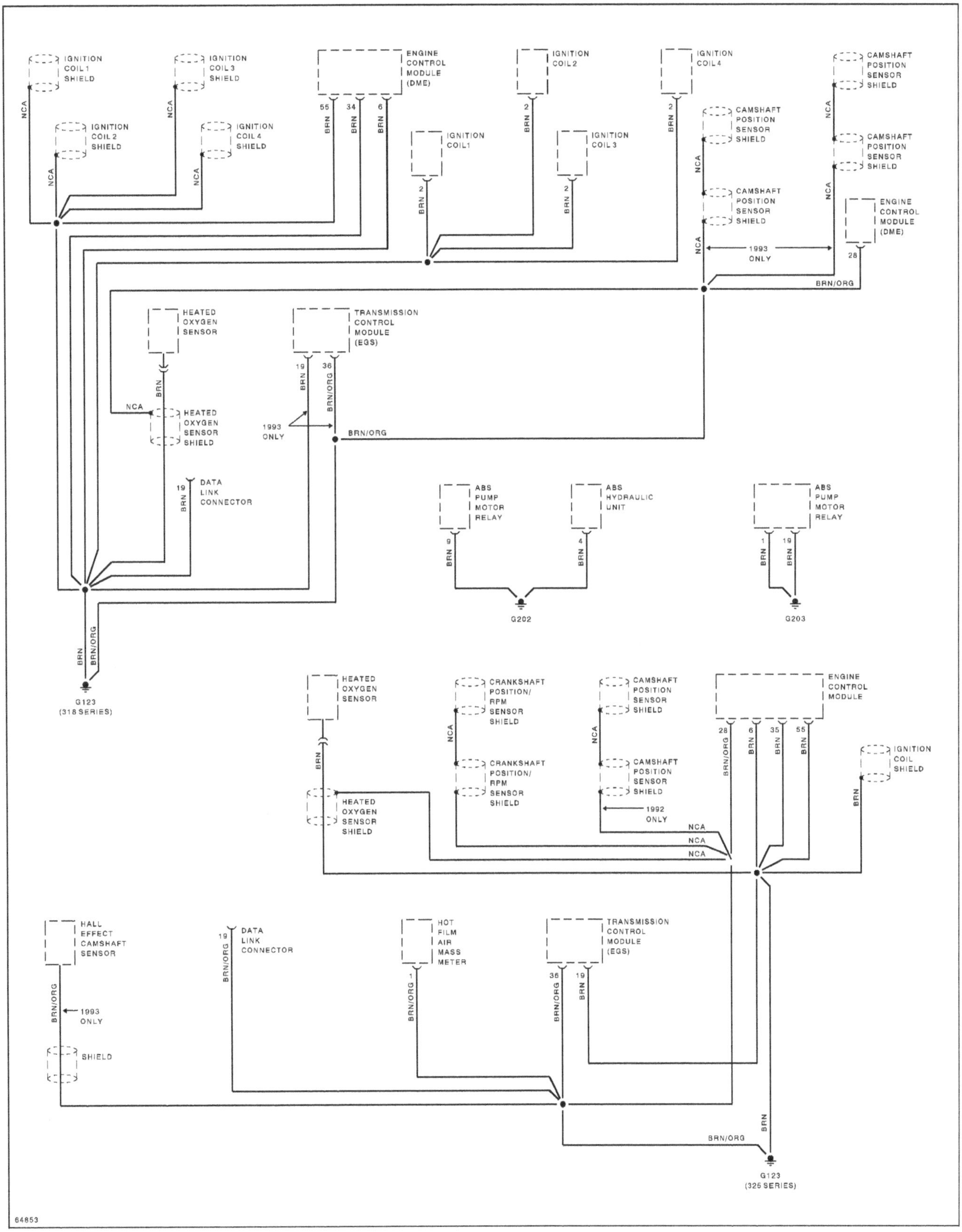

Ground Distribution

1992-93 (4 of 5)

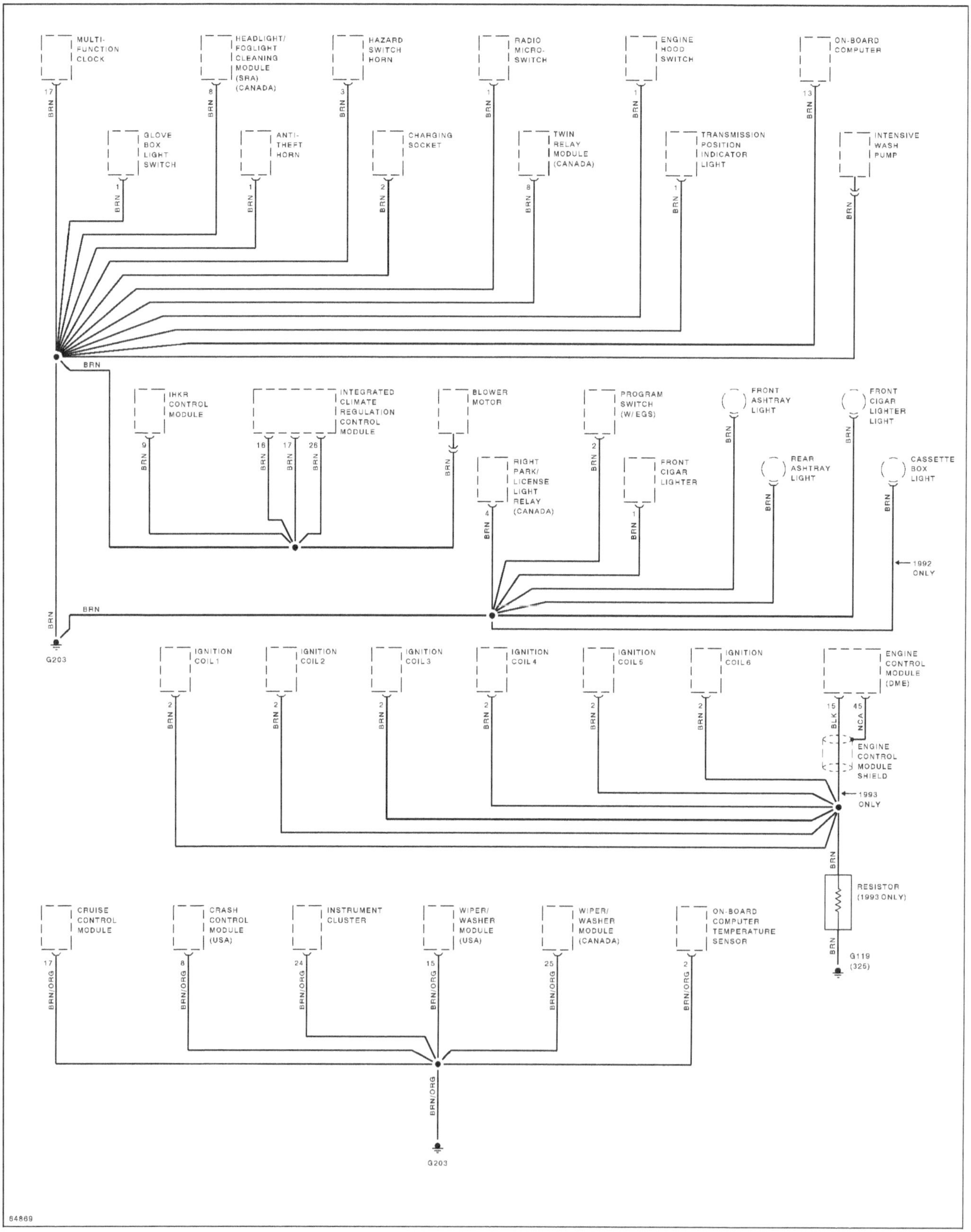

Ground Distribution
1992-93 (5 of 5)

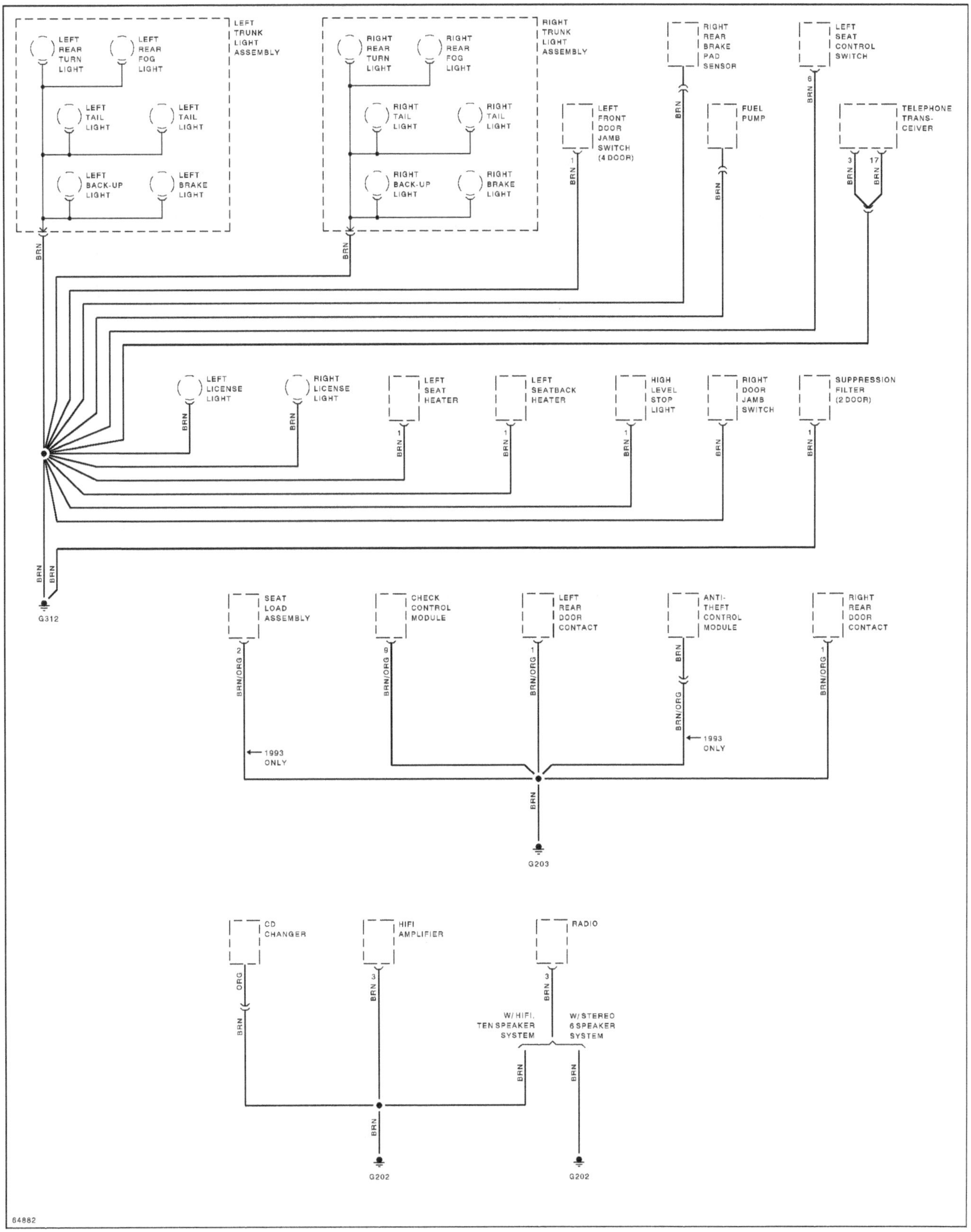

Ground Distribution

1994-95 Except 318ti (1 of 5)

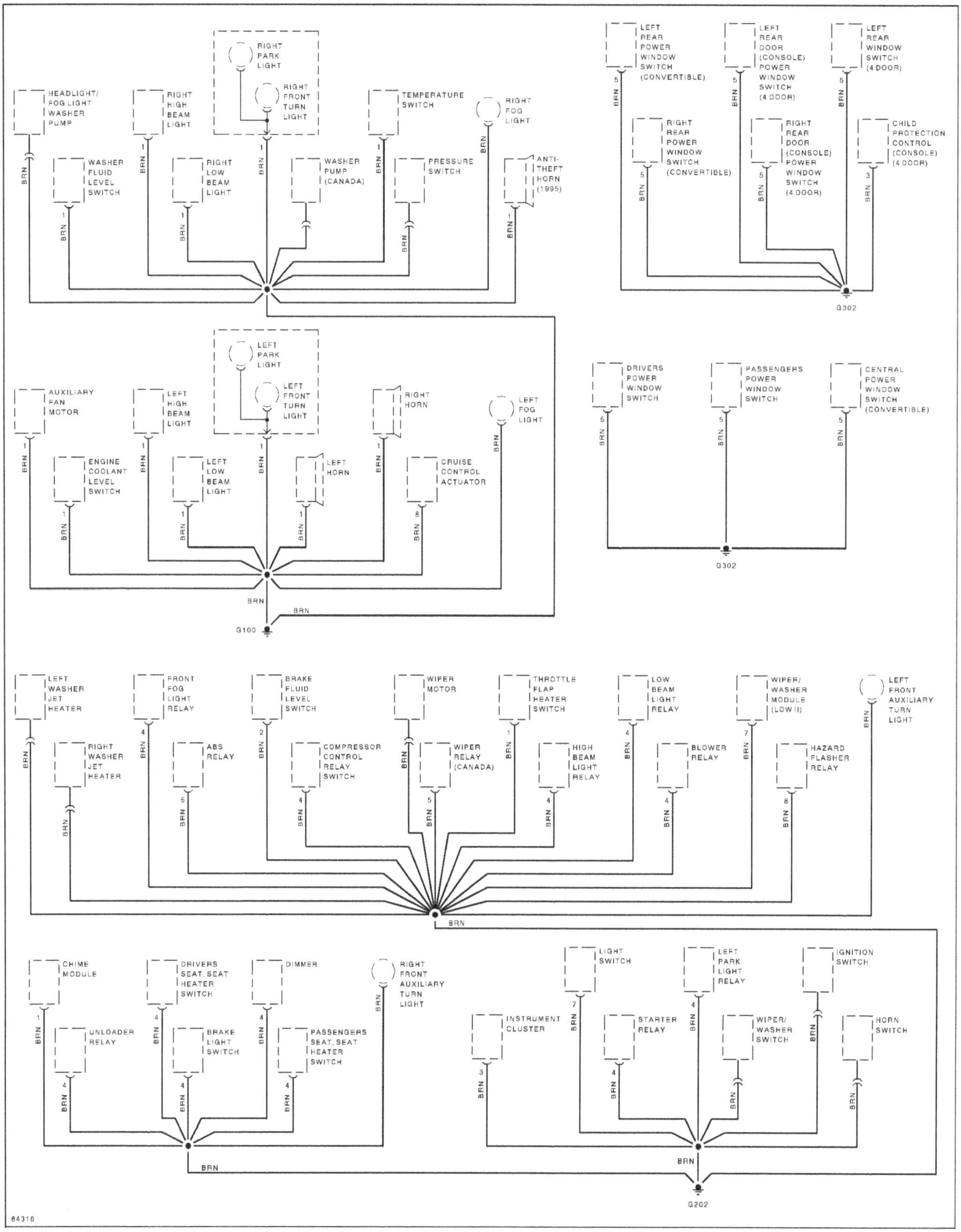

Ground Distribution

1994-95 Except 318ti (2 of 5)

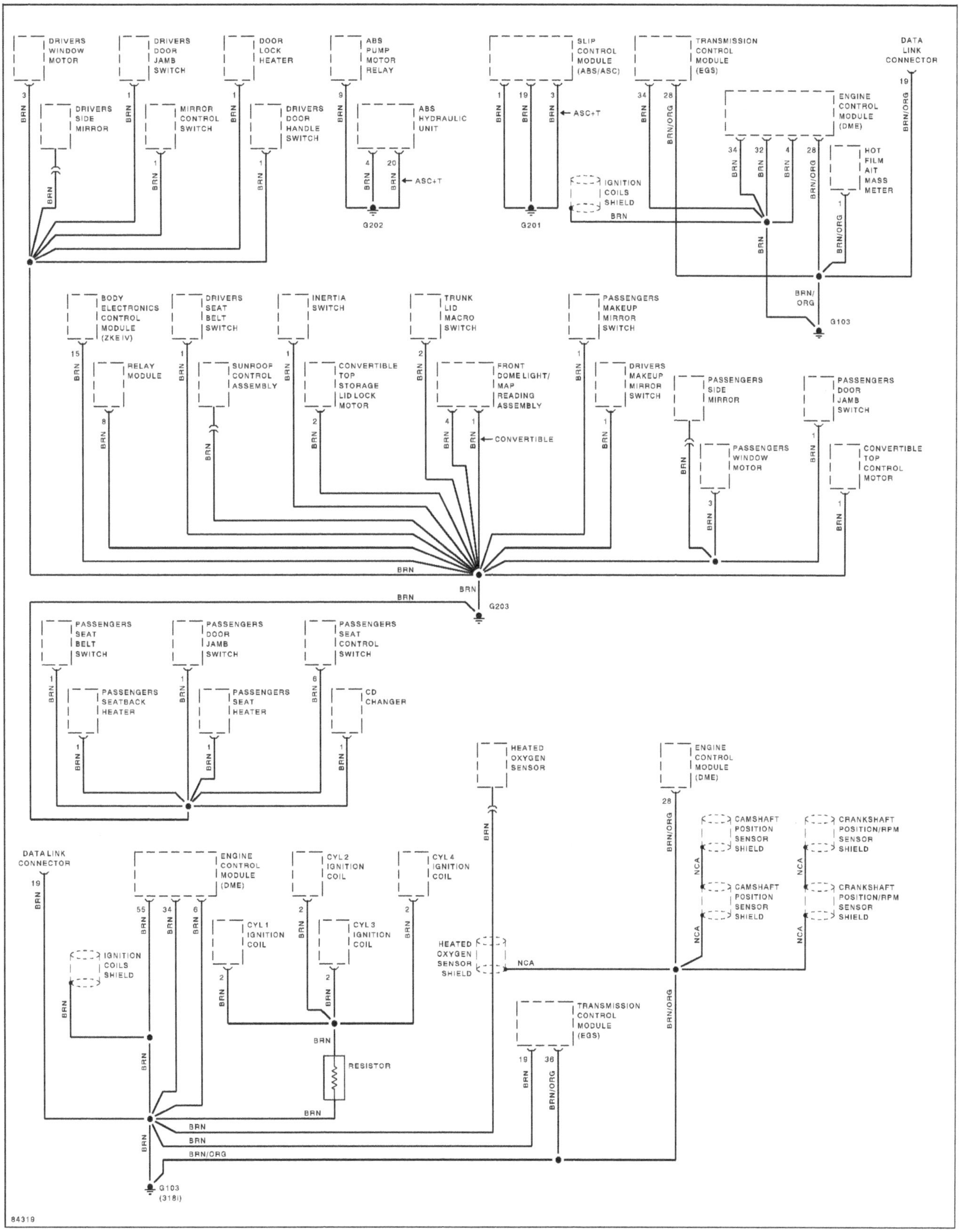

Ground Distribution

1994-95 Except 318ti (3 of 5)

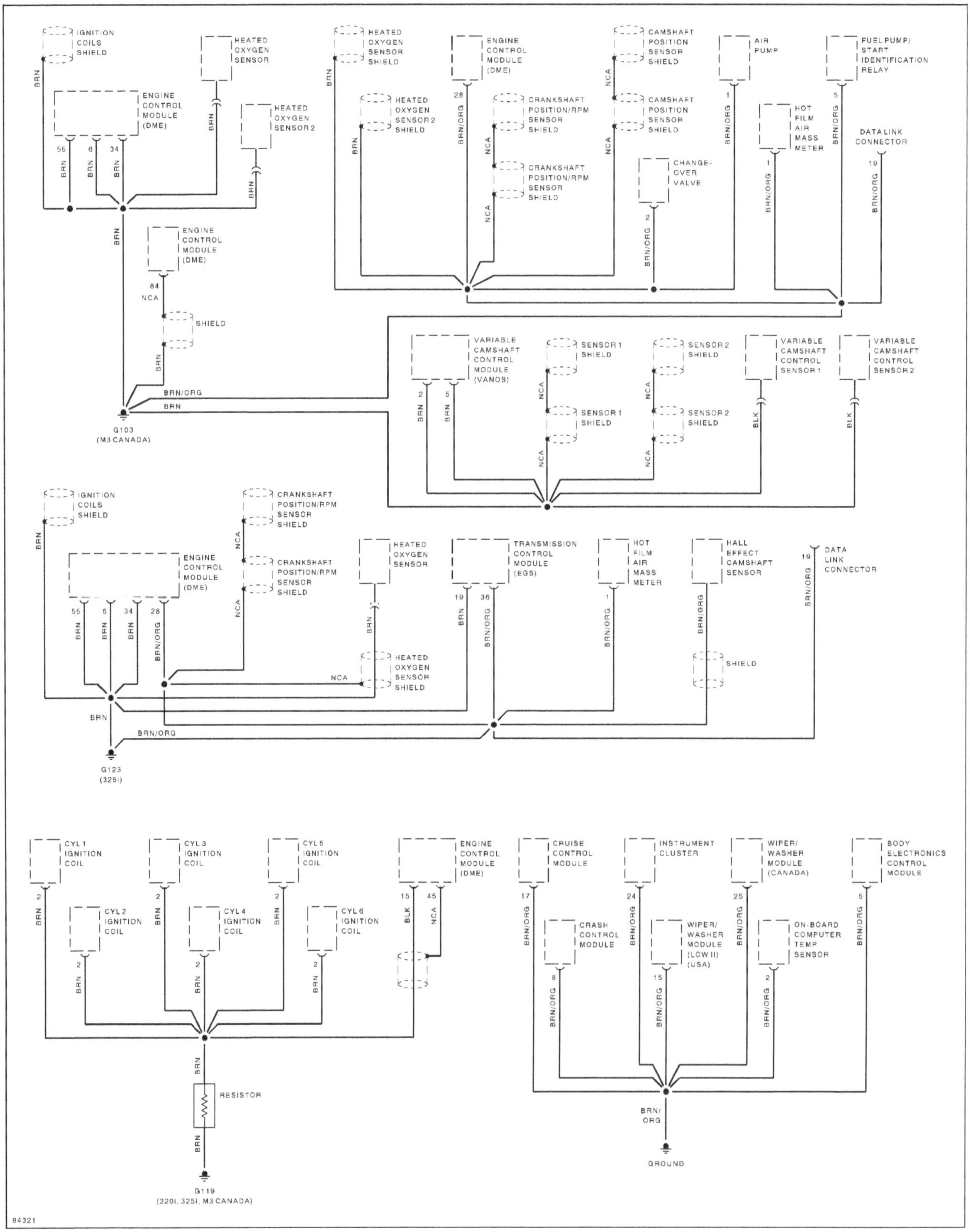

Ground Distribution

1994-95 Except 318ti (4 of 5)

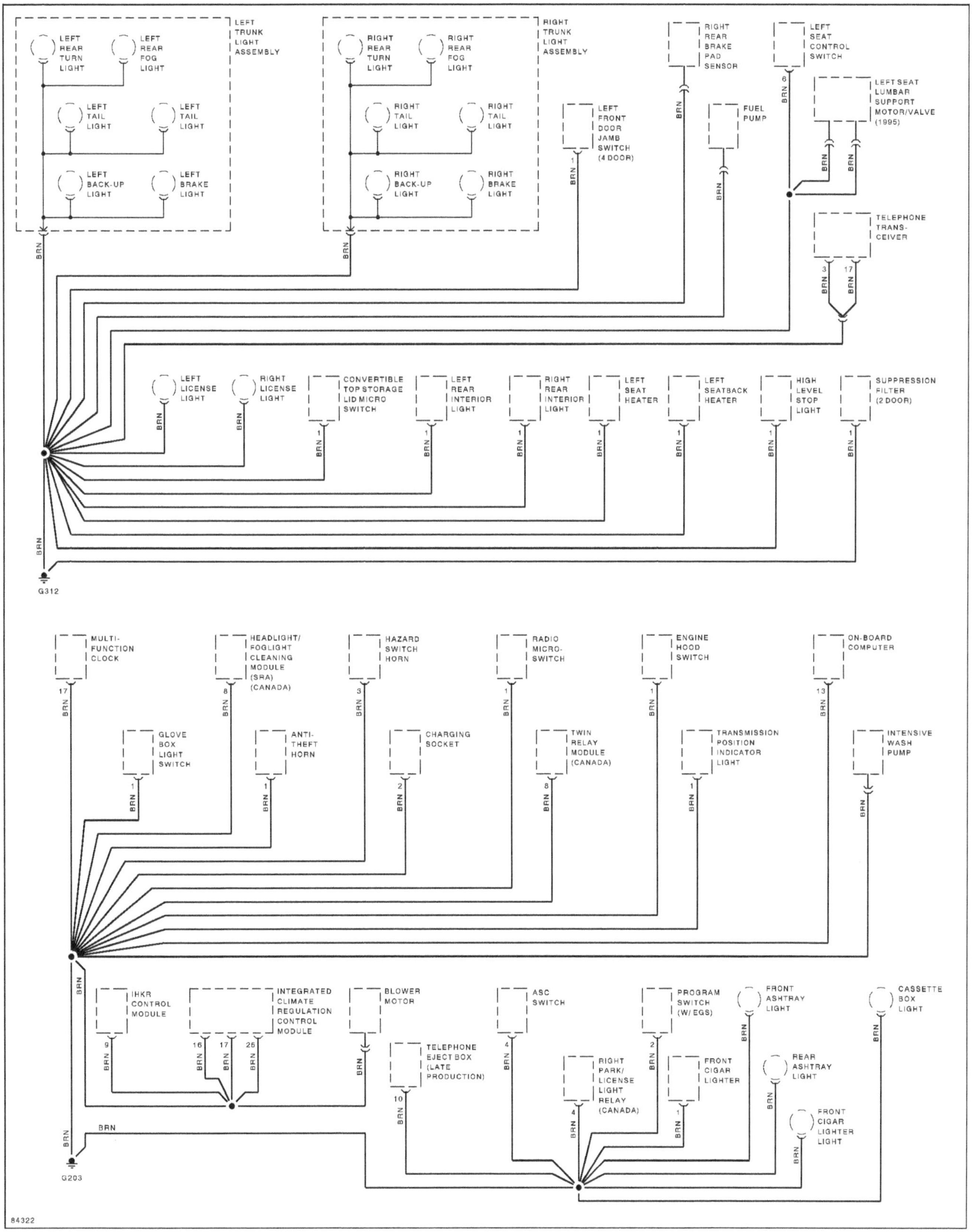

Ground Distribution

1994-95 Except 318ti (5 of 5)

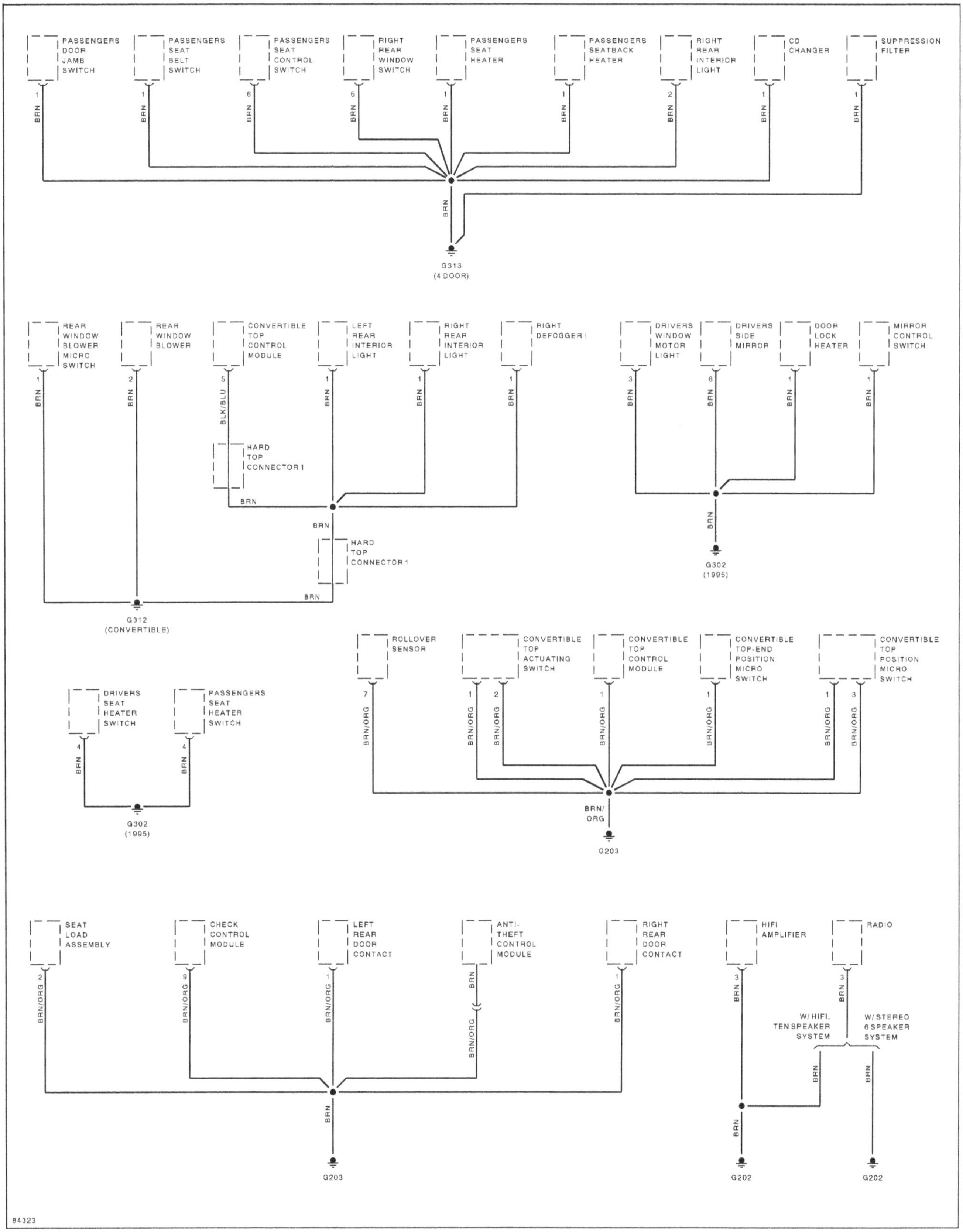

Ground Distribution

1996-97 Except 318ti (1 of 5)

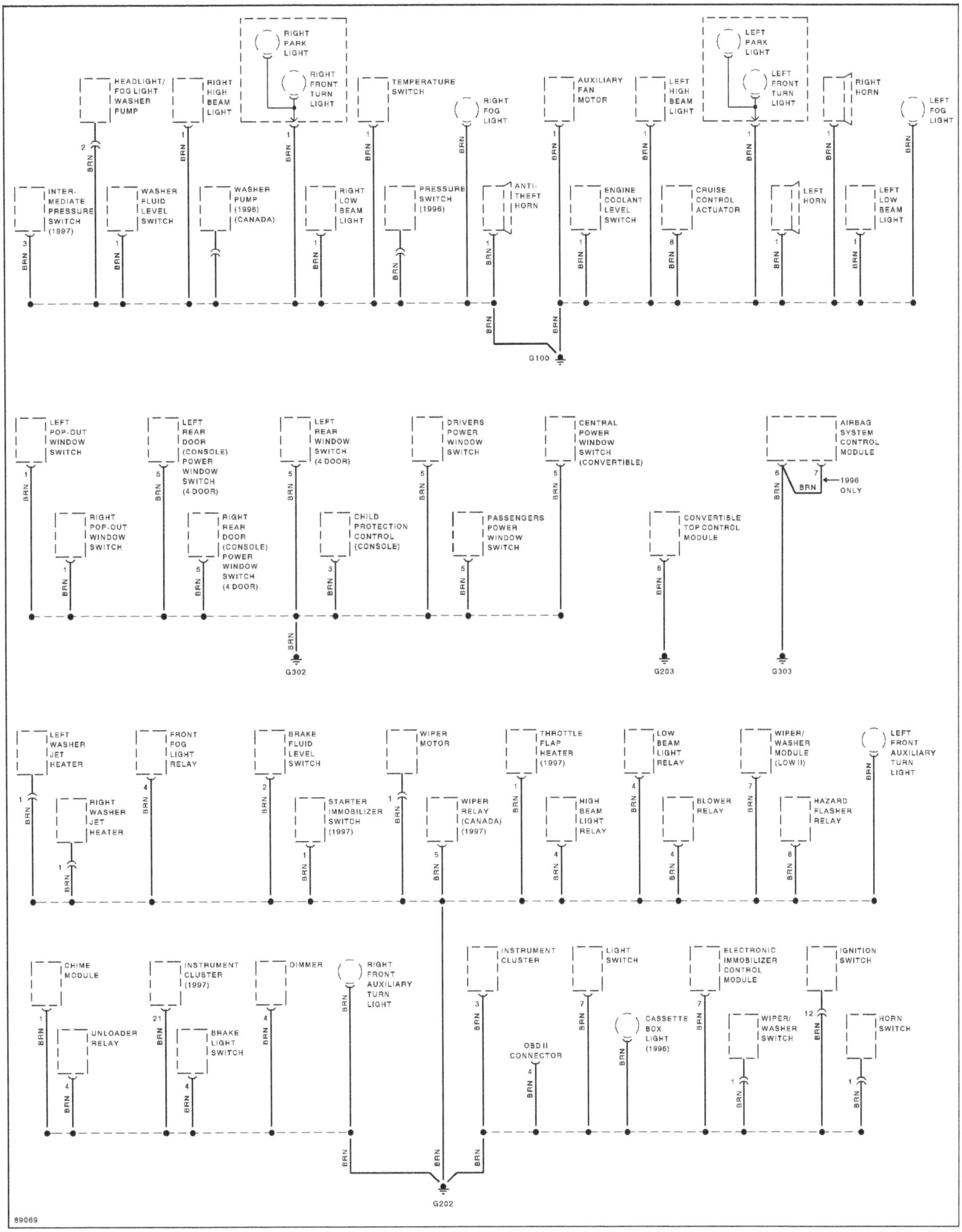

Ground Distribution

1996-97 Except 318ti (2 of 5)

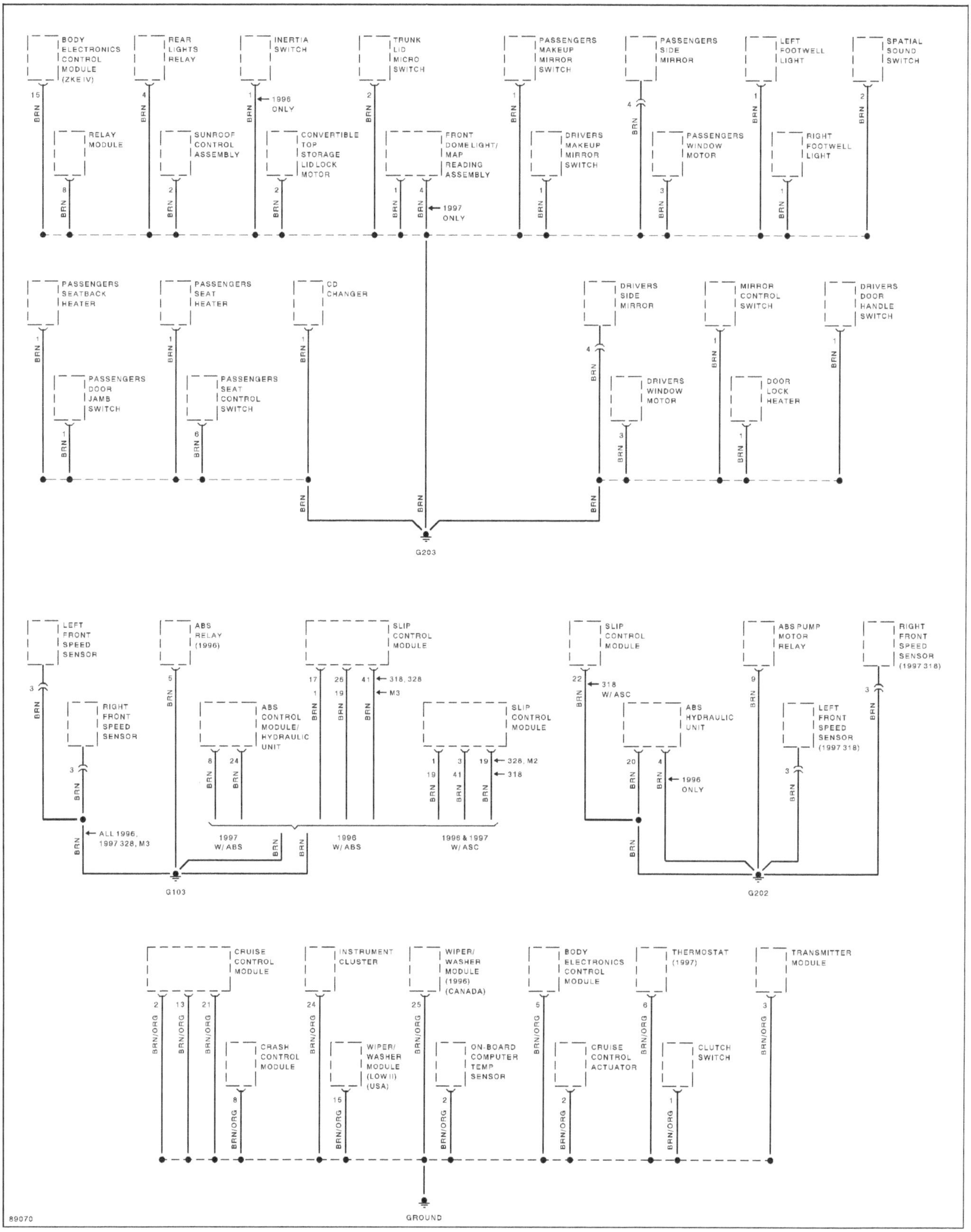

Ground Distribution

1996-97 Except 318ti (3 of 5)

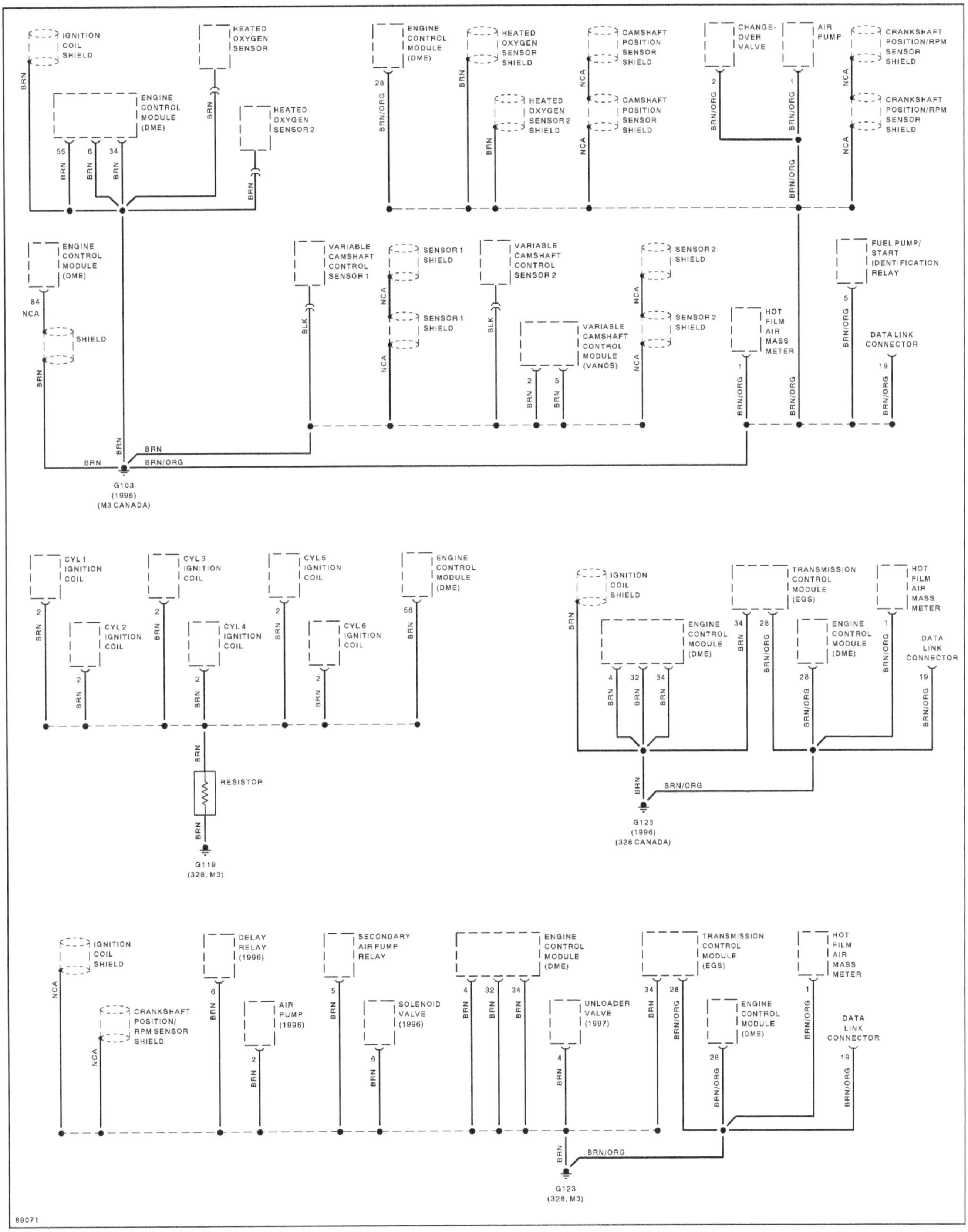

Ground Distribution

1996-97 Except 318ti (4 of 5)

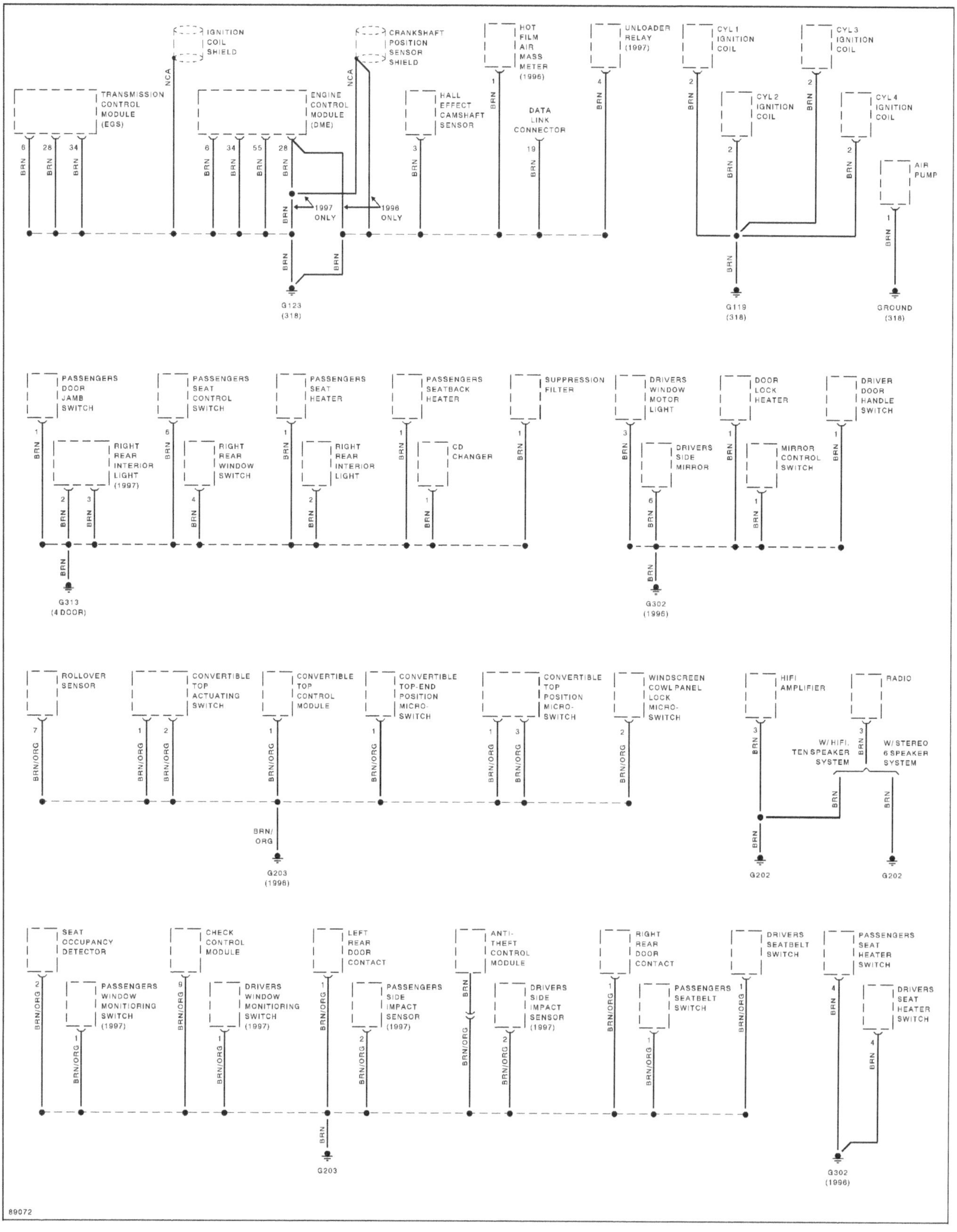

Ground Distribution

1996-97 Except 318ti (5 of 5)

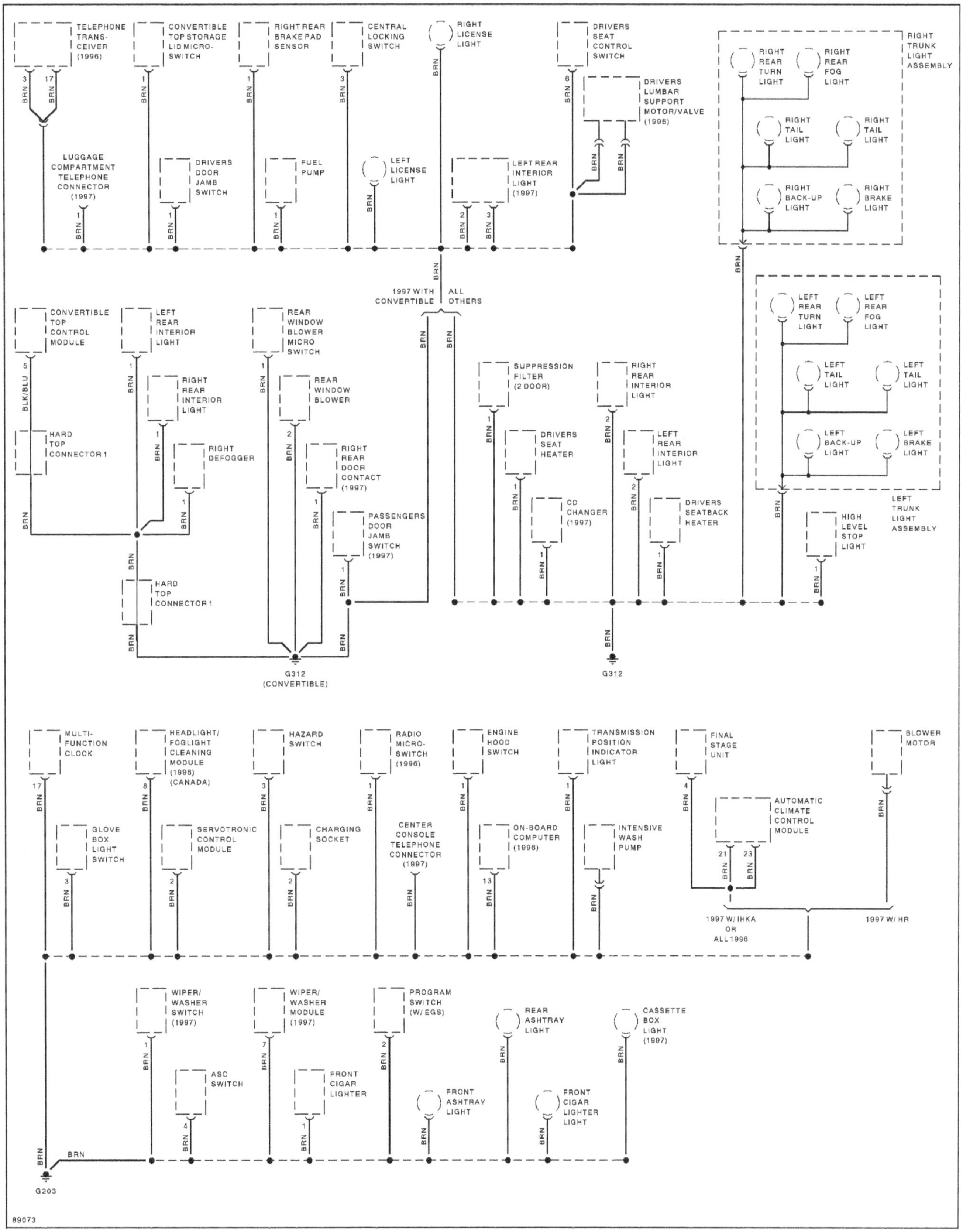

Ground Distribution

1998 Except 318ti (1 of 4)

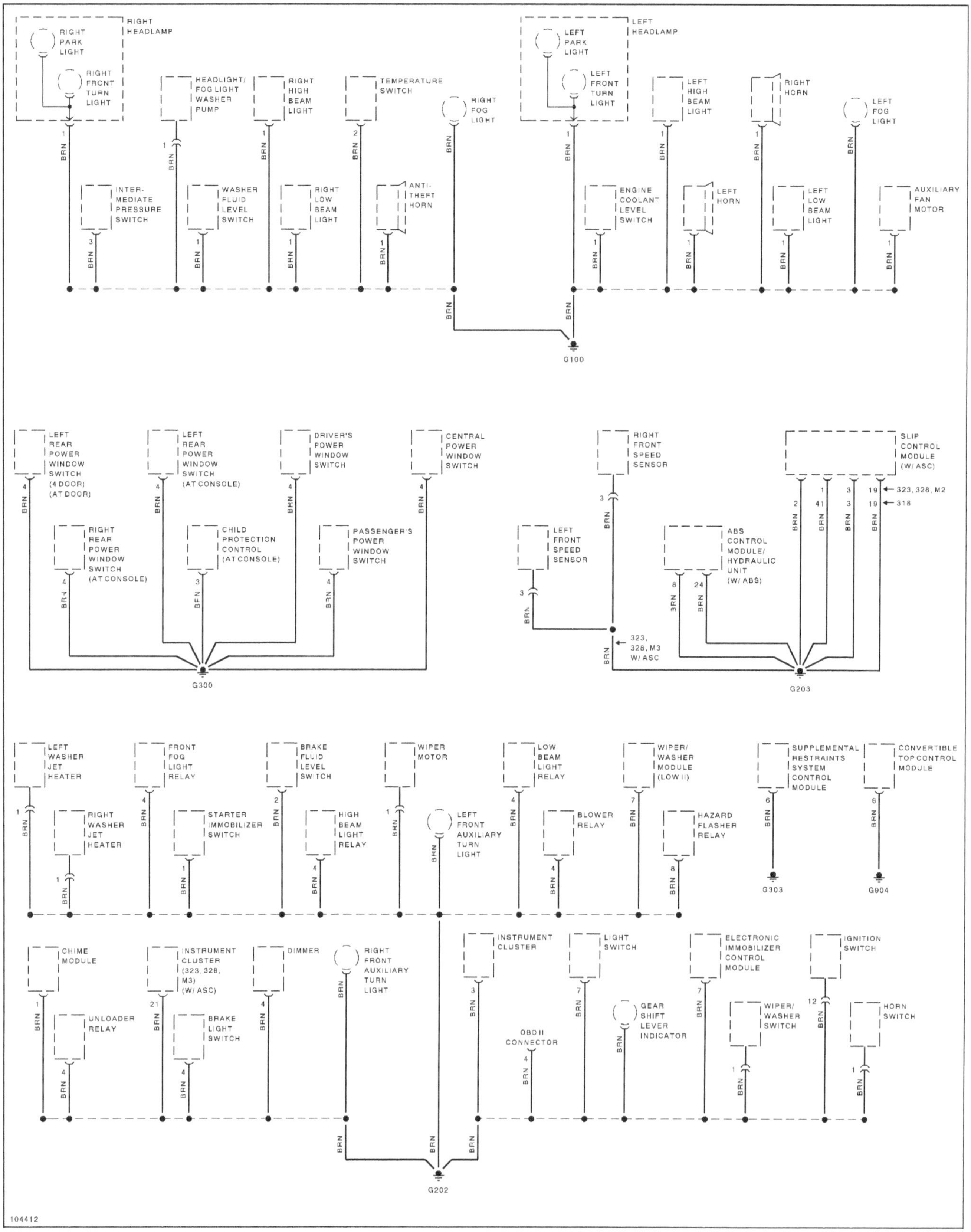

Ground Distribution

1998 Except 318ti (2 of 4)

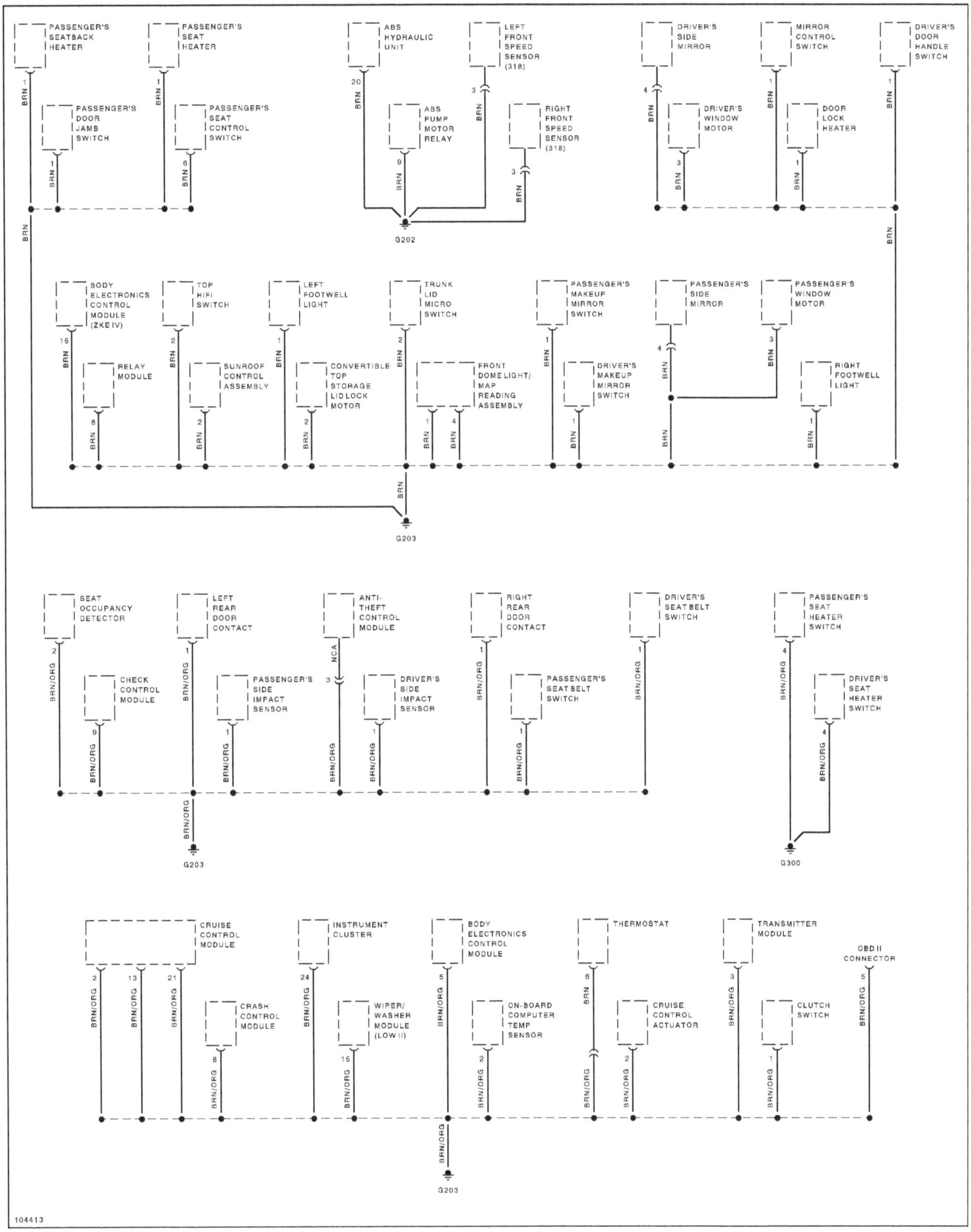

Ground Distribution

1998 Except 318ti (3 of 4)

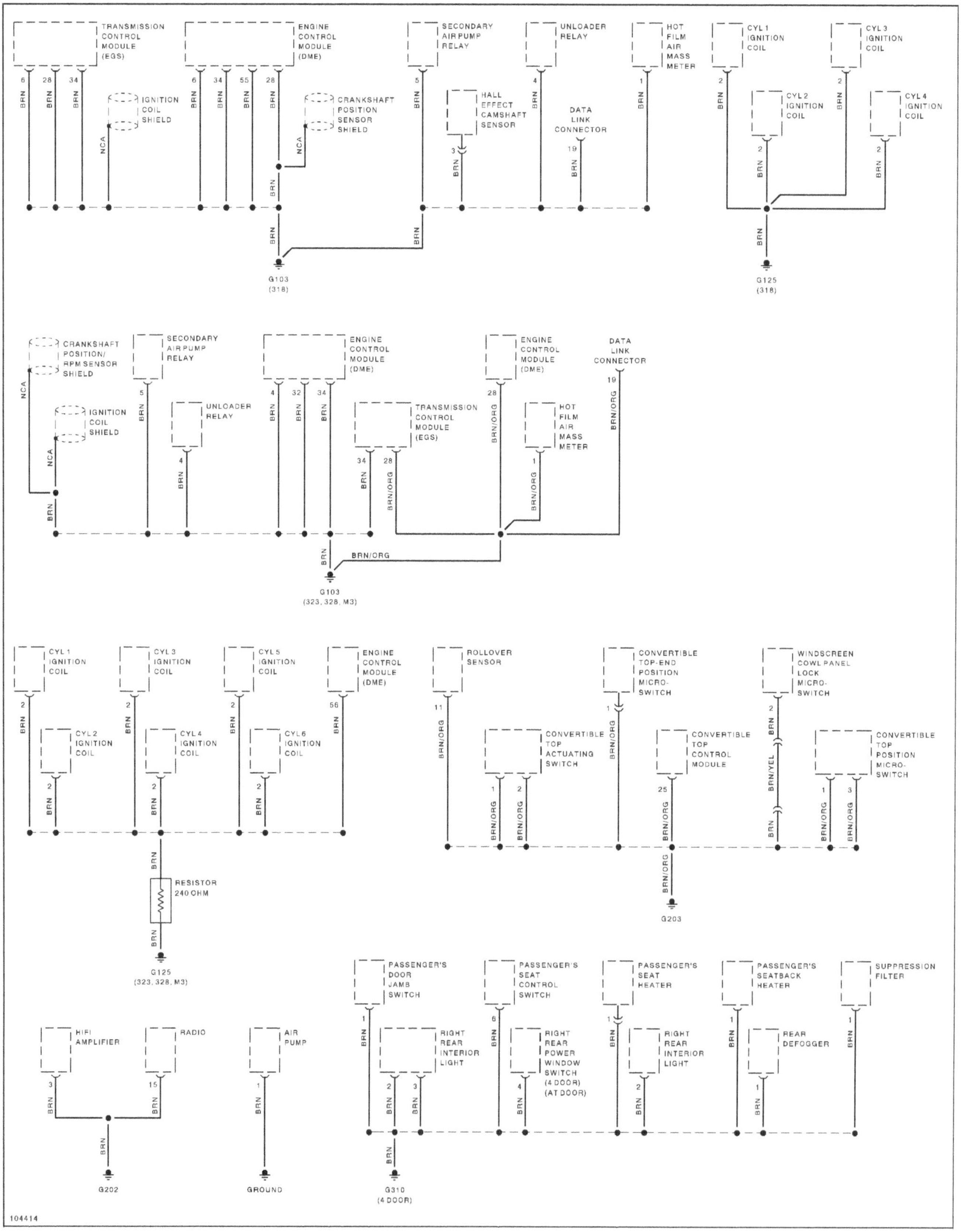

Ground Distribution

1998 Except 318ti (4 of 4)

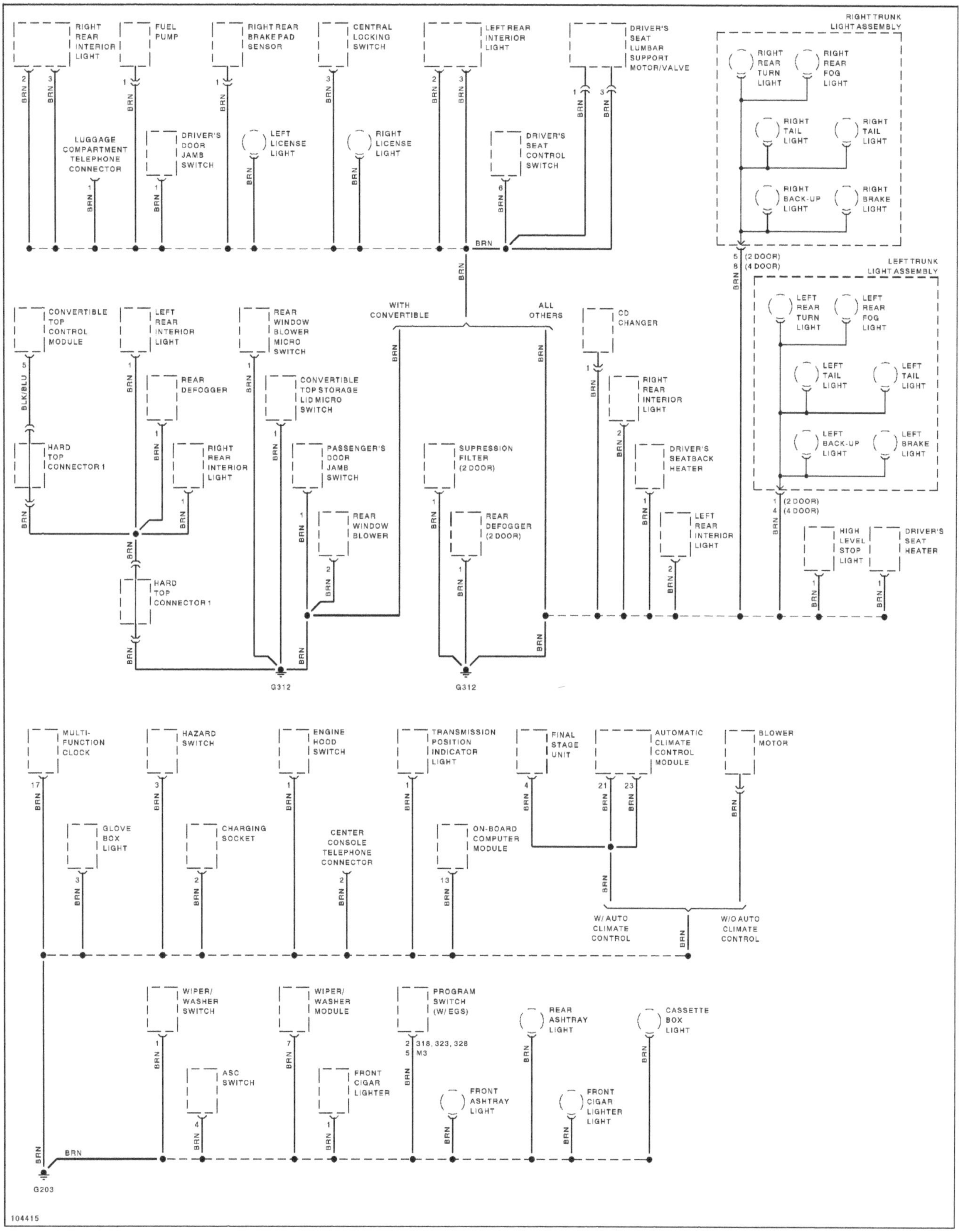

Ground Distribution

1995-96 318ti (1 of 3)

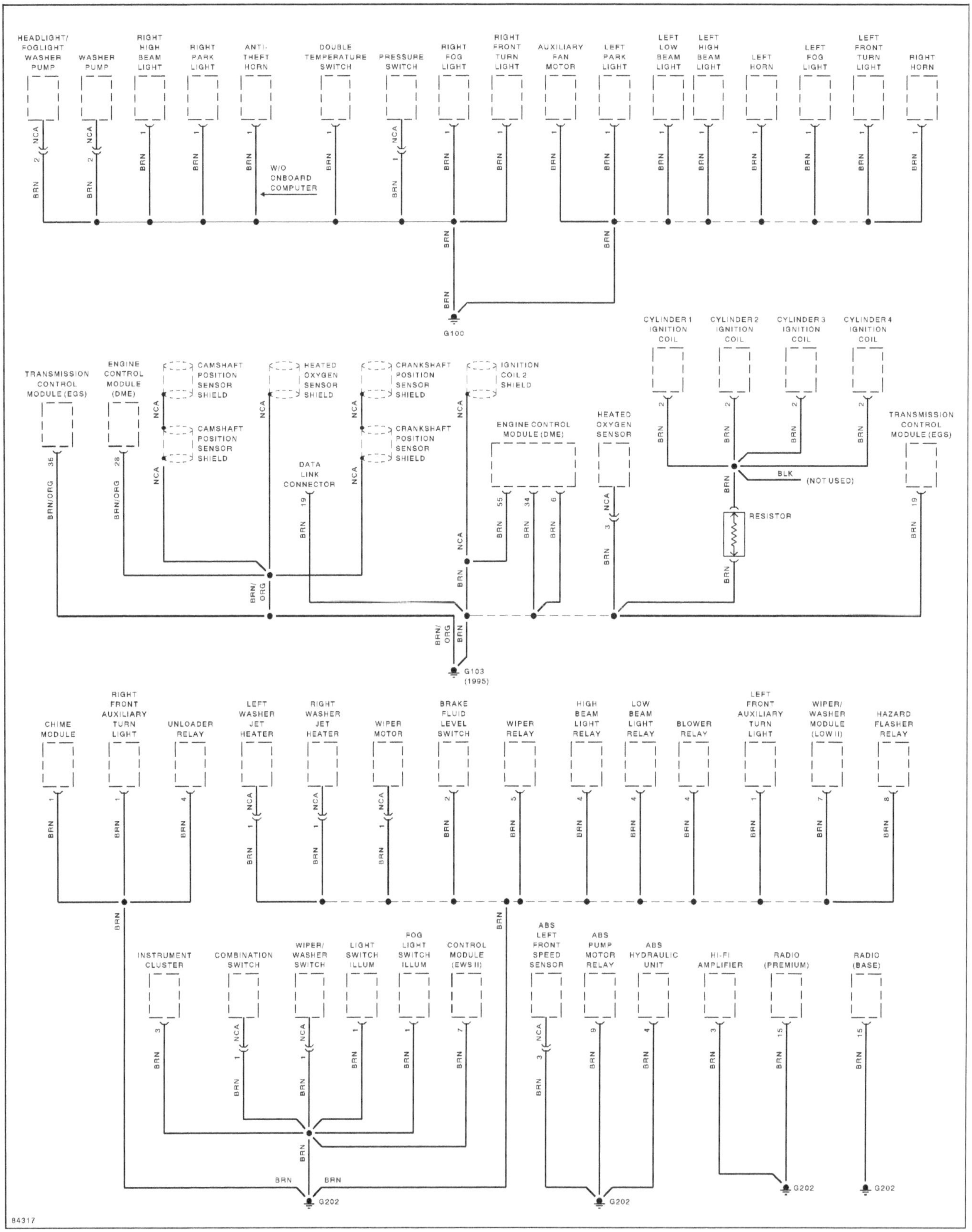

Ground Distribution

1995-96 318ti (2 of 3)

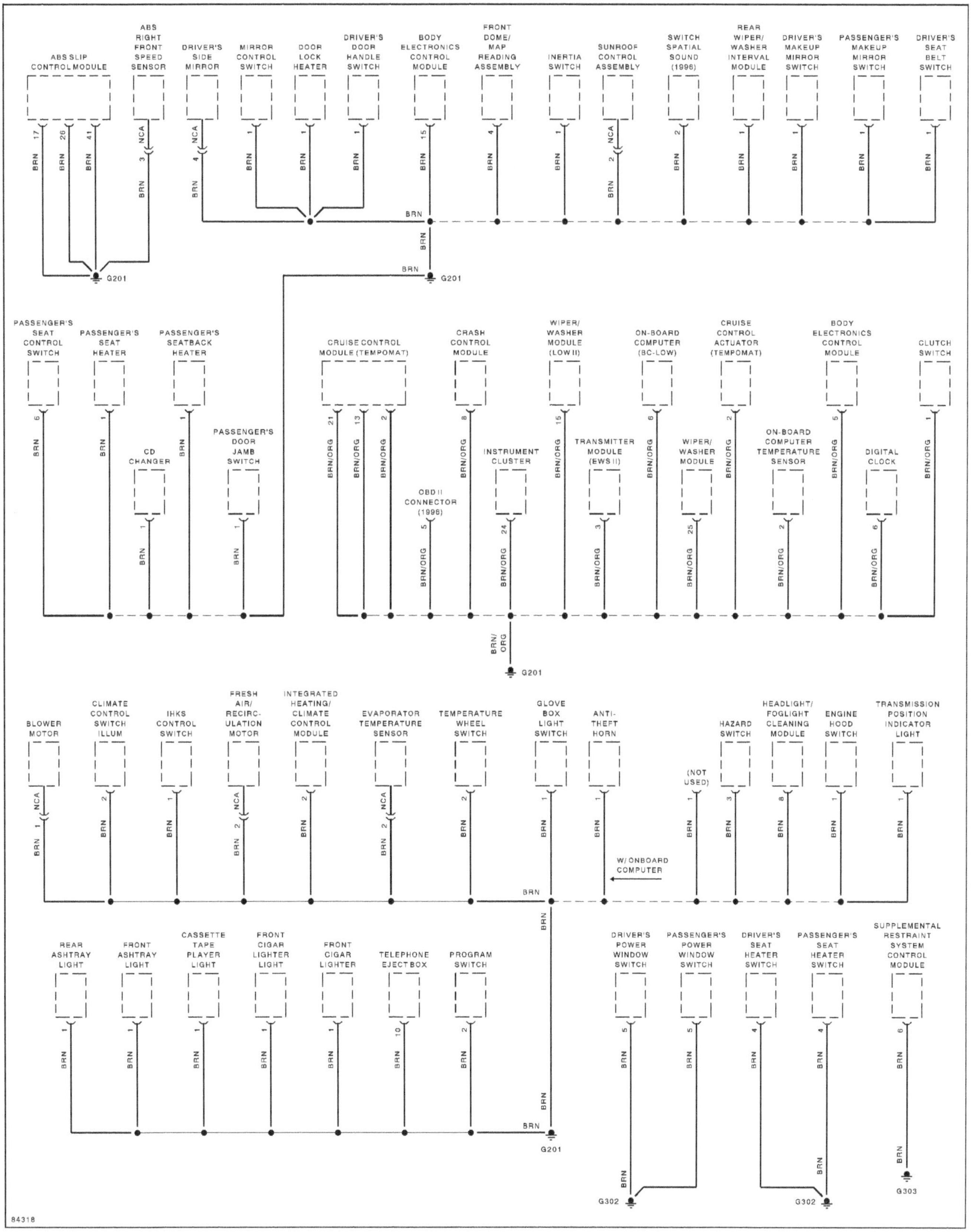

Ground Distribution

1995-96 318ti (3 of 3)

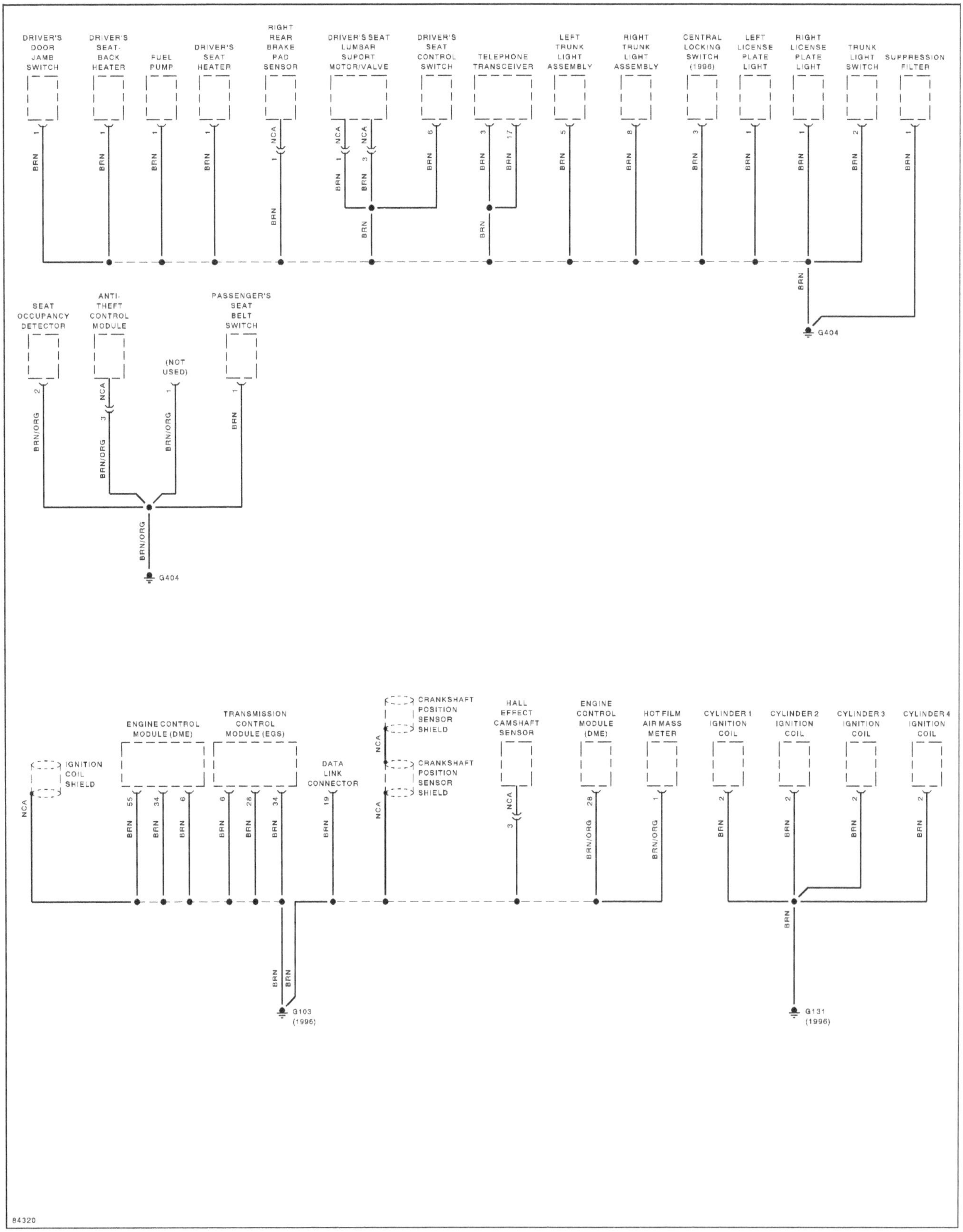

Ground Distribution

1997 318ti (1 of 3)

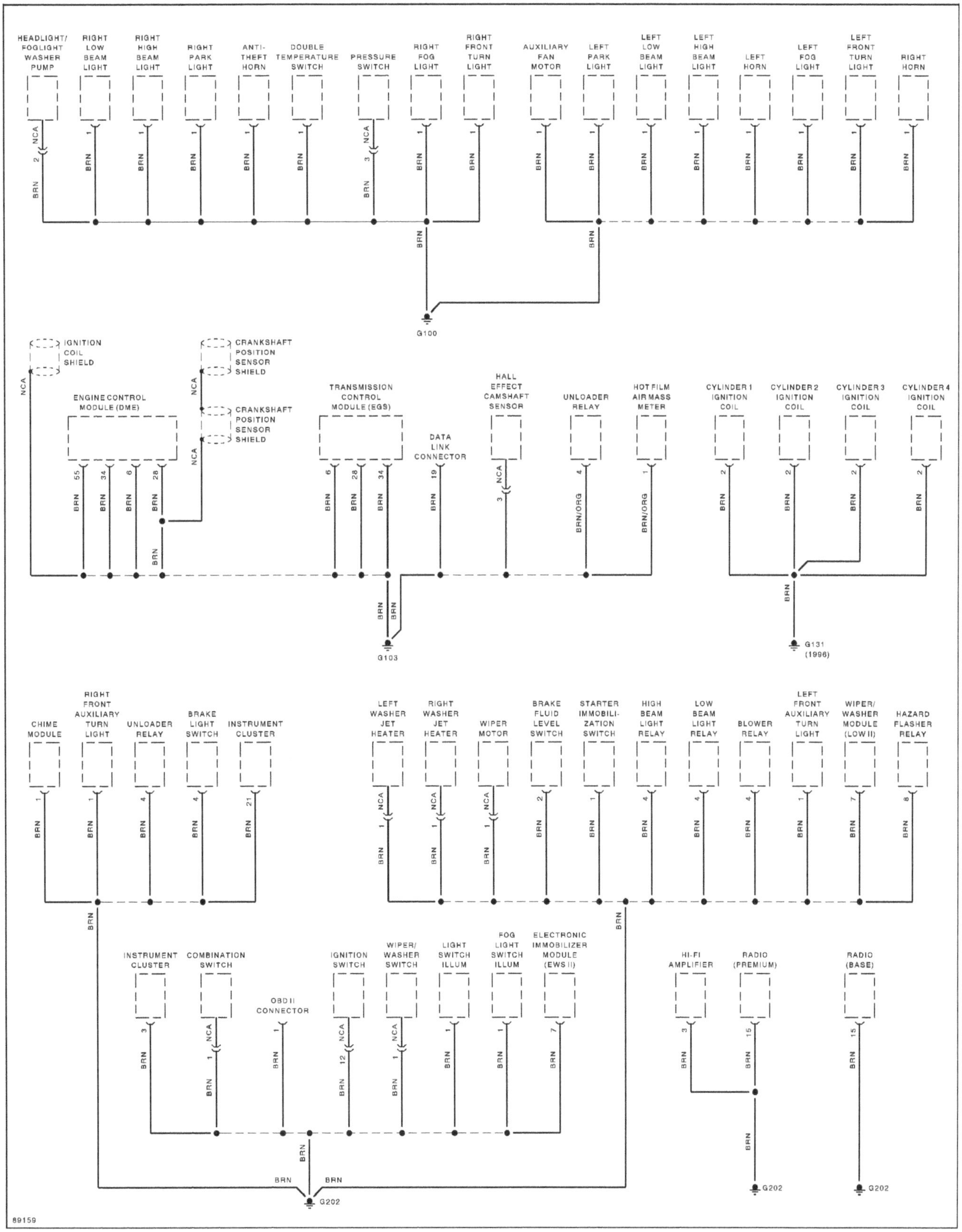

Ground Distribution

1997 318ti (2 of 3)

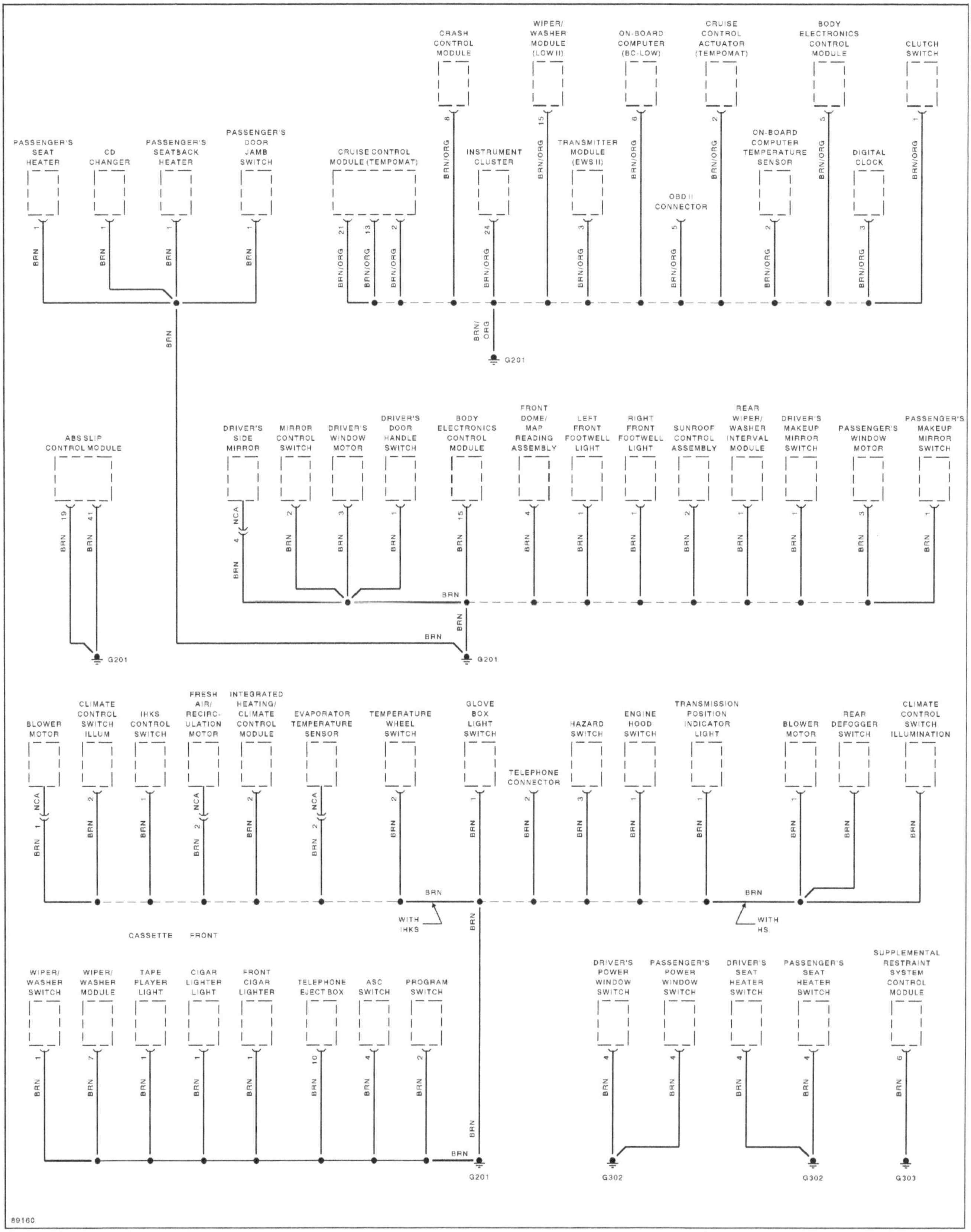

Ground Distribution

1997 318ti (3 of 3)

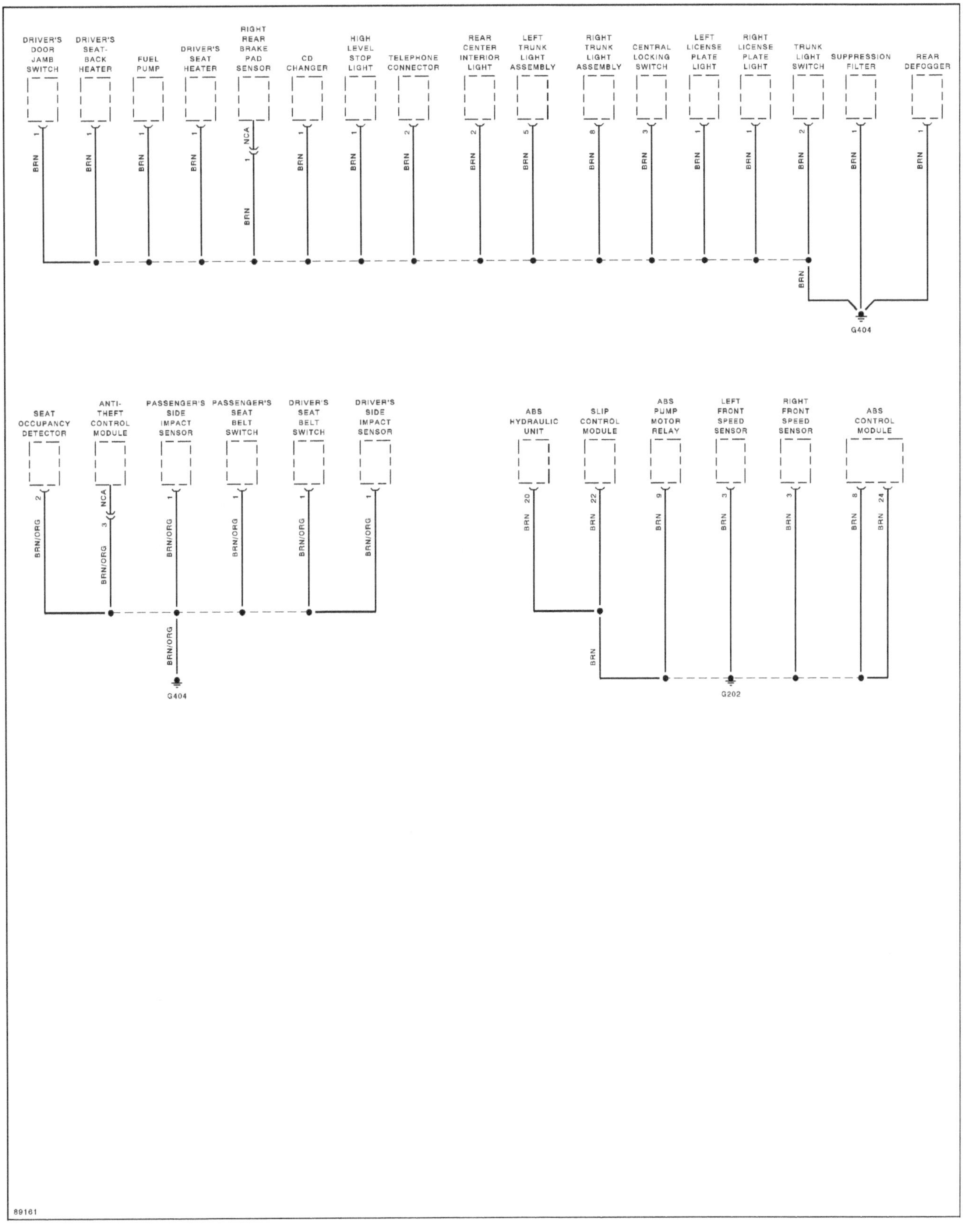

Ground Distribution

1998 318ti (1 of 3)

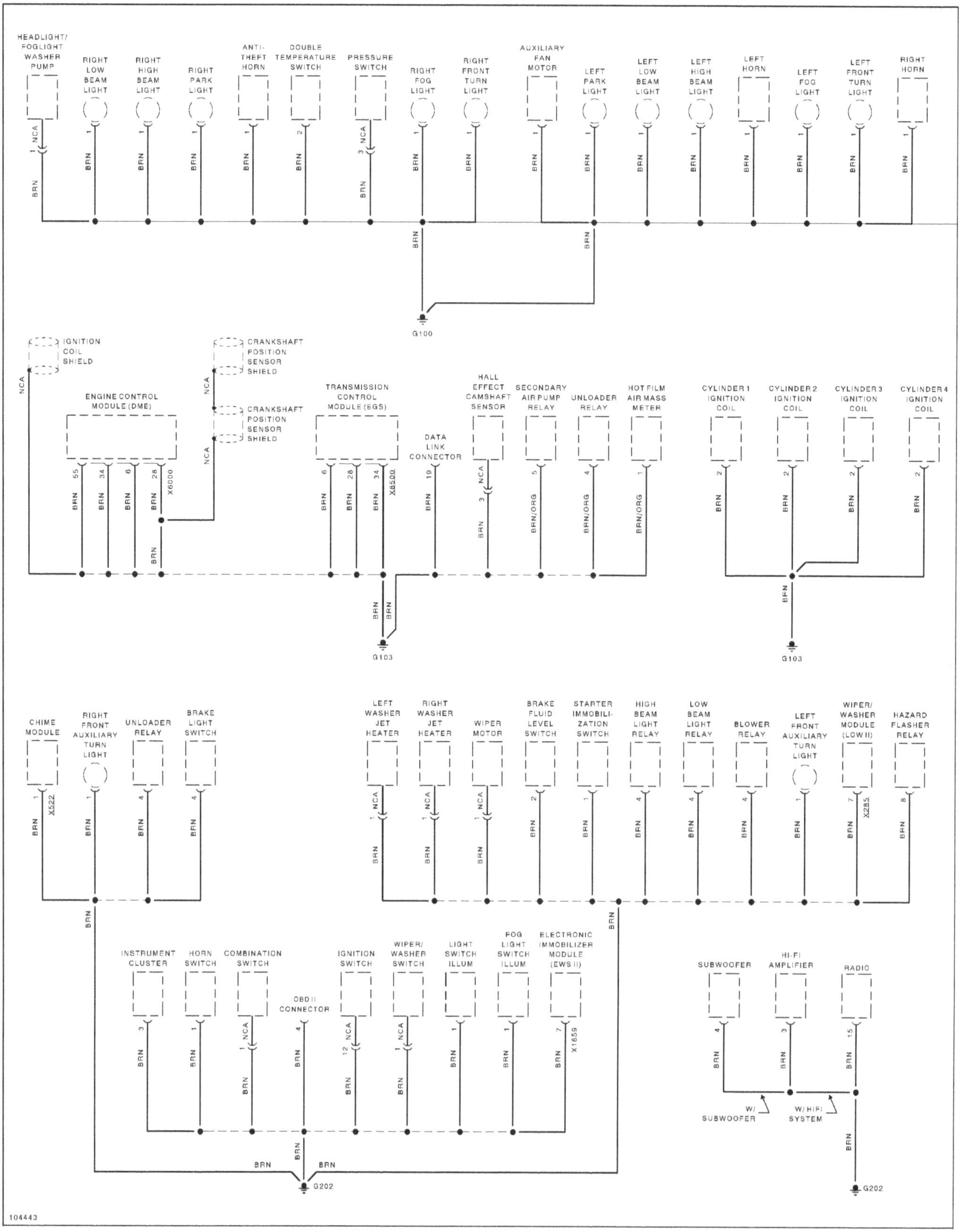

Ground Distribution

1998 318ti (2 of 3)

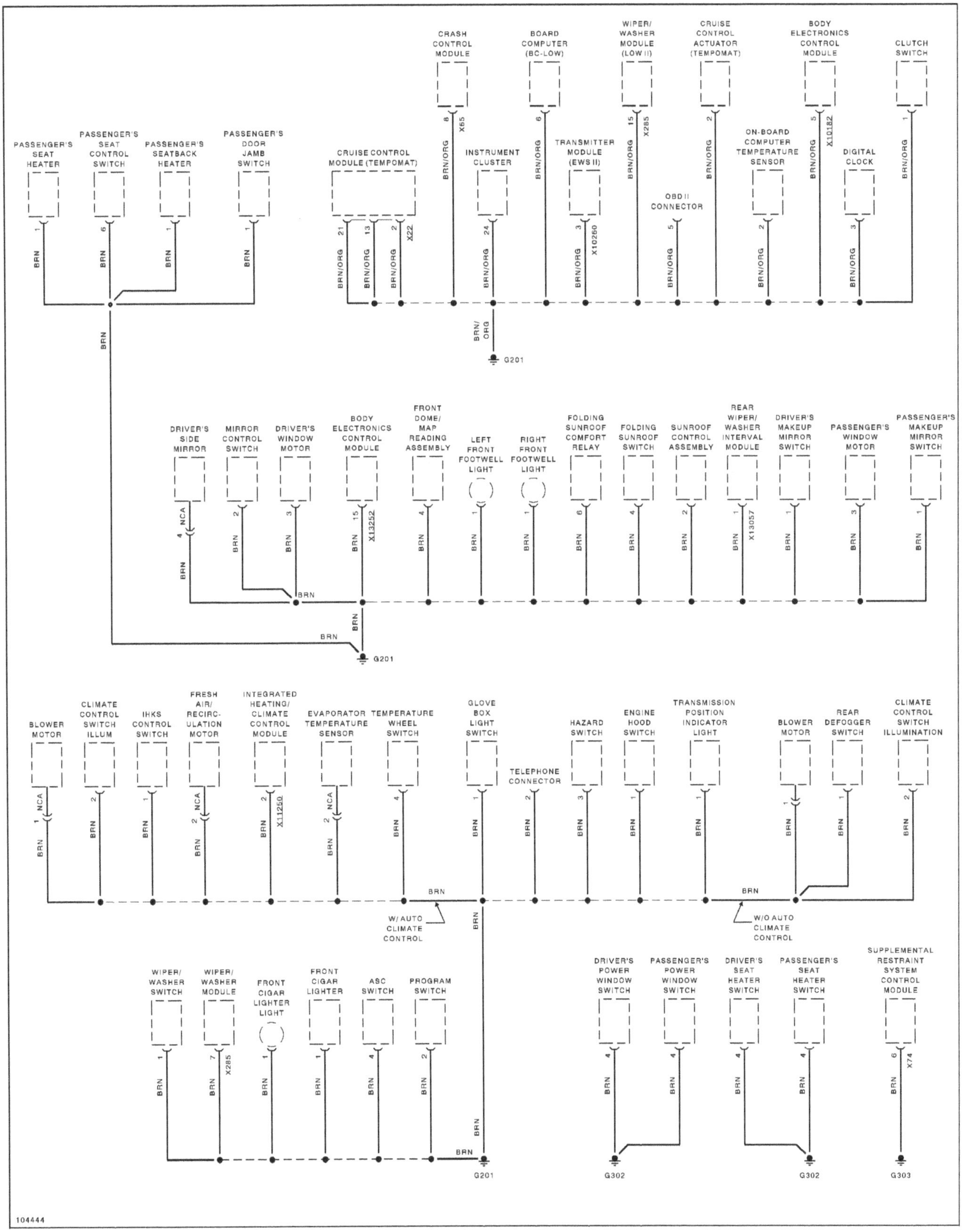

Ground Distribution

1998 318ti (3 of 3)

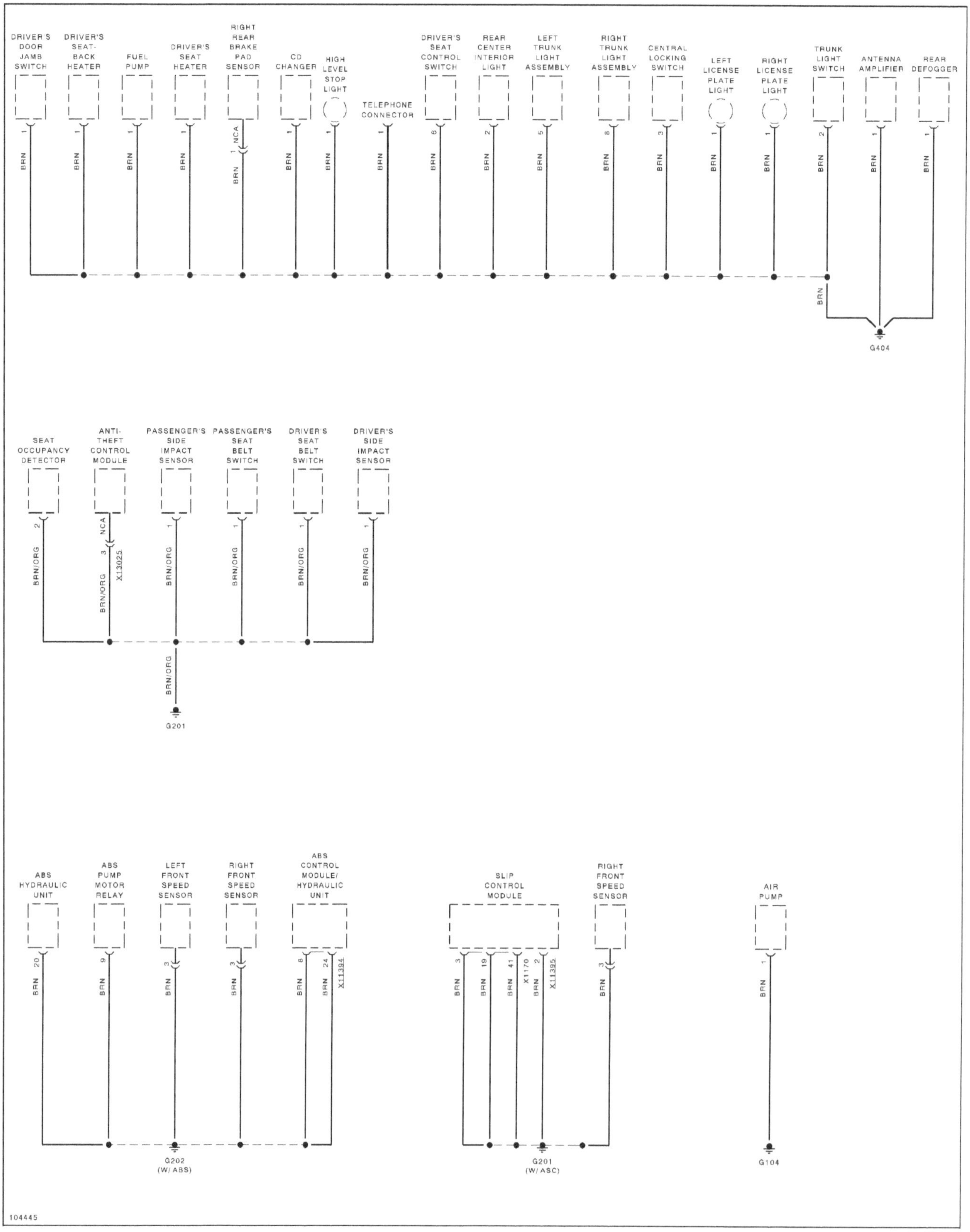

Starting

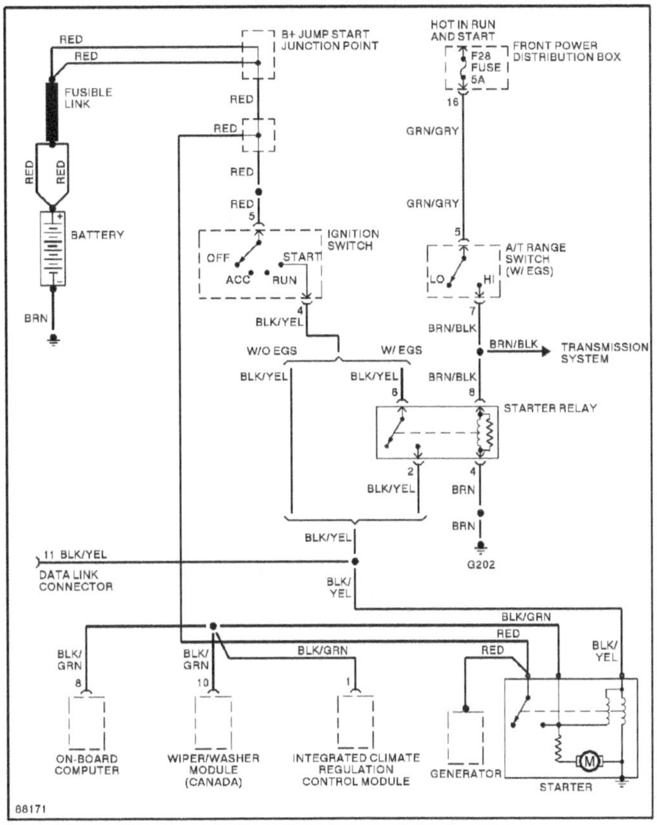

1992
318i & 318is,
1992-93
325i & 325is

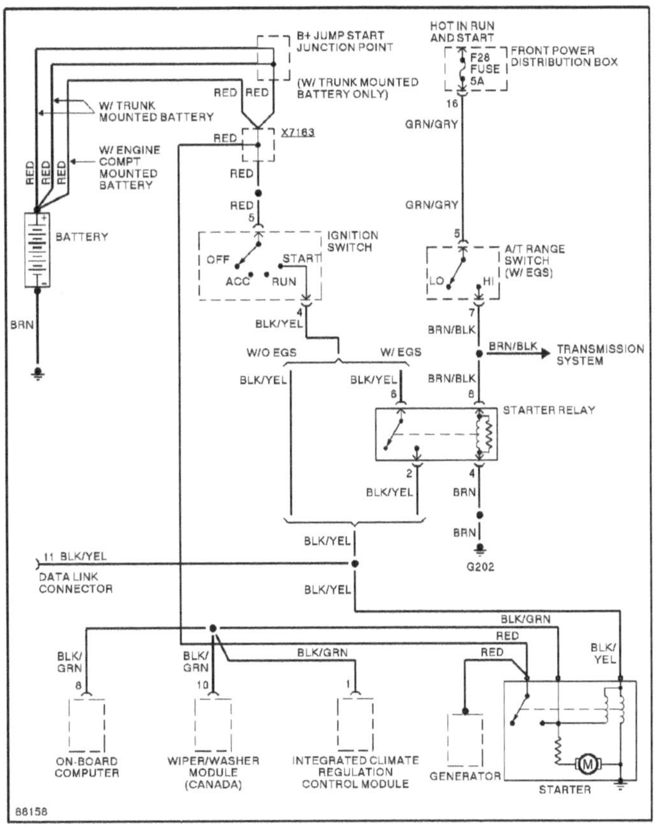

1993
318i & 318is

Starting

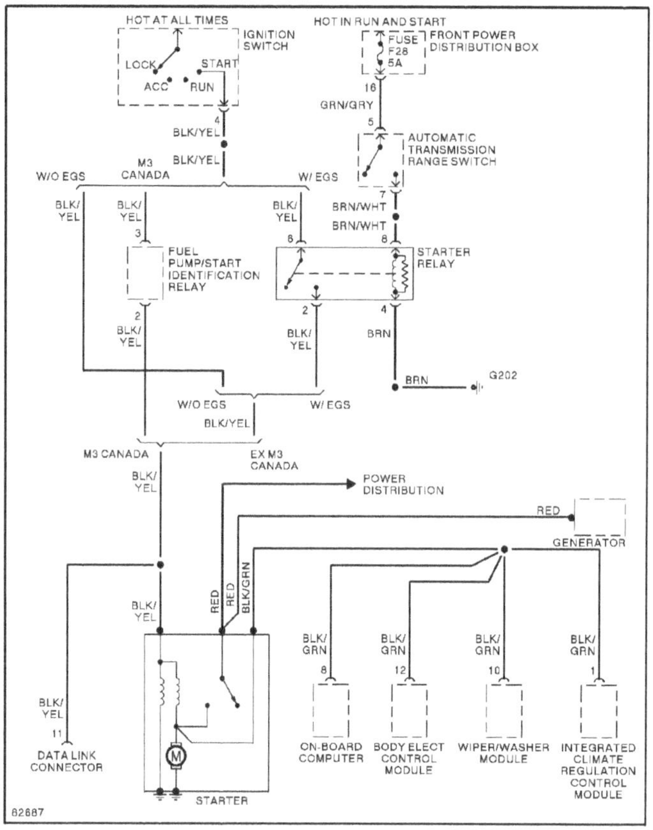

1994 Early Production 318i, 318iC, 318is, 325i, 325iC, 325is & M3

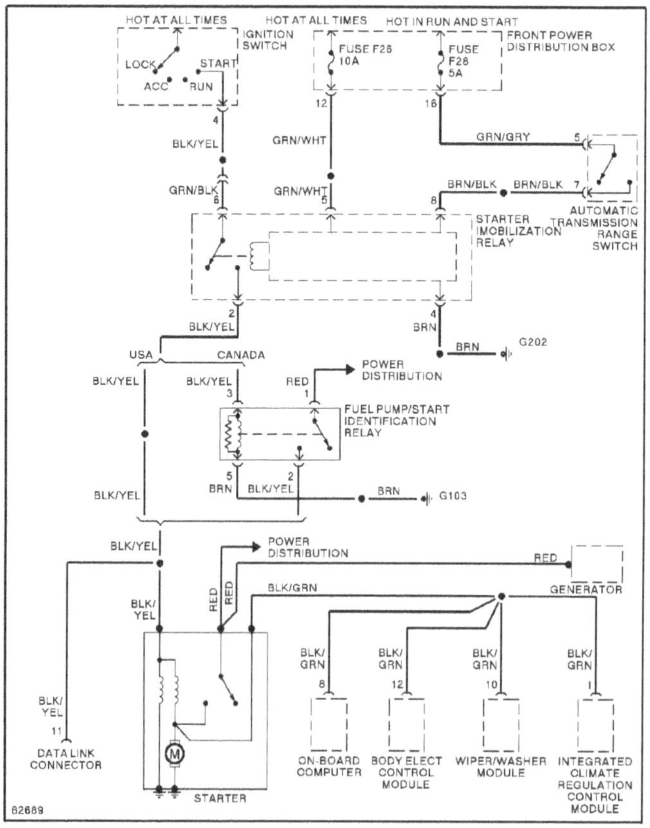

1994 Late Production 318i, 318iC, 318is, 325i, 325iC, 325is & M3

Starting

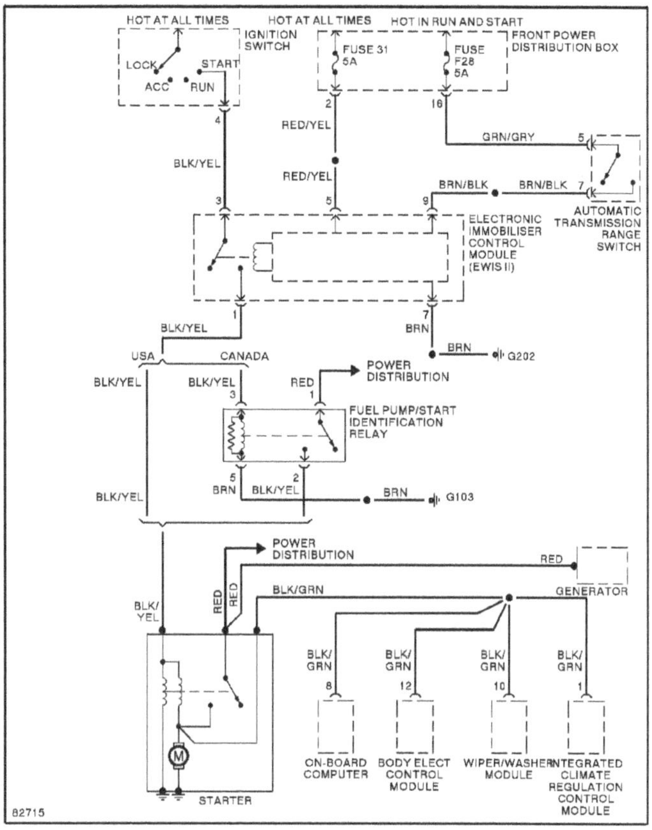

1995-96
318i, 318iC, 318is, 325i, 325iC, 325is, 328i, 328iC, 328is & M3

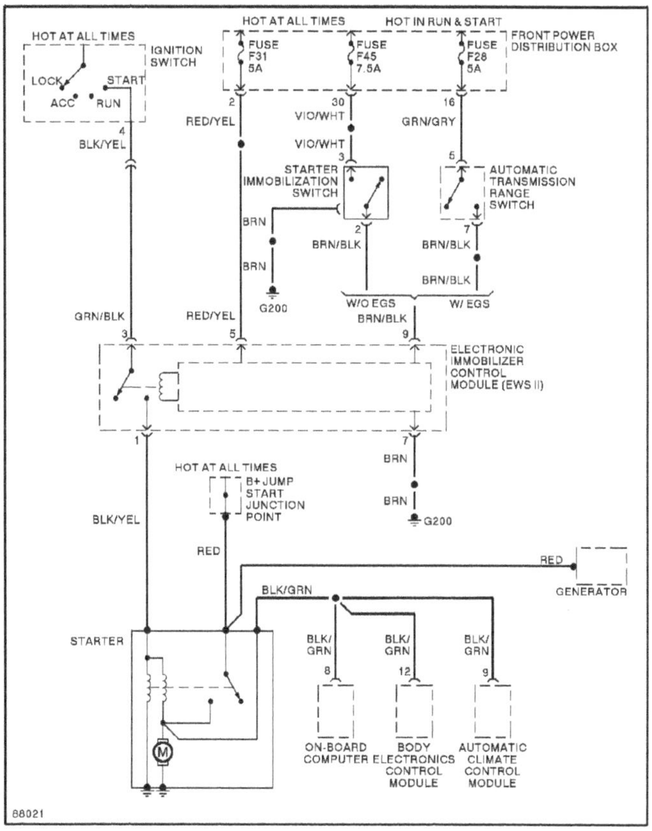

1997-98
318i, 318iC, 318is, 325i, 325iC, 325is, 328i, 328iC, 328is & M3

Starting

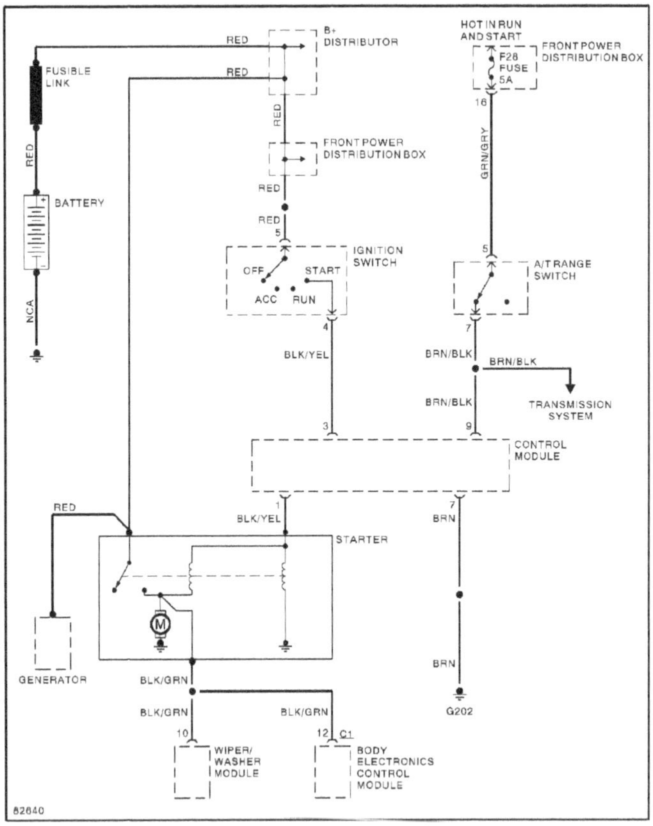

1995-96
318ti

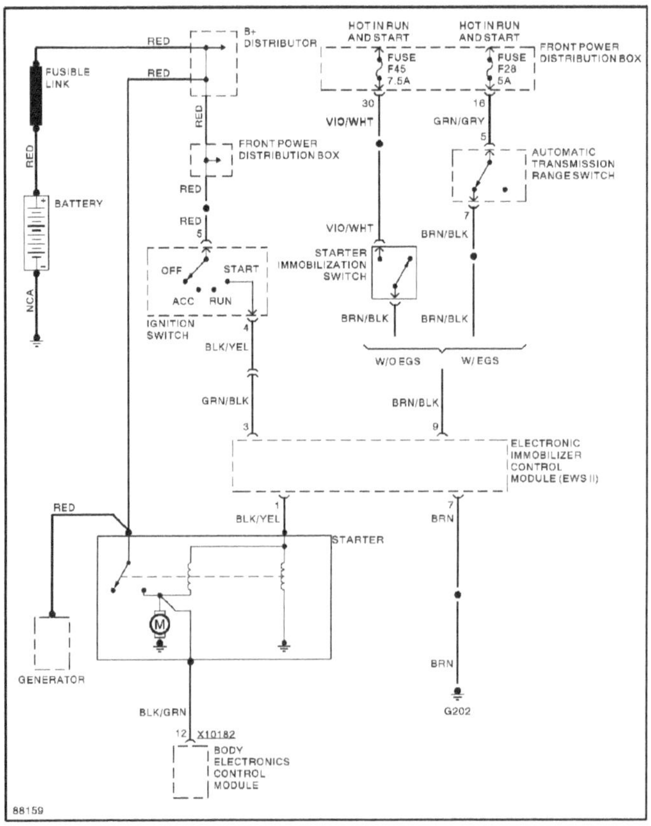

1997-98
318ti

Charging System

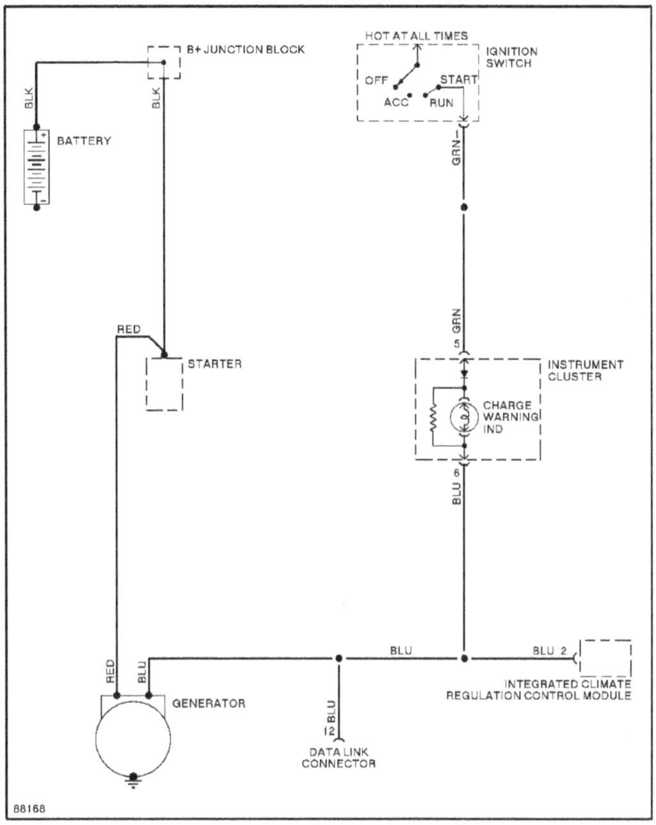

1992-93
318i, 318is, 325i
& 325is

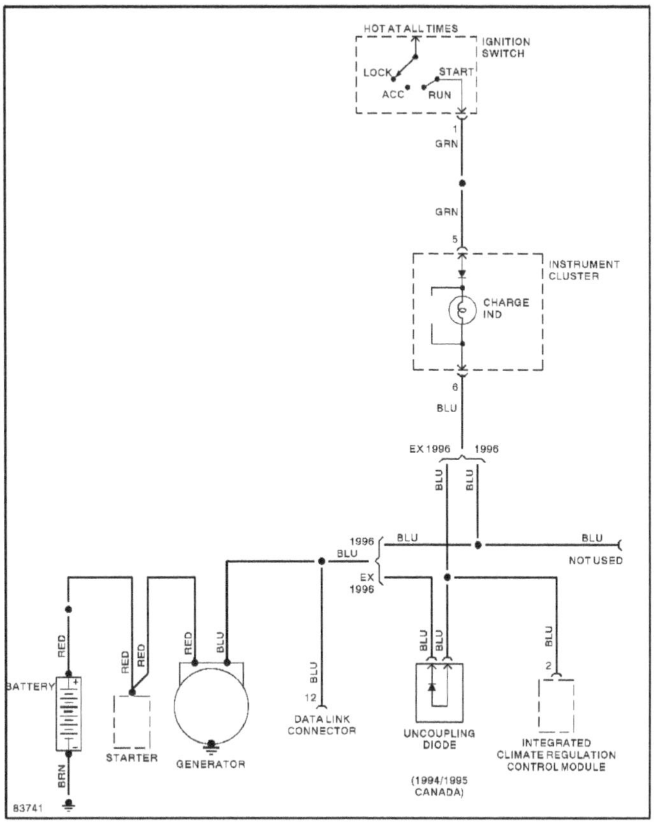

1994-96
318i, 318iC, 318is,
325i, 325iC, 325is,
328i, 328iC, 328is
& M3

Charging System

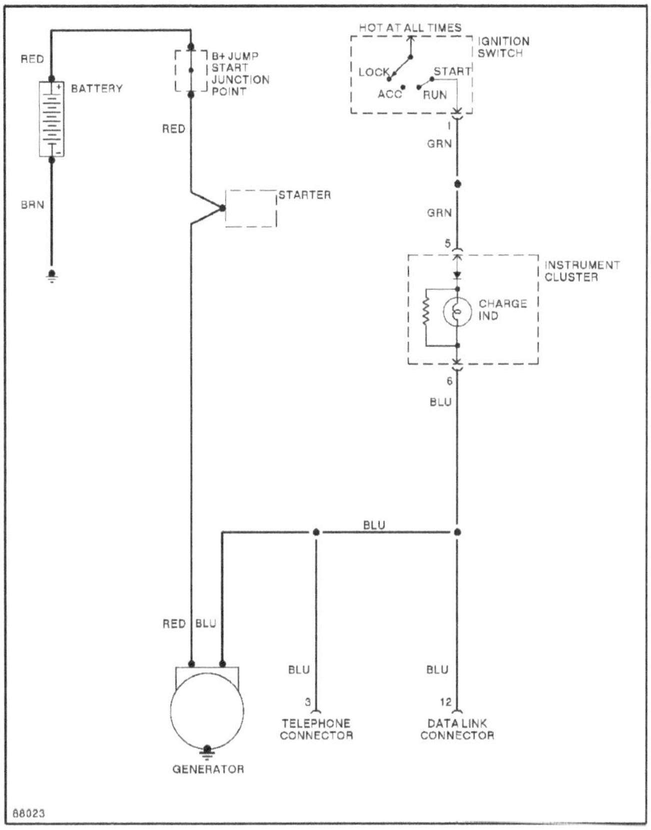

1997-98
318i, 318iC, 318is, 328i, 328iC, 328is & M3

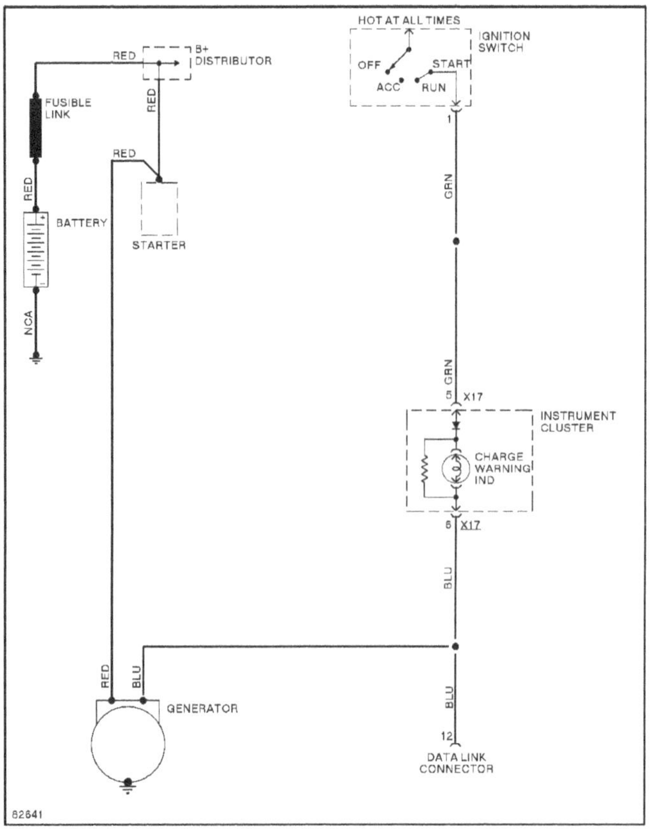

1995-96
318ti

Charging System

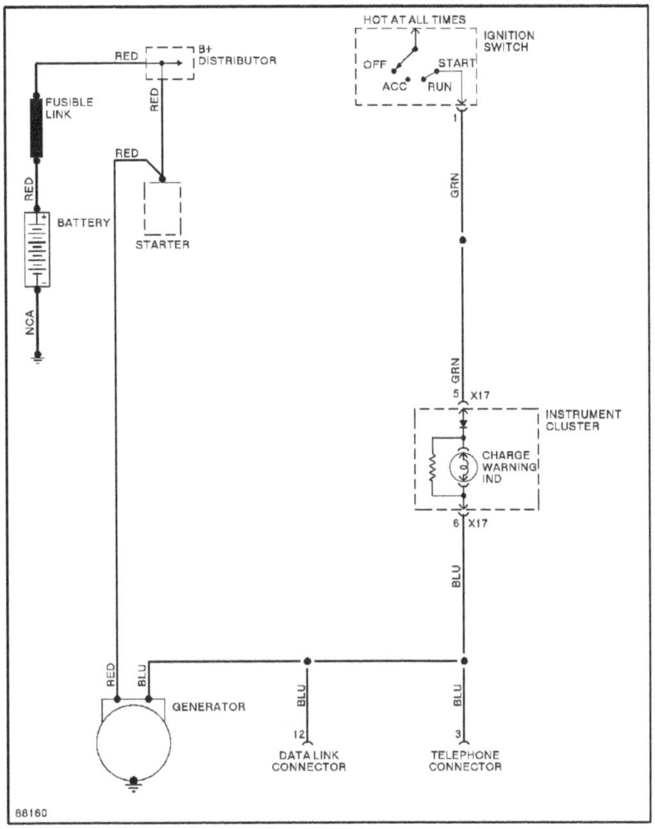

1997-98
318ti

Engine Management

1992 1.8L (1 of 2)

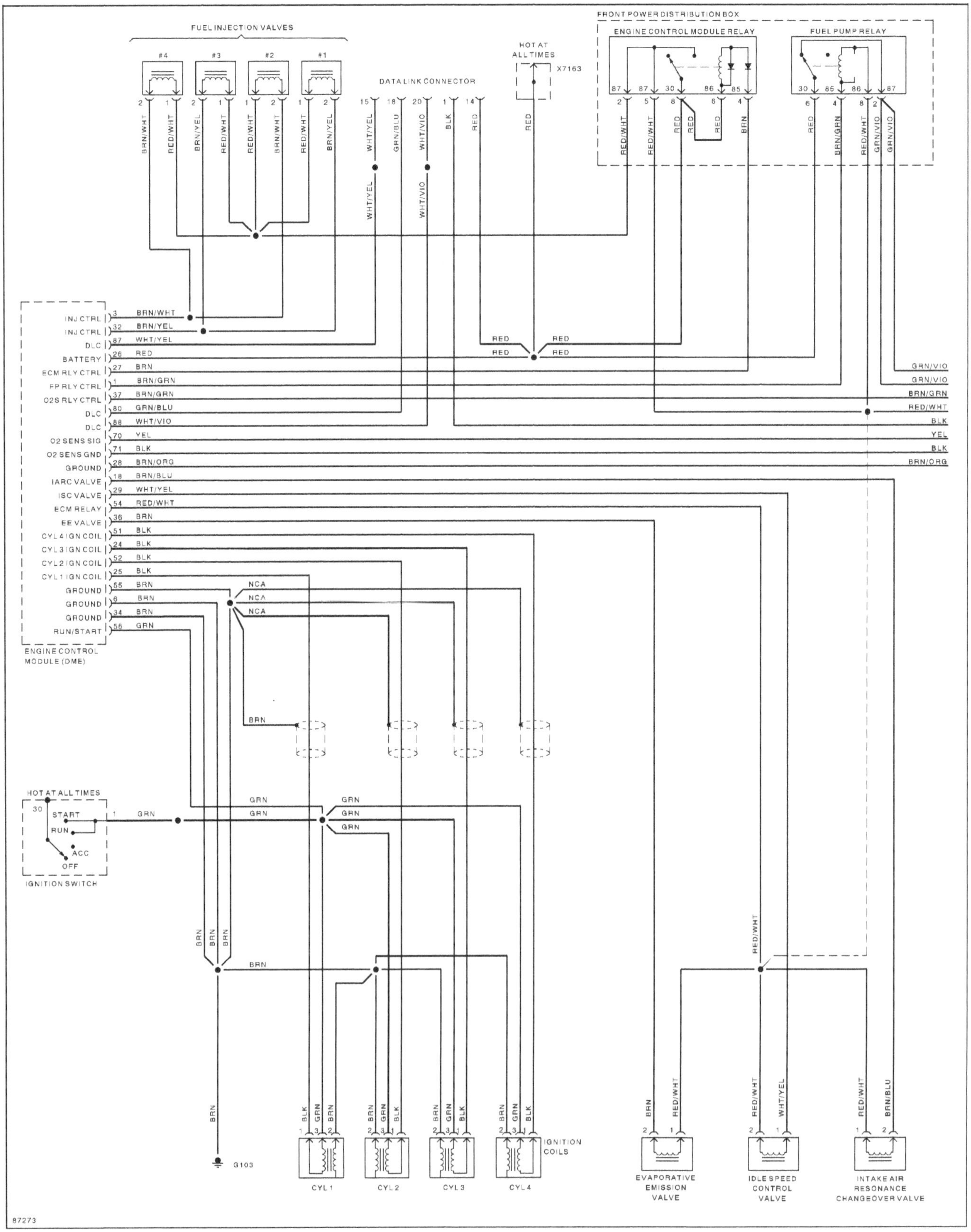

Engine Management

1992 1.8L (2 of 2)

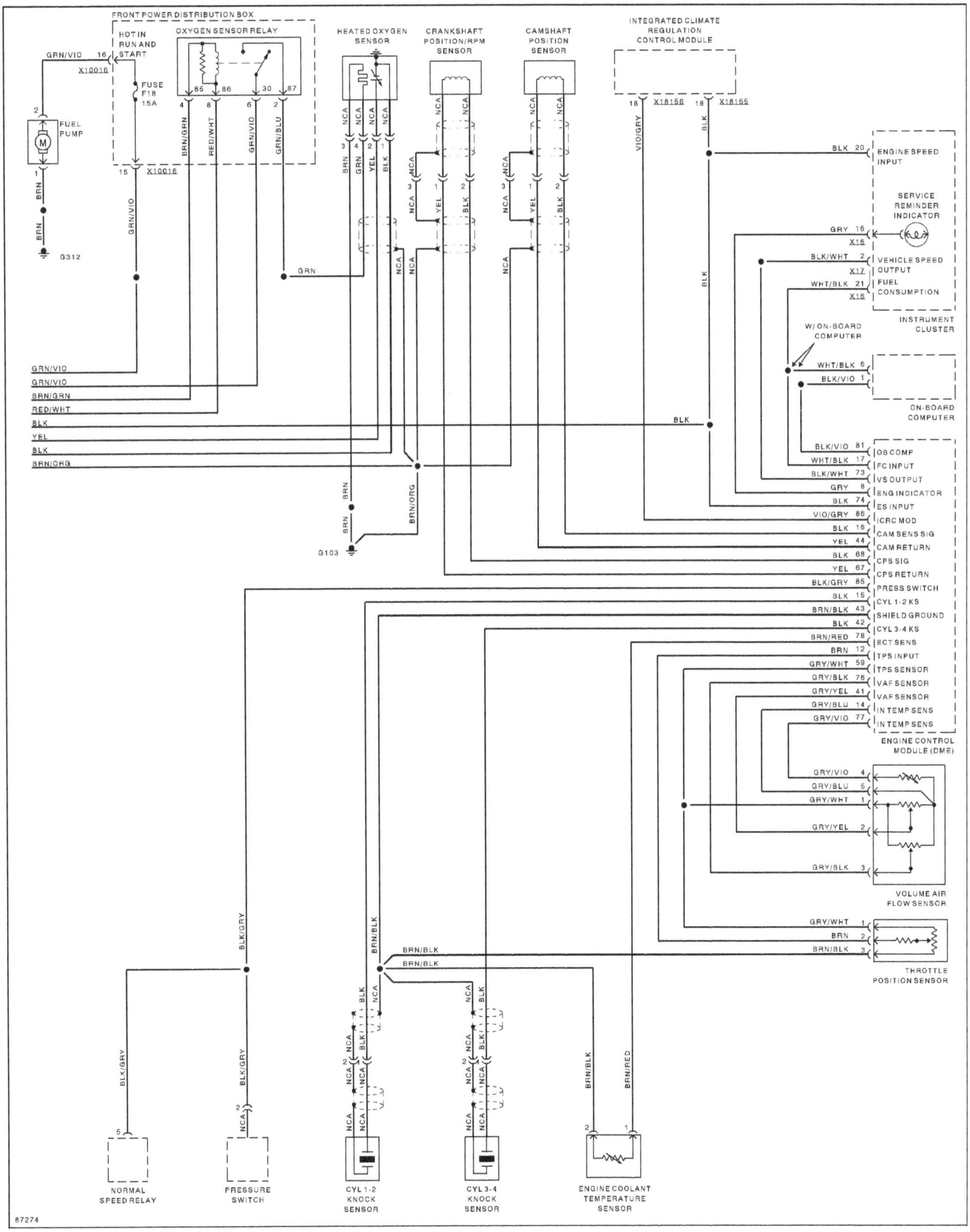

Engine Management

1992 2.5L (1 of 2)

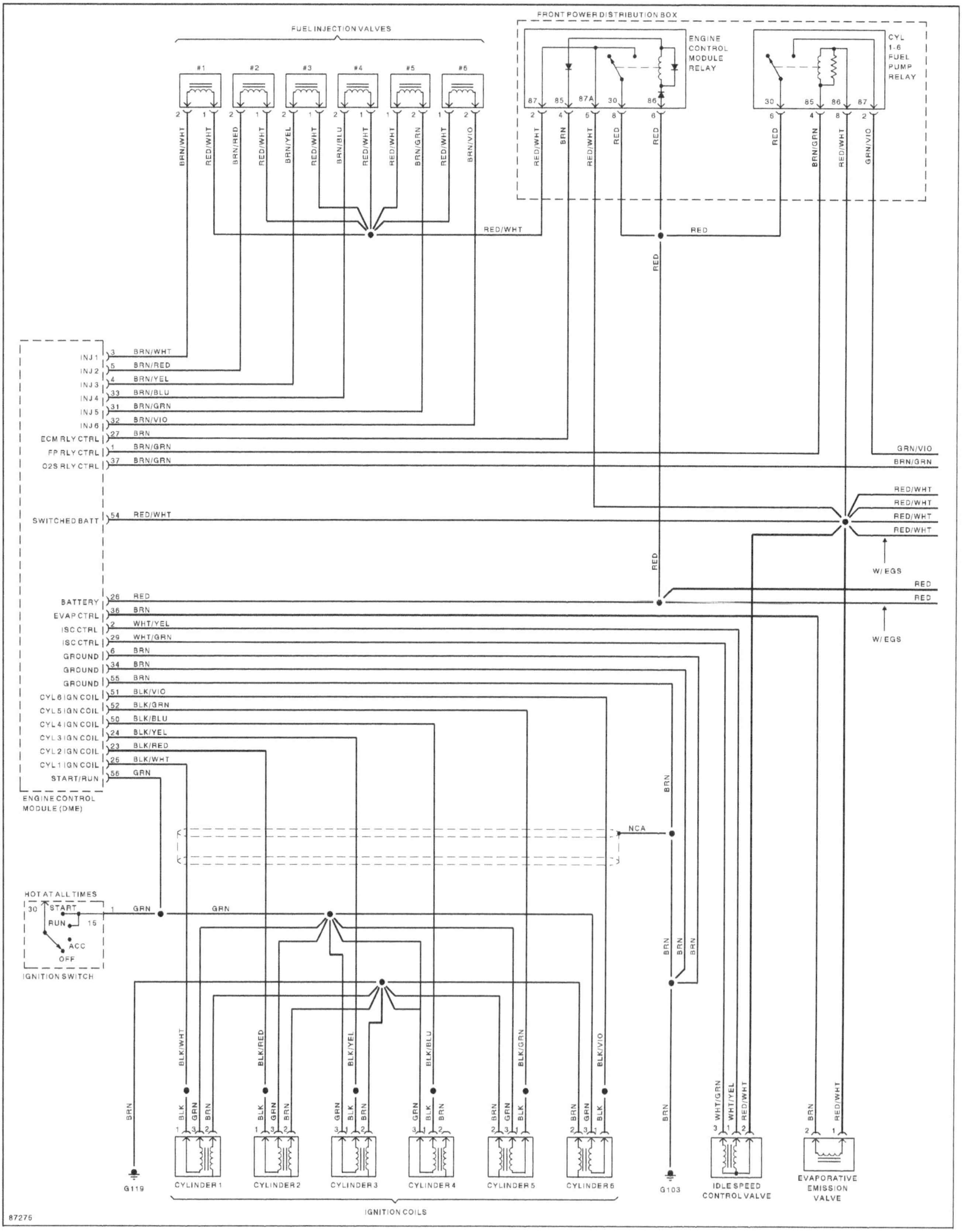

Engine Management

1992 2.5L (2 of 2)

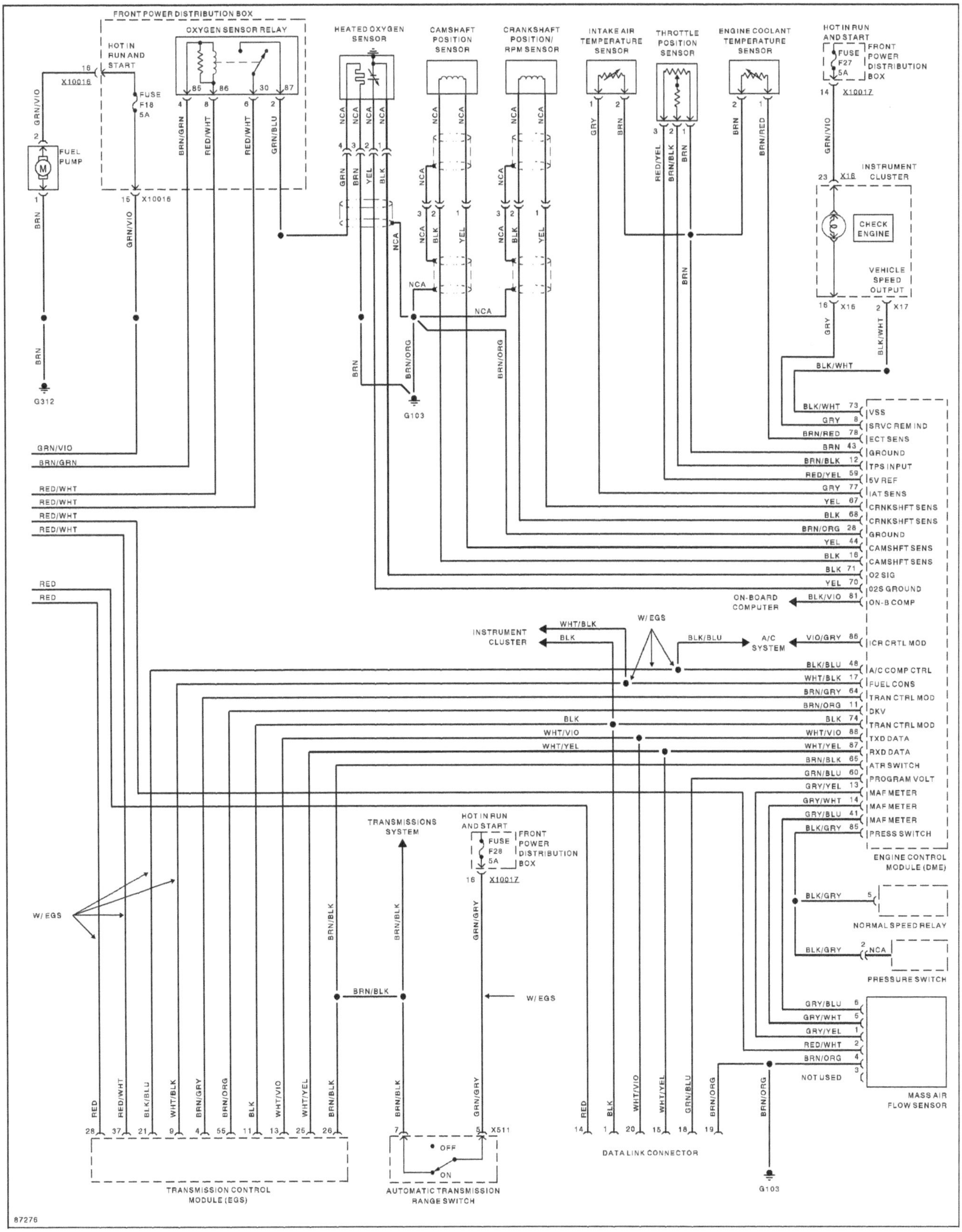

Engine Management

1993-95 1.8L (1 of 2)

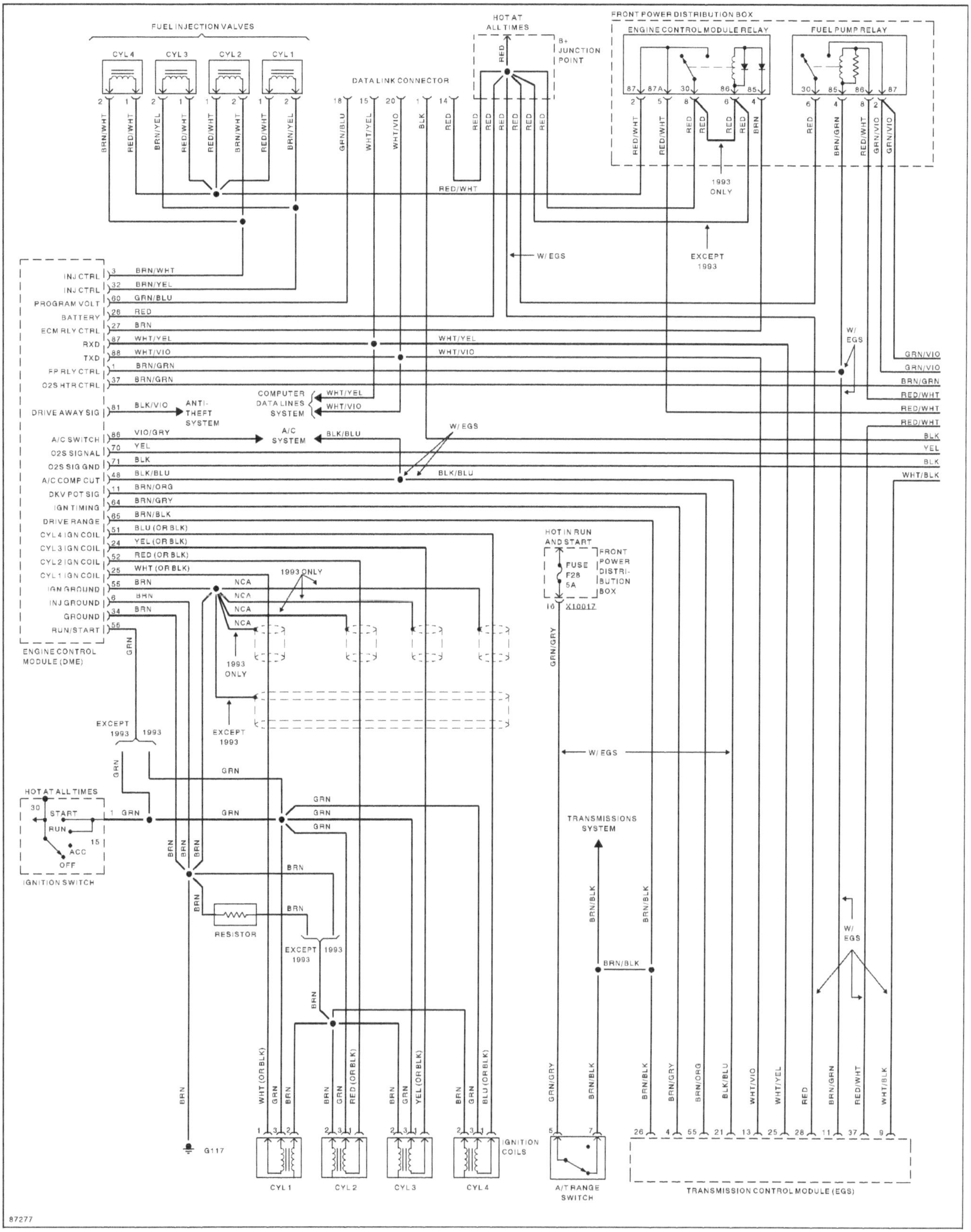

Engine Management

1993-95 1.8L (2 of 2)

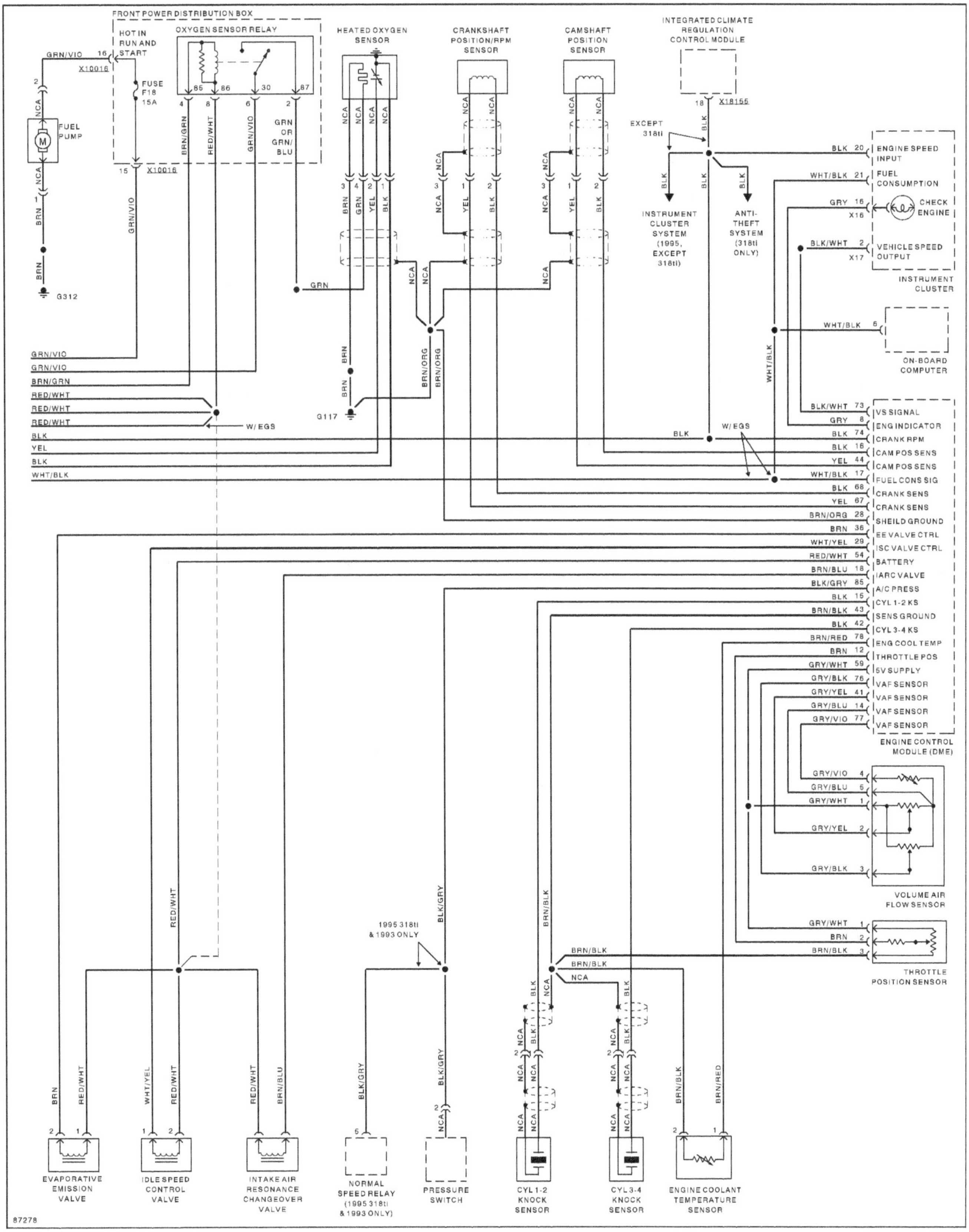

Engine Management

1993-95 2.5L (1 of 2)

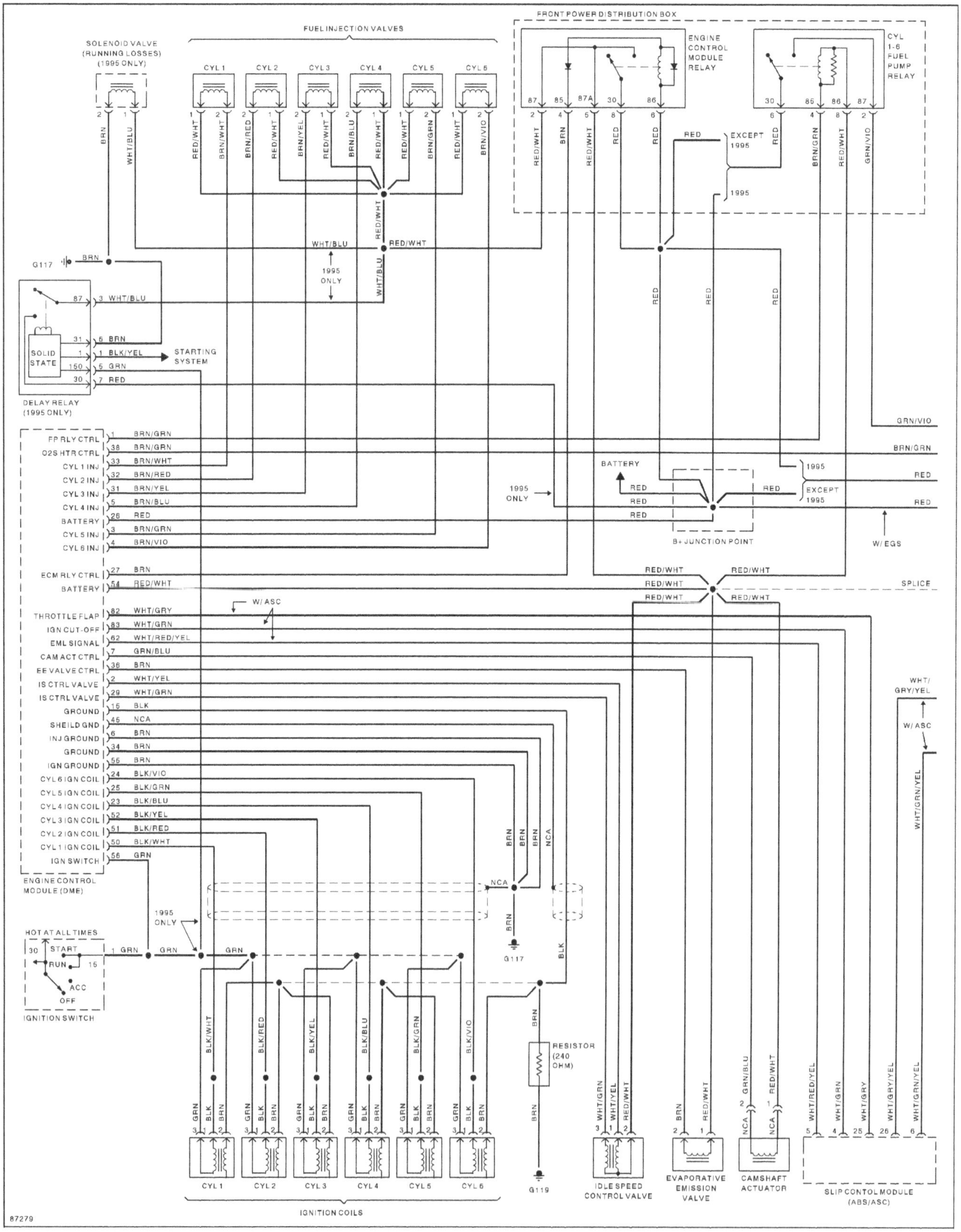

Engine Management

1993-95 2.5L (2 of 2)

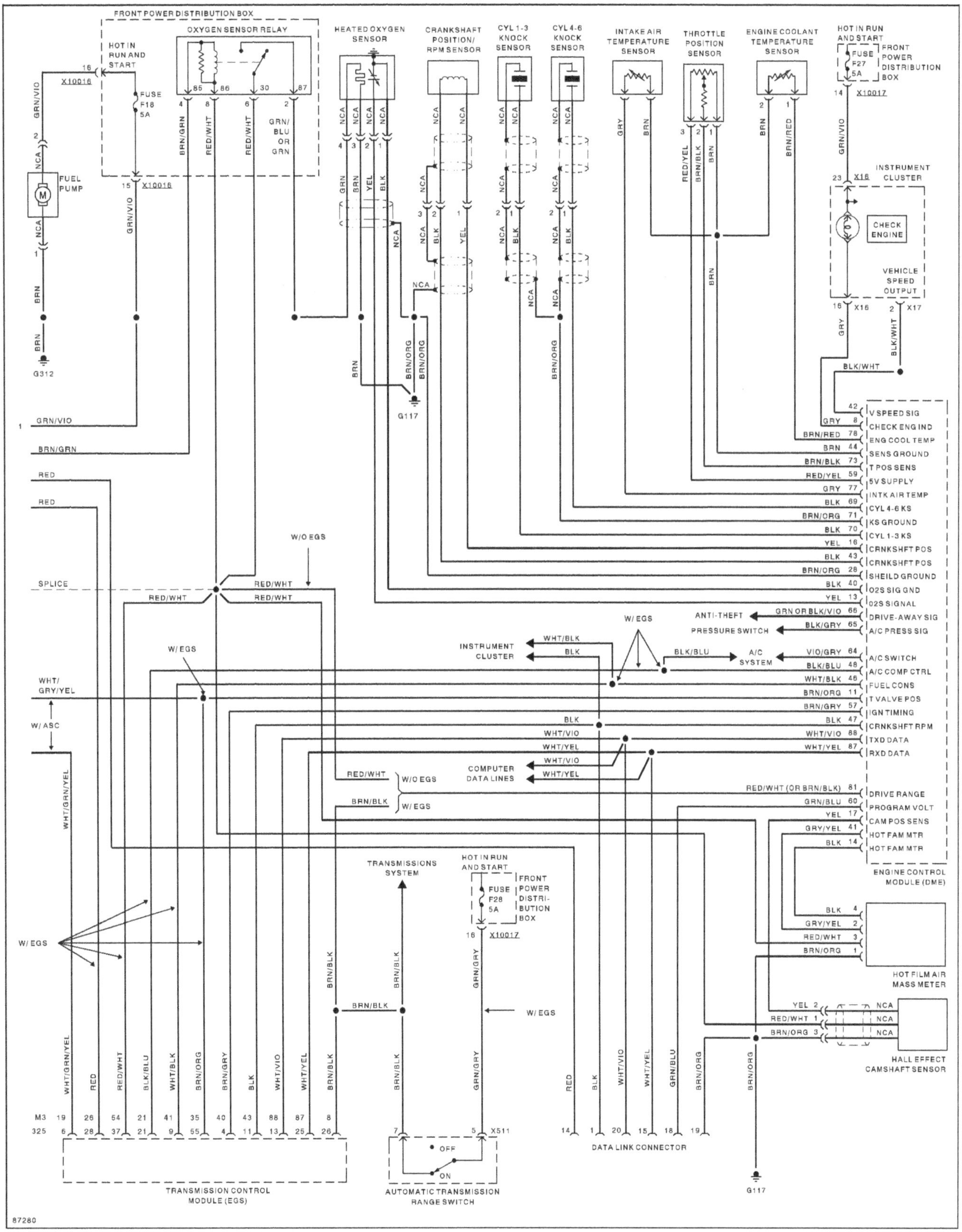

Engine Management

1996-97 1.9L (1 of 2)

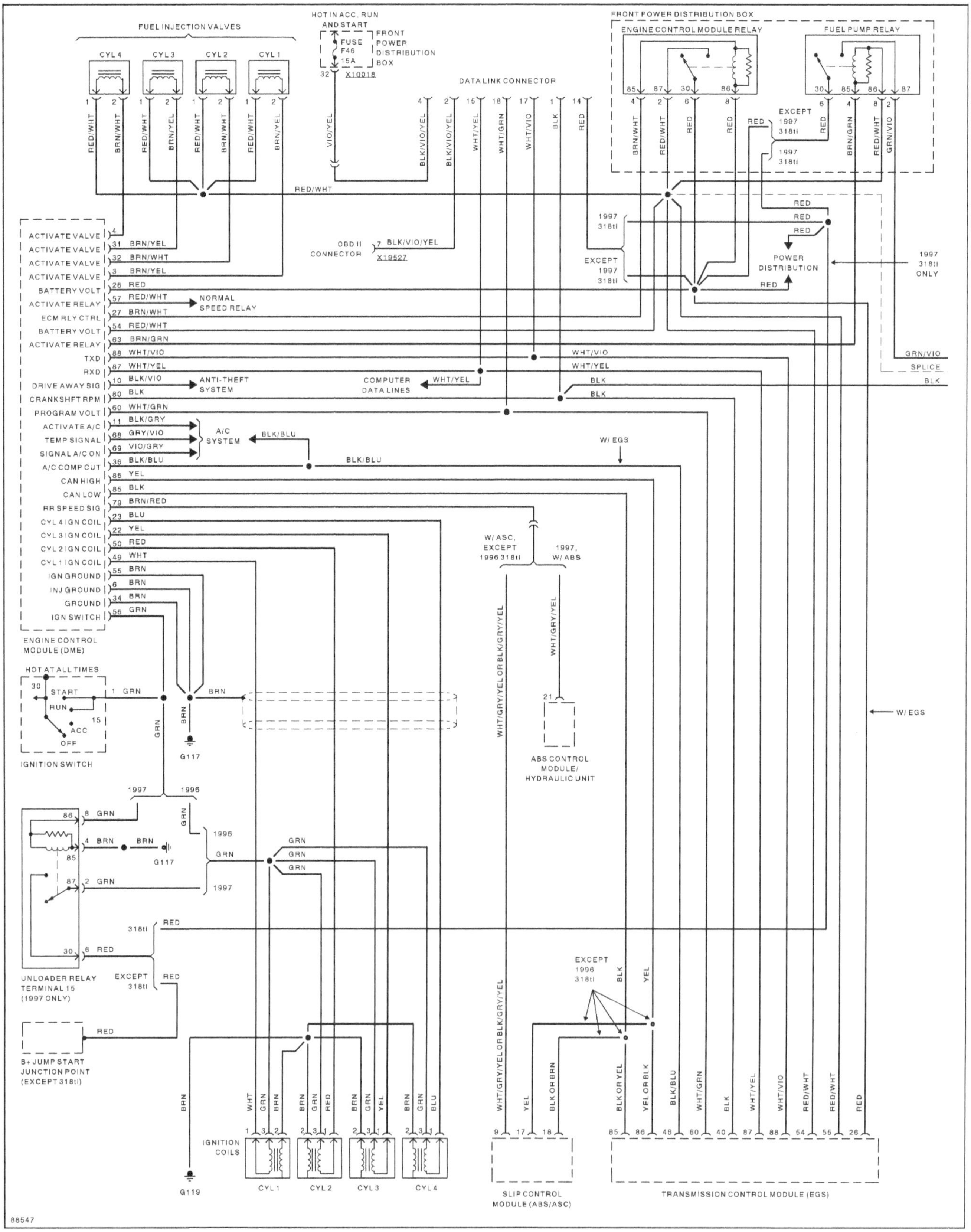

Engine Management

1996-97 1.9L (2 of 2)

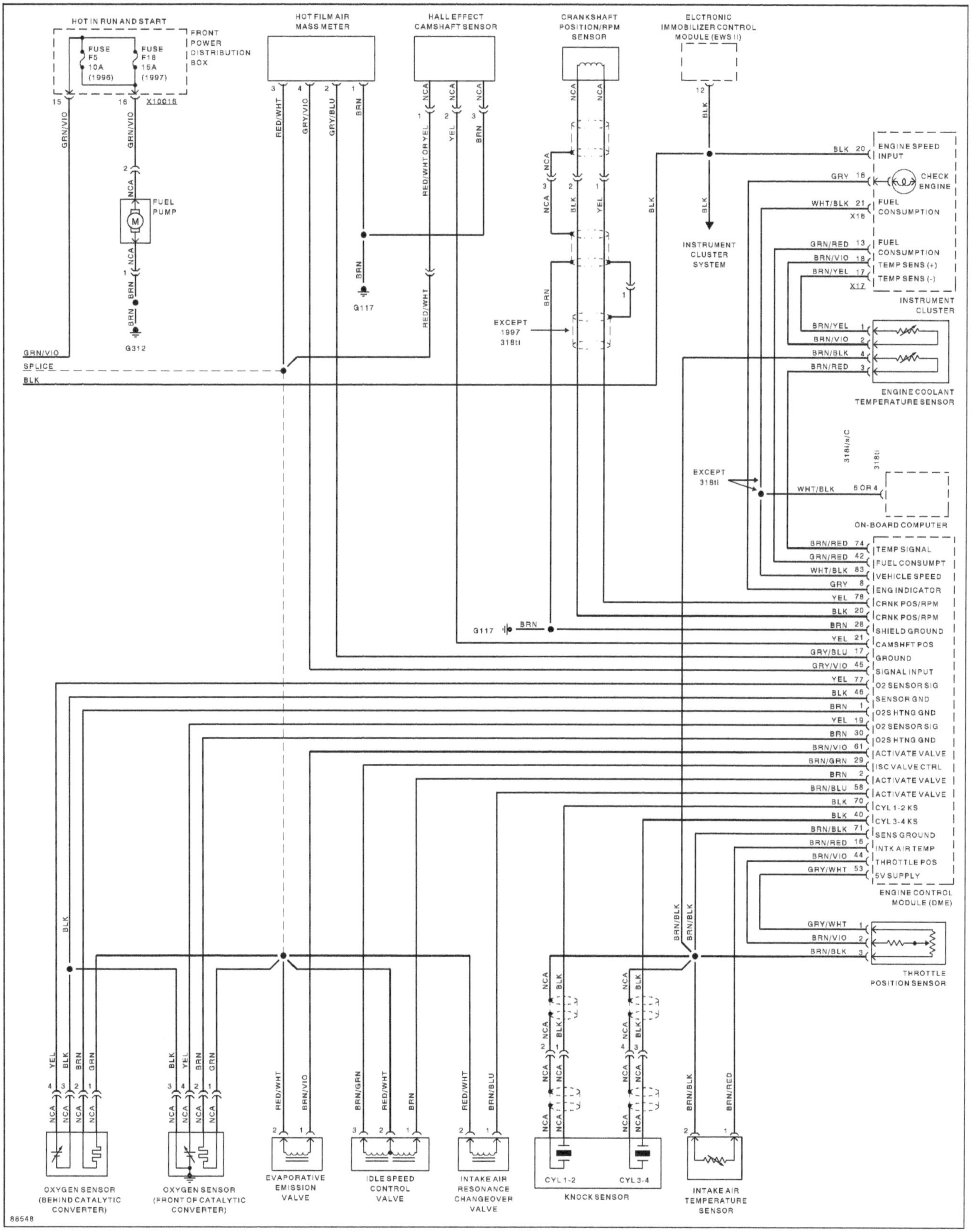

Engine Management

1996-97 2.8L (1 of 2)

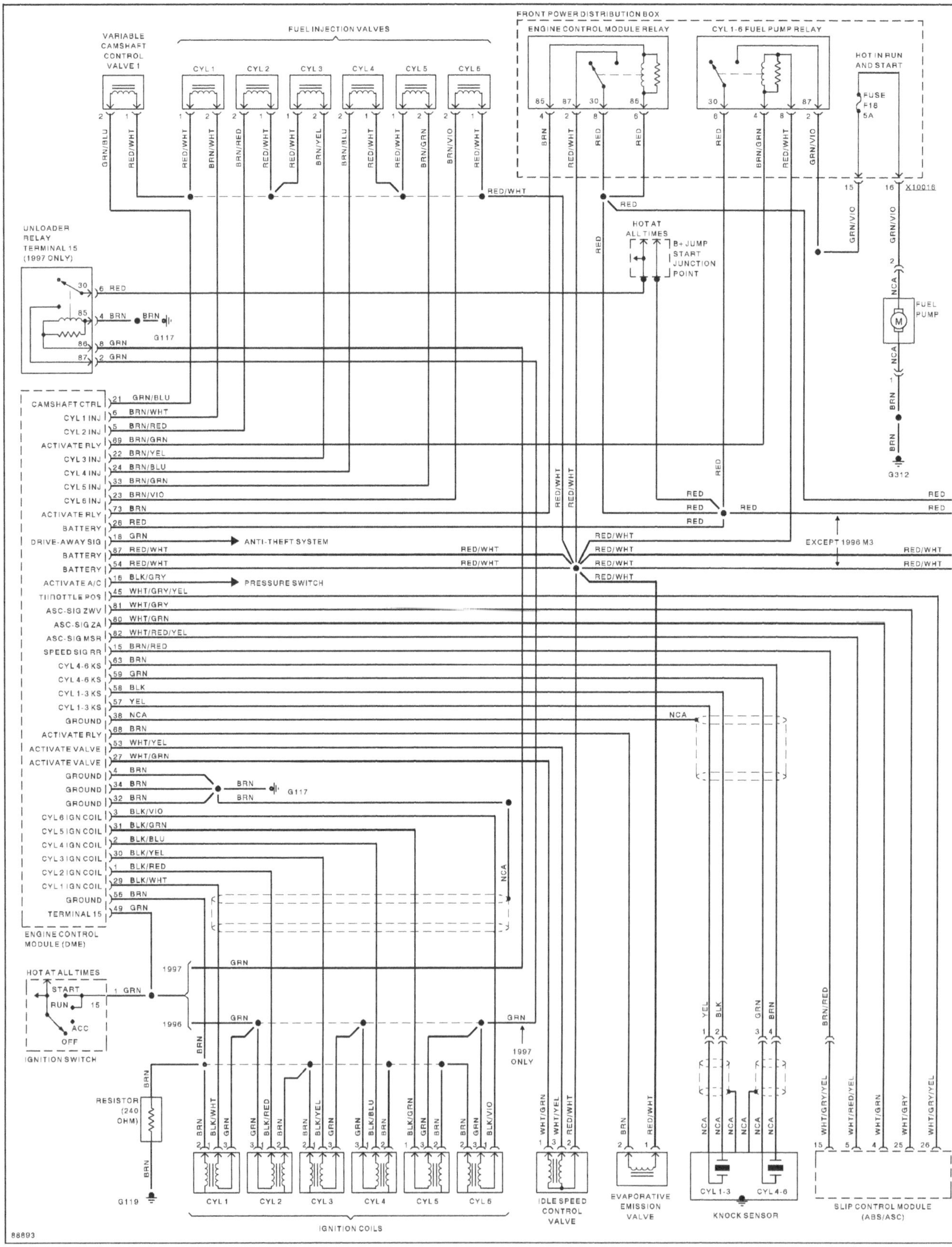

Engine Management

1996-97 2.8L (2 of 2)

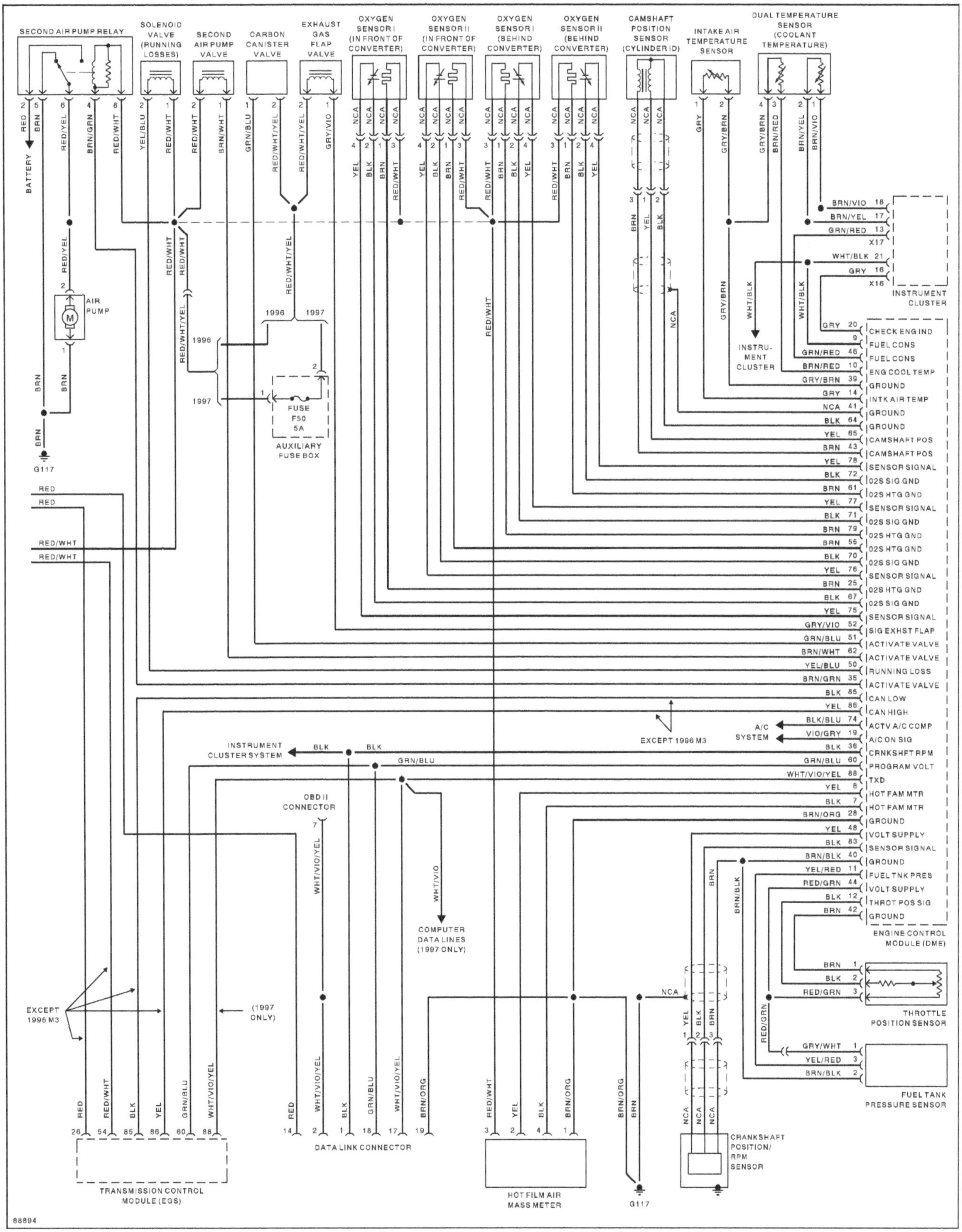

Engine Management

1998 1.9L (1 of 2)

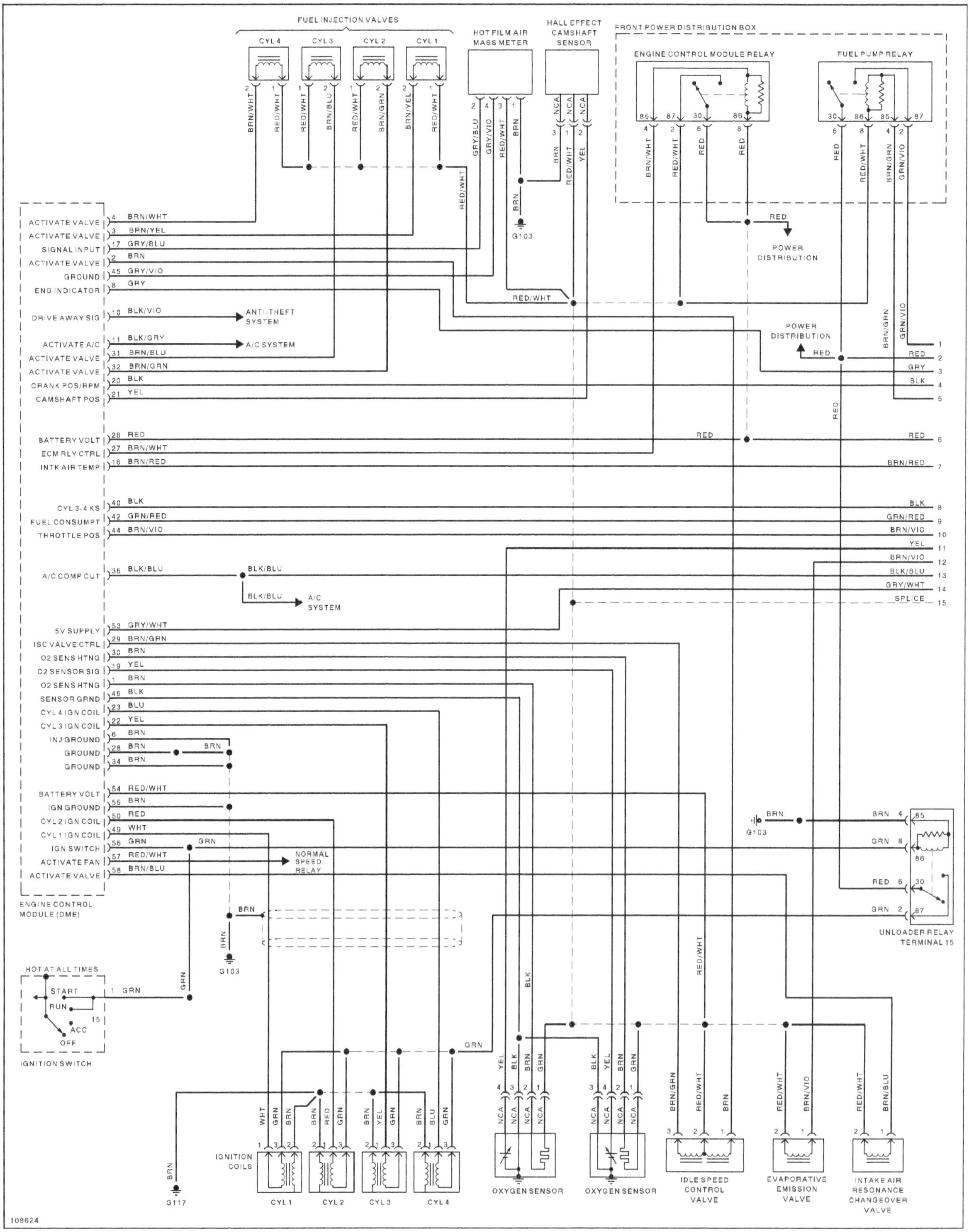

Engine Management

1998 1.9L (2 of 2)

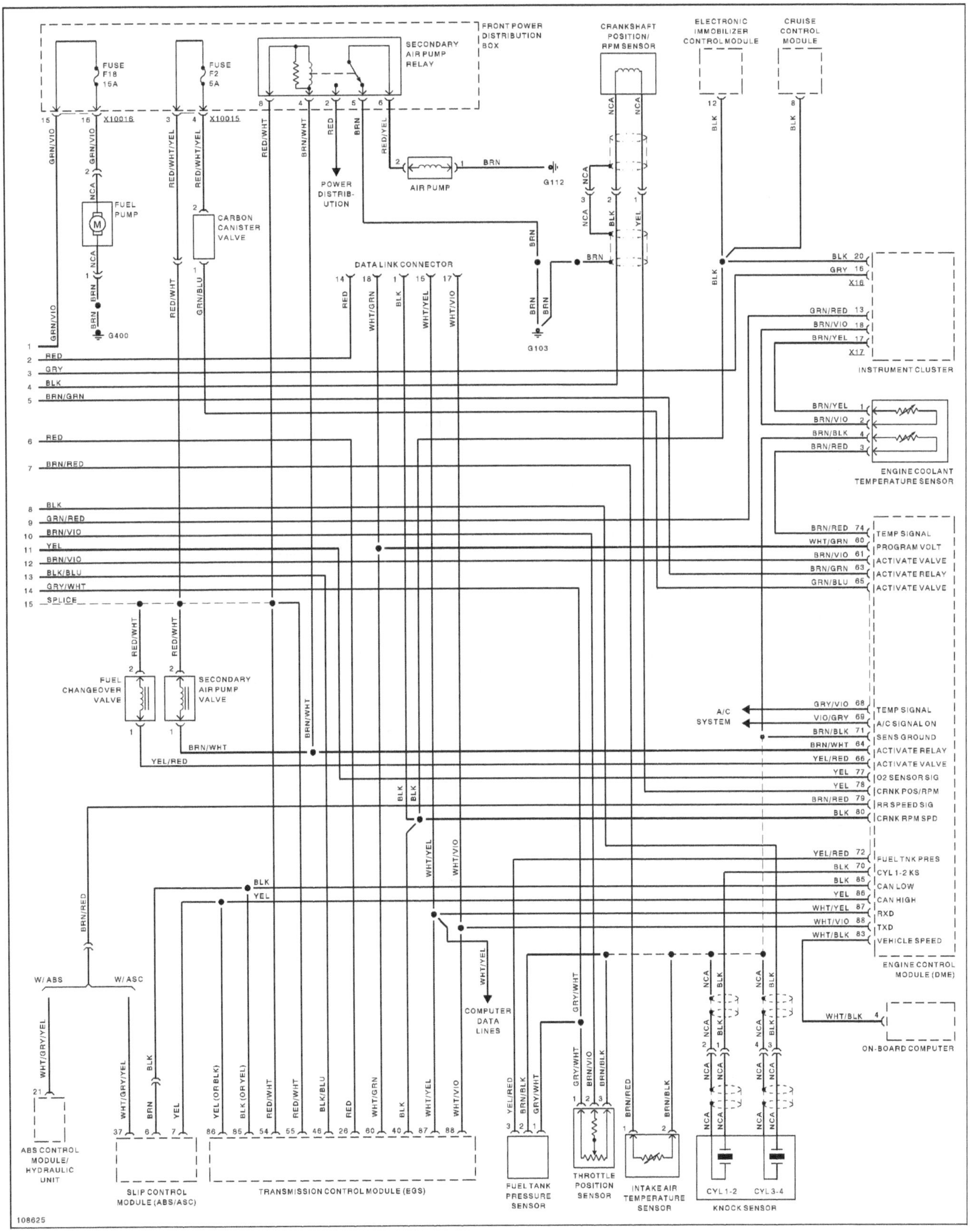

Engine Management

1998 2.5L (323i), 2.8L (328i), 3.2L (M3) (1 of 2)

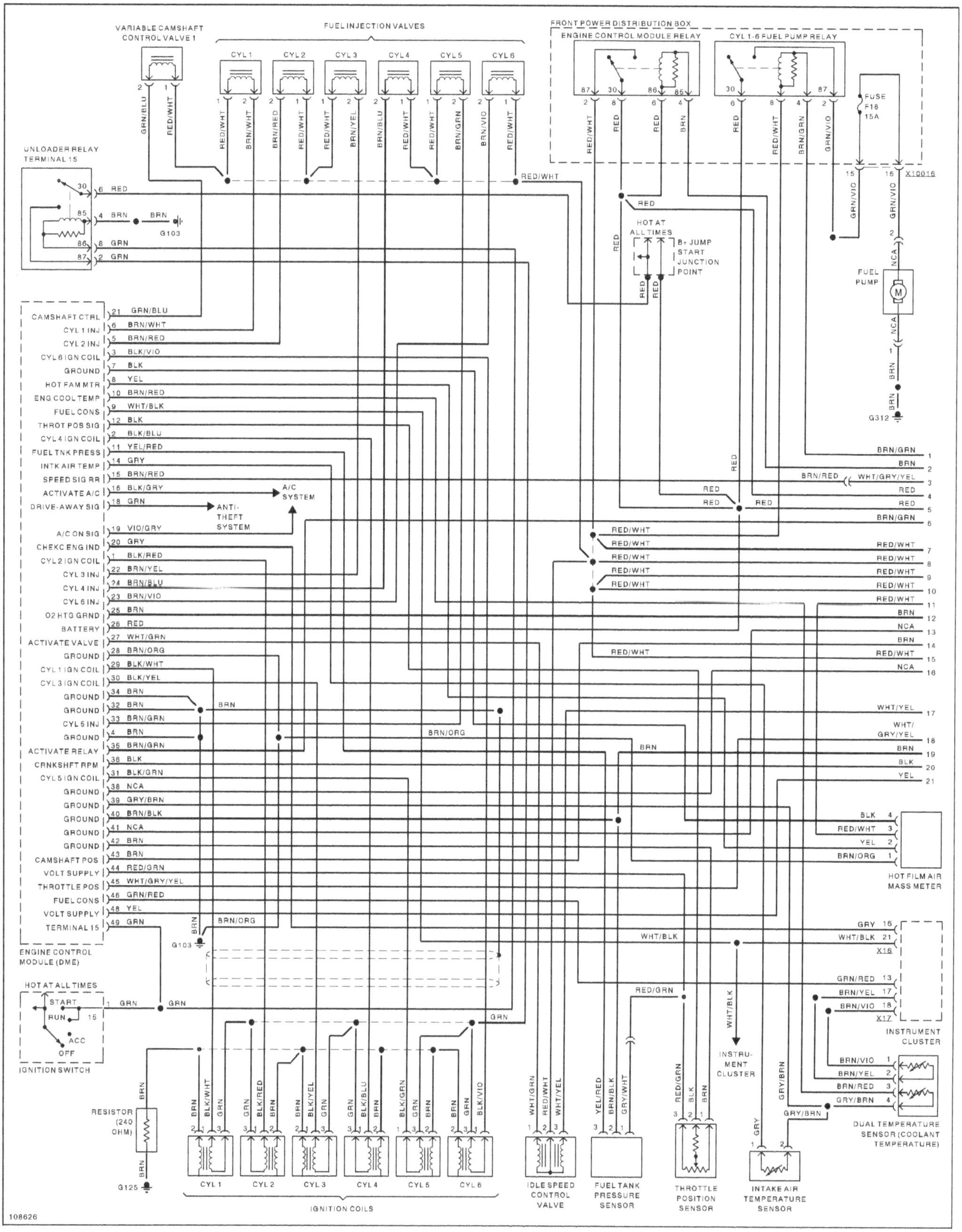

Engine Management

1998 2.5L (323i), 2.8L (328i), 3.2L (M3) (2 of 2)

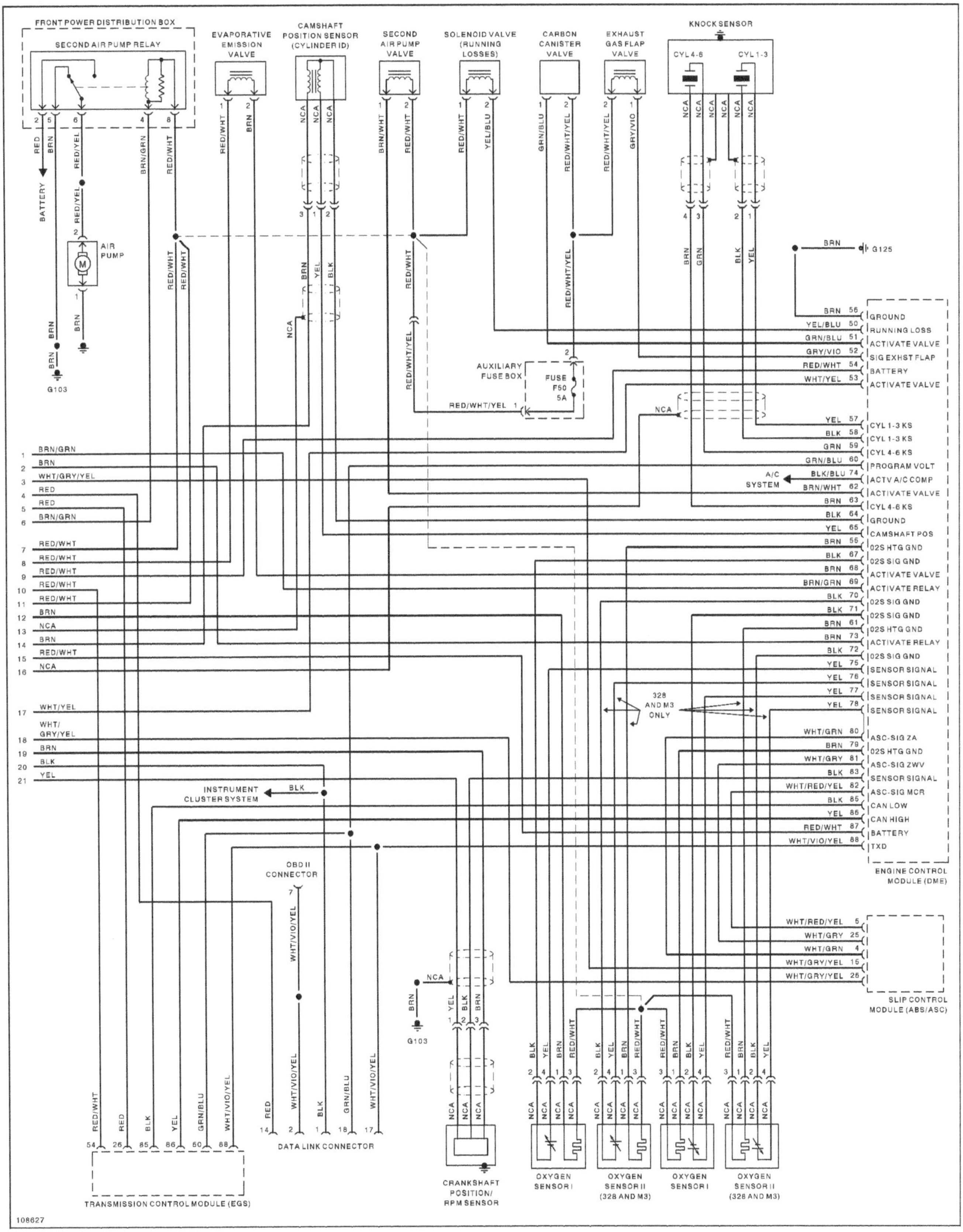

Data Link Connector

1992-93

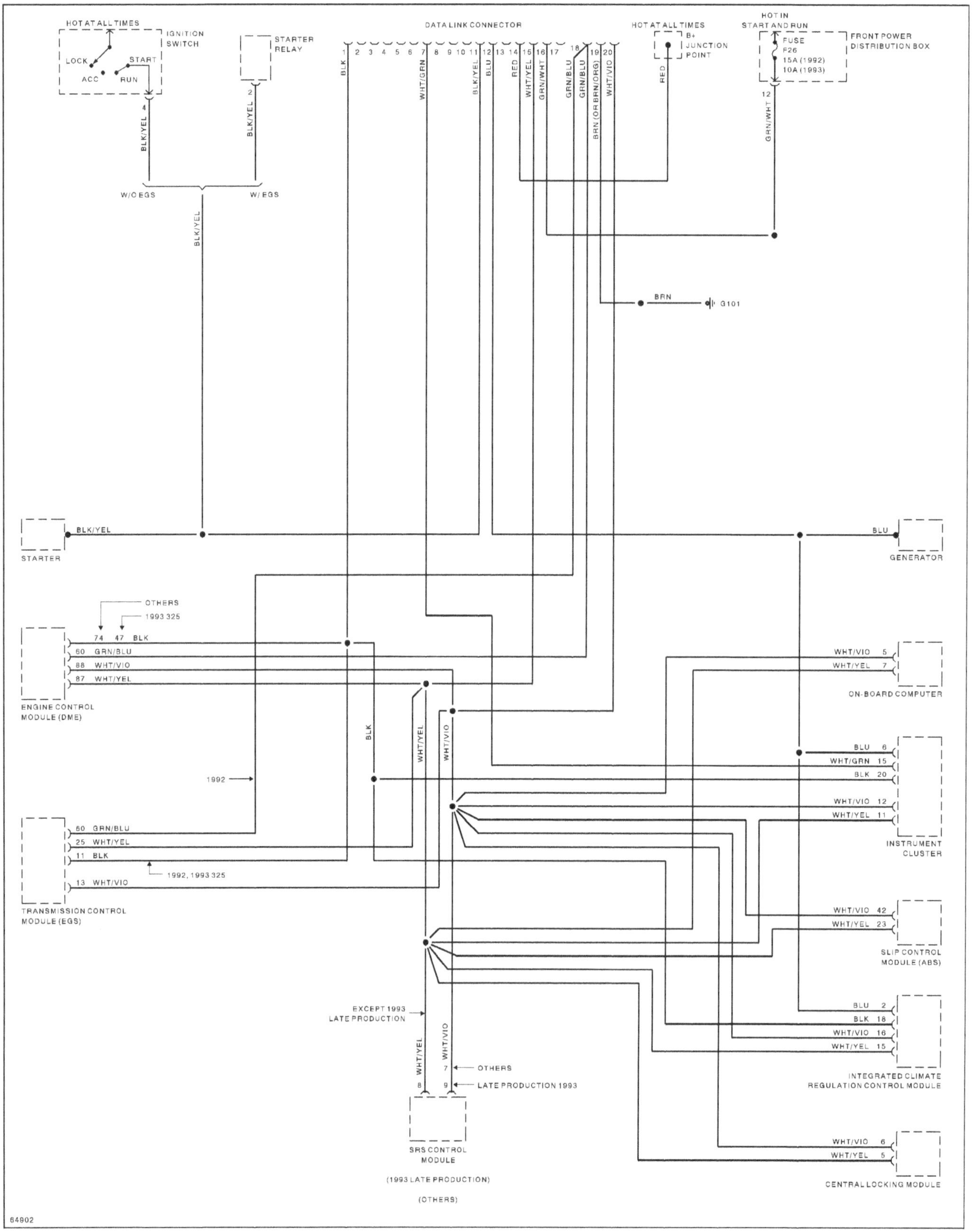

Data Link Connector

1994-95 Early Production Except 318ti

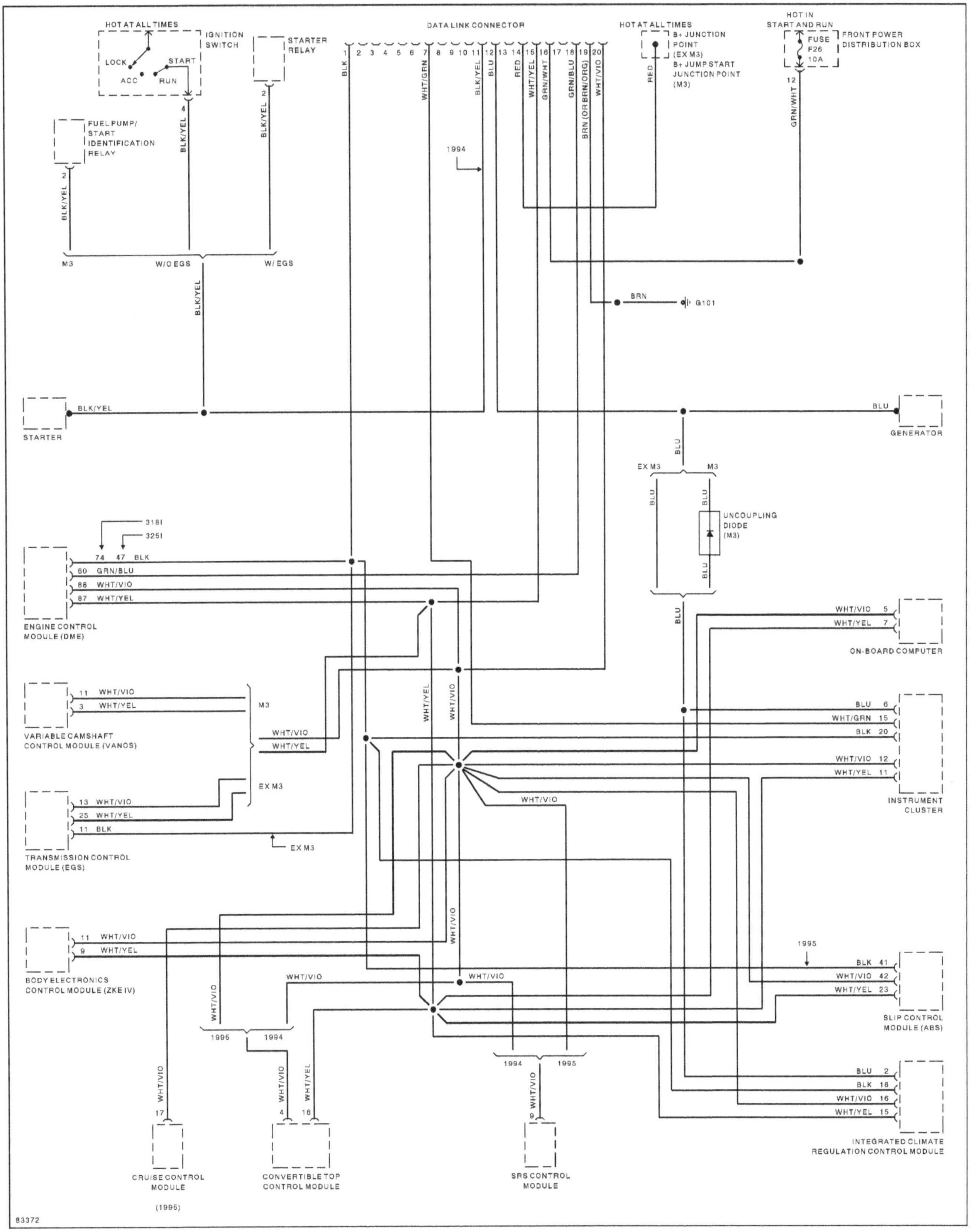

Data Link Connector

1995 Late Production Except 318ti

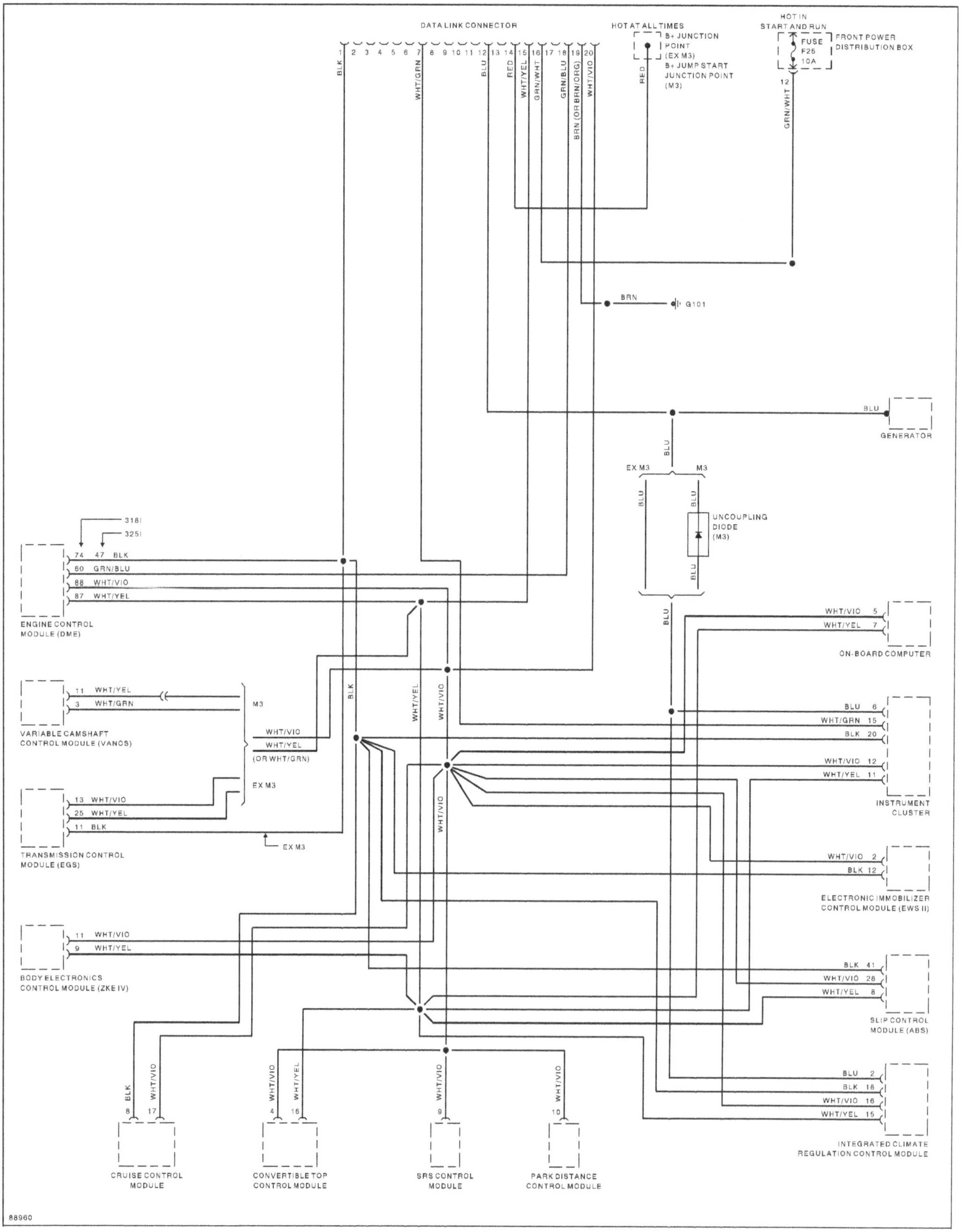

Data Link Connector

1996-97 Except 318ti

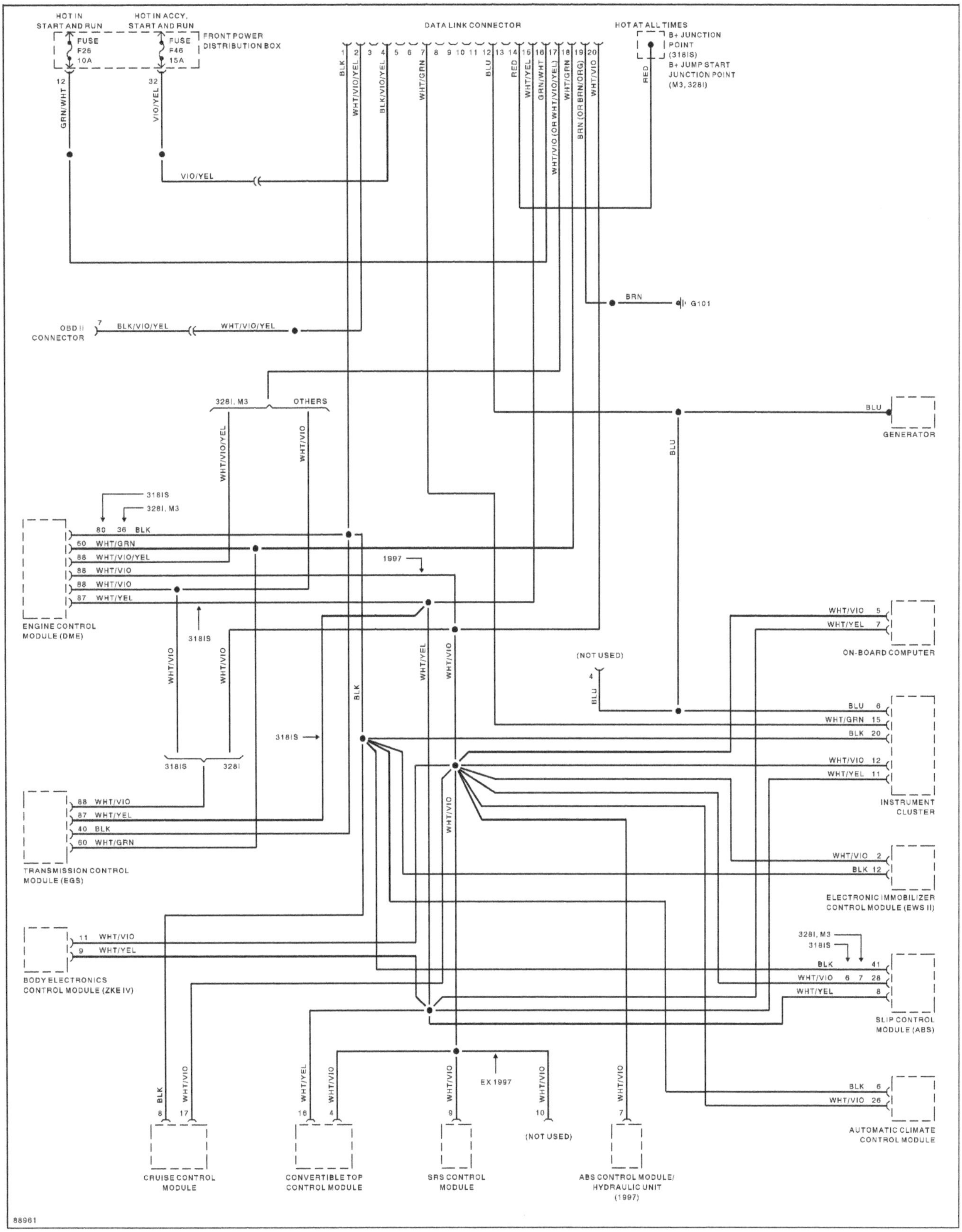

Data Link Connector

1998 Except 318ti

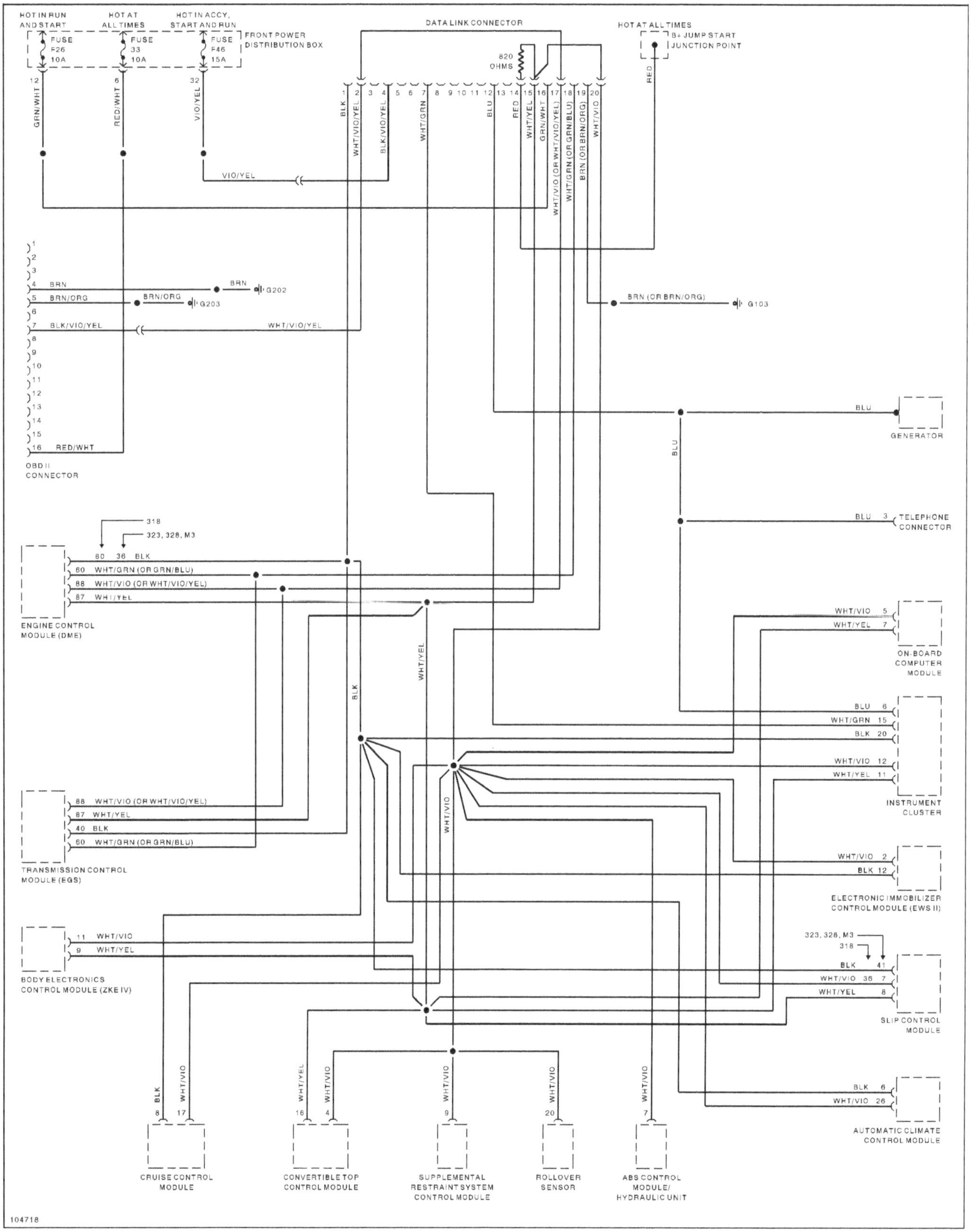

Data Link Connector

1995 318ti

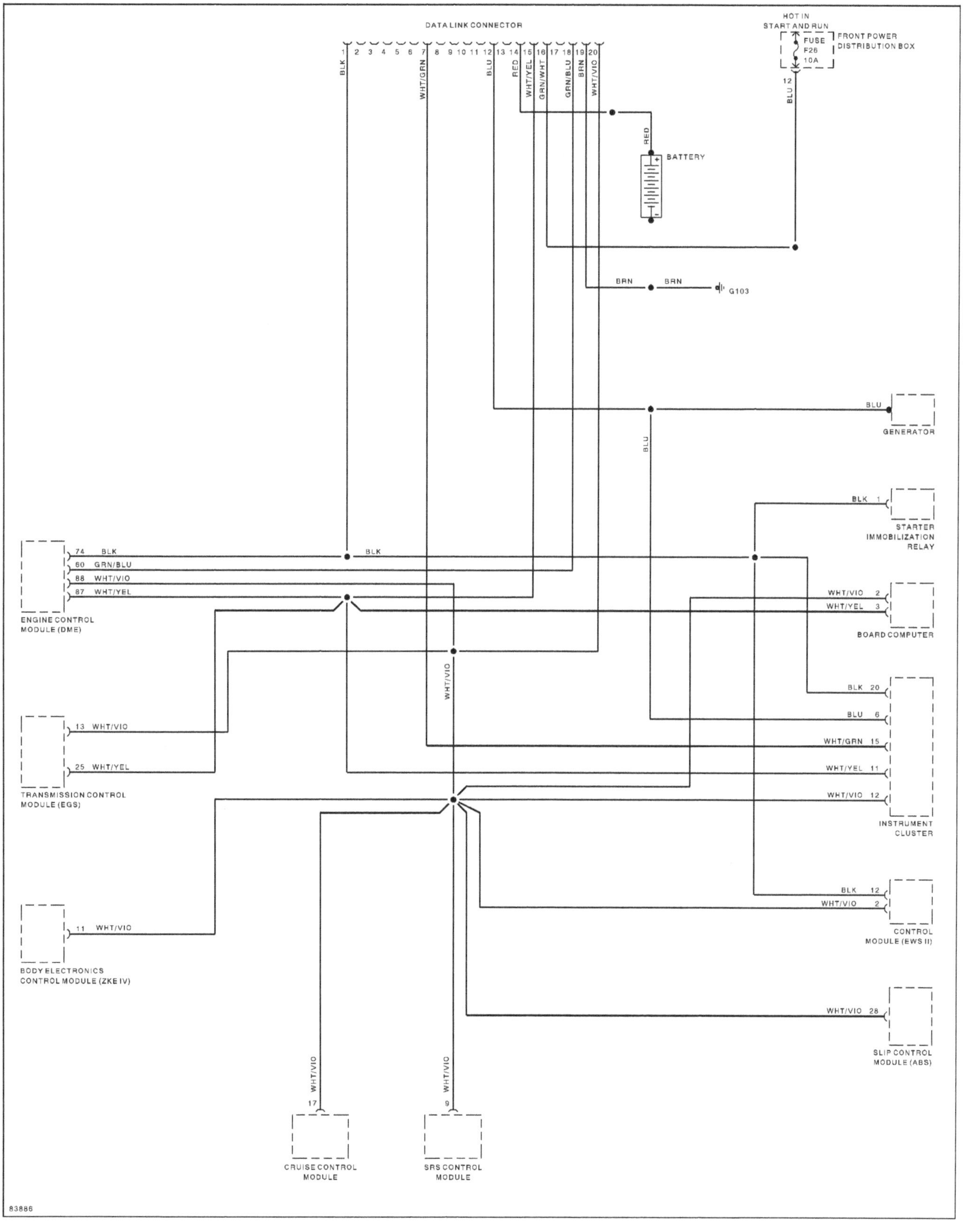

Data Link Connector

1996-97 318ti

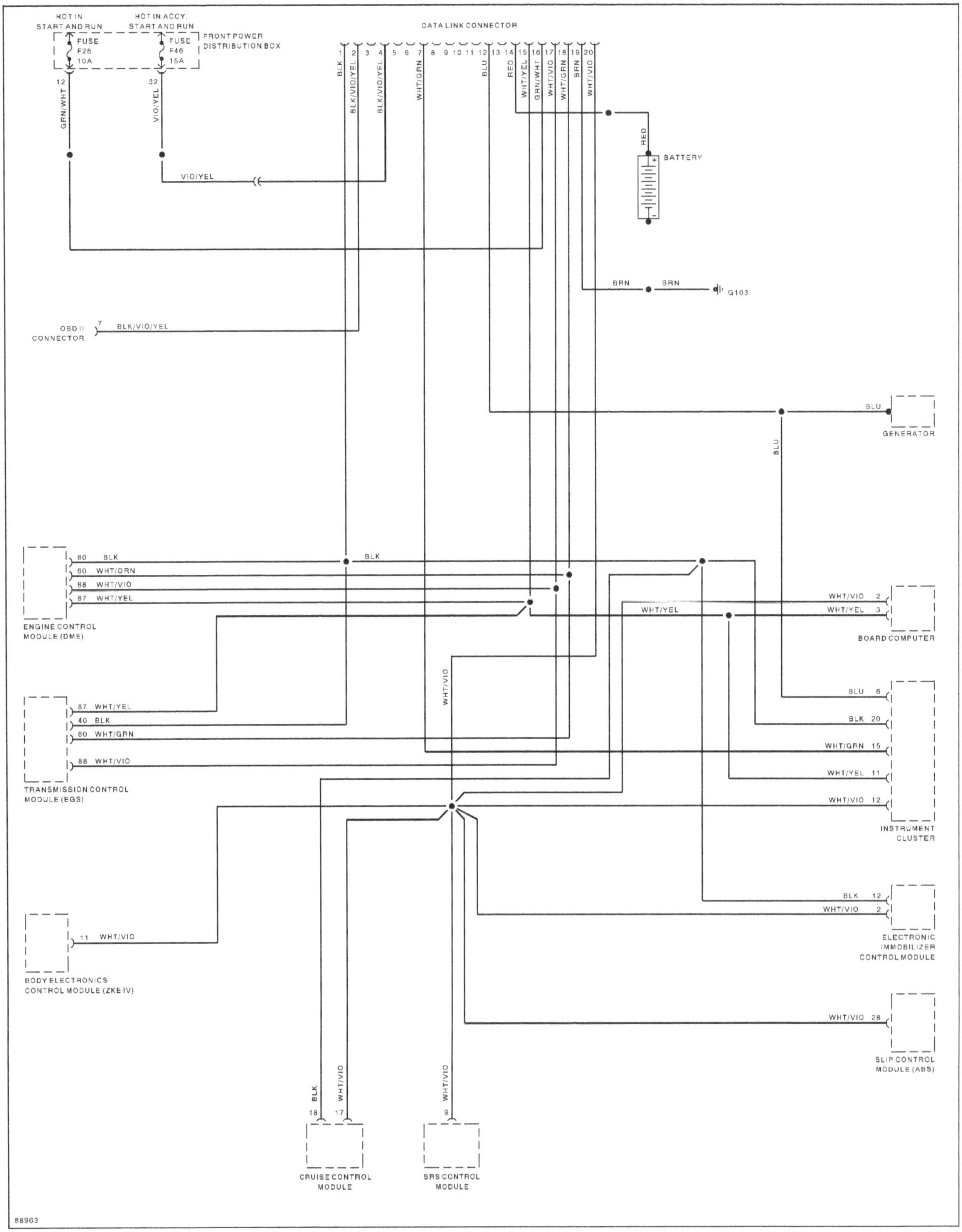

Data Link Connector

1998 318ti

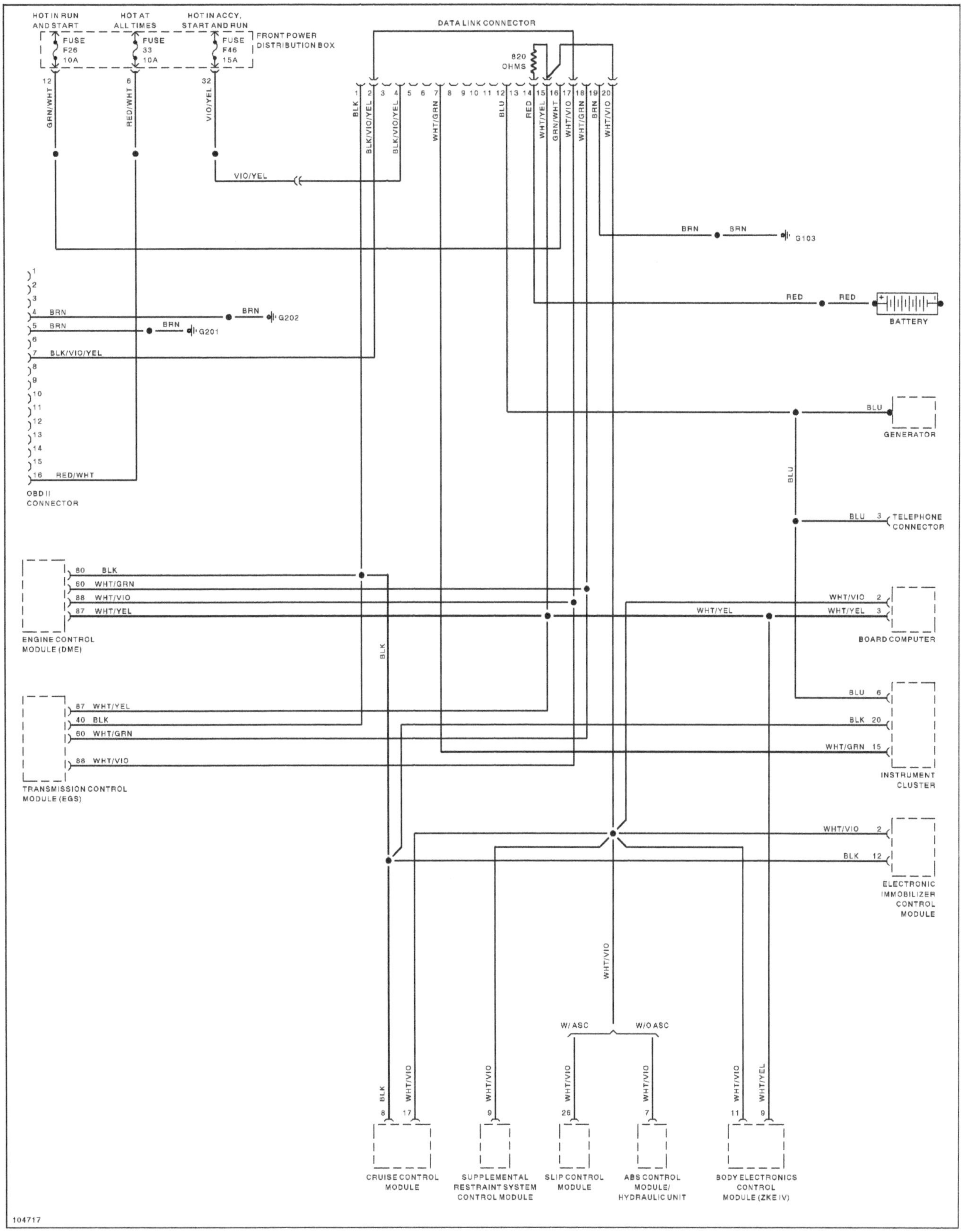

Engine Cooling

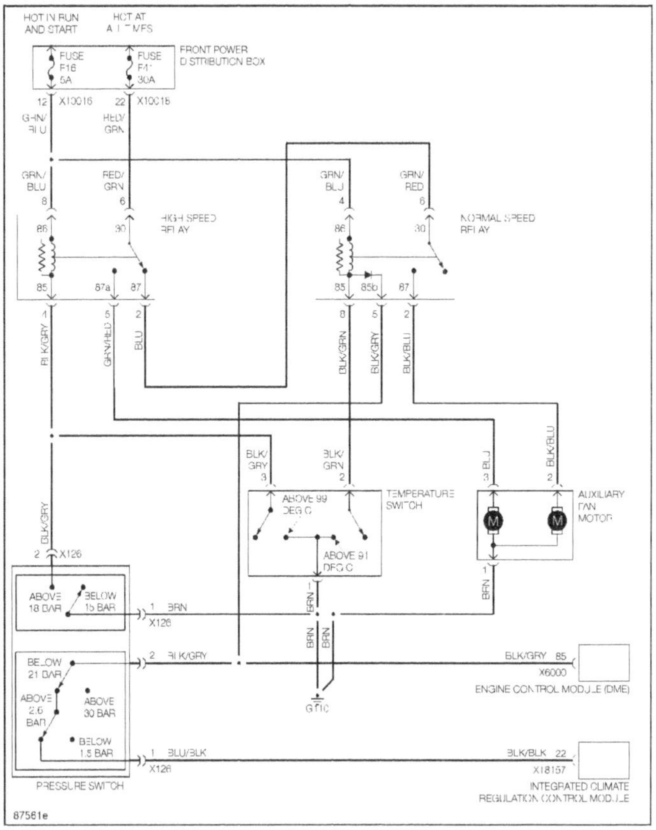

1992

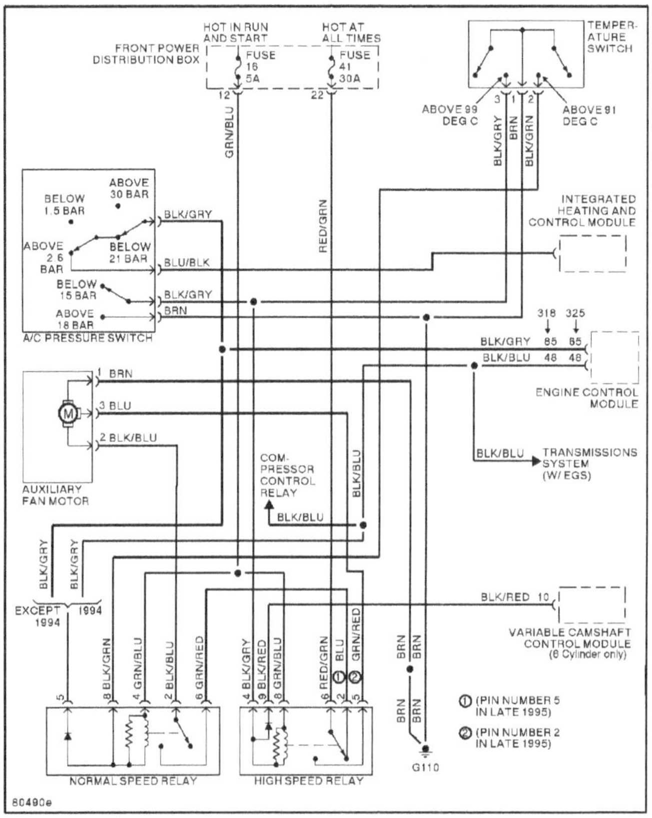

1993-95

Engine Cooling

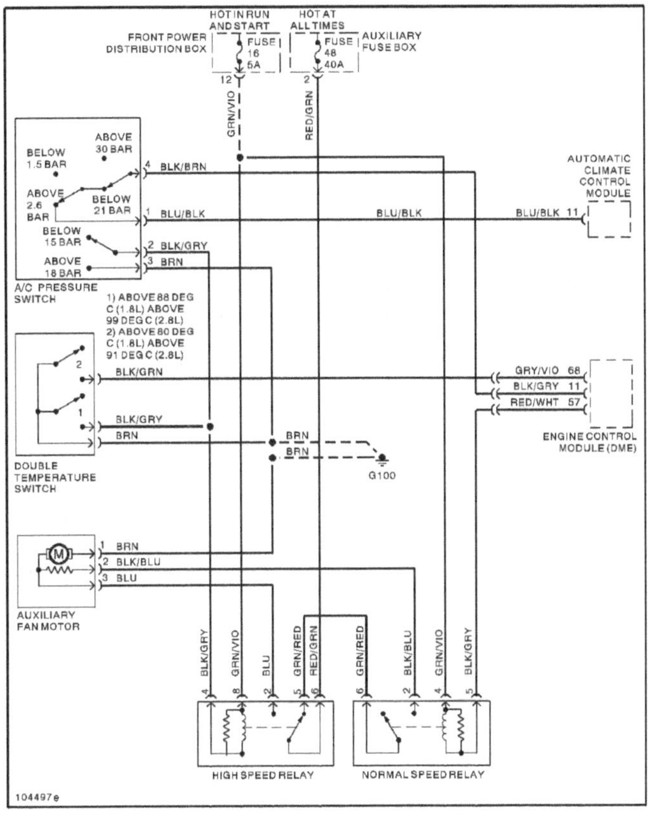

1996-98 4-cylinder

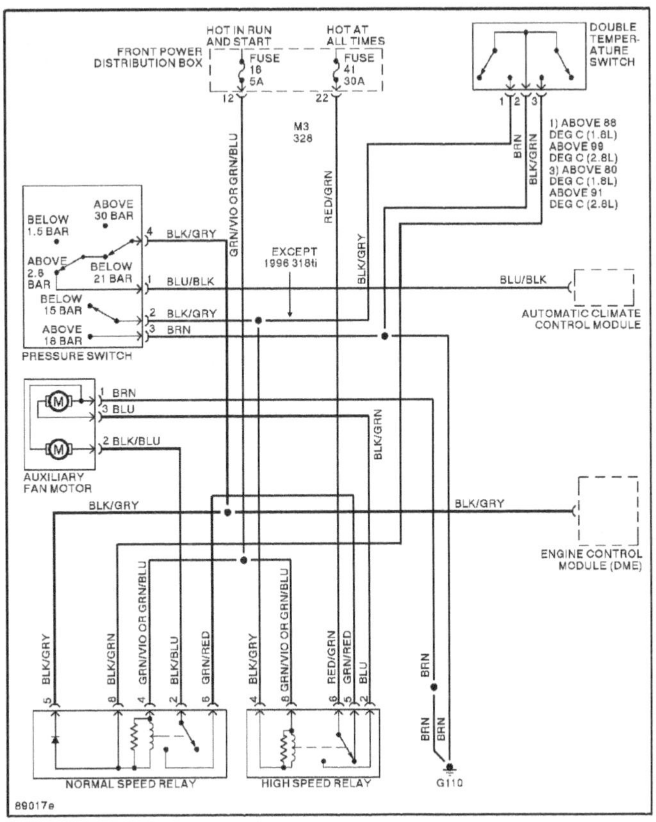

1996-98 6-cylinder

Transmission Electronics

1992 318i, 318is, 325i & 325is

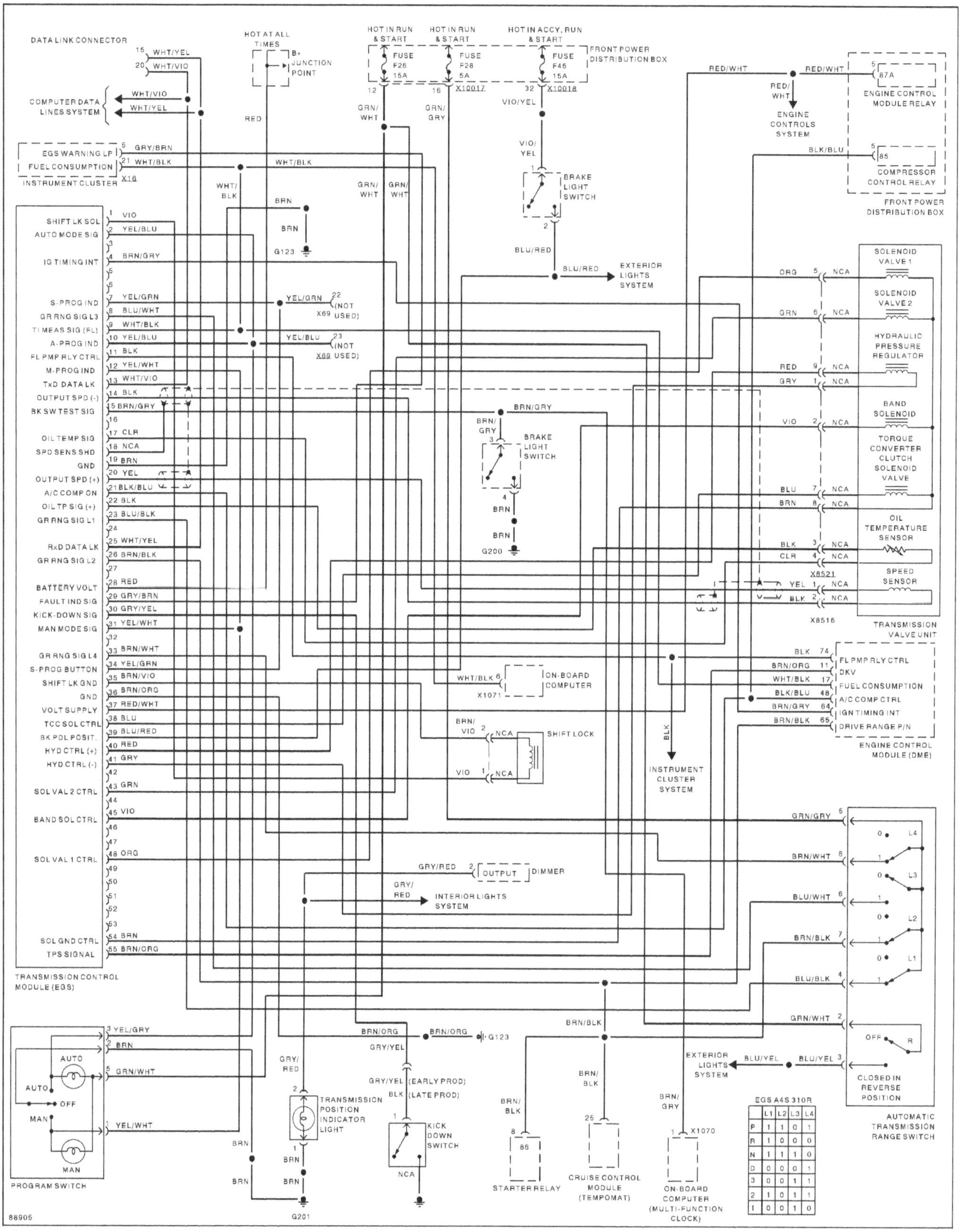

EGS A4S 310R

	L1	L2	L3	L4
P	1	1	0	1
R	1	0	0	0
N	1	1	1	0
D	0	0	0	1
3	0	0	1	1
2	1	0	1	1
1	0	0	1	0

Transmission Electronics

1993-94 318i, 318iC & 318is

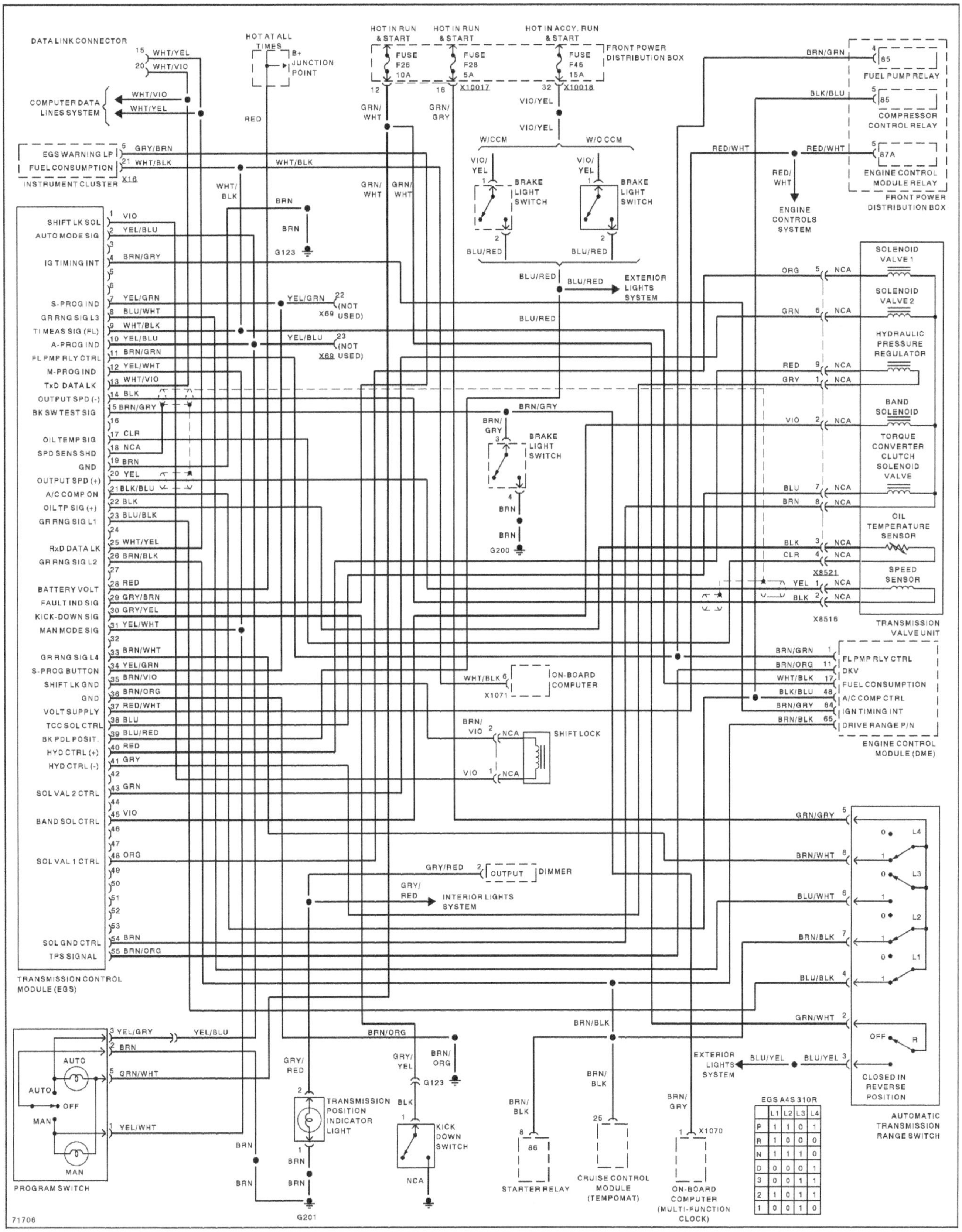

EGS A4S 310R	L1	L2	L3	L4
P	1	1	0	1
R	1	0	0	0
N	1	1	1	0
D	0	0	0	1
3	0	0	1	1
2	1	0	1	1
1	0	0	1	0

Transmission Electronics

1993-94 325i, 325iC & 325is

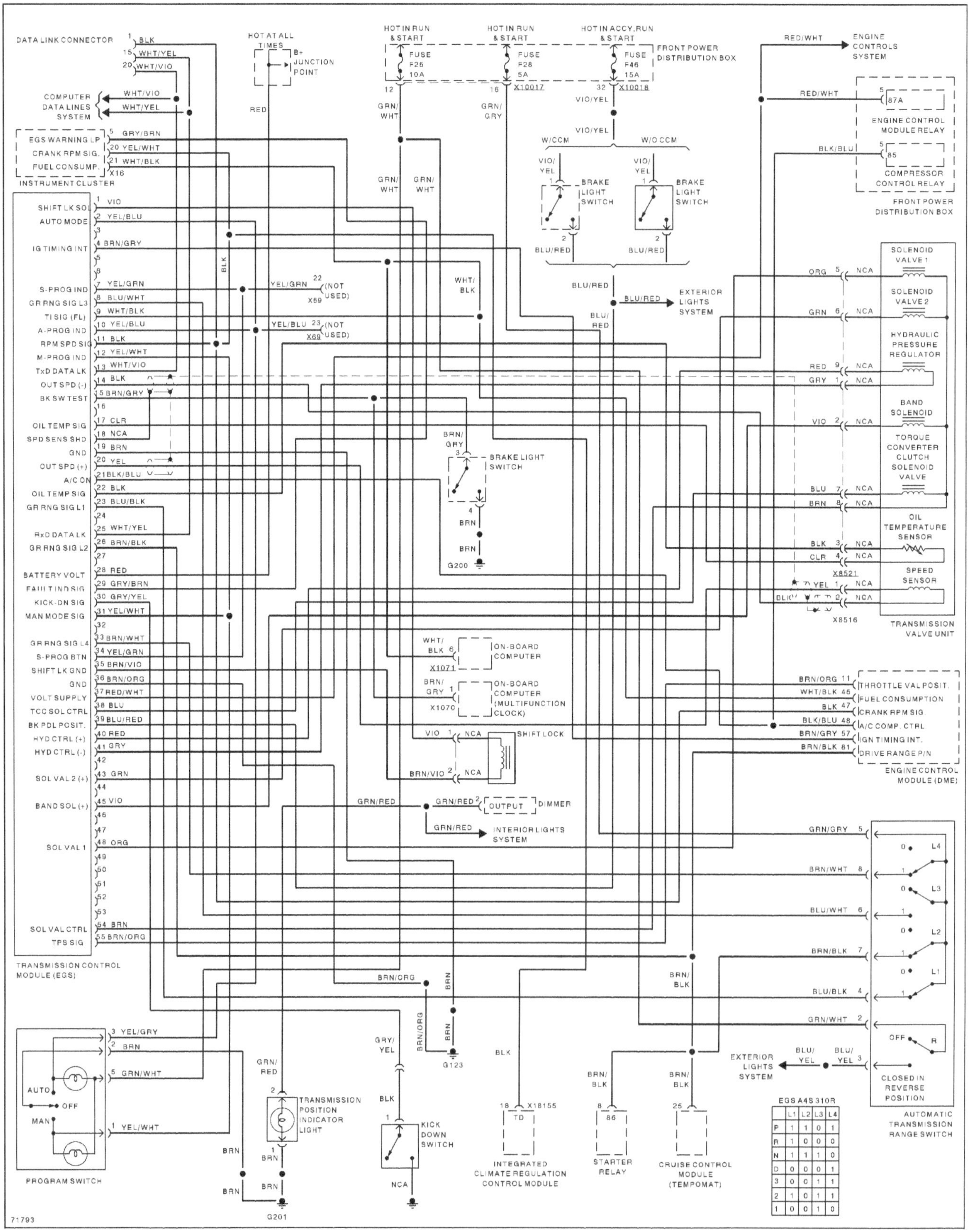

EGS A4S 310R

	L1	L2	L3	L4
P	1	1	0	1
R	1	0	0	0
N	1	1	1	0
D	0	0	0	1
3	0	0	1	1
2	1	0	1	1
1	1	0	0	1

Transmission Electronics

1995 318i, 318iC & 318is

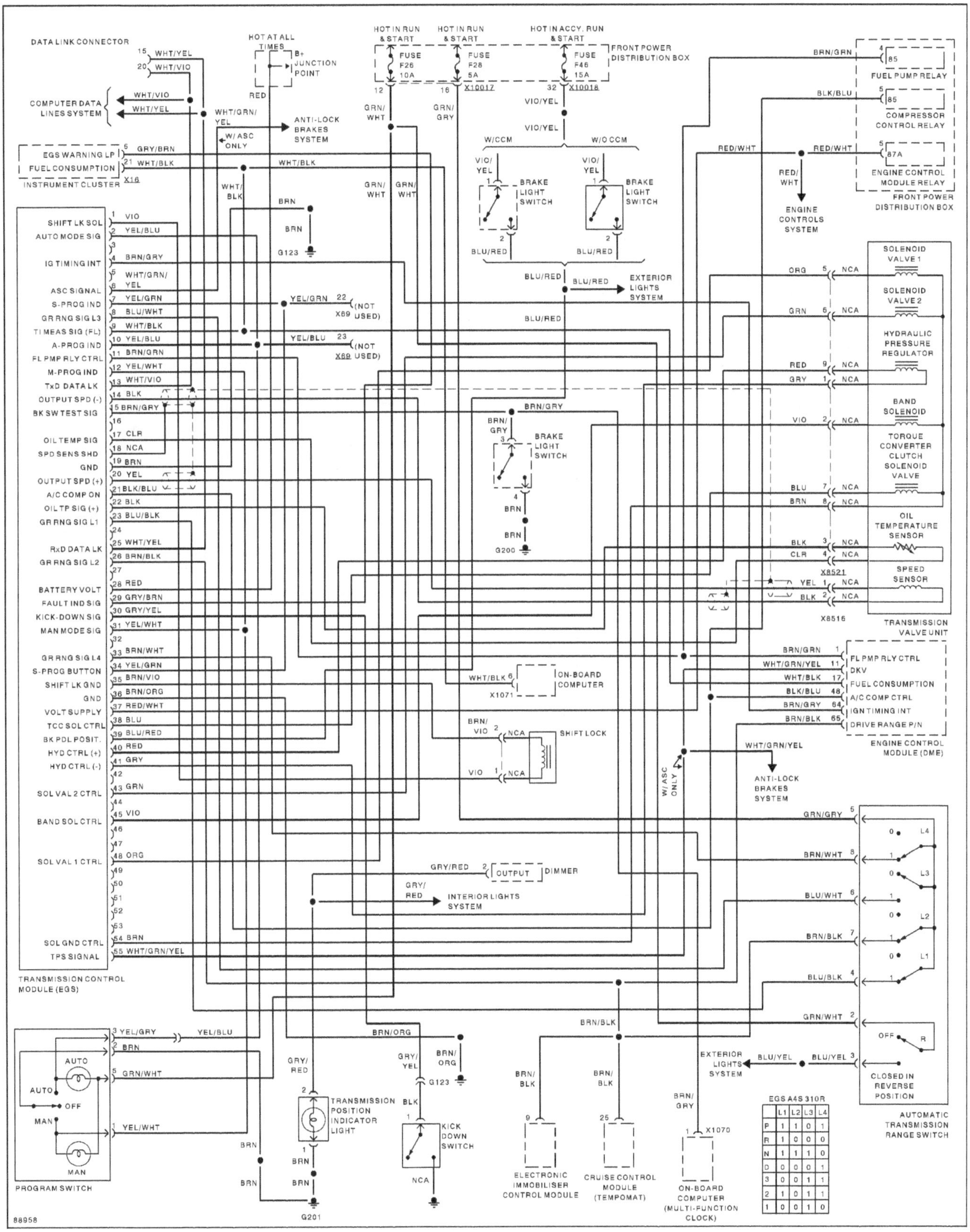

EGS A4S 310R

	L1	L2	L3	L4
P	1	1	0	1
R	1	0	0	0
N	1	1	1	0
D	0	0	0	1
3	0	0	1	1
2	1	0	1	1
1	1	0	0	1	0

Transmission Electronics

1995 325i, 325iC & 325is

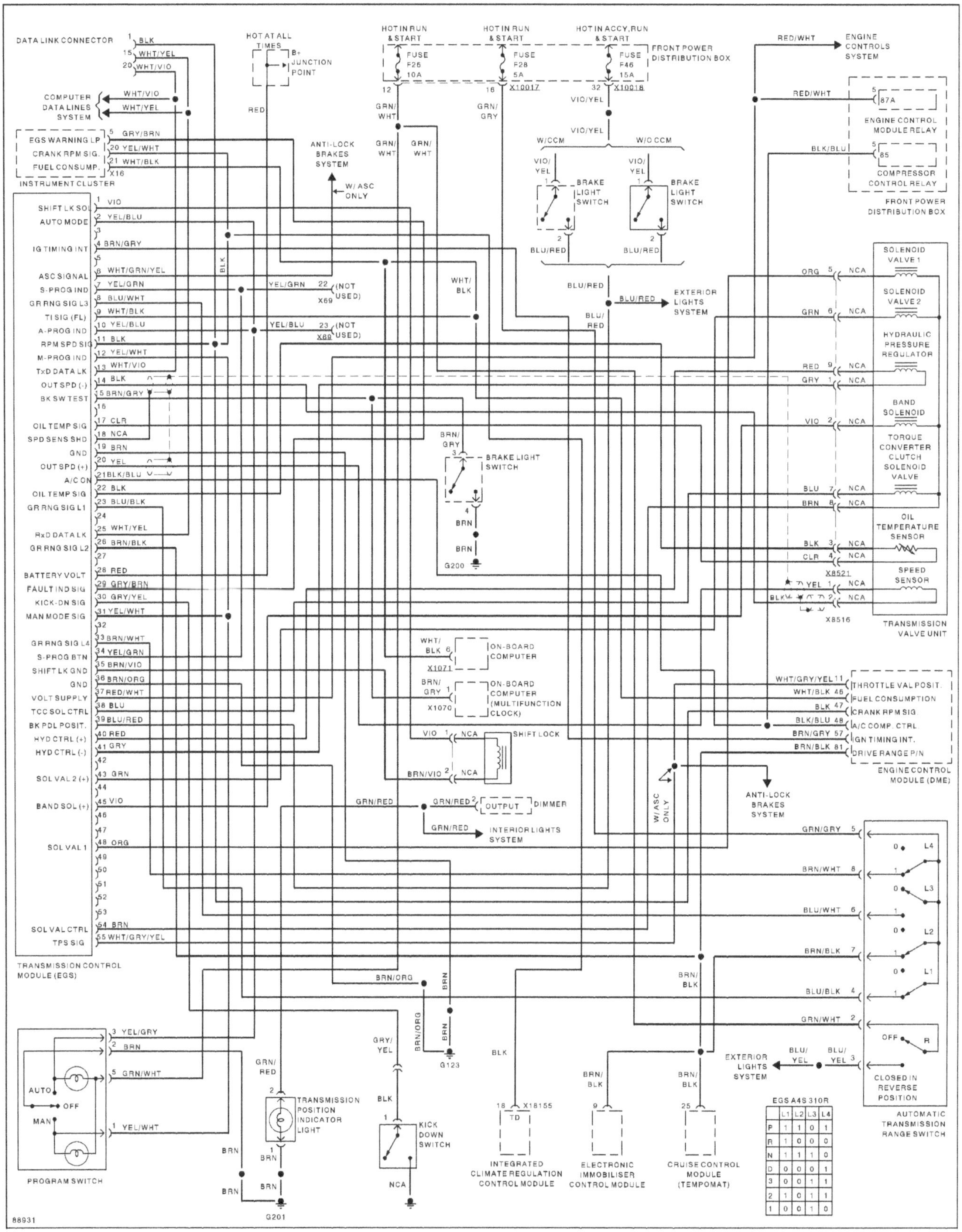

EGS A4S 310R

	L1	L2	L3	L4
P	1	1	0	1
R	1	0	0	0
N	1	1	1	0
D	0	0	0	1
3	0	0	1	1
2	1	0	1	1
1	0	0	1	0

Transmission Electronics

1995 318ti

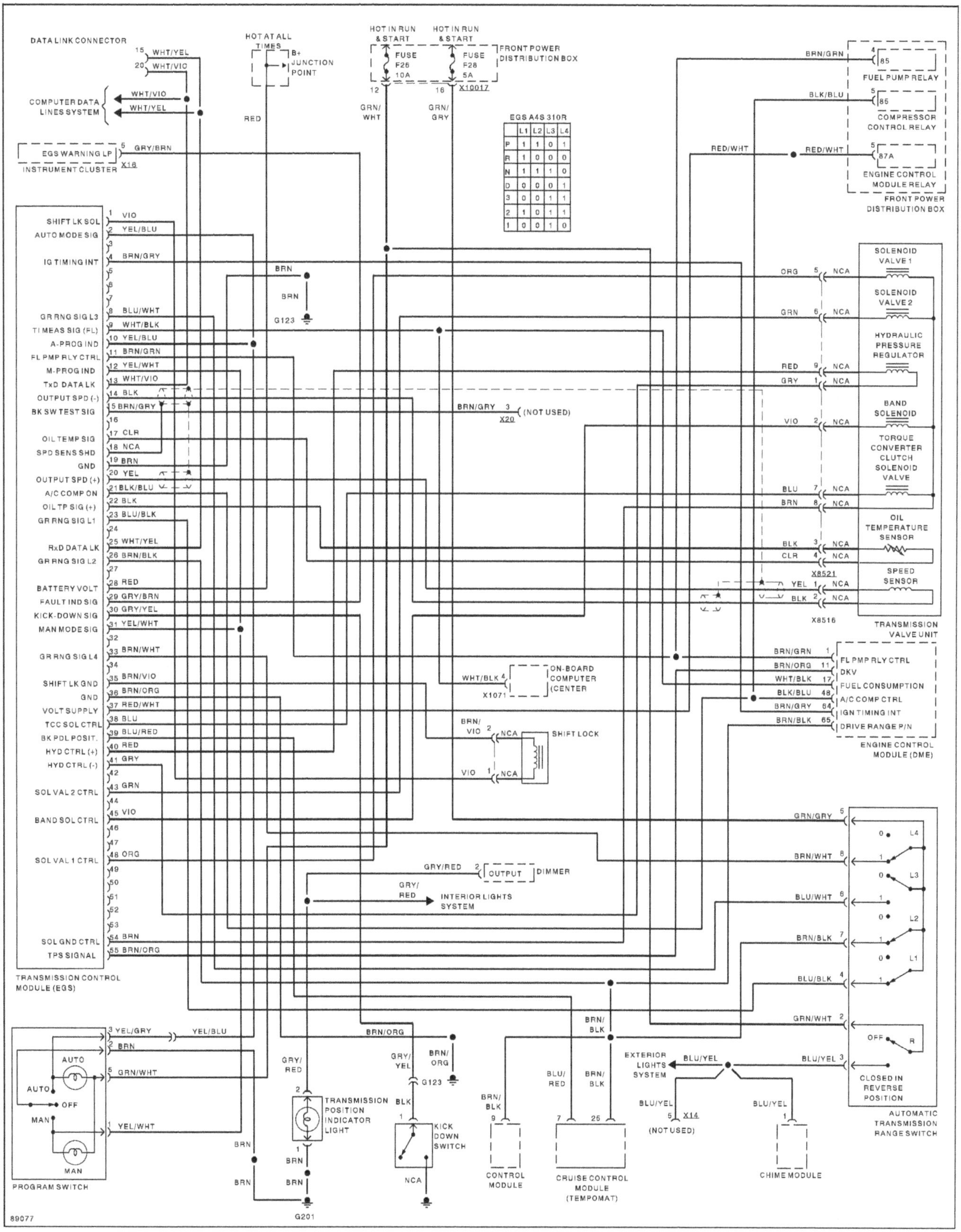

EGS A4S 310R	L1	L2	L3	L4
P	1	1	0	1
R	1	0	0	0
N	1	1	1	0
D	0	0	0	1
3	0	0	1	1
2	1	0	1	1
1	1	0	0	1	0

Transmission Electronics

1996-97 Except M3 (1 of 2)

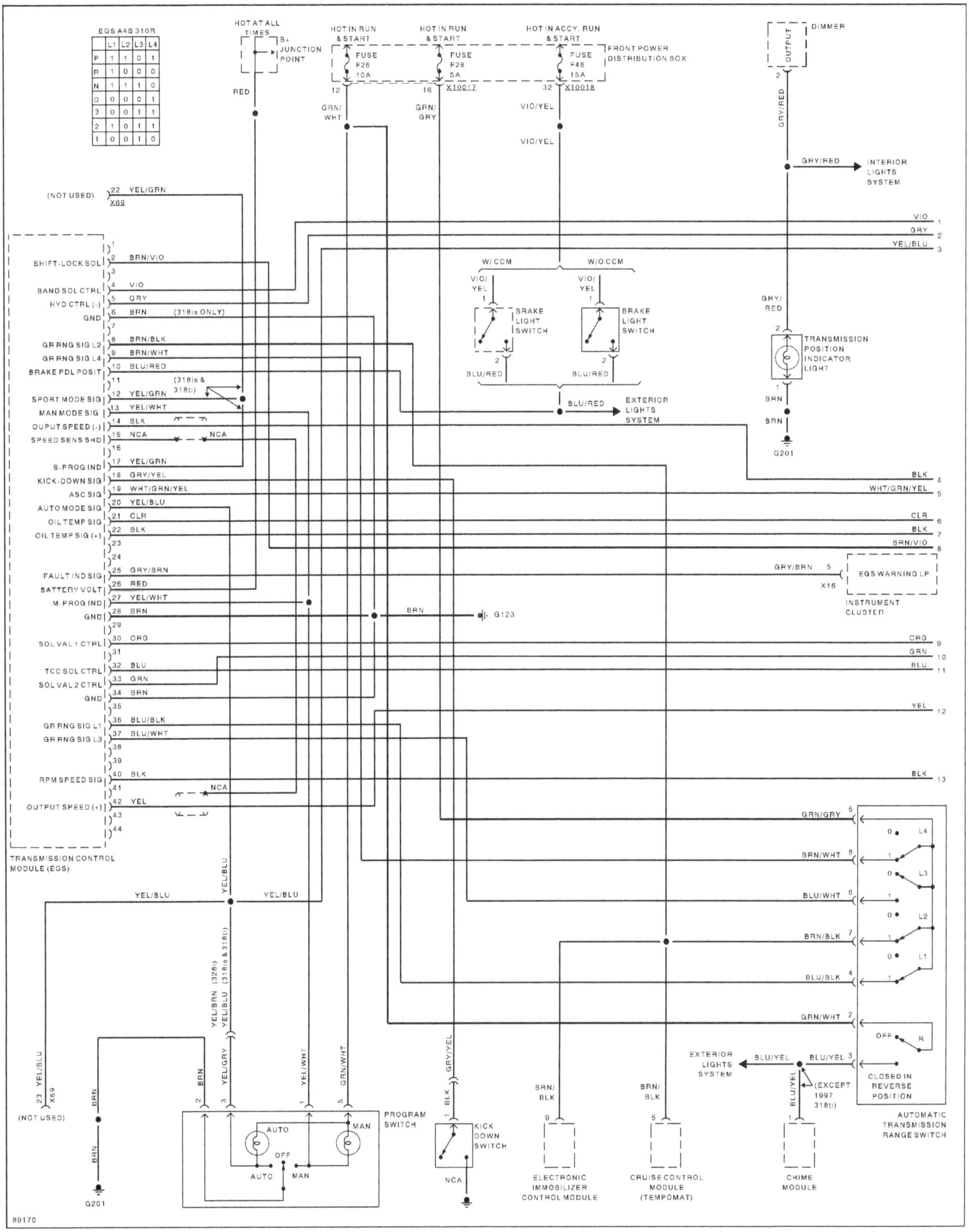

Transmission Electronics

1996-97 Except M3 (2 of 2)

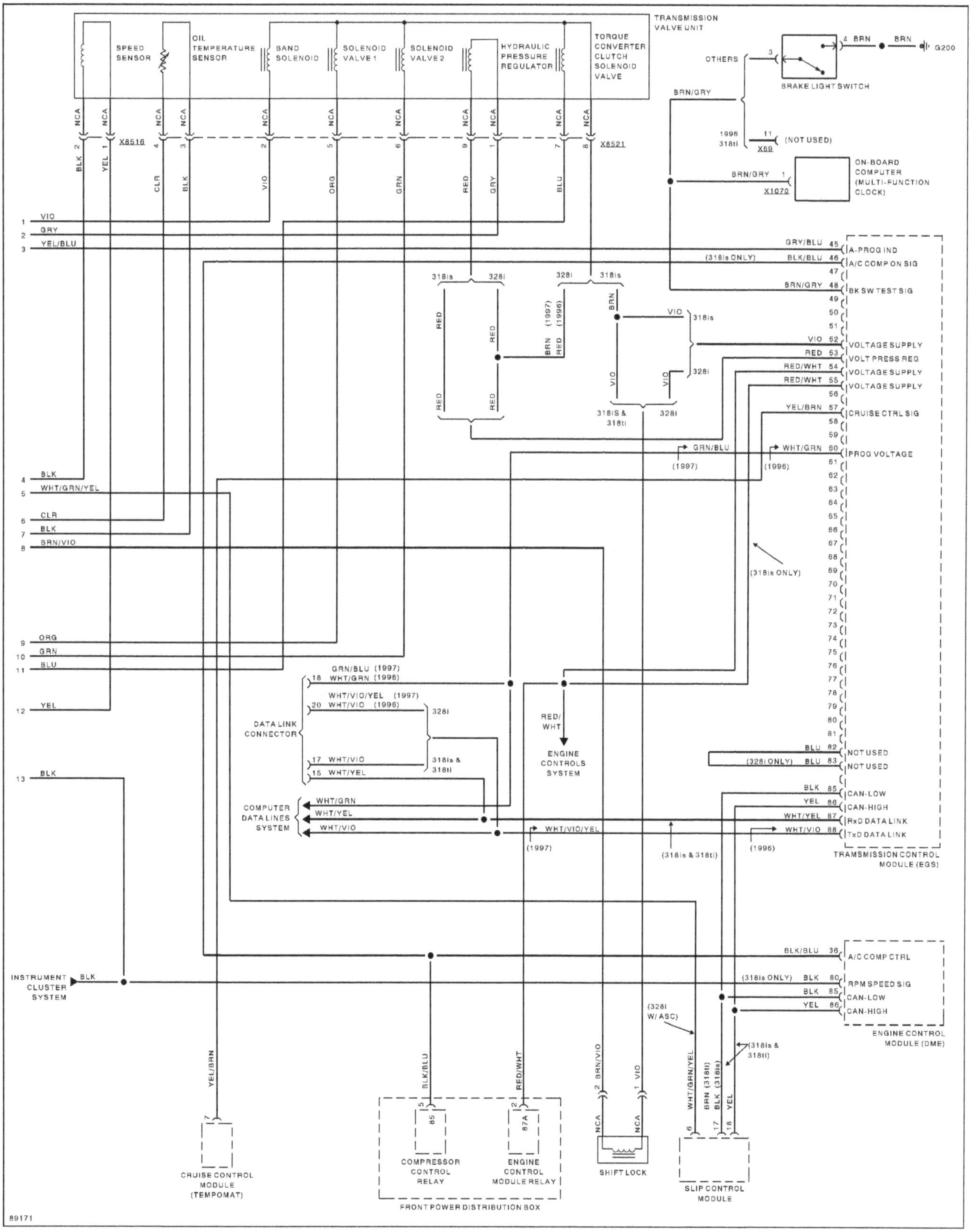

Transmission Electronics

1998 Except M3 (1 of 2)

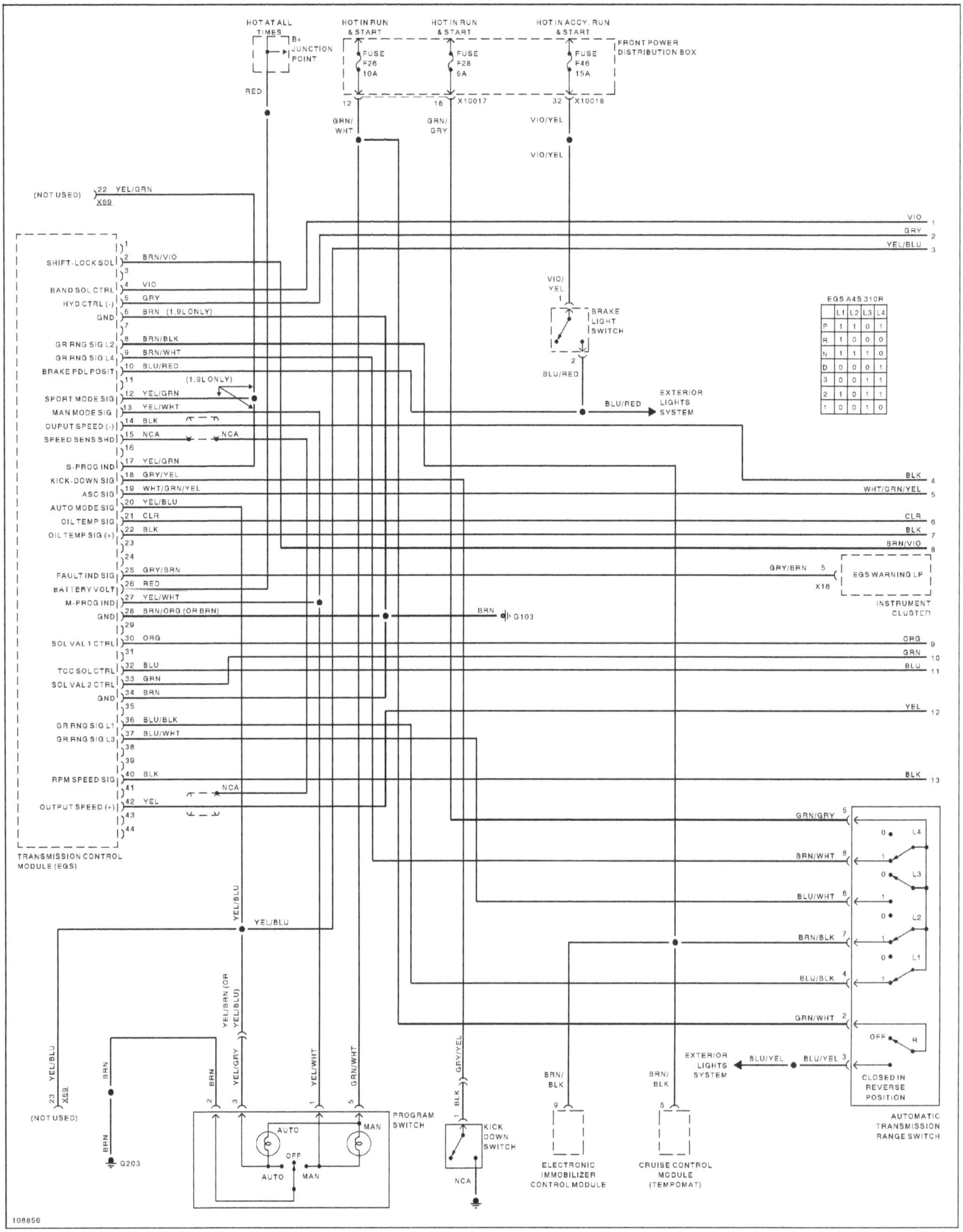

EGS A4S 310R

	L1	L2	L3	L4
P	1	1	0	1
R	1	0	0	0
N	1	1	1	0
D	0	0	0	1
3	0	0	1	1
2	1	0	1	1
1	0	0	1	0

Transmission Electronics

1998 Except M3 (2 of 2)

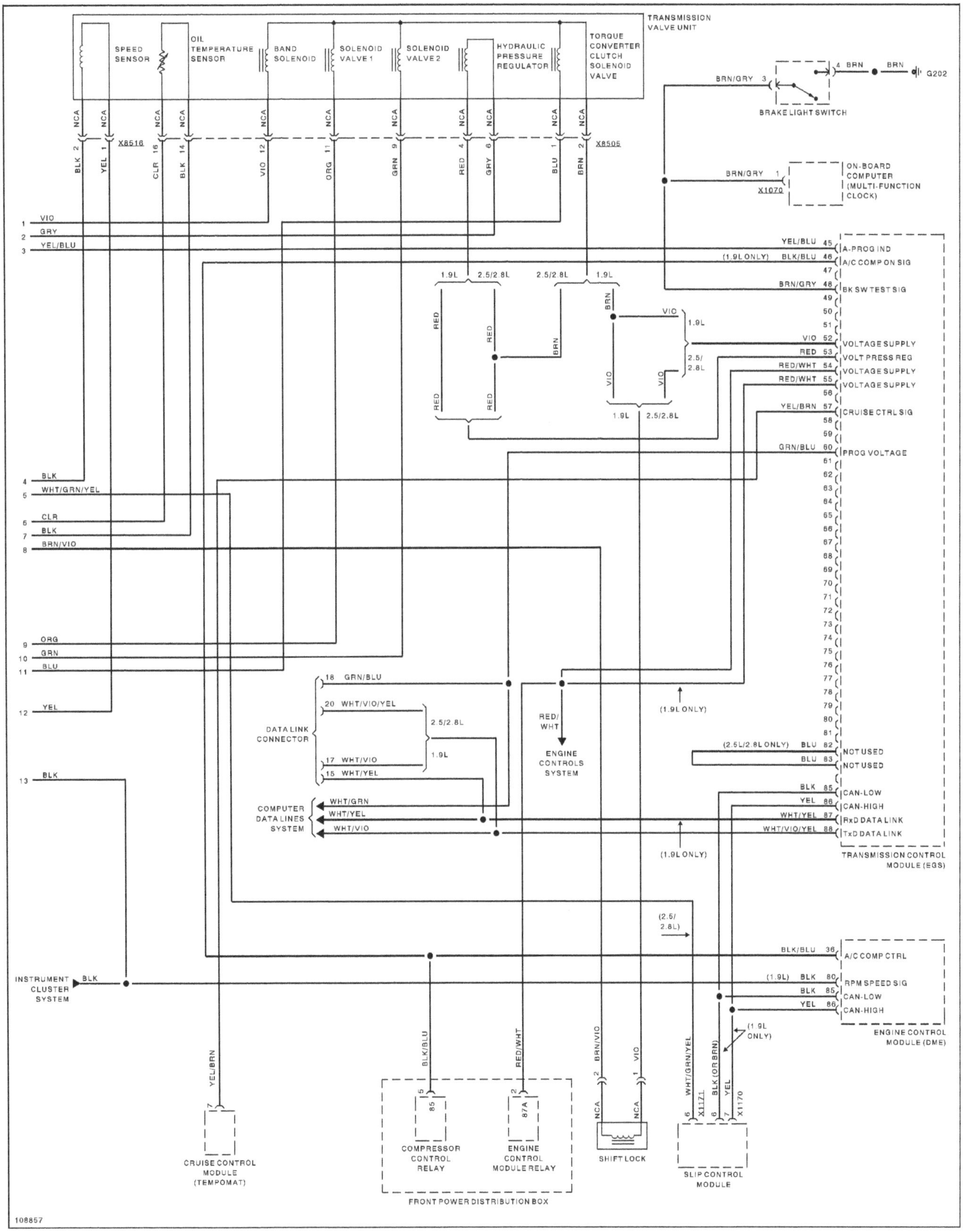

Transmission Electronics

1995-97 M3 (1 of 2)

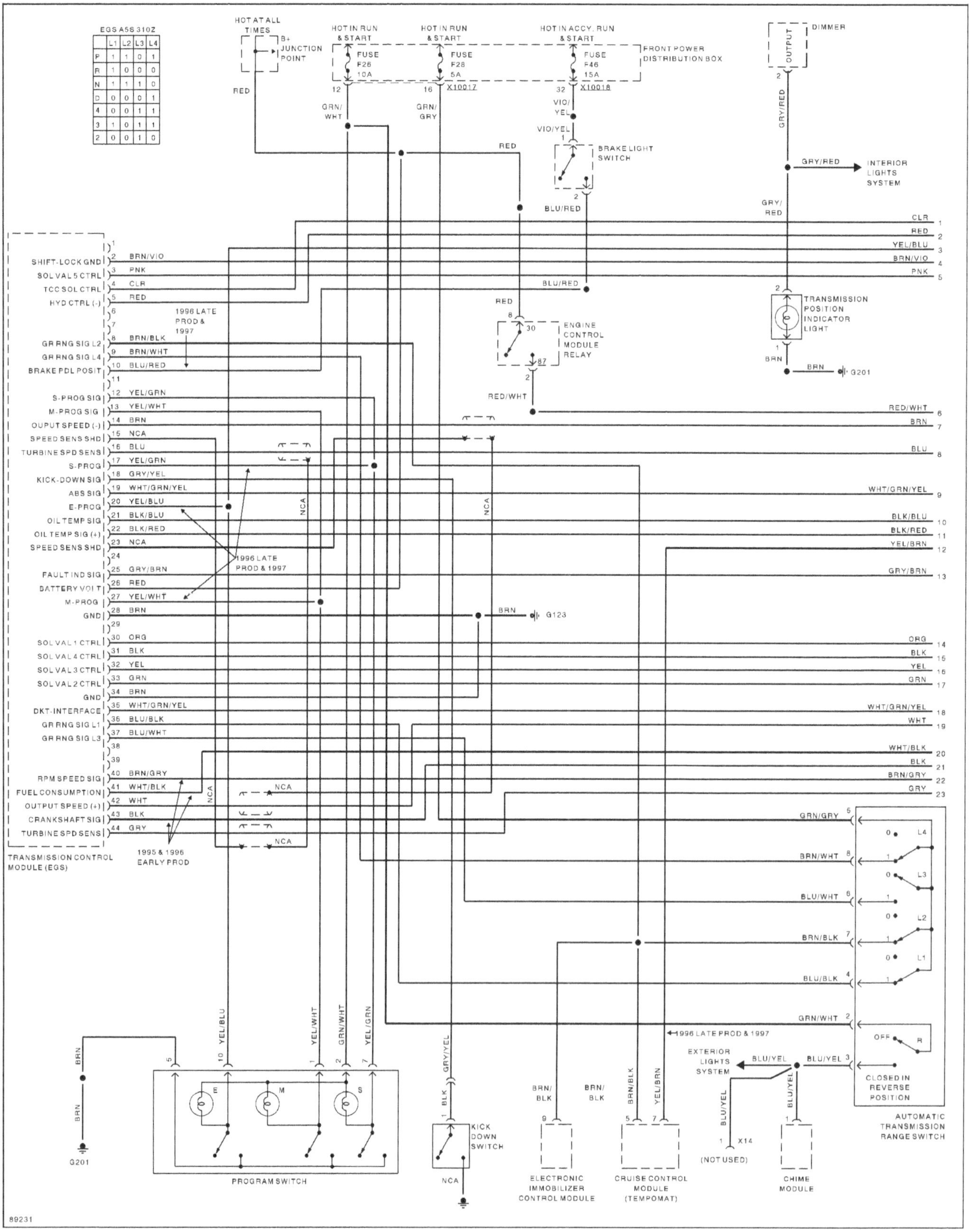

Transmission Electronics

1995-97 M3 (2 of 2)

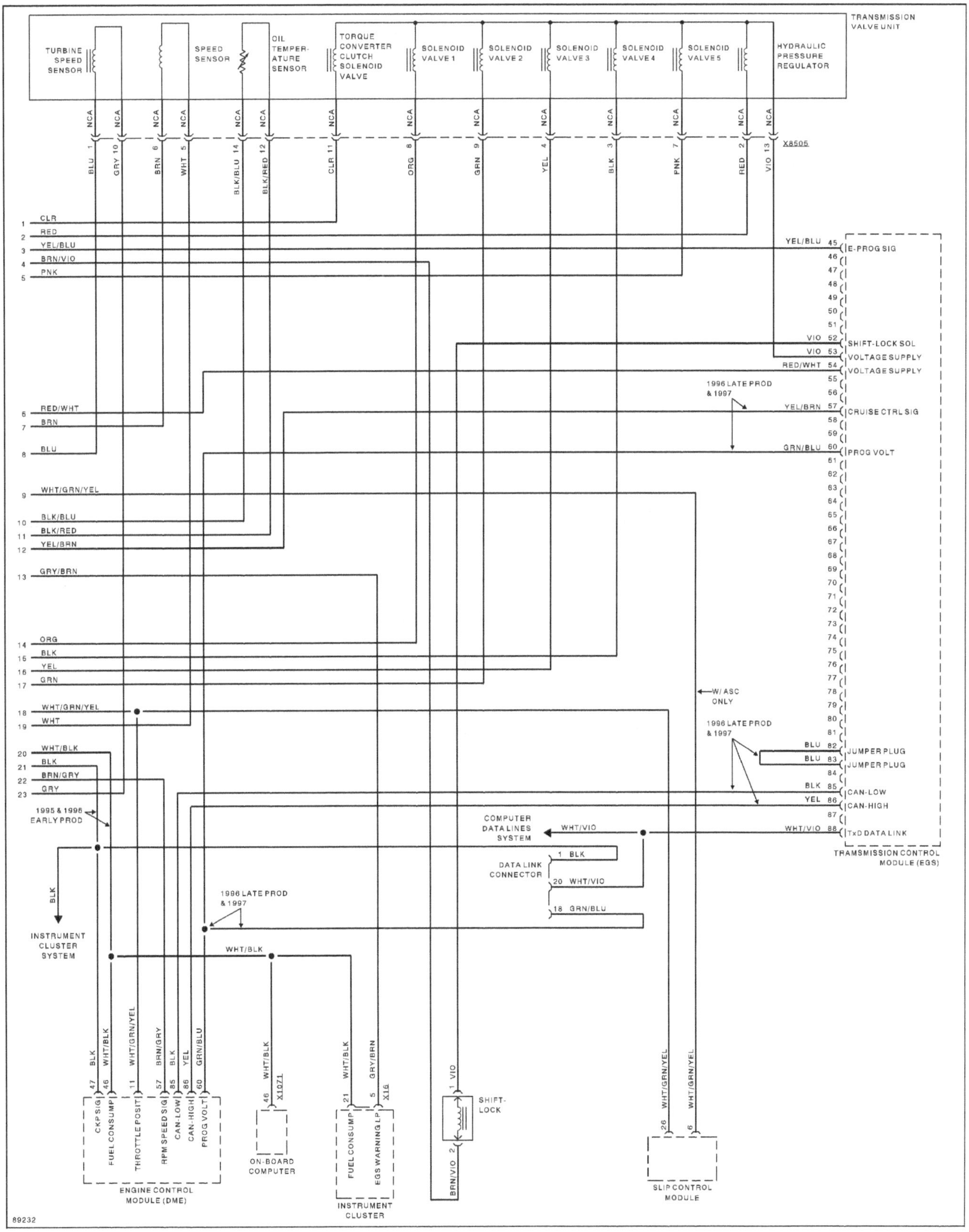

Transmission Electronics

1998 M3 (1 of 2)

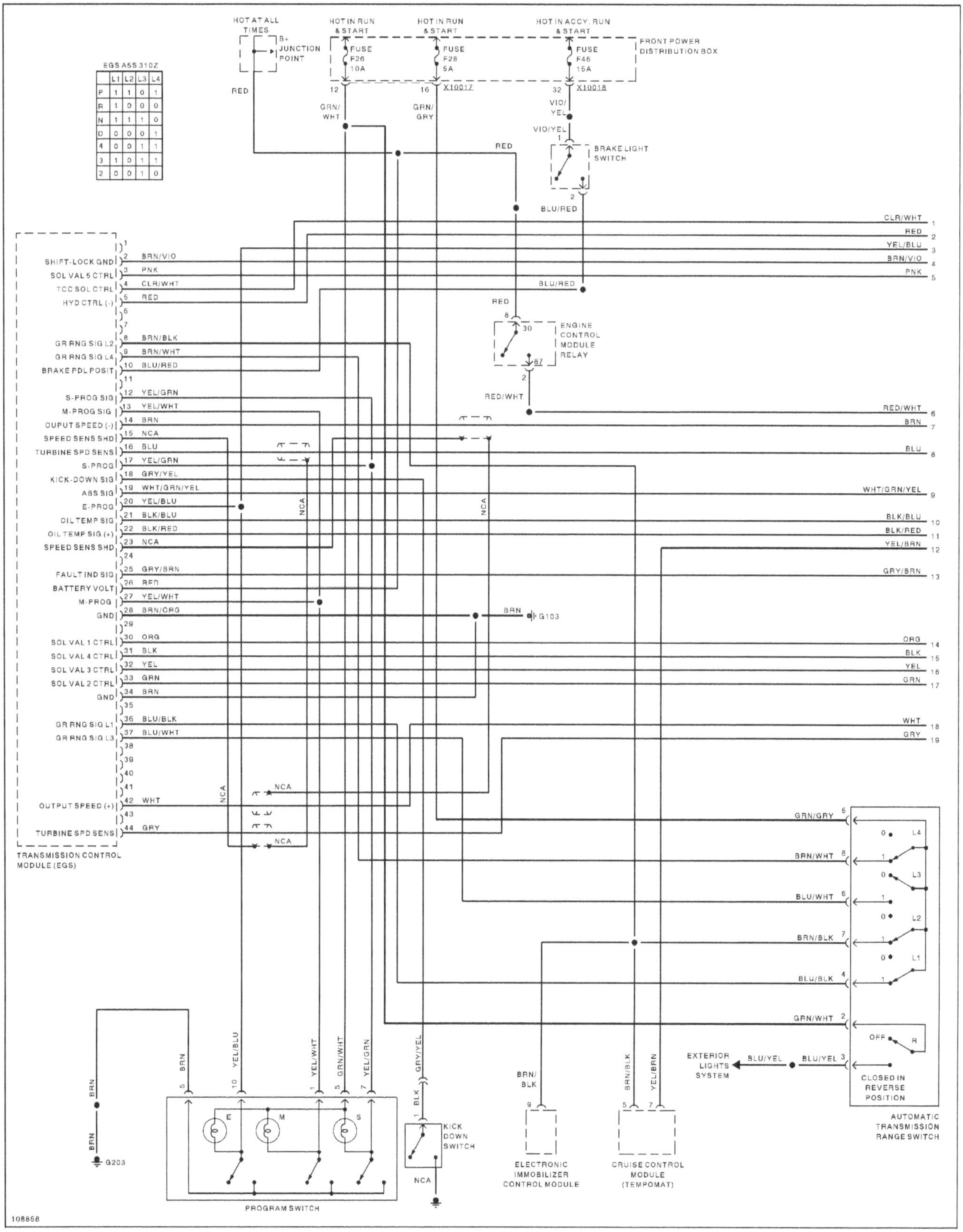

	L1	L2	L3	L4
P	1	1	0	1
R	1	0	0	0
N	1	1	1	0
D	0	0	0	1
4	0	0	1	1
3	1	0	1	1
2	0	0	1	0

Transmission Electronics

1998 M3 (2 of 2)

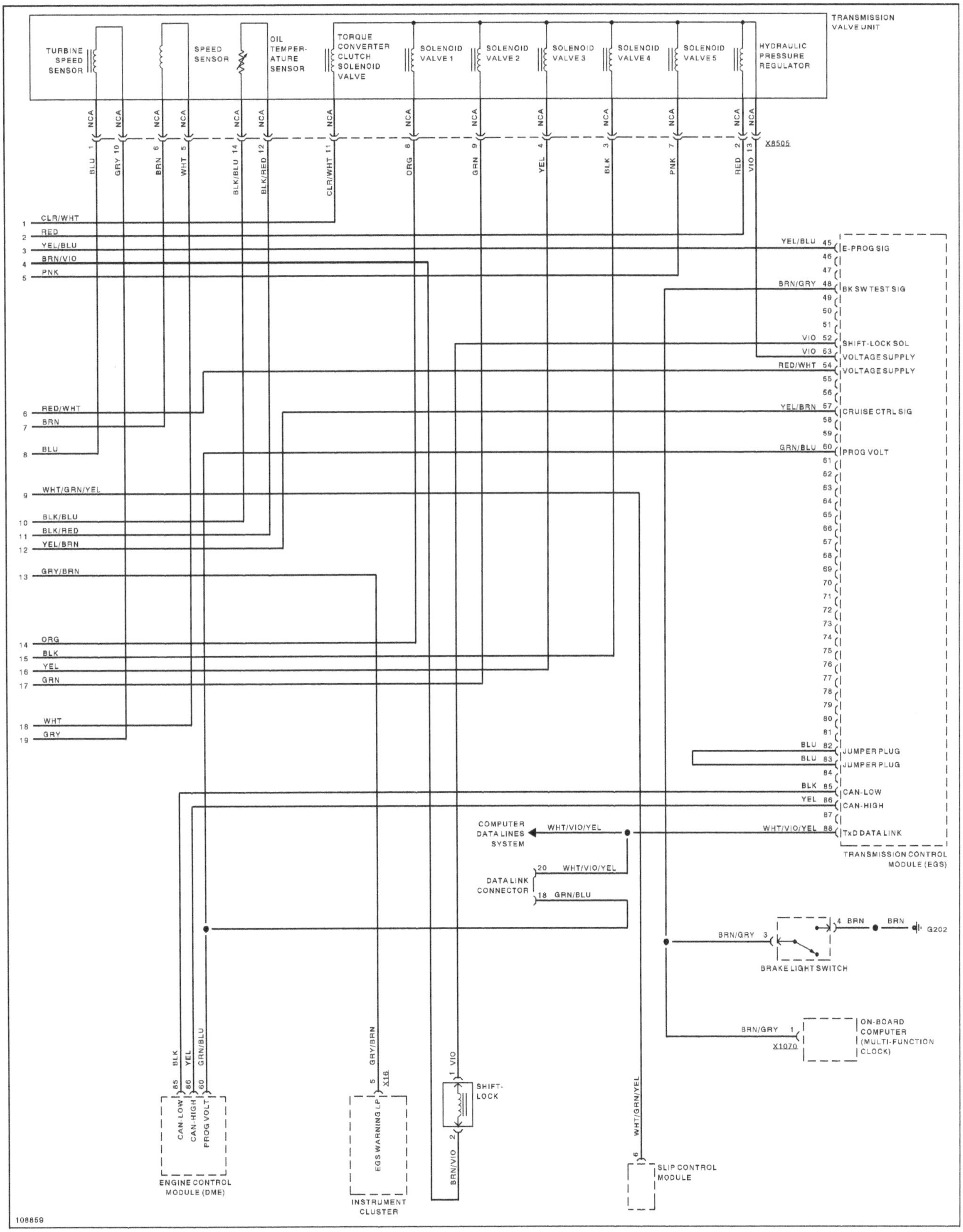

Shift Interlock

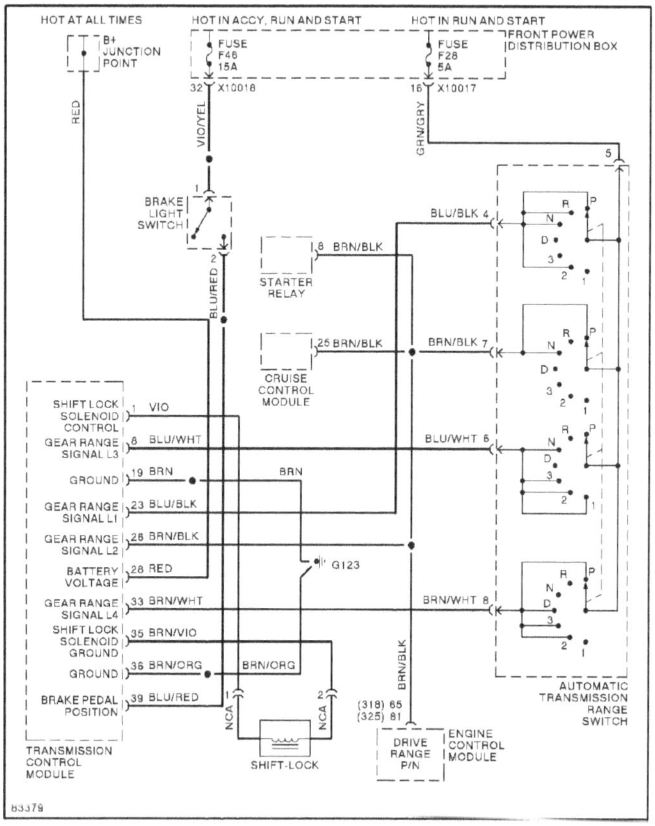

1992-97
Except 318ti

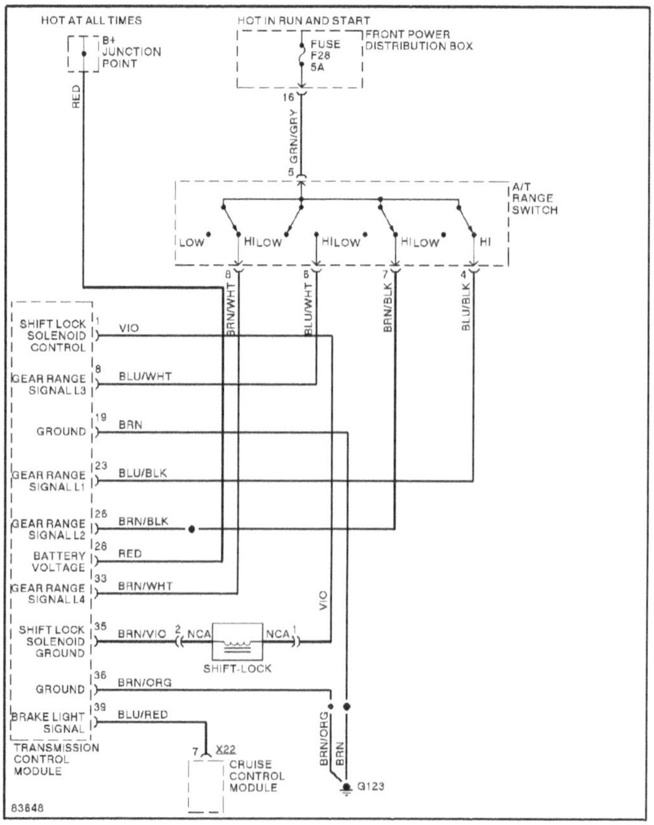

1995-97
318ti

Shift Interlock

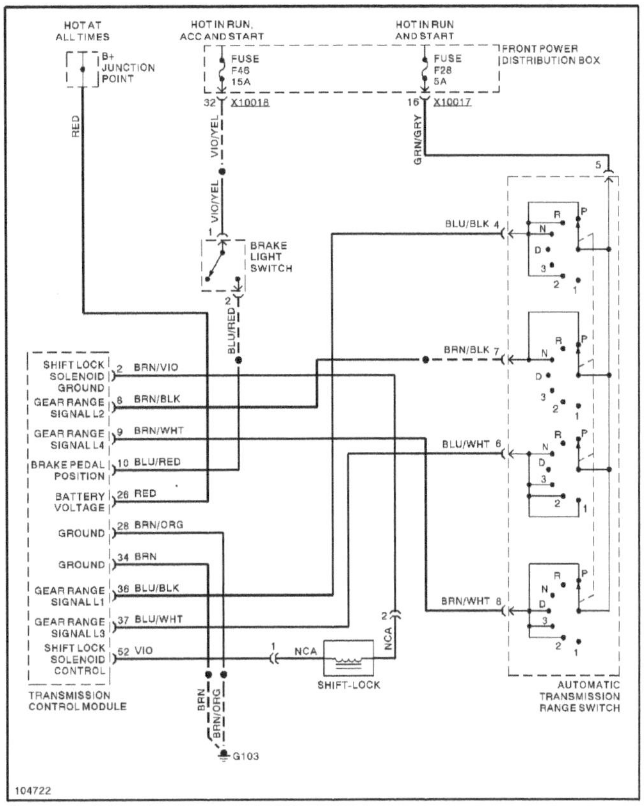

1998
Except 318i, 318ti

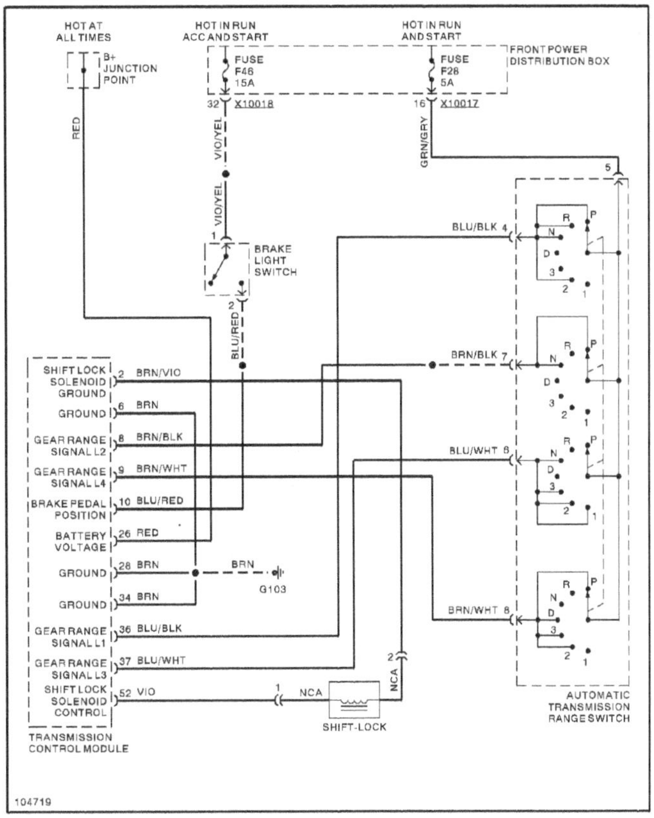

1998
318i, 318ti

Cruise Control

1992-93

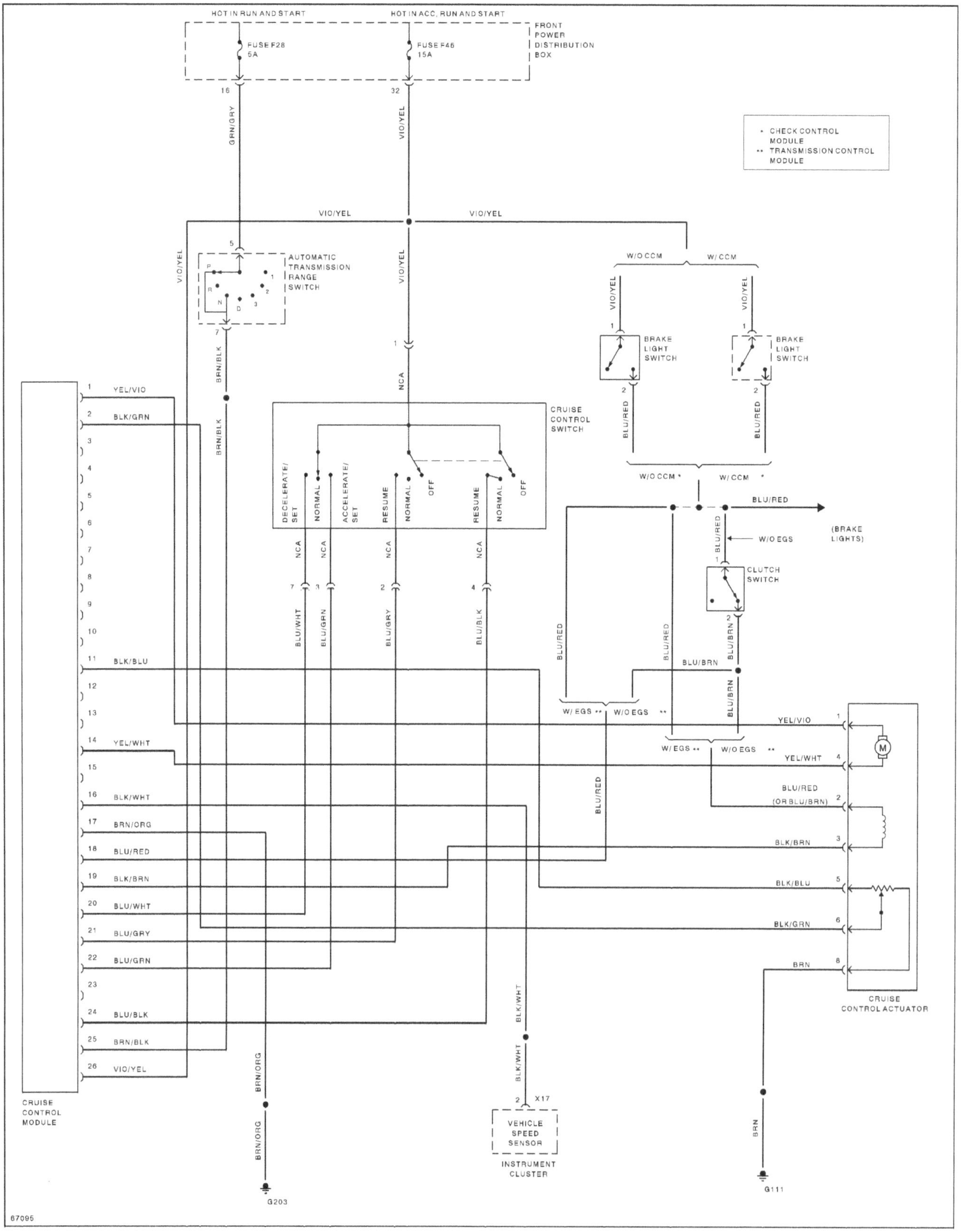

Cruise Control

1994 & 1995 Early Production Except 318ti

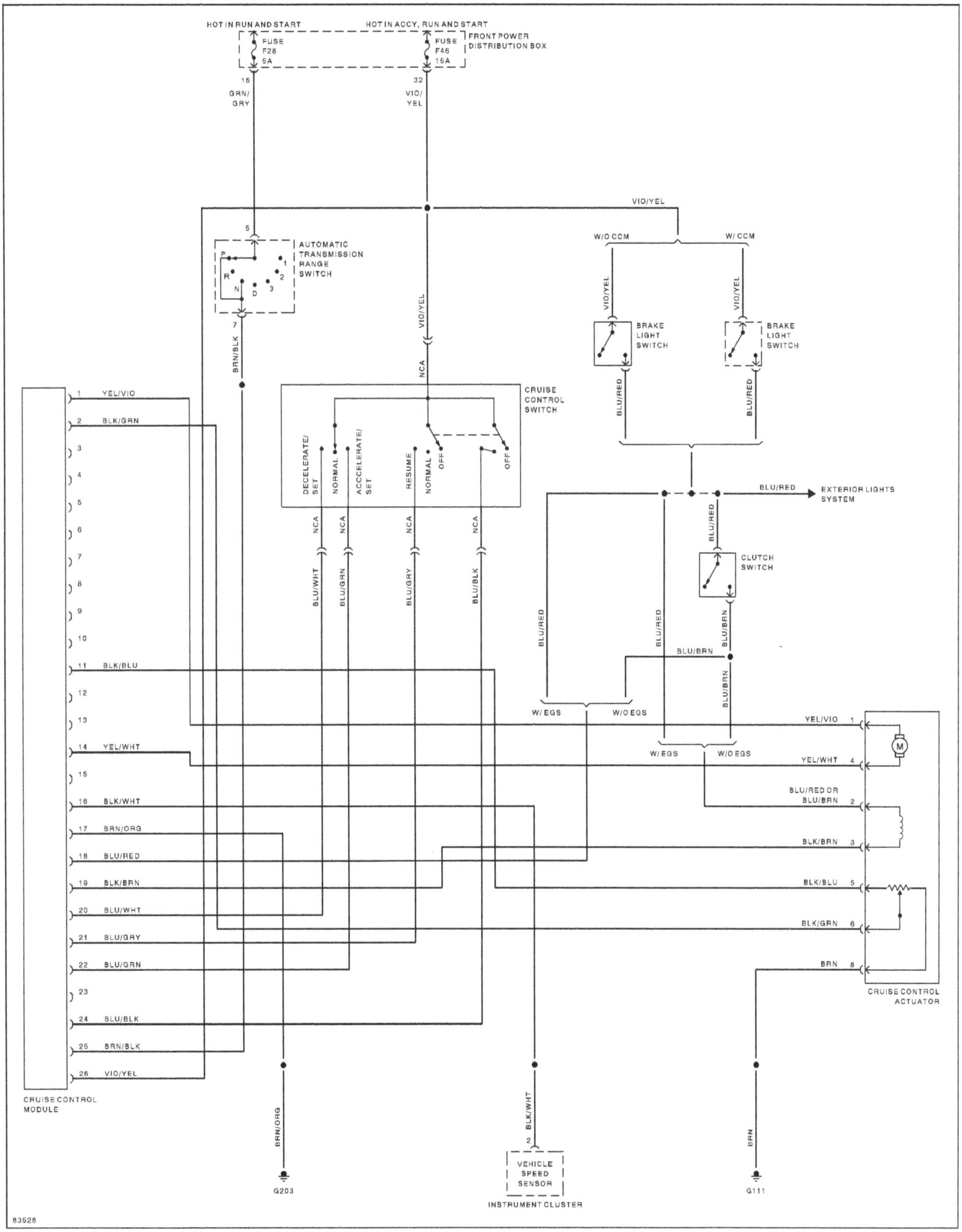

Cruise Control

1995 Late Production & 1996-97 Except 318ti

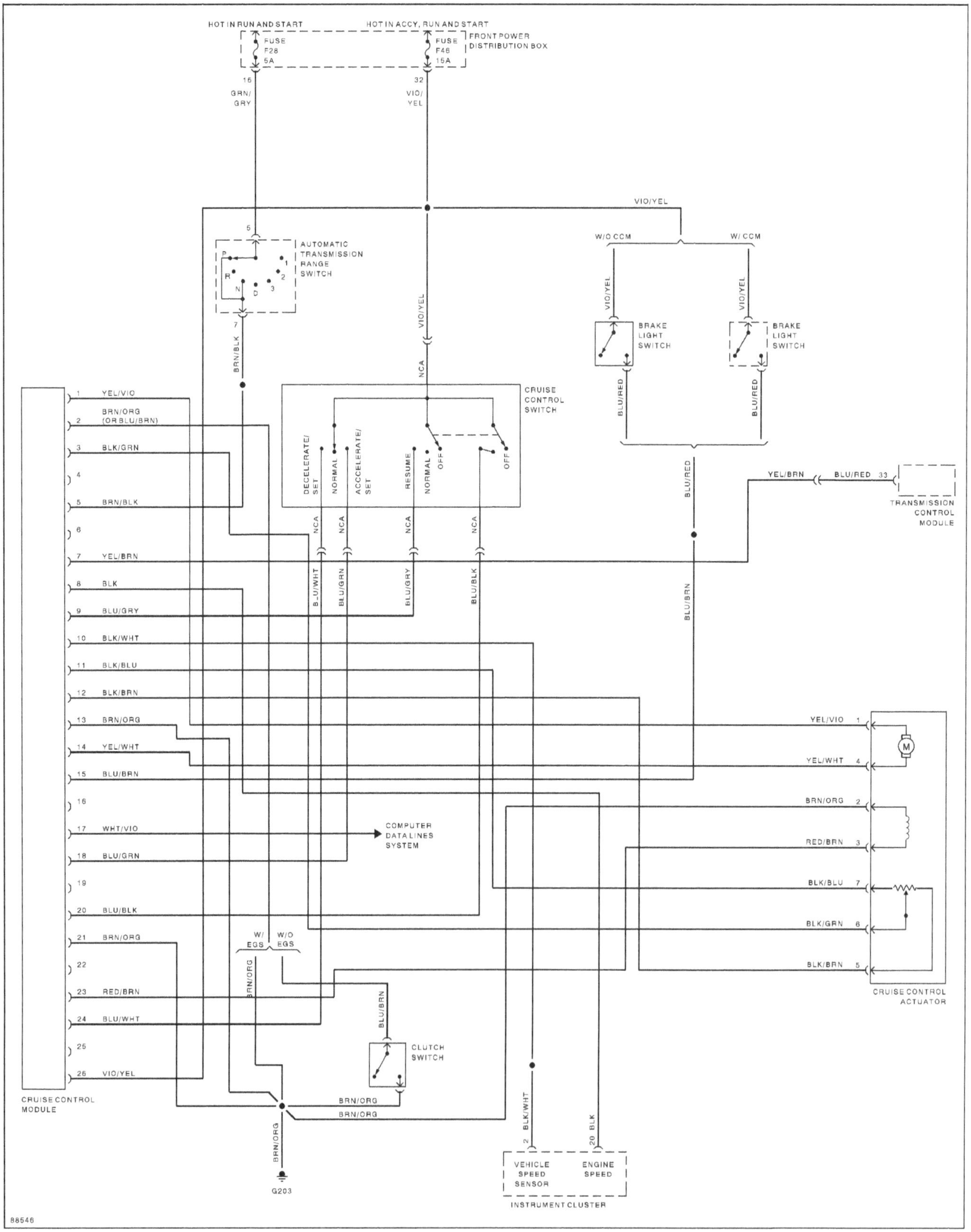

Cruise Control
1995-97 318ti

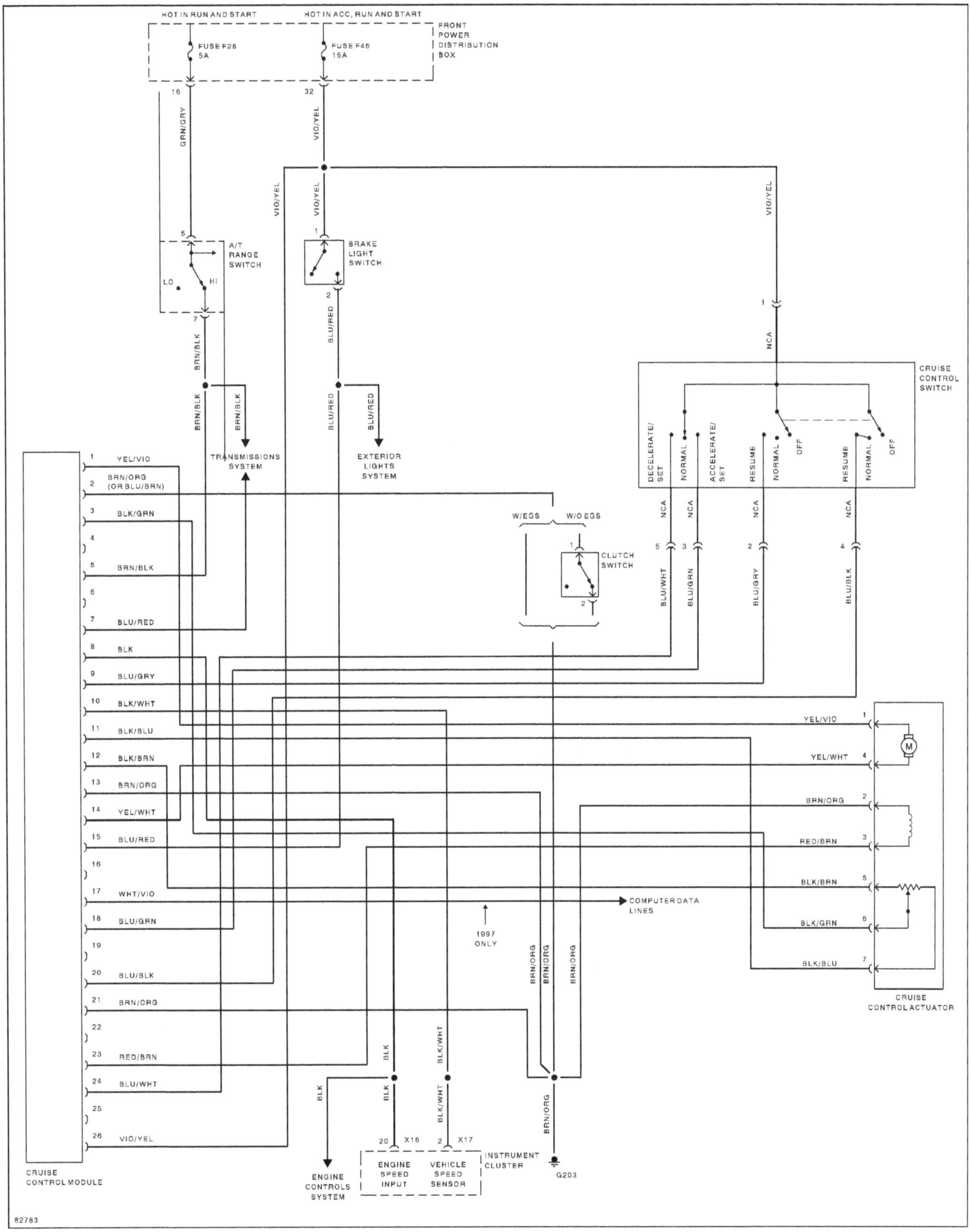

Cruise Control

1998

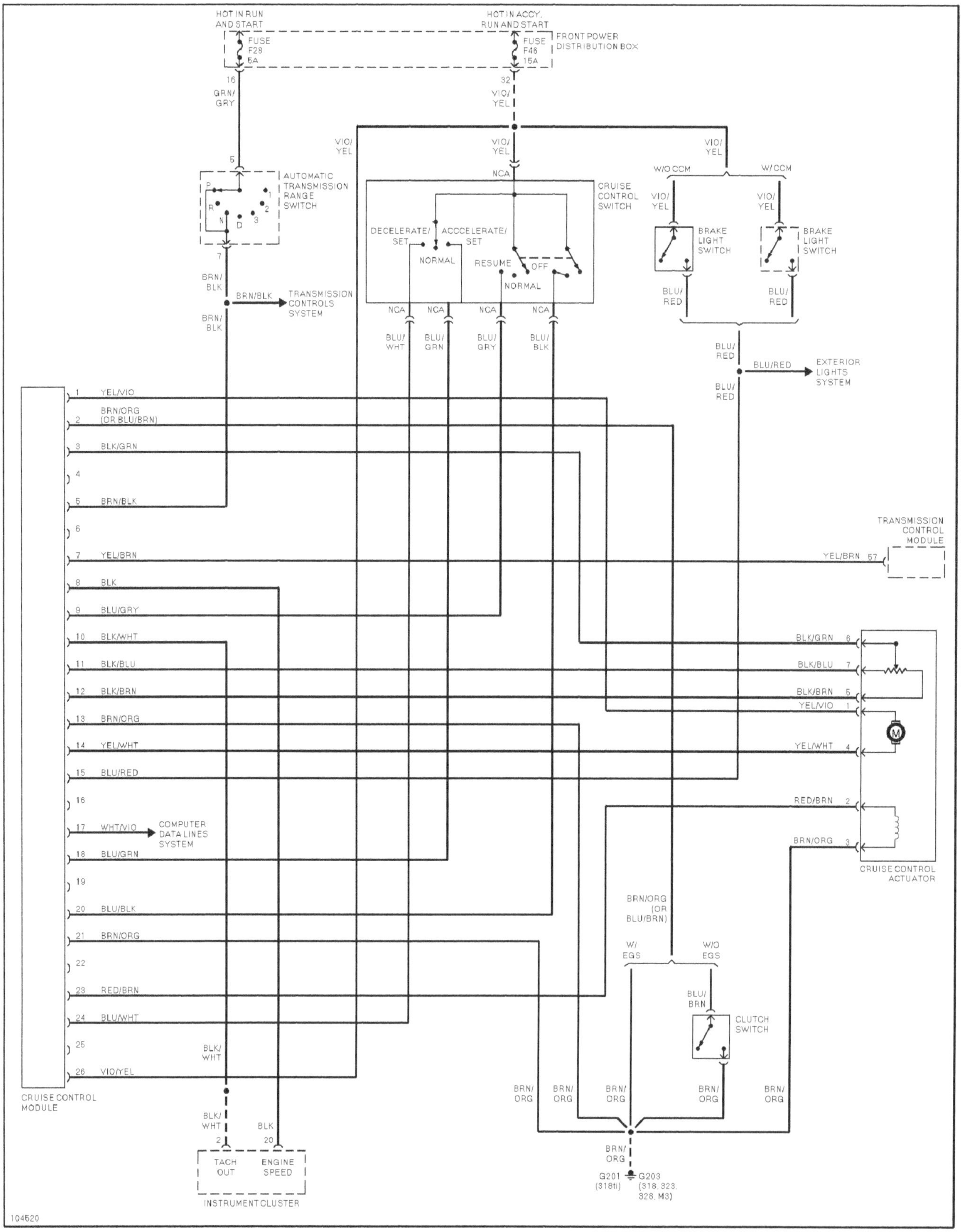

ABS

1992-94 w/o Traction Control

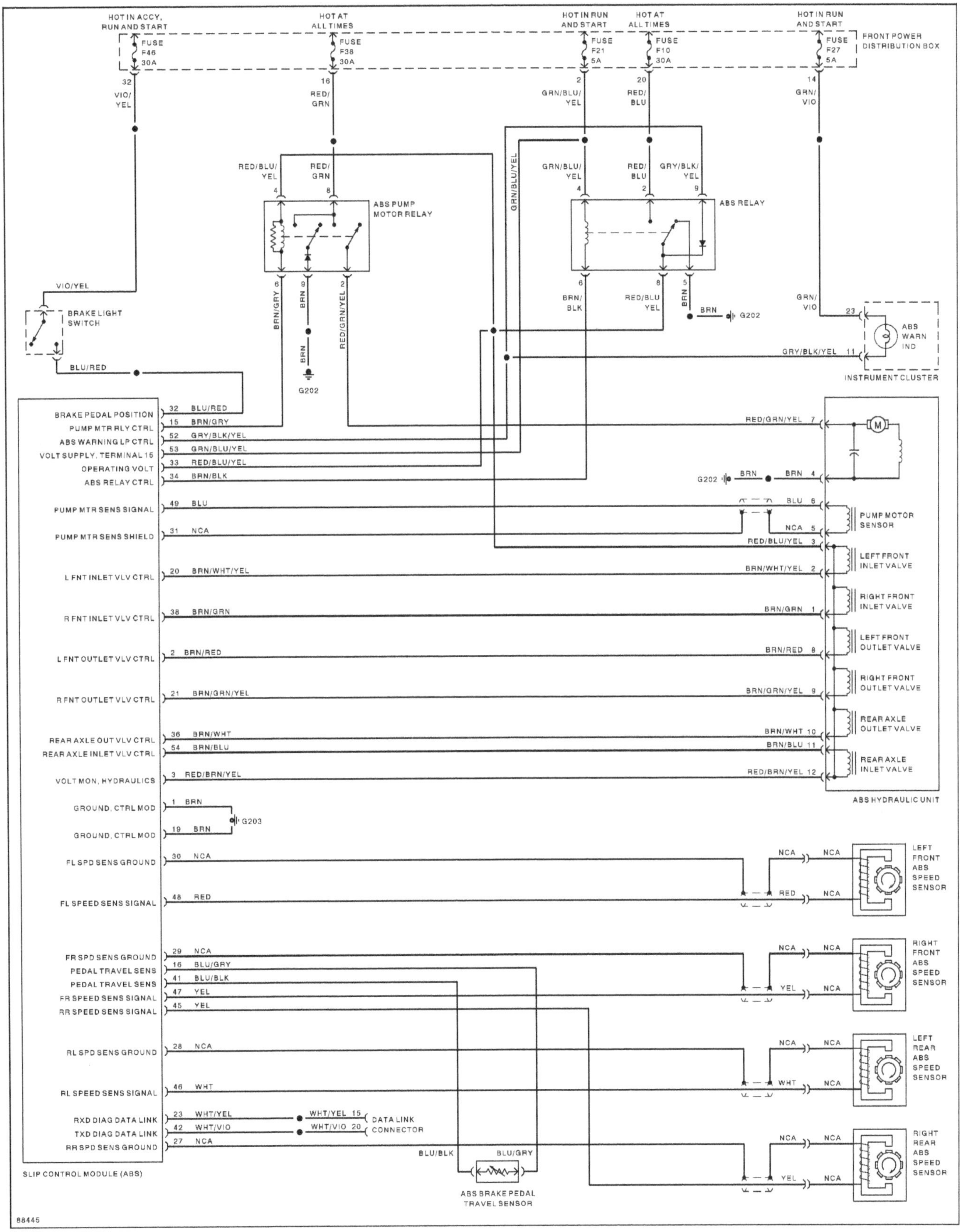

ABS

1995-96 Except 318ti w/o Traction Control

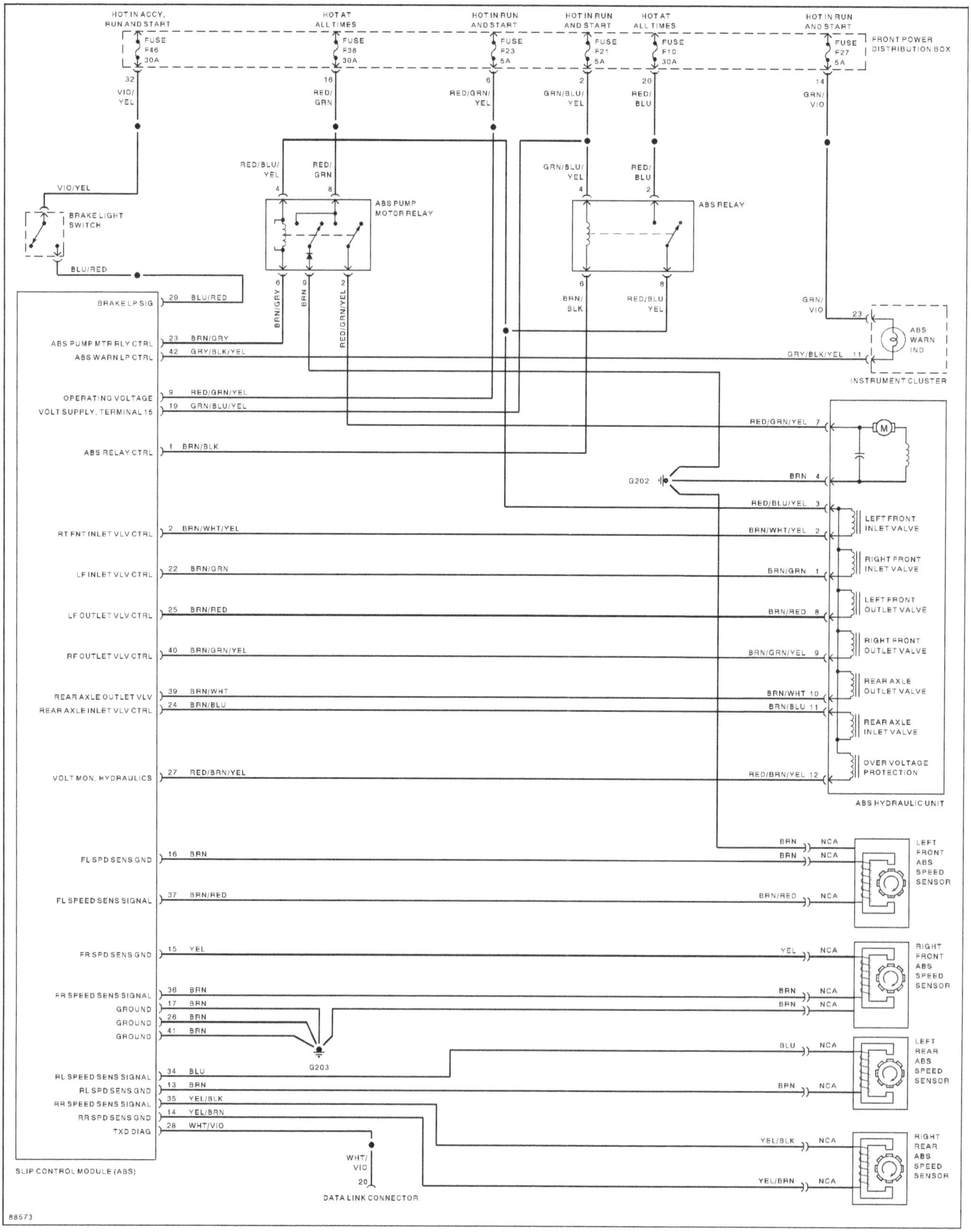

ABS/AST

1995-97 318ti & 1997 318i, 318iC & 318is w/Traction Control

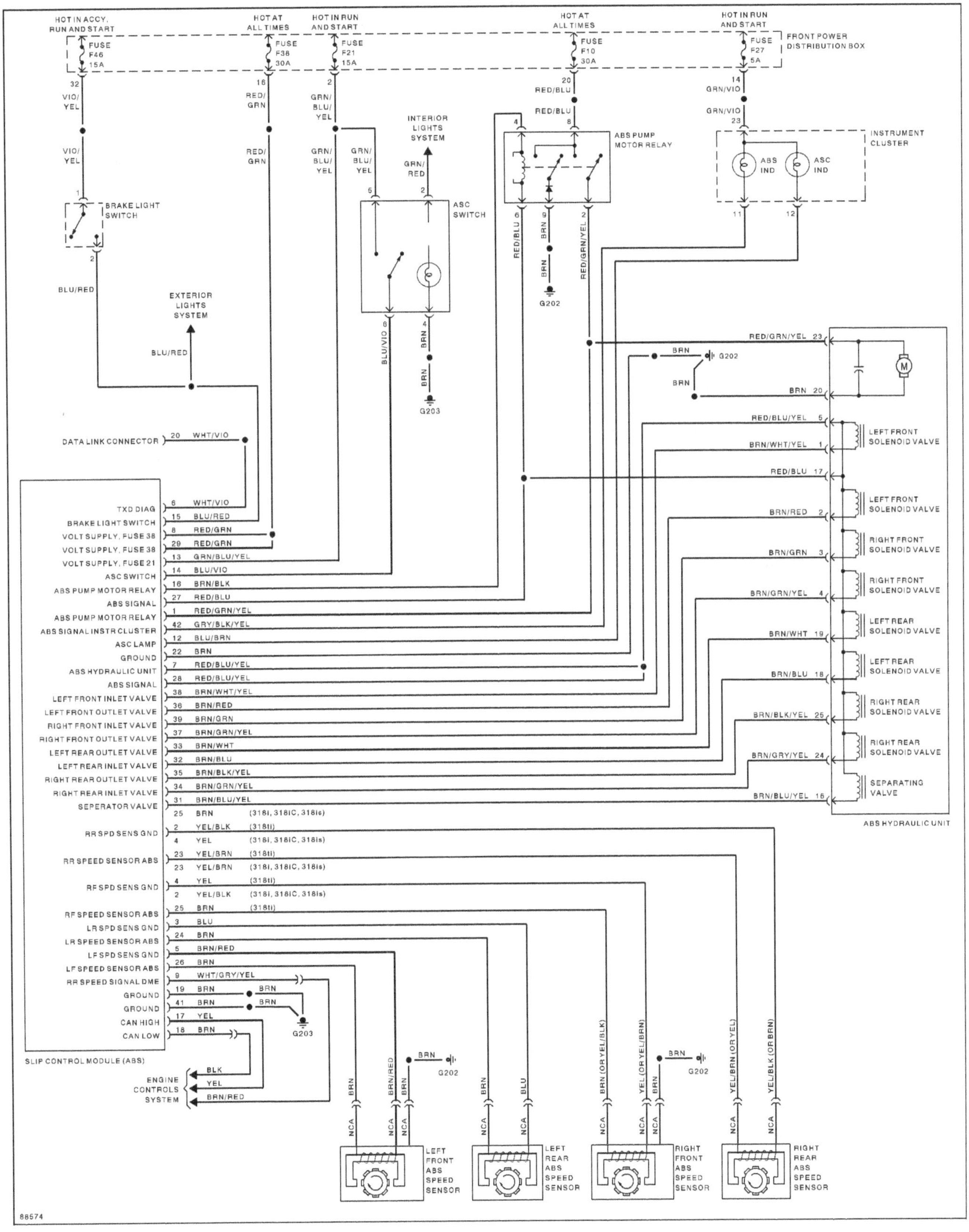

ABS/AST

1994-95 325i, 325iC & 325is, 1996-97 328i, 328iC & 328is, 1995-97 M3 w/ Traction Control (1 of 2)

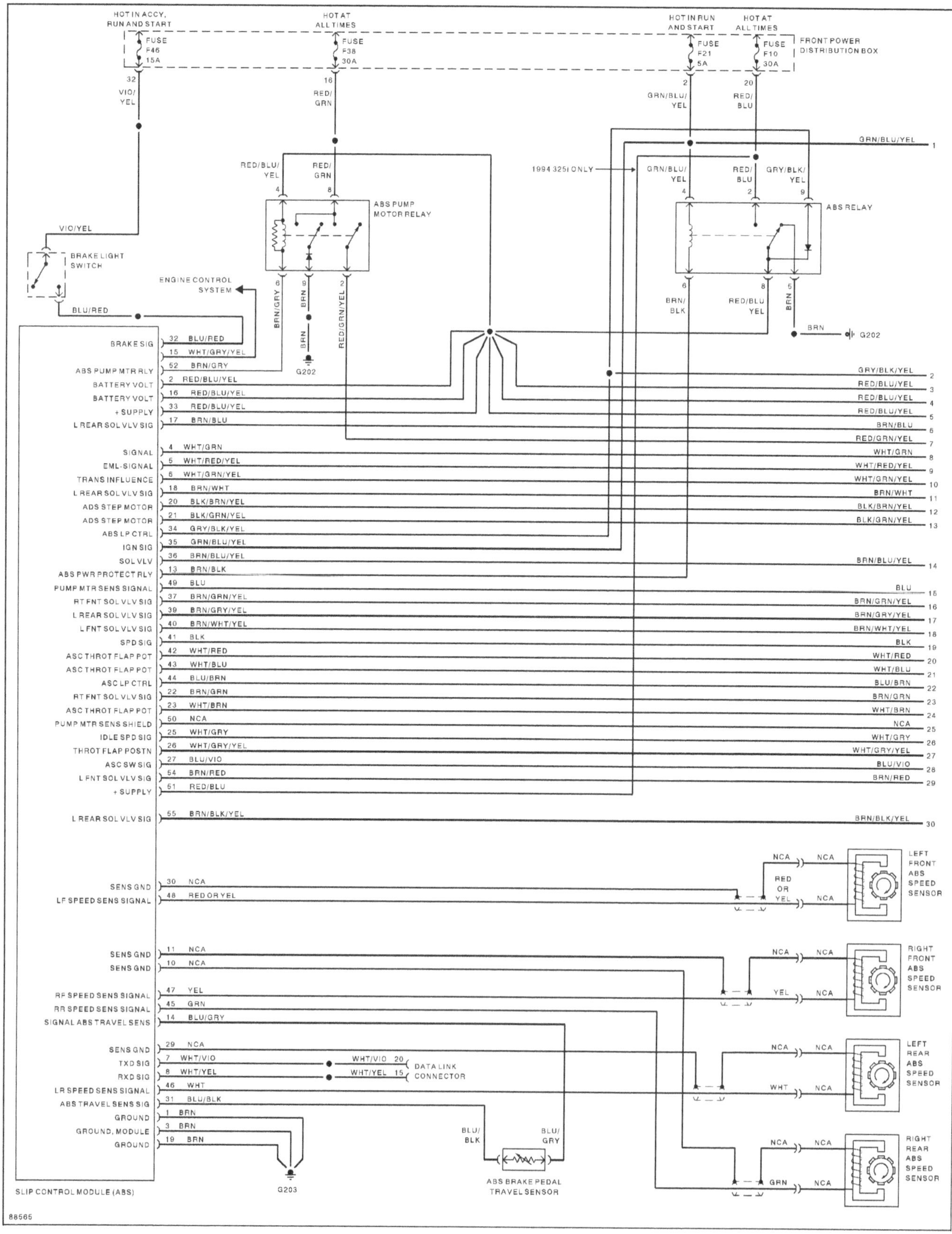

ABS/AST

1994-95 325i, 325iC & 325is, 1996-97 328i, 328iC & 328is, 1995-97 M3 w/ Traction Control (2 of 2)

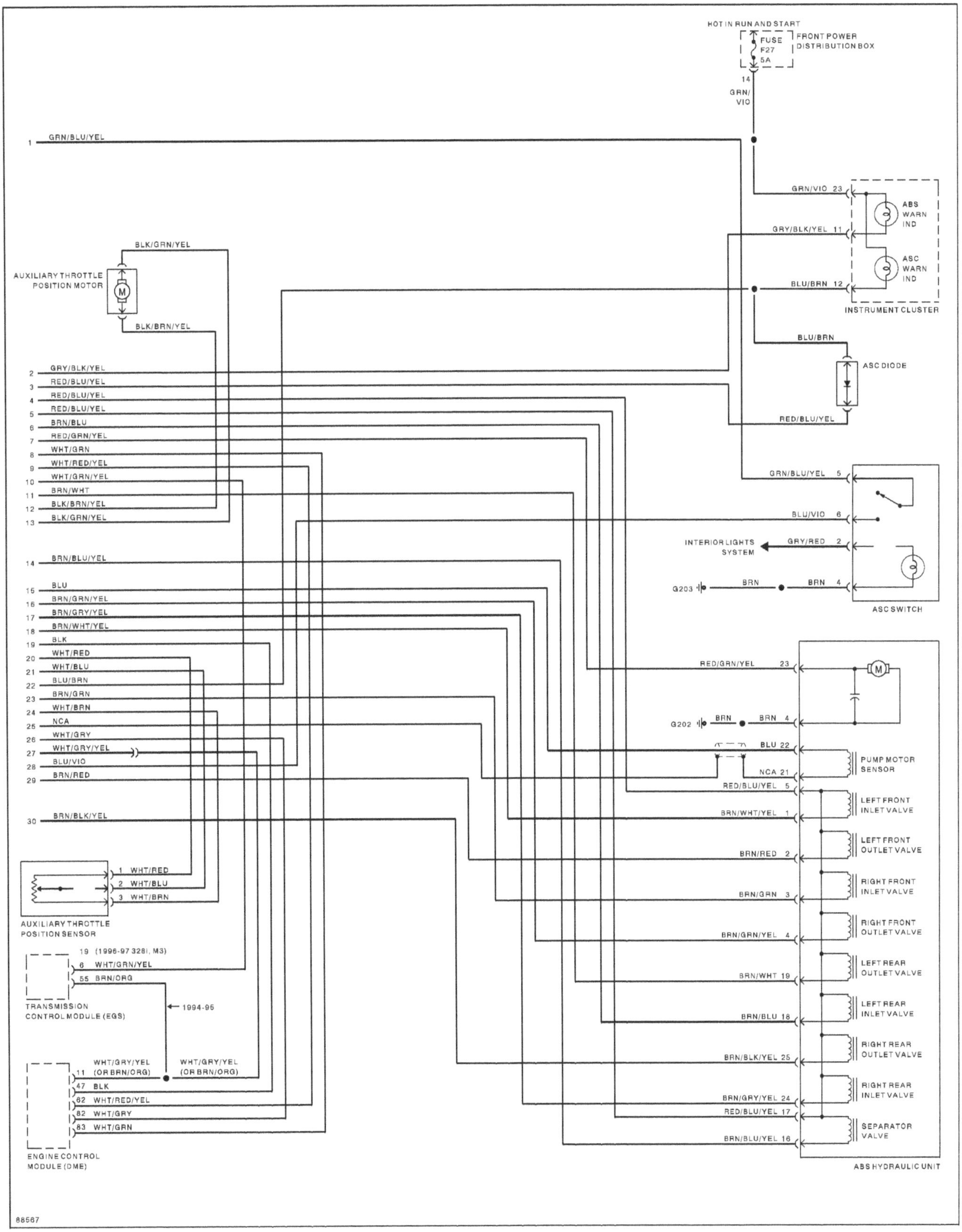

ABS/AST

1996 318i, 318iC & 318is w/Traction Control

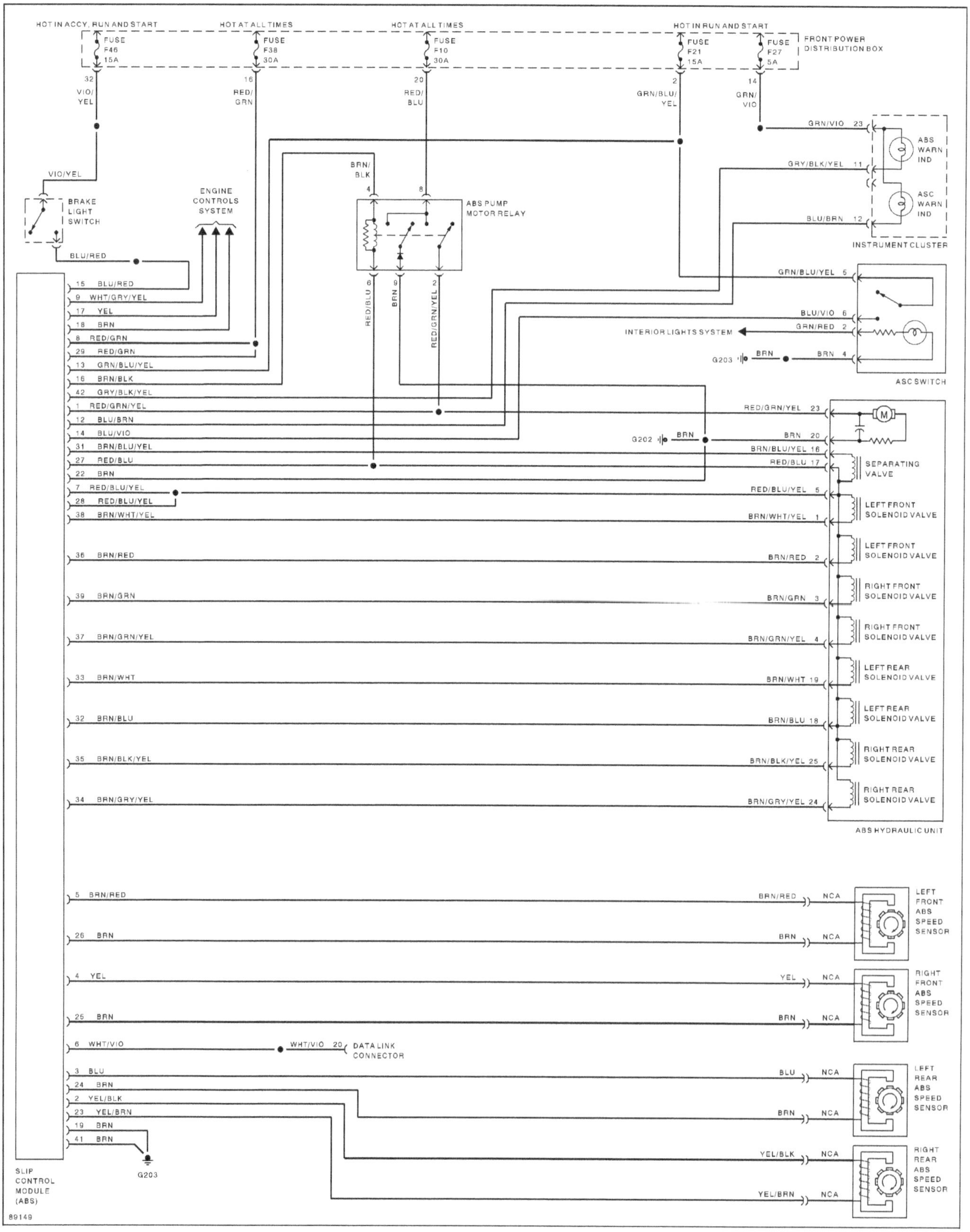

ABS/AST

1998 318i, 318ti w/Traction Control

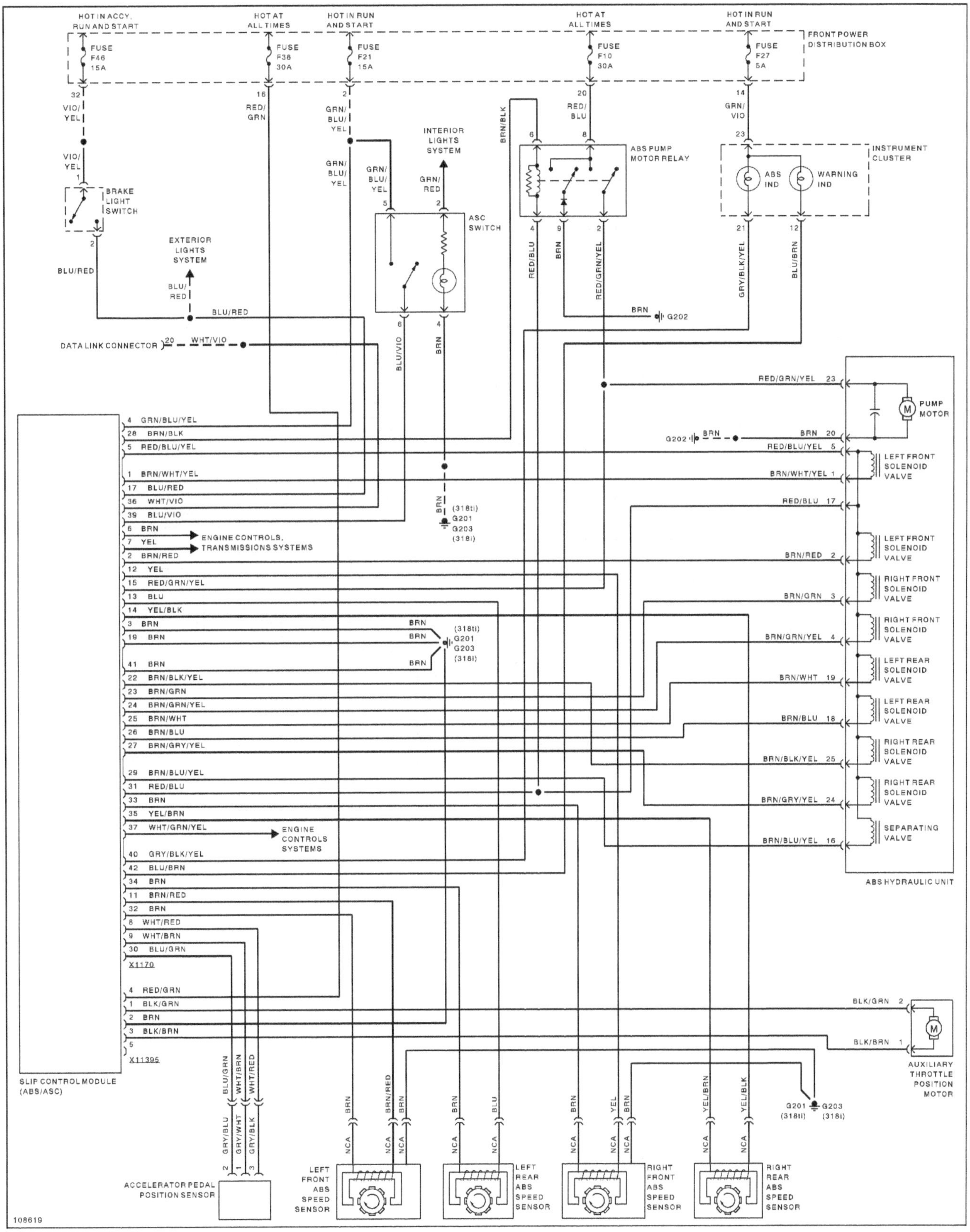

ABS/AST

1998 Except 318i, 318ti w/Traction Control (1 of 2)

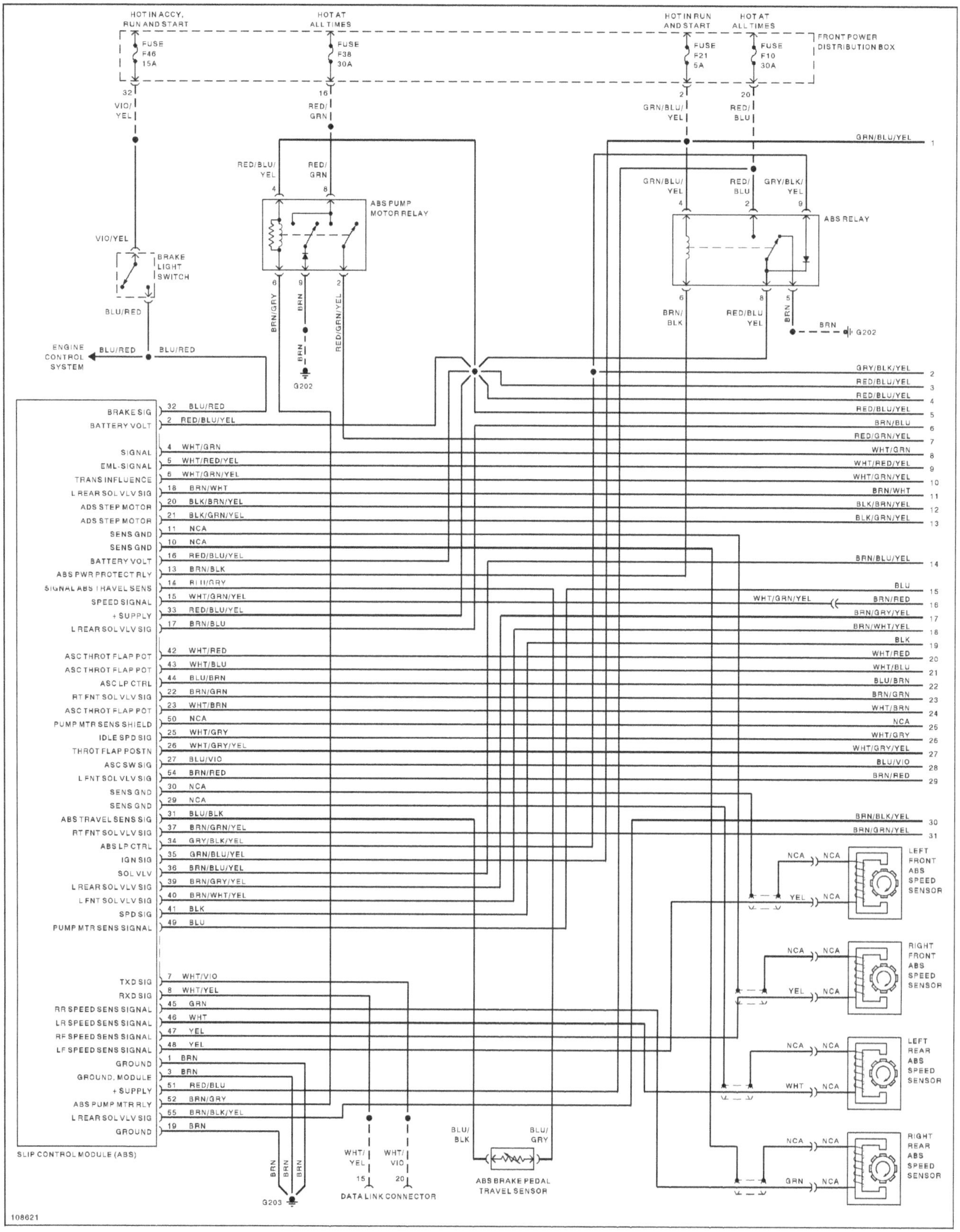

ABS/AST

1998 Except 318i, 318ti w/Traction Control (2 of 2)

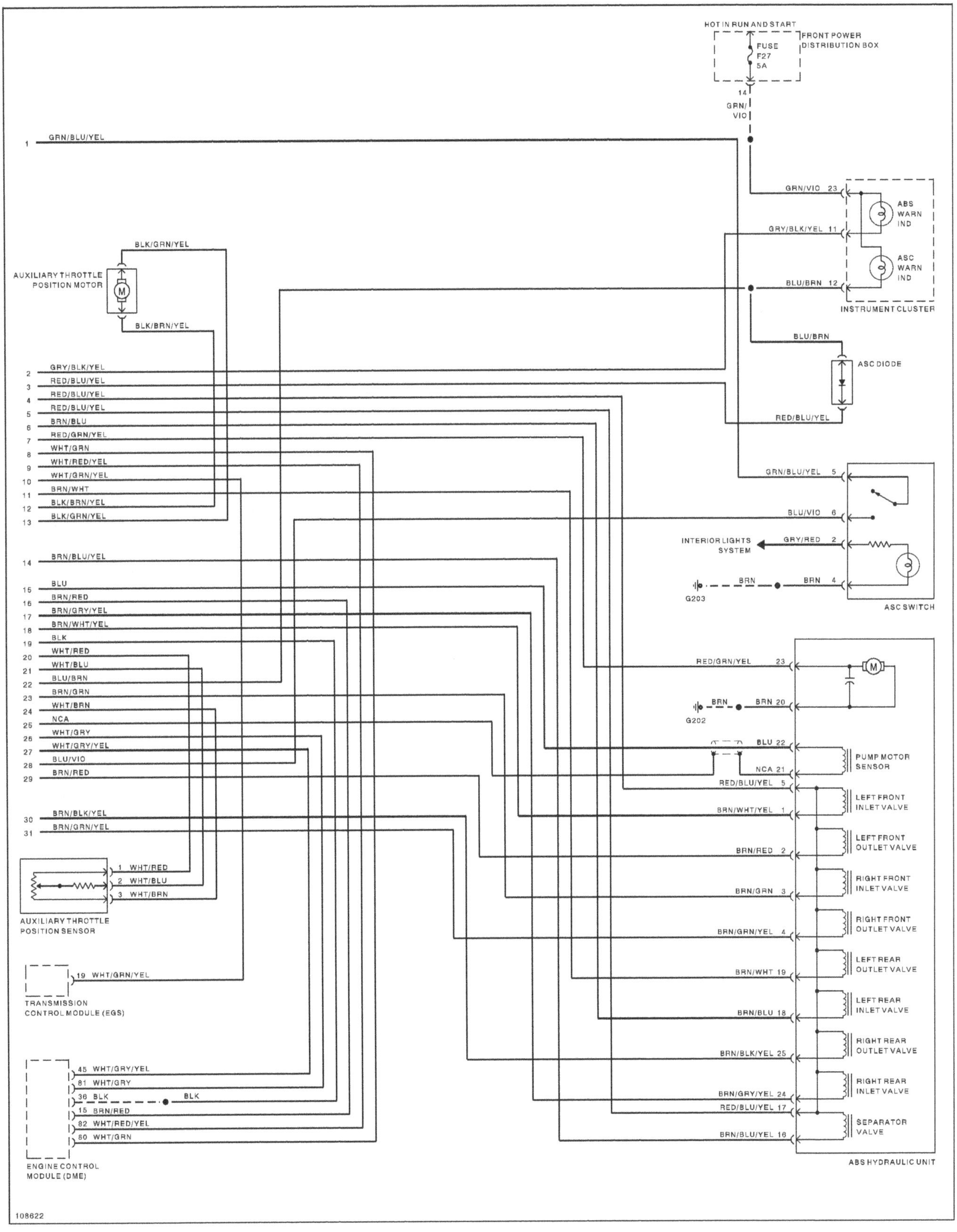

ABS

1995-96 318ti w/o Traction Control

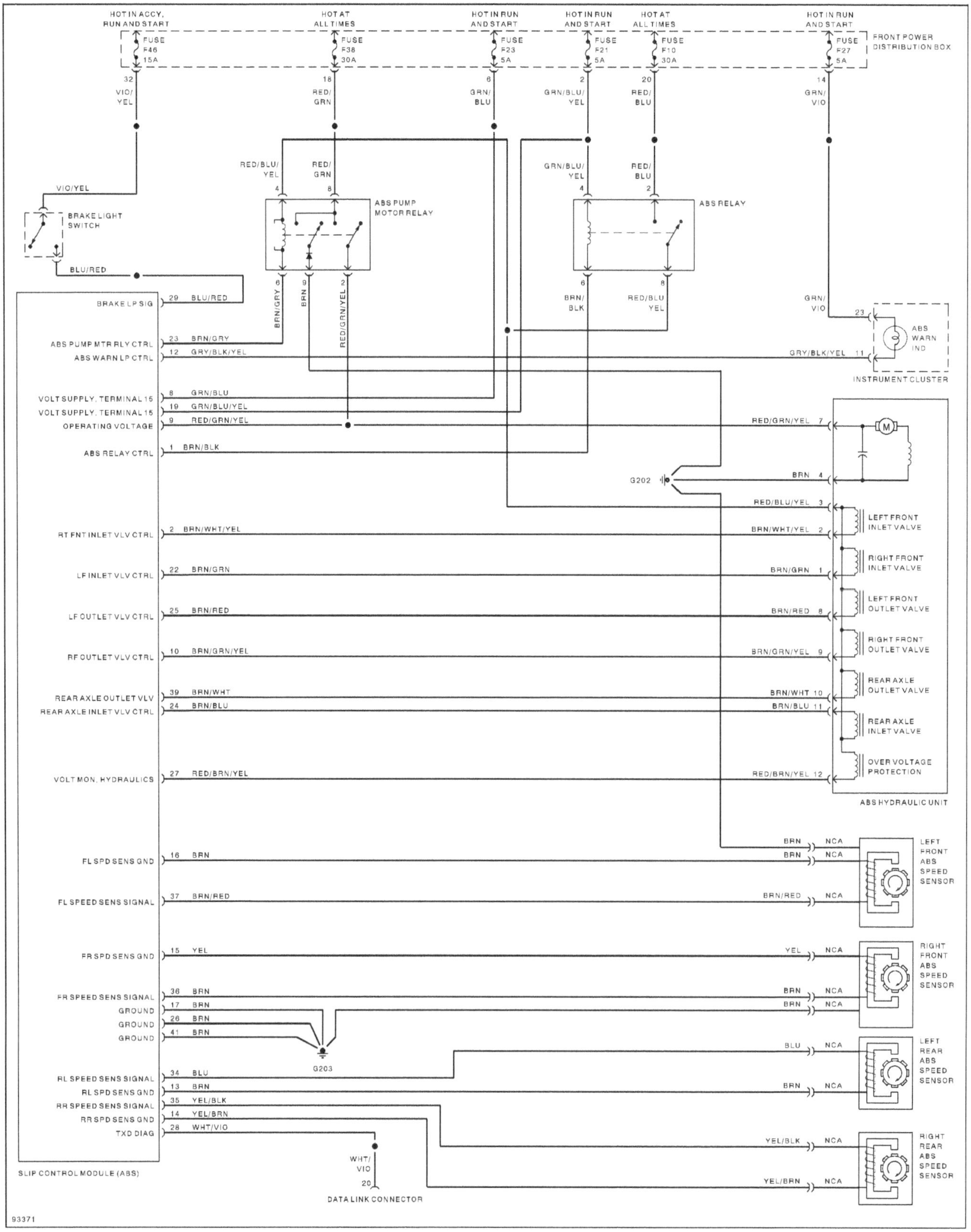

ABS/AST

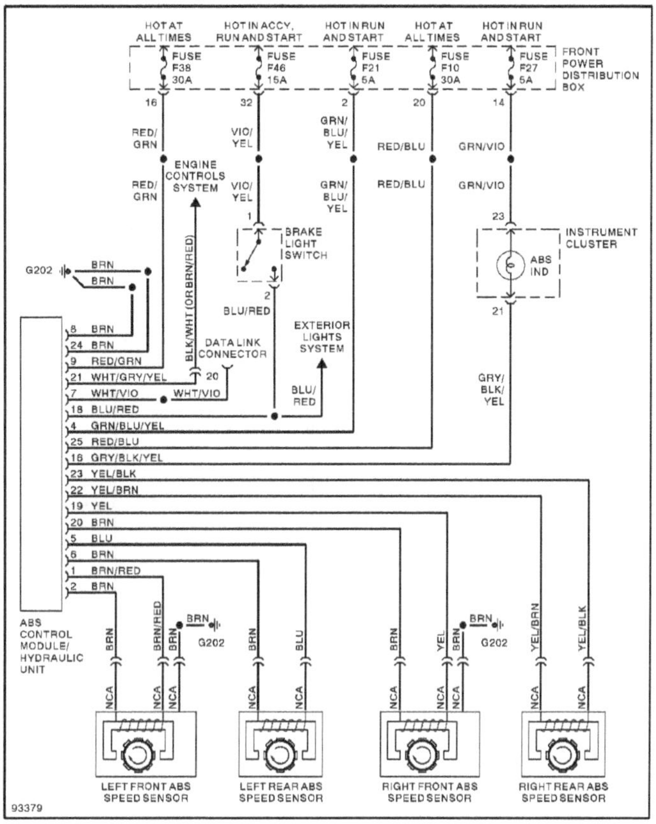

1997 w/ Traction Control

ABS/AST

1998 w/ Traction Control

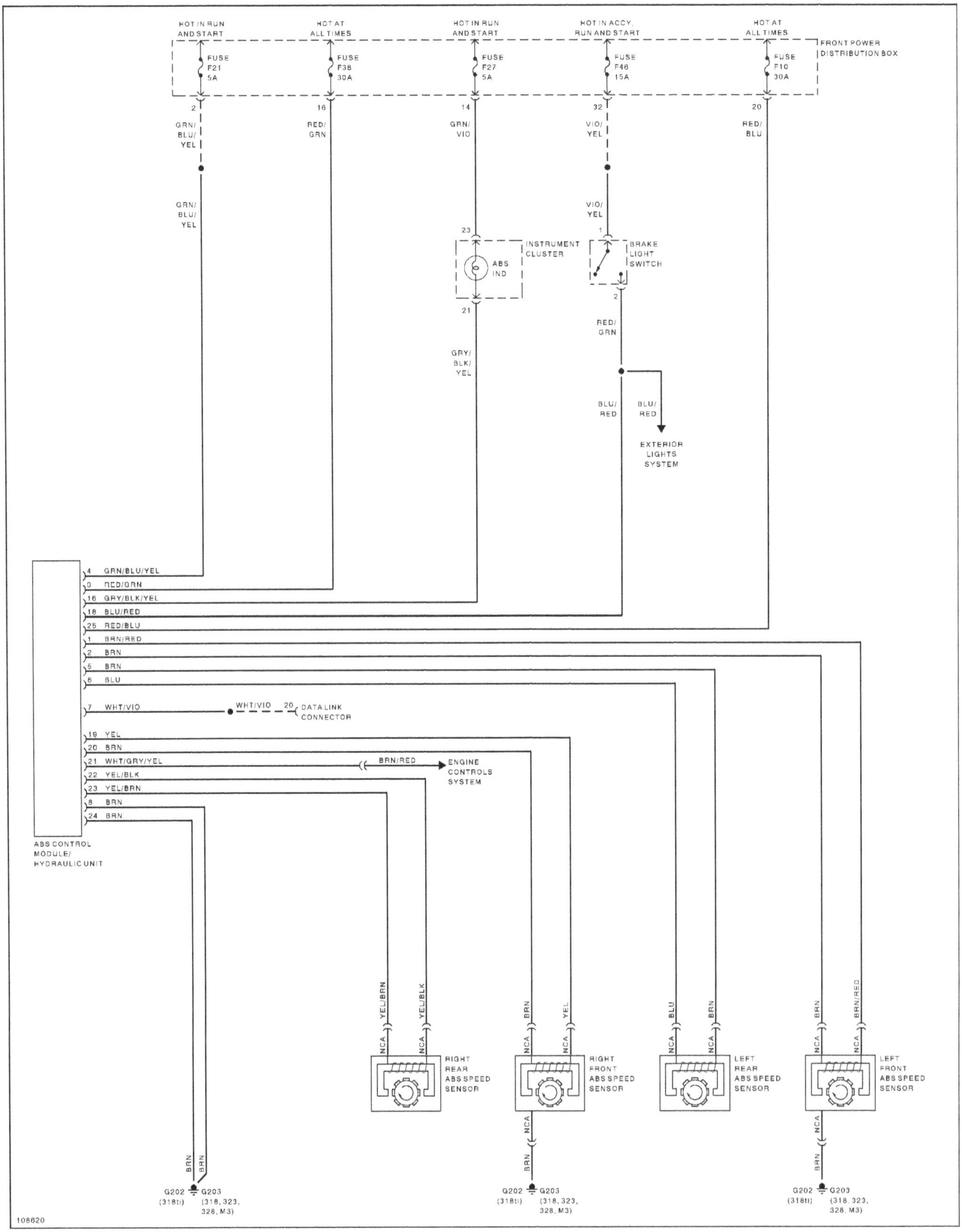

Air Conditioning & Heating

1992 (1 of 2)

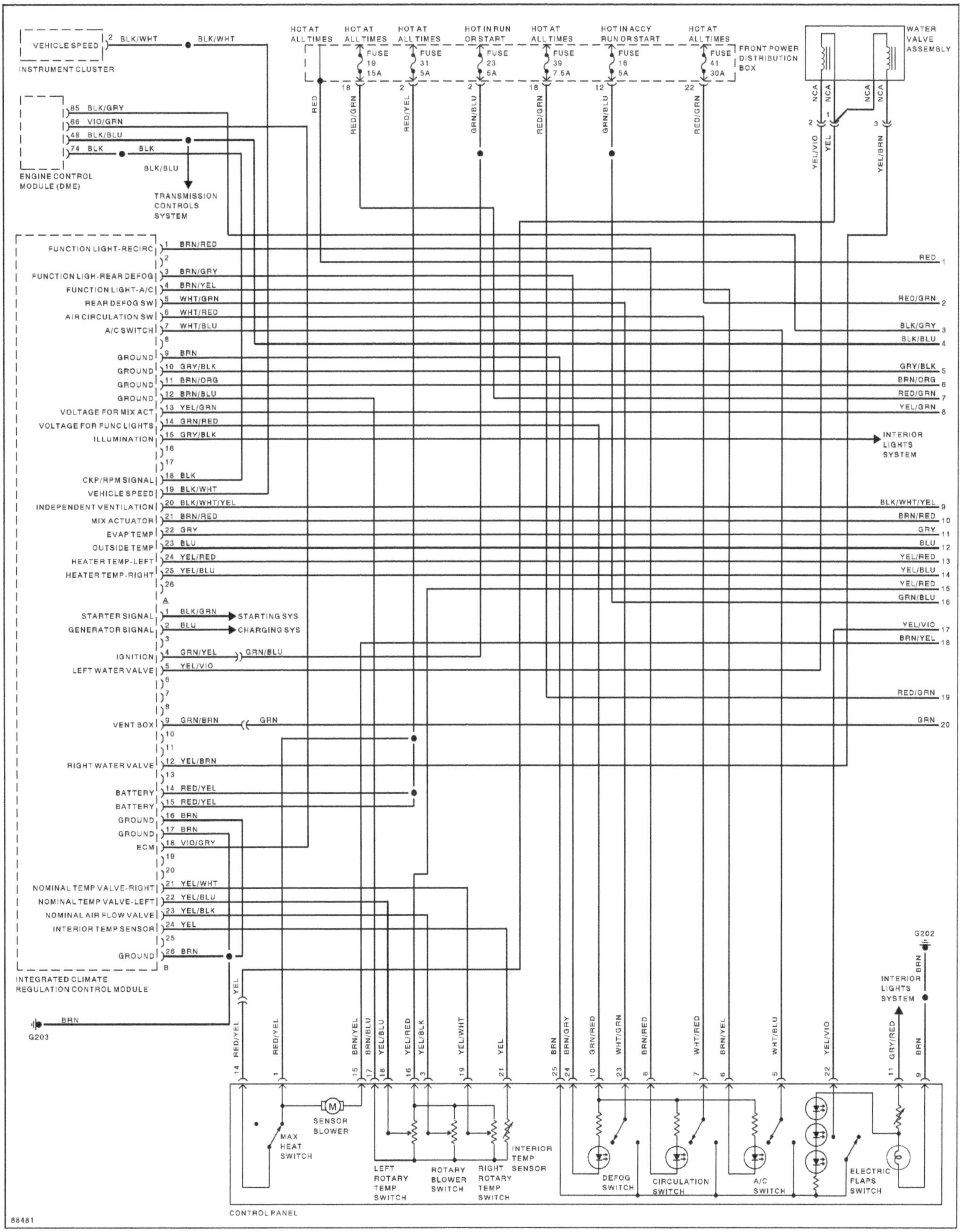

Air Conditioning & Heating

1992 (2 of 2)

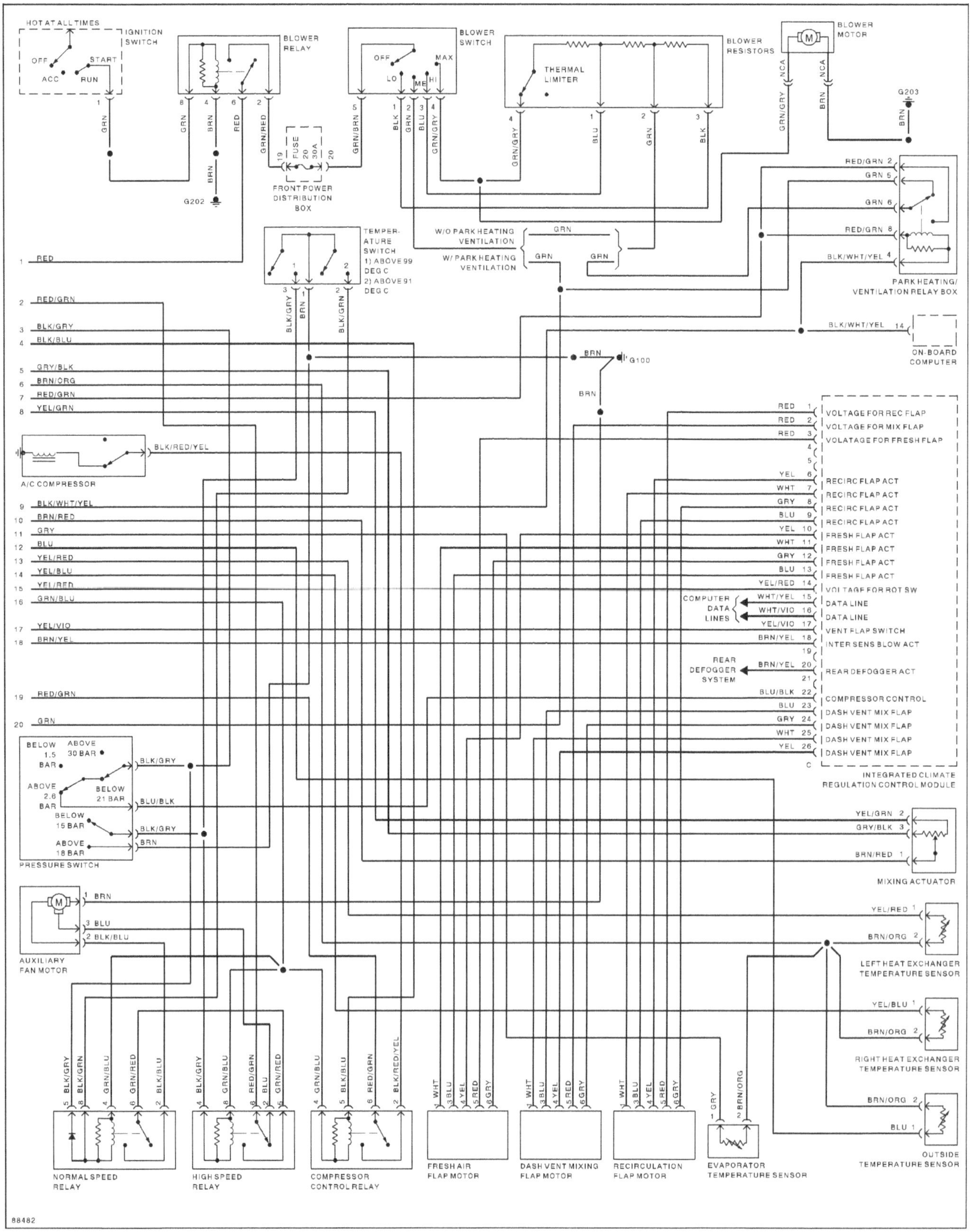

Air Conditioning & Heating

1993-95 Except 318ti (1 of 2)

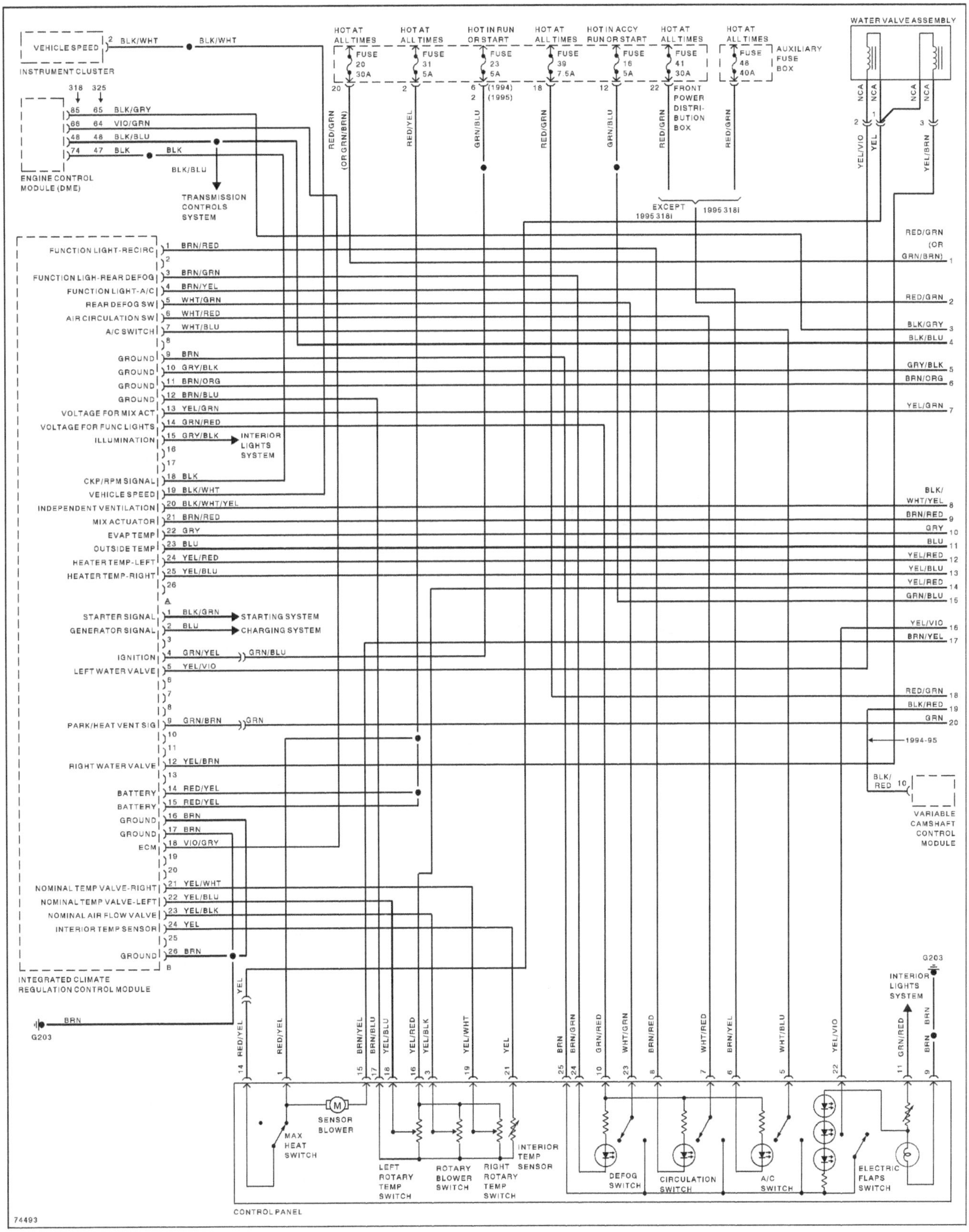

Air Conditioning & Heating

1993-95 Except 318ti (2 of 2)

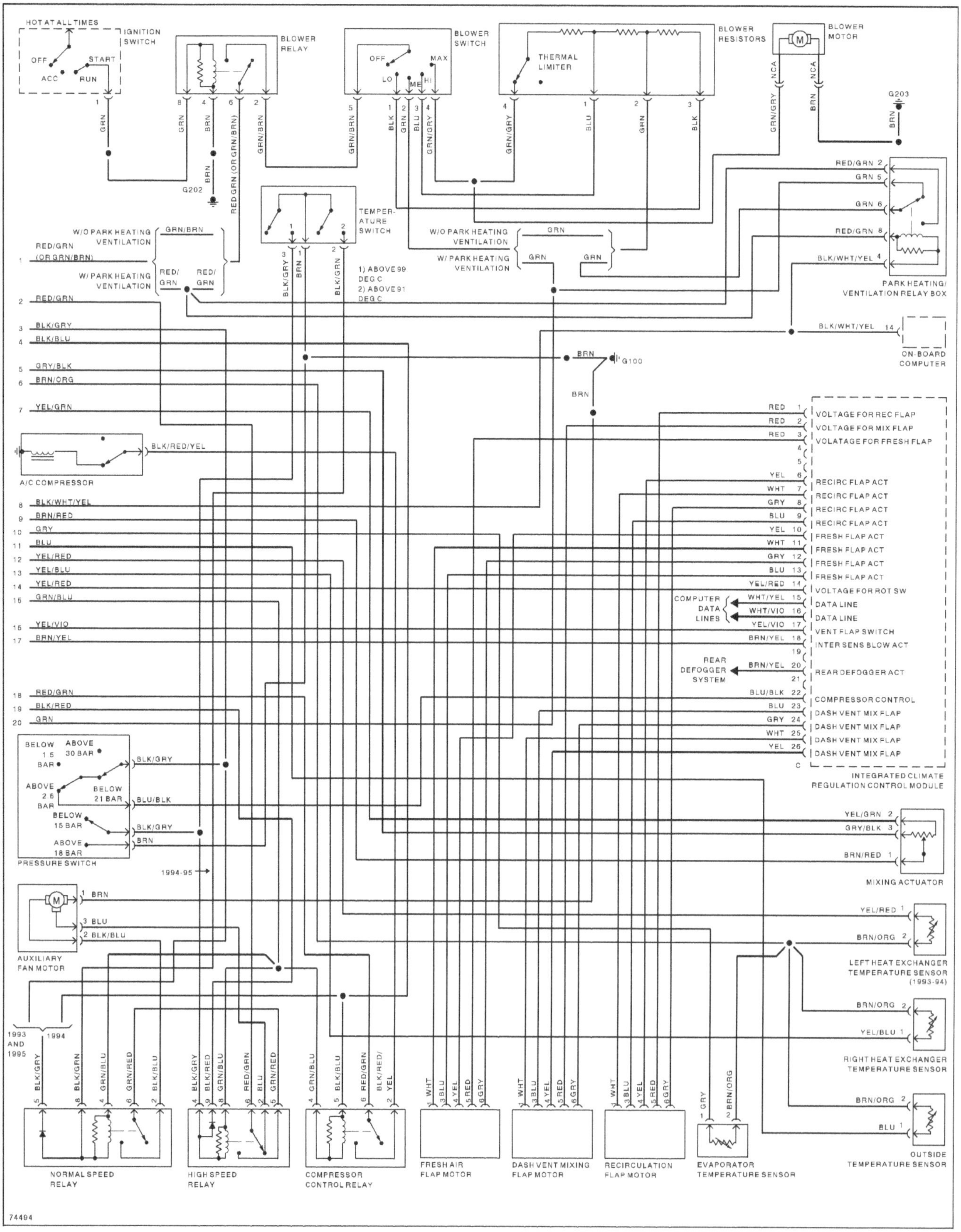

Air Conditioning & Heating

1996-97 Except 318ti

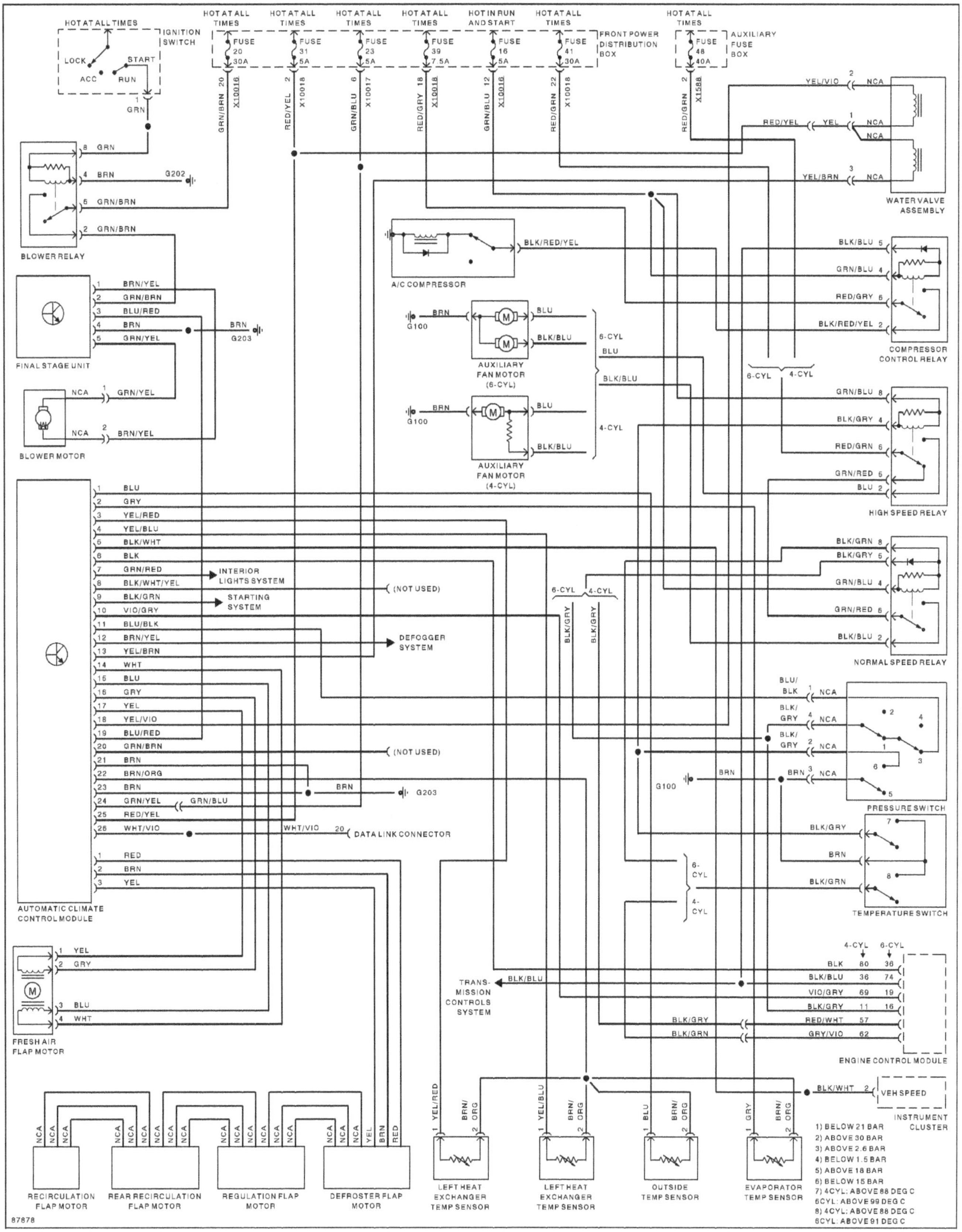

Air Conditioning & Heating

1998 Except 318ti

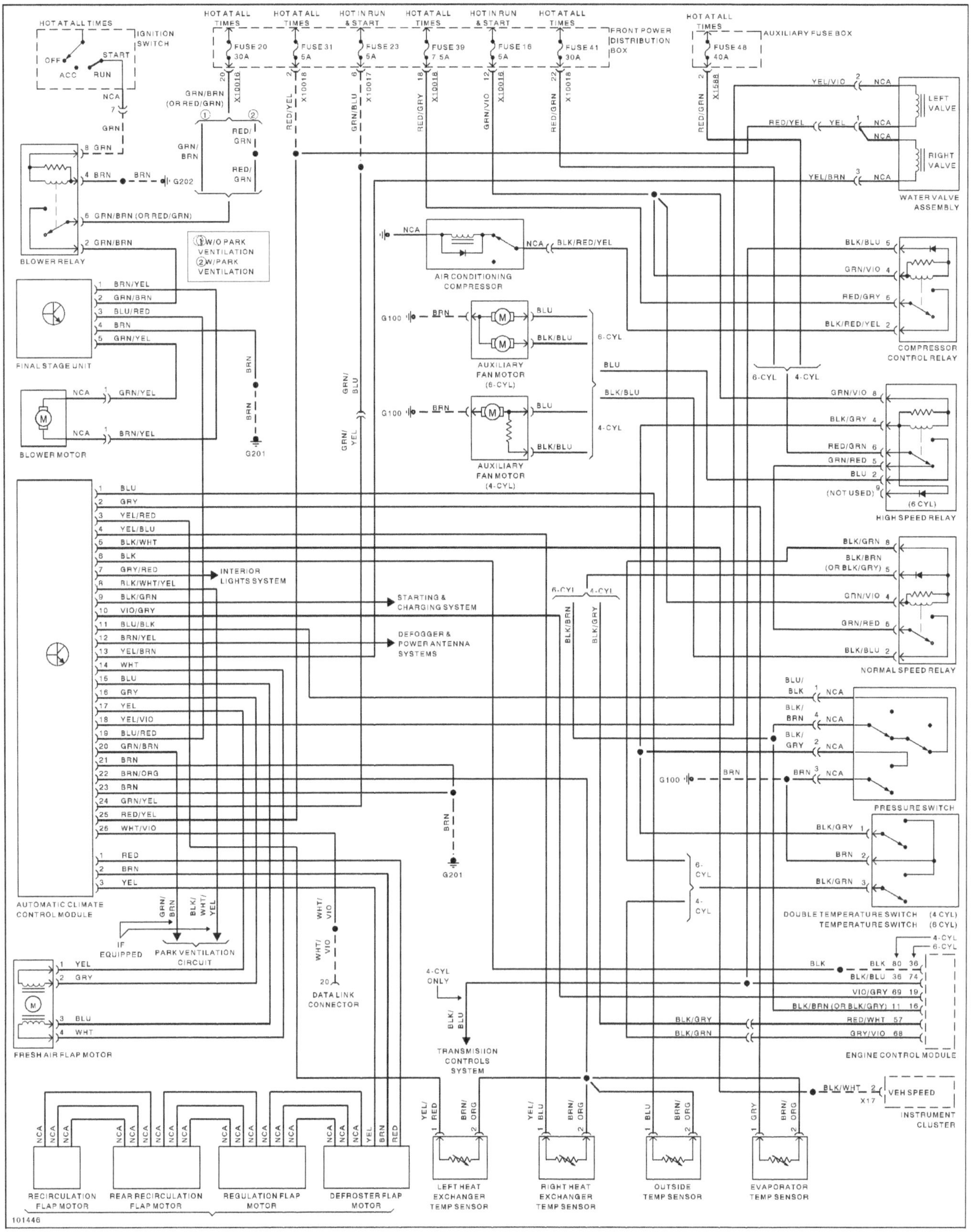

Air Conditioning & Heating

1995 318ti

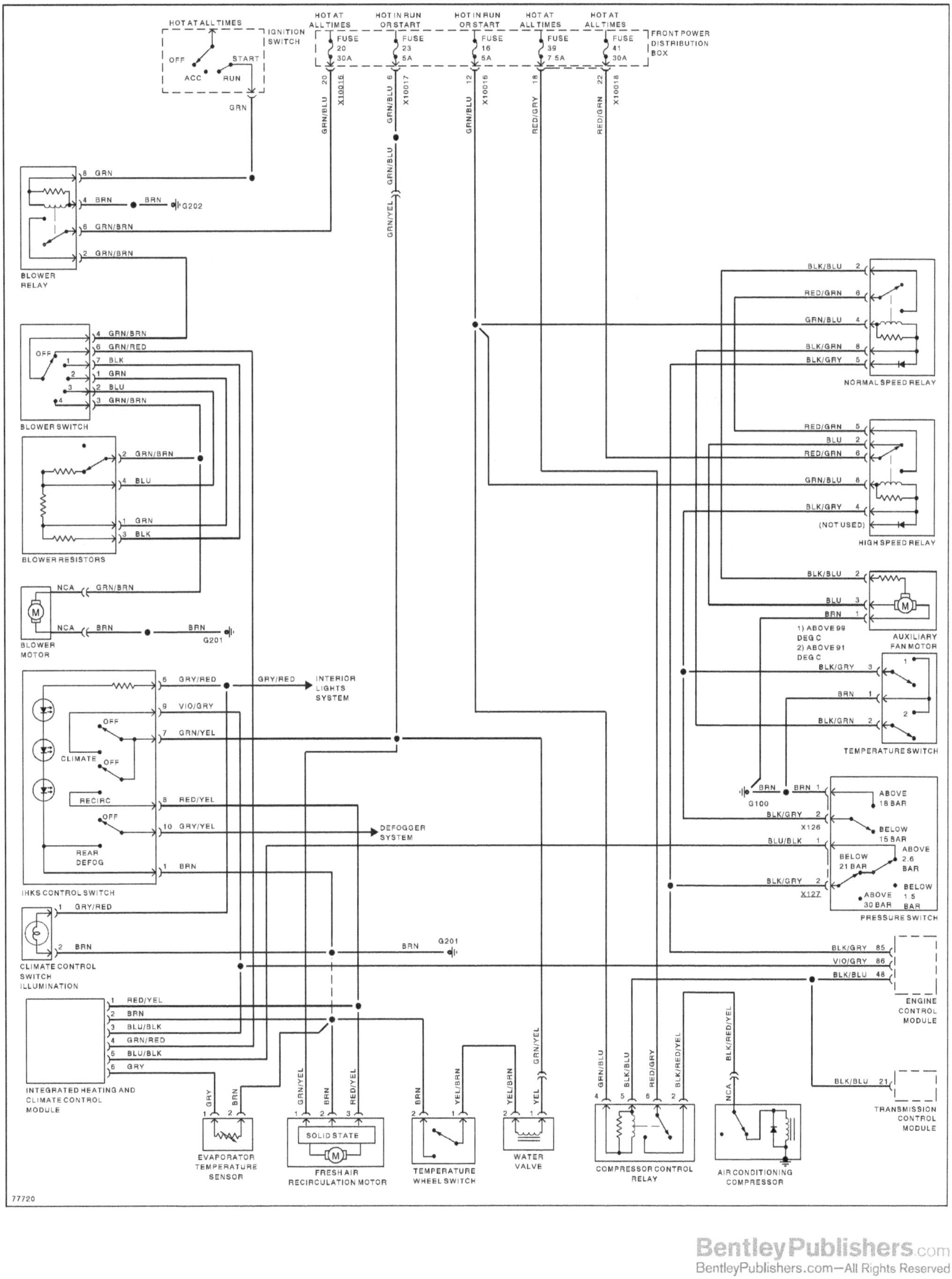

Air Conditioning & Heating

1996 318ti

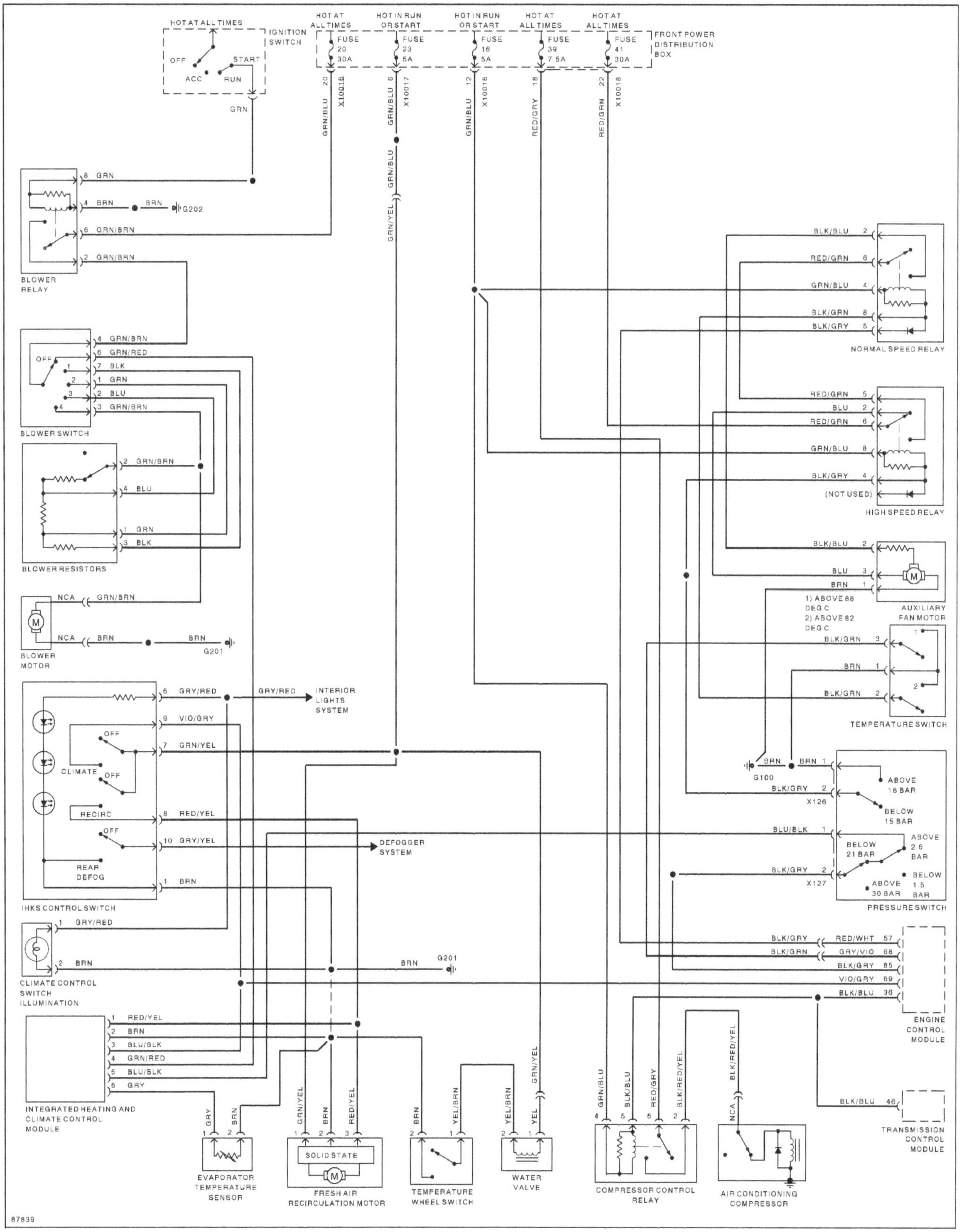

Air Conditioning & Heating

1997 318ti

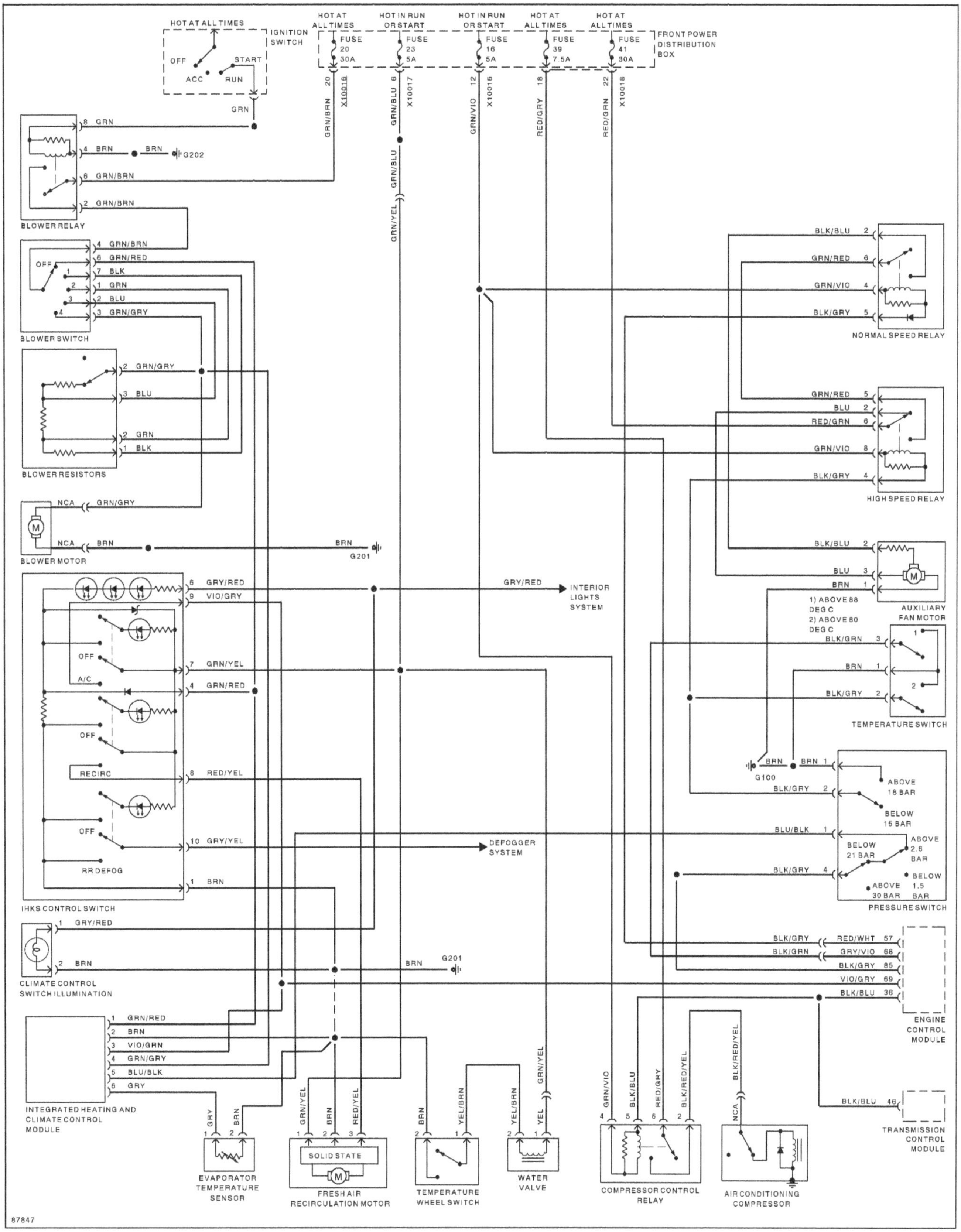

Air Conditioning & Heating

1998 318ti

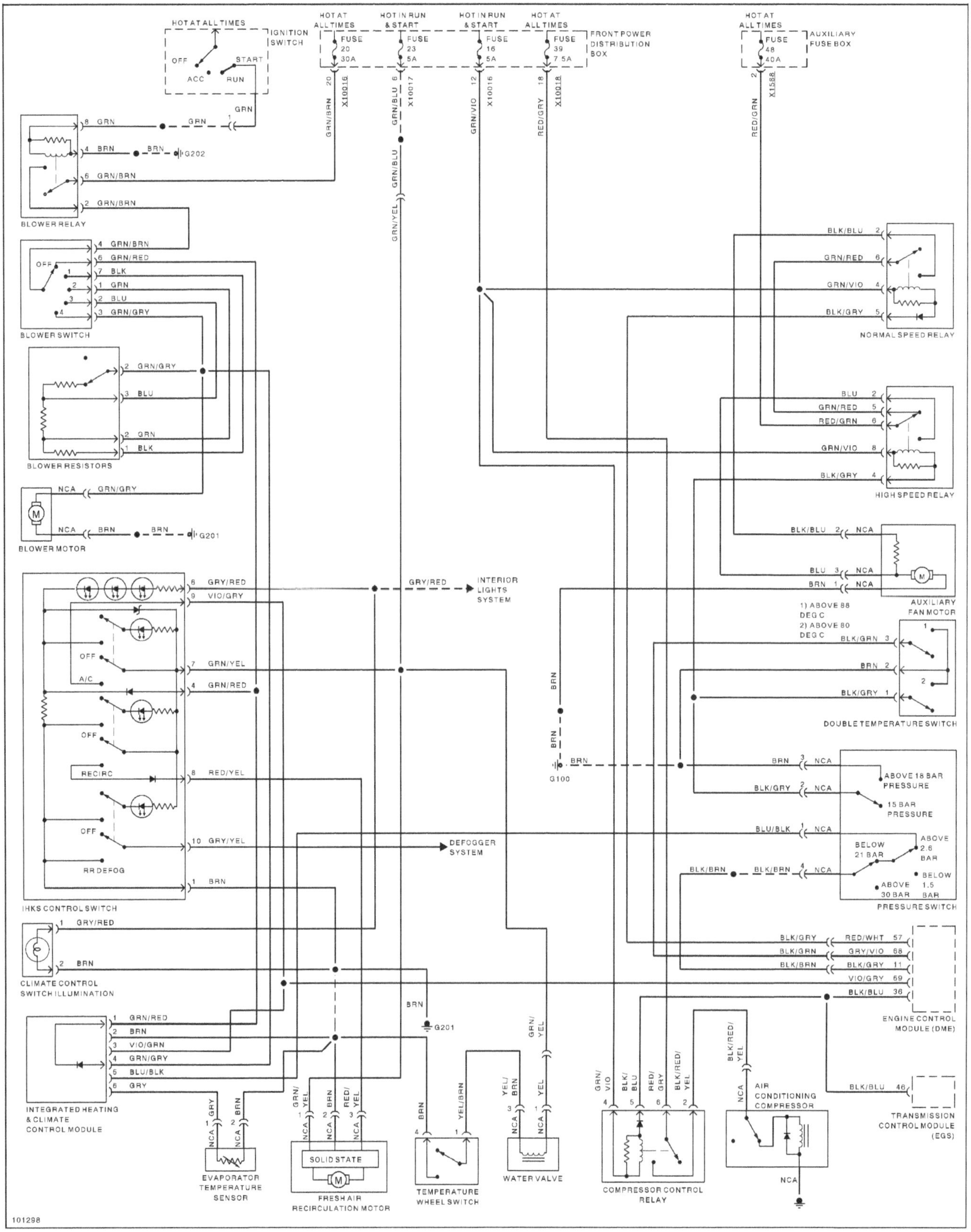

Park Ventilation

1998

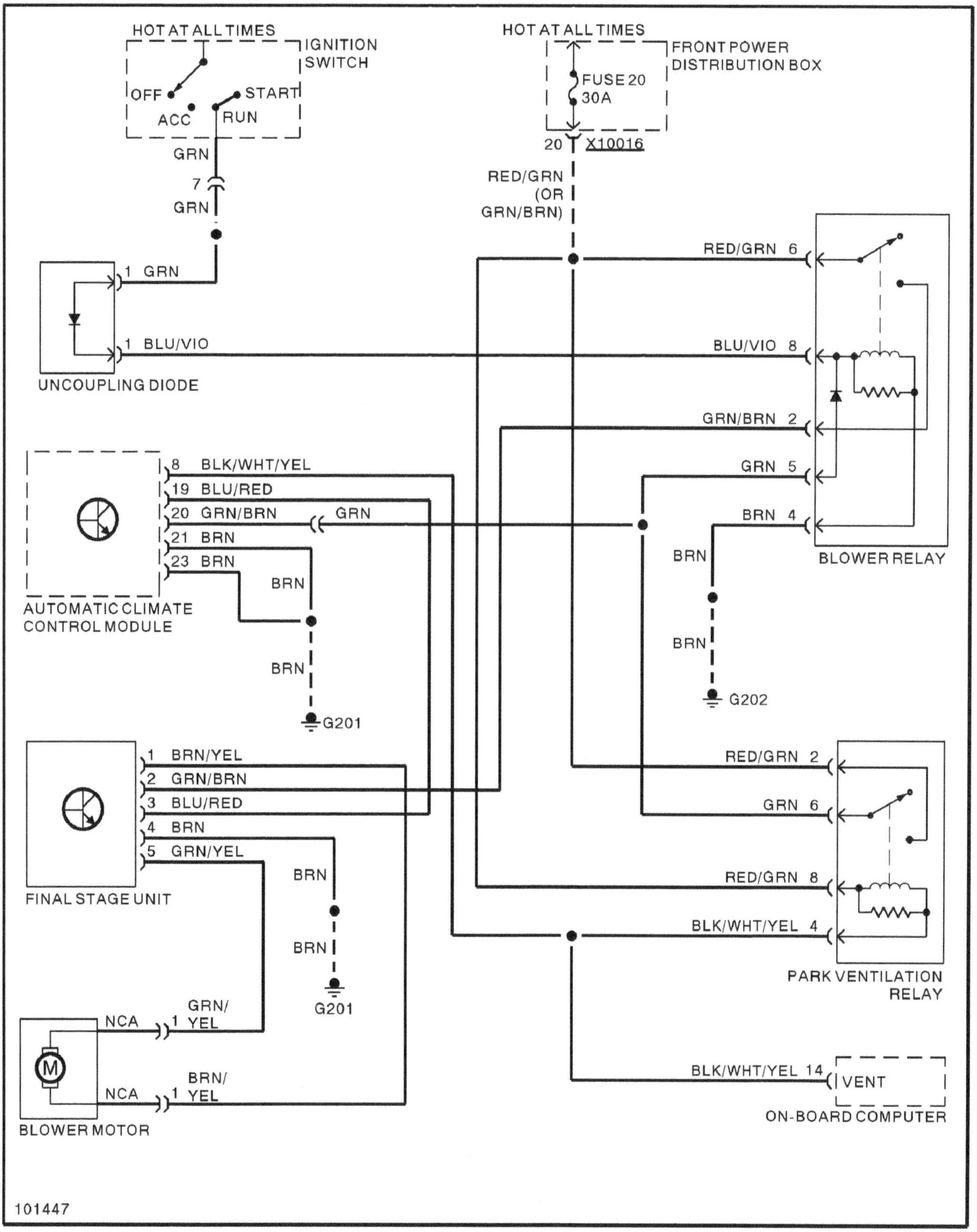

Airbag (Supplemental Restraint System)

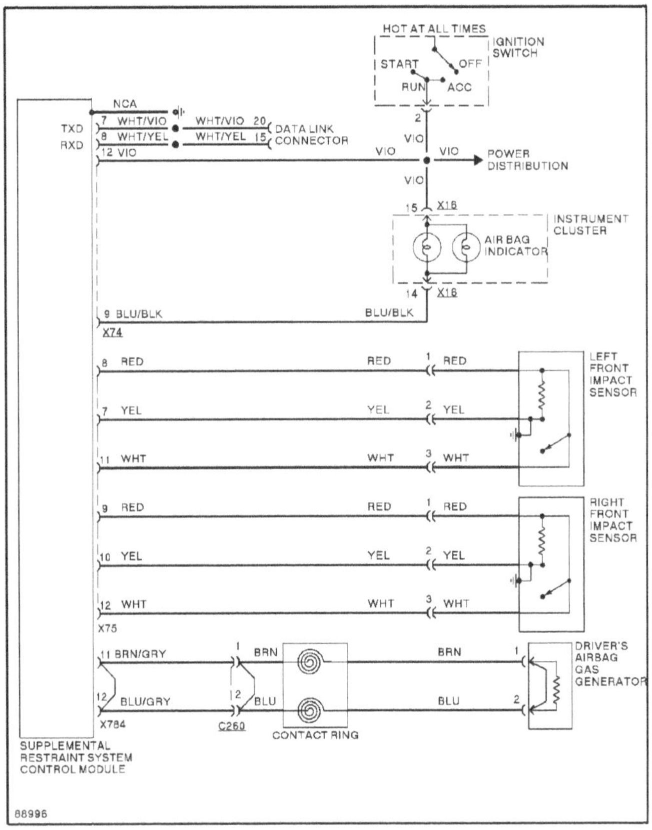

1992-93
Early Production

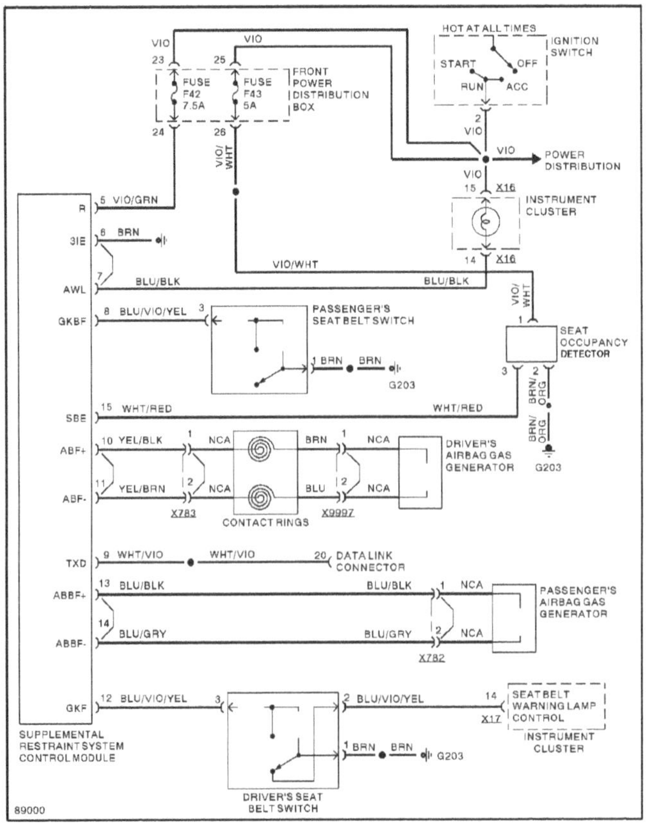

1993
Late Production

Airbag (Supplemental Restraint System)

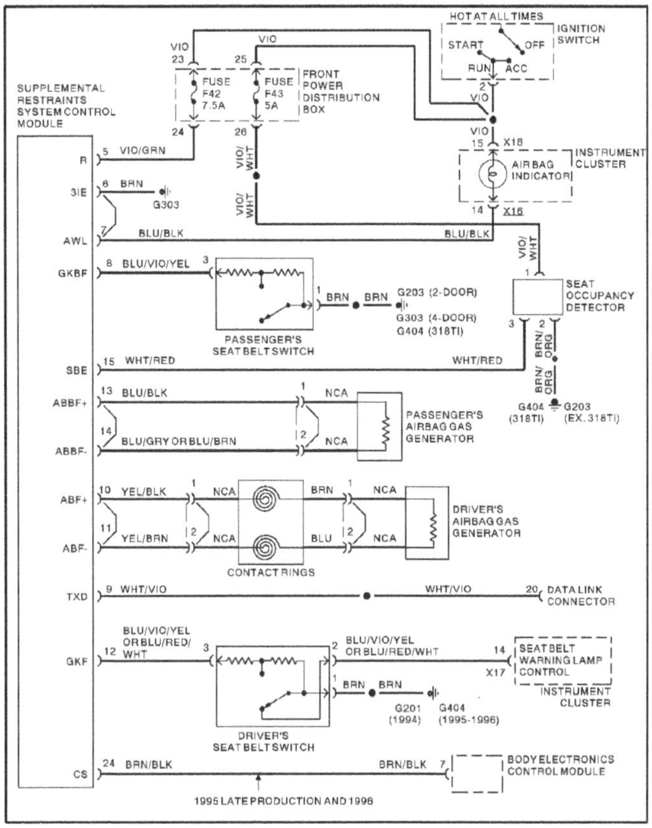

1994-96

Airbag (Supplemental Restraint System)

1997

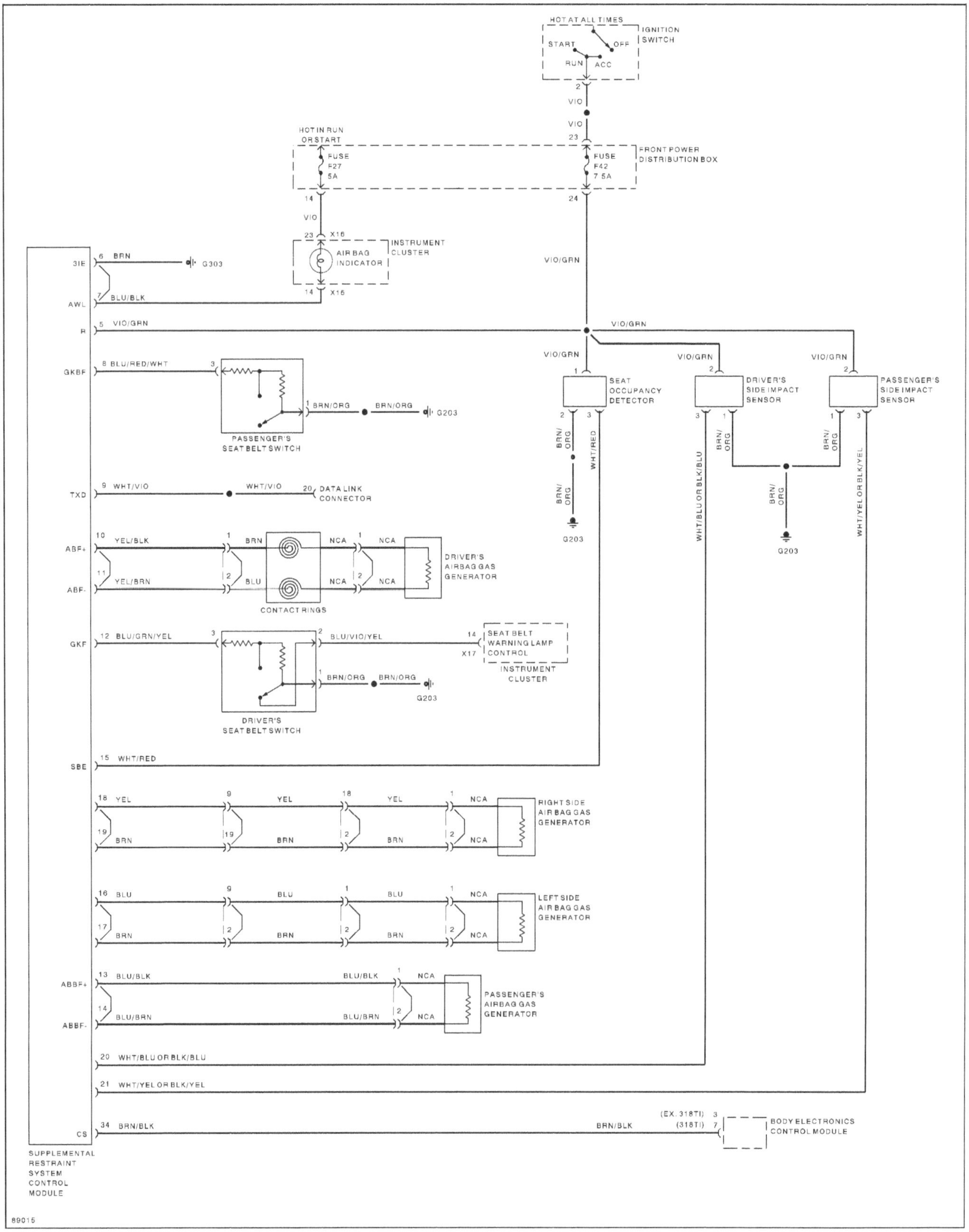

Airbag (Supplemental Restraint System)

1998

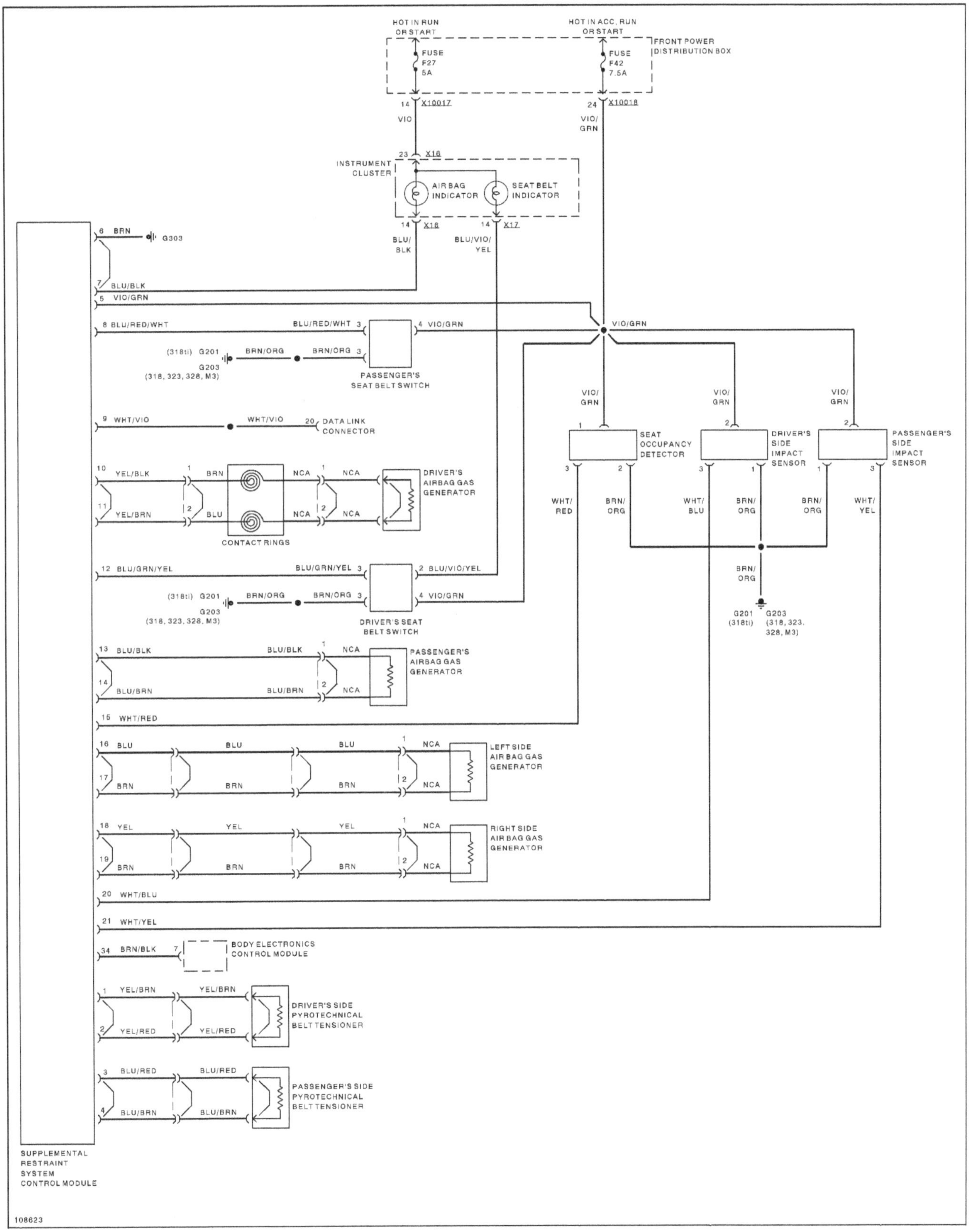

Anti-Theft

1992-93 318i & 325i
Alpine System

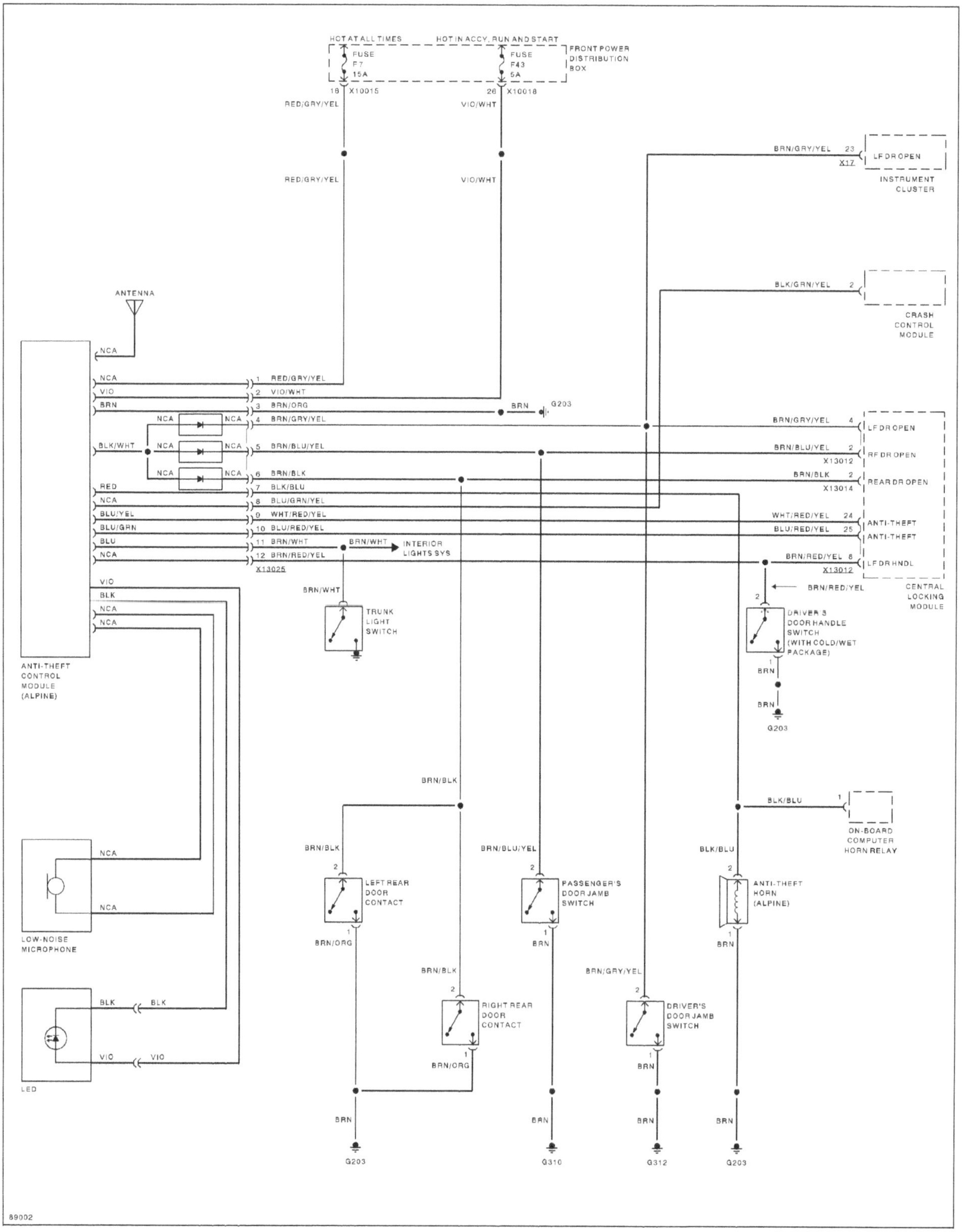

Anti-Theft
1992-93 318is & 325is
Alpine System

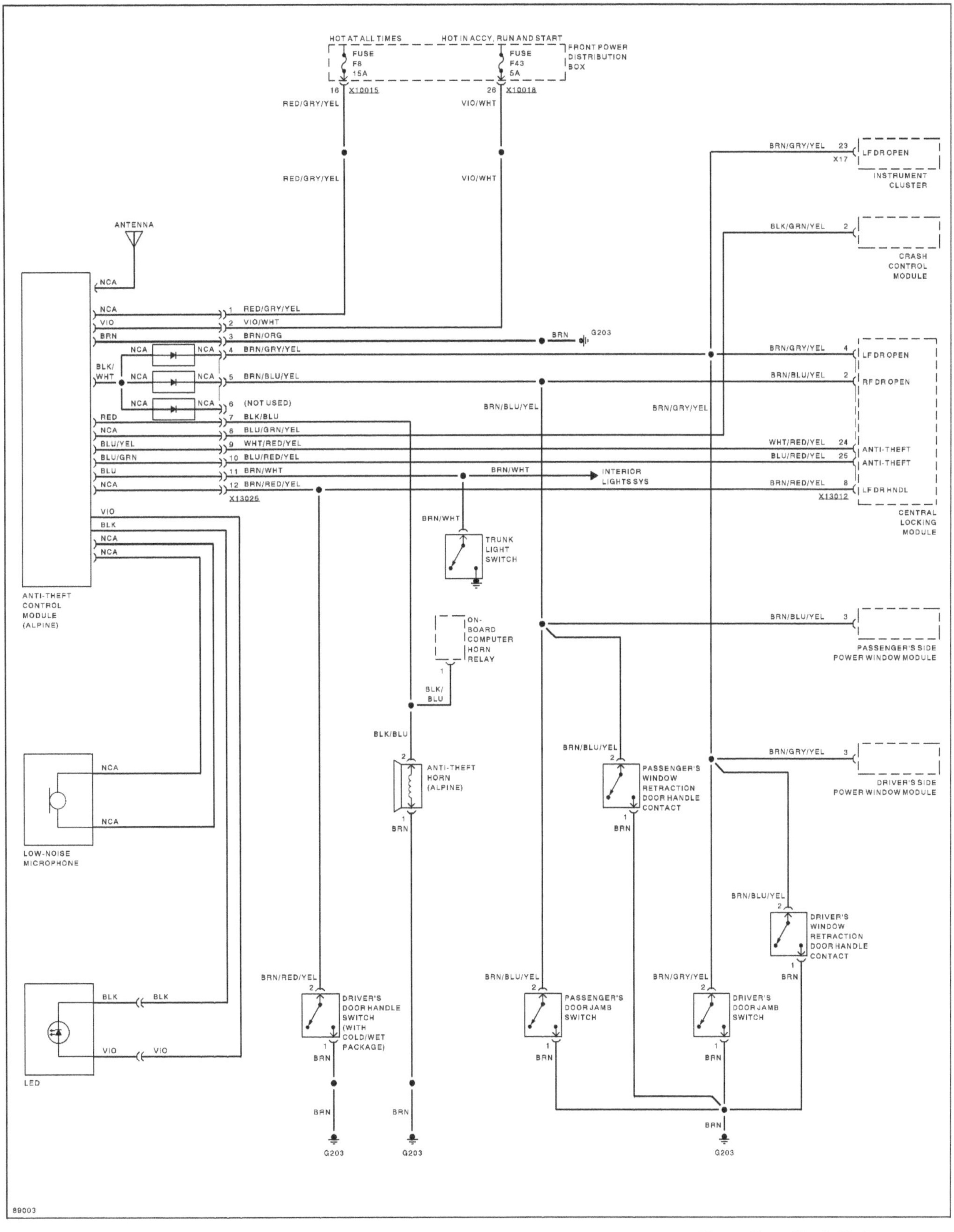

Anti-Theft
1994 Early Production
Alpine System

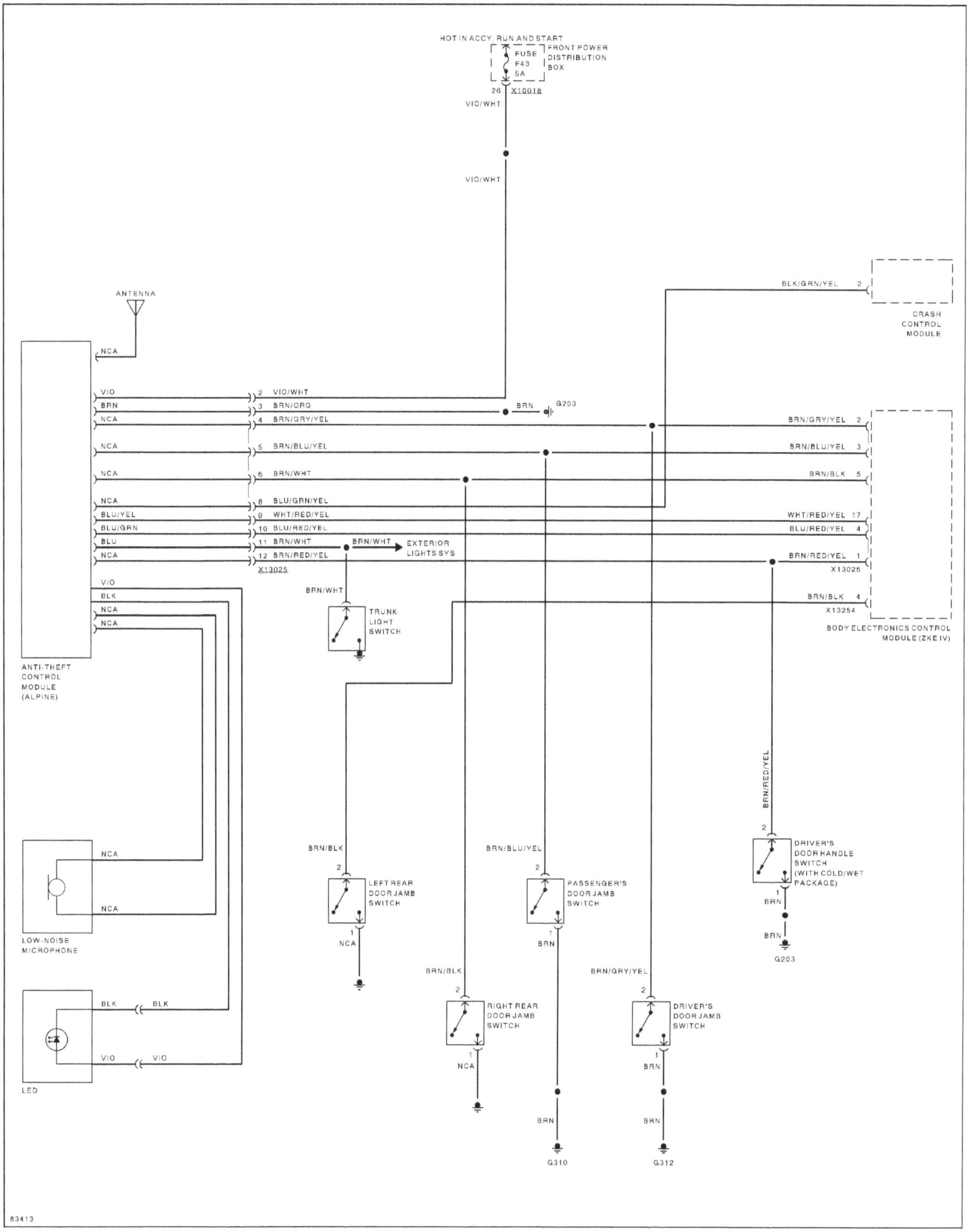

Anti-Theft
1994 Late Production
Alpine System

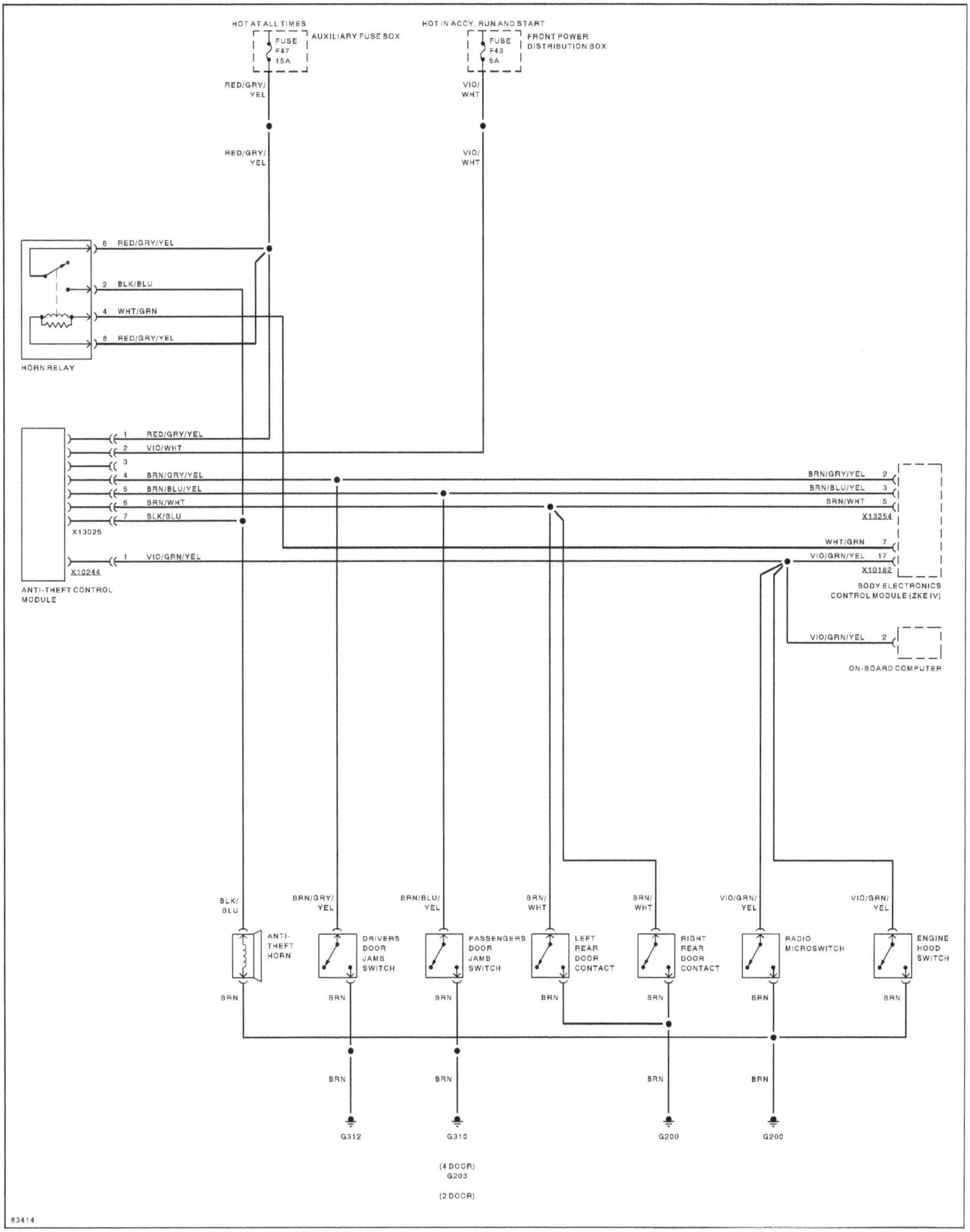

Anti-Theft
1995-97 Except 318ti
Alpine System

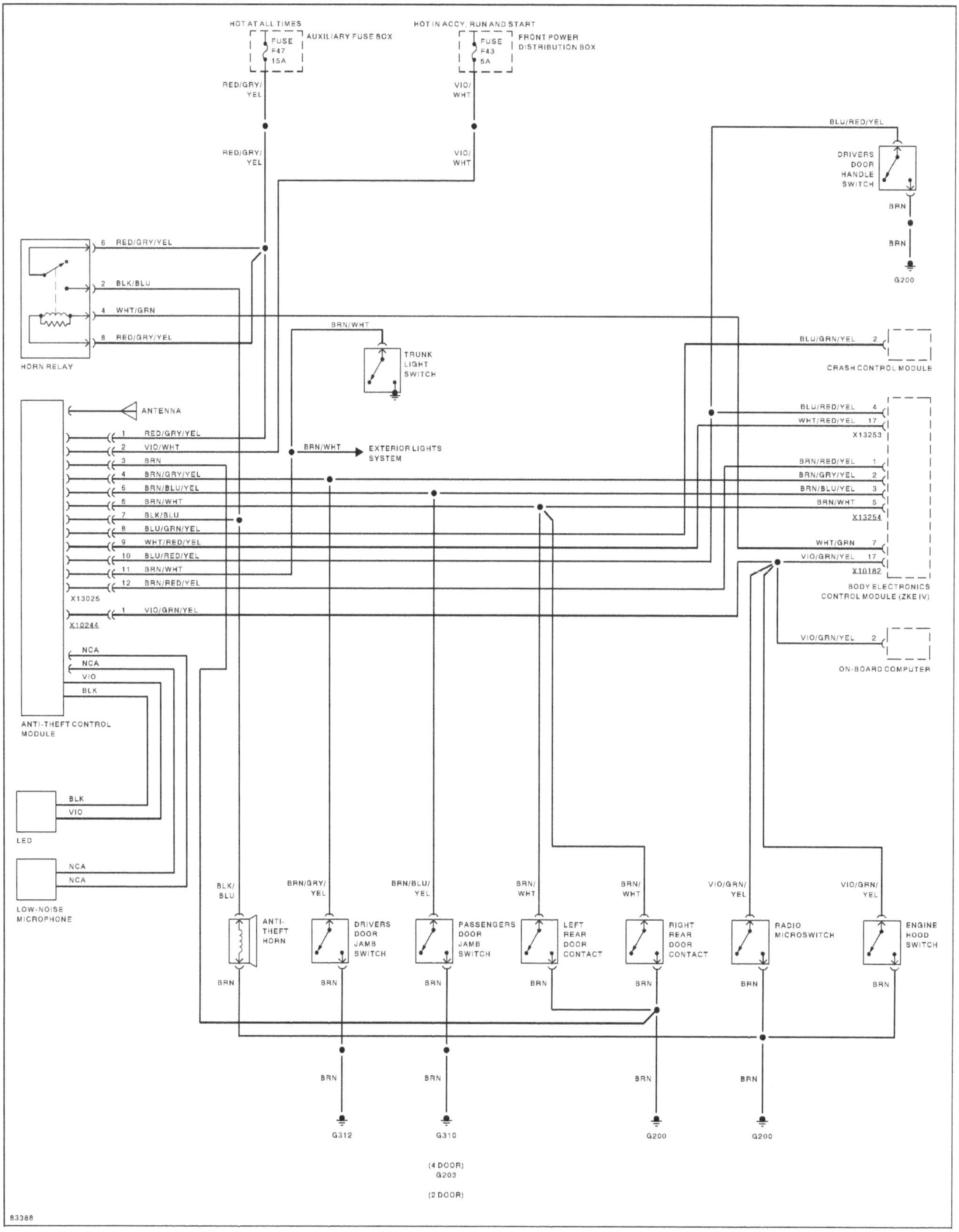

Anti-Theft
1995-97 318ti
Alpine System

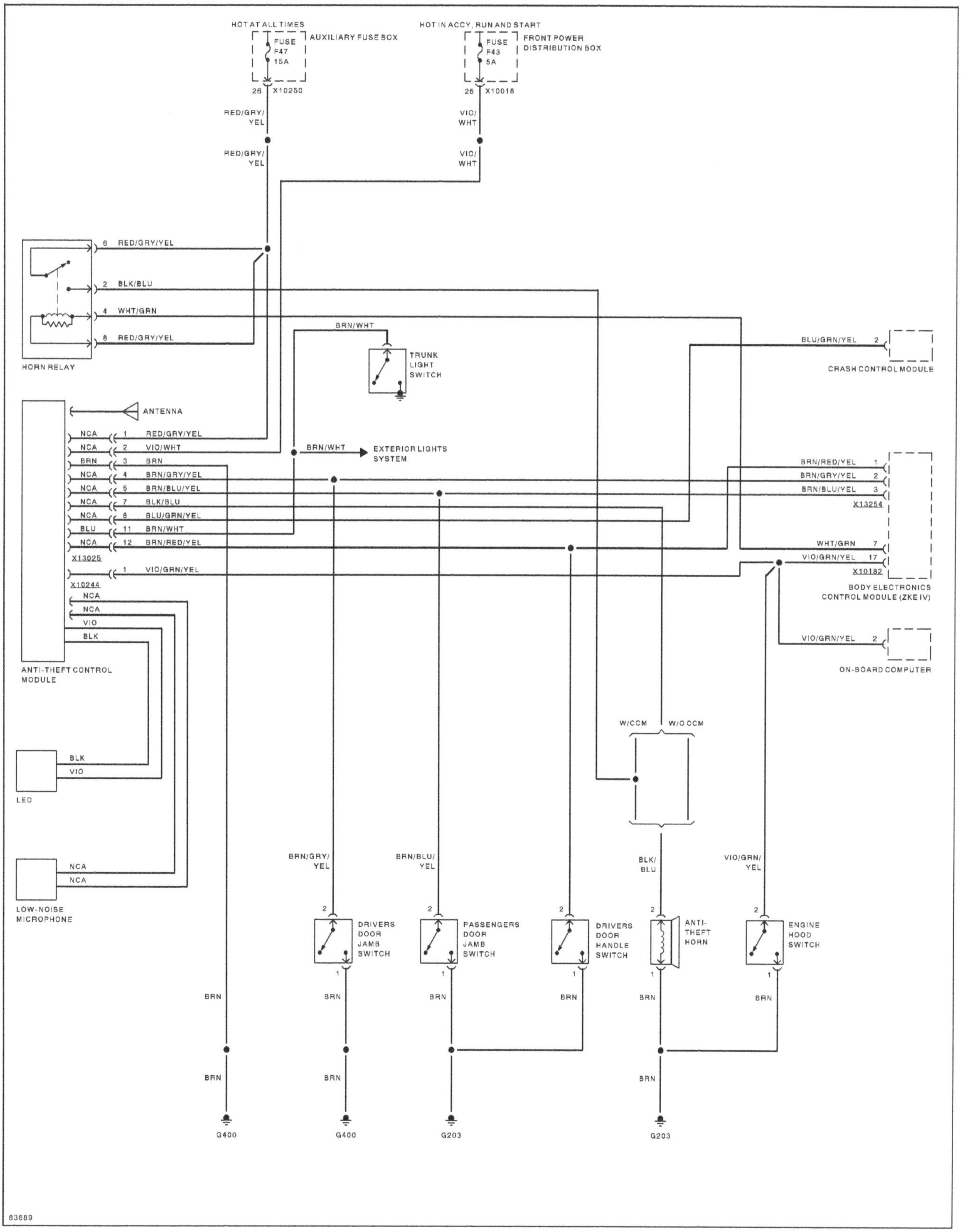

Anti-Theft

1995-98

EWS II

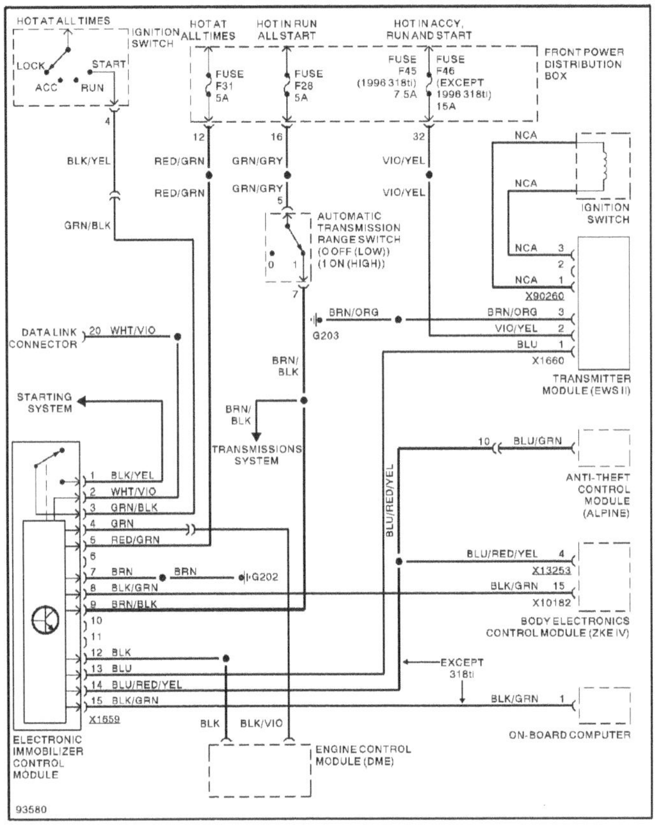

1995-96

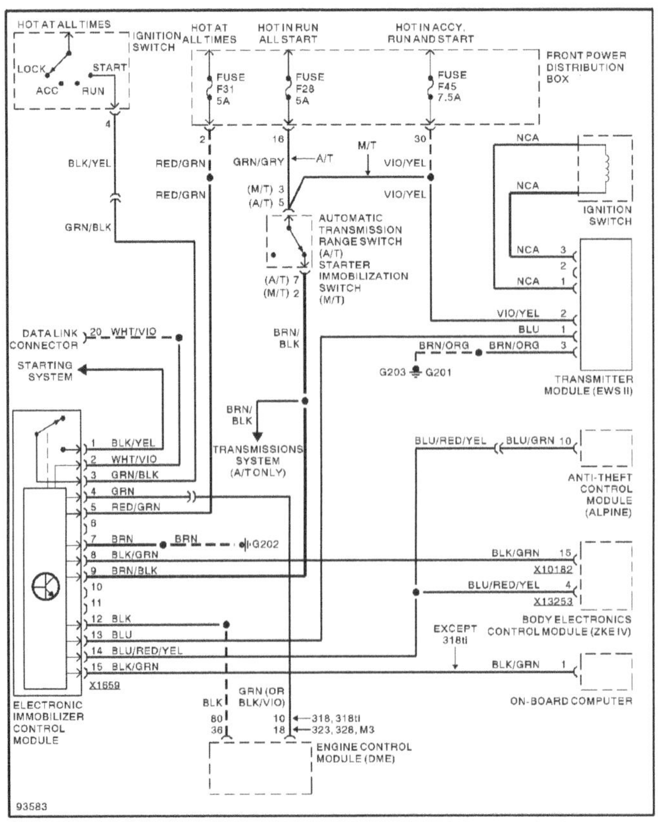

1997-98

Anti-Theft Forced Entry

1998 Except 318ti

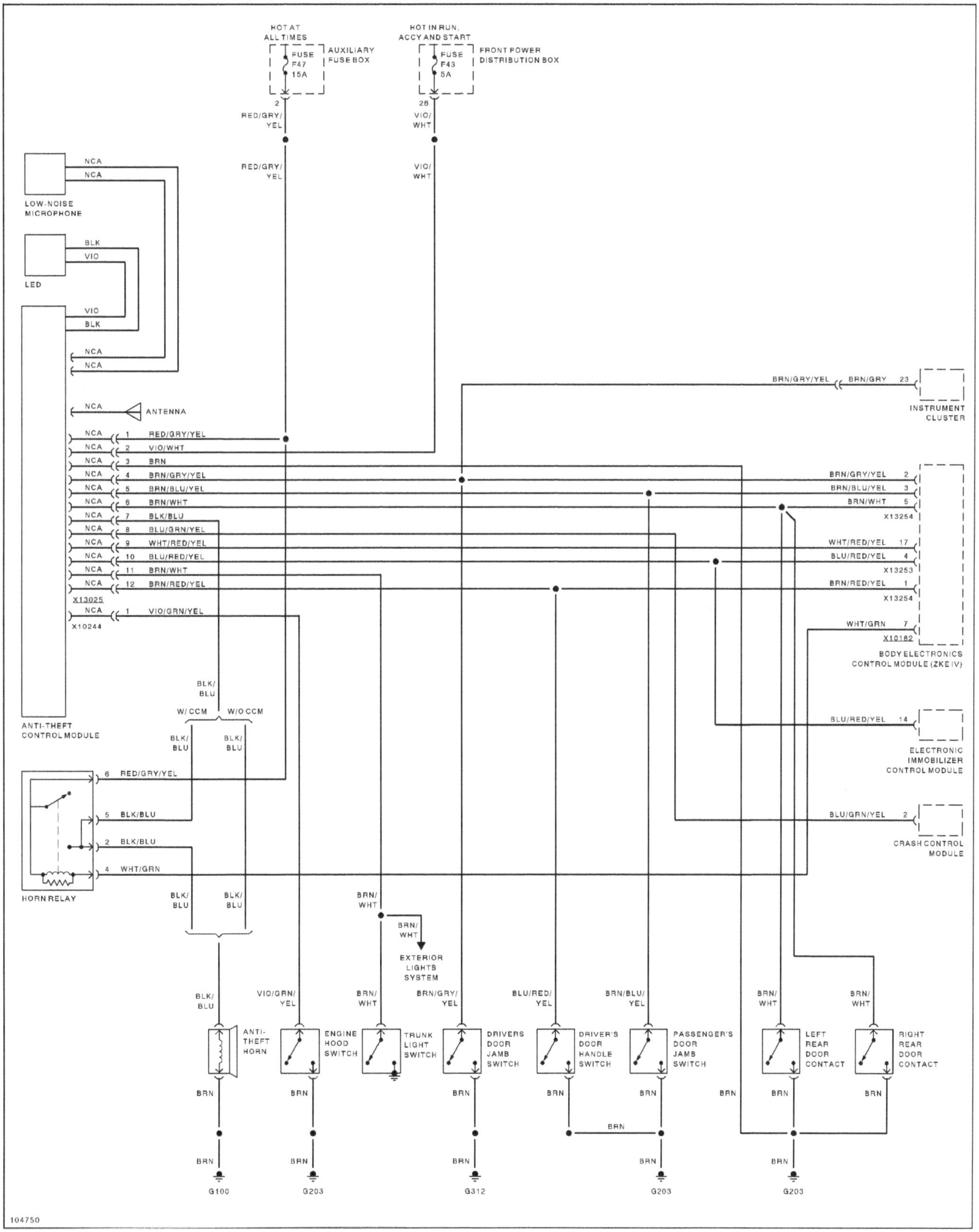

Anti-Theft Forced Entry

1998 318ti

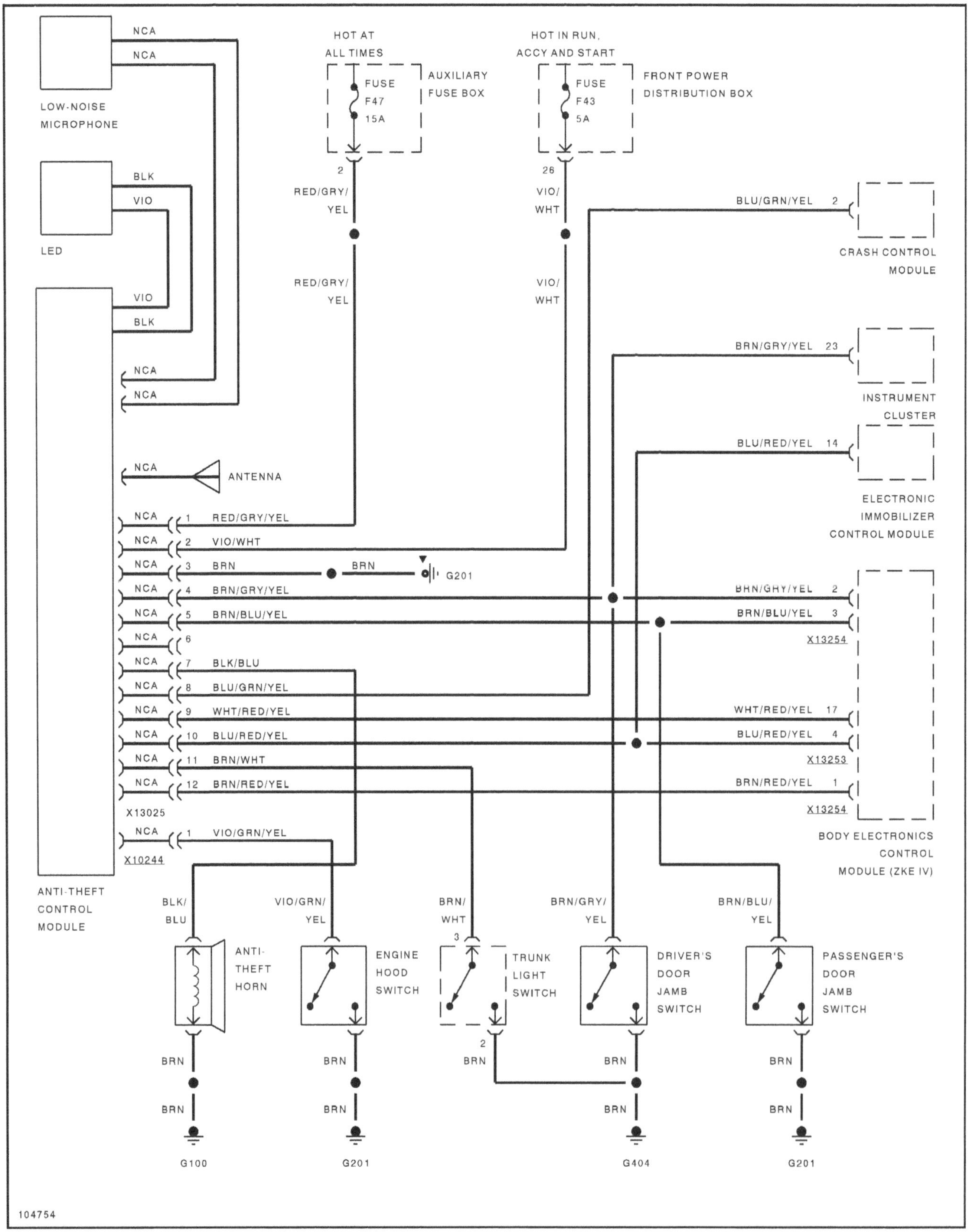

Horns

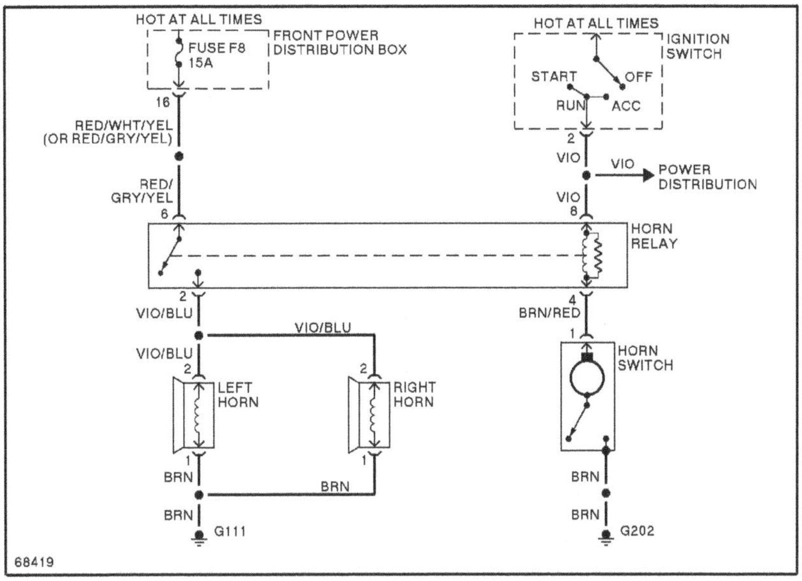

1992-93

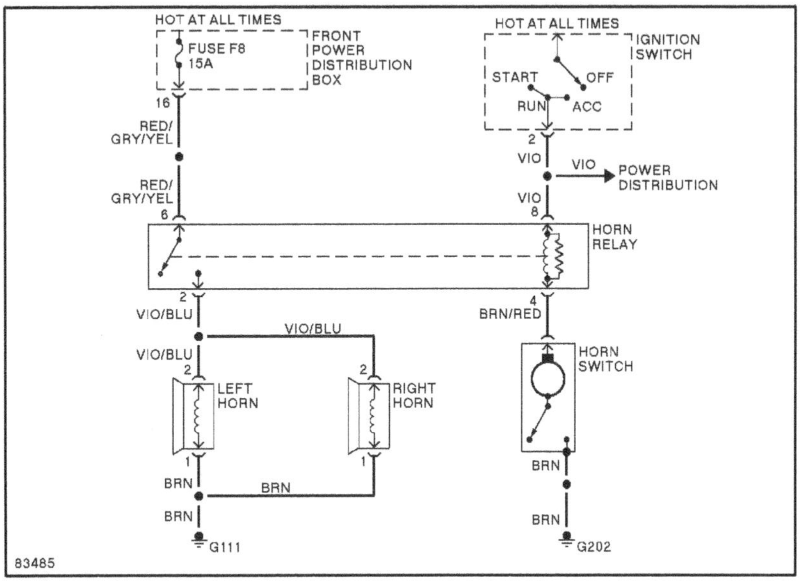

1994-96

Horns

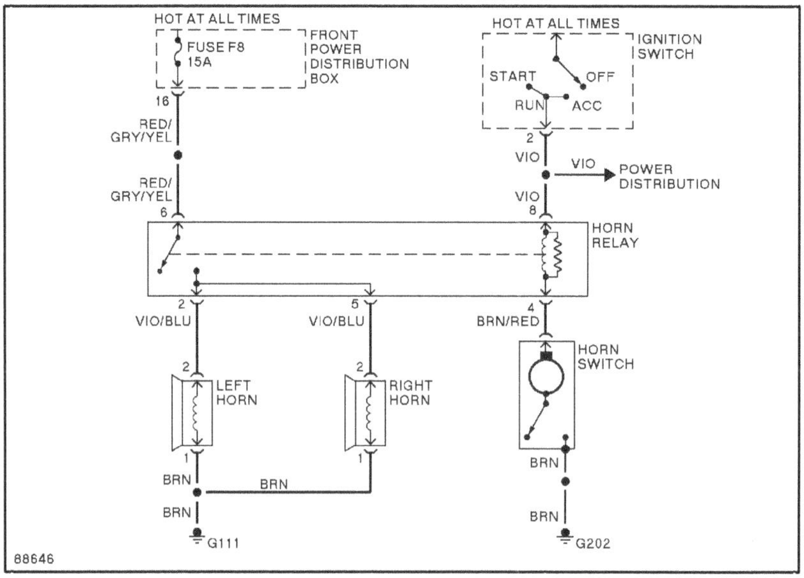

1997

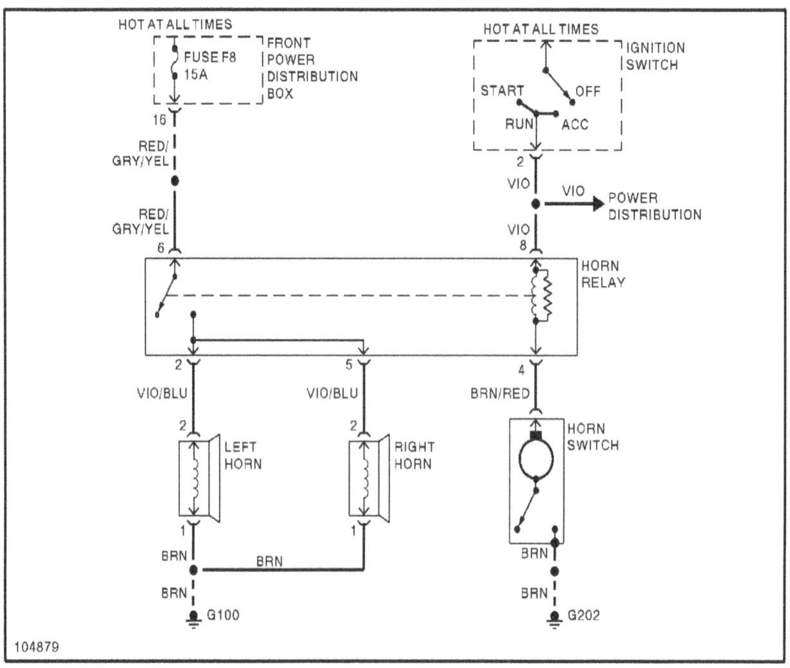

1998

Wiper/Washer

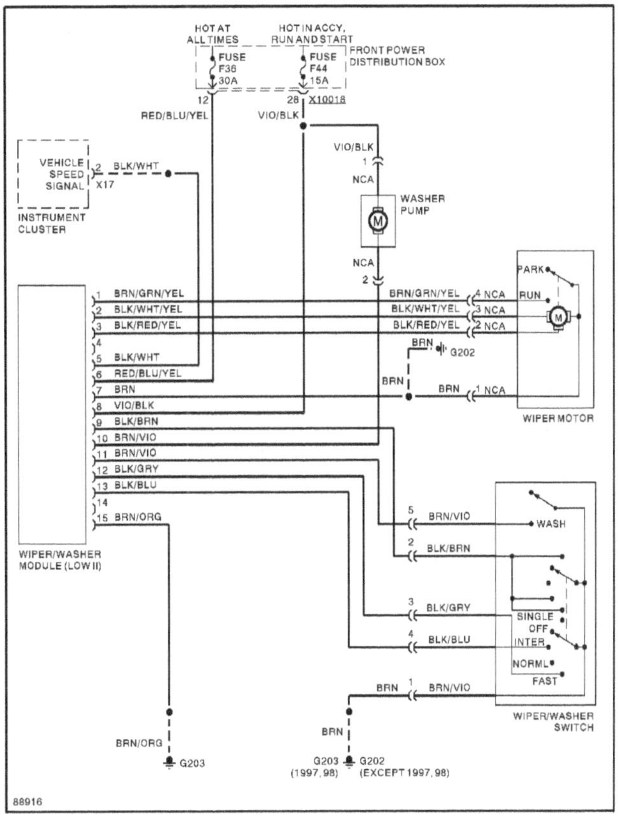

1992-98 USA Except 318ti

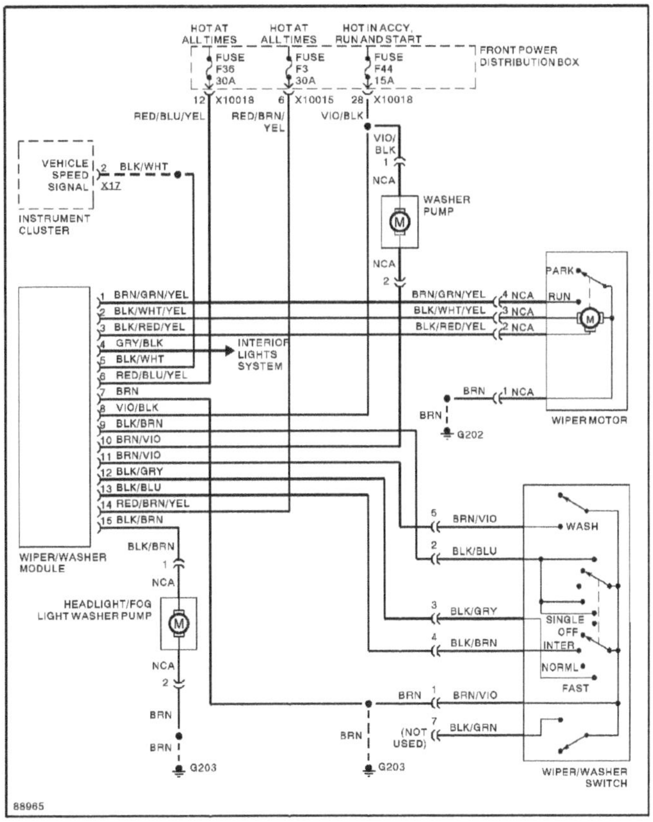

1998 Canada Except 318ti

Wiper/Washer

1992-96 Canada Except 318ti

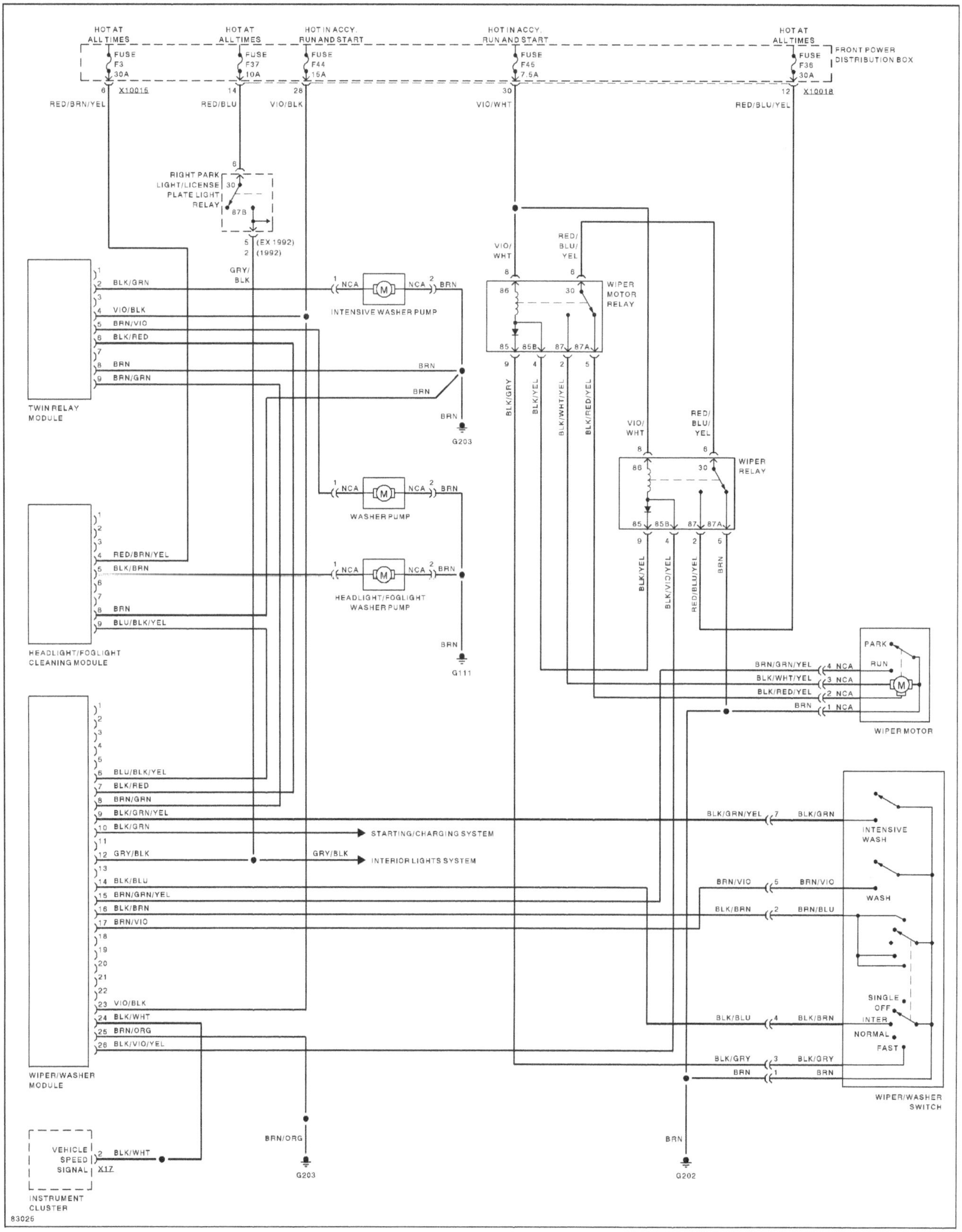

Wiper/Washer

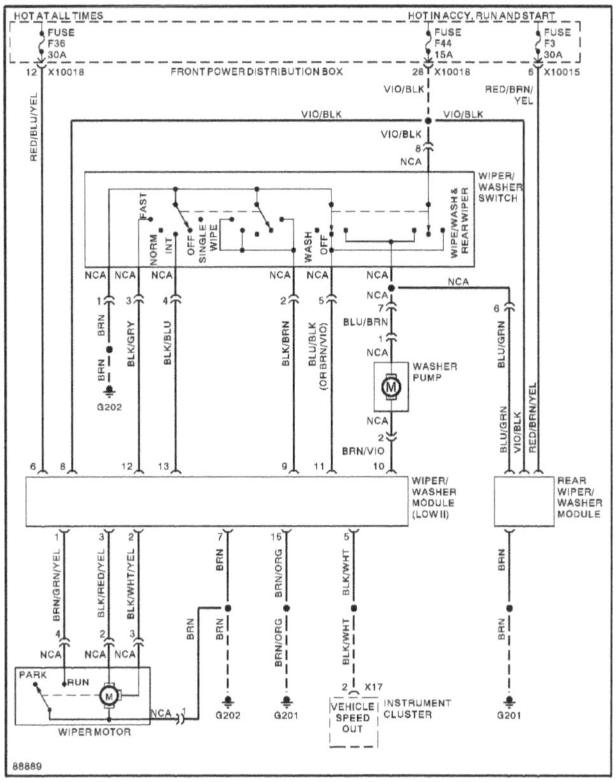

1995-98 USA 318ti

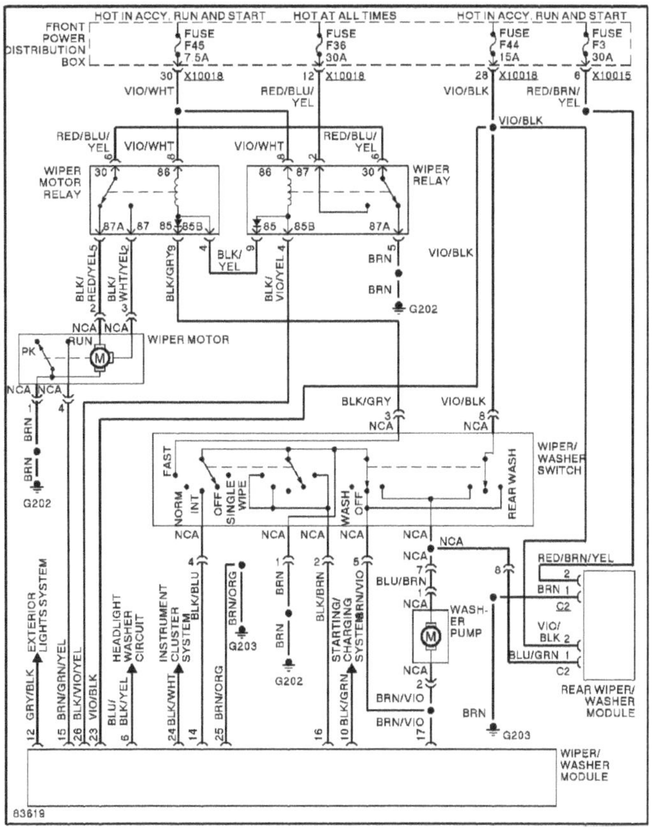

1995-96 Canada 318ti

Wiper/Washer

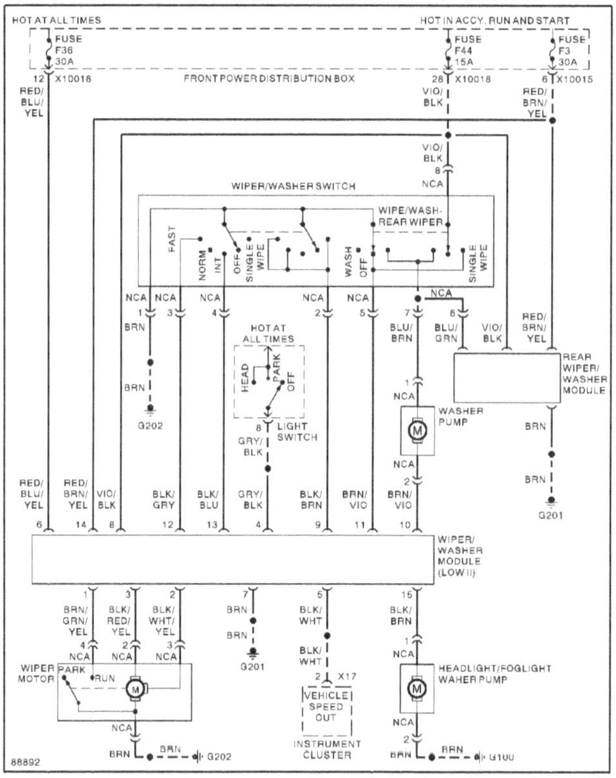

1998 Canada 318ti

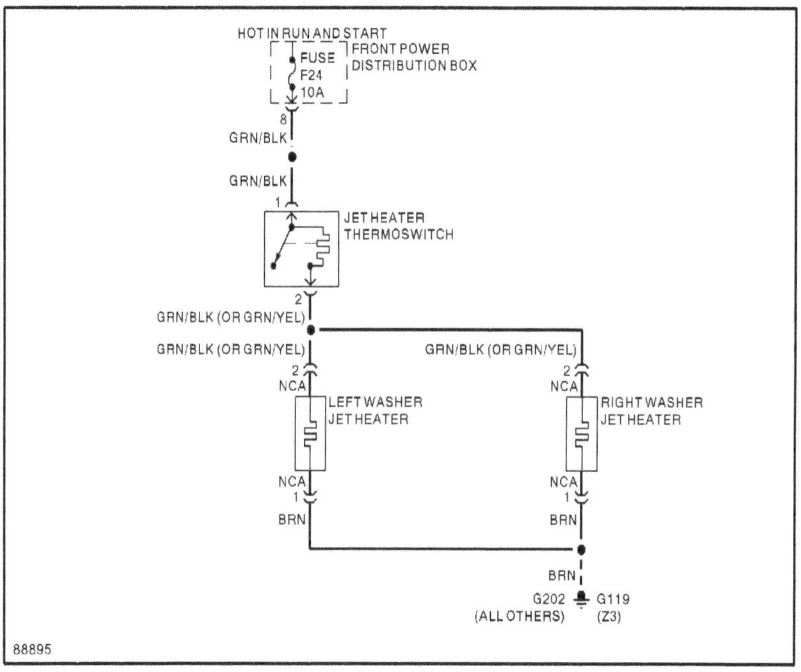

1993-98 Washer Jet Heater

Wiper/Washer

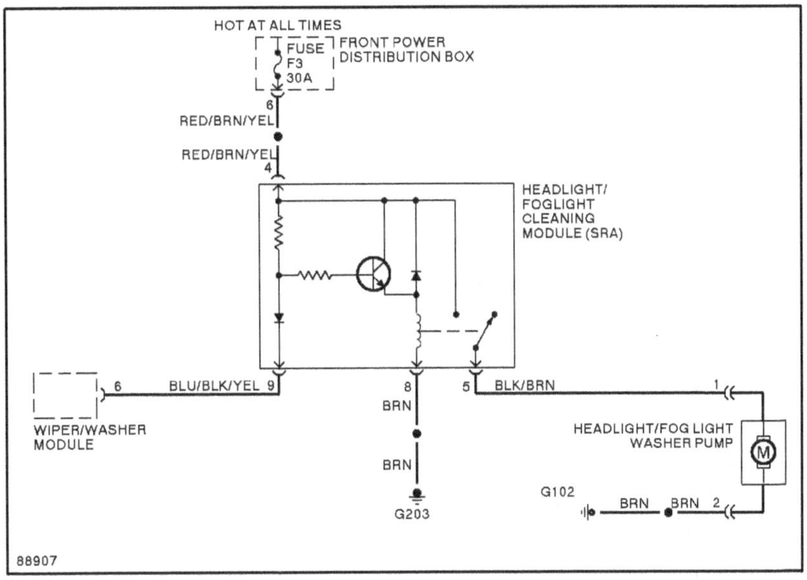

1995-96 Headlight Washer

Rear Window Defogger

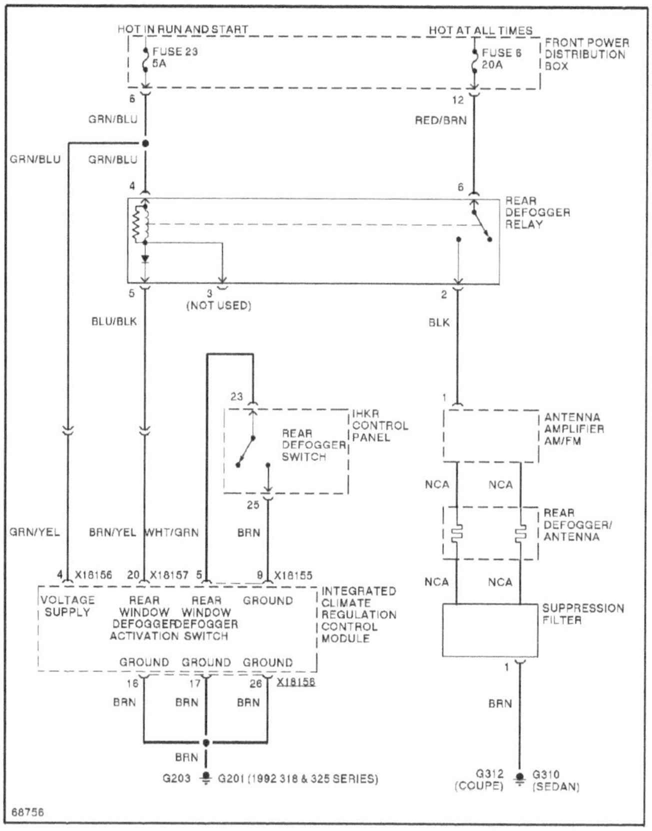

1992-93

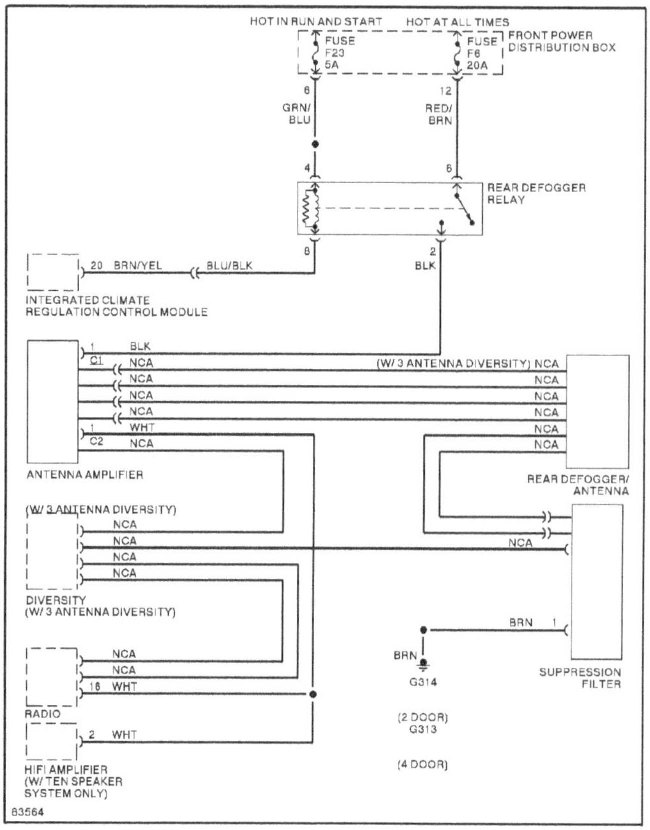

1994-96 Except 318ti & Convertible

Rear Window Defogger

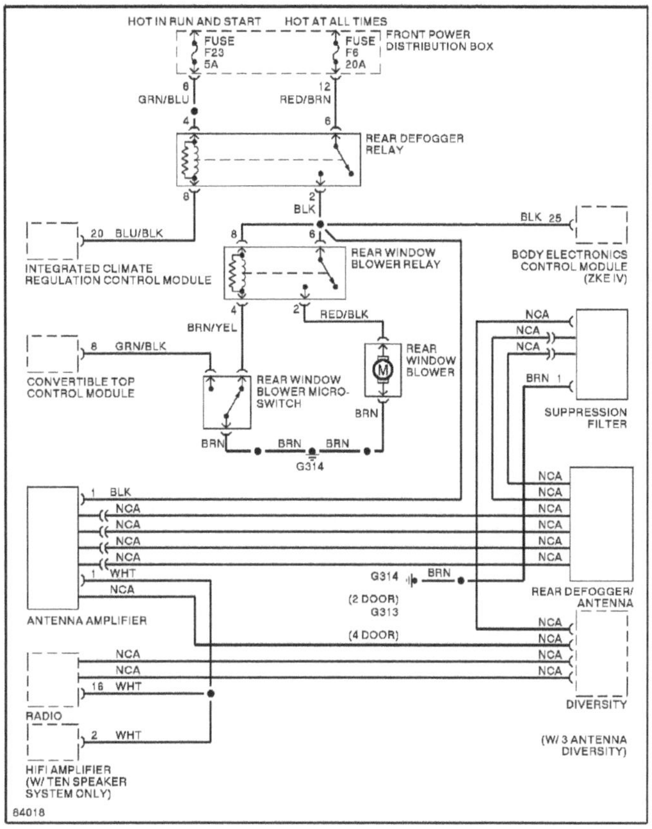

1994-96 Convertible

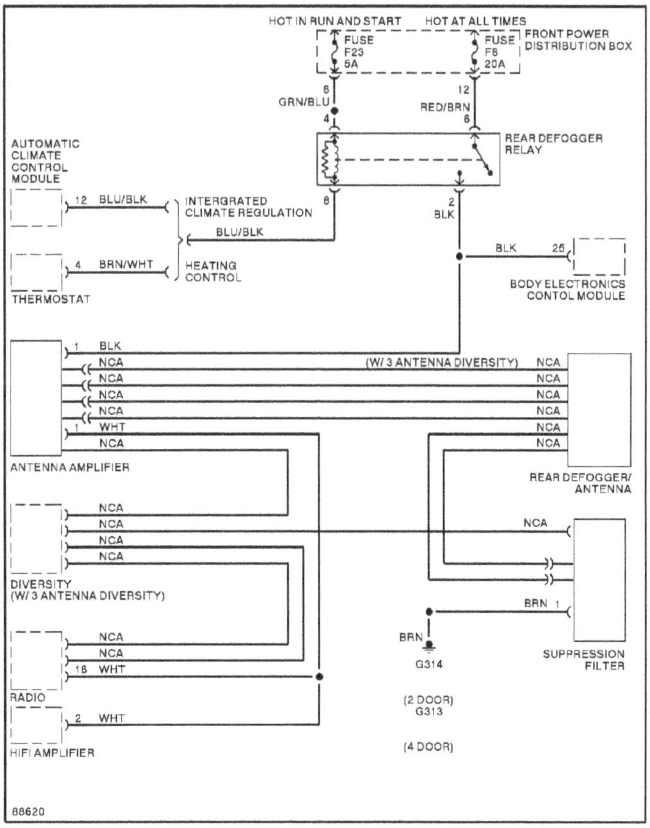

1997 Except 318ti

Rear Window Defogger

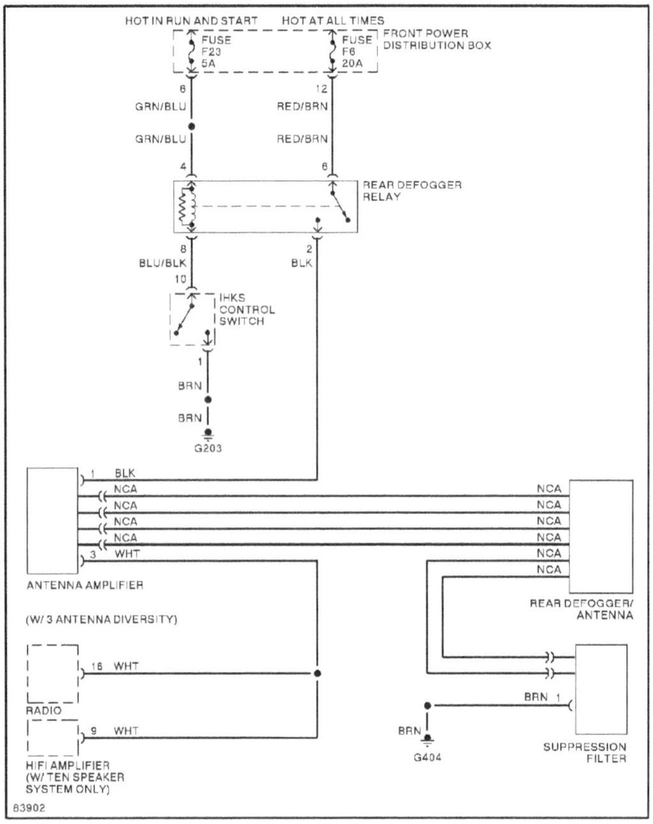

1995-96
318ti

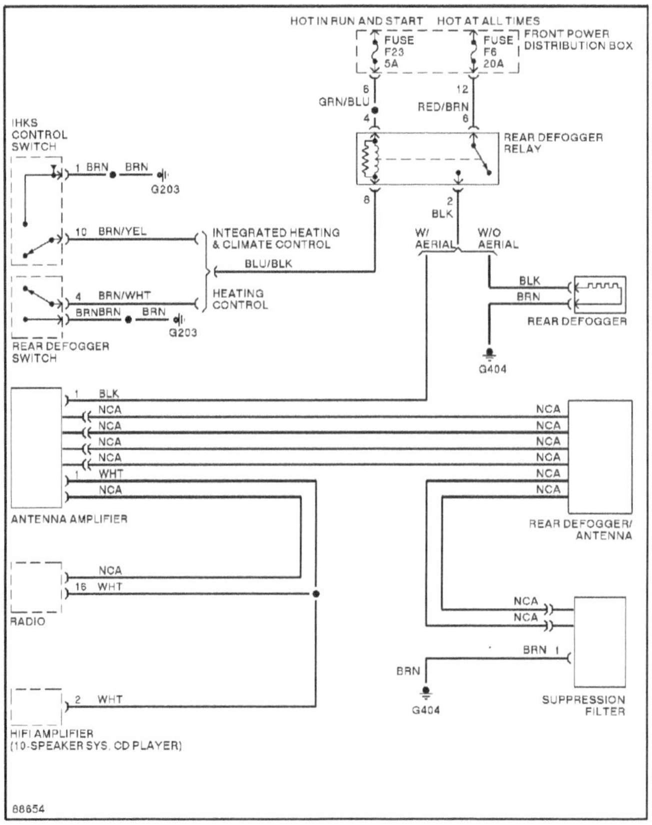

1997
318ti

Rear Window Defogger

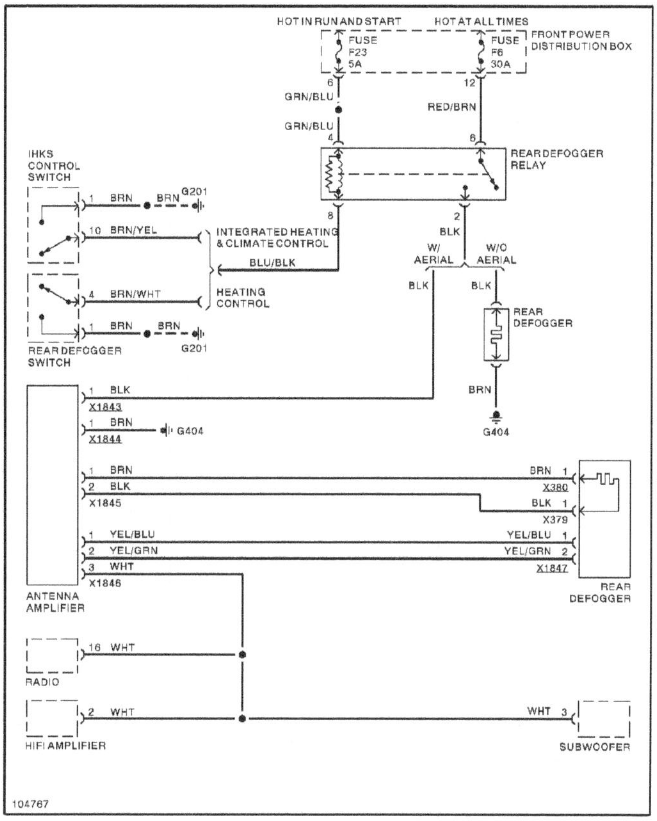

1998 318ti

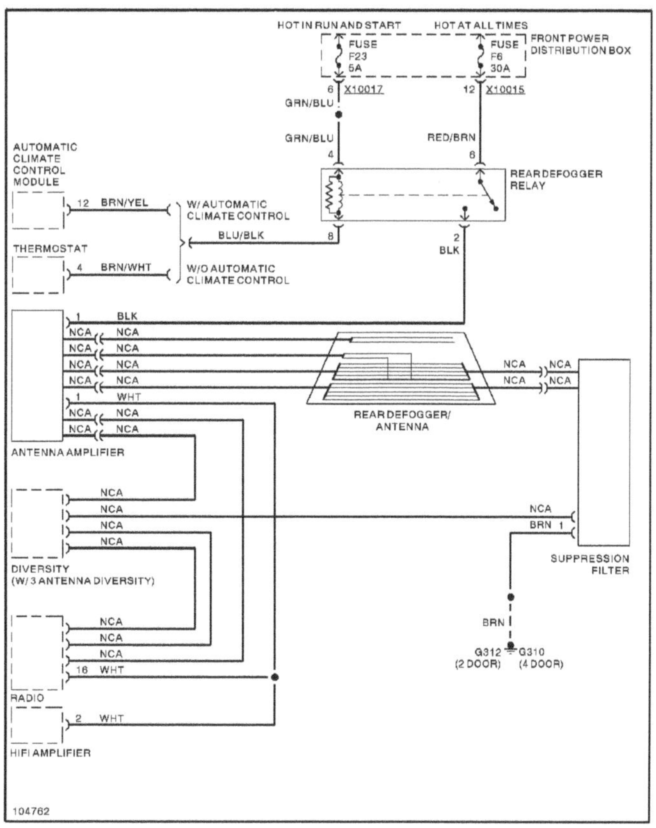

1998 Except 318ti & Convertible

Rear Window Defogger

1998 Convertible

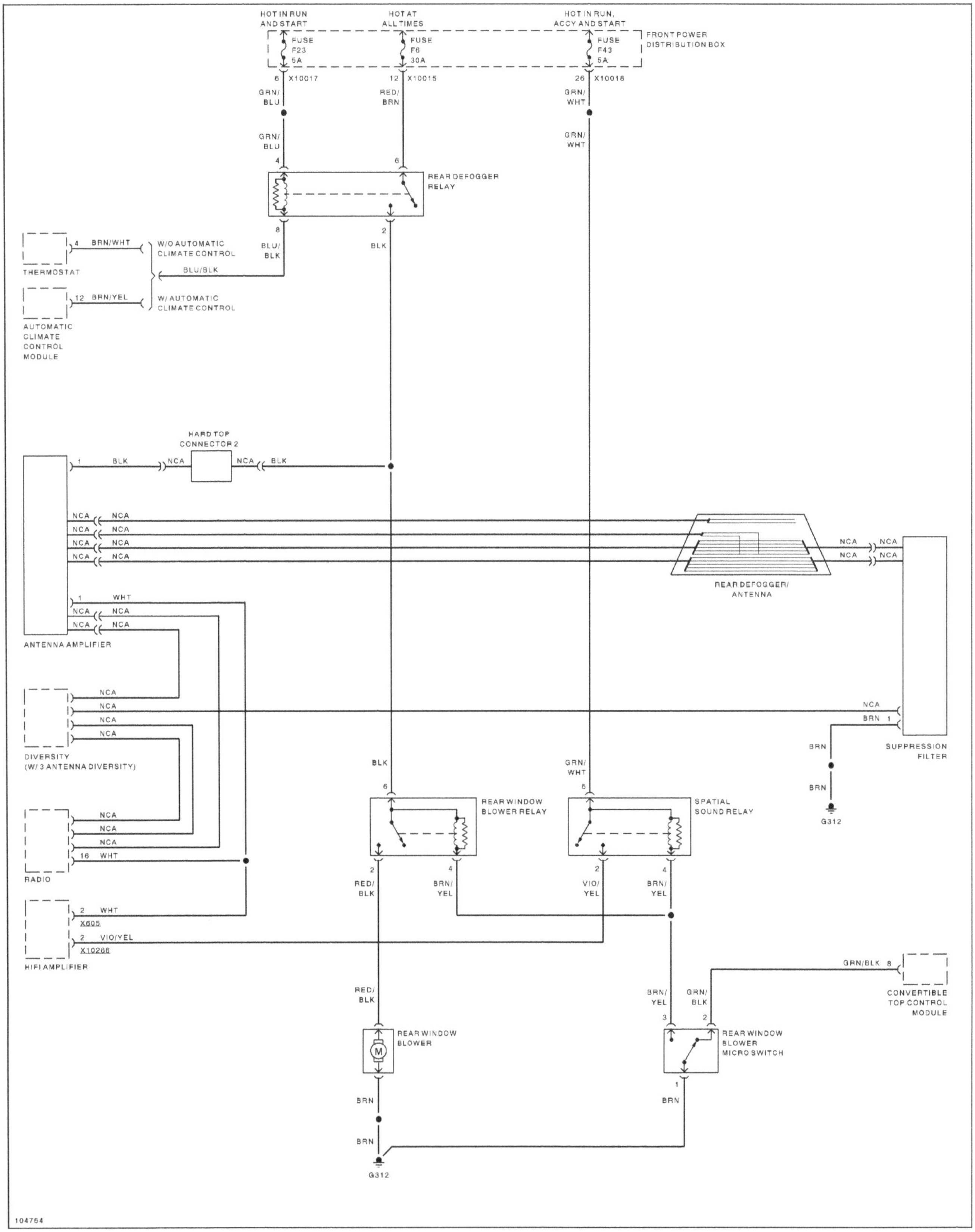

Instrument Panel

1992-97 Except 318ti (1 of 2)

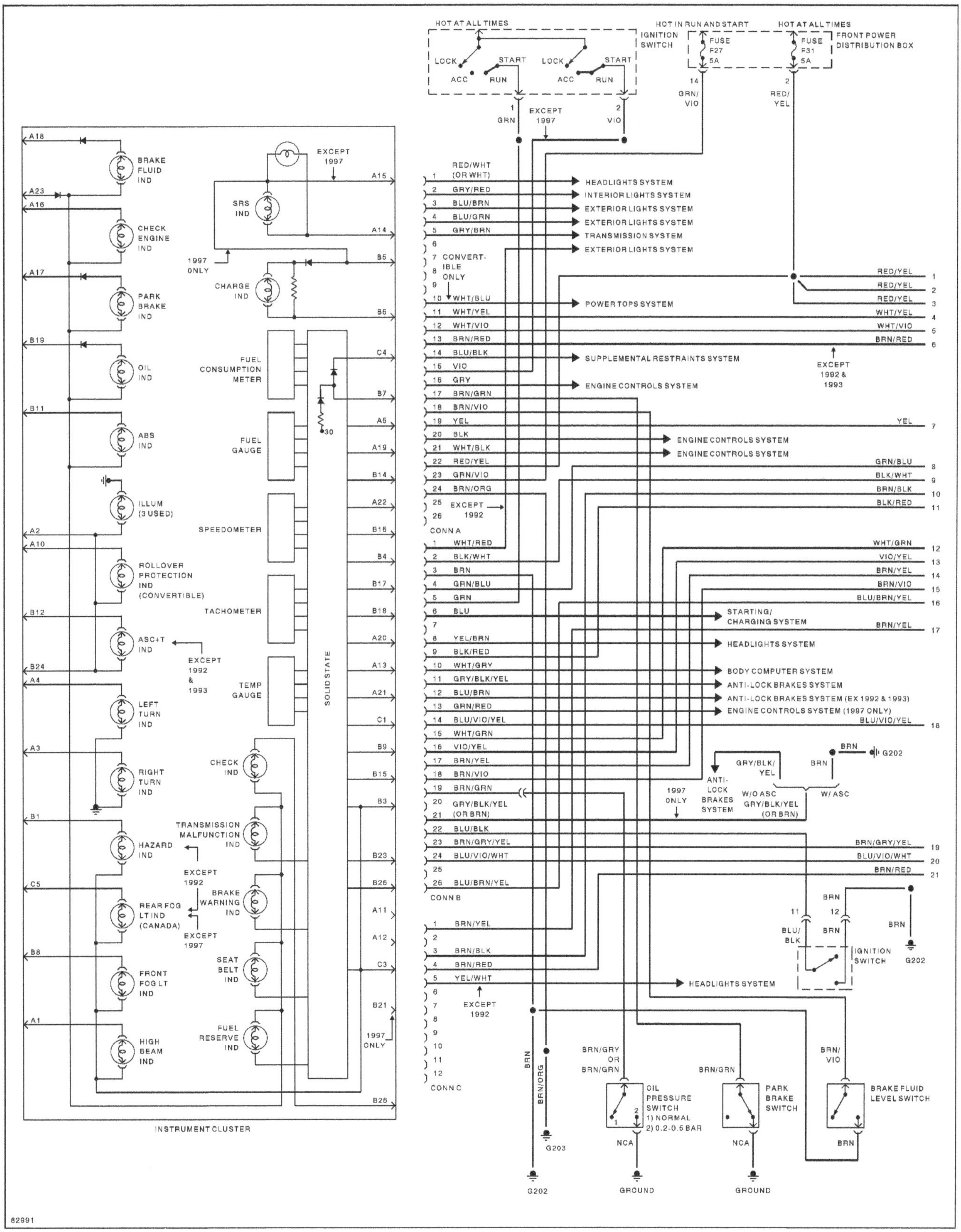

Instrument Panel

1992-97 Except 318ti (2 of 2)

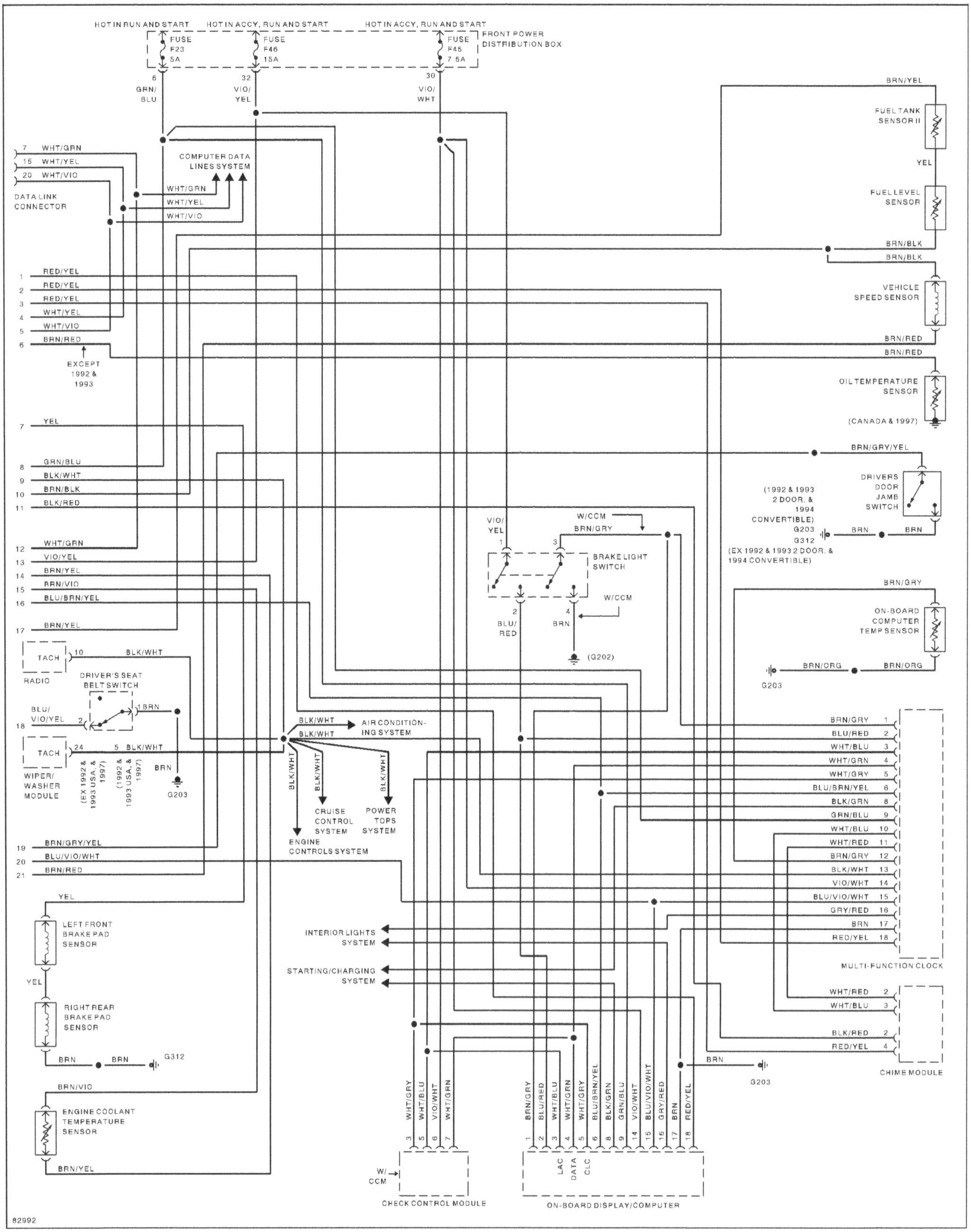

Instrument Panel

1998 Except 318ti (1 of 2)

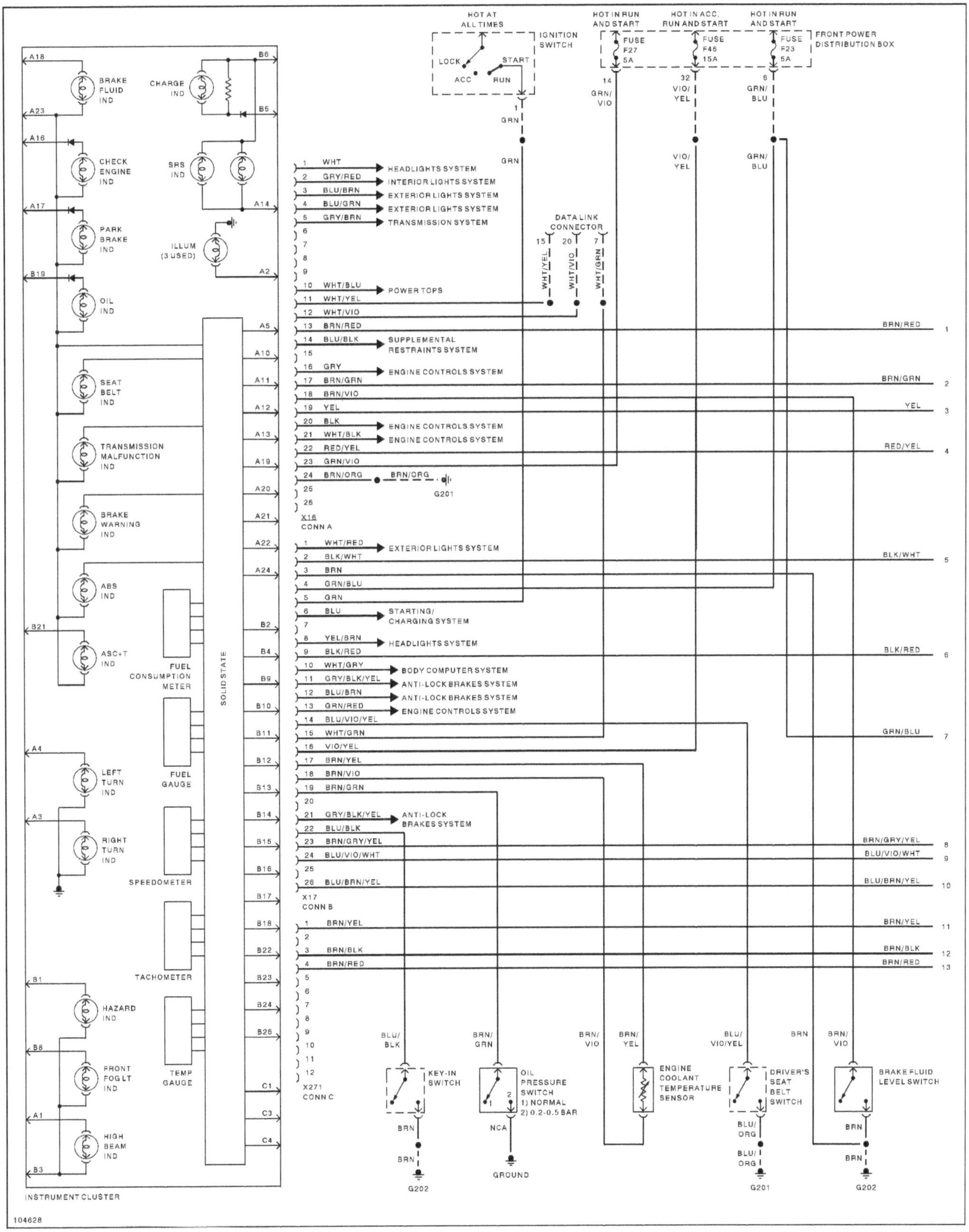

Instrument Panel

1998 Except 318ti (2 of 2)

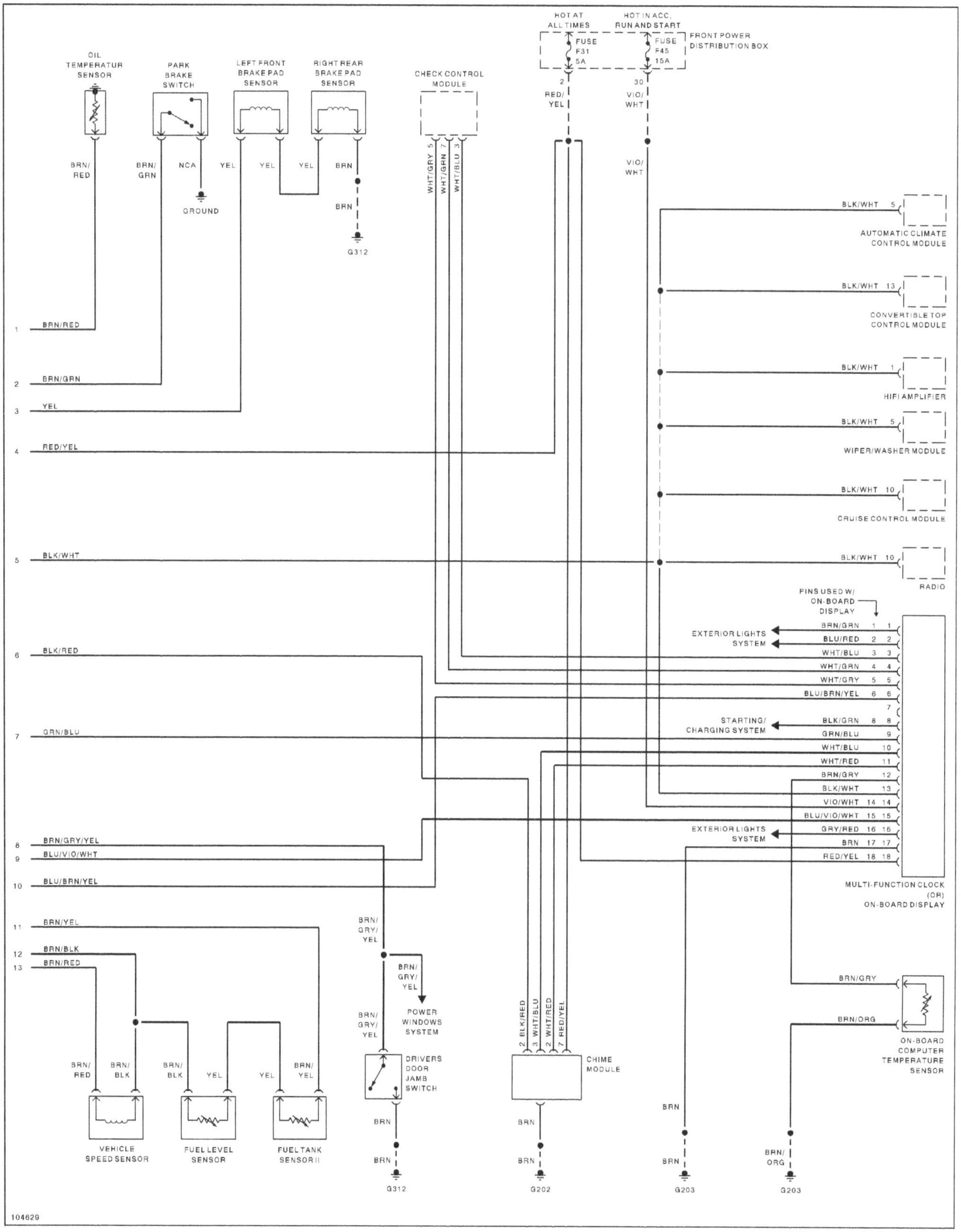

Instrument Panel

1995-97 318ti (1 of 2)

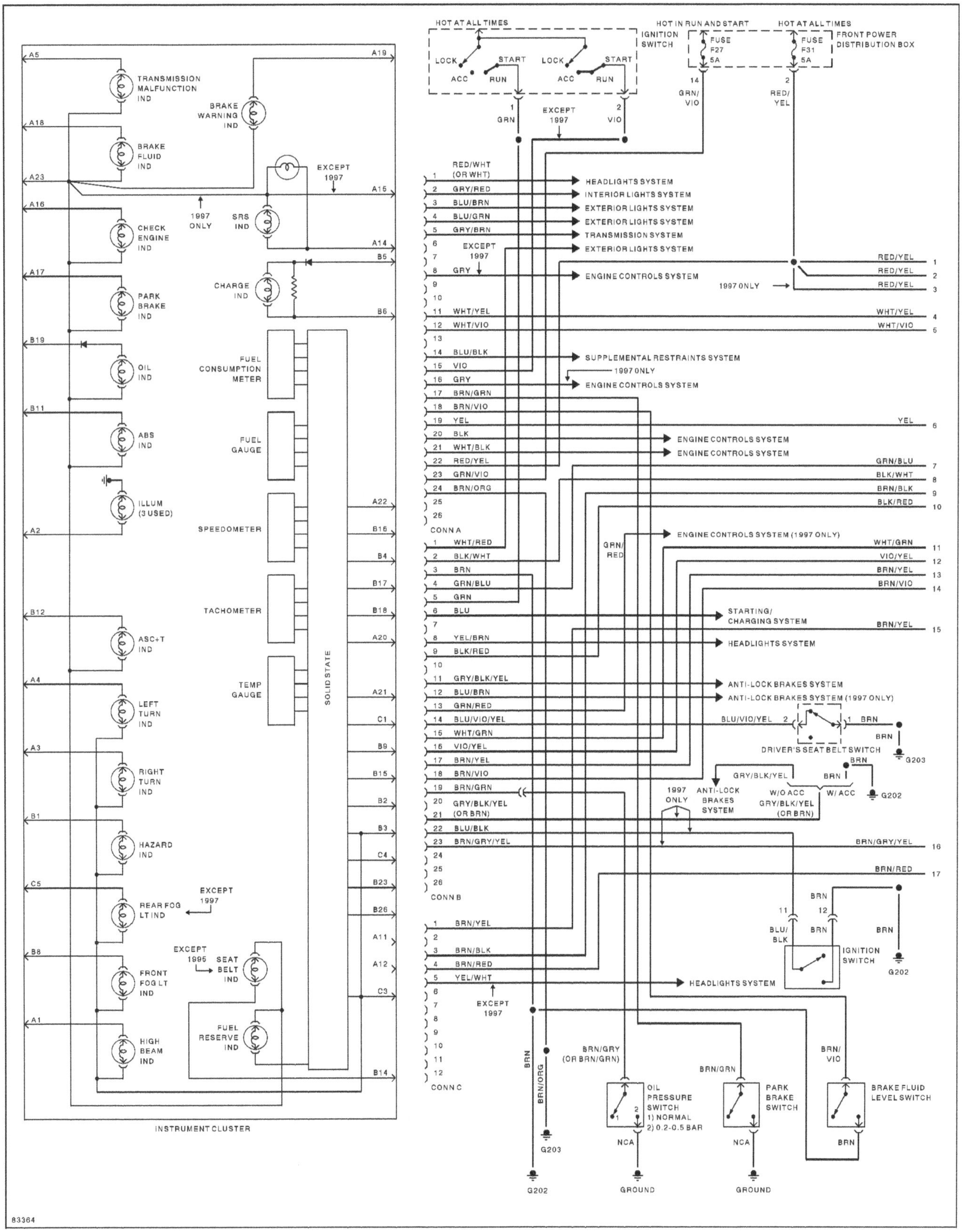

Instrument Panel

1995-97 318ti (2 of 2)

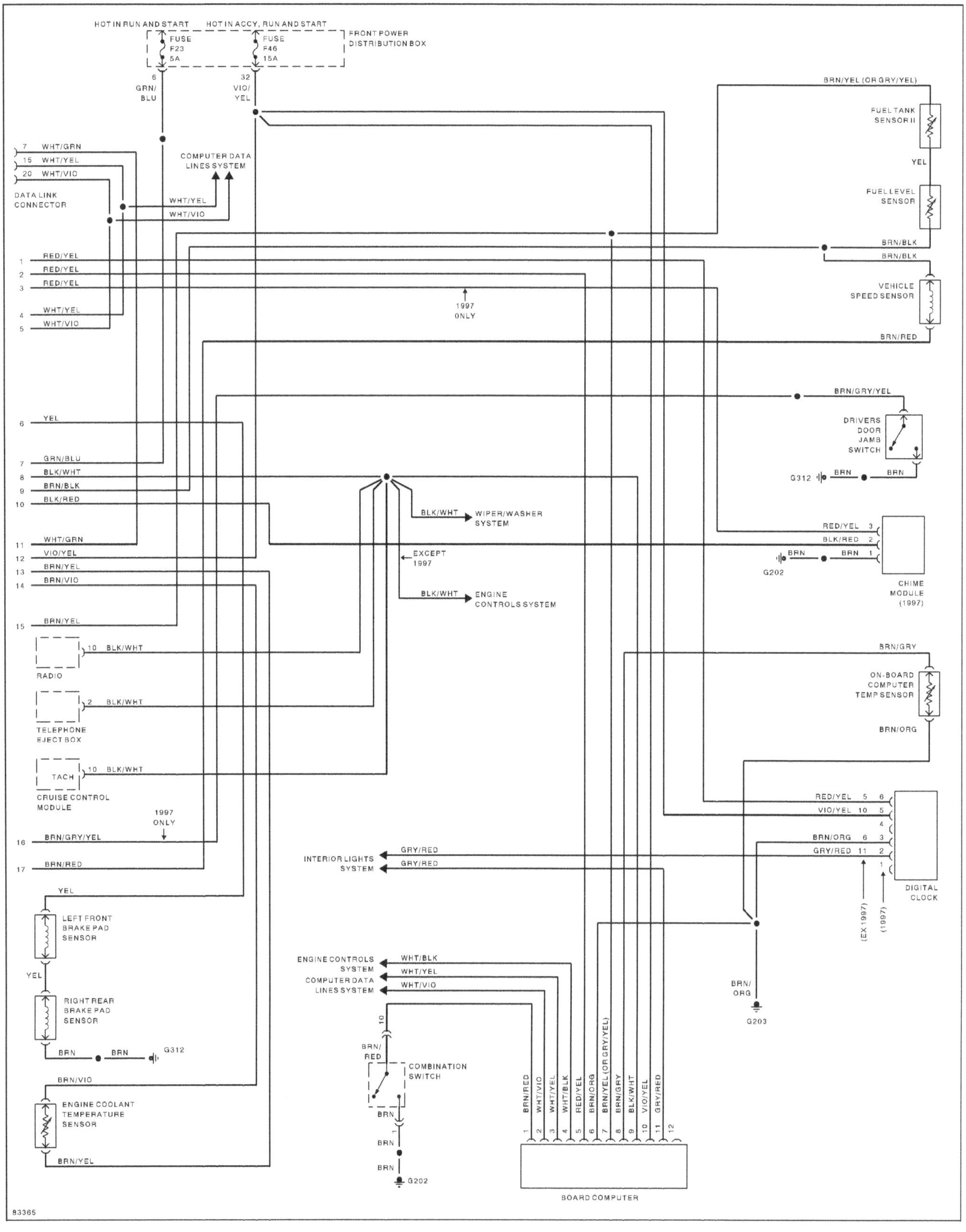

Instrument Panel

1998 318ti (1 of 2)

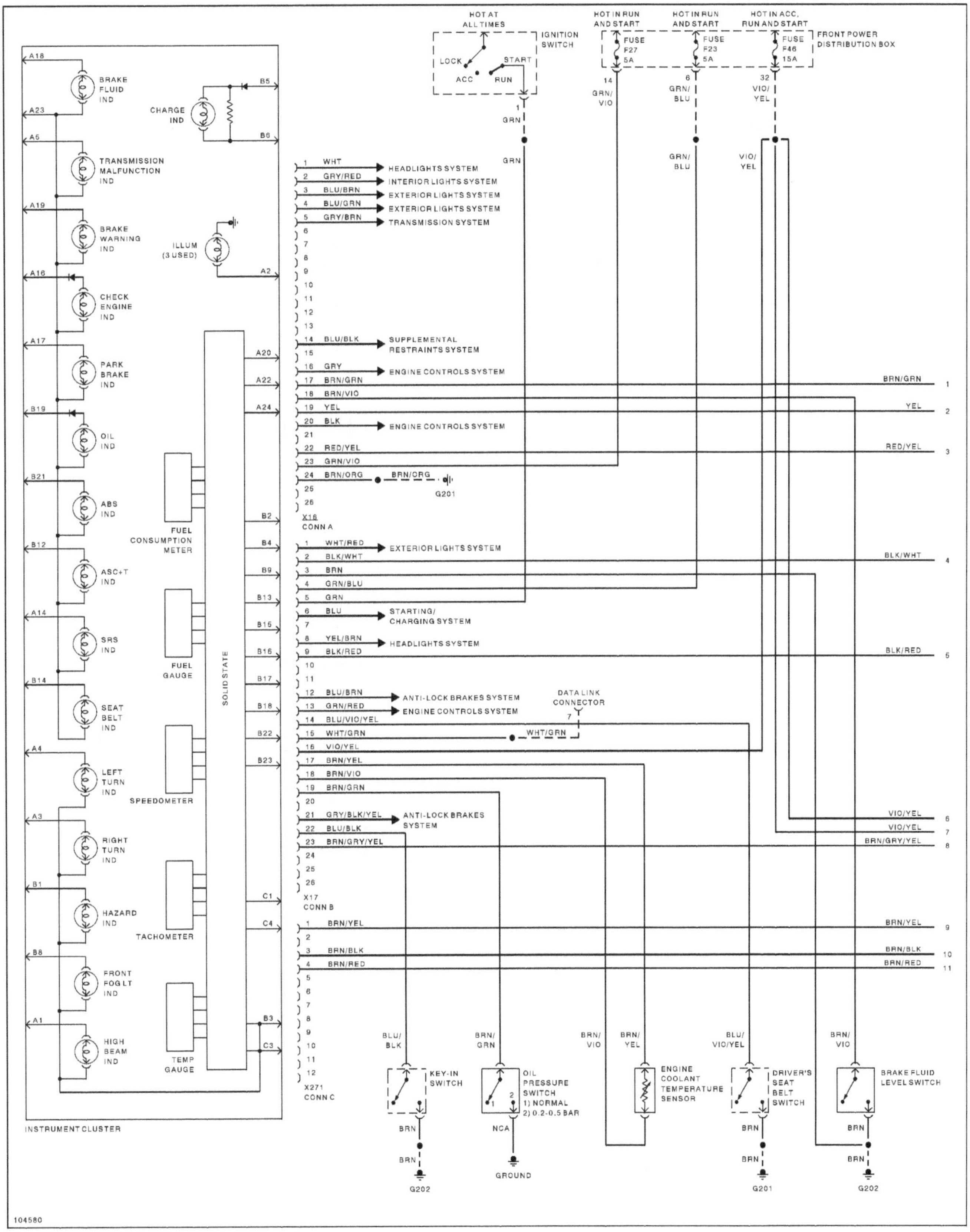

Instrument Panel

1998 318ti (2 of 2)

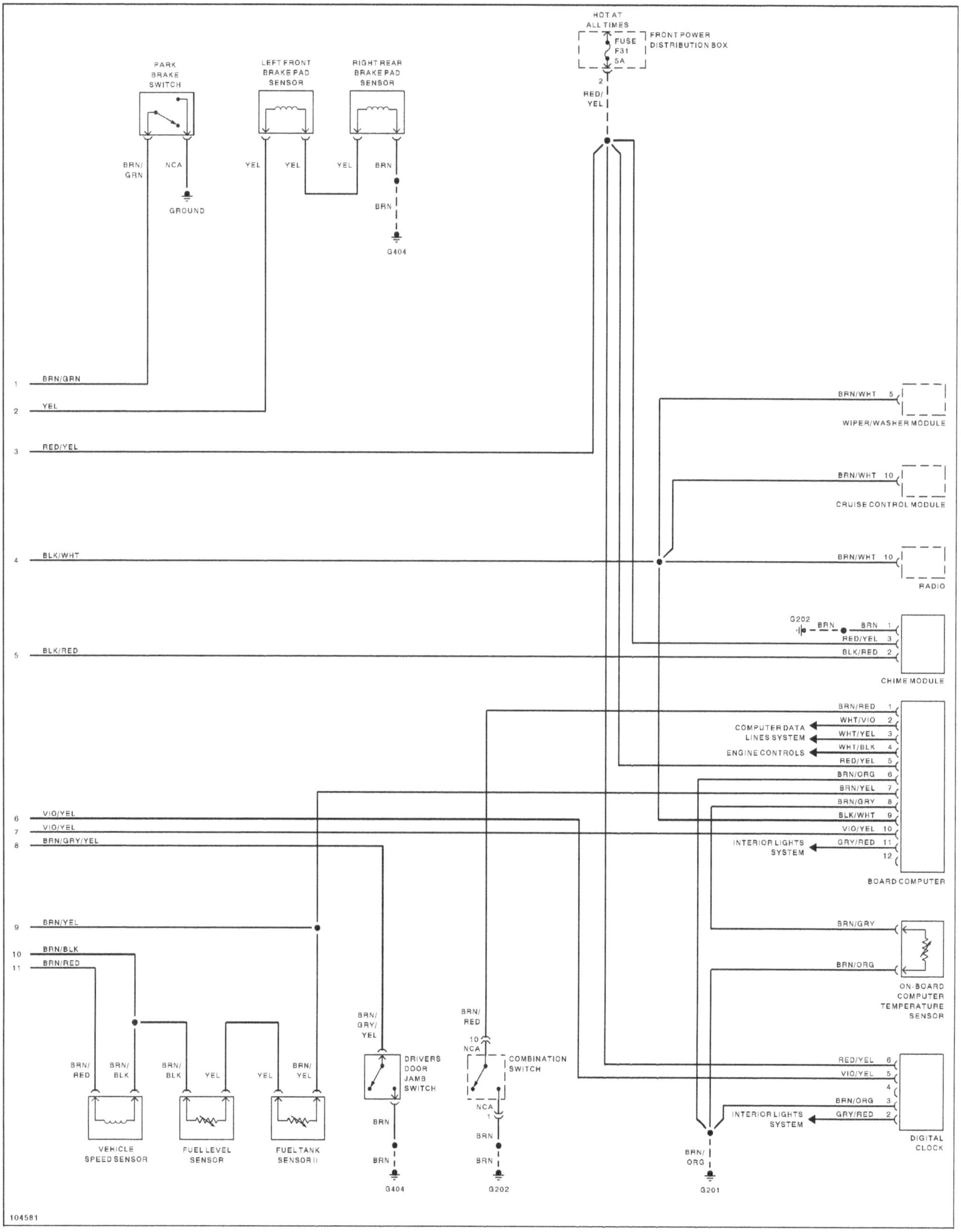

On-Board Computer
1995-98 318ti

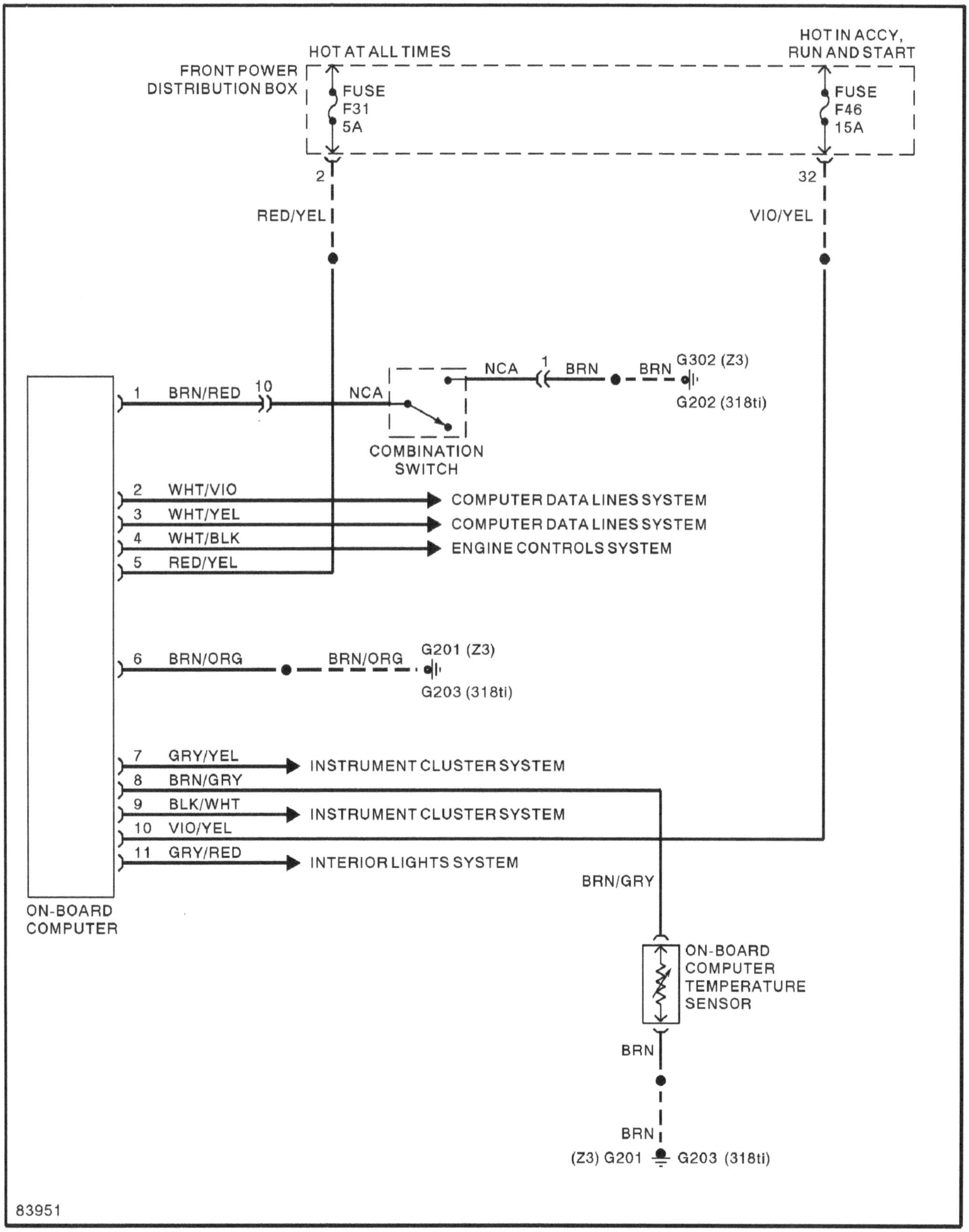

On-Board Computer
1992-97 Except 318ti

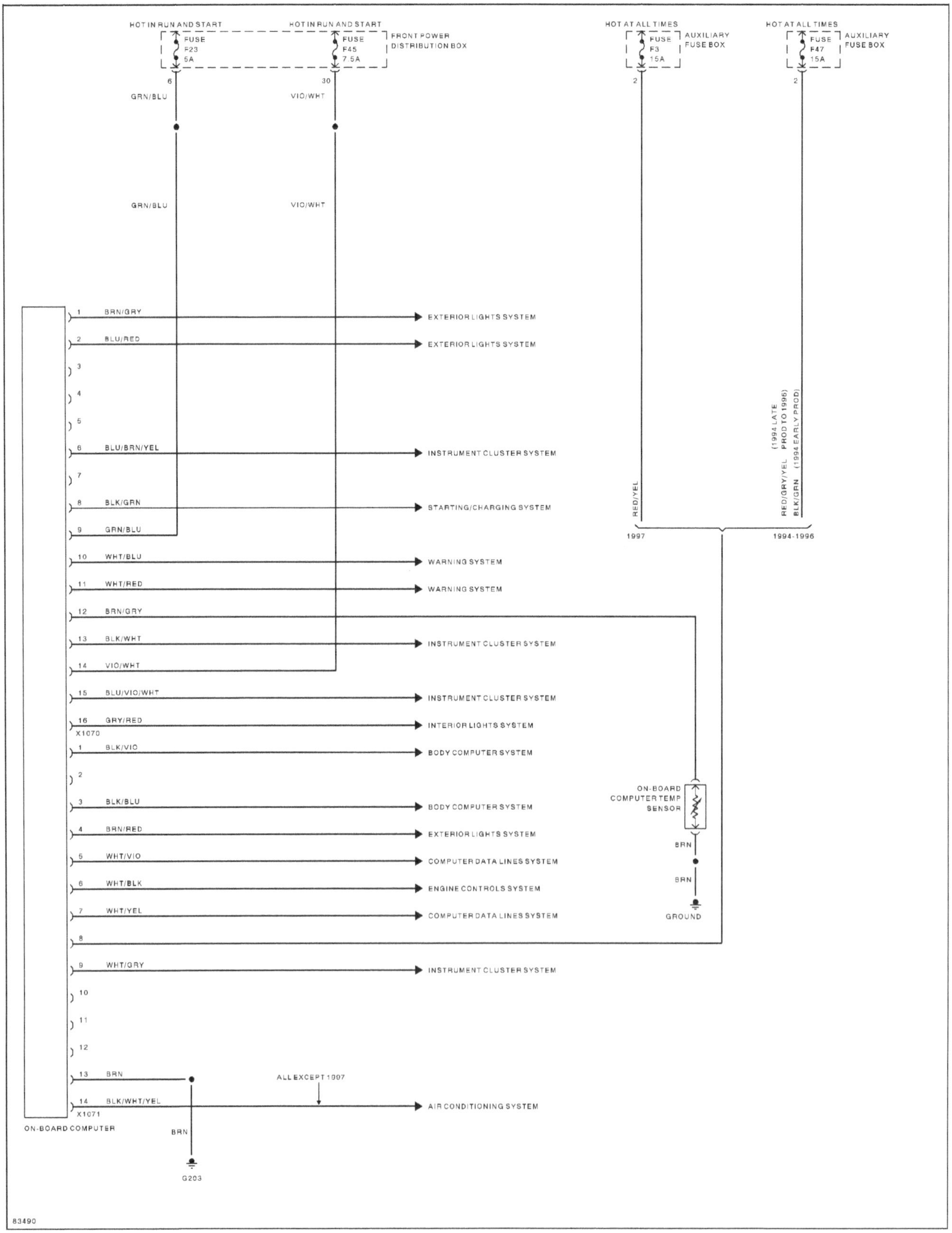

On-Board Computer

1998 Except 318ti

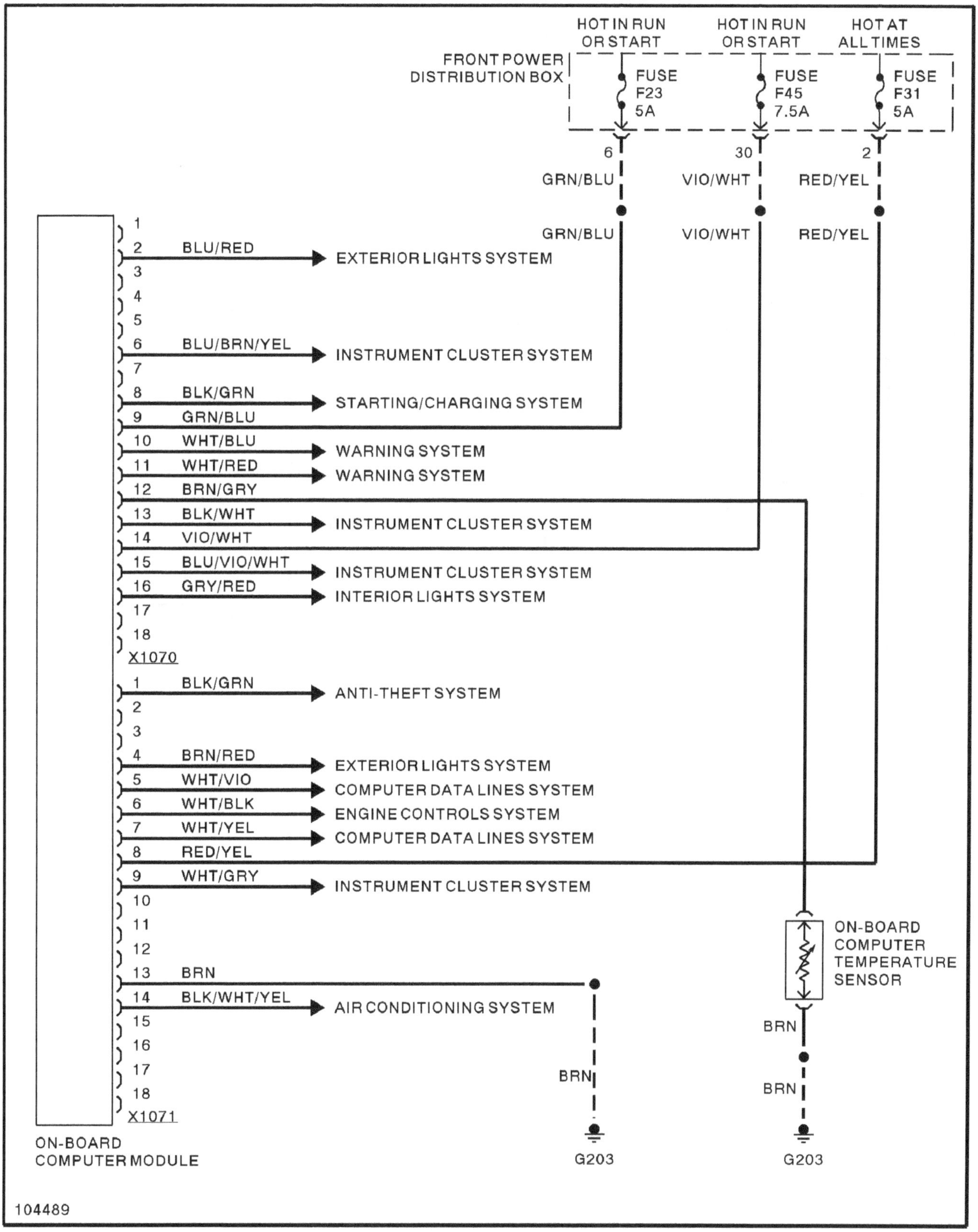

Body Computer Module

1996-98 318ti

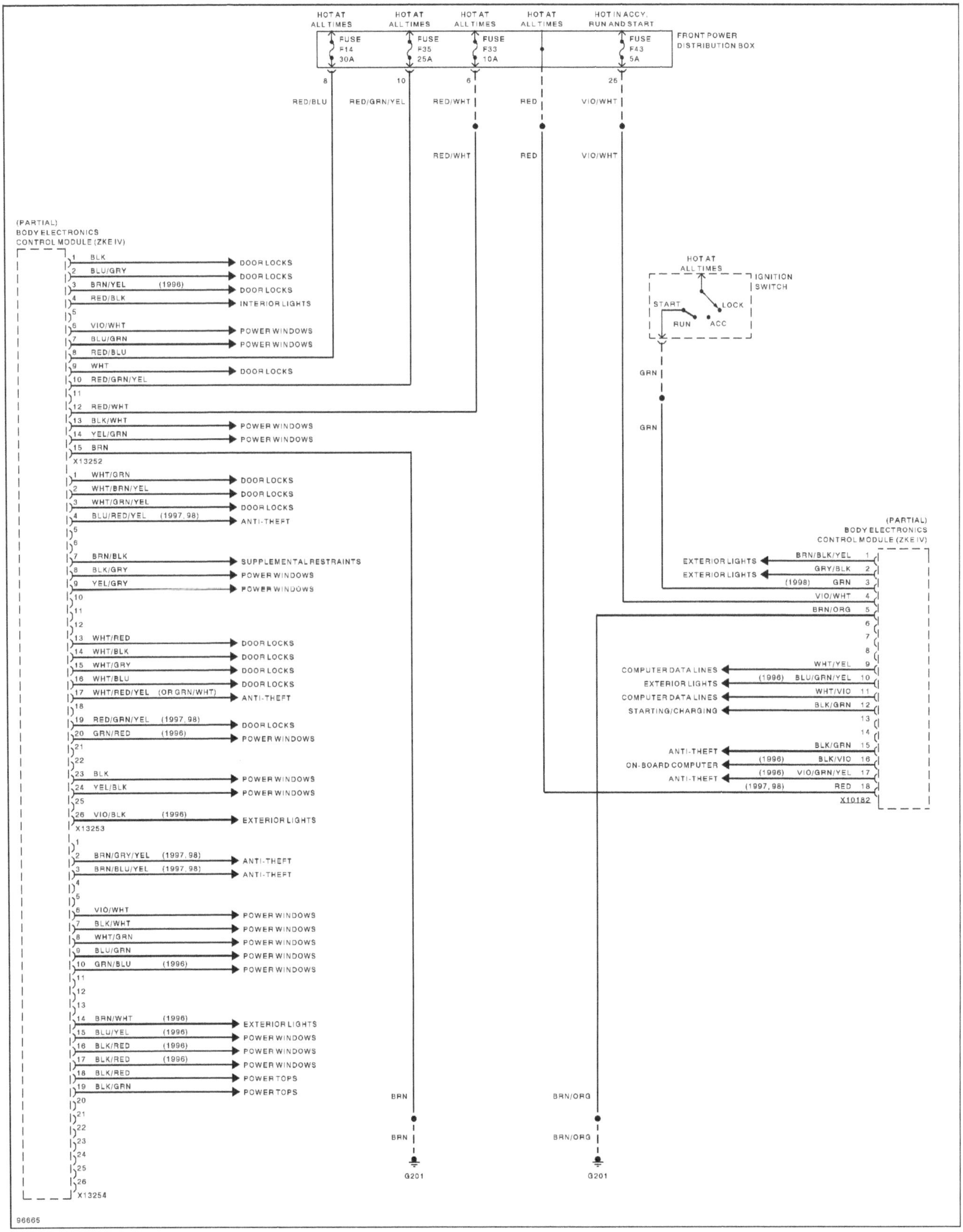

Body Computer Module

1996-98 Except 318ti

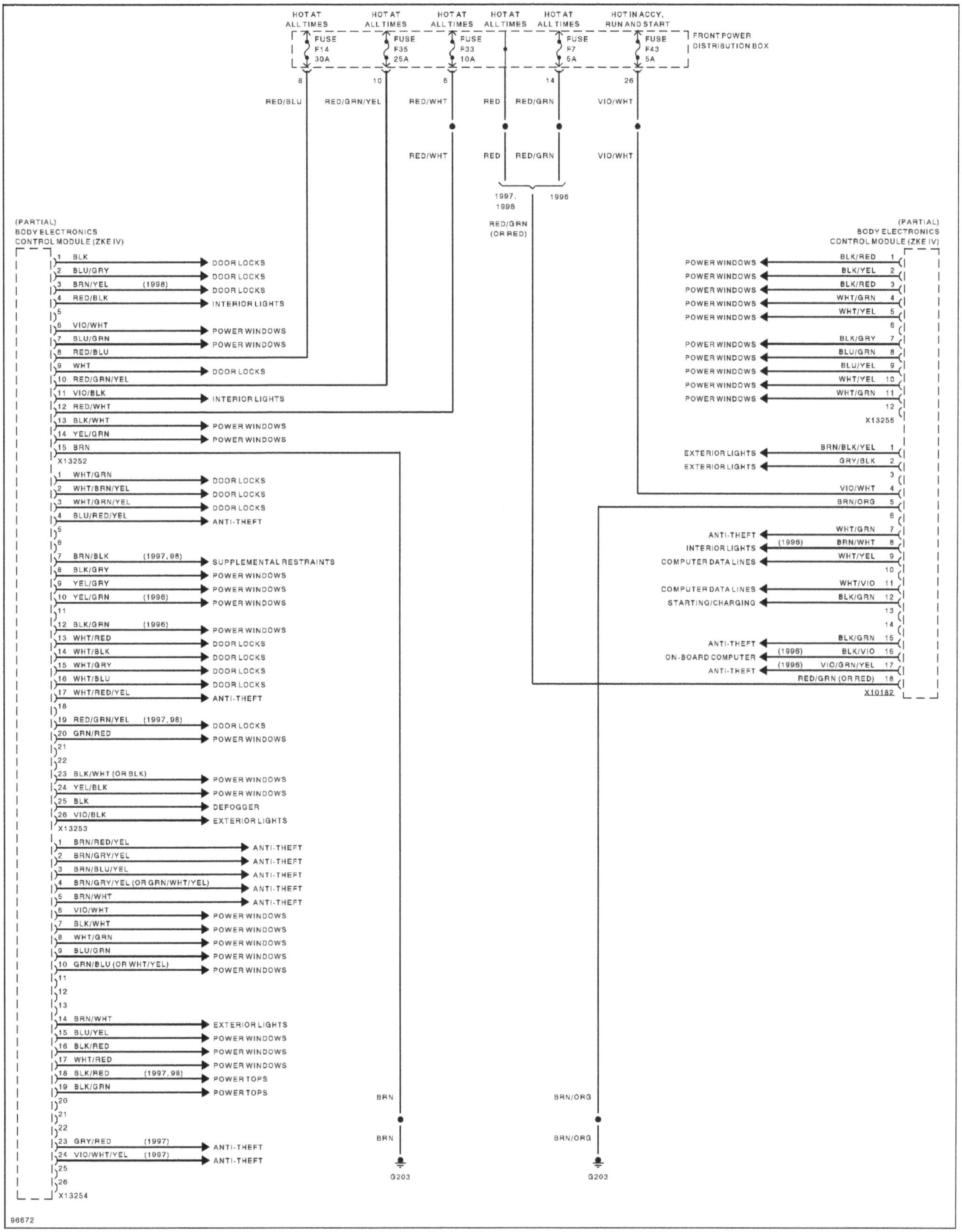

Warning System: Chimes

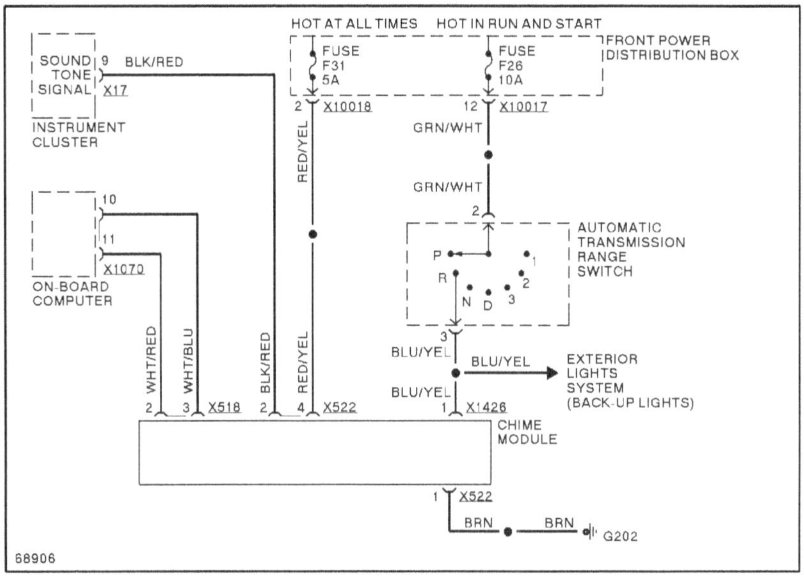

1992-97

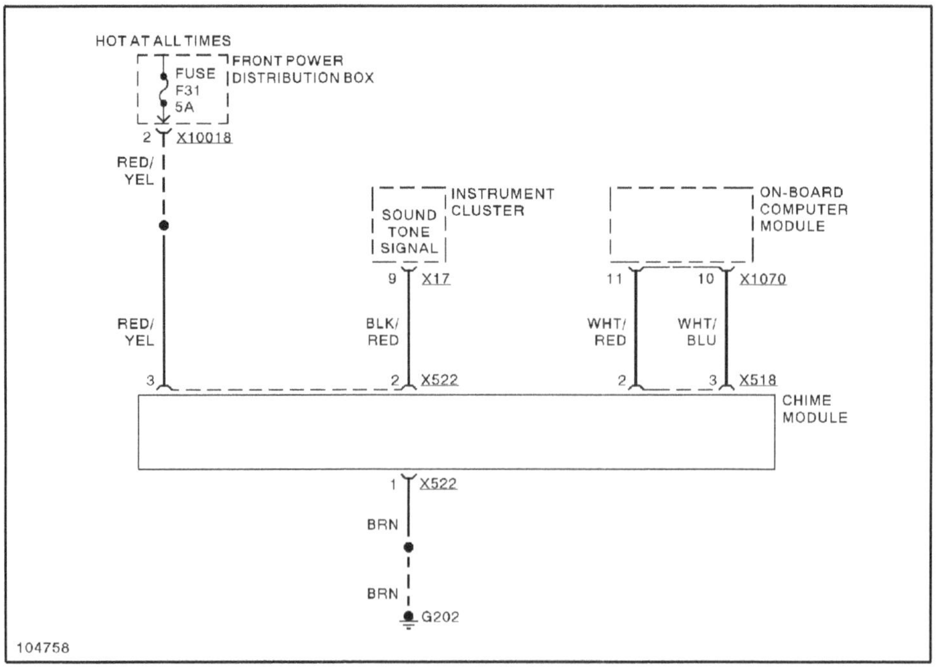

1998

Warning System: Chimes

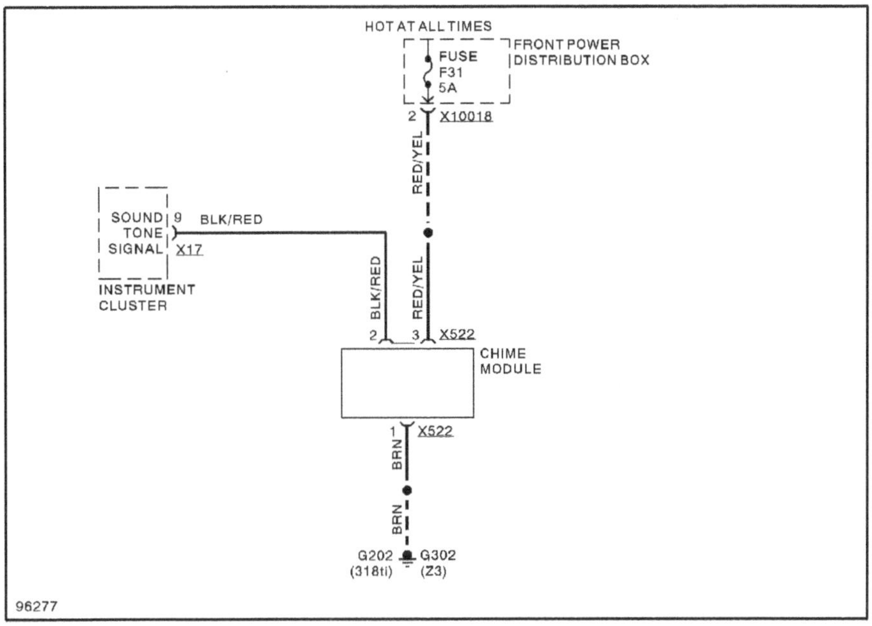

1997-98
318ti

Headlights/Foglights

1992-94 (Canada) (1 of 2)

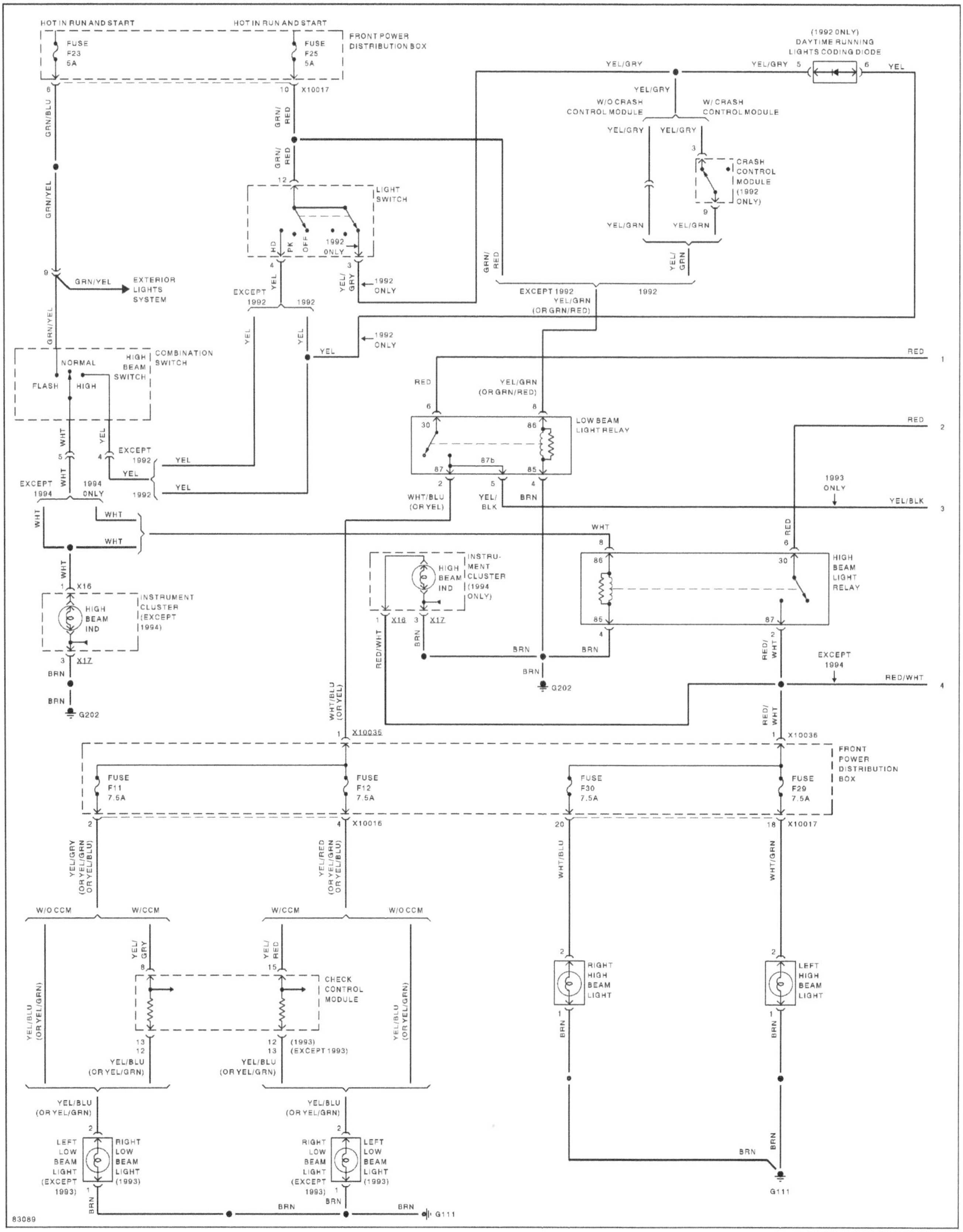

Headlights/Foglights

1992-94 (Canada) (2 of 2)

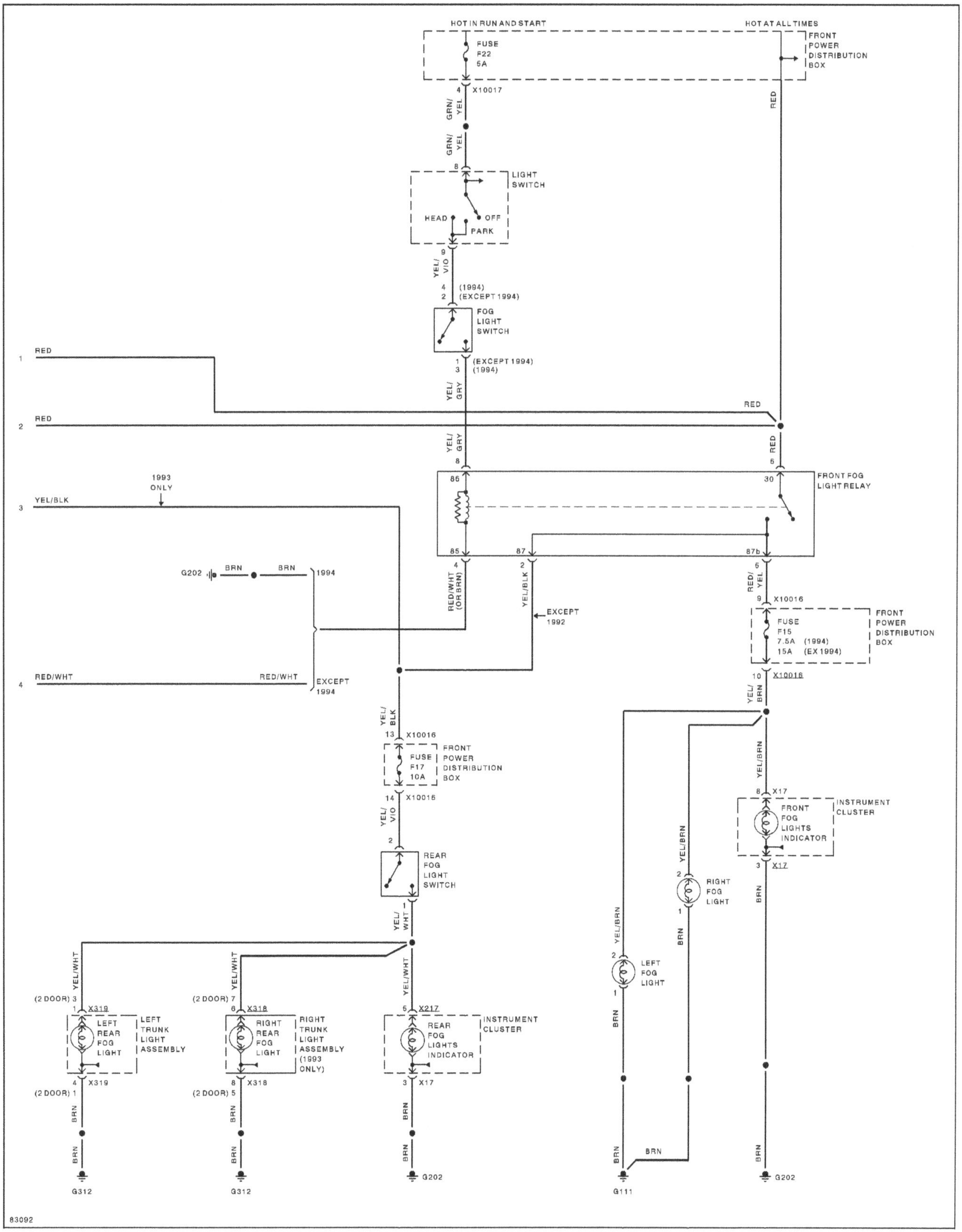

Headlights/Foglights

1992-94 (USA)

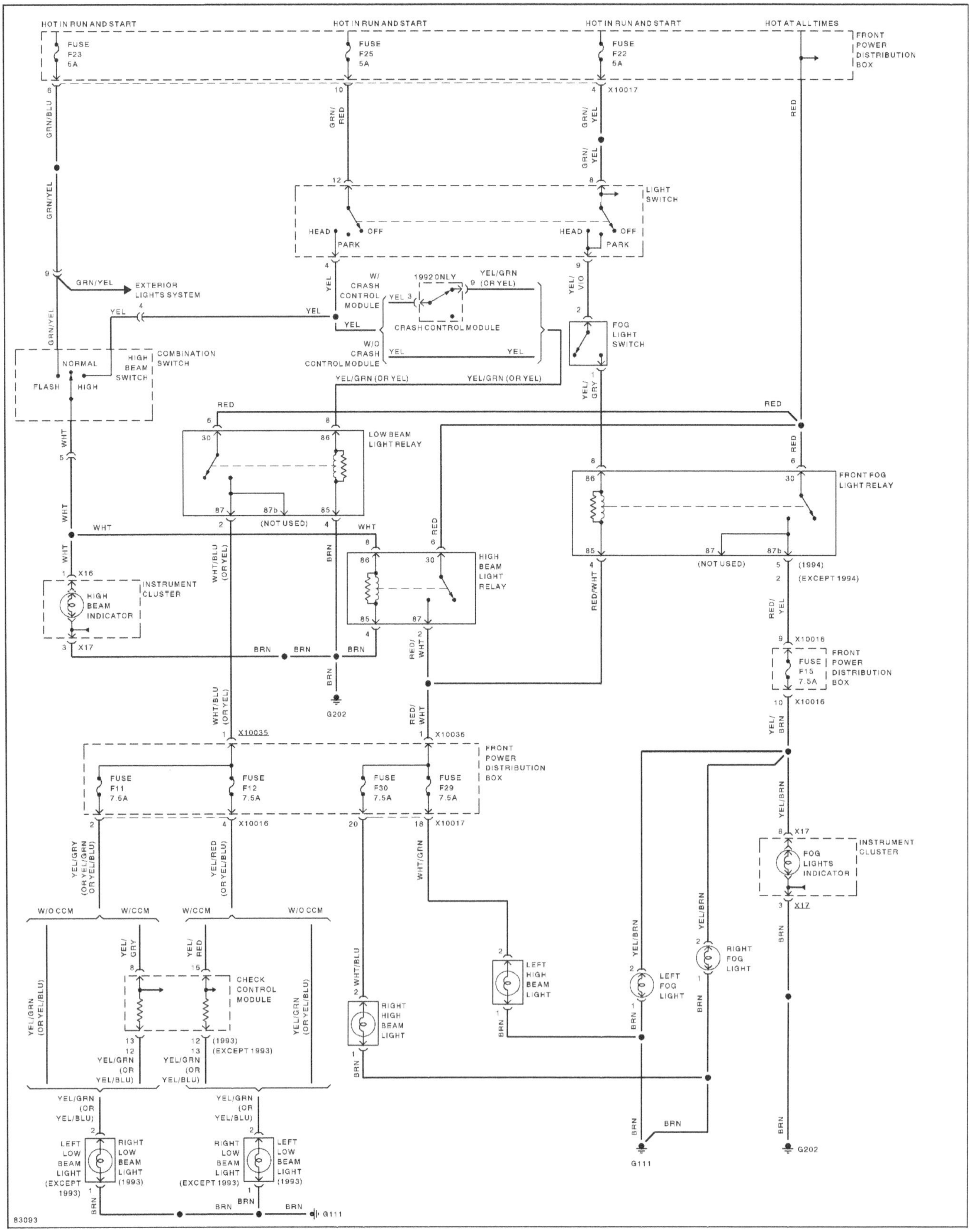

Headlights/Foglights

1995 Except 318ti (1 of 2)

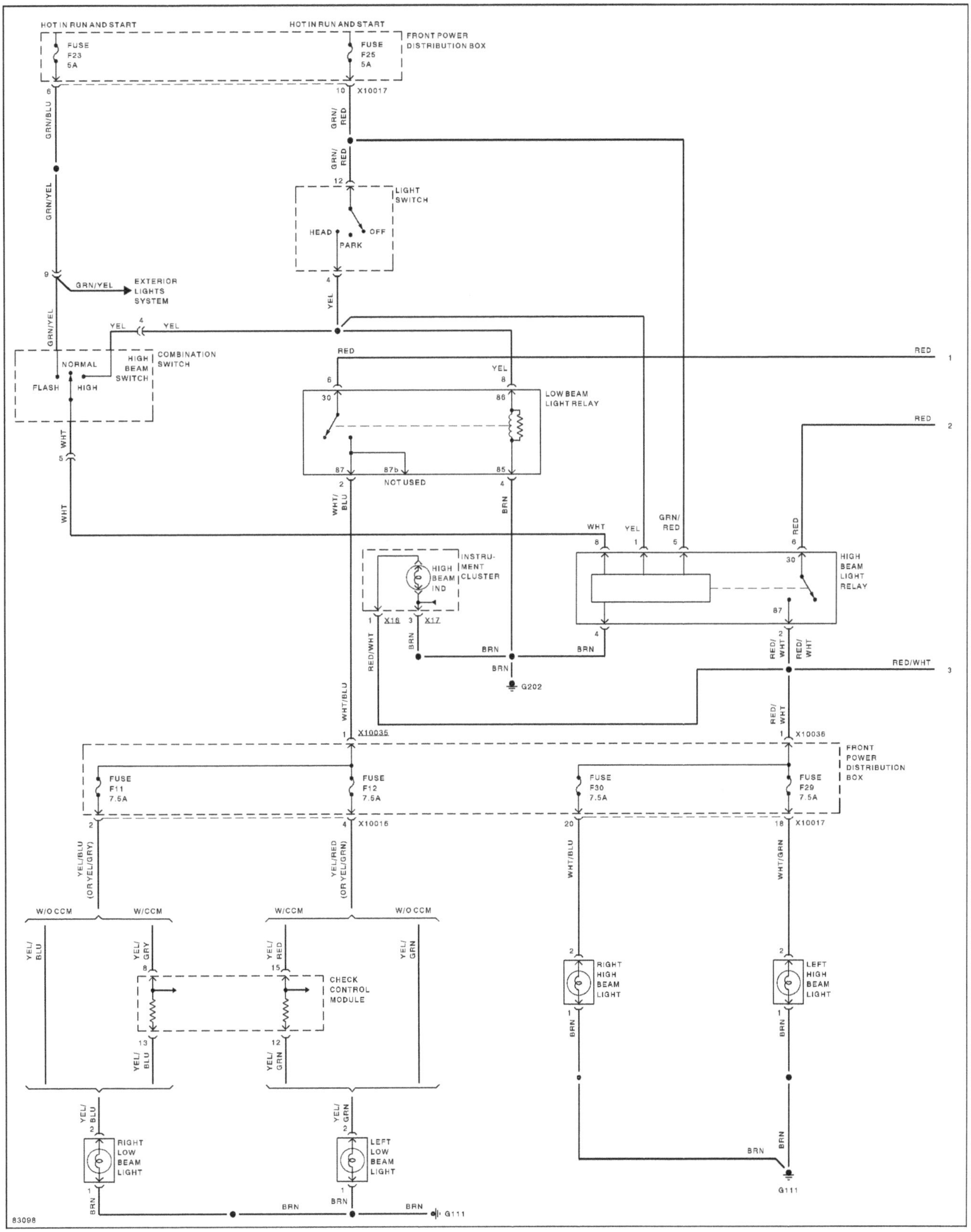

Headlights/Foglights

1995 Except 318ti (2 of 2)

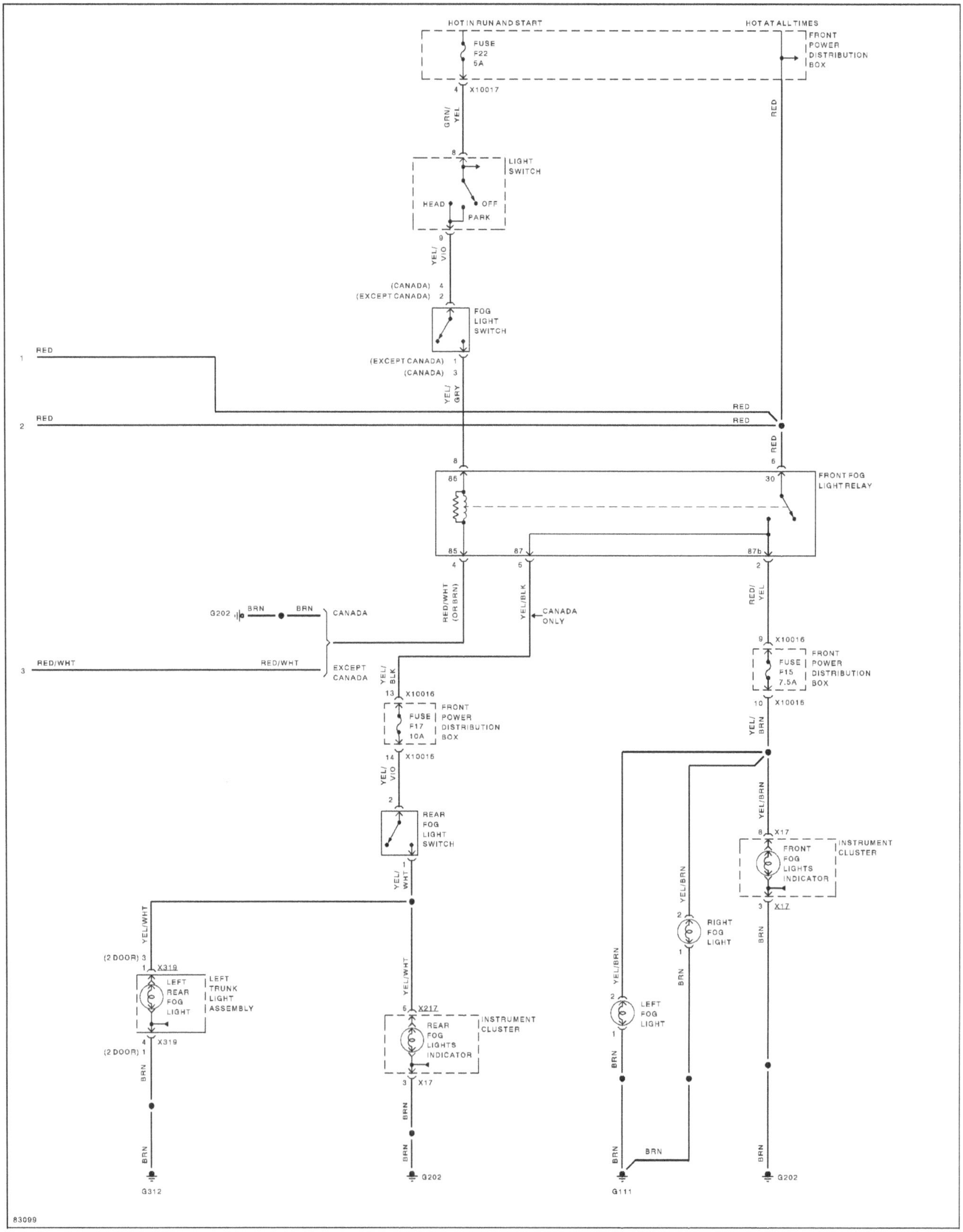

Headlights/Foglights

1996-97 Except 318ti (1 of 2)

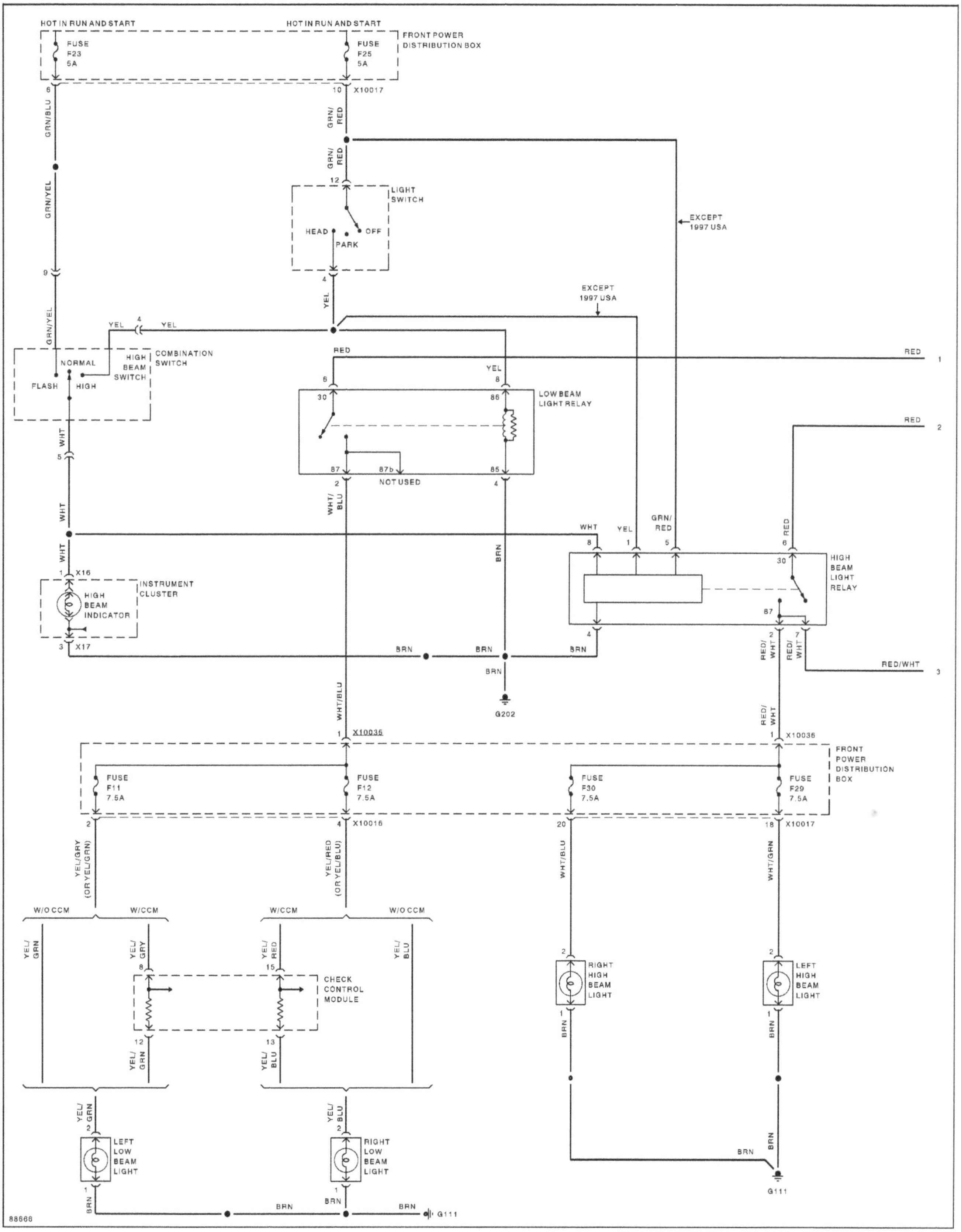

Headlights/Foglights

1996-97 Except 318ti (2 of 2)

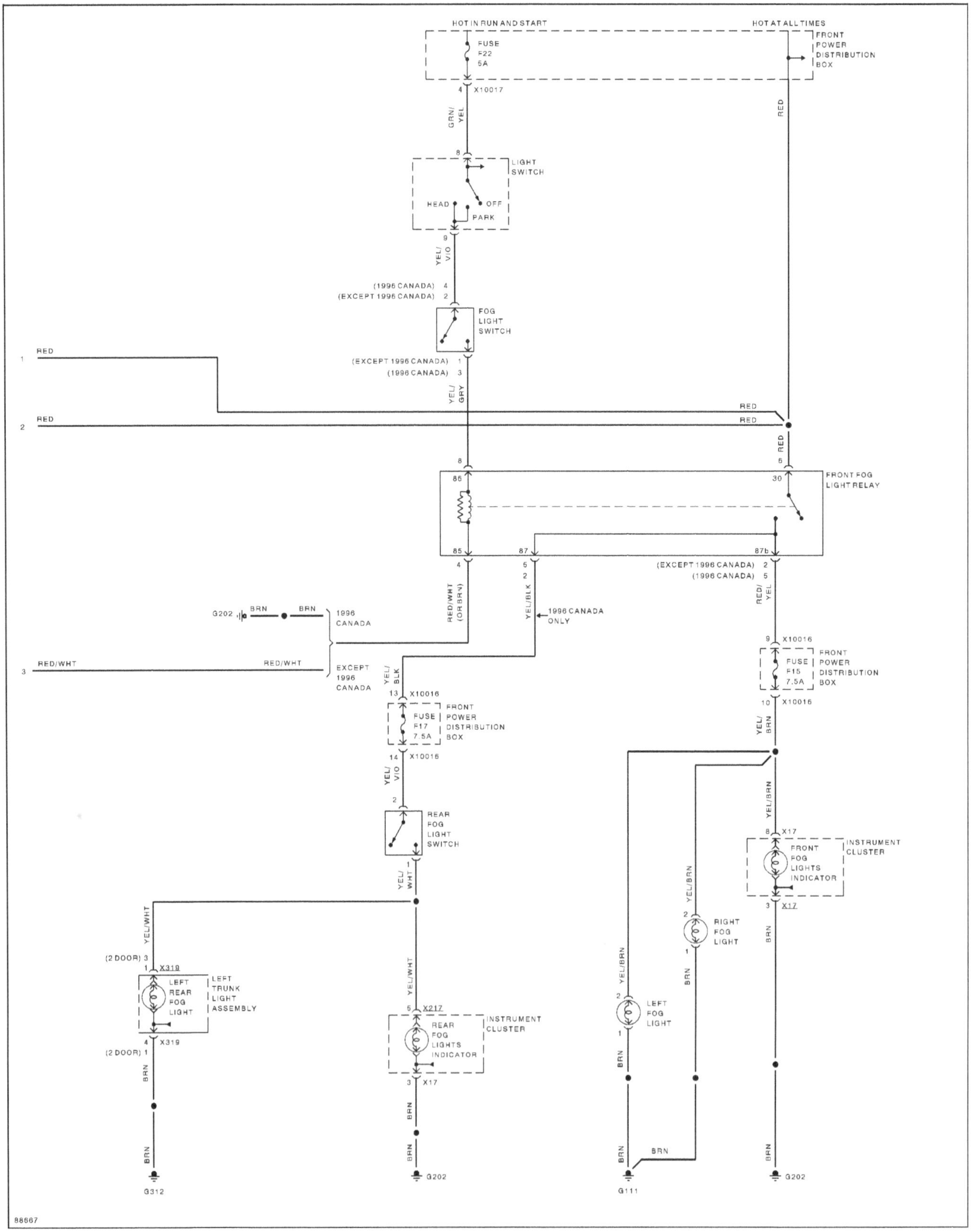

Headlights/Foglights

1998 Except 318ti

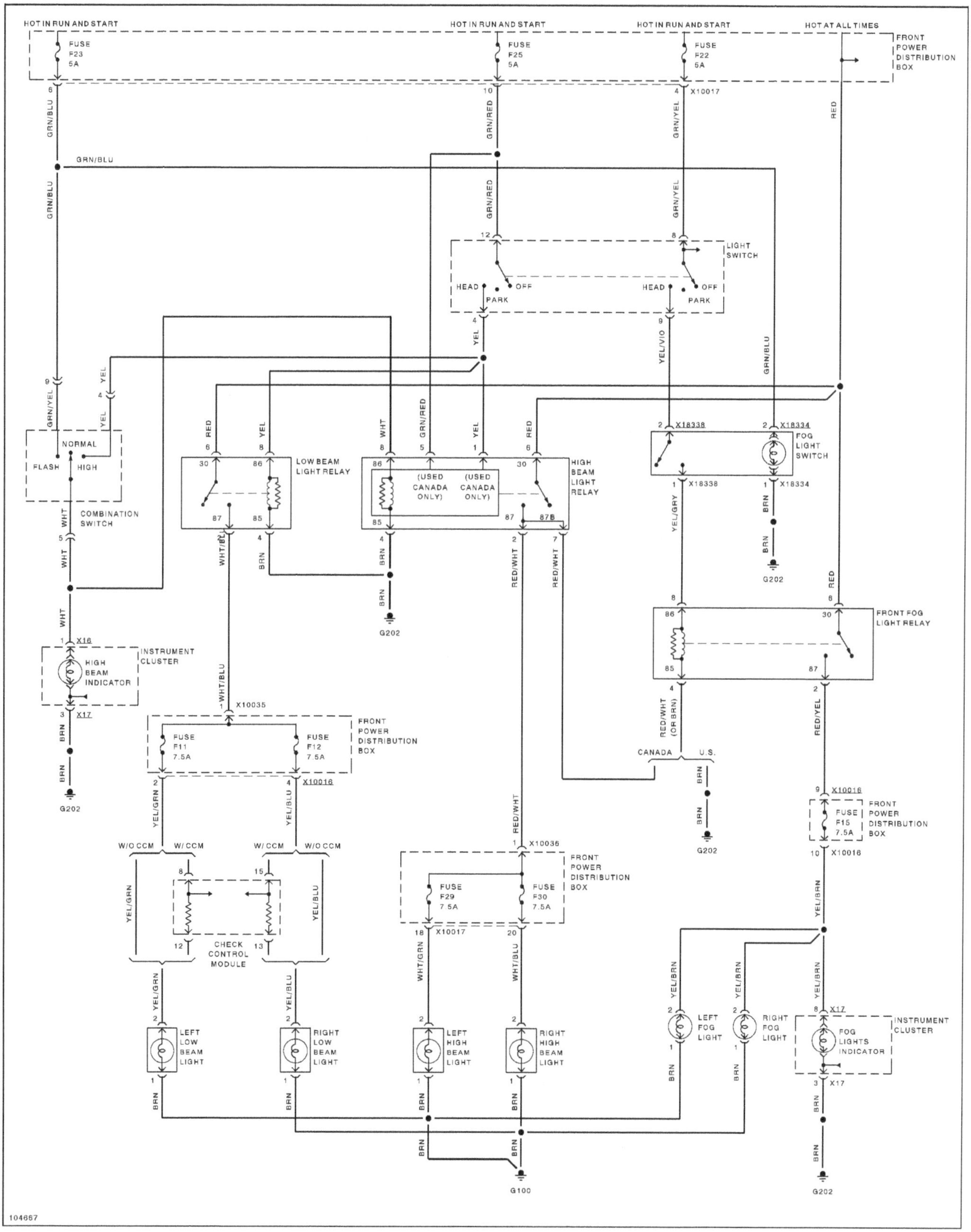

Headlights/Foglights

1995-96 318ti (Canada) (1 of 2)

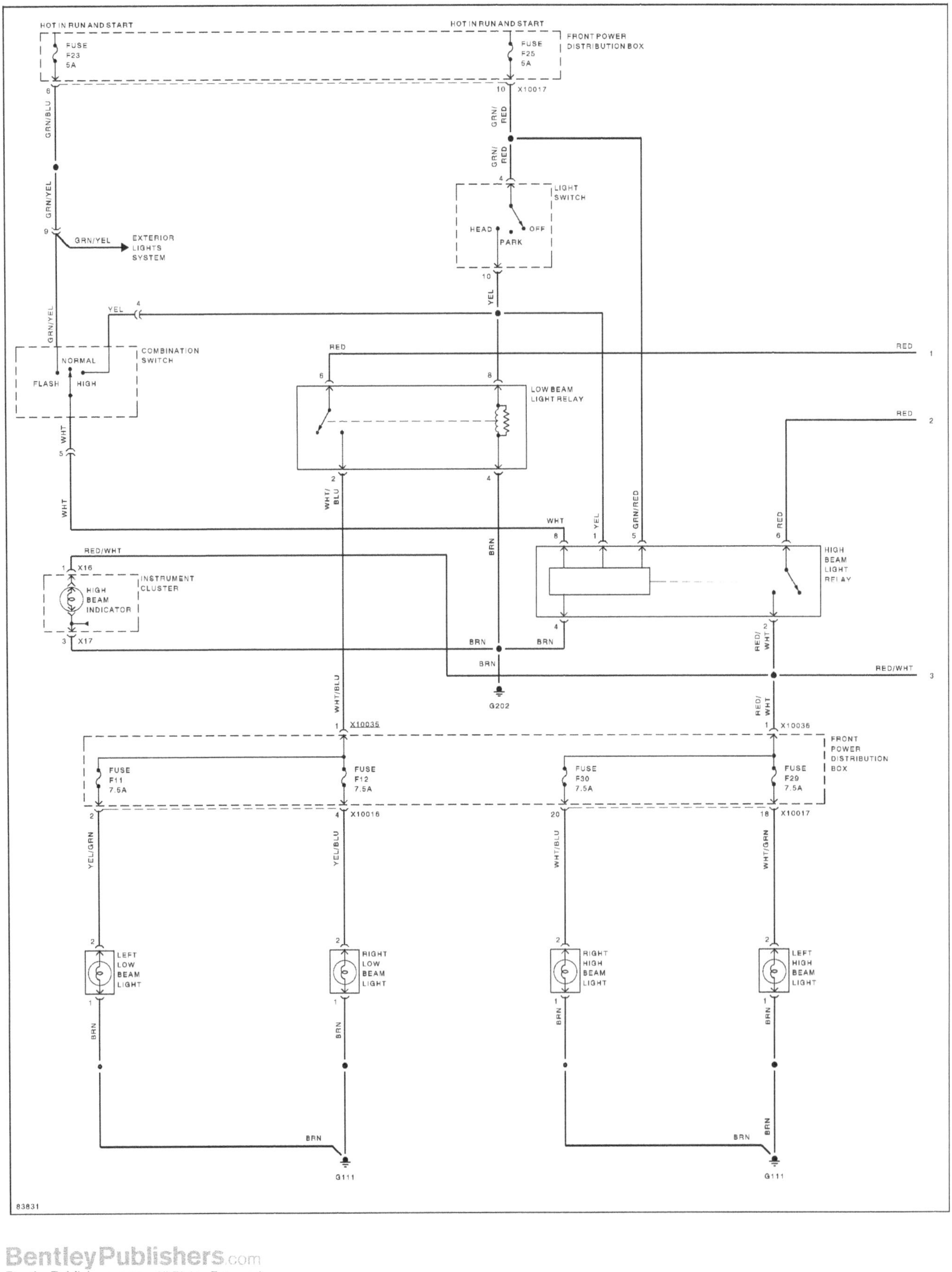

Headlights/Foglights

1995-96 318ti (Canada) (2 of 2)

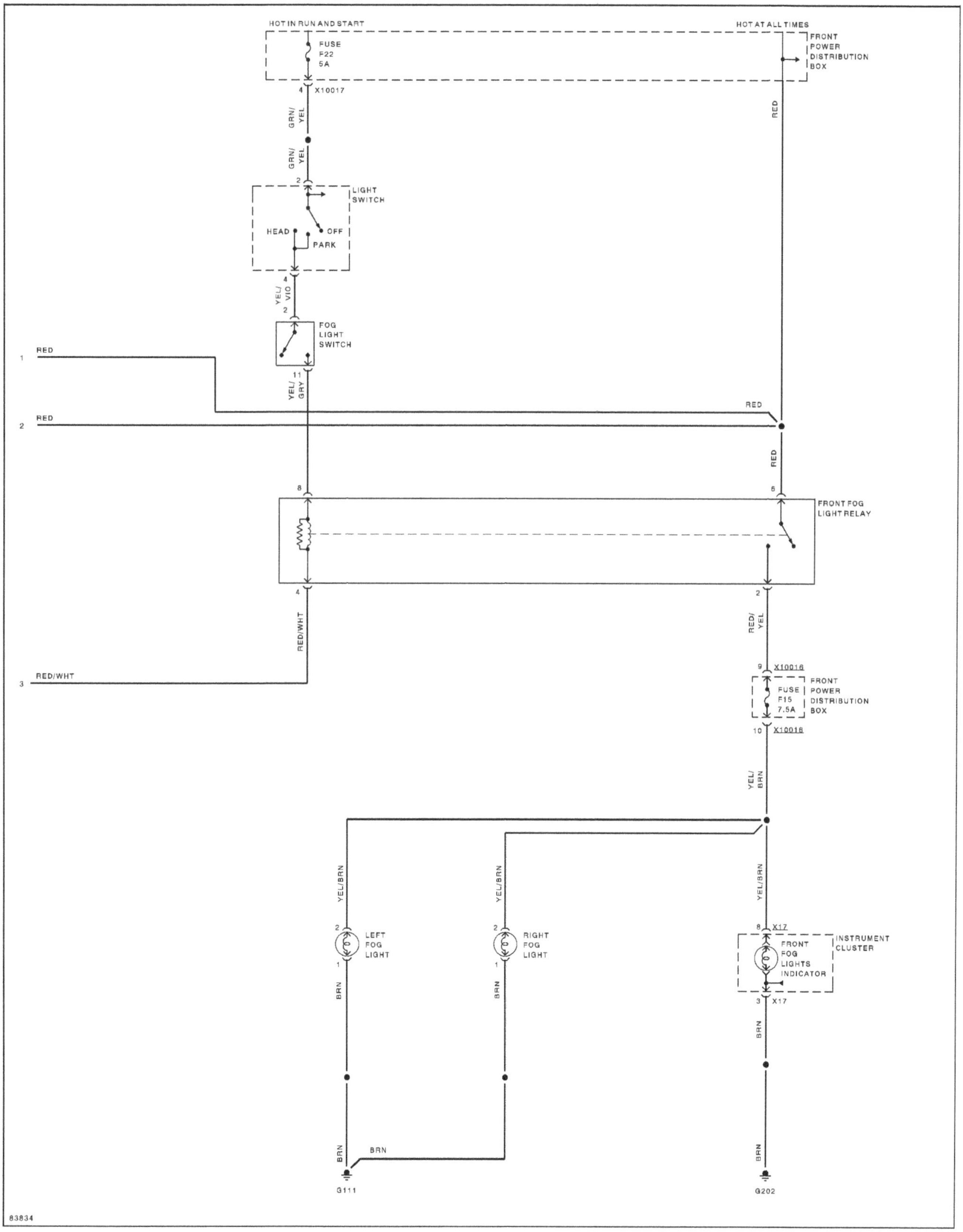

Headlights/Foglights

1995-96 318ti (USA)

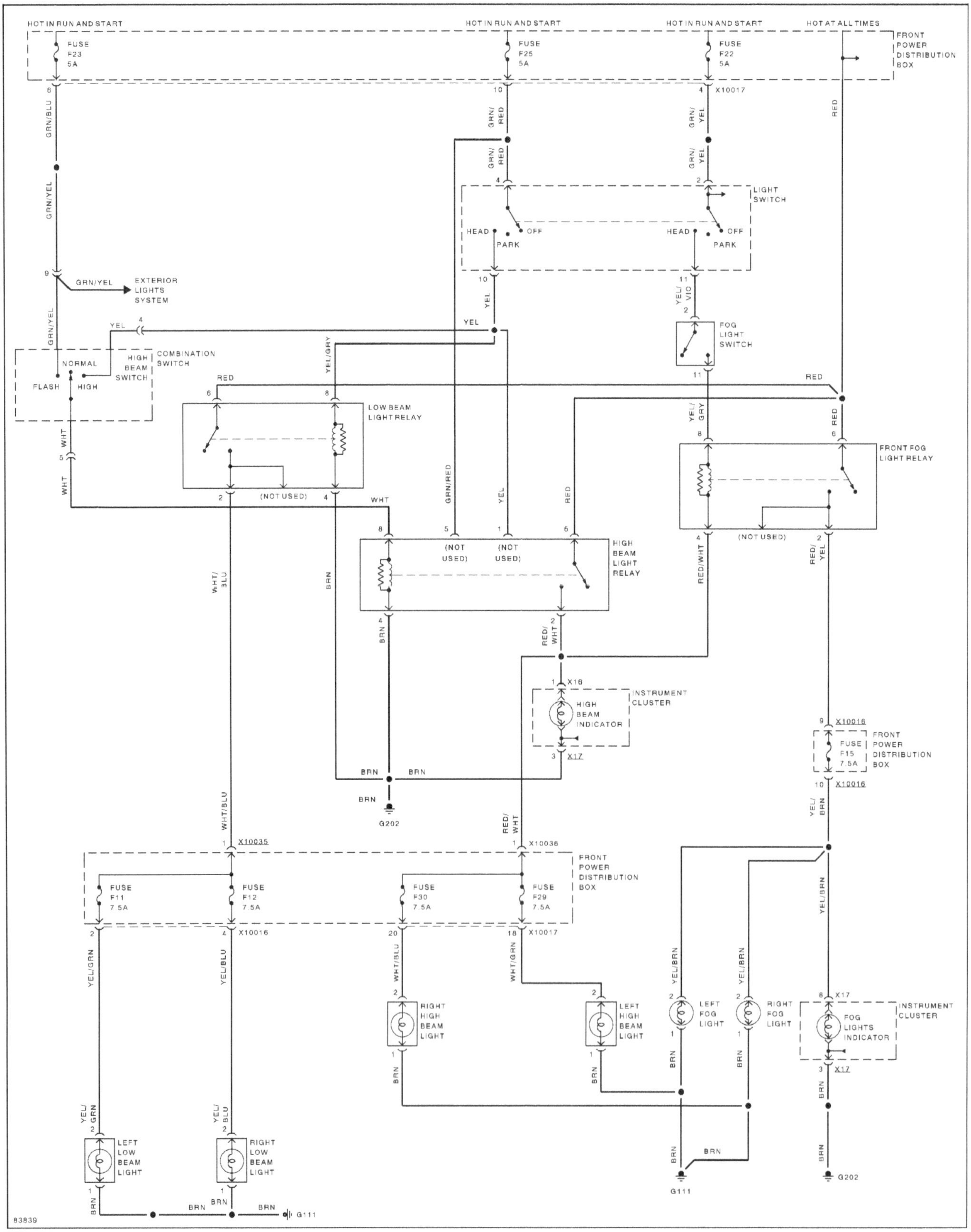

Headlights/Foglights
1997 318ti (Canada) (1 of 2)

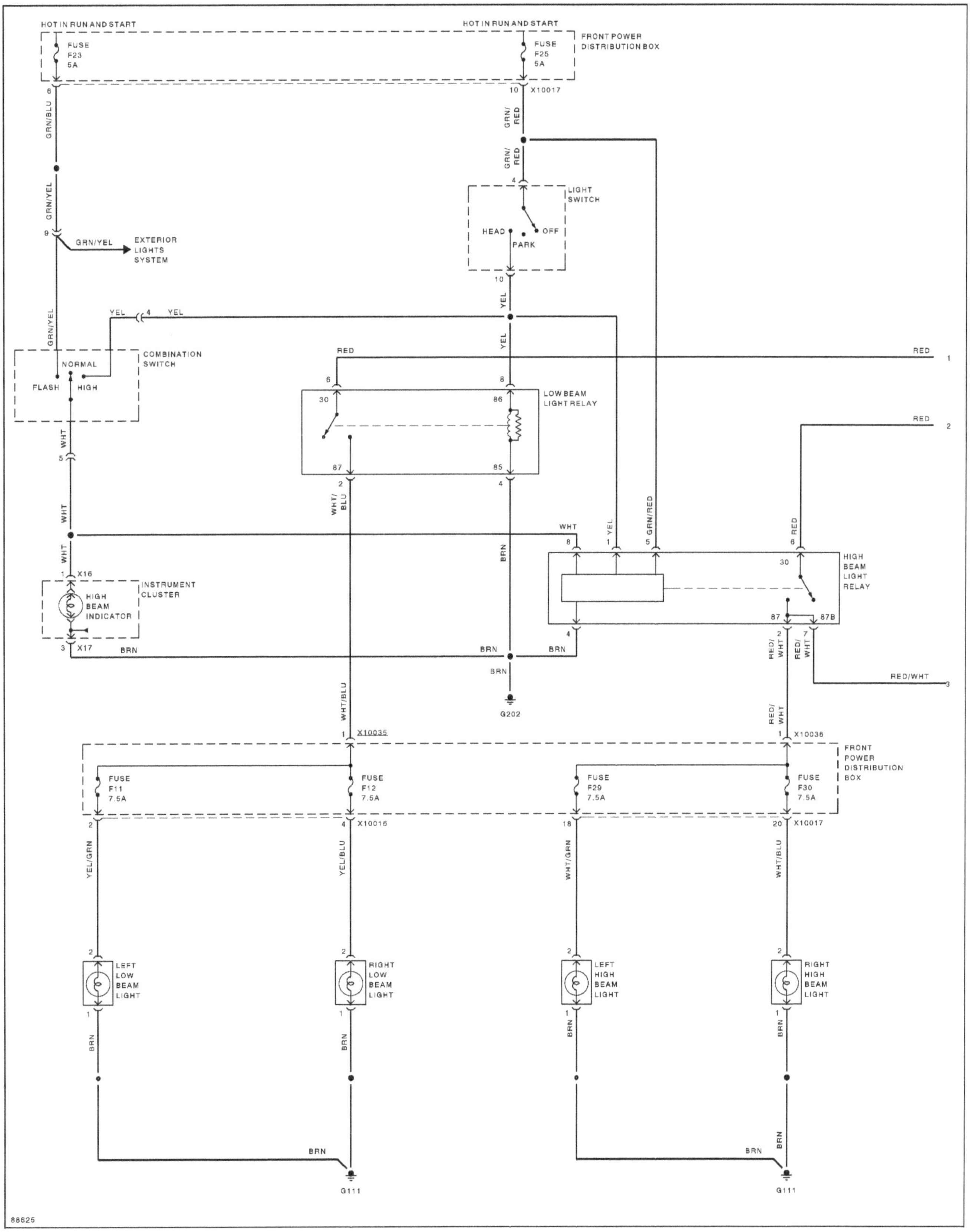

Headlights/Foglights

1997 318ti (Canada) (2 of 2)

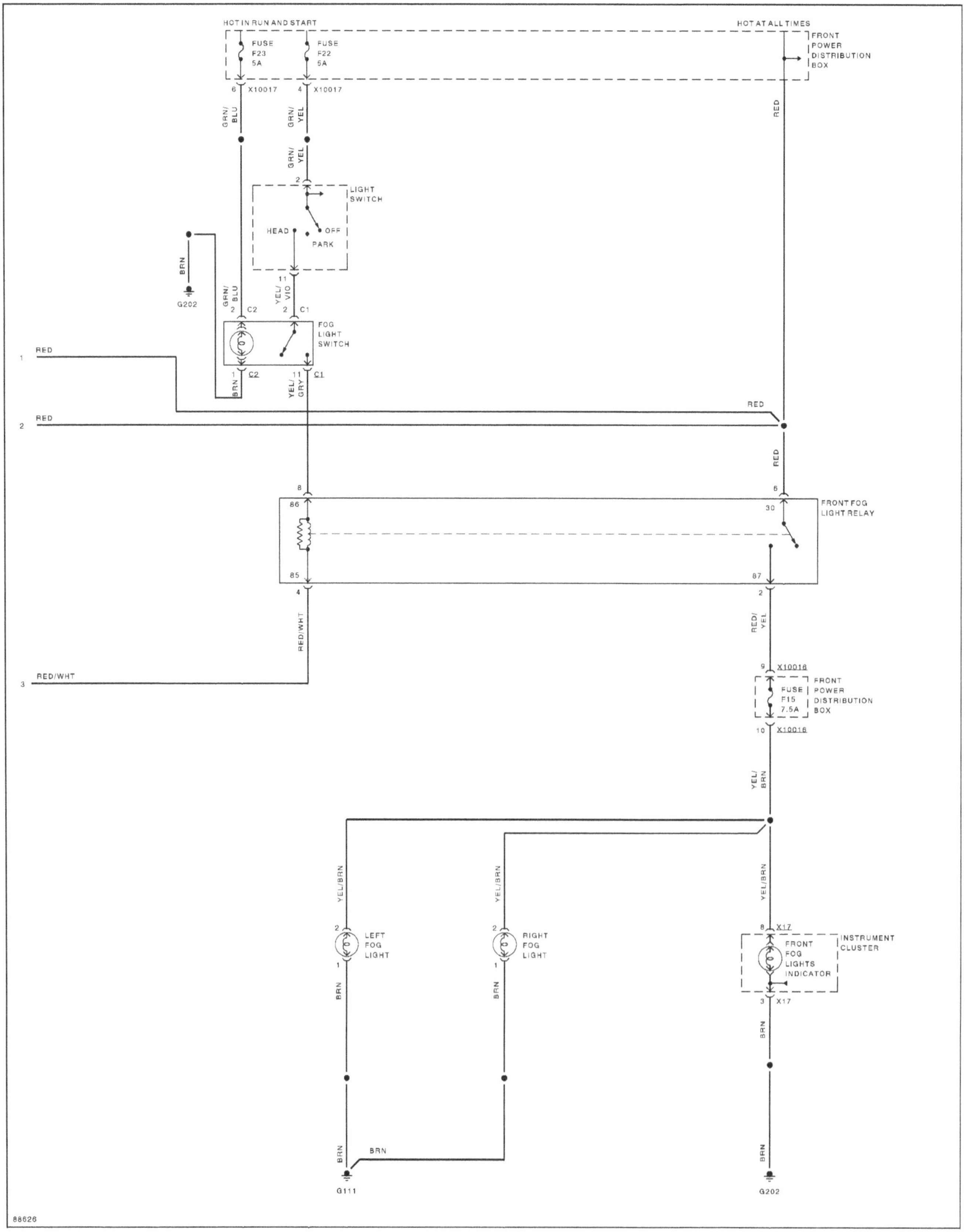

Headlights/Foglights

1997 318ti (USA)

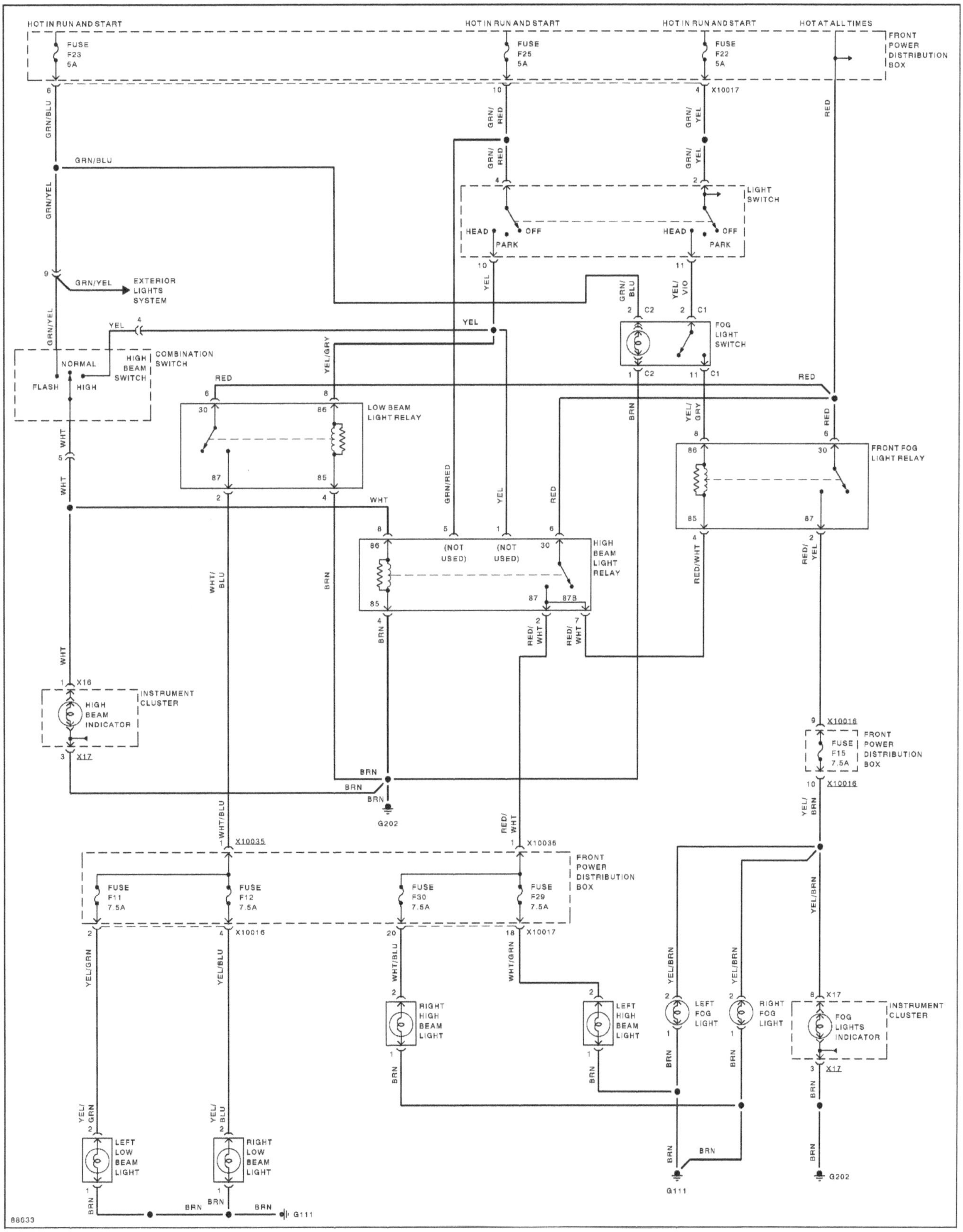

Exterior Lights

1992-93 (Canada) (1 of 3)

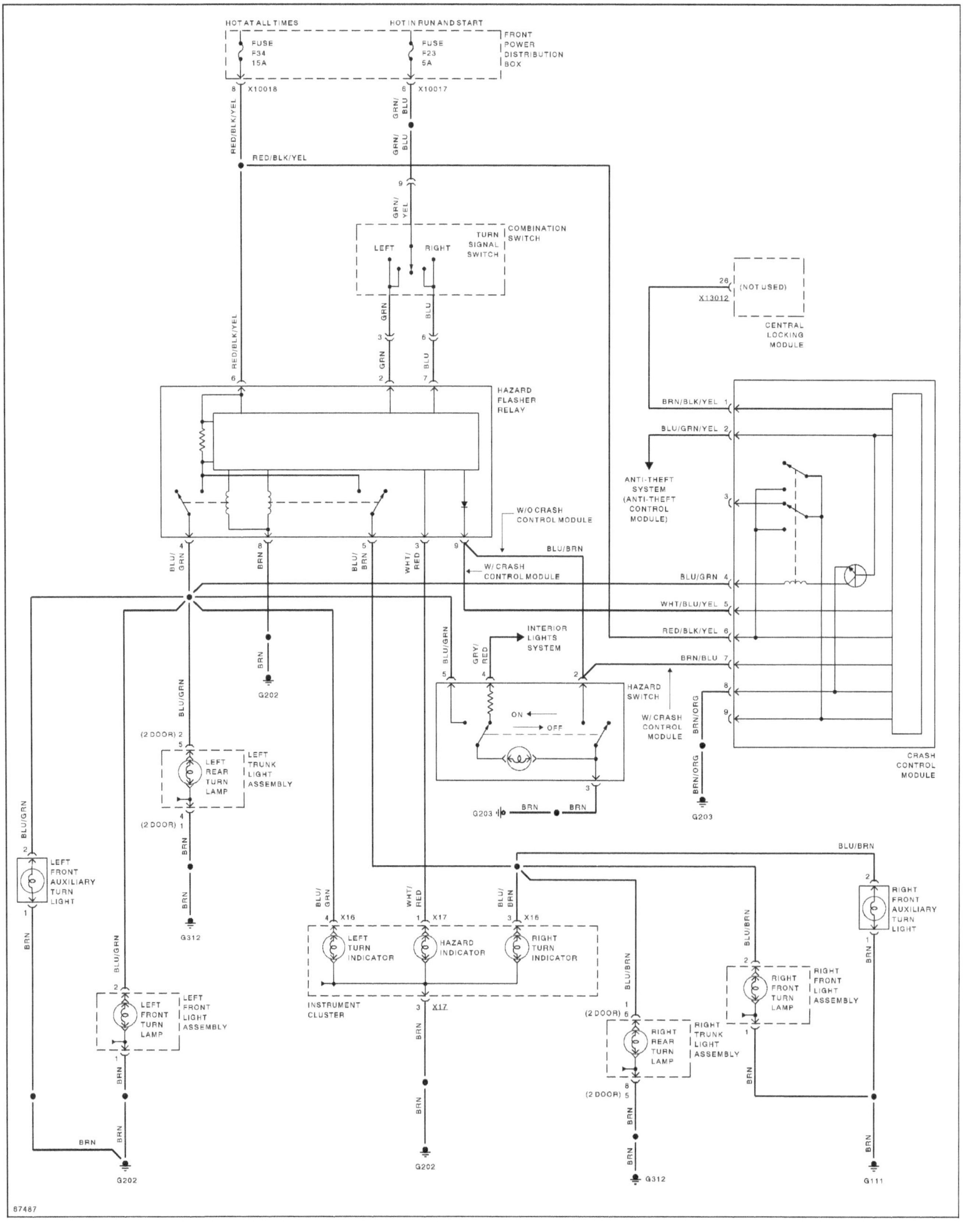

Exterior Lights

1992-93 (Canada) (2 of 3)

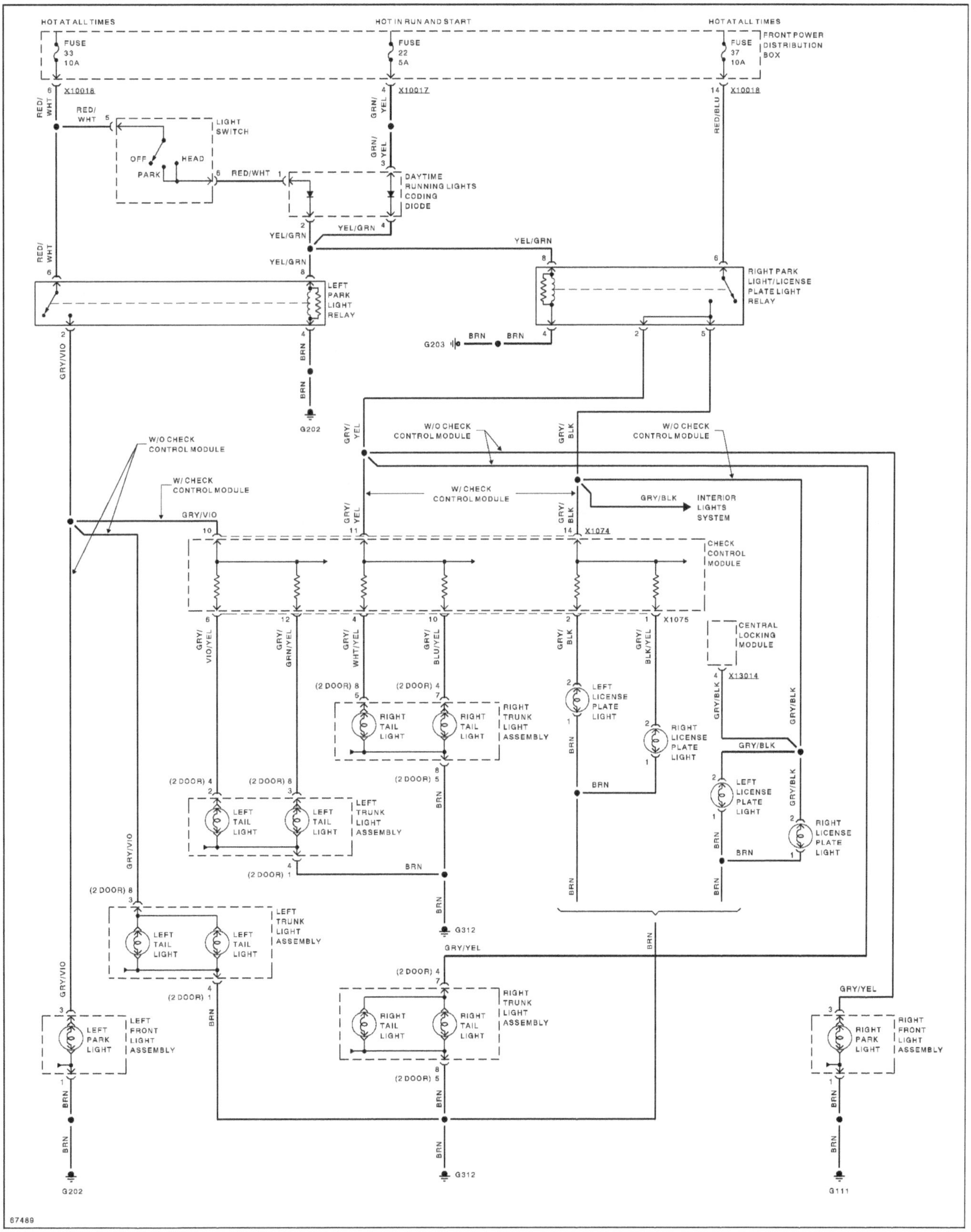

Exterior Lights

1992-93 (Canada) (3 of 3)

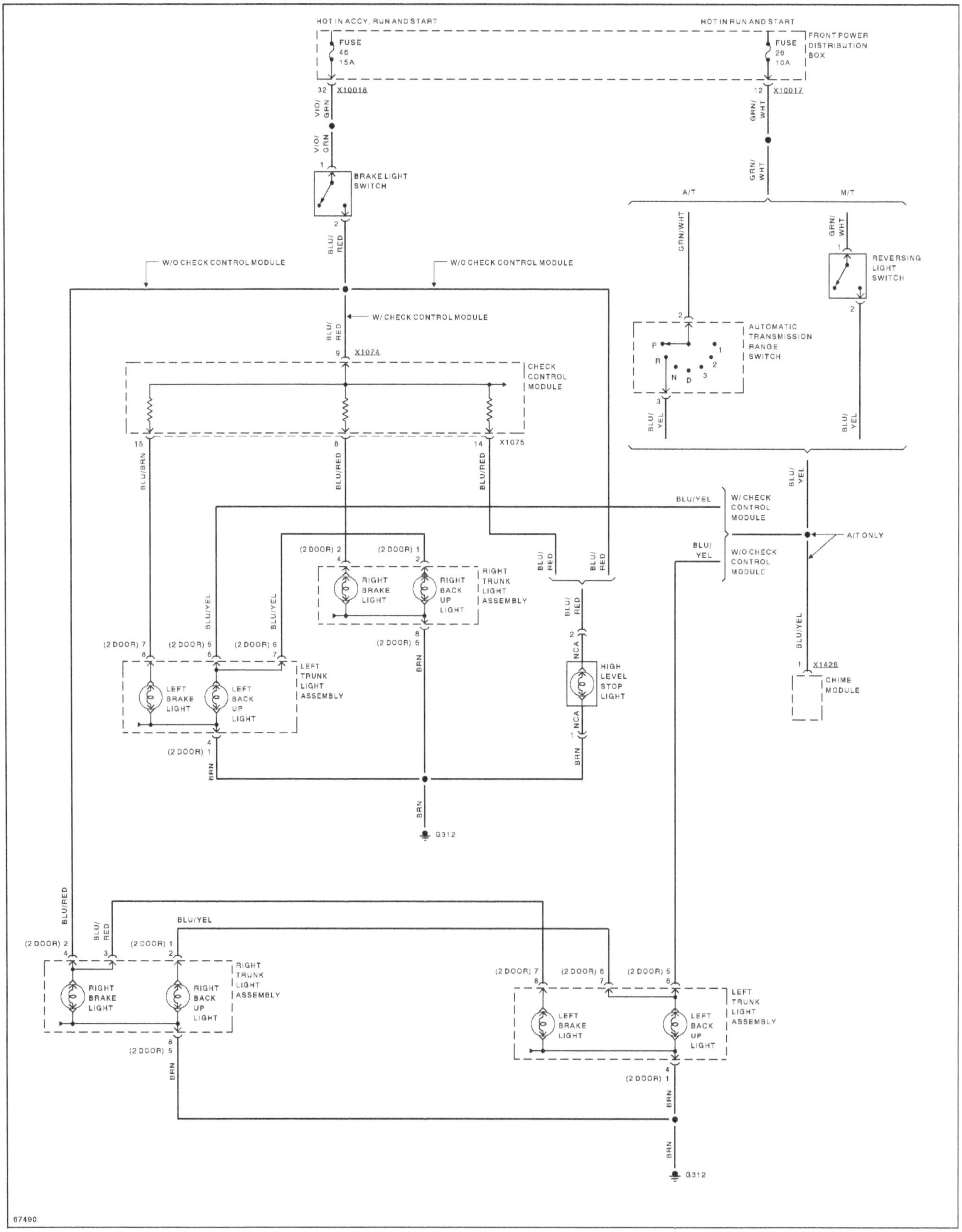

Exterior Lights

1992-93 (USA) (1 of 3)

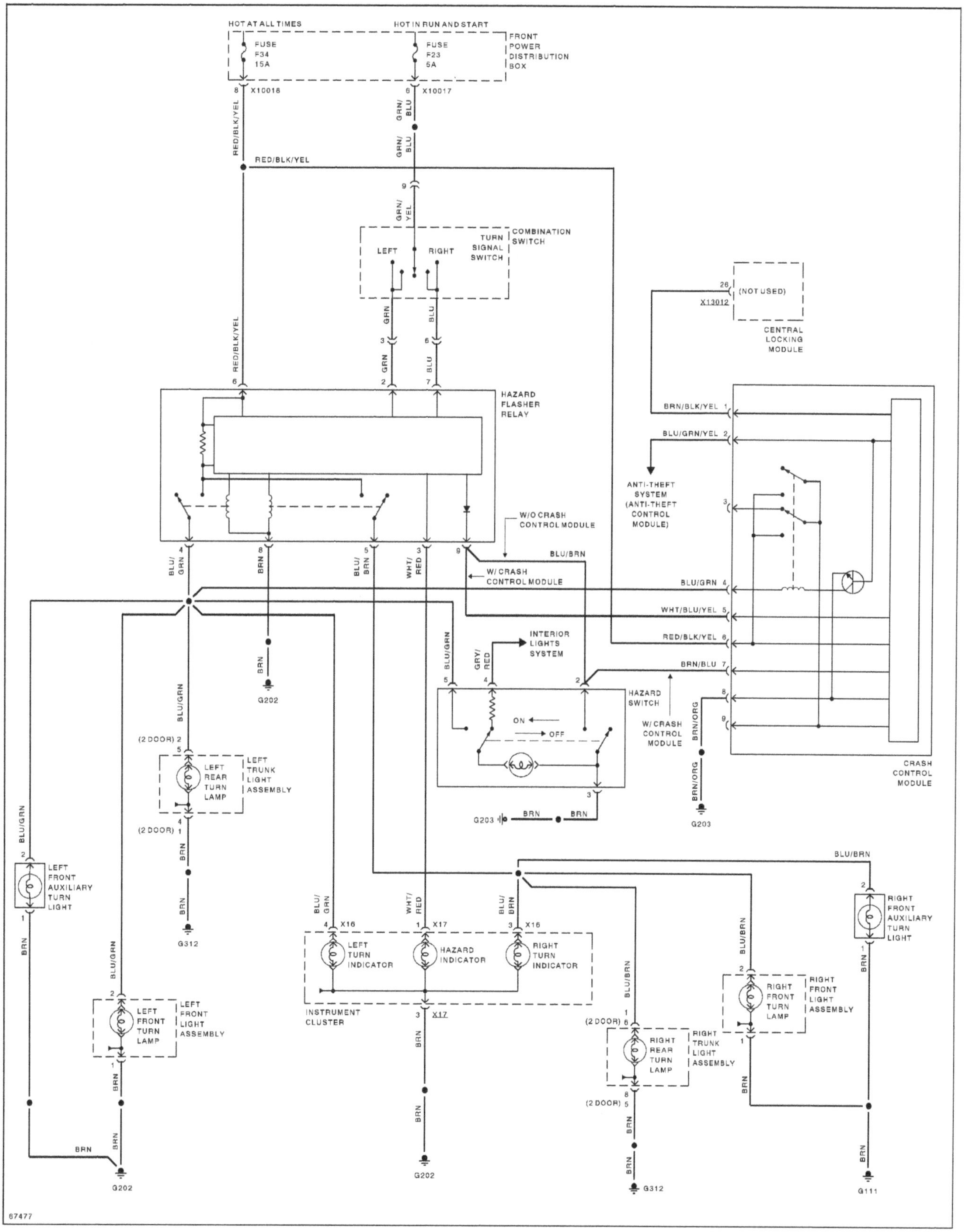

Exterior Lights

1992-93 (USA) (2 of 3)

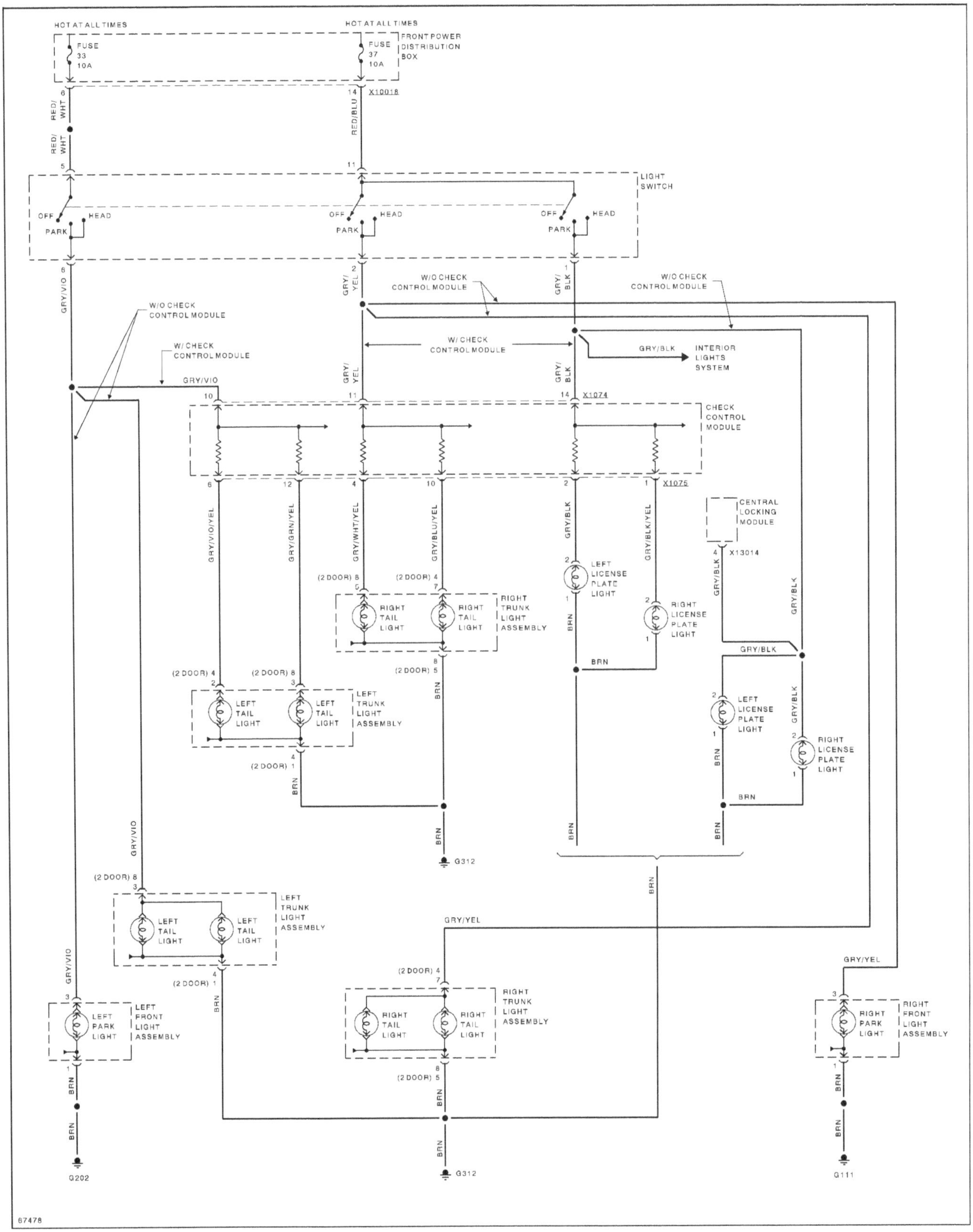

Exterior Lights

1992-93 (USA) (3 of 3)

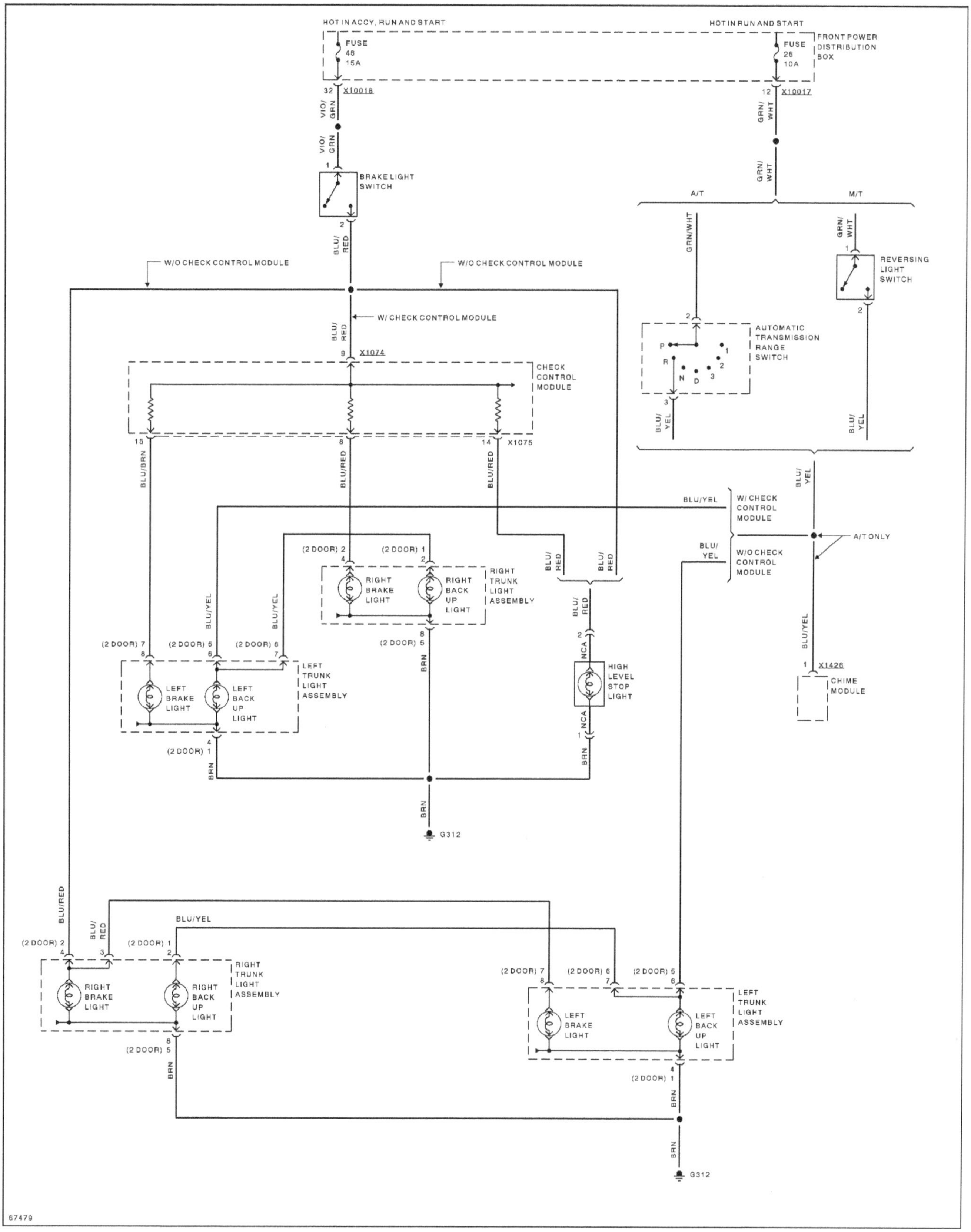

Exterior Lights

1994-96 Except 318ti (Canada) (1 of 3)

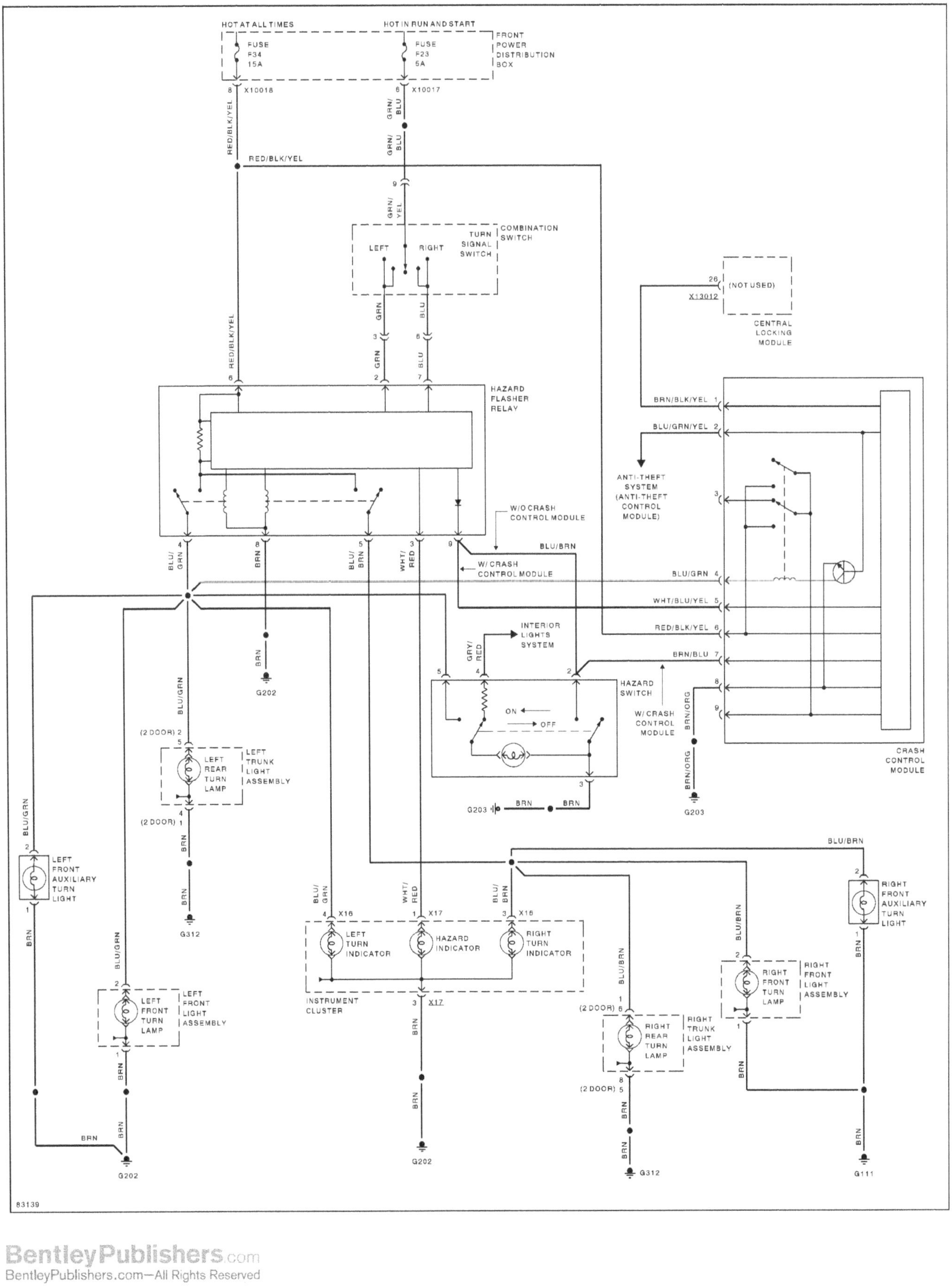

Exterior Lights

1994-96 Except 318ti (Canada) (2 of 3)

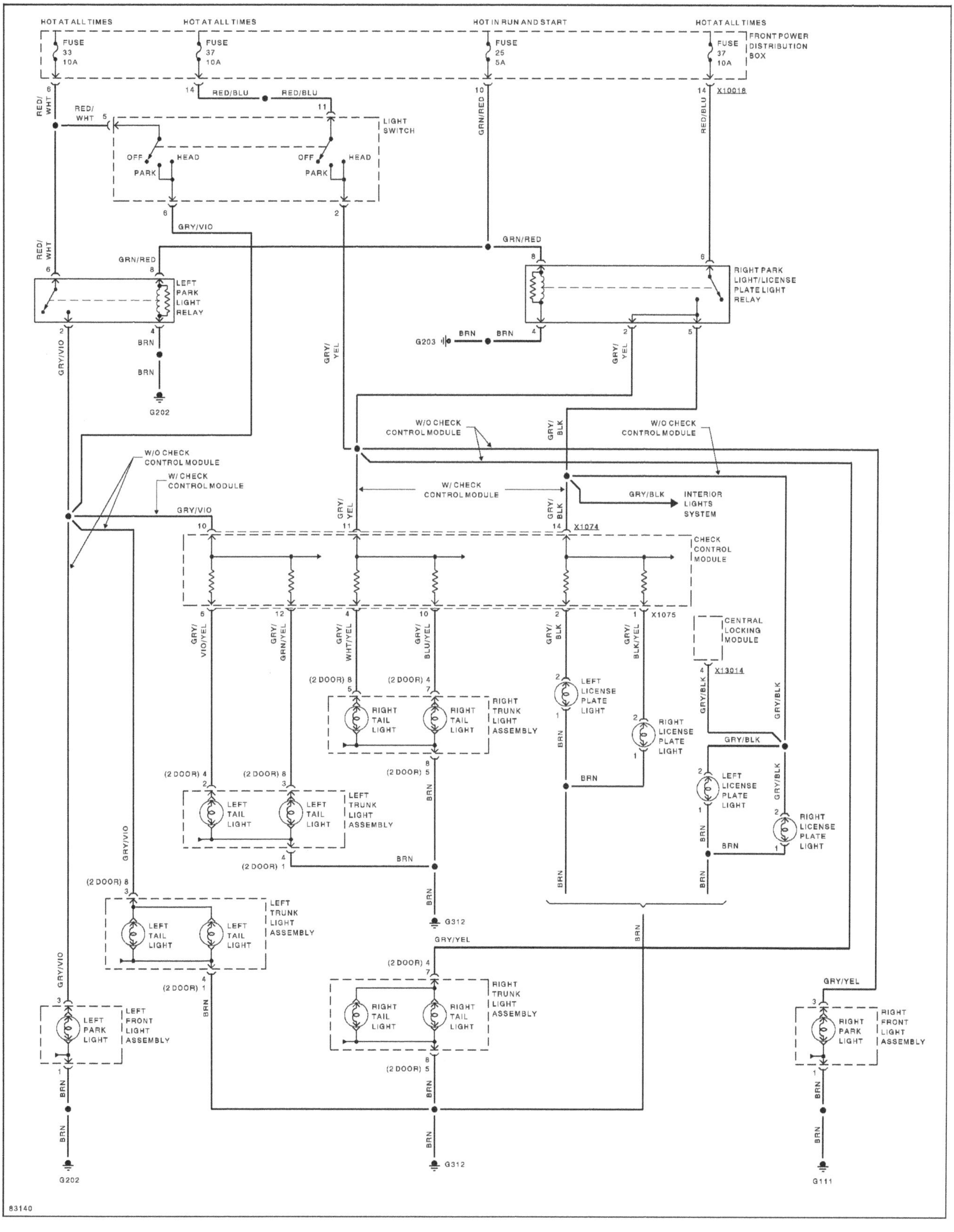

Exterior Lights

1994-96 Except 318ti (Canada) (3 of 3)

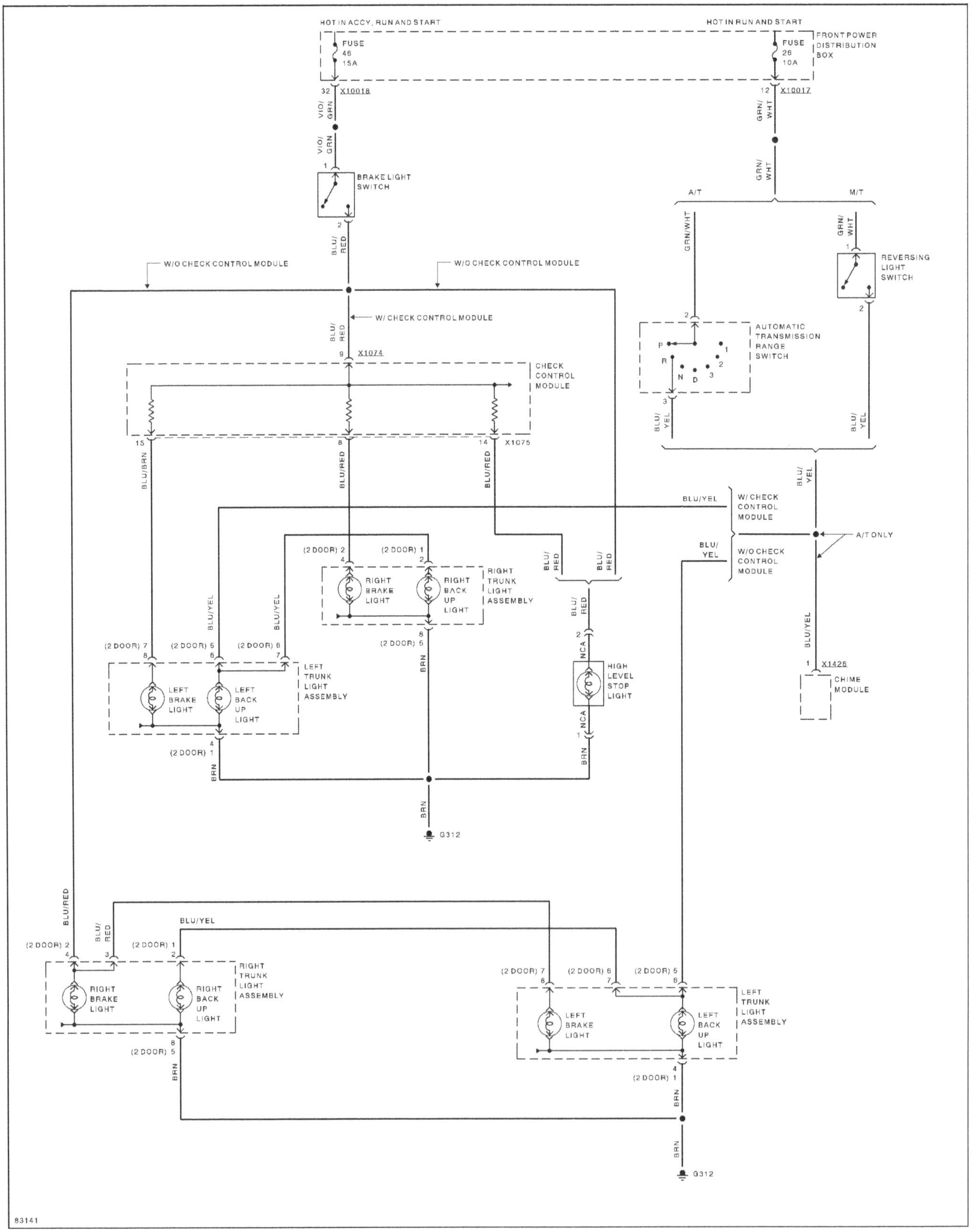

Exterior Lights
1994-96 (USA) & 1997 Except 318ti (1 of 3)

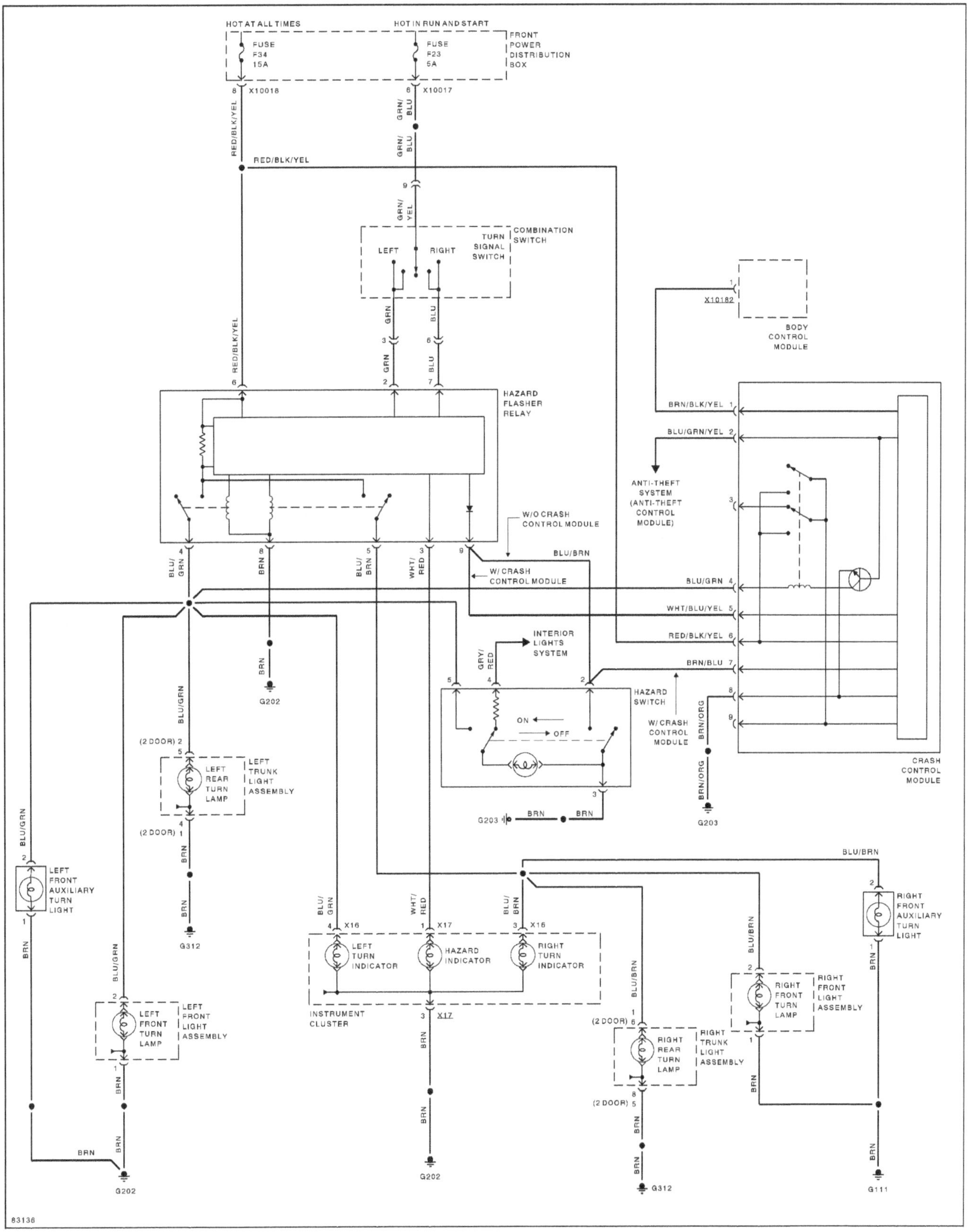

Exterior Lights

1994-96 (USA) & 1997 Except 318ti (2 of 3)

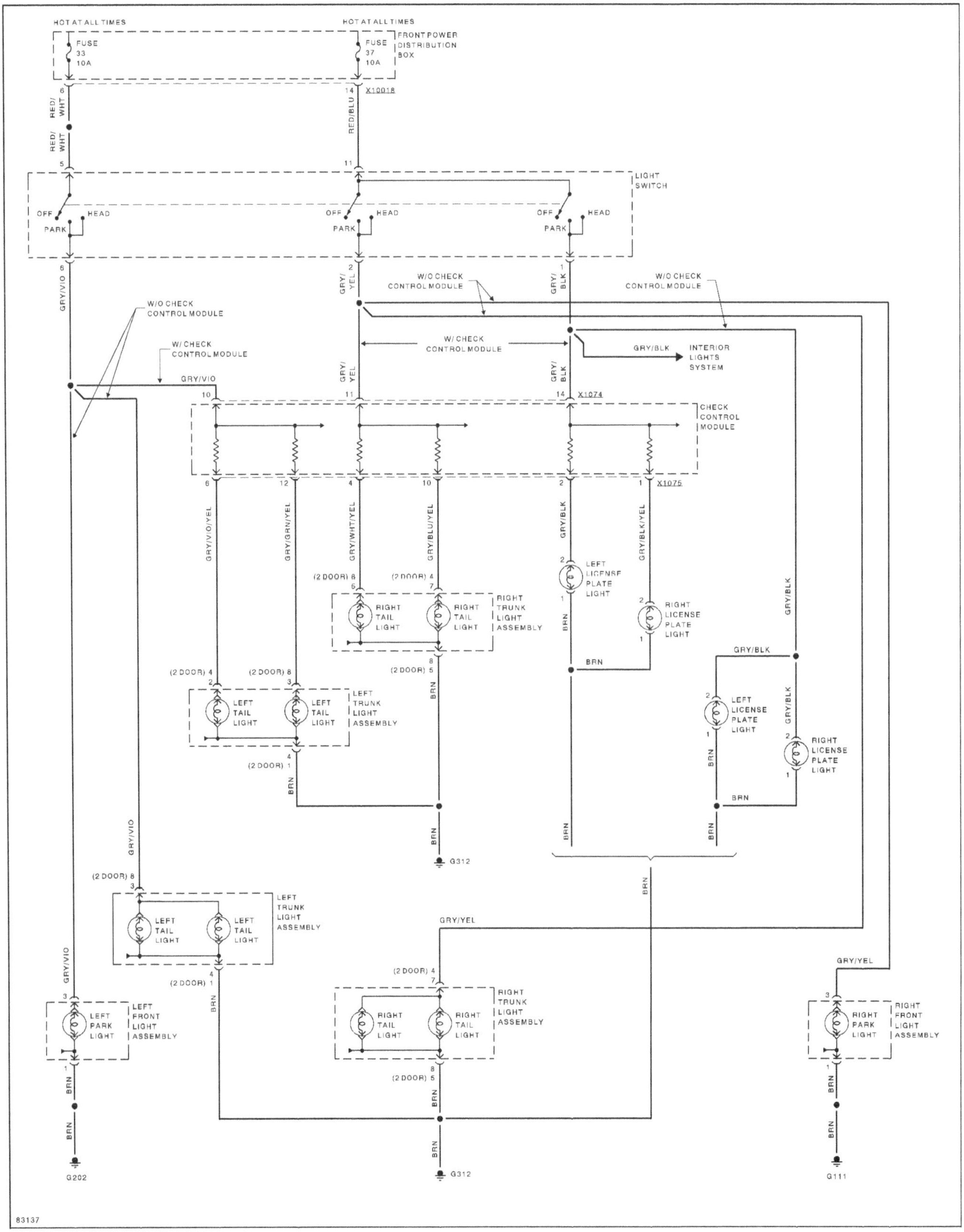

Exterior Lights

1994-96 (USA) & 1997 Except 318ti (3 of 3)

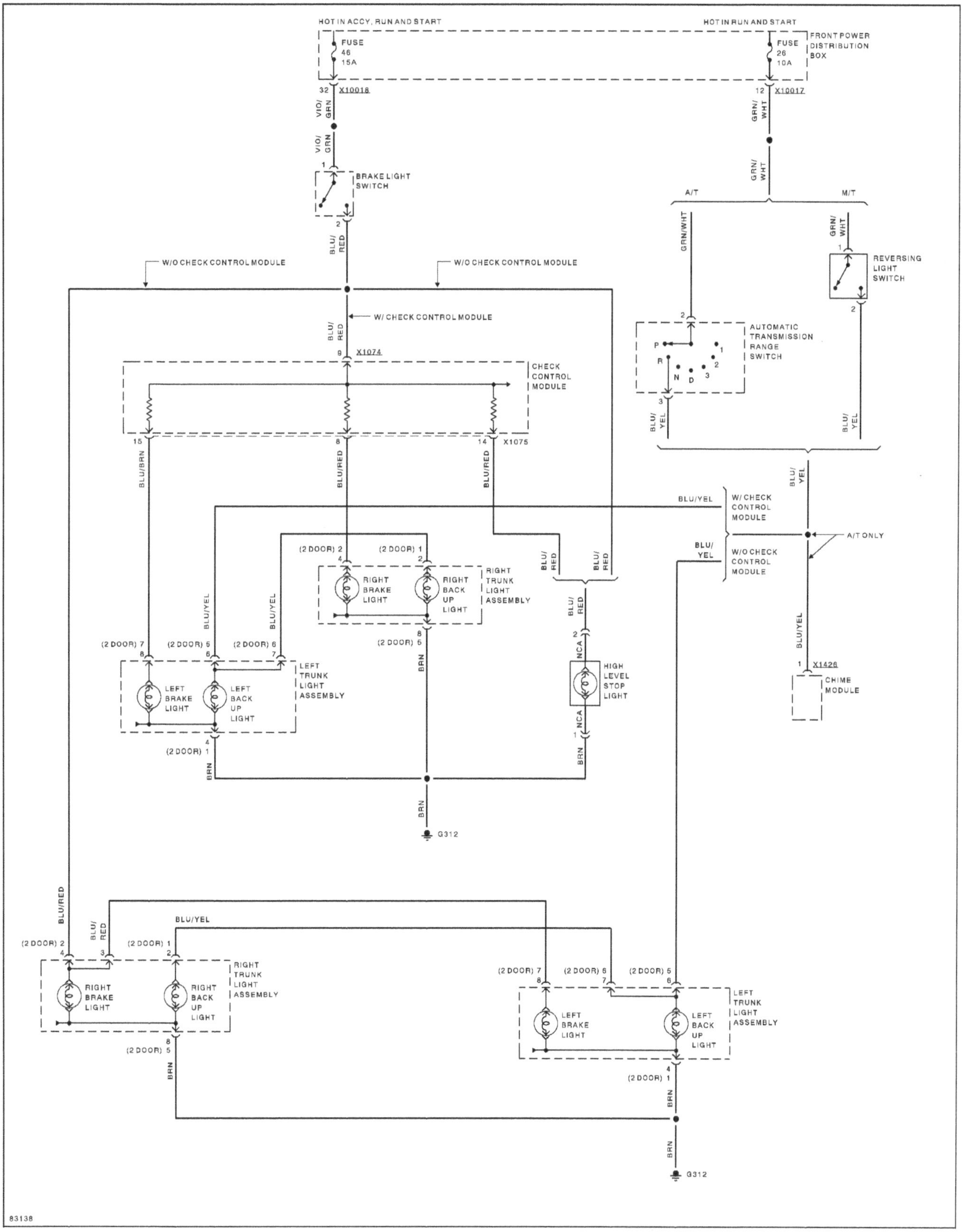

Exterior Lights

1998 Except 318ti (1 of 3)

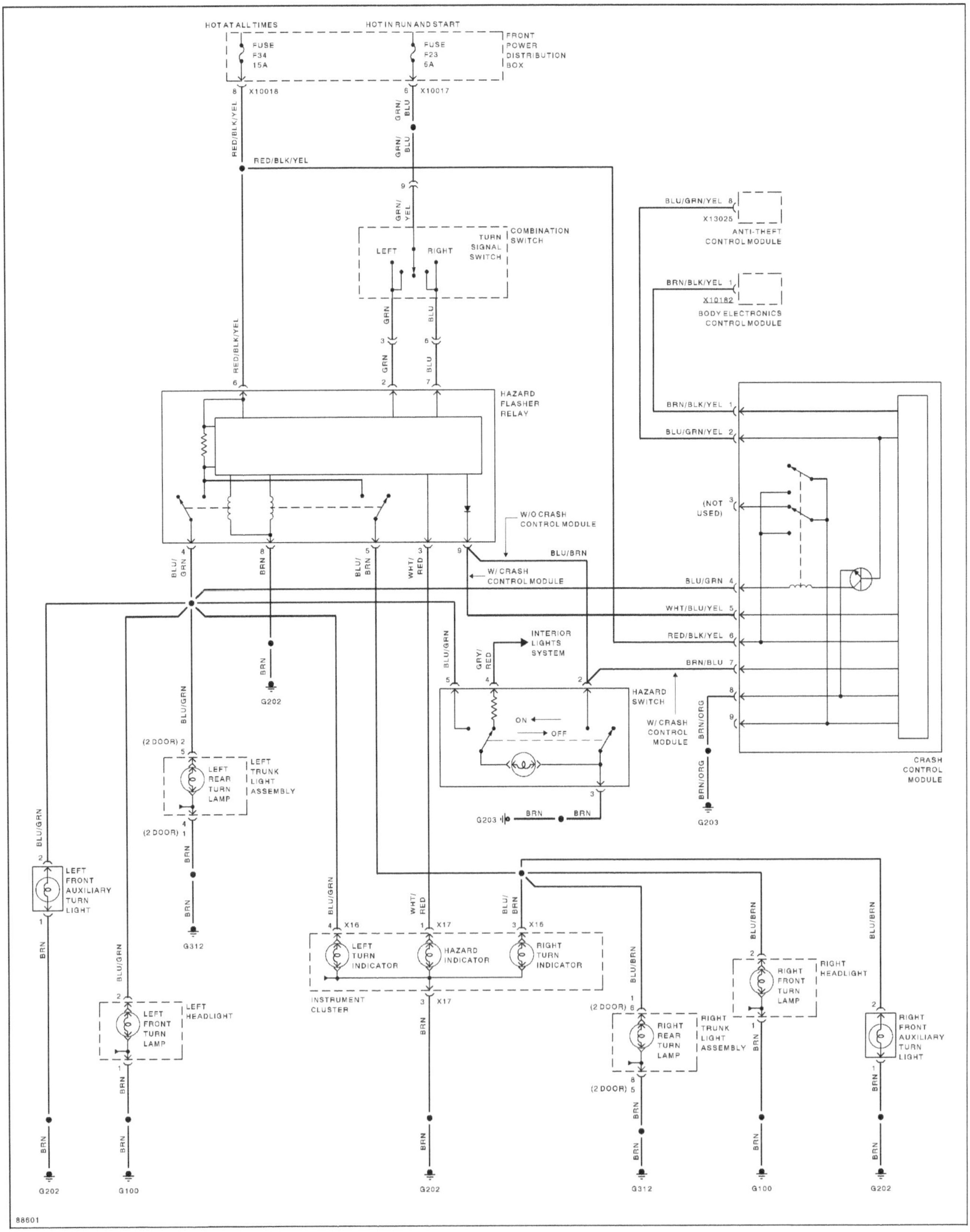

Exterior Lights

1998 Except 318ti (2 of 3)

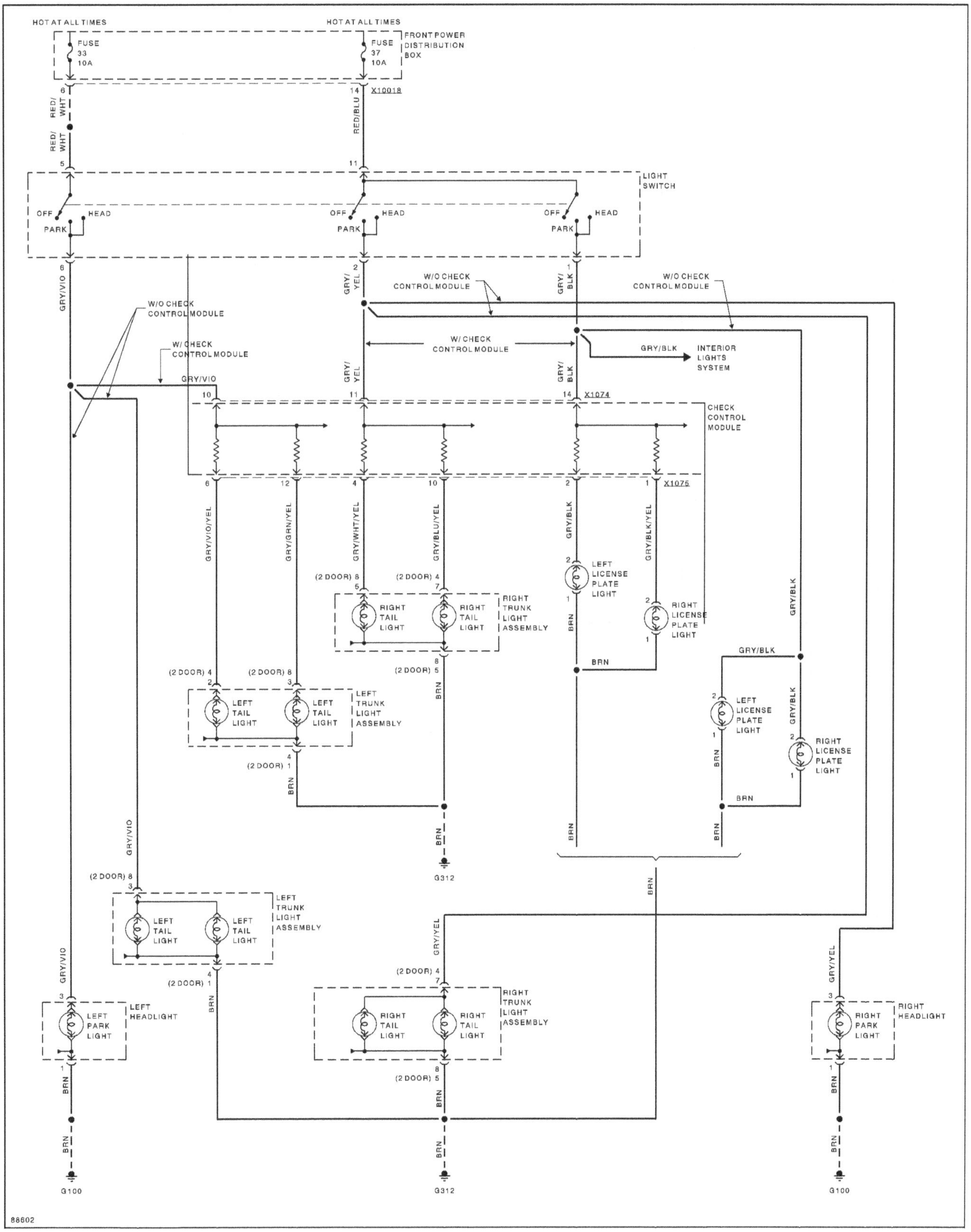

Exterior Lights

1998 Except 318ti (3 of 3)

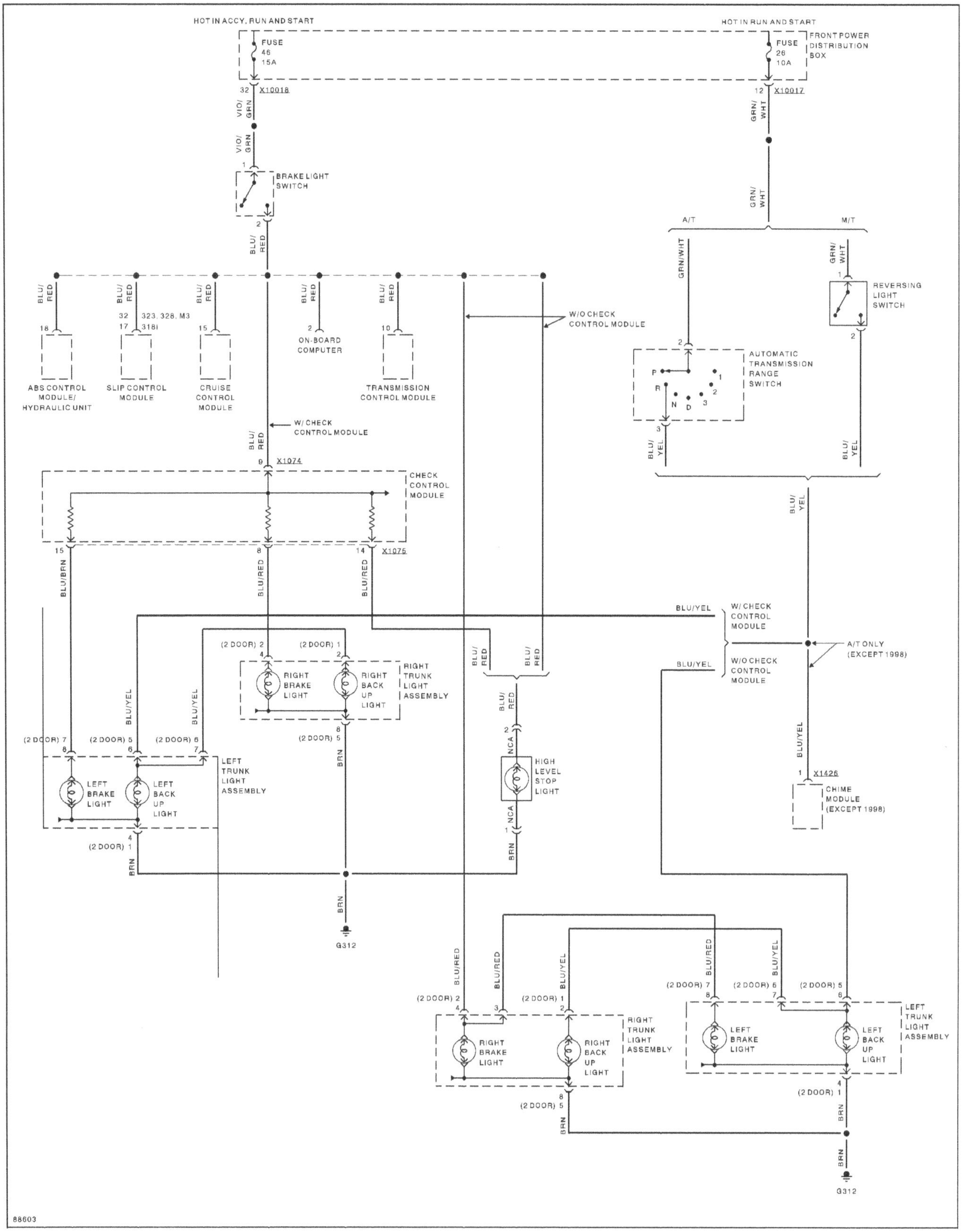

Exterior Lights

1995-98 318ti (1 of 2)

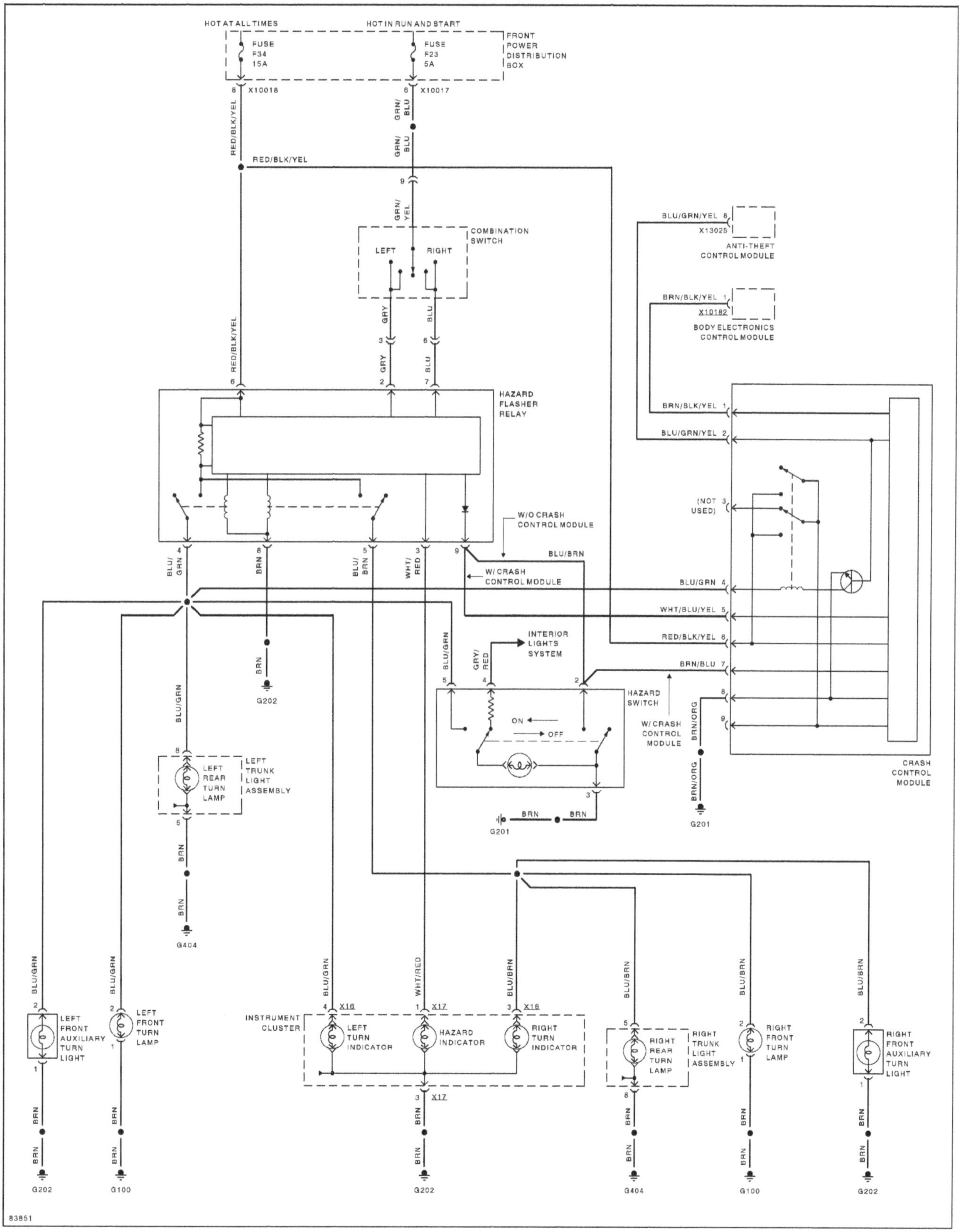

Exterior Lights
1995-98 318ti (2 of 2)

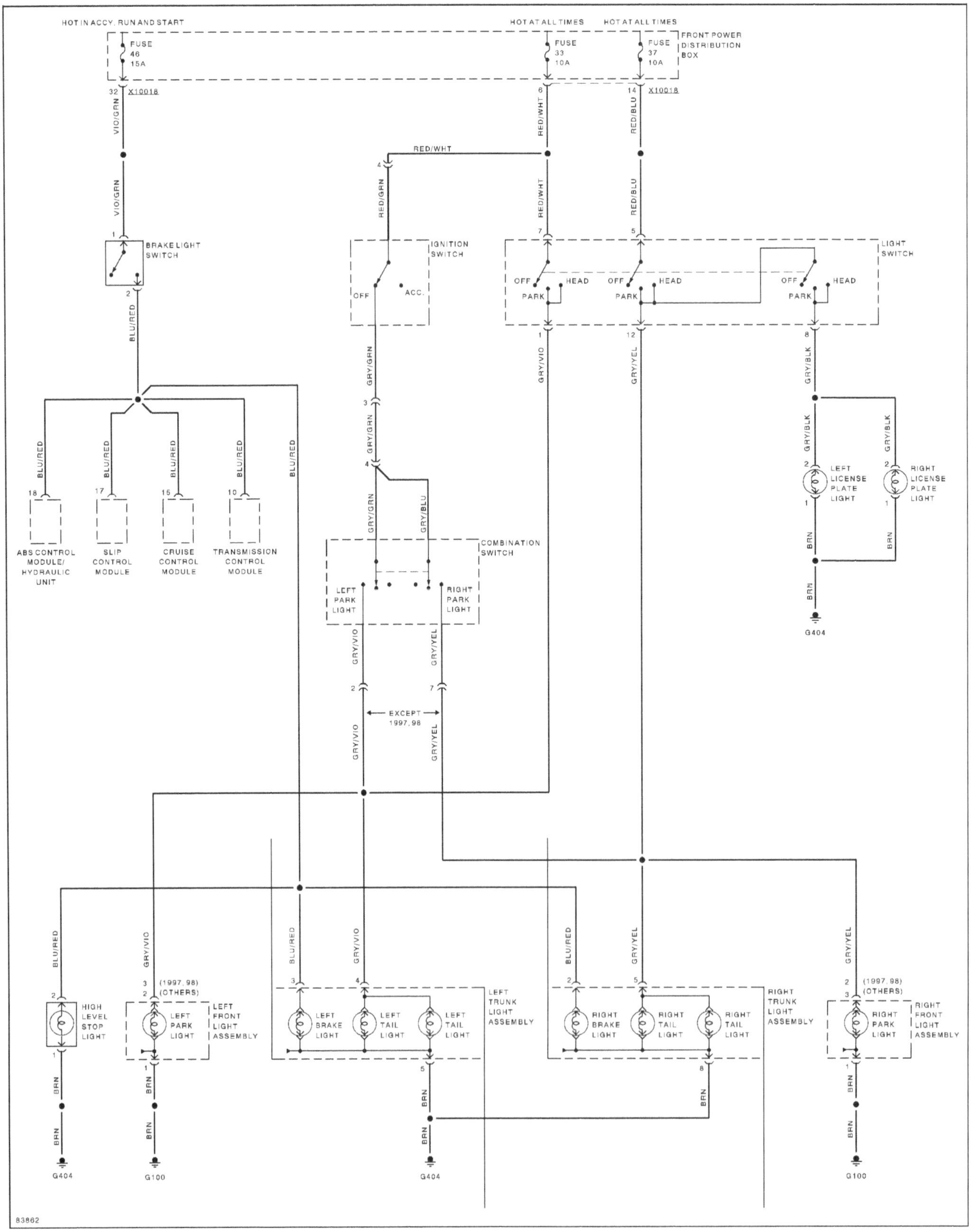

Interior Lights/Illumination

1992-93 (1 of 2)

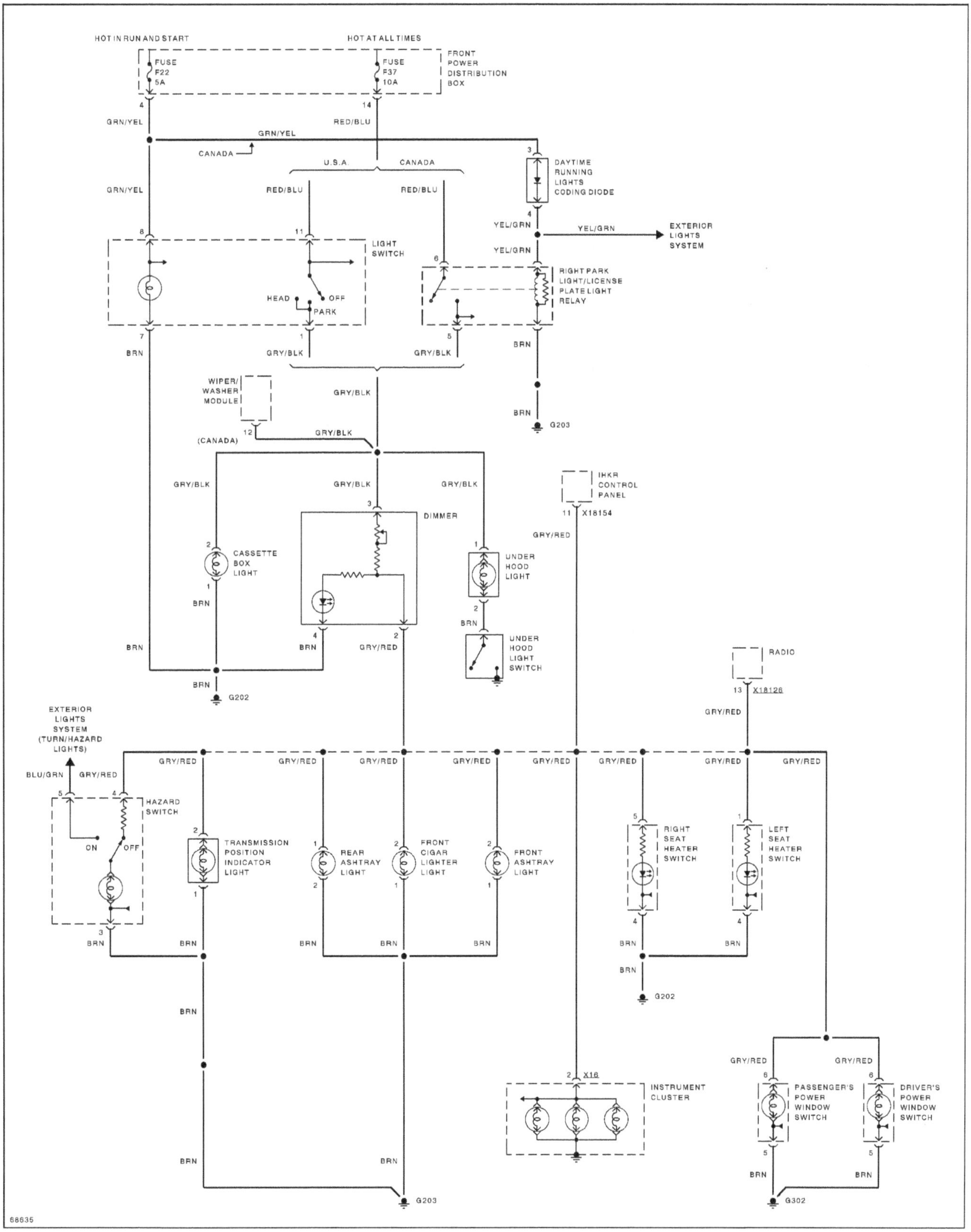

Interior Lights/Illumination

1992-93 (2 of 2)

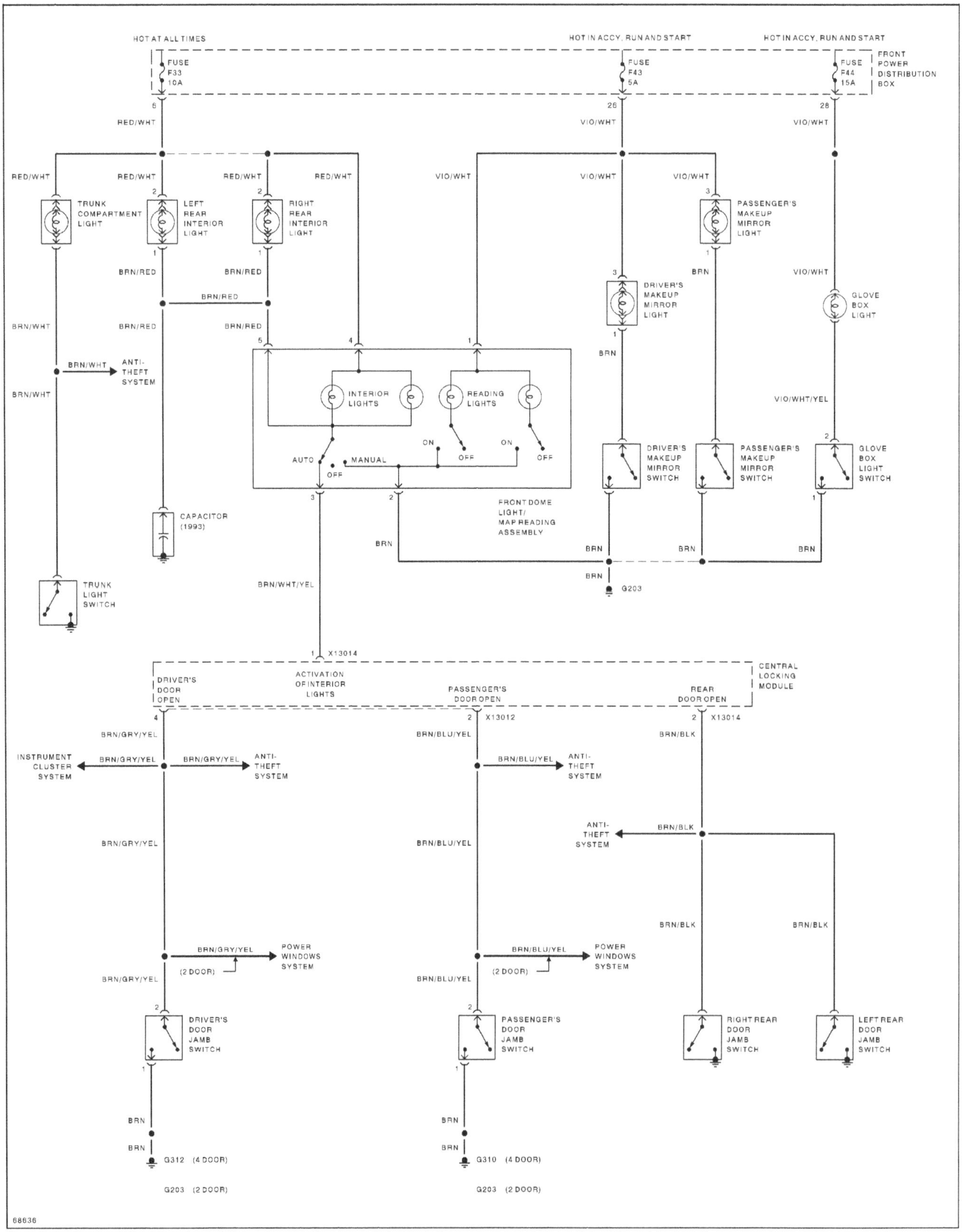

Interior Lights/Illumination

1994-97 Convertible (1 of 2)

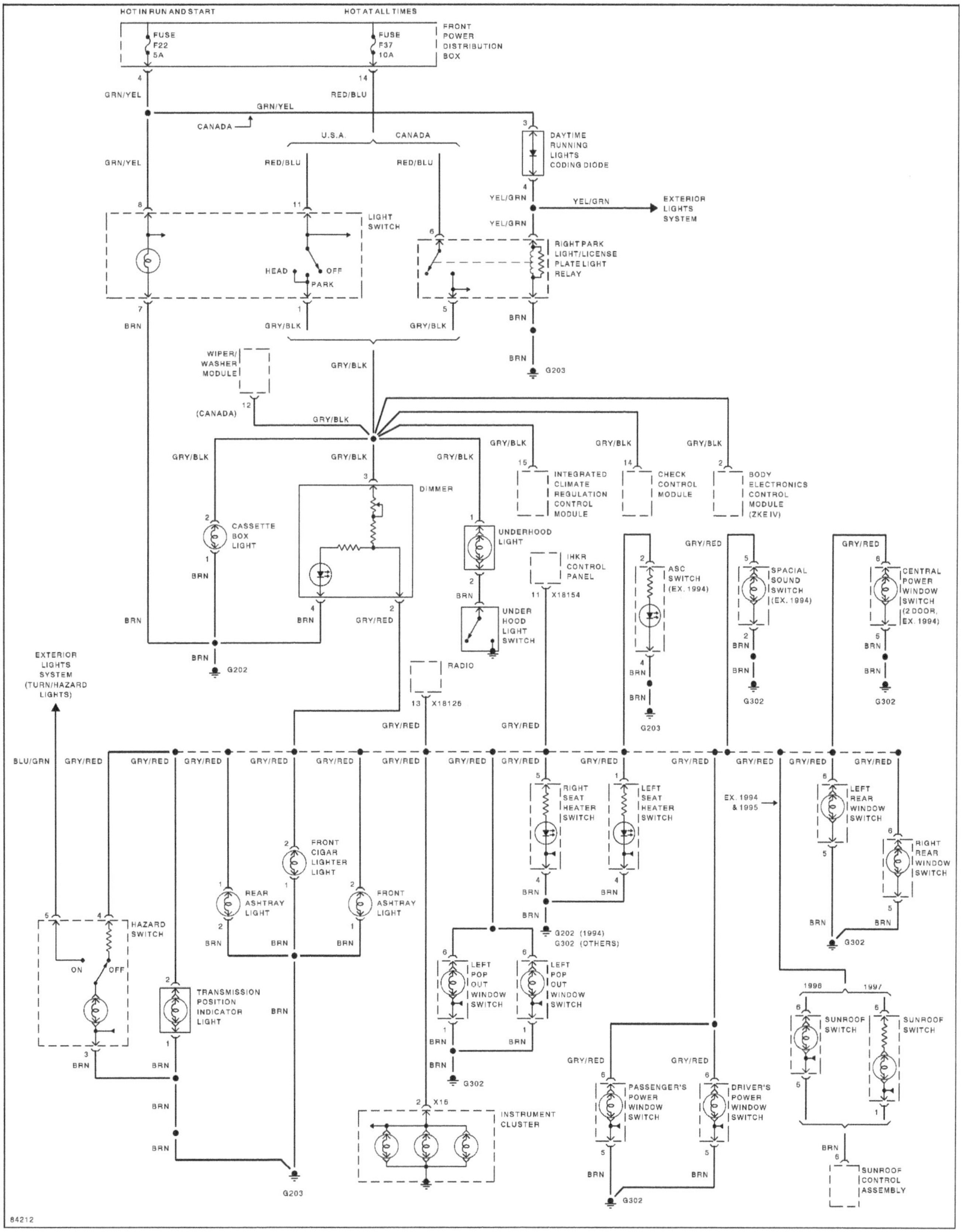

Interior Lights/Illumination

1994-97 Convertible (2 of 2)

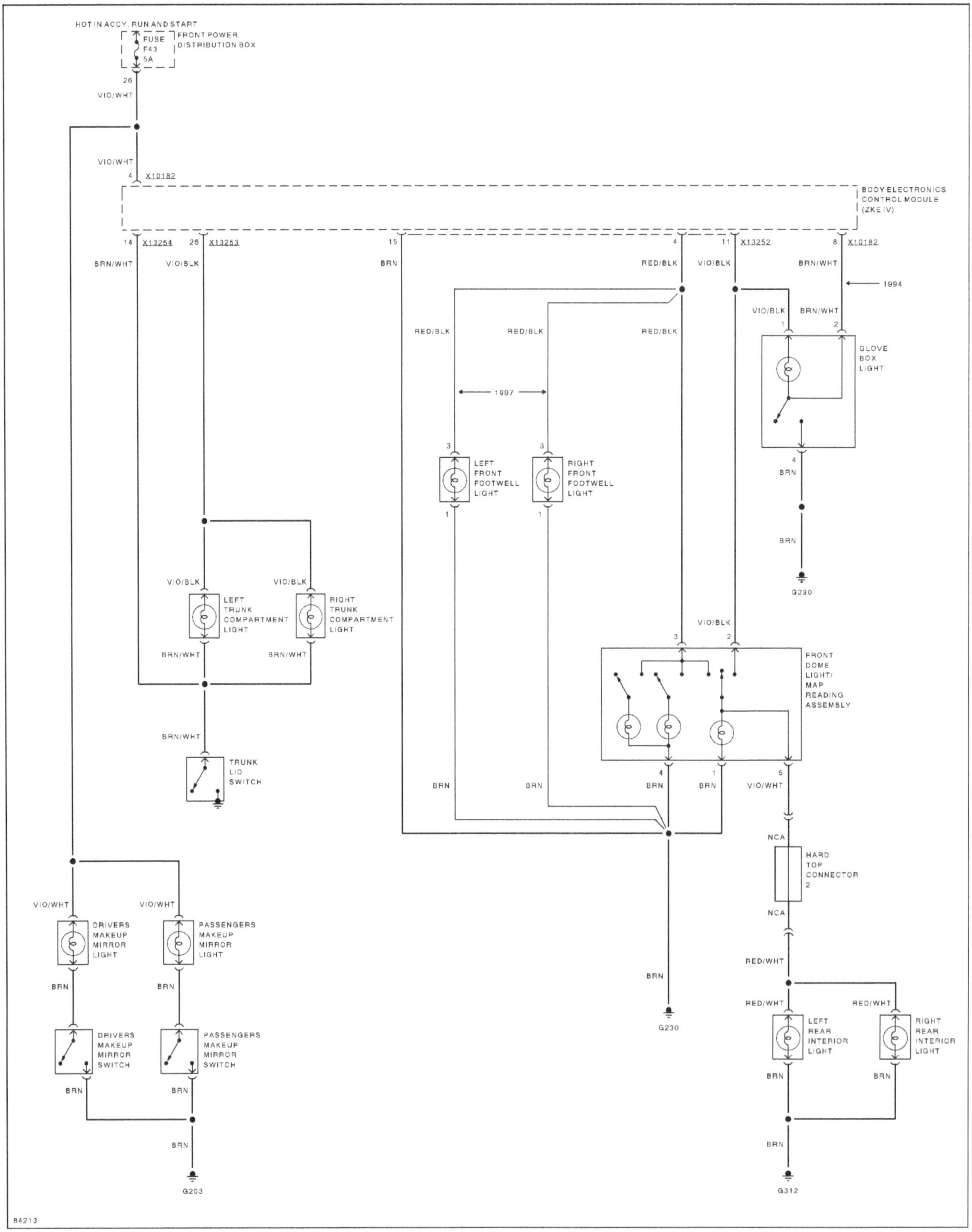

Interior Lights/Illumination

1998 Convertible

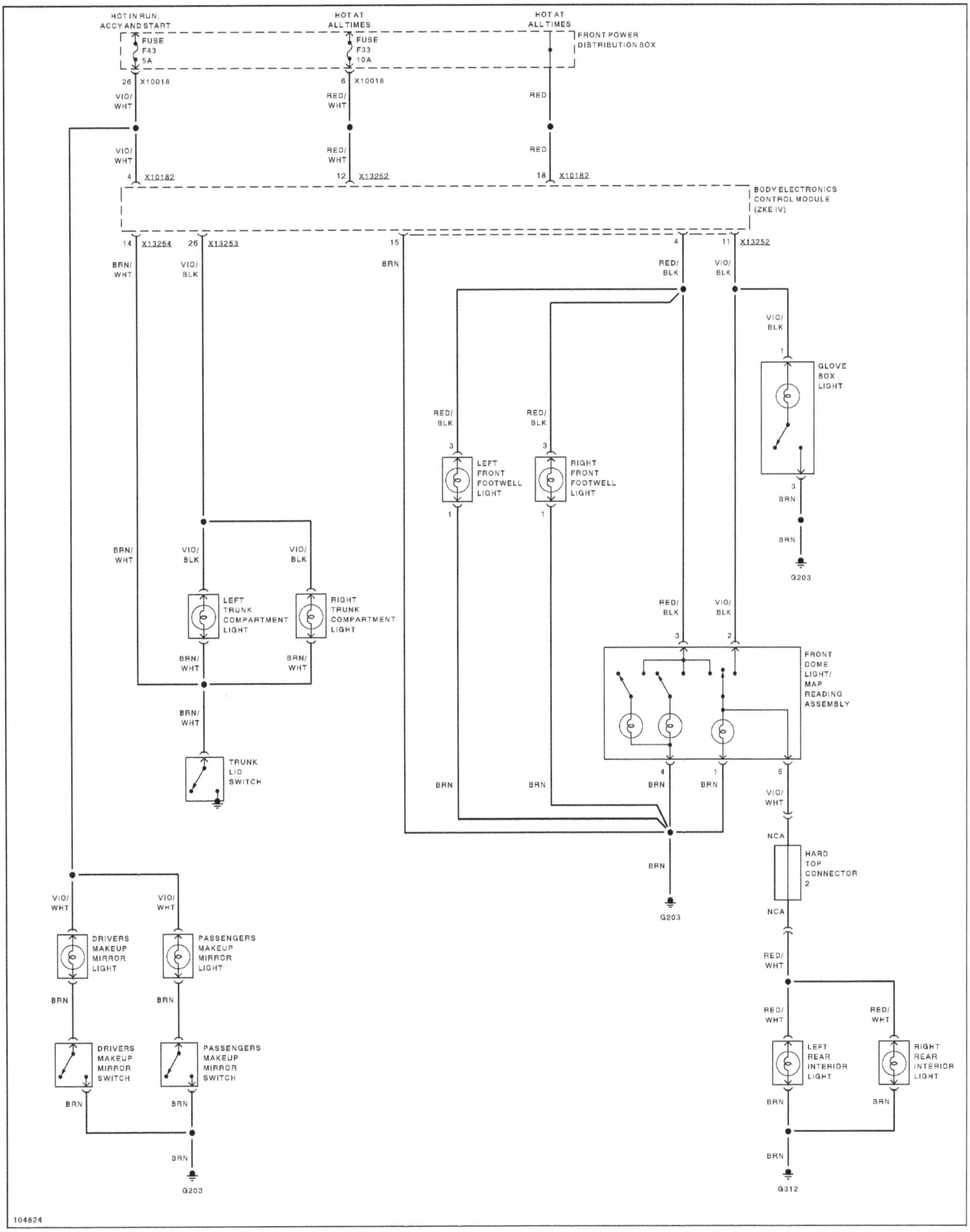

Interior Lights/Illumination

1994-97 Except 318ti & Convertible (1 of 2)

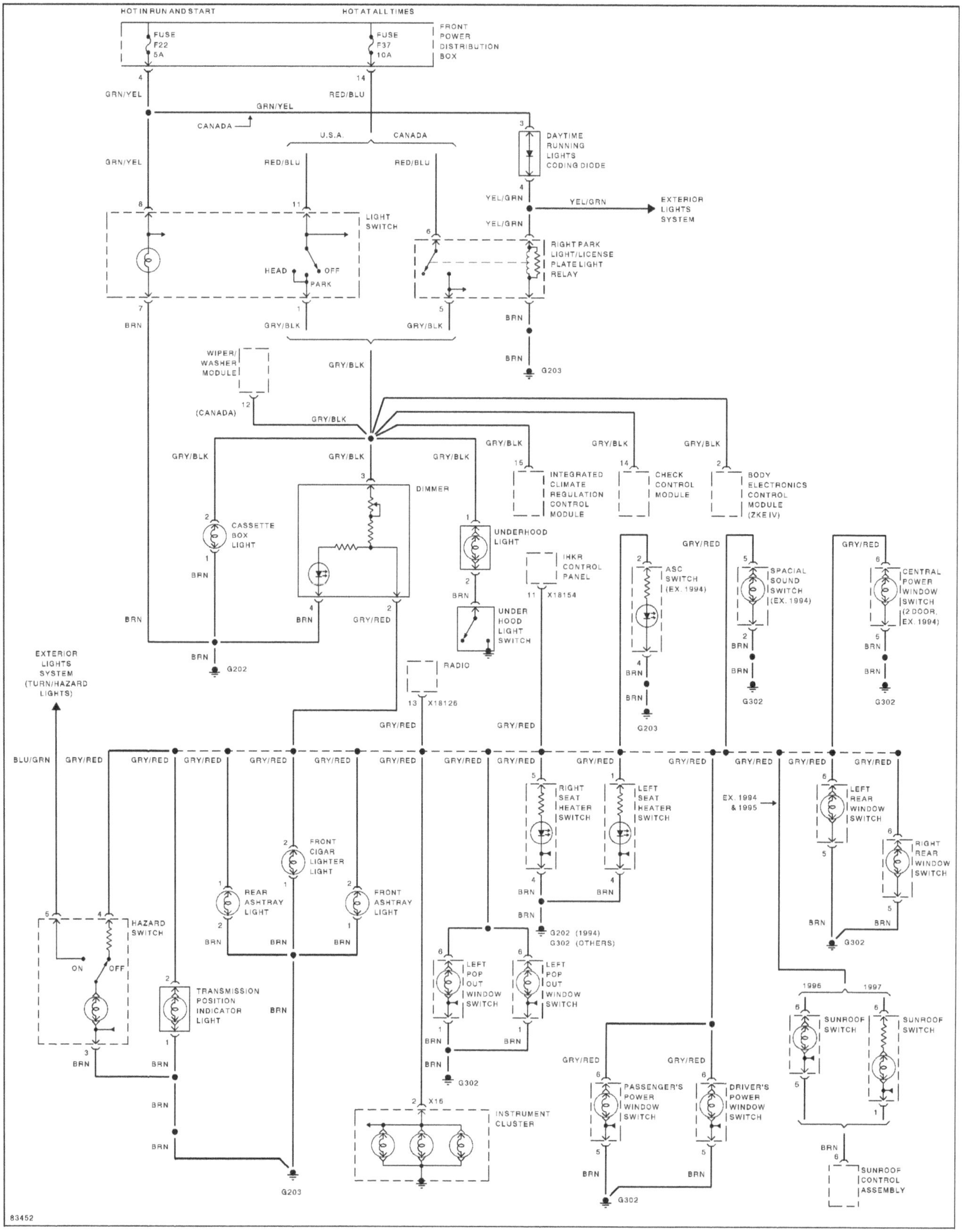

Interior Lights/Illumination

1994-97 Except 318ti & Convertible (2 of 2)

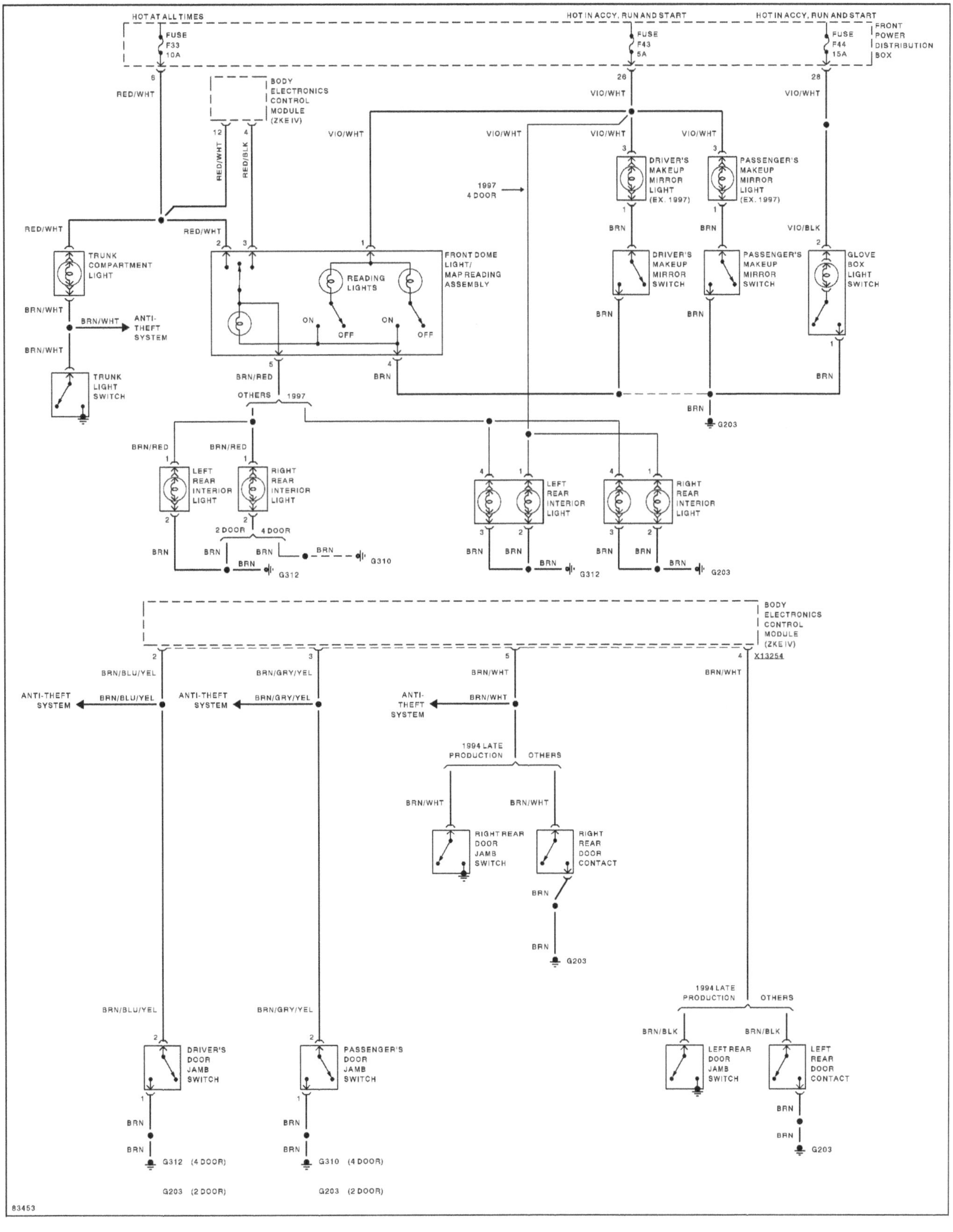

Interior Lights/Illumination

1998 Except 318ti & Convertible

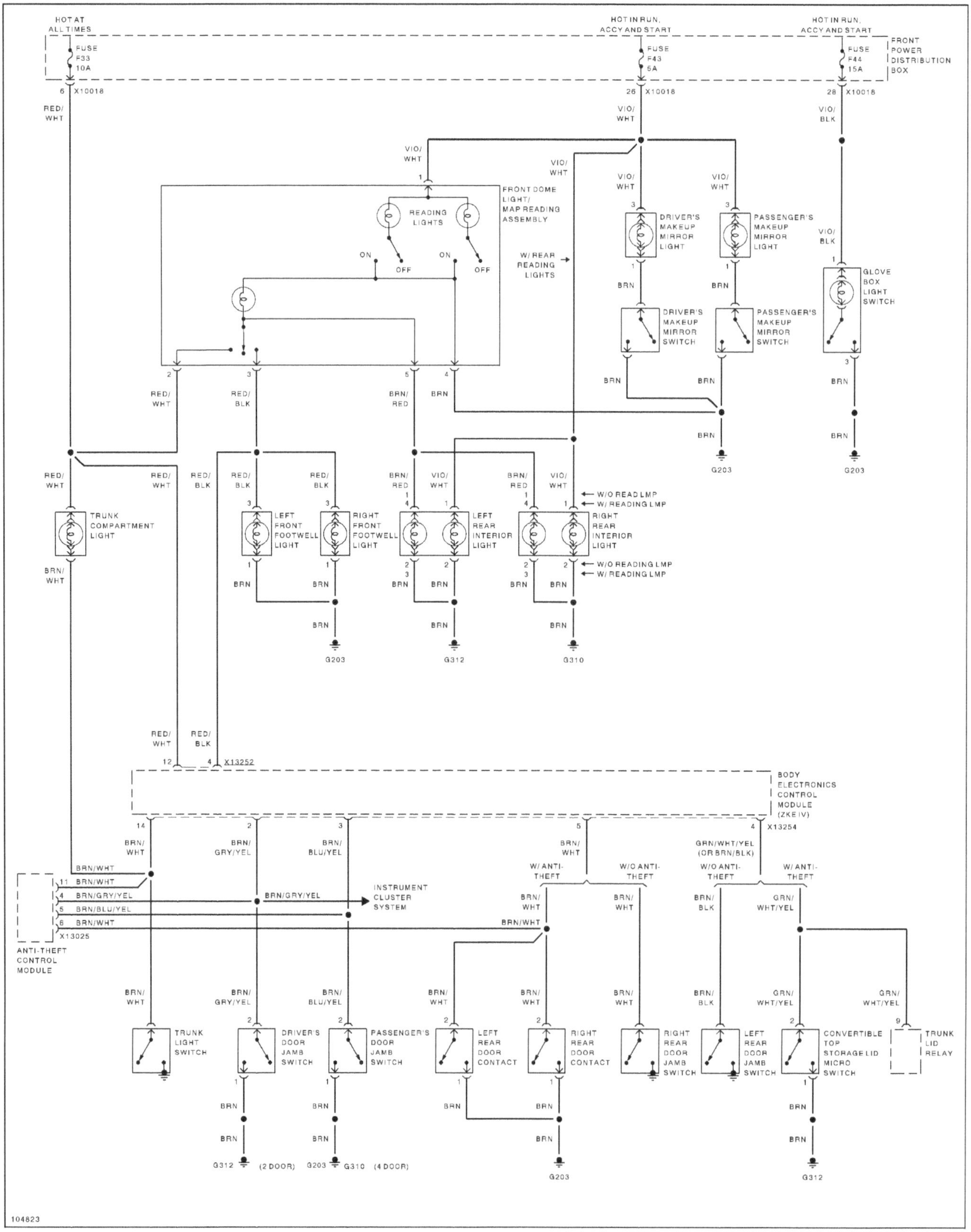

Interior Lights/Instrument Illumination

1998 Except 318ti & Convertible

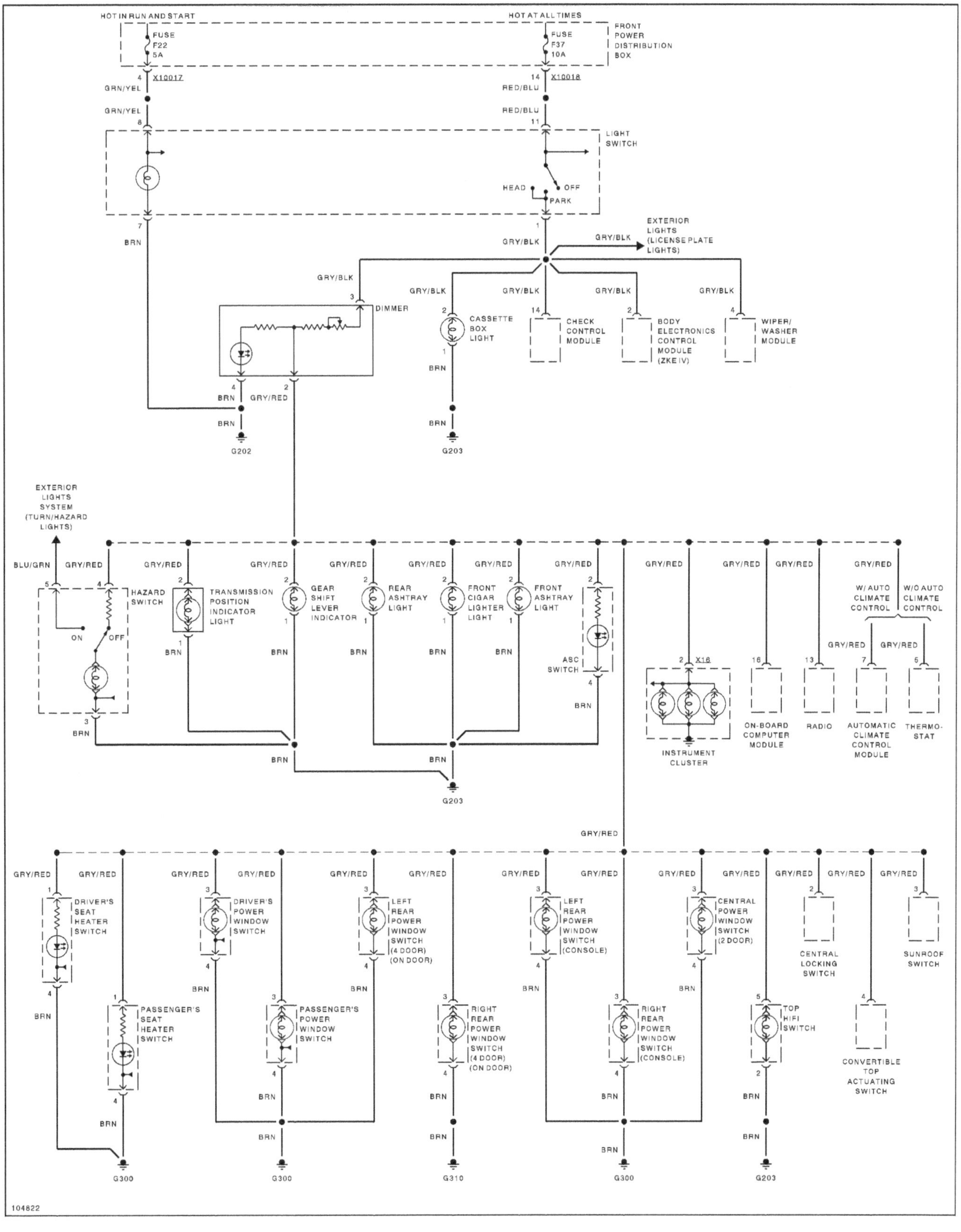

Interior Lights/Illumination

1995-96 318ti (1 of 2)

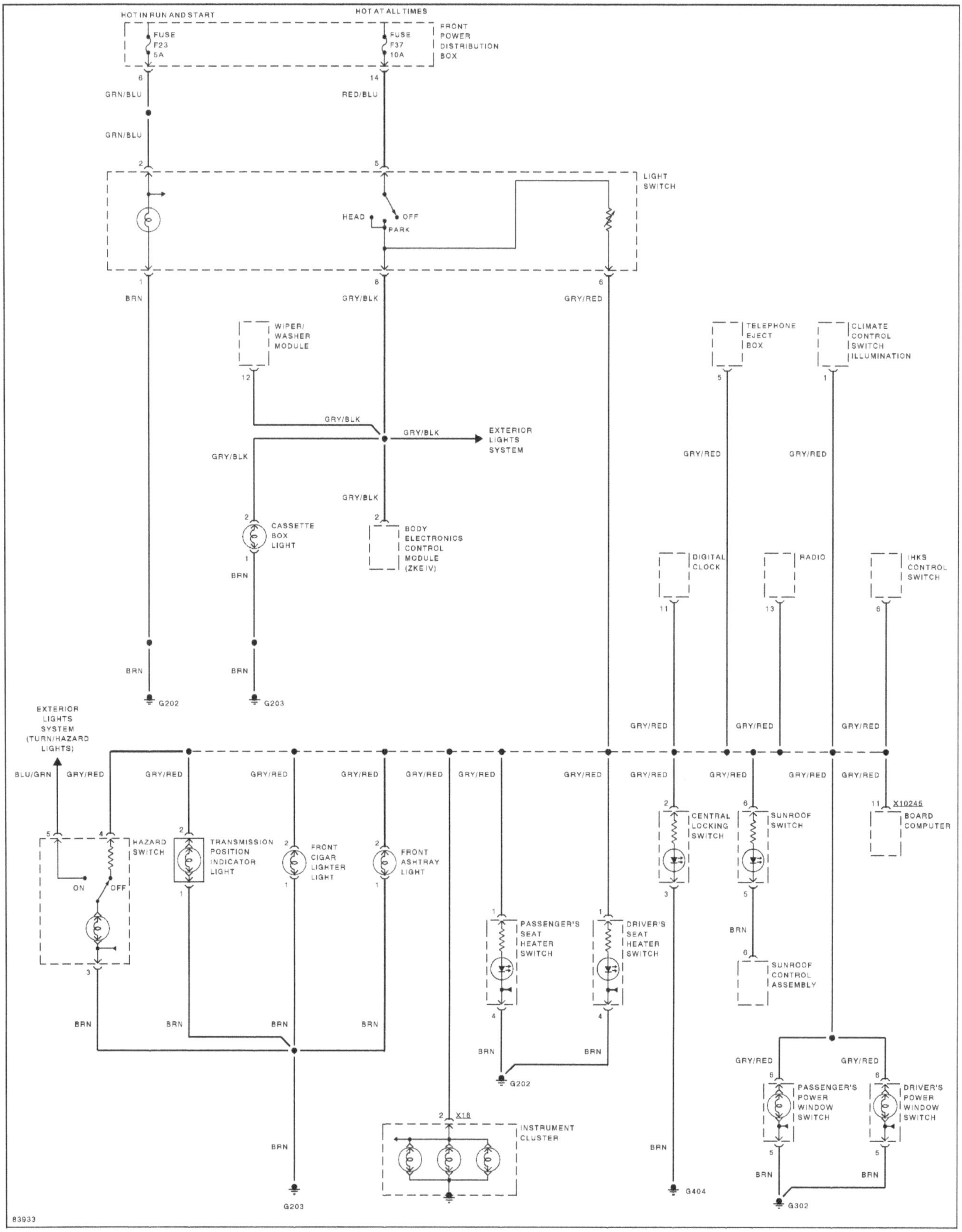

Interior Lights/Illumination

1995-96 318ti (2 of 2)

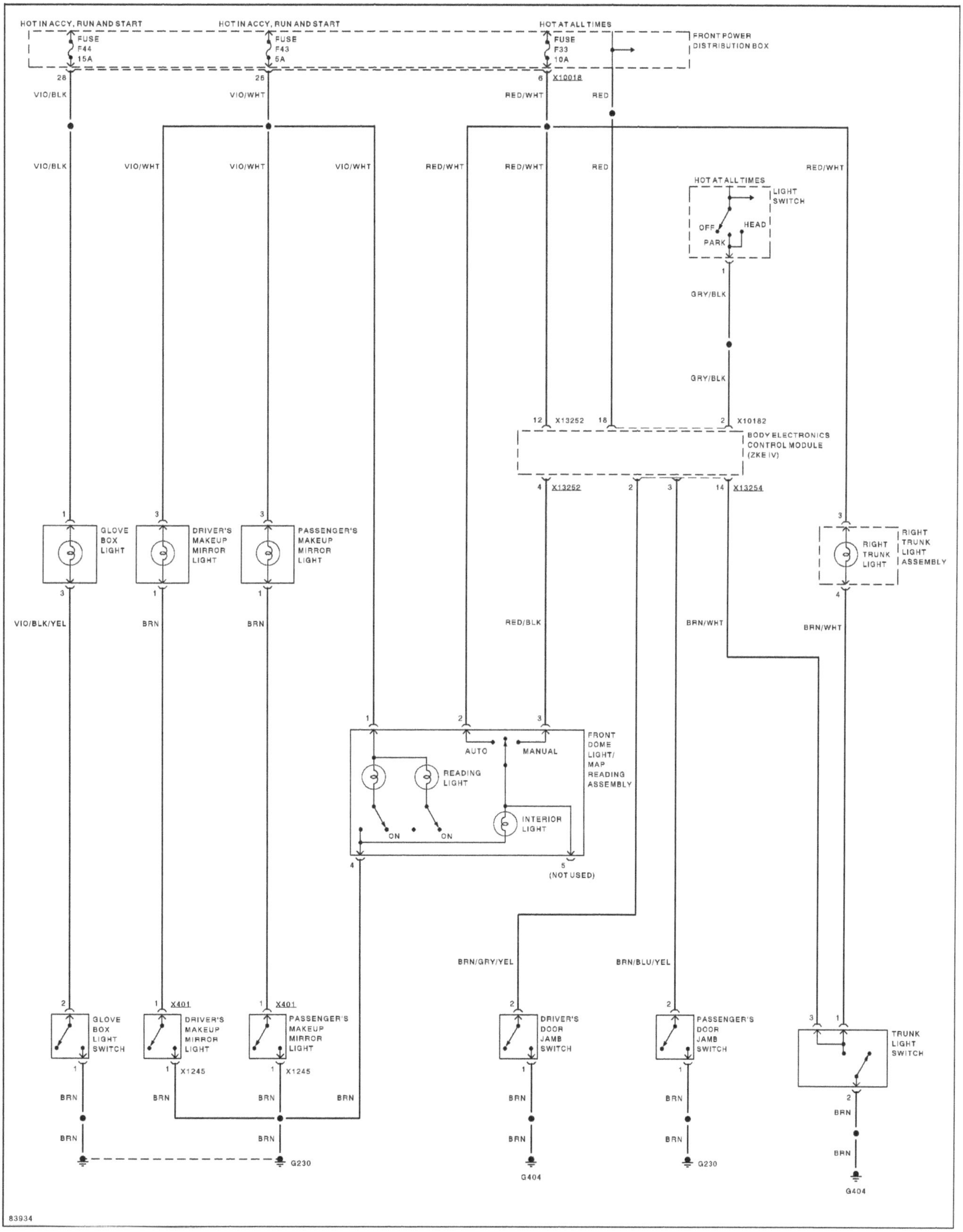

Interior Lights/Illumination

1997 318ti (1 of 2)

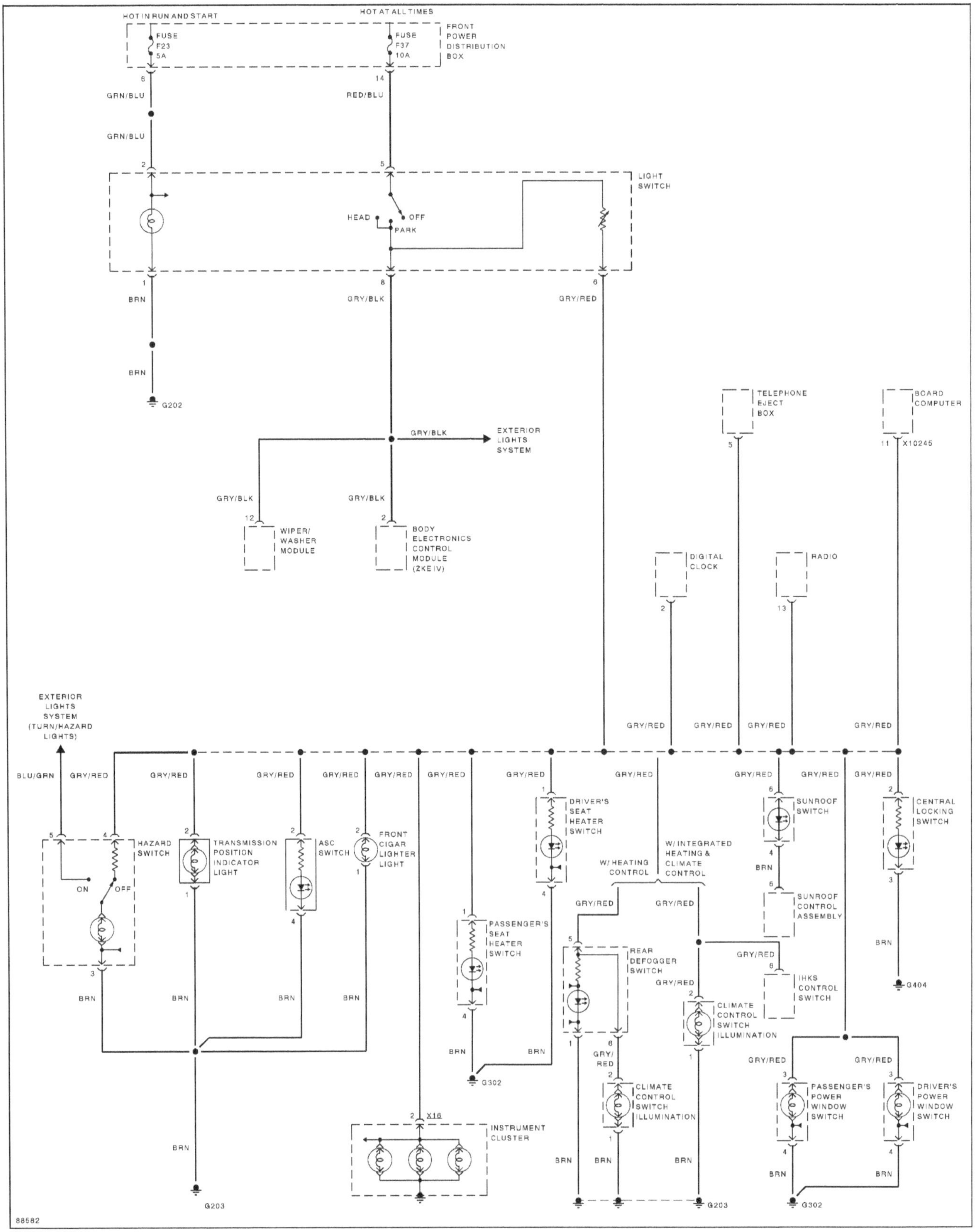

Interior Lights/Illumination

1997 318ti (2 of 2)

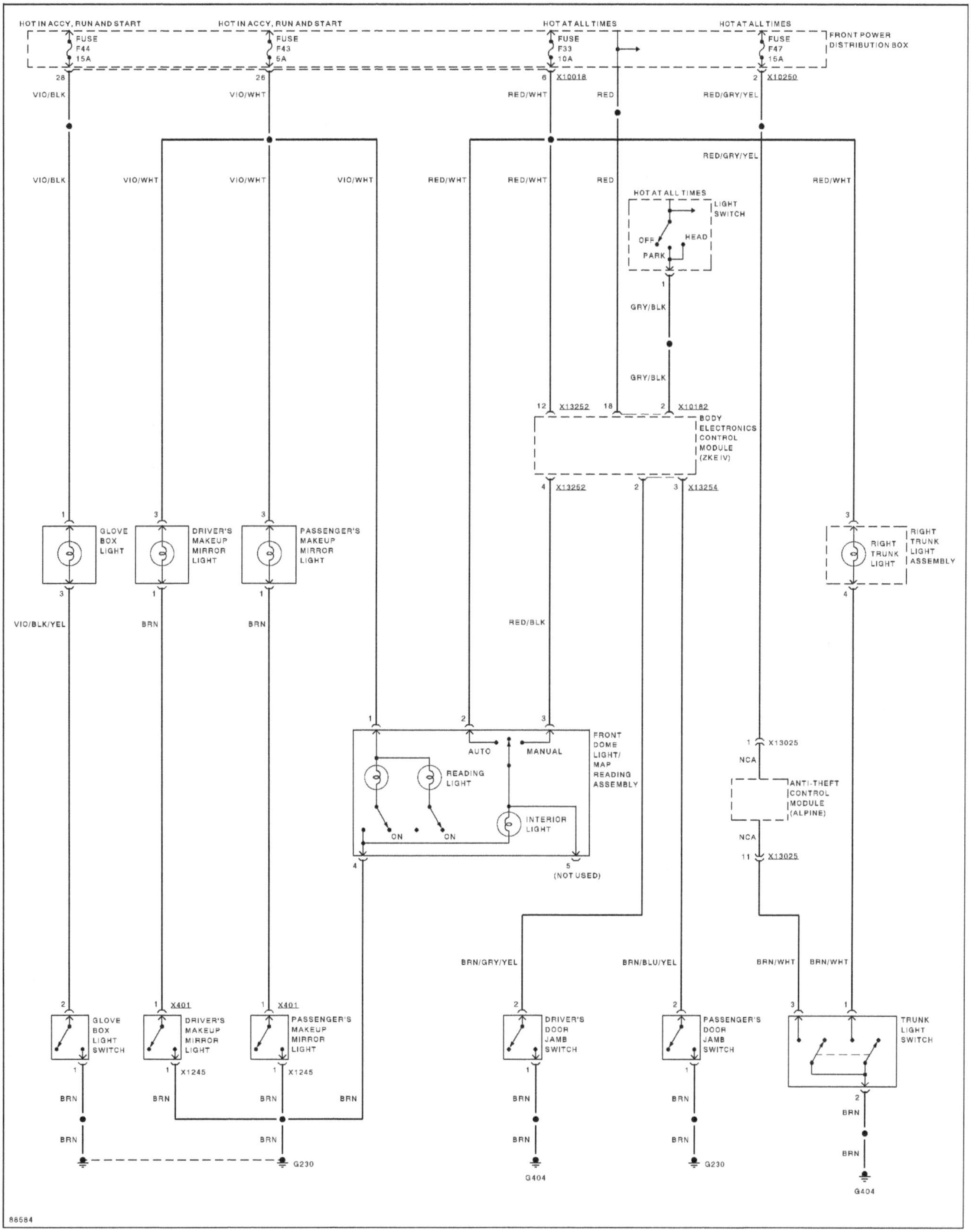

Interior Lights/Illumination

1998 318ti

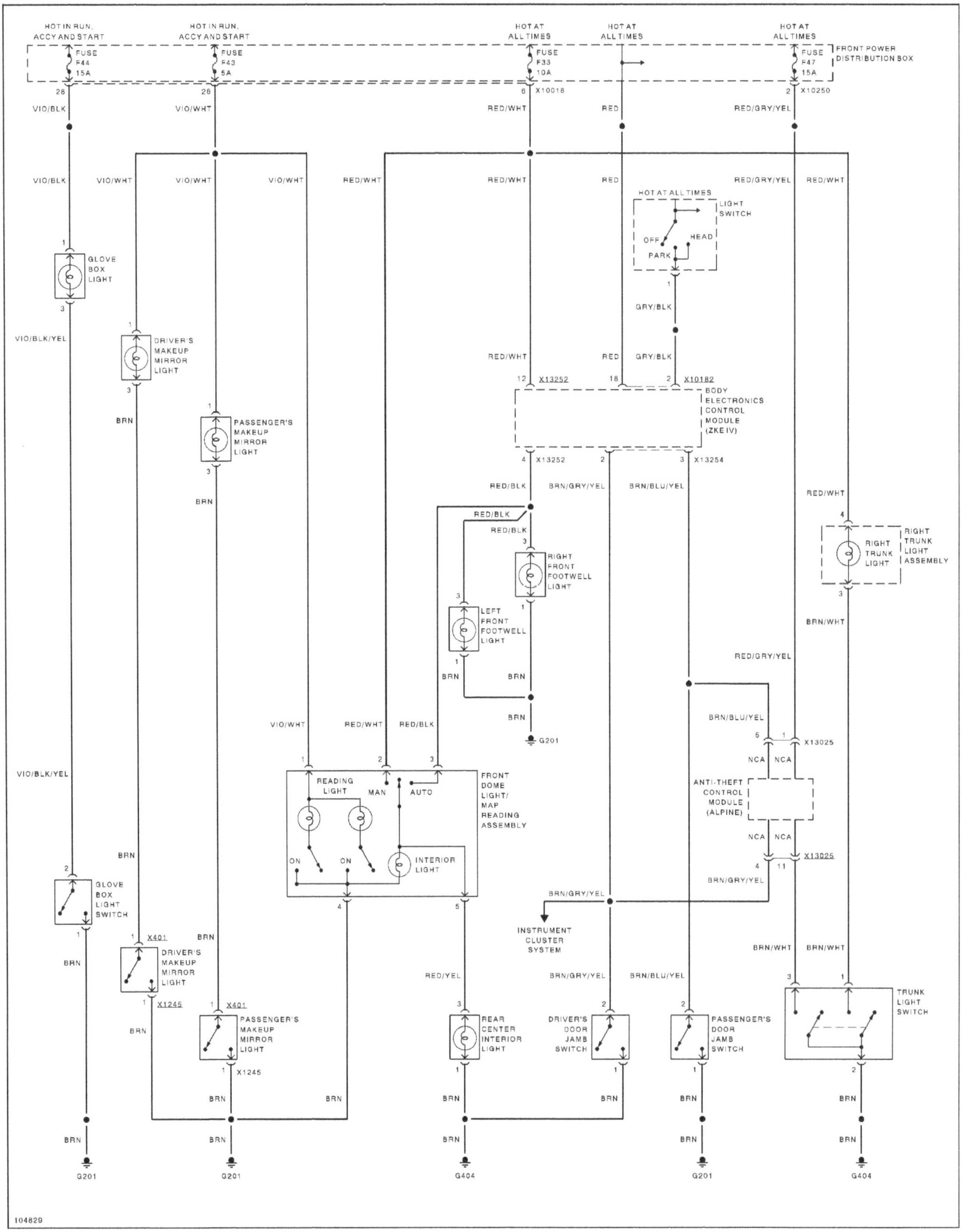

Interior Lights/Instrument Illumination

1998 318ti

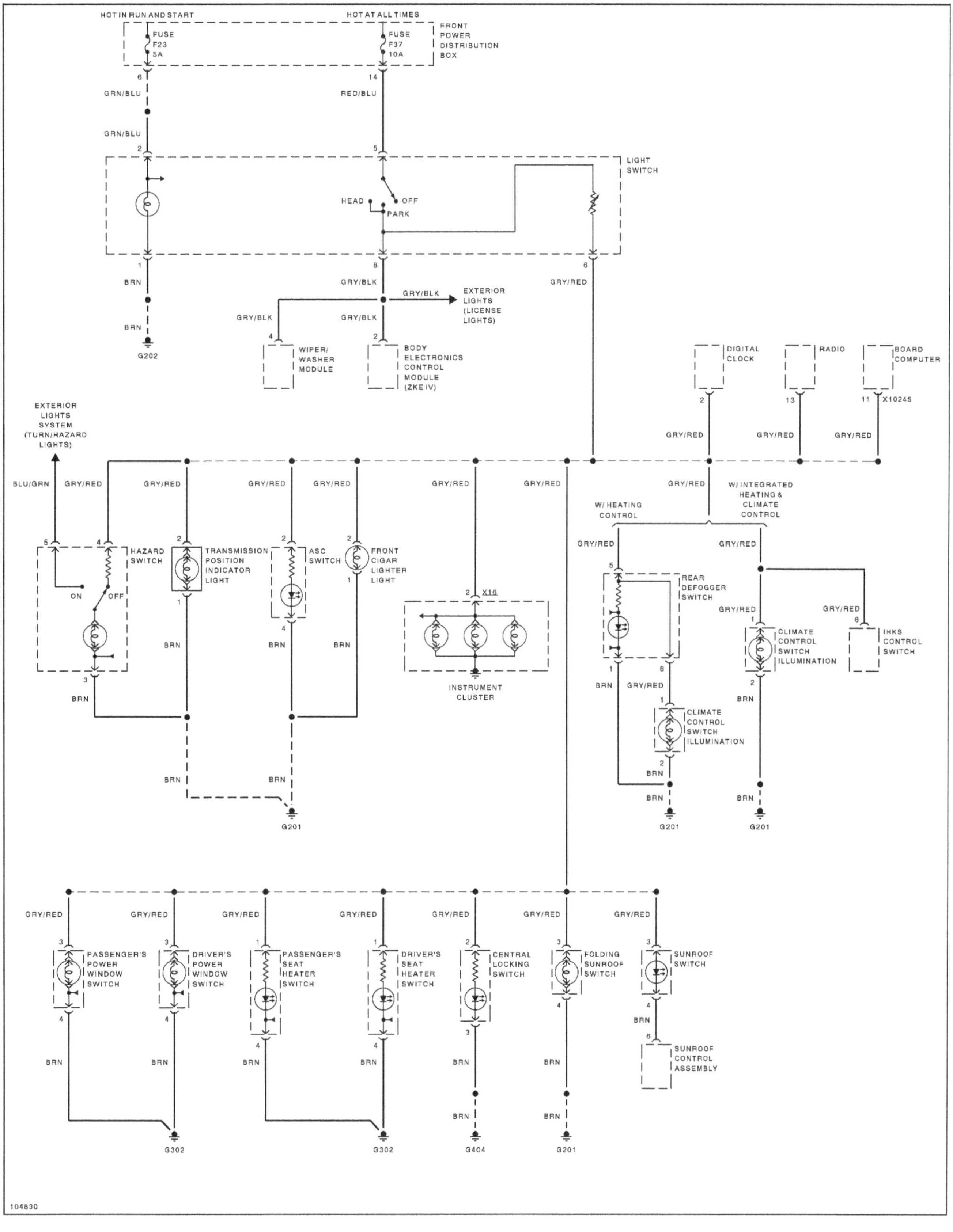

Power Door Locks

1992

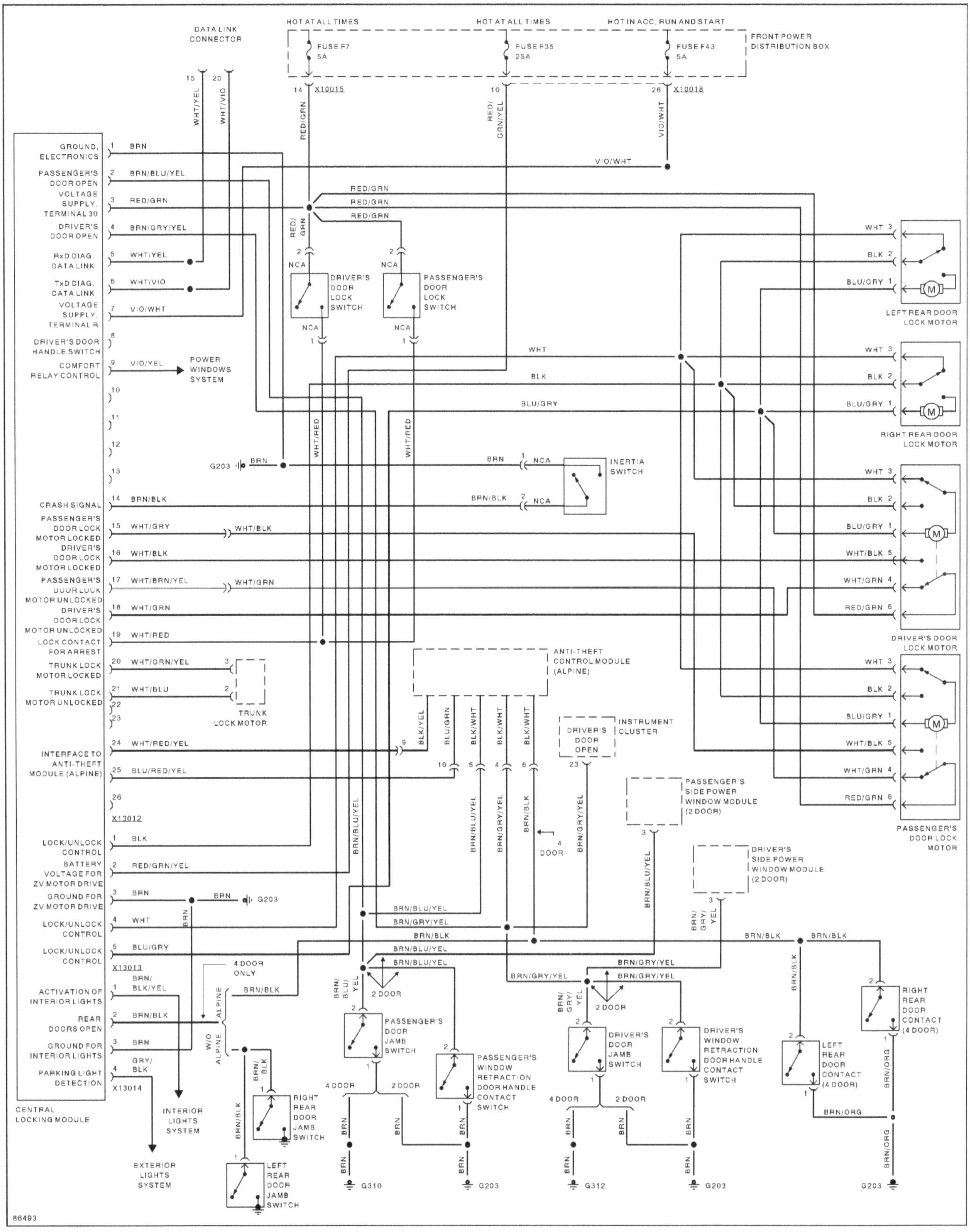

Power Door Locks

1993

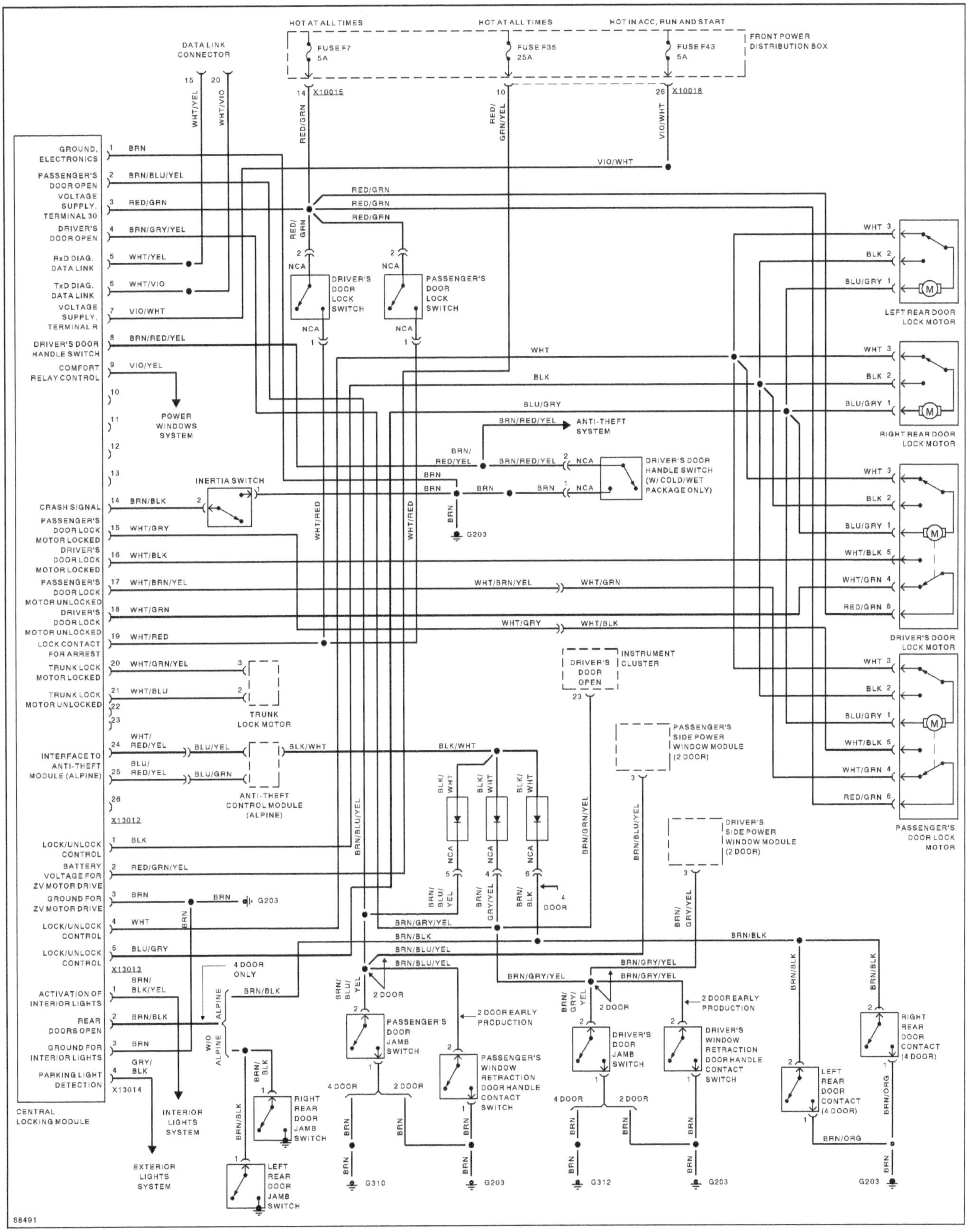

Power Door Locks

1994-97 Except 318ti

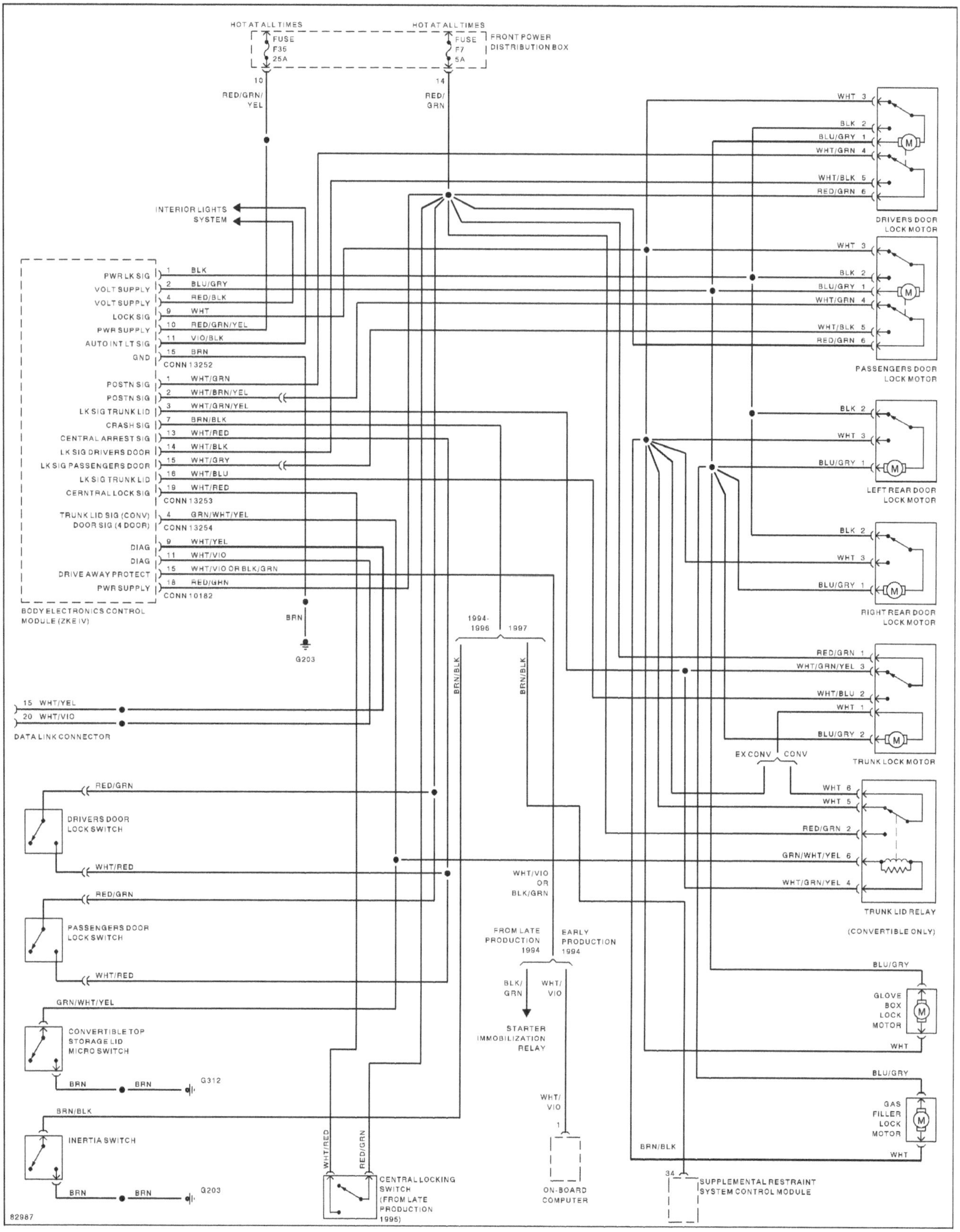

Power Door Locks
1998 Except 318ti

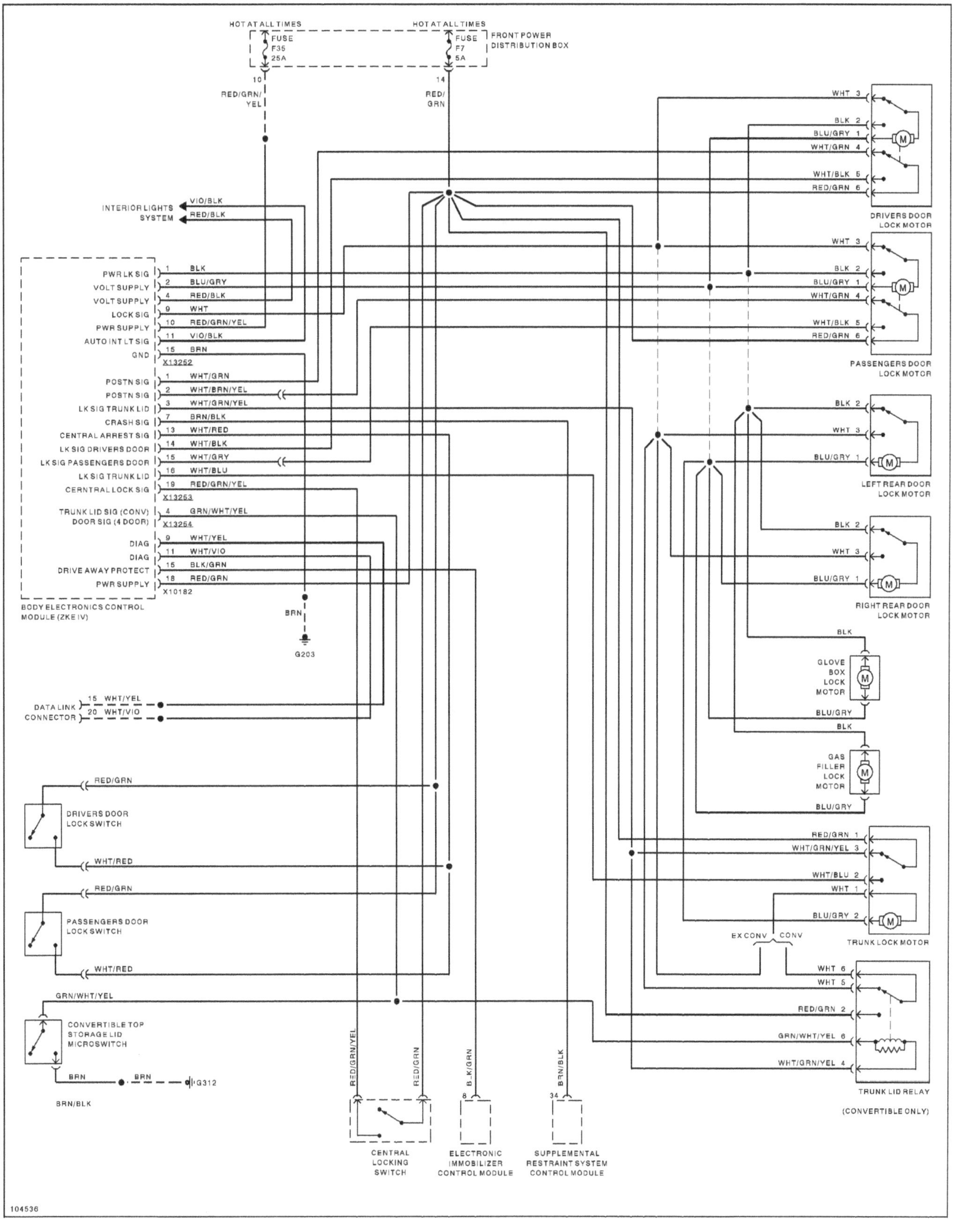

Power Door Locks

1995-1998 318ti

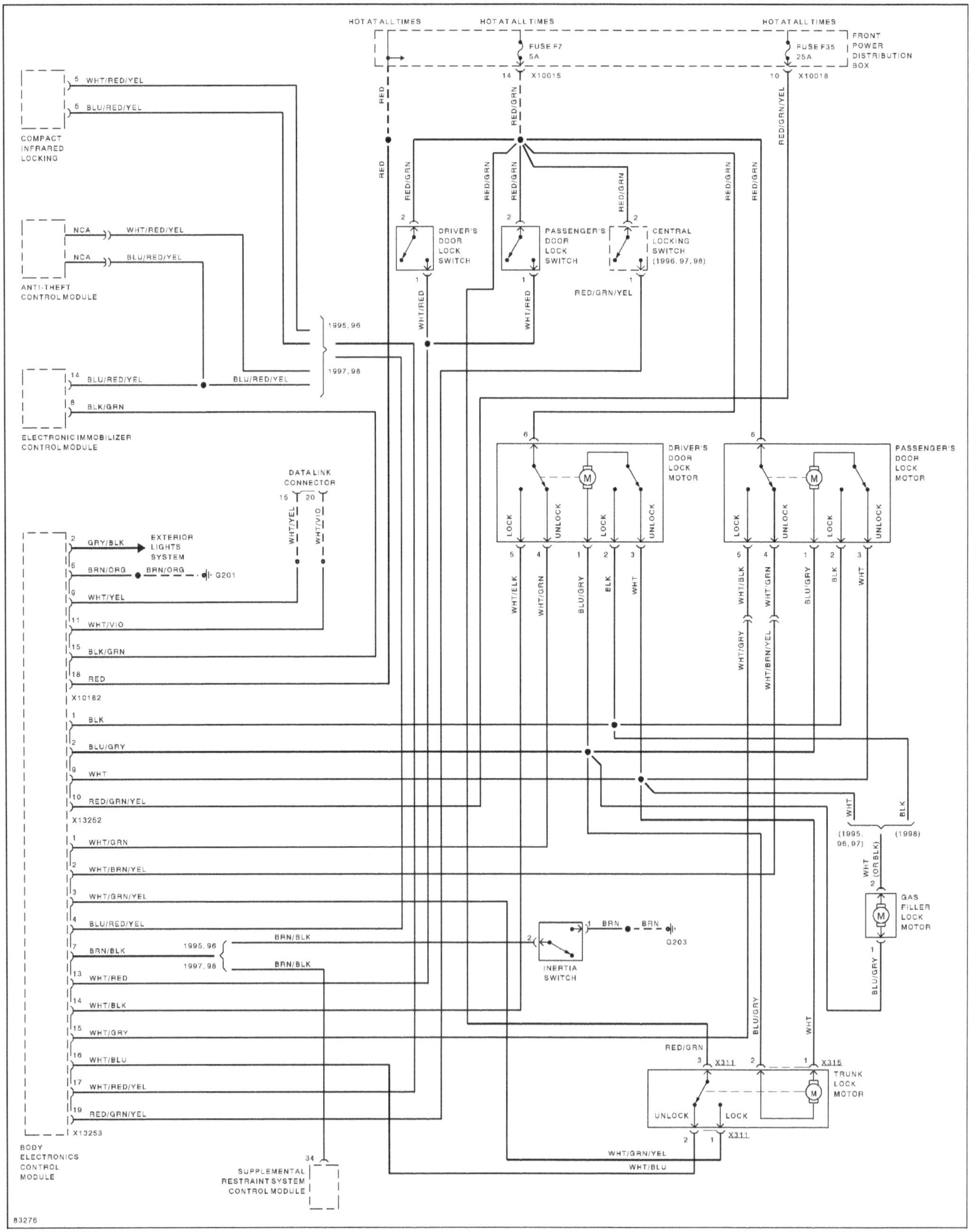

Power Windows

1992 318is & 325is

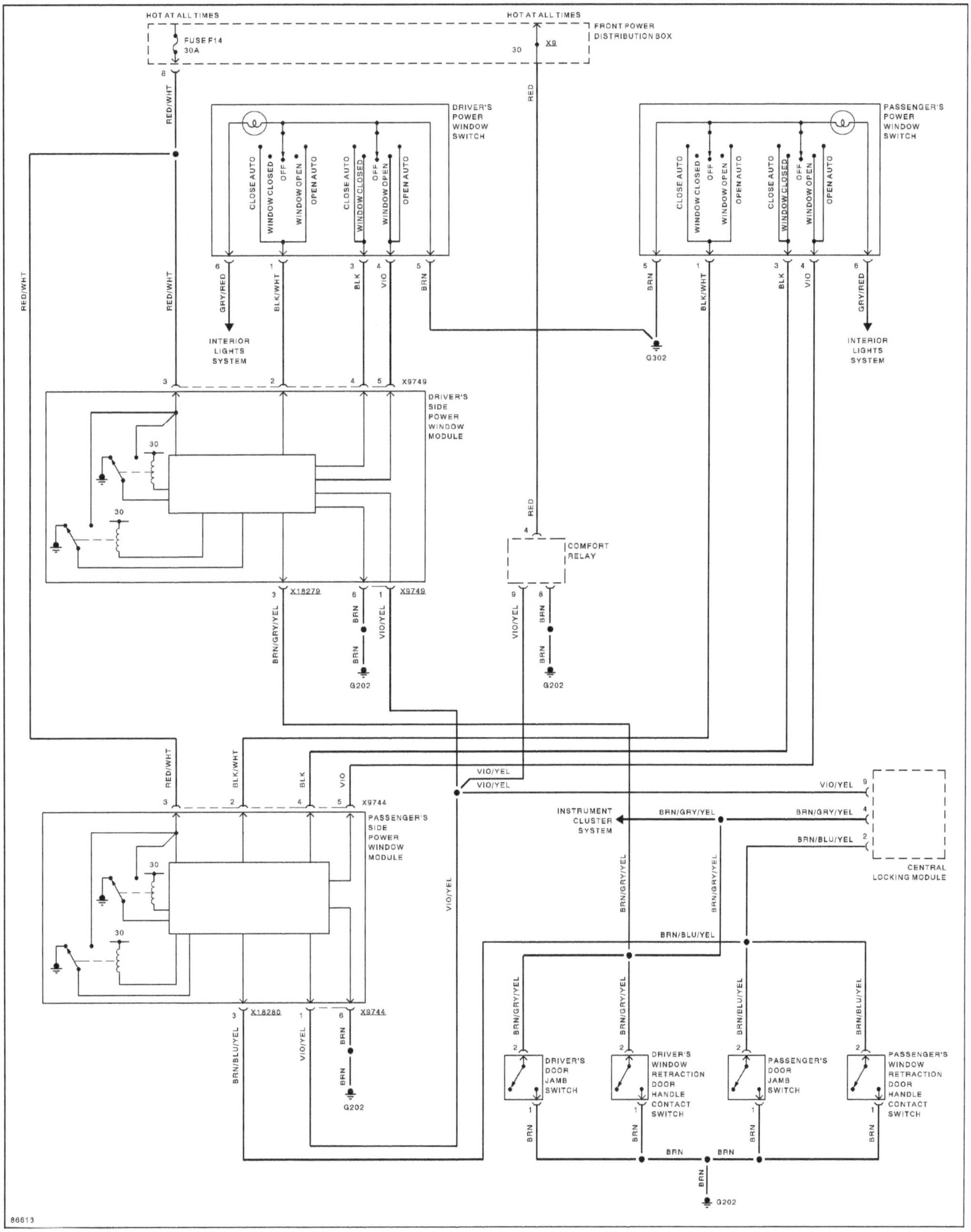

Power Windows

1993 318is & 325is

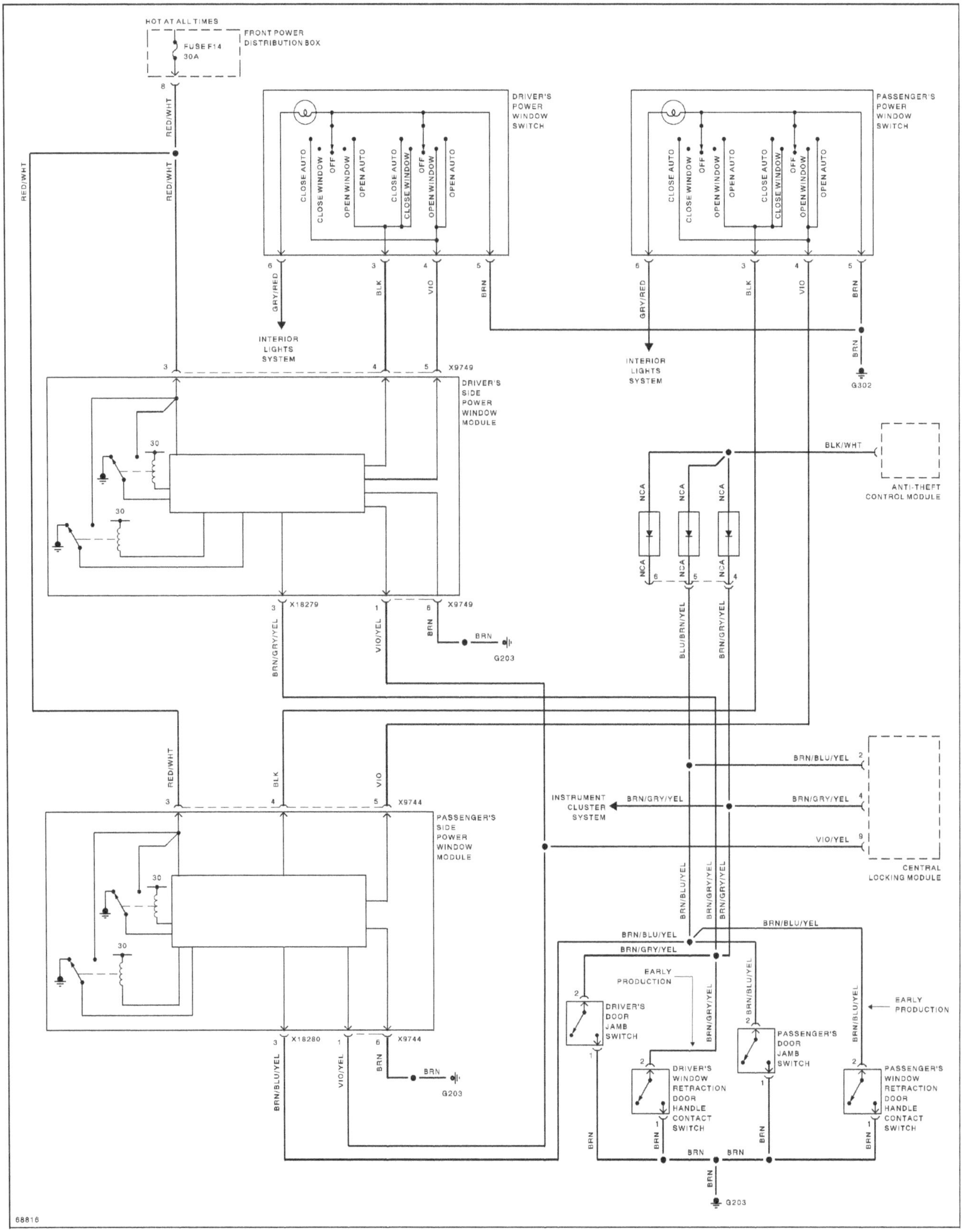

Power Windows
1993 318is & 325is

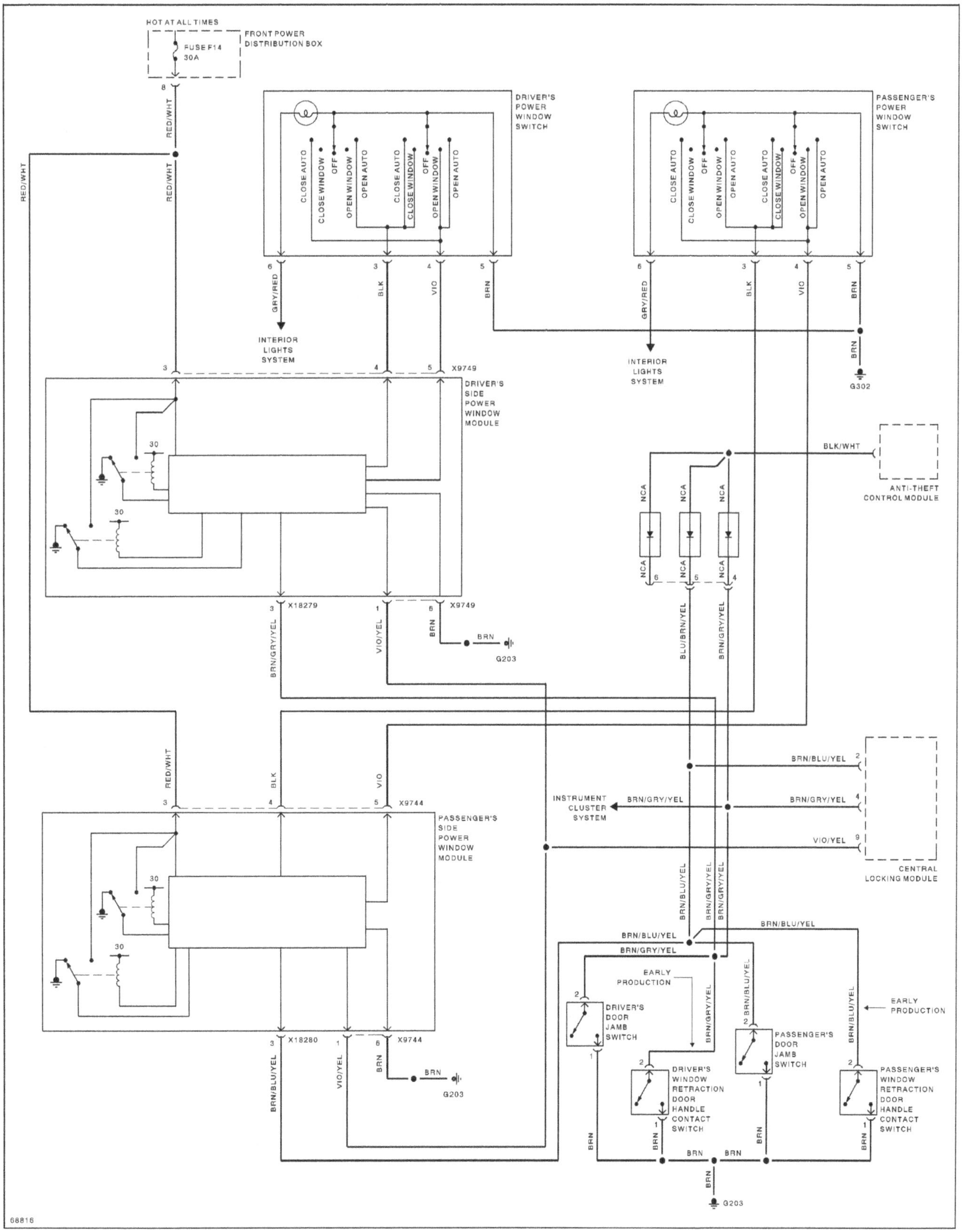

Power Windows

1994-97 Except 318ti (1 of 2)

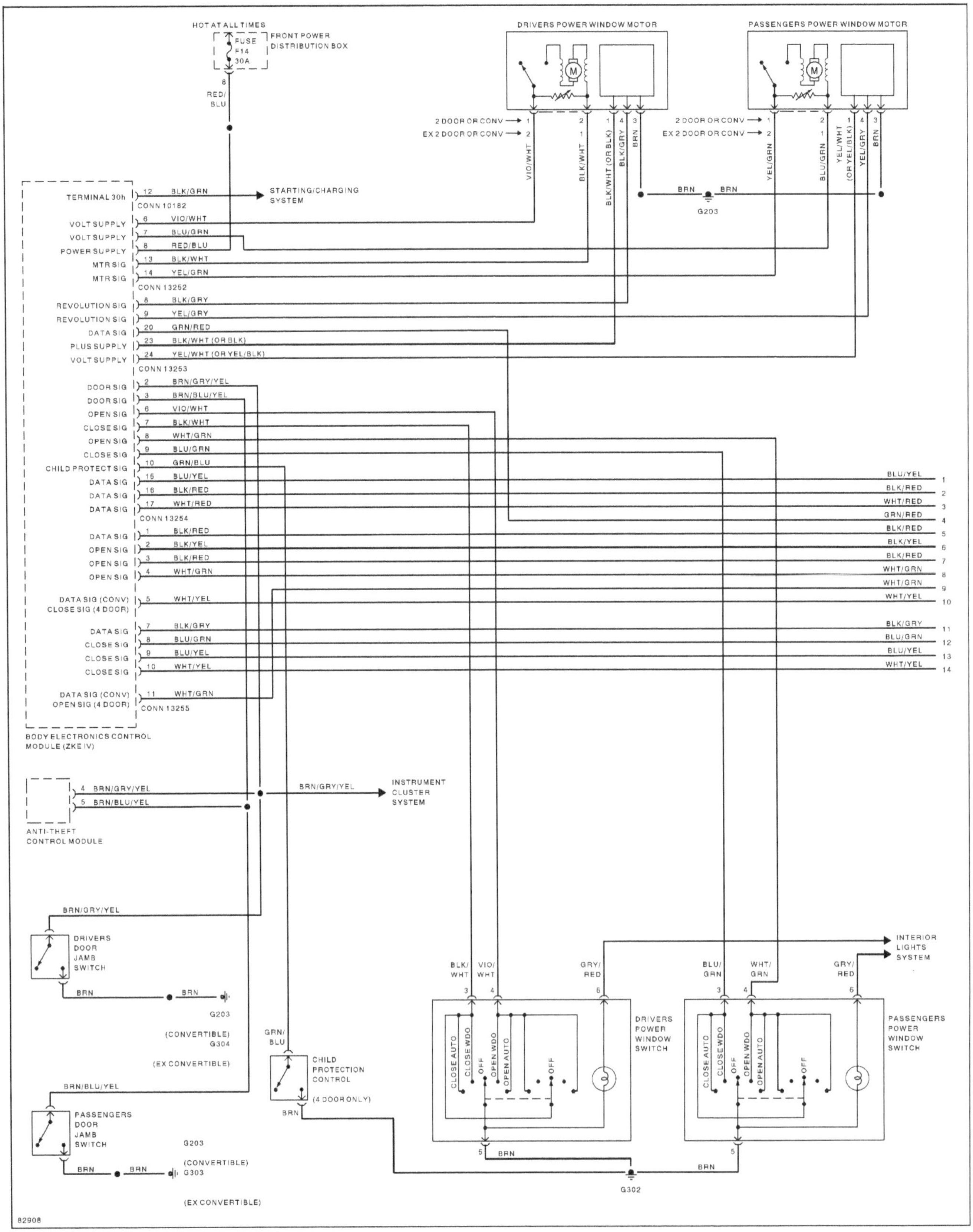

Power Windows

1994-97 Except 318ti (2 of 2)

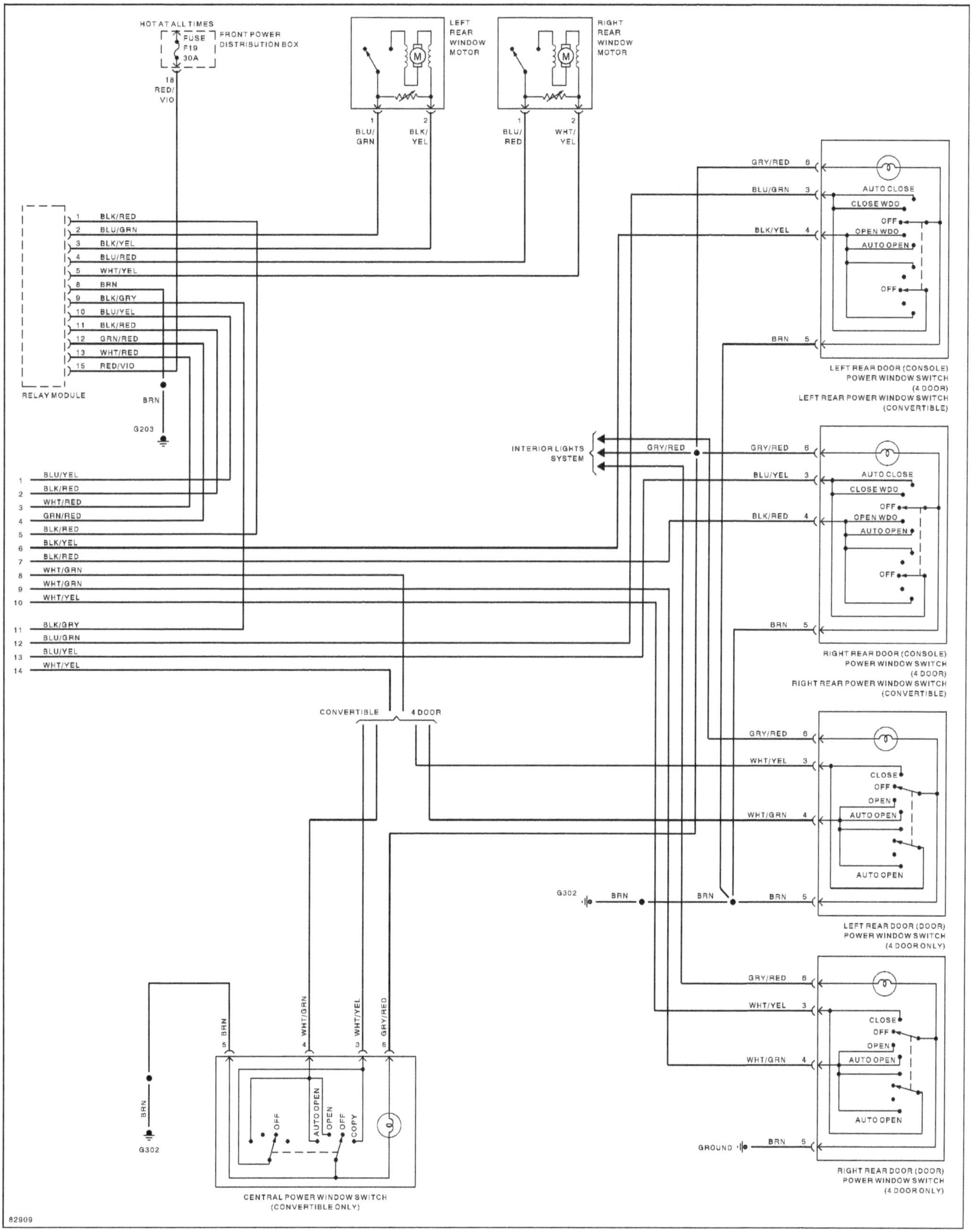

Power Windows

1998 Except 318i, 318ti (1 of 2)

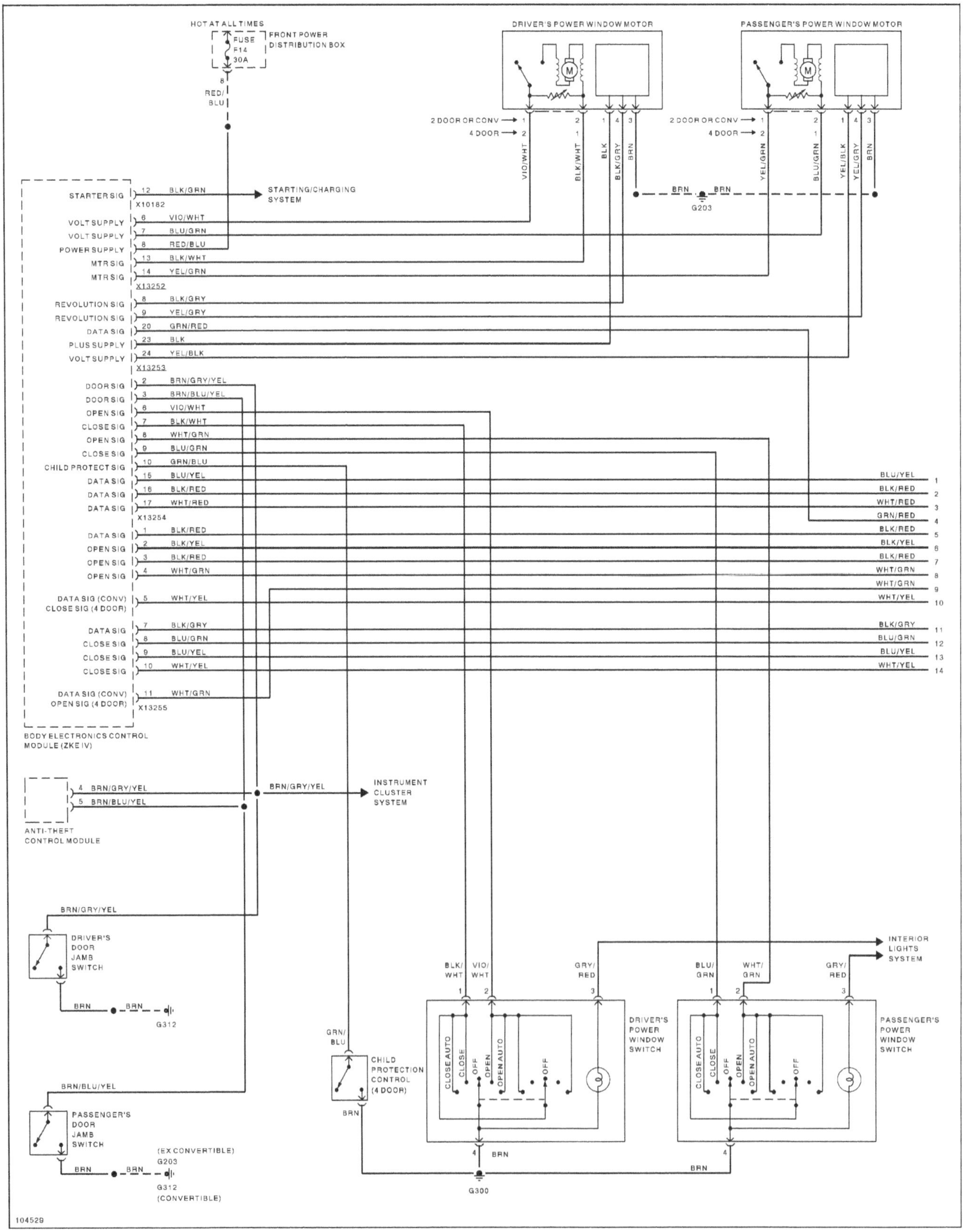

Power Windows

1998 Except 318i, 318ti (2 of 2)

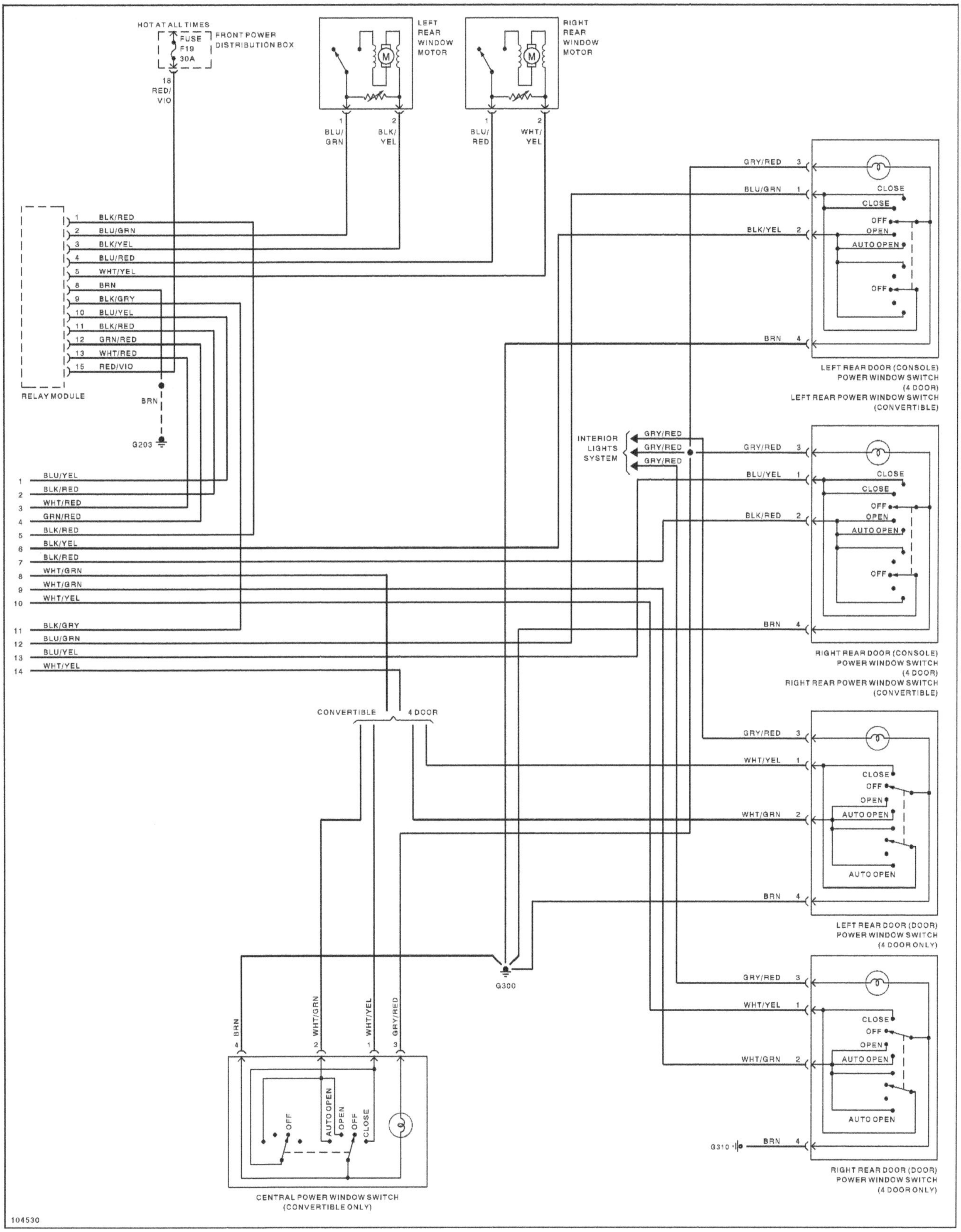

Power Windows

1995-96 318ti

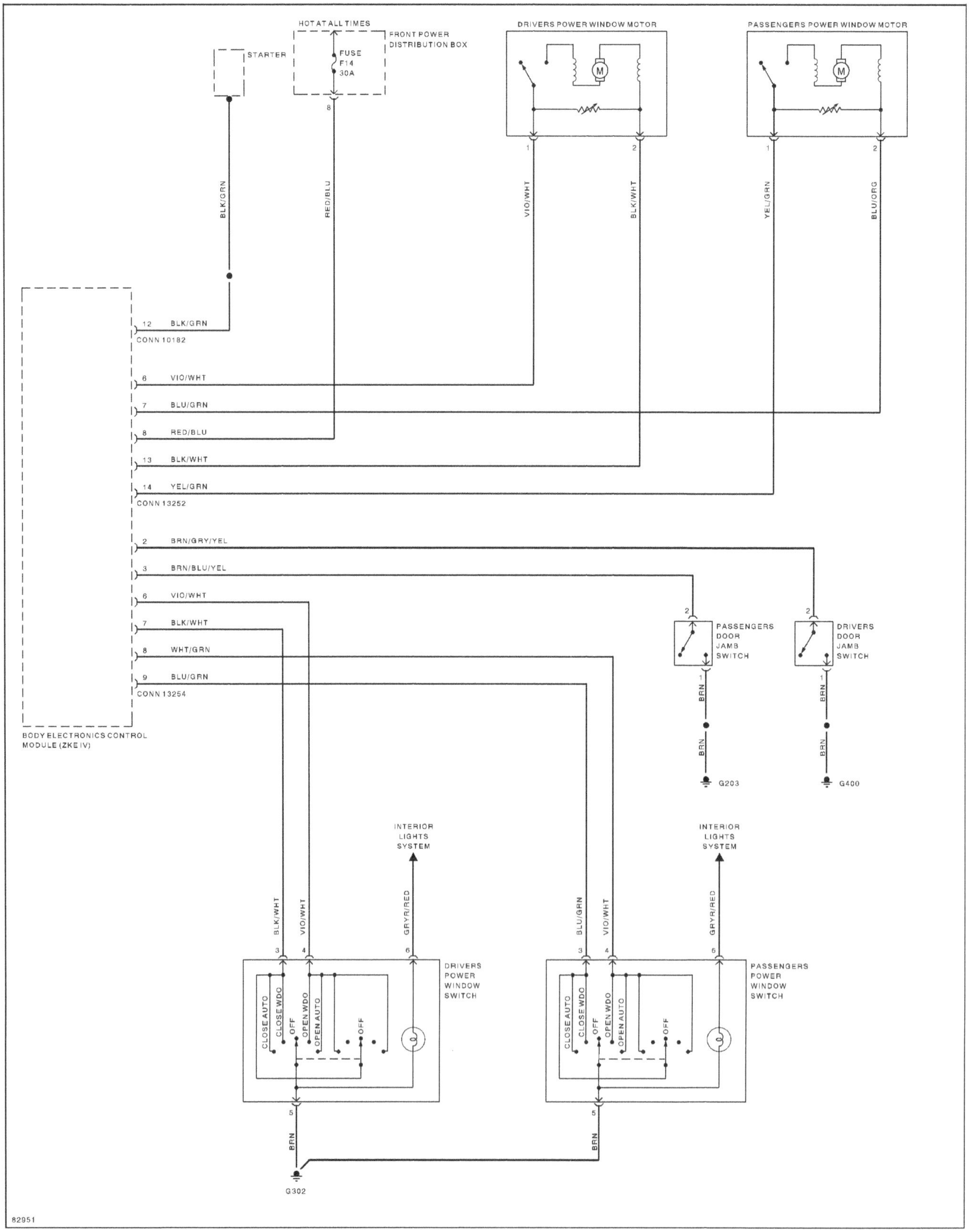

Power Windows

1997 318ti

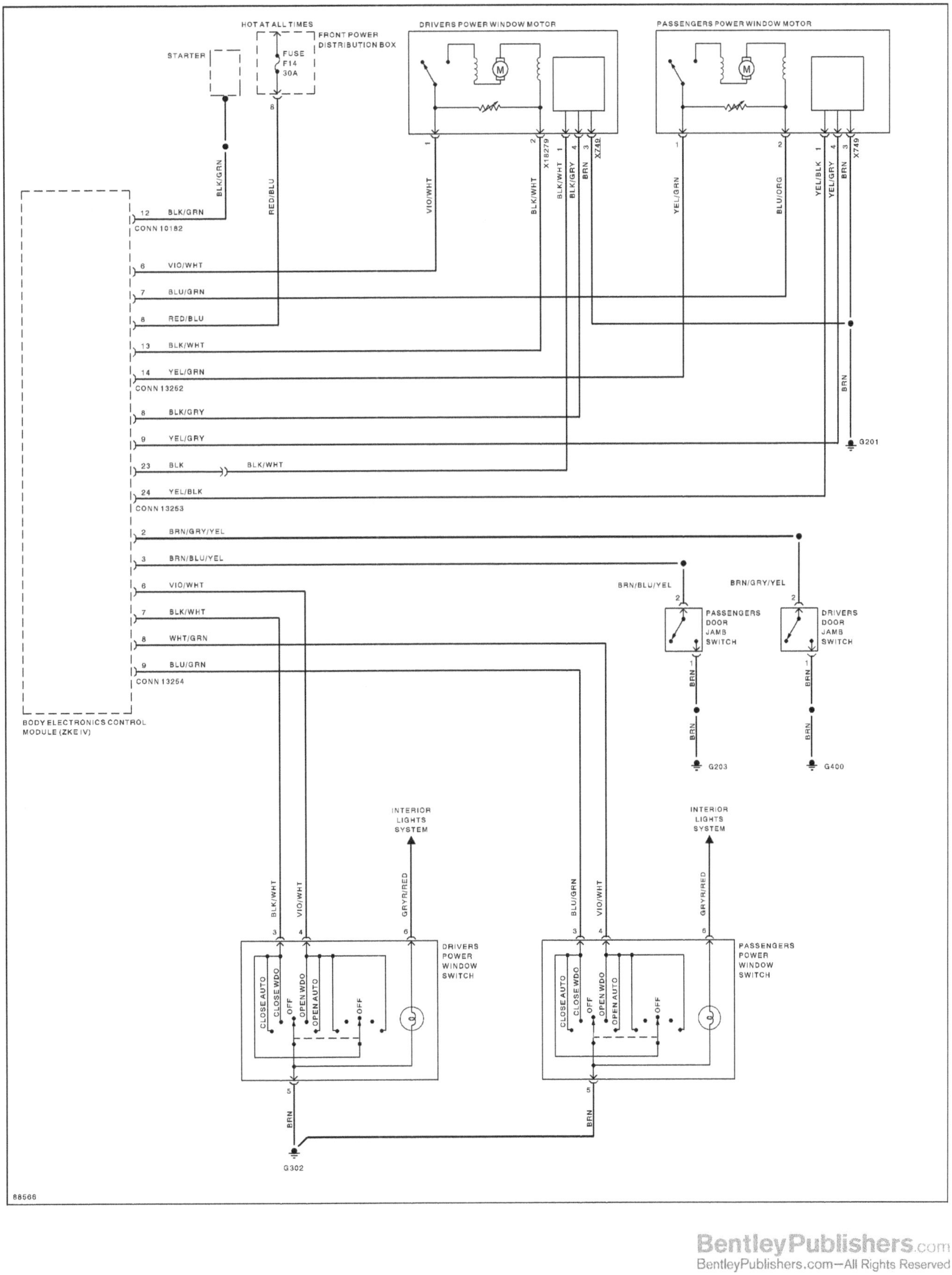

Power Windows

1998 318ti

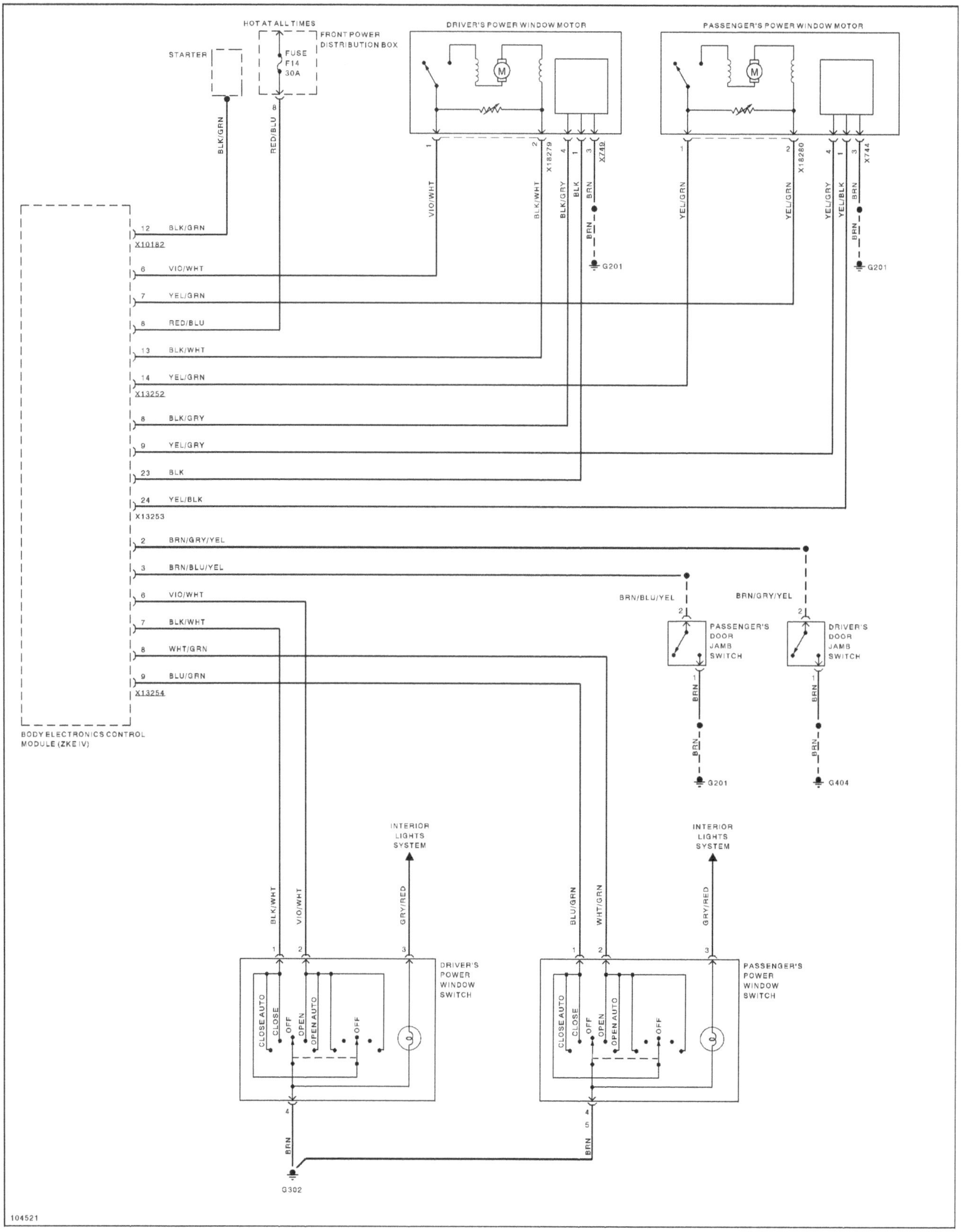

Power Mirrors

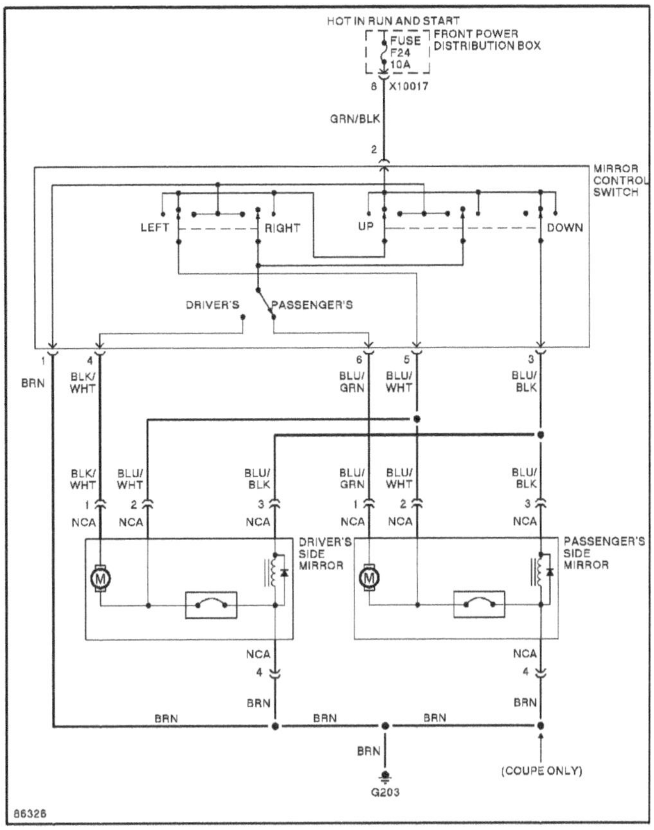

1992

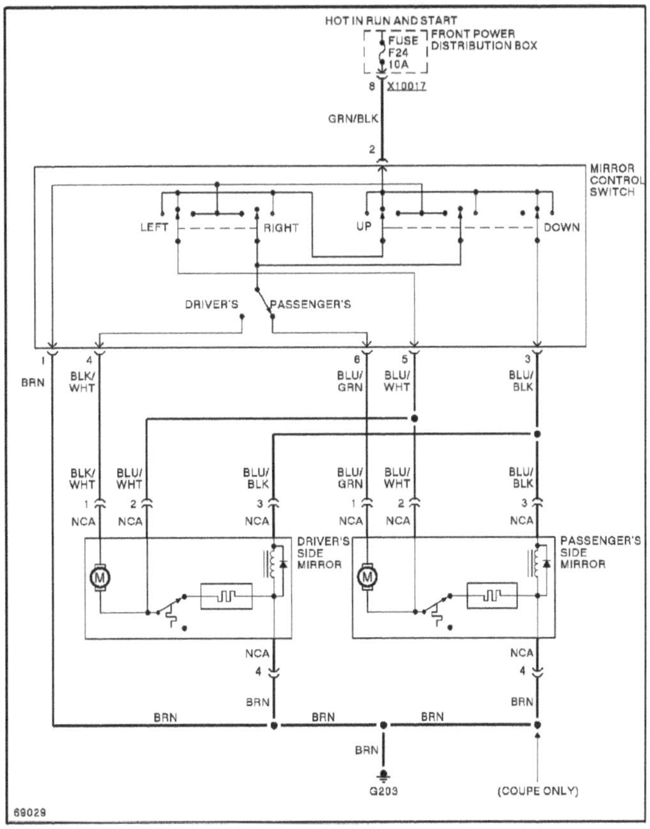

1993

Power Mirrors

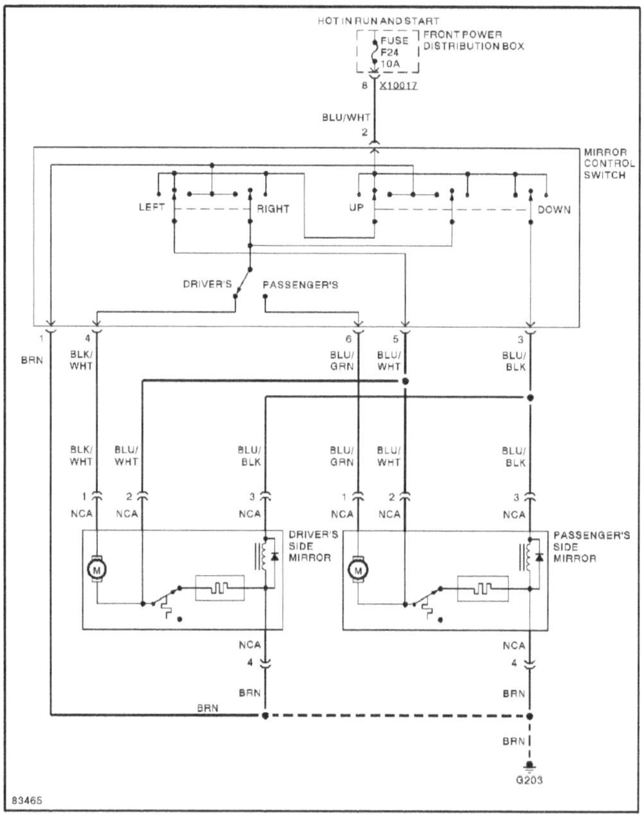

1994-98 Except 318ti

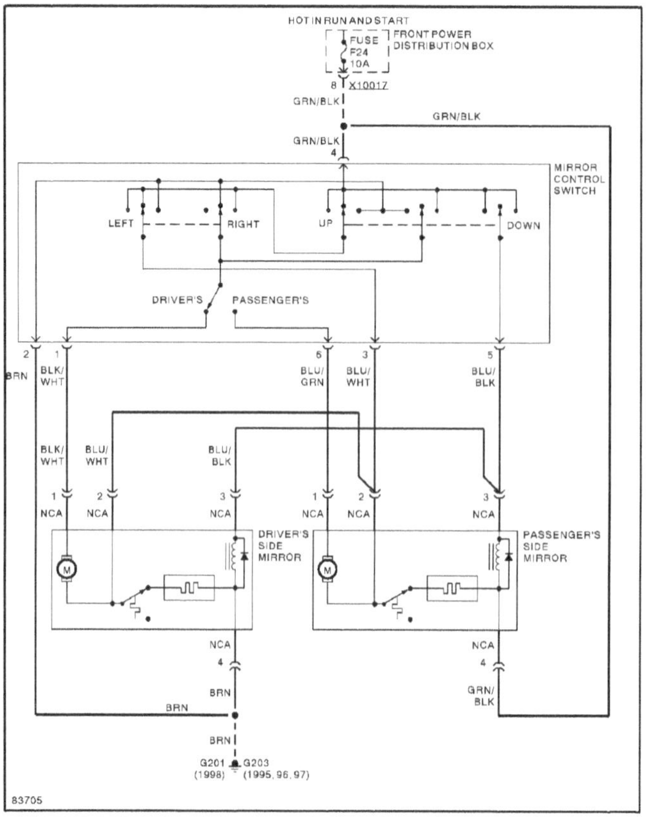

1995-97 318ti

Power Seats

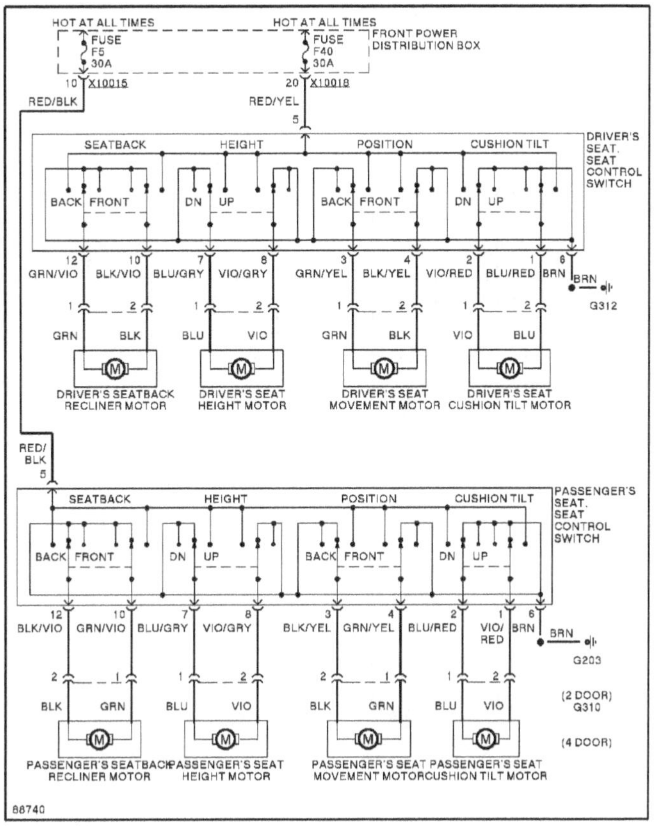

1993

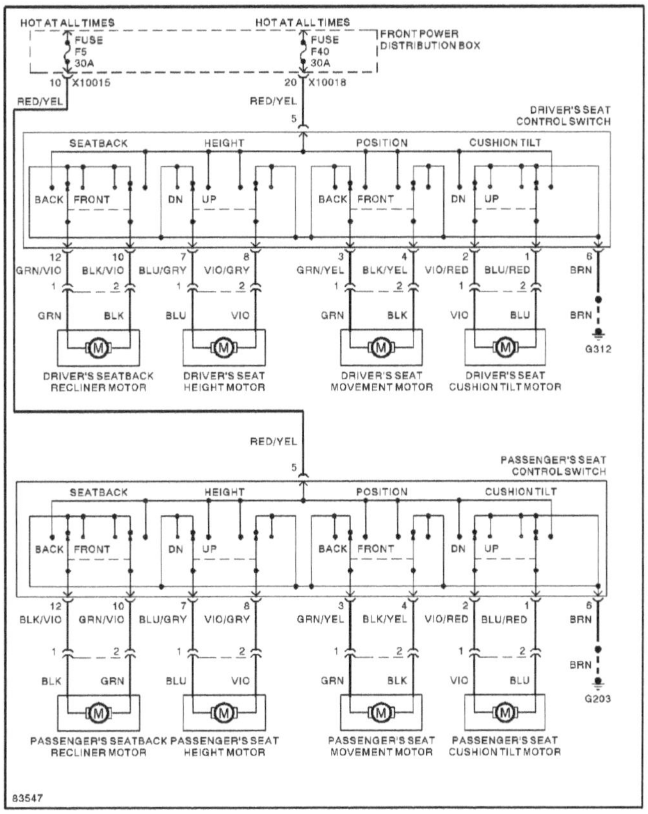

1994-98 Except 318ti

Power Seats

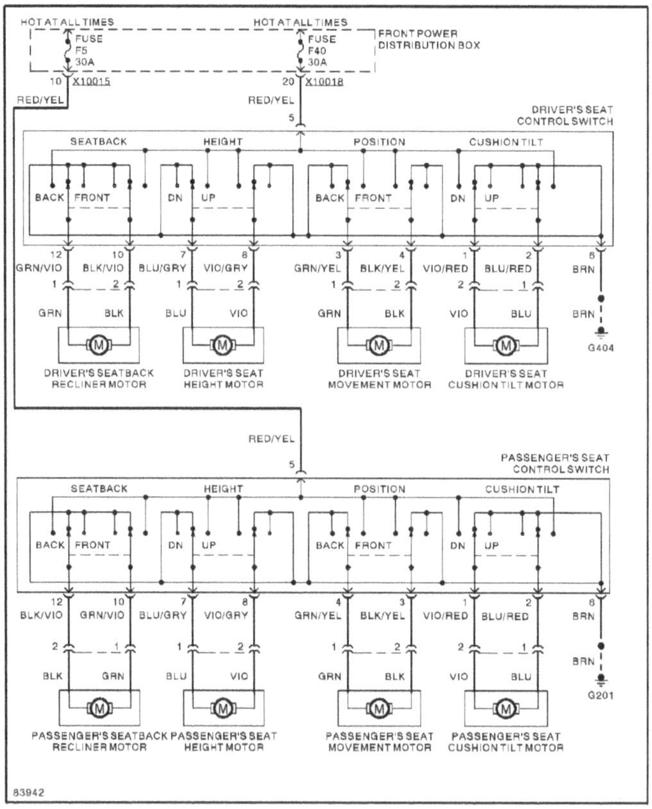

1995-98
318ti

Heated Seats

1992-98 Except 318ti

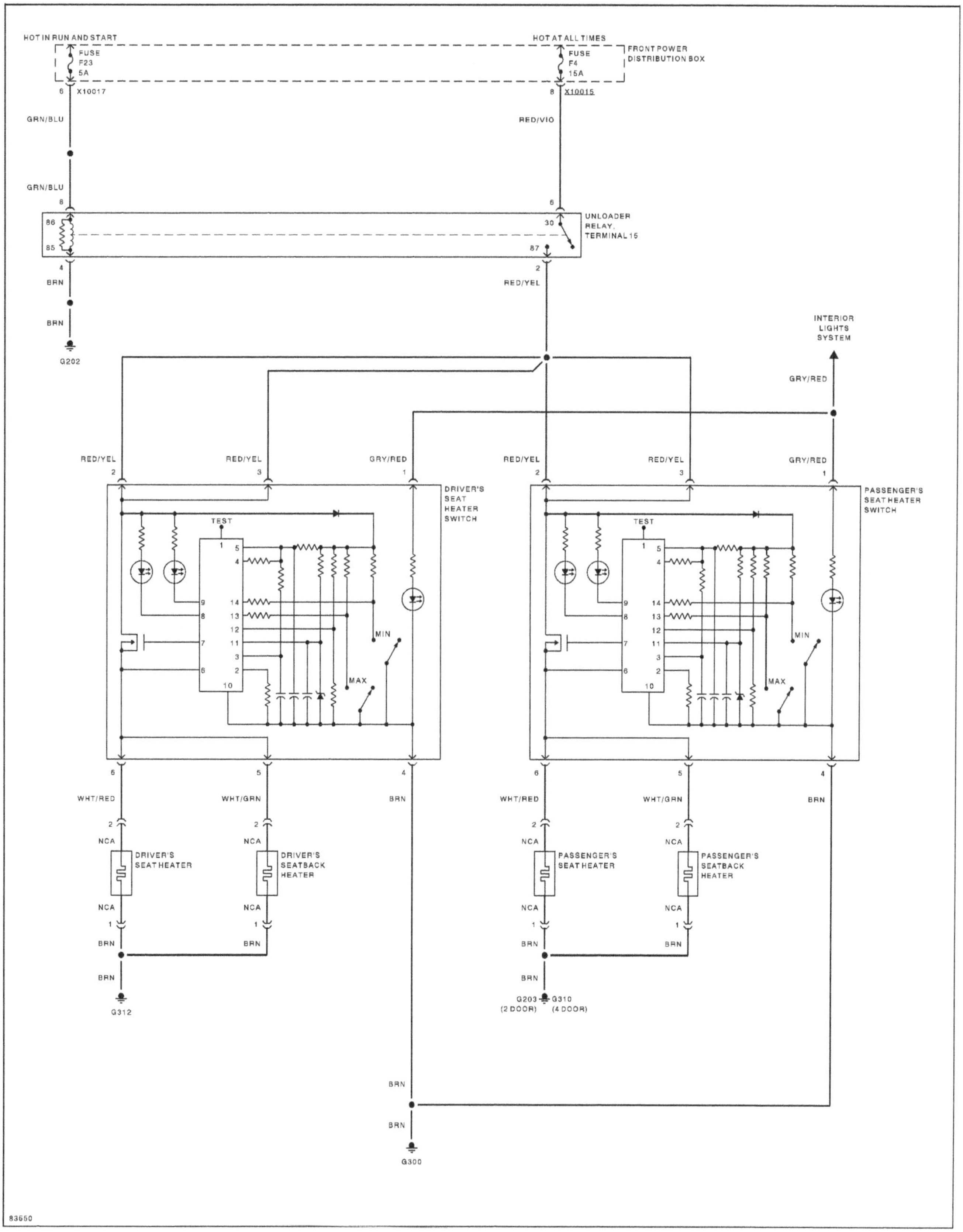

Heated Seats

1995-97 318ti

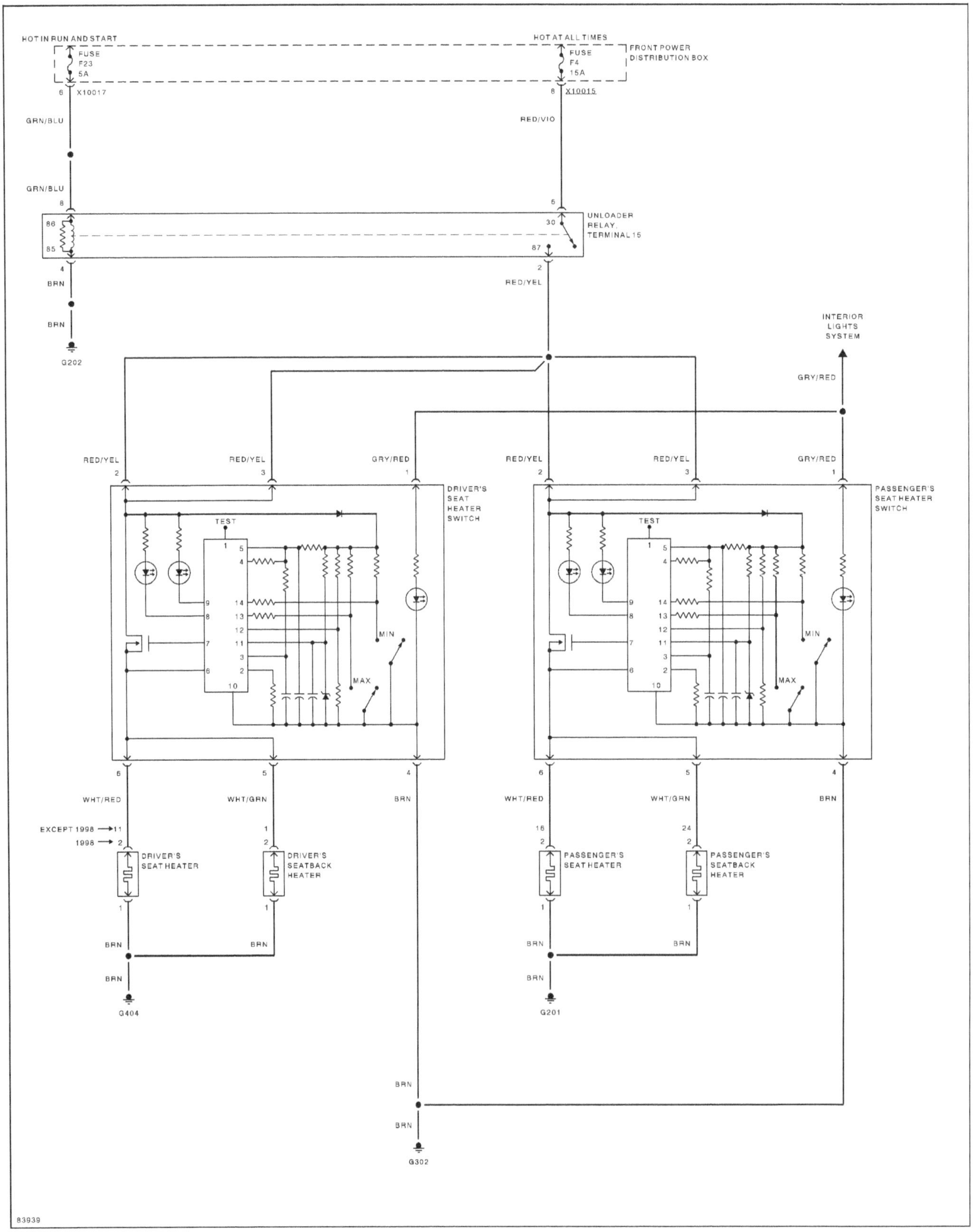

Lumbar Control

1998

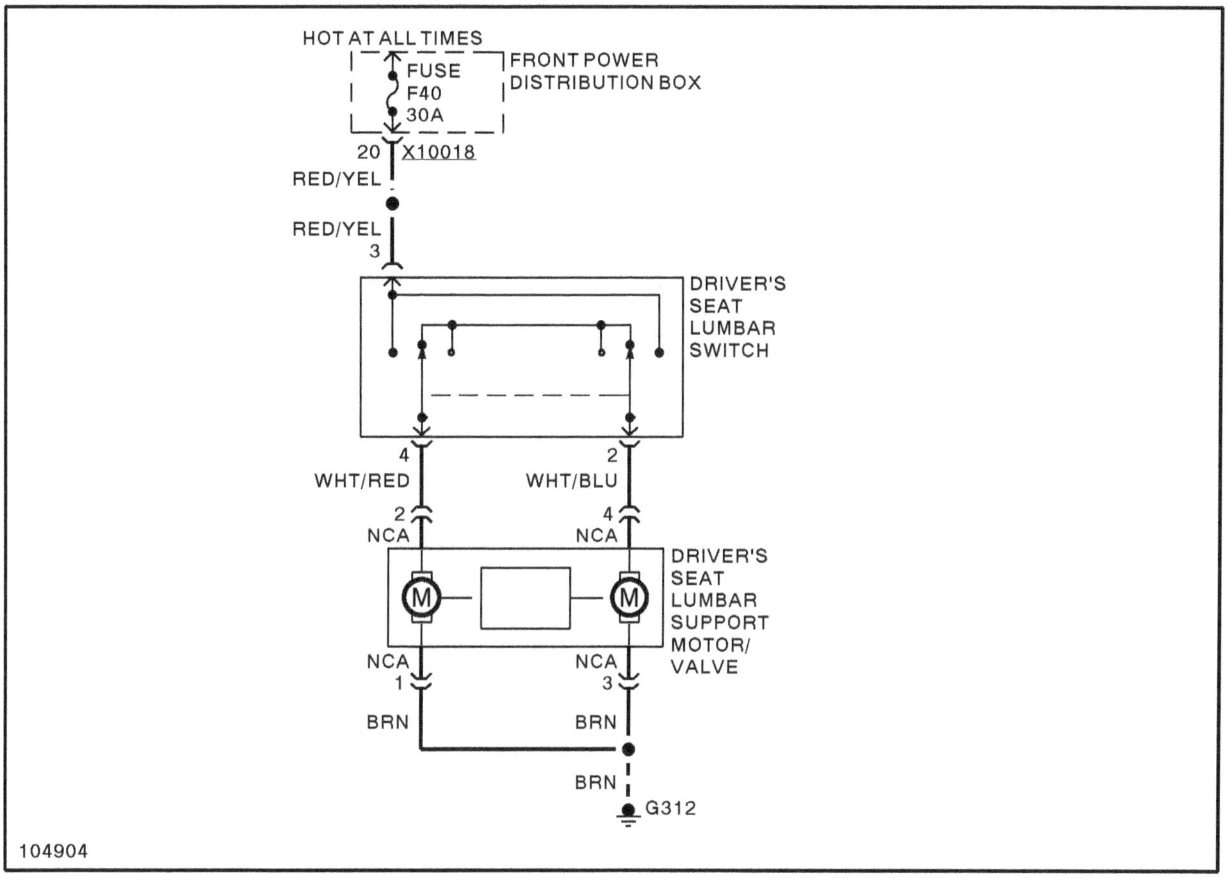

Power Sunroof

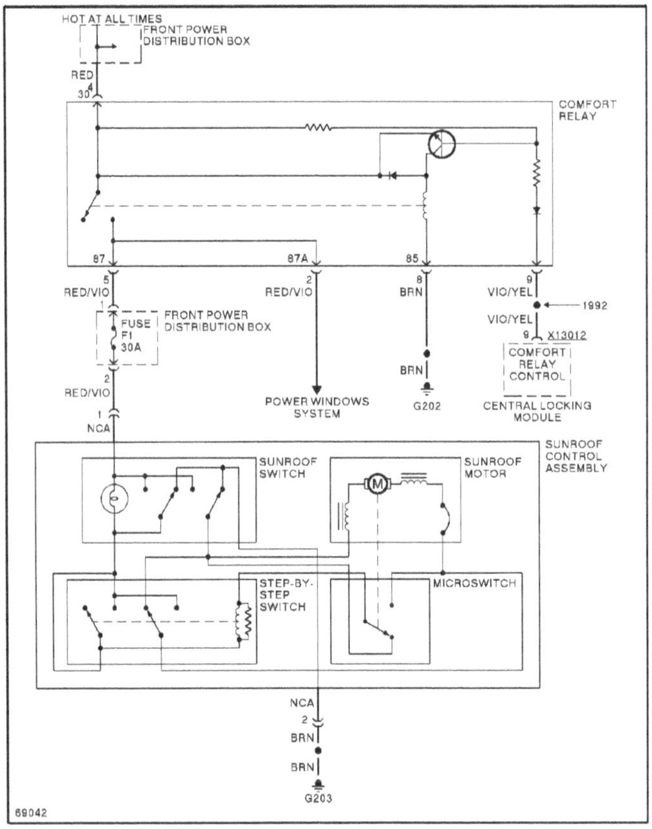

1992-93

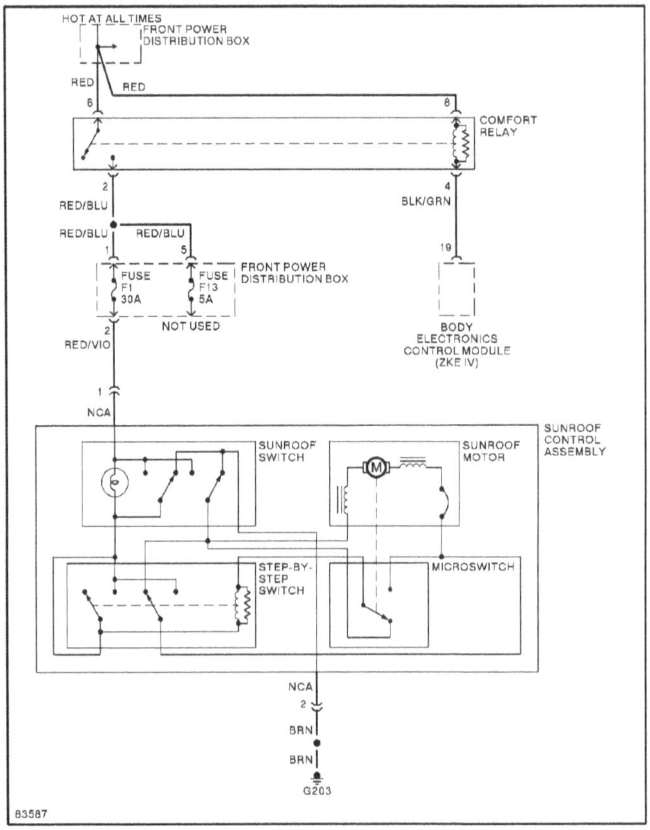

1994

Power Sunroof

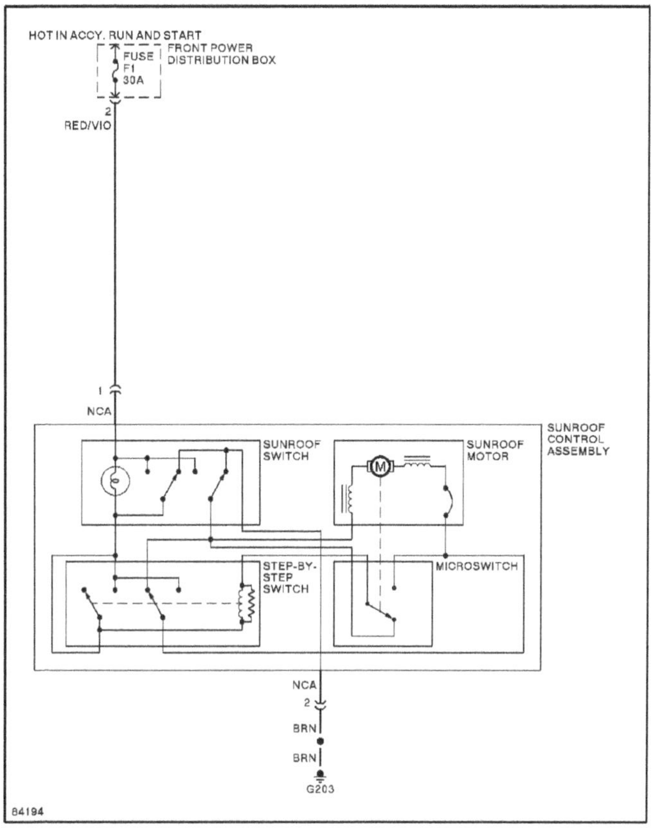

1995
Except 318ti

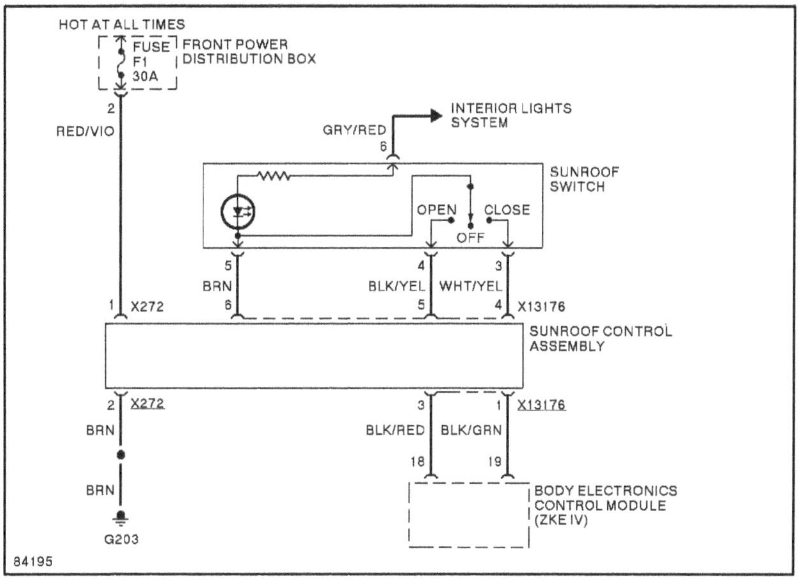

1996-97
Except 318ti

Power Sunroof

1998 Except 318ti

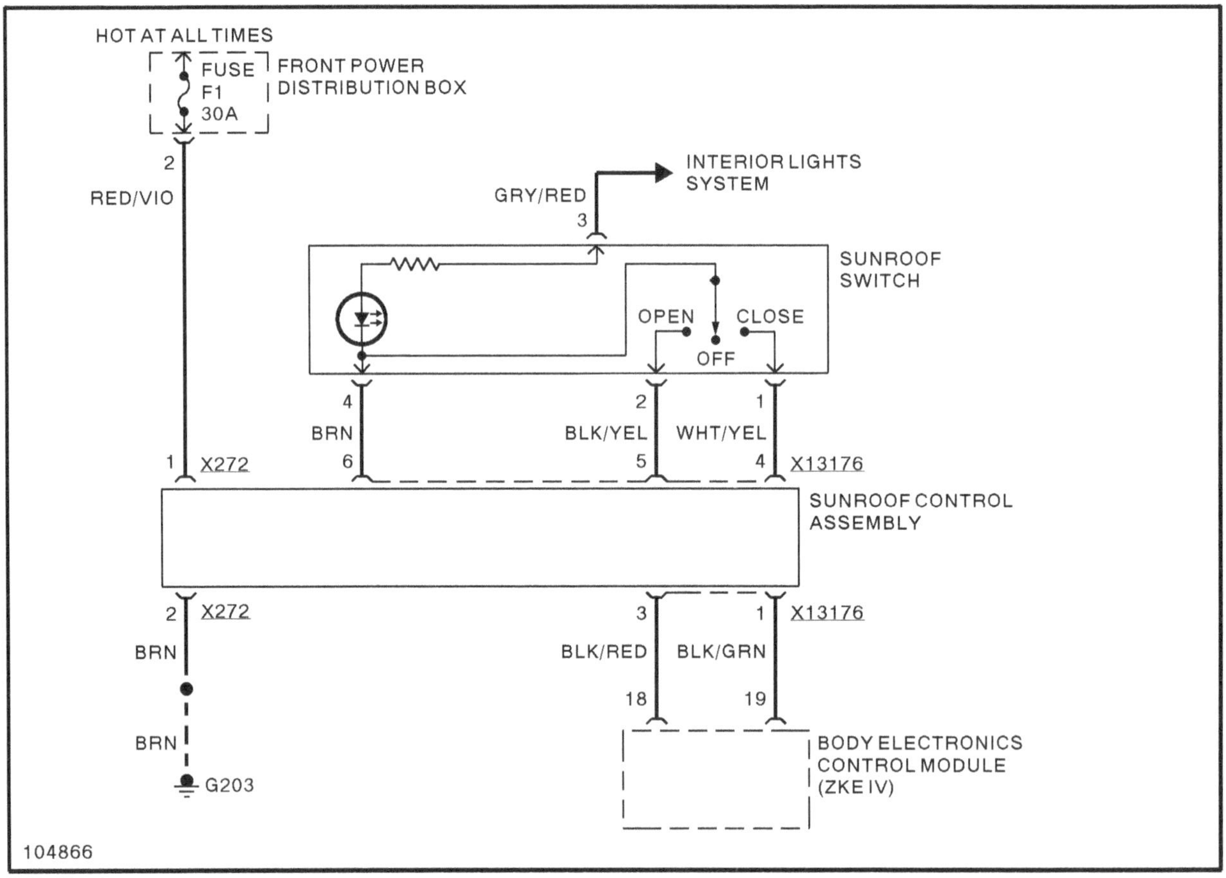

Power Sunroof

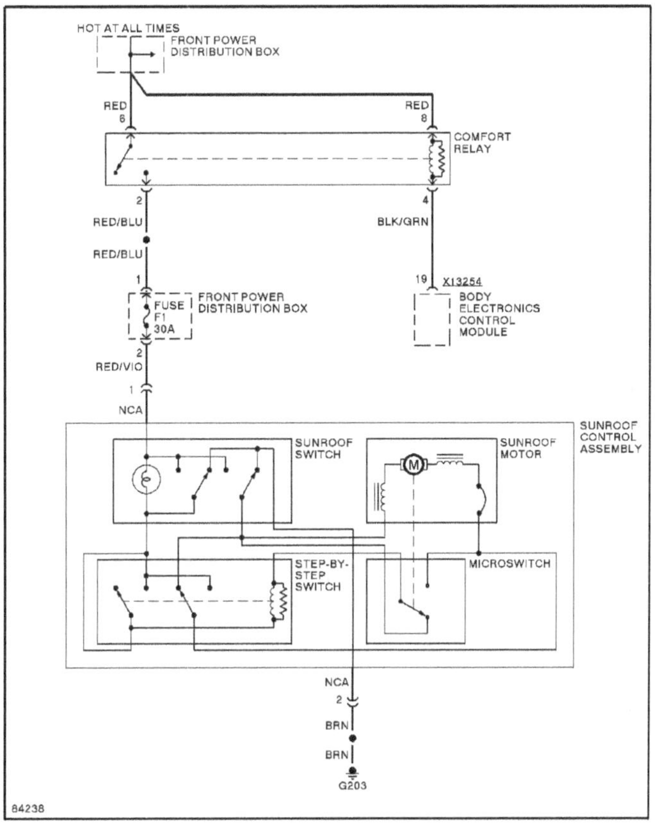

1995 Early Production 318ti

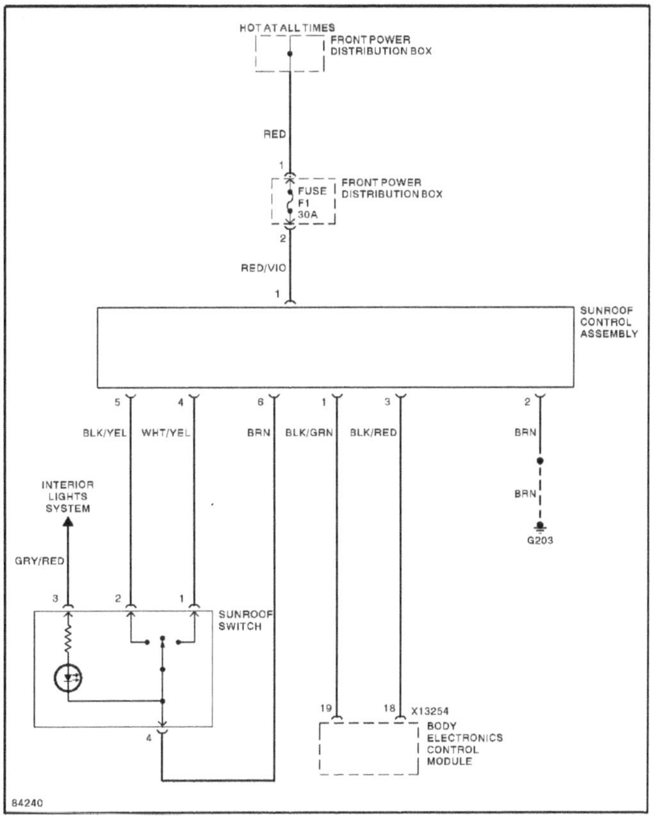

1995 Late Production & 1996-98 318ti Sliding Sunroof

Power Sunroof

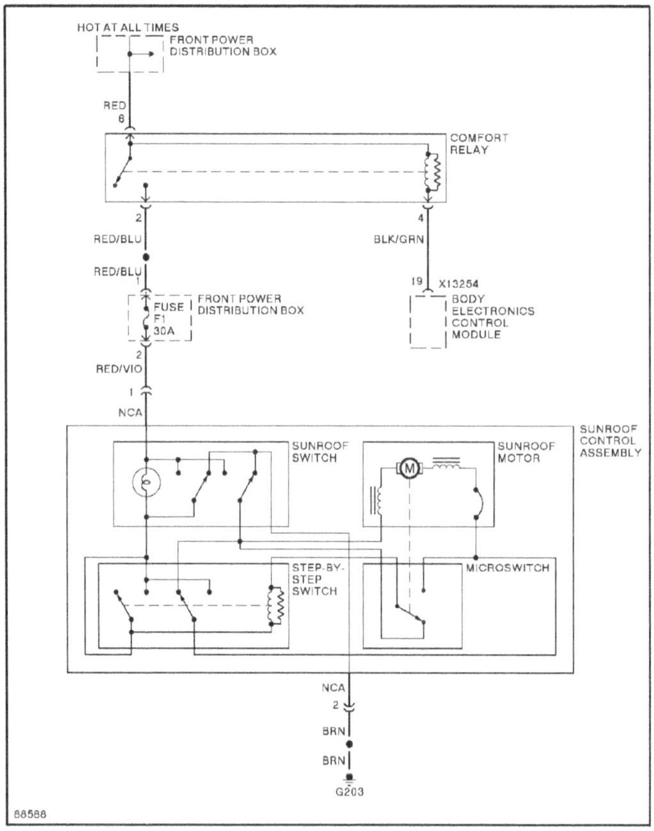

1997 318ti Folding Sunroof

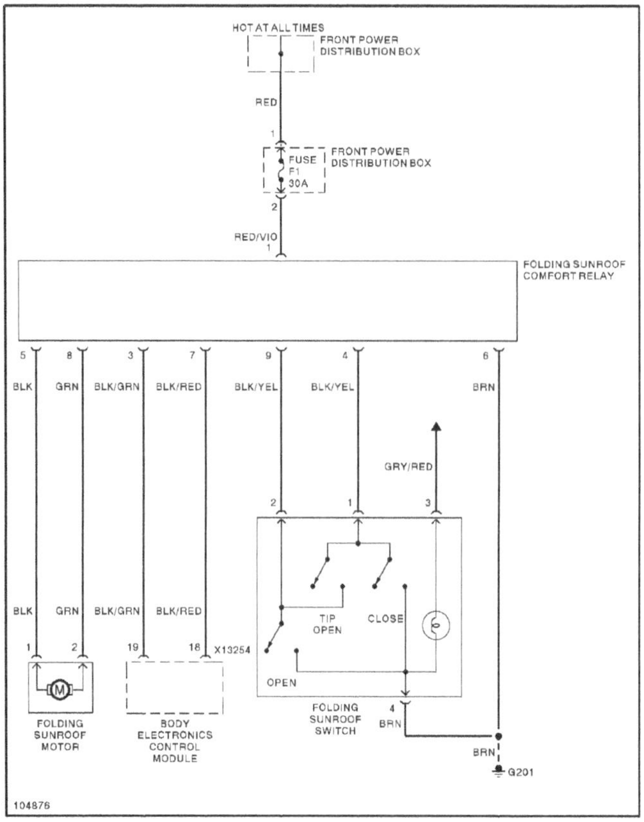

1998 318ti Folding Sunroof

Convertible Top

1994-97

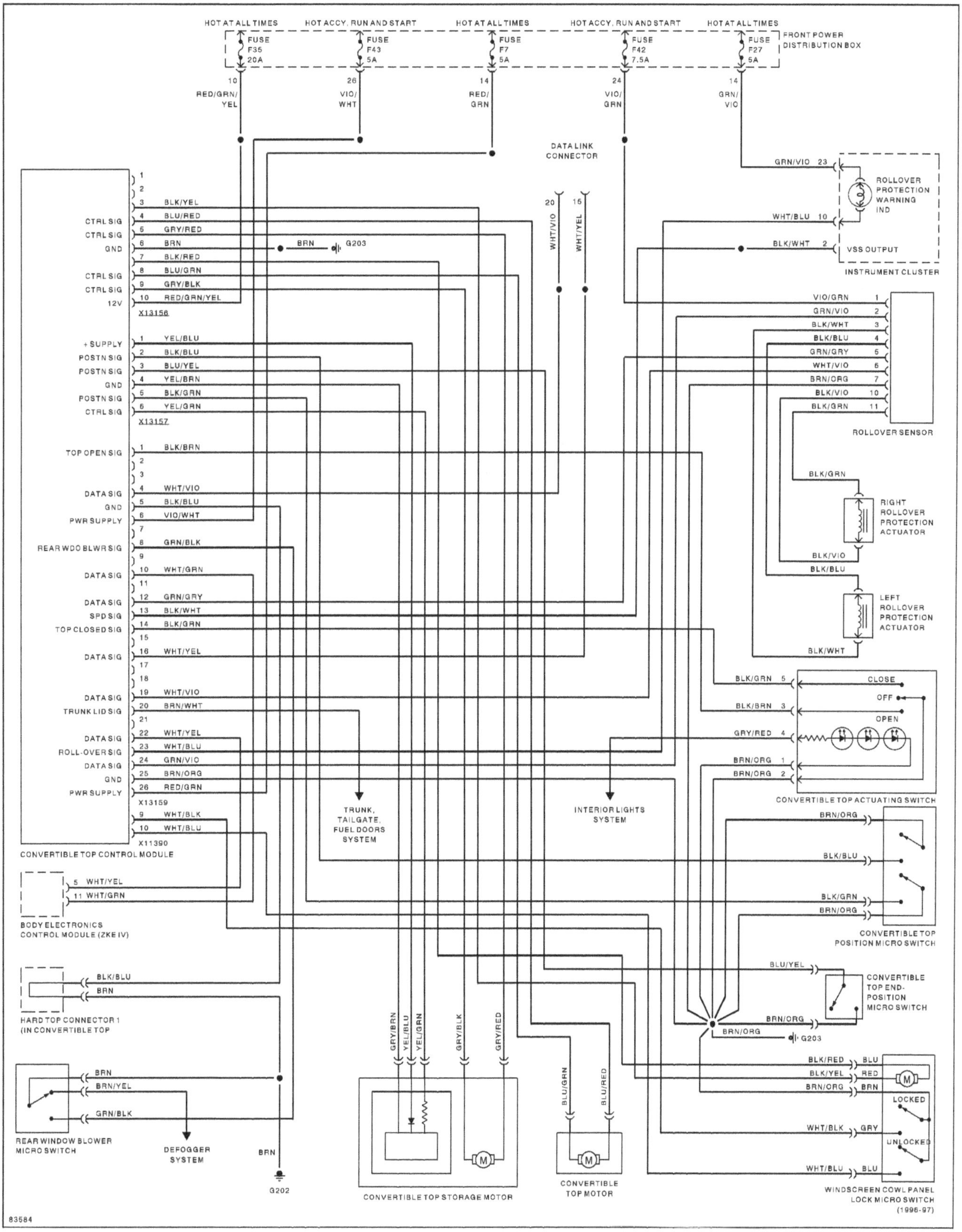

Convertible Top

1998

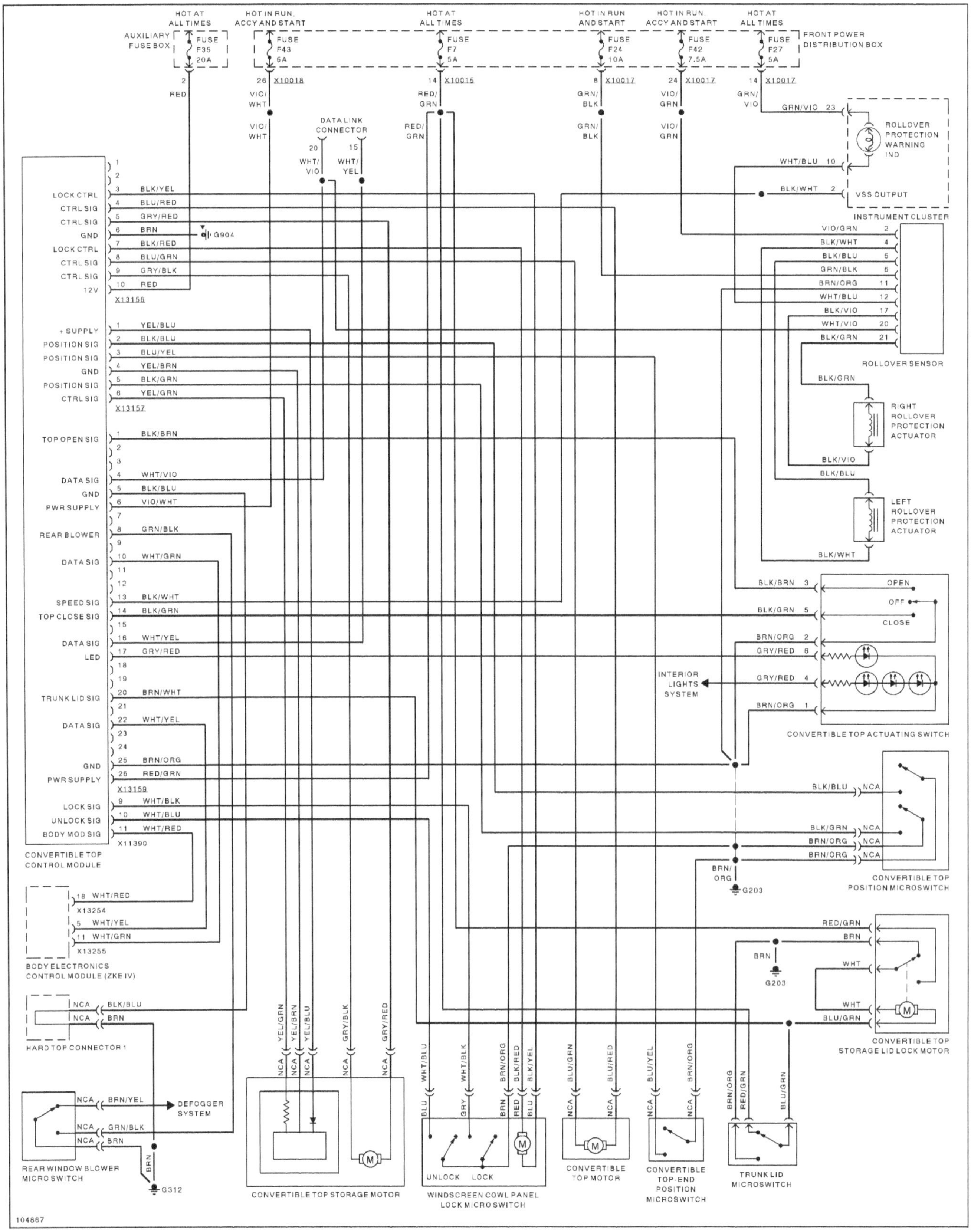

Radio/Cassette/CD/Speakers

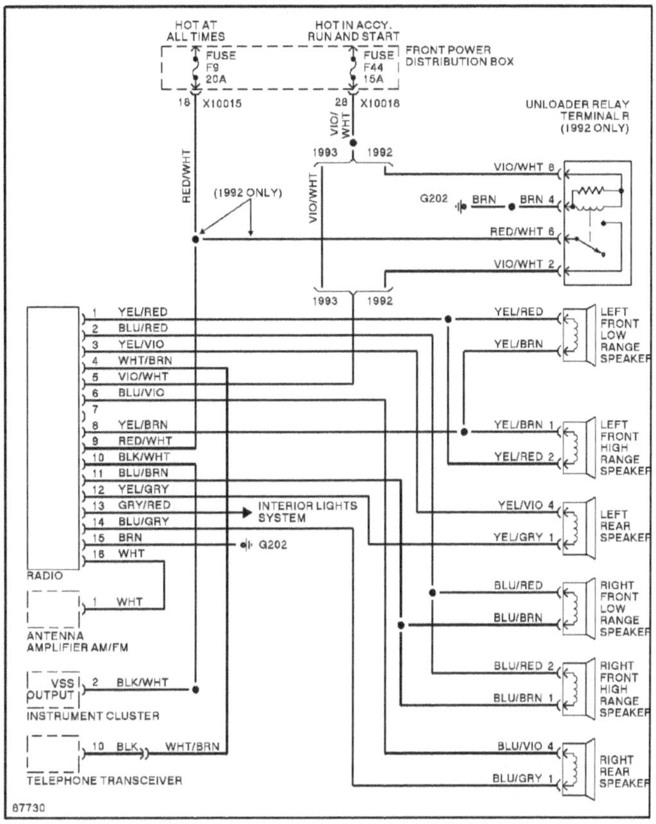

1992-93
6 Speaker
System

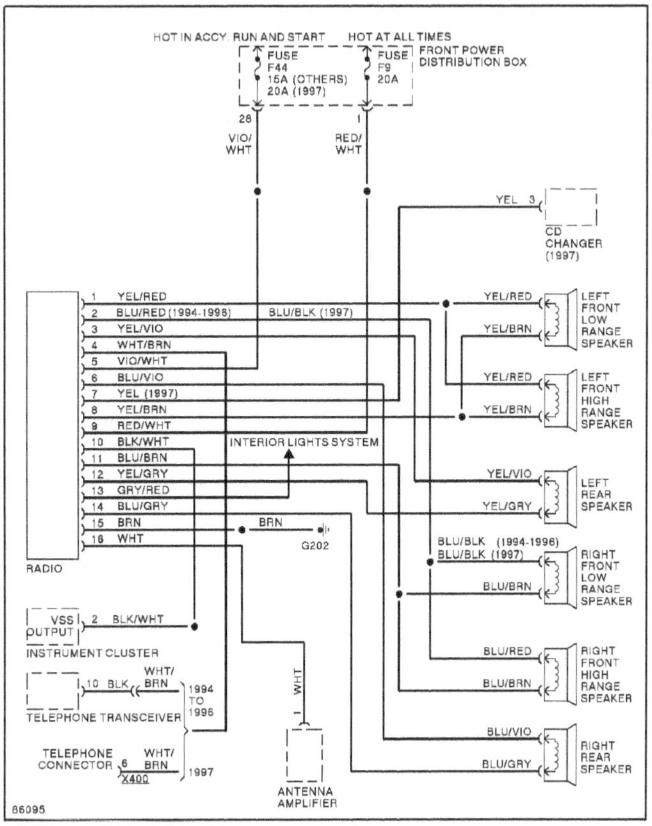

1994-97
Except 318ti
10 Speaker
System

Radio/Cassette/CD/Speakers

1998 Except 318ti 6 Speaker System

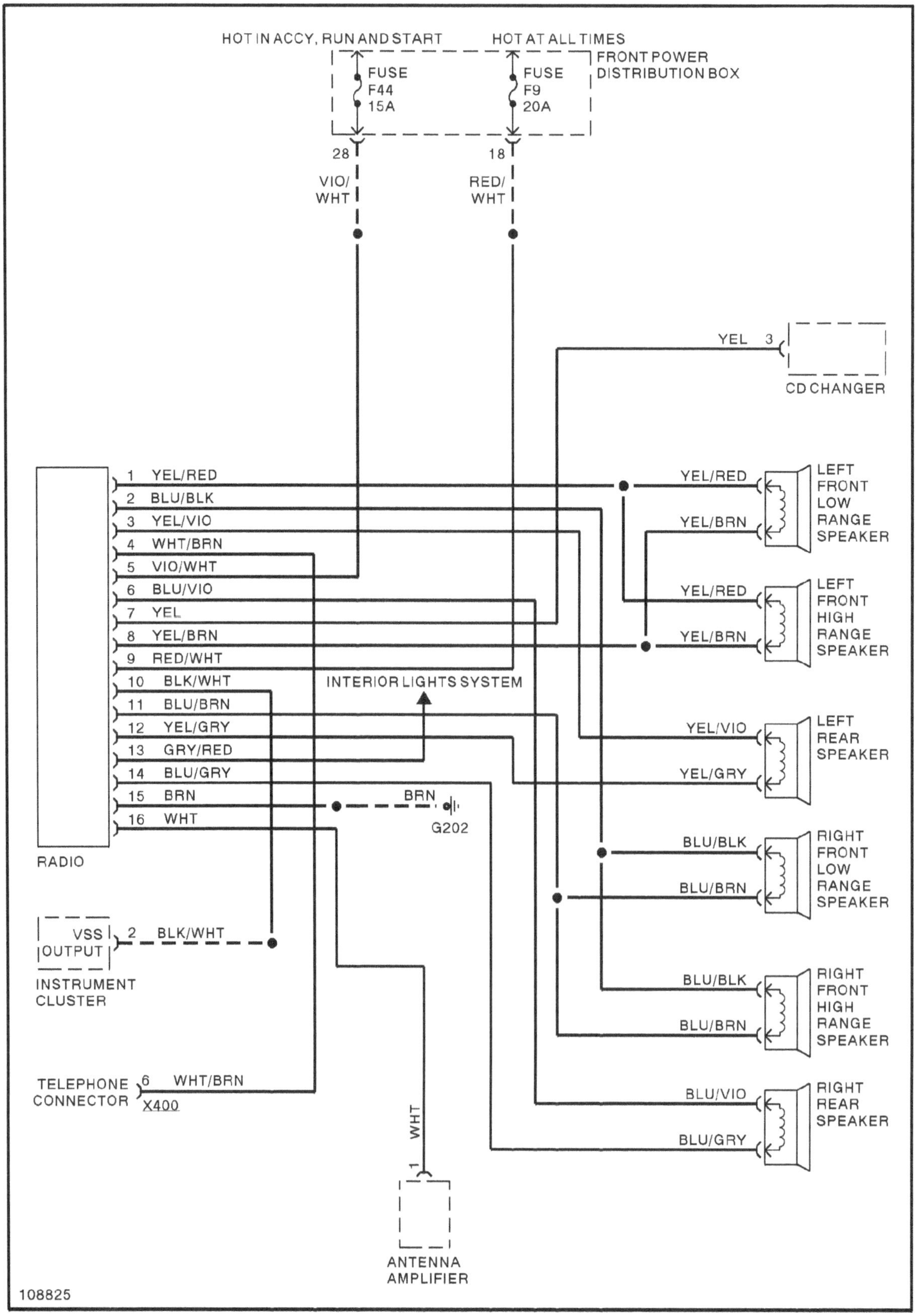

Radio/Cassette/CD/Speakers

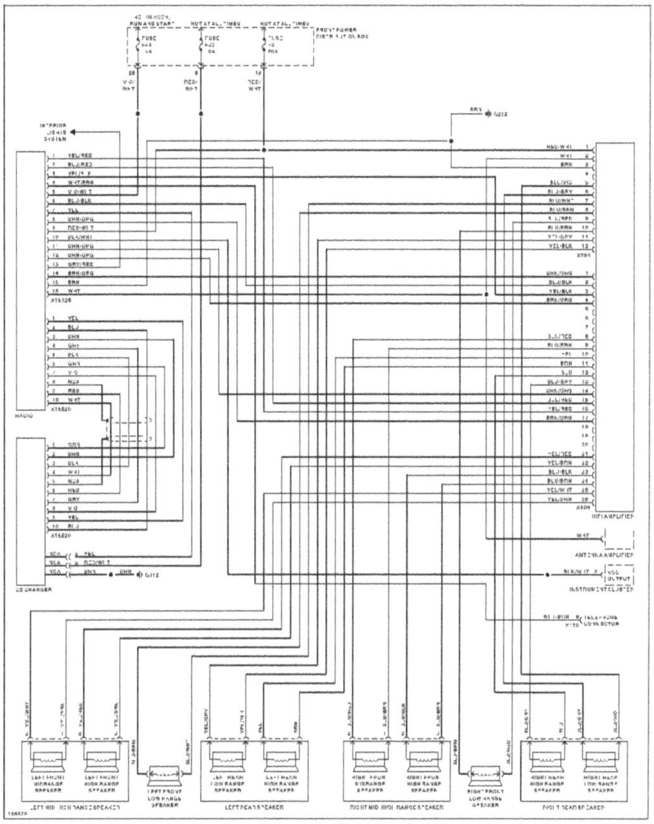

1998 Except 318ti 10 Speaker w/CD Changer System

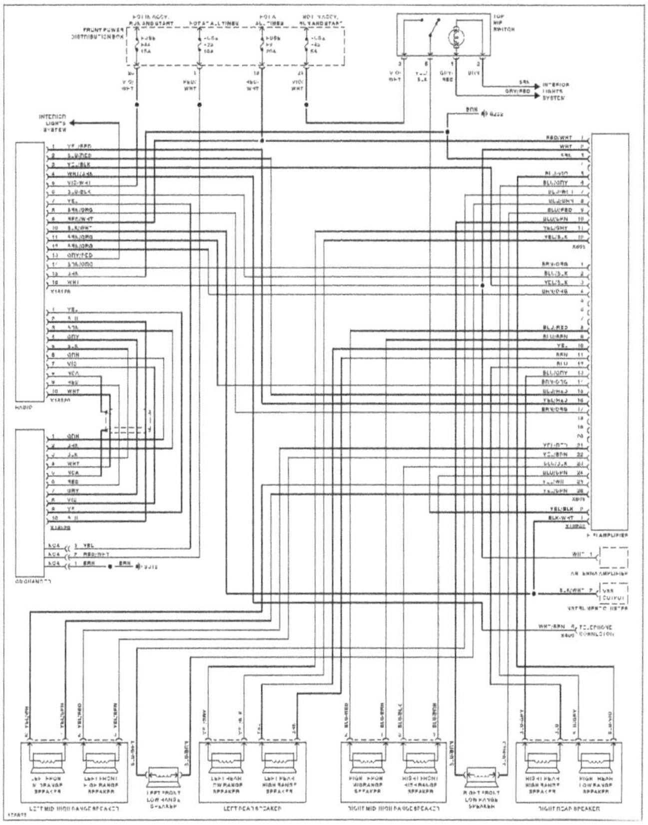

1998 Except convertible 10 Speaker w/CD Changer and Top HiFi System

Radio/Cassette/CD/Speakers

1998 Convertible 10 Speaker w/CD & Top HiFi System

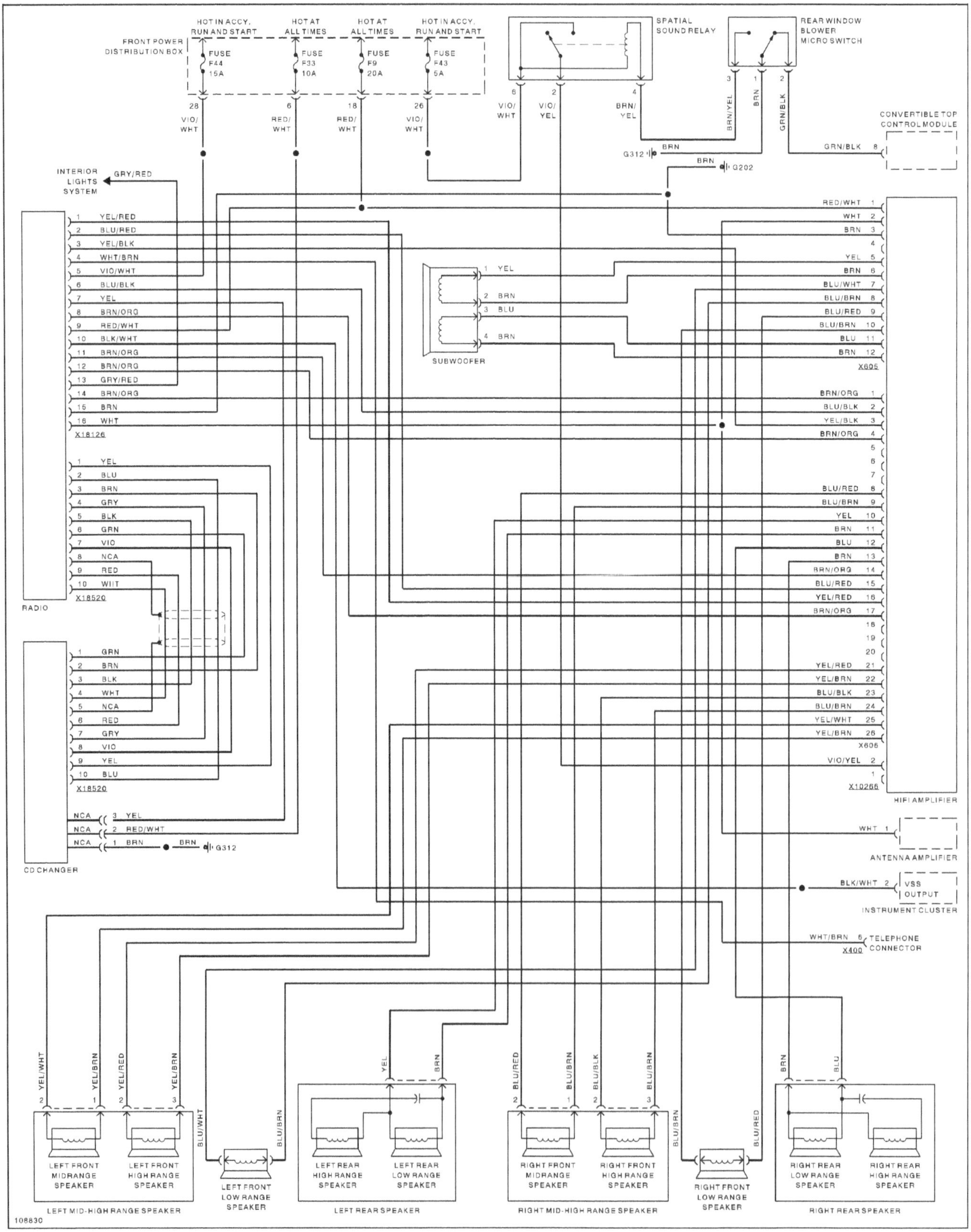

Radio/Cassette/CD/Speakers

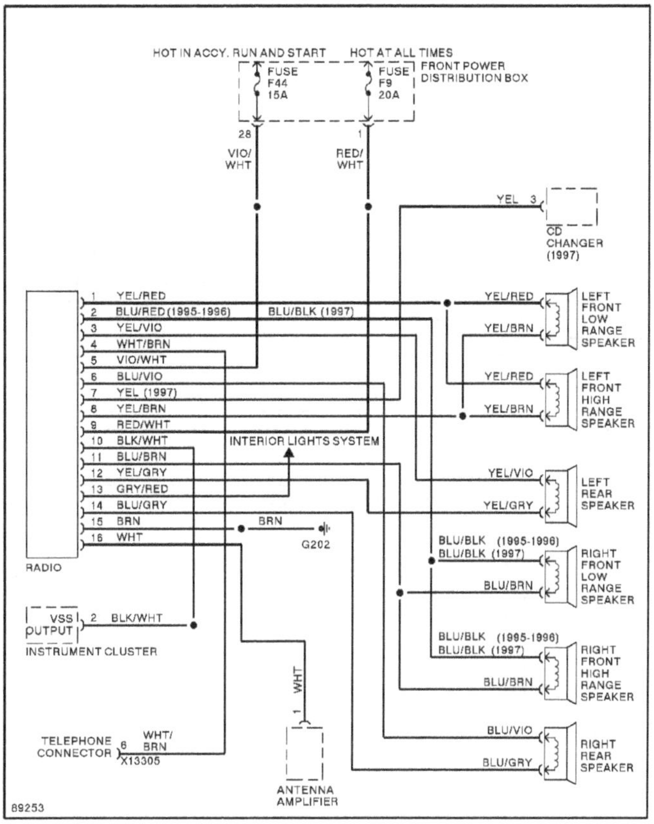

1995-97
318ti
6 Speaker
System

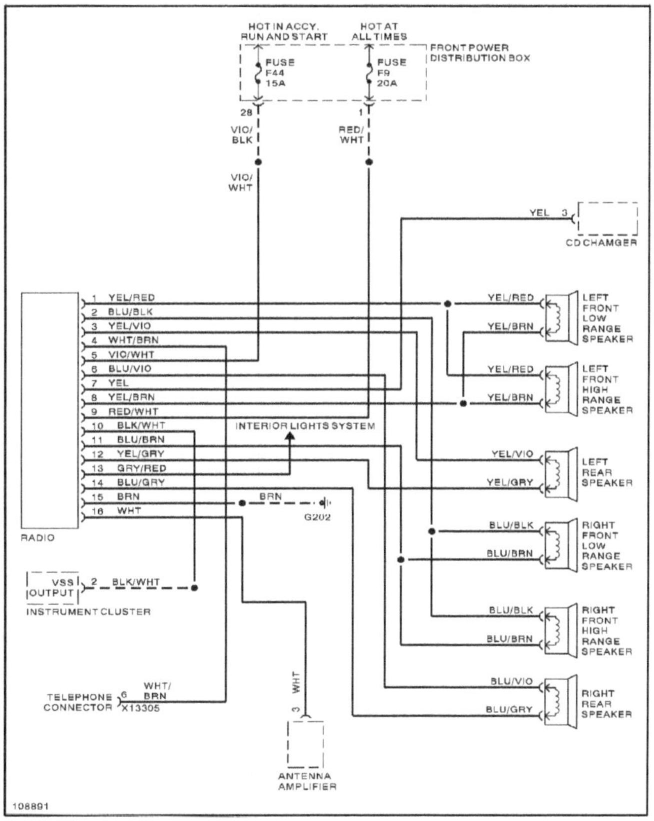

1998
318ti
6 Speaker
System

Radio/Cassette/CD/Speakers

1998 318ti 10 Speaker w/CD Changer System

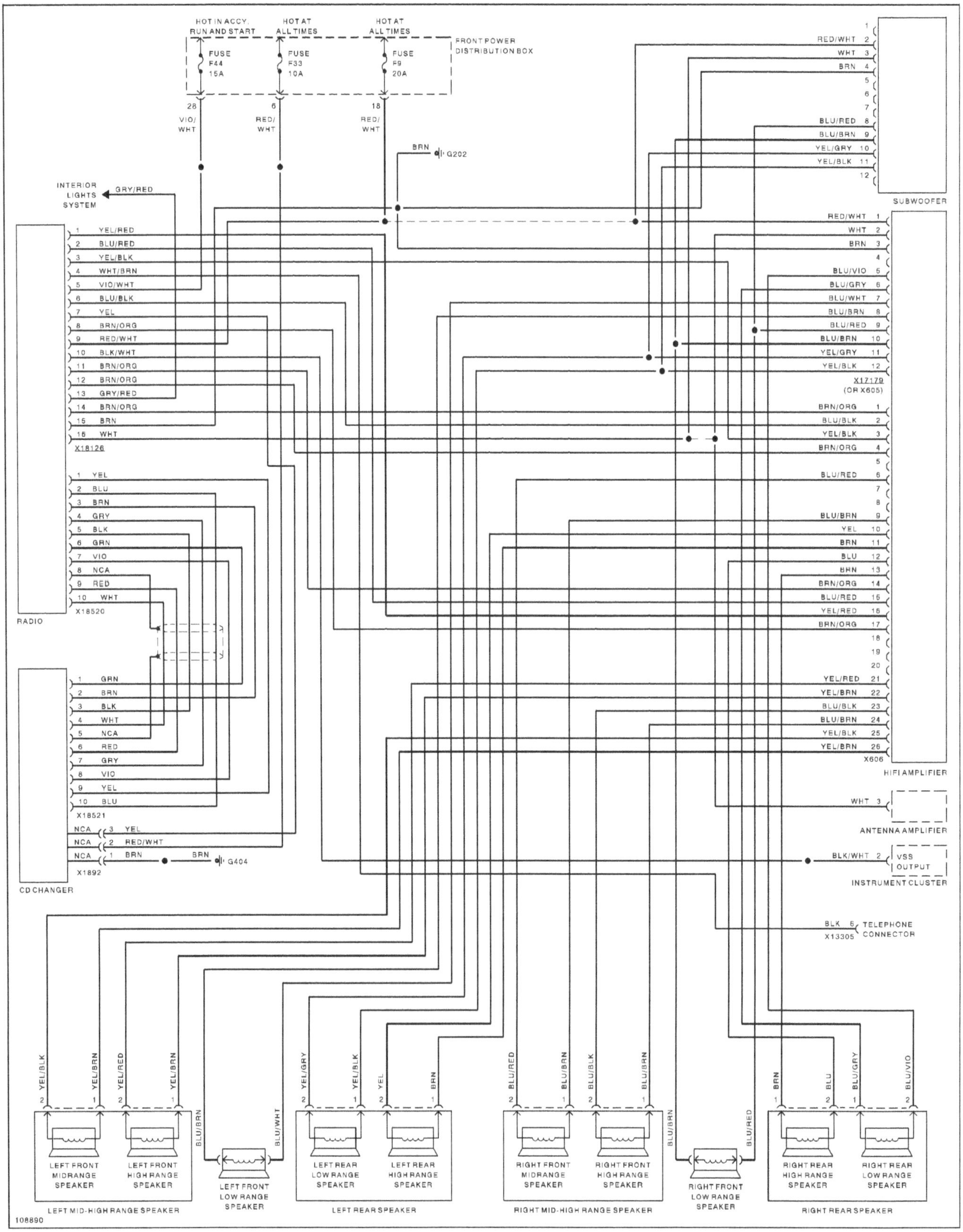

Radio/Cassette/CD/Speakers

1992 10 Speaker System

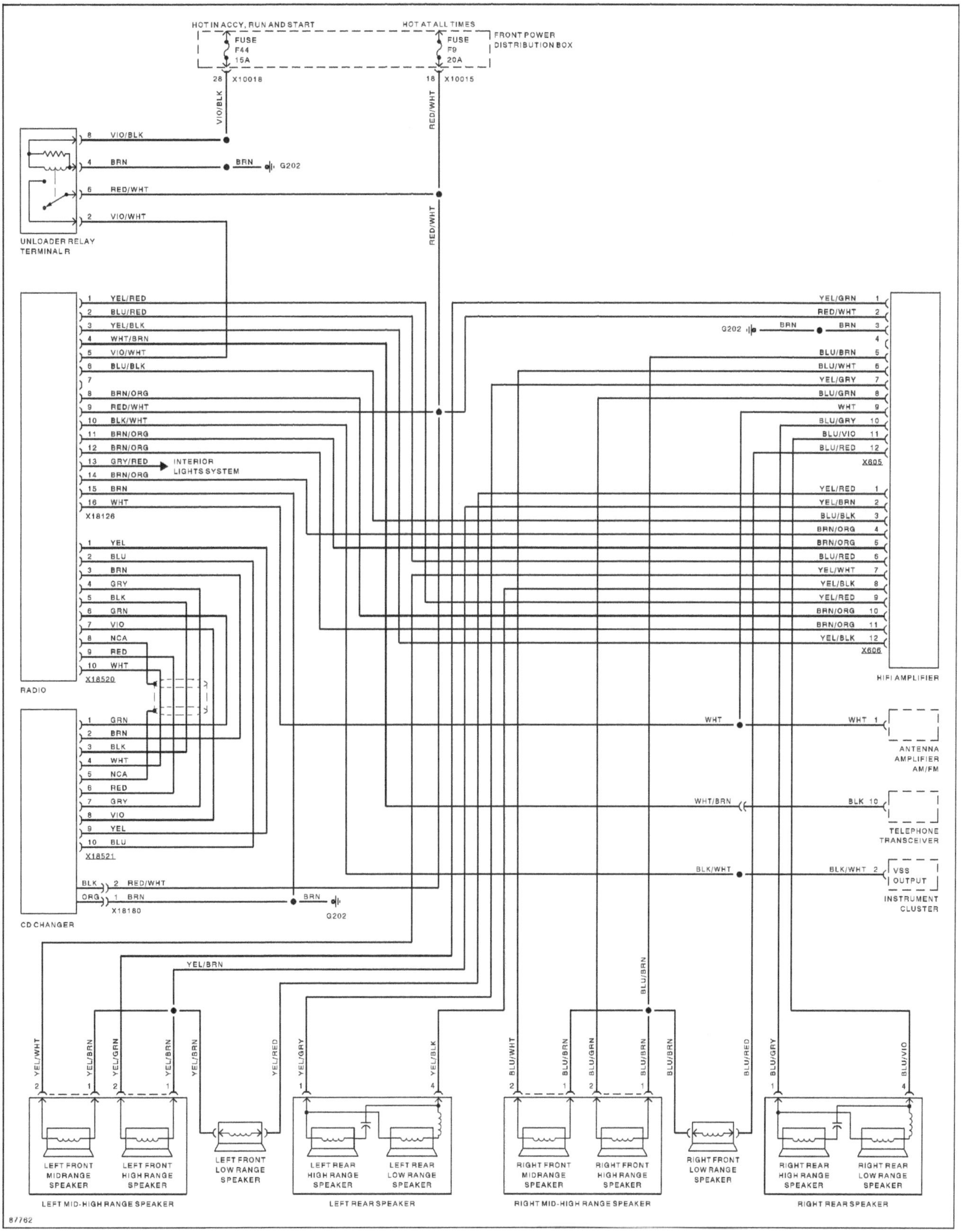

Radio/Cassette/CD/Speakers

1993 Early Production 10 Speaker System

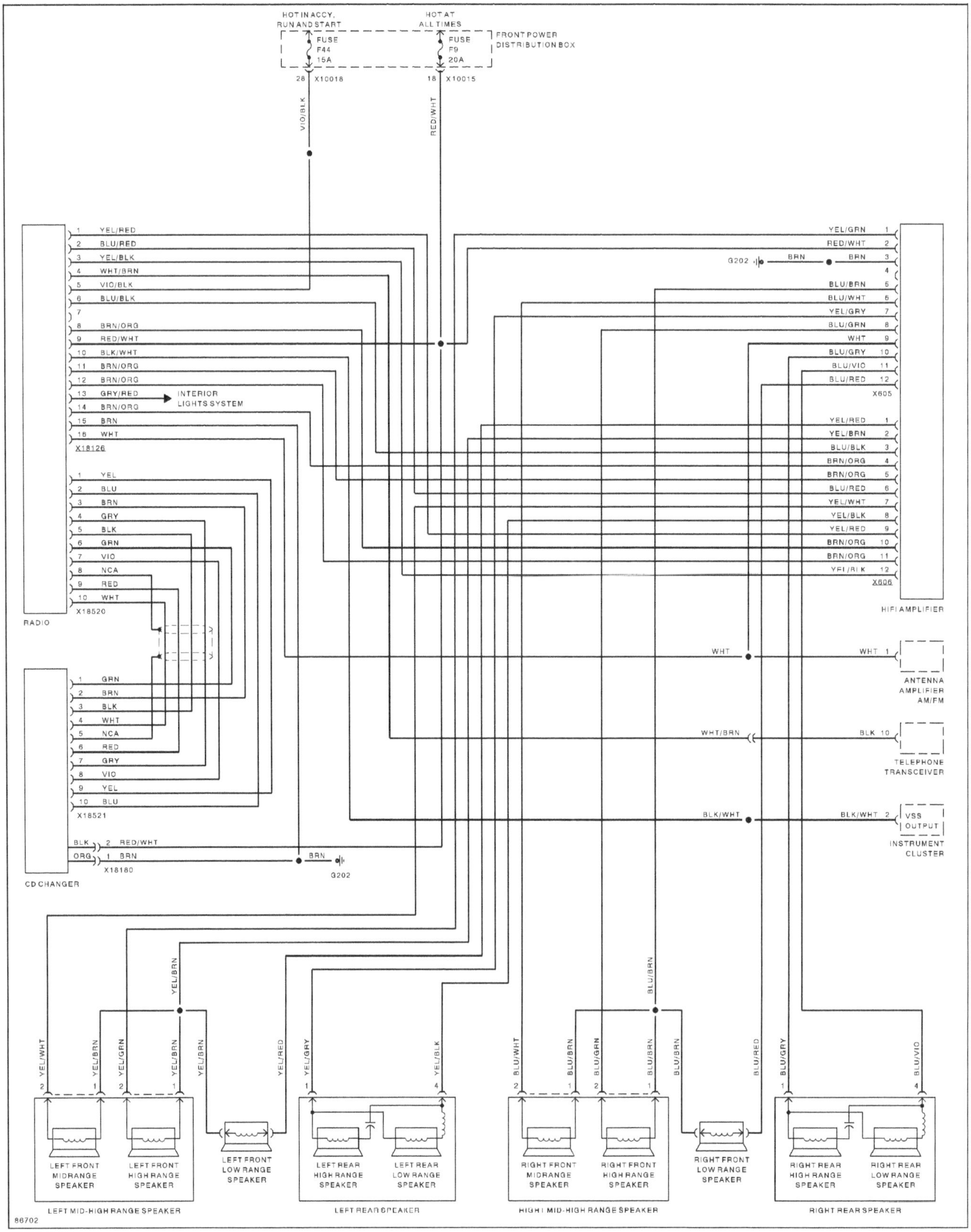

Radio/Cassette/CD/Speakers

1993 Late Production 10 Speaker System

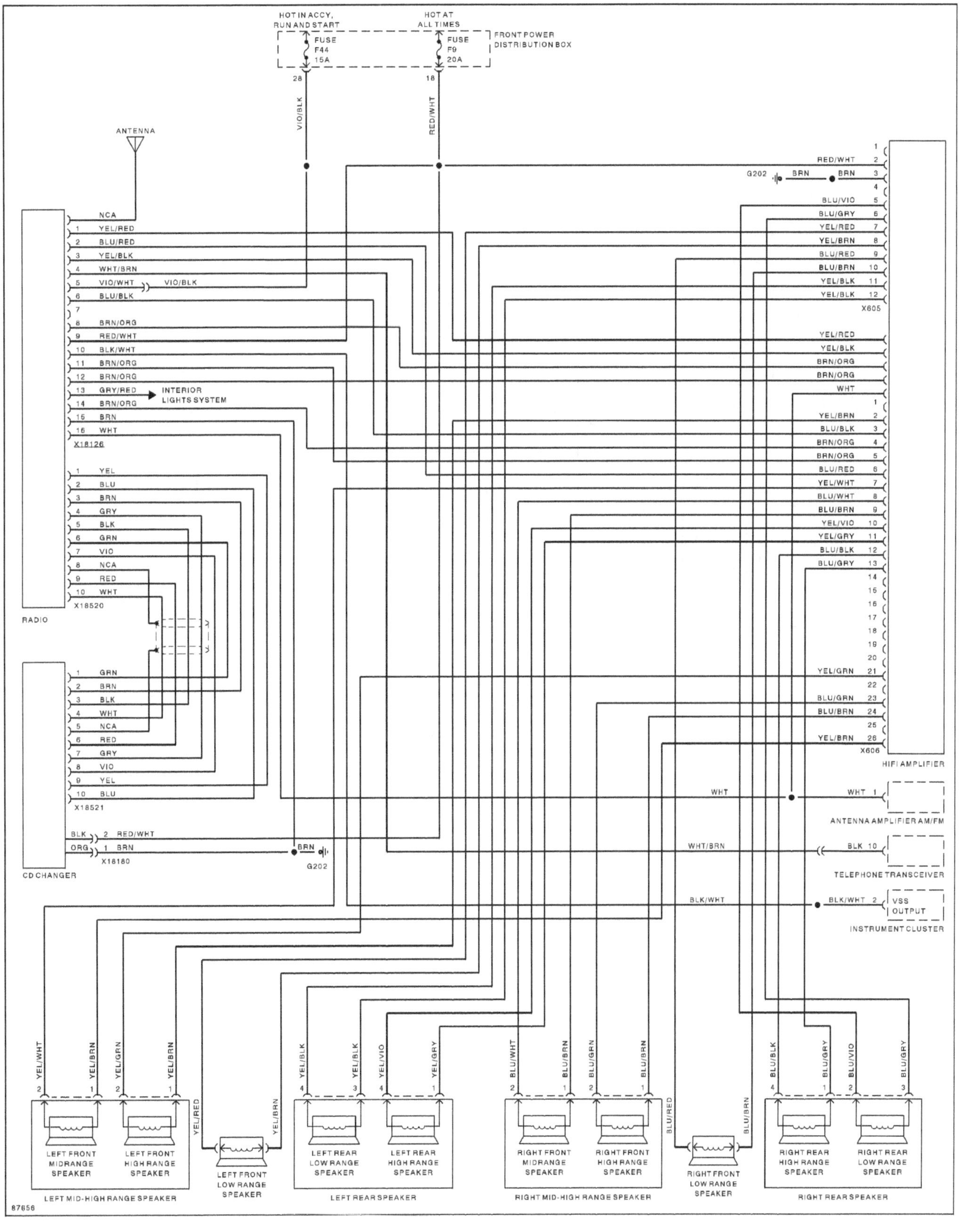

Radio/Cassette/CD/Speakers

1994-97 Except 318ti 10 Speaker System

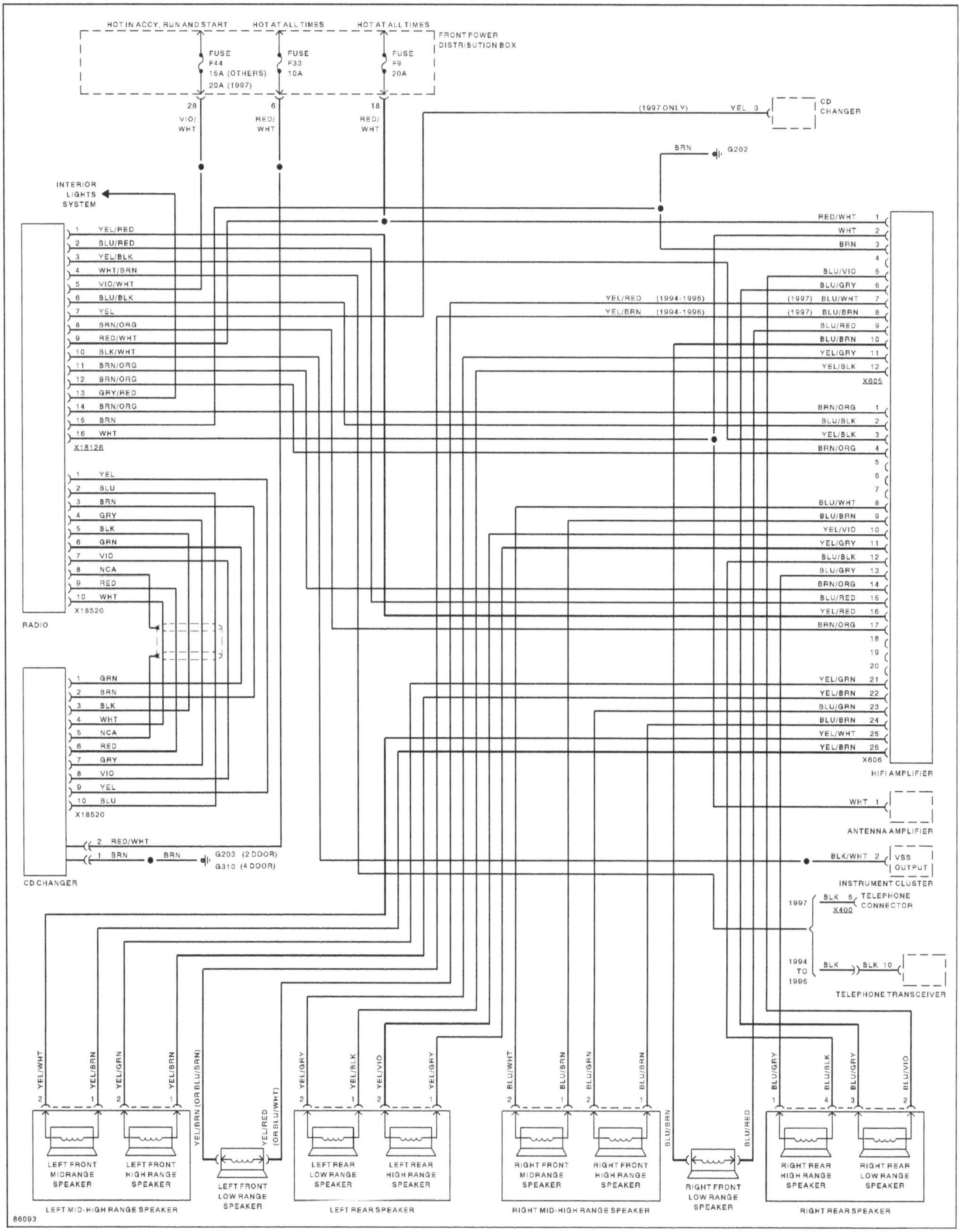

WARNING

Your common sense, good judgement, and general alertness are crucial to safe and successful service work. Before attempting any work on your BMW, read the warnings and cautions on page vii and the copyright page at the front of the manual. Review these warnings and cautions each time you prepare to work on your BMW. Please also read any warnings and cautions that accompany the procedures in the manual.

A

B

WARNING

Your common sense, good judgement, and general alertness are crucial to safe and successful service work. Before attempting any work on your BMW, read the warnings and cautions on page vii and the copyright page at the front of the manual. Review these warnings and cautions each time you prepare to work on your BMW. Please also read any warnings and cautions that accompany the procedures in the manual.

WARNING

Your common sense, good judgement, and general alertness are crucial to safe and successful service work. Before attempting any work on your BMW, read the warnings and cautions on page vii and the copyright page at the front of the manual. Review these warnings and cautions each time you prepare to work on your BMW. Please also read any warnings and cautions that accompany the procedures in the manual.

WARNING

Your common sense, good judgement, and general alertness are crucial to safe and successful service work. Before attempting any work on your BMW, read the warnings and cautions on page vii and the copyright page at the front of the manual. Review these warnings and cautions each time you prepare to work on your BMW. Please also read any warnings and cautions that accompany the procedures in the manual.

N

O

P

R

S

> **WARNING**
>
> Your common sense, good judgement, and general alertness are crucial to safe and successful service work. Before attempting any work on your BMW, read the warnings and cautions on page vii and the copyright page at the front of the manual. Review these warnings and cautions each time you prepare to work on your BMW. Please also read any warnings and cautions that accompany the procedures in the manual.